POLYMERIC BIOMATERIALS

Medicinal and Pharmaceutical Applications

VOLUME 2

Polymeric Biomaterials

Polyméric Biomaterials: Structure and Function, Volume 1

Polymeric Biomaterials: Medicinal and Pharmaceutical Applications, Volume 2

POLYMERIC BIOMATERIALS

Medicinal and Pharmaceutical Applications

VOLUME 2

Founding Editor
Severian Dumitriu

Editor
Valentin Popa

CRC Press
Taylor & Francis Group
Boca Raton   London   New York

CRC Press is an imprint of the
Taylor & Francis Group, an **informa** business

CRC Press
Taylor & Francis Group
6000 Broken Sound Parkway NW, Suite 300
Boca Raton, FL 33487-2742

First issued in paperback 2020

© 2013 by Taylor & Francis Group, LLC
CRC Press is an imprint of Taylor & Francis Group, an Informa business

No claim to original U.S. Government works

Version Date: 20120726

ISBN 13: 978-0-367-38040-3 (pbk)
ISBN 13: 978-1-4200-9468-8 (hbk)

Library of Congress Cataloging-in-Publication Data

Polymeric biomaterials / editors, Severian Dumitriu and Valentin Popa.
    p. ; cm.
    Includes bibliographical references and index.
    ISBN 978-1-4200-9470-1 (v. 1) -- ISBN 978-1-4200-9468-8 (v. 2)
    I. Dumitriu, Severian, 1939- II. Popa, Valentin I.
    [DNLM: 1. Polymers. 2. Biocompatible Materials--therapeutic use. 3. Regenerative Medicine--methods. QT 37.5.P7]

    610.28--dc23   2012029709

Visit the Taylor & Francis Web site at
http://www.taylorandfrancis.com

and the CRC Press Web site at
http://www.crcpress.com

Contents

Supplementary Resources Disclaimer

Additional resources were previously made available for this title on CD. However, as CD has become a less accessible format, all resources have been moved to a more convenient online download option.

You can find these resources available here: https://www.routledge.com/9781420094688

Please note: Where this title mentions the associated disc, please use the downloadable resources instead.

Preface

The field of biomaterials has developed rapidly because of the continuous and ever-expanding practical needs of medicine and health-care practice. There are currently thousands of medical devices, diagnostic products, and disposables on the market, and the range of applications continues to grow. In addition to traditional medical devices, diagnostic products, pharmaceutical preparations, and health-care disposables, the list of biomaterials applications includes smart delivery systems for drugs, tissue cultures, engineered tissues, and hybrid organs.

Undoubtedly, biomaterials have had a major impact on the practice of contemporary medicine and patient care, resulting in both saving and improving the quality of lives of humans and animals. Modern biomaterials practice is continuing to develop into a major interdisciplinary effort involving chemists, biologists, engineers, and physicians. It also takes advantage of developments in the traditional, nonmedical materials field, and much progress has been made since the beginning of the research in biomaterials that made possible the creation of a high-quality and much improved variety of devices, implants (permanent or temporary), and drug carrier devices. All of these now display a greater than ever biocompatibility and biofunctionality. The variety of chemical substances used in these materials is currently very broad, and most biomedical applications are associated with various polymers and materials based on them.

The pace of research in the field of polymeric biomaterials is so fast that two editions of *Polymeric Biomaterials* have already been edited by Severian Dumitriu. Due to the interest generated and the success of these books, Severian was working on a third edition. Unfortunately, he passed away before this could be finalized. Many of the scientists who accepted his invitation to cooperate for this new edition agreed to contribute to the book in memory of the contribution that Severian made to the field of polymeric biomaterials. Together with Daniela, his beloved daughter, and Barbara Glunn and Jessika Vakili from Taylor & Francis Group, we decided to continue the work and finalize this book.

This book is organized in two volumes consisting of 53 chapters that systematically provide the latest developments in different aspects of polymeric biomaterials. Thus, we can mention contributions to the field of synthesis and applications of polymers such as polyesters, poly(vinyl alcohol), polyphosphazenes, elastomers, bioceramics, blends or composites, enzymatic synthesis, along with natural ones such as mucoadhesives, chitin, chitosan, lignin, carbohydrates derivatives, heparin, etc.

Drugs carriers and delivery systems, gene and nucleic acids delivery represent other subjects of some chapters, dealing with both supports (biodegradable and biocompatible) and techniques (nanoparticles, electrospinning, photo- and pH responsive polymers, hydrogels, lipid-core micelles, biomimetic systems, medical devices) aspects. In some cases, biomaterials can be synthesized, modified, and processed by different methods to ensure biocompatibility and biodegradability to be used as membranes, composites, scaffolds, and implants. Some examples of specific utilizations of polymeric biomaterials are presented, such as orthopedic surgery, bone regeneration, wound healing, dental and maxillofacial surgery applications, artificial joints, diabetes, anticancer agents and cancer therapy, modification of living cells, myocardial tissue engineering—repair and reconstruction, and bioartificial organs.

Publishing this book was accomplished with the contributions of renowned scientists from all over the world. They are all experts in their particular field of biomaterials research and have made high-level contributions to various fields of research. We are very grateful to these scientists for their willingness to contribute to this reference work as well as for their engagement. Without their commitment and enthusiasm, it would not have been possible to compile such a book.

I am also grateful to the publisher for recognizing the demand for such a book, for taking the risk to bring out such a book, and for realizing the excellent quality of the publication.

I would like to thank Daniela for her inestimable help and assistance. I dedicate this book to memory of Severian, one of my best friends.

Last but not least, I would like to thank my family for their patience. I sincerely apologize for the many hours I spent in the preparation of this book, which kept me away from them.

This book is a very useful tool for many scientists, physicians, pharmacists, engineers, and other experts in a variety of disciplines, both in academe and industry. It may not only be useful for research and development but may also be suitable for teaching.

This book has a companion CD that contains color figures as noted at the applicable text figures.

Valentin I. Popa

Acknowledgments

My father was passionate about polymeric biomaterials. He was very happy when this project was planned with Taylor & Francis Group. He had worked tirelessly toward this. He would have loved to have seen this book published, but destiny willed otherwise.

I am extremely grateful to Professor Popa for accepting to serve as the editor, to all the authors for their precious contributions, and to the staff at Taylor & Francis Group.

The positive response from the authors to pursue their contribution to this book was amazing and is testimony of their appreciation for the scientific contribution that my father made to the field of polymeric biomaterials.

I trust the book is of great quality and reflects the efforts and dedication that have been put into it by my father and all the contributors.

My small contribution to this book is dedicated to the memory of my parents, Severian and Maria, for their unconditional love and for being the best teachers ever. And to finish on a positive note, I want to cite one quote of Dr. Seuss that I particularly like:

"Don't cry because it's over. Smile because it happened".

Daniela Dumitriu

Editors

Severian Dumitriu (deceased) was a research professor, Department of Chemical Engineering, University of Sherbrooke, Quebec, Canada. He edited several books, including *Polymeric Biomaterials*, second edition, *Polysaccharides in Medicinal Applications*, and *Polysaccharides: Structural Diversity and Functional Versatility* (all three titles were published by Taylor & Francis Group [previously Marcel Dekker]), and authored or coauthored over 190 professional papers and book chapters in the fields of polymer and cellulose chemistry, polyfunctional initiators, and bioactive polymers. He also held 15 international patents. Professor Dumitriu received his BSc (1959) and MS (1961) in chemical engineering and his PhD (1971) in macromolecular chemistry from the Polytechnic Institute of Jassy, Romania. Upon completing his doctorate, he worked with Professor G. Smets at the Catholic University of Louvain, Belgium, and was a research associate at the University of Pisa, Italy; the Hebrew University Medical School, Jerusalem, Israel; and the University of Paris, South France.

Valentin I. Popa earned his BSc and MSc in chemical engineering (1969) and PhD in the field of polysaccharide chemistry (1976) from Polytechnic Institute of Iasi, Romania. He was awarded the Romanian Academy Prize for his contributions in the field of seaweed chemistry (1976). He has published more than 500 papers in the following fields: wood chemistry and biotechnology, biomass complex processing, biosynthesis and biodegradation of natural compounds, allelochemicals, bioadhesives, and bioremediation. He is also the author or coauthor of 37 books or book chapters. Dr. Popa holds six patents and has been involved in many Romanian and European research projects as scientific manager. He was visiting scientist or visiting professor at Academy of Sciences (Seoul, Korea, 1972), Technical University of Helsinki (Finland, 1978), Institute of Biotechnology (Vienna, Austria, 1995), Research Institute for Pulp and Paper (Braila, Romania, 1976), "Petru Poni" Institute of Macromolecular Chemistry (Iasi, Romania, 1985, 1986), Université de Sherbrooke and University McGill (Canada, 2003), STFI–Packforsk (now known as Innventia, Stockholm, Sweden, 2008), and Institute of Wood Chemistry (Riga, Latvia, 2009). Dr. Popa is a member of the International Lignin Institute, International Association of Scientific Papermakers, International Academy of Wood Science, Romanian Academy for Technical Sciences, and American Chemical Society. He is also a professor of wood chemistry and biotechnology in "Gheorghe Asachi" Technical University of Iasi, PhD supervisor (30 students defended their theses), and editor-in-chief of *Cellulose Chemistry and Technology*.

Contributors

M.R. Aguilar
Spanish National Research Council
Institute of Polymer Science and Technology
Madrid, Spain

Mamoru Aizawa
Department of Applied Chemistry
School of Science and Technology
Meiji University
Kawasaki, Japan

N.N. Ali
National Heart and Lung Institute
Imperial College London
London, United Kingdom

Christine Allen
Leslie Dan Faculty of Pharmacy
University of Toronto
Toronto, Ontario, Canada

Carmen Alvarez-Lorenzo
Faculty of Pharmacy
Department of Pharmacy and Pharmaceutical
    Technology
University of Santiago de Compostela
Santiago de Compostela, Spain

Luigi Ambrosio
National Research Council
Institute of Composite and Biomedical
    Materials
Naples, Italy

Karine Andrieux
Faculty of Pharmacy
Institut Galien Paris-Sud
Université Paris-Sud
Chatenay-Malabry, France

Tomohiro Asai
Department of Medical Biochemistry
Graduate School of Pharmaceutical
    Sciences
University of Shizuoka
Shizuoka, Japan

A.K. Bajpai
Bose Memorial Research Laboratory
Department of Chemistry
Government Autonomous Science College
Jabalpur, India

Gillian Barratt
Faculty of Pharmacy
Institut Galien Paris-Sud
Université Paris-Sud
Chatenay-Malabry, France

Nicole A. Beinborn
Aptalis Pharma
Vandalia, Ohio

Leslie R. Berry
Department of Pediatrics
McMaster University
Hamilton, Ontario, Canada

Sujata K. Bhatia
School of Engineering and Applied Sciences
Harvard University
Cambridge, Massachusetts

A.R. Boccaccini
Department of Materials Science
    and Engineering
University of Erlangen-Nuremberg
Erlangen, Germany

Anthony K.C. Chan
Department of Pediatrics
McMaster University
Hamilton, Ontario, Canada

Hao Chen
Department of Reparative Materials
Institute for Frontier Medical Sciences
Kyoto University
Kyoto, Japan

Raje Chouhan
Bose Memorial Research Laboratory
Department of Chemistry
Government Autonomous Science College
Jabalpur, India

Felisa Cilurzo
Department of Health Sciences
University "Magna Græcia" of
    Catanzaro
Germaneto, Italy

Laura Cipolla
Department of Biotechnology
    and Biosciences
University of Milano–Bicocca
Milan, Italy

Angel Concheiro
Faculty of Pharmacy
Department of Pharmacy and Pharmaceutical
    Technology
University of Santiago de Compostela
Santiago de Compostela, Spain

W.F. Daamen
Department of Biochemistry
Nijmegen Centre for Molecular Life
    Sciences
Radboud University Nijmegen
    Medical Centre
Nijmegen, the Netherlands

Gérard Déléris
Department of Materials Engineering
Graduate School of Engineering
The University of Tokyo
Tokyo, Japan

and

Department of Bioengineering
University of California, Los Angeles
Los Angeles, California

Stéphanie Deshayes
Department of Materials Engineering
Graduate School of Engineering
The University of Tokyo
Tokyo, Japan

and

Department of Bioengineering
University of California, Los Angeles
Los Angeles, California

James C. DiNunzio
Pharmaceutical and Analytical R&D
Hoffmann-La Roche, Inc.
Basel, Switzerland

Ryan F. Donnelly
School of Pharmacy
Medical Biology Centre
Queen's University Belfast
Belfast, United Kingdom

K.A. Faraj
Department of Biochemistry
Nijmegen Centre for Molecular Life
    Sciences
Radboud University Nijmegen
    Medical Centre
Nijmegen, the Netherlands

M.M. Fernández
Spanish National Research Council
Institute of Polymer Science and Technology
Madrid, Spain

Jinming Gao
Department of Pharmacology
Simmons Comprehensive Cancer Center
University of Texas Southwestern Medical
    Center at Dallas
Dallas, Texas

Xiang Gao
Department of Pharmaceutical
    Sciences
School of Pharmacy
University of Pittsburgh
Pittsburgh, Pennsylvania

L. García-Fernández
Networking Biomedical Research Centre
    in Bioengineering, Biomaterials and
    Nanomedicine
Madrid, Spain

Karine Gionnet
Chimie et Biologie des Membranes et des
    Nanoobjets
Institut Européen de Chimie et Biologie
Université de Bordeaux
Pessac, France

Antonio Gloria
National Research Council
Institute of Composite and Biomedical
    Materials
Naples, Italy

Shilpi Goswami
Bose Memorial Research Laboratory
Department of Chemistry
Government Autonomous Science College
Jabalpur, India

Thomas Groth
Biomedical Materials Group
Institute of Pharmacy and
    Interdisciplinary Centre for
    Materials Research
Martin Luther University
    Halle-Wittenberg
Halle, Germany

Vincenzo Guarino
National Research Council
Institute of Composite and Biomedical
    Materials
Naples, Italy

I. Sedat Gunes
Department of Polymer Engineering
The University of Akron
Akron, Ohio

S.E. Harding
National Heart and Lung Institute
Imperial College London
London, United Kingdom

T. Hendriks
Department of Surgery
Radboud University Nijmegen Medical
    Centre
Nijmegen, the Netherlands

Hossein Hosseinkhani
Graduate Institute of Biomedical
    Engineering
National Taiwan University of Science
    and Technology
Taipei, Taiwan, Republic of China

Xiao-Jun Huang
Ministry of Education
Key Laboratory of Macromolecular Synthesis
    and Functionalization
Department of Polymer Science and
    Engineering
Zhejiang University
Hangzhou, Zhejiang, People's Republic of
    China

Hiroo Iwata
Department of Reparative Materials
Institute for Frontier Medical Sciences
Kyoto University
Kyoto, Japan

Sadhan C. Jana
Department of Polymer Engineering
The University of Akron
Akron, Ohio

H. Jawad
Department of Materials
Imperial College London
London, United Kingdom

Seong Hoon Jeong
College of Pharmacy
Dongguk University
Goyang, South Korea

Sanjana Kankane
Bose Memorial Research Laboratory
Department of Chemistry
Government Autonomous Science College
Jabalpur, India

M.J.W. Koens
Department of Biochemistry
Nijmegen Centre for Molecular
   Life Sciences
Radboud University Nijmegen
   Medical Centre
Nijmegen, the Netherlands

Hiroyuki Koide
Department of Medical Biochemistry
Graduate School of Pharmaceutical
   Sciences
University of Shizuoka
Shizuoka, Japan

A.G. Krasznai
Department of Surgery
Nijmegen Centre for Molecular Life
   Sciences
Radboud University Nijmegen
   Medical Centre
Nijmegen, the Netherlands

Song Li
Center for Pharmacogenetics
School of Pharmacy
University of Pittsburgh
Pittsburgh, Pennsylvania

M.L. López-Donaire
Spanish National Research Council
Institute of Polymer Science and Technology
Madrid, Spain

Tomokazu Matsuura
Department of Laboratory Medicine
School of Medicine
The Jikei University
Tokyo, Japan

Victor Maurizot
Chimie et Biologie des Membranes et des
   Nanoobjets
Institut Européen de Chimie et Biologie
Université de Bordeaux
Pessac, France

Dusica Maysinger
Department of Pharmacology
   and Therapeutics
McGill University
Montreal, Quebec, Canada

James W. McGinity
Division of Pharmaceutics
College of Pharmacy
The University of Texas at Austin
Austin, Texas

Laurence Moine
Université Paris-Sud
Chatenay-Malabry, France

Tiziana Musacchio
Center for Pharmaceutical Biotechnology
   and Nanomedicine
Northeastern University
Boston, Massachusetts

Julien Nicolas
Institut Galien Paris-Sud
Université Paris-Sud
Chatenay-Malabry, France

Francesco Nicotra
Department of Biotechnology
   and Biosciences
University of Milano–Bicocca
Milan, Italy

Yu Nie
National Engineering Research Center for
   Biomaterials
Sichuan University
Chengdu, Sichuan, People's Republic
   of China

Kyung T. Oh
College of Pharmacy
Chung-Ang University
Seoul, South Korea

Naoto Oku
Department of Medical Biochemistry
Graduate School of Pharmaceutical
   Sciences
University of Shizuoka
Shizuoka, Japan

Donatella Paolino
Department of Health Sciences
University "Magna Græcia" of
   Catanzaro
Germaneto, Italy

Kinam Park
Department of Pharmaceutics and Biomedical
    Engineering
Purdue University
West Lafayette, Indiana

F. Parra
Spanish National Research Council
Institute of Polymer Science and Technology
Madrid, Spain

R. Rai
Department of Materials Science
    and Engineering
University of Erlangen-Nuremberg
Erlangen, Germany

Eva Roblegg
Department of Pharmaceutical Technology
Institute of Pharmaceutical Sciences
University of Graz
Graz, Austria

G. Rodríguez
Spanish National Research Council
Institute of Polymer Science and Technology
Madrid, Spain

J.A. Roether
Department of Materials Science
    and Engineering,
University of Erlangen-Nuremberg
Erlangen, Germany

L. Rojo
Department of Materials and Institute
    of Bioengineering
Imperial College London
London, United Kingdom

J. San Román
Spanish National Research Council
Institute of Polymer Science and Technology
Madrid, Spain

Laura Russo
Department of Biotechnology
    and Biosciences
University of Milano–Bicocca
Milan, Italy

Kengo Sakurai
Department of Reparative Materials
Institute for Frontier Medical Sciences
Kyoto University
Kyoto, Japan

Roberto De Santis
National Research Council
Institute of Composite and Biomedical
    Materials
Naples, Italy

Radoslav Savic
Department of Genetics and Genomic
    Sciences
Mount Sinai School of Medicine
New York, New York

Nasrin Shaikh
Department of Biotechnology
    and Biosciences
University of Milano–Bicocca
Milan, Italy

Kosuke Shimizu
Department of Medical Biochemistry
Graduate School of Pharmaceutical
    Sciences
University of Shizuoka
Shizuoka, Japan

Naohiro Takemoto
Department of Reparative Materials
Institute for Frontier Medical Sciences
Kyoto University
Kyoto, Japan

Yuji Teramura
Radioisotope Research Center
Kyoto University
Kyoto, Japan

Vladimir P. Torchilin
Center for Pharmaceutical Biotechnology
    and Nanomedicine
Northeastern University
Boston, Massachusetts

J.A. van der Vliet
Department of Surgery
Nijmegen Centre for Molecular Life
    Sciences
Radboud University Nijmegen
    Medical Centre
Nijmegen, the Netherlands

T.H. van Kuppevelt
Department of Biochemistry
Nijmegen Centre for Molecular
    Life Sciences
Radboud University Nijmegen
    Medical Centre
Nijmegen, the Netherlands

Regis R. Vollmer
Department of Pharmaceutical Sciences
School of Pharmacy
University of Pittsburgh
Pittsburgh, Pennsylvania

Margherita Vono
Department of Health Sciences
University "Magna Græcia" of
    Catanzaro
Germaneto, Italy

Ernst Wagner
Department of Pharmacy
Center for System-Based Drug Research
and
Center for Nanoscience
Ludwig-Maximilians University
Munich, Germany

Brent D. Weinberg
Department of Radiology
University of Texas Southwestern Medical
    Center at Dallas
Dallas, Texas

Robert O. Williams III
College of Pharmacy
The University of Texas at Austin
Austin, Texas

R.G. Wismans
Department of Biochemistry
Nijmegen Centre for Molecular Life
    Sciences
Radboud University Nijmegen Medical
    Centre
Nijmegen, the Netherlands

A. David Woolfson
School of Pharmacy
Medical Biology Centre
Queen's University Belfast
Belfast, United Kingdom

Zhi-Kang Xu
Ministry of Education
Key Laboratory of Macromolecular Synthesis
    and Functionalization
Department of Polymer Science and
    Engineering
Zhejiang University
Hangzhou, Zhejiang, People's Republic of
    China

Fernando Yañez-Gomez
Faculty of Pharmacy
Department of Pharmacy and Pharmaceutical
    Technology
University of Santiago de
    Compostela
Santiago de Compostela, Spain

Jinzi Zheng
Leslie Dan Faculty
    of Pharmacy
University of Toronto
Toronto, Ontario, Canada

Andreas Zimmer
Department of Pharmaceutical
    Technology
Institute of Pharmaceutical Sciences
University of Graz
Graz, Austria

1 Antithrombin–Heparin Complexes

Leslie R. Berry and Anthony K.C. Chan

CONTENTS

1.1 INTRODUCTION

Antithrombin is a serine protease inhibitor (serpin) that inhibits many plasma proteases. In particular, antithrombin functions as one of the major natural anticoagulants by irreversibly inhibiting enzymes formed during activation of the coagulation cascade in vivo [1]. Antithrombin can form serpin–protease inhibitor complexes with activated factor (F) XII (FXIIa), FXIa, FIXa, FXa, and thrombin [1–3]. Within this group of coagulant proteases, antithrombin has the fastest rate of reaction with thrombin, its preferred reactant [4].

Systemic generation of thrombin in vivo is dependent on a number of factors. These include plasma concentrations of pro- and anticoagulants, the presence of cell surface–associated stimulators (such as phospholipid, tissue factor) and inhibitors (thrombomodulin, tissue factor pathway inhibitor), and interactions with molecules in subendothelial and extravascular spaces [5]. Ultimately, activation or inhibition of proenzymes or activated factors within the coagulation cascade affects generation of thrombin from prothrombin, which, in turn, affects conversion of fibrinogen to fibrin monomer that polymerizes to form a fibrin clot (see Figure 1.1). Thrombin is the pivotal enzyme in the coagulation pathway [6]. Once the initial amounts of thrombin are generated, thrombin causes

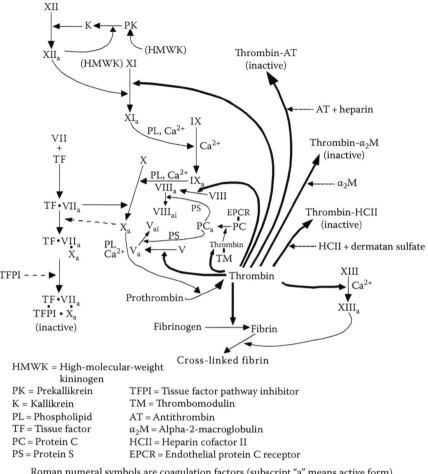

FIGURE 1.1   Plasma coagulation cascade.

feedback activation of its own formation by proteolytic cleavage of FV, FVIII, and FXI (Figure 1.1) to produce FVa, FVIIIa, and FXIa, respectively. Also, thrombin reaction with FXIII leads to the covalent cross-linking of fibrin monomers by FXIIIa. The resultant cross-linked fibrin polymer has more structural stability and integrity within the site of injury [7]. Additionally, thrombin has other procoagulant functions within the vasculature. Thrombin activates platelets, thus facilitating their cell–cell or cell–clot interactions [8], and causes inhibition of fibrinolysis by reaction with thrombin activatable fibrinolysis inhibitor [9–13]. Alternatively, thrombin bound to endothelial cell–associated thrombomodulin activates protein C, which (in association with protein S) then inactivates FVIIIa and FVa, thus limiting thrombin production [14]. In summary, inhibition of thrombin is a critical step in the regulation of coagulant activities in vivo.

Reaction of thrombin with antithrombin can be significantly accelerated by the action of heparin and heparan sulfate glycosaminoglycans (GAGs). In fact, the rate inhibition of thrombin by antithrombin is increased 1000-fold in the presence of native unfractionated heparin (UFH) [15]. The increase in inhibition rate is due to two reasons. First, UFH molecules bind to antithrombin via a specific GAG sequence [16]. The binding of UFH causes an allosteric change in antithrombin that results in a conformation which is more reactive with thrombin. Second, UFH can also bind to thrombin, which allows for a combination of the serpin and protease in a tertiary complex. In effect, UFH can bridge both antithrombin and thrombin. Once an irreversible thrombin–antithrombin (TAT) complex forms, affinity of the antithrombin moiety for UFH decreases, which allows for the GAG to dissociate and repeat the cycle of tertiary complex formation [17]. By increasing the rate of thrombin's reaction with antithrombin, UFH facilitates the inhibition of thrombin formed shortly after activation of coagulation, which in turn decreases generation of thrombin due to feedback activation of the cascade.

Unfortunately, UFH has a number of limitations that are related to its pharmacokinetic and biophysical properties. UFH has a short, dose-dependent intravenous half-life [18], partly due to basic and cell surface proteins that compete with non-covalent binding to antithrombin [19]. The concentrations and distribution of the UFH-binding proteins vary widely between individuals, which results in UFH having an unpredictable anticoagulant effect in vivo [20]. In addition, UFH can readily pass through tissue layers and is lost from the circulation due to its small size, which prevents its sequestration within various vascular spaces [21]. One of the biophysical limitations of UFH relates to its inability to inhibit a number of coagulation factors bound to surfaces. The antithrombin–UFH complex is ineffective at inactivating thrombin bound to fibrin [22] and factor Xa bound to phospholipid [23,24]. Clot propagation is due, in part, to the activity of this clot-bound thrombin [23,24]. The early recurrence of unstable coronary artery syndromes after discontinuation of heparin is likely due to this mechanism [25]. Finally, use of high in vivo concentrations of heparin results in bleeding complications. Previously, it has been shown that UFH anti-factor Xa activity is directly proportional to bleeding time [26]. Attempts to anticoagulate vascular devices by coating surfaces with heparin derivatives have also led to problems. Leaching of weakly attached heparin, resulting in undesired systemic anticoagulation and reduced activity of the modified heparin bound to the surface, is one example of the difficulties involved with heparin coating of biomaterials. Furthermore, approximately only one-third of starting commercial UFH preparations have the anticoagulant high affinity binding sites for antithrombin [27]. Therefore, biomaterials coated with heparin from mixtures derived from UFH would have the majority of their surface area covered by heparin with no antithrombin activity.

In order to address some of the problems associated with the clinical use of heparin, covalent complexes of antithrombin and heparin derivatives have been prepared. The rationale for construction of covalent antithrombin–heparin (ATH) is multifold. First, if heparin was irreversibly bonded to antithrombin, the serpin molecule should be fixed permanently in the active conformation. Second, since the heparin component of ATH cannot dissociate from antithrombin, the intravascular pharmacokinetics would tend to have a longer half-life and anticoagulant effect. Another aspect related to the transport and metabolism of ATH is that of endogenous protein binding in vivo. Because the antithrombin in ATH will interact non-covalently with a significant portion of the heparin moiety (particularly in the case of ATH molecules which contain relatively short heparin chains), binding of plasma or cell surface proteins (or other molecules) with the heparin in ATH should be reduced compared to free heparin. A decreased heparin binding by intra- and extravascular proteins might lead to a more consistent clearance pattern and, consequently, a more predictable anticoagulant response. Furthermore, if ATH binding to cell surface receptors is decreased compared to uncomplexed heparin, risk of hemorrhagic side effects may be reduced. Third, both the technology and resultant polymeric surfaces coated with ATH may be significantly improved compared to that for biomaterials coated with heparin. Attachment of heparin onto a variety of polymers would be facilitated by linking antithrombin to the GAG. Immobilization of ATH could be more readily optimized due to the increased range of functional groups and chemistries found in the amino acid R-groups of the antithrombin. Also, surface linkage of ATH through the antithrombin moiety may allow for a high number of attached molecules in which the heparin chain is oriented outward from the surface into the fluid phase. Again, since all antithrombin species in ATH should be permanently activated, ATH bound to surfaces of biomedical devices may all be active, whereas only a maximum of one-third of GAGs on UFH-coated surfaces would possess anticoagulant activity. Given the potential number of desirable properties of ATH products compared to other UFH derivatives, methods for covalent complexation of antithrombin and heparin and analysis of the covalent conjugates have been investigated.

In this chapter, the development of permanently linked ATH complexes will be assessed. The format for this chapter will be as follows. Previous work on the structure–function relationships and clinical use of antithrombin will be discussed. Heparin's structure, activities, in vivo occurrence, and biochemistry will be covered in light of the long history of medical application. Analysis of the research on antithrombin and heparin will be followed by a broad introduction to the production of covalent ATH complexes. The overview of ATH complexes will include assessment of clinical advantages for conjugation of antithrombin and heparin, novel applications of ATH products, and an introduction to the range of ATH conjugates that are available. A detailed analysis follows for the various types of ATH that have been reported. Each ATH complex will be reviewed according to its synthetic chemistry, physicochemical properties, effect of conjugation on anticoagulant activities, performance in animal models, and particular clinical advantages (including possible use for anticoagulating materials that come into contact with blood). Discussion of the various ATH compounds will appear in the chronological order in which they have been reported. Finally, possible future directions for research on covalent serpin–GAG complexes will be presented.

1.2 ANTITHROMBIN

1.2.1 ANTITHROMBIN CHEMICAL STRUCTURE

Antithrombin is a glycoprotein whose polypeptide moiety shares structural and functional homology with members of a large family of serpins [28–30]. The degree of homology across the various proteins in this family is ~30%. In addition to primary sequence similarities, the serpins have a number of their tertiary structural features in common [31].

Antithrombin is produced within hepatocytes of the liver in mammalians. The gene for antithrombin resides entirely within the long arm of chromosome 1 [32] within a region that extends over an unbroken stretch of ~19 kb, including seven exons and six introns [33,34]. The open reading frame for human antithrombin contains 1396 nucleotides. At the 5′ end of the reading frame there is a section of 96 nucleotides which code for a 32 amino acid segment called the signal peptide [34]. This N-terminal peptide is removed prior to release from the cell. Within mammalian species, 10% to 15% variation in primary sequence has been reported but high conservation has been observed in the critical reactive site regions [35].

The antithrombin molecule is a single-chain plasma glycoprotein with an approximate molecular mass of 60,000 Da [36]. The polypeptide chain in human antithrombin is composed of 432 amino acid residues [37]. Upon folding into its native configuration, three disulfide bonds are formed between three pairs of cysteine residues in the polypeptide chain [37]. Two of the disulfide bonds join a relatively unstructured stretch of 45 amino acid residues at the N-terminus to the third and fourth α-helices of the molecule. Overall, the protein has a neutral to basic pI due to a preponderance of arginyl and lysyl residues, as opposed to the acidic residues. These positively charged amino acids contribute to the UFH-binding regions on the serpin molecule [31,38]. Analysis of the tertiary structure, as determined by peptide modeling of the primary amino acid sequence, has shown that human antithrombin has 31% α-helix, 16% β-sheet, 9% β-turn, and 44% random coil [39,40]. Antithrombin has been crystallized recently and the x-ray crystal structure determined to 3 Å resolution [41]. Conformational studies of the topographic structure of antithrombin using x-ray diffraction show that the serpin can exist in two forms: one active and the other inactive [42]. Results showed that the active antithrombin consists of nine α-helices and three β-sheets. A composite model for the three-dimensional structure of antithrombin, derived from various x-ray diffraction and chemical analyses that have been reported, is shown in Figure 1.2.

N-Linked glycosylation occurs at specific asparagine residues that are flanked by sequences in β-sheet regions that are recognized by a glycosyl transferase during posttranslational modification in vivo. The number of glycan chains that are present on the native antithrombin polypeptide varies from three [43] to four [44], although up to five glycan chains have been reported [45].

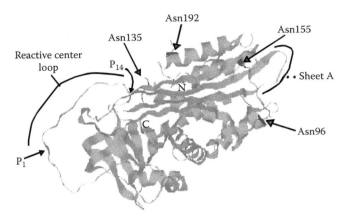

FIGURE 1.2 Antithrombin tertiary structure.

Isoforms of plasma antithrombin, occurring as the four and three glycan chain–containing molecules, have been designated as α- and β-antithrombins, respectively. The fully glycosylated α-antithrombin has glycan chains linked through N-amido glycosidic bonds to the R-groups of asparagine residue numbers 96, 135, 155, and 192 [46]. In the case of β-antithrombin, the glycan at position 135 is missing.

Posttranslational modification of antithrombin occurs in the Golgi apparatus by transfer of tetra-antennary, branched high mannose glycan structures from dolichol phosphate onto the amino acid R-group. These high mannose glycans are further processed by endoglycosidases to a core chitobiose–trimannose stub, which is the substrate for addition of further monosaccharide residues to produce the final "complex-type" glycan structures. In addition to the major heterogeneity of glycan number described earlier, variations in structure within the glycans themselves (termed microheterogeneity) are observed. Complex N-linked glycans have been shown to occur in a number of forms that vary according to the number of branches extending out from the core region of the chain. Antithrombin molecules have been shown to exist as a group of populations in which the branching within glycans at the different glycosylation sites within the same polypeptide chain can range from mono- to bi-, tri-, and tetra-antennary forms [47,48]. The glycans of antithrombin have been shown to vary in level of *N*-acetylneuraminic acid residues present at the termini of the glycan chains [49]. Additional groups may be added to the glycan structures in a heterogeneous fashion. For example, although fucosyl residues are present on the glycans at positions 96 and 192, a portion of glycans at position 155 was not fucosylated [47]. Recently, transgenic animals (goats) have been prepared that contain the human antithrombin gene [50]. Interestingly, the antithrombin expressed in those animals has reduced glycan number and retains high mannose structures [50,51]. Variations due to heterogeneity in antithrombin's carbohydrate moieties have also been identified as having significant effects on its functional activities, as will be outlined later.

1.2.2 ANTITHROMBIN FUNCTIONAL BIOCHEMISTRY

Antithrombin, as well as most of the other serpins, inhibits serine proteases (such as thrombin) by a stress-release mechanism. Within the antithrombin molecule, a reactive center loop exists that is toward the C-terminus of the polypeptide chain. Thrombin interacts non-covalently with antithrombin through the recognition of certain amino acids close to the reactive center. The thrombin-binding region has been mapped out by the natural occurrence of amino acid substitution mutations in the protein, which give rise to expression of molecules with reduced antithrombin activity. The mutations Ala382 → Thr382 (antithrombin Hamilton) [52] and Pro407 → Leu407 (antithrombin Utah) [33] have indicated that the thrombin-binding region extends to at least include amino acid residues 382 and 407. Alternatively, investigations using thrombin mutants have illuminated structural components of the enzyme which are involved in the non-covalent interaction. Gly226 in the thrombin polypeptide has been shown to protrude into the antithrombin specificity pocket, since substitution of a valine at this position removed inhibitory activity [53]. Molecular modeling indicated that the larger valine side chain would not allow enough space for the antithrombin P_1 Arg393 R-group. Once thrombin binding through weak interactions occurs, a more stable covalent bond is formed. Members of the serpin family have specific peptide bonds in the reactive center which are susceptible to enzymatic cleavage by the target protease. Two amino acids make up the reactive center peptide bond and the peptide is designated as $P_1 - P_1'$. The P_1 amino acid residue in antithrombin (Arg393) provides the inhibitor with selectivity toward thrombin and other coagulation factors [30]. Other residues to the N-terminal side of the P_1 Arg393 give further enzyme selectivity and structural elements, such as in the case of the P_2 Gly392, which is preferred for reaction with FXa and is important for preventing release of FXa from the inhibitor complex [54]. The role of the P_1' in antithrombin (Ser394) is not entirely clear.

Thrombin initially treats antithrombin as a substrate by attacking the inhibitor's Arg393 C-terminal amide carbonyl via the protease active serine residue. Thus, thrombin forms a covalent ester bond between the hydroxyl oxygen of the active serine and the Arg393 carboxyl carbonyl group. The active, non-cleaved antithrombin exists in an S-configuration where the two amino acids of the reactive center are in the middle of a stressed loop [55]. Within antithrombin, the reactive center loop connects strand 4 of β-sheet A (central sheet in the serpin ordered numbering system) to strand 1 of β-sheet C [56]. In this stressed configuration, the reactive center loop of antithrombin is under some tension and the resultant surface topology is such that the reactive center is readily available for reaction with thrombin. Once the Arg393–Ser394 has been cleaved, leaving thrombin and antithrombin remaining linked by a covalent ester bond, a radical conformational change occurs leading to a relaxed or R-configuration [55]. Several studies of TAT inhibitor complexes and post-complex cleaved antithrombin have been done, which have helped elucidate the structural properties of the relaxed configuration of the antithrombin within the complex. Antibodies have been produced that do not bind to native antithrombin but recognize either the consumed inhibitor or the inhibitor when it is bound to a tetradecapeptide corresponding to the P_{14} to P_1 residues of the reactive center loop [56–63]. These experiments, in conjunction with x-ray studies, strongly suggested that upon reaction of antithrombin with thrombin, the reactive center loop becomes embedded into the β-sheet A, which exposes new epitopes that are not present at the surface of the intact inhibitor. Model-building studies based on crystal structures and biochemical analyses of protease complexes with normal and mutant antithrombins have suggested that the reactive center loop is inserted into β-sheet A as far as P_{12}, strand s1C is absent from the β-sheet C, and the conformation of the C-terminus has changed so that it interacts with thrombin [64]. In fact, an intact non-inhibitory (latent) conformer of antithrombin exists with a structure where the P_{14} to P_3 residues of the reactive center are completely inserted into β-sheet A [65]. However, experiments with an antibody, that did not bind to native, latent, or reactive center, cleaved antithrombins, recognized antithrombin neoepitopes in either stable TAT complexes or antithrombin complexed to a synthetic peptide corresponding to the P_{14} to P_9 sequence. Therefore, only part of the reactive center may be imbedded in the antithrombin β-sheet A within the inhibitor complex [66]. Regardless, insertion of the reactive center loop positions the thrombin reactive site close to or within the β-sheet A pocket. Thus, the thrombin moiety becomes irreversibly trapped as a reaction intermediate covalently linked to antithrombin [67].

Interaction of antithrombin with heparin and heparan sulfate GAGs accelerates the thrombin inhibition reaction. Activation of antithrombin by heparin binding has been verified by the concomitant increase in protein intrinsic fluorescence, as a result of a conformational change. Research involving antithrombin Trp → Phe mutants has shown that Trp225 and Trp307 each accounts for ~37% of the heparin-induced fluorescence enhancement [68]. A spectral shift in Trp49 toward the blue suggests partial burial due to contact with heparin, while a red shift for Trp225 (along with fluorescence enhancement) indicates the increased access to the solvent due to heparin-induced movement of contact residue Ser380. The main heparin-binding region on anti-thrombin has been shown (according to mutant studies) [69–72] to reside toward the N-terminus of the molecule [31,73]. A second heparin-binding region on antithrombin was proposed to exist at residues from positions 107 to 156 [74]. Ultimately, high resolution of the heparin-binding site was achieved by alanine scanning mutagenesis in a baculovirus expression system, which gives a normal product that is highly similar to plasma β-antithrombin [75]. Mutant antithrombins were screened for heparin affinity by gradient elution from heparin columns. It was determined that only a subset of residues (Lys11, Arg14, Arg24, Arg47, Lys125, Arg129, and Arg145), which line in a 5 nm groove along antithrombin's surface, are critical for heparin binding [75]. A few other residues have been proposed to be significant points for heparin binding [65,67]. Molecular modeling has suggested that interaction of heparin with these amino acid R-groups would induce the breakage of salt bridges between α-helix D and β-sheet B, thus facilitating movement of s123AhDEF to give a species which is conformationally primed for reactive center loop uptake by β-sheet A [75]. X-ray studies have confirmed that binding of heparin pentasaccharide to certain of the key antithrombin residues gives an increased affinity between the GAG and antithrombin, which accompanies the change in conformation [65]. In addition, peptides representing P_{14}–P_3 are able to inhibit forma-tion of the conformation, which would be induced by heparin binding to antithrombin [76]. Upon covalent linkage to thrombin (cleavage), the antithrombin serpin (as part of an inhibitor complex) dissociates from the GAG [17,77]. Undoubtedly, conformational changes resulting from insertion of the reactive loop peptide into the β-sheet A of cleaved antithrombin disturb the heparin-binding site channel.

Variation in glycosylation on the antithrombin molecule has substantial effects on the ability of heparin to catalyze antithrombin's inhibition of thrombin. It has been shown previously that human β-antithrombin, which lacks the carbohydrate side chain at Asn135, has higher affinity for heparin than the fully glycosylated (four glycan) α-antithrombin [78]. Previously, glycoforms of antithrom-bin produced in either BHK or CHO cells that had a 10-fold difference in heparin-binding affinity [48,79] were found to vary in the glycosylation at Asn155 [80]. Rapid kinetic studies of heparin bind-ing to α- and β-antithrombins demonstrated that although the Asn135–glycan moderately interfered with the weak initial binding of heparin, the rate constant for the conformational change induced by heparin was significantly lower for α-antithrombin compared to β-antithrombin [78]. Thus, there is a higher energy required for inducing the activated conformation in α-antithrombin, leading to a decrease in heparin-binding affinity. Since heparin-like GAGs exist in vivo, the variation in bind-ing affinities by the antithrombin glycoforms may have physiological importance. In fact, evidence has been presented showing that β-antithrombin tends to be bound to vessel wall heparin/heparan sulfate while α-antithrombin resides more commonly in the plasma phase [43]. These data suggest specialized functions for the antithrombin glycoforms where β-antithrombin may be vital for con-trolling thrombogenic events arising from vessel wall injury, while α-antithrombin may be largely responsible for inhibition of fluid-phase thrombin.

1.3 HEPARIN

1.3.1 Heparin Chemical Structure

Heparin is a member of the GAG family of molecules that occur not only in mammalians but also in most multicellular, as well as in some single cell, organisms [81]. GAGs are straight-chain poly-saccharides, which are composed of repeating uronic acid–hexosamine disaccharide units [82]. Heparin and heparan sulfate GAGs contain glucosamine derivative residues, whereas dermatan sulfate and the other chondroitins contain galactosamine [82]. Regarding uronic acid content, hepa-rin and heparan sulfate contain both glucuronic and iduronic residues [82]. Saccharides in GAG chains are extensively modified during and after glycosidic polymerization. In the case of hepa-rin and heparan sulfate, glucosamine residues can be N-acetylated or N-sulfated. However, while

>80% of the glucosamines are N-sulfated in heparin, approximately equal amounts of N-sulfated and N-acetylated glucosamines have been detected in various sources of heparan sulfate [83]. Furthermore, whereas there are ≥2 O-sulfates per disaccharide unit in heparin [84], O-sulfation per disaccharide in heparan sulfate has been found to range from 0.2 to 0.75 [83]. Thus, from these and other structural observations, it has been concluded that heparin and heparan sulfate represent separate groups of N-sulfated GAGs.

Heparin, as well as other GAGs, is produced in mast cells by biosynthetic attachment to a protein core. Each core protein may contain as many as 10 heparin chains, of molecular weights ranging from 60,000 to 100,000, linked via O-glycosidic bonds to serine residues in a glycine–serine sequence repeat [85]. Heparin GAG is synthesized by glycosyl transferase addition of monosaccharide residues. An initial xylose–galactose–galactose–glucuronic acid linkage region sequence is built up, with the terminal xylose linked through serine to the polypeptide backbone [86]. Within the growing polysaccharide, uronosyl-β1→4-glucosaminosyl units are joined by α-1→4 bonds [87]. During polysaccharide chain formation, numerous functional group modifications are performed, in a somewhat concerted fashion, on the monosaccharide residues within the nascent oligosaccharide. Glycosyltransferases [88], N-acetylases [89], N-deacetylases [89], O-sulfotranferases [90], N-sulfotransferases [91], and glucuronosyl-5-epimerases [90] have been isolated and characterized, which are involved in anabolism toward the final heparin product in vivo. Segments of glucuronosyl–glucosaminosyl oligosaccharide chain are initially produced, which are partially N-unsubstituted, N-acetylated, or N-sulfated [92]. Following (and during) [92] this stage, the growing chain is acted upon by a combination of a N-deacetylase and a N-sulfotransferase to yield a much greater N-sulfate-containing stretch [89,92]. Shortly after N-sulfation, a glucuronosyl-C5-epimerase causes conversion of the chain-end glucuronosyl residue to an iduronosyl residue [90]. However, the epimerase reaction is one in which an equilibrium exists between glucuronosyl and iduronosyl residues, with the net equilibrium being toward the iduronosyl form. If O-sulfation occurs at either the C6 of the glucosaminosyl or C2 of the iduronosyl residue in question, conversion of the iduronosyl back to a glucuronosyl does not occur [90]. Chronologically, C2-O-sulfation of iduronosyl residues is carried out in the absence of C6-O-sulfation of neighboring glucosaminosyl residues, while C6-O-sulfation of glucosaminosyls occurs readily in the presence of 2-O-sulfates on adjacent iduronic acid units [93]. However, if further glycosyl transfer occurs before C2-O-sulfation of the terminal uronic acid in the nascent chain, then uronic acid remains non-sulfated throughout subsequent modification reactions [93]. Analyses of heparin samples have shown that ~78% of uronic acids are in the form of iduronic acid [87], of which ~75% are 2-O-sulfated [87,94]. No significant amount of 2-O-sulfated glucuronic acid residues has been detected [87]. Termination of the biosynthesis of each polysaccharide on the core protein varies, which yields (as mentioned earlier) side chains ranging in length (molecular weight). Chain size and monosaccharide modifications are not directed by transcriptional expression but are affected by factors within the milieu, such as substrate availability and cell status [95,96].

Studies of heparin metabolism in mastocytoma cells have shown that the newly synthesized heparin proteoglycan chains are partially depolymerized by an endoglucuronidase and stored in cytoplasmic granules [97,98]. Due to this cellular processing, commercial heparin prepared from intestinal mucosa or lung mast cells are free GAG chains ranging in molecular weight from 5,000 to 30,000. Primary source commercial heparin, without isolation of any subpopulation, is called UFH (as discussed earlier). Within the last few decades, low-molecular-weight heparins (LMWHs) have been prepared via various methods [99]. Partial depolymerization of UFH has been carried out by limited treatments with HNO_2, heparinases, heparitinases, and base elimination following partial esterification of uronic acid carboxyls [99,100]. Molecular weights of LMWHs produced vary from 1,800 to 12,000 [99]. Apart from variation in chain size, LMWHs have a number of pharmacokinetic and biological properties, which separate them from their UFH parent compound.

1.3.2 HEPARIN FUNCTIONAL BIOCHEMISTRY

Heparin (both UFH and LMWH) provides two major functions in vivo. First, heparin (and other GAGs), mainly in the proteoglycan form, acts as an extracellular matrix component for structural organization and as a chemoattractant in tissue [101,102]. Second, via particular sequences, heparin can operate as anticoagulant. Heparin's anticoagulant activities are based on its ability to bind to either plasma heparin cofactor II or antithrombin and catalyze their inhibition of thrombin (heparin cofactor II or antithrombin) or other coagulation factors (antithrombin [see earlier]) [103].

Reaction of antithrombin with various coagulation factors is catalyzed by heparin in vivo [104]. Heparin's physiological anticoagulant activity resides particularly in facilitating antithrombin's inhibition of FXa and thrombin [105,106]. However, heparin's augmentation of thrombin inhibition by antithrombin has been shown to be a major basis for the clinical use of heparin (particularly UFH) [107]. Regulation of in vivo FXa and thrombin inhibition by heparin relies on the presence of a particular pentasaccharide sequence in the GAG molecule, which binds to antithrombin [108]. This pentasaccharide antithrombin-binding site has been shown to occur in the GAG chains of the proteoglycan form of heparin obtained from rat skin mast cells (which conserve intact proteoglycan) [109]. Analyses indicated that while most proteoglycans contain no heparin chains with antithrombin-binding sites, a small proportion of proteoglycans had chains with heparin pentasaccharide sequences numbering from 1 to 5 (average of 3) per polysaccharide unit [110]. In the case of commercial UFH, on the average, only about one-third of the molecules have been found to have high affinity antithrombin binding [111]. Interestingly, though, some molecules of commercial UFH have been shown to contain two high affinity antithrombin-binding sites per molecule [112]. It has been shown that the pentasccharide antithrombin-binding sequence occurs somewhat randomly along the chain in UFH molecules [113]. However, reports have indicated that there may be a bias for the antithrombin-binding sites to be located toward the non-aldose half of UFH chains [114,115].

Previously, it was reported that an unusual 3-O-sulfate group was present on an internal glucosamine group within the pentasaccharide sequence [116]. Further work has indicated that the 3-O-sulfated glucosamine residue was critical to the high affinity antithrombin binding and anti-FXa activity [117]. Final characterization and chemical synthesis of the complete pentasaccharide sequence has been accomplished [118] and is shown in Figure 1.3. Important elements of the structure are the appearance of non-sulfated glucuronic acid and 3-O-sulfated glucosamine at positions 2 and 3 from the non-aldose terminus of the sequence. Binding of heparin to antithrombin, through the pentasaccharide, involves a two-stage mechanism, which corresponds to the structural activation events which occur in the serpin. Evidence for the mechanism of activation of antithrombin by heparin pentasaccharide has come from experiments involving binding of various mono-, di-, tri-, and tetrasaccharide derivatives to antithrombin [119–121]. Initially, residues 1, 2, and 3 (Figure 1.3) of the heparin pentasaccharide bind to antithrombin via charge and hydrogen-bond interactions. Binding of the first three residues from the non-aldose end of the pentasaccharide is relatively weak but induces a conformational change in the antithrombin, which is very similar to the heparin activated form [119]. The 2-O-sulfated iduronic acid (residue 4 in Figure 1.3) has been shown to have a flexible capability to convert from a chair to skew boat conformation [121]. Upon binding of pentasaccharide residues 1–3 to antithrombin, movement of

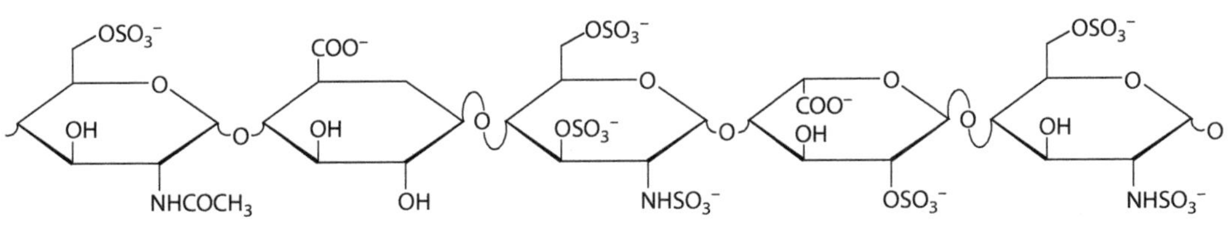

FIGURE 1.3   Heparin pentasaccharide high affinity antithrombin-binding sequence.

the polypeptide due to the trisaccharide-induced conformational change causes amino acid R-groups, particularly Arg47 [122], to come into contact with the remaining pentasaccharide residues **4** and **5**. The skew boat conformation of the 2-O-sulfated iduronic acid also allows the 3-O-sulfate on the central glucosamine and a putative charge cluster on pentasaccharide residues **4** and **5** to interact with the polypeptide after antithrombin's change in conformation [121]. This binding of charge groups from pentasaccharide residues **4** and **5** gives the high affinity binding with antithrombin, which locks the GAG and serpin in place [120,121]. No additional conformational activation of antithrombin occurs due to the locking interactions with residues **4** and **5** of the pentasaccharide [122].

It is now well understood that although activation of antithrombin by heparin pentasaccharide binding is fairly sufficient to accelerate inhibition of FXa, a larger stretch of heparin chain is required for maximal inhibition of thrombin [108]. For heparin catalysis of antithrombin's reaction with thrombin, both the serpin and the enzyme must bind to the GAG. It has been determined that effective ternary complexes of antithrombin, thrombin, and heparin require a heparin pentasaccharide sequence to bind the inhibitor and a total chain length of 18–22 monosaccharide residues in order to also accommodate interaction with thrombin [123]. Heparin binding to thrombin involves an anion-binding exosite on the protease [124]. However, although significant negative heparin charge density is important [125], a specific binding sequence in heparin for thrombin has never been found. Since binding to both antithrombin and thrombin by heparin has a minimum chain length requirement, certain LMWH molecules would be unable to catalyze thrombin inhibition. This has been borne out by the fact that LMWH preparations have a lower antithrombin to anti-FXa activity ratio. The importance of heparin's ability to catalyze thrombin activity has been implicated by reduced in vivo antithrombotic activity of various LMWHs compared to UFH [105]. In fact, the lowest antithrombotic activity (as indicated by in vivo fibrin deposition) has been found with the pentasaccharide [126]. These findings would appear to suggest a reduced clinical usefulness for LMWHs compared to UFH. However, the intravenous half-life of LMWHs is significantly longer than UFH [127]. The difference in pharmacokinetics is due to two main reasons. First, non-specific protein binding by LMWH molecules is reduced compared to UFH [128]. Since LMWH has reduced affinity for other proteins in vivo, there is a reduction in pathways by which LMWH activity can be either pacified or removed from the circulation. Second, although UFH can be either metabolized by uptake into the liver or lost through the kidneys, plasma disappearance of LMWH occurs only via renal elimination [123]. This single phase elimination of LMWH is much slower and less variable than the concave–convex pattern seen with UFH. As a result of the reduction of pathways for LMWH's pharmacokinetics in vivo, LMWH produces a more predictable anticoagulant response than UFH, which gives a decreased risk of bleeding [129,130]. Nevertheless, since different preparations of LMWH exhibit differing amounts of activity against thrombin (possibly due to contaminating higher-molecular-weight chains, varying charge density, etc.) [131], a narrow window of dosage and regimen exists for their application, which must be evaluated for each type of LMWH [100]. Furthermore, unlike UFH, LMWHs cannot be completely neutralized by protamine (administered to prevent bleeding if plasma levels become too high during treatment) [132]. Thus, development of a heparinoid with the thrombin inhibitory potency of UFH and the increased half-life of LMWH would be desirable.

1.4 COVALENT ATH COMPLEXES: OVERVIEW

1.4.1 LIMITATIONS OF CURRENTLY AVAILABLE ANTICOAGULANTS

Limitations in the control of thrombin by heparin administration, as well as major adverse side effects induced by heparinization, have led to the development of new anticoagulants for clinical use. Deficiencies in heparin's efficacy for thrombin regulation stem from either loss of heparin activity within the plasma compartment or the inability of heparin to accelerate inhibition of thrombin bound to different surfaces. Pharmacokinetic loss of UFH is due to binding and uptake of heparin by cells, such as hepatocytes of the liver, or by glomerular filtration through the kidneys [123]. In both elimination mechanisms, removal of UFH from the circulation requires the dissociation of UFH from antithrombin. That antithrombin–UFH complex breaks up prior to UFH elimination in vivo is borne out by the fact that non-covalent mixtures of UFH and antithrombin have significantly different intravenous half-lives when injected as a bolus [111]. It has been determined that the intravenous half-life of UFH in humans is dose-dependent and ranges from 0.3 to 1 h [123]. UFH's short half-life makes it necessary to administer UFH either by intravenous infusion (which assures a constant delivery of UFH) or by subcutaneous injection (which provides a depot of heparin for slow release into the intravascular space) [133]. Subcutaneous UFH injection gives peak plasma concentrations at 4 h [134]. LMWHs have been shown to have longer intravenous half-lives than UFH [135]. Interestingly, LMWH administered subcutaneously in humans at therapeutic doses gave peak plasma levels by 3 h with plasma activity being undetectable after 12 h [129]. Apart from disappearance of heparin from the plasma compartment, heparin's activity can be altered due to binding to other fluid phase or cell surface molecules within the lumen. It has been shown that plasma proteins [136,137], platelets [138], and endothelium [139] can bind UFH. When bound to plasma proteins, UFH exhibits reduced activity [136], which can be regained following displacement from these basic heparin-binding proteins [140]. However, dissociation of UFH from non-anticoagulant plasma proteins (part of the heparin rebound effect) results in variation of heparin activity levels and the regimen used to stop treatment by administration of protamine [140]. As stated earlier, heparin has reduced efficacy at inhibiting fibrin-bound thrombin [22]. In fact, at therapeutic levels, heparin promotes the binding of thrombin to fibrin polymer [141], which, in turn, protects thrombin from inactivation by antithrombin–UFH non-covalent complexes [22]. The mechanism by which fibrin accretion of thrombin protects the protease from inactivation by antithrombin + heparin has been investigated. Fibrin, thrombin, and UFH interact to form a ternary complex [142]. Once in this ternary complex, the action of thrombin against its substrates is altered [143], which causes decreased reaction of the protease with incoming antithrombin-heparin [22]. Although LMWH has been found to have reduced nonselective plasma protein binding [144,145], both UFH and LMWH are unable to inactivate clot-bound thrombin [146]. Numerous adverse side effects have been associated with heparin administration. Approximately 3% of adults receiving heparin develop heparin-induced thrombocytopenia (a condition in which antibodies are developed by the patient against platelet-bound heparin molecules) [147]. Review of heparin use has indicated that the most significant problem involves risk of bleeding [148]. Thus, careful monitoring is required to ensure that UFH's anticoagulant activities remain within the therapeutic range to minimize the risk of recurrent disease while preventing hemorrhagic complications [149]. Since heparin is rapidly cleared from the circulation, constant infusion or frequent subcutaneous injections are required to control the efficacy/bleeding ratio [150]. From data in acute coronary syndrome patients, it has been suggested that LMWH may have a lower risk of heparin-induced thrombocytopenia compared to UFH [148]. However, since LMWHs cross-react with 80% of antibodies generated during UFH exposure, LMWH is not recommended for treatment of established heparin-induced thrombocytopenia [148]. Furthermore, in clinical practice, there is no evidence that bleeding complications are reduced with LMWHs compared to UFH [151,152].

Non-heparin-related anticoagulants have been developed to improve on clinical therapy and to overcome problems with treatments involving heparin. Examples of new and effective anticoagulants include LMWHs (as discussed earlier), hirudin, hirulog, and Phe-Pro-Arg-chloromethyl ketone (PPACK) [153–156]. Hirudin is a small (7000 MW) direct, reversible, thrombin inhibitor produced by the leech [157] and hirulog is the C-terminal portion (residues 53–65) of hirudin covalently linked to a short peptide active site inhibitor [158]. Investigations have determined that hirudin can inhibit clot-bound thrombin since clot growth did not occur even long after hirudin had been cleared from the circulation [159]. However, the intravenous half-lives of hirudin (β-phase ≤1 h, terminal half-life = 2.8 h) [160] and hirulog (36 min) [161] in humans are not significantly longer than that of UFH. Furthermore, it has been shown that persistent formation of thrombin occurs during declining plasma levels of hirudin after administering potent dosages, suggesting that thrombin generation is not blocked [162]. PPACK is a modified tripeptide substrate that reacts selectively and rapidly with thrombin. Again, PPACK is very effective at irreversibly inhibiting clot-bound thrombin, as has been demonstrated in animal models [163]. Given its small size, it is not surprising to note that PPACK's half-life is extremely short [164], which has led to bleeding risk at therapeutic treatment doses [165]. Thus, potency/bioavailability/bleeding issues that are concerns for heparin use, have not been ameliorated by development of alternative anticoagulant agents.

1.4.2 Potential Advantages of Covalent ATH Complexes

To address the aforementioned limitations of UFH, LMWH, and other anticoagulants, researchers have studied the possibility of stabilizing the interaction of antithrombin with heparin by covalent bonds. The rationale for producing a covalent ATH conjugate was several fold. First, if ATH molecules could be produced in which the antithrombin in the complex was able to interact with a pentasaccharide on the heparin component of the same complex, the ATH antithrombin would be activated for rapid reaction with thrombin. Reaction of ATH with thrombin should be faster than non-covalent mixtures of heparin + saturating amounts of antithrombin because the rate-determining step of antithrombin + heparin binding [166] would be eliminated. Second, if antithrombin cannot dissociate from the heparin, the antithrombin would be permanently activated. Thus, wherever and whenever ATH appears at locations in the body, the antithrombin would always be in the conformation of a very potent anticoagulant toward thrombin, FXa, and other coagulant molecules of the cascade. Third, depending on the methodology used, covalent ATH complexes may be prepared in which a selection takes place (prior to covalent linkage) for only the heparin molecules which have pentasaccharide sequences. That is, a mechanism for ATH synthesis may be possible in which antithrombin first binds heparin ionically, through the high affinity sites on the GAG, followed by covalent bonding. In this way, the ATH formed would be a preparation in which all of the molecules would contain serpin species activated by heparin pentasaccharide sites. The heparin in ATH produced by pre-selection of the heparin component by the antithrombin component would have GAG that is a much more potent anticoagulant than the starting heparin, since only approximately one-third of commercial UFH preparations (and even fewer for LMWH) have pentasaccharide-containing molecules [111]. Fourth, ATH may have the capability to inhibit fibrin-bound thrombin. Since the antithrombin and heparin in ATH do not dissociate, ternary complexes of fibrin/heparin/thrombin cannot form. Given that the antithrombin must remain attached to the heparin chain, it is expected that the heparin moiety in ATH will be less likely to also accommodate fibrin and thrombin compared to free heparin. Fifth, if antithrombin was permanently linked to heparin, the heparin moieties would not dissociate and be lost from the circulation through a renal mechanism. Given the increased size of an ATH complex, glomerular filtration of such a compound should be prohibited, or at least vastly reduced compared to the much smaller low-molecular-weight species in polydisperse UFH (even more so for LMWH preparations). Sixth, since the antithrombin in ATH may cover (by non-covalent interaction or steric hindrance) a significant portion of the heparin chain in the complex, binding of the ATH GAG moiety by other plasma and cell surface proteins could be reduced. Seventh, ATH may not promote osteoporosis as readily as UFH. It has been shown previously that LMWH causes less osteopenia than UFH because it only decreases the rate of bone formation and does not increase bone resorption [167]. ATH, again, would only have a portion of its heparin component available (free from the bound antithrombin) and, therefore, may interact poorly with osteoclasts to effect bone loss. Eighth, for the same reasons as those given for increased inhibition of fibrin-bound thrombin, ATH molecules may induce less hemorrhagic side effects compared to either UFH or LMWH. Platelet binding by ATH may be reduced compared to free heparin since the covalently linked antithrombin may not allow heparin to either bridge receptors on the membrane or interact with single protein molecules via the required geometry. Finally, ATH may combine many of the positive attributes exhibited by either UFH or LMWH, without a number of their deficiencies. For example, if ATH is prepared from UFH, the product should have high activity against thrombin (unlike LMWH) but have an increased intravenous half-life (decreased cell surface protein binding compared to UFH). In effect, ATH may be the optimum heparinoid.

1.4.3 Potential Uses of Covalent ATH Complexes

ATH has characteristics which make it attractive for use in a range of clinical indications. The rapid inhibition of thrombin, which should occur if antithrombin is permanently activated by covalently linked heparin, would be highly advantageous for anticoagulant prophylaxis. The low levels of thrombin, which need to be inactivated during ongoing prophylactic treatment, would be quickly complexed by ATH before feedback activation of FV, FVIII, or FXI can occur. Also, since ATH is likely to have a prolonged half-life compared to free heparin, it may be possible to give a single intravenous bolus injection of the conjugate to achieve safe protection against thrombosis. Monitoring of the ATH may be less critical if ATH has a reduced bleeding risk profile. Another possible application of ATH is administration of the complex as an antithrombotic treatment. As suggested earlier, although UFH and, to a lesser extent, LMWH have reduced activity against clot-bound thrombin, ATH may have good reactivity with thrombin associated with fibrin clots. Thus, in addition to an increased half-life for in vivo inhibition of thrombin generation, ATH may be able to pacify the procoagulant activity on thrombin. Rapid inhibition of clot-based procoagulant activity is critical to the prevention of myocardial infarction and stroke, and ATH may be a treatment which could reduce this risk in two ways. By inactivating thrombin on the surface of the polymerized fibrin, clot extension should not occur since there will be no further activation of fluid-phase coagulation by the clot [23,24]. Thus, fibrin accretion would be reduced if not eliminated. In addition, due to its increased potency and half-life compared to UFH and LMWH, ATH could be used at dosages which would not cause significant risk of hemorrhage. Another, novel utilization of ATH involves the likelihood that its permeability from one compartment to another would be reduced compared to free heparin. The molecular size of ATH is much greater than heparin alone, which would prevent its percolation through small pores in cross-linked sections of the extracellular tissue matrix. Therefore, ATH should be retained within the compartment or space in which it is placed. This property can be advantageous in situations where an anticoagulant must have potent direct activity in a confined space, without it being lost from that compartment. An example of a disease which would be best treated by sequestration of covalent serpin–GAG complex is respiratory distress syndrome (RDS). RDS, particularly in neonates, is characterized by intrapulmonary coagulation. Neonatal and adult RDSs are typified by leakage of plasma proteins of varying sizes into the airspace [168–170], which leads to interstitial and intra-alveolar thrombin generation with subsequent fibrin deposition. In fact, extravascular fibrin deposition is a hallmark of RDS [171,172], a common complication of premature birth. Furthermore, the success of surfactant therapy for RDS has not resulted in decreasing either the incidence or severity of chronic bronchopulmonary dysplasia, which continues to be a problem for survivors of prematurity [173–177]. ATH is a strong candidate for RDS treatment and prevention of chronic fibrotic lung disease. Coagulation in the lung space may occur under conditions where there is a paucity of antithrombin for thrombin inhibition. ATH would provide potent direct inhibition of pulmonary thrombin generation, even in the absence of antithrombin from the patient (a requirement for UFH and LMWH). Due to its size, ATH should slowly, if at all, disappear from the lung into the circulation leading to reduced bleeding risk systemically. Finally, once ATH forms a complex with thrombin generated in the alveolar airspace, the serpin–GAG conjugate itself would be neutralized and unable to participate in any further, undesired reactions. In conclusion, prophylaxis, systemic antithrombotic treatment, and anticoagulation of selected compartments by sequestration are significant applications where the particular characteristics of ATH may make it a superior agent compared to other available drug technologies.

1.4.4 Concepts for Producing Covalent ATH Complexes

There are only three general approaches to permanent linkage of antithrombin and heparin. Bonding can occur if heparin is activated to make it reactive, followed by interaction with antithrombin to effect covalent bond formation. Conversely, antithrombin can be pre-activated and then incubated with heparin to obtain a stable complex. Finally, antithrombin and heparin can be conjugated by allowing the two macromolecules to interact non-covalently, followed by addition of a bifunctional reagent, one end of which bonds to the serpin and the other end of which reacts covalently with the GAG. The three synthetic schemes for ATH preparation are outlined in Figure 1.4. In the first two methodologies, it is theoretically possible that the heparin or antithrombin may possess groups that, under the appropriate conditions, are already reactive enough to form a bond with the other macromolecule. Furthermore, in a number of procedures, care must be taken (either by selective chemistry, appropriate reagent stoichiometries, or particular reaction conditions) to prevent linkage of either heparin to itself or antithrombin to itself.

Functional groups on heparin that may be activated (or have active [activatable] groups attached to them) are carboxyl, sulfonyl, hydroxyl, amino, and acetal. Apart from the acetal and amino groups, attachment through heparin's other functional groups would be within the chain of the molecule. Only linkage through acetals, or aldehydes formed at the reducing terminus, would result in end-point attachment of heparin to antithrombin, which more closely resembles the structure of the glycosidically bonded chains in the proteoglycan form of heparin [86]. An example of conjugation which occurs by reaction of antithrombin with a pre-activated heparin follows. Heparin carboxyl groups are reacted with a carbodiimide followed by incubation with a diaminoalkane to form amide-linked groups with a free amino group. Further reaction of the amino-substituted heparin with a diisothiocyanate then gives a heparin derivative with a free isothiocyanate group that can form thiourea bonds with the N-terminal or Lys amino groups of antithrombin. This type of mechanism for joining heparin to antithrombin has been carried out previously and will be discussed further in detail later [178].

Activation of antithrombin for covalent reaction with heparin is more problematic. Not only do proteins have a greater number of different functional R-group types than those on heparin, but also the same type of amino acid R-group can exist in a range of environments, either on the surface or within the tertiary structure. Thus, more finely tuned chemistries may be required to obtain significant yields of purified ATH (containing linkages with heparin that are at the same point on the antithrombin molecule) resulting from initial activation of the antithrombin. One possible example for this type of synthesis would involve activation of antithrombin with N-hydroxysuccinimidyl-4-azidobenzoate (formation of benzoylamide with antithrombin Lys amino group), followed by interaction with heparin in the presence of light to cause photochemical linkage between the benzoyl derivative and heparin. Additionally, the antithrombin carbohydrate residues could be modified to act as linkage points for heparin conjugation. Reaction of antithrombin with $NaIO_4$, under mild conditions, only

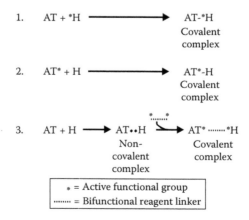

FIGURE 1.4 General approaches for covalent linkage of antithrombin and heparin.

produces aldehydes on the terminal sialic acids of the glycans. Aldehyde-containing antithrombin can be reacted with diaminoalkane + NaBH$_3$CN to form alkyl–amino antithrombin. Reaction of the alkyl–amino groups on the antithrombin with N-hydroxysuccinimidyl-4-azidobenzoate (discussed as earlier) would then be preferred, compared to other amino groups on the antithrombin, since the amino groups are very prominent (essentially on a spacer arm). Conjugation of such a modified antithrombin with heparin should then be possible. The likely advantages for linkage of antithrombin to heparin through the serpin glycan termini are that the number of linkage points on the antithrombin may be more controlled and there would be no perturbation due to modification of the polypeptide.

ATH synthesis involving simultaneous bonding of antithrombin and heparin with an intermediate agent is likely to be more sophisticated than procedures in which initial activation of either heparin or antithrombin is carried out. All of the functional groups that are present on heparin exist on antithrombin as well. Therefore, it would be fortuitous to use heterobifunctional agents that have one functional end, which would only react with a functional group found solely on antithrombin. An example would be use of a linking agent that contains one reactive group which would selectively link with the guanidinyl group of antithrombin Arg residues (vicinal di-one functional group) while the other end of the linker has a photoactivatable group ($-$N$_3$) with a neighboring positively charged group ($-$N(CH$_3$)$_3^+$) to form a covalent bond with the negatively charged heparin chain. One disadvantage of this approach is that more than one link between antithrombin and heparin may occur, which may result in denaturation of the protein. On the other hand, allowing native antithrombin and heparin to interact prior to covalent modification would give the opportunity for the most active non-covalent ATH complexes to form prior to conjugation.

Investigations into ATH synthesis have been reported earlier. The vast majority involves the first method (discussed earlier and illustrated in Figure 1.4) in which heparin is activated prior to interaction with antithrombin. The evolution of ATH compound investigation is reviewed later.

1.5 COVALENT ATH COMPLEXES: DEVELOPMENT

1.5.1 CHEMISTRY, PHYSICOCHEMICAL PROPERTIES, AND IN VITRO ACTIVITY

The chronological development of ATH products shows an evolution over time of more sophisticated, selective chemistries, which are concomitant with higher anticoagulant activities. A number of studies have reported preparations in which non-covalent complexes of antithrombin and heparin have been produced [179–182]. Such complexes are useful for short-term treatment but are not practical for long-term thrombin regulation or anticoagulation at surfaces where binding by the heparin moiety may cause dissociation of the ATH. As the various attempts to join antithrombin and heparin are described, the type of synthetic methods utilized will be related to structure of the final product in order to explain the observed physical properties and anticoagulant activities.

Initial investigations in the construction of a permanently activated antithrombin relied on chemical linkage of GAGs, particularly heparin, to chromatographic matrices and other macroscopic surfaces. Covalent attachment of heparin to other proteins (such as albumin) [183,184], natural polysaccharides [186], and synthetic polysaccharides [185] has added to the knowledge base for coupling of heparin to macromolecules. Some of the earliest attempts to form ATH were carried out by Ceustermans et al. [178]. UFH was activated with CNBr at pH 10 according to the procedure of Cuatrecasas et al. [186], followed by reaction with antithrombin. These experiments failed to produce significant amounts of conjugate. Activation of heparin with CNBr may involve formation of iminocarbonate or cyanate groups with the hydroxyl groups on heparin but, more likely, CNBr reaction requires the presence of amino groups, of which there are fewer than one per heparin molecule on the average [178]. In addition, the amino groups on the heparin molecules may be at locations which are distal to the antithrombin that is interacting with the GAG pentasaccharide-binding site. Further experiments were undertaken to form conjugate through the bifunctional reagent, 1,5-difluoro-2,4-dinitrobenzene [178], according to the methods of Zahn and Meienhofer [187,188]. Again, no ATH was produced. Since 1,5-difluoro-2,4-dinitrobenzene would require the presence of heparin amino groups for nucleophilic substitution on the benzene ring to take place, lack of product was probably due to the same reasons as those for CNBr experiments. Modification of heparin with 4-fluoro-3-nitrophenyl azide or 4-azido-phenacyl bromide, followed by reaction with antithrombin via photoactivation of the $-N_3$ group, has been carried out according to the experimental conditions reported previously [189–192], but with negative results.

Successful linkage of antithrombin and heparin was finally obtained by Ceustermans et al. [178] using tolylene-2,4-diisocyanate according to the procedure of Clyne and coworkers [193–195]. In order to increase reactivity of aminos (or hydroxyls) with the isocyanate groups, reactions were carried out under basic conditions. However, the amount of conjugate obtained was $\leq 5\%$. The yield of ATH decreased rapidly due to the facile hydrolysis of isocyanate groups on the substituted heparin during incubation on ice in pH 9.5 borate buffer, just prior to interaction with the antithrombin. Thus, the more stable tolylene-2,4-diisothiocyanate (TDTC) reagent was substituted in further attempts to couple antithrombin and heparin. Heparin was modified with TDTC using the procedure of Edman and Henschen [196], followed by reaction with antithrombin in pH 8.5 bicarbonate buffer [178]. A yield of 30% was obtained for ATH using the heparin modified with TDTC.

In order to further improve yield and quality of the product with the TDTC reagent, ATH was synthesized following TDTC reaction with heparin that had amino groups introduced it, either by deblocking amino groups on the glucosamine residues or modifying the carboxyl groups with an amino-containing spacer arm [178]. The resultant product was purified, structurally characterized, and functional activities assessed. A scheme for the reaction used to produce ATH with antithrombin linked to amino group–modified heparin is shown as reaction 1 in Figure 1.5. The high affinity

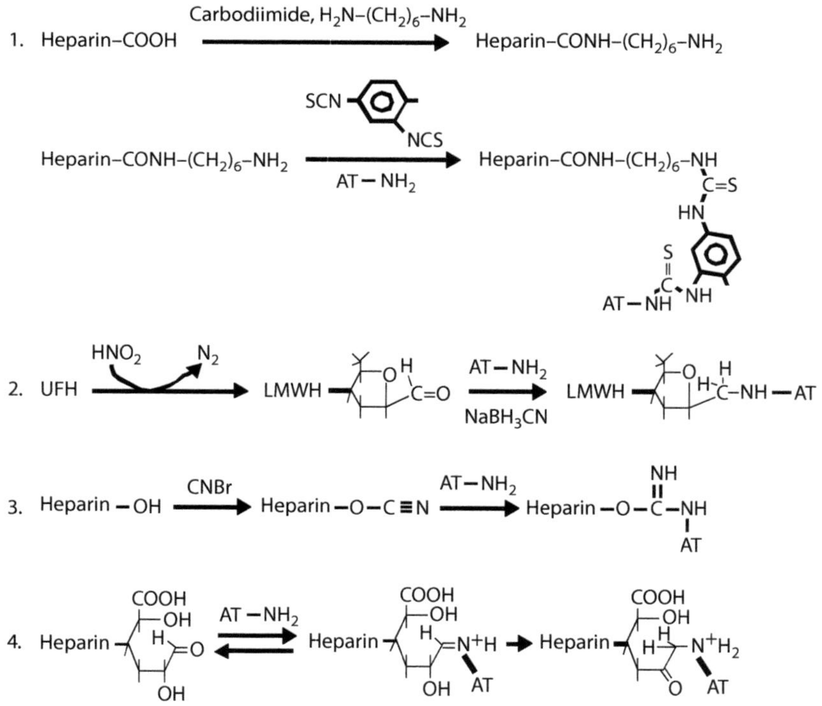

Methods:
1. Ceustermans et al. [178]
2. Björk et al. [201]
3. Mitra and Jordan [204]
4. Chan et al. [111]

FIGURE 1.5   Synthetic methods used to prepare various covalent antithrombin–heparin complexes.

fraction of UFH passed over an immobilized antithrombin column was used as starting material for all of the ATH prepared from amino-substituted heparin. Partial N-desulfation was performed by incubation of the pyridinium salt of the acid form of heparin (prepared by passage of heparin through Dowex 50) in 95% dimethylsulfoxide for 0.5 h at 23°C [197,198], followed by adjustment of the pH to 9.5 and dialysis versus H_2O for 2 days. Introduction of amino groups by carboxyl modification was achieved by incubation of high affinity heparin with 1,6-diaminohexane in the presence of a carbodiimide-condensing agent. Either of the two amino-modified heparins was made to react with excess TDTC (to prevent cross-linkage of the GAG), after which the activated heparin was made to react with antithrombin via the remaining isothiocyanate group. Chemical analyses of the modified heparins and corresponding ATH products are found in Table 1.1. Partial N-desulfation of heparin resulted in an approximate twofold increase in isothiocyanate groups incorporated per activated GAG molecule. Isothiocyanate substitution was improved by a factor of 2.9 in (hexyl-amino)-containing heparin compared to reactions with starting heparin. Yield of ATH (in terms of either starting heparin or antithrombin) was ~30% using either partially N-desulfated heparin or hexyl-amino heparin. Furthermore, up to 25% of the ATH obtained contained two antithrombin molecules per heparin molecule (heparin/antithrombin molar ratio, Table 1.1) [178]. Activity measurements of N-desulfated or hexyl-amino heparin were determined using an activated partial thromboplastin time (APTT) clotting time measurement or anti-FXa inhibition assay (Table 1.1). After partial N-desulfation, heparin APTT activity decreased by 35% and anti-FXa activity was reduced by 24%. Hexyl-amino heparin had 44% and 39% lower APTT and anti-FXa activities, respectively, compared to starting high affinity heparin (Table 1.1). The anti-FXa activities represent the ability of the heparin species to catalyze the inhibition of FXa by exogenous antithrombin (excess added antithrombin that is not covalently bound to heparin). Significant loss of activity occurred in both partially N-desulfated heparin and heparin with diamino-hexyl groups linked to the uronic acid residues, which is not surprising given that it has long been known that the N-sulfate and carboxyl groups are critical for heparin's anticoagulant activity [200]. Anti-FXa measurements of the ATH complexes produced using partially N-desulfated heparin were not reported [179]. Direct, non-catalytic reaction of the hexyl-amino-containing ATH complexes with FXa was studied. Determinations of the reaction rate revealed that the second-order rate constant for this ATH (Table 1.1) was $2.1 \times 10^6 \, M^{-1}s^{-1}$ (~3 times lower than that for non-covalent mixtures of unmodified, high affinity heparin + saturating amounts of antithrombin) [178]. Addition of heparin to the ATH doubled the rate of direct FXa inhibition, suggesting that not all ATH heparin was coupled so that the antithrombin was in the active conformation [178]. Thus, exogenous heparin increased the activity for these non-activated molecules. ATH prepared from the amino-modified heparins had been purified by combinations of anion exchange, gel filtration, and sepharose–antithrombin affinity chromatographies [178]. Further purification of hexyl-amino heparin produced ATH on sepharose–concanavalin A allowed removal of residual-free heparin and possibly some inactive complexes [200]. The highly purified ATH was tested in reactions with thrombin and a second-order rate constant of $6.7 \times 10^8 \, M^{-1}s^{-1}$ was measured (Table 1.1). A bimolecular rate constant for reaction with thrombin of $2.2 \times 10^8 \, M^{-1}s^{-1}$ was determined with high affinity heparin in the presence of saturating amounts of antithrombin (compared to $2.5 \times 10^8 \, M^{-1}s^{-1}$ for inhibition of thrombin by ATH) [200]. Thus, ATH that was an isolate of highly active covalent complexes of one amino-hexyl-thiourea-tolylene-isothiocyanate heparin molecule and one antithrombin molecule had direct inhibitory reactivity with thrombin that was essentially the same as that for inactivation of thrombin with non-covalent mixtures of heparin + saturating amounts of antithrombin. No specific antithrombin activity values have been reported for the hexyl-tolylene linkage-containing ATH. Hoylaerts et al. have also prepared ATH from high affinity hexyl-amino-containing heparin using LMWH obtained by HNO_2 treatment of UFH (followed by size fractionation to give LMWH with chain lengths of 10–14 monosaccharide units $[M_r = 3200]$ or 14–18 monosaccharide units $[M_r = 4300]$) [200]. However, second-order rate constants for reaction with thrombin were significantly reduced for the LMWH-containing ATH ($3 \times 10^5 \, M^{-1}s^{-1}$ and $2 \times 10^7 \, M^{-1}s^{-1}$ for 3200 and 4300 M_r heparin-containing ATH, respectively), compared to the corresponding ATH with standard molecular weight ($M_r = 15,000$) heparin.

TABLE 1.1

Physicochemical Properties and Activities of Heparin and Covalent Antithrombin–Heparin Complexes Prepared by Various Methods

Synthetic Method	Start. Heparin	Activating Groups per Heparin (mole/mole)	ATH H/AT (mole/mole)	Mod. Heparin APTT (% Start. Heparin)	Activity							
					Anti-FXa Catalytic Activity (U/mg)			Anti-IIa Catalytic Activity (U/mg)			ATH + FXa k (M^{-1}s^{-1})	ATH + IIa k (M^{-1}s^{-1})
					Start. Heparin	Mod. Heparin	ATH	Start. Heparin	Mod. Heparin	ATH		
1	High affinity UFH	0.9 ± 0.3	0.9	65 ± 5	250	191 ± 28	ND	ND	ND	ND	ND	ND
2	High affinity UFH	1.9 ± 1.4	0.8	56 ± 7	250	153 ± 34	270	ND	ND	ND	2.1 × 10^6	6.7 × 10^8
3	UFH modified to LMWH	1	0.7	ND	~170 (UFH)	140 (LMWH)	5.2	~170 (UFH)	0 (LMWH)	0	ND	ND
4	UFH	β1	β1	ND	168	ND	Low	168	ND	Low	ND	6.7 × 10^7
5	UFH	0–0.18	1.1	100	209	209	861	198	198	754	3.8 × 10^6	3.1 × 10^9

Definitions: Start. heparin, Starting source heparin used for experiments; ATH, antithrombin–heparin covalent complex; H, heparin; AT, antithrombin; Mod. heparin, Mod. heparin, starting heparin after chemical activation but before conjugation with antithrombin; APTT, activated partial thromboplastin time; anti-FXa, anti-factor Xa heparin activity (catalysis of the FXa + AT reaction); anti-IIa, antithrombin heparin activity; k, rate constant (bimolecular or second order); high affinity, fraction with highest binding strength to antithrombin; UFH, commercial unfractionated heparin; LMWH, low-molecular-weight heparin obtained by partial depolymerization of UFH with HNO$_2$; ND, not determined.

Synthetic Methods:

1. Partially N-desulfated high affinity heparin conjugated to antithrombin with tolylene-2,4-diisothiocyanate. (From Ceustermans, R. et al., *J. Biol. Chem.*, 257, 3401, 1982.)

2. Hexyl-amino substituted high affinity heparin conjugated to antithrombin with tolylene-2,4-diisothiocyanate. (From Ceustermans, R. et al., *J. Biol. Chem.*, 257, 3401, 1982; Hoylaerts, M. et al., *J. Biol. Chem.*, 259, 5670, 1984; Mattsson, C. et al., *J. Clin. Invest.*, 75, 1169, 1985.)

3. Reductive alkylation of antithrombin with high affinity anhydromannose-terminating heparin and NaBH$_3$CN. (From Bjork, I. et al., *FEBS Lett.*, 143, 96, 1982; Dawes, J. et al., *Biochemistry*, 33, 4375, 1994.)

4. Conjugation of antithrombin to CNBr-activated heparin. (From Mitra, G. and Jordan, R.E., Covalently bound heparin–antithrombin III complex. US Patent 4,689,323, Miles Laboratories, 1987; Mitra, G. and Jordan, R.E., Covalently bound heparin–antithrombin III complex, method for its preparation and its use for treating thromboembolism. European Patent 84111048.9 (0 137 356), Miles Laboratories, 1985.)

5. Incubation of antithrombin with aldose-terminating unfractionated heparin to form a Schiff base between an antithrombin lysyl amino group and the heparin aldose aldehyde, which undergoes an Amadori rearrangement. (From Chan, A.K. et al., *J. Biol. Chem.*, 272, 22111, 1997; Berry, L. et al., *J. Biochem.*, 124, 434, 1998; Berry, L. et al., *J. Biol. Chem.*, 273, 34730, 1998.)

Almost coincident to the ATH prepared by Ceustermans et al., Björk et al. produced ATH using end-group attachment to the heparin moiety (synthetic scheme 2 shown in Figure 1.5) [201]. UFH was treated with HNO_2 to cause partial deaminative cleavage. High activity LMWH fragments were obtained by immobilized antithrombin affinity chromatography, followed by covalent linkage to antithrombin by reduction of the Schiff base formed between the aldehyde group of the LMWH anhydromannose termini and lysyl amino groups on the protein (reductive alkylation) [201]. Conjugate was isolated by gel filtration to separate ATH and antithrombin from unreacted LMWH, followed by heparin agarose affinity chromatography to separate ATH from free antithrombin. After completion of the purification procedures, determinations showed that ~40% conversion of the original antithrombin to ATH occurred. The molecular weight of heparin prepared by this partial deaminative cleavage method has been shown to range from 3,700 [201] to 10,000 [202]. Unlike heparin in the hexyl-tolylene-linked ATH, the LMWH generated by HNO_2 treatment of UFH has only one active group per molecule located at the terminal anhydromannose residue (Table 1.1). Therefore, only one type of orientation of the heparin on the antithrombin can occur (end-point attachment). Analyses showed that ATH preparations, formed by reductive alkylation of antithrombin with high affinity anhydromannose-terminating heparin, had an average of 0.7 mol of GAG chain conjugated per antithrombin (Table 1.1). LMWH obtained by HNO_2 for conjugation (prior to isolating the high affinity fraction) was determined to have an anti-FXa activity of 140 units/mg, which is slightly lower than the UFH starting material [203]. Titration with FXa showed that the Björk et al.'s ATH complex had an activity which was 98% of that for the high affinity LMWH material used in the preparation (Table 1.1). Interestingly, although the anti-FXa activity of non-covalent mixtures of LMWH + antithrombin could be neutralized by high NaCl concentrations (1 M) or polybrene, reactivity of the Björk et al.'s ATH with FXa was only slightly decreased by high salt or polybrene [201]. Therefore, it may be concluded that covalent linkage of the HNO_2-produced LMWH to antithrombin prevents removal of the activating heparin species from the serpin by ionic media. Thus, the Björk et al.'s ATH complex is at the same time potent in a variety of conditions and difficult to reverse if used for clinical treatment. Unfortunately, both the LMWH prepared using HNO_2 and the resultant ATH had no measurable antithrombin activity [203]. The lack of activity against thrombin is not surprising, given that short-chain LMWH cannot provide the template required for binding both antithrombin and thrombin [108,123].

Further development of ATH by conjugation of heparin and antithrombin using CNBr was accomplished by Mitra and Jordan (reaction scheme 3 in Figure 1.5) [204]. UFH was activated at pH 10.7 with ~57-fold molar excess of CNBr for 40 min at 23°C, after which excess CNBr was removed by dialysis under basic conditions. Cyanate/iminocarbonate/N-nitrile active group–containing heparin was then incubated with 0.02 mol of antithrombin per mole of GAG at pH 9.4 for 18 h at 5°C. Separation of free heparin from ATH and free antithrombin was accomplished by chromatography on immobilized concanavalin A and isolation of ATH from antithrombin was done using a heparin–Sepharose column [204]. A yield for ATH of 40% (in terms of antithrombin) was estimated. Complexes with antithrombin were also produced using LMWH (obtained by fractionation of UFH on gel filtration columns), but no further characterization of the products was done. Data on the products of this CNBr-based synthetic procedure are shown in Table 1.1. No information on the number of activating groups formed on the heparin molecules by CNBr (cyanate/iminocarbonate/N-nitrile) was given [204]. However, given the molar ratios involved and that Mitra and Jordan suggest the use of glycine to block any excess active groups after conjugation to antithrombin [204,205], it is likely that more than one reactive group per GAG chain was obtained and that ≥1 heparin was conjugated to each antithrombin in the ATH complexes. Titrations of Mitra and Jordan ATH with thrombin and FXa suggested that essentially all of the complexes had rapid, direct inhibitory activity. Also, the increased intrinsic fluorescence of heparin-activated antithrombin was observed with the Mithra et al.'s ATH, which was not further enhanced by addition of UFH. The ability of CNBr-conjugated ATH to catalyze inhibition of either FXa or thrombin by added antithrombin (exogenous to the antithrombin in the covalent complex) was not clearly investigated. However, description of thrombin and

FXa inhibition assays of the complex in plasma suggests that the inhibitory capacity was only that of the covalent complex (no significant catalytic activity) [204,205]. The rate of reaction of the Mitra and Jordan ATH with thrombin was investigated by generating inactivation time curves with equimolar enzyme and inhibitor concentrations (which allow for bimolecular rate constant calculation) [204]. Results showed that the conjugate reacted with thrombin at a bimolecular rate of $6.7 \times 10^7 \, M^{-1} s^{-1}$ (Table 1.1). Interestingly, non-covalent mixtures of antithrombin + high affinity heparin gave a bimolecular rate constant for reaction with thrombin of $2.0 \times 10^7 \, M^{-1} s^{-1}$, which was 3.3 times lower than that for ATH. The authors suggested the possibility that the ATH may react with thrombin at a faster rate than non-covalent antithrombin + heparin because certain equilibrium constraints that apply to the free component system (such as association of the serpin with the GAG) are circumvented by the covalent complex. Unfortunately, since the rate measurements were not done under first-order conditions (reactants not at 5- to 10-fold higher concentrations with respect to the enzyme being inhibited), comparison of absolute values for reaction rates of the non-covalent and covalent complexes cannot be done. No measurement of the rate of inhibition of FXa by the Mitra and Jordan ATH was reported.

More recently, a new approach has been put forward by Chan et al. [111] that utilizes the chemistry of glucose and plasma proteins occurring in diabetics (see Figure 1.5). It has long been known that hemoglobin and various plasma proteins undergo Schiff base formation between the aldehyde function on C_1 of plasma glucose and the amino group of lysyl residues of the protein [206]. This intermediate product is metastable and in equilibrium with the starting materials. However, as shown in Figure 1.5, over time, a tautomeric rearrangement (Amadori) between protons and groups on C_1 and C_2 occurs to form an ene-ol-amine and finally a stable keto-amine [206]. Formation of Amadori rearrangement products is dependent on glucose concentration, pH, temperature, and availability of amino functions. However, once formed, nonenzymatically glycated proteins have been shown to survive in diabetic tissue from weeks to years after hyperglycemic exposure [206]. Reports have indicated that nonenzymatic glycation can have effects on protein function [207]. In the case of antithrombin, brief increases in blood glucose concentration have been shown to result in short alterations of antithrombin activity without loss of protein (likely due to nonenzymatic labile Schiff base formation) [208]. Significant decreases in antithrombin activity due to longer-term glycation have been noted [209], which may explain some of the pathogenesis for thrombosis in severe diabetics [210]. Moreover, nonenzymatic glycation of antithrombin in vitro [211] and in vivo [211] has been found to inhibit heparin-catalyzed antithrombin activity. Even more fascinating was the fact that glycation-induced effects on heparin-catalyzed antithrombin activity were completely reversed by incubation with excess Na heparin before assay [211]. Therefore, reaction of glucose with a particular antithrombin lysyl residue may be occurring to produce Schiff base/Amadori products that can be displaced by a large excess of heparin. Further evidence of selective glycation of antithrombin can be seen by the fact that even in the presence of up to 5 mM glucose for 10 days, only 0.6 mol of glucose can be covalently bound per mole of protein [210]. Although other literature has shown that a number of regions of the antithrombin molecule may form the keto-amine adducts, only one glucose per protein molecule could be yielded during incubations with up to 0.5 M sugar [212]. Thus, once one glycation site is affected, the probability of further glycation is drastically reduced. Given the effects of heparin on glycated antithrombin, the possibility arises that heparin molecules containing aldose residues of their own might be responsible for reversing (competing with) the Schiff base/Amadori structures being formed. Thus, if UFH contained significant numbers of aldose-terminating molecules, it may be possible to form covalent ATH by simple incubation of unmodified UFH and antithrombin under the appropriate conditions. Previous work had shown that Schiff base/Amadori rearrangement of insulin by the disaccharide maltose can proceed [213]. However, numerous workers had concluded that even Schiff base formation between polysaccharide aldoses and amino groups would be extremely limited because the aldose terminus represented a very small part of the sugar residues in the polymer, the equilibrium would be highly in favor of the hemiacetal (masked aldehyde) form and significant steric hindrance for approach of macromolecular amines seemed likely [214].

Chan et al. decided to test the hypothesis of ATH formation via Schiff base/Amadori rearrangement of UFH and antithrombin. It had been determined earlier that in commercial UFH preparations, approximately 10% of all molecules terminated in a free aldose [215]. Since it would be desirable to have ATH complexes in which the antithrombin was activated by the covalently linked heparin, incubation conditions would need to be designed so that antithrombin molecules could have the opportunity to select for heparin with non-covalent pentasaccharide-binding sequences that were close to the aldose end of the chain [111]. Feasibility studies showed that nonenzymatic glycation of polypeptides by polysaccharide aldose was possible; however, yields were greatly variable due to a number of reasons [216]. A format for ATH production via Schiff base/Amadori rearrangement was established. UFH and antithrombin from various sources were mixed in physiological buffer and heated for up to 16 days. Purification of the ATH formed was by a two-step procedure of hydrophobic chromatography (to remove unbound heparin), followed by separation of unbound antithrombin on DEAE Sepharose [111]. Since it had been shown that incubation of antithrombin with monosaccharide aldose could increase the temperature for thermal denaturation (Td) [217], interactions at different temperatures were studied. Yields of conjugate (in terms of antithrombin) were observed to be ~5% at 37°C, while ~50% conversion of antithrombin to complex was observed at 40°C [216]. Properties of the Chan et al.'s ATH are found in Table 1.1. Analyses indicated that ~1 molecule of heparin was combined with each antithrombin in the complex [111]. This would be in agreement with the binding of a heparin chain to the antithrombin-binding site (through the high affinity pentasaccharide), followed by Schiff base/Amadori rearrangement of the aldose terminus with a lysyl residue in the vicinity. Comparison of the full antithrombin amino acid sequence with that of heparin-linked peptides released after partial protease digestion of ATH allowed identification of the antithrombin linkage point for heparin chains in the complexes [218]. Although a small proportion of heparin glucuronic acid termini form a keto-amine [216] with the amino group of Lys139, ~87% of ATH molecules have heparin covalently bonded to the N-terminal His [218]. This result is in concordance with findings stated earlier that the primary non-covalent heparin-binding region is near the antithrombin N-terminus [31,73]. Electrophoretic analysis of the conjugate's antithrombin species after enzymatic release of the heparin chain showed that Chan et al.'s ATH is enriched in the β-isoform [219]. Again, selection of β- over α-antithrombin by heparin during ATH synthesis could be expected with the increased initial binding affinity for the β-species [78]. Similarly, the isolated β-ATH reacts significantly faster with thrombin than the corresponding α-ATH [219].

Since no prior modification of the heparin was used prior to conjugation in Chan et al.'s ATH, the number of activating groups, anti-FXa activity, and anti-IIa activity of the heparin in the coupling reaction were those of the starting UFH. Surprisingly, Chan et al.'s ATH was found to possess catalytic activity for the inhibition of either FXa or thrombin by exogenous antithrombin (Table 1.1) [111]. Furthermore, the catalytic anticoagulant activity of the covalent complex was greater than that for either the starting UFH or heparin with high affinity for antithrombin [111]. This startling result was proposed to be due to the selection, during incubation with antithrombin, of not only heparin chains with a pentasaccharide but also some GAG molecules with more than one pentasaccharide unit. The presence of more than one high affinity antithrombin-binding site in a subpopulation of UFH molecules (1%–3%) has been described earlier [112,220]. Selection of this minor subpopulation of UFH chains by antithrombin was made possible by using a vast molar excess (>200-fold) of UFH to antithrombin during the synthesis [111]. It was posited that since the interaction of incubating antithrombin and UFH would initially be via non-covalent binding to pentasaccharide sequences, heparin chains with more than one high affinity site might be preferred because the intramolecular mean distance of diffusion would be less than the intermolecular mean free distance of diffusion and there is a higher probability that at least one site may be close to the aldose terminus (where covalent linkage would occur). Further studies confirmed that, in fact, Chan et al.'s ATH did contain a significant number of molecules with more than one pentasaccharide [221]. Detailed investigations revealed that pentasaccharides directly interacting with the covalently linked antithrombin moiety were capable of catalyzing thrombin inhibition by exogenous antithrombin, albeit more weakly than distal pentasaccharide sequences [222]. Given that formation of Chan et al.'s ATH involved simple incubation of unmodified heparin with the antithrombin, it was not surprising to find that direct reaction of the inhibitor with thrombin proceeded with a second-order rate constant ($3.1 \times 10^9\,M^{-1}\,s^{-1}$) that was one of the fastest ever recorded [111]. Indeed, second-order rates for reaction of this ATH with most of the activated coagulation factors were significantly faster than that for non-covalent antithrombin + UFH mixtures [223]. The heightened potency of ATH by Chan et al. led to further testing in plasma. ATH outperformed heparin for inhibiting thrombin generation in plasmas from all age groups [224]. Additionally, relative to UFH and LMWH, the ATH exhibited enhanced suppression of thrombin formation in hemophilic patient plasma activated by a variety of stimuli [225].

1.5.2 IN VIVO PERFORMANCE

Further studies have been accomplished that investigated the in vivo properties of many of the ATH compounds described earlier. Investigations ranged from determination of pharmacokinetics to antithrombotic treatment of disease models in animals. Data for the different ATH products are given in Table 1.2.

TABLE 1.2

In Vivo Characteristics of Covalent Antithrombin–Heparin Compounds

Compound	Intravenous Half-Life (h)		Basic Plasma Protein Binding	Bleeding Characteristics	Antithrombotic Activity (Given Equal Mass Concentration Doses)
	α-Phase	β-Phase			
1	<0.5	0.65 ± 0.06	ND	ND	ND
2	0.89 ± 0.26	—	Yes	ND	Same as non-covalent mixtures
3	1	7.85 ± 0.18	ND	ND	Same as non-covalent mixtures
4	1	7.63 ± 0.75	ND	ND	Same as non-covalent mixtures
5	>UFH?	—	ND	ND	ND
6	<3	—	ND	ND	ND
7	2.6	13	Low	Similar to UFH	Efficacy superior to UFH

Definitions: UFH, Commercial unfractionated heparin; ND, not determined.

Synthetic Methods:

1. Partially N-desulfated high affinity heparin conjugated to antithrombin with tolylene-2,4-diisothiocyanate. (From Ceustermans, R. et al., *J. Biol. Chem.*, 257, 3401, 1982.)

2. Hexyl-amino-substituted high affinity heparin conjugated to antithrombin with tolylene-2,4-diisothiocyanate. (From Ceustermans, R. et al., *J. Biol. Chem.*, 257, 3401, 1982; Hoylaerts, M. et al., *J. Biol. Chem.*, 259, 5670, 1984; Mattsson, C. et al., *J. Clin. Invest.*, 75, 1169, 1985.)

3. Hexyl-amino-substituted high affinity low-molecular-weight heparin (MW = 3200) conjugated to antithrombin with tolylene-2,4-diisothiocyanate. (From Hoylaerts, M. et al., *Thromb. Haemost.*, 49, 109, 1983; Collen, D.J., Novel composition of matter of antithrombin III bound to a heparin fragment, US Patent 4,623,718, KabiVitrum AB., 1986.)

4. Hexyl-amino-substituted high affinity low-molecular-weight heparin (MW = 4300) conjugated to antithrombin with tolylene-2,4-diisothiocyanate. (From Hoylaerts, M. et al., *Thromb. Haemost.*, 49, 109, 1983; Collen, D.J., Novel composition of matter of antithrombin III bound to a heparin fragment, US Patent 4,623,718, KabiVitrum AB., 1986.)

5. Antithrombin reductively alkylated with anhydromannose-terminating heparin. (From Bjork, I. et al., *FEBS Lett.*, 143, 96, 1982; Dawes, J. et al., *Biochemistry*, 33, 4375, 1994.)

6. Antithrombin conjugated to CNBr-activated heparin. (From Mitra, G. and Jordan, R.E., Covalently bound heparin–antithrombin III complex. US Patent 4,689,323, Miles Laboratories, 1987; Mitra, G. and Jordan, R.E., Covalently bound heparin–antithrombin III complex, method for its preparation and its use for treating thromboembolism. European Patent 84111048.9 (0 137 356), Miles Laboratories, 1985.)

7. Antithrombin coupled to heparin by Schiff base/Amadori rearrangement. (From Chan, A.K. et al., *J. Biol. Chem.*, 272, 22111, 1997; Berry, L. et al., *J. Biochem.*, 124, 434, 1998; Berry, L. et al., *J. Biol. Chem.*, 273, 34730, 1998.)

ATH produced from the high affinity fraction of UFH, according to the technique of Ceustermans et al. [178], has been tested for a number of characteristics in animals. Initial studies showed that, in rabbits, ATH joined to partially N-desulfated, high affinity heparin disappeared from plasma in a biphasic pattern with a short first phase (α) and a longer second phase ($\beta = 0.65 \pm 0.06$ h) [178]. Further studies were carried out to characterize ATH made by joining hexyl-amino substituted high affinity heparin to antithrombin using diisothiocyanate. Using antithrombin linked to high affinity hexyl-amino-containing heparin (from UFH), disappearance from the circulation in rabbits could be described by a single phase with half-life of 0.89 ± 0.26 h [178]. Neither of these ATH complexes had half-lives that were more than approximately threefold longer than UFH in rabbits (single phase = 0.23 ± 0.03 h) and both were well short of the half-life of antithrombin (11.0 ± 0.4 h) [178]. As with UFH, covalent complexes of the hexyl-amino-containing heparin with antithrombin may be eliminated by binding of the heparin moiety to basic plasma and cell surface proteins since it was shown that these conjugates could also bind to histidine-rich glycoprotein [226]. In order to extend longevity in the circulation, ATH from high affinity hexyl-amino-substituted LMWH (3200 and 4300 MW, produced by partial depolymerization with nitrous acid) was prepared and the turnover parameters in rabbits were measured [227,228]. Both LMWH ATH complexes were lost from plasma with β-phase half-lives of ~7–8 h (Table 1.2). This disappearance rate for the high affinity hexyl-amino LMWH ATH complexes was 30 times longer than that for UFH in rabbits (0.25 ± 0.04 h). However, the half-lives of the unbound 3200 and 4300 starting high affinity heparins were 2.35 ± 0.17 h and 2.50 ± 0.18 h, respectively, only ~3 times shorter than the corresponding conjugates with antithrombin [227]. In addition, it was claimed by the authors that the covalent complexes (Ceustermans et al.'s ATH prepared from high affinity UFH or LMWH) possessed the ability to catalyze reactions with the animal's own antithrombin. Thus, even with an increased half-life, any antithrombotic properties might be determined by their rate of consumption by activated coagulation factors, as opposed to how long the compound or its metabolites remained in the circulation. The antithrombotic properties of ATH prepared from either high affinity hexyl-amino UFH or LMWHs were tested in a Wessler rabbit thrombus prevention model [228,229]. Anesthetized rabbits, injected with test compounds, were injected with glass-activated plasma and stasis of a 2 cm length of isolated jugular vein induced for 10 min, followed by semiquantitative analysis of the number and size of thrombi expressed within the vessel segment [229]. Both effects of dosage and time delay from treatment to insult were examined. Results, up to 45–60 min after administration of anticoagulant, showed that all covalent complexes showed similar antithrombotic effects compared to that for injection of the same mass concentrations of the corresponding non-covalent mixtures of antithrombin + heparinoid. However, antithrombotic effects (reduction in size or number of thrombi) could be detected up to 60–120 min for the conjugates, due to the increased half-lives of ATH compounds compared to the corresponding free heparins [229]. Hemorrhagic properties of Ceustermans et al.'s ATH produced from either UFH or LMWH have not been reported.

Information of in vivo use of Björk et al.'s ATH compounds is scarce. Dawes et al. have prepared ATH according to the method of Björk et al. and injected it, by intravenous bolus, into 12 week old mice [203]. Although no in vivo pharmacokinetic data were reported, administration of only three injections of small amounts of conjugate was necessary to obtain antibodies against the human antithrombin in the complex. Furthermore, interaction of radiolabeled ATH with hybridoma cells in vitro showed very low nonspecific binding to cell surface molecules [203]. These data may suggest that the Björk et al.'s ATH complex has low affinity for binding to proteins in vivo, which would enhance the half-life. Unfortunately, assay of the anti-FXa activity of the conjugate gave a very low activity (5.2 U/mg) and no definite indication exists of its antithrombotic properties.

Mitra and Jordan have tested their ATH in vivo [204,205]. Rabbits were given intravenous infusions of 84 or 200 anti-FXa units/kg using either ATH or UFH, respectively, at rates of 2 mL/kg. On completion of infusion, blood samples were taken over time and the supernatant plasma was tested for anticoagulant activity. It was noted that, compared to the UFH group, animals receiving ATH had a significant prolongation of functional activity (as measured by APTT assays) for up to 3 h after discontinuation of treatment [204]. This result would suggest that, at lower doses, a longer period of anticoagulant prophylaxis can be achieved with the Mitra and Jordan covalent ATH conjugate compared to use of free UFH alone. It is likely that this effect would be due to a longer intravenous half-life of ATH prepared from CNBr-activated heparin, since elimination of UFH from the rabbit circulation at this dosage would be less than 3 h [230] and lower activity of ATH than UFH was administered. Nevertheless, direct determination of the loss of ATH compound by mass analyses of the plasma time samples was not reported and antithrombotic efficacy was not investigated.

A significant number of experiments have been performed to evaluate the in vivo character of ATH produced by Chan et al. Bolus injection of the (keto-amine)-linked ATH into rabbits gave an intravenous α-phase (two-compartment model) half-life of 2.5–2.6 h for both anticoagulant activity and mass (compared to 0.32 and 13 h for UFH and antithrombin, respectively) [111]. Using a three-compartment two-exponential decay model, β-phase half-lives of 13 and 69 h were calculated for loss of the ATH and antithrombin, respectively, from rabbit plasma. This terminal phase half-life was the longest ever reported for any ATH complex in vivo. Prolonged availability of Chan et al.'s ATH in the circulation was likely due to its reduced binding (and removal) by plasma proteins and endothelial surfaces [231]. Single subcutaneous injection of Chan et al.'s ATH resulted in systemic appearance of complex that had peak concentrations at 24–30 h and could be detected up to 96 h post-administration [111]. However, only a small portion of the injected ATH was recovered in the plasma, indicating a very poor bioavailability of the subcutaneously administered Chan et al.'s compound. Given the long intravenous half-life of Chan et al.'s ATH, a likely explanation for the low total plasma levels of the subcutaneously administered adduct is that the large molecular size of ATH drastically reduced its permeability through extravascular tissue. This hypothesis led to the suggestion that Chan et al.'s ATH may be useful as an anticoagulant in sequestered spaces. One possibility is for anticoagulant treatment of RDS in the lung [168,232–235], where administration of heparin is not useful due to ready loss of GAG to the circulation [21]. Intratracheal instillation of keto-amine-conjugated ATH into the lung airspace in rabbits led to significant anticoagulant activities and antithrombin antigen concentrations for up to 48 h after administration [111]. Furthermore, no ATH protein could be measured in the plasma of the treated rabbits, indicating that ATH was retained in the lung without any leakage systemically. Studies of pulmonary ATH in newborn rats confirmed that intact complex could be retained within the alveoli for at least 4 days [236]. Concerns over adverse effects of ATH on lung development during long periods in the airspace may be unwarranted since sophisticated studies demonstrated that, unlike free antithrombin or heparin, ATH actually promotes normal pulmonary angiogenesis and morphology [237]. Functional activity of Chan et al.'s ATH as an anticoagulant on the alveolar surface of the lung is likely. Inhibition of thrombin generation in plasma on the surface of fetal distal lung epithelial cells in vitro was shown to be much more efficient with Chan et al.'s ATH than with similar concentrations of non-covalent mixtures of UFH and antithrombin [236,238]. The superior efficacy of ATH on inhibition of epithelial-based plasma thrombin generation was due, in part, to the very rapid direct inactivation of initial thrombin feedback activation for the conversion of prothrombin to thrombin [238]. Further tests to evaluate the antithrombotic potential of Schiff base/Amadori-prepared ATH were carried out. ATH was compared to non-covalent mixtures of UFH + antithrombin in a rabbit treatment model of preformed jugular vein thrombi. Results showed that administration of a single intravenous bolus of Chan et al.'s ATH caused a reduction in clot mass and fibrin accretion, whereas similar (by mass) injections of non-covalent mixtures of UFH + antithrombin led to an increase in clot size [230]. Parallel experiments using a rabbit bleeding ear model showed no significant difference in cumulative blood loss between covalent complex and non-covalent mixtures of UFH and antithrombin [230]. Superior efficacy of Schiff Base/Amadori ATH over heparin was also illustrated in a rabbit arterial thrombosis prevention model, with a reduction in bleeding risk [239]. Thus, Chan et al.'s

ATH was not only a highly potent antithrombotic but could be used effectively at safe doses. Data from the thrombosis treatment model were suggestive of the possibility that the conjugate tested may function via the direct, non-catalytic inhibition of clot-bound thrombin. As stated earlier, non-covalent antithrombin–UFH has significantly reduced activity against fibrin-bound thrombin compared to that against fluid-phase thrombin. Further investigations with soluble fibrin have shown that the keto-amine adduct of antithrombin and heparin can directly inhibit fibrin-bound thrombin with a similar rate constant to that of thrombin in buffer only [240]. This finding has major implications regarding possible indications for the clinical use of ATH. One rationale for the lack of effect of fibrin on the inhibition of thrombin by Chan et al.'s ATH is that binding of the ATH heparin chain to fibrin is reduced, due to steric hindrance by the covalently linked antithrombin. Moreover, ATH heparin chains cannot dissociate from the antithrombin to form a protective ternary complex with fibrin and thrombin. However, a measurable binding affinity of the heparin component of ATH for fibrin would in fact assist in bringing the antithrombin moiety to the clot-bound thrombin. Non-denaturing gel analysis has confirmed that ATH readily reacts with the surface-bound thrombin by associating with the fibrin through its heparin chain [241]. Resultant thrombin–ATH complexes remain on fibrin [241] and the covalently linked heparin chain retains catalytic activity to render the clot as an anticoagulant surface [242]. Alternatively, ATH prepared with human antithrombin from goats [50] can be selectively directed to anticoagulate the vessel wall, likely through binding of lectins on the vascular endothelium to oligomannose glycans on the recombinant antithrombin [243]. In a broader sense, either antithrombin, heparin, or their combination in ATH can be manipulated to target treatment toward various in vivo compartments [244]. Given the Schiff base/Amadori mechanism by which antithrombin and commercial UFH combine to form Chan et al.'s ATH, a fascinating possibility presented itself that such a reaction may, in fact, occur spontaneously during the clinical use of UFH. Attempts were made to isolate ATH from plasmas of rabbits and a human injected subcutaneously with UFH [216]. Using preparative procedures similar to that employed by Chan et al. for purification of their ATH, covalent adducts of heparin and antithrombin were detected in plasma samples of UFH-treated rabbits and humans [216]. Thus, given that endogenous, nonprotein-linked, heparin chains are known to circulate in humans [245–247], it is apparent that Chan et al.'s ATH may be a natural product.

1.6 COVALENT ATH COMPLEXES: ATTACHMENT TO POLYMER SURFACES

1.6.1 BIOCHEMICAL AND CHEMICAL OVERVIEW

Biomaterials that come into contact with blood have a number of limitations. First, cells such as platelets and polymorphonucleocytes can bind to the foreign surfaces and become activated [248,249]. Second, binding of plasma proteins can cause modification of the surface properties toward other molecules or cells. The non-covalent plasma protein absorption encourages cells and other materials to bind in a layered effect that may even lead to constriction or change in the laminar flow characteristics of the blood. Third, interaction of some biomaterials with the blood cells or surrounding tissue can invoke inflammatory processes [250]. In this case, inflammation can be the result of a plethora of factors such as providing a surface for infectious agents that induce inflammation, damaging blood cells or tissue that respond by arachidonic acid metabolism, and generating products at the biomaterial surface, which are secondary to the inflammatory pathway. Fourth, the polymer composition or functional groups within the matrix of biomaterials may either contain epitopes or react with the blood to produce epitopes that cause an immunogenic response. Induction of complement activation by surface membranes, such as those used in dialysis, may actually exacerbate recovery of the patient [251]. Finally, the surface may be procoagulant. Of all the problems listed earlier, surface-induced activation of the coagulation cascade is probably the most common difficulty encountered with blood-contacting materials. Materials are often coated on the polymeric surface in order to decrease the procoagulant nature of the foreign agent. However, such coatings frequently suffer from either incomplete anticoagulation of the entire surface or long-term instability, whereby the coating becomes modified to render it inactive or some of the coated molecules are lost to the circulation (leaving part of the material unprotected and producing a systemic anticoagulant effect).

Covalently coupled antithrombin and heparin is an agent that may provide a surface covering which could overcome a number of the inherent blood incompatibilities of synthetic products. First, since antithrombin is often found non-covalently bound to the heparan sulfate (heparin-like) proteoglycans on the luminal surface of blood vessels [252], ATH could be a very good model for the native vein or artery. Second, since the heparin portion of ATH already has permanently attached antithrombin, it is less likely that absorption of other proteins in the blood will occur, due to steric reasons. Third, both antithrombin and heparin have been shown to have anti-inflammatory properties. Heparin-coated surfaces have long been known to possess reduced inflammatory effects due to decreased humoral, cellular, and complement activation by this very negative polymer [253]. More recently, however, antithrombin has also been identified as having an anti-inflammatory effect in such conditions as septic shock [254]. Fourth, since both antithrombin and heparin are natural products occurring in the human vascular system, very little immune response should be engendered by their combination. Low antigenicity should certainly be possible for the ATH produced by Chan et al., given that it is produced by spontaneous Schiff base/Amadori rearrangement of unmodified, native antithrombin and UFH. As would be expected, ATH should have potent anticoagulant activity for the direct inhibition of thrombin. However, once the covalently linked antithrombin in ATH is consumed by thrombin, FXa, etc., any further anticoagulant activity of a biomaterial coated with the conjugate would have to come from the heparin moiety. Thus, not all ATH products may be as effective in providing a long-term, highly potent anticoagulant surface. Again, the high catalytic activity of the Chan et al.'s ATH toward inhibition of thrombin by plasma antithrombin in the patient would be a significant advantage for ongoing pacification during insults leading to thrombin generation near or at a biomaterial surface.

One ancillary advantage for the coating of biomaterials with ATH compared to heparin involves technical issues. Since antithrombin has a greater variety of functional groups compared to heparin, there is a wider range of chemistries available for covalent as well as non-covalent attachment of the serpin compared to that of the GAG. Carbamate, guanidinium, imidazole, indole, and phenolic groups on the protein allow for selective linkage of ATH through antithrombin, instead of the heparin chain. In addition, antithrombin's 37 Lys residues (human) provide an enriched amino content compared to heparin [178]. Thus, coupling reactions involving NH_2 functions would be strongly biased toward selection of amino groups on antithrombin, as opposed to the serine amino group or small number of glucosamine residues on heparin. Since the NH_2 group has itself a large number of reactivities, connection of the surface or surface spacer arm to ATH via primary amino functions on antithrombin would be a facile operation with the number of commercial reagents and conditions available. As eluded to earlier, non-covalent binding of biomaterial surfaces to the antithrombin of ATH is feasible. It has been shown that the Chan et al.'s ATH and antithrombin can be separated from free heparin by hydrophobic chromatography on butyl agarose [111], which verifies that hydrophobic surface absorption of ATH would be through the serpin and not the GAG component. Again, this is advantageous in that it is preferable to bind ATH to the blood-contacting surfaces through the antithrombin, thereby allowing the heparin to be directed out into the fluid phase. Stronger attachment of active molecules to biomaterials, using antithrombin as the link point, would result if multiple covalent bonds from the surface to ATH are formed. The number and variety of amino acid R-groups in the ATH protein lend a greater efficacy for joining the conjugate to the surface by several bonds, which would give a more stable coating that is less likely to be removed in vivo. Moreover, if heparin was joined to the biomaterial surface through many link points, the likelihood that the pentasaccharide sequence will be adversely affected would increase significantly. Also, surface binding of heparin by bonds at several positions will almost certainly cause steric problems for approach of either plasma antithrombin or thrombin during catalytic inhibition (orientation of the heparin is improper). In fact, this is the major reason for the significantly superior commercial performance of blood-contacting products that have end-point-linked heparin, compared to those that do not. One other factor that makes coating with ATH advantageous is, simply, the dual functioning of the conjugate. Covering polymer surfaces can be done with either direct (hirudin) or indirect (heparin) thrombin inhibitors. In favor of the direct (non-catalytic) inhibitors is the rapid reaction with thrombin that is not limited by the requirement of a secondary molecule (is not effected by the metabolic environment of the patient to supply an inhibitor or cofactor). On the other hand, surface-bound indirect (catalytic) inhibitors would not be limited to a single knockout of thrombin but could, essentially, enable the inhibition of limitless amounts of ongoing thrombin generation, as long as supplied with plasma-phase inhibitor molecules. Chan et al.'s ATH combines both very rapid non-catalytic and potent catalytic inhibition of thrombin in the same molecule. Therefore, coatings with Chan et al.'s conjugate may, in the short term, quickly eliminate small amounts of thrombin that cause feedback activation when biomaterials are first in contact with blood, as well as catalyze the reaction of plasma antithrombin + thrombin for long-term patency.

1.6.2 CHARACTERISTICS OF COATED SURFACES

At present the only ATH product that has been attached to a biopolymer is the Schiff base/Amadori rearrangement, keto-amine-linked conjugate prepared by Chan et al. Preliminary testing of the material, in vitro and in vivo, has shown promise for the conjugate when covalently linked to grafts composed of synthetic polymers that are found in commercial use.

ATH has been linked to two types of polyurethane (PU). One was a polymer of polycarbonate with urethane extenders and the other material was an ester grade PU. The polycarbonate urethane tubing used was composed of a finely threaded material that in some cases was woven around a flexible stainless-steel wire mesh in order to provide further elasticity for application as an endoluminal vascular graft. This polycarbonate urethane came in several diameters and was produced by Corvita Corporation under the trade name Corethane™. The other material utilized in investigations was composed of a pure PU that contained no plasticizers and low levels of extractables. Pure PU tubing was a clear, flexible, impermeable, product that had resistance to fuels, oils, and a number of hydrophobic solvents. Developed by Nalgene, pure PU conduit came in several wall thicknesses and was ideal for quick assessment of results from the different chemistries tested for graft polymers prepared to covalently link ATH to devices, etc.

Chan et al.'s ATH was attached to polymer surfaces and tested for various properties [255]. Attachment was covalent and involved linkage to a graft poly/oligomer on the two types of PU tubing described earlier. Chemical bonding was performed in three stages and is shown schematically in Figure 1.6. Polycarbonate urethane was activated by reaction with NaOCl, after which, the resultant N-chlorourethane groups were incubated with an initiator (i.e., $Na_2S_2O_4$) and a functional group containing monomer (allyl glycidyl ether). After reaction with the activated urethane surface, the tubing was washed and followed by incubation with ATH for covalent linkage between the graft epoxide groups and lysyl amino groups of antithrombin in the ATH. Under the conditions employed, it has been shown previously that oligomers of the grafted species containing from 1 to 4.6 monomer units per urethane group can be obtained on the average [256,257]. This polymer chemistry is based on the general mechanisms of radical-initiated polymerization on poly(N-chloroamide) donor surfaces [258]. Literature reviews have concluded that surface coating relies on a number of parameters such as surface area and mode of drug attachment [259]. Therefore, all steps in the process must be appraised. Preliminary tests on the pure PU Nalgene surfaces verified that grafting of the epoxide materials had occurred. After reaction had taken place with the allyl glycidyl ether, the pure PU tubing became cloudy and a fine film could be observed that was strongly bound to the surface. Pre-reaction of NaOCl-activated PU followed by addition of the monomer (no initiator) did not result in coated surfaces nor did omission of any one of the reaction components. Epoxide group activity on the pure PU grafted with propyl glycidyl ether was evident by the quenching of colored 2-nitro-5-thio-benzoic acid. Thus active surfaces could be prepared for

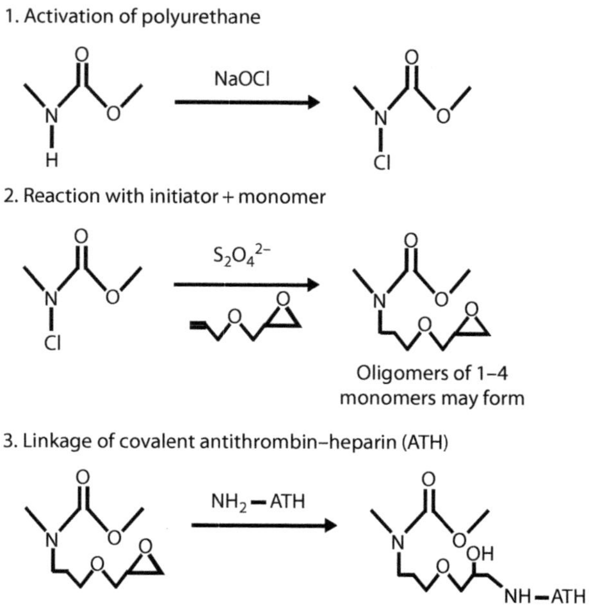

1. Activation of polyurethane

2. Reaction with initiator + monomer

Oligomers of 1–4
monomers may form

3. Linkage of covalent antithrombin–heparin (ATH)

FIGURE 1.6 Covalent coating of antithrombin–heparin conjugates onto polyurethane.

covalent coupling of ATH. Coating of polycarbonate urethane with ATH was verified by staining of the protein or GAG present on the surface [260].

Immobilization of heparin and hirudin (one of the most potent direct thrombin inhibitors known) onto surfaces has been studied for some time. Hirudin has been bound to clinically used biomaterials by a number of methods [261,262]. Significant coating densities with hirudin have been achieved with a surface character that gave efficient thrombin binding and inhibition. However, because hirudin–thrombin complexes are not covalent and only one thrombin can be inactivated per hirudin molecule, attempts have also been made to investigate the properties of heparinized biomaterials. Very ingenious methodologies have been developed to adapt heparin for use in a wide range of synthetic clinical devices. Only one such example is the bonding of heparin to bileaflet valves using a coating of graphite–carbon and benzalkonium chloride [263]. However, the most viable structures for anticoagulating surfaces with heparin have been shown to require end-point attachment of heparin chains [264]. This conclusion is consistent with the concept that active heparin molecules must be directed away from the surface in order to prevent surface activation of coagulants from the blood flow or surrounding tissue. In fact, it has been deduced that the critical action of heparinized biomaterials is to mediate inhibition of the coagulation cascade leading to prothrombin activation (thrombin feedback reactions) [264]. Unfortunately, simple end-point binding of heparin chains to the polymer surface is not sufficient for good thrombin inhibition. A series of experiments have demonstrated that heparin which was directly, covalently linked to polystyrene containing hydrophilic groups could not efficiently catalyze formation of TAT complexes [17]. In order to effect optimum anticoagulant activity of heparin on the surfaces, end-point attachment of the GAG to spacer arms (>2000 MW) is necessary [17]. As a final caveat, in an attempt to further fortify heparinized surfaces, free heparin has either been co-immobilized on the same surface with antithrombin [265], or surfaces with covalently bonded heparin have been pre-saturated with free antithrombin [266]. Although a mild improvement for thrombin inhibition has been noted, optimum arrangement of the antithrombin and heparin has not been achieved. Thus, surface coating with a significant density of antithrombin and heparin that yields rapid, high capacity inhibition of thrombin is desirable. Analysis of the ATH-coated polycarbonate urethanes presented an excellent opportunity for testing the viability of attaching high activity heparin to the surface through antithrombin itself.

Higher substitution of Chan et al.'s ATH on polycarbonate urethane was observed than that with either heparin alone or hirudin [255]. This result verifies that covalent ATH has superior chemical features for attachment of the heparin-containing anticoagulant. Storage properties of the ATH-coated surface are favorable since treated biomaterial could be utilized as an endoluminal graft in vivo after incubation in sterile saline at 4°C for minutes to months, with similar results [255]. Even though ATH was likely covalently bonded through the antithrombin moiety to the propyl glycidyl linker on the surface, the immobilized ATH exhibited significant direct non-catalytic activity against thrombin [255]. Thus, polymer surfaces could be created with ATH coatings that have long-term rapid inhibitory action toward thrombin, which does not require a supply of antithrombin from the patient. However, binding experiments with labeled antithrombin indicated that significant surface concentrations of pentasaccharide sites were readily accessible on polycarbonate urethane grafts coated with the Chan et al.'s ATH [255]. High affinity binding of antithrombin to the surface strongly suggests that ATH-coated material has the capacity to potentiate the reaction of plasma antithrombin with thrombin. The exogenous antithrombin binding results suggested that the ATH-covered polycarbonate urethane product likely possesses the desired catalytic capacity [255].

In vivo experiments were performed to examine if the ATH-coated endoluminal grafts had biological effectiveness that corresponded to the substitution density and antithrombin activity results observed in vitro. Endoluminal tubing (treated and non-treated) was deployed by a wire plunger through a narrow gauge catheter into the jugular veins of anesthetized rabbits. No anticoagulant was administered to the animals and, after 3 h, the tubing segments were recovered and weight of clot adherent to the device determined. A four- to fivefold reduction in weight of clot generated on the tubing was observed with ATH-coated compared to non-treated surfaces [255]. This improvement in antithrombogenicity of the biomaterial was not gained due to the presence of grafted propyl glycidyl groups since tests with these control surfaces (no ATH attached) were negative [255]. Furthermore, the heparin of ATH was critical because in vivo experiments with polycarbonate urethane coated with antithrombin gave results similar to those with surfaces that were reacted with allyl glycidyl ether alone. Finally, hirudin was bonded to polycarbonate urethane in a fashion similar to that of ATH, followed by testing in the rabbit jugular vein. In vitro analyses of the tubing showed that most of the hirudin bound to the endoluminal graft could inhibit thrombin directly. However, significantly more clot was generated in the hirudin tubing than the ATH tubing in vivo. Therefore, ATH-coated surfaces have greater potency than surfaces that are coated only with highly active direct thrombin inhibitors. The dual non-catalytic and catalytic anticoagulant activities of ATH may represent a new generation of pacified biomaterials.

More extensive in vivo experiments have been conducted with PU catheters possessing base coats containing ATH. Coating involved copolymerization of a monomer film with ATH modified by reaction with N-hydroxysuccinimide-PEO-acrylate [267]. Catheters were dipped into a monomer mixture solution containing a small amount of thermolabile 2,2′-azobis(isobutyronitrile), followed by drying and commencement of polymerization by heating at 80°C in an aqueous solution of PEO-acrylate ATH. The resulting catheter was a product covered inside and out with a covalent sheath network containing ATH cross-linked through the antithrombin moiety to the base-coat polymer and other ATH molecules [267,268]. ATH coatings were very robust as they resisted removal of antithrombin protein with general protease treatment under high sheer force [268]. The likelihood that antithrombin in the ATH was cross-linked on the surface during polymerization (more than one PEO-acrylate per ATH molecule was possible) could have greatly hindered protease attack. Immunofluorescence showed coating density to be fairly homogeneous down to $100\,\mu m^2$ of surface area [269]. ATH coated onto Solomon PU catheters were compared to the corresponding commercial uncoated PU and heparin-coated PU (CBAS) catheters [267–269]. Aqueous surface tension testing showed that while both PU and CBAS had hydrophobic character (negative dyne/cm surface tension values), ATH coatings were quite hydrophilic (surface tension = 32 dynes/cm) [267]. Anti-factor Xa heparin activity per cm^2 for ATH surfaces was seven times higher than that for heparin coatings [267]. In addition, specific uptake of antithrombin from human plasma was enhanced and nonselective binding of proteins other than albumin diminished for ATH compared to PU or CBAS [269]. Thus, optimal features for pacification of coagulation seemed to reside with the ATH biomaterial coating. Surfaces were tested for resistance to thrombosis in an acute rabbit model where the catheters were inserted into the jugular vein up to the heart of anesthetized rabbits, followed by syringe removal of blood and reinjection every 5 min up to 4 h or until occlusion occurred. While PU and CBAS catheters occluded (no blood could be drawn) in less than 80 min, no ATH catheters occluded within the 4 h experimental time period [267]. Pretreatment of ATH catheters with $NaIO_4$ abolished the patency characteristics, again affirming the antithrombotic importance of pentasaccharides in the conjugate's heparin chains. Examination of the catheters ex vivo showed that the accretion rate of fibrin (from radiolabeled fibrinogen pre-injected into the animals) was vastly reduced on catheters with ATH coatings compared to uncoated or heparin-coated surfaces [267]. Thus, activation of coagulation by ATH-coated biomaterials was decreased. In view of results from these short-term experiments, chronic testing in rabbits with similar catheter placement was performed. After recovery of the catheterized animals from anesthesia, blood was withdrawn and returned through the catheter every 8–16 h for 3–4 months or until occlusion (blood cannot be withdrawn or fluid injected via the catheter). PU and CBAS catheters had average occlusion times of 7.84 ± 0.96 days and 3.64 ± 0.78 days, respectively [268]. In contrast, catheters with ATH have never occluded, even up to 106 days of use [268]. Although ATH coatings had a much rougher surface topology (atomic force microscopy), scanning electron microscopic observation of catheters after removal from the animal revealed ATH surfaces to be almost pristine [268]. In total, ATH biomaterial coatings display exciting in vivo functionality and the potential to prevent vascular thrombotic activities in the absence of any circulating anticoagulant.

1.7 FUTURE

ATH is an agent that, either in fluid phase or surface immobilized form, pacifies the coagulation system in vivo. The fact that, with optimum methodologies, the highest activity heparin can be selected by antithrombin during complexation, gives a significant advantage for prevention and treatment of coagulation, as well as the construction of highly anticoagulant surfaces for clinical use. ATH preparation might be improved even further if all heparin molecules contained aldose end groups for conjugation. Chan and coworkers have recently reported novel methodology to convert UFH into functionally active, aldose-terminating heparin chains for ATH generation studies [270]. A number of investigations are necessary in order to fully realize the potential applications of ATH. Long-term use of fluid-phase ATH in animals needs to be evaluated for its application as a prophylactic. Furthermore, a number of clinical procedures could be studied to determine the value of ATH treatment during surgery. Recent experiments suggest that ATH may have applications during cardiopulmonary bypass to give improved prevention of brain emboli formation compared to UFH [271]. Other treatment modalities in which ATH may be efficacious are for the treatment of deep vein thrombosis and RDS. Animal models of these pathological states exist which could be used to challenge the viability of ATH as a treatment agent. Linkage of the conjugate to a variety of polymers must be done to determine the optimum type of biomaterials for coating with ATH. Present studies have helped to assess linking agents on gold model surfaces [272] and detailed physicochemical analyses have elucidated some of the mechanisms in ATH conjugation to PU surfaces [273]. In conjunction with the type of surface material is the investigation of other more sophisticated chemistries that could link to the antithrombin in ATH via amino acid R-groups other than that of lysine. Although the lysine amino group is a standard for protein immobilization, the heparin in ATH may tie up some of the key residues. Once the chemical details on ATH attachment to biomaterial surfaces have been determined, in vivo experiments in a number of animal models will be required to evaluate the best indication for use. Finally, clinical trials will be necessary before commercial use of either fluid-phase ATH or ATH-coated biomaterials is possible.

ACKNOWLEDGMENTS

We would like to thank Helen Atkinson for her help in preparation of this manuscript. This work was supported by a Grant-in-Aid (T6208) from the Heart and Stroke Foundation of Ontario. Anthony K. C. Chan is a career investigator of the Heart and Stroke Foundation of Canada. Anthony Chan holds a McMaster Children's Hospital/Hamilton Health Sciences Foundation Chair in Pediatric Thrombosis and Hemostasis.

REFERENCES

1. I. M. Nilson. Coagulation and fibrinolysis. *Scand. J. Gastroenterol.* 137:11 (1987).
2. H. R. Buller and T. Ten Cate. Coagulation and platelet activation pathways. A review of the key components and the way in which these can be manipulated. *Eur. Heart. J.* 16:8 (1995).
3. S. Butenas, C. van't Veer, and K. G. Mann. "Normal" thrombin generation. *Blood* 94:2169 (1999).
4. R. E. Jordan, G. M. Oosta, W. T. Gardner, and R. D. Rosenberg. The kinetics of hemostatic enzyme–antithrombin interactions in the presence of low molecular weight heparin. *J. Biol. Chem.* 255:10081 (1980).
5. B. Risberg, S. Andreasson, and E. Eriksson. Disseminated intravascular coagulation. *Acta Anaesthesiol. Scand.* 95:60 (1991).
6. J. Choay, M. Petitou, J. Lormeau, P. Sinay, B. Casu, and G. Gatti. Structure–activity relationship in heparin: A synthetic pentasaccharide with high affinity for antithrombin III and eliciting high anti factor Xa activity. *Biochem. Biophys. Res. Commun.* 116:492 (1983).
7. L. Muszbek, R. Adany, and H. Mikkola. Novel aspects of blood coagulation factor XIII. I. Structure, distribution, activation, and function. *Crit. Rev. Clin. Lab. Sci.* 33:357 (1996).
8. D. V. Devine and P. D. Bishop. Platelet-associated factor XIII in platelet activation, adhesion, and clot stabilization. *Thromb. Haemost.* 22:409 (1996).
9. W. Wang, M. B. Boffa, L. Bajzar, J. B. Walker, and M. E. Nesheim. A study of the mechanism of inhibition of fibrinolysis by activated thrombin activable fibrinolysis inhibitor. *J. Biol. Chem.* 273:27176 (1998).
10. L. Bajzar, M. Nesheim, J. Morser, and P. B. Tracey. Both cellular and soluble forms of thrombomodulin inhibit fibrinolysis by potentiating the activation of thrombin activable fibrinolysis inhibitor. *J. Biol. Chem.* 273:2792 (1998).
11. M. Nesheim, W. Wang, M. Boffa, M. Nagashima, J. Morser, and L. Bajzar. Thrombin, thrombomodulin, and TAFI in the molecular link between coagulation and fibrinolysis. *Thromb. Haemost.* 78:386 (1997).
12. M. B. Boffa, W. Wang, L. Bajzar, and M. E. Nesheim. Plasma and recombinant thrombin activable fibrinolysis inhibitor (TAFI) and activated TAFI compared with respect to glycosylation, thrombin/thrombomodulin-dependent activation, thermal stability, and enzymatic properties. *J. Biol. Chem.* 273:2127 (1998).
13. K. Kokame, X. Zheng, and J. E. Sadler. Activation of thrombin activable fibrinolysis inhibitor requires epidermal growth factor-like domain 3 of thrombomodulin and is inhibited competitively by protein C. *J. Biol. Chem.* 273:12135 (1998).
14. B. Dahlback. The protein C anticoagulant system: Inherited defects as basis for venous thrombosis. *Thromb. Res.* 77:1 (1995).
15. A. R. Rezaie. Tryptophan 60-D in the B-insertion loop of thrombin modulates the thrombin–antithrombin reaction. *Biochemistry* 35:1918 (1996).
16. B. Mille, J. Watton, T. W. Barrowcliffe, J. Mani, and D. Lane. Role of N- and C-terminal amino acids in antithrombin binding to pentasaccharide. *J. Biol. Chem.* 269:29435 (1994).
17. Y. Byun, H. A. Jacobs, and S. W. Kim. Mechanism of thrombin inactivation by immobilized heparin. *J. Biomed. Mater. Res.* 30:423 (1996).
18. J. W. Estes, E. W. Pelikan, and E. Kruger-Thiemer. A retrospective study of the pharmacokinetics of heparin. *Clin. Pharmacol. Ther.* 10:329 (1969).
19. E. Young, M. Prins, M. N. Levine, and J. Hirsh. Heparin binding to plasma proteins, an important mechanism for heparin resistance. *Thromb. Haemost.* 67:639 (1992).
20. J. Hirsh, W. G. van Aken, A. S. Gallus, C. T. Dollery, J. F. Cade, and W. L. Yung. Heparin kinetics in venous thrombosis and pulmonary embolism. *Circulation* 53:691 (1976).
21. L. B. Jaques, J. Mahadoo, and L. W. Kavanagh. Intrapulmonary heparin: A new procedure for anticoagulant therapy. *Lancet* 7996:1157 (1976).
22. P. J. Hogg and C. M. Jackson. Fibrin monomer protects thrombin from inactivation by heparin–antithrombin III: Implications for heparin efficacy. *Proc. Natl. Acad. Sci. USA* 86:3619 (1989).

23. P. R. Eisenberg, J. E. Siegel, D. R. Abendschein, and J. P. Miletich. Importance of factor Xa in determining the procoagulant activity of whole-blood clots. *J. Clin. Invest.* 91:1877 (1993).

24. N. A. Prager, D. R. Abendschein, C. R. McKenzie, and P. R. Eisenberg. Role of thrombin compared with factor Xa in the procoagulant activity of whole blood clots. *Circulation* 92:962 (1995).

25. P. Theroux, D. Waters, J. Lam, M. Juneau, and J. McCans. Reactivation of unstable angina after the discontinuation of heparin. *N. Engl. J. Med.* 327:192 (1992).

26. M. Palm, C. Mattsson, C. M. Svahn, and M. Weber. Bleeding times in rats treated with heparin, heparin fragments of high and low anticoagulant activity and chemically modified heparin fragments of low anticoagulant activity. *Thromb. Haemost.* 64:127 (1990).

27. J. Langdown, K. J. Belzar, W. J. Savory, T. P. Baglin, and J. A. Huntington. The critical role of hinge-region expulsion in the induced-fit heparin binding mechanism of antithrombin. *J. Mol. Biol.* 386:1278 (2009).

28. L. T. Hunt and M. O. Dayhoff. A surprising new protein superfamily containing ovalbumin, antithrombin-III, and alpha 1-protease inhibitor. *Biochem. Biophys. Res. Commun.* 95:864 (1980).

29. R. W. Carrell, D. R. Boswell, S. O. Brennan, and M. C. Owen. Active site of a1-antitrypsin: Homologous site in antithrombin-III. *Biochem. Biophys. Res. Commun.* 93:399 (1980).

30. T. Chandra, R. Stackhouse, V. Kidd, K. J. Robson, and S. L. Woo. Sequence homology between human alpha-1-antichymotrypsin, alpha-1-antitrypsin and antithrombin III. *Biochemistry* 22:5055 (1983).

31. R. W. Carrell, P. B. Christey, and D. R. Boswell. Serpins: Antithrombin and other inhibitors of coagulation and fibrinolysis: Evidence from amino acid sequences. In: *Thrombosis and Haemostasis* (M. Vertsraete, ed.), Leuven University Press, Leuven, Belgium, 1987, p. 1.

32. T. Takano, Y. Yamanouchi, Y. Mori, S. Kudo, T. Nakayama, M. Sugiura, S. Hashira, and T. Abe. Interstitial deletion of chromosome 1q [del(1)(q24q25.3)] identified by fluorescence in situ hybridization and gene dosage analysis of apolipoprotein A-II, coagulation factor V, and antithrombin III. *Am. J. Med. Genet.* 68:207 (1997).

33. S. C. Bock, J. A. Marrinan, and E. Radziejewska. Antithrombin III Utah: Proline-407 to leucine mutation in a highly conserved region near the inhibitor reactive site. *Biochemistry* 27:6171 (1988).

34. E. V. Prochownik, S. C. Bock, and S. H. Orkin. Intron structure of the human antithrombin III gene differs from that of other members of the serine protease inhibitor superfamily. *J. Biol. Chem.* 260:9608 (1985).

35. R. W. Niessen, A. Sturk, P. L. Hordijk, F. Michiels, and M. Peters. Sequence characterization of a sheep cDNA for antithrombin III. *Biochim. Biophys. Acta* 1171:207 (1992).

36. R. D. Rosenberg and P. S. Damus. The purification and mechanism of action of human antithrombin–heparin cofactor. *J. Biol. Chem.* 248:6490 (1973).

37. H. E. Manson, R. C. Austin, F. Fernandez-Rachubinski, R. A. Rachubinski, and M. A. Blajchman. The molecular pathology of inherited human antithrombin III deficiency. *Transfus. Med. Rev.* 3:264 (1989).

38. M. N. Blackburn, R. L. Smith, J. Carson, and C. C. Sibley. The heparin binding site of antithrombin III. Identification of a critical tryptophan in the amino acid sequence. *J. Biol. Chem.* 259:939 (1984).

39. G. B. Villanueva. Predictions of the secondary structure of antithrombin III and the location of the heparin-binding site. *J. Biol. Chem.* 259:2531 (1984).

40. T. E. Petersen, G. Dudek-Wojciechowska, L. Sottrup-Jensen and S. Magnusson. Primary structure of antithrombin III (heparin cofactor). Partial homology between a1-antitrypsin and antithrombin III. In: *The Physiological Inhibitors of Coagulation and Fibrinolysis* (D. Collen, ed.), Elsevier North-Holland Biomedical Press, Amsterdam, the Netherlands, 1979, p. 43.

41. R. W. Carrell, P. E. Stein, G. Fermi, and M. R. Wardell. Biological implications of a 3 A structure of dimeric antithrombin. *Structure* 2:257 (1994).

42. M. R. Wardell, J. P. Abrahams, D. Bruce, R. Skinner, and A. G. Leslie. Crystallization and preliminary x-ray diffraction analysis of two conformations of intact human antithrombin. *J. Mol. Biol.* 234:1253 (1993).

43. M. R. Witmer and M. W. Hatton. Antithrombin III-beta associates more readily than antithrombin III-alpha with uninjured and de-endothelialized aortic wall in vitro and in vivo. *Arterioscler. Thromb.* 11:530 (1991).

44. T. H. Carlson, M. R. Kolman, and M. Piepkorn. Activation of antithrombin III isoforms by heparan sulphate glycosaminoglycans and other sulphated polysaccharides. *Blood Coagul. Fibrinolysis* 6:474 (1995).

45. S. O. Brennan, J. Y. Borg, P. M. George, C. Soria, J. Soria, J. Caen, and R. W. Carrell. New carbohydrate site in mutant antithrombin (7 Ile–Asn) with decreased heparin affinity. *FEBS Lett.* 237:118 (1988).

46. V. Picard, E. Ersdal-Badju, and S. C. Bock. Partial glycosylation of antithrombin III asparagine-135 is caused by the serine in the third position of its N-glycosylation consensus sequence and is responsible for production of the beta-antithrombin III isoform with enhanced heparin affinity. *Biochemistry* 34:8433 (1995).

47. L. Garone, T. Edmunds, E. Hanson, R. Bernasconi, J. A. Huntington, J. L. Meagher, B. Fan, and P. G. Gettins. Antithrombin–heparin affinity reduced by fucosylation of carbohydrate at asparagine 155. *Biochemistry* 35:8881 (1996).

48. I. Bjork, K. Ylinenjarvi, S. T. Olson, P. Hermentin, H. S. Conradt, and G. Zettlmeissl. Decreased affinity of recombinant antithrombin for heparin due to increased glycosylation. *Biochem. J.* 286:793 (1992).

49. A. Borsodi and T. R. Narasimhan. Microheterogeneity of human antithrombin III. *Br. J. Haematol.* 39:121 (1978).

50. T. Edmunds, S. M. Van Patten, J. Pollock, E. Hanson, R. Bernasconi, E. Higgins, P. Manavalan, C. Ziomek, H. Meade, J. M. McPherson, and E. S. Cole. Transgenically produced human antithrombin: Structural and functional comparison to human plasma-derived antithrombin. *Blood* 91:4561 (1998).

51. Q. Zhou, J. Kyazike, Y. Echelard, H. M. Meade, E. Higgins, E. S. Cole, and T. Edmunds. Effect of genetic background on glycosylation heterogeneity in human antithrombin produced in the mammary gland of transgenic goats. *J. Biotechnol.* 117:57 (2005).

52. R. Devraj-Kizuk, D. H. K. Chui, E. V. Prochownik, C. J. Carter, F. A. Ofosu, and M. A. Blajchman. Antithrombin III-Hamilton: A gene with a point mutation (guanine to adenine) in codon 382 causing impaired serine protease reactivity. *Blood* 72:1518 (1988).

53. H. C. Whinna and F. C. Church. Interaction of thrombin with antithrombin, heparin cofactor II, and protein C inhibitor. *J. Protein Chem.* 12:677 (1993).

54. Y. J. Chuang, P. G. Gettins, and S. T. Olson. Importance of the P2 glycine of antithrombin in target proteinase specificity, heparin activation, and the efficiency of proteinase trapping as revealed by a P2 Gly→Pro mutation. *J. Biol. Chem.* 274:28142 (1999).

55. R. W. Carrell and M. C. Owen. Plakalbumin, a2-antitrypsin, antithrombin and the mechanism of inflammatory thrombosis. *Nature* 317:730 (1985).

56. A. J. Schulze, R. Huber, W. Bode, and R. A. Engh. Structural aspects of serpin inhibition. *FEBS Lett.* 344:117 (1994).

57. K. Skriver, W. R. Wikoff, P. A. Patston, F. Tausk, M. Schapira, A. P. Kaplan, and S. C. Bock. Substrate properties of C1 inhibitor Ma (alanine 434–glutamic acid). Genetic and structural evidence suggesting that the P12-region contains critical determinants of serine protease inhibitor/substrate status. *J. Biol. Chem.* 266:9216 (1991).

58. A. J. Schulze, U. Baumann, S. Knof, E. Jaeger, R. Huber, and C. B. Laurell. Structural transition of alpha-1-antitrypsin by a peptide sequentially similar to beta-strand s4A. *Eur. J. Biochem.* 194:51 (1990).

59. A. J. Schulze, P. W. Frohnert, R. A. Engh, and R. Huber. Evidence for the extent of insertion of the active site loop of intact alpha 1 proteinase inhibitor in beta-sheet A. *Biochemistry* 31:7560 (1992).

60. I. Bjork, K. Nordling, I. Larsson, and S. T. Olson. Kinetic characterization of the substrate reaction between a complex of antithrombin with a synthetic reactive-bond loop tetradecapeptide and four target proteinases of the inhibitor. *J. Biol. Chem.* 267:19047 (1992).

61. I. Bjork, K. Ylinenjarvi, S. T. Olson, and P. E. Bock. Conversion of antithrombin from an inhibitor of thrombin to a substrate with reduced heparin affinity and enhanced conformational stability by binding of a tetradecapeptide corresponding to the P1 to P14 region of the putative reactive bond loop of the inhibitor. *J. Biol. Chem.* 267:1976 (1992).

62. S. Debrock and P. J. Decleck. Characterization of common neoantigenic epitopes generated in plasminogen activator inhibitor-1 after cleavage of the reactive center loop or after complex formation with various serine proteinases. *FEBS Lett.* 376:243 (1995).

63. K. Nordling and I. Bjork. Identification of an epitope in antithrombin appearing on insertion of the reactive-bond loop into the A beta-sheet. *Biochemistry* 35:10436 (1996).

64. J. Whisstock, A. M. Lesk, and R. Carrell. Modeling of serpin–protease complexes: Antithrombin–thrombin, alpha-1-antitrypsin (358Met→Arg)-thrombin, alpha-1-antitrypsin (358Met→Arg)–trypsin, and antitrypsin-elastase. *Proteins* 26:288 (1996).

65. R. Skinner, J.-P. Abrahams, J. C. Whisstock, A. M. Lesk, R. W. Carrell, and M. R. Wardell. The 2.6 A structure of antithrombin indicates a conformational change at the heparin binding site. *J. Mol. Biol.* 266:601 (1997).

66. V. Picard, P.-E. Marque, F. Paolucci, M. Aiach, and B. F. Le Bonniec. Topology of the stable serpin–protease complexes revealed by an autoantibody that fails to react with the monomeric conformers of antithrombin. *J. Biol. Chem.* 274:4586 (1999).

67. L. Jin, J. P. Abrahams, R. Skinner, M. Petitou, R. N. Pike, and R. W. Carrell. The anticoagulant activation of antithrombin by heparin. *Proc. Natl. Acad. Sci. USA* 94:14683 (1997).

68. J. L. Meagher, J. M. Beechem, S. T. Olson, and P. G. Gettins. Deconvolution of the fluorescence emission spectrum of human antithrombin and identification of the tryptophan residues that are responsive to heparin binding. *J. Biol. Chem.* 273:23283 (1998).

69. T. Koide, S. Odani, K. Takahashi, T. Ono, and N. Sakuragawa. Replacement of arginine-47 by cysteine in hereditary abnormal antithrombin III that lacks heparin-binding ability. *Proc. Natl. Acad. Sci. USA* 81:289 (1984).

70. M. C. Owen, J. Y. Borg, C. Soria, J. Soria, J. Caen, and R. W. Carrell. Heparin binding defect in a new antithrombin III variant: Rouen, 47Arg to His. *Blood* 69:1275 (1987).

71. J. Y. Borg, M. C. Owen, C. Soria, J. Soria, J. Caen, and R. W. Carrell. Proposed heparin binding site in antithrombin based on arginine 47: A new variant Rouen-II, 47Arg to Ser. *J. Clin. Invest.* 81:1292 (1988).

72. J. Y. Chang and T. H. Tran. Antithrombin III Basel. Identification of a Pro–Leu substitution in a hereditary abnormal antithrombin with impaired heparin cofactor activity. *J. Biol. Chem.* 261:1174 (1986).

73. H. L. Fitton, R. Skinner, T. R. Dafforn, L. Jin, and R. N. Pike. The N-terminal segment of antithrombin acts as a steric gate for the binding of heparin. *Protein Sci.* 7:782 (1998).

74. J. W. Smith and D. J. Knauer. A heparin binding site in antithrombin III. *J. Biol. Chem.* 262:11964 (1987).

75. E. Ersdal-Badju, A. Lu, Y. Zuo, V. Picard, and S. C. Bock. Identification of the antithrombin III heparin binding site. *J. Biol. Chem.* 272:19393 (1997).

76. R. Skinner, W. S. Chang, L. Jin, X. Pei, J. A. Huntington, J. P. Abrahams, R. W. Carrell, and D. A. Lomas. Implications for function and therapy of a 2.9 A structure of binary-complexed antithrombin. *J. Mol. Biol.* 283:9 (1998).

77. M. W. Hatton, L. R. Berry, and E. Regoeczi. Inhibition of thrombin by antithrombin III in the presence of certain glycosaminoglycans found in the mammalian aorta. *Thromb. Res.* 13:655 (1978).

78. T. B. Brieditis, S. C. Bock, S. T. Olson, and I. Bjork. The oligosaccharide side chain on Asn-135 of alpha-antithrombin, absent in beta-antithrombin, decreases the heparin affinity of the inhibitor by affecting the heparin-induced conformational change. *Biochemistry* 36:6682 (1997).

79. B. Fan, B. C. Crews, I. V. Turko, J. Choay, G. Zettlmeissl, and P. Gettins. Heterogeneity of recombinant human antithrombin III expressed in baby hamster kidney cells. Effect of glycosylation differences on heparin binding and structure. *J. Biol. Chem.* 268:17588 (1993).

80. S. T. Olson, A. M. Frances-Chmura, R. Swanson, I. Bjork, and G. Zettlmeissl. Effect of individual carbohydrate chains of recombinant antithrombin on heparin affinity and on the generation of glycoforms differing in heparin affinity. *Arch. Biochem. Biophys.* 341:212 (1997).

81. K. Lidholt. Biosynthesis of glycosaminoglycans in mammalian cells and in bacteria. *Biochem. Soc. Trans.* 25:866 (1997).

82. S. Ernst, R. Langer, C. L. Cooney, and R. Sasisekharan. Enzymatic degradation of glycosaminoglycans. *Crit. Rev. Biochem. Mol. Biol.* 30:387 (1995).

83. J. T. Gallagher and A. Walker. Molecular distinctions between heparin sulphate and heparin. Analysis of sulphation patterns indicates that heparin sulphate and heparin are separate families of N-sulphated polysaccharides. *Biochem. J.* 230:665 (1985).

84. J. E. Shively and H. E. Conrad. Nearest neighbor analysis of heparin: Identification and quantitation of the products formed by selective depolymerization procedures. *Biochemistry* 15:3943 (1976).

85. H. C. Robinson, A. A. Horner, M. Hook, S. Ogren, and U. Lindahl. A proteoglycan form of heparin and its degradation to single-chain molecules. *J. Biol. Chem.* 253:6687 (1978).

86. M. Iacomini, B. Casu, M. Guerrini, A. Naggi, A. Pirola, and G. Torri. "Linkage Region" sequences of heparins and heparan sulfates: Detection and quantification by nuclear magnetic resonance spectroscopy. *Anal. Biochem.* 274:50 (1999).

87. U. Lindahl and O. Axelsson. Identification of iduronic acid as the major sulfated uronic acid of heparin. *J. Biol. Chem.* 246:74 (1971).

88. T. Lind, U. Lindahl, and K. Lidholt. Biosynthesis of heparin/heparan sulfate. Identification of a 70-kDa protein catalyzing both the D-glucuronosyl- and the *N*-acetyl-D-glucosaminyltransferase reactions. *J. Biol. Chem.* 268:20705 (1993).

89. J. Riesenfeld, M. Hook, and U. Lindahl. Biosynthesis of heparin. Assay and properties of the microsomal *N*-acetyl-D-glucosaminyl-*N*-deacetylase. *J. Biol. Chem.* 255:922 (1980).

90. I. Jacobsson, U. Lindahl, J. W. Jensen, L. Roden, H. Prihar, and D. S. Feingold. Biosynthesis of heparin. Substrate specificity of heparosan *N*-sulfate D-glucuronosyl 5-epimerase. *J. Biol. Chem.* 259:1056 (1984).

91. R. R. Miller and C. J. Waechter. Partial purification and characterization of detergent-solubilized *N*-sulfotransferase activity associated with calf brain microsomes. *J. Neurochem.* 51:87 (1988).
92. K. Lidholt and U. Lindahl. Biosynthesis of heparin. The D-glucuronosyl- and *N*-acetyl-D-glucosaminyltransferase reactions and their relation to polymer modification. *Biochem. J.* 287:21 (1992).
93. I. Jacobsson and U. Lindahl. Biosynthesis of heparin. Concerted action of late polymer-modification reactions. *J. Biol. Chem.* 255:5094 (1980).
94. A. B. Foster, R. Harrison, T. D. Inch, M. Stacey, and J. M. Webber. Amino-sugars and related compounds. Part IX. Periodate oxidation of heparin and some related substances. *J. Am. Chem. Soc.* 85:2279 (1963).
95. K. Lidholt, L. Kjellen, and U. Lindahl. Biosynthesis of heparin. Relationship between the polymerization and sulphation processes. *Biochem. J.* 261:999 (1989).
96. L. Toma, P. Berninsome, and C. B. Hirschberg. The putative heparin specific *N*-acetylglucosaminyl *N*-deacetylase/*N*-sulfotransferase also occurs in non-heparin-producing cells. *J. Biol. Chem.* 273:22458 (1998).
97. S. Ogren and U. Lindahl. Cleavage of macromolecular heparin by an enzyme from mouse mastocytoma. *J. Biol. Chem.* 250:2690 (1975).
98. S. Ogren and U. Lindahl. Metabolism of macromolecular heparin in mouse neoplastic mast cells. *Biochem. J.* 154:605 (1976).
99. J. Fareed, K. Fu, L. H. Yang, and D. A. Hoppensteadt. Pharmacokinetics of low molecular weight heparins in animal models. *Semin. Thromb. Hemost.* 25:51 (1999).
100. J. Fareed, S. Haas, and A. Sasahar. Past, present and future considerations on low molecular weight heparin differentiation: An epilogue. *Semin. Thromb. Hemost.* 25:145 (1999).
101. N. S. Jaikaria, L. Rosenfeld, M. Y. Khan, I. Danishefsky, and S. A. Newman. Interaction of fibronectin with heparin in model extracellular matrices: Role of arginine residues and sulfate groups. *Biochemistry* 30:1538 (1991).
102. T. Zak-Nejmark, M. Krasnowska, R. Jankowska, and M. Jutel. Heparin modulates migration of human peripheral blood mononuclear cells and neutrophils. *Arch. Immunol. Ther. Exp. (Warsz)* 47:245 (1999).
103. W. Jeske and J. Fareed. Antithrombin III- and heparin cofactor II-mediated anticoagulant and antiprotease actions of heparin and its synthetic analogues. *Semin. Thromb Hemost.* 19:241 (1993).
104. R. D. Rosenfeld. Role of heparin and heparin-like molecules in thrombosis and atherosclerosis. *Fed. Proc.* 44:404 (1985).
105. T. W. Barrowcliffe, B. Mulloy, E. A. Johnson, and D. P. Thomas. The anticoagulant activity of heparin: Measurement and relationship to chemical structure. *J. Pharm. Biomed. Anal.* 7:217 (1989).
106. S. Frebelius, U. Hedin, and J. Swedenborg. Thrombogenecity of the injured vessel wall—Role of antithrombin and heparin. *Thromb. Haemost.* 71:147 (1994).
107. L. Liu, L. Dewar, Y. Song, M. Kulczycky, M. Blajchman, J. W. Fenton II, M. Andrew, M. Delorme, J. Ginsberg, K. T. Preissner and F. A. Ofosu. Inhibition of thrombin by antithrombin III and heparin cofactor II in vivo. *Thromb. Haemost.* 73:405 (1995).
108. Y. I. Wu, W. P. Sheffield, and M. Blajchman. Defining the heparin-binding domain of antithrombin. *Blood Coagul. Fibrinolysis* 5:83 (1994).
109. A. A. Horner and E. Young. Asymmetric distribution of sites with high affinity for antithrombin III in rat skin heparin proteoglycans. *J. Biol. Chem.* 257:8749 (1982).
110. K.-G. Jacobsson, U. Lindahl, and A. A. Horner. Location of antithrombin-binding regions in rat skin heparin proteoglycans. *Biochem. J.* 240:625 (1986).
111. A. K. Chan, L. Berry, H. O'Brodovich, P. Klement, L. Mitchell, B. Baranowski, P. Monagle, and M. Andrew. Covalent antithrombin–heparin complexes with high anticoagulant activity: Intravenous, subcutaneous and intratracheal administration. *J. Biol. Chem.* 272:22111 (1997).
112. R. D. Rosenberg, R. E. Jordan, L. V. Favreau, and L. H. Lam. Highly active heparin species with multiple binding sites for antithrombin. *Biochem. Biophys. Res. Commun.* 86:1319 (1979).
113. L. G. Oscarsson, G. Pejler, and U. Lindhahl. Location of the antithrombin-binding sequence in the heparin chain. *J. Biol. Chem.* 264:296 (1989).
114. T. C. Laurent, A. Tengblad, L. Thunberg, M. Hook, and U. Lindahl. The molecular-weight-dependence of the anti-coagulant activity of heparin. *Biochem. J.* 175:691 (1978).
115. L. R. Berry and M. W. Hatton. Controlled depolymerization of heparin: Anticoagulant activity and molecular size of the products. *Biochem. Soc. Trans.* 11:101 (1983).
116. M. Kusche, G. Backstrom, J. Riesenfeld, M. Petitou, J. Choay, and U. Lindahl. Biosynthesis of heparin. O-Sulfation of the antithrombin-binding region. *J. Biol. Chem.* 263:15474 (1988).

117. M. Petitou, P. Duchaussoy, I. Lederman, J. Choay, and P. Sinay. Binding of heparin to antithrombin III: A chemical proof of the critical role played by a 3-sulfated 2-amino-2-deoxy-D-glucose residue. *Carbohydr. Res.* 179:163 (1988).

118. M. Petitou, P. Duchaussoy, I. Lederman, J. Choay, P. Sinay, J. C. Jacquinet, and G. Torri. Synthesis of heparin fragments. A chemical synthesis of the pentasaccharide *O*-(2-deoxy-2-sulfamido-6-*O*-sulfo-alpha-D-glucopyranosyl)-(1–4)-*O*-(beta-D-glucopyranosyluronic   acid)-(1–4)-*O*-(2-deoxy-2-sulfamido-3,6-di-*O*-sulfo-alpha-D-glucopyranosyl)-(1–4)-*O*-(2-*O*-sulfo-alpha-L-idopyranosyluronic acid)-(1–4)-2-deoxy-2-sulfamido-6-*O*-sulfo-D-glucopyranose decasodium salt, a heparin fragment having high affinity for antithrombin III. *Carbohydr. Res.* 147:221 (1986).

119. M. Petitou, T. Barzu, J. P. Herault, and J. M. Herbert. A unique trisaccharide sequence in heparin mediates the early step of antithrombin III activation. *Glycobiology* 7:323 (1997).

120. U. R. Desai, M. Petitou, I. Bjork, and S. T. Olson. Mechanism of heparin activation of antithrombin. Role of individual residues of the pentasaccharide activating sequence in the recognition of native and activated states of antithrombin. *J. Biol. Chem.* 273:7478 (1998).

121. U. R. Desai, M. Petitou, I. Bjork, and S. T. Olson. Mechanism of heparin activation of antithrombin: Evidence for an induced-fit model of allosteric activation involving two interaction subsites. *Biochemistry* 37:13033 (1998).

122. V. Arocas, S. C. Bock, S. T. Olson, and I. Bjork. The role of Arg46 and Arg47 of antithrombin in heparin binding. *Biochemistry* 38:10196 (1999).

123. R. J. Kandrotas. Heparin pharmacokinetics and pharmacodynamics. *Clin. Pharmacokinet.* 22:359 (1992).

124. J. Ye, A. R. Rezaie, and C. T. Esmon. Glycosaminoglycan contributions to both protein C activation and thrombin inhibition involve a common arginine-rich site in thrombin that includes residues arginine 93, 97, and 101. *J. Biol. Chem.* 269:17965 (1994).

125. L. C. Petersen and M. Jorgensen. Electrostatic interactions in the heparin-enhanced reaction between human thrombin and antithrombin. *Biochem. J.* 211:91 (1983).

126. M. Lozano, A. Bos, P. G. de Groot, G. van Willigen, D. G. Meuleman, A. Ordinas, and J. J. Sixma. Suitability of low molecular weight heparin(oid)s and a pentasaccharide for an in vitro human thrombosis model. *Arterioscler. Thromb.* 14:1215 (1994).

127. D. Bergqvist. Low molecular weight heparins. *J. Intern. Med.* 240:63 (1996).

128. E. Young, P. Wells, S. Holloway, J. Weitz, and J. Hirsh. Ex-vivo and in-vitro evidence that low molecular weight heparins exhibit less binding to plasma proteins than unfractionated heparin. *Thromb. Haemost.* 71:300 (1994).

129. L. Bara, E. Billaud, G. Gramond, and A. Kher. Comparative pharmacokinetics (PK 10169) and unfractionated heparin after intravenous and subcutaneous administration. *Thromb. Res.* 39:631 (1985).

130. J. Hirsh and M. N. Levine. Low molecular weight heparin. *Blood* 79:1 (1992).

131. J. Fareed, D. Hoppensteadt, W. Jeske, R. Clarizio, and J. M. Walenga. Low molecular weight heparins: Are they different? *Can. J. Cardiol.* 14:28E (1998).

132. M. Wolzt, A. Weltermann, M. Nieszpaur-Los, B. Schneider, A. Fassolt, K. Lechner, H. G. Eichler, and P. A. Kyrle. Studies on the neutralizing effects of protamine on unfractionated and low molecular weight heparin (Fragmin) at the site of activation of the coagulation system in man. *Thromb. Haemost.* 73:439 (1995).

133. H. F. Schran, D. W. Bitz, F. J. DiSerio, and J. Hirsh. The pharmacokinetics and bioavailability of subcutaneously administered dihydroergotamine, heparin and the dihydroergotamine–heparin combination. *Thromb. Res.* 31:51 (1983).

134. L. Briant, C. Caranobe, S. Saivin, P. Sie, B. Bayrou, G. Houin, and B. Boneu. Unfractionated heparin and CY 216: Pharmacokinetics and bioavailabilities of the antifactor Xa and IIa effects after intravenous and subcutaneous injection in the rabbit. *Thromb. Haemost.* 61:348 (1989).

135. M. D. Laforest, N. Colas-Linhart, S. Guiraud-Vitaux, B. Bok, L. Bara, M. Samama, J. Marin, F. Imbault, and A. Uzan. Pharmacokinetics and biodistribution of technetium 99m labelled standard heparin and a low molecular weight heparin (enoxaparin) after intravenous injection in normal volunteers. *Br. J. Haematol.* 77:201 (1991).

136. L. Manson, J. I. Weitz, T. J. Podor, J. Hirsh, and E. Young. The variable anticoagulant response to unfractionated heparin in vivo reflects binding to plasma proteins rather than clearance. *J. Lab. Clin. Med.* 130:649 (1997).

137. E. Young, T. J. Podor, T. Venner, and J. Hirsh. Induction of the acute-phase reaction increases heparin-binding proteins in plasma. *Arterioscler. Thromb. Vasc. Biol.* 17:1568 (1997).

138. H. L. Messmore, B. Griffin, J. Fareed, E. Coyne, and J. Seghatchian. In vitro studies of the interaction of heparin, low molecular weight heparin and heparinoids with platelets. *N.Y. Acad. Sci.* 556:217 (1989).

139. W. A. Patton, C. A. Granzow, L. A. Getts, S. C. Thomas, L. M. Zotter, K. A. Gunzel, and L. J. Lowe-Krentz. Identification of a heparin-binding protein using monoclonal antibodies that block heparin binding to porcine aortic endothelial cells. *Biochem. J.* 311:461 (1995).
140. K. H. Teoh, E. Young, C. A. Bradley, and J. Hirsh. Heparin binding proteins. Contribution to heparin rebound after cardiopulmonary bypass. *Circulation* 88:420 (1993).
141. P. J. Hogg and C. M. Jackson. Heparin promotes the binding of thrombin to fibrin polymer. Quantitative characterization of a thrombin–fibrin polymer–heparin ternary complex. *J. Biol. Chem.* 265:241 (1990).
142. P. J. Hogg, C. M. Jackson, J. K. Labanowski, and P. E. Bock. Binding of fibrin monomer and heparin to thrombin in a ternary complex alters the environment of the thrombin catalytic site, reduces affinity for hirudin and inhibits cleavage of fibrinogen. *J. Biol. Chem.* 271:26088 (1996).
143. P. J. Hogg and C. M. Jackson. Formation of a ternary complex between thrombin, fibrin monomer, and heparin influences the action of thrombin on its substrates. *J. Biol. Chem.* 265:248 (1990).
144. E. Young, B. Cosmi, J. Weitz, and J. Hirsh. Comparison of the non-specific binding of unfractionated heparin and low molecular weight heparin (enoxaparin) to plasma proteins. *Thromb. Haemost.* 70:625 (1993).
145. B. Cosmi, J. C. Fredenburgh, J. Rischke, J. Hirsh, E. Young, and J. I. Weitz. Effect of nonspecific binding to plasma proteins on the antithrombin activities of unfractionated heparin, low-molecular-weight heparin, and dermatan sulfate. *Circulation* 95:118 (1997).
146. P. Bendayan, H. Boccalon, D. Dupouy, and B. Boneu. Dermatan sulphate is a more potent inhibitor of clot-bound thrombin than unfractionated and low molecular weight heparins. *Thromb. Haemost.* 71:576 (1994).
147. T. E. Warkentin, M. N. Levine, J. Hirsh, P. Horsewood, R. S. Roberts, M. Gent, and J. G. Kelton. Heparin-induced thrombocytopenia in patients treated with low molecular weight heparins or unfractionated heparin. *N. Engl. J. Med.* 332:1330 (1995).
148. M. Cohen. Heparin-induced thrombocytopenia and the clinical use of low molecular weight heparins in acute coronary syndromes. *Semin. Hematol.* 36:33 (1999).
149. R. D. Hull, G. E. Raskob, D. Rosenbloom, J. Lemaire, G. F. Pineo, B. Baylis, J. S. Ginsberg, A. A. Panju, P. Brill-Edwards, and R. Brant. Optimal therapeutic level of heparin therapy in patients with venous thrombosis. *Arch. Intern. Med.* 152:1589 (1992).
150. J. Hirsh, E. W. Salzman, and V. J. Marder. Treatment of venous thromboembolism. In: *Hemostasis and Thrombosis: Basic Principles and Clinical Practice* (R. W. Colman, ed.), Lippincott, New York, 1994, p. 1346.
151. M. M. Koopman, P. Prandoni, F. Piovella, P. A. Ockelford, D. F. Brandjes, J. van der Meer, A. S. Gallus, C. Simoneau, C. H. Chesterman, and M. H. Prins. Treatment of venous thrombosis with intravenous unfractionated heparin administered in the hospital as compared with subcutaneous low molecular weight heparin administered at home. The Tasman Study Group. *N. Engl. J. Med.* 334:682 (1996).
152. M. Levine, M. Gent, J. Hirsh, J. Leclerc, D. Anderson, J. Weitz, J. Ginsberg, A. J. Turpie, C. Demers, and M. Kovaks. A comparison of low molecular weight heparin administered primarily at home with unfractionated heparin administered in the hospital for proximal deep venous thrombosis. *N. Engl. J. Med.* 334:677 (1996).
153. J. Hirsh, S. Siragusa, B. Cosmi, and J. S. Ginsberg. Low molecular weight heparins (LMWH) in the treatment of patients with acute venous thromboembolism. *Thromb. Haemost.* 74:360 (1995).
154. GUSTO IIb investigators. A comparison of recombinant hirudin with heparin for the treatment of acute coronary syndromes. The Global Use of Strategies to Open Occluded Coronary Arteries (GUSTO) IIb investigators. *N. Engl. J. Med.* 335:775 (1996).
155. I. J. Sarembock, S. D. Gertz, L. M. Thome, K. W. McCoy, M. Ragosta, E. R. Powers, J. M. Maraganore, and L. W. Gimple. Effectiveness of hirulog in reducing restenosis after balloon angioplasty of atherosclerotic femoral arteries in rabbits. *J. Vasc. Res.* 33:308 (1996).
156. M. Verstraete. New developments in antiplatelet and antithrombotic therapy. *Eur. Heart J.* 16:16 (1995).
157. W. E. Marki. The anticoagulant and antithrombotic properties of hirudins. *Thromb Haemost.* 64:344 (1991).
158. U. Egner, G. A. Hoyer, and W. D. Schleuning. Rational design of hirulog-type inhibitors of thrombin. *J. Comput. Aided Mol. Des.* 8:479 (1994).
159. G. Agnelli, R. Cinzia, J. I. Weitz, G. G. Nenci, and J. Hirsh. Sustained antithrombotic activity of hirudin after its plasma clearance: Comparison with heparin. *Blood* 80:960 (1992).
160. J. M. Cardot, G. Y. Lefevre, and J. A. Godbillon. Pharmacokinetics of rec-hirudin in healthy volunteers after intravenous administration. *J. Pharmacokinet. Biopharm.* 22:147 (1994).
161. I. Fox, A. Dawson, P. Loynds, J. Eisner, K. Findlen, E. Levin, D. Hanson, T. Mant, J. Wagner, and J. Maraganore. Anticoagulant activity of hirulog, a direct thrombin inhibitor, in humans. *Thromb. Haemost.* 69:157 (1993).

162. P. Zoldhelyi, J. Bichler, W. G. Owen, D. E. Grill, V. Fuster, J. S. Mruk, and J. H. Chesebro. Persistent thrombin generation in humans during specific thrombin inhibition with hirudin. *Circulation* 90:2671 (1994).

163. W. A. Schumacher, T. E. Steinbacher, C. L. Heran, J. R. Megill, and S. K. Durham. Effects of anti-thrombotic drugs in a rat model of aspirin-insensitive arterial thrombosis. *Thromb. Haemost.* 69:509 (1993).

164. N. A. Scott, G. L. Nunes, S. B. King, 3rd, L. A. Harker, and S. R. Hanson. Local delivery of an antithrombin inhibits platelet-dependent thrombosis. *Circulation* 90:1951 (1994).

165. S. Hollenbach, U. Sinha, P. H. Lin, K. Needham, L. Frey, T. Hancock, A. Wong, and D. Wolf. A comparative study of prothrombinase and thrombin inhibitors in a novel rabbit model of non-occlusive deep vein thrombosis. *Thromb. Haemost.* 71:357 (1994).

166. C. H. Pletcher and G. L. Nelsestuen. The rate-determining step of the heparin-catalyzed antithrombin/thrombin reaction is independent of thrombin. *J. Biol. Chem.* 257:5342 (1982).

167. J. M. Muir, J. Hirsh, J. Weitz, M. Andrew, E. Young, and S. G. Shaughnessy. A histomorphometric comparison of the effects of heparin and low molecular weight heparin on cancellous bone in rats. *Blood* 89:3236 (1997).

168. F. Brus, W. van Oeveren, A. Heikamp, A. Okken, and S. B. Oetomo. Leakage of protein into lungs of preterm ventilated rabbits is correlated with activation of clotting, complement, and polymorphonuclear leukocytes in plasma. *Pediatr. Res.* 39:958 (1996).

169. J. Holter, J. Weiland, E. Packt, J. Gadek, and W. Davis. Protein permeability in the adult respiratory distress syndrome. Loss of size selectivity of the alveolar epithelium. *J. Clin. Invest.* 78:1513 (1986).

170. C. L. Sprung, W. M. Long, E. H. Marcial, R. M. H. Schein, R. E. Parker, T. Schomer, and K. L. Brigham. Distribution of proteins in pulmonary edema. *Am. Rev. Respir. Dis.* 136:957 (1987).

171. M. Bachofen and E. R. Weibel. Structural alterations of lung parenchyma in the adult respiratory distress syndrome. *Clin. Chest Med.* 3:35 (1982).

172. D. Gitlin and J. Craig. The nature of the hyaline membrane in asphyxia of the newborn. *Pediatrics* 17:64 (1956).

173. A. H. Jobe. Pulmonary surfactant therapy. *N. Engl. J. Med.* 328:861 (1993).

174. W. Long, T. Thompson, and H. Sundell. Effects of two rescue doses of a synthetic surfactant on mortality rate and survival without bronchopulmonary dysplasia. *J. Pediatr.* 118:595 (1991).

175. The OSIRIS Collaborative Group. Early versus delayed neonatal administration of a synthetic surfactant-the judgement of OSIRIS. *Lancet* 340:1363 (1992).

176. E. A. Liechty, E. Donovan, and D. Purohit. Reduction of neonatal mortality after multiple doses of bovine surfactant in low birth weight neonates with respiratory distress syndrome. *Pediatrics* 88:19 (1991).

177. J. J. Coalson, V. T. Winter, D. R. Gerstmann, S. Idell, R. J. King, and R. A. Delemos. Pathophysiologic, morphometric and biochemical studies of the premature baboon with bronchopulmonary dysplasia. *Am. Rev. Respir. Dis.* 145:872 (1992).

178. R. Ceustermans, M. Hoylaerts, M. DeMol, and D. Collen. Preparation, characterization, and turnover properties of heparin–antithrombin III complexes stabilized by covalent bonds. *J. Biol. Chem.* 257:3401 (1982).

179. R. E. Jordan. Antithrombin–heparin complex and method for its production. US Patent 4,446,126, Cutter Laboratories, 1984.

180. J. Eibl, E. Hetzl, and Y. Linnau. Method of producing an antithrombin III–heparin concentrate or antithrombin III–heparinoid concentrate. Immuno Aktiengesellschaft für chemisch-medizinische Produkte. US Patent 4,510,084, 1985.

181. M. Spannagl, R. Keller, and W. Schramm. A new AT III–heparin–complex preparation: In vitro and in vivo characterisation. *Folia Haematol.* 116:879 (1989).

182. M. Spannagl, H. Hoffman, M. Siebeck, J. Weipert, H. P. Schwarz, and W. Schramm. A purified antithrombin III–heparin complex as a potent inhibitor of thrombin in porcine endotoxin shock. *Thromb. Res.* 61:1 (1991).

183. W. E. Hennink, J. Feijen, C. D. Ebert, and S. W. Kim. Covalently bound conjugates of albumin and heparin: Synthesis, fractionation and characterization. *Thromb. Res.* 29:1 (1983).

184. G. Huhle, J. Harenberg, R. Malsch, and D. L. Heene. Comparison of three heparin bovine serum albumin binding methods for production of antiheparin antibodies. *Semin. Thromb. Hemost.* 20:193 (1994).

185. A. N. Teien, R. Odegard, and T. B. Christensen. Heparin coupled to albumin, dextran and ficoll: Influence on blood coagulation and platelets, and in vivo duration. *Thromb. Res.* 7:273 (1975).

186. P. Cuatrecasas, S. Fuchs, and C. B. Anfinsen. Cross-linking of aminotyrosyl residues in the active site of staphylococcal nuclease. *J. Biol. Chem.* 244:406 (1969).

187. H. Zahn and J. Meienhofer. Reaktionen von 1,5-difluor-2,4-dinitrobenzol mit insulin. 1. Mitt. synthese von modellverbindungen. *Makromol. Chem.* 26:126 (1958).

188. H. Zahn and J. Meienhofer. Reaktionen von 1,5-difluor-2,4-dinitrobenzol mit insulin. 2. Mitt. versuche mit insulin. *Makromol. Chem.* 26:153 (1958).
189. G. W. J. Fleet, R. R. Porter, and J. R. Knowles. Affinity labelling of antibodies with aryl nitrene as reactive group. *Nature* 244:511 (1969).
190. J. V. Staros and F. M. Richards. Photochemical labeling of the surface proteins of human erythrocytes. *Biochemistry* 13:2720 (1974).
191. M. D. Bregman and D. Levy. Labeling of glucagon binding components in hepatocyte plasma membranes. *Biochem. Biophys. Res. Commun.* 78:584 (1977).
192. S. H. Hixson and S. S. Hixson. *P*-Azidophenacyl bromide, a versatile photolabile bifunctional reagent. Reaction with glyceraldehyde-3-phosphate dehydrogenase. *Biochemistry* 14:4251 (1975).
193. D. H. Clyne, S. H. Norris, R. R. Modesto, A. J. Pesce, and V. E. Pollak. The preparation of intermolecular conjugates of horseradish peroxidase and antibody and their use in immunohistology of renal cortex. *J. Histochem. Cytochem.* 21:233 (1973).
194. A. F. Schick and S. J. Singer. On the formation of covalent linkages between two protein molecules. *J. Biol. Chem.* 236:2477 (1961).
195. S. J. Singer and A. F. Schick. The properties of specific stains for electron microscopy prepared by the conjugation of antibody molecules with ferritin. *J. Biophys. Biochem. Cytol.* 9:519 (1961).
196. P. Edman and A. Henschen. Sequence determination. In: *Protein Sequence Determination: A Sourcebook of Methods and Techniques* (S. B. Needleman, ed.), Springer-Verlag, Berlin, Germany, 1975, p. 232.
197. Y. Inque and K. Nagasawa. Selective N-desulfation of heparin with dimethyl sulfoxide containing water or methanol. *Carbohydr. Res.* 46:87 (1976).
198. K. Nagasawa and H. Yoshidome. Solvent catalytic degradation of sulfamic acid and its N-substituted derivatives. *Chem. Pharm. Bull.* 17:1316 (1969).
199. B. Casu. Structure and biological activity of heparin and other glycosaminoglycans. *Pharmacol. Res. Commun.* 11:1 (1979).
200. M. Hoylaerts, W. G. Owen, and D. Collen. Involvement of heparin chain length in the heparin catalyzed inhibition of thrombin antithrombin III. *J. Biol. Chem.* 259:5670 (1984).
201. I. Björk, O. Larm, U. Lindahl, K. Nordling, and M. E. Riquelme. Permanent activation of antithrombin by covalent attachment of heparin oligosaccharides. *FEBS Lett.* 143:96 (1982).
202. A. P. Halluin. Protein heparin conjugates. US Patent 5,308,617, Halzyme, 1994.
203. J. Dawes, K. James, and D. A. Lane. Conformational change in antithrombin induced by heparin probed with a monoclonal antibody against the heparin 1C/4B region. *Biochemistry* 33:4375 (1994).
204. G. Mitra and R. E. Jordan. Covalently bound heparin–antithrombin III complex. US Patent 4,689,323, Miles Laboratories, 1987.
205. G. Mitra and R. E. Jordan. Covalently bound heparin–antithrombin III complex, method for its preparation and its use for treating thromboembolism. European Patent 84111048.9 (0 137 356), Miles Laboratories, 1985.
206. H. Vlassara, M. Brownlee, and A. Cerami. Nonenzymatic glycosylation: Role in the pathogenesis of diabetic complications. *Clin. Chem.* 32:B37 (1986).
207. P. Hall, E. Tryon, T. F. Nikolai, and R. C. Roberts. Functional activities and non-enzymatic glycosylation of plasma proteinase inhibitors in diabetes. *Clin. Chim. Acta* 160:55 (1986).
208. A. Ceriello, D. Giugliano, A. Quatraro, A. Stante, G. Consoli, P. Dello Russo, and F. D'Onofrio. Daily rapid blood glucose variations may condition antithrombin III biologic activity but not its plasma concentration in insulin-dependent diabetes. A possible role for labile non-enzymatic glycation. *Diabetes Metab.* 13:16 (1987).
209. A. Ceriello, P. Dello Russo, C. Zuccotti, A. Florio, S. Nazzaro, C. Pietrantuono, and G. B. Rosato. Decreased antithrombin III activity in diabetes may be due to non-enzymatic glycosylation—A preliminary report. *Thromb. Haemost.* 50:633 (1983).
210. G. B. Villanueva and N. Allen. Demonstration of altered antithrombin III activity due to non-enzymatic glycosylation at glucose concentration expected to be encountered in severely diabetic patients. *Diabetes* 37:1103 (1988).
211. M. Brownlee, H. Vlassara, and A. Cerami. Inhibition of heparin-catalyzed human antithrombin III activity by non-enzymatic glycosylation. Possible role in fibrin deposition in diabetes. *Diabetes* 33:532 (1984).
212. T. Sakurai, J. P. Boissel, and H. F. Bunn. Non-enzymatic glycation of antithrombin III in vitro. *Biochim. Biophys. Acta* 964:340 (1988).
213. M. Brownlee and A. Cerami. A glucose-controlled delivery system: Semi-synthetic insulin bound to lectin. *Science* 206:1190 (1979).

214. J. Hoffman, O. Larm, and E. Scholander. A new method for covalent coupling of heparin and other glycosaminoglycans to substances containing primary amino groups. *Carbohydr. Res.* 117:328 (1983).

215. M. W. Hatton, L. R. Berry, R. Machovich, and E. Regoeczi. Tritiation of commercial heparins by reaction of NaB^3H_4: Chemical analysis and biological properties of the product. *Anal. Biochem.* 106:417 (1980).

216. L. Berry, A. K. C. Chan, and M. Andrew. Polypeptide–polysaccharide conjugates produced by spontaneous non-enzymatic glycation. *J. Biochem.* 124:434 (1998).

217. T. F. Busby and K. C. Ingham. Thermal stabilization of antithrombin III by sugars and sugar derivatives and the effects of non-enzymatic glycosylation. *Biochim. Biophys. Acta* 799:80 (1984).

218. T. A. Mewhort-Buist, M. Junop, L. R. Berry, P. Chindemi, and A. K. C. Chan. Structural effects of a covalent linkage between antithrombin and heparin: Covalent N-terminus attachment of heparin enhances the maintenance of antithrombin's activated state. *J. Biochem.* 140:175 (2006).

219. A. K. C. Chan, L. R. Berry, N. Paredes, and N. Parmar. Isoform composition of antithrombin in a covalent antithrombin–heparin complex. *Biochem. Biophys. Res. Commun.* 309:986 (2003).

220. R. E. Jordan, L. V. Favreau, E. H. Braswell, and R. D. Rosenberg. Heparin with two binding sites for antithrombin or platelet factor 4. *J. Biol. Chem.* 257:400 (1982).

221. L. Berry, A. Stafford, J. Fredenburgh, H. O'Brodovich, L. Mitchell, J. Weitz, M. Andrew, and A. K. Chan. Investigation of the anticoagulant mechanisms of a covalent antithrombin–heparin complex. *J. Biol. Chem.* 273:34730 (1998).

222. N. Paredes, A. Wang, L. R. Berry, L. J. Smith, A. R. Stafford, J. I. Weitz, and A. K. Chan. Mechanisms responsible for catalysis of the inhibition of factor Xa or thrombin by antithrombin using a covalent antithrombin–heparin complex. *J. Biol. Chem.* 278:23398 (2003).

223. S. Patel, L. R. Berry, and A. K. C. Chan. Analysis of inhibition rate enhancement by covalent linkage of antithrombin to heparin as a potential predictor of reaction mechanism. *J. Biochem.* 141:25 (2007).

224. A. K. C. Chan, L. R. Berry, P. T. Monagle, and M. Andrew. Decreased concentrations of heparinoids are required to inhibit thrombin generation in plasma from newborns and children compared to plasma from adults due to reduced thrombin potential. *Thromb. Haemost.* 87:606 (2002).

225. N. Parmar, L. R. Berry, N. Paredes, and A. K. C. Chan. Effect of heparins on thrombin generation in hemophilic plasma supplemented with FVIII, FVIIa, or FEIBA. *Clin. Lab.* 51:157 (2005).

226. H. R. Lijnen, M. Hoylaerts, and D. Collen. Heparin binding properties of human histidine-rich glycoprotein: Mechanism and role in the neutralization of heparin in plasma. *J. Biol. Chem.* 258:3803 (1983).

227. M. Hoylaerts, E. Holmer, M. De Mol, and D. Collen. Covalent complexes between low molecular weight heparin fragments and antithrombin III—Inhibition kinetics and turnover parameters. *Thromb. Haemost.* 49:109 (1983).

228. D. J. Collen. Novel composition of matter of antithrombin III bound to a heparin fragment. US Patent 4623718, KabiVitrum AB., 1986.

229. C. Mattsson, M. Hoylaerts, E. Holmer, T. Uthne, and D. Collen. Antithrombotic properties in rabbits of heparin and heparin fragments covalently coupled to human antithrombin III. *J. Clin. Invest.* 75:1169 (1985).

230. A. K. Chan, L. Berry, P. Klement, J. Julian, L. Mitchell, J. Weitz, J. Hirsh, and M. Andrew. A novel antithrombin–heparin covalent complex: Antithrombotic and bleeding studies in rabbits. *Blood Coagul. Fibrinolysis* 9:587 (1998).

231. A. K. C. Chan, N. Paredes, B. Thong, P. Chindemi, B. Paes, L. R. Berry, and P. Monagle. Binding of heparin to plasma proteins and endothelial surfaces is inhibited by covalent linkage to antithrombin. *Thromb. Haemost.* 91:1009 (2004).

232. K. K. Singhal and L. A. Parton. Plasminogen activator activity in preterm infants with respiratory distress syndrome: Relationship to the development of bronchopulmonary dysplasia. *Pediatr. Res.* 39:229 (1996).

233. S. Idell, A. Kumar, K. B. Koenig, and J. J. Coalson. Pathways of fibrin turnover in lavage of premature baboons with hyperoxic lung injury. *Am. J. Respir. Crit. Care Med.* 149:767 (1994).

234. S. Idell, K. K. James, E. G. Levin, B. S. Schwartz, N. Manchanda, R. J. Maunder, T. R. Martin, J. McLarty, and D. S. Fair. Local abnormalities in coagulation and fibrinolytic pathways predispose to alveolar fibrin deposition in the adult respiratory distress syndrome. *J. Clin. Invest.* 84:695 (1989).

235. P. Bertozzi, B. Astedt, L. Zenzius, K. Lynch, F. Lemaire, W. Zapol, and H. A. Chapman, Jr. Depressed bronchoalveolar urokinase activity in patients with adult respiratory distress syndrome. *N. Engl. J. Med.* 322:890 (1990).

236. L. R. Berry, P. Klement, M. Andrew, and A. K. C. Chan. Effect of covalent serpin–heparinoid complexes on plasma thrombin generation on fetal distal lung epithelium. *Am. J. Respir. Cell. Mol. Biol.* 28:150 (2003).

237. N. Parmar, L. R. Berry, M. Post, and A. K. C. Chan. Effect of covalent antithrombin–heparin complex on developmental mechanisms in the lung. *Am. J. Physiol. Lung Cell. Mol. Physiol.* 296:L394 (2009).

238. A. K. Chan, L. Berry, L. Mitchell, B. Baranowski, H. O'Brodovich, and M. Andrew. Effect of a novel covalent antithrombin–heparin complex on thrombin generation on fetal distal lung epithelium. *Am. J. Physiol.* 274:L914 (1998).

239. A. K. C. Chan, J. Rak, L. R. Berry, P. Liao, M. Vlasin, J. I. Weitz, and P. Klement. Antithrombin–heparin covalent complex: A possible alternative to heparin for arterial thrombosis prevention. *Circulation* 106:261 (2002).

240. D. L. Becker, J. C. Fredenburgh, A. R. Stafford, and J. I. Weitz. Exosites 1 and 2 are essential for protection of fibrin-bound thrombin from heparin-catalyzed inhibition by antithrombin and heparin cofactor II. *J. Biol. Chem.* 274:6226 (1999).

241. L. R. Berry, D. L. Becker, and A. K. Chan. Inhibition of fibrin-bound thrombin by a covalent antithrombin–heparin complex. *J. Biochem. (Tokyo)* 132:167 (2002).

242. L. J. Smith, T. A. Mewhort-Buist, L. R. Berry, and A. K. C. Chan. An antithrombin–heparin complex increases the anticoagulant activity of fibrin clots. *Res. Lett. Biochem.* ID:639829 (2008).

243. P. A. Chindemi, P. Klement, F. Konecny, L. R. Berry, and A. K. C. Chan. Biodistribution of covalent antithrombin–heparin complexes. *Thromb. Haemost.* 95:629 (2006).

244. M. C. van Walderveen, L. R. Berry, and A. K. C. Chan. Intravascular targeting of a new anticoagulant heparin compound. *Cardiovasc. Hematol. Disord. Drug Targets* 9:149 (2009).

245. N. Volpi, M. Cusmano, and T. Venturelli. Qualitative and quantitative studies of heparin and chondroitin sulfates in normal human plasma. *Biochim. Biophys. Acta* 1243:49 (1995).

246. L. Mitchell, R. Superina, M. Delorme, P. Vegh, L. Berry, H. Hoogendoorn, and M. Andrew. Circulating dermatan sulphate/heparin sulfate proteoglycan(s) in children undergoing liver transplantation. *Thromb. Haemost.* 74:859 (1995).

247. M. Andrew, L. Mitchell, L. Berry, B. Paes, M. Delorme, F. Ofosu, R. Burrows, and B. Khambalia. An anticoagulant dermatan sulfate proteoglycan circulates in the pregnant woman and her fetus. *J. Clin. Invest.* 89:321 (1992).

248. C. H. Bamford and K. G. Al-Lamee. Chemical methods for improving the haemocompatibility of synthetic polymers. *Clin. Mater.* 10:243 (1992).

249. C. Karlsson, H. Nygren, and M. Braide. Exposure of blood to biomaterial surfaces liberates substances that activate polymorphonuclear granulocytes. *J. Lab. Clin. Med.* 128:496 (1996).

250. L. Tang and J. W. Eaton. Inflammatory responses to biomaterials. *Am. J. Clin. Pathol.* 103:466 (1995).

251. R. Vanholder and N. Lameire. Does biocompatibility of dialysis membranes affect recovery of renal function and survival? *Lancet* 354:1316 (1999).

252. J. Sanchez and P. Olsson. On the control of the plasma contact activation system on human endothelium: Comparisons with heparin surface. *Thromb. Res.* 93:27 (1999).

253. H. P. Wendel and G. Ziemer. Coating-techniques to improve the hemocompatibility of artificial devices used for extracorporeal circulation. *Eur. J. Cardiothorac. Surg.* 16:342 (1999).

254. G. Dickneite and B. Leithauser. Influence of antithrombin III on coagulation and inflammation in porcine septic shock. *Arterioscler. Thromb. Vasc. Biol.* 19:1566 (1999).

255. P. Klement, Y. J. Du, L. Berry, M. Andrew, and A. K. C. Chan. Blood-compatible biomaterials by surface coating with a novel antithrombin–heparin covalent complex. *Biomaterials* 23:527 (2002).

256. H. H. Hoerl, D. Nussbaumer, and E. Wuenn. Surface grafting of microporous, nitrogen-containing polymer membranes and membranes obtained thereby. German Patent 3,929,648, Sartorius GmbH, 1990.

257. H. H. Heinrich, D. Nussbaumer, and E. Wuenn. Grafting of unsaturated monomers on polymers containing nitrogen. German Patent 4,028,326, Sartorius A-G., 1991.

258. K. van Phung and R. C. Schulz. Pfropfung von vinylverbindugen auf polyamide. *Makromol. Chem.* 180:1825 (1979).

259. V. K. Raman and E. R. Edelman. Coated stents: Local pharmacology. *Semin. Interv. Cardiol.* 3:133 (1998).

260. A. K. C. Chan, L. Berry, M. Andrew, and P. Klement. Antithrombin–heparin covalent complex: A novel approach for improving thromboresistance of cardiovascular devices. *Proceedings International Biomaterial Congress*, Honolulu, HI, 2000.

261. M. D. Phaneuf, S. A. Berceli, M. J. Bide, W. C. Quist, and F. W. LoGerfo. Covalent linkage of recombinant hirudin to poly(ethylene terephthalate) (Dacron): Creation of a novel antithrombin surface. *Biomaterials* 18:755 (1997).

262. M. D. Phaneuf, M. Szycher, S. A. Berceli, D. J. Dempsey, W. C. Quist, and F. W. LoGerfo. Covalent linkage of recombinant hirudin to a novel ionic poly(carbonate) urethane polymer with protein binding sites: Determination of surface antithrombin activity. *Artif. Organs* 22:657 (1998).

263. V. L. Gott and R. L. Daggett. Serendipity and the development of heparin and carbon surfaces. *Ann. Thorac. Surg.* 68:S19 (1999).

264. G. Elgue, M. Blomback, P. Olsson, and J. Riesenfeld. On the mechanism of coagulation inhibition on surfaces with end point immobilized heparin. *Thromb. Haemost.* 70:289 (1993).
265. Y. Miura, S. Aoyagi, F. Ikeda, and K. Miyamoto. Anticoagulant activity of artificial biomedical materials with co-immobilized antithrombin III and heparin. *Biochimie* 62:595 (1980).
266. P. Cahalan, T. Lindhout, B. Fouache, M. Verhoeven, L. Cahalan, M. Hendriks, and R. Blezer. Method for making improved heparinized biomaterials. US Patent 5767108, Medtronic, 1998.
267. Y. J. Du, P. Klement, L. R. Berry, P. Tressel, and A. K. C. Chan. In vivo rabbit acute model tests of polyurethane catheters coated with a novel antithrombin–heparin covalent complex. *Thromb. Haemost.* 94:366 (2005).
268. P. Klement, Y. J. Du, L. R. Berry, P. Tressel, and A. K. C. Chan. Chronic performance of polyurethane catheters covalently coated with ATH complex: A rabbit jugular vein model. *Biomaterials* 27:5107 (2006).
269. Y. J. Du, J. L. Brash, G. McClung, L. R. Berry, P. Klement, and A. K. C. Chan. Protein adsorption on polyurethane catheters modified with a novel antithrombin–heparin covalent complex. *J. Biomed. Mater. Res. A* 80:216 (2007).
270. L. R. Berry, N. Parmar, M. W. Hatton, and A. K. Chan. Selective cleavage of heparin using aqueous 2-hydroxypyridine: Production of an aldose-terminating fragment with high anticoagulant activity. *Biochem. Biophys. Res. Commun.* 346:946 (2006).
271. P. Klement, L. R. Berry, P. Liao, H. Wood, P. Tressel, L. J. Smith, N. Haque, J. I. Weitz, J. Hirsh, N. Paredes, and A. K. Chan. Antithrombin–heparin covalent complex reduces microemboli during cardiopulmonary bypass in a pig model. *Blood* 116:5716 (2010).
272. K. N. Sask, I. Zhitomirsky, L. R. Berry, A. K. Chan, and J. L. Brash. Surface modification with an antithrombin–heparin complex for anticoagulation: Studies on a model surface with gold as substrate. *Acta Biomater.* 6:2911 (2010).
273. Y. J. Du, L. R. Berry, and A. K. Chan. Chemical-physical characterization of polyurethane catheters modified with a novel antithrombin–heparin covalent complex. *J. Biomater. Sci. Polym. Ed.* 22:2277 (2011).

2 Glucose-Sensitive Hydrogels

Seong Hoon Jeong, Kyung T. Oh, and Kinam Park

CONTENTS

2.1 INTRODUCTION

2.1.1 HYDROGELS IN GENERAL

Hydrogels are three-dimensional networks of hydrophilic polymer chains which do not dissolve but can absorb a large amount of water, usually more than 20% of the total weight [1–7]. The hydrophilic properties enable hydrogels to hold a large amount of water within their structure. The high water content together with soft-surface properties can offer good biocompatibility of the hydrogels [8]. Moreover, synthetic or processing methods using various functional polymers can be manipulated to possess various properties including physicochemical, mechanical, and biological properties, as well as new functional properties. For example, hydrogels can be made to respond to environmental stimuli, such as temperature, pH, light, and specific molecules like glucose by employing the functional components. These properties have drawn great attention from academia as well as pharmaceutical and biomedical industries.

Various methods can be used to prepare the three-dimensional hydrogel networks, which are usually formed by chemical or physical cross-linking of hydrophilic polymer chains. In the case

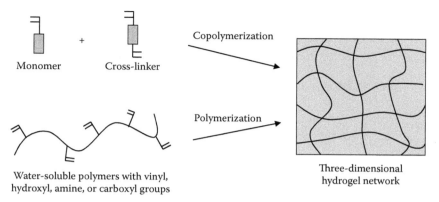

Monomer   Cross-linker

Copolymerization

Polymerization

Water-soluble polymers with vinyl,
hydroxyl, amine, or carboxyl groups

Three-dimensional
hydrogel network

FIGURE 2.1   Schematic view on the chemical method for the preparation of hydrogel networks.

of chemical hydrogels, two different methods can be used (Figure 2.1). One method is polymerization of water-soluble monomers in the presence of bi- or multifunctional cross-linking agents. Typical examples of water-soluble monomers are acrylic acid, acrylamide, hydroxyethyl methacrylate, hydroxypropyl acrylate, and vinylpyrrolidone. The other method is cross-linking water-soluble polymers using chemical reactions that involve functional groups of the polymers. Functional groups for the cross-linking reactions include vinyl, hydroxyl, amine, and carboxyl groups [1,9,10]. In the chemical hydrogels, polymer chains are connected by covalent bonds, so it is difficult to change the shape of the gels (swelling/shrinking behavior). Physical hydrogels are prepared by cross-linking without chemical reactions and they are connected through noncovalent bonds, such as van der Waals interactions, ionic interactions, hydrogen bonding, hydrophobic interactions, etc. Since these bonds are reversible, physical gels possess sol–gel reversibility [11–14]. Table 2.1 lists typical examples of polymers that can be cross-linked to make hydrogels.

The most important property of hydrogels would be the reversible swelling/shrinking in aqueous media. The hydrogels swell in aqueous solution since hydrophilic polymer chains of the cross-linked network try to dissolve in water. The swelling property is usually characterized by measuring their capacity to absorb aqueous media. Measuring the weight of a swollen hydrogel is the simplest way to evaluate the swelling kinetics and equilibrium. The swelling ratio (R_s), which is the most commonly used parameter to express the swelling capacity of hydrogels, is defined as follows:

$$R_s = \frac{(W_s - W_d)}{W_d}$$

where W_s and W_d are the weights of swollen and dried hydrogels, respectively.

There are many factors affecting the swelling properties including the type and composition of monomers, cross-linking density, and other environmental factors such as temperature, pH, and ionic strength. Among the factors, the cross-linking density is inversely related to the swelling property because cross-linking of the network can exert retractive force and also counterbalance the swelling force. If the counteracting forces become equivalent, swelling of the hydrogel reaches its equilibrium state. The degree of cross-linking can be determined by measuring either equilibrium swelling or elastic modulus [1,15]. The cross-linking density is also strongly related to other important properties such as mechanical strength and permeability. When swelled, the glassy polymer network becomes elastic. Due to the high water content of fully swollen hydrogels, they are very weak mechanically. However, the low mechanical strength can be improved by increasing the cross-linking density, making an interpenetrating network (IPN) structure, or copolymerization with hydrophobic comonomers. These approaches are generally accompanied by the decreased swelling property.

TABLE 2.1
Typical Examples of Polymers for Making Hydrogels

Polymer Name	Polymer Structure
Poly(acrylic acid)	
Polyacrylamide	
Poly(N-isopropylacrylamide)	
Poly(ethylene oxide)	
Poly(hydroxyethyl methacrylate)	
Polyvinylpyrrolidone	
Poly(vinyl alcohol)	

2.1.2 ENVIRONMENT-SENSITIVE HYDROGELS

All hydrogels have the same fundamental property of swelling or shrinking in the presence or absence of water. Some hydrogels have additional properties, such as swelling or shrinking, in response to environmental changes, such as temperature, pH, light, and specific molecules like glucose. The presence of functional groups on polymer chains makes the hydrogels sensitive to certain stimulants or environmental factors [16]. These hydrogels can undergo reversible sol–gel phase transition or volume phase transition upon small changes in the surroundings. These hydrogels can be collectively called "environment-sensitive" or "stimuli-responsive" hydrogels. Compared to general ones without such additional properties, environment-sensitive hydrogels are more advanced so they are often called "smart" or "intelligent" hydrogels.

One of the unique properties of these types of hydrogels is that they change their swelling rate rather quickly upon small changes in the environmental factors. This quick volume change is known as "volume transition." If the volume transition happens to make the volume smaller, it is called "volume collapse." The volume transition of smart hydrogels can occur by only a small change in environmental conditions and it is this unique property that makes them very useful in various areas [17–20]. This chapter is more focused on specific hydrogels which can monitor the glucose level in blood and release insulin in a timely manner to maintain optimal blood glucose concentration.

2.1.2.1   pH-Sensitive Hydrogels

Of the many smart hydrogels, pH-sensitive hydrogels have been most frequently used in glucose-sensitive insulin release systems and they can display significant differences in swelling properties depending on pH change. They are cross-linked polyelectrolytes containing acidic or basic pendant groups, which can accept or donate protons upon environmental pH changes. When charged, the hydrogels swell substantially more than at the neutral state due to the electrostatic repulsive interaction among charges present on the polymer chains. Beyond pH, the extent of swelling is also dependent on ionic strength and type of counterions [21]. Figure 2.2 shows examples of pH-sensitive polymers. Hydrogels made of polyanions, such as poly(acrylic acid) (PAA), swell less as pH becomes lower due to the loss of charges. However, polycations, such as poly(N,N'-diethylaminoethyl methacrylate) (PDAEM), swell substantially more at lower pH due to the generation of charges on the polymer chains. Both types of polyelectrolyte hydrogels have been used for preparing glucose-sensitive hydrogels.

2.1.2.2   Glucose-Sensitive Hydrogels

Unlikely to other controlled release drug delivery systems, the insulin delivery systems need to monitor glucose levels and release insulin just enough to reduce the elevated blood glucose level. Since the amount of insulin to be delivered will be different each time and depends on the glucose level, this type of self-regulating insulin delivery system presents the ultimate challenge in controlled drug delivery technologies. Some of the environment-sensitive hydrogels are glucose sensitive and they can be used for glucose sensing and self-regulating insulin delivery. The glucose-sensitive hydrogels undergo changes in the swelling ratio or changes in sol–gel physical states in response to small changes in the glucose concentration in the environment [22]. Development of glucose-sensitive hydrogels is very critical for the development of self-regulating insulin delivery systems.

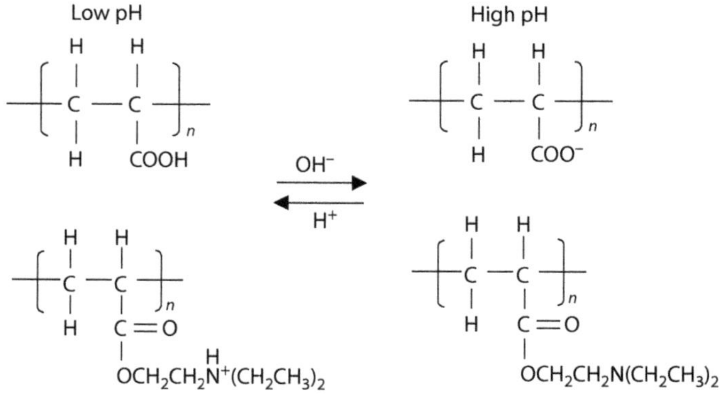

FIGURE 2.2   pH-sensitive (pH-dependent) ionization of polyelectrolytes. Poly(acrylic acid) becomes ionized at high pH, while poly(N,N'-diethylaminoethyl methacrylate) becomes ionized at low pH. Ionized hydrogels have more swelling property due to the presence of charges.

2.2 GLUCOSE-SENSITIVE HYDROGELS FOR INSULIN DELIVERY

In order to maintain relatively constant drug levels in blood, controlled drug delivery systems are generally based on diffusion at a certain rate through polymer membranes or matrices. However, this concept may not be applicable to insulin delivery because the rate of insulin release has to be adjusted depending on the blood glucose level. Daily injections of insulin may not be adequate for maintaining normal blood glucose levels and preventing long-term complications of diabetes. That is why there are increased demands for self-regulated insulin delivery systems.

Since insulin release has to be in harmony with increase in the blood glucose level, the most desirable insulin delivery system requires a glucose-sensing capability and an ability to trigger release of the specific amount of insulin [23]. Many scientists have focused on utilizing glucose-sensitive hydrogels. For the hydrogels to possess glucose sensitivity, they have to contain specific molecules that interact with glucose molecules. Currently used glucose-sensitive molecules are concanavalin A (Con A), phenylboronic acid (PBA), glucose oxidase (GOD), and glucose dehydrogenase (GDH). These molecules have been incorporated into hydrogels for glucose sensing and self-regulated insulin delivery.

2.2.1 LECTIN AS GLUCOSE-RESPONSIVE UNIT

The word "lectin" is based on the Latin word *legere*, meaning "to select" and the lectins were first discovered more than 100 years ago in plants. Lectins can be found easily in nature and most of them are basically nonenzymic in action. Generally, they are sugar-binding proteins which are highly specific for their sugar moieties. Since they may bind to a soluble carbohydrate or to a carbohydrate moiety part of a glycoprotein or glycolipid and also they typically agglutinate certain animal cells and/or precipitate glycoconjugates, they can play a role in biological recognition phenomena involving cells and proteins. The first lectin to be purified on a large scale and available on a commercial basis was Con A, which is now the most used lectin for characterization and purification of sugar-containing molecules and cellular structures [24]. Con A is composed of 237 amino acid residues and has four polypeptide-binding subunits. Each subunit has a molecular weight of 26,000 Da and a dimension of $44 \times 40 \times 39$ Å^3 [25,26].

2.2.1.1 Immobilized Insulin: Glycosylated Insulin Derivatives

Insulin molecules can be attached to a carrier through specific interaction, which can play a role as a functional group. A common example is glycosylated insulin in which insulin is chemically modified to introduce glucose. Glucose-regulated insulin release was first introduced based on competitive binding between glucose and glycosylated insulin for carbohydrate-specific binding sites on Con A. Four subunits of Con A can form a complex with the synthesized glycosylated insulin derivative, which can be released from Con A by external free glucose due to the competitive and complementary binding properties of glycosylated insulin and glucose to Con A (Figure 2.3a) [27–29]. Various insulin derivatives with different binding constants have been investigated for their stability and biological activity [28,30–35].

The Con A-glycosylated insulin complex is further enclosed within a polymer membrane or polymeric microcapsules, which are permeable to glucose and glycosylated insulin but not to Con A [36]. When the blood glucose level increases, glucose enters the reservoir and competitively displaces glycosylated insulin from Con A [28]. The glycosylated insulin is then released by diffusion through the membrane into the body as shown in Figure 2.3b. A system with Con A and succinyl-amidophenyl-glucopyranoside insulin (SAPG-insulin) could control the release of SAPG-insulin fast in response to glucose change [28]. Initially, poly(hydroxyl methacrylate) (poly-HEMA) membrane was utilized. However, due to its weak mechanical strength, it was replaced later with cellulose acetate tubing or thin nylon membrane. Main limitations of the encapsulated system were potential leakage of immunogenic Con A through the polymer membrane and also the slow onset of insulin release as free glucose concentration increases.

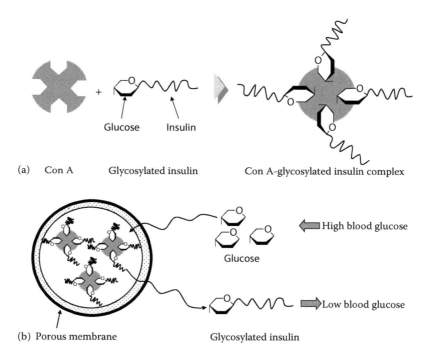

(a)   Con A   Glycosylated insulin   Con A-glycosylated insulin complex

(b) Porous membrane   Glycosylated insulin

FIGURE 2.3 Schematic view on the complex formation between Con A and glycosylated insulin (a) and controlled release of the insulin from the complex as free glucose concentration increases in the environments (b).

2.2.1.2 Complex Formation between Polymers Bearing Saccharide Residues and Con A

Swelling or shrinking of a Con A-loaded hydrogel system is dependent on different saccharides [37]. Hydrogel, consisted of a covalently cross-linked poly(N-isopropyl acrylamide) and physically entrapped Con A associated with dextran sulfate, swelled due to ionic osmotic pressure induced by the anionic inclusion. However, it shrank when the dextran sulfate was replaced from the Con A-binding sites by uncharged saccharides [37].

Various polymers with defined saccharide residues can form a complex with Con A to form precipitates or hydrogels, which can encapsulate insulin [38–40]. The polymers can be natural saccharides as well as synthetic containing saccharide residues [39,41]. For example, complex formation between Con A and polymer with pendant glucose groups, PGEMA, showed that they swelled in the presence of monosaccharide due to the dissociation of the Con A–polymer complex by competitive binding of free glucose [39]. The effect of the monosaccharide on the dissociation extent of the PGEMA–Con A complexes was dependent on the affinity of Con A to the monosaccharide. Turbid PGEMA–Con A solution became transparent as free glucose or mannose was added, but not free galactose. This could provide information that Con A can sense pendant glucose groups of the polymer and the PGEMA–Con A complex is sensitive to monosaccharides. Therefore, the complex can be fabricated as a novel glucose sensor or a glucose-sensitive insulin delivery system. Hydrogels made of Con A and polysucrose complexes were used as rate-determining membranes for the release of a solute in the reservoir as a function of the glucose concentration [41–43]. The release through the membrane was controlled by the viscosity change resulting from the sol–gel transition of Con A–polysucrose complexes.

2.2.1.3 Sol–Gel Phase-Reversible Transitions

Con A and glucose-containing polymers earlier can be utilized to form stable hydrogels capable of undergoing phase transition between sol and gel [44–47]. Con A can be used as a physical

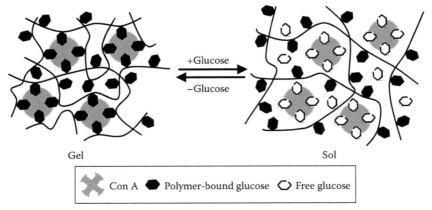

FIGURE 2.4 Schematic view on sol–gel phase transition of glucose-sensitive hydrogel. (Modified from Lee, S.J. and Park, K., *J. Mol. Recogn.*, 9, 549, 1996.)

cross-linking agent for the glucose-containing polymers. Due to the similar complex formation mechanism between pendant glucose groups and Con A, phase-reversible glucose-sensitive hydrogels can be prepared including poly(allyl glucose-*co*-acrylamide) (poly(AG-*co*-AM)), poly(allyl glucose-*co*-vinylpyrrolidone) (poly(AG-*co*-VP)), and poly(allyl glucose-*co*-3-sulfopropylacrylate potassium) (poly(AG-*co*-SPAK)).

As schematically shown in Figure 2.4, Con A interacts with polymer-bound glucose to form a gel state in the absence of free glucose [45]. However, high-level free glucose leads to dissociation of polymer-bound glucose from Con A and hence transition into a sol state [45,47]. The sol–gel phase transition is reversible, so the sol becomes a gel again upon removal of the free glucose. In the sol state, the increased mobility of the polymer chains can give more opportunities for insulin molecules to diffuse out [23]. Therefore, glucose-sensitive phase-reversible hydrogels can be used to regulate the insulin release according to the free glucose concentration in the environment.

2.2.1.4 Entrapment of Con A into Hydrogels

By copolymerization of a monomer with pendant glucose (GEMA) and a divinylmonomer in the presence of Con A, Con A-entrapped PGEMA hydrogels could be obtained [48]. Moreover, Con A played the role of an additional cross-linking agent as the resulting hydrogels' cross-linking density increased as Con A increased [48]. However, the cross-linking density of the hydrogels decreased with increasing glucose concentration. The glucose-sensitive Con A-entrapped PGEMA hydrogels swelled in the presence of free glucose via competitive binding, which resulted in the dissociation of the complex. Mannose caused the swelling of the hydrogels more than glucose did, because mannose inhibited PGEMA–Con A complex formation more effectively than glucose, which could dissociate the complex better. However, galactose did not affect the swelling [40]. Figure 2.5 shows the chemical structure of the three sugars. Since Con A was not able to form complexes with galactose, there was limited change in the swelling of Con A-entrapped PGEMA hydrogels in a galactose

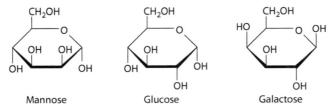

FIGURE 2.5 Molecular structures of mannose, glucose, and galactose.

solution. Therefore, the hydrogels are able to recognize a specific monosaccharide and induce structural changes, which suggest that Con A-entrapped PGEMA hydrogels may have significant applications as a novel glucose-sensitive drug delivery system.

2.2.1.5 Limitations of Con A System

Even though these methods can give unique properties, they still have inherent limitations of their own.

1. *Immunotoxicity of Con A*: Critical not to release Con A from drug delivery system
2. *New chemical entity (NCE) issue*: Each insulin molecule has to be modified with glucose causing complicated approval process
3. *Small reservoir capacity*: Con A is a relatively large molecule with only four glucose-binding sites, poor for long-term delivery
4. *Hypoglycemia issue*: Depending on binding constant between glycosylated insulin and Con A and it needs to be higher than that of glucose

A major limitation to the Con A system is the progressive loss of activity due to Con A leakage through the polymer membrane, which also causes a problem of immunogenic toxicity. To circumvent this problem and obtain a reversible glucose-responsive hydrogels, Con A can be bound covalently. Several strategies were introduced already: A covalent link could be obtained by copolymerizing GEMA with Con A having vinyl groups [48]. Con A was coupled covalently to glycogen with derivatives of Schiff's bases [49,50] or carbodiimide chemistry with carboxylic acid on Carbopol [51,52] or carboxylic acid–modified dextran [53,54]. Even though there are issues of settling by gravity and bulk size of the beads, Con A could be immobilized to Sepharose beads and then enclosed in a macroporous membrane [29].

The long unwanted lag phase for the onset of glycosylated insulin release was partly because of the limited solubility of Con A-glycosylated insulin complex [55]. The limited solubility was due to the tetrameric nature of Con A and the formation of dimers or hexamers by glycosylated insulin derivatives [56,57]. For improved solubility, insulin was derivatized with gylcosyl PEG for improved solubility and solution stability at physiological pH [58]. Moreover, Con A was modified with PEG and the resulting PEGylated Con A improved aqueous solubility and stability, and even higher glucose sensitivity [55]. The hydrogels made of PEG–Con A showed pulsatile release of insulin in response to the changes in glucose concentration [59]. Such pulsatile release of insulin can be maintained as long as the membrane remains intact. To reduce the response time, the surface area can be increased by applying microcapsules and microspheres [29,36,60].

2.2.2 PBA Moiety as Glucose-Responsive Unit

PBA and its derivatives are able to form reversible complexes with compounds bearing dihydroxyls when the two hydroxyl groups are in a coplanar configuration [61–64]. Typical dihydroxyl compounds are poly(vinyl alcohol) (PVA) and carbohydrates including glucose. Since PBA is a synthetic glucose-responsive unit, it may not be immunotoxic compared to Con A.

A polymer bearing PBA moieties (poly(*N*-vinyl-pyrrolidone-*co*-PBA)) can form a gel with a long polyol, for example, PVA, through the complex formation. The PVA is linked to several sites and plays a role as a cross-linker. However, glucose molecules may act as a sort of competing polyol and they will substitute the pendant hydroxyl groups of the polymer. Each glucose molecule is bound to one PBA receptor. This substitution will decrease the cross-linking density and cause the gel to swell and dissolve resulting in increased release of loaded insulin (Figure 2.6). In the absence of free glucose, it may lead to reformation of PBA–polyol cross-linking in the complex gel, which may reduce insulin release. This concept was used to prepare glucose-sensitive insulin delivery system [65–68]. Figure 2.7 shows schematic view on sol–gel phase transition of a PBA polymer

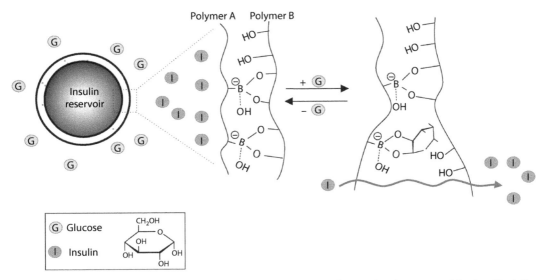

FIGURE 2.6 **(See companion CD for color figure.)** Conceptual view on the glucose-sensitive insulin delivery using PVA/poly(VP-*co*-PBA) complex. Polymer A, poly(VP-*co*-PBA); polymer B, PVA. (Modified from Kitano, S. et al., *J. Control. Release*, 19, 161, 1992.)

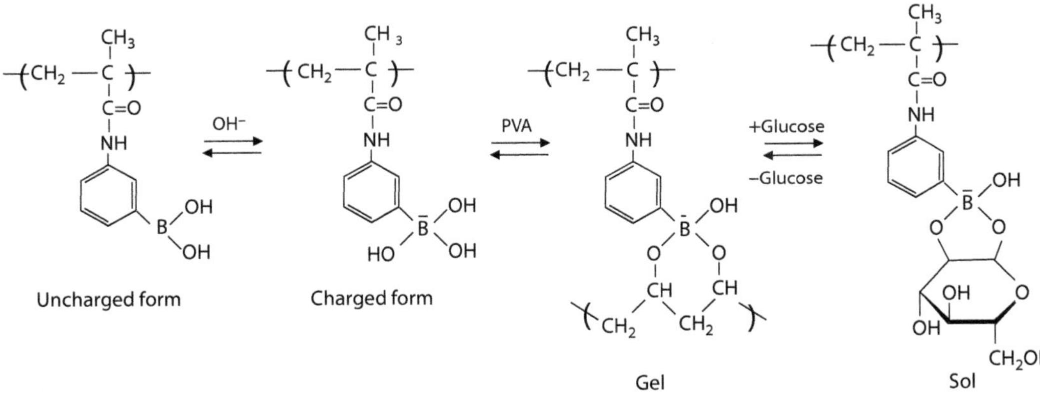

FIGURE 2.7 Schematic view on sol–gel phase transition of a PBA polymer complex together with PBA polymer complex formation. At alkaline pH, the PBA polymer interacts with PVA to form a gel. However, glucose dissociates PVA resulting in transition from a gel to a sol.

complex together with PBA polymer complex formation. At alkaline pH, the PBA polymer interacts with PVA to form a gel. However, glucose dissociates PVA resulting in transition from a gel to a sol. Poly(vinylpyrrolidone-*co-m*-acrylamidophenylboronic acid) is commonly used as a PBA-containing polymer.

Beyond the competitive binding earlier, another swelling mechanism can explain the properties of PBA-derived hydrogels upon glucose addition. Glucose complexation may increase the hydrogel's charge density and the volume of the gel increases when the charge density of a gel increases. Since there are mobile counterions within the gel, they will possess an osmotic pressure [69]. This concept may also be helpful to understand the application of the PBA-derived hydrogels into a glucose-responsive material. PBA exists in equilibrium between charged and uncharged form. Generally, carbohydrates have the cis-diol moiety in their structure and they can form a relatively strong complex with the PBA. However, complexation may change the equilibrium toward the charged form since the charged complex is more stable than the neutral one, which is highly sensitive to hydrolysis

(Figure 2.7) [70]. One of the issues of this system is that the gels are not stable at a physiological pH of 7.4. They are sensitive to glucose only at alkaline conditions (pH 9.0). In order to develop a novel glucose-sensitive hydrogel at physiological pH, amino groups were introduced either into the polymer or in the vicinity of the PBA moiety [65].

The PBA–polyol cross-linked gel was alternatively utilized as a glucose-sensitive erodible matrix system [71]. In this erodible system, a gel is formed between *m*-amino PBA-substituted polyacrylamide (poly(PBA-*co*-AM)) and diglucosyl hexanediamine (DGHDA). Free glucose can disrupt the gel matrix by replacing DGHDA with glucose through competitive binding to PBA, which results in solubilization of the matrix and a significant increase in the insulin release rate. However, in order to be more practical, the glucose specificity of the PBA moiety has to be improved.

Another system was developed to release insulin repeatedly with an "on–off" regulation in response to stepwise changes in the glucose concentration. When combined with thermoresponsive polymers, the PBA derivatives earlier could keep their glucose-responsive properties at physiological salinity. Generally, hydrophilic polymers dissolve more on higher temperature. However, some polymers made of relatively hydrophobic monomers precipitate from aqueous solution upon temperature increase, especially on a minute increase at a certain temperature. The temperature which induces polymer precipitation or phase separation is known as a lower critical solution temperature (LCST). Polymers exist as a homogeneous single phase solution at temperatures below the LCST but phase separation occurs when heated to above the LCST. Thermoresponsive polymers, such as the family of poly(alkylacrylamides), are soluble at low temperatures in water and precipitate when heated above the LCST. When cross-linked, they are swollen at low temperatures and shrink upon heating. The transition occurs at the volume phase transition temperature (VPTT). Copolymers with *N,N*-dimethylacrylamide and PBA were synthesized and they showed that the presence of the PBA moiety decreased the LCST of the polymer, but the LCST increased upon glucose addition [72]. The PBA derivative played the role as a hydrophobic monomer but complexation with glucose imposes hydrophilicity.

2.2.3 GOD AS GLUCOSE-RESPONSIVE UNIT

GOD as a glucose-sensing element was utilized for self-regulated insulin delivery systems. In this approach, glucose-sensitive polymeric hydrogels were prepared using immobilized GOD in pH-sensitive polymers (as shown in Figure 2.2), which can be used as an insulin release controller [73–77]. In such a system, GOD can react with glucose in the presence of oxygen and convert it into gluconic acid.

$$\text{Glucose} + O_2 + H_2O \xrightarrow{\text{GOD}} \text{Gluconate}^- + H^+ + H_2O_2$$

The formation of gluconic acid decreases the microenvironment pH and the hydrogels get the protons. The protonation and deprotonation of the polyelectrolytes in hydrogels lead to change in the structure of hydrogels (swelling or shrinking), resulting in the release of solutes such as insulin molecules (Figure 2.8). Moreover, the charge density of the polymers can be modified resulting in glucose-responsive volume change. The GOD-immobilized hydrogels classified by the utilized polyelectrolytes seem to have different mechanisms of insulin release due to the charge types of polyelectrolytes.

2.2.3.1 Hydrogels Made of Cationic Polyelectrolytes

Common cationic polyelectrolytes are synthesized with cationic monomers, such as *N,N*-dimethylaminoethyl methacrylate (DMAEM) and *N,N*-diethylaminoethyl methacrylate (DEAEM) [73–80]. As shown in Figure 2.9, the deprotonated hydrogels shrink due to hydrophobic interaction and embed the insulin inside. In the presence of glucose, the hydrogels undergo swelling due to the decrease of pH as a result of gluconic acid formation. The increased swelling, in turn, causes more release of insulin. One of the issues of this system is the leveling off of the response to lower than

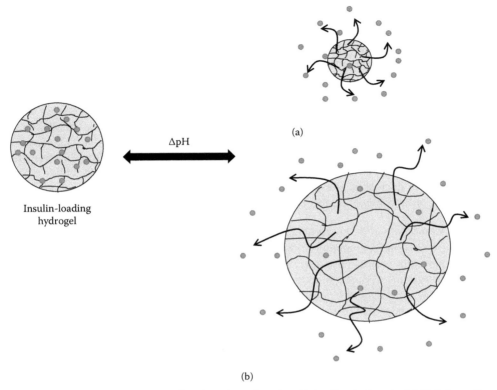

FIGURE 2.8 Schematic mechanism of insulin release from insulin loading hydrogels. (a) Shrinkage: Squeeze release. (b) Swelling: Expansion release.

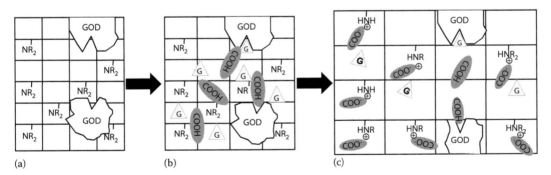

FIGURE 2.9 Schematic representation of mechanism of action of glucose-sensitive membrane. (a) In the absence of glucose, at physiologic pH, few of the amine groups are protonated; (b) in the presence of glucose (G), glucose oxidase (GOD) produces gluconic acid; and (c) gluconic acid can protonate the amine groups. The fixed positive charge on the polymeric network leads to electrostatic repulsion and membrane swelling. (Redrawn from Horbett, T.A. et al., A bioresponsive membrane for insulin delivery. *Recent Advances in Drug Delivery Systems*, Anderson, J.M. and Kim, S.W., eds., Plenum Press, New York, 1984, pp. 209–220.)

0.5 mg/mL glucose concentration below the pathophysiological glucose level (0.5–10 mg/mL) [79]. This limited glucose sensitivity is due to the property of GOD. GOD as a flavoprotein requires an electron acceptor, such as oxygen, to reoxidize the flavin adenine dinucleotide which is reduced as glucose is consumed [79]. Therefore, the sensitivity of this system depends on the presence of oxygen in the medium. The same mechanism has been used to prepare variations of glucose-sensitive insulin delivery systems. Both matrix [81] and reservoir type [82–85] devices have been prepared.

2.2.3.2 Hydrogels Made of Anionic Polyelectrolytes

2.2.3.2.1 "Squeezing" Hydrogels

Glucose-sensitive polyanionic hydrogels, poly(methacrylic acid-*g*-ethylene glycol) (poly(MAA-*g*-EG)), were prepared by copolymerizing with methacrylic acid and PEG monomethacrylate in the presence of GOD [17]. At low pH, hydrogen bonding was formed between the hydrogens of the protonated methacrylic acid units and the oxygens on the ether groups of the PEG chains, resulting in the formation of interpolymer complexes. Such hydrogen bonding results in the collapse of the gel due to increased hydrophobicity in the polymer network. As the carboxylic groups are deprotonated at high pH, the polymer chains in complexes become separated and the gel swells. Therefore, GOD-immobilized hydrogels swell at high pH. However, decrease of pH resulted from the formation of gluconic acid in the presence of free glucose "squeezes" the hydrogels, leading to release of insulin. If the glucose level decreases, slow increase in pH of the environment is occurred by decreased gluconic acid. Consequently, the squeezed hydrogels would swell and be recycled repetitively.

2.2.3.2.2 "Chemical Gate" Hydrogel Layers

Other anionic hydrogels were employed for glucose-sensitive "chemical gate" concept. The chemical gate was closed in normal physiology condition and opened in case of increase of glucose concentration. The systems were prepared from PAA and GOD grafted to porous films such as porous cellulose and poly(vinylidene fluoride) membranes [86,87]. At neutral pH, negatively charged PAA chains are fully extended due to electrostatic repulsion and close up the pores of the membrane for poor diffusion of insulin; conversely, there is a decrement of pH below 5 as a result of GOD action on glucose-protonated PAA chains. The hydrophobicity increase of chains by protonation in carboxyl groups leads to shrinkage of coils on the pore walls, resulting in the opening of the pore channels for the free diffusion of insulin (Figure 2.10). The insulin diffusion through the open gate, however, was only a few times higher than the diffusion in the closed state. In addition, PAA was grafted onto the cellulose membrane by either plasma polymerization or ceric ion-induced radical polymerization, and GOD was immobilized by coupling with L-ethyl-3-(3-dimethylaminopropyl)carbodiimide.

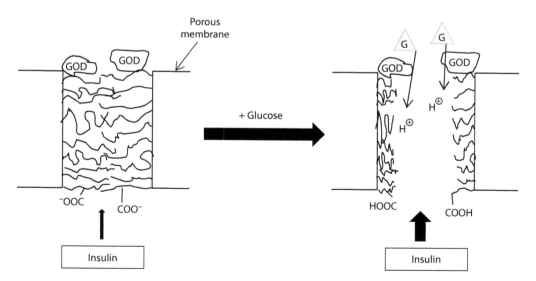

FIGURE 2.10 Principle of controlled release system of insulin. (Left) In the absence of glucose, the chains of poly(acrylic acid) grafts are rodlike, lowering the porosity of the membrane and suppressing insulin permeation. (Right) In the presence of glucose, gluconic acid produced by GOD protonates the poly(acrylic acid), making the graft chains coillike and opening the pores to enhance insulin permeation. (Modified from Ito, Y. et al., *J. Control. Release*, 10, 195, 1989.)

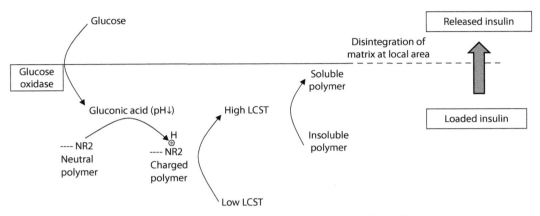

FIGURE 2.11 Schematic representation of glucose-responsive insulin delivery system using poly (DMA-*co*-EMA).

2.2.3.3 Hydrogels Made of pH/Temperature-Sensitive Polymers

pH/temperature-sensitive polymers were prepared by copolymerization with temperature-sensitive poly(ethylacrylamide) (poly(EAM)) blocks and pH-sensitive poly(*N,N*-dimethylaminoethyl methacrylate) (poly(DMAEM)) blocks for glucose-sensitive insulin delivery systems [88,89]. The hydrogels made of poly(DMAEM-*co*-EAM) show different temperature-responsive swelling behaviors at different pH values, for example, pH 4.0 and 7.4. The pH/temperature sensitivity of copolymers resulted from the transition of polymer–water and polymer–polymer interactions through the protonation/deprotonation of amine groups in poly(DMAEM) as the change of pH. The LCST of the copolymer is higher at pH 4.0 than at pH 7.4 due to the disruption of hydrophobic interaction between DMAEM and EAM by protonation of amine groups in DMAEM groups at pH 4.0. This suggests that at pH 7.4, the copolymer does not dissolve at a given temperature, but becomes dissolved at pH 4.0. GOD as glucose response units was incorporated into the poly(DMAEM-*co*-EAM) matrix. As shown in Figure 2.11, the generation of gluconic acid from free glucose through GOD embedded in hydrogels results in decrease of pH inside, and thus induces an increase in the LCST of the surface of the insulin-loaded matrix. The ultimate outcome is the dissolution of the copolymer and subsequently leads to release of the incorporated insulin.

2.2.3.4 Erodible Polymers Surrounded by GOD-Containing Hydrogels

The pH-dependent erodible poly(ortho esters) (POE) were used to develop glucose-sensitive insulin delivery systems [90–94]. pH-sensitive POE was synthesized by incorporating a tertiary amine (e.g., *N*-methyldiethanolamine). The erosion rate of the POE increases as the pH decreases. Insulin is physically dispersed in the POE matrix, which is surrounded by a hydrogel containing immobilized GOD [92]. When external glucose diffuses into the hydrogel to be oxidized by GOD to gluconic acid, the acidic environmental pH within the hydrogel accelerates the erosion of the polymer matrix, and embedded insulin releases. As the concentration of external glucose drops, production of gluconic acid decreases and the pH within the hydrogel rises again as gluconic acid diffuses out from the hydrogel. This process leads to modulated insulin release from the device. One issue related to this device is that the amount of insulin released may not be proportional to the decrease in pH, and the "on-and-off" process of insulin release is not fast enough [92].

2.2.4 GDH as Glucose-Responsive Unit

GDH as a glucose-responsive unit was also developed for glucose-sensitive insulin-releasing systems [86,95]. GDH provided better glucose sensitivity compared to GOD systems. Insulin molecules grafted to the poly(methyl methacrylate) (PMMA) surface through a disulfide bond were

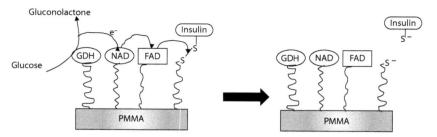

FIGURE 2.12 Design of glucose-sensitive insulin-releasing membrane systems. PMMA, poly(methyl methacrylate); S–S, disulfide bond; GDH, glucose dehydrogenase; NAD, nicotinamide adenine dinucleotide; FAD, flavin adenine dinucleotide. (Modified from Chung, D.-J. et al., *J. Control. Release*, 18, 45, 1992.)

tailored for a higher and faster glucose-sensitive insulin release [95]. Glucoses in solution were oxidized by GDH to produce electrons that broke the disulfide bonds, resulting in the release of the grafted insulin (Figure 2.12). This process requires enzyme cofactors, nicotinamide adenine dinucleotide (NAD), and flavin adenine dinucleotide (FAD) as electron mediators. The glucose sensitivity was increased as NAD and FAD were immobilized to the surface and further enhanced by co-immobilization of GDH (Figure 2.12). While this type of system can provide improved sensitivity to glucose and faster release of the immobilized insulin, the system has a major drawback to graft the limited amount of insulin onto the surface.

2.3 GLUCOSE-SENSITIVE HYDROGELS AS GLUCOSE SENSORS

In order to diagnose and manage diabetes, it is very important to monitor glucose levels in blood. Blood glucose levels keep fluctuating throughout the day depending on food intake, insulin availability, exercise, stress, and illness [95]. Therefore, detection of the accurate blood glucose levels and timely delivery of exact amounts of insulin are critical for the diabetic patients. Currently, diabetic patients determine their blood glucose levels by taking a drop of blood from a fingertip and placing it on a test zone located at one end of a plastic strip. The test zone has chemicals which can react with the glucose in the blood, changing color according to the concentration of glucose. The color change can be read by comparing with a standard chart. Even though this method is accurate, it is painful, inconvenient, and discontinuous with a possibility of infection [96]. If more reliable and noninvasive techniques can be devised, their medical implication can be enormous such as continuous monitoring of glucose levels in the blood.

The in vivo glucose sensors need to be safe, accurate, and reliable clinically, feasible for in vivo recalibration, stable for long periods of time, small in size, easy for insertion and removal, and fast enough to respond while allowing timely intervention [97]. A couple of attempts have been made to prepare noninvasive glucose-sensing devices including near-infrared (NIR) detection, ultrasound, and dielectric spectroscopy. Especially, NIR spectroscopy responds directly to the glucose molecule, unlike electrochemical sensors [98–104]. They can also provide additional information when no finger sticks are available, especially while sleeping. The main limitation of this system would be that the glucose signal is too weak in the midst of much larger signals from proteins, lipids, and scattered light [105]. Table 2.2 summarizes the typical blood glucose measurement techniques currently under evaluation [106].

A more realistic method may involve implanting glucose-sensitive devices, which can be able to sense glucose continuously. The system may consist of a disposable glucose sensor implanted under the skin, a link from the sensor to a non-implanted transmitter which can communicate to a radio receiver, and an electronic receiver like a pager that displays blood glucose levels on a continuous manner [107]. However, there are still limitations of the continuous glucose monitoring systems. Continuous systems must be calibrated with a traditional blood glucose measurement like the finger

TABLE 2.2
Techniques Used for Measuring Glucose Concentrations

Techniques	Methods
1. Invasive techniques	Implanted electrochemical sensors
	Suction blister extraction
	Microdialysis extraction
	Wick extraction
	Competitive fluorescence implants
	Needle puncture and extraction
2. Minimally invasive techniques	Iontophoretic extraction
	Sontophoretic extraction
	Chemically enhanced extraction
3. Noninvasive techniques	Infrared spectroscopy
	Near-infrared spectroscopy
	Raman spectroscopy
	Photoacoustic spectroscopy
	Scatter changes
	Polarization changes

Source: Roe, J.N. and Smoller, B.R., *Crit. Rev. Ther. Drug Carrier Syst.*, 15, 199, 1998.

stick earlier. Moreover, glucose levels in interstitial fluid may not show real blood glucose values. Therefore, patients need to calibrate blood glucose levels using traditional fingertip usually twice a day and are often advised to use the measurements to confirm hypo- or hyperglycemia before any corrective action. Glucose-sensitive hydrogels, such as sol–gel phase-reversible hydrogels, can be implanted, and the glucose level–dependent changes of the hydrogels can be monitored using external spectroscopic devices [108].

On the other hand, the percutaneous sensors are needlelike or inserted through a needle. They are designed to operate for a few days and be replaced by the patient. Those implantable sensors are mainly amperometric enzymatic sensors [109–112]. GOD has been used broadly as a glucose-sensing unit in amperometric glucose sensors since GOD transfers electrons to diffusing and non-diffusing mediators, withstands chemical immobilization techniques, and has a high turnover rate ($\sim 10^3$/s) at ambient temperatures [113]. However, variations in the level of dissolved oxygen may cause fluctuations in the electrode response and the dynamic range of glucose detection can be limited by the lack of dissolved oxygen [114,115].

Diffusion mediators for electron transport such as quinones and ferrocenes have been utilized in enzyme electrodes to overcome oxygen dependence. However, in vivo leaching of the mediator has been a serious issue. To eliminate the leaching, the mediator has been immobilized in different hydrogels. Sensors for in vivo monitoring of glucose also require biocompatible interfaces with the tissue in which they have been implanted [116]. The interface material must be permeable to glucose and the product of its oxidation, gluconolactone. Moreover, it should not be fouled by proteins, cellular attachment, or fibrous encapsulation. The implanted sensors are subject to protein adsorption and cell adhesion, which causes their quick inactivation. Hydrogels can hold a large amount of water and are able to prevent protein adsorption and cell adhesion, which results in excellent biocompatibility. Therefore, coating the surface of implantable glucose sensors with hydrogel layers would be a good strategy for extending the lifetime of the sensors. Table 2.3 summarizes the approaches used for glucose sensing with glucose-sensitive hydrogels.

TABLE 2.3

Glucose-Sensitive Hydrogels Used for Glucose Sensing

Glucose-Sensing Unit	Hydrogel System	Detection Method	References
GOD	Poly(HEMA)	H_2O_2	[120,121]
	Poly(VP) or poly(VP-VI)/OS derivative/ PEG redox hydrogels	Current	[97,113,115,122–126]
	Poly(allylamine)/OS derivative	Current	[127]
	Heparin/PDMAA/PAzSt	Current	[128]
	Poly(DMAA/AzSt/VFe)	Current	[129]
	Poly(PEG diacrylate/VFe) redox hydrogel	Current	[130]
	± Charged hydrogels/PTFE of PC membrane	Current	[131,132]
	Polyacrylamide	Current	[133]
	APEG/SPA hydrogel	Current	[116]
	Gelatin	Photoacoustic wave	[134]
	Poly(carbamoyl sulfonate)	Chemiluminescence	[135]
	Polyacrylamide-based colloidal crystal hydrogel	Optical intensity	[61,136,137]
Con A	FITC-dextran	Fluorescence	[138]
	Poly(GEMA)	Optical intensity	[39]
PBA	Poly(IPPm-*co*-PBA-*co*-DMA-PAA)	Optical intensity	[139]
GDH	Poly(ether amine quinone)	Current	[140]
	Poly(VI)/Os derivative/PEG redox hydrogel	Current	[141]

APEG, 8-armed, amine-terminated PEG; FITC-dextran, fluorescein-labeled dextran; Os, osmium; PAzSt, poly(*m*-azidostyrene); PC, polycarbonate; PDMAA, poly(dimethyl acrylamide-*co*-2-cinnamoylethyl methacrylate); poly(DMAA/AzSt/VFe), poly(dimethylacrylamide-*co*-azidostyrene-covinylferrocene); poly(GEMA), poly(glycosyl-ethyl methacrylate); poly(HEMA), poly(hydroxyethyl methacrylate); poly(IPPm-*co*-PBA-*co*-DMAPAA), poly(*N*-iso-propylacrylamide-*co*-3-acrylamido-phenylboronic acid-*co*-*N*-(3-dimethylaminopropyl)acrylamide; poly(VP), poly(vinylpyrrolidone); poly(VP-VI), poly(vinylpyrrolidone-*co*-1-vinylimidazole); PTFE, polytetrafluoroethylene; SPA, di-succinimidyl ester of PEG α, ω-dipropionic acid.

2.4 PROPERTIES TO IMPROVE FOR PRACTICAL APPLICATIONS

A simple, continuous, and noninvasive glucose sensor and an insulin delivery system mimicking physiological condition were required for treatment of diabetics. Therefore, self-regulated insulin delivery systems have attracted growing interest with respect to detection of increased glucose and delivery of an appropriate amount of insulin with proper kinetics. However, many hurdles to develop clinically useful self-regulated insulin delivery systems based on glucose-sensitive hydrogels were challenged. A main problem is that those systems work very well in the laboratory but rarely work as well when applied in vivo.

A number of progresses are required for practical applications. First, self-regulated insulin delivery devices should clinically release the necessary amount of insulin with kinetically rate-depending changes in the blood glucose concentration. The prompt on-and-off function is critical for adequate control of the dynamically changing glucose levels in the blood. An important parameter on the response time is the structure of hydrogels, which affects the time for diffusion of glucose molecules. Second, more biocompatible glucose-sensitive molecules should be developed for clinical application. Currently used glucose-sensing protein moieties such as Con A, GOD, and GDH are not suitable in long-term implantation. Con A is known to be immunogenic [117,118]. Commonly used GOD and GDH also are not suitable due to loss of enzymatic activity over time.

In addition, the polymers may not be ideal due to their lack of glucose specificity and biocompatibility information [68]. To minimize any side effects resulting from long-term use, glucose-sensing moieties based on nonproteinaceous molecules are preferred. Implantable glucose-sensing moieties should be biocompatible, nontoxic, cost-effective, and independent of environmental factors such as pH, ionic strength, or the presence of divalent cations [119]. Third, the long-term implantable devices tend to be isolated from the body by tissue remodeling around the implants. The formation of new tissues around the implant significantly retards the diffusion of glucose, resulting in delay of the response time and sensitivity. This necessitates frequent recalibration of the implanted devices, since there is no clear understanding of the relationship between glucose levels in the blood and in the tissue [107].

2.5 CONCLUSIONS

The environment-sensitive hydrogels have been intensively investigated for applications in biosensors and bioseparation, as well as drug delivery systems. In this chapter, glucose-sensitive hydrogels focusing on insulin delivery or biosensing were discussed. The glucose-sensitive hydrogels embedded with Con A, PBA, GOD, or GDH as a glucose-responsive unit enable the controlled insulin release from hydrogels, or act as biosensors by changing the physicochemical properties such as sol–gel transition, shrinking–swelling, and oxidation–reduction reaction depending on the glucose. Even though the glucose-sensitive hydrogels are highly promising with their attractive properties, their clinical applications still require many improvements, such as fast kinetic response upon changes in the environmental glucose concentration and capability to go back to their original states rapid enough after the response. Additional constraints can be potential toxicity of the materials, structural fragility, and their complicated biological interaction when applied in the body. Continued research on these hurdles will undoubtedly present answers in the future. The advantages of hydrogels such as high water contents, elastic behavior, ability of protecting drugs from hostile environment, and environmental responsiveness will continuously trigger innovations in hydrogel-based insulin delivery systems and biosensors.

REFERENCES

1. K. Park, S. W. Shalaby, and H. Park. *Biodegradable Hydrogel for Drug Delivery*, Technomic, Lancaster, PA, 1993.
2. V. Kudela. Hydrogels. *Encyclopedia of Polymer Science and Technology* (H. F. Mark and J. I. Kroschwitz, eds.), John Wiley, New York, 1985, pp. 783–807.
3. R. M. Ottenbrite, S. J. Huang, and K. Park. *Hydrogels and Biodegradable Polymers for Bioapplications*, American Chemical Society, Washington, DC, 1996.
4. J. Andrade. *Hydrogels for Medical and Related Applications*, American Chemical Society, Washington, DC, 1976.
5. N. A. Peppas. *Hydrogels in Medicine and Pharmacy*, CRC Press, Boca Raton, FL, 1986.
6. D. DeRossi, K. Kajiwara, Y. Osada, and A. Yamauchi. *Polymer Gels: Fundamentals and Biomedical Applications*, Plenum Press, New York, 1991.
7. J. M. Guenet. *Thermoreversible Gelation of Polymers and Biopolymers*, Academic Press, New York, 1992.
8. Y. H. Bae and S. W. Kim. Hydrogel delivery systems based on polymer blends, block co-polymers or interpenetrating networks. *Adv Drug Deliv Rev* 11: 109–135 (1993).
9. R. Barbucci, G. Leone, and A. Vecchiullo. Novel carboxymethylcellulose-based microporous hydrogels suitable for drug delivery. *J Biomater Sci Polym Ed* 15: 607–619 (2004).
10. J. Berger, M. Reist, J. M. Mayer, O. Felt, N. A. Peppas, and R. Gurny. Structure and interactions in covalently and ionically crosslinked chitosan hydrogels for biomedical applications. *Eur J Pharm Biopharm* 57: 19–34 (2004).
11. T. Miyata, T. Uragami, and K. Nakamae. Biomolecule-sensitive hydrogels. *Adv Drug Deliv Rev* 54: 79–98 (2002).

12. W. E. Hennink and C. F. van Nostrum. Novel crosslinking methods to design hydrogels. *Adv Drug Deliv Rev* 54: 13–36 (2002).

13. K. M. Huh, T. Ooya, W. K. Lee, S. Sasaki, I. C. Kwon, S. Y. Jeong, and N. Yui. Supramolecular-structured hydrogels showing a reversible phase transition by inclusion complexation between poly(ethylene glycol) grafted dextran and α-cyclodextrin. *Macromolecules* 34: 8657–8662 (2001).

14. L. Martin, C. G. Wilson, F. Koosha, and I. F. Uchegbu. Sustained buccal delivery of the hydrophobic drug denbufylline using physically cross-linked palmitoyl glycol chitosan hydrogels. *Eur J Pharm Biopharm* 55: 35–45 (2003).

15. N. A. Peppas, Y. Huang, M. Torres-Lugo, J. H. Ward, and J. Zhang. Physicochemical foundations and structural design of hydrogels in medicine and biology. *Annu Rev Biomed Eng* 2: 9–29 (2000).

16. H. Park and K. Park. Hydrogels in bioapplications. *Hydrogels and Biodegradable Polymers for Bioapplications* (R. M. Ottenbrite, S. J. Huang, and K. Park, eds.), American Chemical Society, Washington, DC, 1996, pp. 2–10.

17. C. M. Hassan, F. J. Doyle, and N. A. Peppas. Dynamic behavior of glucose-responsive poly(methacrylic acid-*g*-ethylene glycol) hydrogels. *Macromolecules* 30: 6166–6173 (1997).

18. A. S. Hoffman. Intelligent polymers. *Controlled Drug Delivery* (K. Park, ed.), American Chemical Society, Washington, DC, 1997, pp. 485–498.

19. J. Kim and K. Park. Smart hydrogels for bioseparation. *Bioseparation* 7: 177–184 (1998).

20. Y. H. Bae. Stimuli-sensitive drug delivery. *Controlled Drug Delivery* (K. Park, ed.), American Chemical Society, Washington, DC, 1997, pp. 147–162.

21. B. A. Firestone and R. A. Siegel. Kinetics and mechanisms of water sorption in hydrophobic, ionizable copolymer gels. *J Appl Polym Sci* 43: 901–914 (1991).

22. J. Heller. Feedback-controlled drug delivery. *Controlled Drug Delivery* (K. Park, ed.), American Chemical Society, Washington, DC, 1997, pp. 127–146.

23. A. A. Obaidat and K. Park. Characterization of protein release through glucose-sensitive hydrogel membranes. *Biomaterials* 18: 801–806 (1997).

24. N. Sharon and H. Lis. Lectins: Cell-agglutinating and sugar-specific proteins. *Science* 177: 949–959 (1972).

25. J. L. Wang, B. A. Cunningham, M. J. Waxdal, and G. M. Edelman. The covalent and three-dimensional structural of concanavalin A. I. Amino acid sequence of cyanogen bromide fragments F1 and F2. *J Biol Chem* 250: 1490–1502 (1975).

26. B. A. Cunningham, J. L. Wang, M. J. Waxdal, and G. M. Edelman. The covalent and three-dimensional structure of concanavalin A. II. Amino acid sequence of cyanogen bromide fragment F3. *J Biol Chem* 250: 1503–1512 (1975).

27. M. Brownlee and A. Cerami. A glucose-controlled insulin-delivery system: Semisynthetic insulin bound to lectin. *Science* 206: 1190–1191 (1979).

28. S. Y. Jeong, S. W. Kim, D. L. Holmberg, and J. C. McRea. Self-regulating insulin delivery systems: III. In vivo studies. *J Control Release* 2: 143–152 (1985).

29. S. W. Kim, C. M. Paii, K. Makino, L. A. Seminoff, D. L. Holmberg, J. M. Gleeson, D. E. Wilson, and E. J. Mack. Self-regulated glycosylated insulin delivery. *J Control Release* 11: 193–201 (1990).

30. S. Y. Jeong, S. W. Kim, M. J. D. Eenink, and J. Feijen. Self-regulating insulin delivery systems I. Synthesis and characterization of glycosylated insulin. *J Control Release* 1: 57–66 (1984).

31. S. W. Kim, S. Y. Jeong, S. Sato, J. C. McRea, and J. Feijen. Self-regulating insulin delivery system—Chemical approach. *Recent Advances in Drug Delivery Systems* (J. M. Anderson, S. W. Kim, eds.), Plenum Press, New York, 1984, pp. 123–136.

32. S. Sato, S. Y. Jeong, J. C. McRea, and S. W. Kim. Self-regulating insulin delivery systems II. In vitro studies. *J Control Release* 1: 67–77 (1984).

33. S. Sato, S. Y. Jeong, J. C. McRea, and S. W. Kim. Glucose stimulated insulin delivery systems. *Pure Appl Chem* 56: 1323–1328 (1984).

34. L. A. Seminoff, J. M. Gleeson, J. Zheng, G. B. Olsen, D. Holmberg, S. F. Mohammad, D. Wilson, and S. W. Kim. A self-regulating insulin delivery system. II. In vivo characteristics of a synthetic glycosylated insulin. *Int J Pharm* 54: 251–257 (1989).

35. L. A. Seminoff, G. B. Olsen, and S. W. Kim. A self-regulating insulin delivery system. I. Characterization of a synthetic glycosylated insulin derivative. *Int J Pharm* 54: 241–249 (1989).

36. K. Makino, E. J. Mack, T. Okano, and S. W. Kim. A microcapsule self-regulating delivery system for insulin. *J Control Release* 12: 235–239 (1990).

37. E. Kokufata, Y.-Q. Zhang, and T. Tanaka. Saccharide-sensitive phase transition of a lectin-loaded gel. *Nature* 351: 302–304 (1991).

38. J. E. Morris, A. S. Hoffman, and R. R. Fisher. Affinity precipitation of proteins by polyligands. *Biotechnol Bioeng* 41: 991–997 (1993).
39. K. Nakamae, T. Miyata, A. Jikihara, and A. S. Hoffman. Formation of poly(glucosyloxyethyl methacrylate)–concanavalin A complex and its glucose-sensitivity. *J Biomater Sci Polym Ed* 6: 79–90 (1994).
40. T. Miyata, A. Jikihara, K. Nakamae, and A. S. Hoffman. Preparation of poly(2-glucosyloxyethyl methacrylate)–concanavalin A complex hydrogel and its glucose-sensitivity. *Macromol Chem Phys* 197: 1135–1146 (1996).
41. M. J. Taylor, S. Tanna, P. M. Taylor, and G. Adams. The delivery of insulin from aqueous and non-aqueous reservoirs governed by a glucose sensitive gel membrane. *J Drug Target* 3: 209–216 (1995).
42. S. Tanna and M. J. Taylor. A self-regulating system using high-molecular weight solutes in glucose-sensitive gel membranes. *J Pharm Pharmacol* 46(Suppl. 2): 1051a (1994).
43. M. J. Taylor, S. Tanna, S. Cockshott, and R. Vaitha. A self regulated delivery system using unmodified solutes in glucose-sensitive gel membranes. *J Pharm Pharmacol* 46(Suppl. 2): 1051b (1994).
44. S. J. Lee and K. Park. Synthesis of sol–gel phase-reversible hydrogels sensitive to glucose. *Proc Int Symp Control Rel Bioact Mater* 21: 93–94 (1994).
45. S. J. Lee and K. Park. Synthesis and characterization of sol–gel phase-reversible hydrogels sensitive to glucose. *J Mol Recogn* 9: 549–557 (1996).
46. S. J. Lee and K. Park. Glucose-sensitive phase-reversible hydrogels. *Hydrogels and Biodegradable Polymers for Bioapplications* (R. M. Ottenbrite, S. J. Huang, and K. Park, eds.), American Chemical Society, Washington, DC, 1996, pp. 11–16.
47. A. A. Obaidat and K. Park. Characterization of glucose dependent gel–sol phase transition of the polymeric glucose–concanavalin A hydrogel system. *Pharm Res* 13: 989–995 (1996).
48. T. Miyata, A. Jikihara, K. Nakamae, and A. S. Hoffman. Preparation of reversibly glucose-responsive hydrogels by covalent immobilization of lectin in polymer networks having pendant glucose. *J Biomater Sci Polym Ed* 15: 1085–1098 (2004).
49. S. Tanna and M. J. Taylor. Characterization of model solute and insulin delivery across covalently modified lectin–polysaccharide gels sensitive to glucose. *Pharm Pharmacol Commun* 4: 117 (1998).
50. S. Tanna, M. J. Taylor, and G. Adams. Insulin delivery governed by covalently modified lectin–glycogen gels sensitive to glucose. *J Pharm Pharmacol* 51: 1093–1098 (1999).
51. S. Tanna, T. Sahota, J. Clark, and M. J. Taylor. Covalent coupling of concanavalin A to a Carbopol 934P and 941P carrier in glucose-sensitive gels for delivery of insulin. *J Pharm Pharmacol* 54: 1461–1469 (2002).
52. S. Tanna, T. Sahota, J. Clark, and M. J. Taylor. A covalently stabilised glucose responsive gel formulation with a Carbopol carrier. *J Drug Target* 10: 411–418 (2002).
53. S. Tanna, M. Joan Taylor, T. S. Sahota, and K. Sawicka. Glucose-responsive UV polymerised dextran–concanavalin A acrylic derivatised mixtures for closed-loop insulin delivery. *Biomaterials* 27: 1586–1597 (2006).
54. R. Zhang, M. Tang, A. Bowyer, R. Eisenthal, and J. Hubble. Synthesis and characterization of a D-glucose sensitive hydrogel based on CM-dextran and concanavalin A. *React Funct Polymers* 66: 757–767 (2006).
55. C. M. Pai, Y. H. Bae, E. J. Mack, D. E. Wilson, and S. W. Kim. Concanavalin A microspheres for a self-regulating insulin delivery system. *J Pharm Sci* 81: 532–536 (1992).
56. M. Baudys, T. Uchio, L. Hovgaard, E. F. Zhu, T. Avramoglou, M. Jozefowicz, B. Rihova, J. Y. Park, H. K. Lee, and S. W. Kim. Glycosylated insulins. *J Control Release* 36: 151–157 (1995).
57. B. A. Cunningham, J. L. Wang, M. N. Pflumm, and G. M. Edelman. Isolation and proteolytic cleavage of the intact subunit of concanavalin A. *Biochemistry* 11: 3233–3239 (1972).
58. F. Liu, S. C. Song, D. Mix, M. Baudys, and S. W. Kim. Glucose-induced release of glycosylpoly(ethylene glycol) insulin bound to a soluble conjugate of concanavalin A. *Bioconjug Chem* 8: 664–672 (1997).
59. J. J. Kim. Phase-reversible glucose-sensitive hydrogels for modulated insulin delivery, Doctor of Philosophy, Purdue University, West Lafayette, 1999.
60. K. Makino, E. J. Mack, T. Okano, and S. W. Kim. Self-regulated delivery of insulin from microcapsules. *Biomater Artif Cells Immobilization Biotechnol* 19: 219–228 (1991).
61. S. Aronoff, T. Chen, and M. Cheveldayoff. Complexation of D-glucose with borate. *Carbohydr Res* 40: 299–309 (1975).
62. J. Boeseken. The use of boric acid for the determination of the configuration of carbohydrates. *Adv Carbohydr Chem* 47: 189–210 (1949).
63. V. Bouriotis, I. J. Galpin, and P. D. G. Dean. Applications of immobilised phenylboronic acids as supports for group specific ligands in the affinity chromatography of enzymes. *J Chromatogr* 210: 267–278 (1981).

64. A. B. Foster. Zone electrophoresis of carbohydrate. *Adv Carbohydr Chem* 12: 81–115 (1957).
65. I. Hisamitsu, K. Kataoka, T. Okano, and Y. Sakurai. Glucose-responsive gel from phenylborate polymer and poly(vinyl alcohol): Prompt response at physiological pH through the interaction of borate with amino group in the gel. *Pharm Res* 14: 289–293 (1997).
66. S. Kitano, K. Kataoka, Y. Koyama, T. Okano, and Y. Sakurai. Glucose-responsive complex formation between poly(vinyl alcohol) and poly(*N*-vinyl-2-pyrrolidone) with pendant phenylboronic acid moieties. *Makromol Chem Rapid Commun* 12: 227–233 (1991).
67. S. Kitano, Y. Koyama, K. Kataoka, T. Okano, and Y. Sakurai. A novel drug delivery system utilizing a glucose responsive polymer complex between poly(vinyl alcohol) and poly(*N*-vinyl-2-pyrrolidone) with a phenylboronic acid moiety. *J Control Release* 19: 161–170 (1992).
68. D. Shiino, Y. Murata, K. Kataoka, Y. Koyama, M. Yokoyama, T. Okano, and Y. Sakurai. Preparation and characterization of a glucose-responsive insulin-releasing polymer device. *Biomaterials* 15: 121–128 (1994).
69. T. Tanaka, D. Fillmore, S. T. Sun, I. Nishio, G. Swislow, and A. Shah. Phase transitions in ionic gels. *Phys Rev Lett* 45: 1636 (1980).
70. J. P. Lorand and J. O. Edwards. Polyol complexes and structure of the benzeneboronate ion. *J Org Chem* 24: 769 (1959).
71. Y. K. Choi, S. Y. Jeong, and Y. H. Kim. A glucose-triggered solubilizable polymer gel matrix for an insulin delivery system. *Int J Pharm* 80: 9–16 (1992).
72. K. Kataoka, H. Miyazaki, T. Okano, and Y. Sakurai. Sensitive glucose-induced change of the lower critical solution temperature of poly[*N,N*-(dimethylacrylamide)-*co*-3-(acrylamido)-phenylboronic acid] in physiological saline. *Macromolecules* 27: 1061–1062 (1994).
73. G. Albin, T. Horbett, and B. Ratner. Glucose-sensitive membranes for controlled release of insulin. *Pulsed and Self-Regulated Drug Delivery* (J. Kost, ed.), CRC Press, Boca Raton, FL, 1990.
74. G. Albin, T. A. Horbett, and B. D. Ratner. Glucose sensitive membranes for controlled delivery of insulin: Insulin transport studies. *J Control Release* 2: 153–164 (1985).
75. G. W. Albin, T. A. Horbett, S. R. Miller, and N. L. Ricker. Theoretical and experimental studies of glucose sensitive membranes. *J Control Release* 6: 267–291 (1987).
76. T. A. Horbett, J. Kost, and B. D. Ratner. Swelling behavior of glucose sensitive membranes. *Polymers as Biomaterials* (S. Shakaby, A. Hoffman, T. Horbett, and B. Ratner, eds.), Plenum Press, New York, 1984, pp. 193–207.
77. J. Kost, T. A. Horbett, B. D. Ratner, and M. Singh. Glucose-sensitive membranes containing glucose oxidase: Activity, swelling, and permeability studies. *J Biomed Mater Res* 19: 1117–1133 (1985).
78. T. A. Horbett, B. D. Ratner, J. Kost, and M. Singh. A bioresponsive membrane for insulin delivery. *Recent Advances in Drug Delivery Systems* (J. M. Anderson and S. W. Kim, eds.), Plenum Press, New York, 1984, pp. 209–220.
79. L. A. Klumb and T. A. Horbett. Design of insulin delivery devices based on glucose sensitive membranes. *J Control Release* 18: 59–80 (1992).
80. L. A. Klumb and T. A. Horbett. The effect of hydronium ion transport on the transient behavior of glucose sensitive membranes. *J Control Release* 27: 95–114 (1993).
81. M. Goldraich and J. Kost. Glucose-sensitive polymeric matrices for controlled drug delivery. *Clin Mater* 13: 135–142 (1993).
82. K. Ishihara, M. Kobayashi, N. Ishimaru, and I. Shinohara. Glucose induced permeation control of insulin through a complex membrane consisting of immobilized glucose oxidase and a poly(amine). *Polymer J* 16: 625–631 (1984).
83. K. Ishihara, M. Kobayashi, and I. Shinohara. Insulin permeation through amphiphilic polymer membranes having 2-hydroxyethyl methacrylate moiety. *Polymer J* 16: 647–651 (1984).
84. K. Ishihara, M. Kobayashi, and I. Shionohara. Control of insulin permeation through a polymer membrane with responsive function for glucose. *Makromol Chem Rapid Commun* 4: 327–331 (1983).
85. K. Ishihara and K. Matsui. Glucose-responsive insulin release from a polymer capsule. *J Polym Sci Polym Lett Edn* 24: 413–417 (1986).
86. Y. Ito, M. Casolaro, K. Kono, and Y. Imanishi. An insulin-releasing system that is responsive to glucose. *J Control Release* 10: 195–203 (1989).
87. H. Iwata and T. Matsuda. Preparation and properties of novel environment-sensitive membranes prepared by graft polymerization onto a porous membrane. *J Memb Sci* 38: 185–199 (1988).
88. S. H. Cho, M. S. Jhon, S. H. Yuk, and H. B. Lee. Temperature-induced phase transition of poly(*N,N*-dimethylaminoethyl methacrylate-*co*-acrylamide). *J Polym Sci Part B Polym Phys* 35: 595–598 (1997).
89. S. H. Yuk, S. H. Cho, and S. H. Lee. pH/temperature-responsive polymer composed of poly((*N,N*-dimethylamino)ethyl methacrylate-*co*-ethylacrylamide). *Macromolecules* 30: 6856–6859 (1997).

90. J. Heller. Chemically self-regulated drug delivery systems. *J Control Release* 8: 111–125 (1988).
91. J. Heller. Use of enzymes and bioerodible polymers in self-regulated and triggered drug delivery systems. *Modulated Controlled Release Systems* (J. Kost, ed.), CRC Press, Boca Raton, FL, 1990, pp. 93–108.
92. J. Heller, A. C. Chang, G. Rood, and G. M. Grodsky. Release of insulin from pH-sensitive poly(ortho esters). *J Control Release* 13: 295–302 (1990).
93. J. Heller, S. H. Pangburn, and D. W. H. Penhale. Use of bioerodible polymers in self-regulated drug delivery systems. *Controlled Release Technology* (P. I. Lee and W. R. Good, eds.), American Chemical Society, Washington, DC, 1987, pp. 172–187.
94. J. Heller, D. W. H. Penhale, and R. F. Helwing. Preparation of poly(ortho esters) by the reaction of ketane acetal and diol. *J Polym Sci Polym Lett Edn* 18: 611–624 (1980).
95. D.-J. Chung, Y. Ito, and Y. Imanishi. An insulin-releasing membrane system on the basis of oxidation reaction of glucose. *J Control Release* 18: 45–53 (1992).
96. R. T. Kurnik, B. Berner, J. Tamada, and R. O. Potts. Design and simulation of a reverse iontophoretic glucose monitoring device. *J Electrochem Soc* 145: 4119–4125 (1998).
97. E. Csoeregi, C. P. Quinn, D. W. Schmidtke, S.-E. Lindquist, M. V. Pishko, L. Ye, I. Katakis, J. A. Hubbell, and A. Heller. Design, characterization, and one-point in vivo calibration of a subcutaneously implanted glucose electrode. *Anal Chem* 66: 3131–3138 (1994).
98. D. M. Back, D. F. Michalska, and P. L. Polavarapu. Fourier transform infrared spectroscopy as a powerful tool for the study of carbohydrates in aqueous solutions. *Appl Spectrosc* 38: 173–180 (1984).
99. G. B. Christison and H. A. MacKenzie. Laser photoacoustic determination of physiological glucose concentrations in human whole blood. *Med Biol Eng Comput* 31: 284–290 (1993).
100. J. W. Hall and A. Pollard. Near-infrared spectrophotometry: A new dimension in clinical chemistry. *Clin Chem* 38: 1623–1631 (1992).
101. H. M. Heise, R. Marbach, T. Koschinsky, and F. A. Gries. Noninvasive blood glucose sensors based on near-infrared spectroscopy. *Artif Organs* 18: 439–447 (1994).
102. L. A. Marquardt, M. A. Arnold, and G. W. Small. Near-infrared spectroscopic measurement of glucose in a protein matrix. *Anal Chem* 65: 3271–3278 (1993).
103. M. R. Robinson, R. P. Eaton, D. M. Haaland, G. W. Koepp, E. V. Thomas, B. R. Stallard, and P. L. Robinson. Noninvasive glucose monitoring in diabetic patients: A preliminary evaluation. *Clin Chem* 38: 1618–1622 (1992).
104. H. Zeller, P. Novak, and R. Landgraf. Blood glucose measurement by infrared spectroscopy. *Int J Artif Organs* 12: 129–135 (1989).
105. A. A. Sharkawy, M. R. Neuman, and W. R. Reichert. Senso-compatibility: Design consideration for biosensor-based drug delivery systems. *Controlled Drug Delivery: Challenges and Strategies* (K. Park, ed.), American Chemical Society, Washington, DC, 1997, pp. 163–181.
106. J. N. Roe and B. R. Smoller. Bloodless glucose measurements. *Crit Rev Ther Drug Carrier Syst* 15: 199–241 (1998).
107. C. Henry. Getting under the skin: Implantable glucose sensors. *Anal Chem* 70: 594A–598A (1998).
108. H. Park and K. Park. Biocompatibility issues of implantable drug delivery systems. *Pharm Res* 13: 1770–1776 (1996).
109. D. Frazer. Biosensing in the body. *Med Device Technol* 5: 24–28 (1994).
110. J. C. Pickup, D. J. Claremont, and G. W. Shaw. Responses and calibration of amperometric glucose sensors implanted in the subcutaneous tissue of man. *Acta Diabetol* 30: 143–148 (1993).
111. F. J. Schmidt, W. J. Sluiter, and A. J. Schoonen. Glucose concentration in subcutaneous extracellular space. *Diabetes Care* 16: 695–700 (1993).
112. P. Vadgama and P. W. Crump. Biosensors: Recent trends. A review. *Analyst* 117: 1657–1670 (1992).
113. M. V. Pishko, A. C. Michael, and A. Heller. Amperometric glucose microelectrodes prepared through immobilization of glucose oxidase in redox hydrogels. *Anal Chem* 63: 2268–2272 (1991).
114. S. Dong, W. Baoxing, and L. Baifeng. Amperometric glucose sensor with ferrocene as an electron transfer mediator. *Biosens Bioelectron* 7: 215–222 (1991).
115. B. Linke, W. Kerner, M. Kiwit, M. Pishko, and A. Heller. Amperometric biosensor for in vivo glucose sensing based on glucose oxidase immobilized in a redox hydrogel. *Biosens Bioelectron* 9: 151–158 (1994).
116. C. A. Quinn, R. E. Connor, and A. Heller. Biocompatible, glucose-permeable hydrogel for in situ coating of implantable biosensors. *Biomaterials* 18: 1665–1670 (1997).
117. E. V. Larsson, M. Gullberg, and A. Coutinho. Heterogeneity of cells and factors participating in the concanavalin A-dependent activation of T lymphocytes with cytotoxic potential. *Immunobiology* 161: 5–20 (1982).

118. A. E. Powell and M. A. Leon. Reversible interaction of human lymphocytes with the mitogen concanavalin A. *Exp Cell Res* 62: 315–325 (1970).
119. T. Li, H. B. Lee, and K. Park. Comparative stereochemical analysis of glucose-binding proteins for rational design of glucose-specific agents. *J Biomater Sci Polym Ed* 9: 327–344 (1998).
120. G. Urban, G. Jobst, E. Aschauer, O. Tilado, P. Svasek, M. Varahram, C. Ritter, and J. Riegebauer. Performance of integrated glucose and lactate thin-film microbiosensors for clinical analysers. *Sens Actuat B Chem* 19: 592–596 (1994).
121. G. Urban, G. Jobst, F. Keplinger, E. Aschauer, O. Tilado, R. Fasching, and F. Kohl. Miniaturized multi-enzyme biosensors integrated with pH sensors on flexible polymer carriers for in vivo applications. *Biosens Bioelectron* 7: 733–739 (1992).
122. T. de Lumley-Woodyear, P. Rocca, J. Lindsay, Y. Dror, A. Freeman, and A. Heller. Polyacrylamide-based redox polymer for connecting redox centers of enzymes to electrodes. *Anal Chem* 67: 1332–1338 (1995).
123. B. A. Gregg and A. Heller. Redox polymer films containing enzymes. 2. Glucose oxidase-containing enzyme electrodes. *J Phys Chem* 95: 5976–5980 (1991).
124. T. J. Ohara, R. Rajagopalan, and A. Heller. Glucose electrodes based on cross-linked bis(2,2′-bipyridine)chloroosmium(+/2+) complexed poly(1-vinylimidazole) films. *Anal Chem* 65: 3512–3517 (1993).
125. T. J. Ohara, R. Rajagopalan, and A. Heller. "Wired" enzyme electrodes for amperometric determination of glucose or lactate in the presence of interfering substances. *Anal Chem* 66: 2451–2457 (1994).
126. D. W. Schmidtke and A. Heller. Accuracy of the one-point in vivo calibration of "wired" glucose oxidase electrodes implanted in jugular veins of rats in periods of rapid rise and decline of the glucose concentration. *Anal Chem* 70: 2149–2155 (1998).
127. C. Danilowicz, E. Cort, F. Battaglini, and E. J. Calvo. An Os(byp)$_2$ClPyCH$_2$NHPoly(allylamine) hydrogel mediator for enzyme wiring at electrodes. *Electrochim Acta* 43: 3525–3531 (1998).
128. Y. Nakayama and T. Matsuda. Surface fixation of hydrogels. Heparin and glucose oxidase hydrogelated surfaces. *Asaio J* 38: M421–M424 (1992).
129. Y. Nakayama, Q. Zheng, J. Nishimura, and T. Matsuda. Design and properties of photocurable electro-conductive polymers for use in biosensors. *Asaio J* 41: M418–M421 (1995).
130. K. Sirkar and M. V. Pishko. Amperometric biosensors based on oxidoreductases immobilized in photopolymerized poly(ethylene glycol) redox polymer hydrogels. *Anal Chem* 70: 2888–2894 (1998).
131. R. Vaidya and E. Wilkins. Application of polytetrafluoroethylene (PTFE) membranes to control interference effects in a glucose biosensor. *Biomed Instrum Technol* 27: 486–494 (1993).
132. R. Vaidya and E. Wilkins. Use of charged membranes to control interference by body chemicals in a glucose biosensor. *Med Eng Phys* 16: 416–421 (1994).
133. C. Jimenez, J. Bartrol, N. F. de Rooij, and M. Koudelka-Hep. Use of photopolymerizable membranes based on polyacrylamide hydrogels for enzymatic microsensor construction. *Anal Chim Acta* 351: 169–176 (1997).
134. K. M. Quan, G. B. Christison, H. A. MacKenzie, and P. Hodgson. Glucose determination by a pulsed photoacoustic technique: An experimental study using a gelatin-based tissue phantom. *Phys Med Biol* 38: 1911–1922 (1993).
135. D. Janasek and U. Spohn. An enzyme-modified chemiluminescence detector for hydrogen peroxide and oxidase substrates. *Sens Actuat B Chem* 39: 291–294 (1997).
136. J. H. Holtz and S. A. Asher. Polymerized colloidal crystal hydrogel films as intelligent chemical sensing materials. *Nature* 389: 829–832 (1997).
137. J. H. Holtz, J. S. W. Holtz, C. H. Munro, and S. A. Asher. Intelligent polymerized crystalline colloidal arrays: Novel chemical sensor materials. *Anal Chem* 70: 780–791 (1998).
138. J. S. Schulz, S. Mansouri, and I. J. Goldstein. Affinity sensor: A new technique for developing implantable sensors for glucose and other metabolites. *Diabetes Care* 5: 245–253 (1982).
139. T. Aoki, Y. Nagao, K. Sanui, N. Ogata, A. Kikuchi, Y. Sakurai, K. Kataoka, and T. Okano. Glucose-sensitive lower critical solution temperature changes of copolymers composed of *N*-isopropylacrylamide and phenylboronic acid moieties. *Polym J* 28: 371–374 (1996).
140. M. Tessema, T. Ruzgas, L. Gorton, and T. Ikeda. Flow injection amperometric determination of glucose and some other low molecular weight saccharides based on oligosaccharide dehydrogenase mediated by benzoquinone systems. *Anal Chim Acta* 310: 161–171 (1995).
141. T. Ruzgas, E. Csoregi, I. Katakis, G. Kenausis, and L. Gorton. Preliminary investigations of an amperometric oligosaccharide dehydrogenase-based electrode for the detection of glucose and some other low molecular weight saccharides. *J Mol Recogn* 9: 480–484 (1996).

3 Advances in Polymeric and Lipid-Core Micelles as Drug Delivery Systems

Tiziana Musacchio and Vladimir P. Torchilin

CONTENTS

3.1 INTRODUCTION

The development of pharmaceutical nanocarriers has been extensively studied over the past decades. Many studies have focused on the use of biodegradable polymers for the manufacture of these drug delivery systems (DDS). The importance of biodegradable nanoparticulate systems is determined by their potential for controlling the release of the drug, the stabilization of labile molecules (such as DNA, RNA, proteins, peptides, etc.) from degradation, and their capacity for site-specific drug targeting when properly surface-modified. In particular, the application of polymer-based DDS in oncology has grown markedly with the advent of the development of biodegradable polymers (Wang et al. 2008). In these polymers, drugs are either physically dissolved, entrapped, encapsulated, or covalently attached to the polymer matrix (Rawat et al. 2006). According to their functions and target of interest in the body, the resulting compounds may have different structures and composition. Thus, the panorama is widely extense and embraces drug carriers like micelles, liposomes, solid–lipid nanoparticles, dendrimers, nanotubes, and polymersomes (Ahmed et al. 2006;

Bae and Kataoka 2009; Chen et al. 2008; Hsu et al. 2009; Kaminskas et al. 2009; Ko et al. 2009; Lalloo et al. 2006; Wong et al. 2007). Microreservoir-type systems, such as liposomes (mainly, for water-soluble drugs) and micelles (mainly, for water-insoluble drugs), have certain advantages over other delivery systems, including a larger load of the drug, a smaller required quantity of a targeting component, and the ease of control of the composition, size, and in vivo stability of the microreservoir. In this chapter, we will be illustrating lipid-core and polymeric micelles as drug carriers since currently they are believed to be able to provide a set of important advantages: they can solubilize poorly soluble drugs and thus increase their bioavailability, they can stay in body (in the blood) long enough to allow accumulation in the pathological areas (tumors) via the enhanced permeability and retention (EPR) effect (Maeda et al. 2001; Yuan et al. 1994), they can be made targeted by attachment of a specific ligand to their surface, they can be easily prepared in reproducibly large quantities. Additionally, when in a micellar form, the drug is protected from possible inactivation by the effects of biological surroundings, it does not provoke undesirable side effects, and its bioavailability is usually enhanced.

3.2 GENERALITIES OF MICELLAR SYSTEMS

Micelles are colloidal dispersions with a particle size usually between 5 and 100 nm. They belong to a group of association or amphiphilic colloids, which form spontaneously under certain conditions of concentration and temperature from amphiphilic or surface-active agents (surfactants), molecules which consist of two distinct segments with opposite affinities for a given solvent (Mittal and Lindman 1991). At low concentrations in an aqueous medium, such amphiphilic molecules exist as unimers. However, as their concentration increases, aggregation takes place within a rather narrow concentration interval. The concentration of a monomeric amphiphile at which micelles appear is called "critical micelle concentration" (CMC), while the temperature, below which amphiphilic molecules exist as unimers, and above as aggregates, is called "critical micellization temperature" (CMT). The formation of micelles is driven by the decrease of free energy in the system associated with the removal of hydrophobic fragments from the aqueous environment and the reestablishment of a hydrogen bond network in water. Hydrophobic fragments of amphiphilic molecules form the core of a micelle, while hydrophilic fragments form the micelle's shell (Attwood and Florence 1983; Elworthy et al. 1968; Lasic 1992; Nishiyama and Kataoka 2006; Torchilin 2007).

Micelles as drug carriers provide a set of clear advantages (Jones and Leroux 1999; Kwon 1998; Torchilin 2001). Thanks to their small size, micelles are capable of an effective "passive targeting" (Maeda et al. 2000; Palmer et al. 1984) via the previously mentioned EPR effect. It has been repeatedly shown that micelle-incorporated anticancer drugs, such as adriamycin (Kwon and Kataoka 1995), accumulate better in tumors than in nontarget tissues, thus reducing the undesired drug toxicity toward normal tissue. In addition, micelles can be made targeted by chemical attachment of target-specific molecules to their surface. In this case, the local release of drug from the micelles within the target organ should lead to the drug's increased efficacy. Additionally, in a micellar form, the drug is better protected from possible inactivation by the effect of biological surroundings, while reducing undesirable side effects on nontarget organs and tissues. At the usual size of a pharmaceutical micelle between 10 and 80 nm, its CMC value is expected to be in a low millimolar region, or even lower, and the loading efficiency toward a hydrophobic drug could ideally be between 5% and 25% wt. Overall, the solubilization of drugs using micelle-forming surfactants results in an increased water solubility of sparingly soluble drug, improved bioavailability, reduction of toxicity and other adverse effects, enhanced permeability across the anatomical and physiological barriers, and substantial favorable changes in drug biodistribution. Micellar compositions of various drugs have been suggested for parenteral (Le Garrec et al. 2004; Shuai et al. 2004; Soga et al. 2005), oral (Mathot et al. 2006; Park et al. 2005), nasal (Gao et al. 2006; Pillion et al. 1996), and ocular (Liaw et al. 2001; Pillion et al. 1996) application.

3.3 POLYMERIC MICELLES

Micelles prepared from amphiphilic copolymers for solubilization of poorly soluble drugs have captured much interest recently (Aliabadi and Lavasanifar 2006; Gaucher et al. 2005; Jones and Leroux 1999; Torchilin 2001). Polymeric micelles consist of hydrophilic and hydrophobic monomer units with the length of a hydrophilic block exceeding to some extent that of a hydrophobic one (Torchilin 2001). When the length of a hydrophilic block is too high, copolymers exist in water as unimers (individual molecules), while molecules with very long hydrophobic blocks form structures with non-micellar morphology, such as rods and lamellae (Zhang and Eisenberg 1995). The major driving force behind self-association of amphiphilic polymers is again the decrease of the free energy of the system, as for surfactants. Polymeric micelles comprise the core of the hydrophobic blocks stabilized by the corona of hydrophilic chains. Polymeric micelles are often more stable compared to micelles prepared from conventional detergents (have lower CMC value), with some amphiphilic copolymers having CMC values as low as 10^{-6} M (Kabanov et al. 2002; La et al. 1996), which is about two orders of magnitude lower than that for surfactants such as Tween 80. The core compartment of a pharmaceutical polymeric micelle should have a high loading capacity, a controlled release profile for the incorporated drug, and good compatibility between the core-forming block and incorporated drug, while the micelle corona should provide an effective steric protection for the micelle and determine the micelle hydrophilicity, charge, length, and surface density of hydrophilic blocks, and serve as site for reactive groups suitable for further micelle derivatization, such as an attachment of targeting moieties (Hagan et al. 1996; Inoue et al. 1998). These properties manage important biological characteristics of a micellar carrier, including its pharmacokinetics, biodistribution, biocompatibility, longevity, surface adsorption of biomacromolecules, adhesion to biosurfaces, and targetability (Hagan et al. 1996; Hunter 1991; Müller 1991). The use of polymeric micelles can allow for achievement of an extended circulation time, a favorable biodistribution, and lower toxicity of a drug (Jones and Leroux 1999; Kwon and Kataoka 1995; Torchilin 2001).

Usually, the amphiphilic micelle-forming unimers include poly(ethylene glycol) (PEG) blocks with a molecular weight of 1–15 kDa to serve as hydrophilic corona-forming blocks (Kwon 2003). This polymer is inexpensive, has low toxicity, serves as an efficient steric shield of various biologically active macromolecules (Abuchowski et al. 1979; Harris et al. 2001; Morcol et al. 2004; Roberts et al. 2002; Veronese and Harris 2002) and particulate delivery systems (Calvo et al. 2001; Klibanov et al. 1990; Lasic and Martin 1995; Moghimi 2002), and has been approved for internal applications by regulatory agencies (Smith and Tanford 1972; Veronese and Harris 2002). Still, other hydrophilic polymers alternatives to PEG may be used as hydrophilic blocks (Torchilin et al. 1995): poly(N-vinyl-2-pyrrolidone) (PVP), poly(vinyl alcohol), and poly(vinylalcohol-co-vinyloleate) copolymer. New materials for pharmaceutical micelles include these new copolymers of PEG (Prompruk et al. 2005) and a variety of completely new macromolecules, such as scorpion-like polymers (Djordjevic et al. 2005; Tao and Uhrich 2006) as well as starlike and core–shell constructs (Arimura et al. 2005). Hydrophobic blocks of polymeric micelles have been prepared with propylene oxide, L-lysine, aspartic acid, β-benzoyl-L-aspartate, γ-benzyl-L-glutamate, caprolactone, D,L-lactic acid, spermine (see the review in references Harada and Kataoka 1998; Jeong et al. 1998; Kabanov and Kabanov 1998; Kabanov et al. 1989; Katayose and Kataoka 1998; Kwon et al. 1995, 1997; Miller et al. 1997; Trubetskoy et al. 1997; Yokoyama et al. 1990).

3.4 LIPID-CORE MICELLES AS PHARMACEUTICAL CARRIERS

In the last decade, phospholipid residues used as hydrophobic core-forming groups (Trubetskoy and Torchilin 1995) have greatly increased in importance. This is attributable to the additional advantages for particle stability when compared with conventional amphiphilic polymer micelles provided by the existence of two fatty acid acyls, which might contribute considerably to an increase in the hydrophobic interactions between the polymeric chains in the micelle's core. Conjugates of lipids with water-soluble polymers are commercially available. The diacyl lipid–PEG molecule represents a characteristic amphiphilic polymer with a bulky hydrophilic (PEG) portion and a very short but very hydrophobic diacyl lipid part. Micelle preparation from the lipid–polymer conjugates is a simple process, since these polymers form micelles spontaneously in an aqueous media (see Figure 3.1). All versions of PEG–phospholipid (PEG–PE) conjugates form micelles with a size of 7–35 nm. Micelles formed from conjugates with polymer (PEG) blocks of higher molecular weight have a slightly larger size. This is a good indicator to suggest that the micelle size may be tailored for a particular application by varying the length of the PEG. Such micelles have a spherical shape and uniform size distribution (Torchilin et al. 2003). From a practical point of view, it is important that micelles prepared from these polymers will stay intact at concentrations much lower than

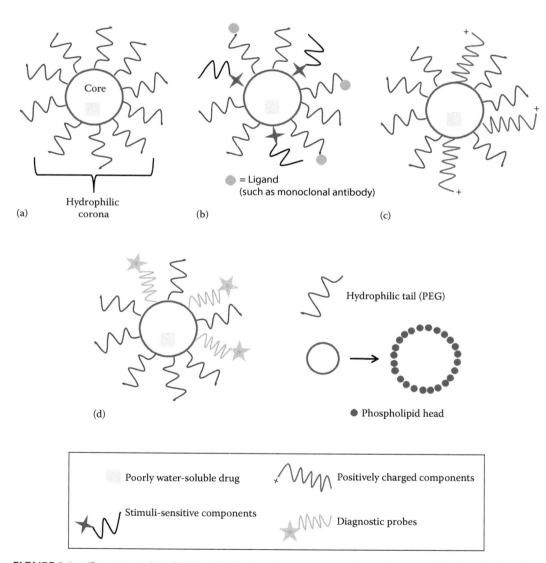

FIGURE 3.1   (See companion CD for color figure.) Lipid-core micelle architecture. (a) Plain micelle with a poorly soluble drug in the hydrophobic core. (b) Targeted micelle surface modified with a ligand (monoclonal antibody, transferrin, etc.) and/or with a stimuli-sensitive component. (c) Positively charged micelle components to improve the intracellular drug delivery. (d) Diagnostic micelle loaded with amphiphilic modified chelating probes (such as gadolinium, manganese, etc.).

that required for drug delivery purposes. Another important useful feature is that PEG_{2000}–PE and PEG_{5000}–PE micelles retain the size characteristic for micelles even after a 48 h incubation in blood plasma (Lukyanov et al. 2002), i.e., the integrity of PEG–PE micelles should not be immediately affected by components of biological fluids upon parenteral administration.

Amphiphilic PVP–lipid conjugates with various polymer lengths have also been prepared by the free-radical polymerization of vinylpyrrolidone and further modified by the attachment of long-chain fatty acid acyls, such as palmityl (P) or stearyl (S) residues, to one of the polymer termini (Torchilin et al. 1994, 1995). Amphiphilic PVPs with a MW of the PVP block between 1500 and 8000 Da readily form micelles in an aqueous environment (Torchilin et al. 1994). CMC values and the size of the micelles formed depend on the length of the PVP block and vary between 10^{-4} and 10^{-6} M and 5 and 20 nm, respectively. As for PEG–PE-based micelles, micelle made of amphiphilic PVP may also be used for the solubilization of poorly water-soluble drugs to yield highly stable biocompatible formulations. The application of micelles prepared from a similar lipidated polymer, polyvinyl alcohol substituted with oleic acid, for transcutaneous delivery of retinyl palmitate has also been proposed (Luppi et al. 2002).

The micelles made of such lipid-containing conjugates can be loaded with various poorly soluble drugs (tamoxifen, paclitaxel, camptothecin, porphyrins, etc.) and demonstrate good stability, longevity, and the ability to accumulate in the areas with a damaged vasculature via the EPR effect such as myocardial infarcts and tumors (Lukyanov et al. 2002, 2004; Trubetskoy and Torchilin 1995). Mixed micelles made of PEG–PE and other micelle-forming components are also very interesting. They provide even better solubilization of certain poorly soluble drugs due to the increase in the capacity of the hydrophobic core for the drug (Krishnadas et al. 2003; Torchilin et al. 2003; Wang et al. 2004). A drug incorporated in lipid-core polymeric micelles is firmly associated with micelles: when PEG–PE micelles loaded with various drugs were dialyzed against aqueous buffer at sink conditions, all tested preparations retained more than 90% of an encapsulated drug for up to 7 h of incubation (Gao et al. 2002).

3.5 SOLUBILIZATION PROCESS OF POORLY WATER-SOLUBLE DRUGS

The process of solubilization of water-insoluble drugs by micelle-forming amphiphilic block copolymers has been investigated in detail (Nagarajan and Ganesh 1989). The mathematical simulation of the solubilization process (Xing and Mattice 1998) indicated that the initial solubilization proceeds via the initial displacement of solvent (water) molecules from the micelle core, and later a solubilized drug begins to accumulate in the very center of the micelle core, "pushing" hydrophobic blocks away from this area. Extensive solubilization may result in some increase of the micelle size due to the expansion of its core with drug. Among other factors influencing the efficacy of drug loading into the micelle, the size of both core-forming and corona-forming blocks is important (Allen et al. 1999). In the former case, the larger the hydrophobic block the bigger the core size and its ability to entrap hydrophobic drugs. In the latter case, the increase in the length of the hydrophilic block results in the increase of an CMC value, i.e., at a given concentration of the amphiphilic polymer in solution, the smaller fraction of this polymer will be present in the micellar form and the quantity of the micelle-associated drug will drop. Drugs, such as diazepam and indomethacin (Lin and Kawashima 1985, 1987), adriamicin (Yokoyama et al. 1990, 1994, 1998), anthracycline antibiotics (Batrakova et al. 1996), polynucleotides (Kabanov and Kabanov 1998; Kabanov et al. 1995), and doxorubicin (Alakhov and Kabanov 1998), were effectively solubilized by various polymeric micelles, including micelles made of Pluronics® (block copolymers of PEG and polypropelene glycol) (Kabanov et al. 2002). Doxorubicin incorporated into Pluronic micelles showed superior properties when compared with free drug in the experimental treatment of murine tumors (leukemia P388, myeloma, Lewis lung carcinoma [LLC]) and human tumors (breast carcinoma MCF-7) in mice (Alakhov and Kabanov 1998). Micellar drugs also show a lower nonspecific toxicity (Matsumura et al. 1999) than free drugs. To prepare drug-loaded micelles by direct entrapment of drug into the micelle core, the whole set of micelle-forming copolymers of PEG with poly(L-amino acids) was used (La et al. 1996). PEG-*b*-poly(caprolactone) copolymer micelles were successfully used as delivery vehicles for dihydrotestosterone (Allen et al. 2000). PEG–PE micelles can efficiently incorporate a variety of poorly soluble and amphiphilic substances including paclitaxel (Huh et al. 2005; Lee et al. 2005; Wang et al. 2005), tamoxifen, camptothecin (Mu et al. 2005; Opanasopit et al. 2004; Watanabe et al. 2006), porphyrine, vitamin K3, and others (Gao et al. 2002; Trubetskoy and Torchilin 1996; Wang et al. 2005).

Numerous studies have dealt with micellar forms of platinum-based anticancer drugs (Cabral et al. 2005; Exner et al. 2005; Xu et al. 2006) and cyclosporin A (Aliabadi and Lavasanifar 2006; Aliabadi et al. 2005). Mixed polymeric micelles made of positively charged polyethyleneimine and pluronic were used as carriers for antisense oligonucleotides (Vinogradov et al. 2004). A typical protocol for the preparation of drug-loaded polymeric micelles from amphiphilic copolymers includes the following steps. Solutions of an amphiphilic polymer and a drug of interest in a miscible volatile organic solvent are mixed, and followed by organic solvent evaporation to form a polymer/drug film. The film obtained is then hydrated in an aqueous buffer, and the micelles are formed by intensive shaking. If the amount of the drug exceeds the solubilization capacity of micelles, the excess drug precipitates in a crystalline form and is removed by filtration. The loading efficiency for different compounds varies from 1.5% to 50% by weight. This value apparently correlates with the hydrophobicity of a drug. In some cases, to improve drug solubilization, additional mixed micelle-forming compounds may be added to polymeric micelles. Thus, to increase the encapsulation efficiency of paclitaxel, egg phosphatidylcholine (PC) was added to the PEG–PE-based micelle composition, which approximately doubled the paclitaxel encapsulation efficiency (from 15 to 33 mg) of the drug per gram of the micelle-forming material (Gao et al. 2002, 2003; Krishnadas et al. 2003).

3.6   MICELLES AS THERAPEUTIC AGENTS

3.6.1   TARGETED MICELLES

Targeted micelles to pathological organs or tissues can further increase pharmaceutical efficiency of a micelle-encapsulated drug. Several approaches have been used to enhance the accumulation of various drug-loaded pharmaceutical nanocarriers, including pharmaceutical micelles, in a pathological area.

3.6.2   "PASSIVE" MICELLE TARGETING

It is based on the already mentioned EPR effect, via the spontaneous penetration of long-circulating macromolecules, particulate drug carriers, and even molecular aggregates into the interstitium through a compromised leaky vasculature, which is characteristic for solid tumors, infarcts, infections, and inflammations (Maeda et al. 2000; Palmer et al. 1984). Of course, the prolonged circulation of drug-loaded micelles facilitates the EPR-mediated target accumulation since it gives a better chance to reach and/or interact with its target. The results of blood clearance studies of various micelles have clearly demonstrated their longevity: micellar formulations, such as PEG–PE-based micelles, had circulation half-lives in mice and rats of around 2 h with certain variations depending on the molecular size of the PEG block (Lukyanov et al. 2002). The increase in the size of a PEG block increases the micelle circulation time in the blood probably by providing a better steric protection against opsonin penetration to the hydrophobic micelle core. Still, circulation times for long-circulating micelles are somewhat shorter compared to those for long-circulating PEG-coated liposomes (Klibanov et al. 1990). Diffusion and accumulation parameters were shown to be strongly dependent on the cutoff size of tumor blood vessel walls, and the cutoff size varies with tumor type (Hobbs et al. 1998; Monsky et al. 1999; Yuan et al. 1995). An increased accumulation of PEG–PE-based micelles in areas with a leaky vasculature including tumors and infarcts has been clearly demonstrated (Lukyanov et al. 2002).

Micelles formed by PEG_{750}–PE, PEG_{2000}–PE, and PEG_{5000}–PE (like long-circulating liposomes; Gabizon 1992, 2001; Papahadjopoulos et al. 1991) accumulate efficiently in tumors via the EPR effect (Mastrobattista et al. 2002). It is worth mentioning that micelles prepared with several different PEG–PE conjugates studied demonstrated much higher accumulation in tumors compared to nontarget tissue (muscle) even with an experimental LLC in mice known to have a relatively small vasculature cutoff size (Hobbs et al. 1998; Weissig et al. 1998). In other words, because of their smaller size, micelles have an additional advantage as a tumor drug delivery system, which utilizes the EPR effect compared to particulate carriers with larger size of individual particles. Thus, the micelle-incorporated model protein (soybean trypsin inhibitor or STI, MW 21.5 kDa) accumulates to a higher extent in subcutaneously established murine LLC than the same protein in larger liposomes (Weissig et al. 1998).

The accumulation pattern of PEG–PE micelles prepared from many versions of PEG–PE conjugates is characterized by peak tumor accumulation times of 3–5 h. The largest total tumor uptake of the injected dose at 5 h postinjection (as the AUC) was for micelles formed by unimers with a relatively large PEG block (PEG_{5000}–PE). This may be explained by the fact that these micelles have the longest circulation time and a lower extravasation into normal tissues compared to micelles prepared from the smaller PEG–PE conjugates. However, micelles prepared from PEG–PE conjugates with shorter versions of PEG may be more efficient carriers of poorly soluble drugs because they have a greater hydrophobic-to-hydrophilic phase ratio and can be loaded with drug more efficiently on a weight-to-weight basis. Similar results have been obtained with another murine tumor model, EL4 T-cell lymphoma (Lukyanov et al. 2002). Other recent data clearly indicates spontaneous targeting of PEG–PE-based micelles into other experimental tumors (Torchilin et al. 2003) in mice as well as into the areas of damaged heart in rabbits with an experimental myocardial infarction (Lukyanov et al. 2004).

3.6.3 pH- AND THERMORESPONSIVE POLYMERIC MICELLES

A different approach to deliver is based on the fact that many pathological processes in various tissues and organs are accompanied by a local temperature increase (by 2°C–5°C) and/or a pH decrease of 1–2.5 units (acidosis) (Helmlinger et al. 1997; Tannock and Rotin 1989). Thus, the efficiency of the micellar carriers for local drug delivery can be improved with micelles that disintegrate and release drug at the increased temperature or decreased pH values in pathological sites, i.e., by combining the EPR effect with stimuli responsiveness. Micelles made of thermo- or pH-sensitive components, such as poly(N-isopropylacrylamide) and its copolymers with poly(D,L-lactide) and other blocks, acquire the ability to disintegrate in targeted areas releasing the micelle-incorporated drug (Meyer et al. 1998; Morcol et al. 2004).

pH-responsive polymeric micelles incapsulating phthalocyanine seem to be promising systems for the photodynamic cancer therapy (Le Garrec et al. 2002), while doxorubicin-loaded polymeric micelles containing acid-cleavable linkages have provided enhanced intracellular drug delivery into tumor cells (Yoo et al. 2002). Similarly, pH-sensitive unimolecular polymeric micelles—star-shaped polymers—have been made of hydrophobic ethyl methacrylate and *t*-butyl methacrylate and hydrophilic poly(ethylene glycol)methacrylate (Jones et al. 2003). With micelles with a size of 10–40 nm, their ionization and possibly drug release should depend on pH. Such micelles are also considered for oral delivery. Micelles based on the poly(2-ethyl-2-oxazoline)-*b*-poly(L-lactide) diblock copolymer have been loaded with doxorubicin and capable of releasing the drug at pH values typical for late endosomes (pH ≈ 5.5) and secondary lysosomes (pH around ≤5.0) (Hsiue et al. 2006; Wang et al. 2005). Phosphorylcholine-based diblock copolymer micelles also demonstrated distinct pH.

Thermoresponsive polymeric micelles showed an increased drug release upon temperature changes (Chung et al. 1999). Micelles combining thermosensitivity and biodegradability have also been suggested (Soga et al. 2005). The penetration of drug-loaded polymeric micelles into cells (tumor cells) as well as drug release from the micelles can also be enhanced by externally applied ultrasound (Gao et al. 2005; Rapoport et al. 2003).

3.6.4 "Active" Micelle Targeting

The drug delivery potential of polymeric micelles may be further enhanced by attaching targeting ligands to the micelle surface (Musacchio et al. 2008), i.e., to the water-exposed termini of hydrophilic blocks (Torchilin 2004). Included among those ligands are antibodies (Chekhonin et al. 1991; Liaw et al. 2001; Torchilin et al. 2003; Vinogradov et al. 1999), sugar moieties (Jule et al. 2003; Nagasaki et al. 2001), transferrin (Dash et al. 2000; Ogris et al. 1999; Vinogradov et al. 1999), and folate residues (Leamon and Low 2001; Leamon et al. 1999). The last two ligands are especially useful for targeting to cancer cells. It was shown that galactose- and lactose-modified micelles made of PEG–polylactide copolymer specifically interact with lectins, thus modeling targeting delivery of the micelles to hepatic sites (Chekhonin et al. 1991; Jule et al. 2003).

Transferrin-modified micelles based on PEG and polyethyleneimine with a size between 70 and 100 nm are expected to target tumors with overexpressed transferrin receptors (Vinogradov et al. 1999). Mixed micelle-like complexes of PEGylated DNA and PEI modified with transferrin (Ogris et al. 1999) have been designed for enhanced DNA delivery into cells overexpressing transferrin receptors. A similar approach was successfully tested with folate-modified micelles in several studies (Jeong et al. 2005; Leamon and Low 2001; Park et al. 2005). Poly(L-histidine)/PEG and poly(L-lactic acid)/PEG block copolymer micelles carrying folate residue on their surface efficiently delivered adriamycin to tumor cells in vitro demonstrating the potential for solid tumor treatment with combined targetability and pH sensitivity (Lee et al. 2003, 2005a,b). Mixed micelles made of folate–PEO-b-poly(D,L-lactic-co-glycolic acid) and PEO-b-PLGA–DOX conjugates showed a superior cell uptake of folate-modified micelles compared to folate-free micelles by human squamous carcinoma cells expressing the folate receptor and better activity against these cells both in vitro and in vivo (Yoo and Park 2004). Folate-targeted for PEO-b-poly(ε-caprolactone) micelles loaded with paclitaxel had significantly higher cytotoxicity against human breast adenocarcinoma MCF-7 and human uterine cervical adenocarcinoma HeLa 229 cells compared to unmodified micelles (Park et al. 2005). Lactose-modified PEO-b-poly(2-(dimethylamino)ethyl methacrylate) that formed an electrostatic micellar complex with plasmid DNA demonstrated a significantly higher transfection efficiency in HepG$_2$ cells (Wakebayashi et al. 2004). Tumor-specific peptide sequences, such as RGD (RGD peptides are small synthetic peptides containing an RGD sequence, Arg-Gly-Asp), have also been used to target drug-loaded micelles to tumors. Thus, tumor endothelial cells have been successfully targeted with doxorubicin-loaded PEO-b-PCL micelles modified with the cyclic pentapeptide C (Arg-Gly-Asp-D-Phe-Lys) by specifically recognizing $\alpha_v\beta_3$ integrins overexpressed in the tumor vasculature (Nasongkla et al. 2006).

3.6.5 Antibody-Targeted Micelles

PEG–PE-based immunomicelles modified with monoclonal antibodies were prepared using PEG–PE conjugates with the free PEG terminus activated with a p-nitrophenyloxycarbonyl (pNP) group (Torchilin et al. 2001a). Diacyl lipid fragments of such bifunctional PEG derivative firmly incorporate within the micelle core, while the water-exposed pNP group, stable at pH values below 6, can interact with amino groups of various ligands (antibodies and their fragments or peptides) at pH values above 7.5 to yield a stable urethane (carbamate) bond. To prepare immunotargeted micelles, the corresponding antibody can simply be incubated with drug-loaded pNP–PEG–PE-containing micelles at a pH ≈ 8.0. Using fluorescent labels, or by SDS-PAGE (Gao et al. 2003; Torchilin et al. 2003), it was calculated that several antibody molecules could be attached to a single 20 nm micelle.

Antibodies attached to the micelle corona (Torchilin et al. 2003) preserve their specific binding ability, and such immunomicelles specifically recognize their target substrates as confirmed by ELISA. For tumor targeting, PEG–PE-based micelles modified with monoclonal 2C5 antibody possessing the nucleosome-restricted specificity (mAb 2C5) were capable of recognition of a broad variety of tumor cells via the tumor cells surface-bound nucleosomes (Iakoubov and Torchilin 1997). Such specific targeting of cancer cells by drug-loaded mAb 2C5-immunomicelles resulted in dramatically improved in vitro cancer cell killing: with human breast cancer MCF-7 cells, paclitaxel-loaded 2C5-immunomicelles showed superior killing efficiency compared to paclitaxel-loaded plain micelles or free drug (Gao et al. 2003). In vivo experiments with LLC-bearing mice revealed an improved tumor uptake of paclitaxel-loaded radiolabeled 2C5-immunomicelles compared to non-targeted micelles (Torchilin et al. 2003). In addition, unlike plain micelles, 2C5-immunomicelles promote delivery of their load not only to tumors with a mature vasculature, but also to tumors at earlier stages of their development and to metastases.

Additionally, 2C5-targeted PEG–PE micelles capable of releasing photosensitizing agents showed an enhanced anticancer activity both in vitro and in vivo. This suggests promise for the photochemical eradication of malignant cells (Roby et al. 2007; Skidan et al. 2008).

3.7   MICELLES AS DIAGNOSTIC AGENTS

The medical diagnostic imagining is an emerging area in the use of micelles as carriers for pharmaceuticals. Medical imagining modalities consist of magnetic resonance (MR), computed tomography (CT), gamma scintigraphy, and ultrasonography. Imaging is greatly important for visualization, localization, and detection in organs and tissues for several pathologies at their onset. Regardless of the specific modality used, medical diagnostic imaging requires a signal of sufficient intensity from the area of interest in order to differentiate this targeted area from the surroundings. To be able to achieve this goal, appropriate contrast agents for specific imaging modalities are needed so that once they are accumulated in the targeted area of interest these sites will be clearly visualized and distinguished by applying the suitable imagining modalities (Torchilin 1995). According to the imaging modality and the signal intensity (intended to provide sensitivity and resolution), the contrast agents will have a different chemical nature and may be required in different amounts. In fact, to be delivered into the area of interest and in the appropriate tissue concentration the quantity varies broadly: quite low for gamma imaging and quite high for MR and CT. Thus, to achieve the optimal local concentration of a labeled diagnostic agent, the use of particulate carriers for a highly efficient and selective delivery was a natural progression. Although micellar transport of contrast agents represents a relatively new field (Torchilin 2002; Trubetskoy and Torchilin 1995), approaches suggesting the use of micellar contrast agents for both pure diagnostic/imaging purposes and for the visual control over the drug delivery are underway.

Chelated paramagnetic metals, such as gadolinium (Gd), manganese (Mn), or dysprosium (Dy), are of the major interest for the design of MR positive (T1) contrast agents. Mixed micelles obtained from monoolein and taurocholate with Mn–mesoproporphyrin have been shown to be a potential oral hepatobiliary imaging agent for T1-weighted MR imaging (MRI) (Schmiedl et al. 1995). Since chelated metal ions possess a hydrophilic character, to be incorporated into micelles, such structures must acquire an amphiphilic nature. Several amphiphilic chelating probes have been developed earlier for liposomes, where a hydrophilic chelating residue is covalently linked to a hydrophobic (lipid) chain, such as diethylene triamine pentaacetic acid (DTPA) conjugated with phosphatidyl ethanolamine (DTPA–PE) (Grant et al. 1989), DTPA–stearylamine, DTPA–SA (Kabalka et al. 1988; Unger et al. 1994), and with amphiphilic acylated paramagnetic complexes of Mn and Gd (Unger 1994). The lipid part of such an amphiphilic chelating molecule can be anchored within the micelle's hydrophobic core while a more hydrophilic chelating group is localized in the hydrophilic shell of the micelle. Amphiphilic chelating probes (paramagnetic Gd–DTPA–PE and radioactive ^{111}In–DTPA–SA) have been incorporated into PEG (5 kDa)–PE micelles and used in vivo for MR and gamma scintigraphy imaging (Trubetskoy et al. 1996). The main feature that makes PEG–lipid micelles attractive for diagnostic imaging applications is their small size which allows for better penetration into the target tissue. In addition, in case of MRI contrast agents, it is especially important that chelated metal atoms are directly exposed to the aqueous environment, to enhance the relaxivity of the paramagnetic ions leading to enhancement of the micelle contrast properties.

CT represents an imaging modality with high spatial and temporal resolution. Diagnostic CT imaging requires an iodine concentration of millimoles per milliliter of tissue (Wolf 1995), so that large doses of low-molecular-weight CT contrast agent (iodine-containing organic molecules) are normally administered to patients. The selective enhancement of blood levels upon such administration is short due to the agent's rapid extravasation and clearance. Micelles to be used as a contrast agent for the blood pool CT imaging were prepared from copolymers of PEG with heavily iodinated, and thus insolubilized, polylysine (Torchilin et al. 1999; Trubetskoy et al. 1997). A micellar iodine-containing CT contrast agent injected intravenously into rats and rabbits showed a three- to fourfold enhancement of the x-ray signal in the visualized blood pool in both animal species for a period of at least 2 h following the injection (Torchilin et al. 1999; Trubetskoy et al. 1997).

Combining both a contrast moiety and therapeutic agent in a single micelle directly connects an enhanced accumulation of drug-loaded micelles in the target (tumor) and increased therapeutic efficiency of such micelles, with radiolabeled tumor-targeted PEG–PE micelles loaded with a porphine derivative and used for photodynamic tumor therapy (Roby et al. 2007).

3.8 FURTHER APPLICATIONS OF POLYMERIC MICELLES

3.8.1 Intracellular Delivery of Micelles

An additional strategy may further improve the efficiency of drug-loaded micelles by enhancing their intracellular delivery to compensate for drug degradation in lysosomes which results from the endocytosis-mediated capture of therapeutic micelles. An attempt to achieve this has been realized by controlling the micelle charge. As known, a net positive charge enhances the uptake of various nanoparticles by cells. Cationic lipid formulations such as Lipofectin® (an equimolar mixture of N-[1-(2,3-dioleyloxy)propyl]-N,N,N-trimethylammonium chloride [DOTMA] and dioleoyl phosphatidylethanolamine [DOPE]), significantly improved the endocytosis-mediated intracellular delivery of various drugs and DNA entrapped into liposomes and other lipid constructs made of these compositions (Almofti et al. 2003; Felgner et al. 1994; Kaiser and Toborek 2001; Ota et al. 2002). Some PEG-based micelles, such as PEG–PE micelles, carry a net negative charge (Lukyanov et al. 2004) that may be expected to hinder their internalization by cells. Compensation for this negative charge by the addition of positively charged lipids to PEG–PE-based micelles could improve their uptake by cancer cells. It is also possible that after the enhanced endocytosis, drug-loaded mixed micelles made of PEG–PE and positively charged lipids could escape from the endosomes and enter the cytoplasm of cancer cells. Considering this, an attempt was made to increase intracellular delivery and the anticancer activity of micellar paclitaxel by preparing paclitaxel-containing micelles from a mixture of PEG–PE and positively charged Lipofectin lipids (LL) (Wang et al. 2005). The cell interaction with breast adenocarcinoma cells BT-20 and intracellular fate of paclitaxel-containing PEG–PE/LL micelles and similar micelles prepared without the addition of the LL were investigated by fluorescence microscopy. It was clearly demonstrated that fluorescently labeled PEG–PE and PEG–PE/LL micelles were both endocytosed by cancer cells. However, in the case of PEG–PE/LL micelles, endosomes were shown to degrade and release drug-loaded micelles into the cell cytoplasm, an apparent result of the destabilizing effect of the LL component on the endosomal membranes (Torchilin et al. 2001b). The in vitro anticancer effects of drug-loaded micelles were significantly improved for intracellularly delivered paclitaxel-containing PEG–PE/LL compared to that of free paclitaxel or paclitaxel delivered using LL-free PEG–PE micelles. In human ovarian carcinoma A2780, the IC_{50} values of free paclitaxel, paclitaxel in PEG–PE micelles, and paclitaxel in PEG–PE/LL micelles were 22.5, 5.8, and 1.2 μM, respectively.

Recently, attempts have been made to prepare pharmaceutical nanocarriers including micelles, which can simultaneously perform targeting to and into tumor cells. To achieve this, micelles have been modified by both cell-penetrating peptides (CPP) and cancer-specific antibodies in such a way that micelle-attached CPP was sterically shielded in the circulatory system by surrounding longer PEG and antibody–PEG moieties. However, after accumulation in the tumor, longer PEG chains conjugated with the carrier via pH-sensitive bonds detached under the action of the lowered intra-tumoral pH, CPP fragments became exposed and facilitated the carrier penetration into cells (Kale and Torchilin 2007; Sawant et al. 2006).

It has been demonstrated that the transactivating transcriptional activator (TAT) protein from HIV-1 enters various cells when added to the surrounding media (Frankel and Pabo 1988). The recent data suggest more than one mechanism for CPP and cell-penetrating proteins (CPP) and CPP-mediated intracellular delivery of various molecules and particles. CPP-mediated intracellular delivery of large molecules and nanoparticles was proven to proceed via energy-dependent macropinocytosis with sub-sequent enhanced escape from endosome into the cell cytoplasm (Wadia et al. 2004), while individual CPPs or CPP-conjugated small molecules penetrate cells via electrostatic interactions and hydrogen bonding and do not seem to depend on the energy (Rothbard et al. 2004). Since traversal through cel-lular membranes represents a major barrier for efficient delivery of macromolecules into cells, CPPs, whatever their mechanism of action is, may serve to transport various drugs and even drug-loaded pharmaceutical carriers into mammalian cells in vitro and in vivo.

Thus, an interesting approach to increase the intracellular micellar delivery is based on the attachment of TAT peptide (TATp) moieties to the surface of the nanocarrier with PEG–PE deriva-tives (Liu et al. 2008; Sawant et al. 2006; Sethuraman and Bae 2007). Recently, it has been demon-strated that TATp-targeted PEG–PE micelles loaded with paclitaxel enhance the interaction with cancer cells compared to non-modified micelles resulting in a significant increase in cytotoxicity to different cancer cells in vitro and in vivo (Sawant and Torchilin 2009).

3.8.2 Micellar Complexes as siRNA Delivery Systems

Lately, small interfering RNA (siRNA) technology has gained great interest for gene therapy given several advantages. A siRNA is a short double-stranded RNA sequence from 21 to 23 nucleotides which in mammalian cells strongly promote gene silencing via an RNA interference (RNAi) mecha-nism (Nawrot and Sipa 2006). Since their discovery, siRNA molecules and their variously modified derivatives have been implemented as potential candidates for therapeutic applications like potential therapeutic agents for different genetic diseases (such as in cancer). siRNA has been chemically conjugated to a variety of bioactive molecules. Previous observations showed that the integrity of the 5′-terminus of the antisense strand, rather than that of the 3′-terminus, is important for the initiation of an RNAi mechanism. Therefore, the 3′- and 5′-terminus of the sense strand and the 3′-terminus of the antisense strand are considered primary potential sites for conjugation with minimal influ-ence on RNAi activity (Jeong et al. 2009). Lipids, polymers, peptides, and inorganic nanostructured materials have been conjugated to the siRNA in attempts to enhance its pharmacokinetic behavior, cellular uptake, target specificity, and safety. Particularly important are siRNA DDS based on the electrostatic interaction between negative charges of siRNA and positive charges of polymers form-ing the nanosystems. This interaction results in the formation of electrostatic complexes in which the oligonucleotides are protected. The most popular polymer used to make polyplexes with siRNAs is polyethylenimine (PEI) (Kim et al. 2008a,b). As an example, a polymeric gene carrier was devel-oped to deliver vascular endothelial growth factor (VEGF) siRNA for prostate cancer cells in a target-specific manner. Prostate cancer–binding peptide (PCP) was conjugated with PEI via a PEG linker (PEI–PEG–PCP). The PEI–PEG–PCP conjugate could effectively condense siRNA to form stable polyelectrolyte complexes. VEGF siRNA/PEI–PEG–PCP polyplexes exhibited significantly higher VEGF inhibition efficiency than PCP-unmodified polycationic carriers (PEI–PEG or PEI) in human prostate carcinoma cells (PC-3) even under serum conditions (Kim et al. 2009).

3.8.3 IMMUNOLOGY AND MICELLES

A very interesting and promising area for the use of polymeric micelles concerns immunology. Nonionic block copolymers, first of all, Pluronics or copolymers of PEG (PEO) and PPG (PPO), are finding application as immunological adjuvants for the modulation of immune response and preparation of new and effective vaccines (Todd et al. 1998). Usually, linear tri-block copolymers with the linear structure PEG–PPG–PEG are used for this purpose. The adjuvant activity of these polymers is strongly influenced by the length of the PPG block. Its increase results in increased adjuvant activity. It is important to mention that Pluronics themselves are able to provoke macrophage activation. Though the exact mechanism of this activation is still under investigation, there are data suggesting that Pluronics actually activate the alternative complement pathway (Hunter and Bennett 1984), and, in turn, certain proteins belonging to the complement system cause macrophage activation.

Pluronics have demonstrated their adjuvant properties both in emulsion and micellar forms. In an emulsion form, they not only activate the alternative complement pathway, but also enhance the binding of protein antigens at the water–oil interface to increase antibody response (Hunter and Bennett 1986; Hunter et al. 1981). Pluronics with higher molecular weights (PPG blocks with MW around 10 kDa with shorter PEG blocks) form micelles able to incorporate various antigens. High adjuvant activity of such micelles was demonstrated with an influenza virus vaccine (Todd et al. 1998). It was also shown that optimization of vaccine properties can be achieved by controlling the size of PPG and PEG blocks. Thus, with ovalbumin as a model antigen, it was shown that the most potent vaccine was obtained with 11 kDa core PPG block copolymer containing between 5% and 10% attached PEG blocks. Naturally, the size of antigen-bearing polymeric micelles also depends on the size of a micelle-incorporated protein antigen (Newman et al. 1997; Todd et al. 1997). The mechanism of protein antigen interaction with polymeric micelles is seen as a hydrogen bonding between the protein antigen molecule and the terminal hydroxyl groups of PEG blocks or with multiple hydrogen bond acceptor sites along the hydrophobic PPG block (Todd et al. 1998).

As noted (Todd et al. 1998), studies on the cellular immune response provoked by ovalbumin in Pluronic micelles demonstrated that a more hydrophilic carrier mainly augments Th2 responses, while more hydrophobic copolymers augment both Th1 and Th2 responses. These data, together with available information on the low toxicity of Pluronic-based composition for vaccination (Triozzi et al. 1997), permit to believe that polymeric micelles may find a real clinical future as adjuvants and vaccine components.

3.9 CONCLUSION

Summing up, polymeric micelles possess an excellent ability to solubilize poorly water-soluble drugs and increase their bioavailability. This has been repeatedly demonstrated for a broad variety of drugs, many of them poorly soluble anticancer drugs, with micelles of a variety of different compositions. In addition, due to their small size micelles demonstrate a very efficient spontaneous accumulation in pathological tissues with increased vascular permeability (tumors, infarcts) in vivo via the EPR effect and can bring increased quantities of drugs to a tissue, thus reducing the whole body toxicity associated with a high drug dose. Micelle-specific targeting to areas of interest can also be achieved by attaching specific targeting ligands (such as target-specific antibodies, transferrin or folate) to the micelle surface. By varying micelle composition and the size of hydrophilic or hydrophobic blocks of the micelle-forming material, one can readily control the properties of micelles, including size, loading capacity, and longevity in the blood. Another interesting option is provided by stimuli-responsive micelles, whose degradation and subsequent drug release proceed at the abnormal pH values or temperatures characteristic of many pathological zones.

Such combinations of micellar properties should lead to the increased practical application of micellar drugs in the next future.

REFERENCES

Abuchowski, A., T. van Es, N. C. Palczuk, J. R. McCoy, and F. F. Davis. 1979. Treatment of L5178Y tumor-bearing BDF1 mice with a nonimmunogenic L-glutaminase–L-asparaginase. *Cancer Treat Rep* 63:1127–1132.

Ahmed, F., R. I. Pakunlu, A. Brannan, F. Bates, T. Minko, and D. E. Discher. 2006. Biodegradable polymer-somes loaded with both paclitaxel and doxorubicin permeate and shrink tumors, inducing apoptosis in proportion to accumulated drug. *J Control Release* 116:150–158.

Alakhov, V. Y. and A. V. Kabanov. 1998. Block copolymeric biotransport carriers as versatile vehicles for drug delivery. *Expert Opin Investig Drugs* 7:1453–1473.

Aliabadi, H. M. and A. Lavasanifar. 2006. Polymeric micelles for drug delivery. *Expert Opin Drug Deliv* 3:139–162.

Aliabadi, H. M., A. Mahmud, A. D. Sharifabadi, and A. Lavasanifar. 2005. Micelles of methoxy poly(ethylene oxide)-*b*-poly(epsilon-caprolactone) as vehicles for the solubilization and controlled delivery of cyclo-sporine A. *J Control Release* 104:301–311.

Allen, C., J. Han, Y. Yu, D. Maysinger, and A. Eisenberg. 2000. Polycaprolactone-*b*-poly(ethylene oxide) copo-lymer micelles as a delivery vehicle for dihydrotestosterone. *J Control Release* 63:275–286.

Allen, C., D. Maysinger, and A. Eisenberg. 1999. Nano-engineering block copolymer aggregates for drug delivery. *Coll Surf B Biointerf* 16:1–35.

Almofti, M. R., H. Harashima, Y. Shinohara, A. Almofti, Y. Baba, and H. Kiwada. 2003. Cationic liposome-mediated gene delivery: Biophysical study and mechanism of internalization. *Arch Biochem Biophys* 410:246–253.

Arimura, H., Y. Ohya, and T. Ouchi. 2005. Formation of core–shell type biodegradable polymeric micelles from amphiphilic poly(aspartic acid)-*block*-polylactide diblock copolymer. *Biomacromolecules* 6:720–725.

Attwood, D. and A. T. Florence (eds.). 1983. *Surfactant System*. London, U.K.: Chapman & Hall.

Bae, Y. and K. Kataoka. 2009. Intelligent polymeric micelles from functional poly(ethylene glycol)–poly(amino acid) block copolymers. *Adv Drug Deliv Rev* 61:768–784.

Batrakova, E. V., T. Y. Dorodnych, E. Y. Klinskii, E. N. Kliushnenkova, O. B. Shemchukova, O. N. Goncharova, S. A. Arjakov, V. Y. Alakhov, and A. V. Kabanov. 1996. Anthracycline antibiotics non-covalently incorporated into the block copolymer micelles: In vivo evaluation of anti-cancer activity. *Br J Cancer* 74:1545–1552.

Cabral, H., N. Nishiyama, S. Okazaki, H. Koyama, and K. Kataoka. 2005. Preparation and biological proper-ties of dichloro(1,2-diaminocyclohexane)platinum(II) (DACHPt)-loaded polymeric micelles. *J Control Release* 101:223–232.

Calvo, P., B. Gouritin, I. Brigger, C. Lasmezas, J. Deslys, A. Williams, J. P. Andreux, D. Dormont, and P. Couvreur. 2001. PEGylated polycyanoacrylate nanoparticles as vector for drug delivery in prion diseases. *J Neurosci Methods* 111:151–155.

Chekhonin, V. P., A. V. Kabanov, Y. A. Zhirkov, and G. V. Morozov. 1991. Fatty acid acylated Fab-fragments of antibodies to neurospecific proteins as carriers for neuroleptic targeted delivery in brain. *FEBS Lett* 287:149–152.

Chen, J., S. Chen, X. Zhao, L. V. Kuznetsova, S. S. Wong, and I. Ojima. 2008. Functionalized single-walled carbon nanotubes as rationally designed vehicles for tumor-targeted drug delivery. *J Am Chem Soc* 130:16778–16785.

Chung, J. E., M. Yokoyama, M. Yamato, T. Aoyagi, Y. Sakurai, and T. Okano. 1999. Thermo-responsive drug delivery from polymeric micelles constructed using block copolymers of poly(*N*-isopropylacrylamide) and poly(butylmethacrylate). *J Control Release* 62:115–127.

Dash, P. R., M. L. Read, K. D. Fisher, K. A. Howard, M. Wolfert, D. Oupicky, V. Subr, J. Strohalm, K. Ulbrich, and L. W. Seymour. 2000. Decreased binding to proteins and cells of polymeric gene delivery vectors surface modified with a multivalent hydrophilic polymer and retargeting through attachment of transfer-rin. *J Biol Chem* 275:3793–3802.

Djordjevic, J., M. Barch, and K. E. Uhrich. 2005. Polymeric micelles based on amphiphilic scorpion-like mac-romolecules: Novel carriers for water-insoluble drugs. *Pharm Res* 22:24–32.

Elworthy, P. H., A. T. Florence, and C. B. Macfarlane (eds.). 1968. *Solubilization by Surface Active Agents*. London, U.K.: Chapman & Hall.

Exner, A. A., T. M. Krupka, K. Scherrer, and J. M. Teets. 2005. Enhancement of carboplatin toxicity by Pluronic block copolymers. *J Control Release* 106:188–197.

Felgner, J. H., R. Kumar, C. N. Sridhar, C. J. Wheeler, Y. J. Tsai, R. Border, P. Ramsey, M. Martin, and P. L. Felgner. 1994. Enhanced gene delivery and mechanism studies with a novel series of cationic lipid for-mulations. *J Biol Chem* 269:2550–2561.

Frankel, A. D. and C. O. Pabo. 1988. Cellular uptake of the tat protein from human immunodeficiency virus. *Cell* 55:1189–1193.

Gabizon, A. A. 1992. Selective tumor localization and improved therapeutic index of anthracyclines encapsulated in long-circulating liposomes. *Cancer Res* 52:891–896.

Gabizon, A. A. 2001. Pegylated liposomal doxorubicin: Metamorphosis of an old drug into a new form of chemotherapy. *Cancer Invest* 19:424–436.

Gao, Z. G., H. D. Fain, and N. Rapoport. 2005. Controlled and targeted tumor chemotherapy by micellar-encapsulated drug and ultrasound. *J Control Release* 102:203–222.

Gao, Z., A. N. Lukyanov, A. R. Chakilam, and V. P. Torchilin. 2003. PEG–PE/phosphatidylcholine mixed immunomicelles specifically deliver encapsulated taxol to tumor cells of different origin and promote their efficient killing. *J Drug Target* 11:87–92.

Gao, Z., A. Lukyanov, A. Singhal, and V. Torchilin. 2002. Diacyllipid–polymer micelles as nanocarriers for poorly soluble anticancer drugs. *Nano Lett* 2:979–982.

Gao, H., Y. W. Yang, Y. G. Fan, and J. B. Ma. 2006. Conjugates of poly(DL-lactic acid) with ethylenediamino or diethylenetriamino bridged bis(beta-cyclodextrin)s and their nanoparticles as protein delivery systems. *J Control Release* 112:301–311.

Gaucher, G., M. H. Dufresne, V. P. Sant, N. Kang, D. Maysinger, and J. C. Leroux. 2005. Block copolymer micelles: Preparation, characterization and application in drug delivery. *J Control Release* 109:169–188.

Grant, C. W., S. Karlik, and E. Florio. 1989. A liposomal MRI contrast agent: Phosphatidylethanolamine–DTPA. *Magn Reson Med* 11:236–243.

Hagan, S. A., A. G. A. Coombes, M. C. Garnett, S. E. Dunn, M. C. Davies, L. Illum, S. S. Davis. 1996. Polylactide–poly(ethylene glycol) copolymers as drug delivery systems. 1. Characterization of water dispersible micelle-forming systems. *Langmuir* 12:2153–2161.

Harada, A. and K. Kataoka. 1998. Novel polyion complex micelles entrapping enzyme molecules in the core: Preparation of narrowly-distributed micelles from lysozyme and poly(ethylene glycol)–poly(aspartic acid) block copolymer in aqueous medium. *Macromolecules* 31:288–294.

Harris, J. M., N. E. Martin, and M. Modi. 2001. Pegylation: A novel process for modifying pharmacokinetics. *Clin Pharmacokinet* 40:539–551.

Helmlinger, G., F. Yuan, M. Dellian, and R. K. Jain. 1997. Interstitial pH and pO$_2$ gradients in solid tumors in vivo: High-resolution measurements reveal a lack of correlation. *Nat Med* 3:177–182.

Hobbs, S. K., W. L. Monsky, F. Yuan, W. G. Roberts, L. Griffith, V. P. Torchilin, and R. K. Jain. 1998. Regulation of transport pathways in tumor vessels: Role of tumor type and microenvironment. *Proc Natl Acad Sci U S A* 95:4607–4612.

Hsiue, G. H., C. H. Wang, C. L. Lo, C. H. Wang, J. P. Li, and J. L. Yang. 2006. Environmental-sensitive micelles based on poly(2-ethyl-2-oxazoline)-*b*-poly(L-lactide) diblock copolymer for application in drug delivery. *Int J Pharm* 317:69–75.

Hsu, S. H., Y. L. Leu, J. W. Hu, and J. Y. Fang. 2009. Physicochemical characterization and drug release of thermosensitive hydrogels composed of a hyaluronic acid/pluronic f127 graft. *Chem Pharm Bull* 57:453–458.

Huh, K. M., S. C. Lee, Y. W. Cho, J. Lee, J. H. Jeong, and K. Park. 2005. Hydrotropic polymer micelle system for delivery of paclitaxel. *J Control Release* 101:59–68.

Hunter, R. J. (ed.). 1991. *Foundations of Colloid Science*. New York: Oxford University Press.

Hunter, R. L. and B. Bennett. 1984. The adjuvant activity of nonionic block polymer surfactants. II. Antibody formation and inflammation related to the structure of triblock and octablock copolymers. *J Immunol* 133:3167–3175.

Hunter, R. L. and B. Bennett. 1986. The adjuvant activity of nonionic block polymer surfactants. III. Characterization of selected biologically active surfaces. *Scand J Immunol* 23:287–300.

Hunter, R., F. Strickland, and F. Kezdy. 1981. The adjuvant activity of nonionic block polymer surfactants. I. The role of hydrophile–lipophile balance. *J Immunol* 127:1244–1250.

Iakoubov, L. Z. and V. P. Torchilin. 1997. A novel class of antitumor antibodies: Nucleosome-restricted antinuclear autoantibodies (ANA) from healthy aged nonautoimmune mice. *Oncol Res* 9:439–446.

Inoue, T., G. Chen, K. Nakamae, and A. S. Hoffman. 1998. An AB block copolymer of oligo(methyl methacrylate) and poly(acrylic acid) for micellar delivery of hydrophobic drugs. *J Control Release* 51:221–229.

Jeong, Y. I., J. B. Cheon, S. H. Kim, J. W. Nah, Y. M. Lee, Y. K. Sung, T. Akaike, and C. S. Cho. 1998. Clonazepam release from core-shell type nanoparticles in vitro. *J Control Release* 51:169–178.

Jeong, J. H., S. H. Kim, S. W. Kim, and T. G. Park. 2005. In vivo tumor targeting of ODN–PEG–folic acid/PEI polyelectrolyte complex micelles. *J Biomater Sci Polym Ed* 16:1409–1419.

Jeong, J. H., H. Mok, Y. K. Oh, and T. G. Park. 2009. siRNA conjugate delivery systems. *Bioconjug Chem* 20:5–14.

Jones, M. and J. Leroux. 1999. Polymeric micelles—A new generation of colloidal drug carriers. *Eur J Pharm Biopharm* 48:101–111.

Jones, M. C., M. Ranger, and J. C. Leroux. 2003. pH-sensitive unimolecular polymeric micelles: Synthesis of a novel drug carrier. *Bioconjug Chem* 14:774–781.

Jule, E., Y. Nagasaki, and K. Kataoka. 2003. Lactose-installed poly(ethylene glycol)–poly(D,L-lactide) block copolymer micelles exhibit fast-rate binding and high affinity toward a protein bed simulating a cell surface. A surface plasmon resonance study. *Bioconjug Chem* 14:177–186.

Kabalka, G. W., E. Buonocore, K. Hubner, M. Davis, and L. Huang. 1988. Gadolinium-labeled liposomes containing paramagnetic amphipathic agents: Targeted MRI contrast agents for the liver. *Magn Reson Med* 8:89–95.

Kabanov, A. V., E. V. Batrakova, and V. Y. Alakhov. 2002. Pluronic block copolymers as novel polymer therapeutics for drug and gene delivery. *J Control Release* 82:189–212.

Kabanov, A. V., V. P. Chekhonin, V. Y. Alakhov, E. V. Batrakova, A. S. Lebedev, N. S. Melik-Nubarov, S. A. Arzhakov, A. V. Levashov, G. V. Morozov, E. S. Severin et al. 1989. The neuroleptic activity of haloperidol increases after its solubilization in surfactant micelles. Micelles as microcontainers for drug targeting. *FEBS Lett* 258:343–345.

Kabanov, V. A. and A. V. Kabanov. 1998. Interpolyelectrolyte and block ionomer complexes for gene delivery: Physico-chemical aspects. *Adv Drug Deliv Rev* 30:49–60.

Kabanov, A. V., I. R. Nazarova, I. V. Astafieva, E.V. Batrakova, V. Y. Alakhov, A. A. Yaroslavov, V. A. Kabanov, and A.V. Kabanov. Micelle formation and solubilization of fluorescent probes in poly(oxyethylene-*b*-oxypropylene-*b*-oxyethylene) solutions. *Macromolecules* 28:2303–2314.

Kaiser, S. and M. Toborek. 2001. Liposome-mediated high-efficiency transfection of human endothelial cells. *J Vasc Res* 38:133–143.

Kale, A. A. and V. P. Torchilin. 2007. Design, synthesis, and characterization of pH-sensitive PEG–PE conjugates for stimuli-sensitive pharmaceutical nanocarriers: The effect of substitutes at the hydrazone linkage on the pH stability of PEG–PE conjugates. *Bioconjug Chem* 18:363–370.

Kaminskas, L. M., B. D. Kelly, V. M. McLeod, B. J. Boyd, G. Y. Krippner, E. D. Williams, and C. J. Porter. 2009. Pharmacokinetics and tumor disposition of PEGylated, methotrexate conjugated poly-L-lysine dendrimers. *Mol Pharm* 6:1190–1204.

Katayose, S. and K. Kataoka. 1998. Remarkable increase in nuclease resistance of plasmid DNA through supramolecular assembly with poly(ethylene glycol)–poly(L-lysine) block copolymer. *J Pharm Sci* 87:160–163.

Kim, S. H., J. H. Jeong, S. H. Lee, S. W. Kim, and T. G. Park. 2008a. LHRH receptor-mediated delivery of siRNA using polyelectrolyte complex micelles self-assembled from siRNA–PEG–LHRH conjugate and PEI. *Bioconjug Chem* 19:2156–2162.

Kim, S. H., J. H. Jeong, S. H. Lee, S. W. Kim, and T. G. Park. 2008b. Local and systemic delivery of VEGF siRNA using polyelectrolyte complex micelles for effective treatment of cancer. *J Control Release* 129:107–116.

Kim, S. H., S. H. Lee, H. Tian, X. Chen, and T.G. Park. 2009. Prostrate cancer cell-specific VEGF siRNA delivery system using cell targeting peptide conjugated polyplexes. *J Drug Target* 17:311–317.

Klibanov, A. L., K. Maruyama, V. P. Torchilin, and L. Huang. 1990. Amphipathic polyethyleneglycols effectively prolong the circulation time of liposomes. *FEBS Lett* 268:235–237.

Ko, Y. T., C. Falcao, and V. P. Torchilin. 2009. Cationic liposomes loaded with proapoptotic peptide D-(KLAKLAK)(2) and Bcl-2 antisense oligodeoxynucleotide G3139 for enhanced anticancer therapy. *Mol Pharm* 6:971–977.

Krishnadas, A., I. Rubinstein, and H. Onyuksel. 2003. Sterically stabilized phospholipid mixed micelles: In vitro evaluation as a novel carrier for water-insoluble drugs. *Pharm Res* 20:297–302.

Kwon, G. S. 1998. Diblock copolymer nanoparticles for drug delivery. *Crit Rev Ther Drug Carrier Syst* 15:481–512.

Kwon, G. S. 2003. Polymeric micelles for delivery of poorly water-soluble compounds. *Crit Rev Ther Drug Carrier Syst* 20:357–403.

Kwon, G. S. and K. Kataoka. 1995. Block copolymer micelles as long-circulating drug vehicles. *Adv Drug Deliv Rev* 16:295–309.

Kwon, G. S., M. Naito, M. Yokoyama, T. Okano, Y. Sakurai, and K. Kataoka. 1995. Physical entrapment of adriamycin in AB block copolymer micelles. *Pharm Res* 12:192–195.

Kwon, G., M. Naito, M. Yokoyama, T. Okano, Y. Sakurai, and K. Kataoka. 1997. Block copolymer micelles for drug delivery: Loading and release of doxorubicin. *J Control Release* 48:195–201.

La, S. B., T. Okano, and K. Kataoka. 1996. Preparation and characterization of the micelle-forming polymeric drug indomethacin-incorporated poly(ethylene oxide)–poly(beta-benzyl L-aspartate) block copolymer micelles. *J Pharm Sci* 85:85–90.

Lalloo, A., P. Chao, P. Hu, S. Stein, and P. J. Sinko. 2006. Pharmacokinetic and pharmacodynamic evaluation of a novel in situ forming poly(ethylene glycol)-based hydrogel for the controlled delivery of the camptothecins. *J Control Release* 112:333–342.

Lasic, D. D. 1992. Mixed micelles in drug delivery. *Nature* 355:279–280.

Lasic, D. D. and F. Martin (eds.). 1995. *Stealth Liposomes*. Boca Raton, FL: CRC Press.

Le Garrec, D., S. Gori, L. Luo, D. Lessard, D. C. Smith, M. A. Yessine, M. Ranger, and J. C. Leroux. 2004. Poly(N-vinylpyrrolidone)-*block*-poly(D,L-lactide) as a new polymeric solubilizer for hydrophobic anticancer drugs: In vitro and in vivo evaluation. *J Control Release* 99:83–101.

Le Garrec, D., J. Taillefer, J. E. Van Lier, V. Lenaerts, and J. C. Leroux. 2002. Optimizing pH-responsive polymeric micelles for drug delivery in a cancer photodynamic therapy model. *J Drug Target* 10:429–437.

Leamon, C. P. and P. S. Low. 2001. Folate-mediated targeting: From diagnostics to drug and gene delivery. *Drug Discov Today* 6:44–51.

Leamon, C. P., D. Weigl, and R. W. Hendren. 1999. Folate copolymer-mediated transfection of cultured cells. *Bioconjug Chem* 10:947–957.

Lee, E. S., K. Na, and Y. H. Bae. 2003. Polymeric micelle for tumor pH and folate-mediated targeting. *J Control Release* 91:103–113.

Lee, H., F. Zeng, M. Dunne, and C. Allen. 2005. Methoxy poly(ethylene glycol)-*block*-poly(delta-valerolactone) copolymer micelles for formulation of hydrophobic drugs. *Biomacromolecules* 6:3119–3128.

Liaw, J., S. F. Chang, and F. C. Hsiao. 2001. In vivo gene delivery into ocular tissues by eye drops of poly(ethylene oxide)–poly(propylene oxide)–poly(ethylene oxide) (PEO–PPO–PEO) polymeric micelles. *Gene Ther* 8:999–1004.

Lin, S. Y. and Y. Kawashima. 1985. The influence of three poly(oxyethylene)poly(oxypropylene) surface-active block copolymers on the solubility behavior of indomethacin. *Pharm Acta Helv* 60:339–344.

Lin, S. Y. and Y. Kawashima. 1987. Pluronic surfactants affecting diazepam solubility, compatibility, and adsorption from i.v. admixture solutions. *J Parenter Sci Technol* 41:83–87.

Liu, L., S. S. Venkatraman, Y. Y. Yang, K. Guo, J. Lu, B. He, S. Moochhala, and L. Kan. 2008. Polymeric micelles anchored with TAT for delivery of antibiotics across the blood–brain barrier. *Biopolymers* 90:617–623.

Lukyanov, A. N., Z. Gao, L. Mazzola, and V. P. Torchilin. 2002. Polyethylene glycol–diacyllipid micelles demonstrate increased accumulation in subcutaneous tumors in mice. *Pharm Res* 19:1424–1429.

Lukyanov, A. N., W. C. Hartner, and V. P. Torchilin. 2004. Increased accumulation of PEG–PE micelles in the area of experimental myocardial infarction in rabbits. *J Control Release* 94:187–193.

Luppi, B., I. Orienti, F. Bigucci, T. Cerchiara, G. Zuccari, S. Fazzi, and V. Zecchi. 2002. Poly(vinylalcohol-*co*-vinyloleate) for the preparation of micelles enhancing retinyl palmitate transcutaneous permeation. *Drug Deliv* 9:147–152.

Maeda, H., T. Sawa, and T. Konno. 2001. Mechanism of tumor-targeted delivery of macromolecular drugs, including the EPR effect in solid tumor and clinical overview of the prototype polymeric drug SMANCS. *J Control Release* 74:47–61.

Maeda, H., J. Wu, T. Sawa, Y. Matsumura, and K. Hori. 2000. Tumor vascular permeability and the EPR effect in macromolecular therapeutics: A review. *J Control Release* 65:271–284.

Mastrobattista, E., G. A. Koning, L. van Bloois, A. C. Filipe, W. Jiskoot, and G. Storm. 2002. Functional characterization of an endosome-disruptive peptide and its application in cytosolic delivery of immunoliposome-entrapped proteins. *J Biol Chem* 277:27135–27143.

Mathot, F., L. van Beijsterveldt, V. Preat, M. Brewster, and A. Arien. 2006. Intestinal uptake and biodistribution of novel polymeric micelles after oral administration. *J Control Release* 111:47–55.

Matsumura, Y., M. Yokoyama, K. Kataoka, T. Okano, Y. Sakurai, T. Kawaguchi, and T. Kakizoe. 1999. Reduction of the side effects of an antitumor agent, KRN5500, by incorporation of the drug into polymeric micelles. *Jpn J Cancer Res* 90:122–128.

Meyer, O., D. Papahadjopoulos, and J. C. Leroux. 1998. Copolymers of N-isopropylacrylamide can trigger pH sensitivity to stable liposomes. *FEBS Lett* 421:61–64.

Miller, D. W., E. V. Batrakova, T. O. Waltner, V. Y. Alakhov, and A. V. Kabanov. 1997. Interactions of Pluronic block copolymers with brain microvessel endothelial cells: Evidence of two potential pathways for drug absorption. *Bioconjug Chem* 8:649–657.

Mittal, K. L. and B. B. Lindman. 1991. *Surfactants in Solution*. New York: Plenum Press.

Moghimi, S. M. 2002. Chemical camouflage of nanospheres with a poorly reactive surface: Towards development of stealth and target-specific nanocarriers. *Biochim Biophys Acta* 1590:131–139.

Monsky, W. L., D. Fukumura, T. Gohongi, M. Ancukiewcz, H. A. Weich, V. P. Torchilin, F. Yuan, and R. K. Jain. 1999. Augmentation of transvascular transport of macromolecules and nanoparticles in tumors using vascular endothelial growth factor. *Cancer Res* 59:4129–4135.

Morcol, T., P. Nagappan, L. Nerenbaum, A. Mitchell, and S. J. Bell. 2004. Calcium phosphate–PEG–insulin–casein (CAPIC) particles as oral delivery systems for insulin. *Int J Pharm* 277:91–97.

Mu, L., T. A. Elbayoumi, and V. P. Torchilin. 2005. Mixed micelles made of poly(ethylene glycol)–phosphatidylethanolamine conjugate and D-alpha-tocopheryl polyethylene glycol 1000 succinate as pharmaceutical nanocarriers for camptothecin. *Int J Pharm* 306:142–149.

Müller, H. 1991. *Colloidal Carriers for Controlled Drug Delivery and Targeting: Modification, Characterization, and In Vivo Distribution*. Boca Raton, FL: CRC Press.

Musacchio, T., V. Laquintana, A. Latrofa, G. Trapani, and V. P. Torchilin. 2008. PEG–PE micelles loaded with paclitaxel and surface-modified by a PBR–ligand: Synergistic anticancer effect. *Mol Pharm* 6:468–479.

Nagarajan, R. and K. Ganesh. 1989. Block copolymer self-assembly in selective solvents: Theory of solubilization in spherical micelles. *Macromolecules* 22:4312–4325.

Nagasaki, Y., K. Yasugi, Y. Yamamoto, A. Harada, and K. Kataoka. 2001. Sugar-installed block copolymer micelles: Their preparation and specific interaction with lectin molecules. *Biomacromolecules* 2:1067–1070.

Nasongkla, N., E. Bey, J. M. Ren, C. Khemtong, and J. S. Guthi. 2006. Multifunctional polymeric micelles as cancer-targeted, MRI-ultrasensitive drug delivery systems. *Nano Lett* 6:2427–2430.

Nawrot, B. and K. Sipa. 2006. Chemical and structural diversity of siRNA molecules. *Curr Top Med Chem* 6:913–925.

Newman, M. J., C. W. Todd, E. M. Lee, M. Balusubramanian, P. J. Didier, and J. M. Katz. 1997. Increasing the immunogenicity of a trivalent influenza virus vaccine with adjuvant-active nonionic block copolymers for potential use in the elderly. *Mech Ageing Dev* 93:189–203.

Nishiyama, N. and K. Kataoka. 2006. Current state, achievements, and future prospects of polymeric micelles as nanocarriers for drug and gene delivery. *Pharmacol Ther* 112:630–648.

Ogris, M., S. Brunner, S. Schuller, R. Kircheis, and E. Wagner. 1999. PEGylated DNA/transferrin–PEI complexes: Reduced interaction with blood components, extended circulation in blood and potential for systemic gene delivery. *Gene Ther* 6:595–605.

Opanasopit, P., M. Yokoyama, M. Watanabe, K. Kawano, Y. Maitani, and T. Okano. 2004. Block copolymer design for camptothecin incorporation into polymeric micelles for passive tumor targeting. *Pharm Res* 21:2001–2008.

Ota, T., M. Maeda, and M. Tatsuka. 2002. Cationic liposomes with plasmid DNA influence cancer metastatic capability. *Anticancer Res* 22:4049–4052.

Palmer, T. N., V. J. Caride, M. A. Caldecourt, J. Twickler, and V. Abdullah. 1984. The mechanism of liposome accumulation in infarction. *Biochim Biophys Acta* 797:363–368.

Papahadjopoulos, D., T. M. Allen, A. Gabizon, E. Mayhew, K. Matthay, S. K. Huang, K. D. Lee, M. C. Woodle, D. D. Lasic, C. Redemann et al. 1991. Sterically stabilized liposomes: Improvements in pharmacokinetics and antitumor therapeutic efficacy. *Proc Natl Acad Sci U S A* 88:11460–11464.

Park, E. K., S. Y. Kim, S. B. Lee, and Y. M. Lee. 2005. Folate-conjugated methoxy poly(ethylene glycol)/poly(epsilon-caprolactone) amphiphilic block copolymeric micelles for tumor-targeted drug delivery. *J Control Release* 109:158–168.

Pillion, D. J., J. A. Amsden, C. R. Kensil, and J. Recchia. 1996. Structure–function relationship among *Quillaja* saponins serving as excipients for nasal and ocular delivery of insulin. *J Pharm Sci* 85:518–524.

Prompruk, K., T. Govender, S. Zhang, C. D. Xiong, and S. Stolnik. 2005. Synthesis of a novel PEG-*block*-poly(aspartic acid-*stat*-phenylalanine) copolymer shows potential for formation of a micellar drug carrier. *Int J Pharm* 297:242–253.

Rapoport, N., W. G. Pitt, H. Sun, and J. L. Nelson. 2003. Drug delivery in polymeric micelles: From in vitro to in vivo. *J Control Release* 91:85–95.

Rawat, M., D. Singh, and S. Saraf. 2006. Nanocarriers: Promising vehicle for bioactive drugs. *Biol Pharm Bull* 29:1790–1798.

Roberts, M. J., M. D. Bentley, and J. M. Harris. 2002. Chemistry for peptide and protein PEGylation. *Adv Drug Deliv Rev* 54:459–476.

Roby, A., S. Erdogan, and V. P. Torchilin. 2007. Enhanced in vivo antitumor efficacy of poorly soluble PDT agent, meso-tetraphenylporphine, in PEG–PE-based tumor-targeted immunomicelles. *Cancer Biol Ther* 6:1136–1142.

Rothbard, J. B., T. C. Jessop, R. S. Lewis, B. A. Murray, and P. A. Wender. 2004. Role of membrane potential and hydrogen bonding in the mechanism of translocation of guanidinium-rich peptides into cells. *J Am Chem Soc* 126:9506–9507.

Sawant, R. M., J. P. Hurley, S. Salmaso, A. Kale, E. Tolcheva, T. S. Levchenko, and V. P. Torchilin. 2006. "SMART" drug delivery systems: Double-targeted pH-responsive pharmaceutical nanocarriers. *Bioconjug Chem* 17:943–949.

Sawant, R. R. and V. P. Torchilin. 2009. Enhanced cytotoxicity of TATp-bearing paclitaxel-loaded micelles in vitro and in vivo. *Int J Pharm* 374:114–118.

Schmiedl, U. P., J. A. Nelson, L. Teng, F. Starr, R. Malek, and R. J. Ho. 1995. Magnetic resonance imaging of the hepatobiliary system: Intestinal absorption studies of manganese mesoporphyrin. *Acad Radiol* 2:994–1001.

Sethuraman, V. A. and Y. H. Bae. 2007. TAT peptide-based micelle system for potential active targeting of anti-cancer agents to acidic solid tumors. *J Control Release* 118:216–224.

Shuai, X., T. Merdan, A. K. Schaper, F. Xi, and T. Kissel. 2004. Core-cross-linked polymeric micelles as paclitaxel carriers. *Bioconjug Chem* 15:441–448.

Skidan, I., P. Dholakia, and V. Torchilin. 2008. Photodynamic therapy of experimental B-16 melanoma in mice with tumor-targeted 5,10,15,20-tetraphenylporphin-loaded PEG–PE micelles. *J Drug Target* 16:486–493.

Smith, R. and C. Tanford. 1972. The critical micelle concentration of L-α-dipalmitoylphosphatidylcholine in water and water–methanol solutions. *J Mol Biol* 67:75–83.

Soga, O., C. F. van Nostrum, M. Fens, C. J. Rijcken, R. M. Schiffelers, G. Storm, and W. E. Hennink. 2005. Thermosensitive and biodegradable polymeric micelles for paclitaxel delivery. *J Control Release* 103:341–353.

Tannock, I. F. and D. Rotin. 1989. Acid pH in tumors and its potential for therapeutic exploitation. *Cancer Res* 49:4373–4384.

Tao, L. and K. E. Uhrich. 2006. Novel amphiphilic macromolecules and their in vitro characterization as stabilized micellar drug delivery systems. *J Colloid Interface Sci* 298:102–110.

Todd, C. W., M. Balusubramanian, and M. J. Newman. 1998. Development of adjuvant-active nonionic block copolymers. *Adv Drug Deliv Rev* 32:199–223.

Todd, C. W., L. A. Pozzi, J. R. Guarnaccia, M. Balasubramanian, W. G. Henk, L. E. Younger, and M. J. Newman. 1997. Development of an adjuvant-active nonionic block copolymer for use in oil-free subunit vaccines formulations. *Vaccine* 15:564–570.

Torchilin, V. P. 1995. *Handbook of Targeted Delivery of Imaging Agents*. Boca Raton, FL: CRC Press.

Torchilin, V. P. 2001. Structure and design of polymeric surfactant-based drug delivery systems. *J Control Release* 73:137–172.

Torchilin, V. P. 2002. PEG-based micelles as carriers of contrast agents for different imaging modalities. *Adv Drug Deliv Rev* 54:235–252.

Torchilin, V. P. 2004. Targeted polymeric micelles for delivery of poorly soluble drugs. *Cell Mol Life Sci* 61:2549–2559.

Torchilin, V. P. 2007. Micellar nanocarriers: Pharmaceutical perspectives. *Pharm Res* 24:1–16.

Torchilin, V. P., M. D. Frank-Kamenetsky, and G. L. Wolf. 1999. CT visualization of blood pool in rats by using long-circulating, iodine-containing micelles. *Acad Radiol* 6:61–65.

Torchilin, V. P., T. S. Levchenko, A. N. Lukyanov, B. A. Khaw, A. L. Klibanov, R. Rammohan, G. P. Samokhin, and K. R. Whiteman. 2001a. *p*-Nitrophenylcarbonyl–PEG–PE–liposomes: Fast and simple attachment of specific ligands, including monoclonal antibodies, to distal ends of PEG chains via *p*-nitrophenylcarbonyl groups. *Biochim Biophys Acta* 1511:397–411.

Torchilin, V. P., T. S. Levchenko, K. R. Whiteman, A. A. Yaroslavov, A. M. Tsatsakis, A. K. Rizos, E. V. Michailova, and M. I. Shtilman. 2001b. Amphiphilic poly-*N*-vinylpyrrolidones: Synthesis, properties and liposome surface modification. *Biomaterials* 22:3035–3044.

Torchilin, V. P., A. N. Lukyanov, Z. Gao, and B. Papahadjopoulos-Sternberg. 2003. Immunomicelles: Targeted pharmaceutical carriers for poorly soluble drugs. *Proc Natl Acad Sci U S A* 100:6039–6044.

Torchilin, V. P., M. I. Shtilman, V. S. Trubetskoy, K. Whiteman, and A. M. Milstein. 1994. Amphiphilic vinyl polymers effectively prolong liposome circulation time in vivo. *Biochim Biophys Acta* 1195:181–184.

Torchilin, V. P., V. S. Trubetskoy, K. R. Whiteman, P. Caliceti, P. Ferruti, and F. M. Veronese. 1995. New synthetic amphiphilic polymers for steric protection of liposomes in vivo. *J Pharm Sci* 84:1049–1053.

Triozzi, P. L., V. C. Stevens, W. Aldrich, J. Powell, C. W. Todd, and M. J. Newman. 1997. Effects of a beta-human chorionic gonadotropin subunit immunogen administered in aqueous solution with a novel nonionic block copolymer adjuvant in patients with advanced cancer. *Clin Cancer Res* 3:2355–2362.

Trubetskoy, V. S., M. D. Frank-Kamenetsky, K. R. Whiteman, G. L. Wolf, and V. P. Torchilin. 1996. Stable polymeric micelles: Lymphangiographic contrast media for gamma scintigraphy and magnetic resonance imaging. *Acad Radiol* 3:232–238.

Trubetskoy, V. S., G. S. Gazelle, G. L. Wolf, and V. P. Torchilin. 1997. Block-copolymer of polyethylene glycol and polylysine as a carrier of organic iodine: Design of long-circulating particulate contrast medium for X-ray computed tomography. *J Drug Target* 4:381–388.

Trubetskoy, V. S. and V. P. Torchilin. 1995. Use of polyoxyethylene–lipid conjugates as long-circulating carriers for delivery of therapeutic and diagnostic agents. *Adv Drug Deliv Rev* 16:311–320.

Trubetskoy, V. S. and V. P. Torchilin. 1996. Polyethyleneglycol based micelles as carriers of therapeutic and diagnostic agents. *STP Pharma Sci* 6:79–86.

Unger, E., T. Fritz, G. Wu et al. 1994. Liposomal MR contrast agents. *J Liposome Res* 4:811–834.

Veronese, F. M. and J. M. Harris. 2002. Introduction and overview of peptide and protein pegylation. *Adv Drug Deliv Rev* 54:453–456.

Vinogradov, S., E. Batrakova, S. Li, and A. Kabanov. 1999. Polyion complex micelles with protein-modified corona for receptor-mediated delivery of oligonucleotides into cells. *Bioconjug Chem* 10:851–860.

Vinogradov, S. V., E. V. Batrakova, S. Li, and A. V. Kabanov. 2004. Mixed polymer micelles of amphiphilic and cationic copolymers for delivery of antisense oligonucleotides. *J Drug Target* 12:517–526.

Wadia, J. S., R. V. Stan, and S. F. Dowdy. 2004. Transducible TAT-HA fusogenic peptide enhances escape of TAT-fusion proteins after lipid raft macropinocytosis. *Nat Med* 10:310–315.

Wakebayashi, D., N. Nishiyama, Y. Yamasaki, K. Itaka, N. Kanayama, A. Harada, Y. Nagasaki, and K. Kataoka. 2004. Lactose-conjugated polyion complex micelles incorporating plasmid DNA as a targetable gene vector system: Their preparation and gene transfecting efficiency against cultured HepG2 cells. *J Control Release* 95:653–664.

Wang, J., D. A. Mongayt, A. N. Lukyanov, T. S. Levchenko, and V. P. Torchilin. 2004. Preparation and in vitro synergistic anticancer effect of vitamin K3 and 1,8-diazabicyclo[5,4,0]undec-7-ene in poly(ethylene glycol)-diacyllipid micelles. *Int J Pharm* 272:129–135.

Wang, J., D. Mongayt, and V. P. Torchilin. 2005. Polymeric micelles for delivery of poorly soluble drugs: Preparation and anticancer activity in vitro of paclitaxel incorporated into mixed micelles based on poly(ethylene glycol)–lipid conjugate and positively charged lipids. *J Drug Target* 13:73–80.

Wang, C. H., C. H. Wang, and G. H. Hsiue. 2005. Polymeric micelles with a pH-responsive structure as intracellular drug carriers. *J Control Release* 108:140–149.

Wang, X., L. Yang, Z. G. Chen, and D. M. Shin. 2008. Application of nanotechnology in cancer therapy and imaging. *CA Cancer J Clin* 58:97–110.

Watanabe, M., K. Kawano, M. Yokoyama, P. Opanasopit, T. Okano, and Y. Maitani. 2006. Preparation of camptothecin-loaded polymeric micelles and evaluation of their incorporation and circulation stability. *Int J Pharm* 308:183–189.

Weissig, V., K. R. Whiteman, and V. P. Torchilin. 1998. Accumulation of protein-loaded long-circulating micelles and liposomes in subcutaneous Lewis lung carcinoma in mice. *Pharm Res* 15:1552–1556.

Wolf, G. 1995. Targeted delivery of imaging agents: An overview. *Handbook of Targeted Delivery of Imaging Agents*, edited by Torchilin, V. P. Boca Raton, FL: CRC Press.

Wong, H. L., R. Bendayan, A. M. Rauth, Y. Li, and X. Y. Wu. 2007. Chemotherapy with anticancer drugs encapsulated in solid lipid nanoparticles. *Adv Drug Deliv Rev* 59:491–504.

Xing L. and W. L. Mattice. 1998. Large internal structures of micelles of triblock copolymers with small insoluble molecules in their cores. *Langmuir* 14:4074–4080.

Xu, P., E. A. Van Kirk, S. Li, W. J. Murdoch, J. Ren, M. D. Hussain, M. Radosz, and Y. Shen. 2006. Highly stable core-surface-crosslinked nanoparticles as cisplatin carriers for cancer chemotherapy. *Colloids Surf B Biointerf* 48:50–57.

Yokoyama, M., S. Fukushima, R. Uehara, K. Okamoto, K. Kataoka, Y. Sakurai, and T. Okano. 1998. Characterization of physical entrapment and chemical conjugation of adriamycin in polymeric micelles and their design for in vivo delivery to a solid tumor. *J Control Release* 50:79–92.

Yokoyama, M., M. Miyauchi, N. Yamada, T. Okano, Y. Sakurai, K. Kataoka, and S. Inoue. 1990. Characterization and anticancer activity of the micelle-forming polymeric anticancer drug adriamycin-conjugated poly(ethylene glycol)–poly(aspartic acid) block copolymer. *Cancer Res* 50:1693–1700.

Yokoyama, M., T. Okano, and K. Kataoka. 1994. Improved synthesis of adriamycin-conjugated poly(ethylene oxide)–poly(aspartic acid) block copolymer and formation of unimodal micellar structure with controlled amount of physically entrapped adriamycin. *J Control Release* 32:269–277.

Yoo, H. S., E. A. Lee, and T. G. Park. 2002. Doxorubicin-conjugated biodegradable polymeric micelles having acid-cleavable linkages. *J Control Release* 82:17–27.

Yoo, H. S. and T. G. Park. 2004. Folate-receptor-targeted delivery of doxorubicin nano-aggregates stabilized by doxorubicin–PEG–folate conjugate. *J Control Release* 100:247–256.

Yuan, F., M. Dellian, D. Fukumura, M. Leunig, D. A. Berk, V. P. Torchilin, and R. K. Jain. 1995. Vascular permeability in a human tumor xenograft: Molecular size dependence and cutoff size. *Cancer Res* 55:3752–3756.

Yuan, F., M. Leunig, S. K. Huang, D. A. Berk, D. Papahadjopoulos, and R. K. Jain. 1994. Microvascular permeability and interstitial penetration of sterically stabilized (stealth) liposomes in a human tumor xenograft. *Cancer Res* 54:3352–3356.

Zhang, L. and A. Eisenberg. 1995. Multiple morphologies of "crew-cut" aggregates of polystyrene-*b*-poly(acrylic acid) block copolymers. *Science* 268:1728–1731.

4 Modular Biomimetic Drug Delivery Systems

Carmen Alvarez-Lorenzo, Fernando Yañez-Gomez, and Angel Concheiro

CONTENTS

4.1 INTRODUCTION

Drug delivery has experienced a remarkable evolution in the last decades (Hoffman 2008). Traditionally, a drug dosage form was considered as just the hardware of the medicine for facilitating the handling and the administration of the drug and for releasing it into the body. The search for better patient compliance and more constant drug levels led to the development of sustained delivery dosage forms (first generation of controlled release systems) that latter made way to devices capable of controlling the place and/or the time at which the delivery, mainly through the gastrointestinal tract, should take place (second generation of controlled release systems). Both approaches have greatly improved the therapeutics and are behind most currently commercialized medicines. Common sustained delivery systems are made with "passive" components with a rigidly predicted behavior. They are mainly expected to dissolve, swell, or erode at a certain rate when entering into contact with the biological fluids and to control the delivery by diffusion or erosion (e.g., hydrophilic or biodegradable matrices) or to sorb water acting as osmotic pumps releasing the drug at a predetermined rate. Therefore, for these systems the incidence of the physiological state of the patient on drug release is minor, but if it happens, the performance of the system may be compromised. The second generation of controlled release systems came up from a better knowledge of the sites through which the delivery system has to pass through in the body and the time involved in the process. Such knowledge is used to select components with physical properties that depend on the "a priori" expected scenario. Examples of these components are some polymers with pH-dependent solubility or time-dependent swelling, or that undergo enzymatic degradation at certain regions of the body, which have found a noticeable position in the design of oral formulations for site-selective delivery (Chien and Lin 2002). Today drug dosage forms are seen as systems capable of delivering the drug at the best conditions as possible (protecting it from adverse agents) to the absorption site and, even, able to modulate drug distribution in the body and its clearance. In such a way, they act as efficient and safe drug delivery systems (DDS).

The new chemical entities candidates to drugs, particularly those from biological and biotechnological sources, are everyday demanding more advanced delivery platforms (Billingsley 2008; Kim et al. 2009). Ideally, drug delivery should resemble the delivery of a packet by a courier (Ranney 2000). As for packet delivery, the item (the drug in our case) has to be conveniently wrapped taking into account its fragility and the route and the conditions of the transport (distance, frontiers, climatic conditions, etc.). Then the item has to be unequivocally directed to the addressee and, finally, the addressee has to accept the packet and open it. If the packet is damaged, lost, or rejected by the addressee, the aim of the delivery fails. Similarly, an advanced DDS may be envisioned as a packet that acts itself as the courier (the pilot plus the cargo) inside the body. Not only biopharmaceuticals but also longtime used drugs could benefit greatly from the development of DDS that enable the achievement of therapeutic levels in specific organs, tissues, or even cellular structures, where and when they are required.

Notable improvement in the chemistry and physics of polymers, but particularly a better knowledge of the transport and the recognition process inside the body, makes nowadays feasible to bring to a reality rudimentary "self-delivery" packets. Biological molecules perform incredibly complex functions adopting architectures which are the result of the correct assembly of components that interact with high specificity in extremely small length scales (Tu and Tirrell 2004). Although human-designed materials are still not as effectively as Nature-designed molecules, in recent years many researchers have sought to find or design synthetic polymeric materials capable of mimicking one or another functional property of biomacromolecules (Kopecek 2003; Hoffman 2006; Bayer and Peppas 2008). Nature only uses 20 building blocks (amino acids) to create molecules with diverse structures and functions through a rigorous control of the sequence of building blocks. Nowadays chemistry offers millions of monomers to be used as components of synthetic molecules. To find the appropriate sequence is the key for generating biomimetic systems (Guan 2008). Biomimetics has been recently defined as an emerging field of science that includes the study of how Nature designs, processes, and assembles/disassembles molecular building blocks (modules) to fabricate high-performance soft materials and mineral–polymer composites, and then applies these designs and processes to engineer new molecules and materials with unique properties (Mano and Reis 2005; Bhushan 2009). Integration of the biomimetic methodologies into tailor-designed drug carriers is being applied to prepare DDS capable of releasing an active molecule at the appropriate site and at a rate that adjusts in response to the progression of the disease or to certain functions/biorhythms of the organism (Dillow and Lowman 2004; Youan 2004; Bayer and Peppas 2008). Mimicking molecule-selective agents, camouflage coatings/shells, or stimuli-sensitive components may render DDS with amazing performances (Vauthier and Labarre 2008).

Biomacromolecules are valuable as models for the design of tailored synthetic materials useful to mimic the behavior of the biological systems. Humans possess their own cellular mechanisms for specific recognition, for selective capture, and for controlled transfer of substances. Biomacromolecules play a fundamental role in all these functions, but their ability to perform functions beyond those required in Nature is constrained by instability and limited versatility (Brudno and Liu 2009). Synthetic polymers can overcome these limitations, since they can be obtained by versatile procedures that lead to tailored structures and features. Furthermore, they can avoid immunogenicity problems and their industrial production is more economic (Roy and Gupta 2003). The elucidation of the structure of numerous biomacromolecules facilitates the understanding of some biological complex mechanisms in which macromolecules play a fundamental role, and enables a quick evolution of the biomedical sciences. The use of such an information in the design of biomimetic materials requires the confluence of polymer science with biomedical sciences as forecasted 30 years ago by Hermann Mark, who stated that only a well-established interdisciplinary team can expect to go beyond a multilingual borderland (Mark 1981). The level of the procedures of synthesis and of the analytical techniques provides a well-characterized variety of polymers with a wide range of molecular compositions (e.g., multiblock, dendritic, hybrid) and structural arrangements (e.g., modular) to be obtained, which can carry out functions that few years ago were difficult to imagine (Qiu and Bae 2006; Kopecek and Yang 2007; Carlmaerk et al. 2009; Jang et al. 2009; Kushner et al. 2009). Biomimetic materials are greatly impacting the therapeutic field as components of targeting and stimuli-responsive DDS, stealth nanocarriers, polymer–drug conjugates, drug-eluting stents, nonviral gene vectors, biohybrid artificial organs, and vaccines (Keegan et al. 2003; Cuchelkar and Kopecek 2006; Lacík 2006; Kukreja et al. 2008; Wang et al. 2009).

The following sections focus on drug carriers that are biomimetic at their surface and those that are biomimetics as a whole. Such an organization is quite artificial since most biomimetic DDS partially fulfill both aspects. Nevertheless, with the aim of simplicity, in each section the systems that predominantly belong to one of the two categories are described and their potential in therapeutics is discussed.

4.2 BIOMIMETICS AT THE SURFACE

The use of nanotechnology for the development of nanosized DDS is perhaps one of the latter revolutionary and more appealing approaches (LaVan et al. 2002; Park 2007; Tong et al. 2009). Nanocarriers seem particularly suitable for formulating injectable biopharmaceuticals and chemotherapy agents, since the carrier can protect them from premature degradation, target localized regions, and release the therapeutic agent in a controlled fashion. Nevertheless, despite the highly promising results obtained in vitro and using cell cultures during the last two decades or so, the practical use of nanoparticles has still to face up many events that occur in vivo and that handicap their performance. Intravenously injected nanocarriers have to deal with mainly three different physiological environments (Ranney 2000):

1. The blood, with the plasma constituents, formed elements and normal endothelia. No intravascular agent has direct access to the endothelium and 95% of drug dose is usually lost by opsonization and precipitation by plasma proteins. The nanocarrier has to avoid these events and has to effectively retain the therapeutic agent inside.
2. The tissue matrix. Normal tissues are rich in lipids, while pathological sites also contain proteolytically cleaved peptides, oligosaccharides, cytokines, and growth factors.
3. The target cells.

A biomimetic-based design of the surface, regarding both the chemical constituents (e.g., stealth components and targeting elements) and the physical properties (e.g., shape and surface texture), may help to a successful dealing with the physiological environment and an efficient operation of the nanocarrier.

4.2.1 BIOADDRESSED NANOCARRIERS

Rationally designed bioaddressins are being explored as a way to overcome the physiological barriers (Ranney 2000). The knowledge of the addresses in the body can be used as a tool for developing "biologically" targeted DDS with sequential bioaddressins complementary to the discontinuous compartments that the carrier will find in its road. Successful targeting to deep tissues and controlled release in vivo require that the transport of the carrier along the different biological compartments has to be adapted to the particular characteristics of each one. This approach, which is termed *discontinuum pharmaceutics*, regards vascular endothelium as the "state address," the

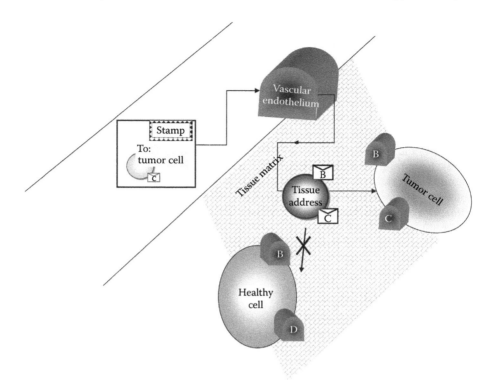

FIGURE 4.1   Drug targeting seen as a postal delivery to the biological addressees. The vascular endothelium is regarded as the "state address," the tissue matrix as the "city street," and the target cell as the "home address."

tissue matrix as the "city street," and the target cell as the "home address," i.e., the street number and floor (Ranney 1986) (Figure 4.1). Anyway, the first problem to be solved is to escape from blood in order to reach the endothelium.

Passive targeting has been mainly achieved by coating the carriers with a hydrophilic layer that generates a water shell at their surface. The nonspecific protein adsorption that is characteristic of hydrophobic materials favors the adsorption of cells on the surface of the foreign-body materials. To overcome this problem, several biomimetic approaches are being tested to regulate cell-surface signaling for suppression of protein adsorption or for selective adhesion of cells (Reintjes et al. 2008). The nonspecific adhesions between tissues and organs are prevented by regions of oligosaccharides that form a dense and confluent hydrophilic layer (glycocalyx) (Holland et al. 1998). Mimicking the glycocalyx properties on particle surface helps to escape it from blood cells. PEG is one of the most used polymers to obtain stealth nanoparticles and liposomes of size below 150 nm. Coating of nanoparticles with copolymers containing PEG blocks and hydrophobic blocks (such as poly(propylene oxide) [PPO] or lipophilic alkyl methacrylates) has attracted much attention in the last two decades (Moghimi and Hunter 2000; Owens and Peppas 2006). The hydrophobic blocks of Pluronic® serve as anchorage points to the nanoparticle surface, while the PEG blocks make the particle surface non-charged and less hydrophobic, providing a mobile stealthy shell that hinders the recognition by RES (Amiji and Park 1992; Göppert and Müller 2005). The structurally related X-shaped copolymers of Tetronic® family possess a central ethylenediamine group that contributes to a more efficient coating and makes the thickness of the coating sensitive to the conditions of the medium (Stolnik et al. 1994; Alvarez-Lorenzo et al. 2010). Immobilization of copolymers on particle surface remarkably modifies the circulation time and the biodistribution when the length of the PEG chains exceeds a minimum. The stealth effect of the coating layer is mainly determined by the length of the PEG chains and the hydrophilic/hydrophobic blocks ratio. In general, the longer the PEG, the thicker the coating is. A thickness of 4 nm has been established as the critical threshold for steric stabilization against phagocytic uptake of large particles (Tröster et al. 1992; Rudt et al. 1993). Interestingly, it has been shown that the coating has not necessarily to occur before administration, but it may also happen in vivo if a Tetronic solution is injected some time before the nanoparticles (Moghimi et al. 2003).

Prolonged circulation time increases the number of passes through target microvessels and, thus, the likelihood of passive extravasation across permeable microvessels. Enhanced permeability and retention (EPR) effect occurs in tumors, granulomas, inflammatory lesions, and infectious sites (Maeda et al. 1999). The ideal size for EPR uptake is 15–20 kDa. EPR does offer access to superficial lesions, but most carriers cannot go deep into the tissue. Furthermore, passive targeting does not enable the carrier to rapidly reach the tissue and, consequently, high drug local concentrations as required in the case of tumors or infections are hard to achieve (Ranney 2000). This causes initial tumor regressions are usually followed by regrowth after a certain time. A way to overcome these limitations is to inject a bolus of free drug followed by a second injection of the drug-loaded nanocarrier. On the other hand, prolonged circulation times (up to 100 h) may result in the uptake at nontarget tissues, leading to toxicity.

Active targeting may be achieved by providing the carrier with binding moieties complementary to either constitutive or induced microvascular surface receptors ("state address") (Ranney 2000). The pharmacokinetics of the uptake of tumor-targeting nanosized DDS has not yet been well tackled (Pirollo and Chang 2008). The use of carriers with antibodies only addressed to the final tumor cells, not to the local endothelial, has a limited success in vivo. This is due to that these carriers initially behave as passive carriers depending on the EPR effect and, once they reach the target tissue, their surface is too specific for recognition of mutated cells. The active recognition of endothelial components mainly depends on (a) the efficiency of the coating of the carrier with binding moieties and stealth components; (b) the capability of the binding moiety to visualize and to interact with the endothelial receptors in vivo, which depends on the layer of camouflaging mucous and coagulant proteins that overcoat endothelial protein targets; and (c) the capability of the carrier to effectively induce extravasation, which is mainly afforded through a multivalent binding surface with individual epitopes of only moderate binding affinity. Elucidation of endothelial integrins and adhesins that correspond to different normal organs (Ruoslahti 1996) and to pathological lesions (Catalina et al. 1999; Binder and Trepel 2009) is essential for a rational design of biomimetics that actively target drug nanocarriers.

The following "state-addressing" substances have been already tested for improving the interaction of nanocarriers with the endothelium (neovascular or organ-selective endothelial determinants):

1. Antibodies (Huang et al. 1997; Byrne et al. 2008)
2. Native and modified tissue pathway factor inhibitors (Hansen et al. 1997; Constantini et al. 1998)
3. RGD phage-display peptides (Ruoslahti 1996; Hui et al. 2008)
4. Sulfated glycosaminoglycans (Ranney 1997; Kovensky 2009)
5. The tetrasaccharides sialyl Lewis A and X, Lewis Y, their synthetic analogs, which are related to E-selectin (ELAM-1) binding (Ball et al. 1992; Pattillo et al. 2009; Theoharis et al. 2009), and some peptidomimetics (Staquicini et al. 2009)

The intense research in this field indicates that carriers with high affinity epitopes (e.g., $K_d > 10^7$), such as antibodies, lectins, and phage-display peptides, are internalized by endothelium but not well released from the abluminal surface, whereas carbohydrate carriers with multiple and more widely spaced epitopes of lower affinity (e.g., $K_d \sim 10^4$) are internalized, transported, and favorably diffused into deep lesional sites. Therefore, leukocytes, bacteria, viruses, and proteases, with their moderately selective but redundant adhesins, selectins, integrins, growth factors, chemokines, and cell-surface carbohydrates, are suitable models for developing biomimetic target drug nanocarriers. Cytolytic adenoviruses and proenzymatic *Salmonella* spp. are being explored as systems capable of active permeation into tumor matrices, preferential replication/proliferation in tumor vs. normal tissues, and propagation throughout the tumor tissue, destroying it and stopping at the tumor boundary (Heise et al. 1997; Bermudes et al. 2002; Chu et al. 2004). Despite promising results already achieved with these biological transporters, important safety concerns indicate that improved knowledge about their in vivo behavior is still required.

Glycosaminoglycan dermatan sulfate can act as a biomimetic polymer capable of adhesion to locally up-regulated vascular adhesins (such as hyaluronan) as leukocytes do. Glycosaminoglycan polymers (ca. 18 kDa) and nanoparticles coated with glycosaminoglycan (50–110 nm in diameter) appear to mimic the local bioadhesion, rolling and stopping of activated leukocytes (Ranney 2000). Covalent or ionic conjugation of drugs to dermatan results in particles with negative zeta potential (–25 to –45 mV) which repels the normal endothelia. The multivalent surface of dermatan-based particles selectively bioadhere to the multiply induced receptors of tumor neovascular endothelium and rapidly activate the transport of the conjugated drug into tumor cells. Therefore, dermatan systems can effectively cross the three key barriers: tumor endothelium ("state address"), the tissue matrix ("city address"), and the tumor cell membrane ("home address"). The preferred receptors at the endothelium are the vascular endothelial growth factor (VEGF), hyaluronan and chondroitin sulfate-like molecules, selectins, and one or more of the induced coagulation factors, including factor VIII antigen (von Willebrand) (Ranney et al. 2005).

Conjugation of nanocarriers with several types of ligands for recognition of overexpressed cell surface receptors can provide more selective cell ("home address") uptake, leading to efficient disease-induced drug targeting (Meijer et al. 2001). Since most receptors are expressed not only on target cells but also on other cells in the body, monoligand-conjugated nanocarriers are also unintendedly taken up by off-target cells resulting in untoward effects (Ghaghada et al. 2005). The dual-ligand approach is based on the fact that tumor cells typically overexpress multiple types of surface receptors. This approach was tested in the human KB cell line, which overexpresses both the folate receptor (FR) and the epidermal growth factor receptor (EGFR). Liposomal nanocarriers loaded with doxorubicin and bearing both folic acid and a monoclonal antibody against the EGFR were designed. Cytotoxicity data demonstrates that nanocarriers can be designed to achieve toxicity only when all targeted receptors are available, providing an approach to improve selectivity over single-ligand approaches (Saul et al. 2006). Mimicking the way that virus use to recognize and to be internalized in cells, polyethylenimine (PEI)-based polyplexes (YC25–PEI–CP9) with two peptides recognizing FGF receptors (peptide YC25) and integrins (peptide CP9) were designed to improve the transfection efficiency. YC25–PEI–CP9 showed a markedly higher transgene efficiency in cell lines that express FGF receptors and integrins, compared with single-peptide-modified PEI or unmodified PEI. In vivo studies with tumor-bearing nude mice demonstrated that the dual-targeting vectors transfect tumor cells possessing FGF receptors and integrins more efficiently (Li et al. 2007). PEI polyplexes conjugated to transferrin (Tf) and transforming growth factor alpha (TGFα) exhibited higher transfection efficiency in A549 cells, owing to receptor-mediated endocytosis, when compared to that by single Tf- or TGFα-introduced polyplexes. No enhancement was observed in CHO-K1 cells. In the presence of excess free Tf or TGFα, the internalization efficiency of the dual-ligand polyplex was strongly inhibited only in A549, but not in CHO-K1. These findings indicate that dual ligands enable cell-selective transfection with enhanced efficiency (Kakimoto et al. 2007).

4.2.2 Biomimicking Shape and Texture

Although less studied, the shape, the mechanical properties, and the surface texture of drug carriers may strongly influence many vital interactions with the body, including phagocytosis (Champion and Mitragotri 2006), circulation (Geng et al. 2007), compartmentalization (Roh et al. 2005), and targeting and adhesion (Tao and Desai 2005). There are many examples in nature where the physical attributes are crucial to biological function (Schulte et al. 2009). For example, macrophages must be able to recognize the enormous diversity in shapes of bacteria (Champion et al. 2008). Platelets and erythrocytes also possess unique shapes and mechanical properties that make their specific functions possible. The mechanical properties of the cellular microenvironment also play an important role in morphogenesis of tissues such as bone, cartilage, and cornea (Engler et al. 2004; Rehfeldt et al. 2007). Macrophages are unable to phagocytose particles that are softer than them (Beningo and Wang 2002). Cells use compartmentalization to control various biochemical reactions in space and time (Nakayama 2004). An excellent review about how the physical properties of the materials determine the biological response has been recently published (Mitragotri and Lahann 2009). The information compiled indicates that precise tailoring of the physical properties of a DDS may notably enhance its performance. Soft and flexible particles are expected to minimize phagocytosis and to circulate for prolonged periods in the blood (Geng et al. 2007). Circulation times, extravasation, targeting, immunogenic response, internalization, intracellular trafficking, degradation, flow properties, clearance, and uptake mechanisms are strongly dependent on the size of the drug carrier. For example, particles in the range of bacteria sizes (>500 nm, preferentially 2–3 μm) can be phagocytosed by macrophages and rapidly cleared from the body (Champion et al. 2008). More desirable for localized delivery is the uptake of smaller particles by non-phagocytic cells. Furthermore, nonspherical drug carriers (mainly oblate or cylindrical) may be preferable to tune phagocytosis and to enhance vascular targeting (Champion and Mitragotri 2006; Decuzzi and Ferrari 2006, 2008; Mitragotri 2009) (Figure 4.2).

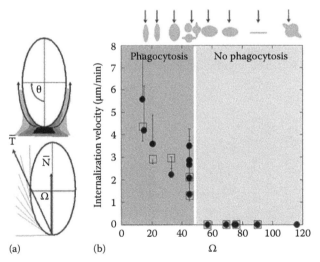

FIGURE 4.2 Definition of Ω and its relationship with internalization velocity (i.e., total distance traveled by macrophage membrane to complete phagocytosis divided by the time involved in the process). (a) A schematic diagram illustrating how membrane progresses tangentially around an elliptical disk. T represents the average of tangential angles from $\theta = 0$ to $\theta = \pi/2$. Ω is the angle between T and membrane normal at the site of attachment, N. (b) Membrane velocity (distance traveled by the membrane divided by time to internalize, $n \geq 3$; error bars represent standard deviation) decreases with increasing Ω for a variety of shapes and sizes of particles. Non-opsonized particles are indicated by filled circles, and IgG-opsonized particles are indicated by open squares. Each data point represents a different shape, size, or aspect ratio particle. The internalization velocity is positive for $\Omega \leq 45°$ ($P < 0.001$). Above a critical value of Ω, $\approx 45°$, the internalization velocity is 0 ($P < 0.001$) and there is only membrane spreading after particle attachment, not internalization. The arrows above the plot indicate the point of attachment for each shape that corresponds to the value of Ω on the x-axis. (Reprinted from Champion, J. and Mitragotri, S., Role of target geometry in phagocytosis, *Proc. Natl. Acad. Sci. USA*, 103, 4930–4934, Copyright 2006. With permission from National Academy of Sciences, USA.)

Biomimetic adhesives based on the capability of bacteria to attach to the human mucosas with the aid of their fimbriae (lectin-like proteins) could be useful for the development of mucoadhesive DDS (Gabor et al. 1997). *Salmonella*-like nanoparticles obtained by associating *Salmonella enteritidis* flagellin (the main component of flagellar filament) to nanoparticles have been shown able to mimic the *Salmonella* invasion and colonization in the gastrointestinal tract. Once orally administered, control nanoparticles displayed a restricted localization in the mucosa, mainly in the outer layer of the ileum (mucus layer), and a low ability to cross this barrier. In contrast, nanoparticles decorated with the flagellar filaments were found broadly distributed in terms of their adhesion to the ileum mucosa. These nanoparticles were able to overcome the mucosa and to permeate deeply in the tissue, internalizing in the enterocytes and Peyer's patches. This new carrier has a great potential for oral targeting and vaccination strategies (Salman et al. 2005).

Recently, gecko-inspired (geometric-based) adhesives have been developed to provide nanoparticles with enhanced mucoadhesion. The toe pads of geckos and other lizards are covered with millions of tiny branching hairs that enable so close a contact with the substrate that intermolecular forces result in excellent adhesion, but are still reversible (Barnes 2007). Reusable tapes that adhere equally well in wet and dry conditions combine the microstructure of gecko pads with the performance of the protein glue of mussels (Lee et al. 2007). Macromolecules can be conjugated to synthetic carriers using the mussel adhesive protein (Lee et al. 2009). A gecko-inspired adhesive based on poly(glycerol sebacate acrylate) (PGSA) with regulable elastic and biodegradable properties has been developed for specific tissue application and for being doped with growth factors or drugs (Mahdavi et al. 2008). This tape-based tissue adhesive platform is claimed to have application in medical therapies ranging from suture/staple replacements, waterproof sealants for hollow organ anatomoses, mesh grafts to treat hernias, ulcers, and burns, and hemostatic wound dressings. On the other hand, silicon nanowire-coated beads can be designed to generate strong bioadhesive forces based on geometric features alone (similar to the interlocking interaction of the Velcro hook-and-loop fasteners). The nanowires were prepared to perfectly fit the space between the microville of Caco-2 cells (Figure 4.3). The interdigitation of the nanowires and the microvilli led to enormous

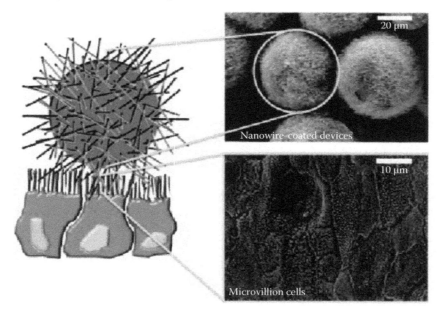

FIGURE 4.3 Beads coated with silicon nanowires that match the space between the microvilli of Caco-2 cells and that show improved bioadhesive properties. (Reprinted with permission from Fischer, K.E., Aleman, B.J., Tao, S.L., et al., Biomimetic nanowire coatings for next generation adhesive drug delivery systems, Nano Lett., 9, 716–720. Copyright 2009 American Chemical Society.)

area of contact between the nanoparticles and the cells, and notably prolonged the retention time of the nanoparticles on the cell culture under flow of mucous (Fischer et al. 2009).

Development of advanced technologies that enable the change of one physical property, while keeping constant all others, should provide valuable information for the optimization of the physical properties of the DDS as a function of the destination of its "cargo" in the body.

4.3 BIOMIMETICS AT THE BULK

The role of a DDS does not finish when it arrives to the adequate body site, as it has to correctly release the therapeutic agent. Both premature release due to leakage en route and too slow release at the target site should be avoided. The triggerable carriers seem to be very suitable for these latter tasks because they could rapidly release the therapeutic agent upon interaction with cell receptors (Kiser et al. 1998) or upon exposure to external or local stimuli (Yoshida et al. 1993; Qiu and Park 2002; Bromberg 2003; Elvira et al. 2005). The DDS that modulate drug release as function of specific stimuli are called "intelligent" or "smart" and can work in an open or closed circuit (Kost and Langer 2001; Sershen and West 2002; Traitel et al. 2008) (Figure 4.4). Open-loop systems release the drug as pulses that are triggered by a specific external stimulus (e.g., applications of ultrasounds or light), disregarding the conditions of the biological environment. Closed-loop or self-regulated systems directly detect certain changes that take place in the biological medium (e.g., in the pH, the temperature, or the concentration of some substances) activating or modulating the response, i.e., switching on/off drug release or automatically adjusting the release rate. Thereby, in these latter systems, a biological variable directly regulates the delivery process (Heller 1997; Ju et al. 2009; Kryscio and Peppas 2009). The development of intelligent DDS demands materials able to react to

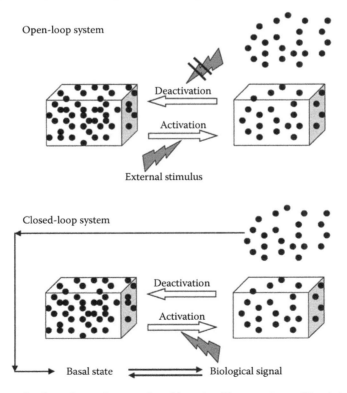

FIGURE 4.4 Schematic view of open-loop or closed-loop intelligent systems. (Reprinted with permission from Alvarez Lorenzo, C. and Concheiro, A., Intelligent drug delivery systems: Polymeric micelles and hydrogels, *Mini-Rev. Med. Chem.*, 8, 1065–1074. Copyright 2008 by Bentham Publishers.)

the stimuli in a predictable, reproducible, proportional to the intensity of the signal, and reversible way (Alvarez-Lorenzo and Concheiro 2008). These approaches are discussed later.

4.3.1 Self-Assembling Peptides and Copolymers

Proteins can serve as models for synthesizing biomimetic materials or be useful themselves as components of drug nanocarriers (Klok 2002), because they offer three unique advantages: molecular recognition, self-assembly, and suitability for genetic manipulation (Sarikaya et al. 2004). Remarkable advances in biotechnology and genomics make the incorporation of peptide sequences responsible for certain functions into drug nanocarriers possible.

A self-assembling peptide or copolymer can be regarded as the simplest brick to construct a system that behaves biomimetic as a whole. One portion of the molecule (the hydrophobic one) directs the assembly with similar molecules into a larger functional structure, while the other (the hydrophilic one) is responsible for an adequate dealing with the biological environment. The design of complex functional structures (e.g., micelles, vesicles, tubules) starting from the assembling of groups of molecularly interlocking parts (modular elements) fits into a framework known as bottom-up design (Shimomura and Sawadaishi 2001; Ossipov et al. 2009). Nature uses the bottom-up design to achieve new functionalities, to control processes with a minimum of resources, and to economize the synthesis of molecules. Proteins are coupled into quaternary structures to perform certain functions. Antibodies perform as dimmers, leucine as zippers, collagen as trimers, and viral capsids as multimeric assemblies. In contrast, dissociation leads to switch the functionality off. Assembly/dissociation depends on subtle conformational changes in certain regions of the natural macromolecules triggered by diverse factors. Trying to emulate Nature, the assembly of synthetic copolymers leads to a local increase in the concentration of functional unimers in an adequate spatial conformation (micelle), which results in features that are not exhibited by the components when isolated (unimers). Namely, compartments to host drug molecules and that possess specific signals at the interface to target the supramolecular entity to certain body sites become feasible (Tu and Tirrell 2004; Branco and Schneider 2009). The relationships between the structure of a copolymer and the function and the toxicity of the supramolecular structure are still only known to a certain extent. Nevertheless, the research about the forces that drive the assembly and the performance of self-assembling systems is beginning to give the first benefit. Physical and chemical tools for controlling the sequence of synthetic polymers and peptides are already available (Barron and Zuckermann 1999; Tu and Tirrell 2004).

Spontaneous micellization in water of amphiphilic molecules, when they reach a certain concentration (critical micellar concentration [CMC]), makes this approach particularly attractive as a green and easy scale-up method to prepare nanocarriers (Kataoka et al. 2001). Traditionally conceived as small compartments inside which the drug is hosted, micelles lead to apparently greater drug solubility in aqueous medium facilitating the pass through biological membranes (Rangel-Yagui et al. 2005). The synthesis of amphiphilic copolymers with low CMC enables the design of nanometric aggregates that possess a greater thermodynamic and kinetic stability (Kwon 2002). The aggregates usually exhibit a core (hydrophobic)–shell (hydrophilic) conformation leading to polymeric micelles or to vesicle-like structures (polymersomes) made of a palisade of alternating layers of water and amphiphilic copolymers (Letchford and Buró 2007; Onaca et al. 2009). Both polymeric micelles and polymersomes are useful as drug carriers since they can host nonpolar substances in the hydrophobic regions and relatively polar substances in the hydrophilic regions. Upon dilution in the physiological fluids, polymeric micelles disassemble very slowly even when the final concentration is below the CMC, enabling longer residence times in the biological environment. In addition, micelles where the hydrophilic blocks are composed of poly(ethylene oxide) (PEO) are sterically stabilized in the aqueous medium, and opsonization and further uptake by macrophages are less feasible (Moghimi et al. 1993; Barratt 2003).

Blood circulation time and control of drug release rate and site can be tuned through an appropriate choice of the copolymer composition and architecture in order to fulfill therapeutic demands

(Kabanov et al. 2002; DesNoyer and McHugh 2003; Aliabadi et al. 2005; Gaucher et al. 2005; Xu et al. 2005a,b). Passive targeting of drug-loaded micelles to pathological sites with affected and leaky vasculature (tumors, inflammations, and infarcted areas) spontaneously occurs via the EPR effect (Ulbrich and Subr 2004; Mondon et al. 2008). The size of polymeric micelles, similar to that of virus, lipoproteins, and other biological systems of transport, provide polymeric micelles to behave as drug carriers toward the interior of the cells. Furthermore, active targeting can be achieved conjugating certain ligand molecules to the shell (Mahmud et al. 2007; Blanco et al. 2009).

An advanced generation of polymeric micelles is envisioned to enable site- and time-selective drug delivery with an architecture that is both stealth and biomimetic (Tu and Tirrell 2004). To bring this into reality, intense attempts for optimizing the structure of block copolymers are being carried out at three levels: cross-linking of the core, cross-linking of the shell, and functionalization of the shell with ligands (Rosler et al. 2001; Kwon 2002). Cross-linking of the core leads to giant wormlike micelles with stable cores from where drug molecules can be slowly released (Won et al. 1999). Cross-linking of the shell reinforces the fragile structure of micelles and results in hollow cages with a shell that controls drug permeability, in some cases as a function of certain stimuli (Huang et al. 1997; Thurmond et al. 1997; Rodríguez-Hernández et al. 2005; Jiang et al. 2007) (Figure 4.5). The cross-linking can be made sensitive to the concentration of certain molecules of cell cytoplasm promoting a fast delivery once the micelles are uptaken (Kakizawa 2001; Matsumoto et al. 2009).

The targeting element should behave as a pilot capable to drive the micellar carrier toward the place where the drug has to act ("home address"). Various pilot molecules have been designed trying to emulate the structure of the molecules that perform such a function in nature, mainly sugars, peptides, and proteins. This approach involves the synthesis of hybrid structures of at least three building blocks: the polymer, a linker, and the bioactive molecule (that acts as the pilot) (Duncan 2003). The polymer has to be synthesized in such a way that an exquisite control of its molecular weight, sequence of blocks, and end functional groups is achieved. Living polymerization may offer such a control (Hadjichristidis et al. 2006). The bioactive molecule may be a peptide with the same amino acids sequence as a fragment of a native protein, which acts as a binding site for specific cells. Solid-phase peptide synthesis offers highly controllable sequences of amino acids (Sabatino and Papini 2008). For example, the protein transduction domain (PTD) peptide from a HIV-1 TAT protein has been synthesized using this approach. Then the bioactive molecule is conjugated to the block copolymer during the synthesis of the components or in a latter step. In the first case, the peptide synthesized on a solid support is functionalized at the N-terminus with an alkoxyamine to initiate the polymerization of the block copolymer. After synthesis, the bioconjugate block copolymer is cleaved from the resin (Becker et al. 2003). The bioactive molecule can also be attached once the micelle is formed. The block copolymers are allowed to self-assemble at concentrations that render individualized spherical micelles. Then the shell is cross-linked using condensation (Huang et al. 1997) or free radical chemistry (Thurmond et al. 1997) resulting in a hydrogel-like shell. The hydrophylic block is functionalized with a carboxylic acid end group and then attached to the N-terminus

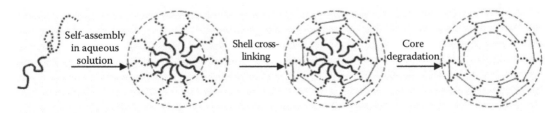

FIGURE 4.5 Preparation of hollow cages via self-assembly of diblock copolymer, stabilization of the shell via cross-linking, and final degradation of the core domain. (Adapted from *Prog. Polym. Sci.*, 30, Rodríguez-Hernández, J., Chécot, F., Gnanou, Y., and Lecommandoux, S., Toward 'smart' nano-objects by self-assembly of block copolymers in solution, *Prog. Polym. Sci.* 30, 691–724. Copyright 2005, with permission from Elsevier.)

of the peptide sequence (Liu et al. 2001). PTD-functionalized micelles selectively interact with CHO and HeLa cells, enabling a site-specific release of small drugs and genes.

It should be noted that the amino acids sequence can be designed in such a way that the peptide itself behaves as a predictable self-assembling structure with a bioactive end. Adequate combinations of hydrophobic (valine, alanine, or leucine), hydrophilic (aspartic acid), and D- and L-amino acids may lead to rodlike micelles, vesicles, hollow tubes, or fibers capable of encapsulating drug molecules or serving as tissue scaffolds (Aggeli et al. 1997; Vauthey et al. 2002). The addition of alkyl tails to the N-terminus of the peptide provides a better control of the architecture of the micelle. Collagen-like peptides containing alkyl tails aggregate resembling a viral envelop that can be loaded with drug molecules and that displays targeting elements (Yu et al. 1998).

The development of synthetic routes for preparing asymmetrically end-functionalized (hetero-bifunctional or heterotelechelic) PEG has opened the way for conjugation at one end of PEG with sugar molecules, in order to target glycoreceptors located on the cell membrane. The other end of PEG can be bound to a block of a biodegradable polymer, such as poly(lactic acid) (PLA), rendering the micelle stealth and biodegradable (Nagasaki et al. 2001). The resultant assemblies are referred as "virus-inspired vehicles." The conjugation with glucose and galactose drives the micelles toward RCA_1 lectins, while the conjugation with lactose leads to specific interaction with RCA_{120} lectins (Toyotama et al. 2001). Lactose-modified poly(ε-caprolactone)–poly(lactobionamidoethyl methacrylate) hybrids (Zhou et al. 2008) and poly(ε-caprolactone)-based polyrotaxanes modified with poly(D-gluconaminoethyl methacrylate) (Dai et al. 2008) render biodegradable and biomimetic micelles and vesicles with specific recognition for lectins.

Micelles made of stimuli-responsive copolymers can specifically release their contents in precise sites at rates finely controlled by an external activator or by self-regulation when the microenvironmental conditions change (Sant et al. 2004; Jones et al. 2006; Shim et al. 2006; Alvarez-Lorenzo et al. 2007, 2009; Rapoport 2007; Alvarez-Lorenzo and Concheiro 2008). Such a gathering of features makes polymeric micelles to be drug carriers capable of modifying drug disposition and of increasing the efficacy and the safety of the treatments (Aliabadi et al. 2008). Comprehensive reviews about the potential of polymeric micelles as components of drug formulations have been recently published, and significant improvements in oral bioavailability of drugs included in polymeric micelles have been reported (Aliabadi et al. 2008; Bromberg 2008; Mondon et al. 2008).

4.3.2 LIPOSOMES AND LIPIDIC BILAYERS

Liposomes are versatile supramolecular structures that resemble the biological membrane in the size scale of biomacromolecules, viruses, and bacteria. Drugs can be hosted in the aqueous compartments or in the hydrophobic bilayers and, as a result, their in vivo pharmacokinetic profile is largely modified. Drug-loaded liposomes are in the pharmaceutical market since more than 10 years, owing to their demonstrated capability to enhance efficiency, safety, and stability of the medicines (Torchilin 2005). Based on liposomes technology, lipidic bilayer deposition onto iron oxide nanoparticles, quantum dots, latex, silica, polystyrene, or even drug particles is an attractive tool for preparing diagnostic and therapeutic agents covered with membrane-like structures that may posses cell receptor function, present antigens to the immunological system, or act as effective antimicrobial agents (Al-Jamal and Kostarelos 2007; Carmona-Ribeiro 2007). The bilayer can overcome the hydrophobicity and the in vivo poor colloidal stability of the nanoparticles and, on the other hand, the nanoparticle enables a narrow size distribution of the whole assembly. The lipid bilayer can resemble the role of the cell membranes regulating the release of substances from the nanoparticle core. Cationic mesoporous silica nanoparticles loaded with calcein rapidly release the drug once placed in CHO cells culture medium, mainly due to a competitive displacement by similar substances. Such premature release hinders the arrival of the drug to the cells. Coating of the nanoparticles with a mix of negatively and positively charged bilayers effectively regulated the delivery (Figure 4.6). The bilayer "protocell" construct has also been shown to reduce premature

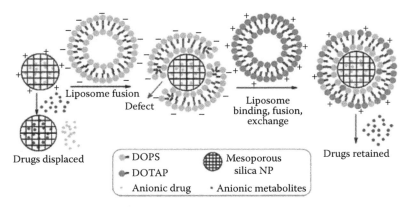

FIGURE 4.6 Lipid bilayers on mesoporous silica nanoparticles can effectively prevent premature drug release, avoiding the competitive displacement of the drug molecules from the nanoparticles by external substances. (Reprinted with permission from Liu, J.W., Jiang, X.M., Ashley, C., and Brinker, C.J., Electrostatically mediated liposome fusion and lipid exchange with a nanoparticle-supported bilayer for control of surface charge, drug containment, and delivery, *J. Am. Chem. Soc.*, 131, 7567–7569. Copyright 2009 American Chemical Society.)

release of cationic drugs (e.g., doxorubicin) from anionic mesoporous silica particles. The protocells are internalized in the cells by endocytosis and once inside, the drug is delivered (Liu et al. 2009).

Antigens are easily adsorbed onto the biomimetic bilayer-covered particles, maintaining the antigenic protein conformation. This results in superior cellular and immune response toward the antigen (Lincopan et al. 2007). Furthermore, the nanoparticles themselves can be not only passive transporters, but also active players in the delivery. Consequently, liposome–nanoparticle hybrids are attracting increasing attention for combining, in a single system, therapeutics and diagnostic agents (Al-Jamal and Kostarelos 2007; Carmona-Ribeiro 2007). For example, superparamagnetic iron oxides encapsulated within phospholipid vesicles render better resolution of magnetic resonance imaging with minor toxicity (Martina et al. 2005). Magnetoliposomes with antigens at the surface and containing drug molecules either in the aqueous or the phospholipidic phase can provide specific targeting and control of the release by applying magnetic forces (Jain et al. 2003). Drug-loaded magnetic liposomes can be designed for selective and preferential presentation to blood monocytes/neutrophils, which result in both drug and magnetite incorporation into these cells. They subsequently become magnetized cells responding to magnetic field. Leukocytes (neutrophils and monocytes) are able to cross the blood–brain barrier in healthy conditions and cause the breakdown of the barrier following brain inflammation. The magnetic leukocytes can then be guided in vivo to the target site, i.e., brain, by applying an external magnetic field of appropriate strength. Negatively charged liposomes and those surface-functionalized with the peptide domain Arg-Gly-Asp (RGD) of the integrin molecule are rapidly taken up by blood monocytes/neutrophils (Senior et al. 1992). Localized delivery of therapeutic substances to inflammatory sites (rich in integrin molecules or RGD domain) has been achieved using these cells as delivery vehicle (Kao et al. 2002). With the aim of achieving noninvasive delivery to the brain, RGD peptide was covalently coupled to negatively charged liposomes composed of Soya lecithin, cholesterol, phosphatidyl serine, and phosphatidyl ethanolamine. In vivo cell sorting studies evidenced a significant increase in relative neutrophils and monocytes count, in blood samples collected from animals treated with uncoated and RGD-coated magnetic liposomal formulations. The cells accumulated at the site of magnet field application. In case of RGD-coated magnetic liposomes loaded with diclofenac, brain levels of the drug were 9.1-fold compared to free drug solution, 6.62-fold compared to nonmagnetic RGD-coated liposomes, and 1.5-fold compared to uncoated magnetic liposomes. Liver uptake was significantly bypassed using RGD-coated magnetic liposomes. Thus, combination of biomimetic nanocarriers and functional cells that can be externally displaced toward certain tissues opens a novel way of understanding active targeting (Jain et al. 2003).

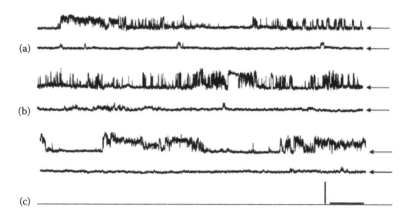

FIGURE 4.7 Continuous 10 s patch-clamp recordings of the gating of the bis-gA channels incorporated in the lipid coating of microcapsules by (a) MBC (10 mM), (b) MTOP (50 mM), and (c) GdCl$_3$ (20 mM). The upper trace in all recordings corresponds to the control condition, where the upward deflections indicate the ion current permeating across the wall of the capsules to the opening of the bis-gA channels. The lower trace was recorded 2 min after addition of the drug to the medium where the microcapsules were immersed. All three drugs blocked the opening of bis-gA and hence the ion permeation across the capsule wall. The horizontal arrow indicates the nonconducting condition of the bis-gA channels where there is zero current recorded. The horizontal scale bar represents 1 s and the vertical scale bar represents 5 pA. (Battle, A.R., Valenzuela, S.M., Mechler, A. et al.: Novel engineered ion channel provides controllable ion permeability for polyelectrolyte microcapsules coated with a lipid membrane. *Adv. Funct. Mater.* 19. 201–208. Copyright 2009 Wiley-VCH Verlag Gmbh & Co. KGaA. Reproduced with permission.)

Magnetoliposomes possessing a cationic surface or antibody conjugation can selectively accumulate in tumor tissues, and act as heat mediators when an alternating magnetic field is applied. The localized hyperthermia results in necrosis and complete tumor regression after multiple exposures to an alternating magnetic field (Yanase et al. 1998; Ito et al. 2005; Kawai et al. 2005). The magnetic field can also regulate drug release rate from temperature-sensitive liposomes (Häfeli 2004). Liposomes and lipidic liquid crystals are also being investigated as responsive to other stimuli for providing *on-demand* drug delivery (Fong et al. 2009; Li and Keller 2009). Polyelectrolyte microcapsules coated with lipidic bilayers that incorporate a functional engineered ion channel protein have been designed in such a way that the bilayer provides stability to the capsule and act as a seal preventing premature delivery. The protein (bis-gA channel) can effectively control the transport of molecules from the microcapsule (Battle et al. 2009). The ability of the lipid-coated microcapsules to respond to a physiologically relevant stimulus has been demonstrated through a pharmacological experiment with drugs that are specific blockers of the bis-gA channels: N,N'-dimethyl-N-(2-[2-(methyl-4-[1,1,3,3-tetramethylbutyl] phenoxy)ethoxy]ethyl)benzylammonium chloride (MBC), GdCl$_3$, and 2-methyl-4-*tert*-octophenol (MTOP). As shown in Figure 4.7, the permeability of the bis-gA channels is blocked by MBC (10 mM), MTOP (50 mM), or GdCl$_3$ (20 mM). This behavior mimics the function of natural cells, whereby an external chemical modulates the gating behavior of ion channels. The gating of incorporated ion channels in the microcapsule system provides a means to control ion composition inside the microcapsule, which can be used as a trigger to release material contained within the microcapsule.

4.3.3 Layer-by-Layer Assemblies

Layer-by-layer (LbL) assembly enables to design the structure of a material at the molecular level in a quite simple and biomimetic way (Tang et al. 2006). The technique consists of successive depositions of monomolecular layers of oppositely charged species, mainly polyelectrolytes, starting from a negatively charged substrate (Figure 4.8). The thickness of each single bilayer (double polyelectrolyte layer) is typically below 1 nm. Thus, the structure and composition of the whole film can be

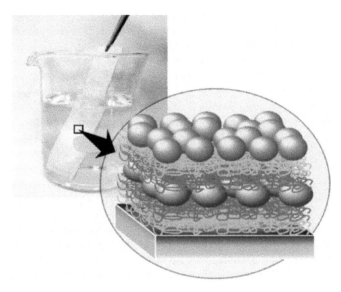

FIGURE 4.8 Preparation and structure of layer-by-layer macromolecular assemblies. (Ariga, K., Hill, J.P., and Ji, Q.M.: Biomaterials and biofunctionality in layered macromolecular assemblies. *Macromol. Biosci.* 8. 981–990. Copyright 2008 Wiley-VCH Verlag Gmbh & Co. KGaA. Reproduced with permission.)

controlled at the nanometer scale. Compared to other techniques to prepare nanostructured films, such as the self-assembled monolayers or the Langmuir–Blodgett deposition, LbL allows for higher loading of therapeutic molecules and leads to more stable films. Versatile combination of interaction forces (i.e., electrostatic, hydrogen bonding, biological recognition, or hydrophobic interactions) and of materials (e.g., polymers, peptides, and nanoparticles) makes LbL an attractive tool to prepare biomimetic DDS (Ariga et al. 2008).

LbL has been intensively evaluated to prepare Nature-inspired superstrong materials (Tang et al. 2003; Luz and Mano 2009), superhydrophobic surfaces (Zhai et al. 2004; Jisr et al. 2005; Sangribsub et al. 2005; Koch and Barthlott 2009), artificial photosynthesis membranes (He et al. 1999; Jussila et al. 2002; Dementiev et al. 2005), biomotors (Jaber et al. 2003), bioreactors (Zu et al. 1999; De Lacey et al. 2000; Xu et al. 2005a,b), or biosensors (Johnson et al. 2009). In the field of drug delivery, LbL allows preparing highly versatile materials with tunable properties owing to the control of the order, location, and concentration of the components with nanometer-scale precision (Bertrand et al. 2000; Tang et al. 2006; Ariga et al. 2008). The release of the therapeutic substances trapped in the layers depends on the permeability or the breakdown of the multilayer structure, and can be programmed to occur when a stimulus disintegrates or triggers the dissolution of the assembly. For example, LbL thin films consisting of PEI and PAA adsorbed a positively charged protein at pH 7.3, where the outermost layer of the film was negatively charged. Complete desorption was obtained at pH 4, where the external layer was neutral (Mueller et al. 2004). pH-dependent degradable LbL films obtained by alternate deposition of a poly(β-amino ester) and polysaccharides have shown pH-induced release (Wood et al. 2005). Delamination of the multilayer and subsequently drug release can also be achieved changing the ionic strength of the medium (Sukhishvili and Granick 2000; Dubas et al. 2001; Izumrudov et al. 2005) or applying UV light (Jensen et al. 2004). Reversible drug loading and release has been achieved with temperature-sensitive components that enable the multilayer to swell or shrink as a function of temperature (Nolan et al. 2004; Quinn and Caruso 2004, 2005). Figure 4.9 shows the strong dependence on the temperature of the loading and release kinetics of rhodamine B from multilayer films of poly(N-isopropylacrylamide) (PNIPA) and poly(acrylic acid) (PAA). This allows the rate of impregnation and release to be easily tuned. Importantly, impregnation and release was also shown to be reversible, with films undergoing several cycles with no loss of loading capacity or temperature-sensitive response (Quinn and Caruso 2004).

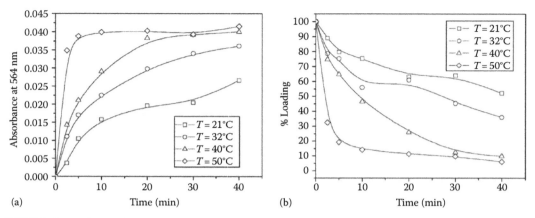

FIGURE 4.9 Rhodamine B impregnation (a) and release (b) profiles at different temperatures for PAA/ PNIPAAm multilayer thin films at pH = 3.0. (Reprinted with permission from Quinn, J.F. and Caruso, F., Facile tailoring of film morphology and release properties using layer-by-layer assembly of thermoresponsive materials, *Langmuir*, 20, 20–22. Copyright 2004 American Chemical Society.)

Biodegradable multilayers consisting of alternating layers of plasmid DNA and a synthetic degradable polyamine have been designed to erode gradually upon incubation in phosphate buffer at 37°C and to sustain the release of incorporated plasmid for several hours. The plasmid was released in a transcriptionally active form and promoted significant levels of gene expression in mammalian cell transfection experiments (Zhang et al. 2004). On the other hand, polymeric prodrugs incorporated into multilayers can deliver the drug at a rate controlled by the hydrolysis of the polymer–drug linkage. Paclitaxel release from such multilayer assembly resulted in rapid death of cancer cells (Thierry et al. 2005).

4.3.4 HYDROGELS

Polymer gels are promising systems for mimicking the behavior of certain macromolecules in the body if their volume phase transitions are adequately designed (Shibayama and Tanaka 1993). Gel collapse can be driven by any one of the four basic types of intermolecular interactions that operate in molecular biological systems (Ilmain et al. 1991); namely, hydrogen bonds, van der Waals, hydrophobic, and Coulomb interactions. According to Flory's theory (Flory 1953), the degree of swelling of a hydrogel is the result of a competition between the entropy due to polymer conformations, which causes rubber elasticity, and the energy associated with internal attractions and repulsions between the monomers themselves and with the solvent. A change in the environmental conditions, such as temperature, pH, or composition of the medium, modifies the balance between the free energy of the internal interactions and the elasticity component, inducing a volume phase transition (Chaterji et al. 2007). The variety of external stimuli that can trigger the phase transition, as well as the present possibilities of modulating the rate and the intensity of the response to the stimulus, guarantee smart gels a wealth of applications, particularly for the controlled release of therapeutic substances (Lin and Metters 2006; Mano 2008; Ju et al. 2009). Furthermore, they can be prepared in a great variety of shapes and sizes and mechanical properties to be adapted to almost all administration routes (Peppas et al. 2000; Raemdonck et al. 2009).

The simplest benefits of pH-responsive micronetworks for oral delivery consist of protecting the drug (mainly peptides or proteins) against the adverse environment of stomach (being collapsed at acid pH) and rapidly releasing it at the gut region (when the hydrogel swells at neutral pH). A correct design of the microhydrogels may even increase the transepithelial paracytosis altering the transmembrane protein networks of the tight junctions. Poly(methacrylic acid)-*grafted* poly(ethylene glycol) [P(MAA-*g*-EG)] hydrogels have been shown to cause a lateral displacement of E-cadherin and caludin-1 enhancing permeation through intestinal mucosa (Fisher and Peppas 2008). Stimuli-responsive networks are also envisioned as advanced DDS capable of using a feedback mechanism to release the drug only where and when is required and to stop the release when back to the normal state. Antipyretics that are only released when the body temperature rises and insulin delivery as a function of glucose concentration are two of the most explored applications. The newest generation of smart DDS tries to mimic the recognition capacity of certain biomacromolecules (e.g., receptors, enzymes, antibodies). The DDS should respond to the concentration of certain molecules, whose presence or absence is responsible for the illness or for the restoring of the health. Advances in this field can lead in a near future to truly intelligent therapeutics (Peppas 2004).

An advanced step in biomimetic systems is not only to imitate the behavior of natural substances, but also to mimic the way that Nature uses to fabricate functional systems (Sarikaya et al. 2003). This is being attempted to be done applying the molecular imprinting technique to the hydrogel synthesis. The unique details of the protein's native state, such as its shape and charge distribution, enable it to recognize and interact with specific molecules. Proteins find their desired conformation out of a nearly infinite number. In contrast, a polymer with a randomly made sequence does not fold in just one way (Tanaka and Annaka 1993; Pande et al. 1997). Therefore, the ability of a polymer (or polymer hydrogel) to always fold back into the same conformation after being stretched and unfolded (i.e., to thermodynamically memorize a conformation) should be related to properly selected or designed nonrandom sequences. To obtain, under proper conditions, synthetic systems with sequences able to adopt conformations with useful functions, the molecular imprinting technology can be applied (Pande et al. 1994a,b, 1995; Wulff 1995). The hydrogels can recognize a substance if they are synthesized in the presence of such a substance (which acts as template) in a conformation that corresponds to the global minimum energy (Figure 4.10). The "memorization" of this conformation, after the swelling of the network and the washing of the template, will only be possible if the network is able to always fold into the conformation upon synthesis that can carry out its designated function (Alvarez-Lorenzo et al. 2000). This revolutionary idea is the basis of new approaches to design imprinted hydrogels and has been developed at different levels, as explained later.

Conventional molecular imprinting technology aims to create tailor-shaped cavities with a high specificity and affinity for a target molecule, inside or at the surface of highly cross-linked polymer networks (Sellergren 2001; Wulff and Biffis 2001; Mayes and Whitcombe 2005; Alvarez-Lorenzo and Concheiro 2006a,b). To do so, the template is added to the monomers and cross-linker solution before polymerization, in order to allow some of the monomers (called "functional") to be arranged in a configuration complementary to the template. The functional monomers are arranged in position through either covalent bonds or non-covalent interactions, such as hydrogen bonds, ionic, hydrophobic, or charge transfer interactions with the template. After synthesis of the network, the bonds are reversibly broken for removal of the template molecules and formation of the imprinted cavities, i.e., pockets of the size and with the most adequate chemical groups to host again the template molecules. Multiple-point interactions between the template molecule and various functional monomers are required for the success of this approach. The non-covalent imprinting protocol allows more versatile combinations of templates and monomers, and provides faster bond association and dissociation kinetics than the covalent imprinting approach (Ansell 2004).

To achieve recognition cavities complementary in shape and functionality with a target molecule, the template should (a) not bear any polymerizable group that could attach itself irreversibly to the polymer network, (b) not interfere in the polymerization process, and (c) be stable at moderately elevated temperatures or upon exposure to UV irradiation. Other key issues are related to the nature and proportion of the monomers and to the synthesis conditions (e.g., solvent or temperature). The cavities should have a structure stable enough to maintain their conformation in the absence of the template and, at the same time, be sufficiently flexible to facilitate the attainment of a fast equilibrium between the release and reuptake of the template in the cavity. Most imprinted systems require around 50%–90% of cross-linker, in order to prevent the polymer network from changing the conformation adopted during synthesis (Sibrian-Vazquez and Spivak 2003). In consequence, the chances to modulate the affinity for the template are very limited,

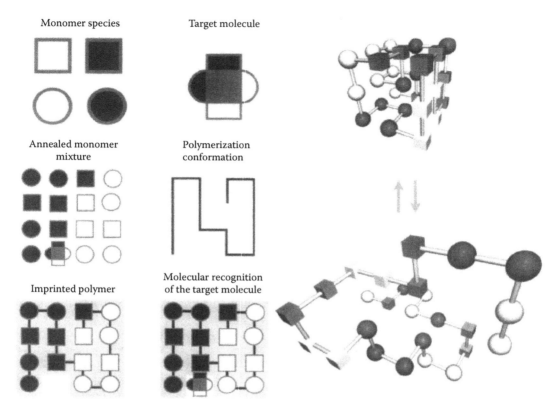

FIGURE 4.10 Design of a heteropolymer using the imprinting approach. Imprinting is depicted in two dimensions for monomers which interact with a template through *p* binding points (there are energetic preferences toward neighbors which have the same shape [square vs. circle] and color [black vs. white]). The target molecule allows different interactions on each side, in this case with the four sides representing all possible monomer species. The target molecule is placed in the presence of monomers prior to polymerization and the "monomer soup" is left to equilibrate, leading to an annealed monomer mixture. Polymerization results in an imprinted polymer with a defined sequence of monomers. The optimization of the monomer arrangement in the monomer soup leads to an imprinted polymer which can renature to the polymerization conformation. Furthermore, the polymerization conformation includes a pocket, or "active site," allowing specific complementary interactions with respect to the target molecule. Thus, imprinted heteropolymers have the protein-like properties of renaturability and specific molecular recognition. On the right hand, the renaturation of an imprinted heteropolymer is shown in three dimensions. (Reprinted with permission from Pande, V.S., Grosberg, A.Y., and Tanaka, T., Phase diagram of heteropolymers with an imprinted conformation, *Macromolecules*, 28, 2218–2227. Copyright 1995 American Chemical Society.)

and it is not foreseeable that the network will have regulatory or switching capabilities. The lack of response to changes in the physical–chemical properties of the medium within the biological range or to the presence of a specific substance notably limits their utility in the biomedical field. Nevertheless, nonresponsive imprinted hydrogels have found attractive application in the field of medicated soft contact lenses. Imprinted cavities in such weakly cross-linked hydrogels notably improved loading and controlled release of ophthalmic drugs, such as timolol, norfloxacin, or ketotifen (Alvarez-Lorenzo et al. 2002, 2006; Hiratani and Alvarez-Lorenzo 2003; Hiratani et al. 2005a,b; Venkatesh et al. 2007).

Stimuli-sensitive imprinted hydrogels are intended to mimic the faculty of biomacromolecules to switch on/off their ability to recognize a certain molecule and to develop a specific function by changing its conformation (Byrne et al. 2002; Miyata et al. 2002; Alvarez-Lorenzo and Concheiro 2004; Cunliffe et al. 2005; Alvarez-Lorenzo and Concheiro 2006a,b). The main difficult to materialize imprinted hydrogel networks able to undergo stimuli-sensitive phase transitions is the regulation of the cross-linking density. It should be low enough to enable conformational changes of the network as a function of environmental stimuli, but high enough to lead to imprinted cavities capable of recognition of the target molecules. Multiple contacts are the key for strong adsorption of the template molecules, because of the larger energy decrease upon adsorption, and also for a high sensitivity due to the greater information provided for recognition. Like in the classical non-covalent approach, the monomers and the template molecules are allowed to settle themselves into a configuration of thermodynamic equilibrium. The monomers are then polymerized in this equilibrium conformation under conditions that render the hydrogel at the collapsed state. As the hydrogel is made from the equilibrium system by freezing the chemical bonds forming the sequence of monomers, the hydrogel should be able to return to its original conformation upon swelling–collapse cycles in which the polymerized sequence remains unchanged. If the memory of the monomer assembly at the template adsorption sites is maintained, truly imprinted hydrogels result. The combination of stimuli-sensitivity and imprinting may have considerable practical advantages: the imprinting provides a high loading capacity of specific molecules, while the ability to respond to stimuli contributes to modulate the affinity of the network for the target molecules, providing regulatory or switching capability of the loading/release processes (Alvarez-Lorenzo and Concheiro 2004, 2008) (Figure 4.11).

Tanaka and coworkers (Alvarez-Lorenzo et al. 2000, 2001; D'Oleo et al. 2001; Moritani and Alvarez-Lorenzo 2001; Stancil et al. 2005) were pioneers in proposing the creation of stimuli-sensitive gels able to recognize and capture target molecules using polymer networks consisting of, at least, two species of monomers, each having a different role. One forms a complex with the template (i.e., the functional or absorbing monomers able to interact with a target molecule), and the other allows the polymers to swell and to shrink reversibly in response to environmental changes (i.e., a smart component such as N-isopropylacrylamide [NIPA]). The gel is synthesized in the collapsed state and, after polymerization, is washed in a swelling medium. The imprinted cavities develop affinity for the template molecules when the functional monomers come into proximity, but when they are separated, the affinity diminishes. The proximity is controlled by the reversible phase transition, which consequently regulates the adsorption/release of the template. A systematic study of the effects of the functional monomer concentration and the cross-linker proportion of the hydrogels (for both imprinted and non-imprinted) and of the ionic strength of the medium on the

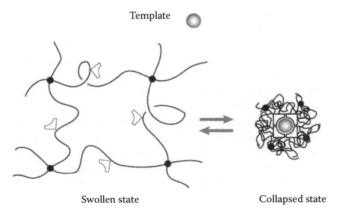

Template

Swollen state   Collapsed state

FIGURE 4.11 **(See companion CD for color figure.)** Diagram of the recognition process of a template by a stimuli sensitive imprinted hydrogel as proposed by Tanaka et al. The volume phase transition of the hydrogel—induced by an external stimuli such as a change in pH, temperature, or electrical field—modifies the relative distance of the functional groups inside the imprinted cavities. This alters their affinity for the template. (Reprinted from *J. Chromatogr. B*, 804, Alvarez-Lorenzo, C. and Concheiro, A., Molecularly imprinted polymers for drug delivery, 231–245. Copyright 2004, with permission from Elsevier.)

affinity of the hydrogels for different templates led to the development of the Tanaka equation (Ito et al. 2003; Alvarez-Lorenzo et al. 2007):

$$\text{Affinity} \approx \frac{[\text{Ad}]^p}{p[\text{Re}]^p} \exp(-p\beta\varepsilon) \exp\left(-(p-1)c\frac{[\text{Xl}]}{[\text{Ad}]^{2/3}}\right) \tag{4.1}$$

where

[Ad] represents the concentration of functional monomers in the gel

[Re] is the concentration of replacement molecules, i.e., ions that are bound to the target molecule when it is not bound to the functional monomers (in the case that they have ionic or protonized groups)

[Xl] is the concentration of cross-linker

p is the number of bonds that each template can establish with the functional monomers

β is the Boltzmann factor ($1/k_B T$)

ε is the difference between the binding energy of an adsorbing monomer to the target molecule and that of a replacement molecule to the target molecule

c is a constant that can be estimated from the persistence length and concentration of the main component of the gel chains (e.g., NIPA)

The main assumption in Equation 4.1 is that the adsorption of target molecules is dominated by one value of p at each state of the gel. The value of p changes from 1 in the swollen state to $p_{\max}$ in the collapsed state; $p_{\max}$ being the number of functional monomers that simultaneously interact with the target molecule.

From a qualitative and very simplified approach, each functional monomer is assumed to be at the end of one effective polymer chain made of n links, where n is inversely proportional to the cross-linking density of the gel. Adsorption of target molecules in the hydrogel deforms the chain network, and the accompanying entropy loss can be analyzed via the entropy of Gaussian chains. This entropic effect is affected not only by the cross-linker density, but also by the density of functional monomers. Note that this density can be adjusted via the gel swelling phase transition. In the swollen state the density decreases, and in the collapsed state it raises. In order for the gel to have a low affinity in the swollen state, the functional monomers should have only a weak attraction to the target molecules, i.e., any adsorption should be single-handed ($p = 1$). To have a high gel affinity, adsorption in the collapsed phase should involve as many functional monomers as possible ($p = p_{\max}$).

Since the Tanaka equation is based on the idea that there are many sets of functional monomers that work together locally to form binding sites for the target molecules and that the probability for chain stretching drops off exponentially with distance, it can also explain the greater affinity of imprinted gels compared to the non-imprinted ones. The synthesis in the presence of the target molecules leads to the distribution of the functional monomers in groups of the p members required for the binding of each target molecule. In the imprinted gels, the p members are closely fixed during polymerization in the collapsed state, due to the template. The template is removed from the gel in the swollen state owing to the deformation of the binding sites. Once again in the collapsed state, the binding sites can be reconstructed, and since each binding site possesses the needed p adsorbing monomers close together, the entropic restrictions to the sorption should be minimized (Alvarez-Lorenzo et al. 2007). However, the design of such hydrogels still requires a significant amount of research, since it is difficult to know a priori the exact value for the binding energy that enables a fine-tuning of the affinity.

According to developments in the statistical mechanics of polymers, to achieve the memory of conformation by flexible polymer chains several requisites must be satisfied (Pande et al. 2000). (1) The polymer must be a heteropolymer, i.e., there should be more than one monomer species, so that some conformations are energetically more favorable than others. (2) There must be frustrations which hinder a typical polymer sequence from being able to freeze to its lowest energy conformation. Such frustrations may be due to the interplay of chain connectivity and excluded volume or may be created by cross-links. For example, a cross-linked polymer chain will not freeze into the same conformation as the uncross-linked polymer chain, at least for most polymer sequences. (3) The sequence of monomers must be selected as to minimize these frustrations (Bryngelson and Wolynes 1987), i.e., a particular polymer sequence should be designed such that the frustrating constraints do not hinder the polymer from reaching its lowest energy conformation. These three conditions allow the polymer to have a global free energy minimum at one designed conformation. Conventional stimuli-responsive hydrogels satisfy the first two conditions, and the application of the molecular imprinting technology enables to satisfy the third one (Alvarez-Lorenzo et al. 2000). The choice of functional monomers and the achievement of an adequate spatial arrangement of functional groups are two of the main factors responsible for specificity and reversibility of molecular recognition. In general, stimuli-sensitive imprinted gels are very weakly cross-linked (less than 2 mol%) and, therefore, the success of the imprinting strongly depends on the stability of the complexes of template/functional monomers during polymerization and after the swelling of the gels. If the molar ratio in the complex is not appropriate or if the complex dissociates to some extent during polymerization, the functional monomers will be far apart from both the template and each other, and the imprinting will be thwarted. Examples of DDS based on molecularly imprinted polymers (MIPs) have been reported for the three main approaches developed to control the moment and the rate of drug release: rate-programmed, activation-modulated, or feedback-regulated drug delivery. Comprehensive reviews about the state of the art of imprinted DDS can be found elsewhere (Alvarez-Lorenzo and Concheiro 2004, 2006a,b; Bayer and Peppas 2008; Byrne and Salian 2008; Kryscio and Peppas 2009).

Molecular imprinting technology is an adequate tool to develop systems with receptors for glucose, which may be the first step in preparing feedback-regulated DDS (Chu 2005). Recognition of glucose moieties through metal coordination (Chen et al. 1997), hydrogen bonding (Mayes et al. 1994; Wizeman and Kofinas 2001; Byrne et al. 2002, 2008; Striegler and Dittel 2003; Oral and Peppas 2004), or interaction with boronic acid (Kataoka et al. 1999) are under investigation. These networks can perform as sensors and actuators undergoing conformational changes proportional to glucose concentration and delivering insulin at a rate proportional to the glucose concentration. Furthermore, they may enable the uptake and the controlled delivery of glycosylated drugs or prodrugs (Byrne et al. 2008).

The synthesis of MIP selective to natural macromolecules, such as peptides and proteins, is still difficult, because a bulky protein cannot easily move in and out through the mesh of a polymer network. The attempts to overcome these limitations have been focused on synthesizing macroporous MIP (Guo et al. 2005; Hawkins et al. 2005; Burova et al. 2011), creating imprinted cavities at the surface of the network (Shnek et al. 1994; Rachkov and Minoura 2001) or developing stimuli-sensitive networks (Demirel et al. 2005; Turan et al. 2009). Imprinted ionic poly(*N-tert*-butylacrylamide-*co*-acrylamide/maleic acid) hydrogels synthesized in the presence of bovine seroalbumin (BSA) showed a remarkably greater affinity for the protein compared to the non-imprinted ones, the adsorption being dependent on both pH and temperature. The hydrogels were synthesized at 22.8°C, at the swollen state. At this temperature, the adsorption is maximal. In contrast, when the gel collapses it is difficult for the protein to diffuse, the imprinted cavities are distorted, and also the nature of the interactions may alter. At low temperature, the interaction between BSA and the hydrogel is based on hydrogen bonds. As the temperature rises, hydrogen bonds become weaker while hydrophobic interactions get stronger (Demirel et al. 2005). These results clearly highlight the relevance of the memorization of the conformation achieved during polymerization for providing the gel with recognition ability for a given template.

Other field where hydrogels are particularly appealing is that of the artificial organs capable of mimicking the production of hormones and growth factors as a healthy body does. Therapeutically active cells can be immobilized within a polymer matrix (commonly alginate) surrounded by a semipermeable membrane that protects the encapsulated cells against immune cell- and antibody-mediated rejection (Lanza et al. 1996; Orive et al. 2003; Chang 2005; Orive et al. 2005; Townsend-Nicholson and Jayasinghe 2006). The cells produce therapeutically active substances that are released through the semipermeable membrane to achieve local or systemic biodistribution (Lörh et al. 2001; De Vos et al. 2006). These "living" DDS act as both drug production and release agents, notably improving the pharmacokinetics of easily degradable peptides and proteins, which often have short half-lives

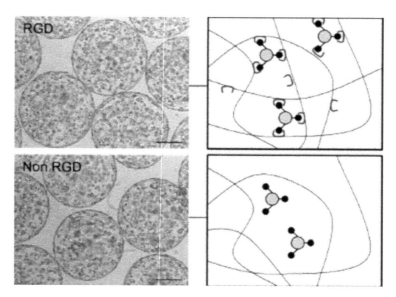

FIGURE 4.12 Biomimetic cell–hydrogel capsules (with peptide arginine–glycine–aspartic acid [RGD] coupled to alginate chains) are mechanically and chemically more resistant and prolong drug release and long-term efficacy of immobilized cells. (Reprinted from *J. Control. Release*, 135, Orive, G., De Castro, M., Kong H.J., et al., Bioactive cell–hydrogel microcapsules for cell-based drug delivery, 135, 203–210. Copyright 2009, with permission from Elsevier.)

in vivo. Biomimetic strategies are being explored for extending the in vivo functionality of the enclosed cells and for improving the mechanical stability of the hydrogel capsules. For example, coupling the adhesion peptide Arg-Gly-Asp (RGD) to alginate polymer chains and using a mixture of alginates of different molecular weights resulted in a cell–hydrogel system capable of sustained delivery of erythropoietin during 300 days without immunosuppressive protocols (Orive et al. 2009) (Figure 4.12).

4.3.5 Cell-Responsive Systems for Tissue Regeneration

Biomimetic systems responsive to cell signals or able to modulate cell behavior are attracting raising interest for improving tissue regeneration (Andreadis and Geer 2006; Little 2009). The release of the therapeutic agents in response to physiological signals could mimic the natural healing process, promoting faster tissue regeneration and reducing scarring (Andreadis and Geer 2006). Although the complex cascade of events that take place during wound healing can be hardly emulated, elastic biomaterials that act as biodegradable scaffolds and release growth factors or genes can, at least partially, cover some events. Importantly, the delivery of growth factors and genes has to be adequately performed to prevent deleterious side effects. Ideally, several growth factors should be released sequentially as occurs in vivo, resulting in wound regeneration rather than repair. In the case of gene therapy, the infiltrating cells should uptake the genes and produce the therapeutic proteins in the wound environment. There are many examples in literature showing the interest of diffusion-controlled and environmental-responsive networks for the sustained delivery of a single growth factor or for adjusting its release rate as a function of the pH, temperature, or ionic strength conditions (Pandit et al. 1998, 2000; Ozeki et al. 2001; Lee et al. 2003; Mellor and Boothman 2003; Obara et al. 2003; Kanematsu et al. 2004). Fewer studies have focused on the delivery of two or more growth factors as occurs during wound healing, cartilage, bone, or nerve regeneration (Desire et al. 2006). Poly(lactide-*co*-glycolide) scaffolds containing VEGF homogeneously dispersed and platelet-derived growth factor (PDGF)-BB inside microspheres have been designed for releasing first VEFG and then PDGF-BB. Such a sequential delivery stimulates the growth of endothelial channels (due to VEGF action) and stabilizes the nascent vessels by recruiting smooth muscle cells (owing to PDGF-BB release) (Richardson et al. 2001). Oligo(poly(ethylene glycol)fumarate) containing transforming growth factor beta-1 (TGF-β1) and gelatin microspheres with insulin-like growth factor 1 (IGF-1) has been shown to promote cartilage repair through a fast release of TGF-β1 followed by a sustained delivery of IGF-1 (Holland et al. 2005). Dual delivery of hepatocyte growth factor (HGF) and fibroblast growth factor 2 (FGF-2) from collagen microspheres enhanced blood vessel formation and produced more mature vasculature in an ischemic mouse hindlimb model at lower doses than either factor alone (Marui et al. 2005) (Figure 4.13). Encapsulation of rabbit marrow mesenchymal stem cells in oligo(poly(ethylene glycol)fumarate) hydrogel composites

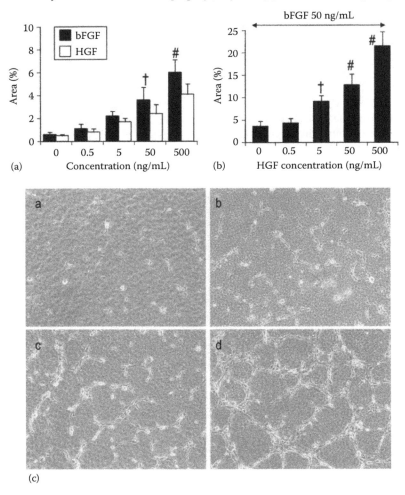

FIGURE 4.13 Capillary network formation assay with human umbilical vein endothelial cells (HUVECs) assessed on Matrigel in the presence of basic fibroblast growth factor (bFGF) or hepatocyte growth factor (HGF), or bFGF and HGF together. (A) Dose–response of capillary network formation area (%) of HUVECs in the presence of bFGF (closed bar) or HGF (open bar). †$P < 0.05$ vs. HGF 50 ng/mL; #$P < 0.01$ vs. HGF 500 ng/mL. (B) Capillary network formation area in the presence of a fixed concentration of bFGF (50 ng/mL) and the indicated concentrations of HGF (0.5–500 ng/mL). †$P < 0.05$, #$P < 0.01$ vs. 0 ng/mL of HGF. (C) Enhancement of capillary network formation of HUVEC in Matrigel by the combination of bFGF and HGF. HUVECs were cultured in Matrigel in the presence of (a) control (no growth factor added); (b) HGF alone (50 ng/mL); (c) bFGF alone (50 ng/mL); or (d) combination of bFGF and HGF (50 ng/mL each). (Reprinted from Marui, A., Kanematsu, A., Yamahara K. et al., Simultaneous application of basic fibroblast growth factor and hepatocyte growth factor to enhance the blood vessels formation, *J. Vasc. Surg.*, 41, 82–90. Copyright 2005 with permission from The Society for Vascular Surgery.)

containing gelatin microparticles loaded with TGF-β1, IGF-1, or both TGF-β1 and IGF-1 revealed that the dual delivery leads to a greater chondrogenic differentiation of the encapsulated cells and thus enables a promising cell therapy to restore cartilage defects. Hydrogels containing micropar- ticles loaded with either TGF-β1 or IGF-1 exhibited 121 ± 20-fold increase of type II collagen gene expression and 71 ± 24-fold increase of aggrecan gene expression after 14 days of in vitro culture, as compared with controls at day 0 (Park et al. 2009). Dual delivery of VEGF and bone morphogenetic protein-2 (BMP-2) for bone regeneration led to similar amounts of bone formation as those achieved by delivery of BMP-2 alone at 12 weeks, but it enhanced bone bridging and union of the critical size defect compared to delivery of BMP-2 alone (Patel et al. 2008).

The third generation of growth factors delivery systems to be explored corresponds to bio- logically inspired DDS that enable cell-triggered drug release and matrix remodeling (Lutolf and Hubbell 2005). These systems use the biochemistry of the wound to promote a cell response or to deliver therapeutic substances upon cellular demand. Namely, they contain biological cues, such as the domains of extracellular matrix molecules, growth factors, or protease substrates, similar to the biological signals that are recognized by the cells that actively repair the injury. Biologically inspired materials are mainly constructed using polymers conjugated to cell recogni- tion domains. For example, hydrogels of polyethylene glycol (PEG) conjugated to VEGF and to the cell adhesion peptide RGD and cross-linked by a peptide sequence, that is recognized by the matrix metalloproteinase 2 (MMP-2), released VEGF upon degradation by MMP-2 secreted by activated endothelial cells (Seliktar et al. 2004). Genetic engineering is being shown particularly useful for immobilizing peptide signals and growth factors into collagen and fibrin hydrogels. Specifically, epithelial growth factor (EGF) was engineered to have high affinity for collagen by fusion of the collagen-binding domain of fibronectin (FNCBD) to its N-terminus (Ishikawa et al. 2001). FNCBD–EGF is a biologically active fusion protein that can stably bind to collagen materials and enhance the effective local concentration of EGF at the site of administration. Application of FNCBD–EGF-containing collagen gels in skin, arterial and hind limb wounds of diabetic mice, accelerated wound closure and improved granulation tissue formation (Ishikawa et al. 2003). Fusion proteins containing the factor XIIIa-recognition domain at the N-terminus were used to successfully deliver nerve growth factor (NGF) and VEGF as a function of cel- lular activity (Zisch et al. 2003). Similarly, the factor XIIIa-recognition domain was covalently attached to keratinocyte growth factor (KGF) and then the complex was conjugated to a fibrin matrix. This enabled a localized delivery at a rate determined by the cell-mediated activation of plasminogen to plasmin, which degraded the fibrin matrix and cleaved the peptides, releasing active KGF to the wound microenvironment (Geer et al. 2005). Fibrin is a particularly attractive scaffold because it is a natural biomaterial that promotes wound healing of epidermal keratino- cytes (Geer et al. 2002; Geer and Andreadis 2003). Fibrin-bound KGF persisted in the wounds for several days and was released gradually as the fibrin matrix was degraded by the infiltrating cells, resulting in significantly enhanced tissue regeneration and wound closure (Geer et al. 2004).

Simultaneous delivery of two genes has also been explored as a way to enhance wound heal- ing. Liposomes containing IGF-1 and KGF (Jeschke and Klein 2004), PGA scaffolds loaded with PDGF-B and VEGF121 (Breitbart et al. 2001, 2003), or fibrin–lipoplex system for topical delivery of multiple genes (Kulkarni et al. 2009) have shown increased rates of wound closure and granula- tion tissue formation, compared with each gene delivered individually.

4.3.6 BIOMIMETIC MOVEMENTS

Movement of drug carriers through the physiological matrix is not easy. Neither the imitation of self-contraction of biological vesicles for pulsate delivery of their contents is. As a way to improve the displacement capability, some researchers are exploring the possibilities of mimicking the movement of living organisms and of certain cells in order to create wireless "swimmers" with a predictable movement and capable of reaching profound tissues (Ghosh and Fischer 2009).

Hydrogels made of PEG and ionic electroactive polymers can be transformed into mini-robots capable of simulating the movement of octopus, myriapod, and spermatozoid under a combination of electric and magnetic signals (Kwon et al. 2008). These hydrogels can also be designed to degrade when a certain substance appears in the medium (e.g., glucose), which prompts the release of the drug. Nanopropellers that combine the oar-like motions of spermatozoa and the cork–screw motion of bacterial flagella have been recently created using functionalized silicon dioxide (Ghosh and Fischer 2009). The potential in therapeutic and diagnostic nanopropellers with fully controlled movement in the three dimensions, using a homogeneous magnetic field, has just begun to be explored.

In addition to externally on–off switchable mobile systems, self-beating materials able to imitate the autonomous oscillation of heartbeat, the brain waves, or the pulsatile secretion of hormones are attracting much attention as unprecedented biomimetic materials with an enormous biomedical potential. Actin- and myosin-based hydrogels can function as artificial muscles in ATP solution (Kakugo et al. 2002). Autonomous swelling–deswelling behavior can be achieved under physiological conditions, as a function of local ATP conversion. ATP can be supplied by the reaction of adenosine diphosphate (ADP) with phosphocreatine in the presence of creatinine kinase (enzyme) and calcium or magnesium ions (coenzyme). This enzymatic reaction is reversible and leads to oscillation of calcium ions concentration in the environment by repeated association with the substrate and dissociation from products. The periodic changes in calcium concentration leads to cyclic changes in the swelling degree of enzyme-immobilized networks made of PNIPA and 2-(methacryloyloxy)ethyl phosphate (Yoshida and Uesusuki 2005) (Figure 4.14). Other hydrogels sensitive to redox reactions similar to those that take place in the Krebs cycle during metabolic process are under study too (Yoshida et al. 2009). In summary, self-oscillating hydrogels may find interest as DDS capable of cyclic release of substances imitating the biorhythms.

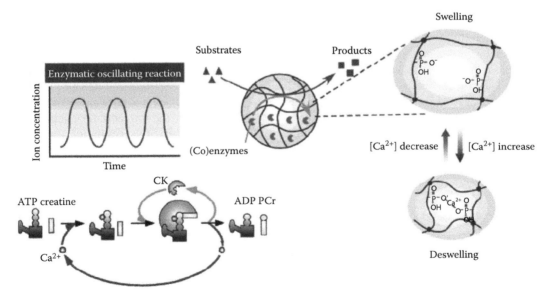

FIGURE 4.14 Mechanism of self-beating of poly(*N*-isopropylacylamide-*co*-2-(methacryloyloxyl)ethyl phosphate) network induced by the reverse enzymatic reaction ATP + creatine → ADP + phosphocreatine (PCr). The ATP-fueled self-oscillating system converts the oscillation in calcium concentration to a beating motion. The periodic changes of calcium concentration are caused by repeated association with substrate and dissociation from products. (Reprinted with permission from Yoshida, R. and Uesusuki, Y., Biomimetic gel exhibiting self-beating motion in ATP solution, *Biomacromolecules*, 6, 2923–2926. Copyright 2005 by American Chemical Society.)

4.3.7 INORGANIC-BASED BIOMIMETIC DDS

Not only organic materials but also inorganic-based composites can greatly benefit of lessons from Nature. There are many examples in which both proteins and inorganics are essential for their assembly and function (Tamerler and Sarikaya 2007). Molecular biomimetics through the selection of inorganic-binding peptides and their tailoring through post-selection engineering have opened the active field of the genetically engineering proteins for inorganics (GEPI). Molecular libraries of peptides are exposed to the material of interest and then eluted for the screening of those peptides that show tighter binding. The information obtained from this first screening can be used for the construction of a second-generation combinatorial library to improve affinity or selectivity (Brown 1997; Whaley et al. 2000; Naik et al. 2002; Thai et al. 2004). Similar to morphogenesis in hard tissues (such as mollusk shells, bones, spicules, and dental tissues) provided by inorganic-binding proteins, GEPI can be used to control the geometrical shapes and sizes of inorganic nanoparticles leading to biomaterials with novel features.

Inorganic biomimetic nanovehicles for drug delivery are prevalently constituted of phosphates or silica, which mimic the composition and structure of bone tissue and silica sponges (Roveri et al. 2008). Hydroxyapatite nanocrystals implanted in bone have been shown, in addition to the implant performance, to be able to release anticancer and antimetastatic drugs locally with controlled release properties (Palazzo et al. 2007). The size, size distribution, and arrangement of nanopores notably determine the amount of drug loaded and the release kinetics while macropores can be utilized for cell growth, the system behaving as active scaffolds (Vallet-Regí 2001; Roveri et al. 2005). Phosphate and silica biomimetic materials are also being designed to release the drug when an external stimulus is applied or an internal chemical factor changes (Liu et al. 2004; Palazzo et al. 2005; Sanchez et al. 2005; Angelos et al. 2007; Slowing et al 2007; Vallet-Regì et al. 2007; Wu et al. 2008). For example, nanoimpeller-controlled mesostructured silica nanoparticles have been recently shown to deliver and to release anticancer drugs into living cells as a function of light pulses (Lu et al. 2008). Experiments carried out with human cancer cell lines showed that once the nanoparticles were taken up by the cells, the anticancer drug camptothecin was only released inside of cells that were illuminated at 413 nm to activate the impellers. The nanoimpellers are azobenzene moieties positioned in the pore interiors with one end attached to the walls and the other end free to undergo photoisomerization. Since *cis*- and *trans*-azobenzene isomers have almost the same extinction coefficient at 413 nm, irradiation at this wavelength causes the azobenzene moieties to move back and forward, driving the drug molecules out of the silica pores. Applying this mechanism, the intracellular release and, consequently, cell apoptosis can be controlled by the light intensity, the irradiation time, and the wavelength. The potential of the fascinating and ambitious stimuli-responsive bioinorganic materials may be enormous both as drug delivery prosthetic materials and as systemic nanovehicles (Roveri et al. 2008; Zhao et al. 2010).

4.4 CONCLUSIONS

Therapy with small organic molecules, peptides, hormones, growth factors, and oligonucleotides has greatly benefited from the rational design of delivery systems adapted to the characteristics of the therapeutic agent, the biological environments through which they have to pass through, and the site of action. Remarkable and even unthinkable improvements in therapy can be achieved if biomimetic approaches are implemented in the design of new carriers. Mimicking the way Nature uses to produce endogenous substances and to deliver them to target places, as well as to generate and to repair specific biological materials and functions, can afford safer and more efficient avant-garde delivery systems. In the last few years, eclosion of biomimetic strategies is leading to highly creative materials and applications, which constitute the first steps of the evolution toward a Nature-guided approach to the delivery of therapeutics and the treatment of illnesses.

ACKNOWLEDGMENTS

This work has been financed by MICINN (SAF2008-01679), FEDER, and Xunta de Galicia (PGIDT07CSA002203PR). F. Yañez is grateful to MEC for a FPI grant.

REFERENCES

Aggeli, A., M. Bell, N. Boden et al. 1997. Responsive gels formed by the spontaneous self-assembly of peptides into polymeric beta-sheet tapes. *Nature* 386:259–262.

Aliabadi, H.M., D.R. Brocks, and A. Lavasanifar. 2005. Polymeric micelles for the solubilization and delivery of cyclosporine A: Pharmacokinetics and biodistribution. *Biomaterials* 26:7251–7259.

Aliabadi, H.M., M. Shahin, D.R. Brocks, and A. Lavasanifar. 2008. Disposition of drugs in block copolymer micelle delivery systems—From discovery to recovery. *Clin. Pharmacokinet.* 47:619–634.

Al-Jamal, W.T. and K. Kostarelos. 2007. Liposome–nanoparticle hybrids for multimodal diagnostic and therapeutic applications. *Nanomedicine—UK* 2:85–98.

Alvarez-Lorenzo, C., L. Bromberg, and A. Concheiro. 2009. Light-sensitive intelligent drug delivery systems. *Photochem. Photobiol.* 85:848–860.

Alvarez-Lorenzo, C. and A. Concheiro. 2004. Molecularly imprinted polymers for drug delivery. *J. Chromatogr. B* 804:231–245.

Alvarez-Lorenzo, C. and A. Concheiro. 2006a. Molecularly imprinted materials as advanced excipients for drug delivery systems. In *Biotechnology Annual Review*, vol. 12, ed. M.R. El-Gewely, pp. 225–264. Amsterdam the Netherlands: Elsevier.

Alvarez-Lorenzo, C. and A. Concheiro. 2006b. Molecularly-imprinted gels, nano- and micro-particles, manufacture and applications. In *Smart Nano and Microparticles*, eds. K. Kono and R. Arshady, pp. 279–336. London U.K.: Kentus Books.

Alvarez-Lorenzo, C. and A. Concheiro. 2008. Intelligent drug delivery systems: Polymeric micelles and hydrogels. *Mini-Rev. Med. Chem.* 8:1065–1074.

Alvarez-Lorenzo, C., A. Concheiro, J. Chuang, and A.Yu. Grosberg. 2007. Imprinting using smart polymers. In *Smart Polymers: Production, Study and Application in Biotechnology and Biomedicine*, eds. I. Galaev and B. Mattiasson, pp. 211–246. Boca Raton, FL: CRC Press.

Alvarez-Lorenzo, C., S. Deshmukh, L. Bromberg et al. 2007. Temperature- and light-responsive blends of Pluronic F127 and poly(N,N-dimethylacrylamide-*co*-methacryloyloxyazobenzene). *Langmuir* 23:11475–11481.

Alvarez-Lorenzo, C., O. Guney, T. Oya et al. 2000. Polymer gels that memorize elements of molecular conformation. *Macromolecules* 33:8693–8697.

Alvarez-Lorenzo, C., O. Guney, T. Oya et al. 2001. Reversible adsorption of calcium ions by imprinted temperature sensitive gels. *J. Chem. Phys.* 114:2812–2816.

Alvarez-Lorenzo, C., H. Hitatani, J.L. Gomez-Amoza, R. Martinez-Pacheco, C. Souto, and A. Concheiro. 2002. Soft contact lenses capable of sustained delivery of timolol. *J. Pharm. Sci.* 91:2182–2192.

Alvarez-Lorenzo, C., A. Rey-Rico, A. Sosnik, P. Taboada, and A. Concheiro. 2010. Poloxamine-based nanomaterials for drug delivery. *Front. Biosci.* E2:424–440.

Alvarez-Lorenzo, C., F. Yanez, R. Barreiro-Iglesias, and A. Concheiro. 2006. Imprinted soft contact lenses as norfloxacin delivery systems. *J. Control. Release* 113:236–244.

Amiji, M. and K. Park. 1992. Prevention of protein adsorption and platelet adhesion on surfaces by PEO/PPO/PEO triblock copolymers. *Biomaterials* 13:682–692.

Andreadis, S.T. and D.J. Geer. 2006. Biomimetic approaches to protein and gene delivery for tissue regeneration. *Trends Biotechnol.* 24:331–337.

Angelos, S., E. Choi, F. Volgtle, L. De Cola, and J.I. Zink. 2007. Photo-driven expulsion of molecules from mesostructured silica nanoparticles. *J. Phys. Chem. C* 111:6589–6592.

Ansell, R.J. 2004. Molecularly imprinted polymers in pseudoimmunoassay. *J. Chromatogr. B* 804:151–165.

Ariga, K., J.P. Hill, and Q.M. Ji. 2008. Biomaterials and biofunctionality in layered macromolecular assemblies. *Macromol. Biosci.* 8:981–990.

Ball, G.E., R.A. O'Neill, J.E. Schultz et al. 1992. Synthesis and structural analysis using 2-D NMR of sialyl Lewis X (Slex) and Lewis X (Lex) oligosaccharides: Ligands related to E-selectin (ELAM-1) binding. *J. Am. Chem. Soc.* 114:5449–5451.

Barnes, W.J.P. 2007. Biomimetic solutions to sticky problems. *Science* 318:203–204.

Barratt, G. 2003. Colloidal drug carriers: Achievements and perspectives. *Cell Mol. Life Sci.* 60:21–37.

Barron, A.E. and R.N. Zuckermann. 1999. Bioinspired polymeric materials: In-between proteins and plastics. *Curr. Opin. Chem. Biol.* 3:681–687.

Battle, A.R., S.M. Valenzuela, A. Mechler et al. 2009. Novel engineered ion channel provides controllable ion permeability for polyelectrolyte microcapsules coated with a lipid membrane. *Adv. Funct. Mater.* 19:201–208.

Bayer, C.L. and N.A. Peppas. 2008. Advances in recognitive, conductive and responsive delivery systems. *J. Control. Release* 132, 216–221.

Becker, M.L., J.Q. Liu, and K.L. Wooley. 2003. Peptide–polymer bioconjugates: Hybrid block copolymers generated via living radical polymerizations from resin-supported peptides. *Chem. Commun.* 180–181.

Beningo, K.A. and Y.L. Wang. 2002. Fc-receptor-mediated phagocytosis is regulated by mechanical properties of the target. *J. Cell Sci.* 115:849–856.

Bermudes, D., L.M. Zheng, and I.C. King. 2002. Live bacteria as anticancer agents and tumor-selective protein delivery vectors. *Curr. Opin. Drug Discov. Devel.* 5:194–199.

Bertrand, P., A. Jonas, A. Laschewsky, and R. Legras. 2000. Ultrathin polymer coatings by complexation of polyelectrolytes at interfaces: Suitable materials, structure and properties. *Macromol. Rapid Commun.* 21:319–348.

Bhushan, B. 2009. Biomimetics: Lessons from nature—An overview. *Phil. Trans. R. Soc. A* 367:1445–1486.

Billingsley, M.L. 2008. Druggable targets and targeted drugs: Enhancing the development of new therapeutics. *Pharmacology* 82:239–244.

Binder, M. and M. Trepel. 2009. Drugs targeting integrins for cancer therapy. *Exp. Opin. Drug Discov.* 4:229–241.

Blanco, E., C.W. Kessinger, B.D. Sumer, and J. Gao. 2009. Multifunctional micellar nanomedicine for cancer therapy. *Exp. Biol. Med.* 234:123–131.

Branco, M.C. and J.P. Schneider. 2009. Self-assembling materials for therapeutic delivery. *Acta Biomater.* 5:817–831.

Breitbart, A.S., D.A. Grande, J. Laser et al. 2001. Treatment of ischemic wounds using cultured dermal fibroblasts transduced retrovirally with *PDGF-B* and *VEGF121* genes. *Ann. Plast. Surg.* 46:555–561.

Breitbart, A.S., J. Laser, B. Parrett et al. 2003. Accelerated diabetic wound healing using cultured dermal fibroblasts retrovirally transduced with the platelet derived growth factor B gene. *Ann. Plast. Surg.* 51:409–414.

Bromberg, L. 2003. Intelligent polyelectrolytes and gels in oral drug delivery. *Curr. Pharm. Biotechnol.* 4:339–349.

Bromberg, L. 2008. Polymeric micelles in oral chemotherapy. *J. Control. Release* 128:99–112.

Brown, S. 1997. Metal recognition by repeating polypeptides. *Nat. Biotechnol.* 15:269–272.

Brudno, Y. and D.R. Liu. 2009. Recent progress toward the templated synthesis and directed evolution of sequence-defined synthetic polymers. *Chem. Biol.* 16:265–276.

Bryngelson, J.D. and P.G. Wolynes. 1987. Spin glasses and the statistical mechanics of protein folding. *Proc. Natl. Acad. Sci. U.S.A.* 2:7524.

Burova, T.V., N.V. Grinberg, E.V. Kalinina et al. 2011. Thermoresponsive copolymer cryogel possessing molecular memory: Synthesis, energetics of collapse and interaction with ligands. *Macromol. Chem. Phys.* 212:72–80.

Byrne, J.D., T. Betancourt, and L. Brannon-Peppas. 2008. Active targeting schemes for nanoparticle systems in cancer therapeutics. *Adv. Drug Del. Rev.* 60:1615–1626.

Byrne, M.E., J.Z. Hilt, and N.A. Peppas. 2008. Recognitive biomimetic networks with moiety imprinting for intelligent drug delivery. *J. Biomed. Mater. Res.* 84A:137–147.

Byrne, M.E., E. Oral, J.Z. Hilt, and N.A. Peppas. 2002. Networks for recognition of biomolecules: Molecular imprinting and micropatterning poly(ethylene glycol)-containing films. *Polym. Adv. Technol.* 13:798–816.

Byrne, M.E., K. Park, and N.A. Peppas. 2002. Molecular imprinting within hydrogels. *Adv. Drug Del. Rev.* 54:149–161.

Byrne, M.E. and V. Salian. 2008. Molecular imprinting within hydrogels II: Progress and analysis of the field. *Int. J. Pharm.* 364:188–212.

Carlmaerk, A., C. Hawker, A. Hult, and M. Malkoch. 2009. New methodologies in the construction of dendritic materials. *Chem. Soc. Rev.* 38:352–362.

Carmona-Ribeiro, A.M. 2007. Biomimetic particles in drug and vaccine delivery. *J. Liposome Res.* 17:165–172.

Catalina, M.D., P. Estess, and M.H. Siegelman. 1999. Selective requirements for leukocyte adhesion molecules in models of acute and chronic cutaneous inflammation: Participation of E- and P- but not L-selectin. *Blood* 93:580–589.

Champion, J. and S. Mitragotri. 2006. Role of target geometry in phagocytosis. *Proc. Natl. Acad. Sci. U.S.A.* 103:4930–4934.

Champion, J., A. Walker, and S. Mitragotri. 2008. Role of particle size in phagocytosis of polymeric microspheres. *Pharm. Res.* 25:1815–1821.

Chang, T.M.S. 2005. Therapeutic applications of polymeric artificial cells. *Nat. Rev. Drug Discov.* 4:221–235.

Chaterji, S., I.K. Kwon, and K. Park. 2007. Smart polymeric gels: Redefining the limits of biomedical devices. *Prog. Polym. Sci.* 32:1083–1122.

Chen, G., Z. Guan, C.T. Chen, L. Fu, V. Sundaresan, and F.H. Arnold. 1997. A glucose-sensing polymer. *Nat. Biotechnol.* 15:354–357.

Chien, Y.W. and S. Lin. 2002. Optimisation of treatment by applying programmable rate-controlled drug delivery technology. *Clin. Pharmacokinet.* 41:1267–1299.

Chu, L.Y. 2005. Controlled release systems for insulin delivery. *Expert Opin. Ther. Pat.* 15:1147–1155.

Chu, R.L., D.E. Post, F.R. Khuri et al. 2004. Use of replicating oncolytic adenoviruses in combination therapy for cancer. *Clin. Cancer Res.* 10:5299–5312.

Constantini, V., P. De Monte, A.O. Cazzato et al. 1998. Systemic thrombin generation in cancer patients is correlated with extrinsic pathway activation. *Blood Coagul. Fibrinolysis* 9:79–84.

Cuchelkar, V. and J. Kopecek. 2006. Polymer–drug conjugates. In *Polymers in Drug Delivery*, eds. I.F. Uchegbu and A.G. Schätzlein, pp. 155–182. Boca Raton, FL: CRC Press LLC.

Cunliffe, D., A. Kirby, and C. Alexander. 2005. Molecularly imprinted drug delivery systems. *Adv. Drug Del. Rev.* 57:1836–1853.

D'Oleo, R., C. Alvarez-Lorenzo, and G. Sun. 2001. A new approach to design imprinted polymer gels without using a template. *Macromolecules* 34:4965–4971.

Dai, X.H., C.M. Dong, and D. Yan. 2008. Supramolecular and biomimetic polypseudorotaxane/glycopolymer biohybrids: Synthesis, glucose-surface nanoparticles, and recognition with lectins. *J. Phys. Chem. B* 112:3644–3652.

De Lacey, A.L., M. Detcheverry, J. Moiroux, and C. Bourdillon. 2000. Construction of multicomponent catalytic films based on avidin–biotin technology for the electroenzymatic oxidation of molecular hydrogen. *Biotechnol. Bioeng.* 68:1–10.

De Vos, P., M.M. Faas, B. Strand, and R. Calafiore. 2006. Alginate-based microcapsules for immunoisolation of pancreatic islets. *Biomaterials* 27:5603–5617.

Decuzzi, P. and M. Ferrari. 2006. The adhesive strength of non-spherical particles mediated by specific interactions. *Biomaterials* 27:5307–5314.

Decuzzi, P. and M. Ferrari. 2008. The receptor-mediated endocytosis of nonspherical particles. *Biophys. J.* 94:3790–3797.

Dementiev, A.A., A.A. Baikov, V.V. Ptushenko, G.B. Khomutov, and A.N. Tikhonov. 2005. Biological and polymeric self-assembled hybrid systems: Structure and properties of thylakoid/polyelectrolyte complexes. *Biochim. Biophys. Acta* 1712:9–16.

Demirel, G., G. Ozcetin, E. Turan, and T. Caykara. 2005. pH/temperature-sensitive imprinted ionic poly(*N-tert*-butylacrylamide-*co*-acrylamide/maleic acid) hydrogels for bovine serum albumin. *Macromol. Biosci.* 5:1032–1037.

Desire, L., E. Mysiakine, D. Bonnafous et al. 2006. Sustained delivery of growth factors from methylidene malonate 2.1.2-based polymers. *Biomaterials* 27:2609–2620.

DesNoyer, J.R. and A.J. McHugh. 2003. The effect of Pluronic on the protein release kinetics of an injectable drug delivery system. *J. Control. Release* 86:15–24.

Dillow, A.K. and A.M. Lowman. 2004. *Biomimetic Materials and Design*. New York: Marcel Dekker.

Dubas, S.T., T.R. Farhat, and J.B. Schlenoff. 2001. Multiple membranes from "true" polyelectrolyte multilayers. *J. Am. Chem. Soc.* 123:5368–5369.

Duncan, R. 2003. The dawning era of polymer therapeutics. *Nat. Rev. Drug Discov.* 2:347–360.

Elvira, C., G.A. Abraham, A. Gallardo, and J. San Roman. 2005. Smart biodegradable hydrogels with applications in drug delivery and tissue engineering. In *Biodegradable Systems in Tissue Engineering and Regenerative Medicine*, eds. R. Reis and J. San Román, pp. 493–508. Boca Raton, FL: CRC Press.

Engler, A., L. Richert, J.Y. Wong, C. Picart, and D. Discher. 2004. Surface probe measurements of the elasticity of sectioned tissue, thin gels and polyelectrolyte multilayer films: Correlations between substrate and cell adhesion. *Surf. Sci.* 570:142–154.

Fischer, K.E., B.J. Aleman, S.L. Tao et al. 2009. Biomimetic nanowire coatings for next generation adhesive drug delivery systems. *Nano Lett.* 9:716–720.

Fisher, O.Z. and N.A. Peppas. 2008. Quantifying tight junction disruption caused by biomimetic pH-sensitive hydrogel drug carriers. *J. Drug Del. Sci. Tech.* 18:47–50.

Flory, P.J. 1953. *Principles of Polymer Chemistry*. New York: Cornell.

Fong, W.K., T. Hanley, and B.J. Boyd. 2009. Stimuli-responsive liquid crystals provide "on-demand" drug delivery in vitro and in vivo. *J. Control. Release* 135:218–226.

Gabor, F., A. Bernkop-Schnurch, and G. Hamilton. 1997. Bioadhesion to the intestine by means of *E. coli* K99-fimbriae: Gastrointestinal stability and specificity of adherence. *Eur. J. Pharm. Sci.* 5:233–242.

Gaucher, G., M.H. Dufresne, V.P. Sant, N. Kang, D. Maysinger, and J.C. Leroux. 2005. Block copolymer micelles: Preparation, characterization and application in drug delivery. *J. Control. Release* 109:169–188.

Geer, D.J. and S.T. Andreadis. 2003. A novel role of fibrin in epidermal healing: Plasminogen-mediated migration and selective detachment of differentiated keratinocytes. *J. Invest. Dermatol.* 121:1210–1216.

Geer, D.J., D.D. Swartz, and S.T. Andreadis. 2002. Fibrin promotes migration in a three-dimensional in vitro model of wound regeneration. *Tissue Eng.* 8:787–798.

Geer, D.J., D.D. Swartz, and S.T. Andreadis. 2004. In vivo model of wound healing based on transplanted tissue-engineered skin. *Tissue Eng.* 10:1006–1017.

Geer, D.J., D.D. Swartz, and S.T. Andreadis. 2005. Biomimetic delivery of keratinocyte growth factor upon cellular demand for accelerated wound healing in vitro and in vivo. *Am. J. Pathol.* 167:1575–1586.

Geng, Y., P. Dalhaimer, S.S. Cai et al. 2007. Shape effects of filaments versus spherical particles in flow and drug delivery. *Nat. Nanotechnol.* 2:249–255.

Ghaghada, K.B., J. Saul, J.V. Natarajan, R.V. Bellamkonda, and A.V. Annapragada. 2005. Folate targeting of drug carriers: A mathematical model. *J. Control. Release* 104:113–128.

Ghosh, A. and P. Fischer. 2009. Controlled propulsion of artificial magnetic nanostructured propellers. *Nano Lett.* 9:2243–2245.

Göppert, T.M. and R.H. Müller. 2005. Protein adsorption patterns on poloxamer- and poloxamine-stabilized solid lipid nanoparticles (SLN). *Eur. J. Pharm. Biopharm.* 60:361–372.

Guan, Z. 2008. Bioinspired supramolecular design in polymers for advanced mechanical properties. In *Molecular Recognition and Polymers: Control of Polymer Structure and Self-assembly*, eds. V. Rotello and S. Thayumanavan, pp. 235–258. Hoboken, NJ: John Wiley & Sons.

Guo, T.Y., Y.Q. Xia, J. Wang, M.D. Song, and B.H. Zhang. 2005. Chitosan beads as molecularly imprinted polymer matrix for selective separation of proteins. *Biomaterials* 26:5737–5745.

Hadjichristidis, N., H. Iatrou, M. Pitsikalis, and J. Mays. 2006. Macromolecular architectures by living and controlled/living polymerizations. *Prog. Polym. Sci.* 31:1068–1132.

Häfeli, U.O. 2004. Magnetically modulated therapeutic systems. *Int. J. Pharm.* 277:19–24.

Hansen, J.B., R. Olsen, and P. Webster. 1997. Association of tissue factor pathway inhibitor with human umbilical vein endothelial cells. *Blood* 90:3567–3578.

Hawkins, D.M., D. Stevenson, and S.M. Reddy. 2005. Investigation of protein imprinting in hydrogel-based molecularly imprinted polymers (HydroMIPs). *Anal. Chim. Acta* 542:61–65.

He, J.E., L. Samuelson, L. Li, J. Kumar, and S.K. Tripathy. 1999. Bacteriorhodopsin thin film assemblies—Immobilization, properties, and applications. *Adv. Mater.* 11:435–446.

Heise, C., A. Sampson-Johannes, A. Williams, F. McCormick, D.D. Von Hoff, and D.H. Kirn. 1997. *ONYX-015*, an E1B gene attenuated adenovirus, causes tumor-specific cytolysis and antitumoral efficacy that can be augmented by standard chemotherapeutic agents. *Nat. Med.* 3:639–645.

Heller, J. 1997. Feedback-controlled drug delivery. *Control. Drug Del.* 127–146.

Hiratani, H. and C. Alvarez-Lorenzo. 2003. The nature of backbone monomers determines the performance of imprinted soft contact lenses as timolol drug delivery systems. *Biomaterials* 25:1105–1113.

Hiratani, H., A. Fujiwara, Y. Tamiya, Y. Mizutani, and C. Alvarez-Lorenzo. 2005a. Ocular release of timolol from molecularly imprinted soft contact lenses. *Biomaterials* 26:1293–1298.

Hiratani, H., Y. Mizutani, and C. Alvarez-Lorenzo. 2005b. Controlling drug release from imprinted hydrogels by modifying the characteristics of the imprinted cavities. *Macromol. Biosci.* 5:728–733.

Hoffman, A.S. 2006. Selecting the right polymer for biomaterial applications. In *Polymers in Drug Delivery*, eds. I.F. Uchegbu and A.G. Schätzlein, pp. 7–22. Boca Raton, FL: CRC Taylor & Francis.

Hoffman, A.S. 2008. The origins and evolution of "controlled" drug delivery systems. *J. Control. Release* 132:153–163.

Holland, N.B., Y. Qiu, M. Ruegsegger, and R.E. Marchant. 1998. Biomimetic engineering of non-adhesive glycocalyx-like surfaces using oligosaccharide surfactant polymers. *Nature* 392:799–801.

Holland, T.A., Y. Tabata, and A.G. Mikos. 2005. Dual growth factor delivery from degradable oligo(poly(ethylene glycol)fumarate) hydrogel scaffolds for cartilage tissue engineering. *J. Control. Release* 101:111–125.

Huang, H.Y., T. Kowalewski, E.E. Remsen, R. Gertzmann, and K.L. Wooley. 1997. Hydrogel-coated glassy nanospheres: A novel method for the synthesis of shell cross-linked knedels. *J. Am. Chem. Soc.* 119:11653–11659.

Huang, X., G. Molema, S. King, L. Watkins, T.S. Edgington, and P.E. Thorpe. 1997. Tumor infarction in mice by antibody-directed targeting of tissue factor to tumor vasculature. *Science* 275:547–550.

Hui, X.L., Y. Han, S.H. Liang et al. 2008. Specific targeting of the vasculature of gastric cancer by a new tumor-homing peptide CGNSNPKSC. *J. Control. Release* 131:86–93.

Ilmain, F., T. Tanaka, and E. Kokufuta. 1991. Volume transition in a gel driven by hydrogen bonding. *Nature* 349, 400–401.

Ishikawa, T., H. Terai, and T. Kitajima. 2001. Production of a biologically active epidermal growth factor fusion protein with high collagen affinity. *J. Biochem.* 129:627–633.

Ishikawa, T., H. Terai, T. Yamamoto, K. Harada, and T. Kitajima. 2003. Delivery of a growth factor fusion protein having collagen-binding activity to wound tissues. *Artif. Organs* 27:147–154.

Ito, K., J. Chuang, C. Alvarez-Lorenzo, T. Watanabe, N. Ando, and A.Yu. Grosberg. 2003. Multiple point adsorption in a heteropolymer gel and the Tanaka approach to imprinting: Experiment and theory. *Prog. Polym. Sci.* 28:1489–1515.

Ito, A., M. Shinkai, H. Honda, and T. Kobayashi. 2005. Medical application of functionalized magnetic nanoparticles. *J. Biosci. Bioeng.* 100:1–11.

Izumrudov, V.A., E. Kharlampieva, and S.A. Sukhishvili. 2005. Multilayers of a globular protein and a weak polyacid: Role of polyacid ionization in growth and decomposition in salt solutions. *Biomacromolecules* 6:1782–1788.

Jaber, J.A., P.B. Chase, and J.B. Schlenoff. 2003. Actomyosin-driven motility on patterned polyelectrolyte mono- and multilayers. *Nano Lett.* 3:1505–1509.

Jain, S., V. Mishra, P. Sing et al. 2003. RGD-anchored magnetic liposomes for monocytes/neutrophils-mediated brain targeting. *Int. J. Pharm.* 261:43–55.

Jang, W.D., K.M.K. Selim, C.H. Lee, and I.K. Kang. 2009. Bioinspired application of dendrimers: From biomimicry to biomedical applications. *Prog. Polym. Sci.* 34:1–23.

Jensen, A.W., N.K. Desai, B.S. Maru, and D.K. Mohanty. 2004. Photohydrolysis of substituted benzyl esters in multilayered polyelectrolyte films. *Macromolecules* 37:4196–4200.

Jeschke, M.G. and D. Klein. 2004. Liposomal gene transfer of multiple genes is more effective than gene transfer of a single gene. *Gene Ther.* 11:847–855.

Jiang, X.Z., Z.S. Ge, J. Xu et al. 2007. Fabrication of multiresponsive shell cross-linked micelles possessing pH-controllable core swellability and thermo-tunable corona permeability. *Biomacromolecules* 8:3184–3192.

Jisr, R.M., H.H. Rmaile, and J.B. Schlenoff. 2005. Hydrophobic and ultrahydrophobic multilayer thin films from perfluorinated polyelectrolytes. *Angew. Chem. Int. Ed.* 44:782–785.

Johnson, E.A.C., R.H.C. Bonser, and G. Jeronimidis. 2009. Recent advances in biomimetic sensing technologies. *Phil. Trans. R. Soc. A* 367:1559–1569.

Jones, M.C., M. Ranger, and J.C. Leroux. 2006. pH-sensitive unimolecular polymeric micelles: Synthesis of a novel drug carrier. *Bioconjug. Chem.* 14:774–781.

Ju, X.J., R. Xie, L. Yang, and L.Y. Chu. 2009. Biodegradable 'intelligent' materials in response to physical stimuli for biomedical applications. *Expert Opin. Ther. Pat.* 19:493–507, 638–696.

Jussila, T., M. Li, N.V. Tkachenko, S. Parkkinen, B. Li, L. Jiang, and H. Lemmetyinen. 2002. Transient absorption and photovoltage study of self-assembled bacteriorhodopsin/polycation multilayer films. *Biosens. Bioelectron.* 17:509–515.

Kabanov, A.V., E.V. Batrakova, and V.Yu. Alakhov. 2002. Pluronic® block copolymers as novel polymer therapeutics for drug and gene delivery. *J. Control. Release* 82:189–212.

Kakimoto, S., T. Moriyama, T. Tanabe, S. Shinkai, and T. Nagasaki. 2007. Dual-ligand effect of transferrin and transforming growth factor alpha on polyethyleneimine-mediated gene delivery. *J. Control. Release* 120:242–249.

Kakizawa, Y. 2001. Glutathione-sensitive stabilization of block copolymer micelles composed of antisense DNA and thiolated poly(ethylene glycol)-*block*-poly(L-lysine): A potential carrier for systemic delivery of antisense DNA. *Biomacromolecules* 2:491–497.

Kakugo, A., S. Sugimoto, J.P. Gong, and Y. Osada. 2002. Gel machines constructed from chemically cross-linked actins and myosins. *Adv. Mater.* 14:1124–1126.

Kanematsu, A., A. Marui, S. Yamamoto et al. 2004. Type I collagen can function as a reservoir of basic fibroblast growth factor. *J. Control. Release* 99:281–292.

Kao, W.J., Y. Liu, R. Gundloori et al. 2002. Engineering endogenous inflammatory cells as delivery vehicle. *J. Control. Release* 78:219–233.

Kataoka, K., A. Harada, and Y. Nagasaki. 2001. Block copolymer micelles for drug delivery: Design, characterization, and biological significance. *Adv. Drug Del. Rev.* 47:113–131.

Kataoka, K., H. Miyazaki, M. Bunya, T. Okano, and Y. Sakurai. 1999. Totally synthetic polymer gels responding to external glucose concentration: Their preparation and application to on–off regulation of insulin release. *J. Am. Chem. Soc.* 120:12694–12695.

Kawai, N., A. Ito, Y. Nakahara et al. 2005. Anticancer effect of hyperthermia on prostate cancer mediated by magnetite cationic liposomes and immune-response induction in transplanted syngeneic rats. *Prostate* 64:373–381.

Keegan, M.E., J.A. Whittum-Hudson, and W.M. Saltzman. 2003. Biomimetic design in microparticulate vaccines. *Biomaterials* 24:4435–4443.

Kim, S., J.H. Kim, O. Jeon, I.C. Kwon, and K. Park. 2009. Engineered polymers for advanced drug delivery. *Eur. J. Pharm. Biopharm.* 71:420–430.

Kiser, P.F., G. Wilson, and D. Needham. 1998. A synthetic mimic of the secretory granule for drug delivery. *Nature* 394:459–461.

Klok, H.A. 2002. Protein-inspired materials: Synthetic concepts and potential applications. *Angew. Chem. Int. Ed.* 41:1509–1513.

Koch, K. and W. Barthlott. 2009. Superhydrophobic and superhydrophilic plant surfaces: An inspiration for biomimetic materials. *Phil. Trans. R. Soc. A* 367:1487–1509.

Kopecek, J. 2003. Smart and genetically engineered biomaterials and drug delivery systems. *Eur. J. Pharm. Sci.* 20:1–16.

Kopecek, J. and J. Yang. 2007. Hydrogels as smart biomaterials. *Polym. Int.* 56:1078–1098.

Kost, J. and R. Langer. 2001. Responsive polymeric delivery systems. *Adv. Drug Deliv. Rev.* 46:125–148.

Kovensky, J. 2009. Sulfated oligosaccharides: New targets for drug development? *Curr. Med. Chem.* 16:2338–2344.

Kryscio, D.R. and N.A. Peppas. 2009. Mimicking biological delivery through feedback-controlled drug release systems based on molecular imprinting. *AIChE J.* 55:1311–1324.

Kukreja, N., Y. Onuma, J. Daemen, and P.W. Serruys. 2008. The future of drug-eluting stents. *Pharmacol. Res.* 57:171–180.

Kulkarni, M., A. Breen, U. Greiser, T. O'Brien, and A. Pandit. 2009. Fibrin–lipoplex system for controlled topical delivery of multiple genes. *Biomacromolecules* 10:1650–1654.

Kushner, A.M., J.D. Vossler, G.A. Williams, and Z. Guan. 2009. A biomimetic modular polymer with tough and adaptive properties. *J. Am. Chem. Soc.* 131:8766–8768.

Kwon, G.S. 2002. Block copolymer micelles as drug delivery systems. *Adv. Drug Del. Rev.* 54:167–167.

Kwon, G.H., J.Y. Park, J.Y. Kim, M.L. Frisk, D.J. Beebe, and S.H. Lee. 2008. Biomimetic soft multifunctional miniature aquabots. *Small* 4:2148–2153.

Lacík, I. 2006. Polymer chemistry in diabetes treatment by encapsulated islets of Langerhans: Review to 2006. *Aust. J. Chem.* 59:508–524.

Lanza, R.P., J.L. Hayes, and W.L. Chick. 1996. Encapsulated cell technology. *Nat. Biotechnol.* 14:1107–1111.

LaVan, D., D. Lynn, and R. Langer. 2002. Moving smaller in drug discovery and delivery. *Nat. Rev. Drug Discov.* 1:77–84.

Lee, H., B.P. Lee, and P.B. Messersmith. 2007. A reversible wet/dry adhesive inspired by mussels and geckos. *Nature* 448:338.

Lee, P.Y., Z.H. Li, and L. Huang. 2003. Thermosensitive hydrogel as a TGF-beta1 gene delivery vehicle enhances diabetic wound healing. *Pharm. Res.* 20:1995–2000.

Lee, H., J. Rho, and P.B. Messersmith. 2009. Facile conjugation of biomolecules onto surfaces via mussel adhesive protein inspired coatings. *Adv. Mater.* 21:431–434.

Letchford, K. and H. Buró. 2007. A review of the formation and classification of amphiphilic block copolymer nanoparticulate structures: Micelles, nanospheres, nanocapsules and polymersomes. *Eur. J. Pharm. Biopharm.* 65:259–269.

Li, M.H. and P. Keller. 2009. Stimuli-responsive polymer vesicles. *Soft Matter* 5:927–937.

Li, D., G.P. Tang, J.Z. Li et al. 2007. Dual-targeting non-viral vector based on polyethylenimine improves gene transfer efficiency. *J. Biomater. Sci. Polym. Ed.* 18:545–560.

Lin, C.C. and A.T. Metters. 2006. Hydrogels in controlled release formulations: Network design and mathematical modelling. *Adv. Drug Del. Rev.* 58:1379–1408.

Lincopan, N., N.M. Espíndola, A.J. Vaz, and A.M. Carmona-Ribeiro. 2007. Cationic supported lipid bilayers for antigen presentation. *Int. J. Pharm.* 340:216–222.

Little, S. 2009. Polymeric microcapsulates as a platform technology for biomimetic drug delivery. Abstracts of Papers, *237th ACS National Meeting*, Salt Lake City, UT, March 22–26, COLL-067.

Liu, N.G., D.R. Dunphy, P. Atanassov et al. 2004. Photoregulation of mass transport through a photoresponsive azobenzene-modified nanoporous membrane. *Nano Lett.* 4:551–554.

Liu, J.W., X.M. Jiang, C. Ashley, and C.J. Brinker. 2009. Electrostatically mediated liposome fusion and lipid exchange with a nanoparticle-supported bilayer for control of surface charge, drug containment, and delivery. *J. Am. Chem. Soc.* 131:7567–7569.

Liu, J.Q., Q. Zhang, E.E. Remsen, and K.L. Wooley. 2001. Nanostructured materials designed for cell binding and transduction. *Biomacromolecules* 2:362–368.

Lörh, M., A. Hoffmeyer, J. Kroger et al. 2001. Microencapsulated cell-mediated treatment of inoperable pancreatic carcinoma. *Lancet* 357:1591–1592.

Lu, J., E. Choi, F. Tamanoi, and J.L. Zink. 2008. Light-activated nanoimpeller-controlled drug release in cancer cells. *Small* 4:421–426.

Lutolf, M.P. and J.A. Hubbell. 2005. Synthetic biomaterials as instructive extracellular microenvironments for morphogenesis in tissue engineering. *Nat. Biotechnol.* 23:47–55.

Luz, G.M. and J.F. Mano. 2009. Biomimetic design of materials and biomaterials inspired by the structure of nacre. *Phil. Trans. R. Soc. A* 367:1587–1605.

Maeda, H., J. Wu, S. Tanaka, and T. Akaike. 1999. Tumor vascular permeability and EPR effect for macromolecular therapeutics. In *Ninth International Symposium on Recent Advances in Drug Delivery Systems*, ed. C. Pitt, pp. 114–117. Shannon Ireland: Elsevier Scientific Publications.

Mahdavi, A., L. Ferreira, C. Sundback et al. 2008. A biodegradable and biocompatible gecko-inspired tissue adhesive. *Proc. Natl Acad. Sci. U.S.A.* 105:2307–2312.

Mahmud, A., X.B. Xiong, H.M. Aliabadi, and A. Lavasanifar. 2007. Polymeric micelles for drug targeting. *J. Drug Target.* 15:553–584.

Mano, J.F. 2008. Stimuli-responsive polymeric systems for biomedical applications. *Adv. Eng. Mater.* 10:515–527.

Mano, J.F. and R.L. Reis. 2005. Some trends on how one can learn from and mimic nature in order to design better biomaterials. *Mater. Sci. Eng. Biomim. Mater. Sens. Syst.* 25:93–95.

Mark, H.F. 1981. Macromolecular chemistry today—Aging roots, sprouting branches. *Angew. Chem. Int. Ed.* 20:303–304.

Martina, M.S., J.P. Fortin, C. Menager et al. 2005. Generation of superparamagnetic liposomes revealed as highly efficient MRI contrast agents for in vivo imaging. *J. Am. Chem. Soc.* 127:10676–10685.

Marui, A., A. Kanematsu, K. Yamahara et al. 2005. Simultaneous application of basic fibroblast growth factor and hepatocyte growth factor to enhance the blood vessels formation. *J. Vasc. Surg.* 41:82–90.

Matsumoto, S., R.J. Christie, N. Nishiyama et al. 2009. Environment-responsive block copolymer micelles with a disulfide cross-linked core for enhanced siRNA delivery. *Biomacromolecules* 10:119–127.

Mayes, A.G., L.I. Andersson, and K. Mosbach. 1994. Sugar binding polymers showing high anomeric and epimeric discrimination obtained by non-covalent molecular imprinting. *Anal. Biochem.* 222:483–488.

Mayes, A.G. and M.J. Whitcombe. 2005. Synthetic strategies for the generation of molecularly imprinted organic polymers. *Adv. Drug Del. Rev.* 57:1742–1778.

Meijer, D.K.F., L. Beljaars, G. Molema, and K. Poelstra. 2001. Disease-induced drug targeting using novel peptide–ligand albumins. *J. Control. Release* 72:157–164.

Mellor, J. and S. Boothman. 2003. TIELLE* hydropolymer dressings: Wound responsive technology. *Br. J. Community Nurs.* 8:14–17.

Mitragotri, S. 2009. In drug delivery, shape does matter. *Pharm. Res.* 26:232–234.

Mitragotri, S. and J. Lahann. 2009. Physical approaches to biomaterial design. *Nat. Mater.* 8:15–23.

Miyata, T., T. Uragami, and K. Nakamae. 2002. Biomolecule-sensitive hydrogels. *Adv. Drug Del. Rev.* 54:79–98.

Moghimi, S.M. and A.C. Hunter. 2000. Poloxamers and poloxamines in nanoparticle engineering and experimental medicine. *TIBTECH* 18:412–420.

Moghimi, S.M., I.S. Muir, L. Illum, S.S. Davis, and V. Kolb-Bachofen. 1993. Coating particles with a block copolymer (poloxamine-908) suppresses opsonization but permits the activity of dysopsonins in the serum. *Biochim. Biophys. Acta* 1179:157–165.

Moghimi, S.M., K.D. Pavey, and A.C. Hunter. 2003. Real-time evidence of surface modification at polystyrene lattices by poloxamine 908 in the presence of serum: In vivo conversion of macrophage-prone nanoparticles to stealth entities by poloxamine 908. *FEBS Lett.* 547:177–182.

Mondon, K., R. Gurny, and M. Moller. 2008. Colloidal drug delivery systems—Recent advances with polymeric micelles. *Chimia* 62:832–840.

Moritani, T. and C. Alvarez-Lorenzo. 2001. Conformational imprinting effect on stimuli-sensitive gels made with an imprinter monomer. *Macromolecules* 34:7796–7803.

Mueller, M., B. Kessler, H.J. Adler, and K. Lunkwitz. 2004. Reversible switching of protein uptake and release at polyelectrolyte multilayers detected by ATR–FTIR spectroscopy. *Macromol. Symp.* 210:157–164.

Nagasaki, Y., K. Yasugi, Y. Yamamoto, A. Harada, and K. Kataoka. 2001. Sugar-installed block copolymer micelles: Their preparation and specific interaction with lectin molecules. *Biomacromolecules* 2:1067–1070.

Naik, R.R., L. Brott, S.J. Carlson, and M.O. Stone. 2002. Silica precipitating peptides isolated from a combinatorial phage display libraries. *J. Nanosci. Nanotechnol.* 2:95–100.

Nakayama, K. 2004. Membrane traffic: Editorial overview. *J. Biochem.* 136:751–753.

Nolan, C.M., M.J. Serpe, and L.A. Lyon. 2004. Thermally modulated insulin release from microgel thin films. *Biomacromolecules* 5:1940–1946.

Obara, K., M. Ishihara, T. Ishizuka et al. 2003. Photocrosslinkable chitosan hydrogel containing fibroblast growth factor-2 stimulates wound healing in healing-impaired db/db mice. *Biomaterials* 24:3437–3444.

Onaca, O., R. Enea, D.W. Hughes, and W. Meier. 2009. Stimuli-responsive polymersomes as nanocarriers for drug and gene delivery. *Macromol. Biosci.* 9:129–139.

Oral, E. and N.A. Peppas, 2004. Responsive and recognitive hydrogels using star polymers. *J. Biomed. Mater. Res.* 68A:439–447.

Orive, G., M. De Castro, H.J. Kong et al. 2009. Bioactive cell–hydrogel microcapsules for cell-based drug delivery. *J. Control. Release* 135:203–210.

Orive, G., R.M. Hernández, A.R. Gascón et al. 2003. Cell encapsulation: Highlights and promise. *Nat. Med.* 9:104–107.

Orive, G., R.M. Hernández, A.R. Gascón, M. Igartua, and J.L. Pedraz. 2005. Cell microencapsulation technology for biomedical purposes: Novel insights and challenges. *Trends Pharm. Sci.* 24:207–210.

Ossipov, D.A., O.P. Varghese, and J. Hilborn. 2009. Modular covalent multifunctionalization of hyaluronic acid for in situ production of biomimetic hydrogels. *Polymer Prepr.* 50 (1).

Owens, D.E. and N.A. Peppas. 2006. Opsonization, biodistribution, and pharmacokinetics of polymeric nanoparticles. *Int. J. Pharm.* 307:93–102.

Ozeki, M., T. Ishii, Y. Hirano, and Y. Tabata. 2001. Controlled release of hepatocyte growth factor from gelatin hydrogels based on hydrogel degradation. *J. Drug Target.* 9:461–471.

Palazzo, B., M. Iafisco, M. La Forgia et al. 2007. Biomimetic hydroxyapatite-drug nanocrystals as potential bone substitutes with antitumor drug delivery properties. *Adv. Funct. Mater.* 17:2180–2188.

Palazzo, B., M.C. Sidoti, N. Roveri et al. 2005. Controlled drug delivery from porous hydroxyapatite grafts: An experimental and theoretical approach. *Mater. Sci. Eng. Biomim. Mater. Sens. Syst.* 25:207–213.

Pande, V.S., A.Yu. Grosberg, and T. Tanaka. 1994a. Folding thermodynamics and kinetics of imprinted renaturable heteropolymers. *J. Chem. Phys.* 101:8246–8257.

Pande, V.S., A.Yu. Grosberg, and T. Tanaka. 1994b. Thermodynamic procedure to synthesize heteropolymers that can renature to recognize a given target molecules. *Proc. Natl. Acad. Sci. U.S.A.* 91:12976–12979.

Pande, V.S., A.Yu. Grosberg, and T. Tanaka. 1995. Phase diagram of heteropolymers with an imprinted conformation. *Macromolecules* 28:2218–2227.

Pande, V.S., A.Yu. Grosberg, and T. Tanaka. 1997. Statistical mechanics of simple models of protein folding and design. *Biophys. J.* 73:3192–3210.

Pande, V.S., A.Yu. Grosberg, and T. Tanaka. 2000. Heteropolymer freezing and design: Towards physical models of protein folding. *Rev. Mod. Phys.* 72:259–314.

Pandit, A.S., D.S. Feldman, J. Caulfield, and A. Thompson. 1998. Stimulation of angiogenesis by FGF-1 delivered through a modified fibrin scaffold. *Growth Factors* 15:113–123.

Pandit, A.S., D.J. Wilson, D.S. Feldman, and J.A. Thompson. 2000. Fibrin scaffold as an effective vehicle for the delivery of acidic fibroblast growth factor (FGF-1). *J. Biomater. Appl.* 14:229–242.

Park, K. 2007. Nanotechnology: What it can do for drug delivery? *J. Control. Release* 120:1–3.

Park, H., J.S. Temenoff, Y. Tabata et al. 2009. Effect of dual growth factor delivery on chondrogenic differentiation of rabbit marrow mesenchymal stem cells encapsulated in injectable hydrogel composites. *J. Biomed. Mater. Res. A* 88A:889–897.

Patel, Z.S., S. Young, Y. Tabata, J.A. Jansen, M.E.K. Wong, and A. Mikos. 2008. Dual delivery of an angiogenic and an osteogenic growth factor for bone regeneration in a critical size defect model. *Bone* 43:931–940.

Pattillo, C.B., B. Venegas, F.J. Donelson et al. 2009. Radiation-guided targeting of combretastatin encapsulated immunoliposomes to mammary tumors. *Pharm. Res.* 26:1093–1100.

Peppas, N.A. 2004. Intelligent therapeutics: Biomimetic systems and nanotechnology in drug delivery. *Adv. Drug Del. Rev.* 56:1529–1531.

Peppas, N.A., P. Bures, W. Leobandung, and H. Ichikawa. 2000. Hydrogels in pharmaceutical formulations. *Eur. J. Pharm. Biopharm.* 50:27–46.

Pirollo, K.F. and E.H. Chang. 2008. Does a targeting ligand influence nanoparticle tumor localization or uptake? *Trends Biotechnol.* 26:552–558.

Qiu, L.Y. and Y.H. Bae. 2006. Polymer architecture and drug delivery. *Pharm. Res.* 23:1–30.

Qiu, Y. and K. Park. 2002. Modulated drug delivery. In *Supramolecular Design for Biological Applications*, ed. N. Yiu, pp. 227–243. Boca Raton, FL: CRC Press.

Quinn, J.F. and F. Caruso. 2004. Facile tailoring of film morphology and release properties using layer-by-layer assembly of thermoresponsive materials. *Langmuir* 20:20–22.

Quinn, J.F. and F. Caruso. 2005. Thermoresponsive nanoassemblies: Layer-by-layer assembly of hydrophilic-hydrophobic alternating copolymers. *Macromolecules* 38:3414–3419.

Rachkov, A. and M. Minoura. 2001. Towards molecularly imprinted polymers selective to peptides and proteins. The epitope approach. *Biochim. Biophys. Acta* 1544:255–266.

Raemdonck, K., J. Demeester, and S. De Smedt. 2009. Advanced nanogel engineering for drug delivery. *Soft Matter* 5:707–715.

Rangel-Yagui, C.O., A. Pessoa, Jr., and L.C.T. Costa-Tavares. 2005. Micellar solubilization of drugs. *J. Pharm. Pharmaceut. Sci.* 8:147–163.

Ranney, D.F. 1986. Drug targeting to the lungs. *Biochem. Pharmacol.* 35:1063–1069.

Ranney, D.F. 1997. In vivo agents comprising cationic metal chelators with acidic saccharides and glycosaminoglycans. US Patent 5,672,334.

Ranney, D.F. 2000. Biomimetic transport and rational drug delivery. *Biochem. Pharmacol.* 59:105–114.

Ranney, D., P. Antich, E. Dadey et al. 2005. Dermatan carriers for neovascular transport targeting, deep tumor penetration and improved therapy. *J. Control. Release* 109:222–235.

Rapoport, N. 2007. Physical stimuli-responsive polymeric micelles for anti-cancer drug delivery. *Prog. Polymer Sci.* 32:962–990.

Rehfeldt, F., A.J. Engler, A. Eckhardt, F. Ahmed, and D.E. Discher. 2007. Cell responses to the mechanochemical microenvironment—Implications for regenerative medicine and drug delivery. *Adv. Drug Deliv. Rev.* 59:1329–1339.

Reintjes, T., J. Tessmar, and A. Göpferich. 2008. Biomimetic polymers to control cell adhesion. *J. Drug Del. Sci. Tech.* 18:15–24.

Richardson, T.P., M.C. Peters, A.B. Ennett, and D.J. Mooney. 2001. Polymeric system for dual growth factor delivery. *Nat. Biotechnol.* 19:1029–1034.

Rodríguez-Hernández, J., F. Chécot, Y. Gnanou, and S. Lecommandoux. 2005. Toward 'smart' nano-objects by self-assembly of block copolymers in solution. *Prog. Polym. Sci.* 30:691–724.

Roh, K.H., D.C. Martin, and J. Lahann. 2005. Biphasic Janus particles with nanoscale anisotropy. *Nat. Mater.* 4:759–763.

Rosler, A., G.W.M. Vandermeulen, and H.A. Klok. 2001. Advanced drug delivery devices via self-assembly of amphiphilic block copolymers. *Adv. Drug Del. Rev.* 53:95–108.

Roveri, N., M. Morpurgo, B. Palazzo et al. 2005. Silica xerogels as a delivery system for the controlled release of different molecular weight heparins. *Anal. Bioanal. Chem.* 381:601–606.

Roveri, N., B. Palazzo, and M. Iafisco. 2008. The role of biomimetism in developing nanostructured inorganic matrices for drug delivery. *Expert. Opin. Drug Deliv.* 5:861–877.

Roy, I. and M.N. Gupta. 2003. Smart polymeric materials: Emerging biochemical applications. *Chem. Biol.* 10:1161–1171.

Rudt, S., H. Wesemeyer, and R.H. Mueller. 1993. In vitro phagocytosis assay of nano- and microparticles by chemiluminescence. IV. Effect of surface modification by coating of particles with Poloxamine and Antarox CO on the phagocytic uptake. *J. Control. Release* 25:123–132.

Ruoslahti, E. 1996. RGD and other recognition sequences for integrins. *Ann. Rev. Cell Dev. Biol.* 12:697–715.

Sabatino, G. and A.M. Papini. 2008. Advances in automatic, manual and microwave-assisted solid-phase peptide synthesis. *Curr. Opin. Drug Discov. Devel.* 11:762–770.

Salman, H.H., C. Gamazo, M.A. Campanero, and J.M. Irache. 2005. *Salmonella*-like bioadhesive nanoparticles. *J. Control. Release* 106:1–13.

Sanchez, C., H. Arribart, and M.M.G. Guille. 2005. Biomimetism and bioinspiration as tools for the design of innovative materials and systems. *Nat. Mater.* 4:277–288.

Sangribsub, S., P. Tangboriboonrat, T. Pith, and G. Decher. 2005. Hydrophobization of multilayered film containing layer-by-layer assembled nanoparticle by Nafion adsorption. *Polym. Bull.* 53:425–434.

Sant, V.P., D. Smith, and J.C. Leroux. 2004. Novel pH-sensitive supramolecular assemblies for oral delivery of poorly water soluble drugs: Preparation and characterization. *J. Control. Release* 97:301–312.

Sarikaya, M., C. Tamerler, A.Y. Jen, K. Schulten, and F. Baneyx. 2003. Molecular biomimetics: Nanotechnology through biology. *Nat. Mater.* 2:577–585.

Sarikaya, M., C. Tamerler, D.T. Schwartz, and F. Baneyx. 2004. Materials assembly and formation using engineered polypeptides. *Ann. Rev. Mater. Res.* 34:373–408.

Saul, J.M., A.V. Annapragada, and R.V. Bellamkonda. 2006. A dual-ligand approach for enhancing targeting selectivity of therapeutic nanocarriers. *J. Control. Release* 114:277–287.

Schulte, A.J., K. Koch, M. Spaeth, and W. Barthlott. 2009. Biomimetic replicas: Transfer of complex architectures with different optical properties from plant surface onto technical materials. *Acta Biomater.* 5:1848–1854.

Seliktar, D., A.H. Zisch, M.P. Lutolf, J.L. Wrana, and J.A. Hubbell. 2004. MMP-2 sensitive, VEGF-bearing bioactive hydrogels for promotion of vascular healing. *J. Biomed. Mater. Res.* A 68:704–716.

Sellergren, B. 2001. The non-covalent approach to molecular imprinting. In *Molecularly Imprinted Polymers*, ed. B. Sellergren, pp. 113–184. Amsterdam the Netherlands: Elsevier.

Senior, R.M., H.D. Gresham, G.L. Griffin, E.J. Brown, and A.E. Chung. 1992. Entactin stimulates neutrophil adhesion and chemotaxis through interactions between its Arg-Gly-Asp (RGD) domain and the leukocyte response integrin. *J. Clin. Invest.* 90:2251–2257.

Sershen, S. and J. West. 2002. Implantable, polymeric systems for modulated drug delivery. *Adv. Drug Del. Rev.* 54:1225–1235.

Shibayama, M. and T. Tanaka. 1993. Volume phase transition and related phenomena of polymer gels. In *Advances in Polymer Science, Responsive Gels: Volume Transitions*, vol. 109, ed. K. Dusek, 1–62. Berlin Germany: Springer.

Shim, S., S.W. Kim, E.K. Choi, H.J. Park, J.S. Kim, and D.S. Lee. 2006. Novel pH sensitive block copolymer micelles for solvent free drug loading. *Macromol. Biosci.* 6:179–186.

Shimomura, M. and T. Sawadaishi. 2001. Bottom-up strategy of materials fabrication: A new trend in nanotechnology of soft materials. *Curr. Opin. Colloid Interf. Sci.* 6:11–16.

Shnek, D.R., D.W. Pack, D.Y. Sasaki, and F.H. Arnold. 1994. Specific protein attachment to artificial membranes via coordination to lipid-bound copper(II). *Langmuir* 10:2382–2388.

Sibrian-Vazquez, M. and D.A. Spivak. 2003. Improving the strategy and performance of molecularly imprinted polymers using cross-linking functional monomers. *J. Org. Chem.* 68:9604–9611.

Slowing, I.I., B.G. Trewyn, S. Giri, and V.S.Y. Lin. 2007. Mesoporous silica nanoparticles for drug delivery and biosensing applications. *Adv. Funct. Mater.* 17:1225–1236.

Stancil, K.A., M.S. Feld, and M. Kardar. 2005. Correlation and cross-linking effects in imprinting sites for divalent adsorption in gels. *J. Phys. Chem. B* 109:6636–6639.

Staquicini, F.I., R. Pasqualini, and W. Arap. 2009. Ligand-directed profiling: Applications to target drug discovery in cancer. *Expert Opin. Drug Discov.* 4:51–59.

Stolnik, S., M.C. Davies, L. Illum, S.S. Davis, M. Boustta, and M. Vert. 1994. The preparation of sub-200 nm biodegradable colloidal particles from poly(β-malic acid-*co*-benzyl malate) copolymers and their surface modification with poloxamer and poloxamine surfactants. *J. Control. Release* 30:57–67.

Striegler, S. and M. Dittel. 2003. Evaluation of new strategies to prepare templated polymers with sufficient oligosaccharide recognition capacity. *Anal. Chim. Acta* 484:53–62.

Sukhishvili, S.A. and S. Granick. 2000. Layered, erasable, ultrathin polymer films. *J. Am. Chem. Soc.* 122:9550–9551.

Tamerler, C. and M. Sarikaya. 2007. Molecular biomimetics: Utilizing Nature's molecular ways in practical engineering. *Acta Biomater.* 3:289–299.

Tanaka, T. and M. Annaka. 1993. Multiple phases of gels and biological implications. *J. Intel. Mat. Syst. Struct.* 4:548–552.

Tang, Z., N.A. Kotov, S. Magonov, and B. Ozturk 2003. Nanostructured artificial nacre. *Nat. Mater.* 2:413–418.

Tang, Z., Y. Wang, P. Podsiadlo, and N.A. Kotov. 2006. Biomedical applications of layer-by-layer assembly: From biomimetics to tissue engineering. *Adv. Mater.* 18:3203–3224.

Tao, S.L. and T.A. Desai. 2005. Micromachined devices: The impact of controlled geometry from cell-targeting to bioavailability. *J. Control. Release* 109:127–138.

Thai, C.K., H.X. Dai, M.S.R. Sastry, M. Sarikaya, D.T. Schwartz, and F. Baneyx. 2004. Identification and characterization of Cu_2O- and ZnO-binding polypeptides by *E. coli* cell surface display: Toward an understanding of metal oxide binding. *Biotechnol. Bioeng.* 87:129–137.

Theoharis, S., U. Krueger, P.H. Tan et al. 2009. Targeting gene delivery to activated vascular endothelium using anti E/P-selectin antibody linked to PAMAM dendrimers. *J. Immunol. Methods* 343:79–90.

Thierry, B., P. Kujawa, C. Tkaczyk, F.M. Winnik, L. Bilodeau, and M. Tabrizian. 2005. Delivery platform for hydrophobic drugs: Prodrug approach combined with self-assembled multilayers. *J. Am. Chem. Soc.* 127:1626–1627.

Thurmond, K.B., T. Kowalewski, and K.L. Wooley. 1997. Shell crosslinked knedels: A synthetic study of the factors affecting the dimensions and properties of amphiphilic core-shell nanospheres. *J. Am. Chem. Soc.* 119:6656–6665.

Tong, R., D.A. Christian, L. Tang et al. 2009. Nanopolymeric therapeutics. *MRS Bull.* 34:422–431.

Torchilin, V.P. 2005. Recent advances with liposomes as pharmaceutical carriers. *Nat. Rev. Drug Discov.* 4:145–160.

Townsend-Nicholson, A. and S.N. Jayasinghe. 2006. Cell electrospinning: A unique biotechnique for encapsulating living organisms for generating active biological microthreads/scaffolds. *Biomacromolecules* 7:3364–3369.

Toyotama, A., S. Kugimiya, J. Yamanaka, and M. Yonese. 2001. Preparation of a novel aggregate like sugar-ball micelle composed of poly(methylglutamate) and poly(ethyleneglycol) modified by lactose and its molecular recognition by lectin. *Chem. Pharm. Bull.* 49:169–172.

Traitel, T., R. Goldbart, and J. Kost. 2008. Smart polymers for responsive drug-delivery systems. *J. Biomater. Sci. Polym. Ed.* 19:755–767.

Tröster, S.D., K.H. Wallis, R.H. Müller, and J. Kreuter. 1992. Correlation of the surface hydrophobicity of ^{14}C-poly(methyl methacrylate) nanoparticles to their body distribution. *J. Control. Release* 20:247–260.

Tu, R.S. and M. Tirrell. 2004. Bottom-up design of biomimetic assemblies. *Adv. Drug Del. Rev.* 56:1537–1563.

Turan, E., G. Ozcetin, and T. Caykara. 2009. Dependence of protein recognition of temperature-sensitive imprinted hydrogels on preparation temperature. *Macromol. Biosci.* 9:421–428.

Ulbrich, K., and V. Subr. 2004. Polymeric anticancer drugs with pH-controlled activation. *Adv. Drug Del. Rev.* 56:1023–1050.

Vallet-Regì, M., F. Balas, and D. Arcos. 2007. Mesoporous materials for drug delivery. *Angew. Chem. Int. Ed.* 46:7548–7558.

Vallet-Regí, M., A. Ramila, R.P. Del Real, and J. Perez-Pariente. 2001. A new property of MCM-41: Drug delivery system. *Chem. Mater.* 13:308–311.

Vauthey, S., S. Santoso, H.Y. Gong, N. Watson, and S.G. Zhang. 2002. Molecular self-assembly of surfactant-like peptides to form nanotubes and nanovesicles. *Proc. Natl. Acad. Sci. U.S.A.* 99:5355–5360.

Vauthier, C. and D. Labarre. 2008. Modular biomimetic drug delivery systems. *J. Drug Del. Sci. Tech.* 18:59–68.

Venkatesh, S., S.P. Sizemore, and M.E. Byrne. 2007. Biomimetic hydrogels for enhanced loading and extended release of ocular therapeutics. *Biomaterials* 28:717–724.

Wang, Y., S.S. Mangipudi, B.F. Canine, and A. Hatefi. 2009. A designer biomimetic vector with a chimeric architecture for targeted gene transfer. *J. Control. Release* 137:46–53.

Whaley, S.R., D.S. English, E.L. Hu, P.F. Barbara, and M.A. Belcher. 2000. Selection of peptides with semiconducting binding specificity for directed nanocrystal assembly. *Nature* 405:665–668.

Wizeman, W.J. and P. Kofinas. 2001. Molecularly imprinted polymer hydrogels displaying isomerically resolved glucose binding. *Biomaterials* 22:1485–1491.

Won, Y.Y., H.T. Davis, and F.S. Bates. 1999. Giant wormlike rubber micelles. *Science* 283:960–963.

Wood, K.C., J.Q. Boedicker, D.M. Lynn, and P.T. Hammond. 2005. Tunable drug release from hydrolytically degradable layer-by-layer thin films. *Langmuir* 21:1603–1609.

Wu, C., C. Chen, J. Lai, J. Chen, X. Mu, J. Zheng, and Y. Zhao. 2008. Molecule-scale controlled-release system based on light-responsive silica nanoparticles. *Chem. Commun.* 2662–2664.

Wulff, G. 1995. Molecular imprinting in cross-linked materials with the aid of molecular templates—A way towards artificial antibodies. *Angew. Chem. Int. Ed. Engl.* 34:1812–1832.

Wulff, G. and A. Biffis. 2001. Molecularly imprinting with covalent or stoichiometric non-covalent interactions. In *Molecularly Imprinted Polymers*, ed. B. Sellergren, pp. 71–111. Amsterdam the Netherlands: Elsevier.

Xu, Z., N. Gao, H. Chen, and S. Dong. 2005b. Biopolymer and carbon nanotubes interface prepared by self-assembly for studying the electrochemistry of microperoxidase-11. *Langmuir* 21:10808–10813.

Xu, J.P., J. Ji, W.D. Chen, and J.C. Shen. 2005a. Novel biomimetic polymersomes as polymer therapeutics for drug delivery. *J. Control. Release* 107:502–512.

Yanase, M., M. Shinkai, H. Honda et al. 1998. Intracellular hyperthermia for cancer using magnetite cationic liposomes: An in vivo study. *Jpn. J. Cancer Res.* 89:463–469.

Yoshida, R., T. Sakai, Y. Hara et al. 2009. Self-oscillating gel as novel biomimetic materials. *J. Control. Release* 140:186–193.

Yoshida, R., K. Sakai, T. Okano, and Y. Sakurai. 1993. Pulsatile drug delivery systems using hydrogels. *Adv. Drug Del. Rev.* 11:85–108.

Yoshida, R. and Y. Uesusuki. 2005. Biomimetic gel exhibiting self-beating motion in ATP solution. *Biomacromolecules* 6:2923–2926.

Youan, B.B.C. 2004. Chronopharmaceutics: Gimmick or clinically relevant approach to drug delivery? *J. Control. Release* 98:337–353.

Yu, Y.C., M. Tirrell, and G.B. Fields. 1998. Minimal lipidation stabilizes protein-like molecular architecture. *J. Am. Chem. Soc.* 120:9979–9987.

Zhai, L., F.C. Cebeci, R.E. Cohen, and M.F. Rubner. 2004. Stable superhydrophobic coatings from polyelectrolyte multilayers. *Nano Lett.* 4:1349–1353.

Zhang, J., L.S. Chua, and D.M. Lynn. 2004. Multilayered thin films that sustain the release of functional DNA under physiological conditions. *Langmuir* 20:8015–8021.

Zhao, Y., J.L. Vivero-Escoto, I.I. Slowing, B.G. Trewyn, and V.S.Y. Lin. 2010. Capped mesoporous silica nanoparticles as stimuli-responsive controlled release systems for intracellular drug/gene delivery. *Expert Opin. Drug Deliv.* 7:1013–1029.

Zhou, W., X.H. Dai, and C.M. Dong. 2008. Biodegradable and biomimetic poly(e-caprolactone)/poly(lactobionamidoethyl methacrylate) biohybrids: Synthesis, lactose-installed nanoparticles and recognition properties. *Macromol. Biosci.* 8:268–278.

Zisch, A.H., M.P. Lutolf, M. Ehrbar et al. 2003. Cell-demanded release of VEGF from synthetic, biointeractive cell ingrowth matrices for vascularized tissue growth. *FASEB J.* 17:2260–2262.

Zu, X., Z. Lu, Z. Zhang, J.B. Schenkman, and J.F. Rusling. 1999. Electroenzyme-catalyzed oxidation of styrene and *cis*-beta-methylstyrene using thin films of cytochrome P450cam and myoglobin. *Langmuir* 15:7372–7377.

5 Polymeric Nanoparticles for Drug Delivery

Karine Andrieux, Julien Nicolas,
Laurence Moine, and Gillian Barratt

CONTENTS

5.1 INTRODUCTION TO COLLOIDAL DRUG CARRIERS

Progress in both synthetic organic chemistry and molecular biology has led to an enormous increase in the number of active molecules which are candidates to become therapeutic or diagnostic agents. However, the physicochemical properties of these molecules often limit their potential. Many molecules derived from combinatorial chemistry libraries are poorly soluble in aqueous media and are therefore difficult to formulate and administer. Thus, although they may show good binding to specific targets in vitro, their bioavailability may be insufficient. On the other hand, molecules such as proteins, peptides, and nucleic acids are water-soluble but are often subject to degradation in biological fluids. Furthermore, because these molecules are hydrophilic, they penetrate poorly through cell membranes, unless specific transporters exist. Appropriate drug carrier systems can help to

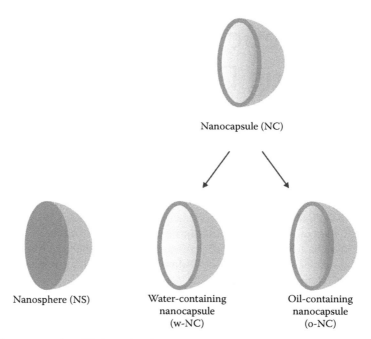

FIGURE 5.1 **(See companion CD for color figure.)** Schematic representation of nanospheres (NS), water-containing nanocapsules (w-NC) and oil-containing nanocapsules (o–NC).

transform candidate molecules into effective drugs. Nanoparticulate systems are particularly useful because, unlike microparticles, they can be administered by parenteral routes and are distributed by normal physiological processes. They present a large surface area, which can facilitate drug dissolution, and can be internalized by the majority of cell types. They differ from prodrugs in that there is usually no covalent linkage between the carrier and the drug.

Over the past few decades, a large amount of work has been carried out on nanoparticulate systems and it is impossible to give a complete review of all the literature in this chapter. Therefore, it will be restricted to nanoparticles (NPs) prepared from synthetic polymers. It will not deal with some interesting results obtained with NPs prepared from natural polymers (proteins and polysaccharides), lipid-based systems (solid–lipid NPs, liposomes, and emulsions) and polymers in the form of micelles or dendrimers. Nanoparticulate systems can be divided into two classes according to their internal structure (Figure 5.1). Nanospheres, which are also often referred to as NPs, are composed of a solid matrix of polymer. Drug molecules may be incorporated into the matrix or adsorbed on the surface. Nanocapsules are a reservoir system in which a liquid core is surrounded by a polymeric shell. The core can be lipophilic (able to solubilize water-insoluble drugs) or aqueous (able to accommodate water-soluble molecules).

The main advantages of the use of nanoparticulate carriers are (i) dispersing a poorly water-insoluble drug to facilitate its administration and dissolution; (ii) protecting a labile drug from degradation; (iii) modifying the distribution of the drug so as to avoid sites of toxicity and metabolism and increase delivery to the site of action, particularly if the NP incorporates specific targeting ligands; and (iv) allowing the drug to penetrate biological membranes. Before describing some results obtained with nanoparticulate systems, the distribution of NPs within the organism will be discussed.

5.1.1 DISTRIBUTION OF NANOCARRIERS AFTER INTRAVENOUS ADMINISTRATION

After intravenous administration, NPs undergo interactions with blood proteins like any "foreign" particle (Leroux et al., 1995, Ameller et al., 2003a). In particular, their hydrophobic surfaces activate the alternative pathway of complement activation. This leads to the deposition of C3b on their surface and uptake by cells of the mononuclear phagocyte system (MPS), mainly Kupffer cells in the liver sinusoids (Alléman et al., 1997). Other plasma proteins, known as opsonins and including immunoglobulin and fibronectin, are also adsorbed onto the particle surface and contribute to the recognition and phagocytosis of the particle by macrophages. The outcome is that NPs are very rapidly removed from the circulation and concentrated in organs of the MPS: principally the liver, spleen, and bone marrow (Moghimi and Szebeni, 2003). Although, as described later, this distribution can provide some therapeutic opportunities, these are quite limited. Therefore, considerable research has been devoted to developing so-called Stealth NPs that are "invisible" to the immune system and can persist longer in the circulation. Based on lessons learned from liposomes (Woodle, 1998), the approach adopted has been to cover the surface of the particle with hydrophilic polymer chains in the brush configuration so that they repel opsonins (Jeon et al., 1991). Poly(ethylene glycol) (PEG) is by far the most popular hydrophilic chain for this application. The introduction of PEG chains can be done in two ways: by coating the surface of the NPs with surfactants containing PEG blocks, such as the poloxamers and poloxamines (Illum et al., 1987), or by forming the NPs from block copolymers combining PEG and a hydrophobic polymer such as poly(D,L-lactide) (PLA) (Bazile et al., 1995, Gref et al., 1995). The first strategy carries the risk of desorption of the surfactant after administration; therefore the second option is generally preferred. The length and density of the PEG chains are both important for "Stealth" properties. On the surface of nanospheres, the average distance between two end-attached PEG chains should be 2.2 nm or less in order to repel proteins effectively (Vittaz et al., 1996). The optimal chain length is 5 kDa (Zambaux et al., 1999). For nanocapsules, where the total surface area per unit weight of polymer is much greater, a lower density of chains can be offset by a higher molecular weight (20 kDa) (Mosqueira et al., 2001).

Long-circulating NPs persist in the circulation but the vascular endothelium remains a barrier for their penetration into tissues. Apart from the sinusoids of the MPS, extravascularization of NPs only occurs in damaged or defective endothelium. One situation in which this can occur is in solid tumors, where the endothelium of the capillaries lacks tight junctions. This combines with a lack of lymphatic drainage to produce the "enhanced permeation and retention" (EPR) effect (Jain, 1987). In this way, long-circulating NPs can be passively accumulated within tumors. Leaky capillary endothelium is also present in sites of inflammation and infection; the accumulation of long-circulating particles in such regions had been demonstrated with radiolabeled liposomes (Oyen et al., 1996). Recently, much attention has been paid to NPs as a means of allowing drugs to cross the blood–brain barrier (BBB). As shown later, there is evidence that long-circulating NPs can enter the brain when a tumor or an inflammatory response is present (Calvo et al., 2001a, Brigger et al., 2002).

5.1.2 Distribution of Nanocarriers after Administration by the Oral Route

Like the vascular endothelium, the intestinal epithelium has tight junctions that prevent the passage of NPs. Small particles may be able to pass the epithelium by transcytosis through enterocytes, but the main route of NP uptake from the digestive tract seems to be via the M cells of Peyer's patches, delivering the particles to the lymphoid follicles (des Rieux et al., 2006). Here, they will be phagocytosed by antigen-presenting cells, but it has been reported that some particles reach the circulation (Florence et al., 1995).

Another phenomenon that occurs after oral dosing of NPs is bioadhesion. Particles are immobilized in the mucus overlying the enterocytes and are eliminated from the digestive tract more slowly than material remaining in the lumen, thus increasing the residence time of the associated drug in the intestine, promoting its dissolution and absorption (Ponchel and Irache, 1998).

5.1.3 Distribution of Nanocarriers after Administration by Other Routes

Colloidal drug carriers administered by the subcutaneous and intramuscular routes are generally retained longer at the site of administration than a free drug and can therefore allow a more sustained action. It has been shown that carriers accumulate in regional lymph nodes after intraperitoneal or subcutaneous administration (Maincent et al., 1992, Hawley et al., 1995, Nishioka and Yoshino, 2001, Liu et al., 2006). In the skin, studies with fluorescent polystyrene NPs showed that these accumulate in hair follicles and skin furrows (Alvarez-Romain et al., 2004). Recently, some interest has been given to the use of colloidal carriers by the ocular route, where they can be retained and release their contents gradually. There is some evidence of NPs penetrating the corneal epithelium (Calvo et al., 1997).

5.2   APPLICATIONS OF "CONVENTIONAL" NPs

For the purposes of this chapter, "conventional" NPs (also called first-generation NPs) are those which are not modified with PEG or other hydrophilic chains on their surface and as a result are rapidly captured by phagocytic cells, especially after administration by the intravenous route.

5.2.1   INTRAVENOUS ROUTE

As might be expected, these NPs have been used most successfully when macrophages are the target of the encapsulated agent. One example is the delivery of antibiotics for the treatment of intracellular infections. Often, the microorganisms are found within the lysosomes, where the acid pH can lead to poor accumulation or inactivation of drug (Pinto-Alphandary et al., 2000). Early work with poly(isohexylcyanoacrylate) (PIHCA) nanospheres loaded with ampicillin showed a large increase in efficacy compared with free antibiotic in mice infected with *Salmonella typhimurium* and *Listeria monocytogenes* (Youssef et al., 1988, Fattal et al., 1989). Both electron microscopy and confocal fluorescence microscope with labeled *S. typhimurium* and nanospheres have shown the carrier system and the bacteria in the same intracellular compartment (Pinto-Alphandary et al., 1994).

Later, a fluoroquinolone antibiotic, ciprofloxacin, was encapsulated within poly(isobutyl cyanoacrylate) (PIBCA) and PIHCA nanospheres in an attempt to kill both dividing and nondividing bacteria; however, the formulation was not effective against persistent *Salmonella* (Page-Clisson et al., 1998). More recently, the same antibiotic was encapsulated in poly(D,L-lactide-*co*-glycolide) (PLGA) nanospheres (Jeong et al., 2008). PLGA nanospheres are also able to deliver gentamicin to the liver and spleen of *Brucella melitensis*–infected mice (Lecaroz et al., 2007). Nanospheres prepared from PLA containing the antiparasitic drug primaquine were also efficient at delivering this drug to the liver (Rodrigues et al., 1994). Co-localization of nanospheres and *Leishmania donovani* parasites in Kupffer cells was observed.

As well as delivering antimicrobial drugs, NPs can also be used to carry immunomodulating substances to macrophages. One such compound is muramyl tripeptide cholesterol (MTP-Chol), which is able to activate macrophages and induce toxicity toward tumor cells, and would therefore be a useful agent to treat metastatic cancer. Work in our laboratory has shown that nanocapsules based on PLA containing MTP-Chol were efficient activators of macrophages in vitro (Morin et al., 1994) and reduced the number of liver metastases in a mouse tumor model (Barratt et al., 1994).

Despite the unfavorable distribution of conventional NPs, some positive results have been achieved in cancer treatment. An example of this is nanospheres prepared from poly(alkyl cyanoacrylate) (PACA) loaded with doxorubicin, a versatile anticancer drug with dose-limiting cardiotoxicity. These NPs were found to be more efficient than free drug against L1210 leukemia and liver metastases of the M5076 reticulosarcoma, both in mice (Brasseur et al., 1986, Chiannikulchai et al., 1989). In the latter model, the mechanism of action was revealed by analyzing the time course of drug accumulation in the tumor and in healthy liver tissue. In the first hours after nanosphere administration, the particles accumulated within Kupffer cells. Later, the drug was found to have been transferred to the tumor cells in the hepatic parenchyma (Chiannikulchai et al., 1990). Despite the increased amount of doxorubicin carried to the liver by the nanospheres, no significant toxicity was observed in mice. Furthermore, the cardiac toxicity was dramatically reduced (Couvreur et al., 1986).

Doxorubicin is one of the drugs whose anticancer activity is reduced by the phenomenon of multidrug resistance. An interesting result obtained in 1992 was that PACA nanospheres, but not those prepared from other polymers such as PLA, were able to overcome resistance to doxorubicin in cultured cells (Cuvier et al., 1992). It was subsequently shown that the formation of an ion pair between positively charged doxorubicin and negatively charged PACA degradation products allows the drug to enter the cells and avoid expulsion by the P-glycoprotein (Colin de Verdière et al., 1997). A similar formulation of doxorubicin associated with PIHCA nanospheres is being developed as Transdrug® by BioAlliance Pharma and has reached Phase III trials in hepatocellular carcinoma (Barraud et al., 2005).

5.2.2   ORAL ROUTE

The results obtained with conventional NPs by the oral route can be resumed as protection of a labile drug from degradation in the gastrointestinal tract, improvement of bioavailability by bioadhesion, and protection of the GI tract from toxicity due to the drug. A major result is that insulin encapsulated in PACA nanocapsules is protected from degradation by gastrointestinal enzymes (Damgé et al., 1997a). Although these nanocapsules did not reduce glycemia in normal rats, they were active in diabetic rats and in normal rats loaded with glucose. After a 2 day lag period, prolonged hypoglycemia (up to 20 days) was observed (Damgé et al., 1990). Microscopic evidence suggested passage of intact nanocapsules across the intestinal mucosa (Aboubakar et al., 2000). The sustained effect could be due to the release of insulin from nanocapsules accumulated at an as yet unidentified site, or a local effect of insulin on intestinal cells. The therapeutic effect of a somatostatin analog given by the oral route was also improved by encapsulation within nanocapsules (Damgé et al., 1997b).

The bioadhesive properties of NPs can be used to improve the absorption of poorly absorbed or unstable drugs. The immobilization of NPs at the mucosal surface can lead to a prolonged residence time at the site of drug action or absorption, localization of the delivery system at a given target site, an increase in the drug concentration gradient due to intense contact of the particles with the mucosal surface and direct contact with intestinal cells, which is the first step before particle absorption (Ponchel and Irache, 1998). For example, AZT-loaded PIHCA nanospheres were able to concentrate the drug in intestinal cells, in particular in Peyer's patches that are sanctuary sites for the HIV virus (Dembri et al., 2001). PLGA nanospheres containing a combination of three antituberculosis drugs were shown to be effective after oral administration in a guinea pig model (Johnson et al., 2005). Oral NP formulations were also found to be more effective than free drug for treating inflammatory bowel disease (Meissner et al., 2006).

Encapsulation in nanocapsules with an oily core is able to protect the gastrointestinal mucosa of rats from the ulcerating effects of non-steroidal anti-inflammatory drugs (diclofenac and indomethacin) after oral administration (Ammoury et al., 1991, Guterres et al., 1995). Despite the reduction in local toxicity, these two drugs exhibited drug concentration–time profiles in the plasma of rats similar to those obtained with the corresponding aqueous solutions. This protection was independent of the nature of the polymer shell: PIBCA produced by interfacial polymerization or preformed PLA (Ammoury et al., 1990). The mechanism of this protection could be a slow release of the drugs in the acidic gastric environment or the reduced toxicity of the acidic form of diclofenac or indomethacin encapsulated in nanocapsules compared with the sodium salt in the aqueous solution. Irritation due to the administration of indomethacin by the rectal route was also reduced by encapsulation in nanocapsules (Fawaz et al., 1996).

5.2.3   OTHER ROUTES

Polymeric NPs have also been used for the local delivery of antisense oligonucleotides (ODN) and small interfering RNA (siRNA) designed to inhibit oncogene expression. These nucleic acid–based agents are sensitive to degradation in biological fluids and, because they are hydrophilic charged molecules, do not penetrate readily into cells. Nanoparticulate delivery systems can help to solve these problems. Early results were obtained with ODN adsorbed onto the surface of PACA NPs by ion pairing with cationic surfactants, which protected the ODN from nucleases and promoted their uptake into cells. Such NPs were able to inhibit PKCα expression in Hep G6 cells and Ha-*ras*-mediated cell growth in vitro and in vivo after subcutaneous administration (Chavany et al., 1992, 1994, Schwab et al., 1994, Lambert et al., 1998).

Despite these results, better protection and cellular penetration of nucleic acids could be achieved by the encapsulation within the core of NPs. It is difficult to incorporate the hydrophilic nucleic acids in the polymeric matrix of nanospheres; therefore, nanocapsules with an aqueous core were developed. Nanocapsules of PIBCA containing both ODN and siRNA have been prepared (Lambert et al., 2000a; Toub et al., 2006a,b). When loaded with an ODN to the fusion gene EWS-Fli-1 expressed in Ewing's sarcoma and injected directly into the tumor, they were effective at reducing the size of xenografts in nude mice (Lambert et al., 2000b).

Subcutaneous administration has been applied to other types of pathology; for example, insulin-loaded nanocapsules similar to those described earlier have been shown to exert hypoglycemic effects in rats by this route (Mesiha et al., 2005). The uptake of NPs by regional lymph nodes after subcutaneous administration has led to their use as carriers for antitumor vaccination (Hamdy et al., 2008).

After intramuscular administration, nanocapsules containing diclofenac were able to reduce inflammation at the site of injection compared with the free drug in solution (Guterres et al., 2000). A similar result was obtained by Hubert et al. (1991) with nanocapsules of darodipine, a poorly soluble antihypertensive.

Application of carrier formulations to the eye retards elimination of drug from the corneal surface. This has been demonstrated for β-blockers (Losa et al., 1993, Marchal-Heussler et al., 1993) and cyclosporin A (Calvo et al., 1996) within nanospheres and nanocapsules. Bioadhesive effects are probably involved.

Some interest has also been devoted to NP administration as aerosols by the pulmonary route. If they can be delivered to the deep lung or alveolar region, small NPs can remain in the lung epithelium and provide sustained release. However, their low inertia means that they are more likely to be exhaled than reach deeper lung tissues. A way of circumventing this problem is the use of "Trojan" particles, in which small NPs are encapsulated within larger porous biodegradable particles with aerodynamic properties allowing them to penetrate deep into the lung (Tsapis et al., 2002).

5.3   APPLICATIONS OF LONG-CIRCULATING NPs

NPs that are able to avoid rapid opsonization and thereby remain longer in the circulation have a wider range of therapeutic applications than conventional ones. After administration by the intravenous route, such NPs (nanospheres or nanocapsules) can act as circulating reservoirs of drug or can deliver their contents to tissues which are accessible because of a discontinuous vascular endothelium. Examples of such tissues are solid tumors and sites of inflammation and infection.

5.3.1   INTRAVASCULAR APPLICATIONS

Nanocapsules prepared from PLA-*b*-PEG block copolymers have been used as a delivery system for halofantrine, a very lipophilic antimalarial drug, allowing it to be administered intravenously. This route of administration must be used in cases of severe malaria, when the patient is unconscious. Halofantrine could be incorporated into the oily core of nanocapsules, and was only released in the presence of plasma lipoproteins (Mosqueira et al., 2006). Apart from other considerations, encapsulation allowed the drug to be administered without the use of irritating solvents and greatly reduced its cardiotoxicity (Leite et al., 2007). Halofantrine pharmacokinetics was determined in mice at an advanced stage of infection with *Plasmodium berghei*, comparing free drug, "conventional" nanocapsules, and long-circulating nanocapsules. The area under the curve for plasma halofantrine was increased sixfold when the drug was presented in nanocapsules, whatever the formulation was. Both nanocapsule formulations also increased the therapeutic efficacy of halofantrine: the "conventional" nanocapsules showed a more rapid onset of effect, whereas that of the PLA-*b*-PEG nanocapsules was more sustained (Mosqueira et al., 2004). The similarity of the behavior of long-circulating and "conventional" nanocapsules might be because in severely infected mice the phagocytic capacity of the liver may be saturated as a result of uptake of hemoglobin breakdown products. Furthermore, serum lipoproteins, which would bind halofantrine released from nanocapsules, are reduced during the disease.

NPs which remain in the circulation can also be loaded with hemoglobin to serve as blood substitutes. Liu and Chang (2008) showed that PLA-*b*-PEG nanocapsules can be used in this way in rats with no adverse side effects. Core–shell NPs with a PIBCA core and a brush-like shell of heparin also function as hemoglobin carriers. The protein is integrated into the surface coating of negatively charged polysaccharide. The hemoglobin retains its oxygen-binding capacity while the heparin retains its antithrombic and complement-inhibiting properties, allowing the particles to remain long-circulating (Chauvierre et al., 2004, Baudin-Creuza et al., 2008).

5.3.2 DELIVERY TO TUMORS

A major application of long-circulating NPs is delivery to solid tumors. In our laboratory, an early focus was on NPs containing antiestrogens for the treatment of hormone-dependent cancers. Both "mixed" antiestrogens such as 4-hydroxytamoxifen and "pure" antiestrogens like RU 58668 benefit from more specific delivery to tumors to avoid side effects in other tissues (Renoir et al., 2006). Attempts to encapsulate 4-hydroxytamoxifen in poly[methoxypoly(ethylene glycol)-*co*-(hexadecyl cyanoacrylate)] (P(MePEGCA-*co*-HDCA)) nanospheres gave only a low level of incorporation (Brigger et al., 2001) that was reflected in weak antitumoral activity (Renoir et al., 2006). On the other hand, RU 58668 was incorporated successfully into nanospheres and nanocapsules of PLA-*b*-PEG that were found to be more effective than the free drug in xenografts of human breast cancer lines growing in nude mice (Ameller et al., 2003a,b, 2004). Interestingly, an antiangiogenic effect was observed, as well as arrest of tumor cell growth. In later work, ferrocenyl derivatives of tamoxifen were encapsulated in PLA-*b*-PEG nanospheres and nanocapsules. The resulting systems showed promising results in cell lines: arresting cell growth and inducing apoptosis in a way that seemed to involve reactive oxygen intermediates (Nguyen et al., 2008).

The siRNA approach has also been investigated. siRNA targeting the estrogen receptor message was incorporated into nanocapsules with a hydrophilic core prepared from amphiphilic PEG-*b*-poly(ε-caprolactone-*co*-dodecyl β-malate) (PEG-*b*-(PCL-*co*-MA)). These nanocapsules inhibited ER expression in cultured MCF-7 cells in vitro and reduced the growth of xenografts in nude mice after intravenous injection (Bouclier et al., 2008).

Another recent approach to cancer therapy is photodynamic therapy (PDT). A photosensitizing compound is allowed to accumulate within tumor cells before irradiation to produce toxic products by photolysis. More specific delivery of photosensitizers to tumor cells would reduce side effects and the delay between administration and irradiation. Moreover, most photosensitizers are lipophilic and therefore difficult to formulate for intravenous injection. Results from Bourdon et al. (2000) using the photosensitizer meta–tetra(hydroxyphenyl)chlorine (mTHPC) encapsulated in PLA-*b*-PEG nanocapsules showed that these problems might be overcome by the use of long-circulating NPs. PLA-*b*-PEG and PLGA-*b*-PEG nanospheres have shown similar results (Konan et al., 2003).

Many more traditional anticancer drugs have been incorporating into long-circulating NPs in order to take advantage of the EPR effect (Avgoustakis, 2004). Only a few examples are given here. Cisplatin loaded into PLGA-*b*-PEG nanospheres was able to delay the growth of HT 29 tumors in SCID mice more effectively than the free drug, while hepatic toxicity was reduced (Mattheolabakis et al., 2009). Much research has been devoted recently to prepare formulations of paclitaxel, to facilitate its intravenous administration, and to concentrate the drug in tumors. Danhier et al. (2009)

loaded paclitaxel into nanospheres of PLGA-*b*-PEG by nanoprecipitation. These were more effective at reducing growth of the transplantable liver tumor TLT in mice than the commercial formulation (Taxol)®. A related compound, docetaxel, has been associated with long-circulating nanospheres and found to accumulate in C26 tumors in mice (Senthilkumar et al., 2008). Doxorubicin has also been encapsulated in PLGA-*b*-PEG nanospheres (Park et al., 2009). In this form, it is effective against subcutaneously implanted A20 murine B-cell lymphoma cells in mice, without producing any evidence of cardiotoxicity.

An interesting observation is that long-circulating NPs seem to be able to penetrate into brain tumors (Calvo et al., 2001a). This is discussed in 5.4.2.2.2.1 on brain targeting.

5.3.3   DELIVERY TO SITES OF INFECTION

The penetration of long-circulating nanocapsules into inflammatory sites has been demonstrated by gamma scintigraphy using nanocapsules labeled with (99m)Tc-HMPAO (Pereira et al., 2009). This suggests that long-circulating NPs be used both as imaging agents and to deliver drugs to these regions. Indeed, PLGA/PLA-*b*-PEG nanospheres loaded with an anti-inflammatory steroid are effective in treating experimental arthritis in rats (Ishihara et al., 2009). Nanocapsules containing liquid perfluorocarbons and modified with PEG-bearing lipids have been proposed as contrast agents for both ultrasound imaging and nuclear magnetic resonance (NMR) (Díaz-Lopez et al., 2009).

Despite some results obtained with long-circulating liposomes (Bakker-Woudenberg et al., 2005), there is not much literature on the use of long-circulating NPs to carry anti-infectious agents into extravascular sites. On the other hand, they have been proposed as carriers for peptides (Mallardé et al., 2003), proteins (Quellec et al., 1999), and nucleic acids (Perez et al., 2001), for both therapeutic and vaccination purposes.

5.3.4   OTHER APPLICATIONS

Despite being designed for administration by the intravenous route, NPs bearing PEG at their surface have also been shown to have advantages for delivery by other routes. PLA-*b*-PEG nanospheres are transported across the nasal mucosa (Vila et al., 2004) and can be used to deliver protein antigens by this route (Vila et al., 2005). The presence of PEG also confers stability in intestinal fluids, giving them potential as protein carriers by this route (Tobío et al., 2000, Garinot et al., 2007). The grafting of polysaccharide chains onto the surface of PACA NPs gives them improved bioadhesive properties in the intestine (Bertholon et al., 2006). Intraperitoneally administered PLA-*b*-PEG nanocapsules containing camptothecin were effective against lung metastases of the B16–F10 melanoma in mice (Loch-Neckel et al., 2007). The coating of PLGA nanospheres with PLA-*b*-PEG was found to increase their drainage into lymph nodes after interstitial administration (Hawley et al., 1997).

5.4   TARGETED NPs

The ultimate goal in the development of NPs for therapeutic and diagnostic purposes is to attach a specific "homing group" which will be recognized by the target cells. Several different types of ligands have been considered. Monoclonal antibodies or fragments such as short-chain Fv are obvious candidates because of their specificity. Other classes of ligand–receptor interactions which could be exploited are sugar–lectin interactions like the mannose/fucose receptor of macrophages and the galactose receptor of hepatocytes, hormone and growth factor receptors, and receptors for cell nutrients such as transferrin and folic acid, which are overexpressed in some tumors. Many of these targeting strategies have already been applied to liposome-based systems; however, as reviewed by Barbet (1995) and by Sapra and Allen (2003), although impressive results may be obtained in vitro, with the exception of targeting to the liver, these are often not confirmed in vivo, since the use of specific ligands cannot overcome some physiological constraints. These observations should guide the development of NP-based systems.

In the first place, the NPs will be recognized and taken up by macrophages if their surfaces are not modified to reduce opsonization. On the other hand, PEG chains could restrict access to a ligand attached directly to the liposome surface. This can be overcome by attaching the targeting group to the end of the PEG chain (Mercadal et al., 1999). The length of the spacer arm between the particle surface and the ligand is an important factor (Ham et al., 2009). However, the presence of ligand at the surface, particularly a large protein such as a whole immunoglobin, can present a new target for opsonization and therefore it is necessary to limit the degree of substitution. Furthermore, ligands that are too closely spaced may hinder correct binding to the receptor. Second, NPs administered by the intravenous route will only be able to pass through the vascular endothelium where this is discontinuous. Third, when intracellular delivery of the encapsulated drug is required for its action (e.g., in the case of ODN or siRNA) the targeting ligand should be one that triggers internalization. In this case, the NPs must be small enough to be taken up by receptor-mediated endocytosis in non-phagocytic cells, that is, 200 nm or less, and it should be remembered that this uptake will usually lead to the lysosomal compartment unless an endosome-disrupting element is present.

Given these restrictions, it is not surprising that most work on targeted NPs has been devoted to delivering drugs and/or imaging agents more specifically to tumor cells while taking advantage of the EPR effect. Although some work has been done with antibodies and fragments directed toward tumor-specific antigens (Brannon-Peppas and Blanchette, 2004), in this chapter we will concentrate on two low-molecular-weight ligands: folate and peptides containing the RGD motif.

Another targeting strategy is to direct the NP to a particular region of capillary endothelium, to concentrate the drug within a particular organ, and to allow it to diffuse from the carrier to the target tissue. The selectin class of receptors, expressed on endothelial cells in response to tissue damage, is a good candidate here. Targeting can be achieved either with antibodies or with the natural ligand or an analog.

A related approach is to prepare NPs which will bind to a receptor which mediates transcytosis across the endothelium. This strategy has been used to promote translocation across the BBB.

In the past, targeting of polymeric NPs has been hampered by the difficulty of performing covalent linkages between the targeting ligand and the polymer making up the NP, particularly when the latter is a polyester with only one reactive group per molecule. However, advances in polymer chemistry and the development of "click chemistry," as described later, could greatly simplify the process of obtaining ligand-bearing NPs.

5.4.1 CHEMICAL STRATEGIES TO OBTAIN TARGETED NPs

This section will mainly focus on two classes of well-established, biodegradable polymers employed for drug delivery purposes: (i) PACA and (ii) polyesters such as PLA, PLGA, and PCL.

5.4.1.1 General Methodologies

With NPs (i.e., nanospheres and nanocapsules), most of the covalent coupling strategies have relied on the synthesis of amphiphilic copolymers in which the hydrophobic part is generally a biodegradable polymer and the hydrophilic part is composed of a PEG chain at the extremity of which is attached the ligand. Two different synthetic approaches are possible: the ligand can be either linked at the surface of preformed NPs (Stella et al., 2000, Lu et al., 2005, 2009, Bae et al., 2007, Stella et al., 2007, Zhang et al., 2007, Chen et al., 2008, Pan and Feng, 2008, 2009, Pang et al., 2008, Wang et al., 2009) or already coupled to the PEG chain prior to synthesis of the hydrophobic block of the copolymer (Salem et al., 2001, Patil et al., 2009, Jubeli et al., 2010).

Other routes have also been investigated. Some examples reported the direct coupling of peptides (with or without a PEG spacer) to the core polymer, followed by self-assembly into the corresponding functionalized NPs (Dawson and Halbert, 2000, Schiffelers et al., 2004, Costantino et al., 2005,

Tosi et al., 2007). The simple adsorption of a PEG-based surfactant at the surface of the NPs and their subsequent coupling with the desired ligand also represent an interesting alternative (Gullberg et al., 2006, Chittasupho et al., 2009). The design of functionalized albumin NPs has also been investigated using PEGylated albumin with activated extremities able to react with a ligand (Michaelis et al., 2006, Mishra et al., 2006, Kreuter et al., 2007, Ulbrich et al., 2009, Zensi et al., 2009).

Non-covalent functionalization to obtain targeted NPs can be achieved using different strategies. The most widely employed uses avidin/streptavidin/neutravidin–biotin couples, which lead to the strongest non-covalent biological interaction known to date. In this case, the biotin-binding protein can be either (i) attached at the surface of the NPs and subsequently incubated with a biotinylated ligand (Michaelis et al., 2006, Haun and Hammer, 2008, Lin et al., 2010) or (ii) functionalized by the desired ligand and linked to preformed biotinylated NPs (Aktaş et al., 2005).

The simple adsorption of the ligand at the surface of NPs has also been investigated (Blackwell et al., 2001, Barbault-Foucher et al., 2002, but its desorption upon in vivo injection could reduce the efficacy of this strategy.

5.4.1.2 Covalent Coupling Strategies

A wide range of ligation strategies has been used so far to prepare targeted NPs. Most of these have employed traditional coupling chemistries that take advantage of functional groups already present in the structures of polymers and biologically active ligands. However, the development of novel coupling approaches has been recently described.

5.4.1.2.1 Traditional Coupling Reactions

On the basis of the methods described earlier, several classical chemistry methods have been developed to introduce bioactive molecules onto the surface of NPs for specific targeting. Among them, a well-established approach consists of the use of N-hydroxysuccinimide (NHS) ester. Due to the high reactivity of NHS at physiological pH, numerous classes of bioactive compounds such as carbohydrates, peptides, nucleic acid, and antibody may be conjugated under mild conditions with preservation of their conformation and activity. This coupling method requires the presence of a coupling agent such as a carbodiimide to activate carboxylic acid groups. This intermediate then reacts with nucleophiles (mainly amines in the case of active targeting) to form an amide bond.

Today, conjugation between molecules and NPs tends to be performed after the formulation step either directly on PLGA (Dawson and Halbert, 2000, Liang et al., 2005, 2006, Nobs et al., 2006) or on the PEG surrounding the NP core (Stella et al., 2000, 2007, Farokhzad et al., 2006, Bae et al., 2007, Garinot et al., 2007, Zhang et al., 2007, Pan and Feng, 2009, Wang et al., 2009, Gan and Feng, 2010). For example, despite its relatively high molecular weight, an aptamer has been successfully inserted onto the PEG corona of PLGA NPs after activation of the carboxy-modified PEG on the NP surface (Farokhzad et al., 2006). This reaction was made possible by the presence of an accessible amine group on the aptamer chemical structure. Similarly, folic acid, which possesses carboxylic acid groups, has often been conjugated onto NH_2 NPs after NHS activation (Stella et al., 2000, 2007, Bae et al., 2007, Pan and Feng, 2009). Recently, Chittasupho et al. (2009) have conjugated a peptide on the surface of the NP in the presence of "functional surfactant." Since it was demonstrated that in the case of hydrophobic NPs, the surfactant is located at the surface, the authors used a modified Pluronic® F-127 bearing activated ester groups in order to conjugate their peptide after formulation of the NP. The relatively high calculated density of peptide at the surface shows the good accessibility of Pluronic-COOH coated on the NP surface.

Rather than coupling the ligand on a preformed NP, another way consists in functionalizing the polymer with a targeting moiety before formulation (Schiffelers et al., 2004, Costantino et al., 2005, Tosi et al., 2007, Patil et al., 2009, Prabaharan et al., 2009, Cao et al., 2010, Tsai et al., 2010). Direct conjugation to pre-synthesized NPs may be hindered by a lack of reactivity or the reaction conditions could be detrimental for the ligand and/or the NPs. To overcome these limitations, Cao et al. (2010) have synthesized a well-defined PLLA star polymer terminated with carboxylic acid groups. These were then activated by NHS in order to conjugate the amine-functionalized folic acids before the formulation step. Similarly, Costantino et al. (2005) have coupled short peptides to the activated carboxy groups of PLGA in solution by means of an amide linkage. After NP formulation, presence of peptide moieties on the NP surface was evidenced by electron spectroscopy for chemical analysis (ESCA).

Another well-established approach to introducing ligands onto NP relies on the high and specific reactivity of sulfydryl groups for maleimides. Introduction of thiol groups either to the NPs or to the ligands opens up many new possibilities for ligand conjugation. Interestingly, thiol groups are naturally present in proteins through cysteine residues. Otherwise, this coupling method generally requires pretreatment of the ligand with 2-iminothiolane (Traut's reagent) in order to incorporate a sulfhydryl group and thereby enable it to attach to the activated NPs. Thus, many ligands such as EGF1 protein for tissue factor targetting (Mei et al., 2010), transferrin (Ulbrich et al., 2009) or lactoferrin (Hu et al., 2009), glycoprotein, apolipoprotein E (Michaelis et al., 2006, Kreuter et al., 2007, Zensi et al., 2009), and antibodies (Cirstoiu-Hapca et al., 2007, Pang et al., 2008) have been successfully conjugated to the surface of NPs and their specificity and sensitivity have been retained.

As an alternative to using maleimide for conjugation with thiol compounds, Gullberg et al. (2006) derivatized a Pluronic with a pyridyl disulfide moiety, a good leaving group for the attachment of thiols. After formulation of the NPs with this activated surfactant, specific attachment of the peptide RGD was carried out using its terminal cysteine residue. Other reactions, such as addition of thiols to carbon double bond (the so-called thiol-ene coupling), have also been investigated, allowing the fixation of various ligands as antibodies and peptides (van der Ende et al., 2009, 2010).

A classical coupling method using standard carbodiimide chemistry for attachment of amines or carboxylic acids has also often been employed. According to the nature of ligands, the carbodiimides were either water-soluble (EDC) or water-insoluble (DCC). As an example, folic acid has been conjugated via its carboxylic acid group to PLA-b-PEG (Zhang et al., 2010) and PCL-b-PEG (Chen et al., 2008) polymers using DCC in anhydrous dimethyl sulfoxide (DMSO). NPs decorated with folate moieties were then prepared and evaluated.

Other less studied routes have been proposed that could offer an interesting alternative to produce functionalized NPs. One of these routes consists in attaching a glycoprotein to the hydroxyl groups of the PVA adsorbed onto the NP surface through a multifunctional epoxy compound (Sahoo and Labhasetwar, 2005). Under mild basic conditions, the amine group of transferring opens the ring to form a stable amide linkage. In 2008, Rao et al. (2008) used the same method to introduce a *trans*-activating transcriptor peptide onto PLA NPs. Recently, Yu et al. (2010) have explored another approach using the N-terminal PEGylation technique. For this purpose, the authors synthesized an amphiphilic copolymer, aldehyde-PEG-b-PLA, to prepare NPs. These were then functionalized with a peptide attached to the aldehyde groups located at the NP surface. The selective N-terminal attachment of ligand to the end group of PEG chain allows a single-positioned PEGylation and usually conserves the peptide's conformation and biological activity.

5.4.1.2.2 Click Chemistry

Copper(I)-catalyzed Huisgen 1,3-dipolar cycloaddition reaction between an azide and an alkyne (CuAAC) (Figure 5.2) belongs to the class of chemical reactions, often referred as "click" chemistry, that share several very important features: (i) a very high efficiency in terms of both conversion

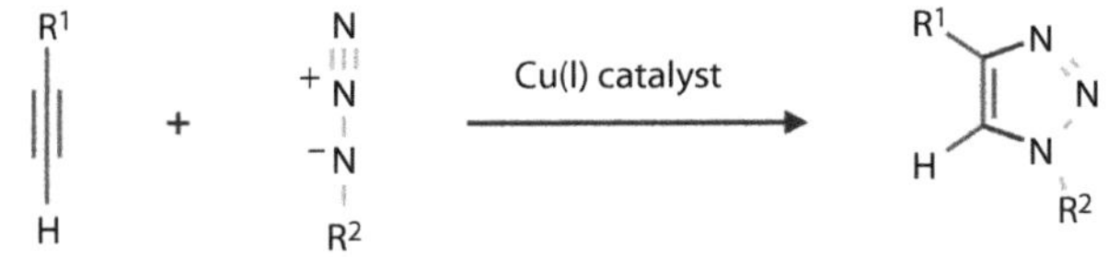

FIGURE 5.2 **(See companion CD for color figure.)** Copper(I)-catalyzed Huisgen 1,3-dipolar cycloaddition reaction between an azide and an alkyne (CuAAC, also termed click chemistry).

and selectivity; (ii) mild experimental conditions; (iii) a simple workup; and (iv) few or no by-products (Kolb et al., 2001, 2004). Click chemistry has recently attracted tremendous interest in many research areas and especially in polymer/material science and bioconjugation (Lutz, 2007, Le Droumaguet and Velonia, 2008, Le Droumaguet and Nicolas, 2010). In addition, CuAAc can be readily performed in aqueous solutions using an appropriate catalyst, thus representing a relevant approach if functionalization of preformed NPs is envisaged. However, in contrast to traditional coupling approaches using ligands with native functional groups (amine, thiol, carboxylic acid, etc.), the click chemistry pathway requires both the NP/copolymer and the ligand to be derivatized with alkyne and azide groups prior to the coupling reaction.

Click chemistry for surface-functionalized NPs was applied to PEGylated PACA copolymers/ NPs carrying azide moieties at the extremity of the PEG chains, acting here as a clickable scaffold. Model and biologically active alkynes derivatives were quantitatively coupled either to the copolymer in homogeneous medium followed by self-assembly in aqueous solution or directly at the surface of the preformed azide-decorated NPs in aqueous dispersed medium, both methods yielding highly functionalized NPs (Nicolas et al., 2008, Le Droumaguet et al. 2011).

A combination of click chemistry and ring-opening polymerization (ROP) has also been reported, allowing clickable polyester homopolymers and copolymers to be synthesized (Riva et al., 2007, Jiang et al., 2008). However, this method requires the specific design of functional monomers and is mainly applied to functionalization of the NP core.

Recently, Lu et al. (2009) derivatized a poly(2-methyl-2-carboxytrimethylene carbonate-*co*-D,L-lactide) (poly(TMCC-*co*-LA)) copolymer with azido-PEG pendant chains via EDC chemistry to form the corresponding amphiphilic copolymer, which was able to self-assemble into well-defined NPs. Alkyne-modified KGRGDS peptides were then synthesized and coupled to the azide-functionalized NPs via click chemistry.

A bifunctional initiator has allowed ROP of lactide and controlled/living radical polymerization of PEG-based macromonomers (bearing glucopyranoside molecules linked by click chemistry) to be sequentially performed by divergent chain growth (Jubeli et al., 2010). The amphiphilic feature of this comb-shaped construct led to well-defined NPs decorated with sugar moieties, the accessibility of which was confirmed by their agglutination by concanavalin A.

5.4.1.3 Non-Covalent Coupling Strategies

5.4.1.3.1 Avidin/Biotin

Avidin, neutravidin, and streptavidin exhibit four biotin-binding sites that can be exploited for conjugation purposes with NPs. To immobilize the biotin-binding protein at their surface, a convenient strategy is to use an heterobifunctional PEG spacer such as NHS–PEG–maleimide (Michaelis et al., 2006) (also widely used to directly link a ligand to NPs), bearing on one side a NHS group for coupling to the NPs and on the other side a maleimide moiety suitable for conjugation to the biotin-binding protein via sulfhydryl modification. Linkage to NPs is then achieved by amidation whereas the protein is linked via thioether formation. The final nanoassemblies are obtained by simple incubation with the desired biotinylated ligand, the synthesis of which is usually achieved by traditional coupling chemistry (biotin carries a primary carboxylic acid function far from its binding site). The coupling of biotin-binding proteins to the NPs can also be directly performed by standard DCC-assisted coupling chemistry without any spacer (Haun and Hammer, 2008, Lin et al., 2010). In this case, NPs are previously activated with DCC and further reacted with the protein. In these examples, NPs must have a suitable functional group at their surface such as NH_2 or COOH for the coupling to take place.

The reverse strategy, which consists in the preparation of biotinylated NPs, requires the coupling of the ligand to the biotin-binding protein (Aktaş et al., 2005). This is achieved by traditional coupling chemistries using cysteine or lysine residues from the protein. Biotin can be inserted at the extremity of PEG chains by conventional coupling, as was exemplified by the design of OX-26 decorated PEGylated chitosan NPs by Aktaş et al. (2005) using avidin as biotin-binding protein.

5.4.1.3.2 Other Non-Covalent Approaches

Although covalent linkages or biotin-based binding are believed to be the most robust approaches with respect to the stability of the corresponding nanocarriers, other non-covalent routes have been used to prepare targeted NPs. For example, Blackwell et al. (2001) reported the simple adsorption of monoclonal antibody recognizing E- and P-selectin adsorbed at the surface of polystyrene NPs.

Interestingly, different routes have been proposed to obtain hyaluronic acid (HA)-coated PCL nanospheres: (i) coating the core by chain entanglement with HA, (ii) coating the nanospheres by a simple HA adsorption, or (iii) coating the nanospheres by electrostatic interactions between negatively charged HA and a cationic surfactant (stearylamine or benzalkonium chloride) (Barbault-Foucher et al., 2002). The best results in terms of HA coating were observed when the positively charged surfactant was used.

5.4.2 Applications of Targeted NPs

5.4.2.1 Targeting to Tumors

5.4.2.1.1 Targeting with Folate

The receptor for folic acid is overexpressed on many tumor cell lines (Zhao et al., 2008). As a result, this small molecule has been coupled to a wide range of active molecules (protein toxins, immunomodulators, anticancer drugs) and particles (liposomes, NPs, imaging agents) for the diagnosis and treatment of cancer (Hildgenbrink and Low, 2005). This section will review some recent results obtained with NPs prepared from synthetic polymers.

Patil et al. (2009) have described a simple method for the preparation of targeted PLA-*b*-PEG nanospheres. Amphiphilic copolymers with ligands on the distal end of the PEG, as described earlier, were assembled by emulsification—evaporation of solvent. Nanospheres bearing two different functional groups could be obtained in this way. Thus, nanospheres loaded with paclitaxel carrying both folic acid and biotin were prepared. These were able to deliver drug to MCF-7 cells, which express receptors for both these ligands, in vitro, and showed a better antitumoral effect than singly targeted nanospheres in nude mice. PCL-*b*-PEG copolymers have also been functionalized with folate (Chen et al., 2008). These were coupled by the PCL end to a highly branched Boltorn H40 polymer to form nanospheres that were able to deliver paclitaxel and 5-fluorouracil to HeLa, A549, and NIH 3T3 cells in vitro. A stronger growth inhibitory effect was noted on the folate receptor–expressing tumor cell lines than on the "normal" NIH 3T3 cells.

PLA and PLGA have also been used as a basis for amphiphilic polymers by conjugation with α-tocopheryl polyethylene glycol succinate (TPGS). Nanospheres formed from these copolymers can be functionalized by adding TPGS folate to the preparation. They have been used to carry doxorubicin (Zhang et al., 2007), paclitaxel (Pan and Feng, 2008), and quantum dots for imaging (Pan and Feng, 2009). The folate targeting strategy has also been applied to long-circulating nanospheres prepared from a cyanoacrylate-based polymer (Stella et al., 2000) (Figure 5.3). Conjugation of folic acid to the distal end of the PEG chain does not affect its ability to bind to its receptor, as determined by surface plasmon resonance. Fluorescent, folate-bearing nanospheres were internalized by folate receptor–positive KB3-1 cells but not by MCF-7 receptor–negative cells (Stella et al., 2007).

Nanocapsules have also been targeted with folate. Bae et al. (2007) prepared PEO-*b*-PPO-*b*-PEO/PEG shell cross-linked nanocapsules encapsulating an oil phase loaded with paclitaxel. By flow cytometry, they showed that targeting increased uptake by KB cells, resulting in the onset of apoptosis. Other colloidal systems based on synthetic polymers, such as polymerosomes (Upadhyay et al., 2010) and copolymer micelles (Le Garrec et al., 2004), have also been targeted to cancer cells by folate.

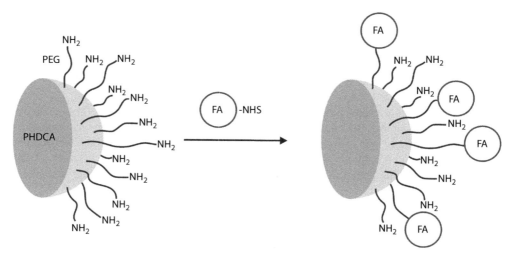

FIGURE 5.3 **(See companion CD for color figure.)** Conjugation of folic acid to poly[(hexadecyl cyanoacrylate)-*co*-aminopoly (ethylene glycol) cyanoacrylate] (P(HDCA-*co*-H₂NPEG CA)) copolymer via an amidation pathway.

Targeting to the folate receptor is a useful strategy for gene and oligonucleotide delivery. In a recent example, folate was conjugated to branched PEI–PEG polymers capable of condensing plasmid DNA. Transfection of cell lines expressing the folate receptor (C6, HEK 293T) was improved in this way (Liang et al., 2008). Folate-linked NPs have also been used to carry a suicide gene (thymidine kinase) (Hattori and Maitani, 2005) and siRNA (Yoshizawa et al., 2008) to prostate cancer cells.

5.4.2.1.2 Targeting with RGD Peptides

Peptides containing the arginine–glycine–aspartic acid (RGD) sequence show strong binding to the αVβ3 integrin, which is frequently expressed by tumor cells and by tumor-associated endothelium. As a result, ligands incorporating this sequence are being investigated for targeting a range of drug carriers and imaging agents to tumors and their vasculature (Temming et al., 2005). Specificity and stability are improved when the RGD sequence is incorporated in a longer, cyclic, peptide. Recently, multimeric structures have been synthesized to improve their affinity (Garanger et al., 2007).

The cRGD peptide has been conjugated to nanospheres based on PLGA. The PLGA was first linked to doxorubicin before forming NPs by the solvent diffusion method. A heterobifunctional PEG was linked to the terminal carboxyl groups of the PLGA and the RGD peptide was coupled to the distal end. The nanosphere diameter was about 400 nm. Despite this large size, uptake was seen in various integrin-expressing cell lines, with induction of apoptosis by the doxorubicin (Wang et al., 2009). A different synthetic approach was adopted by Garinot et al. (2007). Nanospheres were formed by the "double emulsion method" from mixtures of PLGA, PLGA-*b*-PEG, and PCL-*b*-PEG. The RGD-containing peptide was coupled to the surface of the nanospheres by photografting. Ovalbumin was encapsulated within these nanospheres, which were intended for administration by the oral route, to target M cells (see later).

Click chemistry, as described earlier, has been used to couple a hexapeptide containing the RGD sequence to a terminal amino group on PEG carried by nanospheres with a core of poly(2-methyl-2-carboxytrimethylene carbonate-*co*-D,L-lactide) (Lu et al., 2009). The binding of these nanospheres to integrins was confirmed using rabbit corneal epithelial cells. Doxorubicin was again the encapsulated drug in a formulation consisting of an inulin–methacrylate core with a cyclic RGD peptide coupled via PEG-400 (Bibby et al., 2005). The pharmacokinetics and distribution of the drug were determined in mice bearing spontaneous mammary tumors. Although the highest accumulation was in the liver, some accumulation in tumor tissue was observed.

Like folate targeting, the RGD approach has been applied to the delivery of nucleic acids. In this case, the polymer of choice is PEI. In an attempt to target PEI/DNA polyplexes, Ng et al. (2009) attached RGD–gold nanoclusters at the surface and observed greatly increased transfection of HeLa cells with high integrin expression. A similar strategy has been used for siRNA (Schiffelers et al., 2004). PEI conjugated with PEG-bearing RGD in a cyclic decapeptide at the distal end formed nanoplexes spontaneously with siRNA. These provided sequence-specific expression of vascular endothelial growth factor receptor 2 in HUVEC, and inhibition of both tumor growth and angiogenesis in mice.

5.4.2.2 Targeting to Endothelium

5.4.2.2.1 Vascular Endothelium in Disease Sites

The adhesion molecules which are selectively expressed on endothelial cells during infection or inflammation to promote leukocyte arrest and extravasation can be considered as valid targets for drug delivery systems. The first interaction of leukocytes is with the selectins, which allow rolling along the endothelium by multiple, weak interactions. These are succeeded by stronger interactions with cellular adhesion molecules such as ICAM-1. There are several reports in the literature of attempts to target polymeric NPs to these receptors.

A nonselective selectin ligand (a modified sugar) was synthesized and coupled to a PLA homopolymer bearing pendant carboxyl groups and rhodamine as a fluorescent tag by Banquy et al. (2008). Nanospheres prepared from these polymers bound to activated human umbilical vein endothelial cells that expressed both E- and P-selectin. Recently, in our laboratory, amphiphilic copolymers based on D,L-lactide and PEG macromonomer as the hydrophilic part have been prepared and decorated with glucose as a model sugar by click chemistry (Jubeli et al., 2010). These polymers form small nanospheres by nanoprecipitation, exposing the sugar at the surface. The ultimate aim is to attach sialyl Lewis[x], the physiological ligand for E-selectin, in order to target activated endothelium for the treatment of rheumatoid arthritis.

Lin et al. (2010) have used the biotin–avidin system to conjugate glycocalicin, a ligand expressed on activated platelets, to fluorescent polystyrene nanospheres. They were able to demonstrate binding to both P-selectin-coated surfaces and activated endothelial cells, under flow conditions mimicking the physiological situation. Similar observations had been made by Blackwell et al. (2001) using a monoclonal antibody recognizing E- and P-selectin adsorbed onto the surface of polystyrene nanospheres. Haun and Hammer (2008) used a similar approach to study the binding of particles decorated with antibodies against ICAM-1 using the avidin–biotin system to surfaces carrying ICAM. They were able to study the effects of receptor and ligand valency as well as the shear rate.

Chittasupho et al. (2009) attempted to target PLGA nanospheres with a cyclic peptide binding ICAM-1. The peptide was coupled to the surfactant Pluronic 127, which was then used in the formation of the nanospheres. Doxorubicin encapsulated within these nanospheres could be delivered to A549 lung epithelial cells in vitro.

5.4.2.2.2 Targeting to the Blood–Brain Barrier

Drug delivery to the brain is a challenge because of the many mechanisms that protect the brain from the entry of foreign substances. Numerous molecules which could be active against brain disorders are not clinically useful due to the presence of the BBB. NPs can be used to deliver these drugs to the brain (Kreuter, 2001, Garcia-Garcia et al., 2005a, Olivier, 2005, Kim et al., 2007a). Encapsulation within colloidal systems can allow the passage of non-transportable drugs across the BBB by masking their physicochemical properties. It should be noted that the status of the BBB is different depending on the brain disease. In fact, in some pathological situations such as tumors or inflammatory disorders, the permeability of BBB is increased allowing very easy translocation of carriers. However, whatever the status of the BBB is, it is necessary to use long-circulating NPs to reach the brain.

5.4.2.2.2.1 Brain Targeting Using "Stealth" NPs   To deliver drug to the brain tissue, whether or not the BBB is disrupted, two types of "Stealth" polymeric NPs have been designed: surfactant-coated NPs and PEGylated NPs.

5.4.2.2.2.1.1   Surfactant-Coated NPs   In earlier studies from the group of Kreuter, molecules such as dalargin (Kreuter et al., 1995, Schroeder et al., 1998) and loperamide (Alyautdin et al., 1997) have been loaded onto NPs with the aim of brain delivery. After peripheral administration, these molecules themselves do not exhibit any therapeutic effect because they do not diffuse through the BBB. However, when dalargin or loperamide were adsorbed onto the surface of poly(n-butyl cyano-acrylate) (PBCA) NPs further coated with the detergent, polysorbate 80 (PS-80), a pronounced anal-gesic effect was obtained, reaching a maximum 45 min after administration. In vivo experiments in mice have clearly shown that the analgesic effect of dalargin was obtained only when the drug was pre-adsorbed onto the NPs, whereas a simple mixture of dalargin and PBCA NPs did not show any analgesic effect. It is noteworthy that only polysorbate 20, 40, and 80 led to a significant antinocicep-tive effect in mice whereas the dalargin-loaded NPs coated with other types of surfactant did not.

It has been reported that apolipoproteins (APO) could be involved in the brain penetration of PBCA NPs coated with PS-80 (Kreuter et al., 2002). Indeed, a study has been performed using PBCA NPs loaded with dalargin or loperamide and further coated with the APO-A, -B, -C, -E, or -J (with or without precoating with PS-80). The antinociceptive effect was measured in mice by the tail-flick test. In these conditions, the effect was significantly higher after both PS-80 precoat-ing and APO-B or -E coating. No antinociceptive effect was seen after coating with the other APO. Interestingly, in APO-E-deficient mice, the antinociceptive effect was reduced compared with normal mice after injection of the PS-80-coated NPs. This suggests that the PS-80 could act as an anchor for APO-B and -E at the surface of the NPs, which are then able to interact with LDL receptor (LDLR), before being taken up by brain endothelial cells by receptor-mediated endocytosis (Borchard et al., 1994, Kreuter et al., 1995). Thereafter, the drug may be released in these cells and diffused into the brain interior or the particles may be transcytosed (Kreuter, 2001).

These PS-80-coated NPs have been used to deliver other molecules to the brain, such as MRZ 2/576 which is a potent but rather short-acting anticonvulsant drug after intravenous administration. It was observed that the administration of MRZ 2/576 loaded onto PS-80-coated NPs prolonged the duration of the anticonvulsive activity (Friese et al., 2000). Doxorubicin was also adsorbed onto these NPs for the treatment of experimental glioblastoma. Rats treated with this formulation showed significantly higher survival times (Steiniger et al., 2004). Moreover, the acute toxicity of doxoru-bicin was reduced when it was associated with PS-80-coated NPs (Gelperina et al., 2002) and the efficacy of these NPs appeared to be due to antiangiogenic effects (Hekmatara et al., 2009). A recent study examined the antiparkinsonian effect of nerve growth factor adsorbed on the surface of PS-80-coated NPs after IV injection into C57Bl/6 mice with parkinsonian syndrome (Kurakhmaeva et al., 2008). The results showed decreased rigidity and increased locomotor activity compared with con-trol mice without NP treatment and a persistant effect 7 and 21 days after a single injection.

5.4.2.2.2.1.2   PEGylated NPs   PEGylated NPs were prepared using a poly[methoxypoly(ethylene glycol)-co-(hexadecyl cyanoacrylate)] (P(MePEGCA-co-HDCA)) (MePEGCA/HDCA, 1:4) copo-lymer (Peracchia et al., 1998, Nicolas et al., 2008). In this technology, the PEG is covalently attached to the NPs, avoiding the possibility of PEG desorption. These P(MePEGCA-co-HDCA) NPs have shown long-circulating properties in vivo (Peracchia et al., 1999). They have demonstrated their capacity to reach the brain in a model of experimental allergic encephalomyelitis (EAE) rats (Calvo et al., 2002) in which the BBB permeability is increased by inflammation. Two mechanisms have been proposed: passive diffusion as a result of the increase of BBB permeability and transport by NP-containing macrophages infiltrating into these inflammatory tissues. Furthermore, these PEGylated NPs also showed a higher uptake by the brain of scrapie-infected animals, which may be useful for targeting drugs for the treatment of prion disease (Calvo et al., 2001b).

After intravenous administration to rats bearing well-established intracerebral gliosarcoma (Brigger et al., 2002), P(MePEGCA-*co*-HDCA) NPs accumulated preferentially in the tumor tissue, rather than in the peritumoral brain tissue or in the healthy controlateral hemisphere. Interestingly, P(MePEGCA-*co*-HDCA) NPs were concentrated much more in the gliosarcoma than were their non-PEGylated counterparts (PHDCA NPs). P(MePEGCA-*co*-HDCA) NPs did not display any toxicity toward the BBB and penetrated by a mechanism of diffusion/convection, leading to the extravasation of the NPs into the tumor site. Second, in normal brain, an affinity of the P(MePEGCA-*co*-HDCA) NPs for the endothelial cells of the BBB seemed to allow their translocation (Brigger et al., 2002). In order to confirm the influence of hydrophilic PEG chains at the NP surface on their translocation across the BBB, in vitro experiments were performed using primary cultures of rat brain endothelial cells (RBEC) and rat astrocytes in bicompartmental cell culture (Garcia-Garcia et al., 2005b). Experiments studying cell internalization and intracellular distribution of NPs in RBEC have confirmed the in vivo results and suggest that the cellular uptake of P(MePEGCA-*co*-HDCA) NPs was due to specific endocytosis (Garcia-Garcia et al., 2005c). Moreover, several different analytical methods were combined to study the plasma protein adsorption onto the surface of P(MePEGCA-*co*-HDCA) NPs (Kim et al., 2007b). The results suggested that APO-E and/or APO-B-100 is adsorbed onto the surface of P(MePEGCA-*co*-HDCA) NPs after intravenous administration. More recently, it has been shown that NP internalization by RBEC was inhibited by anti-LDLR monoclonal antibody (Kim et al., 2007c), while it was enhanced by the preincubation of these NPs in APO-E or APO-B-100 solutions (Kim et al., 2007d). All these results led to the proposition of the following mechanism: after intravenous administration, the adsorption of APO-E and/or APO-B-100 on the surface of P(MePEGCA-*co*-HDCA) NPs would allow recognition by LDLR on endothelial cells and their receptor-mediated endocytosis. PEG–PHDCA NPs appear to be a potential brain drug delivery system in conditions of both intact and permeabilized BBB.

5.4.2.2.2.2 Active Targeting by Surface-Modified NPs   Active targeting using long-circulating PEGylated nanocarriers can further increase specific tissue accumulation as a result of molecular recognition processes. Therefore, specific antibodies and ligands such as proteins which recognize receptors expressed on endothelial cells of brain capillaries have been conjugated to long-circulating nanocarriers (Olivier, 2005, Kim et al., 2007a). They might control the delivery of drugs both by prolonging drug circulation and by targeting the drug to the site of action in a specific manner. This approach may be the most striking advance in BBB targeting and translocation. Numerous studies have investigated the targeting of transferrin (Tf) and LDLR by coupling antibodies or ligands to NPs.

5.4.2.2.2.2.1   Targeting to the Transferrin Receptor   Active targeting of the transferrin receptor (TfR) has been extensively investigated with NPs coupled to antibody (OX26) or ligand (Tf).

5.4.2.2.2.2.1.1   Monoclonal Antibody: OX26   One active targeting strategy for brain tissue which has been extensively used (especially on liposomes) is based on the use of OX26, a TfR antibody. The TfR is a 180 kDa dimeric transmembrane glycoprotein which is ubiquitously expressed in various tissues including the luminal membrane of the brain capillary endothelium and is overexpressed on tumor and leukemia cells (Qian et al., 2002, Staber et al., 2004). The mechanism of TfR-mediated endocytosis in cells is also well-documented (Pardridge et al., 1987). OX26 is directed against a functional group of TfR which is not the Tf-binding site. Thanks to this characteristic, the efficacy of OX26 is not limited by the competition with endogenous Tf which has high concentration in plasma (Jefferies et al., 1984). This strategy does not allow the receptor-mediated endocytosis but increases the accumulation of NPs at the surface of the endothelial cells. Although antibody conjugation provides nanocarriers with high affinity for the corresponding antigen-expressed cells, in general, immunoparticulate (antibody-conjugated) system has some limitations which hinder their clinical application: (i) the targeting ligand may be immunogenic, (ii) the rigorous procedure of conjugation chemistry could lead to a partial denaturation of the antibody, and (iii) rapid clearance in blood could occur due to Fc receptor–mediated MPS uptake.

Chitosan NPs conjugated with PEG bearing the OX26 have been developed (Aktaş et al., 2005). Streptavidin–OX26 and biotinylated PEG–chitosan NPs were conjugated through avidin–biotin association and the peptide Z-DEVD-FMK, a caspase-3 inhibitor, was encapsulated in these NPs. The conjugation of OX26 facilitated the controlled release of the peptide from the NPs. Moreover, in vivo studies showed that OX26 conjugation allowed the NPs to reach the brain of healthy animals after intravenous administration (Aktaş et al., 2005) and dose dependently decreased the infarct volume, neurological deficit, and ischemia-induced caspase-3 activity in mice subjected to 2 h of MCA occlusion and 24 h of reperfusion, suggesting that they released an amount of peptide sufficient to inhibit caspase activity (Karatas et al., 2009).

More recently, PEG-*b*-PCL NPs were conjugated with OX26 by synthesis of diblock copolymers of PEG-*b*-PCL and maleimide (MAL)–PEG-*b*-PCL (Pan and Feng, 2008). Experiments on brain delivery in rats demonstrated that the increase of surface OX26 density of these NPs decreased the area under the curve of their blood pharmacokinetics, because they were accumulated in the brain. Moreover, when NC-1900 was encapsulated as a model peptide, these optimized NPs improved the scopolamine-induced learning and memory impairments in a water maze task after intravenous administration.

5.4.2.2.2.2.1.2 Transferrin   An alternative strategy to target TfR for brain delivery is the coupling of Tf at the surface of the NPs. In this approach, which leads to specific receptor-mediated endocytosis, the entrapped drug may accumulate more efficiently within the cells. Serum Tf transports ferric ions from the sites of uptake into the systemic circulation to the cells and tissues (Li and Qian, 2002). After binding of Tf to its receptors on the cell surface, the TfR–Tf complexes are internalized to form endosomes through clathrin-coated vesicles by the process of receptor-mediated endocytosis. The iron-loaded Tf releases its iron at the low pH of the endosomes and iron-free Tf remains bound to the receptor. These complexes are further released at the cell surface by means of exocytic vesicles. Tf-conjugated NPs have been developed in order to mimic the Tf cycle (Chiu et al., 2006). Internalization by receptor-mediated endocytosis is observed with Tf-conjugated NPs, but to avoid interference from the PEG chain and to increase the accessibility of Tf at the surface, Tf must be attached on the PEG chain terminus (Kim et al., 2007a). Nevertheless, the main drawback is that high endogenous plasma concentration of Tf may act as a competitor for binding to the receptors. However, the possibility of disruption in iron transport cannot be overlooked. Furthermore, after internalization by receptor-mediated endocytosis, the drug has to escape the endosome/lysosome compartment, a process that may depend on the method of drug encapsulation and NP composition (Andresen et al., 2005).

The strategy of Tf-coupled NPs has been applied with success in brain drug delivery. PEGylated albumin NPs loaded with AZT as a water-soluble antiviral drug were coupled with Tf using chemical cross-linking by glutaraldehyde. A significant enhancement ($21.1\% \pm 1.8\%$) of brain localization of AZT was observed 4 h after intravenous administration in rats compared with nontargeted NPs. This high accumulation could be due to (i) the shielding of MPS uptake by PEG chain, (ii) effective receptor-mediated endocytosis, and (iii) the contribution of the positively charged NP surface (Mishra et al., 2006).

Recently, Tf- and TfR antibody (OX26 or R17217) were covalently coupled to human serum albumin (HSA) NPs using the NHS–PEG–MAL-5000 cross-linker (Ulbrich et al., 2009). Since loperamide does not cross the BBB, it was used as a model drug and was loaded onto the NPs by adsorption. Loperamide-loaded HSA NPs with covalently bound Tf or the OX26 or R17217 antibodies induced significant antinociceptive effects in the tail-flick test in ICR (CD-1) mice after intravenous injection. Control loperamide-loaded HSA NPs with IgG2a antibodies produced only marginal effects.

All these studies have demonstrated that both strategies of targeting the transferrin receptor, using the protein itself or antibodies to the receptor coupled to NPs, are able to transport a drug across the BBB.

5.4.2.2.2.2.2 Targeting to the LDLR   As a result of the observation that brain translocation of "Stealth" NPs (PS-80–PBCA and P(MePEGCA-*co*-HDCA) NPs) was favored by APO-E adsorption at their surface followed by LDLR recognition, some work has been done to investigate the covalent coupling of APO at the surface of NPs.

5.4.2.2.2.2.2.1   APO-E   Michaelis et al. (2006) have reported the preparation of HSA NPs with covalently attached APO-E using a bifunctional MAL-PEGNHS linker reacting on one side with amino groups on the particle surface and on the other with thiol groups of the APO-E. In a recent study (Zensi et al., 2009), HSA NPs with and without APO-E were injected intravenously into SV 129 mice. After 15–30 min, the animals were sacrificed and their brains were examined by transmission electron microscopy. Whereas no uptake into the brain was detectable with NPs without APO-E, the NPs with covalently bound APO-E were detected in brain capillary endothelial cells and neurones. The authors also showed in vitro uptake of these NPs into mouse endothelial (b.End3) cells and their intracellular localization. These results demonstrated that NPs with covalently bound APO-E are taken up into the cerebral endothelium by an endocytic mechanism followed by transcytosis into brain parenchyma. Moreover, the antinociceptive effects of loperamide-loaded HSA NPs that were covalently linked to native APO-E and to variants that do not recognize lipoprotein receptors were compared (Michaelis et al., 2006). Antinociceptive effects in the tail-flick test in ICR mice after intravenous injection were only observed when the APO-E was covalently linked to HSA NPs. The linkage to APO-E variants failed to induce these effects, indicating that APO-E attached to the surface of NPs facilitated the transport of the drug across the BBB, probably after interaction with lipoprotein receptors on the brain capillary endothelial cell membranes.

5.4.2.2.2.2.2.2   Other APO   In another work, the same team has covalently attached APO-E3, A-I, and B-100 to HSA NPs via the MAL–PEG–NHS-3400 linker (Kreuter et al., 2007). The antinociceptive reaction of loperamide-loaded NPs was evaluated after intravenous injection in mice by the tail-flick test. All three NP preparations yielded considerable antinociceptive effects after 15 min that persisted for more than 1 h, whereas the loperamide solution produced no effect. The peak effects of these preparations with the coupled APO-E3, A-I, and B-100 amounted to 95%, 65%, and 50% of the maximum possible effect, respectively. This result suggested that different mechanisms could be involved in the interaction of NPs with the brain endothelial cells and the resulting delivery of drugs to the central nervous system.

5.4.2.2.2.2.3   Other Strategies
5.4.2.2.2.2.3.1   Absorptive-Mediated Transcytosis by Cationic Albumin-Conjugated PEGylated NPs   Cationic bovine serum albumin (CBSA) conjugated with PEG-*b*-PLA NPs (CBSA-NP) has been developed and the effects of CBSA and CBSA-NP were evaluated (Lu et al., 2005). The results in vitro demonstrated a higher uptake of CBSA-NP by rat brain capillary endothelial cells than of native BSA-NP. After CBSA-NP or BSA-NP injection in mice caudal vein, fluorescent microscopy of brain coronal sections showed a higher accumulation of CBSA-NP in the lateral ventricle, third ventricle, and periventricular region than of BSA-NP. By using free CBSA as specific inhibitor, it was shown that CBSA-NP crossed the brain capillary endothelium by absorptive transcytosis (Lu et al., 2007a). Furthermore, a so-called accelerated blood clearance phenomenon was observed when evaluating the blood clearance profiles of CBSA-NP after a single injection or over a period of successive high doses of CBSA-NP. Moreover, aclarubicin (ACL) was incorporated into these NPs to determine its therapeutic potential in rats with intracranially implanted C6 glioma cells (Lu et al., 2007b). A significant increase of median survival time was found in the group of animals treated with these NPs compared with that of saline-treated control animals, animals treated with NPs without ligand and ACL solution, suggesting that CBSA-NP are a promising brain drug delivery carrier with low toxicity.

5.4.2.2.2.2.3.2 Brain Targeting by Synthetic Opioid Peptides Linked to NPs   Costantino et al. (2005) synthesized conjugates of PLGA with five short peptides similar to synthetic opioid peptides, by means of an amidic linkage. Using an in vivo rat brain perfusion technique after femoral vein injection, fluorescent and confocal microscopy studies showed that peptide–PLGA NPs were able to cross the BBB while PLGA NPs could not. More recently, loperamide was loaded into NPs of PLGA derivatized with the peptide H_2N–Gly-L-Phe-D-Thr-Gly-L-Phe-L-Leu-L-Ser(O-β-D-glucose)–$CONH_2$ (g7) (Tosi et al., 2007). A pharmacological study demonstrated that g7 NPs crossed the BBB, were able to reach all the brain areas explored, and ensured a sustained release of the encapsulated drug.

5.4.2.3 Targeting by Other Routes

Although many of the targeted delivery systems described earlier are destined for intravenous administration, the possibility of targeting by other routes has not been ignored. In particular, different targeting systems have been proposed to increase bioadhesion by the oral route. One approach is the use of specific lectins to promote binding to particular sugars in the mucus or on the intestinal cell surface (Lehr and Puzstai, 1995, Irache et al., 1996, Russell-Jones et al., 1999, Clark et al., 2000). The biotin–avidin system is a convenient way of attaching lectins to NPs (Gref et al., 2003, Weiss et al., 2007). Lectin-mediated targeting has been used, for example, to deliver antigens across Peyer's patches for oral immunization (Gupta et al., 2007).

As mentioned earlier, another route of targeting to M cells in Peyer's patches is the use of RGD peptides (Garinot et al., 2007). Non-peptidic analogs of RGD were also effective, and led to a cell-mediated response to encapsulated ovalbumin (Fievez et al., 2009). Uptake of RGD-bearing fluorescent polystyrene nanospheres by follicle-associated epithelium from human Peyer's patches was demonstrated by Gullberg et al. (2006).

A biomimetic approach to intestinal targeting was adopted by Dawson and Halbert (2000), who coated PLGA nanospheres with bacterial invasin by covalent coupling. Association with epithelial cells was increased by coating, and inhibited by free invasin and by the RGD peptide. The intestinal transporter for vitamin B12 has also been used as a target (Chalasani et al., 2007). Vitamin B12 was conjugated to dextran nanospheres with various degrees of cross-linking, which were then swollen in a solution of insulin. Hypoglycemic effects and an increase in plasma insulin were observed in diabetic rats and mice after oral administration of targeted nanospheres, demonstrating that the carrier could promote uptake from the gastrointestinal tract.

As another example of targeting by an extravascular route, ocular delivery is interesting. PCL nanospheres have been coated with HA as a potential ocular delivery system (Barbault-Foucher et al., 2002). HA is a polysaccharide and is both bioadhesive and allows specific interactions with the antigen CD44. Although CD44 is expressed on normal epithelial cells, it is overexpressed in carcinoma cells and therefore also represents a possible target for anticancer therapy (Platt and Szoka, 2008).

5.5 CONCLUSIONS

Nanoparticulate systems prepared from polymers have been studied for several decades. However, no commercial formulations have yet appeared on the market, although doxorubicin-loaded PACA NPs have undergone clinical trials in Europe (Barraud et al., 2005). In order to prepare an effective system for targeting, both the ligand and the polymer making up the core of the NP must be tailored to the application. The ligand should give sufficient specificity to concentrate the encapsulated drug in the target tissue and avoid side effects in other tissues, and promote internalization of the particles if necessary, while the polymer core should be able to carry an adequate amount of the drug and allow its release at the target. Progress in both the methodology of synthesis and modification of polymers and in the understanding of biological recognition mechanisms that can be exploited for targeting should ensure that these goals can be met in the near future.

REFERENCES

Aboubakar, M., Couvreur, P., Pinto-Alphandary, H. et al. 2000. Insulin-loaded nanocapsules for oral administration: In vitro and in vivo investigation. *Drug Dev Res* 49: 109–117.

Aktaş, Y., Yemisci, M., Andrieux, K. et al. 2005. Development and brain delivery of chitosan–PEG nanoparticles functionalized with the monoclonal antibody OX26. *Bioconjug Chem* 16: 1503–1511.

Allémann, E., Gravel, P., Leroux, J.C., Balant, L., and Gurny, R. 1997. Kinetics of blood component adsorption on poly(D,L-lactic acid) nanoparticles: Evidence of complement C3 component involvement. *J Biomed Mater Res* 37: 229–234.

Alvarez-Romain, R., Naik, A., Kalia, Y.N., Guy, R.H., and Fessi, H. 2004. Skin penetration and distribution of polymeric nanoparticles. *J Control Release* 99: 53–62.

Alyautdin, R.N., Petrov, V.E., Langer, K., Berthold, A., Kharkevich, D.A., and Kreuter, J. 1997. Delivery of loperamide across the blood–brain barrier with polysorbate 80-coated polybutylcyanoacrylate nanoparticles. *Pharm Res* 14: 325–328.

Ameller, T., Marsaud, V., Legrand, P., Gref, R., Barratt, G., and Renoir, J.M. 2003a. Polyester–poly(ethylene glycol) nanoparticles loaded with the pure antiestrogen RU 58668: Physicochemical and opsonization properties. *Pharm Res* 20: 1063–1070.

Ameller, T., Marsaud, V., Legrand, P., Gref, R., and Renoir, J.M. 2003b. In vitro and in vivo biologic evaluation of long-circulating biodegradable drug carriers loaded with the pure antiestrogen RU 58668. *Int J Cancer* 106: 446–454.

Ameller, T., Marsaud, V., Legrand, P., Gref, R., and Renoir, J.M. 2004. Pure antiestrogen RU 58668-loaded nanospheres: Morphology, cell activity and toxicity studies. *Eur J Pharm Sci* 21: 361–370.

Ammoury, N., Fessi, H., Devissaguet, J.-P., Allix, M., Plotkine, M., and Boulu, R.G. 1990. Effect on cerebral blood flow of orally administered indomethacin-loaded poly(isobutylcyanoacrylate) and poly(D,L-lactide) nanocapsules. *J Pharm Pharmacol* 42: 558–561.

Ammoury, N., Fessi, H., Devissaguet, J.-P., Dubrasquet, M., and Benita, S. 1991. Jejunal absorption, pharmacological activity and pharmacokinetic evaluation of indomethacin-loaded poly(D,L-lactide) and poly(isobutyl-cyanoacrylate) nanocapsules in rats. *Pharm Res* 8: 101–105.

Andresen, T.L., Jensen, S.S., and Jorgensen, K. 2005. Advanced strategies in liposomal cancer therapy: Problems and prospects of active and tumor specific drug release. *Prog Lipid Res* 44: 68–97.

Avgoustakis, K. 2004. Pegylated poly(lactide) and poly(lactide-*co*-glycolide) nanoparticles: Preparation, properties and possible applications in drug delivery. *Curr Drug Deliv* 1: 321–333.

Bae, K.H., Lee, Y., and Park, T.G. 2007. Oil-encapsulating PEO–PPO–PEO/PEG shell cross-linked nanocapsules for target-specific delivery of paclitaxel. *Biomacromolecules* 8: 650–656.

Bakker-Woudenberg, I.A., Schiffelers, R.M., Storm, G., Becker, M.J., and Guo, L. 2005. Long-circulating sterically stabilized liposomes in the treatment of infections. *Methods Enzymol* 391: 228–260.

Banquy, X., Leclair, G., Rabanel, J.M. et al. 2008. Selectins ligand decorated drug carriers for activated endothelial cell targeting. *Bioconjug Chem* 19: 2030–2039.

Barbault-Foucher, S., Gref, R., Russo, P., Guechot, J., and Bochot, A. 2002. Design of poly-ε-caprolactone nanospheres coated with bioadhesive hyaluronic acid for ocular delivery. *J Control Release* 83: 365–375.

Barbet, J. 1995. Immunoliposomes. In *Liposomes: New Systems and New Trends in their Applications*, eds. F. Puisieux, P. Couvreur, J. Delattre, and J.-Ph. Devissaguet, pp. 159–191. Editions de Santé, Paris France.

Barratt, G., Puisieux, F., Yu, W.P., Foucher, C., Fessi H., and Devissaguet, J.P. 1994. Anti-metastatic activity of MDP-L-alanyl-cholesterol incorporated into various types of nanocapsules. *Int J Immunopharmacol* 16: 457–461.

Barraud, L., Merle, P., Soma, E. et al. 2005. Increase of doxorubicin sensitivity by doxorubicin-loading into nanoparticles for hepatocellular carcinoma cells in vitro and in vivo. *J Hepatol* 42: 736–743.

Baudin-Creuza, V., Chauvierre, C., Domingues, E. et al. 2008. Octamers and nanoparticles as hemoglobin based blood substitutes. *Biochim Biophys Acta* 1784: 1448–1453.

Bazile, D., Prud'Homme, C., Bassoulet, M.-T., Marlard, M., Spenlehauer, G., and Veillard, M. 1995. Stealth Me.PEG–PLA nanoparticles avoid uptake by the mononuclear phagocytes system. *J Pharm Sci* 84: 493–498.

Bertholon, I., Ponchel, G., Labarre, D., Couvreur, P., and Vauthier, C. 2006. Bioadhesive properties of poly(alkylcyanoacrylate) nanoparticles coated with polysaccharide. *J Nanosci Nanotechnol* 6: 3102–3109.

Bibby, D.C., Talmadge, J.E., Dalal, M.K. et al. 2005. Pharmacokinetics and biodistribution of RGD-targeted doxorubicin-loaded nanoparticles in tumor-bearing mice. *Int J Pharm* 293: 281–290.

Blackwell, J.E., Dagia, N.M., Dickerson, J.B., Berg, E.L., and Goetz, D.J. 2001. Ligand coated nanosphere adhesion to E- and P-selectin under static and flow conditions. *Ann Biomed Eng* 29: 523–533.

Borchard, G., Audus, K.L., Shi, F., and Kreuter, J. 1994. Uptake of surfactant-coated poly(methyl methacrylate)-nanoparticles by bovine brain microvessel endothelial cell monolayers. *Int J Pharm* 110: 29–35.

Bouclier, C., Moine, L., Hillaireau, H., Marsaud, V., Connault, E., Opolon, P., Couvreur, P., Fattal, E., and Renoir, J.M. 2008. Physicochemical characteristics and preliminary in vivo biological evaluation of nanocapsules loaded with siRNA targeting estrogen receptor alpha. *Biomacromolecules* 9: 2881–2890.

Bourdon, O., Mosqueira, V., Legrand, P., and Blais, J. 2000. A comparative study of the cellular uptake, localization and phototoxicity of meta–tetra(hydroxyphenyl) chlorine encapsulated in surface-modified submicronic oil/water carriers in HT29 tumor cells. *J Photochem Photobiol B* 55: 164–171.

Brannon-Peppas, L. and Blanchette, J.O. 2004. Nanoparticle and targeted systems for cancer therapy. *Adv Drug Deliv Rev* 56: 1649–1659.

Brasseur, F., Verdun, C., Couvreur, P., Deckers, C., and Roland, M. 1986. Evaluation expérimentale de l'efficacité thérapeutique de la doxorubicine associée aux nanoparticules de polyalkylcyanoacrylate. *Proc 4th Int Conf Pharmaceut Tech* 5: 177–186.

Brigger, I., Chaminade, P. Marsaud, V. et al. 2001. Tamoxifen encapsulation within polyethylene glycol-coated nanospheres. A new antiestrogen formulation. *Int J Pharm* 214: 37–42.

Brigger, I., Morizet, J., Aubert, G. et al. 2002. Poly(ethylene glycol)-coated hexadecylcyanoacrylate nanospheres display a combined effect for brain tumor targeting. *J Pharmacol Exp Ther* 303: 928–936.

Calvo, P., Gouritin, B., Brigger, I. et al. 2001b. PEGylated polycyanoacrylate nanoparticles as vector for drug delivery in prion diseases. *J Neurosci Methods* 111: 151–155.

Calvo, P., Gouritin, B., Chacun, H. et al. 2001a. Long-circulating PEGylated polycyanoacrylate nanoparticles as new drug carrier for brain delivery. *Pharm Res* 18: 1157–1166.

Calvo, P., Gouritin, B., Villarroya, H. et al. 2002. Quantification and localization of PEGylated polycyanoacrylate nanoparticles in brain and spinal cord during experimental allergic encephalomyelitis in the rat. *Eur J Neurosci* 15: 1317–1326.

Calvo, P., Sanchez, A., Martinez, J., Lopez Calonge, M.I., Pastor, J.C., and Alonso, M.J. 1996. Polyester nanocapsules as new topical ocular delivery systems for cyclosporin A. *Pharm Res* 13: 311–315.

Calvo, P., Vila-Jato, J.L., and Alonso, M.J. 1997. Evaluation of cationic polymer-coated nanocapsules as ocular drug carriers. *Int J Pharm* 153: 41–50.

Cao, W., Zhou, J., Wang, Y., and Zhu, L. 2010. Synthesis and in vitro cancer cell targeting of folate-functionalized biodegradable amphiphilic dendrimer-like star polymers. *Biomacromolecules* 11: 3680–3687.

Chalasani, K.B., Russell-Jones, G.J., Jain, A.K., Diwan, P.V., and Jain, S.K. 2007. Effective oral delivery of insulin in animal models using vitamin B12-coated dextran nanoparticles. *J Control Release* 26: 141–150.

Chauvierre, C., Marden, M.C., Vauthier, C., Labarre, D., Couvreur, P., and Leclerc, L. 2004. Heparin coated poly(alkylcyanoacrylate) nanoparticles coupled to hemoglobin: A new oxygen carrier. *Biomaterials* 25: 3081–3086.

Chavany, C., Le Doan, T., Couvreur, P., Puisieux, F., and Helene, C. 1992. Polyalkylcyanoacrylate nanoparticles as polymeric carriers for antisense oligonucleotides. *Pharm Res* 9: 441–449.

Chavany, C., Saison-Behmoaras, T., Le Doan, T., Puisieux, F., Couvreur, P., and Helene, C. 1994. Adsorption of oligonucleotides onto polyisohexylcyanoacrylate nanoparticles protects them against nucleases and increases their cellular uptake. *Pharm Res* 11: 1370–1378.

Chen, S., Zhang, X.Z., Cheng, S.X., Zhuo, R.X., and Gu, Z.W. 2008. Functionalized amphiphilic hyperbranched polymers for targeted drug delivery. *Biomacromolecules* 9: 2578–2585.

Chiannilkulchai, N., Ammoury, N., Caillou, B., Devissaguet J.P., and Couvreur, P. 1990. Hepatic tissue distribution of doxorubicin-loaded nanoparticles after i.v. administration in M5076 metastasis-bearing mice. *Cancer Chemother Pharmacol* 26: 122–126.

Chiannilkulchai, N., Driouich, Z., Benoit, J.P., Parodi A.L., and Couvreur, P. 1989. Doxorubicin-loaded nanoparticles: Increased efficiency in murine hepatic metastases. *Select Cancer Ther* 5: 1–11.

Chittasupho, C., Xie, S.X., Baoum, A., Yakovleva, T., Siahaan, T.J., and Berkland, C.J. 2009. ICAM-1 targeting of doxorubicin-loaded PLGA nanoparticles to lung epithelial cells. *Eur J Pharm Sci* 37: 141–150.

Chiu, S.J., Liu, S., Perrotti, D., Marcucci, G., and Lee, R.J. 2006. Efficient delivery of a Bcl-2-specific antisense oligodeoxyribonucleotide (G3139) via transferrin receptor-targeted liposomes. *J Control Release* 112: 199–207.

Cirstoiu-Hapca, A., Bossy-Nobs, L., Buchegger, F., Gurny, R., and Delie, F. 2007. Differential tumor cell targeting of anti-HER2 (Herceptin (R)) and anti-CD20 (Mabthera (R)) coupled nanoparticles. *Int J Pharm* 331: 190–196.

Clark, M.A., Hirst, B.H., and Jepson, M.A. 2000. Lectin-mediated mucosal delivery of drugs and microparticles. *Adv Drug Deliv Rev* 43: 207–223.

Colin de Verdière, A., Dubernet, C., Nemati, F. et al. 1997. Reversion of multidrug resistance with polyalkylcyanoacrylate nanoparticles: Mechanism of action. *Br J Cancer* 76: 198–205.

Costantino, L., Gandolfi, F., Tosi, G., Rivasi, F., Vandelli, M.A., and Forni, F. 2005. Peptide-derivatized biodegradable nanoparticles able to cross the blood–brain barrier. *J Control Release* 108: 84–96.

Couvreur, P., Grislain, L., Lenaerts, V., Brasseur, F., Guiot, P., and Biornacki, A. 1986. Biodegradable polymeric nanoparticles as drug carrier for antitumor agents. In *Polymeric Nanoparticles and Microspheres*, eds. P. Guiot and P. Couvreur, pp. 27–94. CRC Press, Boca Raton, FL.

Cuvier, C., Roblot-Treupel, L., Chevillard, S. et al. 1992. Doxorubicin-loaded nanoparticles bypass tumor multidrug resistance. *Biochem Pharmacol* 44: 509–519.

Damgé, C., Michel, C., Aprahamiam, M., Couvreur, P., and Devissaguet, J.-P. 1990. Nanocapsules as carriers for oral peptide delivery. *J Control Release* 13: 233–239.

Damgé, C., Vonderscher, J., Marbach, P., and Pinget, M. 1997b. Poly(alkylcyanoacrylate) nanocapsules as a delivery system in the rat for octreotide, a long-acting somatostatin analogue. *J Pharm Pharmacol* 49: 949–954.

Damgé, C., Vranckx, H., Baldschmidt, P., and Couvreur, P. 1997a. Poly(alkylcyanoacrylate) nanospheres for oral administration of insulin. *J Pharm Sci* 86: 1403–1409.

Danhier, F., Lecouturier, N., Vroman, B., Jérôme, C., Marchand-Brynaert, J., Feron, O., and Préat, V. 2009. Paclitaxel-loaded PEGylated PLGA-based nanoparticles: In vitro and in vivo evaluation. *J Control Release* 133: 11–17.

Dawson, G.F. and Halbert, G.W. 2000. The in vitro cell association of invasin coated polylactide-*co*-glycolide nanoparticles. *Pharm Res* 17: 1420–1425.

Dembri, A., Montisci, M.-J., Gantier, J.-C., Chacun, H., and Ponchel, G. 2001. Targeting of 3′ azido 3′ deoxythymidine (AZT) loaded poly(isohexylcyanoacrylate) nanospheres to the gastrointestinal mucosa and associated lymphoid tissues. *Pharm Res* 18: 467–473.

Díaz-López, R., Tsapis, N., Libong, D. et al. 2009. Phospholipid decoration of microcapsules containing perfluorooctyl bromide used as ultrasound contrast agents. *Biomaterials* 30: 1462–1472.

van der Ende, A., Croce, T., Hamilton, S., Sathiyakumar, V., and Harth, E. 2009. Tailored polyester nanoparticles: Post-modification with dendritic transporter and targeting units via reductive amination and thiolene chemistry. *Soft Matter* 5: 1417–1425.

van der Ende, A.E., Sathiyakumar, V., Diaz, R., Hallahan, D.E., and Harth, E. 2010. Linear release nanoparticle devices for advanced targeted cancer therapies with increased efficacy. *Polym Chem* 1: 93–96.

Farokhzad, O.C., Cheng, J.J., Teply, B.A. et al. 2006. Targeted nanoparticle–aptamer bioconjugates for cancer chemotherapy in vivo. *PNAS* 103: 6315–6320.

Fattal, E., Youssef, M., Couvreur P., and Andremont, A. 1989. Treatment of experimental salmonellosis in mice with ampicillin-bound nanoparticles. *Antimicrob Agents Chemother* 33: 1540–1543.

Fawaz, F., Bonini, F., Guyot, M., Lagueny, A.M., Fessi, H., and Devissaguet, J.-P. 1996. Disposition and protective effect against irritation after intravenous and rectal administration of indomethacin loaded nanocapsules to rabbit. *Int J Pharm* 133: 107–115.

Fievez, V., Plapied, L., des Rieux, A. et al. 2009. Targeting nanoparticles to M cells with non-peptidic ligands for oral vaccination. *Eur J Pharm Biopharm* 73: 16–24.

Florence, A.T., Hillery, A.M., Hussain N., and Jani, P.U. 1995. Factors affecting the oral uptake and translocation of polystyrene nanoparticles: Histological and analytical evidence. *J Drug Target* 3: 65–70.

Friese, A., Seiller, E., Quack, G., Lorenz, B., and Kreuter, J. 2000. Increase of the duration of the anticonvulsive activity of a novel NMDA receptor antagonist using poly(butylcyanoacrylate) nanoparticles as a parenteral controlled release system. *Eur J Pharm Biopharm* 49: 103–109.

Gan, C.W. and Feng, S.S. 2010. Transferrin-conjugated nanoparticles of poly(lactide)–D-alpha-tocopheryl polyethylene glycol succinate diblock copolymer for targeted drug delivery across the blood–brain barrier. *Biomaterials* 31: 7748–7757.

Garanger, E., Boturyn, D., and Dumy, P. 2007. Tumor targeting with RGD peptide ligands-design of new molecular conjugates for imaging and therapy of cancers. *Anticancer Agents Med Chem* 7: 552–558.

Garcia-Garcia, E., Andrieux, K., Gil, S., and Couvreur, P. 2005a. Colloidal carriers and blood–brain barrier (BBB) translocation: A way to deliver drugs to the brain? *Int J Pharm* 298: 274–292.

Garcia-Garcia, E., Andrieux, K., Gil, S. et al. 2005c. A methodology to study intracellular distribution of nanoparticles in brain endothelial cells. *Int J Pharm* 298: 310–314.

Garcia-Garcia, E., Gil, S., Andrieux, K. et al. 2005b. A relevant in vitro rat model for the evaluation of blood–brain barrier translocation of nanoparticles. *Cell Mol Life Sci* 62: 1400–1408.

Garinot, M., Fiévez, V., Pourcelle, V. et al. 2007. PEGylated PLGA-based nanoparticles targeting M cells for oral vaccination. *J Control Release* 120: 195–204.

Gelperina, S.E., Khalansky, A.S., Skidan, I.N. et al. 2002. Toxicological studies of doxorubicin bound to polysorbate 80-coated poly(butyl cyanoacrylate) nanoparticles in healthy rats and rats with intracranial glioblastoma. *Toxicol Lett* 126: 131–141.

Gref, R., Couvreur, P., Barratt, G., and Mysiakine, E. 2003. Surface-engineered nanoparticles for multiple ligand coupling. *Biomaterials* 24: 4529–4537.

Gref, R., Domb, A., Quellec, P. et al. 1995. The controlled intravenous delivery of drugs using PEG-coated sterically stabilized nanospheres. *Adv Drug Deliv Rev* 16: 215–233.

Gullberg, E., Keita, A.V., Salim, S.Y. et al. 2006. Identification of cell adhesion molecules in the human follicle-associated epithelium that improve nanoparticle uptake into the Peyer's patches. *J Pharmacol Exp Ther* 319: 632–639.

Gupta, P.N., Khatri, K., Goyal, A.K., Mishra, N., and Vyas, S.P. 2007. M-cell targeted biodegradable PLGA nanoparticles for oral immunization against hepatitis B. *J Drug Target* 15: 701–713.

Guterres, S.S., Fessi, H., Barratt, G., Puisieux, F., and Devissaguet, J.-P. 1995. Poly(D,L-lactide) nanocapsules containing non-steroidal anti-inflammatory drugs: Gastrointestinal tolerance following intravenous and oral administration. *Pharm Res* 12: 1545–1547.

Guterres, S.S., Fessi, H., Barratt, G., Puisieux, F., and Devissaguet, J.P. 2000. Poly(rac-lactide) nanocapsules containing diclofenac: Protection against muscular damage in rats. *J Biomater Sci Polym Ed* 11: 1347–1355.

Ham, A.S., Klibanov, A.L., and Lawrence, M.B. 2009. Action at a distance: Lengthening adhesion bonds with poly(ethylene glycol) spacers enhances mechanically stressed affinity for improved vascular targeting of microparticles. *Langmuir* 25: 10038–10044.

Hamdy, S., Molavi, O., Ma, Z. et al. 2008. Co-delivery of cancer-associated antigen and Toll-like receptor 4 ligand in PLGA nanoparticles induces potent CD8+ T cell-mediated anti-tumor immunity. *Vaccine* 26: 5046–5057.

Hattori, Y. and Maitani, Y. 2005. Folate-linked nanoparticle-mediated suicide gene therapy in human prostate cancer and nasopharyngeal cancer with herpes simplex virus thymidine kinase. *Cancer Gene Ther* 12: 796–809.

Haun, J.B. and Hammer, D.A. 2008. Quantifying nanoparticle adhesion mediated by specific molecular interactions. *Langmuir* 24: 8821–8832.

Hawley, A.E., Davis, S.S., and Illum, L. 1995. Targeting of colloids to lymph nodes: Influence of lymphatic physiology and colloidal characteristics. *Adv Drug Deliv Rev* 17: 129–148.

Hawley, A.E., Illum, L., and Davis, S.S. 1997. Preparation of biodegradable, surface engineered PLGA nanospheres with enhanced lymphatic drainage and lymph node uptake. *Pharm Res* 14: 657–661.

Hekmatara, T., Bernreuther, C., Khalansky, A.S. et al. 2009. Efficient systemic therapy of rat glioblastoma by nanoparticle-bound doxorubicin is due to antiangiogenic effects. *Clin Neuropathol* 28: 153–164.

Hilgenbrink, A.R. and Low, P.S. 2005. Folate receptor-mediated drug targeting: From therapeutics to diagnostics. *J Pharm Sci* 94: 2135–2146.

Hu, K.L., Li, J.W., Shen, Y.H. et al. 2009. Lactoferrin-conjugated PEG–PLA nanoparticles with improved brain delivery: In vitro and in vivo evaluations. *J Control Release* 134: 55–61.

Hubert, B., Atkinson, J., Guerret, M., Hoffman, M., Devissaguet, J.P., and Maincent, P. 1991. The preparation and acute antihypertensive effects of a nanocapsular form of darodipine, a dihydropyridine calcium entry blocker. *Pharm Res* 8: 734–738.

Illum, L., Jacobsen, L.O., Müller, R.H., Mak, R., and Davis, S.S. 1987. Surface characteristics and the interaction of colloidal particles with mouse peritoneal macrophages. *Biomaterials* 8: 113–117.

Irache, J.M., Durrer, C., Duchêne D., and Ponchel, G. 1996. Bioadhesion of lectin–latex conjugates to rat intestinal mucosa. *Pharm Res* 13: 1716–1719.

Ishihara, T., Kubota, T., Choi, T., and Higaki, M. 2009. Treatment of experimental arthritis with Stealth-type polymeric nanoparticles encapsulating betamethasone phosphate. *J Pharmacol Exp Ther* 329: 412–417.

Jain, R.K. 1987. Transport of molecules across tumor vasculature. *Cancer Metastasis Rev* 6: 559–593.

Jefferies, W.A., Brandon, M.R., Hunt, S.V., Williams, A.F., Gatter, K.C., and Mason, D.Y. 1984. Transferrin receptor on endothelium of brain capillaries. *Nature* 312: 162–163.

Jeon, S.I., Lee, J.H., Andrade, J.D., and De Gennes, P.G. 1991. Protein-surface interactions in the presence of polyethylene oxide; 1. Simplified theory. *J Colloid Interf Sci* 142: 149–166.

Jeong, Y.I., Na, H.S., Seo, D.H. et al. 2008. Ciprofloxacin-encapsulated poly(DL-lactide-*co*-glycolide) nanoparticles and its antibacterial activity. *Int J Pharm* 352: 317–323.

Jiang, X., Vogel, E.B., Smith, M.R., and Baker, G.L. 2008. Clickable polyglycolides: Tunable synthons for thermoresponsive, degradable polymers. *Macromolecules* 41: 1937–1944.

Johnson, C.M., Pandey, R., Sharma, S. et al. 2005. Oral therapy using nanoparticle-encapsulated antituberculosis drugs in guinea pigs infected with *Mycobacterium tuberculosis*. *Antimicrob Agents Chemother* 49: 4335–4338.

Jubeli, E., Moine, L., and Barratt, G. 2010. Synthesis, characterization, and molecular recognition of sugar-functionalized nanoparticles prepared by a combination of ROP, ATRP, and click chemistry. *J Polym Sci Part A Polym Chem* 48: 3178–3187.

Karatas, H., Aktas, Y., Gursoy-Ozdemir, Y. et al. 2009. A nanomedicine transports a peptide caspase-3 inhibitor across the blood–brain barrier and provides neuroprotection. *J Neurosci* 29: 13761–13769.

Kim, H.R., Andrieux, K., and Couvreur, P. 2007a. PEGylated polymer-based nanoparticles for drug delivery to the brain. In *Colloids and Interface Science Series, Vol. 3, Colloid Stability and Application in Pharmacy*, ed. T.F. Tadros, pp. 409–427. Wiley-VCH Verlag GmbH and Co KGaA, Weinheim Germany.

Kim, H.R., Andrieux, K., Delomenie, C. et al. 2007b. Analysis of plasma protein adsorption onto PEGylated nanoparticles by complementary methods: 2-DE, CE and protein lab-on-chip system. *Electrophoresis* 28: 2252–2261.

Kim, H.R., Andrieux, K., Gil, S. et al. 2007d. Translocation of poly(ethylene glycol-*co*-hexadecyl)cyanoacrylate nanoparticles into rat brain endothelial cells: Role of apolipoproteins in receptor-mediated endocytosis. *Biomacromolecules* 8: 793–799.

Kim, H.R., Gil, S., Andrieux, K. et al. 2007c. Low-density lipoprotein receptor-mediated endocytosis of PEGylated nanoparticles in rat brain endothelial cells. *Cell Mol Life Sci* 64: 356–364.

Kolb, H.C., Finn, M.G., and Sharpless, K.B. 2001. Click chemistry: Diverse chemical function from a few good reactions. *Angew Chem Int Ed* 40: 2004–2021.

Kolb, H.C. and Sharpless, K.B. 2004. The growing impact of click chemistry on drug discovery. *Drug Discovery Today* 8: 1128–1137.

Konan, Y.N., Berton, M., Gurny, R., and Allémann, E. 2003. Enhanced photodynamic activity of meso–tetra(4-hydroxyphenyl)porphyrin by incorporation into sub-200 nm nanoparticles. *Eur J Pharm Sci* 18: 241–249.

Kreuter, J. 2001. Nanoparticulate systems for brain delivery of drugs. *Adv Drug Deliv Rev* 47: 65–81.

Kreuter, J., Alyautdin, R.N., Kharkevich, D.A., and Ivanov, A.A. 1995. Passage of peptides through the blood–brain barrier with colloidal polymer particles (nanoparticles). *Brain Res* 674: 171–174.

Kreuter, J., Hekmatara, T., Dreis, S., Vogel, T., Gelperina, S., and Langer, K. 2007. Covalent attachment of apolipoprotein A-I and apolipoprotein B-100 to albumin nanoparticles enables drug transport into the brain. *J Control Release* 118: 54–58.

Kreuter, J., Shamenkov, D., Petrov, V. et al. 2002. Apolipoprotein-mediated transport of nanoparticle-bound drugs across the blood–brain barrier. *J Drug Target* 10: 317–325.

Kurakhmaeva, K.B., Voronina, T.A., Kapica, I.G. et al. 2008. Antiparkinsonian effect of nerve growth factor adsorbed on polybutylcyanoacrylate nanoparticles coated with polysorbate-80. *Bull Exp Biol Med* 145: 259–262.

Lambert, G., Bertrand, J.R., Fattal, E. et al. 2000b. EWS fli-1 antisense nanocapsules inhibits Ewing sarco-marelated tumor in mice. *Biochem Biophys Res Commun* 279: 401–406.

Lambert, G., Fattal, E., Brehier, A., Feger, J., and Couvreur, P. 1998. Effect of polyisobutylcyanoacrylate nanoparticles and lipofectin loaded with oligonucleotides on cell viability and PKC alpha neosynthesis in HepG2 cells. *Biochimie* 80: 969–976.

Lambert, G., Fattal, E., Pinto-Alphandary, H., Gulik, A., and Couvreur, P. 2000a. Polyisobutylcyanoacrylate nanocapsules containing an aqueous core as a novel colloidal carrier for the delivery of oligonucleotides. *Pharm Res* 17: 707–714.

Le Droumaguet, B. and Nicolas, J. 2010. Recent advances in the design of bioconjugates from controlled/living radical polymerization *Polym Chem* 1: 563–598.

Le Droumaguet, B., Souguir, H., Brambilla, D. et al. 2011. Selegiline-functionalized, PEGylated poly(alkyl cyanoacrylate) nanoparticles: Investigation of interaction with amyloid-β peptide and surface reorganization. *Int J Pharm* 146: 453–460.

Le Droumaguet, B. and Velonia, K. 2008. Click chemistry: A powerful tool to create polymer-based macromolecular chimeras. *Macromol Rapid Commun* 29: 1073–1089.

Le Garrec, D., Gori, S., Luo, L. et al. 2004. Poly(*N*-vinylpyrrolidone)-*block*-poly(D,L-lactide) as a new polymeric solubilizer for hydrophobic anticancer drugs: In vitro and in vivo evaluation. *J Control Release* 99: 83–101.

Lecaroz, M.C., Blanco-Prieto, M.J., Campanero, M.A., Salman, H., and Gamazo, C. 2007. Poly(D,L-lactide-*co*-glycolide) particles containing gentamicin: Pharmacokinetics and pharmacodynamics in *Brucella melitensis*–infected mice. *Antimicrob Agents Chemother* 51: 1185–1190.

Lehr C.M. and Puzstai A. 1995. The potential of bioadhesive lectins for the delivery of peptide and protein drugs to the gastrointestinal tract. In *Lectins: Biomedical Perspectives*, eds. A. Puzstai and S. Bardocz, pp. 117–140. Taylor & Francis, London U.K.

Leite, E.A., Grabe-Guimarães, A., Guimarães, H.N., Machado-Coelho, G.L., Barratt, G., Mosqueira, V.C. 2007. Cardiotoxicity reduction induced by halofantrine entrapped in nanocapsule devices. *Life Sci* 80: 1327–1334.

Leroux, J.C., De Jaeghere, F., Anner, B., Doelker, E., and Gurny, R. 1995. An investigation on the role of plasma and serum opsonins on the internalization of biodegradable poly(D,L-lactic acid) nanoparticles by human monocytes. *Life Sci* 57: 695–703.

Li, H. and Qian, Z.M. 2002. Transferrin/transferrin receptor-mediated drug delivery. *Med Res Rev* 22: 225–250.

Liang, H.F., Chen, C.T., Chen, S.C. et al. 2006. Paclitaxel-loaded poly(gamma-glutamic acid)–poly(lactide) nanoparticles as a targeted drug delivery system against cultured HepG2 cells. *Bioconjug Chem* 17: 291–299.

Liang, B., He, M.L., Xiao, Z.P. et al. 2008. Synthesis and characterization of folate-PEG-grafted-hyperbranched-PEI for tumor-targeted gene delivery. *Biochem Biophys Res Commun* 367: 874–880.

Liang, H.F., Yang, T.F., Huang, C.T., Chen, M.C., and Sung, H.W. 2005. Preparation of nanoparticles composed of poly(gamma-glutamic acid)–poly(lactide) block copolymers and evaluation of their uptake by HepG2 cells. *J Control Release* 105: 213–225.

Lin, A., Sabnis, A., Kona, S. et al. 2010. Shear-regulated uptake of nanoparticles by endothelial cells and development of endothelial-targeting nanoparticles. *J Biomed Mater Res A* 93: 833–842.

Liu, Z.C. and Chang, T.M. 2008. Effects of PEG–PLA–nano artificial cells containing hemoglobin on kidney function and renal histology in rats. *Artif Cells Blood Substit Immobil Biotechnol* 36: 421–430.

Liu, J., Wong, H.-L., Moselhy, J., Bowen, B., Wu, X.Y., and Johnston, M.R. 2006. Targeting colloidal particulates to thoracic lymph nodes. *Lung Cancer* 51: 377–386.

Loch-Neckel, G., Nemen, D., Puhl, A.C. et al. 2007. Stealth and non-Stealth nanocapsules containing camptothecin: In vitro and in vivo activity on B16-F10 melanoma. *J Pharm Pharmacol* 59: 1359–1364.

Losa, C., Marchal-Heussler, L., Orallo, F., Vila Jato, J.L., and Alonso, M.J. 1993. Design of new formulations for topical ocular administration: Polymeric nanocapsules containing metipranolol. *Pharm Res* 10: 80–87.

Lu, J., Shi, M., and Shoichet, M.S. 2009. Click chemistry functionalized polymeric nanoparticles target corneal epithelial cells through RGD-cell surface receptors. *Bioconjug Chem* 20: 87–94.

Lu, W., Wan, J., She, Z., and Jiang, X. 2007a. Brain delivery property and accelerated blood clearance of cationic albumin conjugated pegylated nanoparticle. *J Control Release* 118: 38–53.

Lu, W., Wan, J., Zhang, Q., She, Z., and Jiang, X. 2007b. Aclarubicin-loaded cationic albumin-conjugated pegylated nanoparticle for glioma chemotherapy in rats. *Int J Cancer* 120: 420–431.

Lu, W., Zhang, Y., Tan, Y.Z., Hu, K.L., Jiang, X.G., and Fu, S.K. 2005. Cationic albumin-conjugated pegylated nanoparticles as novel drug carrier for brain delivery. *J Control Release* 107: 428–448.

Lutz, J.-F. 2007. 1,3-Dipolar cycloadditions of azides and alkynes: A universal ligation tool in polymer and materials science. *Angew Chem Int Ed* 46: 1018–1025.

Maincent, P., Thouvenot, P., Amicabile, C. et al. 1992. Lymphatic targeting of polymeric nanoparticles after intraperitoneal administration in rats. *Pharm Res* 9: 1534–1539.

Mallardé, D., Boutignon, F., Moine, F. et al. 2003. PLGA–PEG microspheres of teverelix: Influence of polymer type on microsphere characteristics and on teverelix in vitro release. *Int J Pharm* 261: 69–80.

Marchal-Heussler, L., Sirbat, D., Hoffman, M., and Maincent, P. 1993. Poly(epsilon-caprolactone) nanocapsules in carteolol ophthalmic delivery. *Pharm Res* 10: 386–390.

Mattheolabakis, G., Taoufik, E., Haralambous, S., Roberts, M.L., and Avgoustakis, K. 2009. In vivo investigation of tolerance and antitumor activity of cisplatin-loaded PLGA–mPEG nanoparticles. *Eur J Pharm Biopharm* 71: 190–195.

Mei, H., Shi, W., Pang, Z.Q. et al. 2010. YEGFP–EGF1 protein-conjugated PEG–PLA nanoparticles for tissue factor targeted drug delivery. *Biomaterials* 31: 5619–5626.

Meissner, Y., Pellequer, Y., and Lamprecht, A. 2006. Nanoparticles in inflammatory bowel disease: Particle targeting versus pH-sensitive delivery. *Int J Pharm* 316: 138–143.

Mercadal, M., Domingo, J.C., Petriz, J.C., Garcia, J., and De Madariaga, M.M. 1999. A novel strategy affords high-yield coupling of antibody to extremities of liposomal surface grafted PEG chains. *Biochim Biophys Acta* 1418: 232–238.

Mesiha, M.S., Sidhom, M.B., and Fasipe, B. 2005. Oral and subcutaneous absorption of insulin poly(isobutylcyanoacrylate) nanoparticles. *Int J Pharm* 288: 289–293.

Michaelis, K., Hoffmann, M.M., Dreis, S. et al. 2006. Covalent linkage of apolipoprotein E to albumin nanoparticles strongly enhances drug transport into the brain. *J Pharmacol Exp Ther* 317: 1246–1253.

Mishra, V., Mahor, S., Rawat, A. et al. 2006. Targeted brain delivery of AZT via transferrin anchored pegylated albumin nanoparticles. *J Drug Target* 14: 45–53.

Moghimi, S.M. and Szebeni, J. 2003. Stealth liposomes and long circulating nanoparticles: Critical issues in pharmacokinetics, opsonization and protein-binding properties. *Prog Lipid Res* 42: 463–478.

Morin, C., Barratt, G., Fessi, H., Devissaguet, J.P., and Puisieux, F. 1994. Improved intracellular delivery of a muramyl dipeptide analog by means of nanocapsules. *Int J Immunopharmacol* 16: 451–456.

Mosqueira, V.C., Legrand, P., and Barratt, G. 2006. Surface-modified and conventional nanocapsules as novel formulations for parenteral delivery of halofantrine. *J Nanosci Nanotechnol* 6: 3193–3202.

Mosqueira, V.C.F., Legrand, P., Gulik, A. et al. 2001. Relationship between complement activation, cellular uptake and surface physicochemical aspects of novel PEG-modified nanocapsules. *Biomaterials* 22: 2967–2979.

Mosqueira, V.C., Loiseau, P.M., Bories, C., Legrand, P., Devissaguet, J.P., and Barratt, G. 2004. Efficacy and pharmacokinetics of intravenous nanocapsule formulations of halofantrine in *Plasmodium berghei*–infected mice. *Antimicrob Agents Chemother* 48: 1222–1228.

Ng, Q.K., Sutton, M.K., Soonsawad, P., Xing, L., Cheng, H., and Segura, T. 2009. Engineering clustered ligand binding into nonviral vectors: Alphavbeta3 targeting as an example. *Mol Ther* 17: 828–836.

Nguyen, A., Marsaud, V., Bouclier, C. et al. 2008. Nanoparticles loaded with ferrocenyl tamoxifen derivatives for breast cancer treatment. *Int J Pharm* 347: 128–135.

Nicolas, J., Bensaid, F., Desmaële, D. et al. 2008. Synthesis of highly functionalized poly(alkyl cyanoacrylate) nanoparticles by means of click chemistry. *Macromolecules* 41: 8418–8428.

Nishioka, Y. and Yoshino, H. 2001. Lymphatic targeting with nanoparticulate system. *Adv Drug Deliv Rev* 47: 55–64.

Nobs, L., Buchegger, F., Gurny, R., and Allemann, E. 2006. Biodegradable nanoparticles for direct or two-step tumor immunotargeting. *Bioconjug Chem* 17: 139–145.

Olivier, J.C. 2005. Drug transport to brain with targeted nanoparticles. *NeuroRx* 2: 108–119.

Oyen, W.J., Boerman, O.C., Storm, G. et al. 1996. Labeled Stealth® liposomes in experimental infection: An alternative to leukocyte scintigraphy? *Nucl Med Commun* 17: 742–748.

Page-Clisson, M.E., Pinto-Alphandary, H., Chachaty, E., Couvreur, P., and Andremont, A. 1998. Drug targeting by polyalkylcyanoacrylate nanoparticles is not efficient against persistent *Salmonella*. *Pharm Res* 15: 544–549.

Pan, J. and Feng, S.S. 2008. Targeted delivery of paclitaxel using folate-decorated poly(lactide)–vitamin E TPGS nanoparticles. *Biomaterials* 29: 2663–2672.

Pan, J. and Feng, S.S. 2009. Targeting and imaging cancer cells by folate-decorated, quantum dots (QDs)-loaded nanoparticles of biodegradable polymers. *Biomaterials* 30: 1176–1183.

Pang, Z., Lu, W., Gao, H. et al. 2008. Preparation and brain delivery property of biodegradable polymersomes conjugated with OX26. *J Control Release* 128: 120–127.

Pardridge, W.M., Eisenberg, J., and Yang, J. 1987. Human blood–brain barrier transferrin receptor. *Metabolism* 36: 892–895.

Park, J., Fong, P.M., Lu, J. et al. 2009. PEGylated PLGA nanoparticles for the improved delivery of doxorubicin. *Nanomedicine* 5: 410–418.

Patil, Y.B., Toti, U.S., Khdair, A., Ma, L., and Panyam, J. 2009. Single-step surface functionalization of polymeric nanoparticles for targeted drug delivery. *Biomaterials* 30: 859–866.

Peracchia, M.T., Harnisch, S., Pinto-Alphandary, H. et al. 1999. Visualization of in vitro protein-rejecting properties of PEGylated Stealth polycyanoacrylate nanoparticles. *Biomaterials* 20: 1269–1275.

Peracchia, M.T., Vauthier, C., Desmaele, D. et al. 1998. Pegylated nanoparticles from a novel methoxypolyethylene glycol cyanoacrylate–hexadecyl cyanoacrylate amphiphilic copolymer. *Pharm Res* 15: 550–556.

Pereira, M.A., Mosqueira, V.C., Carmo, V.A. et al. 2009. Biodistribution study and identification of inflammatory sites using nanocapsules labeled with (99m)Tc-HMPAO. *Nucl Med Commun* 30: 749–755.

Perez, C., Sanchez, A., Putnam, D., Ting, D., Langer, R., and Alonso, M.J. 2001. Poly(lactic acid)–poly(ethylene glycol) nanoparticles as new carriers for the delivery of plasmid DNA. *J Control Release* 75: 211–224.

Pinto-Alphandary, H., Andremont, A., and Couvreur, P. 2000. Targeted delivery of antibiotics using liposomes and nanoparticles: Research and applications. *Int J Antimicrob Agents* 13: 155–168.

Pinto-Alphandary, H., Balland, O., Laurent, M., Andremont, A., Puisieux F., and Couvreur, P. 1994. Intracellular visualization of ampicillin-loaded nanoparticles in peritoneal macrophages infected in vitro with *Salmonella typhimurium*. *Pharm Res* 11: 38–46.

Platt, V.M. and Szoka, F.C. Jr. 2008. Anticancer therapeutics: Targeting macromolecules and nanocarriers to hyaluronan or CD44, a hyaluronan receptor. *Mol Pharm* 5: 474–486.

Ponchel, G. and Irache, J.-M. 1998. Specific and non-specific bioadhesive particulate systems for oral delivery to the gastrointestinal tract. *Adv Drug Deliv Rev* 34: 191–219.

Prabaharan, M., Grailer, J.J., Pilla, S., Steeber, D.A., and Gong, S. 2009. Folate-conjugated amphiphilic hyper-branched block copolymers based on Boltorn (R) H40, poly(L-lactide) and poly(ethylene glycol) for tumor-targeted drug delivery. *Biomaterials* 30: 3009–3019.

Qian, Z.M., Li, H., Sun, H., and Ho, K. 2002. Targeted drug delivery via the transferrin receptor-mediated endocytosis pathway. *Pharmacol Rev* 54: 561–587.

Quellec, P., Gref, R., Dellacherie, E., Sommer, F., Tran, M.D., and Alonso, M.J. 1999. Protein encapsulation within poly(ethylene glycol)-coated nanospheres. II. Controlled release properties. *J Biomed Mater Res* 47: 388–395.

Rao, K.S., Reddy, M.K., Horning, J.L., and Labhasetwar, V. 2008. TAT-conjugated nanoparticles for the CNS delivery of anti-HIV drugs. *Biomaterials* 29: 4429–4438.

Renoir, J.M., Stella, B., Ameller, T., Connault, E., Opolon, P., and Marsaud, V. 2006. Improved anti-tumoral capacity of mixed and pure anti-oestrogens in breast cancer cell xenografts after their administration by entrapment in colloidal nanosystems. *J Steroid Biochem Mol Biol* 102: 114–127.

des Rieux, A., Fievez, V., Garinot, M., Schneider, Y.J., and Préat, V. 2006. Nanoparticles as potential oral delivery systems of proteins and vaccines: A mechanistic approach. *J Control Release* 116: 1–27.

Riva, R., Schmeits, S., Jérôme, C., Jérôme, R., and Lecomte, P. 2007. Combination of ring-opening polymerization and "Click Chemistry": Toward functionalization and grafting of poly(ε-caprolactone). *Macromolecules* 40: 796–803.

Rodrigues, J.M. Jr., Croft, S.L., Fessi, H., Bories, C., and Devissaguet, J.P. 1994. The activity and ultrastructural localization of primaquine-loaded poly(D,L-lactide) nanoparticles in *Leishmania donovani* infected mice. *Trop Med Parasitol* 45: 223–228.

Russell-Jones, G.J., Veitch, H., and Arthur, L. 1999. Lectin-mediated transport of nanoparticles across Caco-2 and OK cells. *Int J Pharm* 190: 165–174.

Sahoo, S.K. and Labhasetwar, V. 2005. Enhanced anti proliferative activity of transferrin-conjugated paclitaxel-loaded nanoparticles is mediated via sustained intracellular drug retention *Mol Pharm* 2: 373–383.

Salem, A.K., Cannizzaro, S.M., Davies, M.C. et al. 2001. Synthesis and characterisation of a degradable poly(lactic acid)–poly-(ethylene glycol) copolymer with biotinylated end groups. *Biomacromolecules* 2: 575–580.

Sapra, P. and Allen, T.M. 2003. Ligand-targeted liposomal anticancer drugs. *Prog Lipid Res* 42: 439–462.

Schiffelers, R.M., Ansari, A., Xu, J. et al. 2004. Cancer siRNA therapy by tumor selective delivery with ligand-targeted sterically stabilized nanoparticle. *Nucleic Acids Res* 32: e149.

Schroeder, U., Sommerfeld, P., and Sabel, B.A. 1998. Efficacy of oral dalargin-loaded nanoparticle delivery across the blood–brain barrier. *Peptides* 19: 777–780.

Schwab, G., Chavany, C., Duroux, I. et al. 1994. Antisense oligonucleotides adsorbed to polyalkylcyanoacrylate nanoparticles specifically inhibit mutated Ha-*ras*-mediated cell proliferation and tumorigenicity in nude mice. *Proc Natl Acad Sci USA* 91: 10460–10464.

Senthilkumar, M., Mishra, P., and Jain, N.K. 2008 Long circulating PEGylated poly(D,L-lactide-*co*-glycolide) nanoparticulate delivery of docetaxel to solid tumors. *J Drug Target* 16: 424–435.

Staber, P.B., Linkesch, W., Zauner, D. et al. 2004. Common alterations in gene expression and increased proliferation in recurrent acute myeloid leukemia. *Oncogene* 23: 894–904.

Steiniger, S.C., Kreuter, J., Khalansky, A.S. et al. 2004. Chemotherapy of glioblastoma in rats using doxorubicin-loaded nanoparticles. *Int J Cancer* 109: 759–767.

Stella, B., Arpicco, S., Peracchia, M.T. et al. 2000. Design of folic acid-conjugated nanoparticles for drug targeting. *J Pharm Sci* 89: 1452–1464.

Stella, B., Marsaud, V., Arpicco, S. et al. 2007. Biological characterization of folic acid-conjugated poly(H2NPEGCA-*co*-HDCA) nanoparticles in cellular models. *J Drug Target* 15: 146–153.

Temming, K., Schiffelers, R.M., Molema, G., and Kok, R.J. 2005. RGD-based strategies for selective delivery of therapeutics and imaging agents to the tumour vasculature. *Drug Resist Updates* 8: 381–402.

Tobío, M. Sánchez, A., Vila A. et al. 2000. The role of PEG on the stability in digestive fluids and in vivo fate of PEG–PLA nanoparticles following oral administration. *Colloids Surf B Biointerf* 18: 315–323.

Tosi, G., Costantino, L., Rivasi, F. et al. 2007. Targeting the central nervous system: In vivo experiments with peptide-derivatized nanoparticles loaded with loperamide and rhodamine-123. *J Control Release* 122: 1–9.

Toub, N., Bertrand, J.R., Malvy, C., Fattal, E., and Couvreur, P. 2006a. Antisense oligonucleotide nanocapsules efficiently inhibit EWS-Fli1 expression in a Ewing's sarcoma model. *Oligonucleotides* 16: 158–168.

Toub, N., Bertrand, J.R., Tamaddon, A. et al. 2006b. Efficacy of siRNA nanocapsules targeted against the EWS-Fli1 oncogene in Ewing sarcoma. *Pharm Res* 23: 892–900.

Tsai, H.C., Chang, W.H., Lo, C.L. et al. 2010. Graft and diblock copolymer multifunctional micelles for cancer chemotherapy and imaging. *Biomaterials* 31: 2293–2301.

Tsapis, N., Bennett, D., Jackson, B., Weitz, D.A., and Edwards, D.A. 2002. Trojan particles: Large porous carriers of nanoparticles for drug delivery. *PNAS* 99: 12001–12005.

Ulbrich, K., Hekmatara, T., Herbert, E., and Kreuter, J. 2009. Transferrin- and transferrin–receptor–antibody-modified nanoparticles enable drug delivery across the blood–brain barrier (BBB). *Eur J Pharm Biopharm* 71: 251–256.

Upadhyay K.K., Bhatt, A.N., Castro, E. et al. 2010. In vitro and in vivo evaluation of docetaxel loaded biodegradables polymerosomes. *Macromol Biosci* 10: 503–512.

Vila, A., Gill, H., McCallion, O., and Alonso, M.J. 2004. Transport of PLA–PEG particles across the nasal mucosa: Effect of particle size and PEG coating density. *J Control Release* 98: 231–244.

Vila, A., Sánchez, A., Evora, C., Soriano, I., McCallion, O., and Alonso, M.J. 2005. PLA–PEG particles as nasal protein carriers: The influence of the particle size. *Int J Pharm* 292: 43–52.

Vittaz, M., Bazile, D., Spenlehauer, G. et al. 1996. Effect of PEO surface density on long-circulating PLA–PEO nanoparticles which are very low complement activators. *Biomaterials* 17: 1575–1581.

Wang, Z., Chui, W.K., and Ho, P.C. 2009. Design of a multifunctional PLGA nanoparticulate drug delivery system: Evaluation of its physicochemical properties and anticancer activity to malignant cancer cells. *Pharm Res* 26: 1162–1171.

Weiss, B., Schneider, M., Muys, L. et al. 2007. Coupling of biotin–(poly(ethylene glycol))amine to poly(D,L-lactide-*co*-glycolide) nanoparticles for versatile surface modification. *Bioconjug Chem* 18: 1087–1094.

Woodle, M.C. 1998. Controlling liposome blood clearance by surface-grafted polymers. *Adv Drug Deliv Rev* 32: 139–152.

Yoshizawa, T., Hattori, Y., Hakoshima, M., Koga, K., and Maitani, Y. 2008. Folate-linked lipid-based nanoparticles for synthetic siRNA delivery in KB tumor xenografts. *Eur J Pharm Biopharm* 70: 718–725.

Youssef, M., Fattal, E., Alonso, M.J. et al. 1988. Effectiveness of nanoparticle-bound ampicillin in the treatment of *Listeria monocytogenes* infections in nude mice. *Antimicrob Agents Chemother* 32: 1204–1207.

Yu, D.H., Lu, Q., Xie, J., Fang, C., and Chen, H.Z. 2010. Peptide-conjugated biodegradable nanoparticles as a carrier to target paclitaxel to tumor neovasculature. *Biomaterials* 31: 2278–2292.

Zambaux, M.F., Bonneaux, F., Gref, R., Dellacherie, E., and Vigneron, C. 1999. MPEO–PLA nanoparticles: Effect of MPEO content on some of their surface properties. *J Biomed Mater Res* 44: 109–115.

Zensi, A., Begley, D., Pontikis, C. et al. 2009. Albumin nanoparticles targeted with Apo E enter the CNS by transcytosis and are delivered to neurones. *J Control Release* 137: 78–86.t

Zhang, L., Hu, C.H., Cheng, S.X., and Zhuo, R.X. 2010. Hyperbranched amphiphilic polymer with folate mediated targeting property. *Colloid Surf B Biointerf* 79: 427–433.

Zhang, Z., Huey Lee, S., and Feng, S.S. 2007. Folate-decorated poly(lactide-*co*-glycolide)–vitamin E TPGS nanoparticles for targeted drug delivery. *Biomaterials* 28: 1889–1899.

Zhao, X., Li, H., and Lee, R.J. 2008. Targeted drug delivery via folate receptors. *Expert Opin Drug Deliv* 5: 309–319.

6 Drug Carrier Systems for Anticancer Agents

Hiroyuki Koide, Tomohiro Asai,
Kosuke Shimizu, and Naoto Oku

CONTENTS

6.1 INTRODUCTION

In the development of novel anticancer drugs including molecular targeting and antibody drugs, unexpected withdraw has been often encountered during phase trials (Kerbel and Folkman 2002). Although these drugs significantly suppressed tumor growth in certain preclinical models, most of them failed to produce significant therapeutic results or demonstrated serious side effects in the human subjects tested. These failed drugs frequently showed fast clearance in the blood or irreversible damage to organs by single or repeated injection. Considering these failures, it is desirable to stringently control the process at the action site for maximizing the positive effects and reducing the side effects of the drugs. However, the control of absorption, distribution, metabolism, and excretion (ADME) of a drug is not so easy depending on its own molecular properties. To overcome these problems, nanocarriers such as liposomes, polymeric micelles, and lipid microspheres have been used as systems for controlling drug delivery. The advantages of using nanocarriers for the desired therapeutics are the following: (1) reduction in nonspecific or undesirable biodistribution of a drug, (2) protection from undesirable metabolism and degradation, (3) enhancement of bioavailability by controlled release, (4) selective delivery to the target tissues, etc. It can be considered that therapeutic effect owing to these advantages is significant.

In this chapter, we will discuss several types of targeting tools for drug delivery, including liposomes, polymeric micelles, and lipid microspheres (Figure 6.1), for application to cancer therapy.

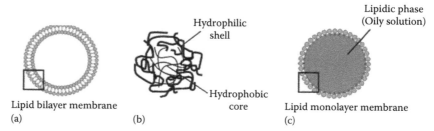

FIGURE 6.1 Structures of (a) liposome, (b) polymeric micelle, and (c) lipid microsphere.

6.2 LIPOSOMES

Liposomes were originally discovered by Bangham in 1964 (Bangham and Horne 1964). They are closed vesicles composed of a lipid bilayer membrane, and contain both hydrophilic and hydrophobic regions. Therefore, liposomes can encapsulate both hydrophilic and hydrophobic drugs, and deliver these drugs with protection against catabolic enzymes in the bloodstream after administration into the body. Furthermore, liposomes are highly biocompatible due to the similarity of their architecture to that of the cell membrane. It is also possible that liposomes can reduce the side effects and enhance the main effect due to the change in the biodistribution of the encapsulated drugs. Up to now, various types of functional liposomes have been developed to deliver certain drugs to the desired target organ.

6.2.1 PEGYLATED LIPOSOMES AND PASSIVE TARGETING

Along with tumor development, a tumor cell acquires a highly malignant phenotype. At this stage, the tumor recruits and requires the induction and maintenance of a dedicated blood supply, and thus secretes a number of growth factors such as vascular endothelial growth factor (VEGF) and fibroblast growth factor (FGF) to produce neovessels. This process is known as tumor angiogenesis (Folkman and Shing 1992, Karamysheva 2008). Angiogenesis was originally discovered by Folkman in 1971 (Folkman 1971). This discovery founded the field of angiogenesis research and opened a new field of investigation worldwide. He hypothesized all cancer growth to be angiogenesis-dependent, and therefore proposed angiogenic vessels to be a relevant target for cancer therapy (Folkman 1974). Angiogenic vessels show a number of specific characteristics in comparison with normal blood vessels, such as the absence of pericytes, a high portion of proliferating endothelial cells, and aberrant vessel formation. One of the most specific characteristics of angiogenic vessels is the weak intercellular connections, with gaps of about 100–200 nm between opposing membranes (Asahara et al. 1997). For this reason, nanocarriers such as liposomes, polymeric micelles, and lipid microspheres can easily penetrate into the interstitial spaces of the tumor after administration of them into the bloodstream. This phenomenon is called the enhanced permeability and retention (EPR) effect (Asahara et al. 1997, Maeda et al. 2009).

To take advantage of this effect, nanocarriers having a long circulation time in the bloodstream accumulate in tumor tissues due to high vascular permeability and rudimentary lymphatic system in the tumor. This accumulation is referred to as passive targeting (Muggia 1999). The most well-known strategy to endow nanocarriers with the property of long circulation is modifying them with polyethylene glycol (PEG). Actually, PEG-modified (PEGylated) nanocarriers have been widely investigated as drug carriers and gene delivery systems. PEG forms a water shell on the carrier surface and provides a steric barrier to the nanocarrier for avoiding interactions with plasma proteins such as γ-globulin and complement components, resulting in escape from trapping by the reticulo-endothelial system (RES) (Lasic et al. 1991, Torchilin et al. 1994).

As a representative example of liposomal drugs, Doxil® has been used clinically. Doxil is the trademark name for PEG-modified liposomes encapsulating doxorubicin (Dox), and it has reduced cardiotoxicity as a side effect than Dox and also enhances the anticancer activity of the drug through the EPR effect of the carrier (Maeda et al. 2000, Muggia 1999).

6.2.2 Antibody-Modified Liposomes for Active Targeting

Since a passive targeting drug delivery system (DDS) is expected to work well for only hypervascular tumors, an active targeting DDS is more attractive. This approach to tumor therapy has been achieved by a modification of nanocarriers with antibodies, peptide probes, or specific molecules such as transferrin, glycoconjugates, and so on. We will discuss antibody-modified liposomes here and transferrin-modified liposomes in the following section.

Membrane type 1 matrix metalloproteinase (MT1-MMP) is a desirable target, since MT1-MMP is expressed in both angiogenic endothelial cells and tumor cells. MT1-MMP, originally identified as a member of the zinc-dependent MMP family, is known to be required for the degradation of the extracellular matrix (ECM), endothelial cell invasion and migration, formation of capillary tubes, and the recruitment of accessory cells (Cao et al. 1995, Sato et al. 1994). This enzyme activates MMP-2 from its precursor, and the active form can break down basement membrane type IV collagen; it is indirectly related to the infiltrative process of cancer cells and metastatic process (Ohuchi et al. 1997, Seandel et al. 2001). Moreover, MT1-MMP itself uses laminin 5 (Giannelli et al. 1997, Koshikawa et al. 2000), CD44 (Mori et al. 2002, Seiki 2002), and lumican (Li et al. 2004) as a substrate, and activates collagenase effective against type IV collagen. Since MT1-MMP is highly expressed in high-grade cancers and angiogenic blood vessels, it can be expected to be a useful target for cancer treatment, as probes against MT1-MMP would bind both cancer and angiogenic endothelial cells (Figure 6.2) (Zucker et al. 1995).

Hatakeyama et al. demonstrated that anti-MT1-MMP antibody-modified PEG liposomes encapsulating Dox were useful for cancer therapy (Hatakeyama et al. 2007). They used Fab′ fragments to prepare immunoliposomes, because Fab′-modified liposomes have a longer circulation time than those modified with whole IgG (Maruyama et al. 1997). Furthermore, Fab′ antibody lacks the Fc

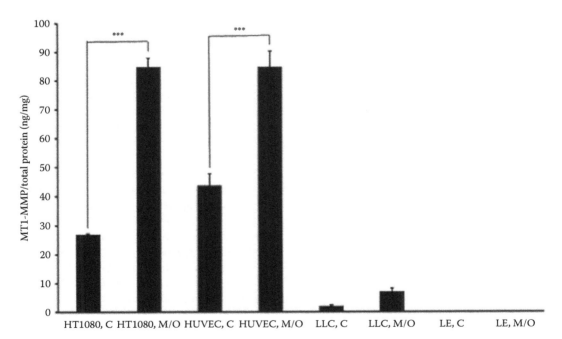

FIGURE 6.2 MT1-MMP protein expression in various cell lines. MT1-MMP protein expression in the cytosol or membranes/organelles of HT1080 fibrosarcoma cells, human umbilical vein endothelial cells (HUVEC), Lewis lung carcinoma (LLC) cells, or lung endothelial (LE) cells was measured by using an ELISA kit for MT1-MMP. The total protein concentration was measured with a Non-Interfering Protein Assay™ Kit, according to manufacturer's instructions. Significant difference is indicated as ***$p < 0.001$. C, cytosol and M/O, membrane/organelle.

fragment and thus can avoid recognition by Fc receptors on macrophages. In their studies, the uptake of anti-MT1-MMP Fab′-modified PEG liposomes into human fibrosarcoma HT1080 cells was tested (Atobe et al. 2007). Three hours after the incubation, the cellular uptake was increased about fivefold by the modification compared with that by the non-modified liposomes. They also examined the survival rate of tumor-bearing mice after the single injection of immunoliposomes encapsulating Dox. More than 60% of the mice were dead after 5 weeks when nontargeted liposomes encapsulating Dox were administered. On the contrary, the treatment with MT1-MMP-targeted immunoliposomes encapsulating Dox prolonged their survival time significantly. These results suggest that antibody-modified immunoliposomes would be a useful DDS tool for cancer therapy. In recent years, antibody preparation for cancer therapy has been developed, because such antibodies have much merit in terms of specificity and avidity.

6.2.3 OXALIPLATIN ENCAPSULATED IN TRANSFERRIN-CONJUGATED LIPOSOMES

Transferrin (TF) is a glycoprotein that transports ferric ion throughout the body. It is well known that the TF receptor is expressed on tumor cells at a significantly higher density than on normal cells. When the TF-conjugated liposomes bind to that receptor, they are internalized by receptor-mediated endocytosis into the cells, where they are present in endosome-like intracellular vesicles (Aisen 1994). This endocytosis is a normal process in iron delivery to certain kinds of cells. Suzuki et al. (2008) prepared TF-conjugated liposomes by coupling TF to the distal terminal of the PEG chains of PEG liposomes. These pendant-type PEG liposomes encapsulating oxaliplatin (OHP), a kind of platinum anticancer agent, showed an anticancer effect. First, they assessed the cytotoxicity of these TF–PEG liposomes encapsulating OHP against colon 26 cells that overexpressed TF receptors.

The TF–PEG liposomes induced dose-dependent cytotoxicity, and their ED_{50} was $8\,\mu g/mL$, or about twofold higher than that for the non-modified PEG liposomes. Moreover, when TF receptor of the cells was blocked with TF, the TF–PEG liposomes were not taken up into colon 26 cells, indicating that the TF–PEG liposomes were taken up into the cells via TF receptor–mediated endocytosis. The biodistribution pattern of TF–PEG liposomes in the plasma, liver, and spleen of tumor-bearing mice was almost the same to that obtained with PEG liposomes. This result indicated that TF–PEG liposomes avoided the RES uptake as well as did the PEG liposomes. Notably, 72 h after their intravenous administration, the TF–PEG liposomes accumulated in the tumor to an approximately twofold higher extent than the PEG liposomes. In addition, these investigators measured serum albumin, total protein, AST, ALT, and BUN after the administration of TF–PEG liposomes encapsulating OHP, and observed that the values of these markers were not different from those in the mice injected with saline. Therefore, TF–PEG liposomes encapsulating OHP would probably not induce significant side effects if used in human patients. When such liposomes were intravenously administered into colon 26-bearing mice on days 9 and 12 after the tumor cell implantation, they suppressed tumor growth significantly in comparison with PEG-liposomal OHP. Based on these preclinical studies, TF–PEG-liposomal OHP was advanced to a phase I study. In that study, the tumor type of the patients was colorectal in 23 patients, pancreatic in 3, and neuroendocrine in 3; most of the patients were pretreated with chemotherapy or chemoradiation. From the dose-limiting toxicity (DLT) and usability of pharmacokinetics, the recommended dose of TF–PEG liposomes encapsulating OHP at phase II was decided to be $226\,mg/m^2$. TF–PEG-liposomal OHP has now been advanced to the phase II trial in the United States.

6.2.4 ACTIVE TARGETING TO ANGIOGENIC VESSELS

Tumor angiogenesis is thought to be regulated by a number of tumor-secreted factors that bind to specific receptors on the angiogenic endothelial cell surface, regulating the formation of the capillary sprouts that grow into the tumor mass and eventually produce a functional vascular system.

Thus, many of the receptors for these factors are expressed on the surface of angiogenic blood vessels, and they could thus be potential target for the drug carriers. The molecules that bind to the receptor of angiogenic blood vessels would be the tool available for delivering the encapsulating agents to the neovessels. Angiogenesis inhibitors do not damage the new blood vessels directly, but inhibit the process of angiogenesis, thus causing tumor dormancy.

On the other hand, we tried to deliver anticancer agents to neovessels by means of active targeting of angiogenic endothelial cells. The advantage of disrupting angiogenic endothelial cells is that such disruption not only damages angiogenic vessels, leading to eradication of the cancer cells, but also inhibits hematogenous metastasis; it also eradicates drug-resistant tumors by the targeting of vascular endothelial cells (Browder et al. 2000). Multivalency, e.g., obtained by displaying ligands on the surface of drug carriers, may further increase the affinity of the interaction with the target cells (Kok et al. 2002). Such a new modality of angiogenic therapy, namely, anti-neovascular therapy (ANET), has been developed in our laboratory (Asai et al. 2002). This therapy aims to damage neovessel endothelial cells by anticancer agents contained in targeting DDS. In the past decade, ANET has revealed to have high therapeutic effect in solid-cancer transplant models.

6.2.5 ANET with Peptide-Modified Liposomes

As described earlier, since endothelial cells in the angiogenic blood vessels within solid tumors express several proteins that are absent or barely detectable in preexisting blood vessels, it is necessary to identify small molecules that may enable the targeting of angiogenic vessels. There are several methods to detect these molecules binding to the proteins specifically presented in angiogenic blood vessels, and in vivo biopanning method using a phage-displayed peptide library is one of them (Figure 6.3) (Oku et al. 2002). This method is useful for discovering novel peptides specific for the angiogenic vessels. A phage-displayed peptide library was injected into angiogenic model mice prepared by the dorsal air sac (DAS) method. The advantage of this method is that the selected phages have the ability to bind to only angiogenic vessels, but not to cancer cells, since the injected phages cannot gain access to the cancer cells (Ishikawa et al. 1998, Takikawa et al. 2000).

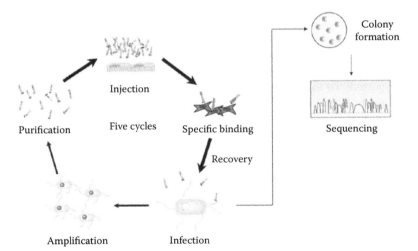

FIGURE 6.3 In vivo biopanning. The phage-displayed peptide library was injected into DAS model mice via a tail vein. After injection, the phages that had accumulated in the target organ were recovered and titrated. After that, recovered phages were amplified and purified. Biopanning steps were repeated for five cycles. After selected phages were cloned, sequences of the peptides presented were determined.

We identified several peptide sequences binding to angiogenic vessels, such as ASSSYPLIHWRPWAR, DRWRPALPVVLFPLH, and PRPGAPLAGSWPGTS, present on the selected phages accumulated in the tumor tissues of two different tumor cell types, and determined their epitope sequences (Oku et al. 2002). For liposomalization of the epitope sequences of peptides, stearoyl 5-mer peptides (APRPG, RWRPA, or HWPRPW) were synthesized by using the DIPCI–HOBt coupling method. For the preparation of peptide-modified liposomes, distearoylphosphatidylcholine (DSPC), cholesterol, and stearoyl 5-mer peptides (10:5:2) were used as components. These peptide-modified liposomes were administered into size-matched Meth A sarcoma-bearing mice via a tail vein. Three hours after the administration, all peptide-modified liposomes showed high accumulation in tumor tissue compared with the non-modified liposomes. In particular, the APRPG peptide-modified liposomes showed the highest accumulation in the tumor. To endow the liposomes with a long circulation time, we attached the APRPG peptide to the PEG termini of PEG–distearoylphosphatidylethanolamine (APRPG–PEG–DSPE) liposomes (Maeda et al. 2004). Both APRPG–PEG-modified liposomes and PEG-modified ones were highly accumulated in the tumor after intravenous injection. However, the intratumoral distribution of the two kinds of liposomes was quite different: the APRPG–PEG-modified liposomes were co-localized with angiogenic endothelial cells, as seen in frozen sections of tumors from colon26 NL-17-bearing mice. In contrast, the PEG-modified ones had accumulated in the interstitial spaces in the tumor tissue. Moreover, the therapeutic efficacy of APRPG–PEG-modified liposomes encapsulating Dox was significantly higher than that of PEG-modified liposomal Dox.

APRPG–PEG-modified liposomal Dox demonstrated therapeutic gain in an orthotopic mouse model of pancreatic cancer (Yonezawa et al. 2007). It is thought that treatment of pancreatic cancer is more difficult than that of other cancer such as breast or colorectal ones because of its hypovasculature nature. Pancreatic cancer is difficult to diagnose, and shows a highly malignant potential. The 5 year survival rate of patients suffering from pancreatic cancer is less than 5% in the United States, Japan, and Europe; its incidence rate is almost equal to the death rate (Jemal et al. 2003). Therefore, existing chemotherapeutics have a limited effect on pancreatic cancer, and thus an effective treatment modality is awaited. In our model, vascular density of orthotropic implanted SUIT-2 tumor cells was significantly low in comparison with that of subcutaneously implanted ones. At 25 days after the implantation, vascular density of s.c. implanted tumor was significantly increased from that at 10 days. In contrast, the vascular density of the orthotropic implanted tumor was not increased. APRPG–PEG-modified liposomes encapsulating Dox showed an efficient anticancer effect against this hypovascular tumor. After liposomal treatment of the mice for four times at days 3, 6, 9, and 12, after tumor implantation the tumor weight was decreased to less than 50% of that of control at day 15.

Since APRPG-modified liposomes accumulate in angiogenic vessels, it is possible to use these liposomes for antiangiogenic therapy. Therefore, the therapeutic effect of APRPG–PEG-modified liposomes encapsulating Z-3-[(2,4-dimethylpyrrol-5-yl)methylidenyl]-2-indolinone (SU5416) (Katanasaka et al. 2008), an inhibitor of VEGF receptor 2 tyrosine kinase, instead of Dox, was examined. SU5416 had shown high therapeutic effect by administration with anticancer drugs in phases I and II clinical trials, although side effects such as anaphylactic shock or hemolysis caused by Cremophor EL which was used for solubilizing the agent (Fong et al. 1999, Gelderblom et al. 2001, Zangari et al. 2004) were observed. APRPG–PEG-modified liposomes encapsulating SU5416 significantly suppressed tumor growth in comparison with control ($p < 0.05$), free SU5416 ($p < 0.05$), and PEG-modified SU5416-encapsulated liposomes ($p < 0.01$) in colon 26 NL-17 carcinoma cell-bearing mice. In particular, the APRPG-modified liposomes encapsulating SU5416 had no hemolytic activity. These results suggest that the antitumor effect obtained by APRPG-modified liposomes is not limited to ANET but may manifest with other modalities such as antiangiogenic therapy and antiangiogenic photodynamic therapy.

6.2.6 ANET Using Various Active Targeting Probes

Besides APRPG, Asn-Gly-Arg (NGR), and Arg-Gly-Asp (RGD) are known to bind to angiogenic blood vessels. These peptides had also been discovered by in vivo biopanning, in tumor-bearing mice by Arap et al. (1998). It is already known that the NGR peptide binds to CD13 (aminopeptidase) and that RGD binds to $\alpha_V\beta_3$ or $\alpha_V\beta_5$ integrin. Aminopeptidase CD13 is a Zn^{2+}-dependent membrane-bound metalloexopeptidase belonging to the M_1 family of aminopeptidases (Luciani et al. 1998). CD13 is a transmembrane protein that cleaves N-terminal, neutral amino acids of various peptides and proteins (Sanderink et al. 1988). It is thought that CD13 plays a key role as a regulator of various hormones, chemokines, protein degradation, cell proliferation, and cancer angiogenesis (Luan and Xu 2007). CD13 is limitedly expressed in normal blood vessels, but highly expressed in cancer angiogenic blood vessels (Curnis et al. 2002). NGR peptide–modified liposomes could thus be target to the angiogenic blood vessels.

Pastorino et al. (2003) prepared NGR-coupled liposomes encapsulating Dox, and used them to treat orthotropic neuroblastoma (NB) xenografts in SCID mice. To confirm whether NGR peptide–modified liposomes would bind to CD13, they investigated the binding using KS1767 cells (Kaposi's sarcoma) and THP-1 cells (acute monocytic leukemia). A monoclonal antibody specific for CD13 was bound to both cells; however, the NGR peptide–modified liposomes did not bind to the THP-1 cells. They next evaluated the tumor accumulation of the NGR or ARA peptide–modified liposomes (control nontargeted liposomes) in xenograft models of orthotropic NB in SCID mice. Both peptide-modified liposomes were taken up by the NB tumor almost the same early after the injection. However, the uptake of the NGR liposomes was increased at least fourfold compared with that of the nontargeted liposomes at 12 h, and about 10-fold at 24 h. Finally, they examined the therapeutic efficiency and metastasis inhibition in SCID mice bearing SH-SY5Y NB cells. The tumor cells were implanted in the left adrenal gland, and the mice were treated with HEPES buffer (control), ARA-modified or NGR-modified liposomes encapsulating Dox for three times in a week (3 mg/kg Dox). Six weeks after the implantation, control or ARA-modified liposomes showed little suppression of the tumor growth; however, the NGR-modified liposomes significantly suppressed the growth. Moreover, the mice treated with NGR-modified liposomes showed little metastasis to the kidney and liver. In addition, NGR-modified liposomal Dox improved the survival of orthotropic NB xenografts beyond 4 months.

Integrins $\alpha_V\beta_3$ and $\alpha_V\beta_5$ are also overexpressed on tumor endothelial cells (Varner and Cheresh 1996). Integrins are composed of 8 different types of α-subunit and 18 types of β-subunit that assemble into at least 24 distinct integrins, and this heterodimer composition generally confers ligand specificity (Johnson 1999, Seftor et al. 1999). Integrins play roles in cell–ECM interaction and in cell–cell adhesion (Van Belle et al. 1999). The importance of these integrins in angiogenesis is remarkable. Inhibition of integrin $\alpha_V\beta_3$ functions such as adhesion and signal transduction by antibody suppressed cancer growth through inhibiting vitronectin binding to endothelial cells during angiogenesis (Brooks et al. 1994, Hodivala-Dilke et al. 2003).

Schiffelers et al. (2003) prepared a cyclic 5-mer peptide derivative containing RGD sequence, c[RGDf(ε-S-acetylthioacetyl)K], and modified PEG liposomes with this peptide. When the cyclic RGD peptide–modified liposomes were incubated with HUVEC cells, the normalized binding increased about fourfold compared with that to non-modified liposomes. In addition, when the RGD peptide–modified liposomes and free RGD peptide were co-incubated with HUVEC cells, specific binding was cancelled. These results indicated that the RGD peptide–modified liposomes and free RGD peptide could bind to the same site. In the therapeutic study, BALB/c mice were subcutaneously implanted with Dox-insensitive murine C26 colon carcinoma cells, and Dox-encapsulated liposomes (10 mg/kg) were administered at day 7. Then, the tumor volume was measured at day 10. RGD peptide–modified liposomal Dox inhibited the tumor growth. These results indicated that ANET is able to inhibit Dox-insensitive tumor growth, since this therapy does not damage cancer cells, but eradicates angiogenic endothelial cells. RGD peptide is considered as a probe not only for liposomes but also for polymeric micelles. Many studies showed that RGD peptide–conjugated block copolymers enhanced drug delivery to the angiogenic blood vessels (Xiong et al. 2007, 2008). These results indicate that the RGD peptide is also useful for the active targeting of nanocarriers.

6.2.7 ACTIVE TARGETING TO DISSEMINATED TUMOR CELLS

As described earlier, angiogenic vessel–specific peptide-modified liposomes showed tumor-specific accumulation, and those encapsulating anticancer drugs were effective for the treatment of solid tumors including orthotropic pancreatic cancer. However, overcoming metastasis is another challenge to conquer cancer. We attempted to treat a peritoneal dissemination of gastric cancer by use of targeted liposomes. Gastric cancer at its early stage is rather easy to treat, and the 5 year survival rate is about 90% in Japan. However, when the cancer has progressed to stage IV, this rate becomes less than 20%. Peritoneal dissemination is the most remarkable form of metastasis in advanced cancers, where cancer cells metastasize diffusely throughout the peritoneum (Kodera et al. 1998, Sakakura et al. 2002). Surgery, chemotherapy, and thermochemotherapy have been used for the treatment, although treatment protocols still remain to be established (Yanagihara et al. 2005).

We attempted to develop an anticancer DDS for treatment of cancer cells that have metastasized to the peritoneum. A novel oligopeptide Ser-Trp-Lys-Leu-Pro-Pro-Ser (SWKLPPS) was identified by in vivo intraperitoneal biopanning, which peptide-bound to peritoneally metastasized cancer (Akita et al. 2006). Thereby the peptide-modified liposomes significantly accumulated in the metastasized cancer in mice compared with the non-peptide-conjugated liposomes, and they showed less accumulation in the spleen and liver at 24 h after administration. Furthermore, SWKLPPS peptide–modified Dox-encapsulated liposomes enhanced anticancer activity in vitro. Liposomes modified with these novel peptides increased the accumulation in the target cells or organ compared with non-modified ones in vivo. Furthermore, these peptide-modified liposomes encapsulating anticancer drugs indicated remarkable anticancer effects. As stated earlier, cancer tissue or angiogenic vessels express a number of molecules that are little expressed in normal or nontarget tissues. These molecules would be suitable for targets, and peptides discovered by in vivo biopanning are considered as useful probes for the active targeting of nanomedicines.

6.2.8 NOVEL DUAL-TARGETING OR MULTI-TARGETING STRATEGY

Active targeting using liposomes modified with ligands such as antibodies and peptides achieves more selective drug delivery to the target tissues. These ligands that recognize target-associated molecules are conjugated to the head of the PEG chains of liposomes. Active targeting using ligand-modified long-circulating PEG liposomes would be more effective, since these liposomes have great opportunity to meet target tissues during their long circulation time. In order to target tumor angiogenesis by the liposomal DDS, peptides that have affinity for molecules expressed on vascular endothelial cells are used to modify the liposomal surface, and there have been a great number of reports on the effectiveness of using such peptides (Chang et al. 2009). In fact, Dox-encapsulating long-circulating liposomes modified with the APRPG, NGR, or RGD sequence indicated a high therapeutic effect on tumor-bearing animals. However, reported studies have only used a single peptide for the modification of nanocarriers. Although cooperative binding of peptides to the target is achieved, the affinity of peptides is far less in comparison with that of specific antibodies.

In light of this background information, we sought to modify liposomes with two different probe peptides to enhance the affinity of nanocarriers to their target. This dual-targeting is a new concept that uses a combination of peptides having different specificities for modification of the liposomal surface. As a result, we found that these "dual-targeting" liposomes showed cooperative binding to the target tissue and that the actual therapeutic effect of such liposomes encapsulating an anticancer drug was synergistically enhanced. In brief, aiming for a higher enhancement of the antitumor effect of ANET, we prepared APRPG- and GNGRG-modified dual-targeting PEG liposomes and evaluated the dual-targeting effect. First, the adhesion of these liposomes to and internalization by proliferating HUVEC cells were examined. These dual-targeting liposomes showed remarkably higher affinity for the proliferating HUVEC cells than the singly modified ones, namely, APRPG–PEG liposomes and GNGRG–PEG liposomes.

Furthermore, these dual-targeting liposomes encapsulating Dox significantly inhibited the growth of HUVEC cells. These results indicated that dual-targeting enhanced the affinity of liposomes for the target cells due to the cooperative effect of APRPG and GNGRG on the liposomal surface. It is known that NGR peptide–modified liposomes tend to accumulate in the spleen after systemic administration (Pastorino et al. 2003). On the other hand, APRPG peptide–modified liposomes tend to avoid capture by the RES in vivo (Oku et al. 2002). However, the dual-targeting liposomes accumulated less in the spleen compared with single GNGRG–PEG-modified liposomes after intravenous injection into mice. In addition, the time in circulation was almost the same between dual-targeting and APRPG–PEG-modified liposomes. To determine the intratumoral distribution of dual-targeting liposomes, we examined tumor slices by confocal laser scanning microscopy after administration of fluorescence-labeled dual-targeting liposomes to tumor-bearing mice. These liposomes became localized over a wide region of the angiogenic blood vessel wall compared with the single-targeting liposomes, indicating that dual-targeting enhanced the affinity of the liposomes for tumor angiogenic endothelial cells. In accordance with this finding, the treatment of tumor-bearing mice with Dox-encapsulated dual-targeting liposomes significantly prolonged the survival of the mice in comparison to that with Dox-encapsulated single-targeting liposomes. These results indicate that dual-targeting enhances the therapeutic effect by augmenting the ability of the liposomes to target the angiogenic blood vessels.

Modification of liposomal surface with two kinds of ligands was also reported by Saul et al. (2006). They indicated that liposomes modified with anti-EGFR antibody and folic acid showed high selectivity for the target cells, although these ligands are quite different from each other, but they indicated that it is not always applicable to modify nanocarriers with two different ligands to increase their selectivity. The dual-targeting strategy shown here, or even a multi-targeting one, should be applicable for molecules that do not have sufficient affinity when used singly as probes, such as peptide ligands. Accordingly, as dual-targeting increases the affinity of liposomes for the target cell, it will be useful for various active targeting strategies.

6.2.9 ORAL ADMINISTRATION OF LIPOSOMES

The history of oral administration of liposomal drugs is quite long. In the early research on liposomes, insulin was encapsulated and orally administered to achieve lowering of the glucose level in the bloodstream. In fact, insulin or calcitonin encapsulation in liposomes increases the absorption of these molecules (Spangler 1990, Takeuchi et al. 1996, 2003). However, theoretically, orally administered liposomes are thought not to be stable in the gastrointestinal (GI) tract. On the contrary, oral administration is a promising route to deliver materials to the intestines.

Recurrence and metastasis after surgical resection of a tumor are two factors associated with the lethality of cancers. Therefore, it is important to overcome cancer metastasis. For the prevention of cancer metastasis, it would be good to have available a means to target the migrating cancer cells. Recently, it was confirmed that natural killer (NK) cells and T cells recognize cancer cells that are migrating in the bloodstream from the primary focus and that these cells have an anticancer effect (Okada et al. 2003). In cancer metastasis, it is empirically known that a certain number of cells are needed to form a metastatic focus at the target organs. If this number is small, tumor cells accumulated at the target site would be cleared (Kikkawa et al. 2000a,b). Immune responses such as macrophage and NK cells are considered as one of the reasons for this clearance (Kikkawa et al. 2000a,b). As the intestines are equipped with an immunological defense mechanism to inhibit invasion by microorganisms or uptake of certain proteins (Holmgren and Czerkinsky 2005, Neurath et al. 2002), oral administration of certain materials may enhance intestinal immunological responses (Van Ginkel et al. 2000).

Phytosterol is known to suppress the absorption of cholesterol from the intestines (Noakes et al. 2002) and also has been demonstrated to have an anti-colonic cancer effect (Awad et al. 1996). Therefore, we speculated that orally administered phytosterol would stimulate intestinal immunity. Since phytosterol is hydrophobic, a liposomal formulation would be desirable, and so we prepared liposomal β-sitosterol, a major component of phytosterol, for the purpose of modulating intestinal immunity. For oral administration, liposomes entrapping β-sitosterol or cholesterol were prepared to have a size of 117 or 121 nm, respectively, since liposomes of about 100 nm had been earlier shown to be well-absorbed (Takeuchi et al. 2003). Although both β-sitosterol- and cholesterol-entrapped liposomes had similar characteristics, the absorption profile of the two sterols determined by radioactivity of radiolabeled sterols was quite different: the plasma level of liposomal cholesterol was greatly increased time-dependently; however, radioactive liposomal β-sitosterol was little detected until 6 h after the oral administration. In the intestines, the production of IL-12 and IL-18 cytokines was increased by the treatment with liposomal β-sitosterol but not by that with liposomal cholesterol. Both of these cytokines are important for immune response. In fact, IL-18 activates interferon gamma (IFN-γ) production in the presence of IL-12 (Vossenkamper et al. 2004). Moreover, IL-18 shows its anticancer effect by enhancing NK cell activity (Micallef et al. 1997). Therefore, the NK cell activity was examined after the oral administration of liposomal β-sitosterol daily for 7 days. As a result, NK cell activity was significantly increased in the spleen by the liposomal β-sitosterol administration compared with that by liposomal cholesterol or phosphate-buffered saline (PBS). These results suggest that enhanced production of IL-12 and IL-18 activates NK cells in the small intestines and that the activated NK cells are carried by the bloodstream to the spleen. In other words, the production of IL-12 or IL-18 could lead not only to an intestinal immune response, but also to whole-body immunity.

Consequently, an antimetastatic effect was observed following the oral administration of liposomal β-sitosterol in an experimental metastatic model. In brief, C57BL6 mice were orally administered liposomal β-sitosterol, cholesterol, or PBS daily for 7 days. The following day, B16BL6 cells were administered via a tail vein, and the numbers of metastatic colonies were measured at 14 days after the injection. The mice administered the liposomal β-sitosterol had a significantly decreased number of metastatic colonies, about one-half compared with the administration of liposomal cholesterol or PBS. It is suggested that liposomal β-sitosterol could be applicable for use as a functional food or as a chemopreventive agent.

6.3 POLYMERIC MICELLES

Polymeric micelles, another type of the DDS nanocarriers, are composed of block copolymers designed to have a hydrophobic, polycationic, or polyanionic core region depending on the encapsulating materials and a hydrophilic, usually PEGylated, surface region. Thus block copolymers form micelles having the drug encapsulated in the interior region and a hydrophilic exterior region (Gaucher et al. 2005). By modification with PEG, polymeric micelles also show the characteristic of long circulation in the bloodstream, as in the case of PEG-modified liposomes. The particle sizes of polymeric micelles can be controlled by the length of both the PEG chain and drug-interacting core chain. Dox-encapsulated polymeric micelles known as NK911 were reported to have a long half-life in the bloodstream and to accumulate in the tumor of tumor-bearing animals due to passive targeting (Yokoyama et al. 1999). In this case, the hydrophobic interaction was used for entrapment of the drug. Polymeric micelles can also encapsulate macromolecules such as DNA or proteins by using electrostatic interaction or hydrogen bonding with the core chain (Kataoka et al. 2005). Preclinical or clinical studies on polymer micelles with entrapped paclitaxel did not show the severe side effects of the drug (Hamaguchi et al. 2007), indicating that the formulation was stable in the bloodstream due to the drug–polymer association. Also, in long-term experiments, polymer micelles would be disassociated into macromolecular chains and excreted. Over all, polymeric micelles can incorporate various kinds of drugs into the inner core by chemical conjugation or physical entrapment, circulate in the bloodstream with relatively high stability, and are expected to accumulate in tumor tissues by the EPR effect.

6.3.1 Paclitaxel-Incorporating Polymeric Micelles

Paclitaxel has been widely used as a standard care for the therapy of cancers such as non-small cell lung, breast, gastric, endometrial, and head and neck ever since its clinical adaptation for the treatment of ovarian cancer was accomplished in the United States in 1992 (Rowinsky and Donehower 1995). However, repeated administration of paclitaxel leads to irreversible neurotoxicity, and the side effects frequently require the cessation of treatment (Formenti 2005). Because paclitaxel is a hydrophobic compound, it can be dissolved in Cremophor EL and alcohol for the administration as an injectable medicine. As mentioned earlier, it is known that Cremophor EL causes hypersensitivity reactions. For this reason, when paclitaxel is administered, steroid or antihistaminic drug treatment is indispensable. Paclitaxel, however, is an attractive medicine with a high anticancer effect even if the paclitaxel has such serious side effects as a disadvantage. It is easy to suppose that overcoming these problems would lead to significant improvement of the patient's quality of life (QOL).

Polymeric micelles encapsulating paclitaxel and known as NK105 (85 nm in size) (Hamaguchi et al. 2005) were developed by facilitating the self-association in an aqueous medium of amphiphilic block copolymers constructed with PEG as the hydrophilic segment and modified polyaspartate as the hydrophobic segment. Owing to the PEG constituting to the outer shell of the micelles, NK105 is injectable as an aqueous solution without Cremophore EL. Moreover, the PEG coating confers a long circulatory half-life to the formulation, and allows the accumulation of the micellar drugs in the tumor tissue by the EPR effect. NK105 demonstrated a remarkable therapeutic gain in preclinical trials using tumor-bearing mice. A single intravenous administration of NK105 or paclitaxel at a dose of 50 or 100 mg/kg, respectively, into colon 26-bearing CDF1 mice resulted in a 50- to 86-fold higher AUC of the drug by NK105 than by the free paclitaxel. Moreover, the half-life in the bloodstream was increased four- to sixfold compared with that of paclitaxel. In addition, the AUC in the tumor tissue was 25 times higher for NK105 than for the free drug, and NK105 continued to reside until 72 h after the injection.

A study was conducted on antitumor activity in mice bearing HT29 cells, a human colonic cell line, implanted subcutaneously into the back skin. The mice were treated with paclitaxel at a dose of 25, 50, or 100 mg/kg or NK105 at the same dose once in a week for 3 weeks. NK105 administered at the dose of 25 mg/kg demonstrated almost the same rate of tumor growth inhibition as the mice treated with 100 mg/kg of paclitaxel. Notably, the tumor completely disappeared in some of the mice after the treatment with 50 or 100 mg/kg NK105. In addition, the NK105-administered mice did not show any body weight change and showed reduced irreversible neurotoxicity. Since NK105 showed usability in the preclinical studies using tumor-bearing mice, NK105 proceeded to the phase I study (Hamaguchi et al. 2007, Matsumura and Kataoka 2009). This study was planned to determine the maximum tolerated dose (MTD), DLTs, and recommended dose for use in the subsequent phase II trial. NK105 was dissolved in 5% glucose solution and administered by intravenous infusion for 1 h every 3 weeks. The injected dose was started from 10 mg/m^2 and was increased up to 180 mg/m^2 without any antiallergic preliminary medication. The toxicity was shown to be comparatively mild; however, grade 4 neutropenia was observed in 42% of the patients and grade 3 and over, in 75%, at a dose of over 150 mg/m^2. The C_{max} and AUC of paclitaxel in the blood increased dose-dependently. The AUC of NK105 at dose of 150 mg/m^2 (as the recommended phase II dose) was about 15-fold higher than that of paclitaxel at dose of 210 mg/m^2 (conventional dose). Moreover, NK105 stabilized the disease for 4 weeks in 6 patients (gastric, bile duct, colon, or pancreatic cancer). Because of these encouraging results, NK105 is now under phase II trial.

6.3.2 SN-38-Conjugated Polymeric Micelles

Polymeric micelles known as NK012 have advanced to clinical trial as well as NK105. NK012 is SN-38 (irinotecan, CPT-11) incorporated into micellar nanoparticles. SN-38 was synthesized and shown to have increased anticancer effects and decreased toxicity compared with camptothecin, a plant alkaloid. Since SN-38 was insoluble in aqueous solution, polymeric micelles were used to allow intravenous administration of the drug. NK012 was constructed with self-assembling PEG–PGlu (SN-38) amphiphilic block copolymers in an aqueous milieu. The average size of NK012 is about 20 nm, smaller than that of other nanocarriers such as liposomes or other anticancer polymeric micelles.

NK012 demonstrated growth inhibition of 5 human renal cell carcinoma (RCC) cell lines (SKRC-49, Caki-1, 769P, 786O, and KU19-20) in vitro. The IC_{50} of NK012 was 96- to 406-fold higher than that of CPT-11. In addition, NK012 significantly inhibited the growth of HT29 tumor xenografts at doses of 15 and 30 mg/kg/day (Koizumi et al. 2006). From these in vitro and in vivo results, it was indicated that NK012 could show stronger anticancer effects than SN-38 by the incorporation of the latter in polymeric micelles; thus NK012 was advanced to clinical trials. In the phase I study on 32 patients, a partial antitumor effect was noted in 4 cases, 16 cases maintained a stable disease, and 9 cases became progressively worse. The last three cases were incapable of being evaluated. Most of the toxicity stemming from the treatment was hematological in nature; i.e., grade 3/4 neutropenia was observed in 15 cases among the 32 cases, and grade 3/4 thrombocytopenia in 3 cases. Because the phase I study showed NK012 to have higher antitumor than CPT-11 without severe side effects, NK012 is expected to have strong therapeutic effects in clinical use. Currently, NK012 is ready to enter the phase II study.

6.3.3 Oral Administration of Polymeric Micelles for Cancer Therapy

TNP-470, a synthetic analog of fumagillin, was discovered in *Aspergillus fumigatus* (Fresenius) (Ingber et al. 1990). This drug has a most potent antiangiogenic effect with reduced side effects in comparison with fumagillin (Kusaka et al. 1991). TNP-470 has already shown an anticancer effect against primary and metastatic tumors in mice and human xenografts of breast cancer, neuroblastoma, ovarian cancer, and prostate cancer (Emoto et al. 2007, Kanamori et al. 2007, Yamaoka et al. 1993, Yanase et al. 1993). Based on these results, TNP-470 was chosen to undergo some clinical trials for cancer treatment as one of the first antiangiogenic medicines (Bhargava et al. 1999, Kruger and Figg 2000, Kudelka et al. 1997). However, TNP-470 was dropped during the phase II trial because of the neurotoxicity and dizziness or mental confusion that manifested (Logothetis et al. 2001). TNP-470 was also difficult for clinical use because of its instability and rapid hydrolysis in vitro and in vivo. Due to its strong potency, poor pharmacokinetic profile, and poor solubility, TNP-470 has been considered a suitable candidate for polymer conjugation or for DDS application (Brannon-Peppas and Blanchette 2004, Kakinoki et al. 2007, Rapoport et al. 2003).

Benny et al. (2008) produced a diblock copolymer, monomethoxy-polyethyleneglycol–polylactic acid (mPEG–PLA), conjugated with TNP-470, with the name of lodamin. TNP-470 was encapsulated in the core of the micelles for protection from the acidic condition after oral administration (Harris and Chess 2003). For study of the intestinal absorption of the polymeric micelles, mice were orally administered fluorescence-labeled micelles and many of fluorescent signals were observed in the stomach, intestine, and liver, but not in the brain, at 3 days after the injection. Fluorescence was first detected in the plasma at 1 h after administration and continued to be detectable at least up to 72 h. The orally administered fluorescence-labeled micelles were highly taken up by the tumors of tumor-bearing mice. Therefore, the tumor growth inhibition by orally administered lodamin as an antiangiogenic agent was evaluated. Oral administration of lodamin inhibited the growth of Lewis lung carcinoma (LLC) in about 87% of tumor-bearing mice about at a dose of 15 mg/kg given every day, in 77% at 30 mg/kg given every other day, and in 74% dosed at 15 mg/kg every other day. This inhibition was first observable at 12 days after the administration. In contrast, administration of free TNP-470 (30 mg/kg every other day) did not show any inhibition of the tumor growth. Moreover, in liver metastatic model mice, where B16–F10 cells were injected into the spleen, ascites accumulated, malignant nodules and cirrhosis appeared in the liver, and 40% of the mice were dead by 20 days. In contrast, such abnormalities were not observed in the lodamin-administered mice, and all of these mice survived. Although TNP-470 is neurotoxic, lodamin did not show any neurotoxicity as evaluated by a sensitive motor-coordination test involving crossing a narrow balance beam (Carter et al. 2001). The reason for this lack of neurotoxicity would be that the polymeric micelles did not penetrate the blood–brain barrier and thus protected against the transfer of TNP-470 to the brain.

6.4 LIPID MICROSPHERES

Lipid microspheres are lipid capsules having a lipid monolayer membrane structure containing neutral lipids in its core. To make lipid microspheres having uniform particle size, phospholipids and neutral lipids are emulsified in a homogenizer. Lipid microspheres had been used as fat emulsions in the clinical setting before being focused on as a drug delivery carrier. Lipid microspheres have gained interest as a drug carrier because they can encapsulate various lipophilic drugs in their core. Lipid microspheres can be prepared from egg PC and soybean oil by an already established large-scale production method. Because of these reasons, clinical applications of lipid microspheres are easily and rapidly made compared with other DDS formulations. Lipid microspheres encapsulating steroidal anti-inflammatory drugs, nonsteroidal anti-inflammatory drugs (NSAIDs), or prostaglandin (PG) have been investigated for the treatment of rheumatic diseases (Igarashi et al. 2001). In recent years, various studies using lipid microsphere composed of various triglycerides or synthesized phosphatidylcholine instead of soybean oil or egg PC have been conducted (Grenha et al. 2008, Ishihara et al. 2009, Nii and Ishii 2005, Suzuki et al. 2008, Zhang et al. 2008). Clinical applications of lipid microspheres are largely dependent on their capacity to entrap hydrophobic materials. These microspheres are phagocytosed by macrophages after injection into bloodstream. Therefore, it would be useful to develop targetable lipid microspheres for general-purpose drug carriers.

6.5 CONCLUSIONS

In this chapter, we introduced background and several recent studies about DDS agents for cancer therapy. Some of them have been already commercialized and some are achieving good results in preclinical and clinical trials. Since chemotherapy with conventional anticancer drugs accompany severe side effects, DDS strategy is useful for this field. DDS could deliver nanocarriers to cancerous tissues due to passive targeting by the EPR effect and reduce the undesirable distribution of the drug to the site of side effects. However, since another advantage of nanocarriers is the ease of modification of them with specific probes, active targeting of DDS would be more beneficial for the development of new anticancer nanomedicines. In this chapter, we introduced some new concepts with supporting evidences such as neovessel-targeted strategy and dual-targeting strategy. The present chapter has also provided important information for the future development of DDS nanocarriers.

REFERENCES

Aisen, P. 1994. The transferrin receptor and the release of iron from transferrin. *Adv Exp Med Biol* 356: 31–40.
Akita, N., Maruta, F., Seymour, L. W. et al. 2006. Identification of oligopeptides binding to peritoneal tumors of gastric cancer. *Cancer Sci* 97: 1075–1081.
Arap, W., Pasqualini, R., and Ruoslahti, E. 1998. Cancer treatment by targeted drug delivery to tumor vasculature in a mouse model. *Science* 279: 377–380.
Asahara, T., Murohara, T., Sullivan, A. et al. 1997. Isolation of putative progenitor endothelial cells for angiogenesis. *Science* 275: 964–967.
Asai, T., Nagatsuka, M., Kuromi, K. et al. 2002. Suppression of tumor growth by novel peptides homing to tumor-derived new blood vessels. *FEBS Lett* 510: 206–210.
Atobe, K., Ishida, T., Ishida, E. et al. 2007. In vitro efficacy of a sterically stabilized immunoliposomes targeted to membrane type 1 matrix metalloproteinase (MT1-MMP). *Biol Pharm Bull* 30: 972–978.
Awad, A. B., Chen, Y. C., Fink, C. S., and Hennessey, T. 1996. Beta-sitosterol inhibits HT-29 human colon cancer cell growth and alters membrane lipids. *Anticancer Res* 16: 2797–2804.

Bangham, A. D. and Horne, R. W. 1964. Negative staining of phospholipids and their structural modification by surface-active agents as observed in the electron microscope. *J Mol Biol* 8: 660–668.

Benny, O., Fainaru, O., Adini, A. et al. 2008. An orally delivered small-molecule formulation with antiangiogenic and anticancer activity. *Nat Biotechnol* 26: 799–807.

Bhargava, P., Marshall, J. L., Rizvi, N. et al. 1999. A Phase I and pharmacokinetic study of TNP-470 administered weekly to patients with advanced cancer. *Clin Cancer Res* 5: 1989–1995.

Brannon-Peppas, L. and Blanchette, J. O. 2004. Nanoparticle and targeted systems for cancer therapy. *Adv Drug Deliv Rev* 56: 1649–1659.

Brooks, P. C., Clark, R. A., and Cheresh, D. A. 1994. Requirement of vascular integrin alpha v beta 3 for angiogenesis. *Science* 264: 569–571.

Browder, T., Butterfield, C. E., Kraling, B. M. et al. 2000. Antiangiogenic scheduling of chemotherapy improves efficacy against experimental drug-resistant cancer. *Cancer Res* 60: 1878–1886.

Cao, J., Sato, H., Takino, T., and Seiki, M. 1995. The C-terminal region of membrane type matrix metalloproteinase is a functional transmembrane domain required for pro-gelatinase A activation. *J Biol Chem* 270: 801–805.

Carter, R. J., Morton, J., and Dunnett, S. B. 2001. Motor coordination and balance in rodents. *Curr Protoc Neurosci* Chapter 8: Unit 8.12.

Chang, D. K., Chiu, C. Y., Kuo, S. Y. et al. 2009. Antiangiogenic targeting liposomes increase therapeutic efficacy for solid tumors. *J Biol Chem* 284: 12905–12916.

Curnis, F., Arrigoni, G., Sacchi, A. et al. 2002. Differential binding of drugs containing the NGR motif to CD13 isoforms in tumor vessels, epithelia, and myeloid cells. *Cancer Res* 62: 867–874.

Emoto, M., Tachibana, K., Iwasaki, H., and Kawarabayashi, T. 2007. Antitumor effect of TNP-470, an angiogenesis inhibitor, combined with ultrasound irradiation for human uterine sarcoma xenografts evaluated using contrast color Doppler ultrasound. *Cancer Sci* 98: 929–935.

Folkman, J. 1971. Tumor angiogenesis: Therapeutic implications. *N Engl J Med* 285: 1182–1186.

Folkman, J. 1974. Tumor angiogenesis: Role in regulation of tumor growth. *Symp Soc Dev Biol* 30: 43–52.

Folkman, J. and Shing, Y. 1992. Angiogenesis. *J Biol Chem* 267: 10931–10934.

Fong, T. A., Shawver, L. K., Sun, L. et al. 1999. SU5416 is a potent and selective inhibitor of the vascular endothelial growth factor receptor (Flk-1/KDR) that inhibits tyrosine kinase catalysis, tumor vascularization, and growth of multiple tumor types. *Cancer Res* 59: 99–106.

Formenti, S. C. 2005. In regard to Kao et al.: Concomitant radiation therapy and paclitaxel for unresectable locally advanced breast cancer: Results from two consecutive phase I/II trials (*Int J Radiat Oncol Biol Phys* 2005;61:1045–1053). *Int J Radiat Oncol Biol Phys* 63: 1275–1276.

Gaucher, G., Dufresne, M. H., Sant, V. P. et al. 2005. Block copolymer micelles: Preparation, characterization and application in drug delivery. *J Control Release* 109: 169–188.

Gelderblom, H., Verweij, J., Nooter, K., and Sparreboom, A. 2001. Cremophor EL: The drawbacks and advantages of vehicle selection for drug formulation. *Eur J Cancer* 37: 1590–1598.

Giannelli, G., Falk-Marzillier, J., Schiraldi, O., Stetler-Stevenson, W. G., and Quaranta, V. 1997. Induction of cell migration by matrix metalloprotease-2 cleavage of laminin-5. *Science* 277: 225–228.

Grenha, A., Remunan-Lopez, C., Carvalho, E. L., and Seijo, B. 2008. Microspheres containing lipid/chitosan nanoparticles complexes for pulmonary delivery of therapeutic proteins. *Eur J Pharm Biopharm* 69: 83–93.

Hamaguchi, T., Kato, K., Yasui, H. et al. 2007. A phase I and pharmacokinetic study of NK105, a paclitaxel-incorporating micellar nanoparticle formulation. *Br J Cancer* 97: 170–176.

Hamaguchi, T., Matsumura, Y., Suzuki, M. et al. 2005. NK105, a paclitaxel-incorporating micellar nanoparticle formulation, can extend in vivo antitumour activity and reduce the neurotoxicity of paclitaxel. *Br J Cancer* 92: 1240–1246.

Harris, J. M. and Chess, R. B. 2003. Effect of pegylation on pharmaceuticals. *Nat Rev Drug Discov* 2: 214–221.

Hatakeyama, H., Akita, H., Ishida, E. et al. 2007. Tumor targeting of doxorubicin by anti-MT1-MMP antibody-modified PEG liposomes. *Int J Pharm* 342: 194–200.

Hodivala-Dilke, K. M., Reynolds, A. R., and Reynolds, L. E. 2003. Integrins in angiogenesis: Multitalented molecules in a balancing act. *Cell Tissue Res* 314: 131–144.

Holmgren, J. and Czerkinsky, C. 2005. Mucosal immunity and vaccines. *Nat Med* 11: S45–S53.

Igarashi, R., Takenaga, M., Takeuchi, J. et al. 2001. Marked hypotensive and blood flow-increasing effects of a new lipo-PGE(1) (lipo-AS013) due to vascular wall targeting. *J Control Release* 71: 157–164.

Ingber, D., Fujita, T., Kishimoto, S. et al. 1990. Synthetic analogues of fumagillin that inhibit angiogenesis and suppress tumour growth. *Nature* 348: 555–557.

Ishihara, T., Tanaka, K., Tasaka, Y. et al. 2009. Therapeutic effect of lecithinized superoxide dismutase against colitis. *J Pharmacol Exp Ther* 328: 152–164.

Ishikawa, D., Kikkawa, H., Ogino, K. et al. 1998. GD1alpha-replica peptides functionally mimic GD1alpha, an adhesion molecule of metastatic tumor cells, and suppress the tumor metastasis. *FEBS Lett* 441: 20–24.

Jemal, A., Murray, T., Samuels, A. et al. 2003. Cancer statistics, 2003. *CA Cancer J Clin* 53: 5–26.

Johnson, J. P. 1999. Cell adhesion molecules in the development and progression of malignant melanoma. *Cancer Metastasis Rev* 18: 345–357.

Kakinoki, S., Taguchi, T., Saito, H., Tanaka, J., and Tateishi, T. 2007. Injectable in situ forming drug delivery system for cancer chemotherapy using a novel tissue adhesive: Characterization and in vitro evaluation. *Eur J Pharm Biopharm* 66: 383–390.

Kanamori, M., Yasuda, T., Ohmori, K., Nogami, S., and Aoki, M. 2007. Genetic analysis of high-metastatic clone of RCT sarcoma in mice, and its growth regression in vivo in response to angiogenesis inhibitor TNP-470. *J Exp Clin Cancer Res* 26: 101–107.

Karamysheva, A. F. 2008. Mechanisms of angiogenesis. *Biochemistry (Mosc)* 73: 751–762.

Katanasaka, Y., Ida, T., Asai, T. et al. 2008. Antiangiogenic cancer therapy using tumor vasculature-targeted liposomes encapsulating 3-(3,5-dimethyl-1*H*-pyrrol-2-ylmethylene)-1,3-dihydro-indol-2-one, SU5416. *Cancer Lett* 270: 260–268.

Kataoka, K., Itaka, K., Nishiyama, N. et al. 2005. Smart polymeric micelles as nanocarriers for oligonucleotides and siRNA delivery. *Nucleic Acids Symp Ser (Oxf)*: 17–18.

Kerbel, R. and Folkman, J. 2002. Clinical translation of angiogenesis inhibitors. *Nat Rev Cancer* 2: 727–739.

Kikkawa, H., Imafuku, H., Tsukada, H., and Oku, N. 2000a. Possible role of immune surveillance at the initial phase of metastasis produced by B16BL6 melanoma cells. *FEBS Lett* 467: 211–216.

Kikkawa, H., Tsukada, H., and Oku, N. 2000b. Usefulness of positron emission tomographic visualization for examination of in vivo susceptibility to metastasis. *Cancer* 89: 1626–1633.

Kodera, Y., Nakanishi, H., Yamamura, Y. et al. 1998. Prognostic value and clinical implications of disseminated cancer cells in the peritoneal cavity detected by reverse transcriptase-polymerase chain reaction and cytology. *Int J Cancer* 79: 429–433.

Koizumi, F., Kitagawa, M., Negishi, T. et al. 2006. Novel SN-38-incorporating polymeric micelles, NK012, eradicate vascular endothelial growth factor-secreting bulky tumors. *Cancer Res* 66: 10048–10056.

Kok, R. J., Schraa, A. J., Bos, E. J. et al. 2002. Preparation and functional evaluation of RGD-modified proteins as alpha(v)beta(3) integrin directed therapeutics. *Bioconjug Chem* 13: 128–135.

Koshikawa, N., Giannelli, G., Cirulli, V., Miyazaki, K., and Quaranta, V. 2000. Role of cell surface metalloprotease MT1-MMP in epithelial cell migration over laminin-5. *J Cell Biol* 148: 615–624.

Kruger, E. A. and Figg, W. D. 2000. TNP-470: An angiogenesis inhibitor in clinical development for cancer. *Expert Opin Investig Drugs* 9: 1383–1396.

Kudelka, A. P., Levy, T., Verschraegen, C. F. et al. 1997. A phase I study of TNP-470 administered to patients with advanced squamous cell cancer of the cervix. *Clin Cancer Res* 3: 1501–1505.

Kusaka, M., Sudo, K., Fujita, T. et al. 1991. Potent anti-angiogenic action of AGM-1470: Comparison to the fumagillin parent. *Biochem Biophys Res Commun* 174: 1070–1076.

Lasic, D. D., Martin, F. J., Gabizon, A., Huang, S. K., and Papahadjopoulos, D. 1991. Sterically stabilized liposomes: A hypothesis on the molecular origin of the extended circulation times. *Biochim Biophys Acta* 1070: 187–192.

Li, Y., Aoki, T., Mori, Y. et al. 2004. Cleavage of lumican by membrane-type matrix metalloproteinase-1 abrogates this proteoglycan-mediated suppression of tumor cell colony formation in soft agar. *Cancer Res* 64: 7058–7064.

Logothetis, C. J., Wu, K. K., Finn, L. D. et al. 2001. Phase I trial of the angiogenesis inhibitor TNP-470 for progressive androgen-independent prostate cancer. *Clin Cancer Res* 7: 1198–1203.

Luan, Y. and Xu, W. 2007. The structure and main functions of aminopeptidase N. *Curr Med Chem* 14: 639–647.

Luciani, N., Marie-Claire, C., Ruffet, E. et al. 1998. Characterization of Glu350 as a critical residue involved in the N-terminal amine binding site of aminopeptidase N (EC 3.4.11.2): Insights into its mechanism of action. *Biochemistry* 37: 686–692.

Maeda, H., Bharate, G. Y., and Daruwalla, J. 2009. Polymeric drugs for efficient tumor-targeted drug delivery based on EPR-effect. *Eur J Pharm Biopharm* 71: 409–419.

Maeda, N., Takeuchi, Y., Takada, M. et al. 2004. Anti-neovascular therapy by use of tumor neovasculaturetargeted long-circulating liposome. *J Control Release* 100: 41–52.

Maeda, H., Wu, J., Sawa, T., Matsumura, Y., and Hori, K. 2000. Tumor vascular permeability and the EPR effect in macromolecular therapeutics: A review. *J Control Release* 65: 271–284.

Maruyama, K., Takahashi, N., Tagawa, T., Nagaike, K., and Iwatsuru, M. 1997. Immunoliposomes bearing polyethyleneglycol-coupled Fab′ fragment show prolonged circulation time and high extravasation into targeted solid tumors in vivo. *FEBS Lett* 413: 177–180.

Matsumura, Y. and Kataoka, K. 2009. Preclinical and clinical studies of anticancer agent-incorporating polymer micelles. *Cancer Sci* 100: 572–579.

Micallef, M. J., Yoshida, K., Kawai, S. et al. 1997. In vivo antitumor effects of murine interferon-gamma-inducing factor/interleukin-18 in mice bearing syngeneic Meth A sarcoma malignant ascites. *Cancer Immunol Immunother* 43: 361–367.

Mori, H., Tomari, T., Koshikawa, N. et al. 2002. CD44 directs membrane-type 1 matrix metalloproteinase to lamellipodia by associating with its hemopexin-like domain. *EMBO J* 21: 3949–3959.

Muggia, F. M. 1999. Doxorubicin–polymer conjugates: Further demonstration of the concept of enhanced permeability and retention. *Clin Cancer Res* 5: 7–8.

Neurath, M. F., Finotto, S., and Glimcher, L. H. 2002. The role of Th1/Th2 polarization in mucosal immunity. *Nat Med* 8: 567–573.

Nii, T. and Ishii, F. 2005. Dialkylphosphatidylcholine and egg yolk lecithin for emulsification of various triglycerides. *Colloids Surf B Biointerf* 41: 305–311.

Noakes, M., Clifton, P., Ntanios, F. et al. 2002. An increase in dietary carotenoids when consuming plant sterols or stanols is effective in maintaining plasma carotenoid concentrations. *Am J Clin Nutr* 75: 79–86.

Ohuchi, E., Imai, K., Fujii, Y. et al. 1997. Membrane type 1 matrix metalloproteinase digests interstitial collagens and other extracellular matrix macromolecules. *J Biol Chem* 272: 2446–2451.

Okada, N., Masunaga, Y., Okada, Y. et al. 2003. Dendritic cells transduced with gp100 gene by RGD fiber-mutant adenovirus vectors are highly efficacious in generating anti-B16BL6 melanoma immunity in mice. *Gene Ther* 10: 1891–1902.

Oku, N., Asai, T., Watanabe, K. et al. 2002. Anti-neovascular therapy using novel peptides homing to angiogenic vessels. *Oncogene* 21: 2662–2669.

Pastorino, F., Brignole, C., Marimpietri, D. et al. 2003. Vascular damage and anti-angiogenic effects of tumor vessel-targeted liposomal chemotherapy. *Cancer Res* 63: 7400–7409.

Rapoport, N., Pitt, W. G., Sun, H., and Nelson, J. L. 2003. Drug delivery in polymeric micelles: From in vitro to in vivo. *J Control Release* 91: 85–95.

Rowinsky, E. K. and Donehower, R. C. 1995. Paclitaxel (taxol). *N Engl J Med* 332: 1004–1014.

Sakakura, C., Hagiwara, A., Nakanishi, M. et al. 2002. Differential gene expression profiles of gastric cancer cells established from primary tumour and malignant ascites. *Br J Cancer* 87: 1153–1161.

Sanderink, G. J., Artur, Y., and Siest, G. 1988. Human aminopeptidases: A review of the literature. *J Clin Chem Clin Biochem* 26: 795–807.

Sato, H., Takino, T., Okada, Y. et al. 1994. A matrix metalloproteinase expressed on the surface of invasive tumour cells. *Nature* 370: 61–65.

Saul, J. M., Annapragada, A. V., and Bellamkonda, R. V. 2006. A dual-ligand approach for enhancing targeting selectivity of therapeutic nanocarriers. *J Control Release* 114: 277–287.

Schiffelers, R. M., Koning, G. A., Ten Hagen, T. L. et al. 2003. Anti-tumor efficacy of tumor vasculature-targeted liposomal doxorubicin. *J Control Release* 91: 115–122.

Seandel, M., Noack-Kunnmann, K., Zhu, D., Aimes, R. T., and Quigley, J. P. 2001. Growth factor-induced angiogenesis in vivo requires specific cleavage of fibrillar type I collagen. *Blood* 97: 2323–2332.

Seftor, R. E., Seftor, E. A., and Hendrix, M. J. 1999. Molecular role(s) for integrins in human melanoma invasion. *Cancer Metastasis Rev* 18: 359–375.

Seiki, M. 2002. The cell surface: The stage for matrix metalloproteinase regulation of migration. *Curr Opin Cell Biol* 14: 624–632.

Spangler, R. S. 1990. Insulin administration via liposomes. *Diabetes Care* 13: 911–922.

Suzuki, Y., Matsumoto, T., Okamoto, S., and Hibi, T. 2008. A lecithinized superoxide dismutase (PC-SOD) improves ulcerative colitis. *Colorectal Dis* 10: 931–934.

Suzuki, R., Takizawa, T., Kuwata, Y. et al. 2008. Effective anti-tumor activity of oxaliplatin encapsulated in transferrin–PEG–liposome. *Int J Pharm* 346: 143–150.

Takeuchi, H., Matsui, Y., Yamamoto, H., and Kawashima, Y. 2003. Mucoadhesive properties of carbopol or chitosan-coated liposomes and their effectiveness in the oral administration of calcitonin to rats. *J Control Release* 86: 235–242.

Takeuchi, H., Yamamoto, H., Niwa, T., Hino, T., and Kawashima, Y. 1996. Enteral absorption of insulin in rats from mucoadhesive chitosan-coated liposomes. *Pharm Res* 13: 896–901.

Takikawa, M., Kikkawa, H., Asai, T. et al. 2000. Suppression of GD1alpha ganglioside-mediated tumor metastasis by liposomalized WHW-peptide. *FEBS Lett* 466: 381–384.

Torchilin, V. P., Omelyanenko, V. G., Papisov, M. I. et al. 1994. Poly(ethylene glycol) on the liposome surface: On the mechanism of polymer-coated liposome longevity. *Biochim Biophys Acta* 1195: 11–20.

Van Belle, P. A., Elenitsas, R., Satyamoorthy, K. et al. 1999. Progression-related expression of beta3 integrin in melanomas and nevi. *Hum Pathol* 30: 562–567.

Van Ginkel, F. W., Nguyen, H. H., and Mcghee, J. R. 2000. Vaccines for mucosal immunity to combat emerging infectious diseases. *Emerg Infect Dis* 6: 123–132.

Varner, J. A. and Cheresh, D. A. 1996. Integrins and cancer. *Curr Opin Cell Biol* 8: 724–730.

Vossenkamper, A., Struck, D., Alvarado-Esquivel, C. et al. 2004. Both IL-12 and IL-18 contribute to small intestinal Th1-type immunopathology following oral infection with *Toxoplasma gondii*, but IL-12 is dominant over IL-18 in parasite control. *Eur J Immunol* 34: 3197–3207.

Xiong, X. B., Mahmud, A., Uludag, H., and Lavasanifar, A. 2007. Conjugation of arginine–glycine–aspartic acid peptides to poly(ethylene oxide)-*b*-poly(epsilon-caprolactone) micelles for enhanced intracellular drug delivery to metastatic tumor cells. *Biomacromolecules* 8: 874–884.

Xiong, X. B., Mahmud, A., Uludag, H., and Lavasanifar, A. 2008. Multifunctional polymeric micelles for enhanced intracellular delivery of doxorubicin to metastatic cancer cells. *Pharm Res* 25: 2555–2566.

Yamaoka, M., Yamamoto, T., Ikeyama, S., Sudo, K., and Fujita, T. 1993. Angiogenesis inhibitor TNP-470 (AGM-1470) potently inhibits the tumor growth of hormone-independent human breast and prostate carcinoma cell lines. *Cancer Res* 53: 5233–5236.

Yanagihara, K., Takigahira, M., Tanaka, H. et al. 2005. Development and biological analysis of peritoneal metastasis mouse models for human scirrhous stomach cancer. *Cancer Sci* 96: 323–332.

Yanase, T., Tamura, M., Fujita, K., Kodama, S., and Tanaka, K. 1993. Inhibitory effect of angiogenesis inhibitor TNP-470 on tumor growth and metastasis of human cell lines in vitro and in vivo. *Cancer Res* 53: 2566–2570.

Yokoyama, M., Okano, T., Sakurai, Y. et al. 1999. Selective delivery of adriamycin to a solid tumor using a polymeric micelle carrier system. *J Drug Target* 7: 171–186.

Yonezawa, S., Asai, T., and Oku, N. 2007. Effective tumor regression by anti-neovascular therapy in hypovascular orthotopic pancreatic tumor model. *J Control Release* 118: 303–309.

Zangari, M., Anaissie, E., Stopeck, A. et al. 2004. Phase II study of SU5416, a small molecule vascular endothelial growth factor tyrosine kinase receptor inhibitor, in patients with refractory multiple myeloma. *Clin Cancer Res* 10: 88–95.

Zhang, H. Y., Tang, X., Li, H. Y., and Liu, X. L. 2008. A lipid microsphere vehicle for vinorelbine: Stability, safety and pharmacokinetics. *Int J Pharm* 348: 70–79.

Zucker, S., Conner, C., Dimassmo, B. I. et al. 1995. Thrombin induces the activation of progelatinase A in vascular endothelial cells. Physiologic regulation of angiogenesis. *J Biol Chem* 270: 23730–23738.

7 Application of Polymer Drugs to Medical Devices and Preparative Medicine

*M.R. Aguilar, L. García-Fernández, M.L. López-Donaire, F. Parra,
L. Rojo, G. Rodríguez, M.M. Fernández, and J. San Román*

CONTENTS

7.1 INTRODUCTION

The "perfect" drug presents the precise therapeutic activity and no side effects. Several efforts have been devoted in order to reach these objectives in the last decades. Currently, two different strategies for improving the therapeutic efficacy of drugs are being followed. The first one is based on the development of new agents that modulate the molecular processes and pathways specifically associated with the disease. Genomics and proteomics play a main role in this strategy. The second one consists in the improvement of existing drugs to make them more effective by using nanocarriers that bring more drug molecules to the desired site.

Polymer drugs or pharmacologically active macromolecules are one of the most promising choices proposed in the controlled drug delivery field for the development of nanocarriers independently of their origin (natural or synthetic), that can be tailored to give the desired physicochemical properties. These polymeric drugs or polymer–drug conjugates show activity, even though the corresponding monomeric species may or may not be biologically active. Although there are infinite possibilities for designing polymer–drug conjugates, the most widely accepted model is the one suggested by Ringsdorf in 1975 (Ringsdorf, 1975), which considers the covalent binding of low-molecular-weight drugs to a polymeric system must be by means of organic functional groups that can be degraded in the physiological medium.

The desired physicochemical properties of a particular polymer drug can be tailored using the numerous tools offered by the organic chemistry and macromolecular sciences.

Many polymer drugs have been developed for the treatment of different diseases in the last years. In this chapter, we report their role in following most important applications:

- Antitumoral polymer drugs
- Antiangiogenic and proangiogenic polymer drugs
- Antibacterial polymer drugs
- Antithrombogenic polymer drugs
- Low friction polymer drugs

7.2 APPLICATIONS OF POLYMER DRUGS TO MEDICAL DEVICES AND REPARATIVE MEDICINE

7.2.1 ANTITUMORAL POLYMER DRUGS

Cancer is considered a leading cause of death worldwide; specifically 7.6 million of deaths were produced in 2008. World Health Organization has estimated to continue rising to 12 million deaths in 2030 (≤Http://Www.Who.Int/Cancer/En≥., W. H. O. C., 2009). The common solid tumors such as breast, prostate, lung, stomach, liver, and colon constitute the most difficult to treat and cause the most cancer deaths.

As mentioned earlier, the efforts devoted to improve anticancer treatment are being concentrated on two different fields: first, by the design and development of new bioactive agents that modulate the molecular processes and pathways associated with the tumor progression and second, by using nanocarriers capable to maintain the anticancer drug in the bloodstream for long periods of time, and to provide a sustained delivery in the tumor tissues, limiting the cytotoxic effect of the drug.

7.2.1.1 Active Tumor Targeting

This therapy is based on the Ringsdorf's (1975) vision about ideal polymer for its use as drug carrier (Figure 7.1). The polymer backbone is attached to solubilizing groups that provide the bioavailability of the carrier system, spacers which are bound to macromolecules and whose chemical stability should be unchanged during transport to the target site, and finally, moieties, such as antibodies, saccharides, proteins, and peptides (Allen, 2002), which are involved in antigens or receptors that are either uniquely expressed or overexpressed on the target cells relative to normal tissues.

One of the first conjugate copolymers bearing a targeting ligand to be tested clinically for cancer treatment was a N-(2-hydroxypropylmethacrylamide) (HPMA) copolymer bearing galactosamine residues, HPMA-Gly-Phe-Leu-Gly-doxorubicin (DOX)-galactosamine (PK$_2$; FCE 28069). In this case, the overhanging galactosamine moieties localize the asialoglycoprotein receptor in hepatocytes (Duncan et al., 1986, Seymour et al., 2002).

Another ligand which is widely used as target for tumor-specific drug delivery is folate (FOL), an anionic form of folic acid. Its use lies on the fact that one of its receptors, glycosyl phosphatidylinositol–anchored glypolypeptide, which presents a dissociation constant in the subnanomolar range, is overexpressed in many cancers such as ovary, lung, breast, kidney, brain, endometrium, colon, and hematopoietic cells of myelogenous origin while its presence in healthy cells is limited on the apical (luminal) surface of polarized epithelial cells where it is inaccessible directly

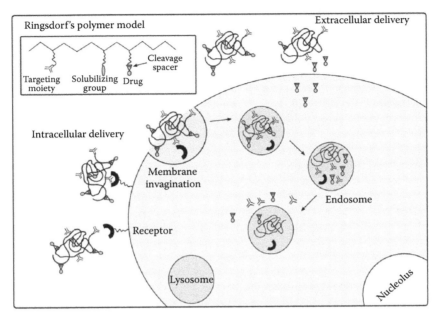

FIGURE 7.1 Endocytic pathway for the cellular uptake of ideal macromolecules carriers defined by Ringsdorf as drug delivery systems. (Modified from Haag, R. and Kratz, F., *Angew. Chem. Int. Ed.*, 45, 1198, 2006; Park, J.H. et al., *Prog. Polym. Sci. (Oxford)*, 33, 113, 2008.)

through the bloodstream (Low et al., 2008, Salazar and Ratnam, 2007). An example of FOL-conjugate systems is based on the amphiphilic block copolymers that self-assemble into spherical micelles FOL–poly(ethylene glycol) (PEG)–poly(aspartate-hydrazone-adriamycin). *Adriamycin* (ADR) was conjugated to the side chains of the core-forming poly(aspartic acid) block through a hydrazone linkage which can be selectively cleaved under the acidic intracellular environment. FOL-conjugated micelles showed lower in vivo toxicity and higher antitumor activity over a broad range of the dosage from 7.5 to 26.21 mg/kg, which was fivefold broader than free drugs (Bae et al., 2007).

Delivery of polymer-conjugated drugs can take place through intra- or extracellular pathways. For intracellular delivery, the lysosometric delivery of the drug defined by De Duve et al. (1974) could be produced by endocytic pathway in the cytoplasm followed by the trafficking through the endosomal compartments to lysosomes. This point is where the linker plays an important role because it should be selectively cleaved under acidic conditions, based on a significant drop in the pH values existing within the endosomes (pH ~ 5–6.5) and in the primary and secondary lysosomes (pH ~ 4) (Bareford and Swaan, 2007).

Table 7.1 collects recent references where drugs were coupled to the suitable carriers through acid-sensitive bonds.

Disulfide constitutes other important linkage that is emerging as a fascinating class of reducing sensitive biodegradable polymers. They are exploited for triggered intracellular delivery of a variety of bioactive molecules such as siRNA, DNA, proteins, and low-molecular-weight drugs based on the difference in reducing potential between the intracellular and extracellular environments (Meng et al., 2009a). In this sense, micelles based on disulfide-linked dextran-*b*-poly(ε-caprolactone) diblock copolymer (Dex-SS-PCL) were prepared for an efficient intracellular release of DOX. In vitro studies revealed the rapid release of DOX to the cytoplasm as well as to the cell nucleus in comparison with that observed from DOX-loaded reduction insensitive Dex–PCL micelles (Sun et al., 2010).

On the other hand, research is also focused on the enzymatic degradation of the linker bond. In this sense, poly(oleyl 2-acetamido-2-deoxy-α-D-glucopyranoside methacrylate-*co-N*-vinyl

TABLE 7.1

Acid-Sensitive Bonds Copolymers

Benzoic imine		Ding (2009) Gu (2008)
Hydrazone		Lee (2006) Etrych (2007) Sirova (2010) Bae (2007)
Ketal	or	Jain (2007) Heffernan (2005) Shim (2008)
Acetal	or	Jain (2007) Chan (2006) Knorr (2007)
cis-Aconityl		Yoo (2002) Lavignac (2009)

pyrrolidone) was developed as an active glycopolymer for brain tumor where the drug delivery occurs via ester hydrolysis by the presence of carboxylesterase enzyme (Lopez Donaire et al., 2009).

7.2.1.2 Passive Tumor Targeting

Different studies have shown that polymer–drug conjugates without targeting ligands in their structures could interact with different (or a broad number of different types of) antibodies and receptors present on the surface of cancer cells and therefore being entrapped or accumulated in solid tumors and retained for prolonged periods (more than 100 h). This passive tumor targeting was identified by Matsumura and Maeda (1986) and is known as *enhanced permeability and retention* (EPR) effect of macromolecules and lipids in solid tumor (Figure 7.2) (Duncan, 1999, 2003, Maeda et al., 2000).

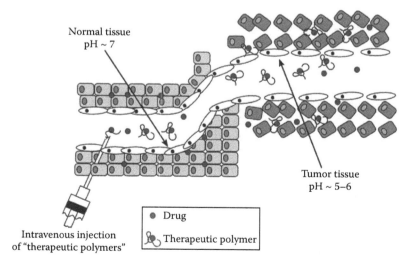

FIGURE 7.2 Passive tumor targeting of polymer therapeutics by the enhanced permeability and retention (EPR) effect. (Modified from Haag, R. and Kratz, F., *Angew. Chem. Int. Ed.*, 45, 1198, 2006; Duncan, R., *Nat. Rev. Drug Discov.*, 2, 347, 2003.)

The EPR effect can be attributed to two factors: the defective vascular architecture based on angiogenesis and hypoxia phenomena that allows macromolecular extravasations not usual in normal tissues, and also, lack of effective tumor lymphatic drainage, preventing clearance of the penetrant macromolecules and promoting their accumulation (Duncan, 2006). The EPR effect is observed for macromolecules with molecular weights greater than 20 kDa, whereas the rate of urinary clearance is inversely related to the tumor uptake. Takakura and Hashida (1996) detailed organ distribution of macromolecules and clearance kinetics with regard to targeting liver and kidney tumors.

This phenomenon was observed with HPMA-Gly-Phe-Leu-Gly-DOX copolymer (PK1; FCE 28068) (Duncan et al., 1992) which was clinically tested and cumulative doses of HPMA copolymer could be administered without signs of immunogenicity or polymer-related toxicity (Duncan, 2009, Vasey et al., 1999). Recently, a new biodegradable graft copolymer–DOX conjugate was designed for passive tumor targeting. HPMA copolymer was grafted to the main chain through degradable, enzymatic, and reductive spacers in order to facilitate the intracellular degradation of the graft polymer carrier after the drug release. The graft polymer–DOX conjugates were tested in mice bearing EL4-T-cells and exhibited a prolongation in the blood circulation and an enhancement in the tumor accumulation compared with its linear analog conjugate (Etrych et al., 2008).

However, tumor-targeting drug delivery cannot be completely achieved only in base of EPR effect especially for tumors that present hypoxic environment, low vascularization, and permeability (Minchinton and Tannock, 2006). The EPR effect can be facilitated by different factor such as nitric oxide (NO), prostaglandin, peroxynitrite, collagenase, bradykinin, and so on. This artificial augmentation of the EPR effect can be achieved by two different strategies (Maeda et al., 2000). The first one is based on *angiotensin II*, which is used to induce hypertension leading to a relative increase in tumor blood flow volume (Figure 7.3a). This method was evaluated with the polymeric drug styrene–maleic acid copolymer-conjugated to the protein neocarzinostatin, SMANCS (Matsumura and Maeda, 1986). Clinical results showed an improvement therapy for various advanced solid tumors such as metastatic liver cancer and pancreas (Nagamitsu et al., 2009).

The second method used to alter the EPR effect is based on the NO generation (a potent mediator of vascular extravasation) from nitroglycerin (NG). NG can be converted into nitrite and then reduced to NO under relatively hypoxic and acidic condition found in cancer tissues (Figure 7.3b) (Maeda et al., 1994). The role of NG in the EPR effect was investigated with PEG-conjugated zinc protoporphyrin IX (PZP) (Fang et al., 2004). The apply of NG exhibited a greater suppression of

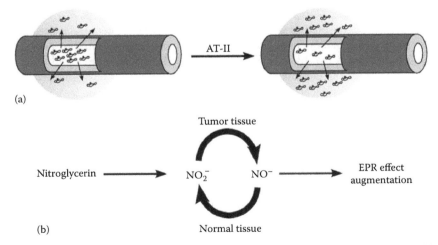

FIGURE 7.3 (a) Blood vasculature in tumor tissue under normotensive (left) and hypertensive (right) conditions. (Modified from Nagamitsu, A. et al., *Jpn. J. Clin. Oncol.*, 39, 756, 2009.) (b) Mechanism of NO generation from nitroglycerin (NG). (Modified from Seki, T. et al., *Cancer Sci.*, 100, 2426, 2009.)

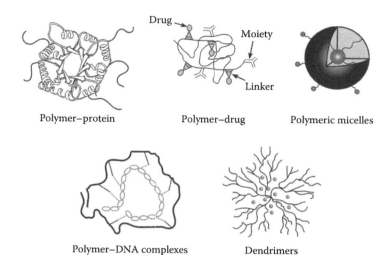

Polymer–protein   Polymer–drug   Polymeric micelles

Polymer–DNA complexes   Dendrimers

FIGURE 7.4   **(See companion CD for color figure.)** Representative therapeutic polymers used in cancer therapy. (Modified from Park, J.H. et al., *Prog. Polym. Sci. (Oxford)*, 33, 113, 2008.)

tumor growth when compared with PZP alone (Seki et al., 2009). However, it should be mentioned that these good results are associated with not only the increase of drug delivery but also a decrease of hypoxic-induced resistance to anticancer drugs in cancer cell lines by means of reduction of hypoxia-inducible factor-1α (HIF-1α) (Yasuda et al., 2006).

At present, polymers used in cancer therapy are defined under the umbrella term of "therapeutic polymers" and include different macromolecular architectures such as polymer–drug and polymer–protein conjugates, DNA–polyplexes, polymeric micelles, and dendrimers (Figure 7.4). All these systems should be water-soluble, non-immunogenic, nontoxic, and degradable in order to be eliminated from the body after their function.

7.2.1.2.1   Polymer–Drugs

Biodistribution, elimination, and metabolism of the polymer–drug conjugate play a crucial role in the design of polymeric systems to be used for drug conjugation, therefore molecular weight and physicochemical properties of the polymeric candidate must be considered.

Despite recent efforts devoted to obtain novel polymer–drug conjugates, at present only three polymers are being tested in clinical phase: PEG, HPMA, and poly(glutamic acid) (PGA).

The most relevant drugs that have been conjugated to HPMA and are under clinical trials are PK_1 and PK_2 (Duncan, 2009, Duncan et al., 1992, Seymour et al., 2009, Vasey et al., 1999). Both systems have a tetrapeptide spacer, Gly-Phe-Leu-Gly that could be cleaved by lysosomal enzymes present in tumor cells. In the case of PK_2, an additional galactosamine group was incorporated onto the main chain as a target ligand in order to promote multivalent targeting of the hepatocyte asialoglycoprotein receptor to treat primary liver cancer (Seymour et al., 2002). The maximum tolerance dose (MTD) of PK_2 in a phase I trial study was 160 mg/m^2 of DOX equivalents, less than the MTD value of PK_1 (320 mg/m^2). It is not clear why the PK_2 presents higher toxicity than its analog without galactosamine. Table 7.2 includes other polymer–drug conjugates based on HPMA, where the anticancer drugs are *Taxol* (Atkins and Gershell, 2002) or *camptothecin* (Bissett et al., 2004, Sarapa et al., 2003). Based on the good results associated with these systems, current approaches are focused on the HPMA carrier based on pH-sensitive linker (Mrkvan et al., 2005, Sirova et al., 2010), enzyme-cleavage linker based on a self-immolative dendritic prodrug with single or multiple triggering modes of action (Erez et al., 2009), and a simultaneous conjugation with multiple chemotherapeutic agents. HPMA was conjugated with DOX through the hydrolytic and degradable

TABLE 7.2
Polymeric Drug and Polymer–Drug Conjugates in Clinical Trials

Name	Drug	Polymer	Indication	References
CT-2103; Xyotax	*Paclitaxel*	PGA (40 kDa)	Ovarian cancer Non-small cell lung cancer	Auzenne et al. (2002) and Duncan (2006)
CT-2106	*Camptothecin*	PGA (50 kDa)	Various cancer	Bhatt (2003)
PK$_1$: FCE 28068	*Doxorubicin*	HPMA (30 kDa)	Lung and breast cancer	Duncan (2009), Duncan et al. (1992), and Vasey et al. (1999)
PK$_2$: FCE 28069	*Doxorubicin*	HPMA-Galactosamine (30 kDa)	Hepatocellular carcinoma	Seymour et al. (2002)
PNU 166945	*Paclitaxel*	HPMA (40 kDa)	Various cancer	Atkins and Gershell (2002)
MAG-CPT	*Camptothecin*	HPMA (40 kDa)	Various cancer	Bissett et al. (2004) and Sarapa et al. (2003)
AP5280	*Carboplatin platinate*	HPMA (28 kDa)	Various cancer	Rademaker-Lakhai (2004)
AP5346 ProLindac	*DACH–platinate*	HPMA	Various cancer	Duncan (2006)
AD-70 DOX-OXD	*Doxorubicin*	Dextran	Various cancer	Danhauser-Riedl (1993)
DE-310	*Camptothecin*	Modified dextran	Various cancer	Kumazawa (2004)
Prothecan	*Camptothecin*	PEG	Various cancer	Rowinsky et al. (2003)
XMT-1001	*Camptothecin*	Polyacetal polymer (PHF or Fleximer®)	Advanced cancer	Yurkovetskiy (2009)
EZN-2208	*SN38*	4 arm PEG (40 kDa)	Advanced cancer	Guo (2008) and Sapra (2008)
NKTR-102	*Irinotecan*	4 arm PEG	Colorectal breast, ovarian and cervical cancer	Pasut and Veronese(2009)
NKTR-105	*Docetaxel*	4 arm PEG	Solid tumors including hormone refractory	Pasut and Veronese (2009)

DACH, Diaminocyclohexane.

hydrazone bond to the drug carrier and additional hydrophobic substituents (oleic acid, cholesterol, and dodecyl moieties) were introduced into the polymer structure in order to get more flexibility favoring the formation of micelles or nanoparticles (NPs) via self-assembly phenomena (Chytil et al., 2008). An example of the mentioned system consist of the HPMA conjugation with two therapeutic agents: gemcitabine and DOX (Lammers et al., 2009).

PGA was the first biodegradable polymer used in polymer–drug conjugates. PGA was linked to *paclitaxel* via an ester bond to the carboxylic acid of PGA, giving a soluble polymer with a high drug-loading efficiency (37 wt%), *PGA–paclitaxel* (CT-2103; *Xyotax*). *Paclitaxel* is released from the polymer backbone by degradation due to the effect of lysosomal cathepsin B after endocytic uptake (Auzenne et al., 2002). In phase I and II clinical trials, *PGA–paclitaxel* was administered every 3 weeks, over 30 min with a MTD of 233 mg/m^2, higher than that observed for free *paclitaxel*, and a partial improvement or disease stabilization was observed (Duncan, 2006). In phase III clinical trial, this system was compared with first-line treatments such as *gemcitabine* and *vinorelbine* (Langer, 2004). The results showed that severe side effects were significantly reduced in comparison with *gemcitabine* and a greater survival was obtained in comparison with that obtained with both components (Duncan, 2006).

Conjugates based on PEG have been obtained with lower drug loads. For example, *Prothecan*, a *PEG–camptothecin* conjugate, contains only 1.7 wt% of *camptothecin* because PEG bears only two hydroxyl terminal groups (C-OH) suitable for conjugation. These terminal groups, specifically C20-OH, favor the desired configuration of *camptothecin* lactone ring. In phase I clinical trial, a prolonged plasma half-life (more than 72 h) and a raise in the activity were observed (Rowinsky et al., 2003).

Despite the desirable properties of PEG, two disadvantages must be overcome such as the low drug payload (two hydroxyl terminal suitable for conjugation) and the nonbiodegradability. In order to overcome these drawbacks, some researchers have focused on the synthesis of new branched PEG through the end-chain groups or coupling on them small dendron structures with the aim to attain higher drug payloads (Choe et al., 2002, Pasut et al., 2008). For the second limitation, researchers developed biodegradable PEGs in response to a specific signal such as changes in pH (Pechar et al., 2005) or reductive conditions (Braunova et al., 2007).

Recently, polymer–drug conjugates carrying the combination of the anticancer agent epirubicin (EPI) and NO functional groups were reported applying PEG as polymer. The combination of NO and EPI in the same carrier presents two advantages: the improvement of the antitumor activity at the tumor site and the reduction of cardiotoxicity (Pasut et al., 2009).

Linear polymer containing cyclodextrins (CDs) as part of the backbone (IT-101) were reported for their conjugation with *camptothecin*. IT-101 is linear, highly biocompatible, nonbiodegradable, and has sufficient size to be cleared renally as a single molecule. Their conjugation with *camptothecin* is currently being investigated in human clinical trials (Davis, 2009).

Several combination therapies have been applied recently with these polymer–drug conjugates. Greco et al. (Greco and Vicent, 2009) described these types of systems namely: (i) polymer–drug conjugate plus free drug (Verschraegen et al., 2009), (ii) polymer–drug conjugate plus polymer–drug conjugate (Hongrapipat et al., 2008), (iii) single polymeric carrier carrying a combination of drug (Chadna et al., 2008), and (iv) polymer-directed enzyme prodrug therapy (PDEPT) (Satchi-Fainaro et al., 2003) and polymer–enzyme liposome therapy (PELT) (Duncan et al., 2001).

PDEPT approach requires initial administration of the polymeric drug in order to allow its arrival at the tumor site before it could be activated for the polymer–enzyme conjugate. One example is the coadministration of both HPMA-DOX (PK$_1$) (copolymer–drug conjugate) and HPMA-cathepsin B (enzyme conjugate) (Satchi et al., 2001). When the PDEPT conjugates were applied in the treatment of B16F10 melanoma tumors, the antimitotic activity was 168%, noticeably higher than that obtained when PK$_1$ was administered alone (152%).

7.2.1.2.2 Polymer–Protein

Polymer–protein conjugates technique is used to reduce protein immunogenicity and increase protein solubility and stability. When the polymer used is PEG, the technique is known as "PEGylation technology" (Davis, 2002).

Linear or branched PEG derivatives are used by coupling to the surface of the proteins. PEGylations produce a drop in the biological activity of proteins. However, they prolong plasma half-life and avoid receptor-mediated protein uptake by cells of the reticuloendothelial system (Caliceti and Veronese, 2003). Furthermore, it is worth noting that the design of the parameters for PEGylation should be based on the type of anticancer proteins that want to be released. Thus, heterologous proteins, whose administration is limited by their immunogenicity, require low-molecular-weight PEGs and random amine coupling while endogenous proteins, which need a prolonged half-life, are conjugated with high-molecular-weight PEGs (Pasut and Veronese, 2009).

PEG-L-asparaginase (ONCASPAR) was the first polymer–protein conjugate to be tested clinically in oncology. The clinical status of other PEGylated conjugates is recorded in recent reviews (Pasut and Veronese, 2009, Vicent and Duncan, 2006). Recently the PEGylation of *recombinant human arginase (rhArg-PEG* 5000 MW), based on the enzyme *rhArg* that can produce the depletion of arginase and consequently the inhibition of the tumor growth, angiogenesis, and NO synthesis have showed similar half maximal inhibitory concentration (IC_{50}) with respect to the native enzyme in several human hepatocellular carcinoma cells line (Cheng et al., 2007).

Despite the great variety of polymer–protein conjugates that are in clinical trials, recent studies have shown the immunogenic response to PEG. It should be suggested that this behavior could be associated with the excessive use of this polymer in the therapeutic applications (Armstrong et al., 2007). The immunogenic response of PEG imposes to develop novel polymers with similar advantages of it/PEG. In this sense, poly(2-ethyl-2-oxazoline) has been suggested as a possible alternative. In fact, this polymer presents similar features than PEG such as high water solubility, amphiphile, flexibility, and nontoxicity being also produced at low polydispersity. Properties of poly (2-ethyl-2-oxazoline) have been determined by its conjugation with a model protein (trypsin) showing a preservation of the enzyme activity and protein rejection properties similar with those of the analogous PEGylated conjugate (Mero et al., 2008).

Additionally, novel dextrin–protein conjugates (Duncan et al., 2008, Ferguson and Duncan, 2009) have been reported in a new concept named "polymer masked–unmasked protein therapy" which is based on the conjugation of a biodegradable polymer in order to mask a protein activity, and subsequently triggered degradation of the polymer is used to regenerate the protein bioactivity in a controlled manner. *Dextrin–phospholipase A$_2$ crotoxin (PLA$_2$)* conjugate is an example of this type of systems. *Dextrin–PLA$_2$* showed a reduction of its enzymatic activity in comparison with free *PLA$_2$*. This activity was restored mostly in the target site by the action of α-amylase, an enzyme present in the extracellular fluid but overexpressed in the tumor site. In this sense, the conjugate displayed enhanced in vitro cytotoxicity in the tumor site (Ferguson and Duncan, 2009).

7.2.1.2.3 Polyplexes

Human Genome Project has received widespread attention in Cancer Gene Therapies during the last decade. This advantageous therapy offers the possibility to kill cancer cells selectively due to the direct action on a deficient gene causing its block or replacement (Merdan et al., 2002, Park et al., 2008). Genes have been applied by direct injection into tumor tissues showing good results (Walther et al., 2002). However, the efficiency is improved when genes are administered by a carrier delivery system avoiding their degradation before reaching the target site (Kawabata et al., 1995).

Until now, viral and nonviral vectors are the two delivery systems used in gene therapy. In spite of the excellent transfection capacity, viral vector (retrovirus, adenovirus, and adeno-associated viruses) therapies have been abandoned due to some disadvantages such as their low ability carrying high-molecular-weight DNA molecules and their associated immunogenicity and oncogenic potential effect in the body. In comparison, nonviral polymeric vector–based therapies show lower side effect risks and high affinity for large DNA molecules although they present low transfection efficiency (Boussif et al., 1995, Collins and Fabre, 2004).

Many nonviral vectors are based on polymer–DNA polyplexes. Optimal transfection efficiencies require positive-charged polymer species in order to promote the electrostatic interactions needed with the negative charges of DNA. Figure 7.5 shows the most common cationic polymers used as nonviral vectors for DNA delivery.

Linear and branched polymers based on polyethylenimine (PEI) constitute the most important nonviral vectors used due to their high affinity for high-molecular-weight DNA molecules (Campeau et al., 2001, Fischer et al., 1999, Marschall et al., 1999). The affinity and transfection efficiency of DNA–PEI polyplexes increase with the PEI molecular weight due to the proton sponge effect associated to the high density of primary, secondary, and tertiary amino groups (Boussif et al., 1995). However, the high number of positively charged amino groups leads to an increased toxicity constituting this fact a real bottleneck of these PEI-based polyplexes.

To overcome the high toxicity of PEI and of PEGylated PEI (PEG–PEI) (Nguyen et al., 2000), FOL was linked on PEG and then grafted the FOL–PEG onto the linear (Benns et al., 2002) and branched PEI 25 kDa (Liang et al., 2008).

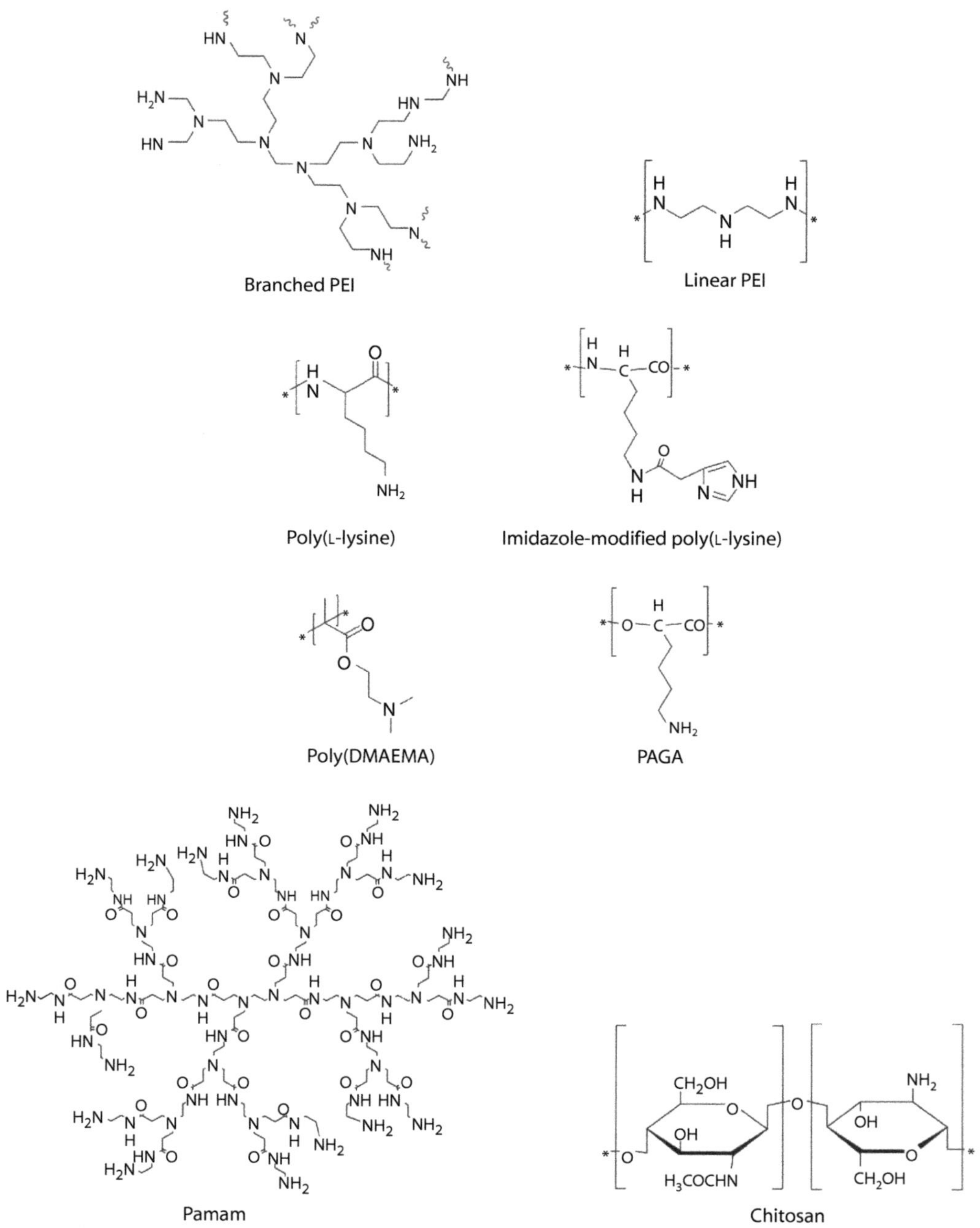

FIGURE 7.5   Nonviral vectors based on cationic polymers for DNA delivery.

Poly(L-lysine) (PLL), a cationic lineal polypeptide, differently from PEI presents an additional desirable feature, its biodegradation. This polymer presents lower toxicity than PEI but also, lower transfection efficiency (Wolfert et al., 1999), which is due to the lack of amino groups present in the polymer structure. However, recent reports using PEGylated PLL (Mannisto et al., 2002) and biocompatible dendrimers such as poly(L-lysine octa(3-aminopropyl)silsesquioxane) dendrimers (Kaneshiro et al., 2007) have shown an increase complex stability and transfection efficiency. These recent studies may lead to a renaissance of PLL as gene delivery agents.

PLL modified with imidazole groups provides systems with enhanced transfection efficiencies, without increasing toxicity, when compared to unmodified PLL (Benns et al., 2000, Yang et al., 2008a) as a consequence of the imidazole heterocycles buffering capacity in the range of endolysosomal pH.

Poly(2-dimethylaminoethyl methacrylate) (PDMAEMA) has also been studied for gene delivery therapies showing good transfection activities. However, these systems also show high cytotoxicity levels and lack of biodegradability limiting their potential in gene delivery (Verbaan et al., 2005). Lately, novel reducible PDMAEMA-based systems combined with biodegradable polymers such as phosphazene (De Wolf et al., 2007) were reported showing minimal cytotoxicity levels and enhanced transfection activity when compared with PDMAEMA.

Different chitosans with molecular weight between 30 and 170 kDa are used in this field, showing transfection efficiency values of the same magnitude as PEI (Koping-Hoggard et al., 2001) and in combination with other polymers (Zhao et al., 2009).

In the case of the dendrimers, the most used one is the polyamidoamine (PAMAM) (Choi et al., 2004, Huang et al., 2007, Lin et al., 2008, Patil et al., 2008) due to their molecular architecture which shows some unique physical and chemical properties and makes them particularly interesting for gene delivery applications (Navarro and Tros De Ilarduya, 2009). Recent studies have shown the use of other type of dendrimers (triazine derivatives) with easier synthesis and higher transfection efficiencies in comparison with PAMAM (Merkel et al., 2009).

Other polymers used are poly(α-(4-aminobutyl)-L-glycolic acid) (Lim et al., 2000, Maheshwari et al., 2000), poly(carbonic acid 2-dimethylamino-ethyl ester 1-methyl-2-(2-methacryloylamino) ethyl ester) (pHPMA-DMAE) (De Wolf et al., 2008, Luten et al., 2006), and CDs (Bellocq et al., 2004, Lake et al., 2006, Li and Loh, 2008, Pun et al., 2004).

The high toxicity of these cationic polymers is associated with their interaction with different plasma proteins such as albumin, fibronectin, immunoglobulin, complement factors, or fibrinogen, through their surface charge (Ogris et al., 1999, Oupicky et al., 1999) so that they can activate the complement system (Plank et al., 1996) leading to removal by the reticuloendothelial system. In order to avoid this phenomenon, the surface of these polymers is modified with hydrophilic polymers like HPMA (Kopecek et al., 2000) or PEG (Kainthan et al., 2006) before or after polyplex formation. In the case of PEG, a half-life rise in the plasma was observed but it was still far from the desired values (Kwoh et al., 1999, Nguyen et al., 2000). These polymer surfaces were also modified with moieties such as transferring for targeting of cancer cells that express transferrin receptor, leading in an increase of the expression levels within the tumor tissues (Bellocq et al., 2003, Kichler, 2004, Kircheis et al., 2001).

Studies based on nonviral DNA gene therapy have been ongoing for years and will continue toward improving systemic delivery and transfection efficiencies to the levels required of in vivo clinical trials. In the meantime, scientists have focused on a novel therapeutic pathway which uses RNA interference (RNAi) by which harmful genes can be "silenced" by delivering complementary short interfering RNA (siRNA) to target cells (Gary et al., 2007).

For this purpose, the same nonviral DNA vectors have been used to deliver siRNA into cancer cells. As examples, linear and branched PEI (Richards Grayson et al., 2006), poly(D,L-lactide-*co*-glycolide) (PLGA) (Khan et al., 2004), PAMAM (Waite and Roth, 2009), poly(isobutyl cyanoacrylate) (Toub et al., 2006), polycations consisting of histidine, and polylysine residues (Read et al., 2005), chitosan (Howard et al., 2006, Pille et al., 2006), PEI–PEG (Mao et al., 2006) were used as polycation-based siRNA delivery systems. However, this novel approach shows the same limitations as DNA-based therapies in terms of toxicity and delivery efficiency.

7.2.1.2.4   Particulate Systems

Particulate system is a term that comprises liposomes, surfactants, polymeric micelles, polymersomes, and NPs based on synthetic polymers. The present section is an introduction of the different advances of the last three systems.

Micellar systems are very interesting in the encapsulation of amphiphilic and low solubility drugs (Lukyanov and Torchilin, 2004, Park et al., 2008). The administration of different drugs in form of micelles offers different advantages such as improvement of drug solubility and it avoids its environmental degradation. Also, due to their low size (10–100 nm) a targeting effect at the tumor site and a high half-life in blood are provided (Kataoka et al., 2001, Maeda et al., 2000, Torchilin, 2001). These kinds of systems are based in liposomes, self-aggregates, microparticles and NPs, and polymer micelles (Park et al., 2008).

Polymer micelles are caused from a spontaneous self-assembling of amphiphilic block copolymers when the critical micellar concentration (CMC) is achieved (Kwon et al., 1994). The size of the micelle depends on the geometry of the constituent monomers, intermolecular interactions, and conditions of the bulk solution (i.e., concentration, ionic strength, pH, and temperature) (Svenson and Tomalia, 2005). The surface properties of polymeric micelles are important factors since they determine their biological fate (Rijcken et al., 2007).

More common polymeric micellar systems are copolymers based on PEG and polyesters such as poly(D,L-lactic acid) (PDLLA), PLGA, and PCL. These systems were widely studied because they do not require removal after their administration due to their biodegradation in the body (Forrest et al., 2006a,b, Lin et al., 2006, Wang et al., 2008, Xiong et al., 2008, Yang et al., 2008b). Most of these systems present a physical encapsulation of the drug. PDMAEMA-*b*-PCL-*b*-PDMAEMA is an example of micellar system loaded with *paclitaxel*. These micelles are based on the self-assembling of these triblock copolymers which present positive charge on their surface (+29.3 to +35.5 mV) allowing the complexation with siRNA. The combinatorial delivery of siRNA and *paclitaxel* showed a reduction on the vascular endothelial growth factor (VEGF) expression (Zhu et al., 2010a).

Polymeric micelles where the drug is conjugated were also reported. For example, a block copolymer PEG-*block*-PCL bearing DOX was synthesized and its efficient control over the rate of DOX release in physiological medium was compared with micelles PEG-*block*-poly(α-benzyl carboxylate-ε-caprolactone) (PEG-*b*-PBCL) where DOX was encapsulated. Both systems could maintain the cytotoxicity effect over cancer cells (Mahmud et al., 2008).

Despite the success of polymeric micelles as drug carriers in vitro and in *animal* studies, clinical trials revealed a number of problems associated with premature drug release from micelles in the circulation or an absence of adequate drug release upon micelle accumulation in the tumor interstitium. The premature drug release can be prevented by appropriate micelle stabilization (Bae et al., 2006) or enhancing drug interaction with the hydrophobic blocks (Forrest et al., 2006a,b). On the other hand, the excessive drug retention in micelle cores can be solved by the application of external stimuli (Rapoport, 2007) that cause micelle destabilization in a specially controlled manner thus increasing the selectivity and efficiency of drug delivery to target cells. The intracellular signals include mainly pH (Bae et al., 2005a,b, Min et al., 2010, Nasongkla et al., 2006, Wu et al., 2010) and glutathione (Oishi et al., 2005).

pH-responsive biodegradable micelles based on a block copolymer consisting of a novel acid-labile hydrophobic polycarbonate and PEG were applied for the encapsulation of *paclitaxel* and *DOX* at the same time (13% and 11.7%, respectively). The release was clearly pH-dependent being higher at acid pH. This phenomenon is associated with the hydrolysis of the acetal groups of the polycarbonate (Chen et al., 2009b).

Two different strategies are proposed for the biodegradation of micelles under reductive environment. One is based on the reduction-sensitive cross-linking of micelles (Xu et al., 2009b) and the other is based on diblock copolymer containing a single disulfide linkage such as PEG-SS-PCL (Sun et al., 2009).

External factors including heat, ultrasound, and light are expected to trigger micellar degradation. In particular, thermoresponsive polymeric micelles that would show a remarkable change of properties responding to minimum changes in temperature are receiving attention, in particular for cancer therapies as the slightly higher temperature shown in tumor tissue (2°C–5°C) compared to healthy tissues (Meng et al., 2009b). In this sense, polymers showing lower and upper critical solution temperatures (LCST and UCST, respectively) such as poly(N-isopropylacrylamide) (PNIPA) (Li et al., 2008, Liu et al., 2008, Wei et al., 2009) were reported for the delivery of different anticancer drugs.

Polymersomes or polymeric vesicles constitute another type of self-assembling nanostructures that, unlike micelles, can encapsulate both hydrophilic and hydrophobic molecules in the hydrophilic fluid–filled core or in the hydrophobic bilayer, respectively (Meng et al., 2009b). Furthermore, unlike liposomes and phospholipids, the features of the polymersome membrane such as thickness, chemical functionality, and stability can be tailored in function of the molecular weight and type of copolymer (Smart et al., 2009). However, the main methods to produce vesicles cannot be efficiently controlled in terms of size and morphology and also the final membrane will always contain an amount of organic solvents. For these reasons, recent studies are based on the design of new methods of preparation of new vesicle systems in order to overcome these drawbacks (Howse et al., 2009). Polymersomes are prepared mainly from amphiphilic diblock, triblock, graft and dendritic polymers, preferentially biodegradable (Wang et al., 2009). Like micelles, recent efforts are focused on the design of stimuli-responsive polymersomes such as redox, pH (Chen et al., 2009b, 2010), and temperature (Xu et al., 2009a).

NPs based on synthetic polymers constitute another type of particulate system which can enhance the intracellular concentration of drugs in cancer cells while avoiding toxicity in normal cells. As examples of these systems, the preparation of Dex–DOX encapsulated in chitosan NPs (Mitra et al., 2001) and the encapsulation of curcumin in biodegradable nanoparticulated formulations based on PLGA were reported in order to improve the bioavailability and the cellular uptake (Anand et al., 2010).

Additionally, new reversible cross-linked dextrans were reported for an efficient intracellular drug delivery within reductive environments that mimic those of the intracellular compartments, showing high drug-loading efficiency and reduction-triggered release of DOX in vitro as well as inside tumor cells, in particular within cell nucleus (Li et al., 2009).

Recently, the oral bioavailability of *paclitaxel* was increased by its encapsulation through complex formation with CDs in poly(anhydride)-based NPs, as a consequence this system produced a synergistic effect associated to the combination of bioadhesive properties of poly(anhydride) and the inhibitory effect of CD (Agueros et al., 2010). Self-assembled NPs based on hydrophobically modified glycol chitosan have also been reported as a carrier for *paclitaxel* and *camptothecin* to display fast cellular and tissue internalization into tumors (Kim et al., 2006, Min et al., 2008).

Like micelles and polymersomes, temperature can also be exploited as an external factor to tailoring the drug release profiles from NPs. In this sense, a novel drug platform comprising a magnetic core and biodegradable thermoresponsive shell of triblock copolymer was synthesized. Oleic acid–coated Fe_3O_4 NPs and *DOX* were encapsulated with PEG-*b*-PLGA-*b*-PEG triblock copolymer (Andhariya et al., 2010).

Furthermore, NPs can be covalently linked to different moieties that are specifically recognized by cancer cell membrane receptors. Farokhzad et al. (2004, 2006) developed a bioconjugate system based on PLGA-*b*-PEG or PLA-*b*-PEG-COOH and a specific aptamer for targeted delivery to prostate cancer cells (Farokhzad et al., 2004, 2006). Galactosyl was also linked to low-molecular-weight chitosan (Gal-LMWC) NPs for hepatocyte targeting and specifically delivering DOX (Jain and Jain, 2010). *Paclitaxel* and DOX-loaded PEGylated PLGA-based NPs were also grafted with RGD or RGD-peptidomimetic moieties in order to target tumor vessels and enhance the antitumor efficacy and tumor growth retardation (Danhier et al., 2009).

7.2.1.2.5 Dendrimers

Dendrimers constitute a family of nanostructured macromolecules with a highly regular branching, globular architecture, multivalency, and well-defined molecular weight charge and surface area. Dendrimeric drug delivery carriers present several advantages compared with linear polymers due to their capacity to cross cell barriers through paracellular and transcellular pathways (Menjoge et al., 2010). On the other hand, they have an exceptionally high drug-loading capacity by chemical bonding or physical encapsulation (Figure 7.6a and b).

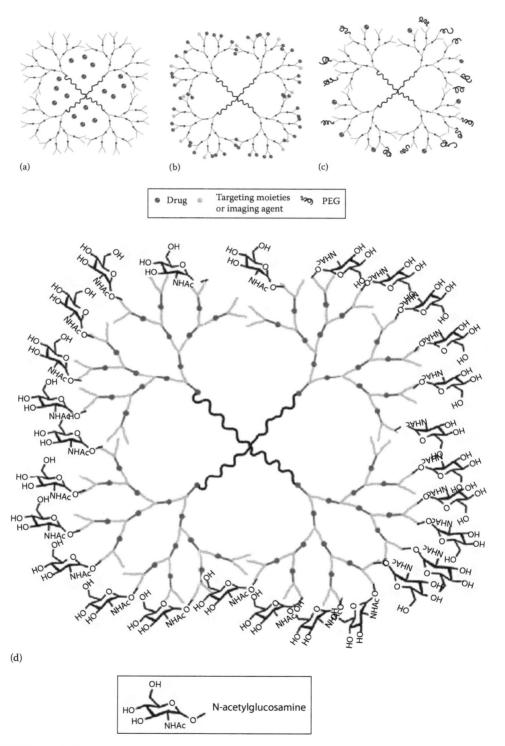

FIGURE 7.6 (a) Encapsulation of the drugs in the dendritic interiors. (b) Dendrimer–drug conjugates, dendrimers linked to targeting moieties and imaging agents. (c) PEGylated dendrimer and (d) glycodendrimer.

On the other hand, dendrimers, unlike polymer micelles based on amphiphilic block copolymers, can maintain their globular structure regardless of their concentration. Their stability is based on the presence of covalent bounds and does not depend on the CMC (Svenson and Tomalia, 2005). Different types of dendrimers such as PAMAM, polypropylene imine (PPI), and poly(ether hydroxylamine) are being commercialized for biomedical applications.

A particular example of dendrimer drug carrier with antitumoral activity consist of the biocompatible polyester composed by glycerol and succinic acid dendrimers where the anticancer drug *camptothecin* was physically encapsulated within intrinsic empty cavities of the dendrimer. The cytotoxicity of this complex was evaluated for several human cancer cell lines showing an increase in cellular uptake and drug retention in human breast adenocarcinoma (MCF-7) (Morgan et al., 2006). A recent publication describes the synthesis of thermoresponsive highly branched PAMAM-PEG-PDLLA dendritic NPs as nanocontainers for *camptothecin*. These constructed NPs show a significantly improved cytocompatibility and encapsulation efficiency, which was significantly improved in comparison with that observed for PAMAM-based dendrimers (Kailasan et al., 2010).

One of the first dendrimer conjugates in which the drug was covalently bonded to the periphery was PAMAM-cisplatin. This system offered excellent water solubility, selective accumulation in tumor tissues, and reduced toxicity in comparison with the native *cisplatin* (Malik et al., 1999).

Different strategies are developed to reduce the cytotoxicity associated to cationic dendrimers based on partial surface derivatization using chemically inert entities. Acetylated PAMAM dendrimer was conjugated to methotrexate or tritium and analyzed in vivo in animal models of human epithelial cancer and the acetylation improved the pharmacokinetics (Kukowska-Latallo et al., 2005). Nevertheless, such modifications may limit the number of targeting molecules and/or drugs. In this sense, PAMAM dendrimers were modified by glutamylation in order to increase the number of available functional groups ($-COOH$ and $-NH_2$) for further biologically active molecules immobilization (Uehara et al., 2010). Moreover, the conversion of primary amine groups to amphoteric ones may prevent toxicity.

Another strategy widely used in order to decrease cytotoxicity and increase half-life of the systems consists of the increase of the molecular weight of the dendrimer by PEGylation (Figure 7.6c) (Ihre et al., 2002, Kaminskas et al., 2008, Lim and Simanek, 2008).

Zhu et al. reported recently the conjugation of DOX to different PEGylated PAMAM dendrimers by acid-sensitive linkage (*cis*-aconityl) or -insensitive linkage (succinic) to produce PPCD and PPSD, respectively. The effect of PEGylation degree and the short of linkage have been investigated against ovarian cancer cell (SKOV-3). Highest PEGylation degree of PPCD conjugate showed greater tumor accumulation in mice inoculated with SKOV-3 cells (Zhu et al., 2010b). Another PEGylate PAMAM dendrimer conjugated with the anticancer drug *adriamycin* by amide or hydrazone bond was also reported (Kono et al., 2008).

As in the linear polymer, dendrimers can also be conjugated with specific ligands for tumor targeting. Based on this concept, PAMAM$_{G5}$ dendrimer was conjugated with biotin-targeting moieties. The good biocompatibility and biodistribution made them interesting for further applications as drug delivery systems (Yang et al., 2009).

FOL and FOL-PEG-PAMAM dendrimers based on the ligand-mediated targeted FOL loading with the anticancer drug *5-fluorouracil* were analyzed and the result showed a reduction in hemolytic toxicity, sustained drug release, as well as a higher accumulation in the tumor area than without PEG-FOL (Singh et al., 2008).

Drug–dendrimer conjugation is based usually on covalent bonds but could also be possible by ionic interaction as is the case of the complexes formed between the anticancer drug *7-ethyl-10-hydroxy-camptothecin* (*SN-38*) and PAMAM. The complexes were stable at pH 7.4 and drug was released at pH 5 (Kolhatkar et al., 2008).

It is worth mentioning the carbohydrate-installed polymers (Figure 7.6d) as targeted anticancer drug carriers, such as glycodendrimers, linear glycopolymers, and spherical glycopolymers. These systems present a specific sugar–protein biomolecular recognition in living systems and the EPR effect. *Methotrexate*-loaded polyether-copolyester dendrimers (PEPE) conjugated with

D-glucosamine as ligand is an example of glycodendrimers. Results showed that glucosamine can be used as an effective ligand not only for targeting glial tumors but also for a permeability enhancement across blood–brain barrier. Thus, glucosylate PEPE dendrimers can serve as potential delivery system for the treatment of gliomas (Dhanikula et al., 2008).

Most of dendrimer vehicles used for anticancer drug delivery are focused on PAMAM but new promising systems are based on triazine dendrimers (Lim and Simanek, 2008).

7.2.2 Antiangiogenic and Proangiogenic Polymer Drugs

Angiogenesis consist of a natural process that occurs in the human body, especially during normal growth of organs and during wound healing and is characterized by the sprouting of new blood vessels from existing ones.

Angiogenesis is regulated by the balance between proangiogenic and antiangiogenic signals. This balance is influenced by the biological function of the heparan sulfate proteoglycans (HSPGs) and their capacity to modulate the biological activity of several growth factors involved in the angiogenesis process, mainly fibroblast growth factors (FGF) and VEGF (Figure 7.7) (Iozzo et al., 2009, Johnson and Williams, 1993). Therefore, the interaction of VEGFs or FGFs with these cell surface HSPGs seems to be an attractive target to modulate angiogenesis.

7.2.2.1 Angiogenesis Inhibitors

Up-regulated angiogenic processes play an important role in the development of various pathological processes such as psoriasis (Arbiser, 1996), rheumatoid arthritis (Paleolog and Miotla, 1998), tumor growth (Folkman, 1990), diabetic retinopathy (Xu et al., 2008), etc. The uncontrolled neoformation of blood vessels contributes to the progression of these diseases, especially in the development of solid malignant tumors. The search of synthetic or natural polymers that interfere on the angiogenic process could be a good approach as angiogenic inhibitors. Figure 7.8 shows different pathways to inhibit the angiogenic process:

Polymer drugs are being used as angiogenic factor (AF) inhibitors, blocking the interaction between the AF and the HSPGs. There are two main strategies in order to avoid the binding between growth factors and HSPG, one consisting on the use of heparin-binding polycations and the other on the use of heparin-like polycations.

Some polycationic compounds, such as protamine, and PLL-based dendrimers are able to compete with growth factors for HSPG interaction. Protamine is a natural polycationic protein that binds to HSPGs and competes with heparin-binding growth factors for its interaction (Neufeld and Gospodarowicz, 1987) (Figure 7.9). The antiangiogenic activity of protamine has been demonstrated

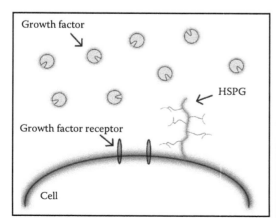
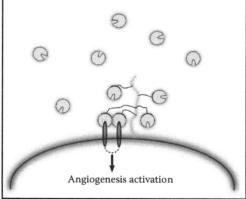

FIGURE 7.7   Interaction between growth factor and its receptors by HSPGs.

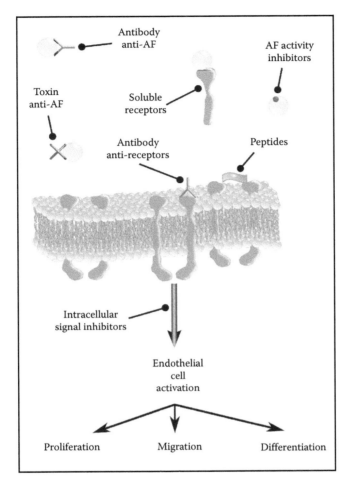

FIGURE 7.8   Mechanisms of inhibition of the HSPG-dependent interaction between growth factors and their receptors.

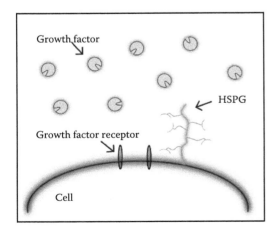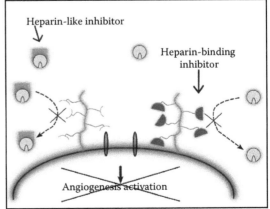

FIGURE 7.9   Interaction between growth factor and growth factor receptors by fucoidan.

in several in vitro tests (Jackson et al., 1994, Kersten et al., 1995). Nowadays, different types of dendrimers have been used as heparin-binding inhibitors such as those synthesized from PLL or arginine showing good inhibition activity of the angiogenic process without toxic effect (Kasai et al., 2002, Al-Jamal et al., 2010).

Polyanionic compounds similar to HSPG compete with HSPGs for the interaction with growth factor (Figure 7.9). In this way, polysulfonate polymers structurally similar to heparin such as sur-amins block the activity of growth factors by inhibiting their binding to HSPGs (Rusnati et al., 1996). The limitation of suramin is due to the toxic side effects at high doses. Similar compounds have been also developed. 5-Amino-2-naphthalenesulfonic acid (ANSA) showed good antian-giogenic activity and lower toxicity (Lange et al., 2007, Lozano et al., 1998). Recent studies on ANSA-based polymer systems showed good results in vitro and in vivo (Garcia-Fernandez et al., 2010a,b). Other polycationic polymers such as poly(2-acrylamido-2-methyl-1-propanesulfonic acid), poly(anetholesulfonic acid), and poly(4-styrenesulfonic acid) have demonstrated a potent inhibition of neovascularization (Liekens et al., 1997, 1999).

In both cases, the binding of AF to endothelial cell surface would be hampered with a conse-quent inhibition of their angiogenic capacity.

Non-polysulfonate-based polymer compounds with different mechanism of actions have also been investigated. Polyacetylene derivatives exhibit significant and potent antiangiogenic activ-ity. This ability is possible through induction of regulators and cell cycle mediator production. Polyacetylenes induced the formation of cyclin-dependent kinase inhibitors that inhibit the forma-tion of tubelike structures (Wu et al., 2004). Other polymers are being used as growth factor recep-tors inhibitors drugs. CEP-7055 (*N,N*-dimethyl glycine ester) or CEP-5210 (a C3-(isopropylmethoxy) fused pyrrolocarbazole) has shown a clear inhibition of cancer angiogenesis.

7.2.2.2 Angiogenesis Inducers

Therapeutic induction of angiogenesis is a potential treatment to induce neovascularization after vessel wall injury (ischemia, angioplasty). As mentioned earlier, HSPGs play an important role by their interactions with proangiogenic growth factors such as VEGF or FGF. Therefore, those molecules that mimic some biological activities of HSPGs can be used to induce angiogenesis (Figure 7.10). In this way, natural sulfated polysaccharides such as fucoidan have been stud-ied as angiogenic modulators (Chabut et al., 2003). Fucoidans are high-molecular-weight sul-fated poly(L-fucopyranose) of marine plants origin. Reduced into low-molecular-weight fraction, the action of fucoidans is similar to HSPGs, binding some proangiogenic growth factors and

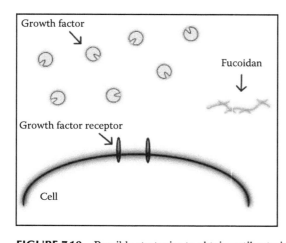
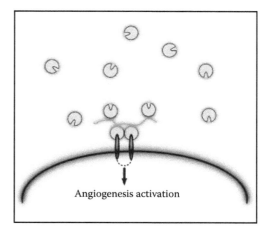

FIGURE 7.10   Possible strategies to obtain antibacterial polymers.

protecting them from proteolysis and enhancing their angiogenic activity (Chabut et al., 2003, Lake et al., 2006, Matou et al., 2002).

The problem of HSPGs-like polymers is that the same polymer can show different behaviors depending on its molecular weight. Numerous studies support the hypothesis that low-molecular-weight species of fucoidan have a generally positive effect on angiogenesis, whereas high-molecular-weight fucoidans present inhibitory effects. (Chabut et al., 2003, Lake et al., 2006, Matou et al., 2002, Soeda et al., 2000).

A similar effect occurs with heparins. Recent studies demonstrated that the angiogenesis capacity also depends on the heparin molecular weight showing inhibition effects for low-molecular-weight fractions and stimulating angiogenesis systemically for the high-molecular-weight fractions (Marchetti et al., 2008, Norrby, 1993, Powers, 2000).

Another way to induce angiogenesis is based on the preparation of scaffolds with bioactive polymers that stimulate the angiogenic process. Generally, the use of tissue scaffold has been limited by the low level of revascularization. The use of polymers that promote angiogenesis is a useful method to obtain scaffolds for the treatment of complicated tissue defects. For example, the preparation of scaffolds based on poly-N-acetyl glucosamine nanofibers showed good results for the treatment of diabetic wound healing (Scherer et al., 2009). This scaffold simulates the extracellular matrix (ECM) and provides to the cell a growth media that enhance the cell metabolism, migration, and therefore the angiogenesis. The problem is the limitation of their use due to the inability to provide sufficient supply of oxygen and nutrients to the inner part of the cell–matrix constructs (Moon and West, 2008). The superficial modification with ECM proteins such as FN, collagen, or sequences corresponding to cell adhesion domains contributes to the angiogenic process (Calonder et al., 2005, Kouvroukoglou et al., 2000, Unger et al., 2004). An example is the preparation of chitosan scaffolds with different degrees of acetylation (DA) coated with FN (Amaral et al., 2009). The DA affects cell adhesion and growth on FN-coated chitosan porous scaffolds due to the influence on the adsorbed FN. The selection of suitable DA will therefore be highly important for further vascularization strategies.

7.2.3 Antibacterial Polymer Drugs

Contamination and infections caused by pathogen microorganisms constitute a serious problem and a great concern in hospitals, healthcare products, medical devices, and many other biomedical applications. In fact, infections are the most common cause of biomaterial implant failure and represent a significant challenge to the more widespread application of biomedical implants (Kurt et al., 2007, Nishi et al., 2007, Waschinski et al., 2008). Generally, an infected implant needs to be removed as the formation of a bacterial biofilm hampers the treatment with antibiotics. Therefore, emphasis has been placed on different ways to prevent the infections caused by bacteria that account for 1 million implant-associated infections and 3 billion dollars of healthcare costs annually (Hetrick and Schoenfisch, 2006). Therefore, there is a definitive need of new materials resistant to microbial colonization and polymeric drugs are good candidates for this purpose.

There are many factors than can affect the biological activity of antibacterial polymers such as hydrophilic–hydrophobic balance, spacer length between active site and polymer, molecular weight or interaction between polymer and pathogen. Consequently, an appropriate balance between all these features is necessary to reach an optimum result.

The basic requirements for antibacterial polymers in biomedical applications are the following:

- Does not decompose to and/or emit toxic products
- Stable in long-term usage and storage at the temperature of its intended application
- Biocidal to a broad spectrum of pathogenic microorganisms in brief times of contact

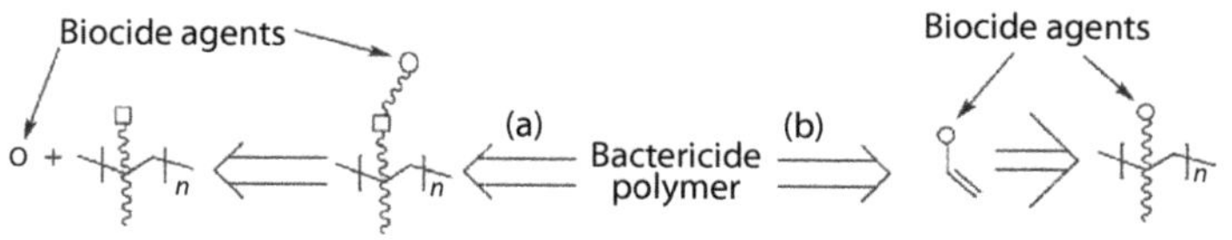

FIGURE 7.11 Antimicrobial polymers obtained by surface functionalization of natural or synthetic polymers (a) or polymerization of pharmaceutical active compounds conjugated to polymerizable groups (b).

From the technological point of view, and without considering the encapsulation of low-molecular-weight active agents to give rise to leachable polymers, the antimicrobial polymers can be obtained following two different and complementary strategies (Figure 7.11):

1. Surface functionalization of natural or synthetic polymers by grafting of antimicrobial agents
2. Polymerization of pharmaceutically active compounds conjugated to polymerizable reactive groups

From the chemical point of view, there are two great groups of bactericidal (kill bacteria) or bacteriostatic (halt bacterial growth) agents that can be included into polymeric systems: cationic biocides and bioactive low-molecular-weight compounds. Cationic biocides are represented to a large extent by guanidines (Wei et al., 2009), biguanidines (Allen et al., 2004), and salts of sulfonium (Kanazawa et al., 1993), phosphonium (Nonaka et al., 2002), pyridinium (Imazato et al., 1999), and ammonium (Kebir et al., 2007), being this last type of salt widely used since the 1930s in domestic and public hygiene for surface disinfection, topical antisepsis, and to control biofouling and microbial contamination in industry. Their mode of action normally involves interaction with the cell envelope, displacing divalent cations. Subsequent interactions with membrane proteins and lipid bilayer depend upon the specific nature of the biocide but generally cationic biocide exposure results in membrane disruption and lethal leakage of cytoplasmic materials (Gilbert and Moore, 2005).

The other great group of substances that can be included into polymeric systems are the low-molecular-weight antibiotic agents widely established in the pharmaceutical industry and the sanitary system from World War II, such as aminoglucosides, ansamicines, carbacephens, carbapenems, cefalosporins, glycopeptides, macrolides, monobactams, penicillins, polypeptides, quinolones, sulfonamides, tetracyclins, and others. Low-molecular-weight antimicrobial agents present some disadvantages, such as short-term antimicrobial activity, toxicity to the environment, and bioresistance. To overcome all these drawbacks, antimicrobial functional groups can be incorporated into polymer molecules if the antimicrobial agent contains reactive functional groups such as amino, carbonyl, or hydroxyl groups that can be covalently linked to a wide variety of polymerizable derivatives (Dizman et al., 2005, Kenawy et al., 2007, Wach et al., 2008).

Considering the effectiveness of bactericides, a distinction needs to be made between two types of bacteria: Gram-positive and Gram-negative. Their essential difference lies in the absence of the cell wall in the Gram-positive bacteria while that the Gram-negative bacteria have an outer membrane structure in the cell wall forming an additional barrier for foreign molecules which determines a lesser sensitivity to chemicals. For the determination of bactericide efficiency, it is important to carry out tests with bacteria representative of both types, being *Staphylococcus aureus* and *Bacillus subtilis* typical representatives for Gram-positive bacteria and *Escherichia coli* and *Pseudomonas aeruginosa* for Gram-negative bacteria.

The membrane of Gram-negative bacteria is composed of 70%–80% of phosphatidylethanolamine (PEA) and 20%–25% of negatively charged lipids, such as phosphatidylglycerol (PG) or cardiolipin (CL) (Glukhov et al., 2005), whereas the membrane of the Gram-positive bacteria are formed mainly by anionic lipids, such as 70% of PG, 12% of PEA, and 4% of CL (Epand et al., 2006).

However, the membrane of eukaryotic cells like human erythrocytes is composed by a lipidic bilayer which has in its internal layer 10% of the negatively charged lipid phosphatidylserine (PS), 25% of PEA, 10% of phosphatidylcholine (PC), and 5% of sphingomyelin (SM), whereas its external part is composed of 25% of cholesterol, 33% of PC, 18% SM, and 9% of PEA (Verkleij et al., 1973). Therefore, there is clear difference between bacterial and eukaryotic cells, being the outer wall of bacteria more negatively charged than eukaryotic membrane.

Biocidal cationic polymers present a clear tendency to interact with the bacterial membrane (negatively charged) instead of the eukaryotic membrane due to ionic affinity (Hancock and Rozek, 2002, Zasloff, 2002). This along with the fact that cholesterol is a membrane-stabilizing agent in mammalian cells but absent in bacterial cell membranes protects the cells from antimicrobial peptides attack, and along with the well-known transmembrane potential that affect peptide–lipid interactions makes bacterial membrane highly vulnerable against positively charged cationic polymers.

In contrast to many conventional antibiotics, the exact mechanism of killing bacteria by cationic polymers is not well-established. These polymers appear to be bactericidal (bacteria killer) instead of bacteriostatic (bacteria growth inhibitor). However, it is known that biocidal cationic polymers interact with the anionic bacteria membrane disrupting the membrane integrity and inducing the cytoplasmic constituents leakage and death of the bacteria (Bagheri et al., 2009, Brogden, 2005).

7.2.3.1 Surface Functionalization of Natural or Synthetic Polymers by Grafting of Antimicrobial Agents

Surface modification by grafting polymerization can be achieved by different coupling mechanisms such as coordination, ionic, or free-radical mechanism (Desmet et al., 2009). However, most recent techniques for surface modification are based on radical living polymerization, reversible addition–fragmentation chain transfer polymerization (RAFT) (Iwasaki et al., 2007, Li et al., 2007, Ranjan and Brittain, 2007), and atom transfer radical polymerization reactions (ATRP) (Edmondson et al., 2004, Huang et al., 2007, Lee et al., 2007).

In this context, Matyjaszewski et al. (Huang et al., 2007) prepared polypropylene (PP)-based surfaces grafted with non-leachable biocide by chemically binding poly(quaternary ammonium) (PQA). The well-defined PDMAEMA, a precursor of PQA, was grown from the surface of PP via ATRP and the tertiary amine groups in PDMAEMA were consequently converted to the quaternary ammonium in the presence of ethyl bromide (Scheme 7.1).

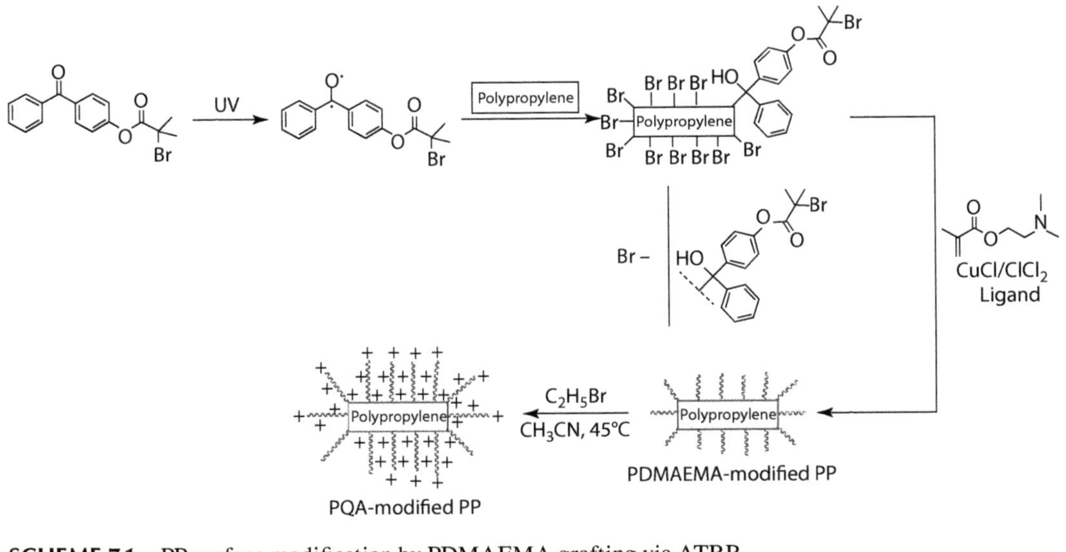

SCHEME 7.1   PP surface modification by PDMAEMA grafting via ATRP.

The biocidal activity test against *E. coli* of the resultant surfaces depended on the number of available quaternary ammonium units (QA density). With the same grafting density, the surface grafted with relatively high-molecular-weight polymers ($M_n > 10,000 \, g/mol$) showed almost 100% killing efficiency, whereas a low biocidal activity (85%) was observed for the surface grafted with shorter PQA chains ($M_n = 1500 \, g/mol$).

Cheng et al. (2008) prepared a switchable polymer surface coating which combined the advantages of both nonfouling and cationic antimicrobial materials overcoming their individual disadvantages. In this system, poly(*N,N*-dimethyl-*N*-(ethoxycarbonylmethyl)-*N*-[2′-(methacryloyloxy) ethyl]-ammonium bromide) was grafted by surface-initiated ATRP onto a gold surface with initiators, giving rise to a surface that killed more than 99.9% of *E. coli* K12 in 1 h and released more than 98% of the dead bacterial cells when the cationic derivative was hydrolyzed to nonfouling zwitterionic polymers.

Natural polymers, such as chitosan and cellulose, offer another source of macromolecules to obtain of antibacterial materials. Nowadays, many attempts have been carried out in order to improve the antibacterial properties of chitosan, a semisynthetic polymer obtained from partial deacetylation of chitin (Cho et al., 1999), and that has many interesting properties, such as antimicrobial activity and nontoxicity (Rabea et al., 2003) (Scheme 7.2).

N-Alkyl chitosan derivatives were prepared by introducing alkyl groups into the amine residue of chitosan, reduction of the intermediate Schiff's base with NaBH$_4$, and subsequent quaternization of the obtained *N*-akyl chitosan derivative with methyl iodide (Kim et al., 1997) (Scheme 7.3).

The chitosan quaternary ammonium salts obtained turned out to be better antibacterial than the original chitosan, and it was found that the antimicrobial activities of the chitosan quaternary

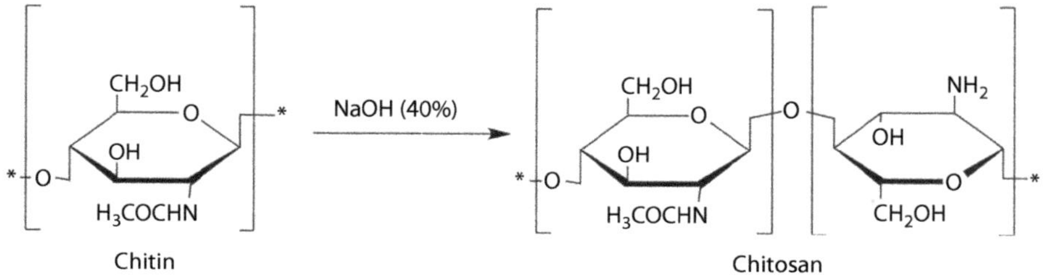

SCHEME 7.2 Transformation of chitin into chitosan.

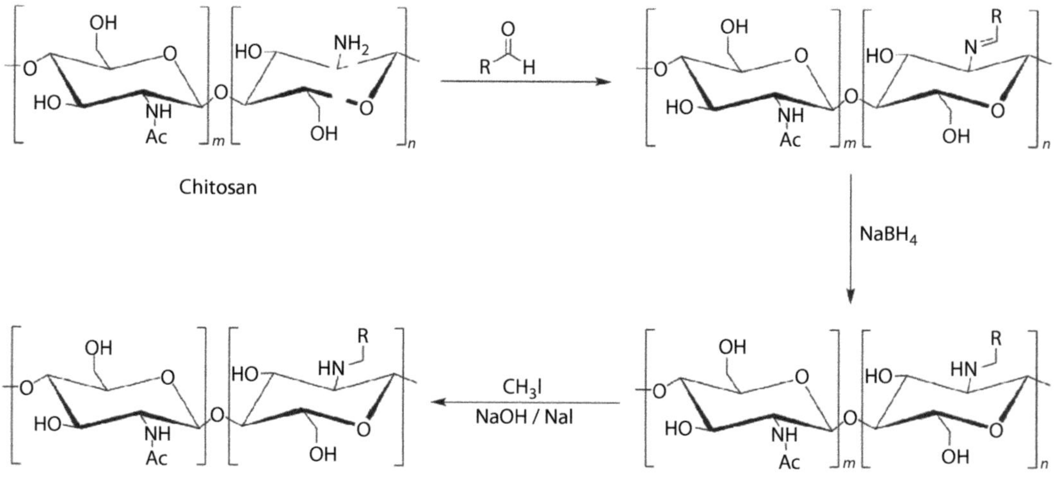

SCHEME 7.3 Synthetic route to *N*-alkyl chitosan derivatives.

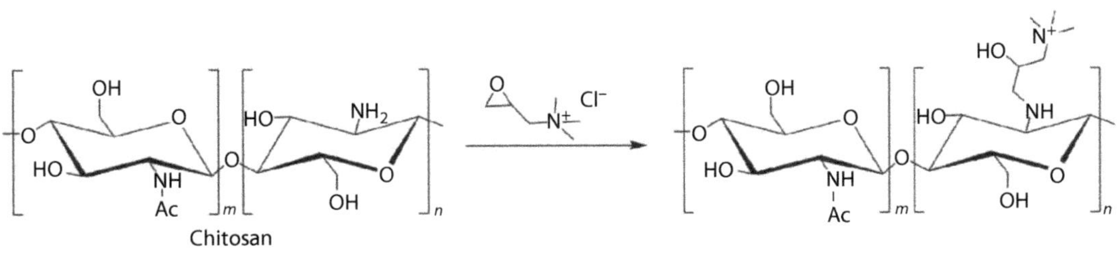

SCHEME 7.4   Reaction of chitosan with glycidyltrimethylammonium chloride.

ammonium salts increased with the increase in the chain length of the alkyl substituent, which was ascribed to the contribution of the increased lipophilic character of the derivatives.

Another way to obtain chitosan quaternary ammonium salts consist of the reaction of chitosan with glycidyltrimethylammonium chloride leading to modified chitosan that showed excellent antimicrobial activity (Nam et al., 1999) (Scheme 7.4).

Phenolic derivatives have antiseptic properties damaging cell membranes and causing release of intracellular constituents, although they also cause intracellular coagulation of cytoplasmic constituents leading to cell death or inhibition of cell growth. Therefore, phenolic derivatives have been conjugated to chitosan through its amino group at C2 leading to chitosan derivatives that were highly active against fungi such as *Aspergillus flavus*, *Candida albicans*, and *Fusarium oxysporium*, as well as against bacteria such as *E. coli*, *S. aureus*, and *B. subtilis* (Kenawy et al., 2005) (Scheme 7.5).

Cellulose constitutes another natural polymer commonly used for the preparation of antibacterial materials. Perrier and coworkers grafted PDMAEMA into cellulosic fibers via RAFT polymerization (Roy et al., 2008) (Scheme 7.6).

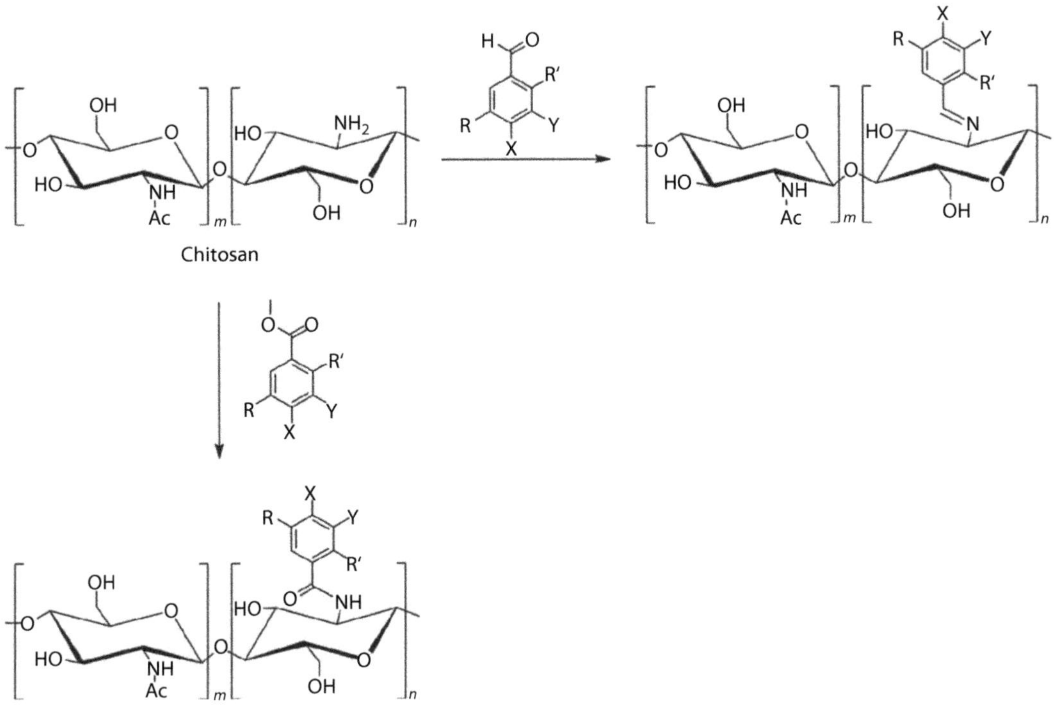

SCHEME 7.5   Phenolic derivatives of chitosan with antimicrobial properties.

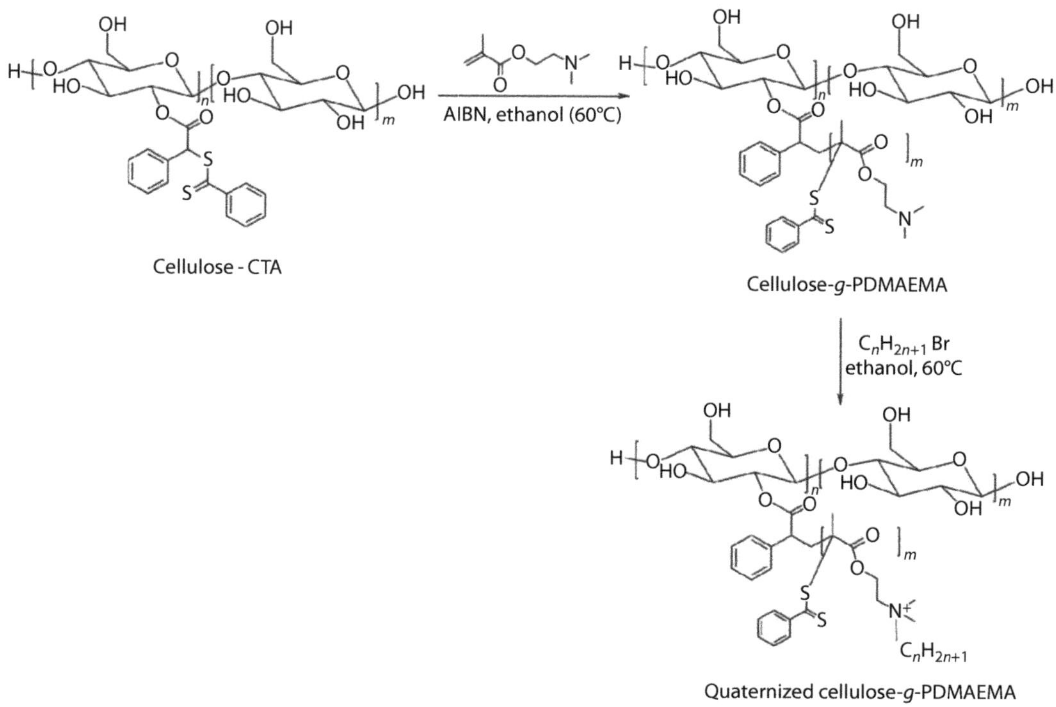

SCHEME 7.6   PDMAEMA grafting into cellulose via RAFT polymerization.

Subsequent quaternization of the tertiary amino groups of the grafted **PDMAEMA** chains with different chain length alkyl bromides (C8–C16) leads to a large concentration of quaternary ammonium groups on the cellulose surface. The antibacterial activity depends on the alkyl chain length and on the degree of quaternization. PDMAEMA-grafted cellulose with the highest degree of quaternization and quaternized with the shortest alkyl chains exhibited the highest activity against *E. coli.*

Gademann and coworkers created antibacterial titanium oxide surfaces allowing the generation of stable, protein-resistant, nonfouling surfaces thanks to a novel biomimetic strategy for surface modification that exploits the evolutionary optimized strong binding affinities of iron chelators such as anachelin, chromophore that contains a catechol moiety as the anchoring group, which is structurally similar to key elements of mussel-adhesive protein sequences that are thought to be responsible for the very strong wet adhesion of mussels to surfaces (Deming, 1999, Gademann et al., 2009, Wach et al., 2008, Zürcher et al., 2006).

In his approximation, Gademann employed an anachelin chromophore–PEG conjugate linked to vancomycin, where the anachelin chromophore enables the immobilization of the hybrid on the surface; the vancomycin interferes with cell-wall biosynthesis and inhibits the growth of bacteria; and the long PEG-3000 linker makes the modified surfaces resistant to proteins and cells, and ensures the optimal positioning of the antibiotic on the surface.

7.2.3.2   Antimicrobial Polymers by Polymerization of Pharmaceutically Active Compounds Conjugated with Polymerizable Groups

The second strategy to overcome the disadvantages of the low-molecular-weight antimicrobial agents consist of their covalently linking onto polymerizable molecules.

Acrylic derivatives bearing pharmaceutically active compounds are one of the most important groups of polymeric drugs. These acrylic drug–conjugated monomers have the advantage that they can be copolymerized with a wide type of compounds and tune the drug concentration or the hydrophilic/hydrophobic functionalities present in the copolymer.

In this context, Rojo et al. (2008) described the preparation of antimicrobial copolymers by radical copolymerization of 2-hydroxyethyl methacrylate (HEMA) with methacryloyl derivatives of eugenol, an essential oil used in medicine as local antiseptic and anesthetic, or mixed with zinc oxide as temporary pulp capping agent and as filling system in root canals. Eugenol derivatives were obtained by reacting methacryloyl chloride and eugenol or 2-eugenyl ethanol in the presence of triethylamine (Scheme 7.7).

The copolymers obtained were active against Gram-negative *E. coli* strains and also against *Streptococcus* mutants, Gram-positive bacteria found in the oral cavity and significant contributor to the tooth decay, and were biocompatible with human fibroblast which makes these kinds of compounds suitable for biomedical systems in the field of dentistry and orthopedic applications (Rojo et al., 2006).

Parra et al. (2009) reported the synthesis of polymerizable quaternary ammonium salts with high refractive index and their copolymerization with different methacrylic monomers for the preparation of bactericide copolymers for ophthalmic applications (Scheme 7.8).

The bactericide quaternary ammonium salts monomers showed inhibition halos of 23–25 mm in antibiogram tests against *Staphylococcus epidermidis* and *P. aeruginosa*, strains found in the ocular cavity and responsible for most postsurgical endolphthalmitis. Biocompatibility of the systems was evaluated in cell cultures using human fibroblasts and in all cases the cellular viability was higher than 90%, and close to 100% in many cases, for the extracts of selected formulations collected at different periods of time. The materials obtained by copolymerization with HEMA

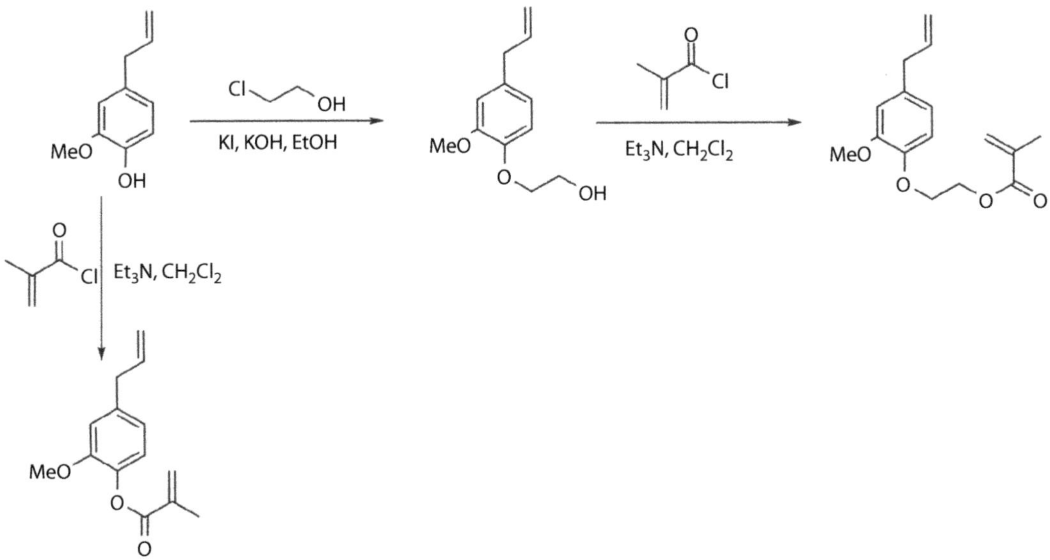

SCHEME 7.7 Synthetic route to polymerizable eugenol derivatives.

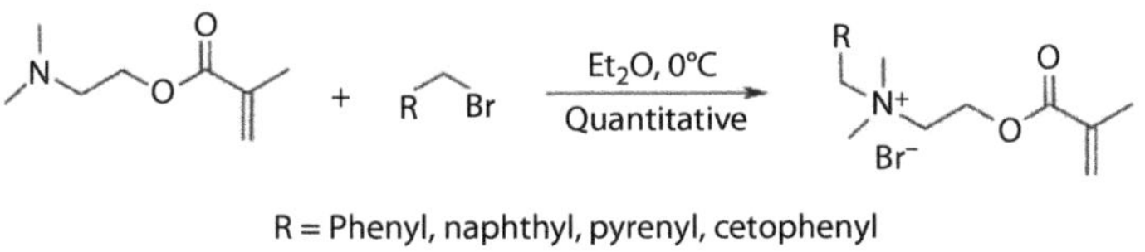

R = Phenyl, naphthyl, pyrenyl, cetophenyl

SCHEME 7.8 Polymerizable quaternary ammonium salts.

and 2-(benzothiazolylthio)ethyl methacrylate presented high refractive index values ($n_D^{20} \approx 1.51$) and good wettability (15%–37%), which make these systems good candidates for the fabrication of bacteriostatic and foldable intraocular lenses (IOLs), and can be considered as a good alternative to the currently used foldable hydrogel IOLs.

It is known that the biocidal cationic polymers can mimic the biological activity of the natural host defense peptides (HDPs) that are an evolutionarily conserved component of the innate immune response and are found among all classes of life. HDPs have demonstrated their broad spectrum antibiotic activity, killing Gram-negative and Gram-positive bacteria (including strains that are resistant to conventional antibiotics), mycobacteria (including *Mycobacterium tuberculosis*), enveloped viruses, fungi, and even transformed or cancerous cells (Gabriel et al., 2007). HDPs are generally between 12 and 50 amino acids and have a large portion of hydrophobic residues, generally >50%, and two or more positively charged residues provided by histidine in acidic environments, or by arginine and lysine. With the aim of mimicking HDPs and taking advantage of the guanidines properties, which are positively charged at physiological pH and have been used widely as antibacterial agents (Broxton et al., 1983), Gabriel et al. synthesized poly(guanidinium oxanorbornene) (PGON) from norbornene monomers via ring-opening metathesis polymerization (Gabriel et al., 2008) (Scheme 7.9).

The synthesized polymer was not membrane-disruptive indicating that it also had properties similar to polyarginine and other cell-penetrating peptides (CPPs) like HIV-TAT (Futaki et al., 2001, Henriques et al., 2006, Mitchell et al., 2000, Miyatake et al., 2006, Rothbard et al., 2002), and represents an exciting and fundamentally novel entry in the expanding field of synthetic mimics of antimicrobial peptides like HDPs. Moreover, PGON possess a remarkable combination of antimicrobial activity against both Gram-negative and Gram-positive bacteria as well as low hemolytic activity against human red blood cells, encouraging the development of powerful, nontoxic, antibacterial materials designed to prevent biofilm formation.

Vancomycin is the prototypical glycopeptide antibiotic, and it remains potent against Gram-positive organisms (e.g., *Staphylococcus* spp.) commonly encountered in association with indwelling medical devices such as orthopedic hardware, although resistance can occur (Boneca and Chiosis, 2003). Its activity derives from binding D-Ala–D-Ala sequences found at the terminal end of peptidoglycan precursors, that blocks the action of both transglycosylases and transpeptidases by complexing with their substrates and preventing proper cross-linking of peptidoglycan structures, reason why the internalization is not required for its activity (Loll and Axelsen, 2000). Following this rationale, Lawson and coworkers (Lawson et al., 2009) synthesized polymerizable vancomycin derivatives bearing either acrylamide or PEG-acrylate and were tethered from a surface pendant to a polyacrylate

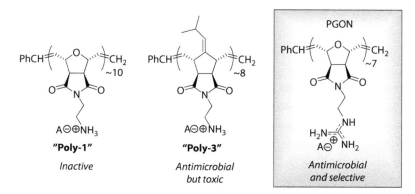

SCHEME 7.9 Synthesis of polyguanidinium oxanorbornene (PGON).

backbone through a living radical polymerization, demonstrating that the vancomycin-PEG-acrylate derivatives showed a significant reduction in bacterial colony-forming units (CFU) with respect to nonfunctionalized control surfaces.

7.2.4 ANTITHROMBOGENIC POLYMER DRUGS

The use of a cardiovascular device represents the introduction of a foreign surface into blood circulation. Under normal conditions (intact blood vessels), blood contacts an endothelium with anticoagulant and antithrombogenic properties. However, when artificial surfaces are placed in contact with blood, the hemostatic mechanism is activated; this is a physiological process that involves a complex set of interdependent reactions between the surface, platelets, and coagulation proteins, resulting in the formation of a clot or thrombus, which can be removed by fibrinolysis (Xue and Greisler, 2003). The first event that occurs when a biomaterial comes in contact with blood is the adsorption of proteins and other molecules; this protein layer has an impact on further biological processes such as cell adhesion or activation of enzyme cascades of coagulation or inflammation (Gorbet and Sefton, 2004).

In this sense, thrombogenicity is defined as the ability of a biomaterial to induce or promote the formation of thromboemboli (Sefton et al., 2000). Thrombogenicity is one aspect of hemocompatibility, which can be defined, according to ISO standards (2002, Seyfert et al., 2002), as the conjunction of thrombosis, coagulation, blood platelets, hematology, and immunology. The first three categories may be summarized as hemostasis.

Hemostasis can be divided into plasmatic (soluble) and cellular aspects. The plasmatic part of coagulation consists of a cascade-like activation of inactive proteases, finally leading to the activation of thrombin, which is the key enzyme for the formation of fibrin. Fibrin spontaneously assembles to fibrils, forming a fibrin clot.

Blood platelets (thrombocytes) present the cellular component of the coagulation/thrombotic system. They are $3\,\mu m$ sized spherical to disk-shaped anuclear cells, which can adhere to surfaces, spread by forming pseudopodia, adhere to each other, release growth factors and cytokines, and enhance the humoral coagulation system.

Immunoreactions to foreign materials are mainly nonspecific. The leading effector of the soluble nonspecific immune system is the complement system. Similar to the coagulation cascade, it consists of a number of proteases, which activate each other in a cascade-like process. Consequences are the opsonization of the target for facilitated phagocytosis, activation of nonspecific inflammatory cells (granulocytes and monocytes), and attraction of these cells (chemotaxis).

Polymers, from natural or synthetic origin, are the biomaterials most widely used for manufacturing medical devices and disposable clinical apparatus which can be used in contact with blood, such as vascular prostheses, artificial kidney, blood pumps, artificial heart, dialyzers, and plasma separators (Mao et al., 2004). The polymeric materials used for these applications are conventional materials like poly(tetrafluoroethylene) (PTFE), poly(vinyl chloride) (PVC), segmented polyetherurethane (SPU), polyethylene (PE), or silicone rubber. These polymers are usually selected because of their technological properties like mechanical and chemical stability, processing and ease sterilization, combined with nontoxicity, and a reasonable hemocompatibility. For this reason, in most of the cases, when the devices are placed in contact with blood they initiate the formation of clots by the activation of platelets and other components of blood coagulation system. These phenomena are harmful for maintaining well-balanced function or even life in patients, being necessary the administration of lifelong antithrombogenic therapy. For example, synthetic vascular grafts are successfully used in the treatment of the pathology of large arteries but when they are used in the replacement of small diameter blood vessels (internal diameter <6mm) are known to be highly thrombogenic and need special treatments to improve their patency after implantation (Boura et al., 2003).

Therefore, the improvement of hemocompatibility of artificial polymer surfaces constitutes one of the main issues in biomaterials science. The main strategies to reduce thrombogenicity in polymeric materials can be resumed in three different approaches (Aldenhoff et al., 1997):

1. Design of bioactive polymeric materials that prevent the reactions involved in blood–material interaction: activation and aggregation of platelets and/or activation of blood coagulation cascade
2. Preparation of inert polymers that do not trigger blood reactions
3. Promotion of endothelial cell growth (re-endothelization)

7.2.4.1 Surface Modification by Bioactive Polymers

The inner wall of natural blood vessel is mainly composed of a monolayer of endothelial cells. These cells posses several active anticoagulant mechanisms, for this reason, blood does not coagulate in normal blood vessels (Werner et al., 2007). As an attempt to reproduce the behavior of this layer, the immobilization of active molecules onto polymeric surfaces represents one of the most popular approaches to improve the hemocompatibility of cardiovascular devices.

Heparin, as the analog to the cell surface anticoagulant heparan sulfate, has been the first anticoagulant successfully immobilized onto materials in order to reduce its thrombogenicity (Gott et al., 1963).

Heparins constitute a family of glycosaminoglycans of various molecular weights with anticoagulant activity due to a high-affinity binding to antithrombin III (AT III) and therefore able to act as indirect thrombin inhibitors. AT III is the molecular target of heparin and, when it is activated, it binds to either thrombin and/or factor X from the coagulation cascade. Many researchers have been able to immobilize heparin by covalent or ionic bonding (Alferiev et al., 2006, Chen et al., 2005b, 2009a, Huang Ly, 2006, Luong-Van et al., 2006).

The main concern of using heparin is based on the fact that, once immobilized, heparin should be able to adopt its native conformation for its interaction with AT III. It has been demonstrated that binding of ATIII is more efficient when heparin is coupled by end-point attachment to a polymeric chain. This strategy was followed by Larm et al. (Larm et al., 1983) for the development of Carmeda® Bioactive Surface, a heparin coating technique which has been licensed for its use on vascular grafts, coronary stents, oxygenation systems, and extracorporeal devices. Heparin is first partially depolymerized by deaminate cleavage to produce heparin fragments terminating in an aldehyde group. These heparin fragments are then covalently linked to the primary amino groups of PEI.

Heparin has also been covalently attached onto poly(carbonate-urea)urethane graft (Myolink®) using spacer arm technology. This is a grafting technique consisting on the use of spacer arms to reduce steric hindrance by the proximity of the ligand to the rigid surface of the polymeric backbone (Krijgsman et al., 2002). Following this rationale, the combination of heparin with an RGD peptide moieties confers antithrombogenic properties to the surface as well as improves endothelial cell adhesion in comparison with native Myolink (Tiwari et al., 2002).

Another approach for surface heparinization consist of the immobilization of heparin macromolecules by ionic bonding onto polymeric surfaces. The strong anionic character of this molecule facilitates the anchorage on surfaces previously treated with a cationic substance. In this sense, Nakayama et al. (2007) described a heparin bioconjugate based on a branched thermoresponsive star-shaped cationic polymer. The system combines a PNIPA-based surface adsorption domain with a cationic polymer as a heparin-binding domain. Upon mixing the polymer with heparin, NPs are formed resulting from the formation of polyionic complexes. These particles can be adsorbed onto hydrophobic polymers commonly used as materials for medical devices, such as silicone, PE, polystyrene, or poly(ethylene terephthalate) (PET).

Another ionic approach has been developed by Hydromer (Vicario et al., 2008) and consist of a polysaccharide-based polymer (F202TM). This polymer has been evaluated as antithrombogenic coating on polyurethanes as well as on medical grade electropolished stainless steel, showing in both cases a reduction in platelet adhesion and thrombus formation.

A commercial procedure for ionically bounding of heparin (Duraflo II, Baxter Bentley Healthcare Systems, Irvine, CA) has been used to coat cardiopulmonary bypass circuits and other medical devices. This coating has proved to reduce surface thrombus formation as well as a significant reduction of C3 and C4 complement activation (Fosse et al., 1997, Mangoush et al., 2007, Wildevuur et al., 1997).

Recent technologies based on the preparation of albumin–heparin coatings have also been described (Brynda et al., 2000, Sperling et al., 2006). These systems are prepared by the layer-by-layer assembly technique consisting on the sequential adsorption of albumin and heparin onto a surface under certain conditions where the constituents bear opposite charges (Brynda and Houska, 1998). Albumin–heparin multilayers have been demonstrated to improve hemocompatibility of cardiovascular devices; moreover, the increasing number of layers enhanced the coating bioactivity, suggesting that heparin maintains its activity inside the assemblies (Houska et al., 2008).

As mentioned earlier, platelet response to a foreign material is one of the most important parameters to take into consideration for the evaluation of blood compatibility of a biomaterial, since coagulation on the surface starts with platelet aggregation and the formation of a fibrin network, giving raise to thrombus formation, from the combination of mutually fused platelets plus the insoluble fibrin and the cells that are trapped from blood (Mao et al., 2004).

There are several drugs that are used to inhibit platelet aggregation; however, only aspirin, dipyridamole, and thienopyridines are currently approved by the Food and Drug Administration (FDA) for use in patients. Attending to the mechanisms of action, these drugs can be classified into the following four groups (Kidane et al., 2004):

1. Cyclooxygenase inhibitors: Aspirin and aspirin-like drugs
2. ADP receptor blockers: Thienopyridine derivatives (ticlopidine and clopidogrel)
3. Adenosine uptake inhibitor: Dipyridamole
4. Inhibitors of platelet glycoprotein IIb/IIIa (GP IIb/IIIa): Abciximab, eptifibatide, tirofiban

Several authors have described the direct immobilization of antiplatelet agents onto polymeric surfaces. For example, Aldenhoff et al. (1997, 2003) described dipyridamole immobilization onto polyurethane surfaces, in which the drug is linked to a photoreactive moiety via an spacer group. These systems showed reduced platelet adhesion in vitro and improved patency in a goat model.

Aspirin and aspirin derivatives have also been extensively incorporated into blood-contacting polymers. In this sense, Paul et al. (Paul and Sharma, 1997) developed poly(vinyl alcohol) (PVA) hemodialysis membranes loaded with acetylsalicylic acid.

San Román et al. (1996) described the application of random copolymers of HEMA and a methacrylic derivative of salicylic acid as coating of Dacron vascular prostheses. The presence of salicylic acid in the coating produced a noticeable decrease in the average number of platelets adhered to the graft.

Polyacrylic derivatives of (2-acetyloxy-4-trifluoromethyl)benzoic acid (Triflusal), an antiplatelet drug with a chemical structure closely related to aspirin, have also been described. These systems are based on a Triflusal derivative synthesized by an esterification reaction with HEMA (THEMA). When the homopolymer bearing Triflusal was applied as coating for commercial vascular prostheses of PTFE, an improvement of the hemocompatibility of these devices was shown, since there was a clear reduction in platelet adhesion when the prostheses were coated with the polymer (Rodriguez et al., 1999).

Copolymerization of THEMA with other hydrophilic biocompatible monomers gave rise to materials that not only present activity in its macromolecular form, but also can be applied as drug delivery systems (Gallardo et al., 2004, Rodriguez et al., 2004). In this sense, copolymers of THEMA and N,N'-dimethylacrylamide (DMAA) were prepared. These systems showed good cell biocompatibility and in vitro antiaggregant behavior. Furthermore, copolymer released Triflusal in a sustained way during several months.

A different family of hydrophilic copolymers was prepared by copolymerization of THEMA with AMPS. These systems combined the presence of the antiaggregant drug with a heparin-like behavior due to the presence of sulfonic group from AMPS monomer. THEMA–AMPS copolymers prevented platelet adhesion and aggregation and also showed a zero-order release for several months (Aguilar et al., 2002, Gallardo et al., 2003).

Recently, new polymeric systems prepared from THEMA and butyl acrylate (BA) have been described (San Roman et al., 2007). These systems have been adequate for the development of stent coatings, since they exhibit good adhesion and crack-bridging properties, requirements that are necessary to withstand the mechanical deformation produced during the application of vascular stents. Furthermore, these polymers present an antithrombogenic behavior as well as good biocompatibility. Poly(THEMA-*co*-BA) copolymers have been loaded with antiproliferative drugs such as simvastatin and paclitaxel in order to develop drug-eluting stents; these coatings simultaneously bear an antiaggregant drug for the prevention of thrombosis and antiproliferative drug to inhibit restenosis postimplantation. The coated stents have been licensed for use in human and commercialized by Iberhospitex (Spain) as IRIST® and ACTIVE®.

Another molecule that prevents blood coagulation and thrombus formation is NO, a signal molecule produced in endothelial cells by the endothelial isoform of nitric oxide synthase (eNOS), a calcium–cadmodulin-sensitive enzyme. In the cardiovascular system, NO triggers a cascade of events leading to smooth muscle relaxation and a subsequent decrease in blood pressure. Furthermore, it prevents platelet adhesion and activation. NO is also involved in the immune response and serves as a potent neurotransmitter at the neuron synapses. In normal physiological conditions, NO has an anti-inflammatory effect, but it could also be a proinflammatory mediator in abnormal situations. On the other hand, NO serves as a central regulator of oxidant reactions and diverse free radical–related disease processes (Moncada et al., 1988).

In recent years, researches in tissue engineering have worked to develop new materials which could release and generate NO. Taite et al. (2008) described a NO-releasing polyurethane–PEG copolymer that incorporates the cell adhesive peptide sequence YIGSR. This system showed a decreasing in platelet adhesion as well as an increasing in endothelial cell proliferation when compared to control polyurethane.

Wu et al. (2007) have described the preparation of multifunctional bilayer polymeric coatings prepared from commercial silicon rubber–polyurethane copolymers in which thrombomodulin and NO are incorporated in separate layers. By this approach, blood-compatible biomedical surfaces are designed by mimicking the highly thromboresistant endothelium layer.

7.2.4.2 Surface Passivation by Inert Polymers

Biomaterials researchers have made a great effort in the last decades for the preparation and development of polymeric systems that are inert or passive with respect to blood reactions, and display thromboresistant behavior by minimizing the interaction with proteins and cells. This strategy is known as surface passivation.

Both synthetic and natural polymers have been investigated in the development of strategies to passivate a biomaterial surface, by reducing or eliminating enthalpic or entropic effects that drive protein and cell absorption on a molecular level (Jordan and Chaikof, 2007). In this sense, brushes of long-chained hydrophilic molecules like poly(ethylene oxide) (PEO) or the related molecules PEG or tetraethylene glycol dimethyl ether (tetraglyme) and others are known to be biologically inert and display excellent biocompatibility, and therefore have been suggested for hemocompatible surface modification (Cao et al., 2007, Chen et al., 2005a, Gorbet and Sefton, 2001, Hansson et al., 2005, Shen et al., 2002). Several methods have been established with this purpose, including bulk modification, covalent grafting, and physical adsorption. These systems has found application as coatings for dialysis membranes (Fushimi et al., 1998, Sirolli et al., 2000) as well as for vascular stents (Okner et al., 2009, Thierry et al., 2005), microencapsulation of cells (Arifin and Palmer, 2005, Haque et al., 2005), or drug delivery systems (Arica et al., 2005, 2008).

Another strategy for the development of inert thromboresistant surfaces consist of the preparation of albumin-coated polymers. Albumin is the most abundant protein in blood plasma. Several studies have demonstrated that this protein induces less platelet adhesion and activation than other plasma protein like fibrinogen or γ-globulin. Surface coating with this molecule is regarded to mask the device from the immune system and coagulation processes; however, this method has limitations due to conformational changes of the adsorbed molecules, exchange processes, physiological degradation of the protein, difficulties during the sterilization process, and risk of transmission of infection diseases due to the human origin of this molecule. Some investigations are directed to the covalent grafting of albumin onto surfaces, while others are based on surface modification with long aliphatic chains in order to enhance the affinity of albumin.

In a similar approach to surface masking with albumin, phosphorylcholine surfaces have been used in order to mimic the cell membrane. Several investigators have hypothesized that a surface similar to the external phospholipid membrane of cells should be nonthrombogenic. Since phosphorylcholine, the main lipid head group present on the external surface of blood cells, is inert in coagulation assays, it has been proposed for incorporation into polymeric surfaces in order to improve its hemocompatibility.

Planar-supported lipid bilayers composed of PC have shown to inhibit protein and cell adhesion in vivo. It has been proposed that this phenomenon is due to the zwitterionic nature of the phosphorylcholine head group that while carrying both positive and negative charges is electrically neutral at physiological pH. Application of supported lipid films as coatings for implantable devices has been limited by the inherent instability of a coating that is formed by individual molecules that self-assemble as a monolayer or bilayer film through relatively weak hydrophobic van der Waals interactions. Therefore, several approaches have been focused on the development of stable membrane-mimetic films through protein anchors, heat stabilization, and in situ polymerization of synthetically modified polymerizable phospholipids (Kazuhiko et al., 2000, Kobayashi et al., 2005, Xu et al., 2003, Yang et al., 2003). The protein and cell-resistant properties of the exposed PC layer were retained in all the studies. Jordan et al. (Jordan et al., 2006) described a polymerized membrane-mimetic film prepared by in situ photopolymerization of an acrylate-functionalized phospholipid assembly at a solid–liquid interface. These systems were applied as coating for small diameter vascular grafts and demonstrated a reduction in platelet adhesion using a baboon femoral arteriovenous shunt model.

A different approach is based on the direct chemical grafting of PC head group to metal or polymer surfaces (Chandy et al., 2000, Chang Chung et al., 2005, Feng et al., 2005, Huang et al., 2005, Nam et al., 2007).

7.2.4.3 Promotion of Endothelial Cells Growth

An intact endothelial cell layer is the most hemocompatible blood-contacting surface; therefore, stimulation of growth of this cell type onto a biomaterial surface allows the preparation of a permanently active hemocompatible surface. The underlying assumption of this strategy is that when cultured on artificial biomaterials, a confluent layer of endothelial cells maintain their nonthrombogenic phenotype. Endothelial cells inhibit thrombosis through three interconnected regulatory systems: at the coagulation cascade level, the cellular components of blood such as leukocytes and platelets, at the complement cascade level, and also through effects on fibrinolysis and vascular tone (Mcguigan and Sefton, 2007).

Combination of endothelial cells with biomaterials have been carried out in a great number of applications in order to prevent thrombotic or inflammatory reactions, or in other words, to improve the integration of the artificial device. For example, the lumen of vascular grafts has been seeded with endothelial cells as a strategy to create an interface that "hides" the biomaterial, enabling blood contact without significant inflammation and thrombosis, and therefore, maintaining graft patency (Heyligers et al., 2005, Kidd et al., 2003). A different approach consist of the fabrication of tissue-engineered constructs, in which blood must be supplied to cells within the construct without contacting the biomaterial scaffold. In this sense, different strategies have been developed in order to encourage blood vessel formation within the biomaterial scaffolds, involving the seeding of the construct with endothelial cells prior to implantation (Feinberg et al., 2009, Lu and Sipehia, 2001, Pawlowski et al., 2004) or the immobilization of biomolecules onto the biomaterial, such as specific antibodies (Aoki et al., 2005) or growth factors (Richardson et al., 2001) in order to encourage endothelial cell attachment.

7.2.5 Low Friction Polymer Drugs

Polymeric materials with slippery or low friction surfaces are valuable in biomedical technologies (Singer and Ollock, 1992). Lubricity is required for medical devices involving moving parts such as artificial joints as well as tubular devices, e.g., catheters or endoscopes that are inserted into blood vessels, urethra, or other parts of the body which have mucous membranes. Therefore, low frictional surface properties of these medical devices are desirable in order to reduce the pain accompanied by their introduction into the body and also to limit the risk of damage to the mucous membranes or the intima of blood vessels, which may lead to infectious diseases or mural thrombus formation (Nagaoka and Akashi, 1990). The use of lubricants to reduce friction and wear between rubbing surfaces has been documented since antiquity (Dowson, 1979, Tabor, 1973). In the last decades, even though many surface modification techniques have been developed in attempt to reduce the friction coefficients (μ) of characteristic medical device surfaces, only a few of them are applicable in the body environment (Figure 7.12) (Table 7.3).

Several methods of decreasing surface friction are known. End-grafting of hydrophilic polymer chains through surface-initiated polymerization of monomers has been extensively investigated as an effective means to impart hydrophilicity and lubricity to various polymer-based medical devices. Nagaoka and Akashi (Nagaoka and Akashi, 1990) described a methodology of binding N-vinyl pyrrolidone (VP) copolymers with epoxy groups, which were covalently coated on polyurethane catheter substrates. Uyama et al. also reported a methodology to prepare catheter devices possessing low μ values (Uyama et al., 1990, 1991). Surface modification of ethylene-vinyl acetate (EVA) copolymers and plasticized PVC, both used as catheter manufacturing polymers, through graft polymerization with nonionic water-soluble monomers such as acrylamide (AAm) and DMAA, was demonstrated useful to reduce their μ values in hydrate states to such an extent as becoming slippery.

2-Methacryloyloxyethyl phosphorylcholine (MPC)-based polymers constitute other example of a common biocompatible and hydrophilic polymers studied so far with high lubricity and low friction and antiprotein adsorption. Recently, Kyomoto and coworkers (Kyomoto et al., 2009) developed an artificial hip joint by using MCP grafted onto the surface of cross-linked PE (PMPC-g-$_{CL}$PE). This device was designed to reduce wear between the ultrahigh-molecular-weight PE component of the prosthesis and the metallic surface that articulates against, suppressing therefore the progressively bone resorption by osteolysis, leading to aseptic loosing of the artificial joint after a number of years, which is recognized as a serious problem (Sochart, 1999).

Ratner and Hoffman (1976) described a process to graft VP-based hydrogels onto organic polymeric substrates using UV radiation for forming biocompatible coatings. Fan and Lawrance (Fan and Marlin, 1994) developed a polymeric complex for forming biocompatible coatings, some of which could render the surface lubricious when exposed to aqueous or body fluids. Hu et al. achieved the same goal by plasma treatment only (Hu et al., 1989, 2007). The lubricous coating is composed of a polyelectrolyte molecular film, along with the hydrophilic, lubricant molecules on a plasma-treated plastic surface. Then the molecular film was further cross-linked with aldehyde-functionalized molecules such as Healon® to form an interpenetrating network, which directly adheres to polymeric matrices fully employed in the development of IOL cartridges (Hu et al., 2007). To minimize friction within the cartridge tip and ease IOL deployment, a lubricious coating is necessary (Kohnen and Kasper, 2005).

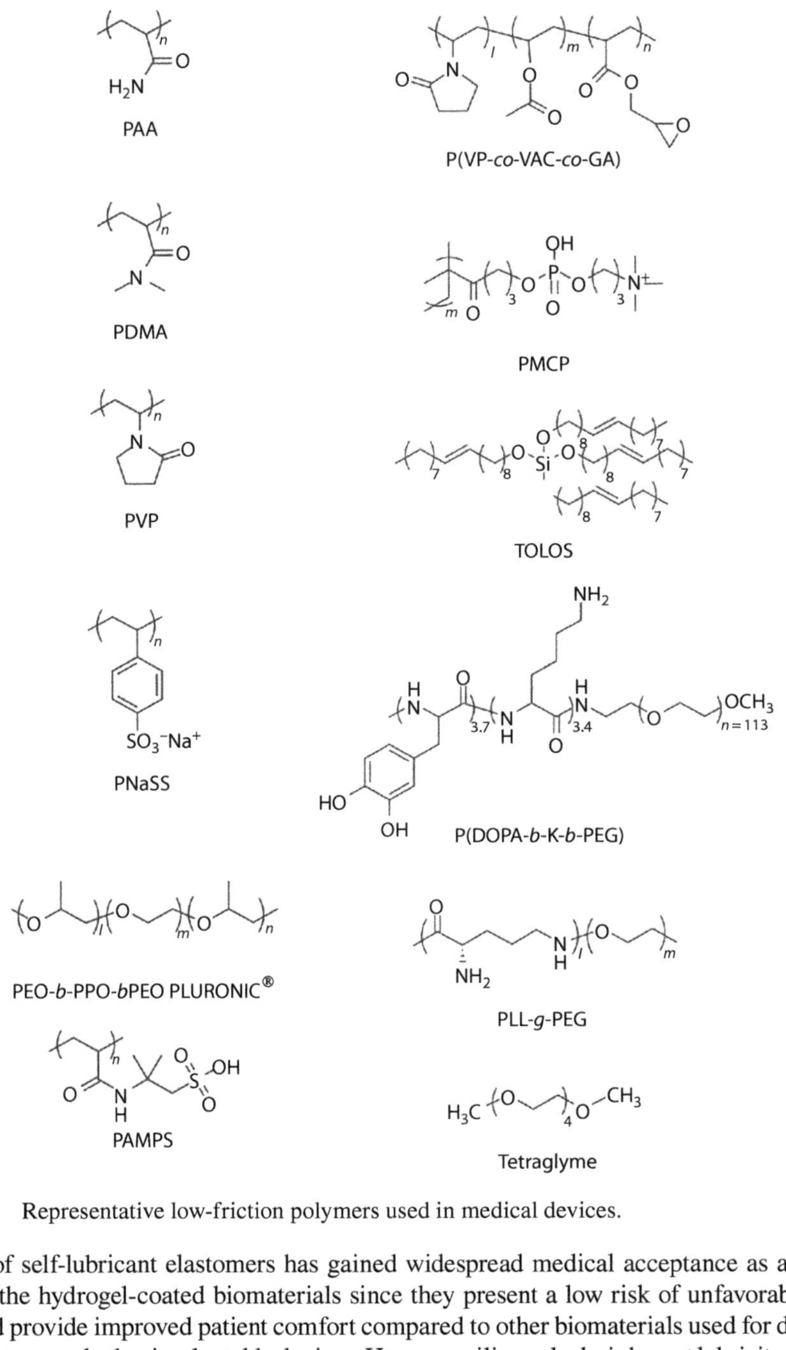

FIGURE 7.12 Representative low-friction polymers used in medical devices.

The use of self-lubricant elastomers has gained widespread medical acceptance as an alternative approach to the hydrogel-coated biomaterials since they present a low risk of unfavorable biological reactions and provide improved patient comfort compared to other biomaterials used for drug delivery, urinary catheters, and other implantable devices. However, silicone lacks inherent lubricity and has a relatively high μ. Woolfson et al. reported on the development of novel self-lubricating silicone elastomers produced by condensation of cure systems employing higher-molecular-weight tetra(alkyloxysilane) cross-linking agents (Woolfson et al., 2003). These silane cross-linkers are simple to synthesize in a one-step process from their low-molecular-weight propoxy analog. The tetraoleyloxysilane (TOLOS) derivative represents a successful example of this low toxicity lubricant agent with a μ near zero to be used in indwelling devices such as prostheses, contact lenses, intravaginal rings, and so on.

TABLE 7.3

Friction Coefficients (μ) Determined for Different Modified Polymeric Surfaces Commonly Used in Medical Devices

Polymeric Coating/ Polymeric Substrate	Coated Surface Friction Coefficient	Substrate Friction Coefficient
PVP/polyurethane	0.035	0.32
PDMA/poly(vinyl acetate)	<0.1	0.7
PMPC/polyethylene	0.026[a]	0.075[a]
TOLOS/tetrapropoxysilane	0.029	0.531
P(DOPA-*co*-K)PEG/PDMS	0.03	0.98
PEO-*b*-PPO-*b*-PEO/PDMS	0.05	0.90
PLL-*g*-PEG/PDMS	0.028	0.90
PNaSS/PDMS	10^{-4}	~1

[a] Dynamic friction coefficient.

Previous studies involving grafting of PEG chains onto elastomeric biomedical devices indicated a dramatic reduction of friction forces under aqueous conditions (Lee et al., 2004, Lee and Spencer, 2008b). Chawla and coworkers achieved the modification of poly(dimethylsiloxane) (PDMS) elastomeric substrates with enhanced tribological properties (Chawla et al., 2009). Synthetic PEG-based polymers containing L-3,4-dihydroxyphenylalanine (DOPA) and lysine (K) (DOPA-*b*-K-*b*-PEG) were grafted through noncovalent interactions between hydrophobic anchoring groups of PEG-based copolymers and the PDMS surfaces in aqueous media.

A very recent alternative to enhance the tribological properties of medical devices consists in tethering polymer chains to the surfaces by one end which act as molecular "brushes" and facilitate sliding when they are swollen by aqueous solvents. This biomimetic approach is based on the architecture of biological lubricant additives, usually glycoproteins, in which large number of sugar chains are bound along a protein backbone (Figure 7.13). For example, mucins are found in most parts of the human body that need lubricating, such as knees and eyes (Bansil et al., 1995). The characteristic bottlebrush structure of these supramolecular complexes plays a crucial role in the mechanism of aid lubrication. The hydrophilic sugars immobilize large amounts of water within the contact region, while the backbone interconnects to other bottlebrushes or to a surface (Lee and Spencer, 2008a). Other important characteristic of natural tribological systems is that

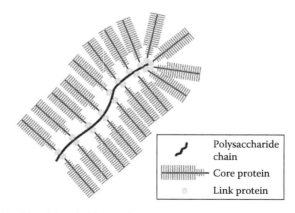

FIGURE 7.13 Hierarchical bottlebrush like architecture of biolubricating systems.

they usually involve soft surfaces and deform elastically in response to external loads, increasing the contact area resulting in a relatively low contact pressure. The synergic combination of hairy polymers and soft surfaces has been extensively studied and had led to applications in biomedical implants. Lee and Spencer recently reported enhanced water lubrication of brush-like grafted PEO-*block*-PEO-*block*-PEO (PEO-*b*-PPO-*b*-PEO, Pluronic®) and PLL-*graft*-PEG (PLL-*g*-PEG) onto silicone–rubber surfaces (Lee and Spencer, 2005), while Gong (2006) have gone a step further in biomimicry by using polyelectrolyte brushes hydrogels based on poly(sodium 4-styrene sulfonate) (PNaSS) for aqueous lubrication. These hieratical systems lead to superior lubrication properties acting in the same way than biolubricating systems (Raviv et al., 2003), where the charged hydrophilic surfaces provide an electrostatic double-layer repulsion, in addition to the steric repulsion of any protruding polyelectrolytes and the hydration layer of tightly bound water molecules (Urbakh et al., 2004).

REFERENCES

2002. TC 194 Biological Evaluation of Medical Devices, ISO 10993-3:2002, Biological Evaluation of Medical Devices—Part 4: Selection of Tests for Interactions with Blood, International Organization for Standardisation.

≤Http://Www.Who.Int/Cancer/En≥., W. H. O. C. 2009.

Agueros, M., Zabaleta, V., Espuelas, S., Campanero, M. A., and Irache, J. M. 2010. Increased oral bioavailability of paclitaxel by its encapsulation through complex formation with cyclodextrins in poly(anhydride) nanoparticles. *Journal of Controlled Release*, 145: 2–8.

Aguilar, M. R., Rodriguez, G., Fernandez, M., Gallardo, A., and San Roman, J. 2002. Polymeric active coatings with functionality in vascular applications. *Journal of Materials Science: Materials in Medicine*, 13: 1099–1104.

Al-Jamal, K. T., Al-Jamal, W. T., Akerman, S. et al. 2010. Systemic antiangiogenic activity of cationic poly-L-lysine dendrimer delays tumor growth. *Proceedings of the National Academy of Sciences of the United States of America*, 107: 3966–3971.

Aldenhoff, Y. B. J., Blezer, R., Lindhout, T., and Koole, L. H. 1997. Photo-immobilization of dipyridamole (Persantin®) at the surface of polyurethane biomaterials: Reduction of in vitro thrombogenicity. *Biomaterials*, 18: 167–172.

Aldenhoff, Y. B. J., Koole, L. H., Curtis, A., and Descouts, P. 2003. Platelet adhesion studies on dipyridamole coated polyurethane surfaces. *European Cells and Materials*, 5: 61–67.

Alferiev, I. S., Connolly, J. M., Stachelek, S. J. et al. 2006. Surface heparinization of polyurethane via bromoalkylation of hard segment nitrogens. *Biomacromolecules*, 7: 317–322.

Allen, T. M. 2002. Ligand-targeted therapeutics in anticancer therapy. *Nature Reviews Cancer*, 2: 750–763.

Allen, M. J., Morby, A. P., and White, G. F. 2004. Cooperativity in the binding of the cationic biocide polyhexamethylene biguanide to nucleic acids. *Biochemical and Biophysical Research Communications*, 318: 397–404.

Amaral, I. F., Unger, R. E., Fuchs, S. et al. 2009. Fibronectin-mediated endothelialisation of chitosan porous matrices. *Biomaterials*, 30: 5465–5475.

Anand, P., Nair, H. B., Sung, B. et al. 2010. Design of curcumin-loaded PLGA nanoparticles formulation with enhanced cellular uptake, and increased bioactivity in vitro and superior bioavailability in vivo. *Biochemical Pharmacology*, 79: 330–338.

Andhariya, N., Chudasama, B., Mehta, R. V., and Upadhyay, R. V. 2010. Biodegradable thermoresponsive polymeric magnetic nanoparticles: A new drug delivery platform for doxorubicin. *Journal of Nanoparticle Research*, 12: 1–12.

Aoki, J., Serruys, P. W., Van Beusekom, H. et al. 2005. Endothelial progenitor cell capture by stents coated with antibody against CD34: The HEALING-FIM (Healthy Endothelial Accelerated Lining Inhibits Neointimal Growth-First in Man) registry. *Journal of the American College of Cardiology*, 45: 1574–1579.

Arbiser, J. L. 1996. Angiogenesis and the skin: A primer. *Journal of the American Academy of Dermatology*, 34: 486–497.

Arica, M. Y., Bayramoglu, G., Arica, B. et al. 2005. Novel hydrogel membrane based on copoly(hydroxyethyl methacrylate/*p*-vinylbenzyl-poly(ethylene oxide)) for biomedical applications: Properties and drug release characteristics. *Macromolecular Bioscience*, 5: 983–992.

Arica, M. Y., Tuglu, D., Basar, M. M. et al. 2008. Preparation and characterization of infection-resistant anti-biotics-releasing hydrogels rods of poly[hydroxyethyl methacrylate-*co*-poly(ethylene glycol)-methacry-late]: Biomedical application in a novel rabbit penile prosthesis model. *Journal of Biomedical Materials Research—Part B Applied Biomaterials*, 86: 18–28.

Arifin, D. R. and Palmer, A. F. 2005. Polymersome encapsulated hemoglobin: A novel type of oxygen carrier. *Biomacromolecules*, 6: 2172–2181.

Armstrong, J. K., Hempel, G., Koling, S. et al. 2007. Antibody against poly(ethylene glycol) adversely affects PEG-asparaginase therapy in acute lymphoblastic leukemia patients. *Cancer*, 110: 103–111.

Atkins, J. H. and Gershell, L. J. 2002. Selective anticancer drugs. Market indicators. *Nature Reviews Drug Discovery*, 1: 491–492.

Auzenne, E., Donato, N. J., Li, C. et al. 2002. Superior therapeutic profile of poly-L-glutamic acid–paclitaxel copolymer compared with Taxol in xenogeneic compartmental models of human ovarian carcinoma. *Clinical Cancer Research*, 8: 573–581.

Bae, K. H., Choi, S. H., Park, S. Y., Lee, Y., and Park, T. G. 2006. Thermosensitive Pluronic micelles stabilized by shell cross-linking with gold nanoparticles. *Langmuir*, 22: 6380–6384.

Bae, Y., Jang, W. D., Nishiyama, N., Fukushima, S., and Kataoka, K. 2005a. Multifunctional polymeric micelles with folate-mediated cancer cell targeting and pH-triggered drug releasing properties for active intracel-lular drug delivery. *Molecular Biosystems*, 1: 242–250.

Bae, Y., Nishiyama, N., Fukushima, S. et al. 2005b. Preparation and biological characterization of polymeric micelle drug carriers with intracellular pH-triggered drug release property: Tumor permeability, con-trolled subcellular drug distribution, and enhanced in vivo antitumor efficacy. *Bioconjugate Chemistry*, 16: 122–130.

Bae, Y., Nishiyama, N., and Kataoka, K. 2007. In vivo antitumor activity of the folate-conjugated pH-sensitive polymeric micelle selectively releasing adriamycin in the intracellular acidic compartments. *Bioconjugate Chemistry*, 18: 1131–1139.

Bagheri, M., Beyermann, M., and Dathe, M. 2009. Immobilization reduces the activity of surface-bound cat-ionic antimicrobial peptides with no influence upon the activity spectrum. *Antimicrobial Agents and Chemotherapy*, 53: 1132–1141.

Bansil, R., Stanley, E., and Lamont, J. T. 1995. Mucin biophysics. *Annual Review of Physiology*, 57: 635–657.

Bareford, L. M. and Swaan, P. W. 2007. Endocytic mechanisms for targeted drug delivery. *Advanced Drug Delivery Reviews*, 59: 748–758.

Bellocq, N. C., Kang, D. W., Wang, X. et al. 2004. Synthetic biocompatible cyclodextrin-based constructs for local gene delivery to improve cutaneous wound healing. *Bioconjugate Chemistry*, 15: 1201–1211.

Bellocq, N. C., Pun, S. H., Jensen, G. S., and Davis, M. E. 2003. Transferrin-containing, cyclodextrin polymer-based particles for tumor-targeted gene delivery. *Bioconjugate Chemistry*, 14: 1122–1132.

Benns, J. M., Choi, J. S., Mahato, R. I., Park, J. S., and Sung Wan, K. 2000. pH-Sensitive cationic polymer gene delivery vehicle: *N*-Ac-poly(L-histidine)-*graft*-poly(L-lysine) comb shaped polymer. *Bioconjugate Chemistry*, 11: 637–645.

Benns, J. M., Mahato, R. I., and Kim, S. W. 2002. Optimization of factors influencing the transfection effi-ciency of folate-PEG-folate-*graft*-polyethylenimine. *Journal of Controlled Release*, 79: 255–269.

Bhatt, R., de Vries, P., Tulinsky, J. et al. 2003. Synthesis and in vivo antitumor activity of poly (l-glutamic acid) conjugates of 20S-camptothecin. *J Med Chem*, 46: 190–193.

Bissett, D., Cassidy, J., De Bono, J. S. et al. 2004. Phase I and pharmacokinetic (PK) study of MAG-CPT (PNU 166148): A polymeric derivative of camptothecin (CPT). *British Journal of Cancer*, 91: 50–55.

Boneca, I. G. and Chiosis, G. 2003. Vancomycin resistance: Occurrence, mechanisms and strategies to combat it. *Expert Opinion on Therapeutic Targets*, 7: 311–328.

Boura, C., Menu, P., Payan, E. et al. 2003. Endothelial cells grown on thin polyelectrolyte multilayered films: An evaluation of a new versatile surface modification. *Biomaterials*, 24: 3521–3530.

Boussif, O., Lezoualc'h, F., Zanta, M. A. et al. 1995. A versatile vector for gene and oligonucleotide transfer into cells in culture and in vivo: Polyethylenimine. *Proceedings of the National Academy of Sciences of the United States of America*, 92: 7297–7301.

Braunova, A., Pechar, M., Laga, R., and Ulbrich, K. 2007. Hydrolytically and reductively degradable high-molecular-weight poly(ethylene glycol)s. *Macromolecular Chemistry and Physics*, 208: 2642–2653.

Brogden, K. A. 2005. Antimicrobial peptides: Pore formers or metabolic inhibitors in bacteria? *Nature Reviews Microbiology*, 3: 238–250.

Broxton, P., Woodcock, P. M., and Gilbert, P. 1983. A study of the antibacterial activity of some polyhexameth-ylene biguanides towards *Escherichia coli* ATCC 8739. *Journal of Applied Bacteriology*, 54: 345–353.

Brynda, E. and Houska, M. 1998. Preparation of organized protein multilayers. *Macromolecular Rapid Communications*, 19: 173–176.

Brynda, E., Houska, M., Jirouskova, M., and Dyr, J. E. 2000. Albumin and heparin multilayer coatings for blood-contacting medical devices. *Journal of Biomedical Materials Research*, 51: 249–257.

Caliceti, P. and Veronese, F. M. 2003. Pharmacokinetic and biodistribution properties of poly(ethylene glycol)–protein conjugates. *Advanced Drug Delivery Reviews*, 55: 1261–1277.

Calonder, C., Matthew, H. W. T., and Van Tassel, P. R. 2005. Adsorbed layers of oriented fibronectin: A strategy to control cell-surface interactions. *Journal of Biomedical Materials Research—Part A*, 75: 316–323.

Campeau, P., Chapdelaine, P., Seigneurin-Venin, S., Massie, B., and Tremblay, J. P. 2001. Transfection of large plasmids in primary human myoblasts. *Gene Therapy*, 8: 1387–1394.

Cao, L., Chang, M., Lee, C. Y. et al. 2007. Plasma-deposited tetraglyme surfaces greatly reduce total blood protein adsorption, contact activation, platelet adhesion, platelet procoagulant activity, and in vitro thrombus deposition. *Journal of Biomedical Materials Research—Part A*, 81: 827–837.

Chabut, D., Fischer, A.-M., Colliec-Jouault, S. et al. 2003. Low molecular weight fucoidan and heparin enhance the basic fibroblast growth factor-induced tube formation of endothelial cells through heparan sulfate-dependent a6 overexpression. *Molecular Pharmacology*, 64: 696–702.

Chadna, P., Saad, M., Wang, Y. et al. 2008. A novel targeted proapoptotic drug delivery system for efficient anticancer therapy. *Proceedings 35th Annual Meeting of the Controlled Release Society*, New York.

Chandy, T., Das, G. S., Wilson, R. F., and Rao, G. H. R. 2000. Use of plasma glow for surface-engineering biomolecules to enhance blood compatibility of Dacron and PTFE vascular prosthesis. *Biomaterials*, 21: 699–712.

Chang Chung, Y., Hong Chiu, Y., Wei Wu, Y., and Tai Tao, Y. 2005. Self-assembled biomimetic monolayers using phospholipid-containing disulfides. *Biomaterials*, 26: 2313–2324.

Chawla, K., Lee, S., Lee, B. P. et al. 2009. A novel low-friction surface for biomedical applications: Modification of poly(dimethylsiloxane) (PDMS) with polyethylene glycol (PEG)–DOPA–lysine. *Journal of Biomedical Materials Research—Part A*, 90A: 742–749.

Chen, H., Chen, Y., Sheardown, H., and Brook, M. A. 2005b. Immobilization of heparin on a silicone surface through a heterobifunctional PEG spacer. *Biomaterials*, 26: 7418–7424.

Chen, W., Meng, F., Cheng, R., and Zhong, Z. 2010. pH-Sensitive degradable polymersomes for triggered release of anticancer drugs: A comparative study with micelles. *Journal of Controlled Release*, 142: 40–46.

Chen, W., Meng, F., Li, F., Ji, S. J., and Zhong, Z. 2009b. pH-responsive biodegradable micelles based on acid-labile polycarbonate hydrophobe: Synthesis and triggered drug release. *Biomacromolecules*, 10: 1727–1735.

Chen, M. C., Wong, H. S., Lin, K. J. et al. 2009a. The characteristics, biodistribution and bioavailability of a chitosan-based nanoparticulate system for the oral delivery of heparin. *Biomaterials*, 30: 6629–6637.

Chen, H., Zhang, Z., Chen, Y., Brook, M. A., and Sheardown, H. 2005a. Protein repellant silicone surfaces by covalent immobilization of poly(ethylene oxide). *Biomaterials*, 26: 2391–2399.

Cheng, P. N. M., Lam, T. L., Lam, W. M. et al. 2007. Pegylated recombinant human arginase (rhArg-peg5,000mw) inhibits the in vitro and in vivo proliferation of human hepatocellular carcinoma through arginine depletion. *Cancer Research*, 67: 309–317.

Cheng, G., Xue, H., Zhang, Z., Chen, S., and Jiang, S. 2008. A switchable biocompatible polymer surface with self-sterilizing and nonfouling capabilities. *Angewandte Chemie—International Edition*, 47: 8831–8834.

Cho, Y. W., Cho, Y. N., Chung, S. H., Yoo, G., and Ko, S. W. 1999. Water-soluble chitin as a wound healing accelerator. *Biomaterials*, 20: 2139–2145.

Choe, Y. H., Conover, C. D., Wu, D. et al. 2002. Anticancer drug delivery systems: Multi-loaded N4-acyl poly(ethylene glycol) prodrugs of ara-C. II. Efficacy in ascites and solid tumors. *Journal of Controlled Release*, 79: 55–70.

Choi, J. S., Nam, K., Park, J. Y. et al. 2004. Enhanced transfection efficiency of PAMAM dendrimer by surface modification with L-arginine. *Journal of Controlled Release*, 99: 445–456.

Chytil, P., Etrych, T., Konak, C. et al. 2008. New HPMA copolymer-based drug carriers with covalently bound hydrophobic substituents for solid tumour targeting. *Journal of Controlled Release*, 127: 121–130.

Collins, L. and Fabre, J. W. 2004. A synthetic peptide vector system for optimal gene delivery to corneal endothelium. *Journal of Gene Medicine*, 6: 185–194.

Danhier, F., Vroman, B., Lecouturier, N. et al. 2009. Targeting of tumor endothelium by RGD-grafted PLGA-nanoparticles loaded with paclitaxel. *Journal of Controlled Release*, 140: 166–173.

Danhauser-Riedl, S., Hausmann, E., Schick, H. et al. 1993. Phase I clinical and pharmacokinetic trial of dextran conjugated doxorubicin (AD-70, doxorubicin-OXD). *Invest New Drugs*, 11: 187–195.

Davis, F. F. 2002. The origin of pegnology. *Advanced Drug Delivery Reviews*, 54: 457–458.

Davis, M. E. 2009. Design and development of IT-101, a cyclodextrin-containing polymer conjugate of camptothecin. *Advanced Drug Delivery Reviews*, 61: 1189–1192.

De Duve, C., De Barsy, T., and Poole, B. 1974. Lysosomotropic agents. *Biochemical Pharmacology*, 23: 2495–2531.

De Wolf, H. K., De Raad, M., Snel, C. et al. 2007. Biodegradable poly(2-dimethylamino ethylamino)phosphazene for in vivo gene delivery to tumor cells. Effect of polymer molecular weight. *Pharmaceutical Research*, 24: 1572–1580.

De Wolf, H. K., Luten, J., Snel, C. J., Storm, G., and Hennink, W. E. 2008. Biodegradable, cationic methacrylamide-based polymers for gene delivery to ovarian cancer cells in mice. *Molecular Pharmaceutics*, 5: 349–357.

Deming, T. J. 1999. Mussel byssus and biomolecular materials. *Current Opinion in Chemical Biology*, 3: 100–105.

Desmet, T., Morent, R., De Geyter, N. et al. 2009. Nonthermal plasma technology as a versatile strategy for polymeric biomaterials surface modification: A review. *Biomacromolecules*, 10: 2351–2378.

Dhanikula, R. S., Argaw, A., Bouchard, J. F., and Hildgen, P. 2008. Methotrexate loaded polyether–copolyester dendrimers for the treatment of gliomas: Enhanced efficacy and intratumoral transport capability. *Molecular Pharmaceutics*, 5: 105–116.

Dizman, B., Elasri, M. O., and Mathias, L. J. 2005. Synthesis, characterization, and antibacterial activities of novel methacrylate polymers containing norfloxacin. *Biomacromolecules*, 6: 514–520.

Dowson, D. (Ed.). 1979. *History of Tribology*. Longmans, London U.K.

Duncan, R. 1999. Polymer conjugates for tumour targeting and intracytoplasmic delivery. The EPR effect as a common gateway? *Pharmaceutical Science and Technology Today*, 2: 441–449.

Duncan, R. 2003. The dawning era of polymer therapeutics. *Nature Reviews Drug Discovery*, 2: 347–360.

Duncan, R. 2006. Polymer conjugates as anticancer nanomedicines. *Nature Reviews Cancer*, 6: 688–701.

Duncan, R. 2009. Development of HPMA copolymer–anticancer conjugates: Clinical experience and lessons learnt. *Advanced Drug Delivery Reviews*, 61: 1131–1148.

Duncan, R., Gac-Breton, S., Keane, R. et al. 2001. Polymer–drug conjugates, PDEPT and PELT: Basic principles for design and transfer from the laboratory to clinic. *Journal of Controlled Release*, 74: 135–146.

Duncan, R., Gilbert, H. R. P., Carbajo, R. J., and Vicent, M. J. 2008. Polymer masked-unmasked protein therapy. 1. Bioresponsive dextrin–trypsin and –melanocyte stimulating hormone conjugates designed for a-amylase activation. *Biomacromolecules*, 9: 1146–1154.

Duncan, R., Seymour, L. W., O'Hare, K. B. et al. 1992. Preclinical evaluation of polymer-bound doxorubicin. *Journal of Controlled Release*, 19: 331–346.

Duncan, R., Seymour, L. C. W., and Scarlett, L. 1986. Fate of *N*-(2-hydroxypropyl)methacrylamide copolymers with pendent galactosamine residues after intravenous administration to rats. *Biochimica et Biophysica Acta—General Subjects*, 880: 62–71.

Edmondson, S., Osborne, V. L., and Huck, W. T. S. 2004. Polymer brushes via surface-initiated polymerizations. *Chemical Society Reviews*, 33: 14–22.

Epand, R. F., Schmitt, M. A., Gellman, S. H., and Epand, R. M. 2006. Role of membrane lipids in the mechanism of bacterial species selective toxicity by two α/β-antimicrobial peptides. *Biochimica et Biophysica Acta—Biomembranes*, 1758: 1343–1350.

Erez, R., Segal, E., Miller, K., Satchi-Fainaro, R., and Shabat, D. 2009. Enhanced cytotoxicity of a polymer–drug conjugate with triple payload of paclitaxel. *Bioorganic and Medicinal Chemistry*, 17: 4327–4335.

Etrych, T., Chytil, P., Mrkvan, T. et al. 2008. Conjugates of doxorubicin with graft HPMA copolymers for passive tumor targeting. *Journal of Controlled Release*, 132: 184–192.

Fan, Y.-L. and Marlin, L. 1994. *Biocompatible Hydrophilic Complexes and Process for Preparation and Use.* Union Carbide Chemicals & Plastics Technology Corporation, Danbury, CT.

Fang, J., Sawa, T., Akaike, T., Greish, K., and Maeda, H. 2004. Enhancement of chemotherapeutic response of tumor cells by a heme oxygenase inhibitor, pegylated zinc protoporphyrin. *International Journal of Cancer*, 109: 1–8.

Farokhzad, O. C., Cheng, J., Teply, B. A. et al. 2006. Targeted nanoparticle–aptamer bioconjugates for cancer chemotherapy in vivo. *Proceedings of the National Academy of Sciences of the United States of America*, 103: 6315–6320.

Farokhzad, O. C., Jon, S., Khademhosseini, A. et al. 2004. Nanoparticle–aptamer bioconjugates: A new approach for targeting prostate cancer cells. *Cancer Research*, 64: 7668–7672.

Feinberg, A. W., Schumacher, J. F., and Brennan, A. B. 2009. Engineering high-density endothelial cell monolayers on soft substrates. *Acta Biomaterialia*, 5: 2013–2024.

Feng, W., Zhu, S., Ishihara, K., and Brash, J. L. 2005. Adsorption of fibrinogen and lysozyme on silicon grafted with poly(2-methacryloyloxyethyl phosphorylcholine) via surface-initiated atom transfer radical polymerization. *Langmuir*, 21: 5980–5987.

Ferguson, E. L. and Duncan, R. 2009. Dextrin–phospholipase A2: Synthesis and evaluation as a bioresponsive anticancer conjugate. *Biomacromolecules*, 10: 1358–1364.

Fischer, D., Bieber, T., Li, Y., Elsasser, H. P., and Kissel, T. 1999. A novel non-viral vector for DNA delivery based on low molecular weight, branched polyethylenimine: Effect of molecular weight on transfection efficiency and cytotoxicity. *Pharmaceutical Research*, 16: 1273–1279.

Folkman, J. 1990. What is the evidence that tumors are angiogenesis dependent? *Journal of the National Cancer Institute*, 82: 4–6.

Forrest, M. L., Won, C. Y., Malick, A. W., and Kwon, G. S. 2006a. In vitro release of the mTOR inhibitor rapamycin from poly(ethylene glycol)-*b*-poly(ε-caprolactone) micelles. *Journal of Controlled Release*, 110: 370–377.

Forrest, M. L., Zhao, A., Won, C. Y., Malick, A. W., and Kwon, G. S. 2006b. Lipophilic prodrugs of Hsp90 inhibitor geldanamycin for nanoencapsulation in poly(ethylene glycol)-*b*-poly(e-caprolactone) micelles. *Journal of Controlled Release*, 116: 139–149.

Fosse, E., Thelin, S., Svennevig, J. L. et al. 1997. Duraflo II coating of cardiopulmonary bypass circuits reduces complement activation, but does not affect the release of granulocyte enzymes in fully heparinized patients: A European multicentre study. *European Journal of Cardio-thoracic Surgery*, 11: 320–327.

Fushimi, F., Nakayama, M., Nishimura, K., and Hiyoshi, T. 1998. Platelet adhesion, contact phase coagulation activation, and C5a generation of polyethylene glycol acid-grafted high flux cellulosic membrane with varieties of grafting amounts. *Artificial Organs*, 22: 821–826.

Futaki, S., Suzuki, T., Ohashi, W. et al. 2001. Arginine-rich peptides. An abundant source of membrane-permeable peptides having potential as carriers for intracellular protein delivery. *Journal of Biological Chemistry*, 276: 5836–5840.

Gabriel, G. J., Pool, J. G., Som, A. et al. 2008. Interactions between antimicrobial polynorbornenes and phospholipid vesicles monitored by light scattering and microcalorimetry. *Langmuir*, 24: 12489–12495.

Gabriel, G. J., Som, A., Madkour, A. E., Eren, T., and Tew, G. N. 2007. Infectious disease: Connecting innate immunity to biocidal polymers. *Materials Science and Engineering R: Reports*, 57: 28–64.

Gademann, K., Kobylinska, J., Wach, J. Y., and Woods, T. M. 2009. Surface modifications based on the cyanobacterial siderophore anachelin: From structure to functional biomaterials design. *BioMetals*, 22: 595–604.

Gallardo, A., Rodriguez, G., Aguilar, M. R., Fernandez, M., and San Roman, J. 2003. A kinetic model to explain the zero-order release of drugs from ionic polymer drug conjugates: Application to AMPS-triflusal-derived polymeric drugs. *Macromolecules*, 36: 8876–8880.

Gallardo, A., Rodriguez, G., Fernandez, M., Aguilar, M. R., and San Roman, J. 2004. Polymeric drugs with prolonged sustained delivery of specific anti-aggregant agents for platelets: Kinetic analysis of the release mechanism. *Journal of Biomaterials Science, Polymer Edition*, 15: 917–928.

Garcia-Fernandez, L., Aguilar, M. R., Fernandez, M. M. et al. 2010a. Structure, morphology, and bioactivity of biocompatible systems derived from functionalized acrylic polymers based on 5-amino-2-naphthalene sulfonic acid. *Biomacromolecules*, 11: 1763–1772.

Garcia-Fernandez, L., Halstenberg, S., Unger, R. E. et al. 2010b. Anti-angiogenic activity of heparin-like poly-sulfonated polymeric drugs in 3D human cell culture. *Biomaterials*, 31: 7863–7872.

Gary, D. J., Puri, N., and Won, Y. Y. 2007. Polymer-based siRNA delivery: Perspectives on the fundamental and phenomenological distinctions from polymer-based DNA delivery. *Journal of Controlled Release*, 121: 64–73.

Gilbert, P. and Moore, L. E. 2005. Cationic antiseptics: Diversity of action under a common epithet. *Journal of Applied Microbiology*, 99: 703–715.

Glukhov, E., Stark, M., Burrows, L. L., and Deber, C. M. 2005. Basis for selectivity of cationic antimicrobial peptides for bacterial versus mammalian membranes. *Journal of Biological Chemistry*, 280: 33960–33967.

Gong, J. P. 2006. Friction and lubrication of hydrogels—Its richness and complexity. *Soft Matter*, 2: 544–552.

Gorbet, M. B. and Sefton, M. V. 2001. Leukocyte activation and leukocyte procoagulant activities after blood contact with polystyrene and polyethylene glycol-immobilized polystyrene beads. *Journal of Laboratory and Clinical Medicine*, 137: 345–355.

Gorbet, M. B. and Sefton, M. V. 2004. Biomaterial-associated thrombosis: Roles of coagulation factors, complement, platelets and leukocytes. *Biomaterials*, 25: 5681–5703.

Gott, V. L., Whiffen, J. D., and Dutton, R. C. 1963. Heparin bonding on colloidal graphite surfaces. *Science*, 142: 1299–1300.

Greco, F. and Vicent, M. J. 2009. Combination therapy: Opportunities and challenges for polymer–drug conjugates as anticancer nanomedicines. *Advanced Drug Delivery Reviews*, 61: 1203–1213.

Guo, K., Li, J., Tang, J. P., Tan, C. P., Wang, H., and Zeng, Q. 2008. Monoclonal antibodies target intracellular PRL phosphatases to inhibit cancer metastases in mice. *Cancer Biology and Therapy*, 7: 752–759.

Haag, R. and Kratz, F. 2006. Polymer therapeutics: Concepts and applications. *Angewandte Chemie—International Edition*, 45: 1198–1215.

Hancock, R. E. W. and Rozek, A. 2002. Role of membranes in the activities of antimicrobial cationic peptides. *FEMS Microbiology Letters*, 206: 143–149.

Hansson, K. M., Tosatti, S., Isaksson, J. et al. 2005. Whole blood coagulation on protein adsorption-resistant PEG and peptide functionalised PEG-coated titanium surfaces. *Biomaterials*, 26: 861–872.

Haque, T., Chen, H., Ouyang, W. et al. 2005. Investigation of a new microcapsule membrane combining alginate, chitosan, polyethylene glycol and poly-L-lysine for cell transplantation applications. *International Journal of Artificial Organs*, 28: 631–637.

Henriques, S. T., Melo, M. N., and Castanho, M. A. R. B. 2006. Cell-penetrating peptides and antimicrobial peptides: How different are they? *Biochemical Journal*, 399: 1–7.

Hetrick, E. M. and Schoenfisch, M. H. 2006. Reducing implant-related infections: Active release strategies. *Chemical Society Reviews*, 35: 780–789.

Heyligers, J. M. M., Arts, C. H. P., Verhagen, H. J. M., De Groot, P. G., and Moll, F. L. 2005. Improving small-diameter vascular grafts: From the application of an endothelial cell lining to the construction of a tissue-engineered blood vessel. *Annals of Vascular Surgery*, 19: 448–456.

Hongrapipat, J., Kopeckova, P., Liu, J., Prakongpan, S., and Kopecek, J. 2008. Combination chemotherapy and photodynamic therapy with Fab′ fragment targeted HPMA copolymer conjugates in human ovarian carcinoma cells. *Molecular Pharmacology*, 5: 696–709.

Houska, M., Brynda, E., Solovyev, A. et al. 2008. Hemocompatible albumin–heparin coatings prepared by the layer-by-layer technique. The effect of layer ordering on thrombin inhibition and platelet adhesion. *Journal of Biomedical Materials Research—Part A*, 86: 769–778.

Howard, K. A., Rahbek, U. L., Liu, X. et al. 2006. RNA interference in vitro and in vivo using a novel chitosan/sirna nanoparticle system. *Molecular Therapy*, 14: 476–484.

Howse, J. R., Jones, R. A. L., Battaglia, G. et al. 2009. Templated formation of giant polymer vesicles with controlled size distributions. *Nature Materials*, 8: 507–511.

Hu, C. B., Gwon, A., Lowery, M., Makker, H., and Gruber, L. 2007. Preparation and evaluation of a lubricious treated cartridge used for implantation of intraocular lenses. *Journal of Biomaterials Science, Polymer Edition*, 18: 179–191.

Hu, C. B., Solomon, D. D., and Williamitis, V. A. 1989. *Method for Preparing Lubricated Surfaces*. Becton, Dickinson and Company, Franklin Lakes, NJ.

Huang, J., Murata, H., Koepsel, R. R., Russell, A. J., and Matyjaszewski, K. 2007. Antibacterial polypropylene via surface-initiated atom transfer radical polymerization. *Biomacromolecules*, 8: 1396–1399.

Huang, X. J., Xu, Z. K., Wan, L. S., Wang, Z. G., and Wang, J. L. 2005. Surface modification of polyacrylonitrile-based membranes by chemical reactions to generate phospholipid moieties. *Langmuir*, 21: 2941–2947.

Huang Ly, Y. M. 2006. Hemocompatibility of layer-by-layer hyaluronic acid/heparin nanostructure coating on stainless steel for cardiovascular stents and its use for drug delivery. *Journal of Nanoscience and Nanotechnology*, 6: 3163–3170.

Ihre, H. R., Padilla De Jesus, O. L., Szoka Jr, F. C., and Frechet, J. M. J. 2002. Polyester dendritic systems for drug delivery applications: Design, synthesis, and characterization. *Bioconjugate Chemistry*, 13: 443–452.

Imazato, S., Ebi, N., Tarumi, H. et al. 1999. Bactericidal activity and cytotoxity of antibacterial monomer MDPB. *Biomaterials*, 20: 899–903.

Iozzo, R. V., Zoeller, J. J., and Nyström, A. 2009. Basement membrane proteoglycans: Modulators Par Excellence of cancer growth and angiogenesis. *Molecules and Cells*, 27: 503–513.

Iwasaki, Y., Takamiya, M., Iwata, R., Yusa, S. I., and Akiyoshi, K. 2007. Surface modification with well-defined biocompatible triblock copolymers. Improvement of biointerfacial phenomena on a poly(dimethylsiloxane) surface. *Colloids and Surfaces B: Biointerfaces*, 57: 226–236.

Jackson, C. J., Giles, I., Knop, A., Nethery, A., and Schrieber, L. 1994. Sulfated polysaccharides are required for collagen-induced vascular tube formation. *Experimental Cell Research*, 215: 294–302.

Jain, N. K. and Jain, S. K. 2010. Development and in vitro characterization of galactosylated low molecular weight chitosan nanoparticles bearing doxorubicin. *AAPS PharmSciTech*, 11: 686–697.

Johnson, D. E. and Williams, L. T. 1993. Structural and functional diversity in the FGF receptor multigene family. *Advances in Cancer Research*, 60: 1–41.

Jordan, S. W. and Chaikof, E. L. 2007. Novel thromboresistant materials. *Journal of Vascular Surgery*, 45: 104–115.

Jordan, S. W., Faucher, K. M., Caves, J. M. et al. 2006. Fabrication of a phospholipid membrane-mimetic film on the luminal surface of an ePTFE vascular graft. *Biomaterials*, 27: 3473–3481.

Kailasan, A., Yuan, Q., and Yang, H. 2010. Synthesis and characterization of thermoresponsive polyamidoamine–polyethylene glycol–poly(D,L-lactide) core–shell nanoparticles. *Acta Biomaterialia*, 6: 1131–1139.

Kainthan, R. K., Gnanamani, M., Ganguli, M. et al. 2006. Blood compatibility of novel water soluble hyperbranched polyglycerol-based multivalent cationic polymers and their interaction with DNA. *Biomaterials*, 27: 5377–5390.

Kaminskas, L. M., Boyd, B. J., Karellas, P. et al. 2008. The impact of molecular weight and PEG chain length on the systemic pharmacokinetics of pegylated poly L-lysine dendrimers. *Molecular Pharmaceutics*, 5: 449–463.

Kanazawa, A., Ikeda, T., and Endo, T. 1993. Antibacterial activity of polymeric sulfonium salts. *Journal of Polymer Science, Part A: Polymer Chemistry*, 31: 2873–2876.

Kaneshiro, T. L., Wang, X., and Lu, Z. R. 2007. Synthesis, characterization, and gene delivery of poly-L-lysine octa(3-aminopropyl)silsesquioxane dendrimers: Nanoglobular drug carriers with precisely defined molecular architectures. *Molecular Pharmaceutics*, 4: 759–768.

Kasai, S., Nagasawa, H., Shimamura, M., Uto, Y., and Hori, H. 2002. Design and synthesis of antiangiogenic/heparin-binding arginine dendrimer mimicking the surface of endostatin. *Bioorganic & Medicinal Chemistry Letters*, 12: 951–954.

Kataoka, K., Harada, A., and Nagasaki, Y. 2001. Block copolymer micelles for drug delivery: Design, characterization and biological significance. *Advanced Drug Delivery Reviews*, 47: 113–131.

Kawabata, K., Takakura, Y., and Hashida, M. 1995. The fate of plasmid DNA after intravenous injection in mice: Involvement of scavenger receptors in its hepatic uptake. *Pharmaceutical Research*, 12: 825–830.

Kazuhiko, I., Hidenori, F., Toshikazu, Y., and Yasuhiko, I. 2000. Antithrombogenic polymer alloy composed of 2-methacryloyloxyethyl phosphorylcholine polymer and segmented polyurethane. *Journal of Biomaterials Science, Polymer Edition*, 11: 1183–1195.

Kebir, N., Campistron, I., Laguerre, A. et al. 2007. Use of telechelic *cis*-1,4-polyisoprene cationomers in the synthesis of antibacterial ionic polyurethanes and copolyurethanes bearing ammonium groups. *Biomaterials*, 28: 4200–4208.

Kenawy, E. R., Abdel-Hay, F. I., El-Magd, A. A., and Mahmoud, Y. 2005. Biologically active polymers: Modification and anti-microbial activity of chitosan derivatives. *Journal of Bioactive and Compatible Polymers*, 20: 95–111.

Kenawy, E. R., Worley, S. D., and Broughton, R. 2007. The chemistry and applications of antimicrobial polymers: A state-of-the-art review. *Biomacromolecules*, 8: 1359–1384.

Kersten, J. R., Pagel, P. S., and Warltier, D. C. 1995. Protamine inhibits coronary collateral development in a canine model of repetitive coronary occlusion. *American Journal of Physiology—Heart and Circulatory Physiology*, 268: H720–H728.

Khan, A., Benboubetra, M., Sayyed, P. Z. et al. 2004. Sustained polymeric delivery of gene silencing antisense ODNs, siRNA, DNAzymes and ribozymes: In vitro and in vivo studies. *Journal of Drug Targeting*, 12: 393–404.

Kichler, A. 2004. Gene transfer with modified polyethylenimines. *Journal of Gene Medicine*, 6(Suppl 1): S3–S10.

Kidane, A. G., Salacinski, H., Tiwari, A., Bruckdorfer, K. R., and Seifalian, A. M. 2004. Anticoagulant and antiplatelet agents: Their clinical and device application(s) together with usages to engineer surfaces. *Biomacromolecules*, 5: 798–813.

Kidd, K. R., Patula, V. B., and Williams, S. K. 2003. Accelerated endothelialization of interpositional 1-mm vascular grafts. *Journal of Surgical Research*, 113: 234–242.

Kim, C. H., Choi, J. W., Chun, H. J., and Choi, K. S. 1997. Synthesis of chitosan derivatives with quaternary ammonium salt and their antibacterial activity. *Polymer Bulletin*, 38: 387–393.

Kim, J. H., Kim, Y. S., Kim, S. et al. 2006. Hydrophobically modified glycol chitosan nanoparticles as carriers for paclitaxel. *Journal of Controlled Release*, 111: 228–234.

Kircheis, R., Wightman, L., Schreiber, A. et al. 2001. Polyethylenimine/DNA complexes shielded by transferrin target gene expression to tumors after systemic application. *Gene Therapy*, 8: 28–40.

Kobayashi, K., Ohuchi, K., Hoshi, H. et al. 2005. Segmented polyurethane modified by photopolymerization and cross-linking with 2-methacryloyloxyethyl phosphorylcholine polymer for blood-contacting surfaces of ventricular assist devices. *Journal of Artificial Organs*, 8: 237–244.

Kohnen, T. and Kasper, T. 2005. Incision sizes before and after implantation of foldable intraocular lenses with 6 mm optic using Monarch and Unfolder injector systems. *Ophthalmology*, 112: 58–66.

Kolhatkar, R. B., Swaan, P., and Ghandehari, H. 2008. Potential oral delivery of 7-ethyl-10-hydroxy-camptothecin (SN-38) using poly(amidoamine) dendrimers. *Pharmaceutical Research*, 25: 1723–1729.

Kono, K., Kojima, C., Hayashi, N. et al. 2008. Preparation and cytotoxic activity of poly(ethylene glycol)-modified poly(amidoamine) dendrimers bearing adriamycin. *Biomaterials*, 29: 1664–1675.

Kopecek, J., Kopeckova, P., Minko, T., and Lu, Z. R. 2000. HPMA copolymer–anticancer drug conjugates: Design, activity, and mechanism of action. *European Journal of Pharmaceutics and Biopharmaceutics*, 50: 61–81.

Koping-Hoggard, M., Tubulekas, I., Guan, H. et al. 2001. Chitosan as a nonviral gene delivery system. Structure–property relationships and characteristics compared with polyethylenimine in vitro and after lung administration in vivo. *Gene Therapy*, 8: 1108–1121.

Kouvroukoglou, S., Dee, K. C., Bizios, R., Mcintire, L. V., and Zygourakis, K. 2000. Endothelial cell migration on surfaces modified with immobilized adhesive peptides. *Biomaterials*, 21: 1725–1733.

Krijgsman, B., Seifalian, A. M., Salacinski, H. J. et al. 2002. An assessment of covalent grafting of RGD peptides to the surface of a compliant poly(carbonate-urea)urethane vascular conduit versus conventional biological coatings: Its role in enhancing cellular retention. *Tissue Engineering*, 8: 673–680.

Kukowska-Latallo, J. F., Candido, K. A., Cao, Z. et al. 2005. Nanoparticle targeting of anticancer drug improves therapeutic response in animal model of human epithelial cancer. *Cancer Research*, 65: 5317–5324.

Kumazawa, E., and Ochi, Y. 2004. DE-310, a novel macromolecular carrier system for the camptothecin analog DX-8951f: Potent antitumor activities in various murine tumor models. *Cancer Sci*, 95: 168–175.

Kurt, P., Wood, L., Ohman, D. E., and Wynne, K. J. 2007. Highly effective contact antimicrobial surfaces via polymer surface modifiers. *Langmuir*, 23: 4719–4723.

Kwoh, D. Y., Coffin, C. C., Lollo, C. P. et al. 1999. Stabilization of poly-L-lysine/DNA polyplexes for in vivo gene delivery to the liver. *Biochimica et Biophysica Acta—Gene Structure and Expression*, 1444: 171–190.

Kwon, G. S., Naito, M., Kataoka, K. et al. 1994. Block copolymer micelles as vehicles for hydrophobic drugs. *Colloids and Surfaces B: Biointerfaces*, 2: 429–434.

Kyomoto, M., Moro, T., Miyaji, F. et al. 2009. Effects of mobility/immobility of surface modification by 2-methacryloyloxyethyl phosphorylcholine polymer on the durability of polyethylene for artificial joints. *Journal of Biomedical Materials Research—Part A*, 90A: 362–371.

Lake, A. C., Vassy, R., Di Benedetto, M. et al. 2006. Low molecular weight fucoidan increases VEGF165-induced endothelial cell migration by enhancing VEGF165 binding to VEGFR-2 and NRP1. *Journal of Biological Chemistry*, 281: 37844–37852.

Lammers, T., Subr, V., Ulbrich, K. et al. 2009. Simultaneous delivery of doxorubicin and gemcitabine to tumors in vivo using prototypic polymeric drug carriers. *Biomaterials*, 30: 3466–3475.

Lange, C., Ehlken, C., Martin, G. et al. 2007. Intravitreal injection of the heparin analog 5-amino-2-naphthalenesulfonate reduces retinal neovascularization in mice. *Experimental Eye Research*, 85: 323–327.

Langer, C. J. 2004. CT-2103: A novel macromolecular taxane with potential advantages compared with conventional taxanes. *Clinical Lung Cancer*, 6: S85–S88.

Larm, O., Larsson, R., and Olsson, P. 1983. A new non-thrombogenic surface prepared by selective covalent binding of heparin via a modified reducing terminal residue. *Biomaterials Medical Devices and Artificial Organs*, 11: 161–173.

Lawson, M. C., Shoemaker, R., Hoth, K. B., Bowman, C. N., and Anseth, K. S. 2009. Polymerizable vancomycin derivatives for bactericidal biomaterial surface modification: Structure–function evaluation. *Biomacromolecules*, 10: 2221–2234.

Lee, S., Iten, R., Muller, M., and Spencer, N. D. 2004. Influence of molecular architecture on the adsorption of poly(ethylene oxide)–poly(propylene oxide)–poly(ethylene oxide) on PDMS surfaces and implications for aqueous lubrication. *Macromolecules*, 37: 8349–8356.

Lee, B. S., Lee, J. K., Kim, W. J. et al. 2007. Surface-initiated, atom transfer radical polymerization of oligo(ethylene glycol) methyl ether methacrylate and subsequent click chemistry for bioconjugation. *Biomacromolecules*, 8: 744–749.

Lee, S. and Spencer, N. D. 2005. Aqueous lubrication of polymers: Influence of surface modification. *Tribology International*, 38: 922–930.

Lee, S. and Spencer, N. D. 2008a. Materials science—Sweet, hairy, soft, and slippery. *Science*, 319: 575–576.

Lee, S. and Spencer, N. D. 2008b. Poly(L-lysine)-*graft*-poly(ethylene glycol): A versatile aqueous lubricant additive for tribosystems involving thermoplastics. *Lubrication Science*, 20: 21–34.

Li, J. and Loh, X. J. 2008. Cyclodextrin-based supramolecular architectures: Syntheses, structures, and applications for drug and gene delivery. *Advanced Drug Delivery Reviews*, 60: 1000–1017.

Li, D., Luo, Y., Zhang, B., Li, B., and Zhu, S. 2007. Raft grafting polymerization of MMA/St from surface of silicon wafer. *Acta Polymerica Sinica*: 699–704.

Li, G., Song, S., Guo, L., and Ma, S. 2008. Self-assembly of thermo- and pH-responsive poly(acrylic acid)-*b*-poly(*N*-isopropylacrylamide) micelles for drug delivery. *Journal of Polymer Science, Part A: Polymer Chemistry*, 46: 5028–5035.

Li, Y. L., Zhu, L., Liu, Z. et al. 2009. Reversibly stabilized multifunctional dextran nanoparticles efficiently deliver doxorubicin into the nuclei of cancer cells. *Angewandte Chemie—International Edition*, 48: 9914–9918.

Liang, B., He, M. L., Xiao, Z. P. et al. 2008. Synthesis and characterization of folate-PEG-grafted-hyperbranched-PEI for tumor-targeted gene delivery. *Biochemical and Biophysical Research Communications*, 367: 874–880.

Liekens, S., Leali, D., Neyts, J. et al. 1999. Modulation of fibroblast growth factor-2 receptor binding, signaling, and mitogenic activity by heparin-mimicking polysulfonated compounds. *Molecular Pharmacology*, 56: 204–213.

Liekens, S., Neyts, J., Degreve, B., and De Clercq, E. 1997. The sulfonic acid polymers PAMPS [poly(2-acrylamido-2-methyl-1-propanesulfonic acid)] and related analogues are highly potent inhibitors of angiogenesis. *Oncology Research*, 9: 173–181.

Lim, Y. B., Han, S. O., Kong, H. U. et al. 2000. Biodegradable polyester, poly[a-(4-aminobutyl)-l-glycolic acid], as a non-toxic gene carrier. *Pharmaceutical Research*, 17: 811–816.

Lim, J. and Simanek, E. E. 2008. Synthesis of water-soluble dendrimers based on melamine bearing 16 paclitaxel groups. *Organic Letters*, 10: 201–204.

Lin, C., Blaauboer, C. J., Timoneda, M. M. et al. 2008. Bioreducible poly(amido amine)s with oligoamine side chains: Synthesis, characterization, and structural effects on gene delivery. *Journal of Controlled Release*, 126: 166–174.

Lin, W. J., Chen, Y. C., Lin, C. C., Chen, C. F., and Chen, J. W. 2006. Characterization of pegylated copolymeric micelles and in vivo pharmacokinetics and biodistribution studies. *Journal of Biomedical Materials Research—Part B Applied Biomaterials*, 77: 188–194.

Liu, B., Yang, M., Li, R. et al. 2008. The antitumor effect of novel docetaxel-loaded thermosensitive micelles. *European Journal of Pharmaceutics and Biopharmaceutics*, 69: 527–534.

Loll, P. J. and Axelsen, P. H. 2000. The structural biology of molecular recognition by vancomycin. *Annual Review of Biophysics and Biomolecular Structure*, 29: 265–289.

Lopez Donaire, M. L., Parra-Caceres, J., Vazquez-Lasa, B. et al. 2009. Polymeric drugs based on bioactive glycosides for the treatment of brain tumours. *Biomaterials*, 30: 1613–1626.

Low, P. S., Henne, W. A., and Doorneweerd, D. D. 2008. Discovery and development of folic-acid-based receptor targeting for imaging and therapy of cancer and inflammatory diseases. *Accounts of Chemical Research*, 41: 120–129.

Lozano, R. M., Jiménez, M. A., Santoro, J., Rico, M., and Giménez-Gallego, G. 1998. Solution structure of acidic fibroblast growth factor bound to 1,3,6-naphthalenetrisulfonate: A minimal model for the antitumoral action of suramins and suradistas. *Journal of Molecular Biology*, 281: 899–915.

Lu, A. and Sipehia, R. 2001. Antithrombotic and fibrinolytic system of human endothelial cells seeded on PTFE: The effects of surface modification of PTFE by ammonia plasma treatment and ECM protein coatings. *Biomaterials*, 22: 1439–1446.

Lukyanov, A. N. and Torchilin, V. P. 2004. Micelles from lipid derivatives of water-soluble polymers as delivery systems for poorly soluble drugs. *Advanced Drug Delivery Reviews*, 56: 1273–1289.

Luong-Van, E., Grøndahl, L., Chua, K. N., Leong, K. W., Nurcombe, V., and Cool, S. M. 2006. Controlled release of heparin from poly(epsilon-caprolactone) electrospun fibers. *Biomaterials*, 27: 2042–2050.

Luten, J., Akeroyd, N., Funhoff, A. et al. 2006. Methacrylamide polymers with hydrolysis-sensitive cationic side groups as degradable gene carriers. *Bioconjugate Chemistry*, 17: 1077–1084.

Maeda, H., Noguchi, Y., Sato, K., and Akaike, T. 1994. Enhanced vascular permeability in solid tumor is mediated by nitric oxide and inhibited by both new nitric oxide scavenger and nitric oxide synthase inhibitor. *Japanese Journal of Cancer Research*, 85: 331–334.

Maeda, H., Wu, J., Sawa, T., Matsumura, Y., and Hori, K. 2000. Tumor vascular permeability and the EPR effect in macromolecular therapeutics: A review. *Journal of Controlled Release*, 65: 271–284.

Maheshwari, A., Mahato, R. I., Mcgregor, J. et al. 2000. Soluble biodegradable polymer-based cytokine gene delivery for cancer treatment. *Molecular Therapy*, 2: 121–130.

Mahmud, A., Xiong, X. B., and Lavasanifar, A. 2008. Development of novel polymeric micellar drug conjugates and nano-containers with hydrolyzable core structure for doxorubicin delivery. *European Journal of Pharmaceutics and Biopharmaceutics*, 69: 923–934.

Malik, N., Evagorou, E. G., and Duncan, R. 1999. Dendrimer–platinate: A novel approach to cancer chemotherapy. *Anti-Cancer Drugs*, 10: 767–776.

Mangoush, O., Purkayastha, S., Haj-Yahia, S. et al. 2007. Heparin-bonded circuits versus nonheparin-bonded circuits: An evaluation of their effect on clinical outcomes. *European Journal of Cardio-Thoracic Surgery*, 31: 1058–1069.

Mannisto, M., Vanderkerken, S., Toncheva, V. et al. 2002. Structure–activity relationships of poly(L-lysines): Effects of pegylation and molecular shape on physicochemical and biological properties in gene delivery. *Journal of Controlled Release*, 83: 169–182.

Mao, S., Neu, M., Germershaus, O. et al. 2006. Influence of polyethylene glycol chain length on the physico-chemical and biological properties of poly(ethylene imine)-*graft*-poly(ethylene glycol) block copolymer/ SiRNA polyplexes. *Bioconjugate Chemistry*, 17: 1209–1218.

Mao, C., Qiu, Y., Sang, H. et al. 2004. Various approaches to modify biomaterial surfaces for improving hemo-compatibility. *Advances in Colloid and Interface Science*, 110: 5–17.

Marchetti, M., Vignoli, A., Russo, L. et al. 2008. Endothelial capillary tube formation and cell proliferation induced by tumor cells are affected by low molecular weight heparins and unfractionated heparin. *Thrombosis Research*, 121: 637–645.

Marschall, P., Malik, N., and Larin, Z. 1999. Transfer of YACs up to 2.3 Mb intact into human cells with poly-ethylenimine. *Gene Therapy*, 6: 1634–1637.

Matou, S., Helley, D., Chabut, D., Bros, A., and Fischer, A.-M. 2002. Effect of fucoidan on fibroblast growth factor-2-induced angiogenesis in vitro. *Thrombosis Research*, 106: 213–221.

Matsumura, Y. and Maeda, H. 1986. A new concept for macromolecular therapeutics in cancer chemother-apy: Mechanism of tumoritropic accumulation of proteins and the antitumor agent SMANCS. *Cancer Research*, 46: 6387–6392.

Mcguigan, A. P. and Sefton, M. V. 2007. The influence of biomaterials on endothelial cell thrombogenicity. *Biomaterials*, 28: 2547–2571.

Meng, F., Hennink, W. E., and Zhong, Z. 2009a. Reduction-sensitive polymers and bioconjugates for biomedi-cal applications. *Biomaterials*, 30: 2180–2198.

Meng, F., Zhong, Z., and Feijen, J. 2009b. Stimuli-responsive polymersomes for programmed drug delivery. *Biomacromolecules*, 10: 197–209.

Menjoge, A. R., Kannan, R. M., and Tomalia, D. A. 2010. Dendrimer-based drug and imaging conjugates: Design considerations for nanomedical applications. *Drug Discovery Today*, 15: 171–185.

Merdan, T., Kopecek, J., and Kissel, T. 2002. Prospects for cationic polymers in gene and oligonucleotide therapy against cancer. *Advanced Drug Delivery Reviews*, 54: 715–758.

Merkel, O. M., Mintzer, M. A., Sitterberg, J. et al. 2009. Triazine dendrimers as nonviral gene delivery systems: Effects of molecular structure on biological activity. *Bioconjugate Chemistry*, 20: 1799–1806.

Mero, A., Pasut, G., Via, L. D. et al. 2008. Synthesis and characterization of poly(2-ethyl 2-oxazoline)-conju-gates with proteins and drugs: Suitable alternatives to PEG-conjugates? *Journal of Controlled Release*, 125: 87–95.

Min, K. H., Kim, J. H., Bae, S. M. et al. 2010. Tumoral acidic pH-responsive MPEG-poly(*b*-amino ester) poly-meric micelles for cancer targeting therapy. *Journal of Controlled Release*, 144: 259–266.

Min, K. H., Park, K., Kim, Y. S. et al. 2008. Hydrophobically modified glycol chitosan nanoparticles-encapsu-lated camptothecin enhance the drug stability and tumor targeting in cancer therapy. *Journal of Controlled Release*, 127: 208–218.

Minchinton, A. I. and Tannock, I. F. 2006. Drug penetration in solid tumours. *Nature Reviews Cancer*, 6: 583–592.

Mitchell, D. J., Steinman, L., Kim, D. T., Fathman, C. G., and Rothbard, J. B. 2000. Polyarginine enters cells more efficiently than other polycationic homopolymers. *Journal of Peptide Research*, 56: 318–325.

Mitra, S., Gaur, U., Ghosh, P. C., and Maitra, A. N. 2001. Tumour targeted delivery of encapsulated dex-tran–doxorubicin conjugate using chitosan nanoparticles as carrier. *Journal of Controlled Release*, 74: 317–323.

Miyatake, T., Nishihara, M., and Matile, S. 2006. A cost-effective method for the optical transduction of chem-ical reactions. Application to hyaluronidase inhibitor screening with polyarginine–counteranion com-plexes in lipid bilayers. *Journal of the American Chemical Society*, 128: 12420–12421.

Moncada, S., Radomski, M. W., and Palmer, R. M. J. 1988. Endothelium-derived relaxing factor. Identification as nitric oxide and role in the control of vascular tone and platelet function. *Biochemical Pharmacology*, 37: 2495–2501.

Moon, J. J. and West, J. L. 2008. Vascularization of engineered tissues: Approaches to promote angiogenesis in biomaterials. *Current Topics in Medicinal Chemistry*, 8: 300–310.

Morgan, M. T., Nakanishi, Y., Kroll, D. J. et al. 2006. Dendrimer-encapsulated camptothecins: Increased solubility, cellular uptake, and cellular retention affords enhanced anticancer activity in vitro. *Cancer Research*, 66: 11913–11921.

Mrkvan, T., Sirova, M., Etrych, T. et al. 2005. Chemotherapy based on HPMA copolymer conjugates with pH-controlled release of doxorubicin triggers anti-tumor immunity. *Journal of Controlled Release*, 110: 119–129.

Nagamitsu, A., Greish, K., and Maeda, H. 2009. Elevating blood pressure as a strategy to increase tumor-targeted delivery of macromolecular drug SMANCS: Cases of advanced solid tumors. *Japanese Journal of Clinical Oncology*, 39: 756–766.

Nagaoka, S. and Akashi, R. 1990. Low-friction hydrophilic surface for medical devices. *Biomaterials*, 11: 419–424.

Nakayama, Y., Okahashi, R., Iwai, R., and Uchida, K. 2007. Heparin bioconjugate with a thermoresponsive cationic branched polymer: A novel aqueous antithrombogenic coating material. *Langmuir*, 23: 8206–8211.

Nam, C. W., Kim, Y. H., and Ko, S. W. 1999. Modification of polyacrylonitrile (PAN) fiber by blending with N-(2-hydroxy)propyl-3-trimethyl-ammonium chitosan chloride. *Journal of Applied Polymer Science*, 74: 2258–2265.

Nam, K., Kimura, T., and Kishida, A. 2007. Physical and biological properties of collagen–phospholipid polymer hybrid gels. *Biomaterials*, 28: 3153–3162.

Nasongkla, N., Bey, E., Ren, J. et al. 2006. Multifunctional polymeric micelles as cancer-targeted, MRI-ultrasensitive drug delivery systems. *Nano Letters*, 6: 2427–2430.

Neufeld, G. and Gospodarowicz, D. 1987. Protamine sulfate inhibits mitogenic activities of the extracellular matrix and fibroblast growth factor, but potentiates that of epidermal growth factor. *Journal of Cellular Physiology*, 132: 287–294.

Navarro, G. and Tros De Ilarduya, C. 2009. Activated and non-activated PAMAM dendrimers for gene delivery in vitro and in vivo. *Nanomedicine: Nanotechnology, Biology, and Medicine*, 5: 287–297.

Nguyen, H. K., Lemieux, P., Vinogradov, S. V. et al. 2000. Evaluation of polyether–polyethyleneimine graft copolymers as gene transfer agents. *Gene Therapy*, 7: 126–138.

Nishi, K. K., Antony, M., Mohanan, P. V. et al. 2007. Amphotericin B–gum Arabic conjugates: Synthesis, toxicity, bioavailability, and activities against *Leishmania* and fungi. *Pharmaceutical Research*, 24: 971–980.

Nonaka, T., Hua, L., Ogata, T., and Kurihara, S. 2002. Synthesis of water-soluble thermosensitive polymers having phosphonium groups from methacryloyloxyethyl trialkyl phosphonium chlorides–N-isopropylacrylamide copolymers and their functions. *Journal of Applied Polymer Science*, 87: 386–393.

Norrby, K. 1993. Heparin and angiogenesis: A low-molecular-weight fraction inhibits and a high-molecular-weight fraction stimulates angiogenesis systemically. *Haemostasis*, 23: 141–149.

Ogris, M., Brunner, S., Schüller, S., Kircheis, R., and Wagner, E. 1999. PEGylated DNA/transferrin–PEI complexes: Reduced interaction with blood components, extended circulation in blood and potential for systemic gene delivery. *Gene Therapy*, 6: 595–605.

Oishi, M., Nagasaki, Y., Itaka, K., Nishiyama, N., and Kataoka, K. 2005. Lactosylated poly(ethylene glycol)–siRNA conjugate through acid-labile β-thiopropionate linkage to construct pH-sensitive polyion complex micelles achieving enhanced gene silencing in hepatoma cells. *Journal of the American Chemical Society*, 127: 1624–1625.

Okner, R., Domb, A. J., and Mandler, D. 2009. Electrochemically deposited poly(ethylene glycol)-based sol–gel thin films on stainless steel stents. *New Journal of Chemistry*, 33: 1596–1604.

Oupicky, D., Konak, C., Dash, P. R., Seymour, L. W., and Ulbrich, K. 1999. Effect of albumin and polyanion on the structure of DNA complexes with polycation containing hydrophilic nonionic block. *Bioconjugate Chemistry*, 10: 764–772.

Paleolog, E. M. and Miotla, J. M. 1998. Angiogenesis in arthritis: Role in disease pathogenesis and as a potential therapeutic target. *Angiogenesis*, 2: 295–307.

Park, J. H., Lee, S., Kim, J. H. et al. 2008. Polymeric nanomedicine for cancer therapy. *Progress in Polymer Science (Oxford)*, 33: 113–137.

Parra, F., Vázquez, B., Benito, L., Barcenilla, J., and San Román, J. 2009. Foldable antibacterial acrylic intraocular lenses of high refractive index. *Biomacromolecules*, 10: 3055–3061.

Pasut, G., Canal, F., Dalla Via, L. et al. 2008. Antitumoral activity of PEG–gemcitabine prodrugs targeted by folic acid. *Journal of Controlled Release*, 127: 239–248.

Pasut, G., Greco, F., Mero, A. et al. 2009. Polymer–drug conjugates for combination anticancer therapy: Investigating the mechanism of action. *Journal of Medicinal Chemistry*, 52: 6499–6502.

Pasut, G. and Veronese, F. M. 2009. PEG conjugates in clinical development or use as anticancer agents: An overview. *Advanced Drug Delivery Reviews*, 61: 1177–1188.

Patil, M. L., Zhang, M., Betigeri, S. et al. 2008. Surface-modified and internally cationic polyamidoamine dendrimers for efficient siRNA delivery. *Bioconjugate Chemistry*, 19: 1396–1403.

Paul, W. and Sharma, C. P. 1997. Acetylsalicylic acid loaded poly(vinyl alcohol) hemodialysis membranes: Effect of drug release on blood compatibility and permeability. *Journal of Biomaterials Science, Polymer Edition*, 8: 755–764.

Pawlowski, K. J., Rittgers, S. E., Schmidt, S. P., and Bowlin, G. L. 2004. Endothelial cell seeding of polymeric vascular grafts. *Frontiers in Bioscience*, 9: 1412–1421.

Pechar, M., Braunová, A., Ulbrich, K., Jelinkova, M., and Rihova, B. 2005. Poly(ethylene glycol)–doxorubicin conjugates with pH-controlled activation. *Journal of Bioactive and Compatible Polymers*, 20: 319–341.

Pille, J. Y., Li, H., Blot, E. et al. 2006. Intravenous delivery of anti-RhoA small interfering RNA loaded in nanoparticles of chitosan in mice: Safety and efficacy in xenografted aggressive breast cancer. *Human Gene Therapy*, 17: 1019–1026.

Plank, C., Mechtler, K., Szoka, F. C. Jr., and Wagner, E. 1996. Activation of the complement system by synthetic DNA complexes: A potential barrier for intravenous gene delivery. *Human Gene Therapy*, 7: 1437–1446.

Powers, C. J. 2000. Fibroblast growth factors, their receptors and signaling. *Endocrine-Related Cancer*, 7: 165–197.

Pun, S. H., Tack, F., Bellocq, N. C. et al. 2004. Targeted delivery of RNA-cleaving DNA enzyme (DNAzyme) to tumor tissue by transferrin-modified, cyclodextrin-based particles. *Cancer Biology and Therapy*, 3: 641–650.

Rabea, E. I., Badawy, M. E. T., Stevens, C. V., Smagghe, G., and Steurbaut, W. 2003. Chitosan as antimicrobial agent: Applications and mode of action. *Biomacromolecules*, 4: 1457–1465.

Rademaker-Lakhai, J. M., Terret, C., Howell, S. B., Baud, C. M., de Boer, R. F., Pluim, D., Beijnen, J. H., Schellens, J. H., and Drozs J.-P. 2004. A phase I and pharmacological study of the platinum polymer AP5280 given as an intravenous infusion once every 3 weeks in patients with solid tumors. *Clinical Cancer Research*, 10: 3386–3395.

Ranjan, R. and Brittain, W. J. 2007. Tandem RAFT polymerization and click chemistry: An efficient approach to surface modification. *Macromolecular Rapid Communications*, 28: 2084–2089.

Rapoport, N. 2007. Physical stimuli-responsive polymeric micelles for anti-cancer drug delivery. *Progress in Polymer Science (Oxford)*, 32: 962–990.

Ratner, B. D. and Hoffman, A. S. 1976. *Process for Radiation Grafting Hydrogels onto Organic Polymeric Substrates*. The United States of America as represented by the United States Energy, Washington, DC.

Raviv, U., Giasson, S., Kampf, N. et al. 2003. Lubrication by charged polymers. *Nature*, 425: 163–165.

Read, M. L., Singh, S., Ahmed, Z. et al. 2005. A versatile reducible polycation-based system for efficient delivery of a broad range of nucleic acids. *Nucleic Acids Research*, 33: 1–16.

Richards Grayson, A. C., Doody, A. M., and Putnam, D. 2006. Biophysical and structural characterization of polyethylenimine-mediated siRNA delivery in vitro. *Pharmaceutical Research*, 23: 1868–1876.

Richardson, T. P., Peters, M. C., Ennett, A. B., and Mooney, D. J. 2001. Polymeric system for dual growth factor delivery. *Nature Biotechnology*, 19: 1029–1034.

Rijcken, C. J. F., Soga, O., Hennink, W. E., and Nostrum, C. F. V. 2007. Triggered destabilisation of polymeric micelles and vesicles by changing polymers polarity: An attractive tool for drug delivery. *Journal of Controlled Release*, 120: 131–148.

Ringsdorf, H. 1975. Structure and properties of pharmacologically active polymers. *Journal of Polymer Science Polymer Symposium*, 51: 135–153.

Rodriguez, G., Gallardo, A., Fernandez, M. et al. 2004. Hydrophilic polymer drug from a derivative of salicylic acid: Synthesis, controlled release studies and biological behavior. *Macromolecular Bioscience*, 4: 579–586.

Rodriguez, G., Gallardo, A., San Roman, J. et al. 1999. New resorbable polymeric systems with antithrombogenic activity. *Journal of Materials Science: Materials in Medicine*, 10: 873–878.

Rojo, L., Barcenilla, J. M., Vázquez, B., González, R., and San Román, J. 2008. Intrinsically antibacterial materials based on polymeric derivatives of eugenol for biomedical applications. *Biomacromolecules*, 9: 2530–2535.

Rojo, L., Vazquez, B., Parra, J. et al. 2006. From natural products to polymeric derivatives of "eugenol": A new approach for preparation of dental composites and orthopedic bone cements. *Biomacromolecules*, 7: 2751–2761.

Rothbard, J. B., Kreider, E., Vandeusen, C. L. et al. 2002. Arginine-rich molecular transporters for drug delivery: Role of backbone spacing in cellular uptake. *Journal of Medicinal Chemistry*, 45: 3612–3618.

Rowinsky, E. K., Rizzo, J., Ochoa, L. et al. 2003. A phase I and pharmacokinetic study of pegylated camptothecin as a 1-hour infusion every 3 weeks in patients with advanced solid malignancies. *Journal of Clinical Oncology: Official Journal of the American Society of Clinical Oncology*, 21: 148–157.

Roy, D., Knapp, J. S., Guthrie, J. T., and Perrier, S. 2008. Antibacterial cellulose fiber via RAFT surface graft polymerization. *Biomacromolecules*, 9: 91–99.

Rusnati, M., Dell'era, P., Urbinati, C. et al. 1996. A distinct basic fibroblast growth factor (FGF-2)/FGF receptor interaction distinguishes urokinase-type plasminogen activator induction from mitogenicity in endothelial cells. *Molecular Biology of the Cell*, 7: 369–381.

Salazar, M. D. and Ratnam, M. 2007. The folate receptor: What does it promise in tissue-targeted therapeutics? *Cancer and Metastasis Reviews*, 26: 141–152.

San Roman, J., Bujan, J., Bellon, J. M. et al. 1996. Experimental study of the antithrombogenic behavior of Dacron vascular grafts coated with hydrophilic acrylic copolymers bearing salicylic acid residues. *Journal of Biomedical Materials Research*, 32: 19–27.

San Roman, J., Rodriguez, G., Fernandez, M. et al. 2007. *Triflusal-Containing Polymers for Stent Coating.* WIPO Patent Application WO/2007/014787; Application Number: EP2006/009156; Publication Date: February 08, 2007, Filing Date: September 20, 2006.

Sapra P., Zhao H., Mehlig M. et al. 2008. Novel delivery of SN38 markedly inhibits tumor growth in xenografts, including a camptothecin-11- refractory model. *Clin Cancer Res*, 14:1888–96.

Sarapa, N., Britto, M. R., Speed, W. et al. 2003. Assessment of normal and tumor tissue uptake of MAG-CPT, a polymer-bound prodrug of camptothecin, in patients undergoing elective surgery for colorectal carcinoma. *Cancer Chemotherapy and Pharmacology*, 52: 424–430.

Satchi, R., Connors, T. A., and Duncan, R. 2001. PDEPT: Polymer-directed enzyme prodrug therapy: I. HPMA copolymer–cathepsin B and PK1 as a model combination. *British Journal of Cancer*, 85: 1070–1076.

Satchi-Fainaro, R., Hailu, H., Davies, J. W., Summerford, C., and Duncan, R. 2003. PDEPT: Polymer-directed enzyme prodrug therapy. 2. HPMA copolymer-*b*-lactamase and HPMA copolymer-C-Dox as a model combination. *Bioconjugate Chemistry*, 14: 797–804.

Scherer, S. S., Pietramaggiori, G., Matthews, J. et al. 2009. Poly-*N*-acetyl glucosamine nanofibers: A new bioactive material to enhance diabetic wound healing by cell migration and angiogenesis. *Annals of Surgery*, 250: 322–330.

Sefton, M. V., Gemmell, C. H., and Gorbet, M. B. 2000. What really is blood compatibility? *Journal of Biomaterials Science, Polymer Edition*, 11: 1165–1182.

Seki, T., Fang, J., and Maeda, H. 2009. Enhanced delivery of macromolecular antitumor drugs to tumors by nitroglycerin application. *Cancer Science*, 100: 2426–2430.

Seyfert, U. T., Biehl, V., and Schenk, J. 2002. In vitro hemocompatibility testing of biomaterials according to the ISO 10993-4. *Biomolecular Engineering*, 19: 91–96.

Seymour, L. W., Ferry, D. R., Anderson, D. et al. 2002. Hepatic drug targeting: Phase I evaluation of polymer-bound doxorubicin. *Journal of Clinical Oncology*, 20: 1668–1676.

Seymour, L. W., Ferry, D. R., Kerr, D. J. et al. 2009. Phase II studies of polymer–doxorubicin (PK1, FCE28068) in the treatment of breast, lung and colorectal cancer. *International Journal of Oncology*, 34: 1629–1636.

Shen, M., Martinson, L., Wagner, M. S. et al. 2002. PEO-like plasma polymerized tetraglyme surface interactions with leukocytes and proteins: In vitro and in vivo studies. *Journal of Biomaterials Science, Polymer Edition*, 13: 367–390.

Singer, I. L. and Ollock, H. M. (Eds.). 1992. *Fundamentals of Friction: Microscopic and Macroscopic Processes*. Kluwer, Dordrecht, the Netherlands.

Singh, P., Gupta, U., Asthana, A., and Jain, N. K. 2008. Folate and folate-PEG-PAMAM dendrimers: Synthesis, characterization, and targeted anticancer drug delivery potential in tumor bearing mice. *Bioconjugate Chemistry*, 19: 2239–2252.

Sirolli, V., Di Stante, S., Stuard, S. et al. 2000. Biocompatibility and functional performance of a polyethylene glycol acid-grafted cellulosic membrane for hemodialysis. *International Journal of Artificial Organs*, 23: 356–364.

Sirova, M., Mrkvan, T., Etrych, T. et al. 2010. Preclinical evaluation of linear HPMA–doxorubicin conjugates with pH-sensitive drug release: Efficacy, safety, and immunomodulating activity in murine model. *Pharmaceutical Research*, 27: 200–208.

Smart, T. P., Mykhaylyk, O. O., Ryan, A. J., and Battaglia, G. 2009. Polymersomes hydrophilic brush scaling relations. *Soft Matter*, 5: 3607–3610.

Sochart, D. H. 1999. Relationship of acetabular wear to osteolysis and loosening in total hip arthroplasty. *Clinical Orthopaedics and Related Research*: 135–150.

Soeda, S., Kozako, T., Iwata, K., and Shimeno, H. 2000. Oversulfated fucoidan inhibits the basic fibroblast growth factor-induced tube formation by human umbilical vein endothelial cells: Its possible mechanism of action. *Biochimica et Biophysica Acta—Molecular Cell Research*, 1497: 127–134.

Sperling, C., Houska, M., Brynda, E., Streller, U., and Werner, C. 2006. In vitro hemocompatibility of albumin–heparin multilayer coatings on polyethersulfone prepared by the layer-by-layer technique. *Journal of Biomedical Materials Research—Part A*, 76: 681–689.

Sun, H., Guo, B., Cheng, R. et al. 2009. Biodegradable micelles with sheddable poly(ethylene glycol) shells for triggered intracellular release of doxorubicin. *Biomaterials*, 30: 6358–6366.

Sun, H., Guo, B., Li, X. et al. 2010. Shell-sheddable micelles based on dextran–SS–poly(ε-caprolactone) diblock copolymer for efficient intracellular release of doxorubicin. *Biomacromolecules*, 11: 848–854.

Svenson, S. and Tomalia, D. A. 2005. Dendrimers in biomedical applications—Reflections on the field. *Advanced Drug Delivery Reviews*, 57: 2106–2129.

Tabor, D. (Ed.). 1973. *Friction*. Doubleday, New York.

Taite, L. J., Yang, P., Jun, H. W., and West, J. L. 2008. Nitric oxide-releasing polyurethane–PEG copolymer containing the YIGSR peptide promotes endothelialization with decreased platelet adhesion. *Journal of Biomedical Materials Research—Part B Applied Biomaterials*, 84: 108–116.

Takakura, Y. and Hashida, M. 1996. Macromolecular carrier systems for targeted drug delivery: Pharmacokinetic considerations on biodistribution. *Pharmaceutical Research*, 13: 820–831.

Thierry, B., Merhi, Y., Silver, J., and Tabrizian, M. 2005. Biodegradable membrane-covered stent from chitosan-based polymers. *Journal of Biomedical Materials Research—Part A*, 75: 556–566.

Tiwari, A., Salacinski, H. J., Punshon, G., Hamilton, G., and Seifalian, A. M. 2002. Development of a hybrid cardiovascular graft using a tissue engineering approach. *FASEB Journal*, 16: 791–796.

Torchilin, V. P. 2001. Structure and design of polymeric surfactant-based drug delivery systems. *Journal of Controlled Release*, 73: 137–172.

Toub, N., Bertrand, J. R., Tamaddon, A. et al. 2006. Efficacy of siRNA nanocapsules targeted against the EWS–Fli1 oncogene in Ewing sarcoma. *Pharmaceutical Research*, 23: 892–900.

Uehara, T., Ishii, D., Uemura, T. et al. 2010. Gamma-glutamyl PAMAM dendrimer as versatile precursor for dendrimer-based targeting devices. *Bioconjugate Chemistry*, 21: 175–181.

Unger, R. E., Peters, K., Wolf, M. et al. 2004. Endothelialization of a non-woven silk fibroin net for use in tissue engineering: Growth and gene regulation of human endothelial cells. *Biomaterials*, 25: 5137–5146.

Urbakh, M., Klafter, J., Gourdon, D., and Israelachvili, J. 2004. The nonlinear nature of friction. *Nature*, 430: 525–528.

Uyama, Y., Tadokoro, H., and Ikada, Y. 1990. Surface lubrication of polymer-films by photoinduced graft polymerization. *Journal of Applied Polymer Science*, 39: 489–498.

Uyama, Y., Tadokoro, H., and Ikada, Y. 1991. Low-frictional catheter materials by photoinduced graft-polymerization. *Biomaterials*, 12: 71–75.

Vasey, P. A., Kaye, S. B., Morrison, R. et al. 1999. Phase I clinical and pharmacokinetic study of PK1 [*N*-(2-hydroxypropyl)methacrylamide copolymer doxorubicin]: First member of a new class of chemotherapeutic agents—Drug–polymer conjugates. *Clinical Cancer Research*, 5: 83–94.

Verbaan, F. J., Klouwenberg, P. K., Van Steenis, J. H. et al. 2005. Application of poly(2-(dimethylamino) ethyl methacrylate)-based polyplexes for gene transfer into human ovarian carcinoma cells. *International Journal of Pharmaceutics*, 304: 185–192.

Verkleij, A. J., Zwaal, R. F. A., and Roelofsen, B. 1973. The asymmetric distribution of phospholipids in the human red cell membrane. A combined study using phospholipases and freeze etch electron microscopy. *Biochimica et Biophysica Acta*, 323: 178–193.

Verschraegen, C. F., Skubitz, K., Daud, A. et al. 2009. A phase I and pharmacokinetic study of paclitaxel poliglumex and cisplatin in patients with advanced solid tumors. *Cancer Chemotherapy and Pharmacology*, 63: 903–910.

Vicario, P. P., Lu, Z., Wang, Z. et al. 2008. Antithrombogenicity of hydromer's polymeric formula F202™ immobilized on polyurethane and electropolished stainless steel. *Journal of Biomedical Materials Research—Part B Applied Biomaterials*, 86: 136–144.

Vicent, M. J. and Duncan, R. 2006. Polymer conjugates: Nanosized medicines for treating cancer. *Trends in Biotechnology*, 24: 39–47.

Wach, J. Y., Malisova, B., Bonazzi, S. et al. 2008. Protein-resistant surfaces through mild dopamine surface functionalization. *Chemistry—A European Journal*, 14: 10579–10584.

Waite, C. L. and Roth, C. M. 2009. PAMAM–RGD conjugates enhance siRNA delivery through a multicellular spheroid model of malignant glioma. *Bioconjugate Chemistry*, 20: 1908–1916.

Walther, W., Stein, U., Fichtner, I. et al. 2002. Intratumoral low-volume jet-injection for efficient nonviral gene transfer. *Applied Biochemistry and Biotechnology—Part B Molecular Biotechnology*, 21: 105–115.

Wang, Y. C., Tang, L. Y., Sun, T. M. et al. 2008. Self-assembled micelles of biodegradable triblock copolymers based on poly(ethyl ethylene phosphate) and poly(ε-caprolactone) as drug carriers. *Biomacromolecules*, 9: 388–395.

Wang, F., Wang, Y. C., Yan, L. F., and Wang, J. 2009. Biodegradable vesicular nanocarriers based on poly(ecaprolactone)-*block*-poly(ethyl ethylene phosphate) for drug delivery. *Polymer*, 50: 5048–5054.

Waschinski, C. J., Zimmermann, J., Salz, U. et al. 2008. Design of contact-active antimicrobial acrylate-based materials using biocidal macromers. *Advanced Materials*, 20: 104–108.

Wei, H., Cheng, S. X., Zhang, X. Z., and Zhuo, R. X. 2009. Thermo-sensitive polymeric micelles based on poly(*N*-isopropylacrylamide) as drug carriers. *Progress in Polymer Science*, 34: 893–910.

Wei, D., Ma, Q., Guan, Y. et al. 2009. Structural characterization and antibacterial activity of oligoguanidine (polyhexamethylene guanidine hydrochloride). *Materials Science and Engineering C*, 29: 1776–1780.

Werner, C., Maitz, M. F., and Sperling, C. 2007. Current strategies towards hemocompatible coatings. *Journal of Materials Chemistry*, 17: 3376–3384.

Wildevuur, C. R. H., Jansen, P. G. M., Bezemer, P. D. et al. 1997. Clinical evaluation of Duraflo II heparin treated extracorporeal circulation circuits (2nd version). The European working group on heparin coated extracorporeal circulation circuits. *European Journal of Cardio-Thoracic Surgery*, 11: 616–623.

Wolfert, M. A., Dash, P. R., Nazarova, O. et al. 1999. Polyelectrolyte vectors for gene delivery: Influence of cationic polymer on biophysical properties of complexes formed with DNA. *Bioconjugate Chemistry*, 10: 993–1004.

Woolfson, A. D., Malcolm, R. K., Gorman, S. P. et al. 2003. Self-lubricating silicone elastomer biomaterials. *Journal of Materials Chemistry*, 13: 2465–2470.

Wu, L. W., Chiang, Y. M., Chuang, H. C. et al. 2004. Polyacetylenes function as anti-angiogenic agents. *Pharmaceutical Research*, 21: 2112–2119.

Wu, X. L., Kim, J. H., Koo, H. et al. 2010. Tumor-targeting peptide conjugated pH-responsive micelles as a potential drug carrier for cancer therapy. *Bioconjugate Chemistry*, 21: 208–213.

Wu, Y., Zhou, Z., and Meyerhoff, M. E. 2007. in vitro platelet adhesion on polymeric surfaces with varying fluxes of continuous nitric oxide release. *Journal of Biomedical Materials Research—Part A*, 81: 956–963.

Xiong, X. B., Mahmud, A., Uludag, H., and Lavasanifar, A. 2008. Multifunctional polymeric micelles for enhanced intracellular delivery of doxorubicin to metastatic cancer cells. *Pharmaceutical Research*, 25: 2555–2566.

Xu, J. H., Li, R. X., Zhang, W., and Qi, F. 2008. Research progress of the mechanism on intraocular neovascularization. *International Journal of Ophthalmology*, 8: 2496–2498.

Xu, Y., Meng, F., Cheng, R., and Zhong, Z. 2009b. Reduction-sensitive reversibly crosslinked biodegradable micelles for triggered release of doxorubicin. *Macromolecular Bioscience*, 9: 1254–1261.

Xu, H., Meng, F., and Zhong, Z. 2009a. Reversibly crosslinked temperature-responsive nano-sized polymersomes: Synthesis and triggered drug release. *Journal of Materials Chemistry*, 19: 4183–4190.

Xu, J., Yuan, Y., Shan, B., Shen, J., and Lin, S. 2003. Ozone-induced grafting phosphorylcholine polymer onto silicone film grafting 2-methacryloyloxyethyl phosphorylcholine onto silicone film to improve hemocompatibility. *Colloids and Surfaces B: Biointerfaces*, 30: 215–223.

Xue, L. and Greisler, H. P. 2003. Biomaterials in the development and future of vascular grafts. *Journal of Vascular Surgery*, 37: 472–480.

Yang, W., Cheng, Y., Xu, T., Wang, X., and Wen, L. P. 2009. Targeting cancer cells with biotin–dendrimer conjugates. *European Journal of Medicinal Chemistry*, 44: 862–868.

Yang, Z. M., Wang, L., Yuan, J., Shen, J., and Lin, S. C. 2003. Synthetic studies on nonthrombogenic biomaterials 14: Synthesis and characterization of poly(ether-urethane) bearing a zwitterionic structure of phosphorylcholine on the surface. *Journal of Biomaterials Science, Polymer Edition*, 14: 707–718.

Yang, Y., Xu, Z., Jiang, J. et al. 2008a. Poly(imidazole/DMAEA)phosphazene/DNA self-assembled nanoparticles for gene delivery: Synthesis and in vitro transfection. *Journal of Controlled Release*, 127: 273–279.

Yang, X., Zhu, B., Dong, T. et al. 2008b. Interactions between an anticancer drug and polymeric micelles based on biodegradable polyesters. *Macromolecular Bioscience*, 8: 1116–1125.

Yasuda, H., Nakayama, K., Watanabe, M. et al. 2006. Nitroglycerin treatment may enhance chemosensitivity to docetaxel and carboplatin in patients with lung adenocarcinoma. *Clinical Cancer Research*, 12: 6748–6757.

Yurkovetskiy, A. V. and Fram, R. J. 2009. XMT-1001, a novel polymeric camptothecin pro-drug in clinical development for patients with advanced cancer. *Adv. Drug Deliv. Rev*, 61: 1193–1202.

Zasloff, M. 2002. Antimicrobial peptides of multicellular organisms. *Nature*, 415: 389–395.

Zhao, J., Duan, J., Zhang, Y. et al. 2009. Cationic polybutyl cyanoacrylate nanoparticles for DNA delivery. *Journal of Biomedicine and Biotechnology,* Hindawi Publishing Corporation, Volume 2009, Article ID 149254, 9 pp. doi: 10.1155/2009/14924.

Zhu, S., Hong, M., Zhang, L. et al. 2010b. PEGylated PAMAM dendrimer–doxorubicin conjugates: In vitro evaluation and in vivo tumor accumulation. *Pharmaceutical Research*, 27: 161–174.

Zhu, C., Jung, S., Luo, S. et al. 2010a. Co-delivery of siRNA and paclitaxel into cancer cells by biodegradable cationic micelles based on PDMAEMA–PCL–PDMAEMA triblock copolymers. *Biomaterials*, 31: 2408–2416.

Zürcher, S., Wäckerlin, D., Bethuel, Y. et al. 2006. Biomimetic surface modifications based on the cyanobacterial iron chelator anachelin. *Journal of the American Chemical Society*, 128: 1064–1065.

8 Polymer Implants for Intratumoral Drug Delivery and Cancer Therapy

Brent D. Weinberg and Jinming Gao

CONTENTS

8.1 INTRODUCTION

Cancer is an enormous health concern in the United States, resulting in almost 1.5 million cases and 550,000 deaths yearly. With only a few exceptions (such as testicular cancer and lymphomas), the definitive treatment of most solid tumors is surgical resection followed by adjuvant chemotherapy or radiation therapy to minimize the risk of recurrence. While some cancers are amenable to resection, other cancer types, such as cancers of the liver, are notoriously difficult to surgically remove. Reasons limiting resection include tumor size, involvement of more than one liver lobe, or a coexisting liver condition (e.g., cirrhosis) (Bentrem et al. 2005; Leung and Johnson 2001). In addition, the overall survival rates for these patients even after surgery are often low (Geller et al. 2006). The situation is even bleaker for cancers that have already spread by the time of their diagnosis. For example, out of 70,000 newly diagnosed colon cancer metastases to the liver in the United States per year, less than 10% of patients are actually eligible for surgery (Bentrem et al. 2005). Other abdominal cancers, such as those of the pancreas and stomach, also have low resection rates and poor overall patient survival, which have only marginally improved over the last decade (Jemal et al. 2009). Intravenously administered chemotherapy for these tumors also has limited usefulness. Since only a small amount of the systemic blood flow is directed to the tumor, a fraction of the total dose reaches the tumor site (Dowell et al. 2000). The remainder of the dose is distributed throughout healthy organs and tissues, leading to a variety of undesirable side effects ranging from neutropenia to cardiomyopathy (Crawford et al. 2004; Wallace 2003). Many chemotherapeutic drugs also have very rapid plasma clearance, meaning what little drug does make it to the tumor is removed relatively rapidly (El-Kareh and Secomb 2000). To improve the outcome of these cancer patients, a new paradigm of minimally invasive and locoregional cancer therapies has rapidly evolved and received considerable attention in the recent years (Gillams 2005).

Image-guided, minimally invasive techniques for therapeutic interventions use regional tumor destruction as an alternative to surgical resection. In each of these techniques, a conduit for administering the therapy, such as a needle or electrode, is inserted with image guidance into the desired treatment region and its position is confirmed. Image guidance can be provided using ultrasound, computed tomography, or magnetic resonance imaging (Clasen and Pereira 2008). Many options for local tumor ablation involve using a localized energy source, such as radiofrequency (RF) (Goldberg 2001), microwave (Martin et al. 2010), laser (Lindner et al. 2010), or ultrasound (Fischer et al. 2010) to heat the tumor to lethal temperatures. Alternative strategies for tumor destruction include cryoablation (Dale et al. 1998; Han and Belldegrun 2004) and chemical injection (Livraghi et al. 1998; Shah et al. 2004). Since they can be applied percutaneously, these minimally invasive treatments typically are viable alternatives to surgery that can be used in patients with poor overall health who might not be able to tolerate a surgical procedure. Additionally, local administration of the treatment maximizes destruction to the tumor target while limiting damage to the surrounding normal tissue. Ablation has been recently described for cancers in a wide variety of organ systems, including the esophagus (Gan and Watson 2010), lung (Nguyen et al. 2006), liver (Shiina 2009), kidney (Joniau et al. 2010), pancreas (Varshney et al. 2006), and prostate (Lindner et al. 2010). The role of local treatment will is different in each of these organ systems and is not likely to immediately supplant well-established treatments. However, focused ablation will provide new options for a subset of patients, and some authors suggest that the impact will be substantial (Bradley 2009).

Other attempts to improve treatment of unresectable tumors have focused on means to increase the tumor specificity of chemotherapeutic drugs through locoregional delivery (Goldberg et al. 2002). Administering an anticancer drug either to the region that contains a tumor or directly within the tumor has the advantage of increasing tumor exposure to a drug while limiting systemic toxicity. One strategy for locoregional chemotherapy is to infuse a solution of a chemotherapeutic agent into the region of the malignancy. Intravesicular instillation of either a drug or Bacillus Calmette–Guerin (BCG) has become a common treatment for superficial bladder tumors, and is associated with reduced tumor recurrence after surgery (Shelley et al. 2010). Local chemotherapy infusion has also been used with some success in the case of advanced nonsmall cell lung cancer when patients have a malignant pleural effusion (Seto et al. 2006). Additionally, several studies have shown significant benefits in treating ovarian cancers with intraperitoneal infusions (Alberts and Delforge 2006). All of these treatments require that malignant cells be in close contact with a potential reservoir to which the chemotherapy can be administered. For tumors that are not externally accessible, localized perfusion, or the administration of a chemotherapeutic agent to a segment of the circulation that preferentially perfuses the tumor, is an alternative. Intra-arterial administration of drugs can maximize drug delivery to blood vessels supplying tumors. For example, transarterial chemoembolization (TACE) benefits from the fact that most hepatocellular carcinomas receive the vast majority of their blood supply from their hepatic artery while normal liver receives its blood supply largely from the portal vein (Ramsey et al. 2002). In this treatment, a catheter is selectively placed in the branches of the hepatic artery that feed the tumor. Once arterial selection has been achieved, a solution of chemotherapeutic agents dissolved in an oily solvents followed by embolic agents is infused into the artery. This approach has been shown to increase concentrations of chemotherapeutic agents in the tumor by 10- to 100-fold (Ramsey et al. 2002) and to improve 1 year survival of patients with unresectable hepatocellular carcinoma (HCC) by as much as 20% (Kokudo and Makuuchi 2004). As a result, TACE has become a commonly administered therapy for unresectable HCC (Tsochatzis et al. 2010). Another strategy to improve delivery is through regional perfusion, in which the portion of the systemic circulation containing a tumor is isolated from the rest of the circulation (Muller and Hilger 2003). Isolated thoracic perfusion (ITP) is achievable by closing off the descending aorta and vena cava with balloon catheters, blocking blood flow to the arms with inflated cuffs, and introducing chemotherapy into the right atrium. This approach has been used to increase the concentration of chemotherapy delivered to lung cancers by 6- to 10-fold (Muller 2002) and an analogous approach exists for isolated abdominal perfusion. Each of these locoregional chemotherapy methods shares the same goal of increasing tumor exposure to drug while reducing systemic toxicity.

Intratumoral cancer treatments extend the locoregional treatment concept by attempting to further limit the scope of chemotherapy exposure. Treatments that have been studied extensively include intratumoral infusions, injections, and implantable devices that deliver either chemotherapeutic drugs or other therapeutic agents (Goldberg et al. 2002). Infusion of chemotherapeutic agents has been heavily studied in the area of brain tumors, where it has spawned a field known as convection-enhanced delivery (CED) (Hall and Sherr 2006). In CED, a microcatheter is inserted into a tumor and the therapeutic agent is slowly administered to the surrounding tissue using positive pressure infusion. Major advantages of CED to brain tumors include bypassing the blood–brain barrier and delivering drugs further from the infusion site due to convection (Raghavan et al. 2006). CED has been used to deliver conventional chemotherapeutic drugs (Mardor et al. 2001) but has shown considerable promise for the delivery of targeted bacterial toxins (Hall and Sherr 2006) and therapeutic antibodies (Sampson et al. 2006). Intratumoral injections of therapeutic solutions have also shown success in treating tumors in locations other than the brain, such as the lung (Celikoglu et al. 2006) and liver (Mok et al. 2001). Several studies have been performed in an attempt to determine optimal parameters for injection and to determine which tumor features, such as collagen content, contribute to the extent of drug delivery (McGuire and Yuan 2001; McGuire et al. 2006). Furthermore, recent studies have shown that using intratumoral injection to deliver viral gene therapy vectors minimizes nonspecific expression of gene products (Wang et al. 2005; Wang and Yuan 2006). Since intratumorally injected liquids may distribute irregularly and be cleared quickly, several investigators have introduced injectable drug depots to prolong the extent of drug release. Examples of intratumoral depots include poly(D,L-lactide-*co*-glycolide) (PLGA) (Emerich et al. 2002; Menei et al. 2004, 2005), alginate (Arica et al. 2002), and albumin (Almond et al. 2003) microsphere formulations as well as injectable gels, which solidify upon intratumoral injection (Jackson et al. 2000; Krupka et al. 2006; Vogl et al. 2002). Injectable depots have the advantage of easy administration and prolonged tumor drug exposure.

Intratumoral, drug-releasing implants are a subset of the intratumoral drug delivery paradigm and have shown increasing promise in recent years. Driven by developments for the treatment of prostate (Grimm and Sylvester 2004; Merrick et al. 2003) and brain cancers (Guerin et al. 2004; Wang et al. 2002), implantable devices containing either radioactive elements or chemotherapeutic drugs have become viable treatment options. The only clinically approved chemotherapeutic implant for cancer treatment is the Gliadel wafer, a carmustine (BCNU)-eluting implant fabricated from a polyanhydride copolymer, 1,3-bis-(*p*-carboxyphenoxy)propane/poly(sebacic acid) (pCPP:SA) (Brem and Gabikian 2001). These implants were designed to treat glioblastoma multiforme, an aggressive brain cancer with extremely limited patient survival, through placement in the surgical cavity after primary surgical resection. After placement, the implants release their drug load over a period of approximately 5 days (Fleming and Saltzman 2002), and drug has been shown to penetrate several millimeters into the brain parenchyma (Strasser et al. 1995). The most extensive long-term study showed that the Gliadel implant placement after surgery increases patient survival to 13.8 months vs. 11.6 months for surgery alone, while maintaining this survival advantage for at least 3 years after initial treatment (Westphal et al. 2006). Despite ongoing work in the development of intratumoral implants over the last 15 years, the use of chemotherapeutic implants has yet to become widespread in the treatment of other cancers, such as those of the pancreas, liver, or lungs. However, it is likely that future chemotherapeutic implants can be optimized for use in a variety of different tumors to maximize patient comfort and survival.

This review article describes work in extending the use of intratumoral implants to treat unresectable liver tumors as part of a combined treatment strategy using ablation and intratumoral implants. However, many of the results and much of the ensuing discussion apply to any intratumoral drug delivery system. The proposed treatment involves the following two steps: (1) Primary treatment of the tumor bulk with radiofrequency (RF) ablation, followed by (2) placement of drug-eluting polymer implants in the ablated tumor region. These biodegradable polymer millirods have been fabricated from PLGA to deliver chemotherapeutic agents through and beyond the RF ablated tumor, thus maximizing tumor destruction and reducing the risk of tumor recurrence. Section 8.2 describes the overall goals that must be considered when developing any local delivery device, including the use of models to predict local drug transport. Section 8.3 describes techniques for measuring local drug concentrations and the use of these measurements to customize drug release. Section 8.4 describes the preliminary results from using these implants to treat a rabbit liver cancer model. Finally, Section 8.5 presents some conclusions drawn from the early use of these implants and some future goals to facilitate using these implants to treat larger tumors similar to unresectable human cancers.

8.2 OVERVIEW OF DRUG DELIVERY GOALS

8.2.1 Definition of Pharmacokinetic Goals for Local Drug Delivery to Unresectable Tumors

In designing an intratumoral implant, it is necessary to consider the generic characteristics that would benefit the device. First, the implant should be able to minimize shortcomings associated with systemically administered chemotherapy. Second, it should provide an optimal drug delivery profile to the tumor, which is to say that it should be able to provide effective drug concentrations to the desired region over a prolonged period of time. Third, the device should be part of a comprehensive and complete treatment strategy that is versatile and applicable in a wide range of realistic situations. Achieving these goals should maximize the treatment success of these intratumoral implants.

When delivering their drug to tumors, intratumoral implants must be able to deliver drug to a large volume, to rapidly reach the therapeutic concentration, and to maintain the therapeutic concentration for an extended time. Previous studies have shown that limited drug penetration distance is one of the major restrictions on the efficacy of intratumoral treatments (Fleming and Saltzman 2002; Strasser et al. 1995). Any successful implant must be designed in such a way that maximizes the distance of drug delivery into tissue. To achieve this goal, the implant must provide drug to the surrounding tissue at an appropriate rate (Qian et al. 2002b). A schematic of ideal drug release rates is shown in Figure 8.1. Consider first an implant A, an implant which releases drug at a constant rate somewhere above the elimination rate. While local drug concentration will slowly rise, it may take too long to reach tissue concentrations that are toxic to the surrounding cancer cells. Alternatively, implant B provides a rapid dose of chemotherapy that will quickly surpass the effective concentration. However, the release rate after the initial burst is insufficient to maintain this concentration for any extended length of time. Such a release rate is undesirable, as it could allow cancer cells to recover, perhaps even with newly acquired drug resistance (Chu 1994). The ideal implant, implant C, combines the best characteristics of both implants: rapid ascent to the effective concentration followed by a maintenance dose to remain at a useful drug level. While this explanation is a simplification (e.g., elimination is almost certainly not constant, etc.), it serves as an example of how different drug release rates might affect local drug concentrations. Additionally, it offers some insight on how the situation can be changed by modifying the elimination rate or therapeutic concentration.

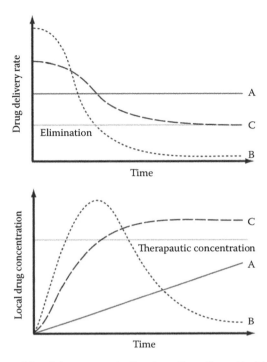

FIGURE 8.1 Drug release and local drug concentration from three theoretical implant types. A zero-order release implant (A) releases drug at a constant rate, but it may take a long period of time to reach the therapeutic concentration. A burst-release implant (B) releases large amounts of drug quickly, but may not have enough continuous release to maintain a therapeutic concentration. A dual-release implant (C) combines an early burst of drug to accelerate the rise to therapeutic concentrations with sustained release to maintain therapeutic concentrations.

These factors are going to be different depending on the type of tissue into which the implants are placed, the type of tumor, and other factors. For this reason, local drug concentrations surrounding implants must be usually be determined experimentally.

The usefulness of chemotherapeutic implants as a clinical treatment also depends on their inclusion in a comprehensive and practical tumor treatment strategy. As an example, with the previously mentioned Gliadel treatment, the tumor is first surgically resected ("debulked") followed by the placement of multiple bis-chloroethylnitrosourea (BCNU)-impregnated implants in the surgical cavity (Guerin 2004). The design of a liver cancer treatment using the polymer millirods discussed in the rest of this chapter proposes a similar strategy to Gliadel treatment: using radiofrequency (RF) ablation to destroy the majority of the tumor mass followed by placement of polymer implants in the tumor to kill any residual cancer cells and limit tumor recurrence. There are several reasons behind the selection of this combined strategy. First, RF ablation is already used clinically to treat liver tumors, but its clinical success has been limited by tumor recurrence, particularly around the ablation boundary (Harrison et al. 2003; Wang et al. 2005). Using chemotherapeutic implants with ablation may also maximize the effects of using the implants themselves. The ablation destroys the majority of the tumor cells, leaving the implants to kill only the remaining cells, thereby reducing the risk of recurrence. Furthermore, tumor ablation may facilitate drug delivery from the implants by changing the fundamental rates governing drug transport in the tumors (Qian et al. 2002b). To develop this strategy, drug-impregnated, PLGA polymer millirods with different release rates were developed and tested in animal models, first in nonablated and ablated liver tissue (Qian et al. 2001, 2002a, 2004; Szymanski-Exner et al. 2003a) and then in nonablated and ablated liver tumors (Weinberg et al. 2006). Results from these studies are described in Sections 8.3 and 8.4.

8.2.2 Interstitial Drug Transport Models in Tumor and Surrounding Tissues

In addition to generic considerations on drug release from implants, the mechanisms of drug transport and elimination in the surrounding tumor tissue have a major effect on how effectively an implant delivers drug to the tumor (Jain 1999). Drug released locally into the tumor has several possible fates that will ultimately affect the outcome of the treatment: (1) drug can be retained in an area and exert its desired effect; (2) drug can move to another location through a transport process; or (3) drug can removed or altered such that it no longer has its desired effect.

Drug can typically be transported by two mechanisms: diffusion and convection (Fleming and Saltzman 2002; Jain 2001; Sinek et al. 2004). In diffusion, the free drug moves from a region of higher drug concentration to an adjacent region of lower drug concentration at a rate proportional to the concentration gradient. Diffusion is the predominant form of transport in tissues that are mostly solid. Convection, on the other hand, is the transport of drug along with the bulk flow of a fluid. In organs that have a high rate of interstitial fluid flow, convection is especially important. Convection also has a significant role in the flow of systemically administered chemotherapeutic agents from the vascular space to the tumor, where it travels along with the interstitial flow that delivers nutrients to the tumor (Jain 1999). The relative importance of diffusion or convection in drug transport depends on the delivery system and tissue type. For instance, in the brain, where cerebrospinal fluid constantly flows from the ventricles to the surrounding parenchyma, convection has a significant effect on the extent of drug penetration (Kalyanasundaram et al. 1997). In situations with small molecular drugs where flow is more limited, diffusion is the predominant mode of drug transport.

Like drug transport, drug elimination can occur through several different mechanisms. One route of drug elimination is through metabolism. Once in a cell, drug can be altered or bound in a variety of ways. Some drug molecules, such as 5-fluorouracil, bind irreversibly to their therapeutic target, after which they are no longer in the population of available drug (Longley et al. 2003). More generally, cells have a variety of nonspecific methods for detoxification, such as organelles for breaking down foreign molecules through enzymatic or pH-mediated degradation. Alternatively, cells contain protective molecules, such as glutathione, which are designed specifically to bind foreign molecules and render them more hydrophilic and less potent (Tanner et al. 1997). Either of these metabolic pathways essentially inactivates the drug. When considering implantable drug delivery systems, another mechanism of drug loss is perfusion away from the target region (Qian et al. 2002b). In this situation, drug is transported by either diffusion or convection into one of two systemic circulations, the blood or the lymph. The vasculature is a fast-moving circulation that rapidly moves drugs away from the target region and into other parts of the body. Since most chemotherapeutic agents have short plasma half-lives, once the drug reaches the plasma it is unlikely to return to the target tumor, and for practical purposes can be considered eliminated. While the lymph is a slower-moving body of fluid that can contribute to drug convection, its effects are probably less influential than those of blood because tumors are known to have limited and poorly organized lymphatic drainage (Jain 2001). Lymphatic elimination may also have the unintended consequence of accidentally treating any isolated tumor cells that may have begun to spread along lymphatic channels. Ultimately, drug that moves into the lymphatic system is dumped into the venous circulation, where it undergoes the same fate as drug that directly diffuses into blood vessels.

Consider the example of drug transport shown in Figure 8.2, in which drug is being delivered to a liver tumor from a cylindrical implant in the center of the tumor. Previous work has shown that transport in liver can be reasonably approximated without including convection because of the reasons described earlier (Qian et al. 2002b). Drug leaving the implant is transported away from the implant into the tumor tissue based on a tumor diffusion rate, D_{tumor}. Once in the tumor, drug can be eliminated in one of two ways, through blood flow and metabolism, proportional to two different constants which sum to contribute to a total elimination, γ_{tumor}. Once drug reaches the outer boundary of the tumor, it can diffuse into the surrounding normal liver tissue, where its fate is again

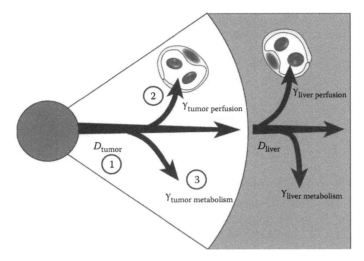

FIGURE 8.2 Simplified scheme of drug transport from an implant centrally placed in a liver tumor. Transport of the drug into the tumor tissue is governed by the diffusion constant of drug in tumor (D_{tumor}), and two simultaneous modes of elimination: metabolism to inactive forms in tumor cells ($\gamma_{tumor\ metabolism}$) or transport into nearby blood vessels which wash drug out of the region ($\gamma_{tumor\ perfusion}$). Once drug reaches the surrounding normal tissue, it continues to diffuse outward into liver tissue (D_{liver}), where it has different rates of elimination by metabolism ($\gamma_{liver\ metabolism}$) or perfusion ($\gamma_{liver\ perfusion}$). Three possibilities for intervening to improve drug penetration are labeled: (1) increasing the diffusion constant in tumor, D_{tumor}, (2) reducing losses to tumor perfusion, and (3) reducing the rate of tumor drug metabolism.

governed by new diffusion and elimination rates. If elimination can be considered approximately first order, the drug transport in each tissue is governed by the following equation:

$$\frac{\partial C}{\partial t} = D\nabla^2 C - \gamma C \tag{8.1}$$

where
  C is the drug concentration
  t is the time
  ∇ is the gradient operator
  D and γ are the tissue-specific rates of diffusion and elimination, respectively

Drug transport properties in each tissue can be estimated by solving this equation computationally and minimizing the error between model output and experimentally collected data.

The use of such a model provides insight into that can facilitate or impede drug transport from a local implant. First, any factor that increases the rate of drug transport or lowers the rate of drug elimination will increase the permeation of drug within the tissue. For instance, some work in ex vivo tumors has shown that prolonged drug exposure raises drug diffusion coefficients, presumably by killing cells and destroying overall structure (Au et al. 2002). Similarly, it can be expected that any factor that reduces elimination will also have a beneficial effect, facilitating deeper drug penetration into the tumor. Other work by Saltzman has shown that including high-molecular-weight molecules, such as dextrans, can increase transport away from implants by increasing the convective fluid flow contribution while decreasing blood perfusion (Strasser et al. 1995). On the other hand, any action that decreases transport or increases elimination will act as a barrier to drug delivery that will effectively reduce the distance over which an implant can be effective. Inflammation, such as that occurring after RF ablation, may raise blood flow and decrease drug diffusion rates as a result of collagen deposition around the wound (Blanco et al. 2004). These side effects could certainly impede successful drug transport from a local tumor treatment. The distance at which an implant can have an effect on a tumor depends on the drug release rate from the implant as well as the balance between local transport and elimination. Unfortunately, past studies have indicated that antitumor implants are likely only effective for a few millimeters away from the implant surface (Wang et al. 1999). Studies into local drug concentration and local transport mechanisms, however, have provided useful information on ways to overcome these limitations.

8.3   MEASURING AND MODULATING LOCAL DRUG PHARMACOKINETICS

8.3.1   Overview of Methods to Investigate Local Drug Release and Tissue Distribution

In developing an intratumoral chemotherapy device, techniques for monitoring local drug concentrations are necessary to optimize implant design. Measuring drug concentration as a function of time provides a quantitative method to compare multiple treatments. Many different techniques can be used to monitor drug release from intratumoral implants. While the gold standard for determining drug concentrations is extraction of tissue and measurement of drug content ex vivo, alternate, noninvasive imaging–based techniques can be used to measure concentrations in vivo. New implant designs or treatment conditions can be tested by creating an implant that has the ideal characteristics described in Section 8.1; a rapid ascent to and prolonged stay above the therapeutic concentration. Additionally, drug concentration information can then be used as input to estimate tissue transport properties. These steps provide a feedback system for continually improving implants through a combination of empirical testing and engineering design.

Considerable information has been obtained by monitoring local drug concentrations using ex vivo analysis of extracted tissues. Two main categories of ex vivo analysis exist: bulk tissue analysis by conventional spectroscopic methods or tissue section analysis by imaging-based methods. The key principle of bulk tissue measurements is the removal of a sizeable piece of tissue followed by the use of a spectroscopic method to determine the average drug concentration in that tissue (Haaga et al. 2005). For targeted drug delivery to tumors, drug concentrations in different tumor regions are the most important, so these tissues are often removed in different sections. To determine tissue drug concentrations, tissues are weighed and either mechanically or chemically homogenized according to the desired detection mechanism. Examples of techniques to measure drug concentration in the extracted tissues include fluorescence detection, high-performance liquid chromatography (HPLC) (Haaga et al. 2005; Zheng et al. 2001), mass spectrometry, and atomic absorption spectroscopy (AAS) (Szymanski-Exner et al. 2003a). By radiolabeling the drug prior to implantation, drug concentrations can also be measured using liquid scintigraphy. Key advantages of measuring drug concentrations in removed tissues include definitive drug detection, high sensitivity, and the ability to detect low or even extremely low drug concentrations. However, these techniques are restricted by low spatial resolution and accuracy, as measurements are an average over an entire piece of tissue. The ability to resolve drug concentrations in different tissue locations depends on the size of pieces cut from the tissue, which usually limits spatial resolution from these techniques to the millimeter range. Imaging of ex vivo tissues can help overcome the spatial resolution limitations of bulk tissue analysis methods. For imaging-based methods, the tissue is removed and sliced into a thin piece followed by drug detection through imaging the slice. At least two techniques have been used for imaging drug detection: autoradiography and fluorescence (Au et al. 2002). For autoradiography, the drug target is radiolabeled, and concentrations of radiolabeled drug in the section of tissue are measured using a flat panel detector or x-ray film (Au et al. 2002; Strasser et al. 1995). The advantages of this technique include high sensitivity, very low detection limits, and high resolution, while the major limitation is working with a radiolabeled drug. Particularly when designing drug containing implants, it can be difficult or even impossible to manufacture implants containing radioactive materials. Moreover, manufacture and usage of the implants then becomes limited by the half-life of the tracer used. Fluorescent detection of drug concentration in tissues is an alternate strategy. In this method, tissue slices are analyzed using a fluorescence scanner or fluorescent microscope to detect drug (Zheng et al. 2001). To use this technique, the drug must be either intrinsically fluorescent or labeled with a fluorescent tag, such as fluorescein isothiocyanate (FITC) (Lu et al. 2004). While also offering low detection limits, reasonable sensitivity, and good resolution, only a few small molecule drugs are fluorescent, and labeling of drugs inevitably modifies their transport and efficacy, unlike radiolabeling methods. For large molecules, such as protein drugs or antibodies, fluorescent labeling may have only a minimal effect on the overall drug properties and may not adversely affect the delivery system, making the approach more tenable. In summary, ex vivo drug detection is a major tool in developing local drug delivery methods, but temporal information is limited because every time point requires animal euthanasia and removal of tissue.

Noninvasive imaging methods for measuring local drug concentrations represent a growing attempt to improve on the temporal limitations of ex vivo approaches. Driven by advances in imaging technology as well as proliferation of scanner availability, noninvasive imaging methods likely hold the future for monitoring drug concentrations from local delivery strategies. With noninvasive imaging, a single subject can be imaged several times throughout the study period, greatly increasing the data available from a smaller number of animal subjects. The most straightforward extension of previous detection technologies is continued use of radiolabeled drugs coupled with positron emission tomography (PET) (Roselt et al. 2004) or single-photon emission tomography (SPECT) for drug detection (Dowell et al. 2000). These detection methods have existed clinically for several years, but recent development of specialized small animal scanners, often coupled with computed tomography (CT) for anatomical information, has improved resolution and usability, making nuclear medicine techniques key for development of targeted therapies. Additionally, these techniques can be easily translated to clinical use for monitoring clinical trials of newly developed devices or treatment strategies. Other imaging techniques, such as in vivo fluorescence imaging, have been specifically developed for use in small animals and can contribute primarily to small animal studies. With in vivo fluorescence imaging, fluorescently labeled molecules are imaged directly in the animal (Moon et al. 2003). Most fluorescent imaging suffers from greater background noise than radiographic imaging and limitation to two dimensions, but developments in tomographic fluorescence have provided noise reduction and are beginning to offer 3D localization of drug (Ntziachristos 2006). The major limitations of in vivo fluorescence are limited by light penetration and high scatter, which prevent its extension into human use. Beyond radiolabeling and fluorescence, magnetic resonance imaging (MRI) detection of drugs or drug carriers labeled with an MRI contrast agent, such as gadolinium or superparamagnetic iron oxide (SPIO), also offers the potential to noninvasively image anatomical detail and drug concentrations simultaneously (Nasongkla et al. 2006; Weinmann et al. 2003; Ye et al. 2006). Recent advances and the benefits afforded by noninvasive imaging make it likely that these techniques will dominate the future landscape of monitoring local drug delivery strategies.

A novel noninvasive method used in the development of polymer millirods for liver cancer treatment is drug detection using x-ray CT (Exner et al. 2004; Szymanski-Exner et al. 2002, 2003b). Polymer implants were loaded with the anticancer drug carboplatin and tested in both normal and ablated rat liver tissue. Carboplatin has a unique property among cancer drugs in that it contains the heavy metal platinum (atomic number 78), which has high x-ray attenuation and provides inherent CT contrast. Polymer millirods containing 10% carboplatin and 90% PLGA were implanted in nonablated or RF ablated rat livers (Exner et al. 2004). To detect the carboplatin, the rats were scanned with CT at multiple time points after the implantation. A representative CT scan of one of these rats is shown in Figure 8.3. Slices perpendicular to the long axis of the implant clearly show the higher absorption of the implant compared to the surrounding tissue. By comparing the intensity of the implant to premeasured implants with known concentrations, the remaining carboplatin in the implant was determined. The drug concentration in each voxel of the surrounding tissue, $C_d(x, y, z)$ was then determined after subtracting the background signal due to ablated liver tissue

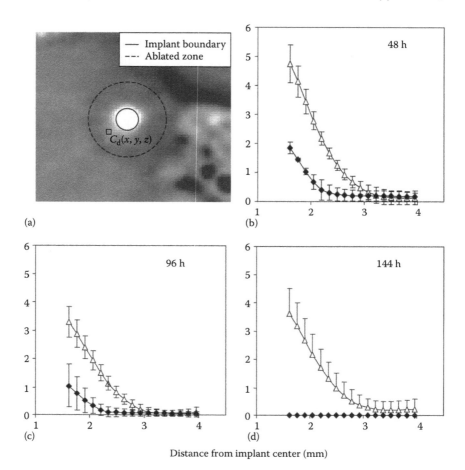

FIGURE 8.3 Computed tomography (CT) scan of rat liver containing a carboplatin-loaded polymer millirod in a region of liver treated by radiofrequency ablation (a). The boundary between the implant and ablated liver (solid line) and ablated and normal liver (dashed line) are denoted. Drug concentrations at a given location, $C_d(x, y, z)$, can be determined from the image intensity contributed by carboplatin in the ablated tissue. Carboplatin distribution in normal (υ) and ablated (ρ) liver tissue measured by CT at 2 days (b), 4 days (c), and 7 days (d) after implant placement. Values reflect mean ± SD between animals.

and accounting for partial volume effects. Tissue concentrations of carboplatin determined from CT images are shown in Figure 8.3. These measurements were then validated using atomic absorption spectroscopy measurements in extracted tissues (not shown). CT provided higher spatial resolution revealed differences in drug distribution in nonablated and ablated tissues not appreciated in comparable extracted tissue measurements. Ablated tissue retained carboplatin for longer times and at greater distances from the implant than nonablated tissue, illustrating a fundamental difference between drug transport in these tissue environments. Additionally, these results demonstrate a novel use of computed tomography to monitor local drug release from implants.

Further studies of local drug concentration around implants using fluorescent imaging allowed for greater quantification of the differences between ablated and normal tissue (Gao et al. 2002). Doxorubicin, a commonly used chemotherapeutic agent (Leung and Johnson 2001), also has the fortuitous property of natural fluorescence with an excitation of 488 nm and an emission of 595 nm. Polymer millirods containing doxorubicin were implanted in nonablated and RF-ablated rat livers (Qian et al. 2003). Liver tissues were removed at various times after implantation, and tissues were sliced into 100 μm sections using a cryostat microtome. Sections were thawed and imaged for

TABLE 8.1

Doxorubicin Transport Properties in Rat Liver

	Normal Liver	Ablated Liver
Apparent diffusion, D^* (cm^2 s^{-1})	6.7×10^{-7}	1.1×10^{-7}
Apparent elimination, γ^* (cm^2 s^{-1})	9.6×10^{-4}	n/a

fluorescence intensity using a fluorescence scanner. Drug distribution profiles were determined by converting net fluorescence intensity to drug tissue concentrations, and average doxorubicin concentrations at each distance from the implant boundary were calculated. These tissue drug concentrations at time points ranging from 1 h to 4 days after implant placement were used to estimate the transport properties of liver tissues within the framework of a theoretical model of drug distribution as described in Section 8.2. The resulting estimates are shown in Table 8.1. These studies established the baseline transport properties of doxorubicin in normal rat liver as well as how ablation modifies them. Ablation reduced the diffusion coefficient, most likely by destroying cell structure and making more sites available for drug binding. Even more notably, drug elimination in the ablated tumor tissue became negligible. This finding might be the expected outcome for two reasons. First, RF ablation kills cells within the central ablation zone, thus stopping active cellular mechanisms for binding and eliminating drug. Second, ablation results in coagulation and destruction of blood cells within the ablated tissue, which should limit perfusion-related losses of drug. This reduction in drug elimination can largely explain why drug penetration distances and retention were higher in ablated liver tissue. In summary, techniques for measuring local drug concentrations surrounding implants are essential for the development of a local drug delivery system for tumors. Many methods exist for measuring drug concentrations, each with advantages and disadvantages, and it is likely that a combination of methods provide the best overall information about drug delivery. After obtaining local drug concentrations, they can be compared empirically to determine qualitative differences in delivery or interpreted through the use of a model to obtain quantitative transport information. Both sets of data can then be used to modify implant properties to provide the best drug coverage to the tumor.

8.3.2 CONTROLLING DRUG RELEASE AND LOCAL PHARMACOKINETICS FROM POLYMER IMPLANTS

Development of techniques to monitor local drug pharmacokinetics allows for the design and assessment of different implant types. As described in Section 8.2, an ideal implant should provide a rapid ascent to the therapeutic concentration and maintenance of this dose for as long as possible. The first generation polymer millirod provided rapid release of a drug mimic, largely within the first few days (Qian et al. 2001). However, modifying implant design makes it possible to customize the drug delivery profile of the implants. Consequently, local drug concentrations arising around the implants can be compared and evaluated based on overall tissue drug exposure.

Several modifications to the initial compression-heat-molded millirods can either prolong drug release or change the timing of the released dose. The first such modification is the addition of a semipermeable membrane around the outside of the implant. Physically, this modification can be made either by wrapping the cylindrical device with a membrane (Qian et al. 2002a) or by dip-coating the implant (Qian et al. 2004). To wrap the implant with a membrane, thin films of PLGA containing NaCl (10%–50% w/w) were solvent cast onto a Teflon dish and then wrapped around the implant. When placed in an aqueous environment, the NaCl component of the outer membrane rapidly dissolved, leaving a semipermeable membrane that could be controlled by modulating the amount of NaCl included in the implant. The addition of such a barrier slowed the overall rate of drug release, allowing for drug release over a period as long as 5 weeks (Qian et al. 2002a).

Dip-coating the implants followed a similar strategy but allowed for a more uniform coating process. To dip-coat the implants, a solution of poly(lactic acid) (PLA) and poly(ethylene oxide) (PEO) was created by dissolving the polymers in methylene chloride. The cylindrical implants were then dipped in the polymer solution and allowed to dry, creating a membrane around the original implant (Qian et al. 2004). Once the implant was exposed to water, the PEO fraction rapidly dissolved, leaving a semipermeable membrane. Again, the release rate could be modified by changing the PEO content (5%–20% w/w) of the layer (Qian et al. 2002a). Either wrapping or dip-coating the implant to create a membrane around it substantially prolonged the drug release from the resulting implants.

By further modifying the polymer millirods, it is possible to create an implant that adds an additional burst dose to the prolonged drug release. As discussed in Section 8.2, the tissue surrounding a sustained release implant may not reach the therapeutic concentration for some time, delaying the onset of action of the drug. To accelerate the rise to the therapeutic concentration, a burst dose can be added to the implant to act as a loading dose. Dual-release implants combining the benefits of a drug burst followed by sustained release were then created by supplementing the implant with two drug coatings (Qian et al. 2002b). Monolithic millirod implants were first created by compression molding of PLGA (60%), NaCl (24%), and doxorubicin (16%). To sustain the release of drug from this implant, it was dip-coated with a layer of PLA/PEG as described earlier. Then, this coated implant underwent further dip-coating with a suspension of doxorubicin (75%) and PEO (25%) in methylene chloride. The total burst amount of drug could then be controlled by applying multiple coatings to increase the thickness of the burst layer. The resulting implants, termed dual-release millirods, released a burst dose of doxorubicin followed by a sustained dose of doxorubicin for as long as 10 days (Qian et al. 2002b). In this manner, polymer millirods that could release doxorubicin into tumors with different dose timings were created.

To evaluate the differences in local drug distribution generated by different implant types, these burst, sustained, and dual-release millirod formulations were tested in vivo (Qian et al. 2004). In the rat model, liver tissue was ablated for 2 min at 90°C to create an ablation region 8–10 mm in diameter. Subsequently, polymer millirods of each type were placed in the ablation needle tract and sutured into place. At specified time points, the rats were euthanized, and the polymer implants and surrounding liver tissue were extracted. Doxorubicin remaining in the implant was quantified by an extraction procedure and used to calculate average release rates, which are shown in Figure 8.4. As expected, the dual-release implants released a higher burst dose of drug in the first 24 h (denoted by the brackets), but after this time the drug release rates were not statistically different. Tissue doxorubicin concentrations were determined using fluorescence scanning of sliced tissues, and the doxorubicin concentration at the outer ablation boundary is shown in Figure 8.4. The dual-release implants provided a more rapid ascent to therapeutically relevant concentrations that was statistically different from the sustained-release implants. The similarity between the experimental results and the theoretical profiles shown in Figure 8.1 (panel 2, curves A and C) is notable. Detailed fluorescent images of tissues that confirm this finding are shown in Figure 8.4. The difference in the sustained release and dual-release implants is emphasized by the differences between tissue drug concentrations on Day 1 (Figure 8.4, panels c and d). Dual-release implants led to local doxorubicin concentrations as high as 1000 μg/g within one day, while it took nearly 4 days for the drug distributions around the sustained implant to reach this extent. This difference was likely provided by the additional burst of drug in the first 24 h of implantation provided by the dual-release implants. This study established that differences in implant formulation could have a substantial impact on local drug concentrations.

8.3.3 Role of the Host Tissue Response in Local Drug Therapy

Histology studies of treated tissue from ablated livers were performed to provide a more detailed understanding of the effect of changing tissue properties on drug transport (Blanco et al. 2004).

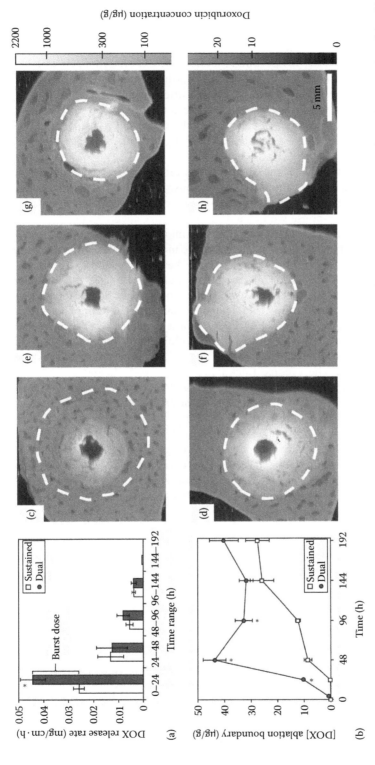

FIGURE 8.4 Average rate of drug release from two implant formulations *in vivo* in ablated rat livers (a). Doxorubicin concentration at the outer edge of the ablated region for the same two implant formulations in ablated rat livers (b). Fluorescence imaging comparing doxorubicin concentrations in ablated liver after placement of sustained release vs. dual-release millirod implants after 1 day (c, sustained vs. d, dual-release), 4 days (e, sustained vs. f, dual-release), and 8 days (g, sustained vs. h, dual-release). The largest difference between the two implant types is on Day 1 (c vs. d) because of the additional burst dose from the dual-release implants. The dotted line is the ablation boundary. (From Weinberg, B. D., et al. *J. Pharm. Sci.*, 97, 1681, 2008. Figures 5 and 6.)

Ablated rat livers were treated with doxorubicin-containing polymer implants, and tissues were subsequently removed at time points ranging from 1 h to 8 days after ablation. Throughout the first 4 days after ablation, an area of coagulation necrosis surrounding the implant was gradually infiltrated by inflammatory cells, particularly neutrophils and monocytes. By 8 days after ablation, fibroblasts and the formation of a dense, collagenous fibrous capsule were evident at the ablation boundary. Furthermore, it was found that doxorubicin concentrations leading up to the fibrous capsule were high but dropped precipitously in the nascent fibrous capsule. From these results, it appeared that the wound healing response after ablation could have a major role in drug diffusion, acting as a barrier to transport outside the ablation region (Blanco et al. 2004). This finding reiterates the importance of considering the tissue surrounding the implant not as a static environment, but instead as a dynamic milieu that can ultimately affect the success of the treatment itself.

Since the tissue surrounding the implant has a large impact on the efficacy of drug therapy, one strategy to overcome this is to modify the response of the surrounding tissue to promote more effective drug delivery. One way to modify the properties of ablated tissue is to moderate the ensuing inflammatory response with an anti-inflammatory agent. To test this hypothesis, the potent corticosteroid dexamethasone (DEX) was loaded into PLGA millirod implants (Blanco et al. 2006). Given that dexamethasone is a very hydrophobic molecule, DEX was first complexed with hydroxypropyl β-cyclodextrin (HPβ-CD) to increase the water solubility before incorporation in the implants. The final formulation of the implants contained 1.7% dexamethasone, 38.3% HPβ-CD, and 60% PLGA. Implants containing the complexed dexamethasone released 95% of their drug over 4 days, compared to 14% of dexamethasone released in implants containing 1.7% uncomplexed dexamethasone (Blanco et al. 2006). These implants were tested for their ability to reduce fibrous capsule formation following liver ablation in rats. Dexamethasone-impregnated implants drastically suppressed the thickness of the collagen fibrous boundary compared to a control ablation treatment. The average thickness of the fibrous boundary in subjects receiving a DEX-containing implant (0.04 ± 0.01 mm) was reduced both compared to a control ablation (0.29 ± 0.08 mm) and ablation followed by an intraperitoneal (IP) DEX injection (0.26 ± 0.07 mm) (Blanco et al. 2006). In addition to reducing fibrous capsule formation, it is speculated that dexamethasone may have other beneficial effects when administered after ablation. Chemokine release, growth factor production, and angiogenic processes during wound healing have been implicated in tumor growth and recurrence (Coussens and Werb 2002; Raz et al. 2000; Wang et al. 1998). At least one report has suggested that liver tumor recurrence after ablation is potentiated by inflammation (Harrison et al. 2003). For this reason, including dexamethasone in the implants may improve the primary outcome of ablation while facilitating drug delivery. Future polymer implants which concomitantly release dexamethasone and an anticancer agent may maximize the therapeutic benefits of using polymer millirod devices after RF ablation.

8.3.4 Summary of Factors Influencing Implant Design and Drug Release

Several different implant types were developed and evaluated in thermal ablation models of rat and rabbit livers. Two factors appeared to have a role in the extent of drug delivery: the rate of drug release and the properties of the surrounding tissue. Dual-release implants, consisting of two dip-coated layers, provided the fastest ascent to therapeutic concentrations and maintained local concentrations for at least 8 days. RF ablation, by destroying the surrounding vasculature, potentiated drug release into the surrounding tissue but may have ultimately restricted it by instigating the encapsulation of the ablated region within a thick fibrous shell. One approach, including dexamethasone complexed with β-cyclodextrin in the implants, showed the potential to overcome this limitation. Overall, studies of drug release from implants demonstrated that PLGA implants are a versatile platform for drug delivery that is capable of different release kinetics and local pharmacokinetics following RF ablation.

8.4   TREATMENT OF ANIMAL TUMOR MODELS

8.4.1   DRUG DISTRIBUTION AND ANTITUMOR EFFICACY FROM LIVER TUMOR TREATMENT WITH POLYMER IMPLANTS

After extensive pharmacokinetic study of polymer millirods in normal livers, studies of drug distribution and treatment efficacy in tumor tissue were performed. One study assessed the use of implants alone for treatment and local control of small liver tumors (Weinberg et al. 2006); the second study explored drug distribution and therapeutic effects of an approach combining RF ablation followed by implant placement (Weinberg et al. 2006). Both studies were performed using the rabbit VX2 model of liver carcinoma, which has been widely used as a model of human hepatocellular carcinoma for assessment of new interventional therapies (Boehm et al. 2002; Ramirez et al. 1996; Yoon et al. 2003). Together, these studies provide insight on the efficacy of liver tumor treatment with implantable polymer devices.

The first study demonstrates the first use of polymer millirods as a standalone strategy for the treatment of small liver cancers (Weinberg et al. 2006). The primary goal of this work was to determine the drug distribution and the resulting treatment efficacy from using polymer implants to treat tumors approximately 1 cm in diameter. Such a scenario might be encountered in hepatocellular carcinoma, when multiple small tumors in various stages might be found throughout the liver. In this case, surgery is often excluded because of insufficient liver function or the involvement of both liver lobes (Kashef and Roberts 2001). VX2 carcinoma cells were implanted in the livers of New Zealand white rabbits and allowed to grow for 12 days to an approximate diameter of 8 mm. The tumors were then treated with the implantation of either a control (0% doxorubicin) or burst-release (13.5% doxorubicin) polymer millirod into the center of the tumor (Weinberg et al. 2006). Tumors were evaluated at 4 and 8 days after the treatment using gross pathology, fluorescent drug concentration measurements, and tissue histology. On a gross level, the implants demonstrated considerable tumor control at both time points, as tumors were 50% and 90% smaller than their respective controls. The treated tumors had a substantially different morphology than controls, showing considerable necrosis and cell damage. Effective drug concentrations were seen penetrating tumor tissue to a distance of 2.8 mm (day 4) and 1.3 mm (day 8) from the implant boundary. Outside the tumor, drug concentrations dropped sharply and were below the detectable levels of drug. Furthermore, untreated tumor cells, which are likely to grow into recurrent tumors over time, were found outside the main tumor boundary. This study established that the polymer implants could be used to treat small tumors in a palliative or neoadjuvant role.

8.4.2   DRUG DISTRIBUTION AND ANTITUMOR EFFICACY FROM COMBINED LIVER TUMOR TREATMENT WITH RADIOFREQUENCY ABLATION AND POLYMER IMPLANTS

After demonstrating the drug coverage and treatment effects provided by the implants alone, polymer millirods containing doxorubicin were tested as part of a combined liver cancer treatment (Weinberg et al. 2006). The combined treatment consisted of RF ablation of the center of the tumor followed by the placement of a doxorubicin-containing implant. For liver tumor treatment, this approach has two distinct advantages. First, the RF treatment destroys the majority of the tumor mass, leading to a considerable reduction of viable tumor. Remnant tumor cells may also have increased susceptibility to drug because of sublethal damage from RF-induced heating. Second, RF ablation of tissue facilitates drug distribution to greater distances from the implant, an effect established in earlier studies in normal liver (Qian et al. 2003; Szymanski-Exner et al. 2003a). This property should provide greater drug exposure to the tumor, and hence, a greater degree of success, than either treatment alone. Additionally, the combined treatment may be more clinically relevant, as combined treatments are frequently applied in the treatment of many human cancers (Ashby and Ryken 2006).

RF ablation followed by polymer implant placement was tested in the same VX2 carcinoma model in rabbits (Weinberg et al. 2006). VX2 tumors were implanted in the middle lobe of the rabbit liver and allowed to grow for 18 days, until they reached an approximate size of 1.1 cm in diameter. These tumors were then treated with RF ablation to a probe temperature of 80°C for 2 min, which was shown in test ablations to generate a coagulated region measuring 8 mm in diameter. These conditions were chosen to create a situation in which a partially ablated rim of tumor persisted around the coagulated region, mimicking a scenario in which a tumor is not completely treated. Following ablation, polymer millirods of the same composition tested without ablation (65% PLGA, 21.5% NaCl, and 13.5% doxorubicin) or control implants (100% PLGA) were implanted into the ablation tract in the center of the tumors. The rabbits were subsequently monitored for 4 and 8 days after the tumor treatment, at which point the tumors were removed and assessed for gross tumor appearance, histological cell morphology, and local doxorubicin concentration. Results from the combined treatment allowed for comparison of drug distribution properties between nonablated and ablated tumors. In both cases, drug concentrations in the center of the tumor were over 1 mg/g at the implant tissue interface. However, the drug penetrated more deeply into ablated tumor tissues, providing greater drug coverage to the tumor 4 mm away from the implant boundary. Doxorubicin penetration distances in the ablated tumor tissue were found to be 3.7 mm on day 4 and 2.1 mm on day 8 (compared to 2.8 and 1.3 mm on day 4 and day 8 in nonablated tumors, respectively). Ablation almost tripled the total mass drug estimated to be in the tumor on day 4, increasing the value from 210 ± 120 μg without ablation to 590 ± 300 μg with ablation. The overall half-life of drug removal from the tumor volume was found to be 2.0 ± 0.1 days, slower than the 1.6 ± 0.2 days seen in nonablated tumor. These results confirm findings from normal liver, demonstrating that ablating tumor reduces drug elimination and leads to greater penetration into the tumor tissue. Ultimately this provides greater coverage of the tumor with therapeutic drug values. Such data was qualitatively confirmed in images of drug distribution, which showed greater amounts of drug further from the tumor.

Gross pathological and histological observations after the combined treatment also allowed for preliminary assessment of the success of the combined treatment. The total area of coagulation necrosis and inflammatory tissue surrounding the ablated area was similar regardless of which type of implant was used. Totaling both time points, 2 out of 7 animals treated with ablation and a control implant were found to have significant regions of residual tumor; similarly, 2 out of 7 animals treated with ablation and a doxorubicin implant had residual tumor. None of the animals (0/3) in the 8 day ablation plus implant group had areas of residual tumor, indicating a possible, although not statistically significant improvement in this group. Considerable knowledge on why the residual tumors remain untreated was determined from histological assessment. In the two residual tumors in the treatment group, areas of residual tumor began on average 4.1 mm away from the implant location, with the median distance between the implants and viable tumor measuring 7.9 mm. Some residual tumor cells were found as far as 12 mm away from the implant. Comparison of tumor histology with fluorescence microscopy images provided further insight into why the treatment did not reach the entire tumor. Starting at day 4 and more considerable by day 8, fibrous capsule formation around the coagulated zone was evident. Fluorescence attributed to doxorubicin was seen up to the fibrous boundary on day 8, but little fluorescence was seen beyond this barrier, suggesting that collagen deposition in the boundary may have inhibited drug transport to untreated regions outside the ablated region. The two main barriers to treatment success revealed by this study were drug penetration distance from the millirod implant and the formation of a fibrous barrier to drug transport (Weinberg et al. 2006).

Together, the two studies of liver tumor treatment with polymer millirods provide several interesting findings about the probability of success for tumor treatment with implantable polymer devices. With doxorubicin containing millirods alone, relatively small lesions (<1.0 cm diameter) were controlled in terms of tumor size but may not have been ultimately cured because of the presence of residual cells around the periphery (Weinberg et al. 2007a). However, these results suggest that these implants may be successful in reducing tumor load and could serve as a palliative and life-prolonging strategy in patients who are not good candidates for surgical resection. For instance, patients with multiple small metastatic lesions to the liver could be treated with percutaneous, image-guided placement of an implant in each lesion. The study of the combined treatment was unable to elicit statistically significant values in likelihood of remnant viable tumor, but did reveal that fibrous capsule formation and therapeutic distance from the implant limit treatment of residual tumor (Weinberg et al. 2006). Subsequent studies address some of these limitations by using similar implants to treat a greater number of subjects in a mouse model and by using a multiple implant strategy to treat larger tumors in rabbits.

8.4.3 ANTITUMOR EFFICACY IN A MOUSE MODEL

Building on the previous studies of polymer implants in rabbit tumors, chemotherapeutic polymer implants were further tested using a mouse model of prostate cancer (Dong et al. 2009). To perform this study, a polymer millirod incorporating β-lapachone was formulated. β-Lapachone is a novel chemotherapeutic agent derived from the bark of the South American Lapacho tree (Planchon et al. 2001). β-Lapachone kills cells using a mechanism involving inducing futile cycling of the metabolic enzyme NAD(P)H:quinone oxidoreductase-1 (NQO1), an enzyme which is upregulated in many types of cancers. One unique property of this antitumor mechanism is that it is independent of many antiapoptotic mechanisms, such as caspases, P53 status, and cell cycle stage (Reinicke et al. 2005). Unfortunately, β-lapachone is very insoluble in water and its systemic uses have been limited. However, complexing β-lapachone with molecules from a class of cyclodextrins drastically increases water solubility and allows for it to be incorporated in polymer implants (Wang et al. 2006). Varying the composition of the implants allows for modulation of the release kinetics of β-lapachone. Using this strategy, implants for use in an animal study were fabricated using 31% β-lapachone-hydroxypropyl-β-cyclodextrin complex, 19% free β-lapachone, and 50% PLGA using a similar compression molding technique described for making the implants described throughout this chapter. In vitro, these implants provided a burst release of approximately 0.5 mg of β-lapachone in the first 12 h followed by a relatively constant release of 0.05 mg/day over the next three days. The burst dose was designed to provide maximum exposure to the tumors at the beginning of the study to provide the greatest tumor killing effect while sparing surrounding normal tissues (Dong et al. 2009). β-Lapachone polymer millirods were tested against control polymer implants in a mouse model of prostate cancer (Dong et al. 2009). Control implants were manufactured using 24% hydroxypropyl-β-cyclodextrin and 76% PLGA to ensure that no antitumor effects were provided by the cyclodextrin molecule. The tumor model was a prostate cancer cell line, PC-3, which was injected into the flanks of athymic nude mice and allowed to grow until the tumors reached 300 mm³ in volume. At that time, treatment or control implants were placed into the tumors ($n = 10$). Tumor volumes were calculated every other day, and mice were euthanized if their tumor volumes exceeded 10% of their body weight. Results from this study are shown in Figure 8.5. In the first 3 days after implantation, the β-lapachone millirod tumors had decreased in size to 200 ± 10 mm³ while control group tumors increased dramatically in size 550 ± 40 mm³. In terms of tumor size, the β-lapachone millirods controlled tumor size for the first 17 days, after which tumor growth returned to approximately the control rate. Mice treated with β-lapachone implants also had a survival advantage, with 50% survival of 25 days in the control group vs. 35 days in the treated group. Two mice in the treatment group survived for longer than 40 days. The treated animals also had no adverse effects in terms of weight loss or histological damage to liver or lungs.

These results display the potential benefit of using drug containing polymer implants to treat inoperable tumors. These implants provided an immediate regression in tumor size that was maintained for at least 2 weeks as well as a 40% increase in survival time, both of which were statistically significant. Although the implants did not provide a cure from the tumor, such results strongly support further development of these types of implants for use in a palliative and life-prolonging role. Use of the implants in a combined treatment may further increase these benefits.

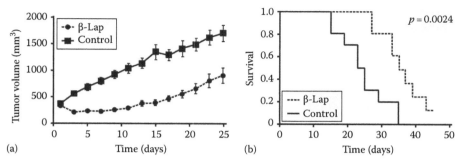

FIGURE 8.5 Comparison of PC-3 tumor volume in mice treated with β-lapachone or control implants over time (a). Kaplan-Meier survival curves for the same animals treated with β-lapachone demonstrate increased survival of the treated animals (b).

8.4.4 THREE-DIMENSIONAL MODELING OF INTRATUMORAL DRUG DELIVERY

Because of limitations in treating larger tumors, additional studies were performed to develop and validate a three-dimensional (3-D) finite element model to evaluate the effects of different implant designs and treatment strategies. Such a model can be used to evaluate changes in therapy strategy, such as incorporating an anti-inflammatory agent in the implant and the use of multiple millirods in a single tumor. This approach limits the number of animal experiments and allows for rapid prototyping of different treatment strategies. To model drug distribution into ablated tissues, transport properties of nonablated and ablated tumor were estimated. The resulting properties were then used to simulate drug distribution to ablated tumors with different types or arrangement of implants. Finally, these estimates were validated using measurements of drug concentrations in a tumor model.

Tumor drug transport parameters were estimated by minimizing the error between a finite element solution to the transport mass balance equation and experimental data (Weinberg et al. 2007c). Nonablated tumor was found to have drug diffusion slightly less than normal liver tissue and elimination considerably less than normal liver tissue. Ablated tumor, on the other hand, had a doxorubicin diffusion rate higher than either normal liver or nonablated tumor. As was found in normal liver, ablation completely stopped drug elimination for 4 days, but due to reperfusion of the ablated tumor areas elimination returned to values similar to that in normal tumor between day 4 and day 8. This finding indicates that, much as in normal liver tissue, ablation provides a window for improved drug delivery that lasts 4–8 days. The estimated tissue parameters were then used to simulate 3-D drug distribution profiles using multiple, peripherally placed implants to treat larger tumors (Weinberg et al. 2008). Scenarios for treating a 2 cm tumor with an RF-ablated core and 1 mm rim of viable tissue around the periphery were designed and simulated. Implants configurations included using a single central implant, multiple peripherally placed implants, or multiple peripherally placed implants around a central implant. Drug transport was then simulated using a finite element solution to determine the drug concentrations throughout the tumor over an 8 day period. Simply increasing the number of implants (and thereby the doxorubicin dose) increased the average drug exposure in the entire tumor independent of configuration. However, placement of implants around the tumor periphery provided more effective delivery of drug to tumor where drug was needed the most (the unablated tumor periphery). Four peripherally placed implants maintained the entire tumor volume at greater than two times the therapeutic concentration (>12.8 μg/g) for 74 h, while a single central implant never reached 100% tumor coverage during the simulated 8 days. These findings emphasized the value of using drug distribution simulations to rapidly compare different treatments and determine which has the maximum likelihood of success. To confirm the validity of the simulations performed earlier, experimental validation using four equally spaced peripheral implants in a rabbit tumor model was then performed (Weinberg et al. 2008). To achieve this, VX-2 tumors were

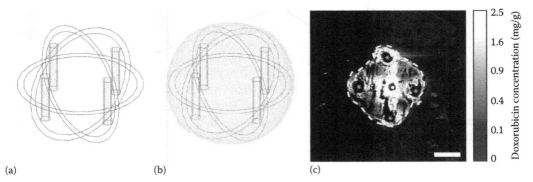

(a)   (b)   (c)

FIGURE 8.6 Simulated three-dimensional (3-D) geometry of a 2 cm diameter tumor containing a central RF-ablated region of diameter 1.8 cm and treated peripheral implants at a radius of 0.7 cm from the tumor center (a). Region predicted to be covered by therapeutic concentrations of doxorubicin on day 4 incorporate the entire tumor volume (b). Experimentally measured doxorubicin concentrations from an ablated rabbit tumor on day 4 show drug extending to the outer ablation boundary, denoted by the dashed line (c). Scale bar is 5 mm.

allowed to grow in rabbit livers for 28 days until they reached approximately 2 cm in diameter. On day 28, the tumors were treated with RF ablation to a temperature of 90°C for 9 min to create an ablation size of roughly 1.8 cm. Subjects were then euthanized at day 4 and day 8 after ablation and tissues removed for determining drug concentrations in the tissues. The experimental setup and drug concentrations from extracted tissue are shown in Figure 8.6. Experimentally calculated concentration values vs. simulated values are shown in Table 8.2. While simulations predicted 100% coverage of the nonablated tumor rim with therapeutic doxorubicin concentrations on day 4, experimentally 87% of the tumor volume was covered by therapeutic doxorubicin concentrations. This finding is likely because there was a greater area of untreated tumor than anticipated in the experimental group. However, considerable overlap between experimental and simulated drug concentrations was shown. Despite the difference, the average drug concentration values in the nonablated tumor overlapped considerably between simulated and experimental data. On day 8, coverage of the nonablated tumor correlated much more closely between the model and experimental data, which have values of 61% and 65%, respectively. Overall, for these relatively large tumors, the doxorubicin concentration exceeded therapeutic concentrations in a large fraction of the tumor rim on both days. The success of achieving this therapeutic drug level with polymer millirods in such a large tumor is a major step toward the use of these implants in practical clinical situations.

Future extensions to this model can be used to predict how implants with different release profiles, such as dual-release doxorubicin implants, could improve tumor coverage. Additionally, the

TABLE 8.2
Doxorubicin Coverage of Simulated and Experimental Tissues

	Model		Experimental	
	Day 4	Day 8	Day 4	Day 8
Area of ablated tumor (mm³)	254	254	343	418
Area of nonablated tumor (mm³)	60	60	195	90
Nonablated tumor over [DOX*] (%)	100%	61%	87%	65%
[DOX] in nonablated tumor (μg/g)[a]	140 [78–232]	20 [8–56]	79 [30–173]	26 [7–87]

[a] Values shown are the median concentration. Brackets contain the 25th and 75th percentiles.

model will be able to evaluate how other changes to implant design that modify tissue properties, such as including dexamethasone within the implants, can affect drug distribution in the tumors. Ultimately, a drug transport model may be used as part of a comprehensive treatment planning tool. Image based data obtained from CT or MRI could be used to determine tumor geometry. Subsequently, ablation treatment could be planned using a thermal damage model (Johnson and Saidel 2002), after which drug coverage in the ablated tumor could be predicted using this 3-D finite element model. Using a computational tool to plan combined treatment would allow assessment of the best ways to treat complex lesions and extension of the findings from smaller tumors already reported into larger, more clinically relevant tumor models.

8.5 CONCLUSIONS AND FUTURE OUTLOOK

Conventional systemic chemotherapy for tumors is restricted by lack of tumor specificity and severe side effects associated with intrinsic drug toxicity (Crawford et al. 2004; Dowell et al. 2000; Wallace 2003). With the emergence of minimally invasive, image-guided interventional technology, tumor chemotherapy is at the threshold of a major breakthrough because of such technological advances in targeting strategies that overcome the previous limitations. Tumor-directed therapies, such as focal ablation and locoregional chemotherapy, are being developed to increase the specificity of tumor destruction and reduce undesired side effects. Intratumoral implants reduce systemic drug exposure by using image-guided placement directly into the target region, thus delivering the entire drug dose with reduced systemic exposure. The key consideration with intratumoral chemotherapy is to design an implant system that provides optimal drug distribution and therefore maximal tumor destruction.

Polymer millirod implants were specifically designed to treat unresectable liver tumors in conjunction with RF ablation. Implants with different drug release rates have been developed and extensively studied in both nonablated and ablated liver tissues, where they effectively delivered drugs into the surrounding tissue. Particularly, dual-release implants combining a burst of drug release with sustained drug release maximized drug coverage in the ablated region (Qian et al. 2004). Modeling of tissue properties using a pharmacokinetic transport model emphasized the importance of tumor pretreatment with RF ablation, which facilitated drug delivery to tissues further away from the implant by preventing drug elimination. Polymer millirods were further tested in a rabbit model of hepatocellular carcinoma and mouse model of prostate cancer. Tumor control was achieved within a limited distance from the implant over a limited amount of time. The success of the tumor treatment appeared to be limited by two factors: drug transport distance from the implant and the formation of a fibrous capsule that restricted drug transport. Limitations in drug transport distance were partially overcome by using a drug delivery model to develop a strategy for using multiple implants in a single tumor. In limited experimental validation, this strategy provided excellent coverage of a realistically sized tumor which can conceivably be extrapolated to human use in future studies.

The future of polymer millirods for tumor treatment depends on optimizing drug delivery efficiency to the tumor periphery. One way of achieving this goal is to include both dexamethasone and doxorubicin in a single implant. Dexamethasone-containing polymer millirods have already been shown to prevent fibrous capsule formation and decrease new blood vessel formation after ablation (Blanco et al. 2006). These effects should promote drug delivery by increasing doxorubicin diffusion and reducing elimination. Further developments in the most recent studies using multiple implants around the periphery of a larger tumor may also improve the probability of treatment success. Placing implants closer to the boundary increases the likelihood of drug exposure at and beyond the ablation periphery, where recurrence is most likely to occur. As these developments progress, three-dimensional modeling can be used as a tool to rapidly evaluating different treatment protocols and gain mechanistic insights to optimize dosage regimen design. The integrated modeling and experimental approach should greatly assist the clinical translation of polymer implants as a viable option for locoregional chemotherapy of unresectable tumors.

REFERENCES

Alberts, D.S. and A. Delforge. 2006. Maximizing the delivery of intraperitoneal therapy while minimizing drug toxicity and maintaining quality of life. *Semin Oncol* 33 (6 Suppl 12):8–17.

Almond, B.A., A.R. Hadba, S.T. Freeman et al. 2003. Efficacy of mitoxantrone-loaded albumin microspheres for intratumoral chemotherapy of breast cancer. *J Control Release* 91 (1–2):147–155.

Arica, B., S. Calis, H. Kas, M. Sargon, and A. Hincal. 2002. 5-Fluorouracil encapsulated alginate beads for the treatment of breast cancer. *Int J Pharm* 242 (1–2):267–269.

Ashby, L.S. and T.C. Ryken. 2006. Management of malignant glioma: steady progress with multimodal approaches. *Neurosurg Focus* 20 (4):E3.

Au, J.L., S.H. Jang, and M.G. Wientjes. 2002. Clinical aspects of drug delivery to tumors. *J Control Release* 78 (1–3):81–95.

Bentrem, D.J., R.P. Dematteo, and L.H. Blumgart. 2005. Surgical therapy for metastatic disease to the liver. *Annu Rev Med* 56:139–156.

Blanco, E., F. Qian, B. Weinberg, N. Stowe, J.M. Anderson, and J. Gao. 2004. Effect of fibrous capsule formation on doxorubicin distribution in radiofrequency ablated rat livers. *J Biomed Mater Res A* 69 (3):398–406.

Blanco, E., B.D. Weinberg, N.T. Stowe, J.M. Anderson, and J. Gao. 2006. Local release of dexamethasone from polymer millirods effectively prevents fibrosis after radiofrequency ablation. *J Biomed Mater Res A* 76 (1):174–182.

Boehm, T., A. Malich, S.N. Goldberg et al. 2002. Radio-frequency ablation of VX2 rabbit tumors: assessment of completeness of treatment by using contrast-enhanced harmonic power Doppler US. *Radiology* 225 (3):815–821.

Bradley, W.G., Jr. 2009. MR-guided focused ultrasound: a potentially disruptive technology. *J Am Coll Radiol* 6 (7):510–513.

Brem, H. and P. Gabikian. 2001. Biodegradable polymer implants to treat brain tumors. *J Control Release* 74 (1–3):63–67.

Celikoglu, S.I., F. Celikoglu, and E.P. Goldberg. 2006. Endobronchial intratumoral chemotherapy (EITC) followed by surgery in early non-small cell lung cancer with polypoid growth causing erroneous impression of advanced disease. *Lung Cancer* 54 (3):339–346.

Chu, G. 1994. Cellular responses to cisplatin. The roles of DNA-binding proteins and DNA repair. *J Biol Chem* 269 (2):787–790.

Clasen, S. and P.L. Pereira. 2008. Magnetic resonance guidance for radiofrequency ablation of liver tumors. *J Magn Reson Imaging* 27 (2):421–433.

Coussens, L.M. and Z. Werb. 2002. Inflammation and cancer. *Nature* 420 (6917):860–867.

Crawford, J., D.C. Dale, and G.H. Lyman. 2004. Chemotherapy-induced neutropenia: risks, consequences, and new directions for its management. *Cancer* 100 (2):228–237.

Dale, P.S., J.W. Souza, and D.A. Brewer. 1998. Cryosurgical ablation of unresectable hepatic metastases. *J Surg Oncol* 68 (4):242–245.

Dong, Y., S.F. Chin, E. Blanco et al. 2009. Intratumoral delivery of beta-lapachone via polymer implants for prostate cancer therapy. *Clin Cancer Res* 15 (1):131–139.

Dowell, J.A., A.R. Sancho, D. Anand, and W. Wolf. 2000. Noninvasive measurements for studying the tumoral pharmacokinetics of platinum anticancer drugs in solid tumors. *Adv Drug Deliv Rev* 41 (1):111–126.

El-Kareh, A.W. and T.W. Secomb. 2000. A mathematical model for comparison of bolus injection, continuous infusion, and liposomal delivery of doxorubicin to tumor cells. *Neoplasia* 2 (4):325–338.

Emerich, D.F., P. Snodgrass, D. Lafreniere et al. 2002. Sustained release chemotherapeutic microspheres provide superior efficacy over systemic therapy and local bolus infusions. *Pharm Res* 19 (7):1052–1060.

Exner, A.A., B.D. Weinberg, N.T. Stowe et al. 2004. Quantitative computed tomography analysis of local chemotherapy in liver tissue after radiofrequency ablation. *Acad Radiol* 11 (12):1326–1336.

Fischer, K., W. Gedroyc, and F.A. Jolesz. 2010. Focused ultrasound as a local therapy for liver cancer. *Cancer J* 16 (2):118–124.

Fleming, A.B. and W.M. Saltzman. 2002. Pharmacokinetics of the carmustine implant. *Clin Pharmacokinet* 41 (6):403–419.

Gan, S. and D.I. Watson. 2010. New endoscopic and surgical treatment options for early esophageal adenocarcinoma. *J Gastroenterol Hepatol* 25 (9):1478–1484.

Gao, J., F. Qian, A. Szymanski-Exner, N. Stowe, and J. Haaga. 2002. In vivo drug distribution dynamics in thermoablated and normal rabbit livers from biodegradable polymers. *J Biomed Mater Res* 62 (2):308–314.

Geller, D.A., A. Tsung, J.W. Marsh, I. Dvorchik, T.C. Gamblin, and B.I. Carr. 2006. Outcome of 1000 liver cancer patients evaluated at the UPMC Liver Cancer Center. *J Gastrointest Surg* 10 (1):63–68.

Gillams, A.R. 2005. Image guided tumour ablation. *Cancer Imaging* 5:103–109.

Goldberg, S.N. 2001. Radiofrequency tumor ablation: principles and techniques. *Eur J Ultrasound* 13 (2):129–147.

Goldberg, E.P., A.R. Hadba, B.A. Almond, and J.S. Marotta. 2002. Intratumoral cancer chemotherapy and immunotherapy: opportunities for nonsystemic preoperative drug delivery. *J Pharm Pharmacol* 54 (2):159–180.

Grimm, P. and J. Sylvester. 2004. Advances in brachytherapy. *Rev Urol* 6 (S4):S37–S48.

Guerin, C., A. Olivi, J.D. Weingart, H.C. Lawson, and H. Brem. 2004. Recent advances in brain tumor therapy: local intracerebral drug delivery by polymers. *Invest New Drugs* 22 (1):27–37.

Haaga, J.R., A.A. Exner, Y. Wang, N.T. Stowe, and P.J. Tarcha. 2005. Combined tumor therapy by using radiofrequency ablation and 5-FU-laden polymer implants: evaluation in rats and rabbits. *Radiology* 237 (3):911–918.

Hall, W.A. and G.T. Sherr. 2006. Convection-enhanced delivery: targeted toxin treatment of malignant glioma. *Neurosurg Focus* 20 (4):E10.

Han, K.R. and A.S. Belldegrun. 2004. Third-generation cryosurgery for primary and recurrent prostate cancer. *BJU Int* 93 (1):14–18.

Harrison, L.E., B. Koneru, P. Baramipour et al. 2003. Locoregional recurrences are frequent after radiofrequency ablation for hepatocellular carcinoma. *J Am Coll Surg* 197 (5):759–764.

Jackson, J.K., M.E. Gleave, V. Yago, E. Beraldi, W.L. Hunter, and H.M. Burt. 2000. The suppression of human prostate tumor growth in mice by the intratumoral injection of a slow-release polymeric paste formulation of paclitaxel. *Cancer Res* 60 (15):4146–4151.

Jain, R.K. 1999. Transport of molecules, particles, and cells in solid tumors. *Annu Rev Biomed Eng* 1:241–263.

Jain, R.K. 2001. Delivery of molecular and cellular medicine to solid tumors. *Adv Drug Deliv Rev* 46 (1–3):149–168.

Jemal, A., R. Siegel, E. Ward, Y. Hao, J. Xu, and M.J. Thun. 2009. Cancer statistics, 2009. *CA Cancer J Clin* 59 (4):225–249.

Johnson, P.C. and G.M. Saidel. 2002. Thermal model for fast simulation during magnetic resonance imaging guidance of radio frequency tumor ablation. *Ann Biomed Eng* 30 (9):1152–1161.

Joniau, S., T. Tailly, L. Goeman, W. Blyweert, P. Gontero, and A. Joyce. 2010. Kidney radiofrequency ablation for small renal tumors: oncologic efficacy. *J Endourol* 24 (5):721–728.

Kalyanasundaram, S., V.D. Calhoun, and K.W. Leong. 1997. A finite element model for predicting the distribution of drugs delivered intracranially to the brain. *Am J Physiol* 273 (5 Pt 2):R1810–R1821.

Kashef, E. and J.P. Roberts. 2001. Transplantation for hepatocellular carcinoma. *Semin Oncol* 28 (5):497–502.

Kokudo, N. and M. Makuuchi. 2004. Current role of portal vein embolization/hepatic artery chemoembolization. *Surg Clin North Am* 84 (2):643–657.

Krupka, T.M., B.D. Weinberg, N.P. Ziats, J.R. Haaga, and A.A. Exner. 2006. Injectable polymer depot combined with radiofrequency ablation for treatment of experimental carcinoma in rat. *Invest Radiol* 41 (12):890–897.

Leung, T.W. and P.J. Johnson. 2001. Systemic therapy for hepatocellular carcinoma. *Semin Oncol* 28 (5):514–520.

Lindner, U., N. Lawrentschuk, and J. Trachtenberg. 2010. Focal laser ablation for localized prostate cancer. *J Endourol* 24 (5):791–797.

Livraghi, T., V. Benedini, S. Lazzaroni, G. Meloni, G. Torzilli, and C. Vettori. 1998. Long term results of single session percutaneous ethanol injection in patients with large hepatocellular carcinoma. *Cancer* 83 (1):48–57.

Longley, D.B., D.P. Harkin, and P.G. Johnston. 2003. 5-Fluorouracil: mechanisms of action and clinical strategies. *Nat Rev Cancer* 3 (5):330–338.

Lu, Y., E. Sega, C.P. Leamon, and P.S. Low. 2004. Folate receptor-targeted immunotherapy of cancer: mechanism and therapeutic potential. *Adv Drug Deliv Rev* 56 (8):1161–1176.

Mardor, Y., Y. Roth, Z. Lidar et al. 2001. Monitoring response to convection-enhanced taxol delivery in brain tumor patients using diffusion-weighted magnetic resonance imaging. *Cancer Res* 61 (13):4971–4973.

Martin, R.C., C.R. Scoggins, and K.M. McMasters. 2010. Safety and efficacy of microwave ablation of hepatic tumors: a prospective review of a 5-year experience. *Ann Surg Oncol* 17 (1):171–178.

McGuire, S. and F. Yuan. 2001. Quantitative analysis of intratumoral infusion of color molecules. *Am J Physiol Heart Circ Physiol* 281 (2):H715–H721.

McGuire, S., D. Zaharoff, and F. Yuan. 2006. Nonlinear dependence of hydraulic conductivity on tissue deformation during intratumoral infusion. *Ann Biomed Eng* 34 (7):1173–1181.

Menei, P., L. Capelle, J. Guyotat et al. 2005. Local and sustained delivery of 5-fluorouracil from biodegradable microspheres for the radiosensitization of malignant glioma: a randomized phase II trial. *Neurosurgery* 56 (2):242–248; discussion 242–248.

Menei, P., E. Jadaud, N. Faisant et al. 2004. Stereotaxic implantation of 5-fluorouracil-releasing microspheres in malignant glioma. *Cancer* 100 (2):405–410.

Merrick, G.S., K.E. Wallner, and W.M. Butler. 2003. Permanent interstitial brachytherapy for the management of carcinoma of the prostate gland. *J Urol* 169 (5):1643–1652.

Mok, T.S., S. Kanekal, X.R. Lin et al. 2001. Pharmacokinetic study of intralesional cisplatin for the treatment of hepatocellular carcinoma. *Cancer* 91 (12):2369–2377.

Moon, W.K., Y. Lin, T. O'Loughlin et al. 2003. Enhanced tumor detection using a folate receptor-targeted near-infrared fluorochrome conjugate. *Bioconjug Chem* 14 (3):539–545.

Muller, H. 2002. Combined regional and systemic chemotherapy for advanced and inoperable non-small cell lung cancer. *Eur J Surg Oncol* 28 (2):165–171.

Muller, H. and R. Hilger. 2003. Curative and palliative aspects of regional chemotherapy in combination with surgery. *Support Care Cancer* 11 (1):1–10.

Nasongkla, N., E. Bey, J. Ren et al. 2006. Multifunctional polymeric micelles as cancer-targeted, MRI-ultrasensitive drug delivery systems. *Nano Lett* 6 (11):2427–2430.

Nguyen, C.L., W.J. Scott, and M. Goldberg. 2006. Radiofrequency ablation of lung malignancies. *Ann Thorac Surg* 82 (1):365–371.

Ntziachristos, V. 2006. Fluorescence molecular imaging. *Annu Rev Biomed Eng* 8:1–33.

Planchon, S.M., J.J. Pink, C. Tagliarino, W.G. Bornmann, M.E. Varnes, and D.A. Boothman. 2001. beta-Lapachone-induced apoptosis in human prostate cancer cells: involvement of NQO1/xip3. *Exp Cell Res* 267 (1):95–106.

Qian, F., N. Nasongkla, and J. Gao. 2002a. Membrane-encased polymer millirods for sustained release of 5-fluorouracil. *J Biomed Mater Res* 61 (2):203–211.

Qian, F., G.M. Saidel, D.M. Sutton, A. Exner, and J. Gao. 2002b. Combined modeling and experimental approach for the development of dual-release polymer millirods. *J Control Release* 83 (3):427–435.

Qian, F., N. Stowe, E.H. Liu, G.M. Saidel, and J. Gao. 2003. Quantification of in vivo doxorubicin transport from PLGA millirods in thermoablated rat livers. *J Control Release* 91 (1–2):157–166.

Qian, F., N. Stowe, G.M. Saidel, and J. Gao. 2004. Comparison of doxorubicin concentration profiles in radiofrequency-ablated rat livers from sustained- and dual-release PLGA millirods. *Pharm Res* 21 (3):394–399.

Qian, F., A. Szymanski, and J. Gao. 2001. Fabrication and characterization of controlled release poly(D,L-lactide-co-glycolide) millirods. *J Biomed Mater Res* 55 (4):512–522.

Raghavan, R., M.L. Brady, M.I. Rodriguez-Ponce, A. Hartlep, C. Pedain, and J.H. Sampson. 2006. Convection-enhanced delivery of therapeutics for brain disease, and its optimization. *Neurosurg Focus* 20 (4):E12.

Ramirez, L.H., Z. Zhao, P. Rougier et al. 1996. Pharmacokinetics and antitumor effects of mitoxantrone after intratumoral or intraarterial hepatic administration in rabbits. *Cancer Chemother Pharmacol* 37 (4):371–376.

Ramsey, D.E., L.Y. Kernagis, M.C. Soulen, and J.F. Geschwind. 2002. Chemoembolization of hepatocellular carcinoma. *J Vasc Interv Radiol* 13 (9 Pt 2):S211–S221.

Raz, A., G. Levine, and Y. Khomiak. 2000. Acute local inflammation potentiates tumor growth in mice. *Cancer Lett* 148 (2):115–120.

Reinicke, K.E., E.A. Bey, M.S. Bentle et al. 2005. Development of beta-lapachone prodrugs for therapy against human cancer cells with elevated NAD(P)H:quinone oxidoreductase 1 levels. *Clin Cancer Res* 11 (8):3055–3064.

Roselt, P., S. Meikle, and M. Kassiou. 2004. The role of positron emission tomography in the discovery and development of new drugs; as studied in laboratory animals. *Eur J Drug Metab Pharmacokinet* 29 (1):1–6.

Sampson, J.H., G. Akabani, A.H. Friedman et al. 2006. Comparison of intratumoral bolus injection and convection-enhanced delivery of radiolabeled antitenascin monoclonal antibodies. *Neurosurg Focus* 20 (4):E14.

Seto, T., S. Ushijima, H. Yamamoto et al. 2006. Intrapleural hypotonic cisplatin treatment for malignant pleural effusion in 80 patients with non-small-cell lung cancer: a multi-institutional phase II trial. *Br J Cancer* 95 (6):717–721.

Shah, S.S., D.L. Jacobs, A.M. Krasinkas, E.E. Furth, M. Itkin, and T.W. Clark. 2004. Percutaneous ablation of VX2 carcinoma-induced liver tumors with use of ethanol versus acetic acid: pilot study in a rabbit model. *J Vasc Interv Radiol* 15 (1 Pt 1):63–67.

Shelley, M.D., M.D. Mason, and H. Kynaston. 2010. Intravesical therapy for superficial bladder cancer: a systematic review of randomised trials and meta-analyses. *Cancer Treat Rev* 36 (3):195–205.

Shiina, S. 2009. Image-guided percutaneous ablation therapies for hepatocellular carcinoma. *J Gastroenterol* 44 Suppl 19:122–131.

Sinek, J., H. Frieboes, X. Zheng, and V. Cristini. 2004. Two-dimensional chemotherapy simulations demonstrate fundamental transport and tumor response limitations involving nanoparticles. *Biomed Microdevices* 6 (4):297–309.

Strasser, J.F., L.K. Fung, S. Eller, S.A. Grossman, and W.M. Saltzman. 1995. Distribution of 1,3-bis(2-chloroethyl)-1-nitrosourea and tracers in the rabbit brain after interstitial delivery by biodegradable polymer implants. *J Pharmacol Exp Ther* 275 (3):1647–1655.

Szymanski-Exner, A., A. Gallacher, N.T. Stowe, B. Weinberg, J.R. Haaga, and J. Gao. 2003a. Local carboplatin delivery and tissue distribution in livers after radiofrequency ablation. *J Biomed Mat Res* 67A (2):510–516.

Szymanski-Exner, A., N.T. Stowe, R.S. Lazebnik et al. 2002. Noninvasive monitoring of local drug release in a rabbit radiofrequency (RF) ablation model using X-ray computed tomography. *J Control Release* 83 (3):415–425.

Szymanski-Exner, A., N.T. Stowe, K. Salem et al. 2003b. Noninvasive monitoring of local drug release using X-ray computed tomography: optimization and in vitro/in vivo validation. *J Pharm Sci* 92 (2):289–296.

Tanner, B., J.G. Hengstler, B. Dietrich et al. 1997. Glutathione, glutathione S-transferase alpha and pi, and aldehyde dehydrogenase content in relationship to drug resistance in ovarian cancer. *Gynecol Oncol* 65 (1):54–62.

Tsochatzis, E.A., G. Germani, and A.K. Burroughs. 2010. Transarterial chemoembolization, transarterial chemotherapy, and intra-arterial chemotherapy for hepatocellular carcinoma treatment. *Semin Oncol* 37 (2):89–93.

Varshney, S., A. Sewkani, S. Sharma et al. 2006. Radiofrequency ablation of unresectable pancreatic carcinoma: feasibility, efficacy and safety. *Jop* 7 (1):74–78.

Vogl, T.J., K. Engelmann, M.G. Mack et al. 2002. CT-guided intratumoural administration of cisplatin/epinephrine gel for treatment of malignant liver tumours. *Br J Cancer* 86 (4):524–529.

Wallace, K.B. 2003. Doxorubicin-induced cardiac mitochondrionopathy. *Pharmacol Toxicol* 93 (3):105–115.

Wang, F., E. Blanco, H. Ai, D.A. Boothman, and J. Gao. 2006. Modulating beta-lapachone release from polymer millirods through cyclodextrin complexation. *J Pharm Sci* 95 (10):2309–2319.

Wang, J.M., X. Deng, W. Gong, and S. Su. 1998. Chemokines and their role in tumor growth and metastasis. *J Immunol Methods* 220 (1–2):1–17.

Wang, P.P., J. Frazier, and H. Brem. 2002. Local drug delivery to the brain. *Adv Drug Deliv Rev* 54 (7):987–1013.

Wang, C.C., J. Li, C.S. Teo, and T. Lee. 1999. The delivery of BCNU to brain tumors. *J Control Release* 61 (1–2):21–41.

Wang, Y., S. Liu, C.Y. Li, and F. Yuan. 2005. A novel method for viral gene delivery in solid tumors. *Cancer Res* 65 (17):7541–7545.

Wang, Y. and F. Yuan. 2006. Delivery of viral vectors to tumor cells: extracellular transport, systemic distribution, and strategies for improvement. *Ann Biomed Eng* 34 (1):114–127.

Weinberg, B.D., H. Ai, E. Blanco, J.M. Anderson, and J. Gao. 2007a. Antitumor efficacy and local distribution of doxorubicin via intratumoral delivery from polymer millirods. *J Biomed Mater Res A* 81A (1):161–170.

Weinberg, B.D., E. Blanco, S.F. Lempka, J.M. Anderson, A.A. Exner, and J. Gao. 2006. Combined radiofrequency ablation and doxorubicin-eluting polymer implants for liver cancer treatment. *J Biomed Mater Res A* 81A (1):205–213.

Weinberg, B.D., R.B. Patel, A.A. Exner, G.M. Saidel, and J. Gao. 2007c. Modeling doxorubicin transport to improve intratumoral drug delivery to RF ablated tumors. *J Control Release* 124 (1–2):11–19.

Weinberg, B.D., R.B. Patel, H. Wu et al. 2008. Model simulation and experimental validation of intratumoral chemotherapy using multiple polymer implants. *Med Biol Eng Comput* 46 (10):1039–1049.

Weinmann, H.J., W. Ebert, B. Misselwitz, and H. Schmitt-Willich. 2003. Tissue-specific MR contrast agents. *Eur J Radiol* 46 (1):33–44.

Westphal, M., Z. Ram, V. Riddle, D. Hilt, and E. Bortey. 2006. Gliadel wafer in initial surgery for malignant glioma: long-term follow-up of a multicenter controlled trial. *Acta Neurochir (Wien)* 148 (3):269–275; discussion 275.

Ye, F., T. Ke, E.K. Jeong et al. 2006. Noninvasive visualization of in vivo drug delivery of poly(L-glutamic acid) using contrast-enhanced MRI. *Mol Pharm* 3 (5):507–515.

Yoon, C.J., J.W. Chung, J.H. Park et al. 2003. Transcatheter arterial chemoembolization with paclitaxel-lipiodol solution in rabbit VX2 liver tumor. *Radiology* 229 (1):126–131.

Zheng, J.H., C.T. Chen, J.L. Au, and M.G. Wientjes. 2001. Time- and concentration-dependent penetration of doxorubicin in prostate tumors. *AAPS PharmSci* 3 (2):E15.

9 Biological Stimulus-Responsive Hydrogels

A.K. Bajpai, Sanjana Kankane, Raje Chouhan,
and Shilpi Goswami

CONTENTS

9.1 INTRODUCTION

The field of controlled drug delivery has led to the development of smart drug delivery systems (SDDSs), which are also known as stimuli-sensitive delivery systems. The concept of SDDS is based on rapid transitions of a physicochemical property of polymer systems upon imposition of an environmental stimulus, which include physical (temperature, mechanical stress, ultrasound, electricity, light), chemical, (pH, ionic, strength), or biological (enzymes or biomolecules) signals. Such stimuli can be either "internal" signals, resulting from changes in the physiological condition of a living subject, or "external" signals artificially induced to provoke desired events. As illustrated in Figure 9.1 SDDS provides a programmable and predictable drug release profile in response to various

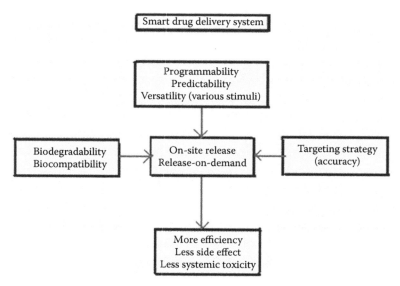

FIGURE 9.1 Various advantages of the smart drug delivery systems.

stimulation sources (De Villiers et al., 2009). Depending on the applications, one can design on–off system, pulsatile/sustained drug release, and closed loop drug delivery systems, for enhanced therapeutic efficiency with low systemic toxicity and side effects. SDDS provides various advantages over conventional drug delivery systems. The conventional controlled release systems are based on the predetermined drug release rate irrespective of the environmental condition at the time of the application. On the other hand, SDDS is based on the release-on-demand strategy, allowing a drug carrier to liberate a therapeutic drug only when it is required in response to a specific stimulation. The best example of SDDS has been self-regulated insulin delivery system that can respond to changes in the environmental glucose level (Chu et al., 2004).

Stimuli-sensitive variable architecture polymeric materials that can respond with a change in conformation to relevant stimuli action such as temperature, pH, ionic strength, and biological molecules are of considerable interest for a wide range of applications because of their responsivity to the modification of environmental conditions. Sensing capabilities are attractive in many biomedical applications. By combining a temperature-sensitive polymer/monomer usually N-isopropylacrylamide with a pH-sensitive one in different architectures, dual/multiple-responsive materials can be obtained (Vasile et al., 2009).

The use of hydrogels for controlled release of drugs has been well established. Hydrogels are particularly suited to drug delivery applications because they can be tailored to produce release characteristics desired for a particular drug. In the simplest case, drug release can occur through swelling and diffusion processes. Hydrogels may also be designed to incorporate environmentally responsive pendant groups along the hydrogel chain backbone. These environmentally responsive groups then have the ability to change the degree of swelling of the hydrogel carrier, and therefore change the diffusion and release profile of the drug. This method has been successfully demonstrated for a variety of drugs incorporating pH, ionic, or temperature change as the release triggers (Bayer and Peppas, 2008).

One of the most remarkable and useful features of a polymer's swelling ability manifests itself when that swelling can be triggered by a change in the environment surrounding the delivery system. Depending upon the polymer, the environmental change can involve pH, temperature, or ionic strength, and the system can either shrink or swell upon a change in any of these environmental factors (Aminabhavi et al., 2004). A number of these environmentally sensitive or "intelligent" hydrogel materials (Del Valle et al., 2009) are listed in Table 9.1.

TABLE 9.1
Swelling Stimuli and Mechanisms

Stimulus	Hydrogel	Mechanism
pH	Acidic or basic hydrogel	Change in pH-swelling-release of drug
Ionic strength	Ionic hydrogel	Change in ionic strength-change in concentration of ions inside gel-change in swelling-release of drug
Chemical species	Hydrogel containing electron-accepting groups	Electron-donating compounds—formation of charge/transfer complex—change in swelling-release of drug
Enzyme–substrate	Hydrogel containing immobilized enzymes	Substrate present—enzymatic conversion—product change—swelling of gel—release of drug
Magnetic	Magnetic particles dispersed in alginate microspheres	Applied magnetic field—change in pores in gel—change in swelling—release of drug
Thermal	Thermoresponsive hydrogel poly(N-isopropylacrylamide)	Change in temperature—change in polymer—polymer and water–polymer interactions—change in swelling—release of drug
Electrical	Polyelectrolyte hydrogel	Applied electric field—membrane charging—electrophoresis of charged drug-change in swelling—release of drug
Ultrasound irradiation	Ethylene-vinyl alcohol hydrogel	Ultrasound irradiation—temperature increase—release of drug

9.1.1 BIOLOGICAL STIMULUS-RESPONSIVE DRUG DELIVERY SYSTEMS

The hydrogel materials with biological responsiveness change properties in response to selective biological recognition events. When exposed to a biological target (nutrient, growth factor, receptor, antibody, enzyme, or whole cell), molecular recognition events trigger changes in molecular interactions that translate into macroscopic responses, such as swelling/collapse or solution-to-gel transitions (Ulijn et al., 2007). These bioresponsive or biointeractive materials contain receptors for biomolecules that, when stimulated, cause localized or bulk changes in the material properties. This is of special interest in developing autonomous systems that can detect disease markers and respond to them to repair the diseased area (Langer and Tirrell, 2004). Three types of stimuli for bioresponsive hydrogel systems can be distinguished. First, hydrogel materials can be modified to contain small biomolecules that selectively bind to biomacromolecules, including protein receptors or antibodies. Upon binding, a macroscopic transition follows (Figure 9.2, system I).

Second, systems may be modified with enzyme-sensitive substrates, such as short peptides (Figure 9.2, system II). Here, the initial molecular recognition event is similar to that in the first category (enzyme protein binds to substrate), but it is followed by a chemical event involving the making or breaking of bonds within the enzyme-sensitive substrate. Since enzymes are highly selective, materials can be programmed to respond to a specific enzyme by incorporation of the specific substrate (or a substrate mimic). This concept is especially attractive because the distribution of enzymes can differ between healthy and diseased cells, between different cell types, and during cellular migration, differentiation, and cell division (Ulijn, 2006; Yang and Xu, 2006). Third, systems may have biomacromolecules, such as enzymes, incorporated into their structures that recognize small biomolecules (Figure 9.2, system III). Enzymatic conversion of these biomolecules into molecules with different physical properties (e.g., an acid or basic compound) then triggers hydrogel swelling or collapse. One of the major treatments for insulin-dependent diabetes is to control glycemia by the administration of insulin. The ideal delivery systems should be the continuous, noninvasive, or minimally invasive close-loop systems, and they should attempt to deliver insulin in direct response to the levels of blood glucose, and to achieve the feedback-controlled release of insulin (Guiseppi-Elie et al., 2005). Potential benefits from such a system include the improved disease

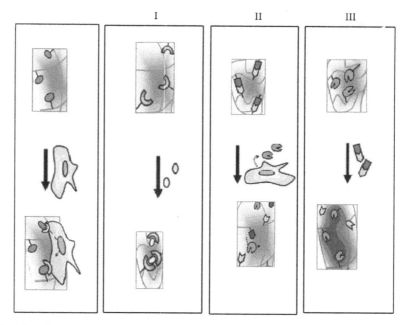

FIGURE 9.2   Bioactive hydrogel (top) and three different types of bioresponsive hydrogels (bottom) that change properties in response to: (I) small molecules via receptor/ligand interactions; (II) (cell-secreted) enzymes via cleavable linkers, and (III) small molecules that are converted by immobilized enzymes. The macroscopic response (swelling/collapse of the hydrogel) is shown.

management, enhanced patient compliance, and reduction of long-term complications of diabetes, which will improve significantly the lives of diabetic patients (Calceti et al., 2004). Because of outstanding mechanical swelling properties, the glucose-sensitive hydrogels are very suitable biomaterials for development of smart insulin delivery systems (Mahkam, 2007).

9.1.1.1   Glucose-Sensitive Gels

The measurement of glucose is extremely important in the treatment of diabetes and is also of value in monitoring cell growth, since glucose is the primary carbon source in most fermentation processes (Kabilan et al., 2004). Therefore, precisely engineered glucose-sensitive gels have huge potential in the quest to generate self-regulated modes of insulin delivery and to facilitate the construction of an artificial pancreas that would function in a manner similar to the β-cells of the pancreas. Among glucose-sensitive gels, immobilized enzymes, specifically gels with glucose oxidase conjugated to them have been extensively investigated (Ghanem and Ghaly, 2004; Srivastava et al., 2005). The glucose-responsiveness of the poly (vinylidine floride) (PVDF) membranes was found to be strongly influenced by the functional poly(acrylic acid) (PAA) gating. In fact, for an ideal gating response a proper grafting yield of PAA is critical. In addition, glucose-sensitive microcapsules, with covalently bound GOx and constituted of PAA-gated porous polyamide membranes have been prepared (Chu et al., 2004). The proposed microcapsules show promise as self-regulated injectable drug delivery systems having the ability of adapting the release rate of drugs such as insulin in response to changes in glucose concentration. Such closed feedback loops are highly attractive for diabetes therapy. Further, sulfonamide-based hydrogels with decreased swelling capacity at high glucose concentration were fabricated (Kang and Bae, 2003).

Delivery of insulin is different from delivery of other drugs, since insulin has to be delivered in an exact amount at the exact time of need. Thus, self-regulated insulin delivery systems require the glucose-sensing ability and an automatic shut-off mechanism. Many hydrogel systems have been developed modulating insulin delivery, and all of them have a glucose sensor built into the system. Several existing strategies that may be feasible for glucose-responsive drug delivery are discussed next.

9.1.1.2 pH-Sensitive Membranes

Self-regulating insulin delivery devices depend on the concentration of glucose in the blood to control the release of insulin. Glucose oxidase is probably the most widely used enzyme in glucose sensing. Incorporating glucose oxidase into a polymer will result in a decrease in the pH of the environment immediately surrounding the polymer in the presence of glucose as a result of the enzymatic conversion of glucose to gluconic acid. This decrease in pH would cause the membrane to swell, forcing the insulin out of the device (Bryers and Ratner, 2006). This makes it possible to use different types of pH-sensitive hydrogels for modulated insulin delivery.

The reaction of glucose oxidase and glucose to form gluconic acid can also be used to drive the swelling of pH-dependent membrane, reducing the barrier to insulin diffusion across the membrane in the presence of glucose. A dual membrane system consisting of glucose oxidase immobilized on cross-linked polyacrylamide and copolymer membrane composed of *N,N*-diethylaminoethyl methacrylate (NNDEAEMA) and 2-hydroxypropyl methacrylate (DEA-HPMA) has demonstrated the promise of this idea. The glucose oxidase is placed in contact with the release media, while the barrier membrane is worked as an interface between insulin reservoir and sensing membrane (Karathnasis et al., 2007).

As shown in Figure 9.3, gluconic acid formed by the interaction of glucose and glucose oxidase caused the tertiary amine groups in the barrier membrane to protonate and induce a swelling response in the membrane. Insulin in the reservoir was able to diffuse across the swollen barrier membrane. When the glucose concentration decreased, the pH of the barrier membrane increased and it returned to a more collapsed and impermeable state (Lalwani and Santani, 2007).

Glucose-sensitive P (MAA-*g*-EG) gels were synthesized by copolymerizing methacrylic acid and poly(ethylene glycol) monomethacrylate in the presence of activated glucose oxidase (Dai et al., 2008). The surface of the polymer contained a series of molecular "entrances," which opened and released insulin dependent on glucose concentration. Thus, when these gels are placed in contact with a glucose solution, it results in a pH drop due to oxidation of glucose into gluconic acid. Due to this pH drop, the released protons caused the pendent poly (methacrylic acid) (PMAA) chains of the hydrogel to contract, thus opening the gates to allow insulin transport (see Figure 9.4).

A hydrogel-based photonic crystal acts as a glucose sensor for patients with *diabetes mellitus*. Glucose oxidase is attached to arrays of polystyrene nanospheres, which are then polymerized within a hydrogel matrix. The resulting material reversibly swells in the presence of glucose, similar to the glucose-responsive systems. The swelling event increases the mean separation between the immobilized nanospheres, shifting the Bragg peak of diffracted light to longer wavelengths and

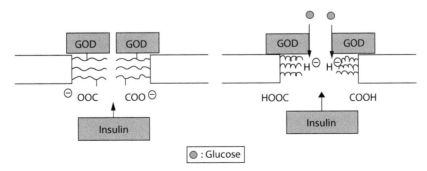

FIGURE 9.3 Principles of glucose-responsive insulin release system.

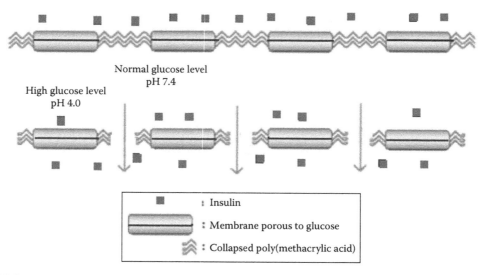

FIGURE 9.4 P(MAA-*g*-EG) responsive hydrogel systems for controlled release of insulin.

producing a red-shift in the optical properties (i.e., a readily observed color change) of the polymer. This system can be implanted as contact lenses or ocular inserts to detect small changes in blood glucose levels indirectly via tear fluid. Glucose binds to the derivatives, producing cross-links that shrink the hydrogel and cause a blue-shift. The patient is then able to determine their blood glucose levels via a color chart (Ben-Moshe, 2006).

9.1.1.3 Con A-Immobilized Systems

Concanavalin A (Con A) has also been frequently used in modulated insulin delivery. Con A is a glucose-binding protein obtained from the jack bean plant, *Canavalia ensiformis*. In this type of system, insulin molecules are attached to a support or carrier through specific interactions, which can be interrupted by glucose itself. In one approach, insulin was chemically modified to introduce glucose, which itself binds especially to Con A. Glycosylated insulin(G-insulin) bound to Con A can be displaced by glucose, thus releasing it from the drug delivery system.

Tanna et al. (2006a,b) had developed a chemical self-regulated insulin delivery system based on a polymeric glucose-responsive biomaterial. In this approach, the novel glucose-responsive biomaterial responds reversibly to glucose triggers and is being developed for use as part of an implantable self-regulated insulin delivery system. These materials show a graded decrease in complex viscosity when tested across a glucose concentration range of 0%–5% w/w and were stable against degradation and component loss and also able to deliver insulin, reversibly in response to glucose, when tested in in vitro diffusion experiments.

A number of biological or artificial receptors for glucose have been described, which can transduce glucose concentrations into changes in fluorescence, including lectins, enzymes, bacterial-binding proteins, and boronic acid derivatives, and which might be engineered as nano sensors. For example, a glucose assay in which Con A was covalently labeled with the highly near infrared (NIR) fluorescent protein allophycocyanin (donor), and dextran was labeled with the nonfluorescent dye, malachite green (acceptor). Addition of glucose displaces dextran from Con A, thereby reducing fluorescence resonance energy transfer (FRET) between the donor and acceptor and the measured fluorescence lifetime (Pickup et al., 2008).

A second sensing strategy with glucose-galactose binding protein (GBP) is to monitor glucose-dependent changes in fluorescence of an environmentally sensitive dye (i.e., one where fluorescence is low in a polar environment and high in a nonpolar one) linked to a suitable site in the protein.

For example, we found that with the environmentally sensitive dye, badan (ex 400 nm, em 550 nm), covalently attached to a cysteine residue introduced by site-directed mutagenesis at position 152 of GBP, glucose addition caused a 300% increase in fluorescence, likely due to folding of the protein around the dye producing a less polar environment (Khan et al., 2008).

9.1.1.4 Sol–Gel Phase Reversible Hydrogel Systems

Hydrogels can be made to undergo sol–gel phase transformations depending on the glucose concentration in the environment. Reversible sol–gel phase transformations require glucose-responsive cross-linking. A highly specific interaction between glucose and Con A was used to form a cross-link between glucose-containing polymer chains. Con A exists as a tetramer at physiological pH and each subunit has a glucose-binding site. It functions as a cross-linking agent for glucose-containing polymer chains. For example, when Con A is mixed with dextran derivatives, the four-valent Con A acts as a cross-linker for insulin incorporation into the gel. When such a composite insulin-containing gel is exposed to an increased glucose concentration, there is competition between dextran and free glucose for the Con A-binding sites. This results in the rupture of gel cross-links and hence insulin releases (Malmsten, 2006).

Glucose-sensitive phase reversible hydrogels can also be prepared without using Con A. Glucose-responsive hollow polyelectrolyte multilayer capsules (approximately 10 μm) based on poly(3-acryl amido phenyl boronic acid)-*co*-poly(dimethyl amino ethyl acrylate) copolymers were recently prepared. The phenylboronic acid units are partially uncharged in an aqueous medium. Through complexation with glucose the equilibrium is shifted toward the charged state, thereby rendering the entire polymer more hydrophilic. Only at glucose concentrations higher than 2.5 mg/mL (which is above healthy levels), the complexation between the charged phenyl borates and glucose was great enough to result in disassembly of the capsules within 5 min (Geest et al., 2006).

9.1.2 GLUTATHIONE-SENSITIVE GELS

Glutathione (g-ECG; i.e., g-glutamate–cysteine–glycine), a g-glutamyl tripeptide, is a highly distinctive amino acid derivative with several important roles. For example, glutathione, present at high levels (5 mM) in animal cells, protects red blood cells from oxidative damage precipitating from reactive oxygen species. Glutathione (G) also facilitates maintenance of overall cellular redox homeostasis by transitioning between a reduced thiol form (GSH with a free sulfhydryl group) and an oxidized disulfide form (GSSG with a disulfide linkage). GSSG is reduced to GSH by glutathione reductase, a flavor protein that uses NADPH as the electron source. The ratio of GSH to GSSG in most cells is greater than 500 (Alberts et al., 2002). From the importance of GSH in cells as a powerful metabolite and a sulfhydryl buffer, it is logical to infer that a delivery system sensitive to levels of glutathione can be beneficial in delivering therapeutics or drugs to intracellular compartments of cells. This rationale has prompted several researchers to develop innovative glutathione-sensitive forms of delivery systems. For example, Kataoka et al. developed a novel cytoplasmic delivery system for delivering antisense oligodeoxynucleotides (asODN). This delivery system was engineered by assembling a PEG–asODN conjugates with a disulfide-based smart linkage, PEG-SS-asODN. This conjugate was complexed with branched polyethylenimine (B-PEI) to form polyion complex (PIC) micelles. Results indicated that the fabricated smart micelles demonstrated significant antisense effect via disulfide bond cleavage in the cellular interior. This was because the delivery system cross-linked using disulfide linkages "sensed" high glutathione concentrations in the cellular cytoplasmic compartment (Oishi et al., 2005).

A novel class of biochemically degradable ABA triblock copolymer where A is 2-(methacryloyloxy)ethylphosphorylcholine (MPC) and B is *N*-isopropylacrylamide (NIPAM) was fabricated to form "flower micelles" under physiologically relevant conditions at a copolymer concentration above approximately 8% w/v (Li et al., 2006). Glutathionemediated degradation of these flower micelles resulted in alterations of the rheological properties of the gel due to the breakdown of

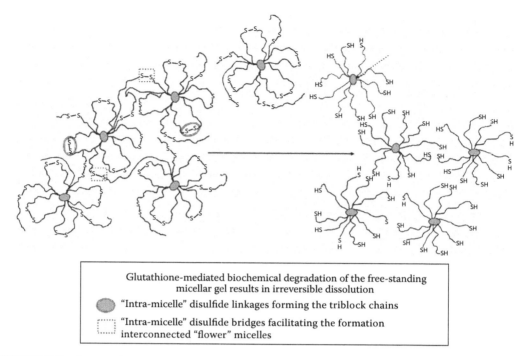

Glutathione-mediated biochemical degradation of the free-standing
micellar gel results in irreversible dissolution

"Intra-micelle" disulfide linkages forming the triblock chains

"Intra-micelle" disulfide bridges facilitating the formation
interconnected "flower" micelles

FIGURE 9.5   Strategy for the fabrication of glutathione-responsive and thermally responsive "doubly smart gels." Irreversible biochemical degradation of novel PNIPAM-PMPC-S-S-PMPC–PNIPAM "flower micelles" to free-flowing sulfhydryl-terminating micelles by the cleavage of intermicellar bridges is mediated by the tripeptide, glutathione. The reductive degradation of the interacting micelles to free-flowing micelles results in approximate halving of the original copolymer molecular weight.

the "flower structure." Changing the block polymer architecture from triblock disulfide-linked "flower" micelles to diblock sulfhydryl-terminated conventional micelles resulted in "degelation" of the system (Figure 9.5). The resulting sulfhydryl-terminated micelles still exhibited temperature-sensitive micellar self-assembly. This strategy illustrates a promising paradigm for altering the rheology of polymeric delivery systems using biochemical stimuli that are known to be ubiquitous in intracellular compartments. Further, important design parameters in engineering dissolvable glutathione-sensitive hydrogels include the disulfide cross-linking density, the dimensions of the hydrogel, and the distribution of the disulfide linkages. Aluru et al. (Chatterjee et al., 2003) modeled the transduction of biochemical signals in dissolvable hydrogels to optimize the design parameters in hydrogels with disulfide chemistries. Once the design parameters of such hydrogels are optimized, precisely engineered microfluidic devices using such biochemically modulated hydrogels can be devised.

9.1.3   ANTIGEN-SENSITIVE GELS

To induce reversible antigen-responsiveness, Miyata et al. (Miyata et al., 1999 a,b) synthesized an antigen–antibody semi-interpenetrating (SIPN) hydrogel network. Antigen (rabbit immunoglobulin G (IgG)) and antibody (goat anti-rabbit IgG) were chemically modified by coupling respective antigen and antibody with N-succinimidylacrylate (NSA). The modified antibody monomers were copolymerized with acrylamide. Following this, the modified antigen monomers were then copolymerized with acrylamide and N,N-methylene bisacrylamide (MBAAm) as a cross-linker in the presence of the polymer bearing the antibody, resulting in semi-IPNs containing antigen and corresponding antibody. Additional cross-links were introduced by the binding between antigen and

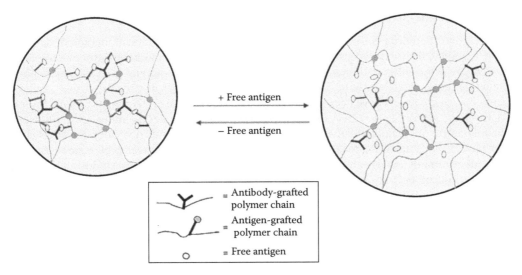

FIGURE 9.6 Strategy for the fabrication of a reversibly antigen-responsive gel. *Left*: The gel is in its collapsed state due to cross-linking via grafted antigens and antibodies. *Right*: The addition of free antigen to the system results in competitive binding interaction between the grafted antigen and the free antigen resulting in gel swelling due to decreased extent of noncovalent cross-linking. Here, the antigen is rabbit IgG and the antibody is GAR IgG.

antibody in the network. The polymerized antibody showed a higher binding constant to the native antigen than to the polymerized antigen. Therefore, the presence of a free antigen caused swelling of the hydrogels by dissociating the noncovalent cross-links induced by the intrachain antigen–antibody binding, because the free antigen replaced the binding of antibody from the polymerized antigen to native antigen. The antigen-responsive property could thus be specifically utilized in a drug delivery system (Figure 9.6) (Miyata et al., 1999a).

9.1.4 GELS SENSITIVE TO OTHER ANALYTES

The responsiveness of hydrogels to signal biomolecules can translate to potential applications in smart devices. A hydrogel membrane sensitive to the metabolite nicotinamide adenine dinucleotide (NAD) and containing immobilized ligands and receptors was investigated for the controlled diffusion of model proteins (Tang et al., 2004). Both cibacron blue (ligand) and lysozyme (receptor) were covalently linked to dextran. NAD serves as a competing ligand and competes with cibacron blue in its interaction with lysozyme. Using cytochrome c and hemoglobin as model proteins to examine diffusion across the hydrogel membrane in response to differential concentrations of NAD, the approach of sensing ambient levels of NAD can be generalized to diagnose the levels of different analytes by a suitable selection of a competing ligand–receptor interaction, thereby affecting the permeability of the polymer membrane. Further, in this kind of ligand-modulated release, it is important to maximize selective transport and minimize intrinsic diffusion (Chaterji et al., 2007). This can be achieved by optimizing the affinity kinetics between the analyte and the receptors as well as engineering the cross-linking density to attain a desired distribution and geometry of pores in the network. Effective modeling of the transport kinetics of the analyte of interest with differentially engineered cross-linking parameters can facilitate selective transport through the polymer matrix. Another point of concern raised by the authors is the hysteresis associated with the transition of the network properties of the gel with the waxing and waning of the stimuli strength (i.e., antigen concentration); this hysteresis has been found to be an artifact of the polymer chain relaxation time.

9.2 MAGNETIC-SENSITIVE DRUG DELIVERY SYSTEMS

Magnetic-controlled drug delivery by particulate carriers is a very efficient method of delivering drug to a localized disease site. Very high concentrations of chemotherapeutic or radiological agents can be achieved near the target site, such as a tumor, without any toxic effects to normal surrounding tissue or to the whole. Biological polymers, liposomes, hydrogels, viruses, and other controlled release carriers have been widely investigated (Figure 9.7) (Dijkmans, 2004; Hofheinz, 2005; Lesniak, 2005). These vehicles release therapeutic agents under the influence of ultrasound, pH, temperature, or chemical interaction, but in many cases, however, the drug delivery vehicles do not have a mechanism for localization where it is possible to deliver high concentrations of drugs with minimally invasive techniques. This is especially true when repeat dosing is required. The magnetic drug delivery system proposed herein overcomes many of these difficulties, and provides a method for concentrating drugs at selected sites in the body with minimal stress on the patient.

Ferrogels consisting of magnetic nanoparticles embedded in polymer gels are an important category of stimuli-sensitive hydrogels that respond to external magnetic stimuli (Collin et al., 2003; Lattermann and Krekhova, 2006). Their unique magnetoelastic property endows ferrogels with great potential for magnetic controllable drug delivery systems. However, compared to drug release studies of other stimuli-sensitive hydrogels, about which many papers have been published (Grassi et al., 2005). Recent studies by several groups on various types of ferrogels for in vitro release have implemented magnetically controlled drug release (Liu et al., 2006, 2008b; Francois et al., 2007; Hu et al., 2007). Nevertheless, nearly all reported studies are focusing only on water-soluble drugs, due to the overall hydrophilic nature of ferrogels. Effective incorporation of important hydrophobic drugs in ferrogels still remains a challenge.

A U.S. patent (Handy et al., 2006) focuses on a treatment method that involves the administration of a magnetic material composition containing single-domain magnetic particles. The application of an alternating magnetic field to inductively heat the magnetic material composition causes the triggered release of therapeutic agents at target tumor or cancerous cells. Pankhurst et al. (2003) studied the physical principles underlying some current biomedical applications of magnetic nanoparticles, that is, the relevant physics of magnetic materials and their responses to applied magnetic fields. They studied the way these properties are controlled with reference to (a) magnetic separation of labeled cells and other biological entities, (b) therapeutic drug, gene, and radionuclide delivery, (c) radio frequency methods for catabolism of tumors via hyperthermia, and (d) enhancement agents for magnetic resonance imaging applications. Tertaj et al. (2003) presented various synthetic routes for the preparation of magnetic nanoparticles useful for biomedical applications.

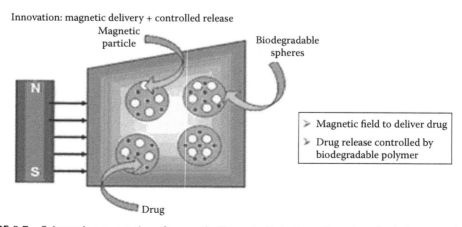

FIGURE 9.7 Schematic presentation of magnetically controlled release from drug-loaded nanoparticles.

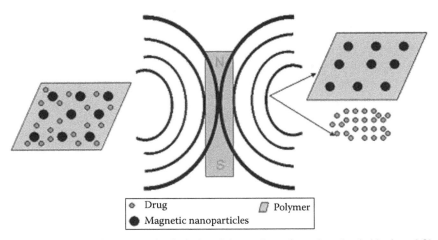

FIGURE 9.8   Diagram showing magnetically induced drug release from drug-loaded hydrogel film.

The process of drug localization using magnetic delivery systems is based on the competition between forces exerted on the particles by blood compartment and magnetic forces generated from the magnet, that is, applied field. When the magnetic forces exceed the linear blood flow rates in arteries (10 cm s^{-1}) or capillaries (0.05 cm s^{-1}), the magnetic particles are retained at the target site and may be internalized by the endothelial cells of the target tissue (Hafeli, 2004).

Magnetic particles have been shown to aid in the release of encapsulated molecules from inside of a polymer matrix when magnetically induced by an oscillating or alternating current (AC) magnetic field (Lu et al. 2005; De Ikehara et al., 2006; Paoli et al., 2006) (Figure 9.8). Magnetic release can control the activity of a drug delivery system in several ways, including decreasing the amount of unnecessary damage to healthy tissue, increasing the efficacy of the drug, and treating ailments in a minimally invasive way. Unnecessary damage to healthy tissue can be decreased by using an external magnet to guide polymeric particles containing magnetic material and drugs to cells that need to be treated (Kumar et al., 2004); an external magnetic field can then be used to release the drugs at the target site (Kim and Park, 2005).

9.3   ELECTRIC FIELD–RESPONSIVE DRUG DELIVERY SYSTEMS

Electric field–responsive polymers have been investigated as a form of hydrogels to have swelling, shrinking, or bending behavior in response to an external field. This property of such polymers has been applied for bio-related applications such as drug delivery systems, artificial muscle, or biomimetic actuator. The electromechanical behavior of cross-linked strong acid hydrogels was investigated in different salt solutions and in a 1.6 V/cm dc electric field (Yao and Krause, 2003). The sulfonated polystyrene-based copolymer networks were observed to bend toward the cathode when they were preequilibrated in the corresponding salt solutions. A polythiophene-based conductive polymer gel actuator was reported to show expansion/contraction behavior in response to an applied potential (Irvin et al., 2001). The axial pressure generated by the expansion of the gel against a fixed wall was measured, which demonstrated that the generated closure pressures could be utilized as a small actuator valve. In most cases, electric field responsive polymer networks are based on electrically driven motility. On the other hand, some neutral polymer gels were reported to show electric field sensitivity. Weakly cross-linked poly (dimethyl siloxane) gels containing finely distributed electric field–sensitive particles (TiO$_2$ particle) showed a significant and quick bending in silicon oil (Filipcsei et al., 2000). The mechanism of this responsiveness was that the forces acting on the particles, which could not leave the gel networks, directly caused the locomotion or the deformation of the gel.

In a recent study by Bajpai et al. (2008), polyaniline (PANI) was impregnated into a macromolecular matrix of poly(vinyl alcohol)-*g*-PAA and electrical conductivity and electroactive behavior of the resulting nanocomposite was studied.

9.4   REDOX POTENTIAL–SENSITIVE DRUG DELIVERY SYSTEMS

Naturally occurring redox potentials within the body can be utilized as a stimulus to trigger the release of encapsulated molecules from nanocarrier systems. Both oxidative conditions existing physiologically in extracellular fluids and pathophysiologically in, for instance, inflamed or tumor tissues, as well as reductive environments within the cell, can be exploited to destabilize carrier systems. Hubbell and coworkers developed oxidation-responsive polymersomes based on triblock copolymers PEG-poly(propylene sulfide)-PEG (PEG-PPS-PEG) (Napoli et al., 2004). Upon exposure to either aqueous H_2O_2 or H_2O_2 from a glucose-oxidase/glucose/oxygen system, the hydrophobic PPS core was oxidized and transformed within 2 h into hydrophilic poly(sulfoxides) and poly(sulfones), thereby destabilizing the vesicular structure. The metal-containing polydimethylsiloxane-*b*-polyferrocenylsilane (PDMS-*b*-PFS) diblock copolymer was utilized to construct redox-active organometallic vesicles offering a potential for redox-tunable encapsulation (Power-Billard et al., 2004).

The existence of a high difference in redox potential between the mildly oxidizing extracellular space and the reducing intracellular space, due to the presence of endogenous thiols, including glutathione (GSH, ca.10 μM in human plasma), free reduced homocysteine (0.1–0.35 μM), and free cysteine (ca. 5 μM), can offer a valuable alternative to design polymeric release systems. Disulfide bonds can respond to this reductive condition via reversible cleavage into free thiols. For example, disulfides have been used in bioreducible poly(amido amine)s-based DNA delivery systems (Lin et al., 2007).

The response of the polymer system to thiol was sensitive enough so that a single reduction per polymer chain triggered the cleavage of the macroamphiphile to a hydrophobic and a hydrophilic homopolymer. In cellular experiments, uptake, disruption, and release were observed within 10 min of exposure of the polymer system to cells within the time frame of the early endosome of endolysosomal processing (ca. 30 min). The reduction-sensitive polymer system can protect biomolecules in the extracellular environment and suddenly burst releasing their contents within the early endosome prior to exposure to the harsh conditions encountered after lysosomal fusion. Therefore, they may be suitable for the cytoplasmic delivery of bimolecular drugs such as peptides, proteins, oligonucleotides, and DNA (Meng et al., 2009).

9.5   pH-RESPONSIVE DRUG DELIVERY SYSTEMS

pH is another important environmental parameter for drug delivery systems, because the pH change occurs at many specific or pathological body sites, such as stomach, intestine, endosome, lysosome, blood vessels, vagina, and tumor extracellular sites. Therefore, the pH-triggered drug delivery systems have attracted considerable interest. pH-sensitive drug delivery is an self-regulated drug delivery system in which the release rates are adjusted by the system, in response to feedback information, without any external intervention. It is also known as closed-loop delivery systems.

Ionizable polymers with a pK_a value in between 3 and 10 are included under pH-responsive materials. Weak acids and bases like carboxylic acids, phosphoric acids, and amines, respectively, exhibit a change in the ionization state upon variation of the pH. This leads to the conformational change for the soluble polymers and a change in the swelling behavior of the hydrogels when these ionizable groups are linked to the polymer structure.

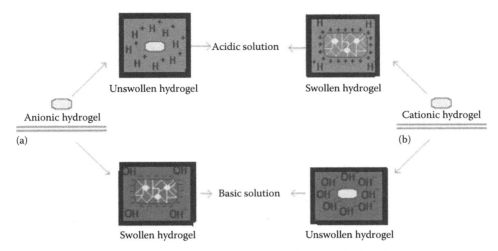

FIGURE 9.9 The pH-responsive swelling of (a) anionic and (b) cationic hydrogels.

All pH-sensitive polymers contain pendant acidic or basic groups that either accept or donate protons in response to the environmental pH (Bartil et al., 2007). Swelling of a hydrogel increases as the external pH increases in the case of weakly acidic (anionic) groups, but decreases if the polymer contains weakly basic (cationic) groups (Guo et al., 2007). This mechanism is shown in Figure 9.9. When the ionic strength of the solution is increased, the hydrogel can exchange ions with the solution. By doing so, the hydrogel maintains charge neutrality and the concentration of free counter ions inside the hydrogel increases. pH-sensitive hydrogels found wide application in the biomedical field (Pal et al., 2007) including control drug release (Xiao et al., 2009), for oral drug delivery (Kumar et al., 2007), and bio-sensor (Bhardwaj et al., 2008), etc. pH-responsive hydrogels may be grouped into two main classes: (a) cationic hydrogels and (b) anionic hydrogels according to the method of ionization that is, donating or accepting protons.

9.5.1 Cationic Hydrogels

Cationic hydrogels swell and release a drug in the low pH environment of the stomach. When external pH is lower than the pK_b of the ionizable groups, cationic hydrogels protonate and swell more (Lin and Metters, 2006). Typical examples of the basic polyelectrolytes include poly(tertiary amine methacrylate) such as poly(2-(dimethylamino)ethylmethacrylate) (Dieudonne et al., 2006) and poly(2-(diethylamino)ethylmethacrylate), poly(2-vinylpyridine), and biodegradable poly(β-amino ester).

Cationic polymer, for example, poly(ethyleneimine), forms an ionic complex with therapeutic genes. After cellular uptake, the complex is located in endosomes derived by fusion between lysosomes and endocytotic vesicles. Many primary, secondary, and tertiary amino groups in the cationic polymer induce osmotic unbalance, thanks to proton buffering effect, which finally results in the dissociation of therapeutic genes and endosome rupture (Kim et al., 2009).

9.5.2 Anionic Hydrogels

Hydrogels of PAA or PMA can be used to develop formulations that release drugs in a neutral pH environment. Typical acidic pH-sensitive polymers for drug delivery are based on the polymers containing carboxylic groups, such as poly(acrylic acid) (Bai et al., 2008), poly(methacrylic acid), poly(L-glutamic acid), and polymers containing sulfonamide groups.

The synthesis and swelling behavior of a superabsorbent hydrogel based on starch (St) and poly-acrylonitrile (PAN) are investigated (Sadeghi and Hosseinzadeh, 2008). The absorbency of the hydrogels indicated that the swelling ratios decreased with increasing ionic strength. The hydrogels exhibited a pH-responsive swelling–deswelling behavior at pH 2 and 8. This on–off switching behavior provides the hydrogel with the potential to control delivery of bioactive agents. Release profiles of ibuprofen from the hydrogels were studied under both simulated gastric and intestinal pH conditions. The release was much quicker at pH 7.4 than at pH 1.2. The swelling rates of the hydrogels with various particle sizes were investigated as well.

A biodegradable pH-sensitive hydrogel for potential colon-specific drug delivery was synthesized by Casadei et al. (2008). Their composite hydrogel, based on a methacrylate and succinic derivative of dextran, and a methacrylate and succinic derivative of poly (N-2-hydroxyethyl)-DL-aspartamide was produced by photocross-linking polymerization. In vitro drug release studies, performed using 2-methoxyestradiol as a model drug, show that obtained hydrogel is able to release the drug in simu-lated intestinal fluid, due to its pH-sensitive swelling and enzymatic degradability.

Wang and his coworkers (Yue et al., 2009) synthesize novel microstructure and pH-sensitive poly(acrylic acid-*co*-2-hydroxyethyl methacrylate)/poly(vinyl alcohol) (P(AA-*co*-HEMA)/PVA) interpenetrating network (IPN) hydrogel films by radical precipitation copolymerization and sequential IPN technology. The IPN films were studied as controlled drug delivery material in dif-ferent pH buffer solutions using cationic compound crystal violet as a model drug. The drug release followed different release mechanism at pH 4.0 and 7.4, respectively.

Nugent et al. (McGann et al., 2009) developed novel pH-sensitive hydrogel composite for the delivery of aspirin to wounds using a freeze–thaw process. Physically cross-linked hydrogels com-posed of poly(vinyl alcohol) (PVA) and poly(acrylic acid) were prepared by a freeze–thaw treatment of aqueous solutions and aspirin was incorporated into the systems. The pH-sensitive nature of the hydrogels was apparent from solvent uptake studies carried out. Increasing alkaline media led to a greater degree of swelling due to increased ionization of PAA. The release rates shown by prepared hydrogels were relatively slow and exhibited non-Fickian release kinetics.

Richter et al. (2008) prepared hydrogel-based pH sensors and microsensors. These sensors are directly working devices that can have a high selectivity and sensitivity. The pH-sensitive nano-vehicles demonstrated increased anticancer activity following enhanced drug release triggered by acidic tumor extracellular pH or by the cleavage of chemical bonds between the carrier and drug. The acidity or pH primarily depends on the tumor's histology and volume. Recently, a pH-sensitive polymeric micelle system has been potentially used for targeted antitumor drug delivery (Lee et al., 2008) based on the observation that one of the consistent differences between various solid tumors and the surrounding normal tissue is the acidity of the surrounding tissue (Lee et al., 2008).

9.6 TEMPERATURE-SENSITIVE DRUG DELIVERY SYSTEMS

Temperature is the most widely utilized triggering signal for a variety of triggered or pulsatile drug delivery systems. The use of temperature as a signal has been justified by the fact that the body temperature often deviates from the physiological temperature (37°C) in the presence of pathogens or pyrogens. This deviation sometimes can be a useful stimulus that activates the release of thera-peutic agents from various temperature-responsive drug delivery systems for diseases accompany-ing fever. The drug delivery systems that are responsive to temperature utilize various polymer properties, including the thermally reversible coil/globule transition of polymer molecules, swelling change of networks, glass transition, and crystalline melting (Anal, 2007).

Thermosensitive polymer hydrogels have been extensively studied because of their valuable applications in drug delivery system, cell encapsulation, and tissue engineering. These polymeric hydrogels are formed from aqueous polymer solutions with temperature changes, which come mainly from packing of polymeric micelles or physical associations between polymer segments in aqueous solution. Therefore, thermosensitive polymer hydrogels can avoid toxic organic cross-linker usually employed to form hydrogel. In this system, various drugs can be incorporated by a simple mixing and the solution containing drugs is locally injected to specific body site (Kang et al., 2006). As a result, the solution is instantly converted to hydrogel at the injected site and drugs are slowly released through three-dimensional networks of the hydrogel for a long period.

Temperature-sensitive hydrogels undergo a volume phase-transition or a sol–gel phase-transition at a critical temperature, namely, cloud point or lower critical solution temperature (LCST) and upper critical solution temperature (UCST). The LCST polymers exhibit a hydrophilic-to-hydrophobic transition with increasing temperature (Aqil et al., 2008), whereas the UCST systems undergo the opposite transition. In contrast to the UCST systems, the LCST systems have received more attention for drug delivery because mixing of the UCST systems and drugs needs to be performed at relatively high temperature, which may be harmful to some unstable drugs or biomolecules and bring inconvenience into the drug formulation (He et al., 2008).

Phase transition of these polymers is controlled by a delicate balance between hydrophobic and hydrophilic conditions. The phase transitions in terms of the biologically relevant intermolecular forces can rely on several different interactions.

1. *Van-der-Waals interaction*: Van-der-Waals interaction causes a phase transition in hydrophilic gels in mixed solvents, such as an acrylamide gel in an acetone–water mixture. The nonpolar solvent is needed to decrease the dielectric constant of the solvent.
2. *Hydrophobic interaction*: Hydrophobic gels, such as N-isopropylacrylamide (NIPAM) gels, undergo a phase transition in pure water, from a swollen state at low temperature to a collapsed state at high temperature.
3. *Hydrogen bonding with change in ionic interaction*: Gels with cooperative hydrogen bonding, such as an IPN of poly(acrylic acid) and poly(acrylamide), undergo a phase transition in pure water (the swollen state at high temperatures). The repulsive ionic interaction determines the transition temperature and the volume change at the transition.
4. *Attractive ionic interaction*: The attractive ionic interaction is responsible for the pH-driven phase transition, such as in acrylamide–sodium acrylate/methacrylamido propyltrimethyl ammonium chloride gels (Schmaljohann, 2006).

Temperature-sensitive hydrogels can be classified as negatively thermosensitive, positively thermosensitive, and thermally reversible hydrogels.

9.6.1 Negatively Thermosensitive Hydrogels

Negatively thermosensitive hydrogels tend to shrink or collapse as the temperature is increased above the LCST, and swell upon lowering the temperature below the LCST. The change in the hydration state is due to the volume phase transition, Thermodynamics can explain this with a balance between entropic effects due to the dissolution process itself and due to the ordered state of water molecules in the vicinity of the polymer. Enthalpy effects are due to the balance between intra- and intermolecular forces and due to salvation, for example, hydrogen bonding and hydrophobic interaction. The transition is then accompanied by coil-to-globule transition (Liu et al., 2008). By controlling the polymer composition and topology, the coil-to-globule transition could be kinetically and thermodynamically controlled.

Typical LCST polymers include poly(methylvinylether) (PMVE)-19, *N-tert*-butyl acrylamide) (*N*-tBAAm), poly(*N*-(3-(dimethylamino)propyl)methyacrylamide) (PDMAPMA), Poly(*N*-vinyl-*n*-butyramide) (PNVIBAM), poly(*N*-vinylcaprolactam) (PNVCa), poly(dimethylamino ethylmethacrylate), and elastin-like polypeptide (ELPs).

A well-known polymer with LCST at 32°C is poly(*N*-isopropyl acrylamide) (NIPAAm), which has been extensively employed as a negative thermosensitive hydrogel because it exhibits a coil–globule transition upon exterior temperature changes. Polymers with LCST have been tested in controlled drug delivery matrices and in on–off release profiles in response to a stepwise temperature change. The on–off release profile of drugs are due to the formation of a dense, less permeable surface layer of gel, described as a skin-type barrier. The skin barrier was formed upon a sudden temperature change, due to the faster collapse of the gel surface than the interior (Satish et al., 2006). The surface shrinking process was found to be regulated by the length of the methacrylate alkyl side chain, that is, the hydrophobicity of the comonomers.

Cellulose nitrate and cellulose acetate monolayer membranes containing *n*-heptyl-cyanobiphenyl were developed as thermoresponsive barriers for drug permeation (Atyabi et al., 2007). Methimazole and paracetamol as hydrophilic and hydrophobic drug models were used, respectively. Atyabi's group found that upon changing the temperature of the system around 41.5°C, both cellulose membranes without cyanobiphenyl showed no temperature sensitivity to drug permeation, whereas the results for cyanobiphenyl-entrapped membranes exhibited a distinct jump in permeability when temperature was raised to above 41.5°C for both drug models. Drug permeation through the membranes was reversible, reproducible, and followed zero-order kinetics. The pattern of on–off permeation through these membranes was more distinguished for methimazole compared to that of paracetamol, seemingly due to its lower molecular weight.

Coughlan et al. (2006) evaluated the swelling and release profile of cross-linked pNiPAAm hydrogels as a function of the physicochemical properties of the loaded drugs. Dried hydrogel disks were loaded by sorption of a drug solution, the solvent was removed and the hydrogels were allowed to swell in a buffer solution. Hydrogel swelling was decreased in the presence of hydrophobic drugs and the opposite effect was observed for hydrophilic drugs.

9.6.2   Positively Thermosensitive Hydrogels

Certain hydrogels formed by IPNs show swelling at high temperature, and shrinking at low temperature. Such types of hydrogels are called positively thermosensitive hydrogels. A positive temperature-sensitive hydrogel has an UCST. The hydrogel contracts upon cooling below the UCST. Polymer networks of PAA and polyacrylamide (PAAm) (Kimura et al., 2007) or poly(acrylamide-*co*-butylmethacrylate) exhibit positive temperature dependence of swelling.

The obtained P(AAm-AAc) IPNs were very stable at 70°C in an aqueous solution. Dissociation temperatures of the hydrogels shifted to higher values with increasing AAm content. These IPNs showed limited swelling ratios between their swelling transition temperatures and lower swelling ratios above these transition temperatures. Transition temperatures shift to higher values with increasing AAm content.

9.6.3   Thermo-Reversible Hydrogels

Aqueous solutions of some polymers undergo sol-to-gel transition in response to a certain stimulus (Van-Tomme et al., 2008). Polymers with hydrophobic domains can cross-link in aqueous environments via reverse thermal gelation. The hydrophobic segment is coupled to a hydrophilic polymer segment by postpolymerization grafting or by directly synthesizing a block copolymer to create a polymer amphiphile. Such amphiphiles are water-soluble at low temperature. As the temperature is increased, hydrophobic domains aggregate to minimize the hydrophobic surface area contacting the bulk water, reducing the amount of structured water surrounding the hydrophobic domains and maximizing the solvent entropy. The temperature at which gelation occurs depends on the concentration of the polymer, the length of the hydrophobic block, and the chemical structure of the polymer (Hoare and Kohane, 2008).

The most widely used thermally reversible hydrogels are triblock copolymers of poly(ethylene oxide)-poly(propylene oxide)-poly(ethylene oxide) (PEO-PPO-PEO), the poloxamers/pluronics (Xiong et al., 2006). Aqueous solution of poloxamers demonstrates phase transitions from sol to gel at 5°C–30°C and gel to sol at 35°C–50°C with the temperature increasing monotonically over the polymer concentration range of 20–30 wt%.

Stereocomplexed hydrogels based on PEG–(PLLA) and PEG–(PDLA) star block copolymers were reported to show the in vitro protein release profiles for 10–16 days and retarded therapeutic effect in vivo (Hiemstra et al., 2007). Very recently, a novel pH and temperature-sensitive hydrogel based on pH-sensitive sulfonamide oligomers and thermosensitive PCLA–PEG–PCLA block copolymers were reported to show the antitumor effect for 2 weeks (Shim et al., 2007).

Recently, the use of Pluronic F127 (synonymous to poloxamer 407) was reported for tissue engineering applications. This polymer has been found to have a rapid gelation at 37°C (after 1 min incubation, 30% solution in cell culture medium) (Weinand et al., 2006).

Some natural polymers also undergo reverse thermal gelation. Chitosan-glycerol phosphate-water system is an interesting example, which has being investigated for protein delivery, gene delivery, and tissue engineering applications. Recently, novel di-block copolymers of chitosan and PEG were synthesized by block copolymerization of monomethoxy-PEG onto chitosan backbone, using potassium persulfonate as a free radical initiator (Ganji and Abdekhodaie, 2008). The obtained hydrogels undergo a thermosensitive transition from a free-flowing solution at room temperature to a gel around body temperature. Their gelation time varied from 6 to 11 min. Their gelation time varied from 6 to 11 min. A sharp increase in viscosity around 35°C indicates the beginning of the gelation process. It is also found that solutions with high polymer concentrations and low PEG content gel faster than those with low polymer concentrations or high PEG content.

9.7 ULTRASOUND-SENSITIVE DRUG DELIVERY SYSTEMS

Ultrasound is a safe, portable, and low-cost imaging modality that shows excellent potential for applications in molecular imaging and targeted drug delivery (Borden et al., 2006). Ultrasound (US) has an ever-increasing role in the delivery of therapeutic agents including genetic material, proteins, and chemotherapeutic agents. Just as audio sound is the transmission of pressure waves through a medium such as air or water, ultrasound is the same type of transmission of pressure waves, but at frequencies above human hearing, or above 20,000 Hz.

As with light waves, these ultrasonic waves can be reflected, refracted (bent), focused, and absorbed. Unlike light waves, ultrasonic waves are very physical in nature; they are actual movement of molecules as the medium is compressed (at high pressure) and expanded (at low pressure), and thus ultrasound can act physically upon biomolecules and cells. Most importantly, unlike visible light waves, ultrasonic waves are absorbed relatively little by water, flesh, and other tissues (Pitt et al., 2004). Therefore, ultrasound can "see" into the body (e.g., diagnostic ultrasound) and can be used to transmit energy into the body at precise locations. This safe, noninvasive, and painless transmission of energy into the body is the key to ultrasonic-activated drug delivery. Major mechanisms that are involved in ultrasound-triggered drug release are local temperature increase [hyperthermia]; cavitation, which increases the permeability of cell membranes; and the production of highly reactive free radical species, which can accelerate polymer degradation (Blanco et al., 2009).

The first contribution of ultrasound is the disruption of drug carriers. Drug-loaded vesicles (or micelles), which are denser than the surrounding liquid, will be absorbed into the shear field surrounding an oscillating bubble. If the shear stress exceeds the strength of the drug carrier, it will rupture and release its contents. The second mechanism arises from collapse cavitation, which is produced when the bubble collapses during the contraction cycle by the high acoustic intensity. The collapse produces a shock wave-like a spike of dense fluid and when it passes over a polymeric carrier, the backbone of the polymers can be ruptured if the critical stress is exceeded. Consequently, the drug might be sheared from the polymer backbone to achieve controlled release (Oh et al., 2007).

Ultrasonication and cavitation (Haar, 2007) can be involved in drug delivery by several mechanisms. The simplest mechanism derives from the oscillatory motion of the insonated fluid and can occur in the absence of cavitation. The oscillating fluid increases the effective diffusivity of molecules; thus the transport of any drug, whether free or bound to a carrier, will be augmented by the oscillatory motion of the fluid. Such ultrasonic-enhanced transport may occur within blood, cells, or extracellular fluids.

Ultrasound is mostly used as an enhancer for the improvement of drug permeation through biological barriers, such as skin, lung, intestinal wall, and blood vessels. There are several reports describing the effect of ultrasound on controlled drug delivery (Hernot and Klibanov, 2008). Kost at al. described an ultrasound-enhanced polymer degradation system. During polymer degradation, incorporated drug molecules were released by repeated ultrasonic exposure. As degradation of biodegradable matrix was enhanced by ultrasonic exposure, the rate of drug release also increased. Thus, pulsed drug delivery was achieved by the on–off application of ultrasound.

Ultrasound involves the use of ultrasonic energy to enhance the transdermal delivery of solutes either simultaneously or via pretreatment and is frequently referred to as sonophoresis or phonophoresis. The proposed mechanism behind the increase in skin permeability is attributed to the formation of gaseous cavities within the intercellular lipids on exposure to ultrasound, resulting in disruption of the stratum corneum (Brown et al., 2008). Ultrasound parameters such as treatment duration, intensity, and frequency are all known to affect percutaneous absorption, with the latter being the most important.

Ultrasound-activated drug release from Pluronic [PEG–poly(propylene oxide) (PPO)–PEG] micelle solutions has been systematically investigated by Rapoport's and Pitt's groups (Husseini and Pitt, 2008). Gao et al. evaluated the effects of ultrasound on drug targeting and release using Pluronic P-105 micelles, stabilized micelles using PEG-diacylphospholipid, and DOX-loaded micelles (Gao et al., 2005).

Myhr and Moan (2006) used Plurogel to encapsulate a chemotherapy drug (fluorouracil). The authors then applied ultrasound to mice inoculated with a human colon cancer cell line. Ultrasound significantly reduced the tumor volume compared to the control group. The authors also concluded that more significant tumor reductions were observed when higher drug concentrations were administered.

It has been reported that collapse cavitation was implicated in colon cancer cell membrane disruption. Their study showed that calcein uptake was reduced by increasing the hydrostatic pressure (up to 3 atm) at constant ultrasonic intensity and frequency. Increasing hydrostatic pressure reduced acoustic bubble collapse associated with collapse cavitation, although stable cavitation still occurs. In these experiments, lower membrane permeability correlated directly with higher hydrostatic pressure and cavitation suppression.

Howard et al. (2006) studied the effect of ultrasound on paclitaxel in micelles of methyl capped poly(ethylene oxide)-*co*-poly-(L-lactide)-tocopherol on a breast cancer drug-resistant cell line. Their study used 1 MHz ultrasound at a power density of $1.7\,W/cm^2$ and duty cycle of 33%. They showed that upon the application of ultrasound, drug accumulation from encapsulated Paclitaxel drastically increased and surpassed the amount of drug found in noninsonated cells.

Juffermans et al. (2009) studied ultrasound and microbubble-targeted delivery, which provides opportunities for new therapies due to its low toxicity, low immunogenicity, noninvasive nature, local application, and its cost effectiveness. The bio-effects found in their studies provide important new insights into the mechanisms of ultrasound and microbubble-targeted delivery of therapeutic compounds and will lead to the rational design of new drug or gene therapies involving ultrasound and microbubbles (see Figure 9.10).

Killiany et al. developed a 512-channel transducer and driving system and tested for the normal brain in vivo (Hynynen et al., 2006). Focused ultrasound has also been shown to have the ability to induce selective opening of the blood–brain barrier (BBB) without damaging normal neuronal tissue (Kim et al., 2008). This enables ultrasound-enhanced drug delivery to specific areas of the diseased brain.

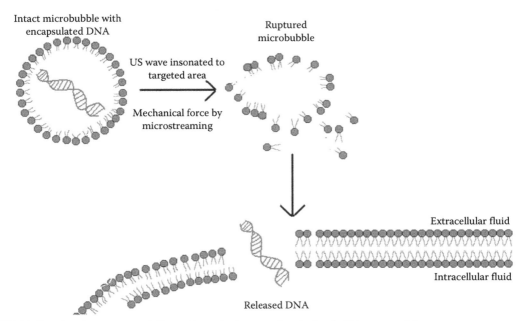

FIGURE 9.10 Drug or gene delivery through ultrasound and microbubble-targeted therapy.

9.8 LIGHT-SENSITIVE DRUG DELIVERY SYSTEMS

The interaction between light and a material can also be used to modulate drug delivery. This can be accomplished by combining a material that absorbs light at a desired wavelength and a material that uses energy from the absorbed light to modulate drug delivery (Sershen and West, 2002). Light responsiveness is receiving increasing attention owing to the possibility of developing materials sensitive to innocuous electromagnetic radiation (mainly in the UV, visible, and NIR range), which can be applied on demand at well delimited sites of the body. Some light-responsive DDS are of a single use (i.e., the light triggers an irreversible structural change that provokes the delivery of the entire dose) while others are able to undergo reversible structural changes when cycles of light–dark are applied, behave as multiswitchable carriers (releasing the drug in a pulsatile manner). Polymeric micelles, gels, liposomes, and nanocomposites with light-sensitiveness have been recently analyzed (Carmen et al., 2009). Research into light-responsive DDS has been focused mainly on self-assembled colloids such as copolymer micelles and liposomes, although other photoresponsive supramolecular architectures are also under study. Modern laser systems enable a precise control of light wavelength, duration, intensity, and diameter of the beam and offer a wide range of possibilities for biomedical applications.

NIR light has been used to modulate the release of various proteins from a composite material fabricated from gold nanoshells and poly(NIPAAm-*co*-AAm) (Sershen et al., 2000). Gold nanoshells are a new class of optically active nanoparticles that consist of a thin layer of gold surrounding a dielectric core (Averitt et al., 1999). Varying the shell thickness, core diameter, and the total nanoparticles diameter allows the optical properties of the nanoshells to be tuned over the visible and NIR spectrum. Since the core and shell sizes can be easily manipulated, the optical extinction profiles of the nanoshells can be modified to optimally absorb light emitted from various lasers. Embedding the nanoshells in an NIPAAm-*co*-AAm hydrogel formed a composite material that possessed the absorption spectrum of the nanoshells and the phase transition characteristics of an NIPAAm-*co*-AAm copolymer with an LCST of 400°C. When exposed to NIR light, the nanoshells absorb the light and convert it to heat, raising the temperature of the composite hydrogel above its LCST. This in turn initiates the thermoresponsive collapse of the hydrogel, resulting in an increased rate of release of soluble drug held within the polymer matrix.

Zhao (2009) developed photocontrollable block copolymer micelles. Photoisomerization, photocleavage, and photodimerization, and various types of control over polymer micelles can be achieved using light. These include either reversible or irreversible dissociation, reversible cross-linking, and reversible morphological transition of polymer micelles in response to light exposure. The design rationale for light-responsive block copolymers, the underlying mechanisms of their photocontrol, and the possible future development in exploring light as a powerful stimulus to control the structures or functions of block copolymer micelles. Although light has long been recognized as an external stimulus and used for small-molecule (surfactant-type) micelles (Wang et al., 2007).

9.8.1 Reversible Dissociation and Formation

The general approach of designing a diblock copolymer whose micelles can be reversibly dissociated and formed upon illumination at two different wavelengths (λ_1 and λ_2): one block contains a photochromic moiety in the pendant group, which can be photoswitched between two isomers that can change the block's solubility in a given solvent. For amphiphilic diblock copolymers, the hydrophobic block should bear the photochromic groups and form the micelle core with one isomeric form (the more stable one). Upon absorption of photons at λ_1, the photochromic groups should be converted to an isomeric form that increases significantly the polarity of the hydrophobic block to shift the hydrophilic–hydrophobic balance toward the destabilization of the micelles. The reverse isomerization upon absorption of photons at λ_2 brings the photochromic groups back to the initial isomeric form and the restored hydrophilic hydrophobic balance allows micelles to be reformed. This mechanism is schematically depicted in Figure 9.11a.

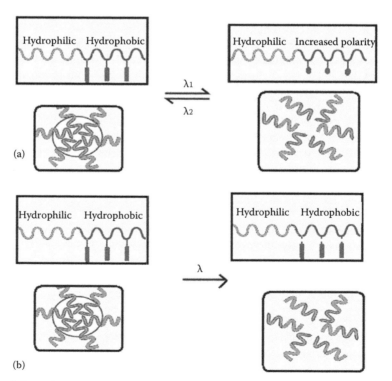

FIGURE 9.11 Schematic presentations of (a) reversible dissociation and formation, and (b) irreversible dissociation and formation.

9.8.2 IRREVERSIBLE DISSOCIATION

Photochromic molecules can be used to design block copolymers (BCPs) whose micelles dissociate irreversibly as a result of the photoreaction. To achieve irreversible photocontrolled dissociation of BCP micelles, the photoreaction of the used chromophore should result in a permanent structural change for the hydrophobic block that triggers the process that is schematically depicted in Figure 9.11b; this can be accomplished through a photoinduced cleavage reaction of pendant photochromic groups that transforms the hydrophobic block into a hydrophilic one. It is easy to understand the dissociation of micelles when the initially amphiphilic BCP becomes a double-hydrophilic BCP.

Rational design of light-controllable polymer micelles was developed by Zhao (2007). Its principle is based on light-changeable or light-switchable amphiphilicity is general and can be applied to many polymer/chromophore combinations. BCP micelles have some important advantages over micelles formed by small-molecule surfactants. They can be more stable due to a much lower critical micellization concentration and macromolecular nature. Their sizes can easily be adjusted to allow their preferential accumulation in porous cancer tissues through the enhanced permeation and retention mechanism. In simplistic terms, an ideal controlled delivery consists of three steps. The first step is a stable encapsulation of the drug by BCP micelles. It means that after being administrated in the body, the micelle protects the drug and prevents it from leaking out quickly. The second step is a site-specific transportation of the drug. It means that the drug-loaded BCP micelle should be selectively captured by the target (pathological sites). The third step is that once arrived on the target, the BCP micelle "opens the door" to release the drug.

Liposomes have been evaluated as drug delivery vehicles for decades (Peer et al., 2007), but their clinical significance has been limited by slow release or poor availability of the encapsulated drug. Liposomal release can be initiated within seconds by irradiating hollow gold nano shells (HGNs) with a NIR pulsed laser (Wu et al., 2008). The use of light to stimulate the release of encapsulated compounds from liposomes is attractive, because it is possible to control the spatial and temporal delivery of the radiation.

9.9 MECHANICAL FORCE–SENSITIVE DRUG DELIVERY SYSTEMS

Mechanical stimulation is an important signal that could be readily exploited. We hypothesize those polymer matrices, which release growth factors in response to mechanical stimulation, could provide a novel approach to engineer tissues in mechanically stressed environments. Drug delivery can also be initiated by the mechanical stimulation of an implant.

Lee and coworkers have developed alginate hydrogels that release vascular endothelial growth factor (VEGF) in response to compressive forces of varying strain amplitudes. Free drug that is held within the polymer matrix is released during compression; once the strain is removed the hydrogel returns to its initial volume. This concept is essentially similar to squeezing the drug out of a sponge. VEGF released from this system under mechanical stimulation induced a significant increase in the amount of new blood vessel formation in nonobese diabetic mice as compared to nonstimulated controls (Lee et al., 2000). Pulsatile delivery of growth factors can prevent saturation of receptors specific for a given molecule, and may extend the effective lifetime of a growth factor-based therapy.

The provision of mechanical stimulation is believed to be necessary for the functional assembly of skeletal tissues, which are normally exposed to a variety of biomechanical signals in vivo (Mitrakovic et al., 2009). Mechanical stimulation can prompt healing of bone fractures. However, it is largely unknown how osteogenesis is promoted by mechanical stimulation. In this study, it has been found that mechanical strain induced proliferation of osteoblastic cells (MC3T3-E1) accompanied increased levels of platelet-derived growth factor-A (PDGF-A) mRNA, determined by quantitative reverse polymerase chain reaction. The effects of mechanical stimulation on hemodynamics, such as due to mechanical transduction in vascular endothelial cells, have been widely discussed recently. In this study, Hsiu et al. (2006) monitored the effects of external mechanical stimulation at a frequency of double the heart rate (HR) on BPW (blood pressure waveform), HRV (HR variability), and BPHV (blood-pressure-harmonics variability) in rats.

The importance of mechanical stimulation for the development of tissue-engineered cartilage is evident. Chondrocyte metabolism and synthesized matrix proteins can be modulated with mechanical loading (Kuo et al., 2006; Lima et al., 2007). Similar results are found with bone-marrow-derived mesenchymal stem cells from various species that undergo increased chondrogenesis in three-dimensional cultures when appropriate mechanical stimulation is applied (Mauck et al., 2006, 2007). In addition, dynamically loaded tissue-engineered cartilage has superior properties compared to unloaded tissue with the same amount of matrix (Kelly et al., 2006). Tissue engineering is a rapidly growing field that utilizes cell–scaffolds constructs with chemical signaling molecules as potential therapeutic products for tissue regeneration. Mesenchymal stem cells (MSCs) provide excellent novel strategies for tissue engineering application. Recently, it has been recognized that understanding mechanical stimulation is an important key to the development of efficient and controllable methods as well as chemical signaling for differentiation of MSCs and tissue engineering application. Mechanical stimuli can enhance the synergy effects for differentiation of MSCs and tissue formation or regeneration (Kim et al., 2009).

Zhang et al. (2008) examined how BMSC and their cytoskeleton responded to mechanical stimulation. They investigated their collagen synthesis and F-actin expression through mechanical stimuli. This study showed that the cyclic stretching favored the synthesis of collagen types I and III, but decreased the amount of F-actin in the BMSC. Therefore, these studies indicate that mechanical stimulation induce synergy effects for differentiation of MSCs into specific cells and could be important role for tissue engineering application.

9.10   FUTURE CHALLENGES AND SCOPE

The concept of delivery of drugs from a suitable device induced by external signals is not as simple as it looks. There are many considerations that have to be taken into considerations while designing a desired drug carrier. It is, therefore, a joint responsibility of a polymer chemist and pharmaceutical expert to think of certain factors prior to building up a scenario for responsive drug delivery systems. These factors may be, for example, biocompatibility of the device, cytotoxicity, in vivo studies, Food and Drug Administration (FDA) approval, efficiency, inconvenience caused to patients, cost effectiveness, etc. Only after these factors are examined, can a drug delivery system be therapeutically acceptable. The delivery of a drug at a predetermined rate over a specified time to a selected target organ has been the ideal requisite in drug delivery technology and pharmacokinetics. Moreover, the need for carriers that exhibit oscillatory behavior of the releasing bioactive agent has also emerged as a significant problem of drug design and formulation in recent years. The traditional methods of drug administration in conventional forms, such as pills and subcutaneous or intravenous injection, are still the predominant routes for drug administration. But pills and injections offer limited control over the rate of drug release into the body; usually they are associated with an immediate release of the drug. Consequently, to achieve therapeutic levels that extend over time, the initial concentration of the drug in the body must be high, causing peaks that gradually diminish over time to an ineffective level. In this mode of delivery, the duration of the therapeutic effect depends on the frequency of dose administration and the half-life of the drug. This peak-and-valley delivery is known to cause toxicity in certain cases, most frequently with chemotherapy for cancer. Thus, the design of a drug delivery system with optimum performance in specific circumstances poses challenges. In an overview of the whole scenario, the field of drug delivery systems has to confront the following challenges:

- Improved efficacy
- Targeted delivery and reduced side effects
- Optimum performance

- Interfacing and pacing with modern methodologies
- Guarantees of safe environment
- Ease of fabrication and application in reality

These benefits may be realized by adopting approaches that basically involve judicious combination of highly specific monomers and polymers of both synthetic and natural origin. The use of smart materials in drug delivery technologies has not only to focus on the possible medical benefits but also must consider economic aspects of the developed materials and/or technology. Furthermore, huge effort on synthetic polymer chemistry must be undertaken to design tailor-made macromolecular systems that will offer novelty in their operation and performance. Above all, the systems developed must be acceptable to the patient community who are the end-users of any successful research and technology. A logical consideration of the possibilities about bright prospects for controlled drug delivery gives rise to positive signals and, therefore, more effort deserves to be put into its growth and expansion. Since smart materials have specific mode of operability and are prone to typical experimental conditions, there is large scope for synthetic polymer chemistry to design multiresponsive delivery systems. Despite the tremendous research input that has been applied to achieve high-performance technologies, a number of aspects still remain to be worked on

- Designing of drug delivery systems with multistimuli responsive potential
- More precise synthetic routes for making responsive materials with greater responsive sensitivity
- Assurance of economic viability so as to popularize devices on a large commercial and population scale
- Design of more localized drug delivery systems
- Oral delivery of insulin using body-friendly natural polymers with enhanced absorption in blood

Thus, it may be concluded that although much advancement has been demonstrated by untiring efforts of researchers worldwide, still there exist numerous challenges that have to be addressed. However, in spite of these challenges the field of controlled drug delivery offers a wide scope and future prospects to build up a high performing, economically viable, and potentially efficient technology.

REFERENCES

Alberts, B.J.A., Lewis, J., Raff, M., Roberts, K., and Walter, P. 2002. *Molecular Biology of the Cell.* Garland Science, New York.

Aminabhavi, T.M., Kulkarni, R.V., and Kulkarni, A.R. 2004. Polymers in drug delivery: Polymeric transdermal drug delivery systems. *Polym News* 29(7): 214–218.

Anal, A.K. 2007. Stimuli-induced pulsatile or triggered release delivery systems for bioactive compounds. *Recent Patents on Endocrine, Metab Immun Drug Discov* 1: 83–90.

Aqil, A., Qiu, H., Greisch, J.F., Jerome, R., Pauw, E.D., and Jerome, C. 2008. Coating of gold nanoparticles by thermosensitive poly(*N*-isopropylacrylamide) end-capped by biotin. *Polymer* 49: 1145–1153.

Atyabi, F., Khodaverdi, E., and Dinarvand, R. 2007. Temperature modulated drug permeation through liquid crystal embedded cellulose membranes. *Int J Pharm* 339: 213–221.

Averitt, R.D., Westcott, S.L., and Halas, N.J. 1999. Linear optical properties of gold nanoshells. *J Opt Soc Am B* 16: 1824–1832.

Bai, L., Gu, F., Feng, Y., and Liu, Y. 2008. Synthesis of microporous pH-sensitive polyacrylic acid/poly (ethylene glycol) hydrogels initiated by potassium diperiodatocuprate (III). *Iran Polym J* 17: 325–332.

Bajpai, A.K., Bajpai, J., and Soni, S.N. 2008. Preparation and characterization of electrically conducted composites of poly(vinyl alcohol)-g-poly(acrylic acid) hydrogels impregnated with polyaniline (PANI). *Exp Polym Lett* 7: 26–9.

Bartil, T., Bounekhel, M., Cedric, C., and Jerome, R. 2007. Swelling behavior and release properties of pH-sensitive hydrogels based on methacrylic derivatives. *Acta Pharm* 57: 301–314.

Bayer, C.L. and Peppas, N.A. 2008. Advances in recognitive, conductive and responsive delivery systems. *J Control Release* 132: 216–221.

Ben-Moshe, M. 2006. Fast responsive crystalline colloidal array photonic crystal glucose sensors. *Anal Chem* 78(14): 5149–5157.

Bhardwaj, U., Papadimitrakopoulos, F., and Burgess, D.J. 2008. A review of the development of a vehicle for localized and controlled drug delivery for implantable biosensors. *J Diabetes Sci Technol* 2(6): 1016–1029.

Blanco, E., Kessinger, C.W., Sumer, B.D., and Gao, J. 2009. Multifunctional micellar nanomedicine for cancer therapy. *Exp Biol Med* 234: 123–131.

Borden, M.A. et al. 2006. Ultrasound radiation force modulates ligand availability on targeted contrast agents. *Mol Imaging* 5(3): 139–147.

Brown, M.B. et al. 2008. Transdermal drug delivery systems: Skin perturbation devices. *Methods Mol Biol* 437: 119–139.

Bryers, J.D. and Ratner, B.D. 2006. Biomaterials approaches to combating oral biofilms and dental disease. *BMC Oral Health* 6(Suppl 1): S15.

Calceti, P., Salmaso, S., Walker, G., and Bernkop-Schnurch, A. 2004. Development and in vivo evaluation of an oral insulin-PEG delivery system. *Eur J Pharm Sci* 22: 315–323.

Carmen, A.L., Lev, B., and Angel, C. 2009. Light-sensitive intelligent drug delivery systems. *Photochem Photobiol* 85: 848–860.

Casadei, M.A., Pitarresi, G., Calabrese, R., Paolicelli, P., and Giammona, G. 2008. Biodegradable and pH-sensitive hydrogels for potential colon-specific drug delivery: Characterization and in-vitro release studies. *Biomacromolecules* 9: 43–49.

Chaterji, S., Kwon, I.K., and Park, K. 2007. Smart polymeric gels: Redefining the limits of biomedical devices. *Prog Polym Sci* 32: 1083–1122.

Chatterjee, A.N., Moore, J.S., Beebe, D.J., and Aluru, N.R. 2003. Dissolvable hydrogels as transducers of bio-chemical signals: Models and simulations. In: *IEEE International Conference on Transducers, Solid-State Sensors, Actuators, and Microsystems*, Vol. 1, pp. 734–737.

Chu, L.Y., Liang, Y.J., Chen, W.M., Ju, X.J., and Wang, H.D. 2004. Preparation of glucose-sensitive microcapsules with a porous membrane and functional gates. *Colloids Surf. B: Biointerfaces* 37: 9–14.

Collin, D., Auernhammer, G.K., Gavat, O., Martinoty, P., and Brand, H.R. 2003. Frozen-in magnetic order in uniaxial magnetic gels: Preparation and physical properties. *Macromol Macromol Rapid Commun* 24(12): 737–741.

Coughlan, D.C., Quilty, F.P., and Corrigan, O.I. 2006. Effect of drug physicochemical properties on swelling/deswelling kinetics and pulsatile drug release from thermoresponsive poly(*N*-isopropylacrylamide) hydrogels. *J Control Release* 98: 97–114.

Dai, S., Ravi, P., and Tam, K.C. 2008. pH-Responsive polymers: Synthesis, properties and applications. *Soft Matter* 4: 435–449.

De Paoli, V.M., De Paoli Lacerda, S.H., Spinu, L., Ingber, B., Rosenzweig, Z., and Rosenzweig, N. 2006. Effect of an oscillating magnetic field on the release properties of magnetic collagen gels. *Langmuir* 22(13): 5894–5899.

De Villiers, M.M. et al. 2009. *Nanotechnology in Drug Delivery*. American Association of Pharmaceutical Scientist, Springer, New York, DOI; 10.1007/978-0-387-77667-5-19.

Del Valle, E.M.M., Galan, M.A., and Carbonell, R.G. 2009. Drug delivery technologies: The way forward in the new decade. *Ind Eng Chem Res* 48: 2475–2486.

Dieudonne, L., Pohl, V., Jones, A., and Schmaljohann, D. 2006. Controlled synthesis strategies for the preparation of tailor-made polymer therapeutics. *Proceeding of the International Symposium of Controlled Release of Bioactive Materials* 33: 907.

Dijkmans, P.A. 2004. Microbubbles and ultrasound: From diagnosis to therapy. *Eur J Echocardiogr* 5(4): 245–256.

Filipcsei, G., Feher, J., and Zrinyi, M. 2000. Electric field sensitive neutral polymer gels. *J Mol Struct* 554: 109–117.

Francois, N.J., Allo, S., Jacobo, S.E., and Daraio, M.E. 2007. Composites of polymeric gels and magnetic nanoparticles: Preparation and drug release behavior. *J Appl Polym Sci* 105: 647.

Ganji, F. and Abdekhodaie, M.J. 2008. Synthesis and characterization of a new thermoreversible chitosan-PEG diblock copolymer. *Carbohydr Polym* 74: 435–441.

Gao, Z.G., Fain, H.D., and Rapoport, N. 2005. Controlled and targeted tumor chemotherapy by micellar-encapsulated drug and ultrasound. *J Control Release* 102: 203–222.

Geest, B.G.D., Jonas, A.M., Demeester, J., and Smedt, S.C.D. 2006. Glucoseresponsive polyelectrolyte capsules. *Langmuir* 22: 5070–5074.

Ghanem, A. and Ghaly, A. 2004. Immobilization of glucose oxidase in chitosan gel beads. *J Appl Polym Sci* 91: 861–866.

Gil, E.S. and Hudson, S.M. 2004. Stimuli-reponsive polymers and their bioconjugates. *Prog Polym Sci* 29: 1173–1222.

Grassi, G., Farra, R., Caliceti, P., Guarnieri, G., Salmaso, S., Carenza, M., and Grassi, M. 2005. Temperature-sensitive hydrogels: potential therapeutic applications. *Am J Drug Deliv* 3(4): 239–251.

Guiseppi-Elie, A., Brahim, S., Slaughter, G., and Ward, K.R. 2005. Design of a subcutaneous implantable bio-chip for monitoring of glucose and lactate. *IEEE Sens J* 5: 345–355.

Guo, J., Li, L., Ti, Y., and Zhu, J. 2007. Synthesis and properties of a novel pH sensitive poly(*N*-vinyl-pyrrol-idone-*co*-sulfadiazine) hydrogel. *Exp Polym Lett* 1(3): 166–172.

Haar, G.T. 2007. Therapeutic application of ultrasound. *Prog Biophys Mol Biol* 93: 111–129.

Hafeli, U.O. 2004. Magnetically modulated therapeutic systems. *Int J Pharma* 277: 19–24.

Handy, E.S. et al. 2006. Thermotherapy via targeted delivery of nanoscale magnetic 2608 particles. US Patent 20066997863B2.

He, C., Kim, S.W., and Lee, D.S. 2008. In situ gelling stimuli-sensitive block copolymer hydrogels for drug delivery. *J Control Release* 127: 189–207.

Hernot, S. and Klibanov, A.L. 2008. Microbubbles in ultrasound-triggered drug and gene delivery. *Adv Drug Deliv Rev* 60(10): 1153–1166.

Hiemstra, C. et al. 2007. In vitro and in vivo protein delivery from in situ forming poly (ethylene glycol)–poly(lactide) hydrogels. *J Control Release* 119: 320–327.

Hoare, T.R. and Kohane, D.S. 2008. Hydrogels in drug delivery: Progress and challenges. *Polymer* 49: 1993–2007.

Hofheinz, R.D. 2005. Liposomal encapsulated anti-cancer drugs. *Anticancer Drugs* 16(7): 691–707.

Howard, B., Gao, A., Lee, S.W., Seo, M.H., and Rapoport, N. 2006. Ultrasound-enhanced chemotherapy of drug resistant breast cancer tumors by micellar encapsulated paclitaxel. *Am J Drug Deliv* 4(2): 97–104.

Hsiu, H., Jan, M.Y., Wang, W.K., and Wang, Y.Y.L. 2006. Effects of whole-body mechanical stimulation at double the heart rate on the blood pressure waveform in rats. *Physiol Meas* 27: 131–144.

Hu, S.H., Liu, T.Y., Liu, D.M., and Chen, S.Y. 2007. Controlled pulsatile drug release from a ferrogel by a high-frequency magnetic field. *Macromolecules* 40(19): 6786–6788.

Husseini, G.A. and Pitt, W.G. 2008. The use of ultrasound and micelles in cancer treatment. *J Nanosci Nanotechnol* 8(5): 2205–2215.

Hynynen, K. et al. 2006. Pre-clinical testing of a phased array ultrasound system for MRI-guided noninvasive surgery of the brain: A primate study. *Eur J Radiol* 59: 149–156.

Ikehara, Y., Niwa, T., Biao, L., Ikehara, S.K., Ohashi, N., Kobayashi, T., Shimizu, Y., Kojima, N., and Nakanishi, H. 2006. A carbohydrate recognition-based drug delivery and controlled release system using intraperito-neal macrophages as a cellular vehicle. *Cancer Res* 66(17): 8740–8748.

Irvin, D.J. et al. 2001. Direct measurement of extension and force in a conductive polymer gel actuator. *Chem Mater* 13: 1143–1145.

Juffermans, L.J.M. et al. 2009. Ultrasound and microbubble-targeted delivery of therapeutic compounds. *Neth Heart J* 17: 82–86.

Kabilan, S. et al. 2004. Glucose-sensitive holographic sensors. *J Mol Recognition* 17: 162–166.

Kang, S.I. and Bae, Y.H. 2003. A sulfonamide based glucose-responsive hydrogel with covalently immobilized glucose oxidase and catalase. *J Control Release* 86: 115–121.

Kang, G.D., Cheon, S.H., Khang, G., and Song, S.C. 2006. Thermosensitive poly(organophosphazene) hydro-gels for a controlled drug delivery. *Eur J Pharmaceut Biopharmaceut* 63: 340–346.

Karathnasis, E., Bhavane, R., and Annapragada, A.V. 2007. Glucose-sensing pulmonary delivery of human insulin to the systemic circulation of rats. *Int J Nanomedicine* 2(3): 501–513.

Kelly, T.A., Ng, K.W., Wang, C.C.B., Ateshian, G.A., and Hung, C.T. 2006. Spatial and temporal development of chondrocyte-seeded agarose constructs in free-swelling and dynamically loaded cultures. *J Biomech* 39(8): 1489–1497.

Khan, F., Gnudi, L., and Pickup, J.C. 2008. Fluorescence-based sensing of glucose using engineered glucose/galactose-binding protein: a comparison of fluorescence resonance energy transfer and environmentally sensitive dye labelling strategies. *Biochem Biophys Res Commun* 365: 102–106.

Kim, Y.S. et al. 2008. High-intensity focused ultrasound therapy: An overview for radiologists. *Korean J Radiol* 9(4): 291–302.

Kim, J.H. et al. 2009. Mechanical stimulation of mesenchymal stem cells for tissue engineering. *Tissue Eng Regenerative Med* 6: 199–206.

Kim, S., Kim, J., Kwon, I.C., and Park, K. 2009. Engineered polymers for advanced drug delivery. *Eur J Pharmaceut Biopharmaceut* 71: 420–430.

Kim, K.S. and Park, J.K. 2005. Magnetic force-based multiplexed immunoassay using superparamagnetic nanoparticles in microfluidic channel. *Lab Chip* 5(6): 657–664.

Kimura, M., Takai, M., and Ishihara, K. 2007. Biocompatibility and drug release behavior of spontaneously formed phospholipid polymer hydrogels. *J Biomed Mater Res A* 80: 45–54.

Kumar, A., Lahiri, S.S., Punyani, S., and Singh, H. 2007. Synthesis and characterization of pH sensitive poly(PEGDMA-MAA) copolymeric microparticles for oral insulin delivery. *J Appl Polym Sci* 107(2): 863–871.

Kumar, C.S., Leuschner, C., Doomes, E.E., Henry, L., Juban, M., and Hormes, J. 2004. Efficacy of lytic peptide-bound magnetite nanoparticles in destroying breast cancer cells. *J Nanosci Nanotechnol* 4(3): 245–249.

Kuo, C.K., Li, W.J., Mauck, R.L., and Tuan, R.S. 2006. Cartilage tissue engineering: Its potential and uses. *Curr Opin Rheumatol* 18(1): 64–73.

Lalwani, A. and Santani, D.D. 2007. Pulsatile drug delivery systems. *Int J Nanomedicine* 2(3): 501–513.

Langer, R. and Tirrell, D.A. 2004. Designing materials for biology and medicine. *Nature* 428(6982): 487–492.

Lattermann, G. and Krekhova, M. 2006. Thermoreversible ferrogels. *Macromol Rapid Commun* 27(16): 1373–1379.

Lee, E.S., Kim, D., Youn, Y.S., Oh, K.T., and Bae, Y.H. 2008. A virus mimetic nanogel vehicle. *Angew Chem Int Ed* 47(13): 2418–2421.

Lee, K.Y., Peters, M.C., Anderson, K.W., and Mooney, D.J. 2000. Controlled growth factor release from synthetic extracellular matrices. *Nature* 408: 998–1000.

Lee, E.S. and Youn, Y.S. 2008. Poly(benzyl-L-histidine)-b-poly(ethylene glycol) micelle engineered for tumor acidic pH-targeting, in-vitro evaluation. *Bull Korean Chem Soc* 29(8): 1539–1544.

Lesniak, M.S. 2005. Novel advances in drug delivery to brain cancer. *Cancer Res Treat* 4(4): 417–428.

Li, C., Madsen, J., Armes, S.P., and Lewis, A.L. 2006. A new class of biochemically degradable, stimulus-responsive triblock copolymer gelators. *Angew Chem Int Ed Engl* 45: 3510–3513.

Lima, E.G. et al. 2007. The beneficial effect of delayed compressive loading on tissue-engineered cartilage constructs cultured with TGF-β3. *Proceedings of the Annual Meeting of the Orthopaedic Research Society*, San Diego, CA.

Lin, C.C. and Metters, A.T. 2006. Hydrogels in controlled release formulations: Network design and mathematical modeling. *Adv Drug Delivery Rev* 58: 1379–1408.

Lin, C., Zhong, Z.Y., Lok, M.C., Jiang, X.L., Hennink, W.E., Feijen, J., and Engbersen, J.F.J. 2007. Novel bioreducible poly(amido amine)s for highly efficient gene delivery. *Bioconjugate Chem* 18: 138–145.

Liu, T.Y., Hu, S.H., Liu, D.M., Chen, S.Y., and Chen, I.W. 2008a. Biomedical nanoparticle carriers with combined thermal and magnetic responses. *Nano Today* 7: 1–14.

Liu, T.Y., Hu, S.H., Liu, K.H., Liu, D.M., and Chen, S.Y. 2006. Preparation and characterization of smart hydrogels and its use for drug release. *J Magn Magn Mater* 304: e397–e399.

Liu, T.Y., Hu, S.H., Liu, K.H., Liu, D.M., and Chen, S.Y. 2008b. Study on controlled drug permeation of magnetic-sensitive ferrogels: Effect of Fe_3O_4 and PVA. *J Control Release* 126(3): 228–236.

Lu, Z.H., Prouty, M.D., Guo, Z.H., Golub, V.O., Kumar, C.S.S.R., and Lvov, Y.M. 2005. Magnetic switch of permeability for polyelectrolyte microcapsules embedded with Co@Au nanoparticles. *Langmuir* 21(5): 2042–2050.

Mahkam, M. 2007. New pH-sensitive glycopolymers for colon-specific drug delivery. *Drug Deliv* 14: 147–153.

Malmsten, M. 2006. Soft drug delivery systems. *Soft Matter* 2: 760–769.

Mauck, R.L., Byers, B.A., Yuan, X., and Tuan, R.S. 2007. Regulation of cartilaginous ECM gene transcription by chondrocytes and MSCs in 3D culture in response to dynamic loading. *Biomech Model Mechanobiol* 6(1–2): 113–125.

Mauck, R.L., Yuan, X., and Tuan, R.S. 2006. Chondrogenic differentiation and functional maturation of bovine mesenchymal stem cells in longterm agarose culture. *Osteoarthr Cartil* 14(2): 179–189.

Mc Gann, M.J., Higginbotham, C.L., Geever, L.M., and Nugent, M.J.D. 2009. The synthesis of novel pH-sensitive poly (vinyl alcohol) composite hydrogels using a freeze/thaw process for biomedical applications. *Int J Pharmaceutics* 372(1–2): 154–161.

Meng, F., Zhong, Z., and Feijen, J. 2009. Stimuli-responsive polymersomes for programmed drug delivery. *Biomacromolecules* 10(2): 197–209.

Mitrakovic, D., Bugarski, B., Vonwil, D., Martin, I., and Obradovic, B. 2009. A novel bioreactor with mechanical stimation for skeletal tissue engineering. *Chem Ind Chem Eng Quar* 15(1): 41–44.

Miyata, T., Asami, N., and Uragami, T. 1999a. Preparation of an antigensensitive hydrogel using antigen–antibody bindings. *Macromolecules* 32: 2082–2084.

Miyata, T., Asami, N., and Uragami, T. 1999b. A reversibly antigenresponsive hydrogel. *Nature* 399: 766–769.

Myhr, G. and Moan, J. 2006. Synergistic and tumour selective effects of chemotherapy and ultrasound treatment. *Cancer Lett* 232: 206–213.

Napoli, A., Boerakker, M.J., Tirelli, N., Nolte, R.J.M., Sommerdijk, N., and Hubbell, J.A. 2004. Glucose-oxidase based self-destructing vesicles. *Langmuir* 20: 3487–3491.

Oh, K.T., Yin, H., Lee, E.S., and Bae, Y.H. 2007. Polymeric nanovehicles for anticancer drugs with triggering release mechanisms. *J Mater Chem* 17: 3987–4001.

Oishi, M. et al. 2005. Supramolecular assemblies for the cytoplasmic delivery of antisense oligodeoxynucleotide: polyion complex (PIC) micelles based on poly(ethylene glycol)-SS-oligodeoxynucleotide conjugate. *Biomacromolecules* 6: 2449–2454.

Pal, K., Banthia, A.K., and Majumdar, D.K. 2007. Preparation and characterization of polyvinyl alcohol–gelatin hydrogel membranes for biomedical applications. *AAPS PharmSciTech* 8(1): 21.

Pankhurst, Q.A., Connolly, J., Jones, S.K., and Dobson, J. 2003. Applications of magnetic nanoparticles in biomedicine. *J Phys D: Appl Phys* 36: 167–181.

Peer, D., Karp, J.M., Hong, S., Farokhzad, O.C., Margalit, R., and Langer, R. 2007. Nanocarriers as an emerging platform for cancer therapy. *Nat Nanotechnol* 2: 751–760.

Pickup, J.C., Zhi, Z.L., Khan, F., Saxl, T., and Birch, D.J.S. 2008. Nanomedicine and its potential in diabetes research and practice. *Diabetes Metab Res Rev* 24: 604–610.

Pitt, W.G., Husseini, G.A., and Staples, B.J. 2004. Ultrasonic drug delivery—A general review. *Expert Opin Drug Deliv* 1(1): 37–56.

Power-Billard, K.N., Spontak, R.J., and Manners, I. 2004. Redox-active organometallic vesicles: Aqueous self-assembly of a diblock copolymer with a hydrophilic polyferrocenylsilane polyelectrolyte block. *Angew Chem Int Ed* 43: 1260–1264.

Richter, A., Paschew, G., Klatt, S., Lienig, J., Arndt, K.F., and Adler, H.J.P. 2008. Review on hydrogel-based pH sensors and microsensors. *Sensors* 8: 561–581.

Sadeghi, M. and Hosseinzadeh, H. 2008. Synthesis of starch-poly(sodium acrylate-*co*-acrylamide) superabsorbent hydrogel with salt and pH-responsiveness properties as a drug delivery system. *J Bioact Compact Polym* 23: 381–404.

Satish, C.S., Satish, K.P., and Shivkumar, H.G. 2006. Hydrogels as controlled drug delivery systems: Synthesis, crosslinking, water and drug transport mechanism. *Indian J Pharm Sci* 68(2): 133–140.

Schmaljohann, D. 2006. Thermo- and pH-responsive polymers in drug delivery. *Adv Drug Delivery Rev* 58: 1655–1670.

Sershen, S. and West, J. 2002. Implantable, polymeric systems for modulated drug delivery. *Adv Drug Deliv Rev* 54: 1225–1235.

Sershen, S.R., Westcott, S.L., Halas, N.J., and West, J.L. 2000. Temperature-sensitive polymer-nanoshell composites for photothermally modulated drug delivery. *J Biomed Mater Res* 51: 293–298.

Shim, W.S. et al. 2007. pH- and temperature-sensitive, injectable, biodegradable block copolymer hydrogels as carriers for paclitaxel. *Int J Pharm* 331: 11–18.

Srivastava, R., Brown, J.Q., Zhu, H., and McShane, M.J. 2005. Stabilization of glucose oxidase in alginate microspheres with photoreactive diazoresin nanofilm coatings. *Biotechnol Bioeng* 91: 124–131.

Tang, M., Zhang, R., Bowyer, A., Eisenthal, R., and Hubble, J. 2004. NAD-sensitive hydrogel for the release of macromolecules. *Biotechnol Bioeng* 87: 791–796.

Tanna, S., Sahota, T.S., Sawicka, K., and Taylor, M.J. 2006a. The effect of degree of acrylic derivatisation on dextran and concanavalin-A glucose-responsive materials for closed-loop insulin delivery. *Biomaterials* 27: 4498–4507.

Tanna, S., Taylor, M.J., Sahota, T.S., and Sawicka, K. 2006b. Glucose-responsive UV polymerised dextran–concanavalin A acrylic derivatised mixtures for closed-loop insulin delivery. *Biomaterials* 27(8): 1586–1597.

Tertaj, P., Morales, M., Verdoguer, S., Carreno, T., and Serna, C.J. 2003. The preparation of magnetic nanoparticles for applications in biomedicine. *J Phys D: Appl Phys* 36: 182–197.

Ulijn, R.V. 2006. Enzyme-responsive materials: A new class of smart biomaterials. *J Mater Chem* 16: 2217–2225.

Ulijn, R.V., Bibi, N., Jayawarna, V., Thornton, P.D., Todd, S.J., Mart, R.J., Smith, A.M., and Gough, J.E. 2007. Bioresponsive hydrogel. *Mater Today* 10(4): 40–48.

Van-Tomme, S.R., Storm, G., and Hennink, W.E. 2008. In situ gelling hydrogels for pharmaceutical and biomedical applications. *Int J Pharm* 355: 1–18.

Vasile, C., Dumitriu, R.P., Cheaburu, C.N., and Oprea, A.M. 2009. Architecture and composition influence on the properties of some smart polymeric materials designed as matrices in drug delivery systems. A comparative study. *Appl Surf Sci* 256: S65–S71.

Wang, Y., Ma, N., Wang, Z., and Zhang, X. 2007. Photocontrolled reversible supramolecular assemblies of an azobenzene-containing surfactant with α-cyclodextrin. *Angew Chem Int Ed* 46: 2823.

Weinand, C. et al. 2006. Hydrogel-b-TCP scaffolds and stem cells for tissue engineering bone. *Bone* 38: 555–563.

Wu, G., Mikhailovsky, A., Khant, H.A., Fu, C., Chiu, W., and Zasadzinski, J.A. 2008. Remotely triggered liposomal release by near-infrared light absorption via hollow gold nanoshells. *J Am Chem Soc* 130(26): 8175–8177.

Xiao, Y. et al. 2009. Preparation and characterization of a novel pachyman-based pharmaceutical aid. II: A pH-sensitive, biodegradable and biocompatible hydrogel for controlled release of protein drugs. *Carbohydrate Polym* 77: 612–620.

Xiong, X.Y., Tam, K.C., and Gan, L.H. 2006. Polymeric nanostructures for drug delivery applications based on pluronic copolymer systems. *J Nanosci Nanotechnol* 6: 2638–2650.

Yang, Z. and Xu, B. 2006. Using enzymes to control molecular hydrogelation. *Adv Mater* 18: 3043–3046.

Yao, L. and Krause, S. 2003. Electromechanical responses of strong acid polymer gels in DC electric fields. *Macromolecules* 36: 2055–2065.

Yue, Y., Sheng, X., and Wang, P. 2009. Fabrication and characterization of microstructured and pH sensitive interpenetrating networks hydrogel films and application in drug delivery field. *Eur Polym J* 45(2): 309–315.

Zhang, L. et al. 2008. Cyclic stretching promotes collagen synthesis and affects F-actin distribution in rat mesenchymal stem cells. *Biomed Mater Eng* 18: 205.

Zhao, Y. 2007. Rational design of light-controllable polymer micelles. *Chem Rec* 7: 286–294.

Zhao, Y. 2009. Photocontrollable block copolymer micelles: What can we control? *J Mater Chem* 19: 4887–4895, DOI: 10.1039/b819968j.

10 Polymeric Materials for Surface Modification of Living Cells

Yuji Teramura, Hao Chen, Naohiro Takemoto,
Kengo Sakurai, and Hiroo Iwata

CONTENTS

10.1 INTRODUCTION

Various studies have investigated the interactions of polymeric materials with living cells, focusing on gene (or plasmid DNA) transfer, drug delivery systems (DDS), cell fusion, and cell immobilization onto biomaterials. For example, cationic poly(ethyleneimine) (PEI) and poly-L-lysine have been used to carry plasmid DNA into cells (Boussif et al. 1995, Chanana et al. 2005), and poly(ethylene glycol) (PEG) has been used to induce cell fusion in the preparation of hybridomas for the production of monoclonal antibodies (Honda et al. 1981). Recently, cell surface modification has attracted much attention in the biomedical fields. One example is the transplantation of islets of Langerhans, an insulin-releasing tissue isolated from the pancreas, to treat type 1 diabetes (Shapiro et al. 2000, Ryan et al. 2001). The islet surface is modified to increase blood compatibility or reduce their antigenic properties in order to enhance the clinical outcome (Miura et al. 2006, Cabric et al. 2007, Stabler et al. 2007, Teramura et al. 2007, 2008, 2010a,b, Teramura and Iwata 2008, 2009a,b, 2010a,b, Totani et al. 2008, Inui et al. 2010). Cell transplantation has attracted much attention as a promising method of treating serious diseases because various kinds of pluripotent stem cells, such as embryonic stem (ES) cells, induced pluripotent stem (iPS) cells, and mesenchymal stem cells, have been developed or identified, and their differentiation to functional cells has been extensively studied. The modification of the surface of these cells is expected to play an important role in their transplantation and reducing unfavorable recipient reactions, such as graft rejection and recurrence of an autoimmune reaction. In addition, modification of the surfaces of living cells with polymeric materials allows for new opportunities in biomedical engineering and science because a variety of functional groups and bioactive substances can be introduced onto the cell surface, adding new functionality to the cell. These modifications can be applied in various biomedical studies, such as development, regenerative medicine, and the immune system, some of which will be discussed in some detail.

In this chapter, we provide an overview of polymeric materials that have been developed and used for modifying the surfaces of living animal cells and their applications in biomedical engineering and science.

10.2 CELL SURFACE MODIFICATION WITH SYNTHETIC POLYMERS

Polymers that have been used for surface modification in living animal cells are listed in Figure 10.1. The methods of cell surface modification can be classified into three categories (Figure 10.2a): covalent conjugation of polymers to the amino groups of membrane proteins, incorporation of amphiphilic polymers into the lipid bilayer of the cell membrane by hydrophobic interaction, and electrostatic interaction between cationic polymers and a negatively charged cell surface. Using these methods, various functional groups and bioactive substances can be introduced onto the cell surface to generate new cell functions.

10.2.1 COVALENT BONDING

Polymeric substances are covalently conjugated to the amino groups of membrane proteins by chemical modification using reactive groups, such as N-hydroxyl-succinimidyl ester (NHS) and cyanuric chloride (Scott et al. 1997, Murad et al. 1999, Chen and Scott 2003, Contreras et al. 2004, Nacharaju et al. 2005, Hashemi-Najafabadi et al. 2006, Lee et al. 2006, 2007, Cabric et al. 2007,

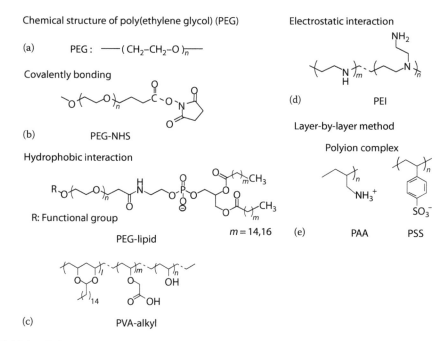

FIGURE 10.1 Cell surface modification with synthetic polymers. Chemical structure of (a) poly(ethylene glycol) (PEG), (b) PEG carrying N-hydroxyl-succinimidyl ester (PEG-NHS), (c) PEG-conjugated phospholipid (PEG-lipid) and poly(vinyl alcohol) carrying side alkyl chains (PVA-alkyl), (d) poly(ethyleneimine) (PEI), and (e) poly(allylamine) (PAA) and poly(styrene)sulfate (PSS).

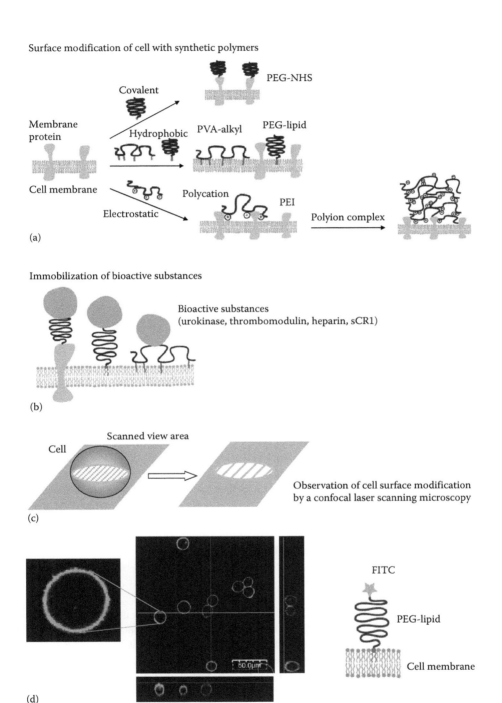

FIGURE 10.2 (See companion CD for color figure.) (a) Schematic illustration of cell surface modification with synthetic polymers by covalent bonding, hydrophobic interactions, electrostatic interactions, and the layer-by-layer method. (b) Schematic illustration of the immobilization of bioactive substances to the cell surface via polymers. (c) Cells modified with polymers observed by confocal laser scanning microscopy. (d) Confocal laser scanning microscope photos of CCRF-CEM cell surface modification with FITC-labeled PEG-lipid.

Stabler et al. 2007). These methods are well-established for modifying biomaterials and proteins. PEG-NHS (Figure 10.1b) can react with membrane proteins of living animal cells by simply adding it to the cell suspension in aqueous solution (Figure 10.2a), resulting in a cell surface coated with PEG chains. This chemical modification randomly occurs with membrane proteins and is difficult to control. A deterioration of membrane protein function is expected due to unselective modification, and examining the function of the cells after treatment is necessary. In addition, an NHS group is easily hydrolyzed in aqueous solution, making the conjugation reaction competitive with its inactivation.

Bertozzi et al. reported a cell surface modification method utilizing azido sugars. Azide sugars fed to living cells are integrated by the glycan biosynthetic machinery into various glycoconjugates that are expressed on the cell surface (Saxon and Bertozzi 2000, Prescher et al. 2004). The azide sugars are used to modify some polymers via an azide-specific reaction, such as the Staudinger ligation reaction. The methods utilizing modified sugars can be used to introduce various functional groups, such as biotin, azide, and ketones, to living cell surfaces. The technologies, however, are limited to the introduction of small functional groups to cells and might perturb cell functions (Paulick et al. 2007, Rabuka et al. 2008). Although covalently immobilized polymers exist for a long period of time due to the direct conjugation to the membrane components, they disappear from the cell surface more rapidly than expected (Cabric et al. 2007, Teramura et al. 2008, Inui et al. 2010).

10.2.2 Hydrophobic Interaction

Hydrophobic interactions between amphiphilic polymers and the lipid bilayer of the cell membrane are frequently employed for cell surface modification, and some amphiphilic polymers such as PEG-lipid and PVA-alkyl (Figure 10.1c) have been used (Miura et al. 2006, Teramura et al. 2007, 2008, 2010a,b, Teramura and Iwata 2008, 2009a,b, 2010a,b, Totani et al. 2008, Inui et al. 2010). When a polymer solution is added to a cell suspension, the hydrophobic alkyl chains of the amphiphilic polymer spontaneously anchor to the lipid bilayer of the cell membrane through the hydrophobic interactions in aqueous solution. When fluorescein isothiocyanate (FITC)-labeled PEG-lipid is added to CCRF-CEM, a human cell line derived from T-cell leukemia, and incubated for 30 min, the fluorescence of FITC is observed at the periphery of all cells by confocal microscopy (Figure 10.2c and d). This finding indicates that PEG-lipids exist on the cell surface, that is, the cell surface can be coated with PEG-lipid. These amphiphilic polymers tend to gradually dissociate from the cell surface when cultured in fresh medium. The retention time of PEG-lipid on cell membranes can be controlled by the lipid alkyl chain length. The dissociation rate of PEG-lipid with longer hydrophobic domains is much lower than that of PEG-lipid with shorter hydrophobic domains.

10.2.3 Electrostatic Interaction

This method is based on the multiple electrostatic interactions between cationic polymers, such as PEI and PAA (Figure 10.1d and e), and the negatively charged cell surface (Figure 10.2a). For further modification, the layer-by-layer method between anionic and cationic polymers is used. Polyion complexes have been formed using polycations, poly(allylamine hydrochloride), poly-L-lysine, and PEI, and the polyanion complex poly(styrene) sulfate (Elbert et al. 1999, Germain et al. 2006, Krol et al. 2006, Veerabadran et al. 2007). The layer-by-layer method is a simple and easy method for modifying the cell surface (Decher 1997). The surface properties are controlled by the outermost polymer layer. The thickness of the membrane is also controlled by exposure to polymer solutions. However, most cationic polymers, including PLL and PEI, are cytotoxic and exert severe damage to cells. The cell membrane tends to be destroyed by exposure to most cationic polymers. Although there have been some successful attempts with *Escherichia coli* cells and yeast cells, which have rigid cell walls (Diaspro et al. 2002, Hillberg and Tabrizian 2006), it is difficult to use cationic polymers for modifying the surface of animal cells.

10.3 IMMOBILIZATION OF BIOACTIVE SUBSTANCES ON THE CELL SURFACE

The immobilization of bioactive substances on the cell surface can be achieved with a polymer layer on the cell surface (Figure 10.2b). Polymers carrying the various functional groups, such as amine, thiol, maleimide, biotin, and DNA, can be introduced to the cell surface through the methods previously described. Bioactive substances are immobilized by conjugation to the functional groups via covalent conjugation to polymers that are covalently immobilized to membrane proteins, or covalent conjugation to polymers that are immobilized on membrane proteins through hydrophobic interactions.

10.3.1 COVALENTLY CONJUGATED POLYMERS

In order to immobilize bioactive substances on a cell surface, hetero- or homo-polyfunctional cross-linkers must be used. An NHS group is generally used for covalent bonding to the cell membrane as one group, and NHS, biotin, or phosphine groups are used as a second group for conjugation with bioactive substances. Cabric et al. immobilized heparin on islets using biotin molecules covalently bound to the amino groups of membrane proteins through cross-linkers carrying NHS. The surface was then sequentially treated with avidin. Finally, anionic polysaccharide, heparin, was immobilized to the surface through an electrostatic interaction with avidin (isoelectric point of avidin, pI = 10) (Cabric et al. 2007). Stabler et al. (2007) used phosphine covalently conjugated to amino groups of the membrane proteins, and the recombinant soluble domain of thrombomodulin was immobilized by the Staudinger ligation reaction. Contreras et al. immobilized albumin, which is not a bioactive substance itself, on islets to hide the islet surface antigens from the recipient's immune system. The serum albumins were immobilized using PEG carrying NHS groups on both ends; one NHS was used for anchoring to the amino group of membrane proteins and the other for the immobilization of serum albumin (Contreras et al. 2004).

10.3.2 AMPHIPHILIC POLYMERS

Bioactive substances can also be immobilized by utilizing amphiphilic polymers, such as PEG-lipid and PVA-alkyl, carrying various functional groups, such as thiol, maleimide, biotin, and DNA, without reducing cell viability (Miura et al. 2006, Teramura et al. 2007, 2008, 2010a,b, Teramura and Iwata 2008, 2009a,b, 2010a,b, Totani et al. 2008, Inui et al. 2010). After the introduction of these functional groups via the incorporation of amphiphilic polymers, bioactive substances can be conjugated onto the polymers through the functional groups. Most of the work in this area has been done by our groups (Miura et al. 2006, Teramura et al. 2007, 2008, 2009a,b, 2010a,b, Teramura and Iwata 2008, 2009a,b, 2010a,b, Totani et al. 2008, Inui et al. 2010). Figure 10.3 summarizes the amphiphilic polymers and methods used to immobilize bioactive substances.

When an amphiphilic PVA derivative carrying long alkyl side chains and thiol groups (SH-PVA-alkyl) is used for surface modification (Figure 10.3a and b), thiol groups can easily be introduced to the cell surface (Totani et al. 2008). A bioactive substance carrying maleimide groups, which can react with thiol groups under physiological conditions, is prepared using a heterobifunctional cross-linker, such as N-(6-maleimidocaproyloxy)sulfosuccinimide) (sulfo-EMCS). Then, the bioactive substance can be added to the cell suspension to be immobilized on the cell surface by the maleimide/thiol reaction. In the case of PEG-lipid carrying maleimide groups at the end of PEG (Mal-PEG-lipid, Figure 10.3c), bioactive substances carrying thiol groups can also be immobilized on the cell surface via the thiol/maleimide reaction (Teramura et al. 2007). Maleimide groups are introduced onto the cell surface using Mal-PEG-lipid, and thiol groups are introduced to bioactive substance using 2-iminothiolane hydrochloride, also known as Traut's reagent (Figure 10.3c and d). When bioactive substances have a free cysteine, the free SH groups can be used for immobilization.

Biotin can easily be introduced on the cell surface using biotin-PEG-lipid (Figure 10.3e) (Teramura and Iwata 2008). Biotin and streptavidin form a stable conjugate (association constant ~10^{15} M^{-1})

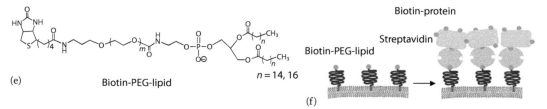

FIGURE 10.3 **(See companion CD for color figure.)** Chemical structure of amphiphilic polymers and cross-linkers for the immobilization of bioactive substances on the cell surface. (a) PVA-alkyl carrying SH groups (SH-PVA-alkyl). (b) Introduction of maleimide groups on bioactive substances using sulfo-EMCS. (c) PEG-lipid carrying a maleimide group at the end of a PEG chain (Mal-PEG-lipid). (d) Introduction of SH groups on bioactive substances using 2-iminothiolane (Traut's reagent). (e) PEG-lipid carrying biotin group at the end of a PEG chain (biotin-PEG-lipid). (f) Schematic illustration of the immobilization of biotinylated bioactive substances. (g) PEG-lipid carrying a polyDNA segment at the end of a PEG chain (DNA-PEG-lipid). (h) Introduction of polyDNA on bioactive substances using sulfo-EMCS and polyDNA-SH.

under physiological conditions, and they can be used for immobilizing bioactive substances on the cell surface. After cells have been treated with biotin-PEG-lipid, streptavidin can be immobilized on the cell surface. Because streptavidin is a tetrameric protein with four binding sites for biotin, unoccupied binding sites on streptavidin can be used for further reaction as shown schematically in Figure 10.3f. Bioactive substances carrying biotin, which can be prepared with commercially available reagents

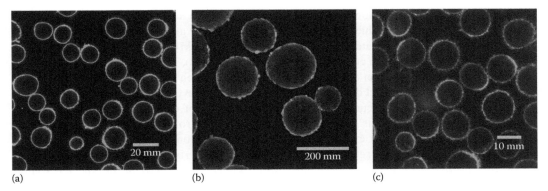

FIGURE 10.4 **(See companion CD for color figure.)** Confocal laser scanning microscope photographs of the surface modification of CCRF-CEM cells and islets. (a) Immobilization of FITC–streptavidin on cells through biotin-PEG-lipid. (b) Islets modified with polyA-PEG-lipid treated with FITC-labeled polyT. (c) Immobilization of FITC–albumin on cells through hybridization between polyA on albumin and polyT on cells.

such as sulfo-succinimidyl biotin (sulfo-NHS-biotin), can be immobilized on the cell surface through the biotin–streptavidin conjugation. For example, CCRF-CEM cells can be treated with biotin-PEG-lipid and then reacted with FITC-labeled streptavidin (Figure 10.4a). Clear fluorescence is seen on the periphery of the cells, indicating the immobilization of streptavidin on the cell surface through biotin-PEG-lipid. Streptavidin is a protein purified from the bacterium *Streptomyces avidinii*. The protein's antigenic property is problematic for use in cell surface modification for transplantation.

We introduced DNA hybridization as an alternative to biotin–avidin conjugation. PEG-lipid carrying polyDNA at the end of the PEG chain (polyDNA-PEG-lipid, Figure 10.3g) is used for surface modification (Teramura et al. 2010a,b). The polyDNA-PEG-lipid is synthesized from Mal-PEG-lipid and polyDNA-SH, which is commercially available. PolyDNA can be immobilized on the cell surface by exposing the cell to the polyDNA-PEG-lipid. Complementary polyDNA-SH can be introduced on a bioactive substance that carries maleimide groups (Figure 10.3h). An example of cell surface modification with polyDNA-PEG-lipid using the hybridization between polyA and polyT is shown in Figure 10.4b and c. The polyT segment is introduced onto the surface of islets using polyT-PEG-lipid. FITC-labeled polyA is then added to the islet suspension. Fluorescence from FITC-labeled polyA can clearly be seen at the periphery of each islet, indicating that FITC-labeled polyA is immobilized on the cell surface through polyT/polyA hybridization. Similarly, no fluorescence is observed on islets not treated with polyT-PEG-lipid. In addition, when proteins carrying polyT are added to the cells treated with polyA-PEG-lipid, proteins can be immobilized on the cell surface. Figure 10.4c shows the immobilization of a FITC-labeled albumin–polyT conjugate on the cell surface using polyA/polyT hybridization. Clear fluorescence can be seen on the periphery of each cell, indicating the immobilization of FITC–albumin. This method is versatile and applicable to the immobilization of various bioactive substances onto the cell surface.

10.4 BIOMEDICAL APPLICATION OF CELL SURFACE MODIFICATIONS

Various applications of cell surface modification with polymers have been used in biomedical studies. Two topics have been selected to demonstrate the usefulness of cell surface modification.

10.4.1 ISLET TRANSPLANTATION

10.4.1.1 Inhibition of Instant Blood-Mediated Inflammatory Reactions

First, we introduce the immobilization of bioactive substances with amphiphilic polymers as applied to islet transplantation. Islet transplantation in the liver has been proposed as a safe and effective method of treating patients with insulin-dependent diabetes mellitus (type I diabetes).

However, multiple pancreas donors are still needed to obtain sufficient numbers of islets for blood glucose normalization due to islet loss in the early phase after intraportal transplantation caused by instant blood-mediated inflammatory reactions (IBMIR) (Bennet et al. 1999, 2000, Moberg et al. 2002, Johansson et al. 2005, Korsgren et al. 2008). The exposure of islets to blood activates the blood coagulation and complement systems, which induce nonspecific inflammatory reactions, resulting in islet destruction. If these initial unfavorable reactions are controlled, most of the islets can be rescued. Systemic administration of drugs, such as anticoagulants and complement inhibitors, has been examined as way to control IBMIR, and some promising results have been obtained in experimental islet transplantation. However, the application of these methods in the clinical setting is difficult due to a high risk of bleeding. The immobilization of bioactive substances, such as heparin, urokinase, thrombomodulin, and the soluble domain of human complement receptor 1 (sCR1), to the islet surface is one way to overcome these issues and suppress IBMIR (Cabric et al. 2007, Stabler et al. 2007, Teramura and Iwata 2008, Totani et al. 2008).

PVA-alkyl carrying long alkyl side chains and thiol carboxyl groups has been used for immobilizing urokinase on the surface of islets (Figure 10.5a) (Totani et al. 2008). Urokinase is a serine protease that activates plasminogen, which triggers a proteolytic cascade that participates in thrombolysis. The immobilization of urokinase on the islet surface is expected to help dissolve small

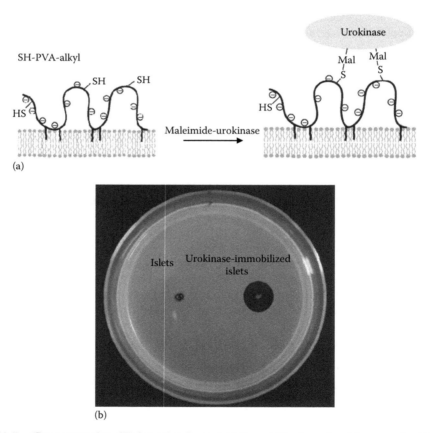

FIGURE 10.5 **(See companion CD for color figure.)** (a) Immobilization of urokinase on the islet surface using amphiphilic PVA-alkyl. Maleimide groups are introduced onto urokinase using sulfo-EMCS. (b) Fibrin plate assay. Urokinase-immobilized islets and nontreated islets ($n = 100$ each) were placed on a fibrin gel plate. A large transparent area formed around the urokinase-immobilized islets due to the dissolution of the fibrin gel by plasmin, which is produced from plasminogen by urokinase.

blood clots that may form on islets. As described in the previous section, we can introduce thiol groups onto the islet surface using PVA-alkyl, and maleimide groups on urokinase using sulfo-EMCS (Figure 10.3a).

Urokinase can be immobilized on the islet surface by the maleimide/thiol reaction. A fibrin plate–based assay can then be performed to assess the enzyme's function on the islets as shown in Figure 10.5b. Urokinase islets and naïve islets ($n = 100$ islets each) were spotted on a fibrin gel. After 24 h, a large transparent area (1.8 cm in diameter) of dissolved fibrin was observed around the urokinase islets, indicating the presence of fibrinolytic activity as expected with active urokinase. As expected, the transparent area was small around the naïve islet spot. These results suggest that the immobilization of urokinase on the islet surface is a promising method for controlling IBMIR and improving islet graft survival following intraportal transplantation. Clinical application is expected in the near future.

10.4.1.2 Coverage of an Islet with Living Cells

If the surface of islets can be covered with a patient's own vascular endothelial cells, histocompatibility and blood compatibility will be significantly improved, controlling graft rejection and the destruction caused by IBMIR. This modification of islets has been achieved by DNA hybridization as schematically shown in Figure 10.6a (Teramura et al. 2010a). PolyT was introduced onto the surface of HEK cells using polyT-PEG-lipid, and polyA was introduced on the surface of islets using polyA-PEG-lipid. The polyA-PEG-lipid-modified islets were mixed with the polyT-PEG-lipid-treated cells, immobilizing the HEK cells on the islet surface through hybridization between polyA and polyT

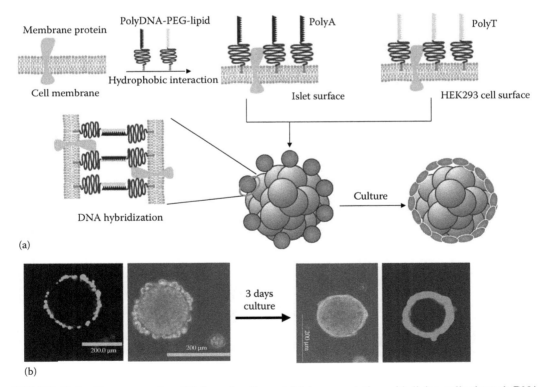

FIGURE 10.6 **(See companion CD for color figure.)** Islet encapsulation with living cells through DNA hybridization. (a) Schematic illustration of islet enclosure in living cells by utilizing surface modifications with polyDNA. (b) Phase contrast and confocal microscope images of islets carrying HEK293 cells in culture. PolyA-PEG-lipid-treated HEK293 cells (GFP expressing) are immobilized on the surface of islets modified with polyT-PEG-lipid. Islets are completely encapsulated with GFP-expressing HEK293 after 3 days in culture.

(Figure 10.6b). Although the cells exist as single cells on the islet just after immobilization, the surface of the islets were completely covered with a cell layer after 3–5 days in culture. No central necrosis of the islet cells was observed. Insulin secretion upon glucose stimulation was well-maintained after cell encapsulation. Using our technique, it is possible to microencapsulate islets with cells derived from recipients with type 1 diabetes. We expect that this novel method will enable the preparation of self-adjusting bioartificial pancreases to control host reactions.

10.4.2 CELL PRINTING

The hybridization of polyDNA-PEG-lipid can be applied to cell alignment on a substrate (Teramura et al. 2010b). PolyDNA carrying a thiol group at one end (polyDNA-SH) can be immobilized on a substrate with a thin layer of gold through the SH/Au reaction, and cells treated with polyT-PEG-lipid applied to the surface. Figure 10.7a shows a fluorescence microscope image of CCRF-CEM cells labeled with fluorescent dye. The cells with polyT are immobilized at the spots with polyA, but not at the spots with polyT, through hybridization between polyA and polyT.

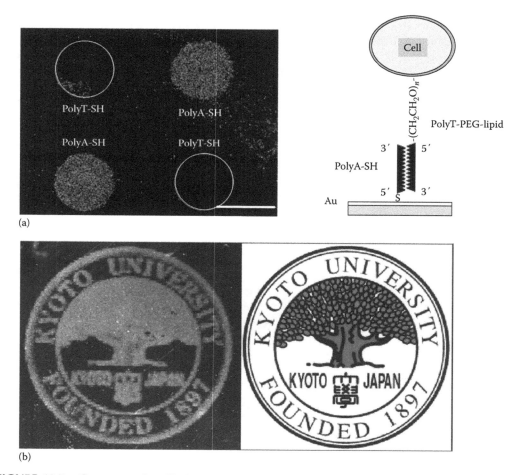

FIGURE 10.7 **(See companion CD for color figure.)** Cell–substrate attachment via DNA hybridization. (a) Stereomicroscope images of the attachment of polyT20-PEG-lipid-modified CCRF-CEM cells to spots with immobilized polyA-SH and polyT-SH (solid lines) on Au surface. (b) Cell patterning using ink-jet printing technology. A solution of polyA-SH was spotted on Au surface by an ink-jet printer, and then the surface was exposed to the polyT-PEG-lipid-treated cell suspension.

A bare glass surface is first treated with a silane-coupling reagent, (3-aminopropyl)triethoxysilane, to introduce amine groups. After modification with the silane-coupling reagent, the surface is sequentially treated with sulfo-EMCS and polyDNA-SH, which can be immobilized on the glass surface. Cells treated with the polyT-PEG-lipid are then applied to the polyA surface.

If we want to align cells in a complicated form, polyDNA must be accurately printed as the complicated form on a substrate. An ink-jet printer can be used for this purpose, as it can freely draw a pattern with a polyDNA solution. Figure 10.7b shows this process with the logo of Kyoto University, which was printed onto a gold substrate using a polyA-SH solution in an ink-jet printer. PolyT-PEG-lipid-treated cells were then applied and incubated, making the logo of Kyoto University as printed by cells. The technique described here is quite simple and, thus, versatile for preparing any pattern of cells. This method can find various applications in cell-based arrays and studies of cell–cell interactions.

10.5   SUMMARY

Three methods have been developed for the cell surface modification: (1) covalent conjugation of polymers with cell membrane proteins, (2) incorporation of amphiphilic polymers into the lipid bilayer of the cell membrane by hydrophobic interaction, and (3) electrostatic interaction between cationic polymers and a negatively charged cell surface. The use of amphiphilic polymers is most promising in these methods because they do not decrease cell viability and protein denaturation. These methods are applied in many biomedical settings, and we selected two topics to demonstrate the usefulness of cell surface modification: (1) islet transplantation to treat type I diabetes and (2) cell printing on a substrate.

Various kinds of pluripotent stem cells, such as ES cells, iPS cells, and mesenchymal stem cells, have been developed or identified. These cells can open a door to regenerative medicine. High-throughput methods are needed to develop methods for their differentiation to functional cells and control host reactions when they are transplanted. Cell surface modification is one of the key technologies for this research.

REFERENCES

Bennet, W., Sundberg, B., Groth, C.G. et al. 1999. Incompatibility between human blood and isolated islets of Langerhans: A finding with implications for clinical intraportal islet transplantation? *Diabetes* 48: 1907–1914.

Bennet, W., Sundberg, B., Lundgren, T. et al. 2000. Damage to porcine islets of Langerhans after exposure to human blood in vitro, or after intraportal transplantation to cynomolgus monkeys: Protective effects of sCR1 and heparin. *Transplantation* 69: 711–719.

Boussif, O., Lezoualc'h, F., Zanta, M.A. et al. 1995. A versatile vector for gene and oligonucleotide transfer into cells in culture and in vivo: Polyethylenimine. *Proc. Natl. Acad. Sci. U.S.A.* 92: 7297–7301.

Cabric, S., Sanchez, J., Lundgren, T. et al. 2007. Islet surface heparinization prevents the instant blood-mediated inflammatory reaction in islet transplantation. *Diabetes* 56: 2008–2015.

Chanana, M., Gliozzi, A., Diaspro, A. et al. 2005. Interaction of polyelectrolytes and their composites with living cells. *Nano Lett.* 5: 2605–2612.

Chen, A.M. and Scott, M.D. 2003. Immunocamouflage: Prevention of transfusion induced graft-versus-host disease via polymer grafting of donor cells. *J. Biomed. Mater. Res. A* 67: 626–636.

Contreras, J.L., Xie, D., Mays, J. et al. 2004. A novel approach to xenotransplantation combining surface engineering and genetic modification of isolated adult porcine islets. *Surgery* 136: 537–547.

Decher, G. 1997. Fuzzy nanoassemblies: Toward layered polymeric multicomposites. *Science* 277: 1232–1237.

Diaspro, A., Silvano, D., Krol, S. et al. 2002. Single living cell encapsulation in nano-organized polyelectrolyte shells. *Langmuir* 18: 5047–5050.

Elbert, D.L., Herbert, C.B., and Hubbell, J.A. 1999. Thin polymer layers formed by polyelectrolyte multilayer techniques on biological surfaces. *Langmuir* 15: 5355–5362.

Germain, M., Balaguer, P., Nicolas, J.C. et al. 2006. Protection of mammalian cell used in biosensors by coating with a polyelectrolyte shell. *Biosens. Bioelectron.* 21: 1566–1573.

Hashemi-Najafabadi, S., Vasheghani-Farahani, E., Shojaosadati, S.A. et al. 2006. A method to optimize PEG-coating of red blood cells. *Bioconjug. Chem.* 17: 1288–1293.

Hillberg, A.L. and Tabrizian, M. 2006. Biorecognition through layer-by-layer polyelectrolyte assembly: In-situ hybridization on living cells. *Biomacromolecules* 7: 2742–2750.

Honda, K., Maeda, Y., Sasakawa, S. et al. 1981. The components contained in polyethylene glycol of commercial grade (PEG-6000) as cell fusogen. *Biochim. Biophys. Res. Commun.* 101: 165–171.

Inui, O., Teramura, Y., and Iwata, H. 2010. Retention dynamics of amphiphilic polymers PEG-lipids and PVA-alkyl on the cell surface. *ACS Appl. Mater. Interf.* 2: 1514–1520.

Johansson, H., Lukinius, A., Moberg, L. et al. 2005. Tissue factor produced by the endocrine cells of the islets of Langerhans is associated with a negative outcome of clinical islet transplantation. *Diabetes* 54: 1755–1762.

Korsgren, O., Lundgren, T., Felldin, M. et al. 2008. Optimising islet engraftment is critical for successful clinical islet transplantation. *Diabetologia* 51: 227–232.

Krol, S., del Guerra, S., Grupillo, M. et al. 2006. Multilayer nanoencapsulation. New approach for immune protection of human pancreatic islets. *Nano Lett.* 6: 1933–1939.

Lee, D.Y., Lee, S., and Nam, J.H. 2006. Minimization of immunosuppressive therapy after islet transplantation: Combined action of heme oxygenase-1 and PEGylation to islet. *Am. J. Transplant.* 6: 1820–1828.

Lee, D.Y., Nam, J.H., and Byun, Y. 2007. Functional and histological evaluation of transplanted pancreatic islets immunoprotected by PEGylation and cyclosporine for 1 year. *Biomaterials* 28: 957–1966.

Miura, S., Teramura, Y., and Iwata, H. 2006. Encapsulation of islets with ultra-thin polyion complex membrane through poly(ethylene glycol)–phospholipids anchored to cell membrane. *Biomaterials* 27: 5828–5835.

Moberg, L., Johansson, H., Lukinius, A. et al. 2002. Production of tissue factor by pancreatic islet cells as a trigger of detrimental thrombotic reactions in clinical islet transplantation. *Lancet* 360: 2039–2045.

Murad, K.L., Gosselin, E.J., Eaton, J.W. et al. 1999. Stealth cells: Prevention of major histocompatibility complex class II-mediated T-cell activation by cell surface modification. *Blood* 94: 2135–2141.

Nacharaju, P., Boctor, F.N., Manjula, B.N. et al. 2005. Surface decoration of red blood cells with maleimido-phenyl-polyethylene glycol facilitated by thiolation with iminothiolane: An approach to mask A, B, and D antigens to generate universal red blood cells. *Transfusion* 45: 374–383.

Paulick, M.G., Forstner, M.B., Groves, J.T. et al. 2007. A chemical approach to unraveling the biological function of the glycosylphosphatidylinositol anchor. *Proc. Natl. Acad. Sci. U.S.A.* 104: 20332–20337.

Prescher, J.A. Dube, D.H., and Bertozzi, C.R. 2004. Chemical remodelling of cell surfaces in living animals. *Nature* 430: 873–877.

Rabuka, D., Forstner, M.B., Groves, J.T. et al. 2008. Noncovalent cell surface engineering: Incorporation of bioactive synthetic glycopolymers into cellular membranes. *J. Am. Chem. Soc.* 130: 5947–5953.

Ryan, E.A., Lakey, J.R., Rajotte, R.V. et al. 2001. Clinical outcomes and insulin secretion after islet transplantation with the Edmonton protocol. *Diabetes* 50: 710–719.

Saxon, E. and Bertozzi, C.R. 2000. Cell surface engineering by a modified Staudinger reaction. *Science* 287: 2007–2010.

Scott, M.D., Murad, K.L., Koumpouras, F. et al. 1997. Chemical camouflage of antigenic determinants: Stealth erythrocytes. *Proc. Natl. Acad. Sci. U.S.A.* 94: 7566–7571.

Shapiro, A.M., Lakey, J.R., Ryan, E.A. et al. 2000. Islet transplantation in seven patients with type 1 diabetes mellitus using a glucocorticoid-free immunosuppressive regimen. *N. Engl. J. Med.* 343: 230–238.

Stabler, C.L., Sun, X.L., Cui, W. et al. 2007. Surface re-engineering of pancreatic islets with recombinant azido-thrombomodulin. *Bioconjug. Chem.* 18: 1713–1715.

Teramura, Y., Chen, H., Kawamoto, T. et al. 2010b. Control of cell attachment through polyDNA hybridization. *Biomaterials* 31: 2229–2235.

Teramura, Y. and Iwata, H. 2008. Islets surface modification prevents blood-mediated inflammatory responses. *Bioconjug. Chem.* 19: 1389–1395.

Teramura, Y. and Iwata, H. 2009a. Islet encapsulation with living cells for improvement of biocompatibility. *Biomaterials* 30: 2270–2275.

Teramura, Y. and Iwata, H. 2009b. Surface modification of islets with PEG-lipid for improvement of graft survival in intraportal transplantation. *Transplantation* 88: 624–630.

Teramura, Y. and Iwata, H. 2010a. Cell surface modification with polymers for biomedical studies. *Soft Matter* 6: 1081–1091.

Teramura, Y. and Iwata, H. 2010b. Bioartificial pancreas microencapsulation and conformal coating of islet of Langerhans. *Adv. Drug Deliv. Rev.* 62: 825–838.

Teramura, Y., Kaneda, Y., and Iwata, H. 2007. Islet-encapsulation in ultra-thin layer-by-layer membranes of poly(vinyl alcohol) anchored to poly(ethylene glycol)-lipids in the cell membrane. *Biomaterials* 28: 4818–4825.

Teramura, Y., Kaneda, Y., Totani, T. et al. 2008. Behavior of synthetic polymers immobilized on cell membrane. *Biomaterials* 29: 1345–1355.

Teramura, Y., Minh, L.N., Kawamoto, T. et al. 2010a. Microencapsulation of islets with living cells using polyDNA-PEG-lipid conjugate. *Bioconjug. Chem.* 21:792–796.

Totani, T., Teramura, Y., and Iwata, H. 2008. Immobilization of urokinase to islet surface by amphiphilic poly (vinyl alcohol) carrying alkyl side chains. *Biomaterials* 29: 2878–2883.

Veerabadran, N.G., Goli, P.L., Stewart-Clark, S.S. et al. 2007. Nanoencapsulation of stem cells within polyelectrolyte multilayer shells. *Macromol. Biosci.* 7: 877–882.

11 Biomedical Applications of Shape Memory Polymers and Their Nanocomposites

I. Sedat Gunes and Sadhan C. Jana

CONTENTS

11.1 INTRODUCTION

Shape memory polymers (SMPs) are attractive in biomedical applications due to their ability to change shape upon application of appropriate external stimuli. The ability of shape change could be useful in various applications, such as in minimally invasive surgical procedures and in implants, to name a few. First, however, let us present a classification of SMPs, so that we can identify the general scope of them, before discussing the applications of SMPs in the biomedical field.

11.1.1 CLASSIFICATION OF SHAPE MEMORY POLYMERS

SMPs are a class of polymers that can "remember" the original, undeformed shape. The original shape is recovered from a deformed, strained state with the application of proper stimuli. In view of the broad definition presented earlier, many polymer systems can be labeled as SMPs. It is well known that many polymeric materials show some degree of shape recovery. The extent of recovery depends on the deformation temperature—for example, whether above or below the glass transition (T_g) or crystalline melting temperature (T_m)—and the total strain originally imposed on the materials in the deformation step. Examples include recovery of cold-stretched specimen of polyethylene, although polyethylene is not considered as an SMP. Thus, SM property of polymers should not be inferred simply only on the basis that some strain is recovered, for example, upon heating of deformed polymer. A new definition should be based on the extent of elastically recoverable total strain. For SM materials, the ratio of elastically recoverable total strain and the originally introduced total strain approaches unity. A large number of polymeric systems also fulfill this criterion, such as electroactive polymers,[1-4] electrostrictive polymers,[5-7] and certain liquid crystalline (LC) elastomers,[8] to name a few.

In view of various modes of deformation, for example, tension, compression, bending, etc., and the varied nature of stimuli, for example, heat, electricity, magnetic field, light, etc., it is imperative to present a classification of different SMP materials reported in literature. Other researchers[9-11] also presented definitions and classifications of SMPs by identifying unique characteristics, such as morphology—amorphous or semicrystalline,[11] nature of cross-links—chemical vs. physical,[10] and segmental vs. main chain relaxation.[10] Here, we present a classification of SMPs based on the underlying physical mechanism responsible for the SM properties. In this classification, SMPs are divided into two main groups:

1. *Rubberlike* SMPs
2. *Mesomorphic* SMPs

Rubberlike SMPs: These materials show an irreversible SM, whereby materials assume a deformed shape under the effect of a mechanical stress field and the deformed shape is preserved usually by cooling the materials below T_g or T_m. Upon reheating, the materials recover the original shape.

Mesomorphic SMPs: Mesomorphic SMPs exhibit an intermediate form, which is an anisotropic, stimulus-induced morphology and which returns to the initial state as soon as the stimulus is removed. These materials respond to external stimuli, such as electric field or light, by creating strain and stress responses. The applied stimulus induces a *non-permanent* intermediate shape in the materials. The original shape is recovered as soon as the external stimulus is removed. Thus, unlike *rubberlike* SMPs, a second, stable deformed shape at room temperature and in the absence of stimuli does not exist.

The stability of the deformed shape at room temperature raises another fundamental question. How stable is stable? Generally, the stability of the shape of polymer articles cannot be assessed absolutely, since it changes gradually as function of both temperature and elapsed time. In *mesomorphic* SMP systems, the time scale for shape change after removal of the stimulus is a characteristic unique to the material system, and varies from a few seconds to a few hours. In contrast, a typical *rubberlike* SMP may keep the deformed shape at room temperature without any appreciable recovery for long periods of time[12]—a time scale of the order of a few months or even years.[13]

Mesomorphic SMP systems, such as electroactive polymers, electrostrictive polymers, and certain LC elastomers, are not discussed in this chapter, as a detailed account on all these systems is not possible within the scope of this discussion. Excellent reviews[1,8] are available on more specific systems. In the following, we present a brief summary of SM behavior of some LC elastomers in order to elaborate the definition of *mesomorphic* SMP presented earlier.

Some LC polymers have been known for their ability to change orientation upon application of external stimuli, such as electrical field.[14–16] In addition, LC elastomers can *remember* the initial orientation, for example, before they are cross-linked in the presence of a mechanical stress field[17–23] and can significantly change their shape reversibly over a narrow temperature range.[24–29] In these systems, the intermediate shape induced by external field is not stable—the material returns to its original shape as soon as the stimulus is removed, for example, due to lowering of the temperature. Another interesting type of LC elastomers show SM effects induced by light. The deformation is actuated by photoisomerization of certain LC moieties, such as azobenzene. Light-induced photoisomerization and resulting orientation of azobenzene containing LC polymers have been known for some time.[30–33] Recently, this photoisomerization property was combined with new chemical structures to obtain polymers with the capability to remarkably change their shape upon photon absorption.[34–36] In the case of some LC elastomers,[37] however, the distinction between *rubberlike* and *mesomorphic* SMP systems is much less clear. We limit the discussion in this chapter mostly to *rubberlike* SMP materials. Hence, if not otherwise stated, the term SMP refers to *rubberlike* SMP.

11.1.2 Scope of Shape Memory Polymers

The presence of a cross-linked polymer network and thermally stable secondary structures, such as crystals or glassy domains, are the basic requirements of SMPs. Note that a cross-linked polymer network would create an instantaneous recovery force upon deformation due to entropic elasticity of cross-linked polymer chains. This is the main driving force for shape recovery. Crystals and glassy domains are needed in order to preserve the residual orientation of polymer chains after deformation by crystallizing or vitrifying them. Note that in the absence of the thermally stable secondary structures, chains will shrink immediately upon removal of the load, similar to conventional elastomers, and the deformed shape cannot be preserved. Hence, most SMPs show structures similar to elastomeric networks,[38] such as vulcanized rubber, phase-separated triblock copolymers, highly entangled systems, interpenetrating polymer networks, to name a few. In this case, the chains relax upon application of the stimuli and, therefore, SM properties of SMPs originate from entropy elasticity of the polymer chains. Figure 11.1 presents an illustration of shape recovery of shape memory polyurethane (SMPU) from deformed states under tension and bending. Shape recovery from bending deformation can be exploited in closing an electrical circuit.

For decades, shape memory metal alloys (SMA) received considerable attention as stimuli-responsive materials.[39] In contrast, SMPs are relatively new materials, compared to metal alloys, however attracted significant attention from scientific community as evident from a number of recent review papers.[9–11] This growing interest in SMPs originates from unique advantages of SMPs over other SM materials. Examples include inherently high recoverable strain of several hundred percents in SMPs compared to only a few percent for metals,[40] much lower density, and much lower cost. SMPs also provide latitude in conveniently adjusting material properties. For instance, one can alter the recovery time and even the recovery temperature by simple adjustment of chemical formulations. The SMP articles can be easily produced and shaped into complex forms by conventional processing methods, such as injection molding, film casting, fiber spinning, profile extrusion, and foaming. Another promising property of SMP was described recently[41]—the ability to show not only two stable shapes such as the *original* and the *recovered final* shapes, but also a third shape called an *intermediate* shape depending upon the temperature. The existence of multiple intermediate shapes makes SMPs amenable to exhibit morphing characteristics, such as those applicable in unmanned air vehicles.

Despite increasing current research activities on SMPs[11] and a great future outlook, some current scientific and technological barriers prevent widespread applications of SMPs. For instance, SMPs have relatively low recovery stress, which is usually 1–5 MPa, compared to 0.5–1 GPa for SMAs. This becomes a limiting factor especially in cases where SMP articles should overcome a large resisting stress from its surroundings; examples include stents, implants, hinges, etc.

A common avenue to improve mechanical properties is to mix with various solid, particulate fillers with high modulus. However, there is usually an inverse relationship between modulus enhancement

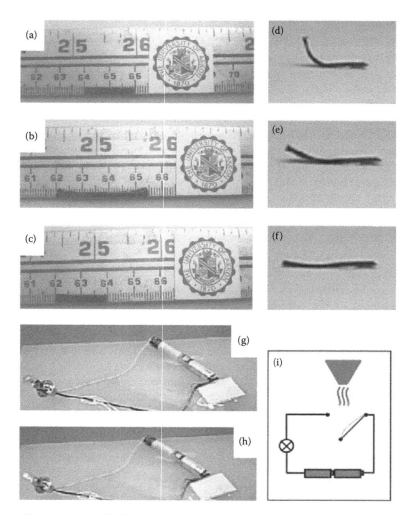

FIGURE 11.1 **(See companion CD for color figure.)** An illustration of shape recovery of a shape memory polyurethane (SMPU) from two different deformed states. All images are taken at room temperature. (a–c) Example from tensile deformation: (a) original shape as thin strip, (b) deformed by tensile elongation, and (c) shape recovery upon heating. (d–f) Example from bending deformation: (d) deformed shape and (e and f) shape recovery upon heating. (g and h) Shape recovery from bending deformation can be exploited in closing an electrical circuit: (g) specimen is in the deformed state, circuit is open. (h) Specimen recovers its original shape and closes the circuit. (i) A schematic of the electric circuit.

and recoverable strain ratio. A general observation is that recoverable strains decrease in the presence of fillers. This can be attributed to filler particle size, much higher stiffness of fillers compared to the matrix polymer, or potential infringement with the polymer networks especially at high filler loadings. Many of these issues can be alleviated by reducing filler loading, for example, by incorporation of small fraction of nanoscopic fillers, or by designing fillers appropriate for the polymer system. Nevertheless, research efforts on development of SMP nanocomposites are scarce and the field is still evolving.

In the following section, we discuss the underlying physical principles of shape recovery process. A description of the current state of knowledge on SMP and its nanocomposites is then presented, with emphasis on selected biomedical applications reported in patent literature and extensive review of journal articles. In each case, we attempt to identify the observed trends and present commentary on future outlook.

11.2 BASIC MOLECULAR MECHANISM OF SHAPE MEMORY FUNCTION

A brief description of the underlying physical concepts responsible for SM function is presented. A more detailed description of all relevant phenomena is beyond the scope of this chapter. Interested readers are referred to fundamental studies reported elsewhere.[38,42–44]

The driving force for strain recovery of SMP is of entropic origin as ramified by elasticity of polymer chains.[45] SMPs possess structures similar to elastomeric networks. These structures are usually of multiphase nature—materials that contain *fixed* (or *hard*) phase and *reversible* (or *soft*) phase. The *fixed* phase acts in the same manner as thermally stable cross-link points. An analog of fixed phase in the vulcanized rubber is the conventional sulfur cross-links and is in the case of amine-cured epoxies three-dimensional cross-link sites. "Cross-links" in general can be crystals, glassy domains, or other forms of entangled chains that prevent unobstructed or "free" flow of at least a part of the materials upon application of stress (Figure 11.2). The *reversible* phase, on the other hand, is the major constituent of SMP and is responsible for elasticity during deformation and undergoes strain recovery. The *reversible* phases become "fluid" above a predefined temperature and, hence, move freely. They also assume either glassy or crystalline forms upon cooling below the predefined temperature. This predefined temperature could be called the *actuation tempera-ture*. The glass transition temperature (T_g) or the melting temperature (T_m) of the reversible phase can be thought of as the actuation temperature. As the SMP is deformed, for example, stretched at above the actuation temperature, the chains undergo orientation and the conformational entropy of the system is reduced. This condition describes an entropically unfavorable state. The result-ing orientation can be preserved—except for a small instantaneous recovery—by quickly freezing the material below the actuation temperature with the applied stress in place. The conformational rearrangement of the reversible phase induced by stretching deformation is severely restricted at temperatures below the actuation temperature; hence, chains cannot recover. Note, however, that not all molecular motions are frozen at temperatures below the actuation temperature. For example, crankshaft-like motions of chain segments and side group rotations are still possible at or below the actuation temperature.[43] However, these molecular motions do not exert much effect on the main chain orientation and, therefore, on shape recovery process. Finally, as the stretched material is heated above the actuation temperature, the chain backbone experiences large-scale movement leading to an increase of entropy and attains statistically more probable conformations. Thus, the recovery of the macroscopic strain has its origin in entropy-driven chain relaxation.[38] However, a majority of SMP systems, such as phase-separated, physically cross-linked polymers and polyure-thanes (PU), are not Gaussian networks.[46–48] Hence, many considerations and mathematical expres-sions derived for entropy elasticity of Gaussian networks are not directly applicable to analysis of these systems.

Some critical parameters are now defined. These are useful in the analysis and evaluation of SM performance. A typical SM testing cycle is first explained with the aid of illustration presented in Figure 11.3. A sample initially undeformed and usually at room temperature (T_{cold}) is heated to an elevated temperature (T_{hot}), which is higher than the actuation temperature. After reaching this temperature, the sample is deformed from length l_o to l_s and cooled down quickly to a temperature below the actuation temperature. The sample undergoes instantaneous recovery during the cooling step and the length reduces from l_s to l_D. The sample can retain its length l_D for long periods of time at below the actuation temperature. As the sample is reheated to above the actuation temperature, it undergoes shape recovery and its length reduces to l_f. Note that the recovered length l_f is a strong function of the constraining force. The value of l_f attains a maximum in the absence of any resisting force from the surroundings of the SMP (unconstrained recovery condition).

The performance parameters of typical SMP are now defined. These parameters were already used in describing the behavior of heat-shrinkable polymers.[49] The total deformation of SMPs can have three contributions such as (i) instantaneous elastic component related to shape fixity, (ii) a viscoelastic part related to recovery ratio, and (iii) a plastic component related to permanent set.

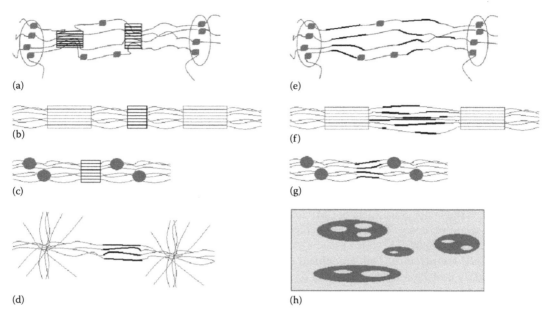

(a)

(e)

(b)

(f)

(c)

(g)

(d)

(h)

FIGURE 11.2 **(See companion CD for color figure.)** Schematic of various SMP systems with different *fixed* and *reversible* phases. (a) Phase-separated, ordered domains form the *fixed* phase and the crystalline soft segments constitute the *reversible* phase. Shape recovery is triggered by melting of crystals. An example is a polyurethane system with crystallizable soft segments. (b) Crystals with high melting point form the *fixed* phase and the crystals with lower melting point constitute the *reversible* phase. Shape recovery is triggered by melting of crystals with lower melting point. An example is polyamide–polyethylene copolymer. (c) Covalent cross-links form the *fixed* phase and the crystals constitute the *reversible* phase. Shape recovery is triggered by melting of crystals. An example is heat-shrinkable polyethylene. (d) Chain entanglements form the fixed phase and glassy chains constitute the reversible phase. Shape recovery is triggered by glass transition. An example is polynorbornene. (e) Phase separated, ordered domains form the fixed phase and the glassy soft segments constitute the reversible phase. Shape recovery is triggered by glass transition. An example is a polyurethane system with glassy soft segments. (f) Crystals with high melting point form the fixed phase and the glassy chain segments constitute the reversible phase. Shape recovery is triggered by glass transition. (g) Covalent cross-links form the fixed phase and glassy chains form the reversible phase. Shape recovery is triggered by glass transition. An example is epoxy. (h) Glassy matrix forms the fixed phase and phase separated rubber particles constitute the reversible phase. Shape recovery is triggered by glass transition of the matrix. An example is given in Ref. [55].

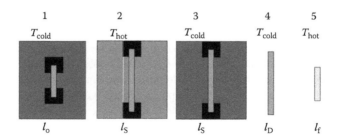

1	2	3	4	5
T_{cold}	T_{hot}	T_{cold}	T_{cold}	T_{hot}
l_o	l_s	l_s	l_D	l_f

FIGURE 11.3 **(See companion CD for color figure.)** Schematic of SMP cycle. Step 1: Undeformed specimen is at *cold temperature* (T_{cold}). Specimen length is l_o. Step 2: Specimen is heated to *hot temperature* (T_{hot}) and deformed. Specimen length is increased to l_s. Step 3: Specimen is cooled down to T_{cold} while keeping the constraining stress in place. Specimen length is still l_s. Step 4: Constraining stress is removed after finishing cooling down to T_{cold}. Specimen length decreased to l_D due to instantaneous recovery. Step 5: Specimen is heated to T_{hot} and hence it recovers. Specimen length is l_f, which is close to l_o.

Shape fixity is defined as the level of deformation that may be fixed upon rapid cooling of the deformed material to room temperature. Although the word "fixity" is rare in English dictionaries,[50] it appears to be first suggested by Mitsubishi researchers and is now widely accepted by others working on SMPs. The value of fixity is computed from the sample lengths as follows (see also Figure 11.3):

$$\text{Shape fixity (SF)} = \frac{l_D - l_o}{l_s - l_o} \tag{11.1}$$

The recovery ratio is the level of deformation that is recovered upon heating and is defined as

$$\text{Recovery ratio (RR)} = \frac{l_D - l_f}{l_D - l_o} \tag{11.2}$$

Recovery stress (σ_R) is designated as the highest level of stress that develops in the materials undergoing shape recovery. The stress develops as the chains "shrink" or recover upon heating. Figure 11.4 shows schematically the evolution of stress in polymer specimen during heating. Note that σ_R in Figure 11.4 is the stress needed to keep the specimen length constant at l_f during heating. The temperature (T_{max}, Figure 11.4) at which the recovery stress reaches maximum is usually close to the actuation temperature of the system. The value of T_{max} depends on heating rate during shape recovery and the rate of deformation.

The steps of SM cycle presented in Figure 11.3 include the initial heating step before stretching and the fast cooling step after stretching. The initial heating step is needed to convert the stiff polymer chains to an easily stretchable elastomer and to prevent appreciable plastic deformation. Hansen et al.[51] observed small recovery ratio results if the specimens are cold drawn. Thus, cold drawings are associated with permanent set as was seen for heat-shrinkable polyolefin systems.[51]

Fast cooling of the stretched specimen is needed for storage of energy and for fixing orientation of the reversible phase. For instance if the cooling rate is not high enough, oriented chains can relax easily and the material cannot fix orientation.

Major factors affecting the SMP parameters such as σ_R, RR, and SF are the number of *locked* chains, the level of orientation of the *locked* chains, and the moduli of SMP at *cold* and *hot*

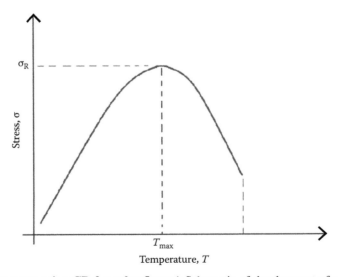

FIGURE 11.4 (See companion CD for color figure.) Schematic of development of recovery stress with temperature. T_{max} is the temperature which corresponds to the maximum recovery stress.

temperatures. In this context, *locked* chains or the chain segments are the ones oriented by stretching and kept in highly oriented state by crystallization or vitrification at T_{cold}. The SM performance is affected by the deformation parameters, such as *hot* and *cold* temperatures, rate of cooling, rate of stretching, the rate of crystallization, total crystallinity, and strain-induced crystallinity.

In light of the role of crystallization presented earlier, SMPs exhibit many similarities with commercial polymeric fibers. The primary production method of fibers consists of spinning and drawing,[52] whereby a polymer melt undergoes uniaxial stretching to create highly oriented chains. Subsequent crystallization of the elongated chains locks in the state of orientation and contributes to higher mechanical properties of the fibers.

A parallel to fiber spinning is found in the stretching of SMP materials. If affine deformation is invoked, all chains experience the same degree of deformation during stretching. The orientation of chains produced by stretching is preserved due to glass transition or crystallization as materials are cooled. In this context, the rate of cooling is critically important as significant relaxation of oriented chains may occur if cooling rate is slow. Some chains and chain segments may remain mobile, for instance, amorphous chain segments not yet crystallized or low-molecular-weight chains not undergone glass transition. Owing to entropic elasticity, these mobile chains create an instantaneous retractive force upon unloading of force. This retractive force, however, is usually small and alone cannot contribute to instantaneous shape recovery. The second factor is the modulus of the sample at the "*cold*" temperature, i.e., the temperature at which the specimens are cooled before the load is removed (Figure 11.3). It was mentioned earlier that room temperature is usually accepted as the "*cold*" temperature. A material with high modulus resists the instantaneous retractive stress and retains the deformed state. On the other hand, instantaneous recovery from deformed state is promoted by a low value of modulus which results in a low value of SF. The stress exerted by shrinking chains during heating is phenomenologically similar to the retractive force of an extended spring. The rate of heating in this step is also crucial, since a low heating rate prolongs the duration of shrinkage. As a consequence, this may result in lower values of RR and σ_R. Thus, different steps of a SM cycle depicted in Figure 11.3 are directly correlated. A low level of SF in stretched specimen results from a higher number of "*unlocked*" chains and chain segments which may shrink immediately upon unloading. These chains cannot participate in recovery process since they do not possess any residual orientation and do not contribute to entropic elasticity during heating. Consequently, low SF causes lower RR and also lower σ_R.

The SMPs and heat-shrinkable polymers may appear to represent the same class of polymeric materials. The molecular mechanism responsible for shape recovery, for example, shrinkage of oriented chains upon heating, applies to both materials. In addition, the underlying physical phenomenon for recovery—i.e., entropic elasticity of a polymeric network—is independent of the nature of the polymer, for example, amorphous, semicrystalline, multiphase, thermoset, and thermoplastic, and also independent of the specific ramifications, for example, heat shrinkable, SM, elastic memory, and thermoresponsive. Thus, one might argue if there is any distinction between heat-shrinkable polymers and SMPs. In our present discussion, we designate chemically cross-linked polyolefins and polyesters as heat-shrinkable polymers, which has been a widely accepted custom for over a century.[53] Block copolymers, PU, thermoset networks, etc., and highly entangled polynorbornene, are considered as examples of SMPs. Note that this designation is based on chemical structures; however, it is evident that a more clear distinction is impossible. In current scientific literature, distinction between these systems is not clear either, and sometimes the terms "heat-shrinkable" and "SM" are used interchangeably.[54]

11.3   SHAPE MEMORY POLYMERS

In this section, we present a brief review on pristine SMPs. The SM behavior of SMP nanocomposites is governed primarily by the SMP matrix and the nanosize fillers exert synergistic effects. Thus, a review of the current state of knowledge of SMP is also central to a subsequent discussion on nanocomposites. Three generic SMPs are primarily used in current research—PU, selected thermoset polymers, such as epoxy, and custom-synthesized block copolymers. In view of this, the review presented later focused primarily on the aforementioned SMPs.

11.3.1 Materials

Although research interests in SMPs grew more rapidly in last decades, technological promises of SMP systems were recognized much earlier. For instance, a US patent[55] was awarded in 1949 for a new kind of rivet and associated production method from a two-phase mixture of a vulcanized rubber and a thermoplastic polymer with structure similar to rubber-toughened glassy polymers, such as acrylonitrile–butadiene–styrene (ABS) terpolymer. A pioneering work on potentials and limitations of SMP and on actuating SM by water uptake was published by Ward.[56] An array of SMPs with different chemical structures was published.[57–89]

PU has been a popular SMP material due to its versatility of formulation and unique morphological features. The relationship between formulation and morphology of PU is presented elsewhere.[90] In 1972, Morbitzer and Hespe[91] synthesized a series of PU which could undergo strain-induced crystallinity and hence possessed high levels of SF and RR. First patent on SMPU foam was issued in 1966.[92] A recent patent describes SMP and SMPU actuated by electromagnetic radiation.[93] Mitsubishi Heavy Industries (MHI) was first to commercialize SMPU in a series of applications, ranging from engine valves for car engines to breathable fabric for sportswear. SMPU of MHI is usually actuated by T_g of soft segments. Hayashi and coworkers described the basic properties of SMPU formulation in a series of technical publications on SM articles such as thin films and foams.[94–98] Other researchers also studied SM behavior of foams and films under various conditions.[99–104] A series of articles also articulated the importance of formulation and specific structures on SM behavior of PU.[105–133] A reflection into the findings of these studies reveals a clear trend in the selection of polyols, which have been predominantly poly(caprolactone)diol (PCL diol) and poly(tetramethyleneglycol)diol (PTMG diol)—the former was used for its ability to crystallize[134] and the latter provided amorphous soft segment phase. PTMG diol also offers long relaxation time[135] that helps preserve orientation in the deformed state as discussed earlier. Lendlein and Kelch[9] presented a comprehensive list of polyols and diisocyanates used in the context of SMPU exhibiting thermal induction. Besides the nature of polyols, the ratio of elastic moduli below and above the actuation temperature and hard segment content play crucial role on SM behavior. A high value of the ratio of elastic moduli is essential for good SM properties, for example, to fix the deformed shape easily and to recover the original shape rapidly. The elastic modulus ratio is dictated by the hard segment content, which in turn influences the extent of cross-links. The hard segment content in SMPs is usually varied between 30 and 60 wt%. The system may not show shape recovery if the extent of cross-links is inadequate and the hard segment content is less than a critical value. On the other hand, the system may be too brittle for practical applications and the maximum level of attainable deformation may be limited if the hard segment content is greater than an optimal value.

Thermoset SMPs have been popular especially in aerospace applications for their stronger mechanical properties and higher thermal stability compared to PU. In particular, the use of epoxy as a common SMP for aerospace applications is not surprising in light of the long history of epoxy polymers as matrix materials.[136] Although some pristine epoxies show SM functions,[137,138] they are almost always used in the form of composites. A major drawback of most thermoset SMPs is low recoverable strain level, ~10%, although a thermoset SMP with recoverable elongations of up to 200%[139] has been reported in literature.

Custom-synthesized block copolymers also exhibit SM functions and find usage in biomedical applications. In this case, biocompatibility and biodegradability are the primary attributes.[140] Note also that PU represent a class of biocompatible materials,[141] although they must meet certain structural requirements,[142,143] such as UV stability. A detailed account on SMPs for biomedical applications can be found elsewhere.[144,145] Liu et al.[11] classified SMPs into four categories and presented a comprehensive list of polymers exhibiting SM behavior.

11.3.2 PREFERRED DEFORMATION MODES

Two primary deformation modes appear to be dominant in the testing of both pristine and filled SMP systems. These are bending and uniaxial elongation, apparently due to convenience of experimentation with these deformation methods. The advantage of bending mode is evidently its ability to create high levels of deformation, as ramified by the radius of curvature, at a relatively low strain. Bending deformation is very common in applications, such as hinges and actuators. On the other hand, uniaxial elongation is usually chosen for testing of thermoplastic SMPs with high recoverable strain level and has ramification in packaging applications or in biomedical applications such as stents. A high level of chain orientation in a short period of time can be achieved in this deformation mode compared to other forms of stretching, for example, biaxial stretching. In uniaxial stretching, the distance between two material points change exponentially with time as follows:

$$l = l_0 e^{\dot{\varepsilon}_0 t} \tag{11.3}$$

where
  l is the time-dependent distance between two particles
  l_0 is the initial distance
  t is time
  $\dot{\varepsilon}_0$ is the rate of elongation

Two additional modes of deformation have been studied—dilatation and compression/biaxial elongation, although researchers did not pay as much attention to these modes. Deformation by dilatation is achieved by uniaxially compressing a specimen in a closed piston–cylinder assembly. Under these conditions, achievable deformation is limited by the incompressibility of the materials. A high level of deformation is possible, if the material is compressible. Thus, this deformation mode finds limited practical applications, except in the cases of highly compressible SMP composites,[10] as will be discussed later.

Biaxial elongation is achievable by either simultaneously or sequentially stretching the specimen—similar to the production of biaxially oriented polyolefin films—or by compression. In this deformation mode, the distance between two adjacent material points change somewhere between exponentially with time and linearly with time, depending on the material properties, such as Poisson's ratio, and the testing conditions. Note that biaxial deformation mode is of central importance, especially in the production of SMP articles with hollow, globular shapes, for example, the articles shaped by blow molding or stents which were deployed in coronary arteries, such as during a balloon angioplasty procedure.

It is evident that the selection of deformation mode is usually dictated by the application. However, one may argue that uniaxial tensile deformation is the most advantageous one to realize the best potential SM actions. Note that the distance between adjacent material points changes exponentially with time (Equation 11.1) which in turn induces highest level of orientation, and hence potentially the highest value of σ_R. Another advantage of uniaxial tensile deformation is strain-induced crystallization which can further contribute to augmented SF and σ_R.

11.4 RATIONALE FOR BIOMEDICAL APPLICATIONS OF SHAPE MEMORY POLYMERS AND THEIR NANOCOMPOSITES

11.4.1 PROMISE OF SMPs IN BIOMEDICAL APPLICATIONS

Both the *mesomorphic* and *rubberlike* SMPs have promises and limitations in biomedical applications. The major potential advantage of *mesomorphic* SMPs is their reversible shape change upon application of the external stimuli. These polymers obviously could be attractive as artificial substitutes for human muscles due to their reversible shape change. However, the majority of research on the *mesomorphic* SMPs has been limited to thin films and their mechanical properties, whereby especially the recovery stress needs to be analyzed in order to asses their viability as artificial muscles. Another potential application field of the *mesomorphic* SMPs could be anticipated as artificial valves to be used in human body, such as heart valves and artificial sphincters, for example for urinary incontinence.

The *rubberlike* SMPs, on the other hand, have been considered for the applications in which a one-time recovery is desired, such as surgical products, stents, sutures, and implants. They are especially promising in performing minimally invasive surgical procedures. Two major advantages of SMPs in biomedical applications might be anticipated as supplying a certain degree of remoteness and securing minimal invasion during surgical. A SMP surgical tool which would be deployed to the target area using the existing biological transport channels of human body, such as blood veins or ureters, and which would be actuated in the target area, could reduce both the direct and collateral tissue damage during surgery. This approach could simplify the procedure, reduce the associated risks, and shorten the healing duration.

11.4.2   Limitations of SMPs in Biomedical Applications

Relatively low recovery stress and the usual necessity of thermal actuation are two major barriers which limit the utility of SMPs in biomedical applications. The inherently low values of stiffness and mechanical strength compared to other SM materials, especially metals and ceramics, are practical issues which limit widespread applications of SMPs, as alluded to earlier. Due to organic nature, the modulus of SMPs is orders of magnitude lower than that of an SMA. In addition, the stiffness of SMP decreases while that of SMA increases with temperature, thus limiting the scope for high temperature applications of the former. Moreover, the relatively lower value of σ_R for SMPs compared to SM metals or ceramics does not qualify SMPs for applications where the materials must undergo shape recovery under a resisting stress. To illustrate this point further, the case of catheters and stents can be considered. Numerous US patents have been awarded for this application of SMPs[146–152] especially on their ability to change stiffness over a limited temperature interval. The value of σ_R of SMP catheter placed in a blood vein should exceed the resisting stress of surrounding tissues for full recovery of its shape. Otherwise, the catheter recovers its shape slowly and obstructs the blood flow.[86] As pointed out by Lendlein and Langer,[80] typical value of σ_R of SMPs and stresses in soft tissue in human body are of the same order of magnitudes, about 1–3 MPa. Thus, any avenue for augmentation of σ_R is of central importance in such applications. We will return to the discussion of recovery stresses of SMPs in the conjunction with SMP nanocomposites.

The rate of heat transfer is another dominant factor which could influence the rate of recovery, recovery ratio, and recovery stress, since most SMPs are thermally actuated materials. Note, however, that SMPs, like other organic polymers, and human body tissues are almost perfect thermal insulators due to their organic nature. Thus, providing efficient heat transfer is a fundamental problem in design and operation of SMP articles. Let us now elaborate the issue of heat transfer in more detail in conjunction with biomedical applications.

The human body itself can be the first choice for the heat source to actuate SMPs in biomedical applications, by selecting the actuation temperature of SMP as about 37°C which is close to body temperature. However, drawbacks of this approach are evident. The actuation temperature of about 37°C may cause immediate relaxation of SMP articles, such as stents and other implants upon insertion into the body.[76,153] Note that very rapid relaxation of stents is not desirable in order to provide proper window of time for completion of surgical operations required for insertion of the stent. Rapid relaxation may cause premature loss of residual orientation of chains and thus SMP article might lose its functionality before actuation. There are two potential avenues to alleviate the possibility of almost immediate relaxation of SMPs. First, amorphous SMP systems can be used. In this case, the glass transition temperature may be chosen such that it is higher than the body temperature and the effective actuation temperature is reduced due to plasticization by absorbed water.[76] Note the rate of water absorption and the maximum amount of absorbed water in common polymers are usually limited. Hence a potential drawback of such an avenue may be relatively longer time needed for sufficient fluid uptake,[154,155] unless a hydrogel was used. Hydrogels can supply sufficient and rapid water absorption; however, their mechanical performances were inherently inferior compared to those of polymers. A second avenue is the possibility of external heating after insertion of an SMP article into the body. This external heating might be direct, such as Joule heating by low electric current or indirect heating, for example, by photon absorption.[156] However, the extent of collateral damage on the surrounding tissues due to external heating limits the scope of external heating. One may argue that the upper limit of the available temperature interval for actuation of SMPs in human body could not exceed 40°C–42°C which could imitate the level for a high-grade fever. Let us return the discussion of external heating in conjunction with nanocomposites.

11.4.3 RATIONALE OF SMP NANOCOMPOSITES

11.4.3.1 Improving the Recovery Stress

One method to obtain enhanced value of σ_R is to fill the pristine polymers with high modulus, inorganic fillers. The size, shape, aspect ratio of the extent of dispersion, the presence of favorable filler–matrix interactions could all influence the mechanical properties and hence the recovery stress of filled SMPs. The reports on SMPs filled with micrometer size fillers, such as glass fiber, carbon fiber, silica, and carbon black, established a trade-off between enhancement in mechanical properties and SM performance. Usually, incorporation of the conventional fillers reinforce the SMPs; however, the resulting composites could not recover their original shapes.[157–160]

The research efforts on filled SMPs have been more focused on the nanoparticulate-filled systems. Nanocomposites combine the mechanical reinforcement by preservation of SM properties. For example, Cao and Jana[161,162] reported SM properties of an SMPU/organoclay nanocomposite based on PCL diol soft segments, prepared by melt mixing and bulk polymerization. They achieved good dispersion of organoclay following a methodology developed by Pattanayak and Jana.[163–166] It was seen that a nanocomposite containing 1 wt% organoclay offered about 30% higher σ_R compared to the unfilled system. Cao and Jana[167] further prepared a glassy SMPU/organoclay nanocomposite and their analysis established the importance of matrix formulation and filler–matrix interactions for SMP nanocomposites. It was observed that the presence of 5 wt% organoclay augmented σ_R by more than 40%[167] and the resulting SMPU/organoclay nanocomposites exhibited a recovery stress of about 20 MPa which was about an order of magnitude higher than the usual SMPs[162] (Figure 11.5).

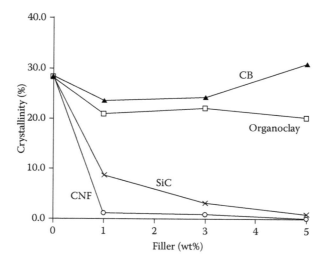

FIGURE 11.5 Soft segment crystallinity of melt-processed semicrystalline shape memory polyurethane composites, containing organoclay, carbon nanofiber (CNF), silicon carbide (SiC), and carbon black (CB). (Reprinted from Gunes, I.S., Cao, F., and Jana, S.C., *Polymer*, 49, 2223. Copyright 2008, with permission from Elsevier.)

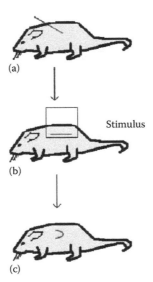

FIGURE 11.6 **(See companion CD for color figure.)** Schematic of potential application of shape memory polymers in biomedical applications. The line insert represents a stent or a catheter. (a) The stent is inserted to the body, (b) external stimulus is applied, and (c) shape recovers in the body.

Gunes et al.[168] compared the performance of various nanofillers, such as organoclay, carbon nanofiber (CNF), 30 nm silicon carbide (SiC), and conductive CB in SMPU formulations based on PCL diol. All composites were prepared by melt mixing of SMPU with the nanofillers. It was observed that SM properties, such as SF, RR, and σ_R reduced significantly in the presence of CNF and SiC, even at a low filler loading, for example, only 1 wt%. These were attributed to reduction in soft segment crystallinity in the presence of CNF and SiC particles (Figure 11.6). Note that chain orientation cannot be preserved effectively in the absence of substantial quantities of crystals.

Jimenez[169] studied the effects of filler surface properties on soft segment crystallinity and SM performance of SMPU composites of PCL diol filled with CNF. He showed that oxygen-rich surface-modified carbon nanofiber (ox-CNF) offered better dispersion, higher crystallinity, and better SM properties, such as higher σ_R, than their counterparts with untreated CNF. Mondal and Hu[170] prepared SMPU/functionalized MWNT system by solution mixing and reported an improvement in RR at filler content of 2.5 wt%. More recently, considerably higher recovery stress values of about 150 MPa were reported for SMP/SWNT nanocomposite fibers recovering from a strain of about 800%.[171]

11.4.3.2 Supplying Multifunctionality

Nanoparticles can also act as agents to contribute multifunctionality to SMPs (Figure 11.7). The choice of particulate fillers in filled SMP systems may also aid in actuating SM function by mechanisms other than application of heat. For example, SMP functions can be actuated by electric and magnetic fields and light if appropriate choice of fillers is taken, even though the matrix polymer does not respond to such stimuli. This added functionality can have profound importance in biomedical applications. As we will discuss later, many creative SMP composite designs have been developed to enable SM functions utilizing different indirect heating methods.

SMPs actuated by resistive heating could be instrumental in analyzing the promise of nanofillers in preparation of multifunctional SMP nanocomposites. The most popular nanofiller for resistive heating actuated SMP matrices has been carbon nanotubes (CNT), both to reinforce and to induce electrical and infrared light-assisted heating to actuate shape recovery. Vaia and coworkers[172,173] prepared SMPU filled with multiwall CNT (MWNT) that respond to infrared radiation and electric

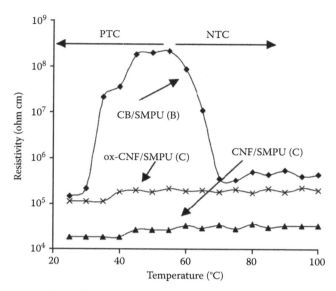

FIGURE 11.7 Temperature-dependent volume resistivity of CNF/SMPU and CB/SMPU composites, each with 3 wt% filler content. (Reprinted from Gunes, I.S., Jimenez, G.A., and Jana, S.C., *Carbon*, 47, 981. Copyright 2009, with permission from Elsevier.)

current. Cho et al.[174] developed SMPU/modified MWNT nanocomposites by solution mixing and sonication and actuated SM action by resistive heating. Jimenez[170] showed that the presence of 5 wt% CNF augmented the recovery stress, and made the composite electrically conductive so as to actuate SM action by resistive heating.

There are some basic design and performance criteria for conductive SMP nanocomposites which have also been of major interest in the context of conductive polymer composites.[175] The first one is the minimum filler loading which ensures a high level of electrical conductivity and which also does not deteriorate the SM properties of SMP. Gunes et al.[168] observed that at about 5 wt% CB content, SMPU/CB composite was extremely brittle and lost all SM properties, although the composite exhibited high enough electrical conductivity for voltage actuating.[168] Other design parameters are the stability of the electrical conductivity under strain and under temperature gradients. Note that SMPs may experience strains on the order of several hundred percents during deformation and large temperature gradients during cooling and heating (Figure 11.3). In this context, the knowledge of temperature- and strain-dependent electrical resistivity of SMP composites actuated by resistive heating is of central importance. For instance, a major challenge in the design and functioning of SMP composite actuators with a semicrystalline matrix is the positive temperature coefficient (PTC) of resistivity, due to which the conductive compound transforms into an insulator with an increase of the temperature; further resistive heating by the continued application of the electrical voltage is not possible. An avenue to alleviate PTC effect may be to use higher filler content in the composite formulation, such that enough electrically conductive networks survive after melting of crystals. However, this may severely deteriorate the SM properties of the composites.[168] Another critical aspect of functioning of SMP composites actuated by resistive heating is the stability of electrical conductivity when materials undergo strain in SM cycles. It is imperative, therefore, that the electrical conductivity should be insensitive to large strains in the materials. Gunes et al.[176] found that filler loading, crystallinity of the polymer, filler–polymer interactions, and the temperature, all influenced PTC effects (Figure 11.8). In addition, the resistivity of CNF/SMPU and ox-CNF/SMPU composites showed weak dependence on strain, while the resistivity of CB/SMPU composites increased several orders of magnitude with imposed strain on specimen.

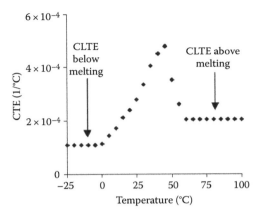

FIGURE 11.8 The change in soft segment crystallinity and coefficient of thermal expansion of pristine SMPU as function of temperature. Note the overlap between the temperature interval of crystal melting and that of changes in nonlinear thermal expansion coefficient. (Reprinted from Gunes, I.S., Jimenez, G.A., and Jana, S.C., *Carbon*, 47, 981. Copyright 2009, with permission from Elsevier.)

11.4.3.3 Improving Thermal Properties

Fillers also improve thermal conductivity, in addition to assisting direct and indirect heating and increasing σ_R, as addressed by Liu and Mather.[177] As polymers are of organic nature and are insulators, heat transfer has been a major issue in manufacturing of SMP articles, actuating SMP functions, and testing. Note that heating is the initial step of typical SMP programming cycle (Figure 11.3) and application of indirect heating is a useful remedy to overcome the limitations imposed by low thermal conductivity of the polymer. In view of this and as pointed out earlier, cooling and freezing of morphology of oriented polymer sample might be a challenge in unfilled polymers, since fast cooling is essential for keeping chains in oriented state without relaxation. In this context, addition of inorganic fillers enhances thermal conductivity, and may cause better SM performance.

Another potential contribution of fillers is on the value of thermal expansion coefficient which has been mostly ignored in technical literature. Most SMPs are multiphase materials and at least a portion of the materials usually undergoes a thermodynamic change during shape recovery, such as glass transition or crystal melting. These thermodynamic changes may actuate abrupt and significant thermal expansion of the SMPs (Figure 11.9). Hence, an understanding of thermal expansion is essential to identify the basic mechanisms of SM functions and to develop more realistic models of SM behavior. It is also worthy to note that the thermal expansion and strain recovery processes during heating

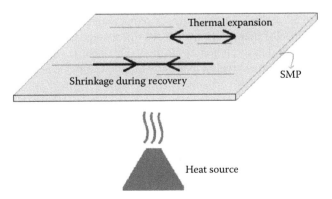

FIGURE 11.9 **(See companion CD for color figure.)** Relationship between thermal expansion and shape recovery during heating step.

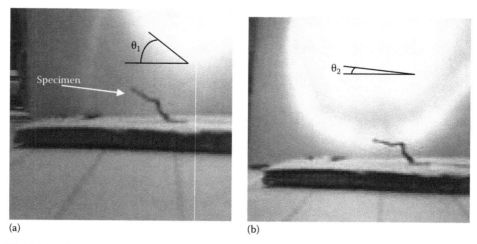

(a)   (b)

FIGURE 11.10 **(See companion CD for color figure.)** Laser triggered shape recovery of shape memory polyurethane/carbon black composite film with 1 wt% filler content. (a) The deformed shape and (b) the recovered shape. The values of θ_1 and θ_2 were 32° and 9°. The excitation source was an argon laser (Lexel 95, Cambridge Lasers Laboratories, Inc., Fremont, CA) with a wavelength of 514.53 nm. The focusing spot was a circular area with a diameter of about 2 μm and the power density was 2.5 mW.

step occur in opposite directions, as shown in Figure 11.10. The impact of thermal expansion on SM properties may be further reduced in composites, especially in view of reduction of the coefficient of linear thermal expansion (CLTE) of polymers in the presence of inorganic fillers.[178,179] Gunes et al.[180] evaluated the effects of thermal expansion on SM performance of a series of SMPU nanocomposites and composites, filled with organoclay, carbon nanofiber, nanosize SiC, and CB. They observed that thermal expansion reduces the recovered strain, the extent of which depends on the magnitude of temperature gradient, coefficient of thermal expansion, and the level of tensile strain.

11.5 BIOMEDICAL APPLICATIONS OF SMPS AND SMP NANOCOMPOSITES

11.5.1 ARTIFICIAL MUSCLES

Mesomorphic SMPs, including LC elastomers, electroactive elastomers, have been primarily considered for artificial muscle applications due to their ability of reversible shape recovery. These SMPs have been shown to exhibit recovery strains on the order of 100% and could be actuated by the application of direct heating, voltage, and light. Yang et al.[181] demonstrated actuation by photon absorption also in a SWNT/LC-elastomer nanocomposite. This concept is illustrated in Figure 11.11 with a SMPU/CB composite with 1 wt% filler content. In SMPU/CB composites, CB particles underwent heating due to photon absorption. This heat is transferred to the surrounding polymer chains which in turn are actuated thermally and induce macroscopic shape change.

11.5.2 SURGICAL APPLICATIONS

A number of US patents have been awarded for the proposed application of SMPs in catheters. These applications mainly recognized the SMPs as rigid polymers at room temperature, that would allow puncture of the human skin at ambient temperature, which, however, would turn to a rubbery medium upon insertion to the human body and hence could prevent any damage on the surrounding tissues.[182,183]

Lendlein and Langer[80] illustrated that SMPs could be used as sutures which could close a wound remotely upon heating. Light-actuated SMP and SMP nanocomposites could also be especially

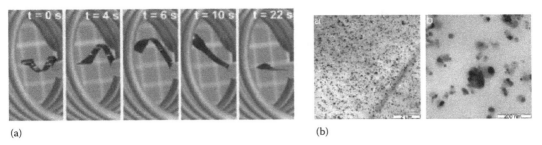

(a)   (b)

FIGURE 11.11 **(See companion CD for color figure.)** (a) Series of photographs showing the shape memory effect of composite with 10 wt% particle content. The permanent shape is a plane stripe of composite material, and the temporary shape is a corkscrew-like spiral. (b) TEM images of composite with 10 wt% particle content. Scale bars: (a) 2 μm; (b) 200 nm. (Reprinted from Mohr, R., Kratz, K., Weigel, T., Lucka-Gabor, M., Moneke, M., and Lendlein, A., *Proc. Natl. Acad. Sci. USA*. 103, 3540. Copyright 2006 by National Academy of Sciences, USA.)

useful in surgical applications which require truly remote actuation. These SMP systems can be actuated over long distances with a laser source which emits highly directional rays. A major critical design and performance parameters are dependent on the thickness of the specimen.[184] Another critical design parameter of the light-actuated SMPs is the size of the specimen. The laser sources commonly employed in actuation of SMPs have limited focusing spots with diameters of a couple of microns which in turn limit the dimensions of the SMP article. Maitland et al.[185] doped SMPU with a green dye and successfully actuated shape recovery by laser absorption-induced heating. The resulting SMPU was proposed to be useful as microactuators and grippers for treating stroke.

SMPs which could be actuated by the application of magnetic field were other promising candidates in surgical applications. Note that magnetic field–generating equipments, such as magnetic resonance imaging (MRI), have been commonly and safely used in the medical field. Mohr et al.[186] presented that application of magnetic field to these materials can induce heating and actuate SMPU in the presence of 20–30 nm iron(III) oxide nanoparticles (Figure 11.12). Filled SMPs with 10 nm iron(II, III) oxide nanoparticles[187] and 9 μm iron(II, III) oxide particles[188] were also shown to respond to magnetic field.

11.5.3 CASE STUDY OF SMP APPLICATIONS IN SURGERY: SMPs FOR CATARACT SURGERY

In this section, we will analyze the potential application of SMPs in surgical applications in more detail by analyzing the cataract surgery as an example. The cataract surgery was selected as the targeted application in which the substitution of conventional procedure was aimed with one which uses SMPs. Note that cataract is a relatively common disease. World Health Organization (WHO) estimated that as of 2001 around 20 million people around the world suffered severely reduced vision due to cataract and that figure was expected to be doubled by the year 2020.[189] Note also that cataract is the most common ophthalmic surgery in the United States, whereby the median age for the development of cataract is 70 years.[190] The usually high average age of patients necessitates a minimally invasive procedure.

Cataract surgery and various other refractive surgery procedures require the substitution of the original lens, with a synthetic, intraocular lens (IOL).[191,192] Foldable IOLs are developed for minimally invasive refractive surgeries[193] which are usually applied to the eye in a folded state in order to facilitate insertion through small incisions and unfolded in the eye after insertion. IOLs made of SMPs were suggested to be used in the surgical procedures as a more minimally invasive and less complicated alternative to foldable IOLs.[194] The suggested SMP was a thermally actuated semicrystalline SMP with an actuation temperature equals the human body temperature. The crystals melt at the actuation temperature, and SMP recovers its original shape. However, the suggested SMP required a cooling jacket to prevent immediate relaxation upon insertion, and the cooling jacket

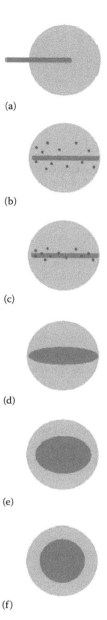

(a)

(b)

(c)

(d)

(e)

(f)

FIGURE 11.12 **(See companion CD for color figure.)** Schematic intraocular lens (IOL) made of a glassy shape memory polymer (SMP) which has a glass transition temperature (actuation temperature) slightly higher than the body temperature. (a) Insertion of glassy IOL to the eye through a small incision. (b and c) IOL experiences simultaneous heat and mass transfer. Its temperature is elevated to that of the human body and started to absorb water which would plasticize the glassy IOL and which in turn would reduce the glass transition temperature. (d–f) IOL fully recovers its original shape.

had to be kept at place until the end of the surgery. In this context, one may argue that an initially glassy SMP hydrogel which could be actuated by water uptake could function as an IOL which did not require an external cooling jacket (Figure 11.12). It should also be mentioned that hydrogels have long been used as contact lens and IOL materials,[195,196] whereas these materials absorb significant amounts of water up to three to five times of their dry weights, in a relatively short time span, which is usually on the order of tens of minutes.[197,198] Note also that SMP is a stiff polymer before actuation.

Thus, one may also argue that a surgical procedure of inserting a stiff polymer may damage the cell which would contact the surface of SMP. However, previous works[199–201] suggest that the surface of stiff IOLs could be relatively easily modified by using hydrophilic polymers and their solutions in order to minimize the cell damage. Consequently, IOLs made of SMPs are anticipated to be used in the surgical procedures without exerting a significant damage to the surrounding cells.

The analysis of application of SMPs to cataract surgery illustrated the potential advantages of them in surgical procedures. SMPs posses the ability to supply minimally invasive and less complicated surgical procedures.

11.5.4 OTHER BIOMEDICAL APPLICATIONS

A current commercial application of SMPs targets the patients who receive oncological treatment.[202] SMPs were used to prepare custom-made face masks which would fix the head of the patient in a given position during radiotherapy. Another current application area of SMPs is the consumer cosmetics industry. One example of such an application is the SMP hair fixation products.[203] SMPs have also been shown to be used in dental applications, such as guiding wires in orthodontic brace applications.[204]

SMPs are also anticipated to be utilized soon in field of drug delivery and targeted release agents, although no current-related report is available in public domain. However, an analogy of this anticipated application was demonstrated with thermoset polymers which soften upon heating.[205]

11.6 CONCLUSIONS AND FUTURE OUTLOOK

SMPs and SMP nanocomposites have the potential to change the current paradigm in surgical procedures and in the field of implants and artificial muscles. However, the knowledge of SMPs are still evolving and it might be safe to claim that only a fraction of the significant potential of SMPs have been utilized in actual biomedical applications. Obtaining deeper fundamental understanding of SMPs and SMP nanocomposites is essential in achieving the ability to design working products and processes which could replace the existing ones. Deformation, relaxation, and recovery behavior of polymer networks, achieving an acceptable level of filler dispersion at high loading, application of industrially viable processing methods in nanocomposite preparation, and obtaining satisfactory filler–matrix bonding are among the fields where deeper understanding is required. It is accepted that research efforts regarding SMP nanocomposites and the area of polymer nanocomposites are interconnected, in that developments in one field will affect the other and evidently will contribute to the advancement of general knowledge on polymers, nanocomposites, and stimuli-responsive materials.

ACKNOWLEDGMENT

ISG would like to thank Bausch & Lomb, Inc. for the 2008 Bausch & Lomb Student Innovation Award. SCJ acknowledges financial support from National Science Foundation CAREER Program.

REFERENCES

1. Y. Bar-Cohen, in *Electroactive Polymer Actuators as Artificial Muscles: Reality, Potential and Challenges*, 2nd ed., Y. Bar-Cohen, Ed., SPIE Press, Bellingham, WA (2004), pp. 4–52.
2. J. Su, Q. M. Zhang, and R. Y. Ting, *Appl. Phys. Lett.* 71, 386 (1997).
3. R. Pelrine, R. Kornbluh, Q. Pei, and J. Joseph, *Science* 287, 836 (2000).
4. S. Zhang, C. Huang, R. J. Klein, F. Xia, Q. M. Zhang, and Z.-Y. Cheng, *J. Intel. Mat. Syst. Str.* 18, 133 (2007).

5. M. Zhenyi, J. I. Scheinbeim, J. W. Lee, and B. A. Newman, *J. Polym. Sci. Part B: Polym. Phys.* 32, 2721 (1994).
6. R. E. Pelrine, R. D. Kornbluh, and J. P. Joseph, *Sensors Actuat. A: Phys.* 64, 77 (1998).
7. W. Lehmann, H. Skupin, C. Tolksdorf, E. Gebhard, R. Zentel, P. Kruger, M. Losche, and F. Kremer, *Nature* 410, 447 (2001).
8. M. Warner and E. M. Terentjev, *Liquid Crystal Elastomers*, Oxford University Press, New York (2003), pp. 95–119.
9. A. Lendlein and S. Kelch, *Angew. Chem. Int. Ed.* 41, 2034 (2002).
10. V. A. Beloshenko, V. N. Varyukhin, and Yu. V. Voznyak, *Russ. Chem. Rev.* 74, 265 (2005).
11. C. Liu, H. Qin, and P. T. Mather, *J. Mater. Chem.* 17, 1543 (2007).
12. S. Hayashi, S. Kondo, P. Kapadia, and E. Ushioda, *Plast. Eng.* 51, 29 (1995).
13. L. C. E. Struik, *Polymer* 28, 1521 (1987).
14. P. G. de Gennes and J. Prost, *The Physics of Liquid Crystals*, 2nd ed., Oxford University Press, New York (1993), pp. 98–162.
15. J. D. Margerum and L. J. Miller, *J. Colloid. Interf. Sci.* 58, 559 (1977).
16. H. Finkelmann, D. Naegele, and H. Ringsdorf, *Makromol. Chem.* 180, 803 (1979).
17. J. Kupfer and H. Finkelmann, *Makromol. Chem. Rapid Commun.* 12, 717 (1991).
18. C. H. Legge, F. J. Davis, and G. R. Mitchell, *J. Phys. II* 1, 1253 (1991).
19. G. R. Mitchell, F. J. Davis, and W. Guo, *Phys. Rev. Lett.* 71, 2947 (1993).
20. P. Bladon, M. Warner, and E. M. Terentjev, *Macromolecules* 27, 7067 (1994).
21. M. Camacho-Lopez, H. Finkelmann, P. Palffy-Muhoray, and M. Shelley, *Nat. Mater.* 3, 307 (2004).
22. A. I. Leonov and V. S. Volkov, *J. Eng. Phys. Thermophys.* 77, 717 (2004).
23. M. Warner and E. M. Terentjev, *Prog. Polym. Sci.* 21, 853 (1996).
24. A. Kaiser, M. Winkler, S. Krause, H. Finkelmann, and A. M. Schmidt, *J. Mater. Chem.* 19, 538 (2009).
25. A. R. Tajbakhsh and E. M. Terentjev, *Eur. Phys. J. E* 6, 181 (2001).
26. D. L. Thomsen III, P. Keller, J. Naciri, R. Pink, H. Jeon, D. Shenoy, and B. R. Ratna, *Macromolecules* 34, 5868 (2001).
27. K. Hiraoka, W. Sagano, T. Nose, and H. Finkelmann, *Macromolecules* 38, 7352 (2005).
28. Z. Yang, W. T. S. Huck, S. M. Clarke, A. R. Tajbakhsh, and E. M. Terentjev, *Nat. Mater.* 4, 486 (2005).
29. S. V. Ahir, A. R. Tajbakhsh, and E. M. Terentjev, *Adv. Funct. Mater.* 16, 556 (2006).
30. K. Anderle, R. Birenheide, M. Eich, and J. H. Wendorff, *Makromol. Chem. Rapid Commun.* 10, 477 (1989).
31. U. Wiesner, M. Antonietti, C. Boeffel, and H. W. Spiess, *Makromol. Chem.* 191, 2133 (1990).
32. J. Stumpe, L. Muller, D. Kreysig, G. Hauck, H. D. Koswig, R. Ruhmann, and J. Rubner, *Makromol. Chem. Rapid Commun.* 12, 81 (1991).
33. S. Ivanov, I. Yakovlev, S. Kostromin, V. Shibaev, L. Läsker, J. Stumpe, and D. Kreysig, *Makromol. Chem. Rapid Commun.* 12, 709 (1991).
34. H. Finkelmann, E. Nishikawa, G. G. Pereira, and M. Warner, *Phys. Rev. Lett.* 87, 15501-1 (2001).
35. Y. Yu, M. Nakano, and T. Ikeda, *Nature* 425, 145 (2003).
36. N. Tabiryan, S. Serak, X.-M. Dai, and T. Bunning, *Opt. Express* 13, 7442 (2005).
37. I. A. Rousseau and P. T. Mather, *J. Am. Chem. Soc.* 125, 15300 (2003).
38. B. Erman and J. E. Mark, *Structures and Properties of Rubberlike Networks*, Oxford University Press, New York (1997), pp. 7–21.
39. A. V. Srinivasan and D. M. McFarland, *Smart Structures: Analysis and Design*, Cambridge University Press, New York (2001), pp. 26–72.
40. Z. G. Wei, R. Sandstrom, and S. Miyazaki, *J. Mater. Sci.* 33, 3743 (1998).
41. I. Bellin, S. Kelch, R. Langer, and A. Lendlein, *Proc. Natl. Acad. Sci. U.S.A.* 103, 18043 (2006).
42. L. R. G. Treloar, *The Physics of Rubber Elasticity*, 2nd ed., Oxford University Press, London, U.K. (1967), pp. 59–79.
43. N. G. McCrum, B. E. Read, and G. Williams, *An Elastic and Dielectric Effects in Polymeric Solids*, Wiley, New York (1967), pp. 180–182.
44. A. I. Leonov, *Polymer* 46, 5596 (2005).
45. J. E. Mark, *J. Chem. Educ.* 58, 898 (1981).
46. G. M. Estes, R. W. Seymour, D. S. Huh, and S. L. Cooper, *Polym. Eng. Sci.* 9, 383 (1969).
47. K. Dusek and W. Prins, *Adv. Polym. Sci.* 6, 1 (1969).
48. R. Bonart, *Polymer* 20, 1389 (1979).
49. A. Ram, Z. Tadmor, and M. Schwartz, *Int. J. Polym. Mater.* 6, 57 (1977).

50. *Webster's Third New International Dictionary of the English Language (Unabridged) with Seven Language Dictionary*, Encyclopedia Britannica, Chicago, IL (1966), pp. 952.
51. D. Hansen, W. F. Kracke, and J. R. Falender, *J. Macromol. Sci. Phys.* 4, 583 (1970).
52. F. W. Billmeyer, Jr., *Textbook of Polymer Science*, 2nd ed., Wiley, New York (1971), pp. 513–532.
53. M. Engels, US Patent 825,116 (1906).
54. H. A. Khonakdar, S. H. Jafari, S. Rasouli, J. Morshedian, and H. Abedini, *Macromol. Theory Simul.* 16, 43 (2007).
55. E. A. Eakins, US Patent 2,458,152 (1949).
56. R. S. Ward, *Med. Device Diagn. Ind.*, August, 24 (1985).
57. J. R. Lin and L. W. Chen, *J. Appl. Polym. Sci.* 69, 1563 (1998).
58. J. R. Lin and L. W. Chen, *J. Appl. Polym. Sci.* 69, 1575 (1998).
59. F. Li, Y. Chen, W. Xu, X. Zhang, and M. Xu, *Polymer* 39, 6929 (1998).
60. F. Li and R. C. Larock, *J. Appl. Polym. Sci.* 84, 1533 (2002).
61. H. M. Jeong, J. H. Song, K. W. Chi, I. Kim, and K. T. Kim, *Polym. Int.* 51, 275 (2002).
62. D. Perez-Foullerat, S. Hild, A. Mucke, and B. Rieger, *Macromol. Chem. Phys.* 205, 374 (2004).
63. K. Gall, C. M. Yakacki, Y. Liu, R. Shandas, N. Willett, and K. S. Anseth, *J. Biomed. Mater. Res.* 73A, 339 (2005).
64. K. Inoue, M. Yamashiro, and M. Iji, *Polym. Mater. Sci. Eng.* 93, 967 (2005).
65. C. Liu and P. T. Mather, *J. Appl. Med. Polym.* 6, 47 (2002).
66. C. Liu and P. T. Mather, *Society of Plastics Engineers Annual Technical Conference*, Nashville, TN, May 4–8, 1962 (2003).
67. C. J. Campo and P. T. Mather, *Society of Plastics Engineers Annual Technical Conference*, Charlotte, NC, May 7–11, 1510 (2006).
68. M. Wang, X. Luo, X. Zhang, and D. Ma, *Polym. Adv. Technol.* 8, 136 (1997).
69. X. Luo, X. Zhang, M. Wang, D. Ma, M. Xu, and F. Li, *J. Appl. Polym. Sci.* 64, 2433 (1997).
70. M. Wang, X. Luo, and D. Ma, *Eur. Polym. J.* 34, 1 (1998).
71. D. Ma, M. Wang, M. Wang, X. Zhang, and X. Luo, *J. Appl. Polym. Sci.* 69, 947 (1998).
72. M. Wang and L. Zhang, *J. Polym. Sci. B: Polym. Phys.* 37, 101 (1999).
73. B. C. Chun, S. H. Cha, Y.-C. Chung, and J. W. Cho, *J. Appl. Polym. Sci.* 83, 27 (2002).
74. B. C. Chun, S. H. Cha, C. Park, Y.-C. Chung, M. J. Park, and J. W. Cho, *J. Appl. Polym. Sci.* 90, 3141 (2003).
75. C. Park, J. Y. Lee, B. C. Chun, Y.-C. Chung, J. W. Cho, and B. G. Cho, *J. Appl. Polym. Sci.* 94, 308 (2004).
76. Y. Cao, Y. Guan, J. Du, J. Luo, Y. Peng, C. W. Yip, and A. S. C. Chan, *J. Mater. Chem.* 12, 2957 (2002).
77. G. Liu, X. Ding, Y. Cao, Z. Zheng, and Y. Peng, *Macromolecules* 37, 2228 (2004).
78. A. Lendlein, A. M. Schmidt, and R. Langer, *Proc. Natl. Acad. Sci. U.S.A.* 98, 842 (2001).
79. C. Liu, S. B. Chun, P. T. Mather, L. Zheng, E. H. Haley, and E. B. Coughlin, *Macromolecules* 35, 9868 (2002).
80. A. Lendlein and R. Langer, *Science* 296, 1673 (2002).
81. M. Bertmer, A. Buda, I. Blomenkamp-Hofges, S. Kelch, and A. Lendlein, *Macromolecules* 38, 3793 (2005).
82. C. Min, W. Cui, J. Bei, and S. Wang, *Polym. Adv. Technol.* 16, 608 (2005).
83. G. M. Zhu, Q. Y. Xu, G. Z. Liang, and H. F. Zhou, *J. Appl. Polym. Sci.* 95, 634 (2005).
84. A. Kraft, G. Rabani, C. Schuh, K. Muller, and M. C. Lechmann, *Polym. Mater. Sci. Eng.* 93, 935 (2005).
85. S. Kelch, S. Steuer, A. M. Schmidt, and A. Lendlein, *Biomacromolecules* 8, 1018 (2007).
86. C. M. Yakacki, R. Shandas, C. Lanning, B. Rech, A. Eckstein, and K. Gall, *Biomaterials* 28, 2255 (2007).
87. Y.-W. Chang, J. K. Mishra, J.-H. Cheong, and D.-K. Kim, *Polym. Int.* 56, 694 (2007).
88. X. Zheng, S. Zhou, X. Li, and J. Weng, *Biomaterials* 27, 4288 (2006).
89. E. L. Kamienski, S. Mandelbaum, P. Vemuri, and R. A. Weiss, *Society of Plastics Engineers Annual Technical Conference*, Cincinnati, OH, May 6–10, 2714 (2007).
90. G. Woods, *The ICI Polyurethanes Book*, ICI Polyurethanes and Wiley, The Netherlands (1987), pp. 27–54.
91. L. Morbitzer and H. Hespe, *J. Appl. Polym. Sci.* 16, 2697 (1972).
92. G. R. Nelson, US Patent 3,284,275 (1966).
93. V. A. Topolkaraev, T. W. Odorzynski, D. A. Soerens, M. J. Garvey, and D. G. Uitenbroek, US Patent 7,074,484 (2006).
94. S. Hayashi, *Int. Prog. Urethanes* 6, 90 (1993).
95. T. Takahashi, N. Hayashi, and S. Hayashi, *J. Appl. Polym. Sci.* 60, 1061 (1996).
96. K. Ito, K. Abe, H.-L. Li, Y. Ujihira, N. Ishikawa, and S. Hayashi, *J. Radioanal. Nucl. Ch.* 211, 53 (1996).

97. C. Poilane, P. Delobelle, C. Lexcellent, S. Hayashi, and H. Tobushi, *Thin Solid Films* 379, 156 (2000).
98. H. Tobushi, R. Matsui, S. Hayashi, and D. Shimada, *Smart Mater. Struct.* 13, 881 (2004).
99. S. J. Tey, W. M. Huang, and W. M. Sokolowski, *Smart Mater. Struct.* 10, 321 (2001).
100. A. Metcalfe, A.-C. Desfaits, I. Salazkin, L. Yahia, W. M. Sokolowski, and J. Raymond, *Biomaterials* 24, 491 (2003).
101. J. L. Hu, Y. M. Zeng, and H. J. Yan, *Textile Res. J.* 73, 172 (2003).
102. J. L. Hu, F. L. Ji, and Y. W. Wong, *Polym. Int.* 54, 600 (2005).
103. B. Yang, W. M. Huang, C. Li, C. M. Lee, and L. Li, *Smart Mater. Struct.* 13, 191 (2004).
104. J. Kaursoin and A. K. Agrawal, *J. Appl. Polym. Sci.* 103, 2172 (2007).
105. B. K. Kim, S. Y. Lee, and M. Xu, *Polymer* 37, 5781 (1996).
106. F. Li, X. Zhang, J. Hou, M. Xu, X. Luo, D. Ma, and B. K. Kim, *J. Appl. Polym. Sci.* 64, 1511 (1997).
107. B. K. Kim, S. Y. Lee, J. S. Lee, S. H. Baek, Y. J. Choi, J. O. Lee, and M. Xu, *Polymer* 39, 2803 (1998).
108. H. M. Jeong, B. K. Ahn, S. M. Cho, and B. K. Kim, *J. Polym. Sci. B: Polym. Phys.* 38, 3009 (2000).
109. H. M. Jeong, B. K. Kim, and Y. J. Choi, *Polymer* 41, 1849 (2000).
110. B. K. Kim, Y. J. Shin, S. M. Cho, and H. M. Jeong, *J. Polym. Sci. B: Polym. Phys.* 38, 2652 (2000).
111. H. M. Jeong, B. K. Ahn, and B. K. Kim, *Polym. Int.* 49, 1714 (2000).
112. H. M. Jeong, S. Y. Lee, and B. K. Kim, *J. Mater. Sci.* 35, 1579 (2000).
113. H. M. Jeong, B. K. Ahn, and B. K. Kim, *Eur. Polym. J.* 37, 2245 (2001).
114. B. K. Kim, J. S. Lee, Y. M. Lee, J. H. Shin, and S. H. Park, *J. Macromol. Sci. Phys.* B40, 1179 (2001).
115. S. H. Park, J. W. Kim, S. H. Lee, and B. K. Kim, *J. Macromol. Sci. Phys.* B43, 447 (2004).
116. S. H. Lee, J. W. Kim, and B. K. Kim, *Smart Mater. Struct.* 13, 1345 (2004).
117. B. S. Lee, B. C. Chun, Y.-C. Chung, K. I. Sul, and J. W. Cho, *Macromolecules* 34, 6431 (2001).
118. J. W. Cho, Y. C. Jung, S. H. Lee, B. C. Chun, and Y.-C. Chung, *Fiber Polym.* 4, 114 (2003).
119. J. H. Yang, B. C. Chun, Y.-C. Chung, and J. H. Cho, *Polymer* 44, 3251 (2003).
120. J. W. Cho, Y. C. Jung, B. C. Chun, and Y.-C. Chung, *J. Appl. Polym. Sci.* 92, 2812 (2004).
121. J. W. Cho, Y. C. Jung, Y.-C. Chung, and B. C. Chun, *J. Appl. Polym. Sci.* 93, 2410 (2004).
122. W. Chen, C. Zhu, and X. Gu, *J. Appl. Polym. Sci.* 84, 1504 (2002).
123. H. M. Wache, D. J. Tartakowska, A. Hentrich, and M. H. Wagner, *J. Mater. Sci. Mater. Med.* 14, 109 (2003).
124. H.-H. Wang and U.-E. Yuen, *J. Appl. Polym. Sci.* 102, 607 (2006).
125. P. Ping, W. Wang, X. Chen, and X. Jing, *J. Polym. Sci. Part B: Polym. Phys.* 45, 557 (2007).
126. C. P. Buckley, C. Prisacariu, and A. Caraculacu, *Polymer* 48, 1388 (2007).
127. G. Baer, T. S. Wilson, D. L. Matthews, and D. J. Maitland, *J. Appl. Polym. Sci.* 103, 3882 (2007).
128. P. Ping, W. Wang, X. Chen, and X. Jing, *Biomacromolecules* 6, 587 (2005).
129. W. M. Huang, C. W. Lee, and H. P. Teo, *J. Intel. Mat. Syst. Str.* 17, 753 (2006).
130. S. Chen, Q. Cao, B. Jing, Y. Cai, P. Liu, and J. Hu, *J. Appl. Polym. Sci.* 102, 5224 (2006).
131. Y. Zhu, J. Hu, K.-W. Yeung, K.-F. Choi, Y. Liu, and H. Liem, *J. Appl. Polym. Sci.* 103, 545 (2007).
132. S. Mondal and J. L. Hu, *J. Elastom. Plast.* 39, 81 (2007).
133. B. C. Chun, T. K. Cho, and Y.-C. Chung, *Eur. Polym. J.* 42, 3367 (2006).
134. F. Li, J. Hou, W. Zhu, X. Zhang, M. Xu, X. Luo, D. Ma, and B. K. Kim, *J. Appl. Polym. Sci.* 62, 631 (1996).
135. H.-J. Kim, D. C. Worley, II, and R. S. Benson, *Polymer* 38, 2609 (1997).
136. G. Dorey, *J. Phys. D: Appl. Phys.* 20, 245 (1987).
137. Y. Liu, K. Gall, M. L. Dunn, and P. McCluskey, *Smart Mater. Struct.* 12, 947 (2003).
138. E. R. Abrahamson, M. S. Lake, N. A. Munshi, and K. Gall, *J. Intel. Mat. Syst. Str.* 14, 623 (2003).
139. S. Black, *High-Performance Compos.* 14(5), 52 (2006).
140. J. Black, *Biological Performance of Materials*, Marcel Dekker, New York (1981), pp. 3–28.
141. L. G. Griffith, *Acta Mater.* 48, 263 (2000).
142. R. J. Zdrahala and I. J. Zdrahala, *J. Biomater. Appl.* 14, 67 (1999).
143. G. L. Wilkes, in *Polymers in Medicine and Surgery*, R. L. Kronenthal, Z. Oser, and E. Martin, Eds., Plenum Press, New York (1975).
144. F. El Feninat, G. Laroche, M. Fiset, and D. Mantovani, *Adv. Eng. Mater.* 4, 91 (2002).
145. W. Sokolowski, A. Metcalfe, S. Hayashi, L. Yahia, and J. Raymond, *Biomed. Mater.* 2, S23 (2007).
146. J. M. Walker, R. C. Brown, and J. R. Thomas, US Patent 4,846,812 (1989).
147. J. M. Walker and N. J. Sheehan, US Patent 4,955,863 (1990).
148. J. M. Walker and J. R. Thomas, US Patent 4,994,047 (1991).
149. A. Utsumi, Y. Morita, T. Kaide, K. Onishi, and S. Hayashi, US Patent 5,441,489 (1995).

150. W. J. Benett, P. A. Krulevitch, A. P. Lee, M. A. Northrup, and J. A. Folta, US Patent 5,609,608 (1997).
151. F. Stinger, US Patent 5,634,913 (1997).
152. P. A. Gunatillake, S. J. McCarthy, G. F. Meijs, and R. Adhikari, US Patent 6,858,680 (2005).
153. V. Tucci, B. Gibson, D. DiBiasio, S. Shivkumar, and G. Borah, *Society of Plastics Engineers Annual Technical Conference*, Boston, MA, May 8–11, 2102 (1995).
154. B. Yang, W. M. Huang, C. Li, and J. H. Chor, *Eur. Polym. J.* 41, 1123 (2005).
155. W. M. Huang, B. Yang, L. An, C. Li, and Y. S. Chan, *Appl. Phys. Lett.* 86, 114105-1 (2005).
156. A. Lendlein, H. Jiang, O. Junger, and R. Langer, *Nature* 434, 879 (2005).
157. C. Liang, C. A. Rogers, and E. Malafeew, *J. Intel. Mat. Syst. Str.* 8, 380 (1997).
158. T. Ohki, Q.-Q. Ni, N. Ohsako, and M. Iwamoto, *Compos. Part A: Appl. Sci.* 35, 1065 (2004).
159. F. Li, L. Qi, J. Yang, M. Xu, X. Luo, and D. Ma, *J. Appl. Polym. Sci.* 75, 68 (2000).
160. K. Gall, M. L. Dunn, Y. Liu, D. Finch, M. Lake, and N. A. Munshi, *Acta. Mater.* 50, 5115 (2002).
161. F. Cao and S. C. Jana, *Society of Plastics Engineers Annual Technical Conference*, Charlotte, NC, May 7–11, 646 (2006).
162. F. Cao and S. C. Jana, *Polymer* 48, 3790 (2007).
163. A. Pattanayak and S. C. Jana, *Polymer* 46, 3275 (2005).
164. A. Pattanayak and S. C. Jana, *Polymer* 46, 3394 (2005).
165. A. Pattanayak and S. C. Jana, *Polymer* 46, 5183 (2005).
166. A. Pattanayak and S. C. Jana, *Polym. Eng. Sci.* 45, 1532 (2005).
167. F. Cao and S. C. Jana, *Society of Plastics Engineers Annual Technical Conference*, Milwaukee, WI, May 4–8, 172 (2008).
168. I. S. Gunes, F. Cao, and S. C. Jana, *Polymer* 49, 2223 (2008).
169. G. A. Jimenez, Ph.D. Thesis, The University of Akron (2007).
170. S. Mondal and J. L. Hu, *Iran. Polym. J.* 15, 135 (2006).
171. P. Miaudet, A. Derre, M. Maugey, C. Zakri, P. M. Piccione, R. Inoubli, and P. Poulin, *Science* 318, 1294 (2007).
172. H. Koerner, G. Price, N. A. Pearce, M. Alexander, and R. A. Vaia, *Nat. Mater.* 3, 115 (2004).
173. R. A. Vaia, H. Koerner, D. Powers, P. Mirau, M. Alexander, and M. Arlen, *Polymer* Preprints 46, 542 (2005).
174. J. W. Cho, J. W. Kim, Y. C. Jung, and N. S. Goo, *Macromol. Rapid Commun.* 26, 412 (2005).
175. G. Kraus, *Rubber Chem. Technol.* 38, 1070 (1965).
176. I. S. Gunes, G. A. Jimenez, and S. C. Jana, *Carbon* 47, 981 (2009).
177. C. Liu and P. T. Mather, *Society of Plastics Engineers Annual Technical Conference*, Chicago, IL, May 16–20, 3080 (2004).
178. L. Holliday and J. Robinson, *J. Mater. Sci.* 8, 301 (1973).
179. R. S. Raghava, *Polym. Compos.* 9, 1 (1988).
180. I. S. Gunes, F. Cao, and S. C. Jana, *J. Polym. Sci. Polym. Phys.* 46, 1437 (2008).
181. L. Yang, K. Setyowati, A. Li, S. Gong, and J. S. Chen, *Adv. Mater.* 20, 2271 (2008).
182. J. M. Lambert, D. D. Solomon, and D. R. Rhodes, US Patent 5,102,401 (1992).
183. A. P. Lee, M. A. Northrup, D. R. Ciarlo, P. A. Krulevitch, and W. J. Benett, US Patent 5,911,737 (1999).
184. N. B. Colthup, L. H. Daly, and S. E. Wiberley, *Introduction to Infrared and Raman Spectroscopy*, Academic Press, New York (1964), pp. 2–29.
185. D. J. Maitland, M. F. Metzger, D. Schumann, A. Lee, and T. S. Wilson, *Lasers Surg. Med.* 30, 1 (2002).
186. R. Mohr, K. Kratz, T. Weigel, M. Lucka-Gabor, M. Moneke, and A. Lendlein, *Proc. Natl. Acad. Sci. U.S.A.* 103, 3540 (2006).
187. A. M. Schmidt, *Macromol. Rapid Commun.* 27, 1168 (2006).
188. M. Y. Razzaq, M. Anhalt, L. Frormann, and B. Weidenfeller, *Mater. Sci. Eng. A-Struct.* 444, 227 (2007).
189. G. Brian and H. Taylor, *Bull. World Health Organ.* 79, 249 (2001).
190. C. A. McCarty and H. R. Taylor, *Invest. Ophthalmol. Vis. Sci.* 42, 1677 (2001).
191. I. H. Fine, M. Packer, and R. S. Hoffman, in *Cataract and Refractive Surgery*, T. Kohnen and D. D. Koch, Eds., Springer, New York, 2005 (chapter 2).
192. J. G. F. Worst, US Patent 5,192,319 (1993).
193. L. Werner and N. Mamalis, in *Cataract and Refractive Surgery*, T. Kohnen and D. D. Koch, Eds., Springer, New York, 2005 (chapter 4).
194. S. Q. Zhou, C. D. Wilcox, C. Liau, and I. Valyunin. US patent 6,679,605 (2004).
195. O. Wichterle, in *Soft Contact Lenses: Clinical and Applied Technology*, M. Ruben, Ed., John Wiley & Sons, New York, 1978 (chapter 1).

196. B. J. Tighe, in *Hydrogels in Medicine and Pharmacy, Vol. 3: Properties and Applications*, N. A. Peppas, Ed., CRC Press, Boca Raton, FL, 1987 (chapter 3).

197. S. H. Gehrke, D. Biren, and J. J. Hopkins, *J. Biomater. Sci. Polym. Ed.* 6, 375 (1994).

198. E. H. Immergut and H. F. Mark, in *Plasticization and Plasticizer Processes*, N. A. J. Platzer, Ed., American Chemical Society, Washington, DC, 1965 (chapter 1).

199. H. E. Kaufman, J. Katz, J. Valenti, and J. W. Sheets, *Science* 198, 525 (1977).

200. J. Katz, H. E. Kaufman, E. P. Goldberg, and J. W. Sheets, *Trans. Am. Acad. Ophthalmol. Otolaryngol.* 83, 204 (1977).

201. S. Reich, M. Levy, A. Meshorer, M. Blumental, M. Yalon, J. W. Sheets, and E. P. Goldberg, *J. Biomed. Mater. Res.* 18, 737 (1984).

202. R. Stewart, *Plast. Eng.* 65, 27 (2009).

203. G. Martino, M. Vitale, and P. Vanemon, *Cosmet. Toiletries* 118, 49 (2003).

204. C. Liu, P. T. Mather, and C. Burstone, *Society of Plastics Engineers Annual Technical Conference*, Charlotte, NC, May 7–11, 1356 (2006).

205. T. G. Leong, C. L. Randall, B. R. Benson, N. Bassik, G. M. Stern, and D. H. Gracias, *Proc. Natl. Acad. Sci. U.S.A.* 106, 703 (2009).

12 Bioadhesive Drug Delivery Systems

Ryan F. Donnelly and A. David Woolfson

CONTENTS

12.1 THEORIES OF BIOADHESION

Adhesion can be defined as the bond produced by contact between a pressure-sensitive adhesive (PSA) and a surface (Jiménez-Castellanos et al., 1993). The American Society for Testing and Materials (1984) has defined it as the state in which two surfaces are held together by interfacial forces, which may consist of valence forces, interlocking action, or both. Good (1983) defined bioadhesion as the state in which two materials, at least one biological in nature, are held together for an extended period of time by interfacial forces. It is also defined as the ability of a material, synthetic or biological, to adhere to a biological tissue for an extended period of time. In biological systems, three main types of bioadhesion can be distinguished:

Type 1: Adhesion between two biological phases, for example, cell fusion, platelet aggregation, wound healing, adhesion between a normal cell and a foreign substance or a pathological cell.

Type 2: Adhesion of a biological phase to an artificial substrate, for example, cell adhesion to culture dishes, platelet adhesion to biomaterials, microbial fouling and barnacle adhesion to ships, biofilm formation on prosthetic devices and inserts.

Type 3: Adhesion of an artificial material to a biological substrate, for example, adhesion of synthetic hydrogels to soft tissues (Henriksen et al., 1996) and adhesion of sealants to dental enamel.

Cell-to-cell adhesion can be explained as a balance between nonspecific, repulsive, and attractive physical forces and macromolecular bridges (Bell et al., 1984). In contrast to Type 1 adhesion, Types 2 and 3 involve one synthetic phase. When many hard synthetic materials contact a biological fluid, there is formation of an interfacial film of biomolecules. This Type 2 bioadhesion is usually irreversible (Larsson, 1980) and has important implications in areas such as artificial kidney hemodialysis where there exists a risk of blood damage. The adhesion of human endothelial cells to polymeric surfaces is dependent on surface wettability (Van Wachem et al., 1985). Optimum adhesion occurs when surfaces show moderate wettability, with more hydrophilic and more hydrophobic polymers showing less, or even no, adhesion. Polymers of both natural and synthetic origin comprise the majority of examples that display Type 3 bioadhesion. Polymeric matrices are, of course, the basis of many novel drug delivery systems, including adhesive patches for transdermal and topical indications (Govil, 1988; Woolfson et al., 1998).

For drug delivery purposes, the term bioadhesion implies attachment of a drug carrier system to a specified biological location. The biological surface can be epithelial tissue or the mucus coat on the surface of a tissue. If adhesive attachment is to a mucus coat, the phenomenon is referred to as mucoadhesion. Leung and Robinson (1988) described mucoadhesion as the interaction between a mucin surface and a synthetic or natural polymer.

For bioadhesion to occur, a succession of phenomena is required. The first stage involves an intimate contact between a bioadhesive and a biological substrate, either from wetting of the bioadhesive surface or from swelling of the bioadhesive. In the second stage, after contact is established, penetration of the bioadhesive into the crevice of the tissue surface or interpenetration of the chains of the bioadhesive with those of the substrate takes place. Secondary chemical bonds can then establish themselves (Duchêne et al., 1988).

One of the most important factors for bioadhesion is tissue surface roughness. Tissue surfaces often display undulations and crevices that provide both attachment points and a key for the deposition and inclusion of polymeric adhesives. These surface irregularities have been analyzed by Merrill (1977), who stated that the surface geography can be expressed in terms of an aspect ratio of maximum depth, d, to maximum width, h, as shown in Figure 12.1.

Surfaces that have a ratio of $d/h < 1/20$ are considered too smooth for satisfactory adhesion (Duchêne et al., 1988). For higher ratio values, only adhesives of low viscosity can penetrate into the tissue topography to form mechanical or physical bonds.

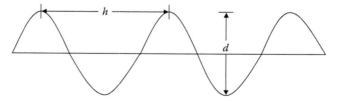

FIGURE 12.1   Schematic representation of the surface roughness of a soft tissue.

Viscosity and wetting are also important factors in bioadhesion (Van Wachem et al., 1985). Bioadhesion can be considered to be a wet-stick adhesion process (Woolfson et al., 1998). During wet-stick, polymer chains are released from restraining dry lattice forces by hydration. These chains then move and entangle into the matrix of the substrate. Interaction may then occur on a molecular level between bioadhesive and substrate. The interaction between two molecules is composed of attraction and repulsion. Attractive interactions arise from van der Waal's forces, electrostatic attraction, hydrogen bonding, and hydrophobic interaction. Repulsive interactions occur because of electrostatic and steric repulsion. Attractive and, or, repulsive interactions may occur between the bioadhesive and the lipids, glycolipids, proteins, glycoproteins, and polysaccharides found on epithelial cell membranes. Alternatively the bioadhesive may interact with the glycocalyx, a structure containing polysaccharides on the cell surface. In regions of the body where mucus overlies the epithelial surface, such as the gastrointestinal tract (GIT), the interaction may largely be between bioadhesive and the mucin glycoproteins rather than the epithelial cell itself (Yang and Robinson, 1988). For bioadhesion to occur, the attractive interaction should be larger than nonspecific repulsion (Kamath and Park, 1992).

Various theories exist to explain at least some of the experimental observations made during the bioadhesion process. Unfortunately, each theoretical model can only explain a limited number of the diverse range of interactions that constitute the bioadhesive bond (Longer and Robinson, 1986). However, four main theories can be distinguished.

12.1.1 WETTING THEORY OF BIOADHESION

The mechanical or wetting theory is perhaps the oldest established theory of adhesion. It explains adhesion as an embedding process, whereby adhesive molecules penetrate into surface irregularities of the substrate and ultimately harden, producing many adhesive anchors. Free movement of the adhesive on the surface of the substrate means that it must overcome any surface tension effects present at the interface (McBain and Hopkins, 1925). The wetting theory is best applied to liquid bioadhesives, describing the interactions between the bioadhesive, its angle of contact, and the thermodynamic work of adhesion.

The work done is related to the surface tension of both the adhesive and the substrate, as described by Equation 12.1, the Dupré equation (Pritchard, 1971):

$$\omega_A = \gamma_b + \gamma_t - \gamma_{bt} \qquad (12.1)$$

where
ω_A is the specific thermodynamic work of adhesion
γ_b, γ_t, and γ_{bt} represent the surface tensions of the bioadhesive polymer, the substrate, and the interfacial tension, respectively

The adhesive work done is a sum of the surface tensions of the two adherent phases, less the interfacial tensions apparent between both phases (Wake, 1982). Figure 12.2 shows a drop of liquid bioadhesive spreading over a soft tissue surface.

Horizontal resolution of the forces gives the Young equation (Equation 12.2):

$$\gamma_{ta} = \gamma_{bt} + \gamma_{ba} \cos\theta \qquad (12.2)$$

where
θ is the angle of contact
γ_{bt} is the surface tension between the tissue and polymer
γ_{ba} is the surface tension between polymer and air
γ_{ta} is the surface tension between tissue and air

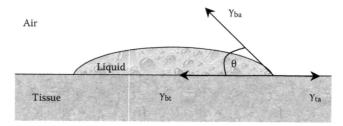

FIGURE 12.2   (See companion CD for color figure.) Liquid bioadhesive spreading over a typical soft tissue surface.

Equation 12.3 states that if the angle of contact, θ, is greater than zero, the wetting will be incomplete. If the vector γ_{ta} greatly exceeds $\gamma_{bt} + \gamma_{ba}$, that is

$$\gamma_{ta} \geq \gamma_{bt} + \gamma_{ba} \tag{12.3}$$

then θ will approach zero and wetting will be complete. If a bioadhesive material is to successfully adhere to a biological surface, it must first dispel barrier substances and then spontaneously spread across the underlying substrate, either tissue or mucus. The spreading coefficient, S_b, can be defined as shown in Equation 12.4:

$$S_b = \gamma_{ta} - \gamma_{bt} - \gamma_{ba} > 0 \tag{12.4}$$

which states that bioadhesion is successful if S_b is positive, thereby setting the criteria for the surface tension vectors, in other words, bioadhesion is favored by large values of γ_{ta} or by small values of γ_{bt} and γ_{ba} (Wake, 1982).

12.1.2 ELECTROSTATIC THEORY OF BIOADHESION

Separation of two distinct bodies is achieved by overcoming electrostatic forces originating from the establishment of an electrical double layer between an adhesive and a substrate. The electrostatic theory of repulsion proposes the transfer of electrons across the adhesive interface, establishing the electrical double layer and a series of attractive forces responsible for maintaining contact between the two layers (Deraguin and Smilga, 1969). The adhesive interface is considered to be analogous to a parallel-plate condenser, such that work must be done against any electrical charges before separation is achieved. The work of adhesion can be equated to the energy of the condenser, provided that no work is done in overcoming van der Waal's forces.

12.1.3 DIFFUSION THEORY OF BIOADHESION

The interpenetration or diffusion theory is currently the most widely accepted physical theory and was first proposed by Voyutskii (1963), who studied the autodiffusion of two contacting, identical polymers. The term autodiffusion was used to describe the diffusion of long-chain molecules across the interface from one polymer into an identical one and is the most likely explanation for the self-tact of rubbers. Many adhesives are applied to the adherents as a solvent or aqueous solution and, on evaporation of the vehicle, the polymer residues are combined. Failure of the joint is not normally expected along the plane of combination because diffusion has made the two layers into one, with a loss of the interface (Wake, 1982).

While the concept of autodiffusion can be used to explain the autodiffusion of polymers and the heat sealing of thermoplastics, the universality of this theory is not widely accepted. For example, adhesion between different polymers and polymer-to-metal adhesion are not believed to occur as a result of diffusion (Wake, 1982).

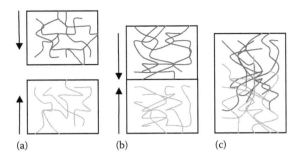

(a)   (b)   (c)

FIGURE 12.3   (See companion CD for color figure.) Schematic representation of the diffusion theory of bioadhesion. Blue polymer layer and red mucus layer before contact (a), upon contact (b), the interface becomes diffuse after contact for a period of time (c).

With regard to bioadhesion, polymeric chains from the bioadhesive and the biological substrate, for example the glycoprotein mucin chains found in mucus, intermingle and reach a sufficient depth within the opposite matrix to allow formation of a semipermanent bond (Jiménez-Castellanos et al., 1993). The process can be visualized from the point of initial contact. The existence of concentration gradients will drive the polymer chains of the bioadhesive into the mucus network and the glycoprotein mucin chains into the bioadhesive matrix until an equilibrium penetration depth is achieved as shown in Figure 12.3.

The exact depth needed for good bioadhesive bonds is unclear, but is estimated to be in 0.2–0.5 µm range (Peppas and Buri, 1985). The mean diffusional depth of the bioadhesive polymer segments, s, may be represented by Equation 12.5:

$$s = \sqrt{2tD} \tag{12.5}$$

where
  D is the diffusion coefficient
  t is the contact time

Duchêne et al. (1988) adapted Equation 12.5 to give Equation 12.6, which can be used to determine the time, t, to bioadhesion of a particular polymer:

$$t = \frac{l^2}{D_b} \tag{12.6}$$

where
  l represents the interpenetrating depth
  D_b the diffusion coefficient of a bioadhesive through the substrate

Once intimate contact is achieved, the substrate and adhesive chains move along their respective concentration gradients into the opposite phases. Depth of diffusion is obviously dependent on the diffusion characteristics of both phases. Reinhart and Peppas (1984) reported that the diffusion coefficient depended on the molecular weight of the polymer strand and that it decreased with increasing cross-linking density. This process of simple physical entanglement has been described as analogous to forcing two pieces of steel wool to intermingle upon contact (Yang and Robinson, 1988).

12.1.4   Adsorption Theory of Bioadhesion

According to the adsorption theory, after an initial contact between two surfaces, the materials adhere because of surface forces acting between the chemical structures at the two surfaces

(Ahuja et al., 1997). When polar molecules or groups are present, they reorientate at the interface (Wake, 1982). Chemisorption can occur when adhesion is particularly strong. The theory maintains that adherence to tissue is due to the net result of one or more secondary forces (Huntsberger, 1967; Kinloch, 1980; Yang and Robinson, 1988). The three types of secondary forces are

1. van der Waal's forces
2. Hydrogen bonding
3. Hydrophobic interaction

The formation of primary chemical bonds, such as those formed between adhesives used in dentistry and surgery and tissue, is undesirable in bioadhesion. This is because the high strength of ionic or covalent bonds may result in permanent adhesion (Ahuja et al., 1997).

12.2 BIOADHESIVE MATERIALS

Bioadhesive polymers have numerous hydrophilic groups, such as hydroxyl, carboxyl, amide, and sulfate. These groups attach to mucus or the cell membrane by various interactions such as hydrogen bonding and hydrophobic or electrostatic interactions. These hydrophilic groups also cause polymers to swell in water and, thus, expose the maximum number of adhesive sites (Yang and Robinson, 1988).

An ideal polymer for a bioadhesive drug delivery system should have the following characteristics (Jiménez-Castellanos et al., 1993; Ahuja et al., 1997):

1. The polymer and its degradation products should be nontoxic and nonabsorbable.
2. It should be nonirritant.
3. It should preferably form a strong non-covalent bond with the mucus or epithelial cell surface.
4. It should adhere quickly to moist tissue and possess some site specificity.
5. It should allow easy incorporation of the drug and offer no hindrance to its release.
6. The polymer must not decompose on storage or during the shelf life of the dosage form.
7. The cost of the polymer should not be high, so that the prepared dosage form remains competitive.

Polymers that adhere to biological surfaces can be divided into three broad categories (Jiménez-Castellanos et al., 1993; Ahuja et al., 1997):

1. Polymers that adhere through nonspecific, non-covalent interactions which are primarily electrostatic in nature
2. Polymers possessing hydrophilic functional groups that hydrogen bond with similar groups on biological substrates
3. Polymers that bind to specific receptor sites on the cell or mucus surface

The latter category includes lectins, which are generally defined as proteins or glycoprotein complexes of nonimmune origin that are able to bind sugars selectively in a non-covalent manner (Smart et al., 1999). Lectins are capable of attaching themselves to carbohydrates on the mucus or epithelial cell surface and have been extensively studied, notably for drug targeting applications (Naisbett and Woodley, 1994; Nicholls et al., 1996). These second-generation bioadhesives provide not only for cellular binding but also for subsequent endo- and transcytosis. However, lectins have proved problematic, with recent literature indicating variability in drug absorption associated with their use (Yang and Robinson, 1988; Lehr, 1996), as well as with their potential toxicity (Banchonglikitkul et al., 2002; Smart et al., 2003). Table 12.1 shows the chemical structures of several bioadhesive polymers commonly used in modern drug delivery.

TABLE 12.1
Chemical Structures of Some Bioadhesive Polymers Used in Drug Delivery

Chemical Name (Abbreviation)	Chemical Structure
Poly(ethylene glycol) (PEG)	$-[CH_2-CH_2O]_n-$
Poly(vinyl alcohol) (PVA)	$-[CH_2-\underset{OH}{\overset{H}{C}}]_n-$
Poly(vinyl pyrrolidone) (PVP)	$[CH_2-\overset{H}{\underset{|}{C}}]_n$ with pyrrolidone ring (N–C=O)
Poly(acrylic acid) (PAA or Carbopol)	$[CH_2-\overset{H}{\underset{|}{C}}]_n$ with $C(=O)OH$
Poly(hydroxyethyl methacrylate) (PHEMA)	$[CH_2-\underset{}{\overset{CH_3}{C}}]_n$ with $C(=O)O-CH_2-CH_2OH$
Chitosan	CH_2OH, OH, HO, NH_2 glucosamine ring repeating unit n
Hydroxyethylcellulose (HEC): R = H and CH_2CH_2OH Hydroxypropylcellulose (HPC): R = H and $CH_2CH(OH)CH_3$ Hydroxypropylmethylcellulose (HPMC): R = H, CH_3, and $CH_2CH(OH)CH_3$ Methylcellulose: R = H and CH_3 Sodium carboxymethylcellulose (NaCMC): R = H and CH_2COONa	cellulose backbone with CH_2OR and OR groups, repeating unit n

12.3 FACTORS AFFECTING BIOADHESION

Bioadhesion may be affected by a number of factors, including hydrophilicity, molecular weight, cross-linking, swelling, pH, and the concentration of the active polymer (Jiménez-Castellanos et al., 1993; Ahuja et al., 1997; Peppas et al., 2000).

12.3.1 Hydrophilicity

Bioadhesive polymers possess numerous hydrophilic functional groups, such as hydroxyl and carboxyl. These groups allow hydrogen bonding with the substrate, swelling in aqueous media, and allowing maximal exposure of potential anchor sites. Hydrophilic groups also allow polymers to swell in water. In addition, swollen polymers have the maximum distance between their chains leading to increased chain flexibility and efficient penetration of the substrate.

12.3.2 Molecular Weight

There is a certain molecular weight (mw) at which bioadhesion is at a maximum. The interpenetration of polymer molecules is favored by low-molecular-weight polymers, whereas entanglements are favored at higher molecular weights. The optimum molecular weight for the maximum bioadhesion depends on the type of polymer, with bioadhesive forces increasing with the molecular weight of the polymer up to 100,000. Beyond this level, there is no further gain (Gurny et al., 1984).

12.3.3 Cross-Linking and Swelling

Cross-link density is inversely proportional to the degree of swelling (Gudeman and Peppas, 1995). The lower the cross-link density, the higher the flexibility and hydration rate; the larger the surface area of the polymer, the better the bioadhesion. Interpenetration of chains is easier as polymer chains are disentangled and free of interactions. The force generated by pulling water from the underlying substrate into the swelling polymer also helps polymer strands to penetrate deep into the substrate. To achieve a high degree of swelling, a lightly cross-linked polymer is favored. However, if too much moisture is present and the degree of swelling is too great a slippy mucilage results and this is easily removed from the substrate (McCarron et al., 2004). The bioadhesion of cross-linked polymers can be enhanced by the inclusion in the formulation of adhesion promoters, such as free polymer chains and polymers grafted onto the preformed network (Peppas et al., 2000).

12.3.4 Spatial Conformation

Besides molecular weight or chain length, spatial conformation of a polymer is also important. Despite a high molecular weight of 19,500,000 for dextrans, they have adhesive strength similar to that of polyethylene glycol (PEG), with a molecular weight of 200,000. The helical conformation of dextran may shield many adhesively active groups, primarily responsible for adhesion, unlike PEG polymers, which have a linear conformation (Jiménez-Castellanos et al., 1993).

12.3.5 pH

The pH at the bioadhesive to substrate interface can influence the adhesion of bioadhesives possessing ionizable groups. Many bioadhesives used in drug delivery are polyanions possessing carboxylic acid functionalities. If the local pH is above the pK_a of the polymer, it will be largely ionized; if the pH is below the pK_a of the polymer, it will be largely unionized. The approximate pK_a for the poly(acrylic acid) (PAA) family of polymers is between 4 and 5. The maximum adhesive strength of these polymers is observed around pH 4–5 and decreases gradually above the pH of 6. A systematic investigation of the mechanisms of bioadhesion clearly showed that the protonated carboxyl groups rather than the ionized carboxyl groups react with mucin molecules, presumably by the simultaneous formation of numerous hydrogen bonds (Park and Robinson, 1985). Fully ionized anionic polymers, despite showing maximal swelling in water, will tend to be repelled by mucus and epithelial surfaces, both of which carry a net negative charge. The pH at the bioadhesive to substrate interface can be influenced by the nature of the tissue or mucus substrate and the composition of the bioadhesive formulation. Polycations, such as poly(lysine), can bind to negatively charged mucus at pH 7.4, presumably by electrostatic interactions. However, cationic polymers tend not to adhere as strongly as anionic ones and the former may cause cell aggregation and other toxic reactions (Yang and Robinson, 1988; Jiménez-Castellanos et al., 1993).

12.3.6 Concentration of Active Polymer

Ahuja et al. (1997) stated that there is an optimum concentration of polymer corresponding to the best bioadhesion. In highly concentrated systems, the adhesive strength drops significantly. In concentrated solutions, the coiled molecules become solvent-poor and the chains available for interpenetration are not numerous. This result seems to be of interest only for more or less liquid bioadhesive formulations. It was shown by Duchêne et al. (1988) that, for solid dosage forms, such as tablets, the higher the polymer concentration, the stronger the bioadhesion.

12.3.7 Drug/Excipient Concentration

BlancoFuente et al. (1996) showed that the addition of propranolol hydrochloride to Carbopol® (a lightly cross-linked PAA polymer) hydrogels increased adhesion when water was limited in the system, due to an increase in the elasticity caused by complex formation between drug and polymer. However, when large quantities of water were present, the complex precipitated out, leading to a slight decrease in adhesive character. Increasing toluidine blue O concentration in mucoadhesive patches based on Gantrez® (poly(methylvinylether/maleic acid)) significantly increased mucoadhesion to porcine cheek tissue (Donnelly et al., 2007). This was attributed to increased internal cohesion within the patches due to electrostatic interactions between the cationic drug and anionic copolymer. By contrast, miconazole nitrate did not influence bioadhesion up to a concentration of 30% in PAA-based tablets (Bouckaert and Remon, 1993). Voorspoels et al. (1996) demonstrated that increasing concentrations of testosterone and its esters produced a decrease in adhesive characteristics of buccal tablets.

For ionic bioadhesive polymers, dissolved salts can have a significant effect on bioadhesion by a variety of mechanisms, including salting out of polymeric components and charge neutralization, preventing expansion of coiled polymer chains and, thus, entanglement with epithelial mucin. Woolfson et al. (1995a,b) added sodium chloride to polymer blends containing Gantrez and found no effect on bioadhesive character, indicating that the sensitivity of bioadhesive polymers to the presence of dissolved salts varies considerably, depending on structure. However, sodium chloride can affect bioadhesion through a direct action on the substrate, Thus, sodium chloride, when added to several mucoadhesive polymers, was observed to reduce the stiffness of mucus gel (Mortazavi and Smart, 1994).

12.3.8 Other Factors Affecting Bioadhesion

Bioadhesion may be affected by the initial force of application (Smart, 1991). Higher forces lead to enhanced interpenetration and high bioadhesive strength (Duchêne et al., 1988). In addition, the greater the initial contact time between bioadhesive and substrate, the greater the swelling and interpenetration of polymer chains (Kamath and Park, 1992). Physiological variables can also affect bioadhesion. The rate of mucus turnover can be affected by disease states and by the presence of a bioadhesive device (Lehr and Poelma, 1991). In addition, the nature of the surface presented to the bioadhesive can vary significantly depending on the body site and the presence of local or systemic disease (Kamath and Park, 1992).

12.4 DETERMINATION OF BIOADHESIVE FORCE OF ATTACHMENT

The evaluation of bioadhesive properties is fundamental to the development of novel bioadhesive delivery systems. Measurement of the mechanical properties of a bioadhesive material after interaction with a substrate is one of the most direct ways to quantify the bioadhesive performance. To measure the force of bioadhesive attachment to tissue involves the application of a stress to the bonding interface. Numerous designs of apparatus have been proposed for this purpose. Since no standard apparatus is available for testing bioadhesive strength, an inevitable lack of uniformity between test methods has arisen. Nevertheless, three main testing modes are recognized: tensile, shear, and peel tests.

When the force of separation is applied perpendicularly to the tissue/adhesive interface, a state of tensile stress is set up. This is, perhaps, the most common configuration used in bioadhesive testing. During shear stress, the direction of the forces is reoriented so that it acts along the joint interface. In both tensile and shear modes, an equal pressure is distributed over the contact area (Park and Park, 1990).

The peel test is most applicable to systems involving adhesive tape, where removal of the device is an important parameter. By pulling the two interfaces apart at an acute angle, the force is focused along a single line of contact. This effectively concentrates the applied force at the point of separation and removal of the tape can be achieved easily. The peel test is of limited use in most bioadhesive systems. However, it is of value when the bioadhesive system is formulated as a patch (McCarron et al., 2005).

Once formed, the interface of the bioadhesive bond is likely to exist in a microenvironment that will inevitably influence its further performance. Factors such as pH, temperature, ionic strength, and water content will all affect bond durability. Experimental rigs have been devised to investigate these important considerations. Although the force used to break the bond can be applied in one of three fashions, as described, the majority of researchers prefer to use the tensile stress method. Irrespective of which configuration is used to apply the force, the measurement of the performance of the bioadhesive bond is generally performed in one of three environments.

Most in vitro methods involve measurement of shear or tensile stress. Smart et al. (1984) described a method to study bioadhesion whereby a polymer-coated glass slide was withdrawn from a mucus solution. The force required to accomplish this was equated to mucoadhesion. A fluorescence probe technique was developed by Park and Robinson (1985) to determine bonding between epithelial cells and a test polymer. Other methods have involved measurement of the force required to detach polymeric materials from excised rabbit corneal endothelium. Mikos and Peppas (1986) described a method whereby a polymer particle was blown across a mucus-filled channel. The motion, recorded photographically, gave details regarding the adhesion process.

McCarron et al. (2004, 2005, 2006) and Donnelly et al. (2006) have reported extensively on the use of a commercial apparatus, in the form of a texture profile analyzer (Figure 12.4) operating in bioadhesive test mode, to measure the force required to remove bioadhesive films from excised tissue in vitro.

The Texture Analyser, operating in tensile test mode and coupled with a sliding lower platform, was also used to determine peel strength of similar formulations (McCarron et al., 2005) (Figure 12.5).

The ultimate destination for successful bioadhesive devices is a tissue surface such as the GIT or the buccal cavity. Some investigators have used previous results obtained from their own in vitro work to predict and, subsequently test for, in vivo performance. Examples of in vivo studies in the literature are not as plentiful as in vitro work because of the amount and expense of animal trials

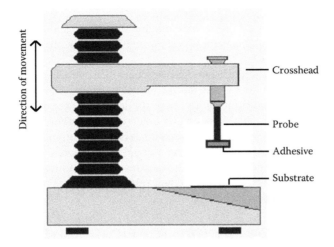

FIGURE 12.4   Texture profile analyzer in bioadhesion test mode.

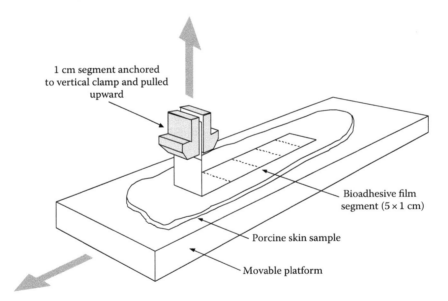

FIGURE 12.5   **(See companion CD for color figure.)** Simplified representation of a typical test setup used to determine peel strength of bioadhesive films.

and the difficulty in maintaining experimental consistency. Commonly measured variables include gastrointestinal transit times of bioadhesive-coated particles and drug release from in situ bioadhesive devices.

Ch'ng et al. (1985) studied the in vivo transit time for bioadhesive beads in the rat. A ^{51}Cr-radiolabeled bioadhesive was carefully introduced into the stomach. At selected time intervals, the GITs were removed, cut into 20 equal segments, and the radioactivity in each was measured.

Reich et al. (1984) described an instrument for measuring the force of adhesion between intraocular lens materials and the endothelium of excised rabbit corneas. It used a metal or glass fiber deflection technique for measuring pressures that were usually lower than $1 \, \text{g/cm}^2$. The contacting surfaces were submerged in saline to eliminate surface tension effects. In contrast to many bioadhesive testing situations, the authors were attempting to find a polymer with a lower adhesion than poly(methylmethacrylate) (PMMA), the material most commonly used to make intraocular lenses.

A direct relationship had been shown between the adhesive forces of a polymer to the endothelium and the extent of cell damage. Poly(hydroxyethylmethacrylate) and Duragel®, both hydrophilic materials used to make soft lenses, gave less adhesion and subsequently less cell damage than the hydrophobic PMMA.

Davis (1985) described a method to study gastric-controlled release systems. It can be advantageous in certain diseased states for a single dosage unit to be retained in the stomach. This enables released drug to empty from the stomach and have the length of the small intestine available for absorption. Mucoadhesives can help achieve this objective. Therefore, as an alternative to using invasive in vivo techniques, a formulation was used containing a gamma-emitting radionuclide. The release characteristics and the position of the device could be monitored using gamma scintigraphy. This technique could measure two different radionuclides simultaneously so that a subject could be given two different formulations together, with their release and transit times then being studied in a single occasion crossover study.

12.5   CHARACTERIZATION METHODS FOR BIOADHESIVE DRUG DELIVERY SYSTEMS

A range of techniques are available for the study of bioadhesive delivery systems, with the choice being influenced by site of attachment and substrate tissue, together with the design of the dosage form. Most viable bioadhesive drug delivery applications involve mucoadhesion to accessible (topical) epithelia. Certain tests will be required for regulatory purposes, whilst others, although optional, often constitute an important part of the development of a pharmaceutics package for a novel bioadhesive carrier. Bioadhesive (mucoadhesive) dosage forms vary in bioadhesive performance, with water content being a primary determinant of bioadhesion. Delivery platforms include liquids and gels (low adhesion), viscoelastic semisolids (moderate adhesion), flexible hydrogel films, particulates, and compacts (strong adhesion). Applicable characterization methodologies include

- Drug release studies
- Drug diffusion (membrane penetration) studies
- Examination of mechanical and textural properties
- Examination of continuous shear properties
- Examination of structural (viscoelastic) properties at defined temperatures
- Evaluation of adhesion to model substrates
- Examination of product/packaging interactions

Texture profile analysis may be used to determine the following characteristics of semisolid bioadhesive delivery systems (Jones and Woolfson, 1997; McCarron et al., 2005).

- Hardness/compressibility—A measure of the resistance of the formulation to probe depression, which can be used to characterize product spreadability
- Cohesiveness—A measure of the effects of successive deformations on the structural properties of a product
- Adhesiveness—A measure of the work required to remove the probe from the sample, a property related to bioadhesion

Flow rheometry may be used to obtain the following information on bioadhesive semisolids:

- Effects of successive shearing stresses on the rate of deformation of a product
- Information concerning product viscosity which, in turn, affects drug release and ease of application at the site

- Information concerning the rate of structural recovery following deformation (thixotropy) which, in turn, affects product retention at the application site
- Drug release
- Manufacture

Oscillatory rheometry, a nondestructive test (unlike flow rheometry), may be used to obtain the information on bioadhesive semisolids used at sites subjected to variable stresses, for example, in the oral cavity, where chewing, swallowing, and talking all affect the structural rheology of bioadhesive systems (Andrews et al., 2009; Jones et al., 2009). Oscillatory rheometry determines strain in the system, resolved into two components: the storage modulus (G') is the part of the strain which is in phase with the stress and is a measure of the solid character and the loss modulus (G") is the part of the strain which lags the stress by 90° and is a measure of the liquid character.

Dielectric spectroscopy is an analytical technique which involves the application of an oscillating electric held to a sample and the measurement of the corresponding response over a range of frequencies, from which information on sample structure and behavior may be extrapolated. Craig and Tamburic (1997) described studies on sodium alginate gels whereby a model was proposed in order to relate the low frequency response to the gel structure. They also reported on the application of dielectric spectroscopy to the study of cross-linked PAA with respect to the effects of additives such as propylene glycol and chlorhexidine gluconate, the influence of the choice of neutralizing agent, and the effects of ageing on the gel structure.

12.6 BIOADHESIVE DRUG DELIVERY SYSTEMS

There has been an increasing interest in the use of bioadhesive polymers in the design of drug delivery systems. One of the advantages of using these materials is that they can maintain contact with mucosal surfaces for much longer periods of time than non-bioadhesive polymers. Since polymers possessing bioadhesive properties can retain drugs in close proximity to membranes rich in underlying vasculature, they may offer a solution to the poor bioavailability of some drugs and a method to avoid enzymatic degradation of others.

Bioadhesive drug delivery systems have in the past been formulated as powders, compacts, sprays, semisolids, or films. For example, compacts have been used for drug delivery to the oral cavity (Ponchel et al., 1987) and powders and nanoparticles (NP) have been used to facilitate drug administration to the nasal mucosa (Nagai and Konishi, 1987; Sayin et al., 2008). Bioadhesive films are generally prepared from solutions of film-forming bioadhesive polymers in appropriate solvents. The final film is traditionally produced by casting of the solution into a mould of defined dimensions. A casting knife may be used to spread the solution and remove excess (Radebaugh, 1992). Film-forming bioadhesive polymers used in the production of bioadhesive films include the cellulose derivatives (Anders and Merkle, 1989), PAA such as Carbopol (Woolfson et al., 1995a,b), and Gantrez copolymers such as poly(methylvinylether/maleic anhydride) (Woolfson et al., 1998).

12.6.1 Bioadhesive Devices for the Oral Cavity

The oral cavity is a convenient and accessible area, ideally suited to bioadhesive drug delivery. The most commonly used areas are the buccal and sublingual areas. The non-keratinized regions in the oral cavity, such as the soft palate, the mouth floor, the ventral side of the tongue, and the buccal mucosa also offer least resistance to drug absorption (Leung and Robinson, 1992). In many instances, these areas are most suitable for locating bioadhesives. Molecular transport across the barrier membrane is achieved by simple diffusion along a concentration gradient from carrier to tissue, possibly aided by some form of penetration enhancement. Larger, hydrophilic molecules encounter greater resistance to diffusion and are believed to cross the oral epithelium by intercellular pathways. The advantages of drug delivery through the oral mucosa have been detailed by Veillard et al. (1987) and include bypassing hepatic first-pass metabolism, excellent accessibility, unidirectional drug flux, and improved barrier permeability compared, for example, to intact skin.

Many drugs, such as glyceryl trinitrate, testosterone, and buprenorphine (Nagai and Machida, 1985), have been delivered via the buccal route. Absorption is rapid and drains into the reticulated vein. This avoids hepatic first-pass metabolism and intestinal enzymatic attack on the drug species. Because of its accessibility, the buccal area offers excellent patient compliance in comparison to epithelial routes that involve insertion of oleaginous devices, as sometimes found in rectal and vaginal delivery.

The process of drug absorption from conventional oral delivery systems normally occurs from saliva after dissolution from the formulation. Bioadhesive devices differ in that drug diffuses directly through the swollen polymer and then into the membrane. The intimacy of contact concentrates the drug on the epithelial surface. At other mucosal surfaces, the activity of goblet cells builds up a diffusion-limiting mucus layer that can impede drug absorption. This potential problem does not manifest itself at the oral mucosa, which contains no goblet cells, the mouth receiving its mucus primarily from the parotid, submaxillary, and sublingual salivary glands (Veillard et al., 1987).

Perioli et al. (2008) studied the influence of compression force tablet behavior and drug release rate for mucoadhesive buccal tablets. Several tablet batches were produced by varying the compression force and by using hydroxyethyl cellulose (HEC) and Carbopol 940 in a 1:1 ratio as matrix-forming polymers. All the tablets hydrated quickly and their high hydration percentage showed that the compression forces used did not significantly affect water penetration and polymer chain stretching. Mucoadhesion performance and drug release were mainly influenced by compression force; its increase produced higher ex vivo and in vivo mucoadhesion. In vitro and in vivo drug release were both observed to decrease with increase of compression force. However, tablets fabricated by using the lowest compression force showed the best in vivo mucoadhesive time and hydrated faster when compared to the others. Tablets prepared with the highest forces caused pain during in vivo application and gave rise to irritation, needing to be detached by human volunteers. Tablets prepared with the lowest force gave the best results, as they were able to produce the highest drug salivary concentration and no pain. All tablets exhibited an anomalous release mechanism.

The buccal route may be an attractive site for peptide delivery because it is deficient in enzymatic degradation pathways. The cavity is lined with a relatively thick mucous membrane which is highly vascularized and approximately $100 \, cm^2$ in area (Leung and Robinson, 1992). Prompted by studies showing rectal absorption of insulin in oleaginous vehicles, various workers tried, largely unsuccessfully, to achieve pharmacologically active blood levels of insulin using bioadhesive buccal patches (Ishida et al., 1981; Nagai, 1985, 1986a,b; Nagai and Machida, 1985). Recently, Cui et al. (2009) have described a delivery system based on a mucoadhesive layer (chitosan [CA]–ethylenediaminetetraacetic acid) containing insulin and an impermeable protective layer made of ethylcellulose. In vitro mucoadhesion studies showed that the mucoadhesive force of the hydrogel remained over $17,000 \, N/m^2$ during 4 h in a simulated oral cavity. The insulin-loaded bilaminate film showed a pronounced hypoglycemic effect following buccal administration to healthy rats, despite only achieving a 17% bioavailability compared with a subcutaneous insulin injection.

Oral mucosal ulceration is a common condition with up to 50% of healthy adults suffering from recurrent minor mouth ulcers (aphthous stomatitis). Shemer et al. (2008) evaluated the efficacy and tolerability of a mucoadhesive patch compared with a pain-relieving oral solution for the treatment of aphthous stomatitis. Patients with active aphthous stomatitis were randomly treated either once a day with a mucoadhesive patch containing citrus oil and magnesium salts ($n = 26$) or three times a day with an oral solution containing benzocaine and compound benzoin tincture ($n = 22$). All patients were instructed to apply the medication until pain had resolved, and completed a questionnaire detailing multiple clinical parameters followed by an evaluation of the treatment. The mucoadhesive patch was found to be more effective than the oral solution in terms of healing time and pain intensity after 12 and 24 h. Local adverse effects 1 h after treatment were significantly less frequent among the mucoadhesive patch patients compared with the oral solution patients.

Donnelly et al. (2007) reported on a mucoadhesive patch containing toluidine blue O (TBO), as a potential delivery system for use in photodynamic antimicrobial chemotherapy (PACT) of oropharyngeal candidiasis. Patches prepared from aqueous blends of poly(methyl vinyl ether/maleic anhydride) and tripropylene glycol methyl ether possessed suitable properties for use as mucoadhesive drug delivery systems and were capable of resisting dissolution when immersed in artificial saliva. When releasing directly into an aqueous sink, patches containing 50 and 100 mg TBO/cm^2 both generated receiver compartment concentrations exceeding the concentration (2.0–5.0 mg/mL) required to produce high levels of kill (>90%) of both planktonic and biofilm-grown *Candida albicans* upon illumination. However, the concentrations of TBO in the receiver compartments separated from patches by membranes intended to mimic biofilm structures were an order of magnitude below those inducing high levels of kill, even after 6 h release. Therefore, the authors concluded that short application times of TBO-containing mucoadhesive patches should allow treatment of recently acquired oropharyngeal candidiasis, caused solely by planktonic cells. Longer patch application times may be required for persistent disease where biofilms are implicated.

Diseases of the oral cavity may be broadly differentiated into two categories, namely, inflammatory and infective conditions. In many cases, the demarcation between these two categories is unclear as some inflammatory diseases may be microbiological in origin (Addy, 1994). Most frequently, therapeutic agents are delivered into the oral cavity in the form of solutions and gels, as this ensures direct access of the specific agent to the required site in concentrations that vastly exceed those that may be achieved using systemic administration (Addy, 1994; Jones et al., 1997). However, the retention of such formulations within the oral cavity is poor, due, primarily, to their inability to interact with the hard and soft tissues and, additionally, to overcome the flushing actions of saliva. Therefore, the use of bioadhesive formulations has been promoted by several authors to overcome these problems and hence improve the clinical resolution of superficial diseases of the oral cavity (Gandhi and Robinson, 1994; Jones et al., 1997). In particular, within the oral cavity, bioadhesive formulations have been reported for the treatment of periodontal diseases and superficial oral infection, in which specific interactions between the bioadhesive formulations and the oral mucosa may be utilized to "anchor" the formulations to the site of application.

Periodontitis is an inflammatory disease of the oral cavity which results in the destruction of the supporting structures of the teeth (Medlicott et al., 1994). It is characterized by the formation of pockets between the soft tissue of the gingiva and the tooth and which, if untreated, may result in tooth loss (Medlicott et al., 1994; Jones et al., 1996). Drug delivery problems associated with the periodontal pocket may be overcome by the use of novel, bioadhesive, syringeable semisolid systems (Jones et al., 1996, 1997). Such systems may be formulated to exhibit requisitory flow properties (and hence may be easily administered into the periodontal pocket using a syringe), mucoadhesive properties (ensuring prolonged retention within the pocket), and sustained release of therapeutic agent within this environment. In a series of papers, Andrews et al., (2004); Bruschi et al., (2008); Jones et al. (2004, 2008, 2009) described the formulation and physicochemical characterization of syringeable semisolid, bioadhesive networks (containing tetracycline, metronidazole, or model protein drugs). Upon contact with mucus, water diffuses into these formulations and controls swelling of the bioadhesive polymers which, in turn, allows interpenetration of the fluidized polymer chains with mucus and, hence, ensures physical and chemical adhesion. The authors concluded that, when used in combination with mechanical treatments, antimicrobial-containing bioadhesive semisolid systems described in these studies would augment periodontal therapy by improving the removal of pathogens and, hence, enhance periodontal health.

12.6.2 BIOADHESIVE DEVICES FOR THE GIT

Bioadhesive polymers may provide useful delivery systems for drugs that have limited bioavailability from more conventional dosage forms. To attain a once-daily dosing strategy using peroral bioadhesion, it is desirable for a dosage form to attach itself to the mucosa of the GIT. If attachment is successful in the gastric region, then a steady supply of drug is available to the intestinal tract for absorption. However, gastric motility and muscular contractions will tend to dislodge any such device. Motility in the GIT during the fasted state is known as the interdigestive migrating motor complex (IMMC) (Leung and Robinson, 1992). During certain phases of the IMMC, a "house-keeper wave" migrates from the foregut to the terminal ileum and is intended to clear all nondigestible items from the gut. The strength of this "housekeeper wave" is such that poorly adhered dosage forms are easily removed and only strong bioadhesives will be of any practical use.

In addition to physical abrasion and erosion exerted on a bioadhesive dosage form, the mucin to which it is attached turns over quickly, especially in the gastric region. Any device that has lodged to surface mucus will be dislodged as newly synthesized mucus displaces the older surface layers. The acid environment will also affect bioadhesion, especially with polyacid polymers, such as PAA, where the mechanism of bond formation is thought to occur chiefly through hydrogen bonding and electrostatic interactions.

Liu et al. (2005) prepared amoxicillin mucoadhesive microspheres using ethylcellulose as matrix and Carbopol 934P (C934P) as mucoadhesive polymer for the potential use of treating gastric and duodenal ulcers, which were associated with *Helicobacter pylori*. It was found that amoxicillin stability at low pH was enhanced when entrapped within the microspheres. In vitro and in vivo mucoadhesive tests showed that the amoxicillin-containing mucoadhesive microspheres adhered more strongly to the gastric mucous layer than nonadhesive amoxicillin microspheres did and could be retained in the GIT for an extended period of time. Amoxicillin-containing mucoadhesive microspheres and amoxicillin powder were orally administered to rats. The amoxicillin concentration in gastric tissue was higher in the mucoadhesive microspheres group. In vivo *H. pylori* clearance tests were also carried out by administering, respectively, amoxicillin-containing mucoadhesive microspheres or amoxicillin powder to *H. pylori*–infected BALB/c mice under fed conditions at single or multiple oral dose(s). The results showed that amoxicillin-containing mucoadhesive microspheres had a better clearance effect than amoxicillin powder did. The authors concluded that the prolonged gastrointestinal residence time and enhanced amoxicillin stability observed with the mucoadhesive microspheres of amoxicillin might make a useful contribution to *H. pylori* clearance.

Any swallowed dosage form will be subjected to shear forces as the intestinal contents move. This motility will either prevent the attachment of the bioadhesive or attempt to remove it if bonding to the gut wall has occurred. Previous attempts to reduce the GIT transit time of normal dosage forms have included devices that swelled and floated on the stomach contents, their increasing size preventing them from passing through the pylorus.

Ahmed and Ayres (2007) studied gastric retention formulations made of naturally occurring carbohydrate polymers and containing riboflavin in vitro for swelling and dissolution characteristics as well as in fasting dogs for gastric retention. The bioavailability of riboflavin, from the gastric retention formulations (GRFs), was studied in fasted healthy humans and compared to an immediate release formulation. It was found that when the GRFs were dried and immersed in gastric juice, they swelled rapidly and released their drug payload in a zero-order fashion for a period of 24 h. In vivo studies in dogs showed that a rectangular shaped GRF stayed in the stomach of fasted dogs for more than 9 h, then disintegrated and reached the colon in 24 h. Endoscopic studies in dogs showed that the GRF hydrates and swells back to about 75% of its original size in 30 min. Pharmacokinetic parameters, determined from urinary excretion data from six human subjects under fasting conditions, showed that bioavailability depended on the size of the GRF. Bioavailability of riboflavin from a large size GRF was more than triple than that measured after administration of an immediate release formulation. In vivo studies suggested that the large size GRF stayed in the stomach for about 15 h.

Li et al. (2007) investigated distribution, transition, bioadhesion, and release behavior of insulin-loaded pH-sensitive NP in the gut of rats, as well as the effects of a viscosity-enhancing agent. Insulin was labeled with fluorescein isothiocyanate (FITC). The FITC–insulin solution and FITC–insulin nanoparticulate aqueous dispersions, with or without hydropropylmethylcellulose (HPMC, 0.2%,

0.4%, or 0.8% [w/v]), were orally administered to rats. The amounts of FITC–insulin in both the lumen content and the intestinal mucosa were quantified. The release profiles in the gut were plotted by the percentages of FITC–insulin released versus time. FITC–insulin NP aqueous dispersion showed similar stomach, but lower intestinal, empty rates and enhanced intestinal mucosal adhesion in comparison with FITC–insulin solution. Addition of HPMC reduced the stomach and intestinal empty rates and enhanced adhesion of FITC–insulin to the intestinal mucosa. Release of FITC–insulin from NP in the gut showed an S-shaped profile and addition of HPMC prolonged the release half-life from 0.77 to 1.51 h. It was concluded that the behaviors of pH-sensitive NP tested in the GIT of rats and the addition of HPMC were favorable to the absorption of the incorporated insulin.

The oral route constitutes the preferred route for drug delivery. However, numerous drugs remain poorly available when administered orally. Drugs associated with bioadhesive polymeric nanoparticulates or small particles in the micrometer size range may be advantageous in this respect due to their ability to interact with the mucosal surface. Targeting applications are possible by this method if there are specific interactions occurring when a ligand attached to the particle is used for the recognition and attachment to a specific site at the mucosal surface (Ponchel and Irache, 1998).

Salman et al. (2007) aimed to develop polymeric nanoparticulate carriers with bioadhesive properties and to evaluate their adjuvant potential for oral vaccination. Thiamine was used as a specific ligand–nanoparticle conjugate (TNP) to target specific sites within the GIT, namely enterocytes and Peyer's patches. The affinity of NP to the gut mucosa was studied in orally inoculated rats. In contrast to conventional noncoated NP, higher levels of TNP were found in the ileum tissue, showing a strong capacity to be captured by Peyer's patches. The adhesion of TNP was found to be three times higher than for control NP. To investigate the adjuvant capacity of TNP, ovalbumin (OVA) was used as a standard antigen. Oral immunization of BALB/c mice with OVA-TNP induced higher serum titers of specific IgG2a and IgG1 and mucosal IgA compared to OVA-NR. This mucosal immune response (IgA) was about four titers higher than that elicited by OVA-NP. The authors concluded that thiamine-coated NP showed promise as particulate vectors for oral vaccination and immunotherapy.

Rastogi et al. (2007) formulated spherical microspheres able to prolong the release of isoniazid (INH) by a modified emulsification method, using sodium alginate as the hydrophilic carrier. The release profiles of INH from the microspheres were examined in simulated gastric fluid (SGF, pH 1.2) and simulated intestinal fluid (SIF, pH 7.4). Gamma scintigraphic studies were carried out to determine the location of microspheres on oral administration to rats and the extent of transit through the GIT. The microspheres had smooth surfaces and were found to be discreet and spherical in shape. The particles were heterogeneous with the largest average diameter of 3.7 μm. Results indicated that the mean particle size of the microspheres increased with an increase in the concentration of polymer and cross-linker, as well as cross-linking time. The entrapment efficiency was found to be in the range of 40%–91%. Concentrations of the cross-linker up to 7.5% w/w increased entrapment efficiency and the extent of drug release. Optimized INH-alginate microspheres were found to possess good bioadhesion, which resulted in prolonged retention in the small intestine. Microspheres could be observed in the intestinal lumen at 4 h and were still detectable at lower levels in the intestine 24 h post-oral administration. Approximately 26% of the INH loading was released in SGIF pH 1.2 in 6 h and 71.25% in SIF pH 7.4 in 30 h.

In general, oral applications of bioadhesive technology have been less successful than where the delivery system can be applied directly to an accessible site. Issues such as overhydration, attachment to gastrointestinal mucus, and subsequent shedding and poor resistance to mechanical forces in the GIT have tended to limit the utility of bioadhesive gastric retentive systems to date.

12.6.3 BIOADHESIVE DEVICES FOR RECTAL DRUG DELIVERY

The function of the rectum is mostly concerned with removing water. It is only 10 cm in length, with no villi, giving it a relatively small surface area for drug absorption (Leung and Robinson, 1992).

Drug permeability differs from that found in both the oral cavity and intestinal regions. Most rectal absorption of drugs is achieved by a simple diffusion process through the lipid membrane. However, the rectal route is readily accessible, penetration enhancers can be used, and there is access to the lymphatic system. In contrast, drugs absorbed via the small intestine are transported mostly to the blood with only a small proportion entering the lymphatic system.

Drugs that are liable to extensive first-pass metabolism can benefit greatly if delivered to the rectal area, especially if they are targeted to areas close to the anus. This is because the blood from the lower rectum drains directly into the systemic circulation, whereas blood from the upper regions drain into the portal systems via the superior hemorrhoidal vein and the inferior mesenteric vein. Drug absorbed from this upper site is subjected to liver metabolism. A bioadhesive suppository will attach to the lower rectal area and once inserted will reduce the tendency for migration upward to the upper rectum; this migration can occur with conventional suppositories.

Kim et al. (2009) aimed to develop a thermoreversible flurbiprofen liquid suppository base composed of poloxamer and sodium alginate for improvement of rectal bioavailability of flurbiprofen. Cyclodextrin derivatives, such as α-, β-, γ-cyclodextrin and hydroxypropyl-β-cyclodextrin (HP-β-CD), were used to enhance the aqueous solubility of flurbiprofen. The effects of HP-β-CD and flurbiprofen on the physicochemical properties of liquid suppository were then investigated. Pharmacokinetic studies were performed after rectal administration of flurbiprofen liquid suppositories with and without HP-β-CD or after intravenous administration of a commercially available product (Lipfen®, flurbiprofen axetil-loaded emulsion) to rats, and their pharmacokinetic parameters were compared. HP-β-CD decreased the gelation temperature and reinforced the gel strength and bioadhesive force of liquid suppositories, while the opposite was true for flurbiprofen. Thermoreversible flurbiprofen liquid suppositories demonstrated physicochemical properties suitable for rectal administration. The flurbiprofen liquid suppository with HP-β-CD showed significantly higher plasma levels, AUC, and C_{max} for flurbiprofen than those of the liquid suppository without HP-β-CD, indicating that flurbiprofen could be well-absorbed, due to the enhanced solubility by formation of an inclusion complex. Moreover, the flurbiprofen liquid suppository containing HP-β-CD showed an excellent bioavailability in that the AUC of flurbiprofen after its rectal administration was not significantly different from that after intravenous administration of Lipfen. The authors concluded that HP-β-CD could be a preferable solubility enhancer for the development of liquid suppositories containing poorly water-soluble drugs.

Uchida et al. (2001) prepared insulin-loaded acrylic hydrogel formulations containing various absorption enhancers, performed in vitro and in vivo characterization of these formulations, and evaluated the factors affecting insulin availability upon rectal delivery. The acrylic block copolymer of methacrylic acid and methacrylate, Eudispert®, was used to make the hydrogel formulations. As absorption enhancers, 2,6-di-O-methyl-β-cyclodextrin (DM-β-CyD), lauric acid (C_{12}), or the sodium salt of C_{12} (C_{12}Na) was incorporated into the hydrogels. The in vitro release rate of insulin from the hydrogels decreased as polymer concentration was increased. The addition of C_{12}Na further increased release rate, which was greater at higher concentrations of the enhancer. Serum insulin levels were determined at various time points after the administration of insulin solution or insulin-loaded (50 units/kg body weight) Eudispert hydrogels containing 5% w/w of C_{12}, C_{12}Na, or DM-β-CyD to in situ loops in various regions of the rat intestine. The most effective enhancement of insulin release was observed with formulations containing C_{12}Na. The bioavailability of insulin from the hydrogels was lower than that from the insulin solutions, however. Hydrogel formulations containing 7% or 10% w/w Eudispert remained in the rectum for 5 h after rectal administration. However, the 5% w/w C_{12}Na solution stained with Evan's blue had diffused out and the dye had reached the upper intestinal tract within 2 h. Finally, the rectal administration of insulin-loaded hydrogels containing 4%, 7%, or 10% w/w Eudispert and 5% w/w enhancer (C_{12}, C_{12}Na, or DM-β-CyD) to normal rats was shown to decrease serum glucose concentrations. However, despite such promising results, bioadhesives have not been extensively employed for rectal delivery.

12.6.4 BIOADHESIVE DEVICES FOR CERVICAL AND VULVAL DRUG DELIVERY

Nagai (1986a,b) has investigated various bioadhesive forms to treat uterine and cervical cancers. Uterine cancers comprise about 25% of all malignant tumors in Japan, among which carcinoma colli accounts for 95% of this figure (Machida et al., 1979). The target cells remain at, or near, the cervical epithelium and can be readily targeted with an appropriate dosage form. Pessaries have been of limited use because drug release is rapid and leakage into the surrounding tissue causes inflammation of vaginal mucosa. In order to meet three important criteria, drug release, swelling of the preparation, and adhesion to the diseased tissue, a bioadhesive disk was prepared. Bleomycin was incorporated into a hydroxypropylcellulose/Carbopol mix and molded into a suitable shape. This could be placed on or into the cervical canal, where it adhered and released the cytotoxic drug. After treatment and following colposcopic examination, areas of necrosis on the lesion were observed, with surrounding normal mucosal cells unaffected. With pessaries, however, this was not the case. In approximately 33% of cases, cancerous foci had completely disappeared.

Woolfson et al. (1995a,b) described a novel bioadhesive cervical patch containing 5-fluorouracil for the treatment of cervical intraepithelial neoplasia (CIN). The patch was of bilaminar design, with a drug-loaded bioadhesive film cast from a gel containing 2% w/w Carbopol 981 plasticized with 1%w/w glycerin. The casting solvent was ethanol/water 30:70, chosen to give a nonfissuring film with an even particle size distribution. The film, which was mechanically stable on storage under ambient conditions, was bonded directly to a backing layer formed from thermally cured poly(vinyl chloride) emulsion. Bioadhesive strength was independent of drug loading in the bioadhesive matrix over the range investigated but was influenced by both the plasticizer concentration in the casting gel and the thickness of the final film. Release of 5-fluorouracil from the bioadhesive layer into an aqueous sink was rapid but was controlled down to an undetectable level through the backing layer. The latter characteristic was desirable to prevent drug spill from the device onto vaginal epithelium in vivo. Despite the relatively hydrophilic nature of 5-fluorouracil, substantial drug release through human cervical tissue samples was observed over approximately 20 h. Drug release, which was clearly tissue rather than devicedependent, may have been aided by a shunt diffusion route through aqueous pores in the tissue. The bioadhesive and drug release characteristics of the 5-fluorouracil cervical patch indicated that it would be suitable for further clinical investigation as a drug treatment for CIN (Sidhu et al., 1997).

Donnelly et al. (2009) described the design, physicochemical characterization, and clinical evaluation of bioadhesive drug delivery systems for photodynamic therapy of difficult-to-manage vulval neoplasias and dysplasias. In photodynamic therapy (PDT), a combination of visible light and a sensitizing drug causes the destruction of selected cells. Aminolevulic acid (ALA) is commonly delivered to the vulva using creams or solutions, which are covered with an occlusive dressing to aid retention and enhance drug absorption. Such dressings are poor at staying in place at the vulva, where shear forces are high in mobile patients. To overcome the problems associated with delivery of ALA to the vulva, the authors produced a bioadhesive patch by a novel laminating procedure. The ALA loading was 38 mg/cm^2. Patches were shown to release more ALA over 6 h than the proprietary cream (Porphin®, 20% w/w ALA). The ALA concentration in excised tissue at a depth of 2.375 mm following application of the patch was an order of magnitude greater than that found to be cytotoxic to HeLa cells in vitro. Clinically, the patch was extensively used in successful PDT of vulval intraepithelial neoplasia, lichen sclerosus, squamous hyperplasia, Paget's disease, and vulvodynia.

12.6.5 BIOADHESIVE DEVICES FOR VAGINAL DRUG DELIVERY

Bioadhesives can control the rate of drug release from, and extend the residence time of, vaginal formulations. These formulations may contain drug or, quite simply, act in conjunction with moisturizing agents as a control for vaginal dryness.

Alam et al. (2007) developed an acid-buffering bioadhesive vaginal tablet for the treatment of genitourinary tract infections. From bioadhesion experiment and release studies, it was found that polycarbophil and sodium carboxymethylcellulose was a good combination for an acid-buffering bioadhesive vaginal tablet. Sodium monocitrate was used as a buffering agent to provide an acidic pH (4.4), which is an attribute of a healthy vagina. The effervescent mixture (citric acid and sodium bicarbonate) along with a super disintegrant (Ac-Di-sol) was used to enhance the swellability of the bioadhesive tablet. The drugs clotrimazole (antifungal) and metronidazole (antiprotozoal and antibacterial) were used in the formulation along with *Lactobacillus acidophilus* spores to treat mixed vaginal infections. From ex vivo retention studies, it was found that the bioadhesive polymers held the tablet for more than 24 h inside the vagina. The hardness of the acid-buffering bioadhesive vaginal tablet was optimized, at 4 to 5 kg hardness, the swelling was found to be good and the cumulative release profile of the developed tablet was matched with a marketed conventional tablet (Infa-V®). The in vitro spreadability of the swelled tablet was comparable to the marketed gel. In the in vitro antimicrobial study, it was found that the acid-buffering bioadhesive tablet produced better antimicrobial action than marketed intravaginal drug delivery systems (Infa-V, Candid-V®, and Canesten®1).

Cevher et al. (2008) aimed to prepare clomiphene citrate (CLM) gel formulations possessing appropriate mechanical properties, exhibiting good vaginal retention, and providing sustained drug release for the local treatment of human papilloma virus infections. In this respect, 1% w/w CLM gels including PAA polymers such as C934P, Carbopol 971P (C971P), Carbopol 974P (C974P) in various concentrations, and their conjugates containing thiol groups, were prepared. Based on obtained data, gel formulations containing C934P and its conjugate had appropriate hardness and compressibility to be applied to the vaginal mucosa and showed the highest elasticity and good spreadability. Such gels also exhibited the highest cohesion. The mucoadhesion of the gels changed significantly depending on the polymer type and concentration. Addition of conjugates containing thiol groups caused an increase in mucoadhesion. Gels containing C934P–Cys showed the highest adhesiveness and mucoadhesion. A significant decrease was observed in drug release from gel formulations as the polymer concentration increased.

12.6.6 Bioadhesive Devices for Nasal Drug Delivery

The area of the normal human nasal mucosa is approximately 150 cm². The nasal mucosa and submucosa are liberally populated with goblet cells along with numerous mucous and serous glands. These keep the nasal mucosal surfaces moist. Drug administration to this region is normally reserved for local treatment, such as nasal allergy or inflammation. The nasal mucosa is thin and incorporates a dense vascular network, indicating that drug absorption may be good from this site (Igawa et al., 1990). Absorbed drugs avoid first-pass metabolism and lumenal degradation associated with the oral route (Leung and Robinson, 1992). The nasal mucosa itself is sensitive to drug molecules and to surfactants which are often used to enhance drug absorption. Moreover, the mucociliary escalator travels at 5 mm/min as it drags mucous fluid backward toward the throat. Therefore, if a drug is applied in either a simple powder or liquid formulation it will be quickly cleared from this site of absorption. Thus, any useful dosage form must be nonirritant to the nasal mucosa and be retained for extended periods of time.

Charlton et al. (2007) studied the effect of bioadhesive formulations on the direct transport of an angiotensin antagonist drug from the nasal cavity to the central nervous system in a rat model. Three different bioadhesive polymer formulations (3% w/w pectin, 1.0% w/w pectin, and 0.5% w/w CA) containing the drug were administered nasally to rats by inserting a dosing cannula 7 mm into the nasal cavity after which the plasma and brain tissue levels were measured. It was found that the polymer formulations provided significantly higher plasma levels and significantly lower brain tissue levels of drug than a control, in the form of a simple drug solution. Changing the depth of insertion of the cannula from 7 to 15 mm in order to reach the olfactory region in the nasal cavity significantly decreased plasma levels and significantly increased brain tissue levels of drug for the two formulations studied (1.0% w/w pectin and a simple drug solution). There was no significant difference between the drug availability for the bioadhesive formulation and the control in the brain when the longer cannula was used for administration. The authors suggested that the conventional rat model is not suitable for evaluation of the effects of bioadhesive formulations in nose-to-brain delivery.

Nasal delivery of protein and peptide therapeutics can be compromised by the brief residence time at this mucosal surface. Some bioadhesive polymers have been suggested to extend residence time and improve protein uptake across the nasal mucosa. McInnes et al. (2007) quantified nasal residence of bioadhesive formulations using gamma scintigraphy and investigated absorption of insulin. A four-way crossover study was conducted in six healthy male volunteers, comparing a conventional nasal spray solution with three lyophilized nasal insert formulations (1%–3% w/w hydroxypropylmethylcellulose [HPMC]). The conventional nasal spray deposited in the posterior nasal cavity in only one instance, with a rapid clearance half-life of 9.2 min. The nasal insert formulations did not enhance nasal absorption of insulin. However, an extended nasal residence time of 4–5 h was observed for the 2% w/w HPMC formulation. The 1% w/w HPMC insert initially showed good spreading behavior. However, clearance was faster than for the 2% w/w formulation. The 3% w/w HPMC nasal insert showed no spreading and was usually cleared intact from the nasal cavity within 90 min. The authors concluded that the 2% w/w HPMC lyophilized insert formulation achieved extended nasal residence, demonstrating an optimum combination of rapid adhesion without overhydration.

Coucke et al. (2009) studied viscosity-enhancing mucosal delivery systems for the induction of an adaptive immune response against viral antigen. Powder formulations based on spray-dried mixtures of starch (Amioca®) and PAA (C974P) in different ratios were used as carriers of the viral antigen. A comparison of these formulations for intranasal delivery of heat-inactivated influenza virus combined with LTR 192G adjuvant was made in vivo in a rabbit model. Individual rabbit sera were tested for seroconversion against hemagglutinin (HA), the major surface antigen of influenza. The powder vaccine formulations were able to induce systemic anti-HA IgG responses. The presence of C974P improved the kinetics of the immune responses and the level of IgG titers in a dose-dependent way which was correlated with moderately irritating capacities of the formulation. In contrast, mucosal IgA responses were not detected. The authors concluded that the use of bioadhesive carriers based on starch and PAA facilitates the induction of a systemic anti-HA antibody response after intranasal vaccination with a whole virus influenza vaccine.

Overall, the nasal route remains highly promising, particularly for macromolecular absorption, but there are problems regarding effects of bioadhesive delivery systems on nasal cilial beat.

12.6.7 BIOADHESIVE DEVICES FOR OCULAR DRUG DELIVERY

Extended drug delivery to the eye is difficult for several reasons. Systemically administered drugs must cross the blood–aqueous humor barrier and to achieve local therapeutic concentrations of drug, high levels of drug in the blood are needed. In addition, lacrimation, blinking, and tear turnover will all reduce the bioavailability of topically administered drug to approximately 1%–10% (Robinson, 1989). Conventional delivery methods are not ideal. Solutions and suspensions are readily washed from the cornea and ointments alter the tear refractive index and blur vision.

Sensoy et al. (2009) aimed to prepare bioadhesive sulfacetamide sodium microspheres to increase residence time on the ocular surface and to enhance treatment efficacy of ocular keratitis. Microspheres were fabricated by a spray-drying method using a mixture of polymers, such as pectin, polycarbophil, and HPMC at different ratios. A sulfacetamide sodium-loaded polycarbophil microsphere formulation with a polymer/drug ratio of 2:1 was found to be the most suitable for ocular application and used in in vivo studies on New Zealand male rabbit eyes with keratitis caused by *Pseudomonas aeruginosa* and *Staphylococcus aureus*. Sterile microsphere suspension in light mineral oil was applied to infected eyes twice a day. Plain sulfacetamide sodium suspension was used as a positive control. On the third and sixth days, the eyes were examined in respect to clinical signs of infection (blepharitis, conjunctivitis, iritis, corneal edema, and corneal infiltrates) and then cornea samples were counted microbiologically. The rabbit eyes treated with microspheres demonstrated significantly lower clinical scores than those treated with sulfacetamide sodium alone. A significant decrease in the number of viable bacteria in eyes treated with microspheres was observed in both infection models when compared to those treated with sulfacetamide sodium alone.

Gene transfer is considered to be a promising alternative for the treatment of several chronic diseases that affect the ocular surface. de la Fuente et al. (2009) investigated the efficacy and mechanism of action of a bioadhesive DNA nanocarrier made of hyaluronan (HA) and CS, specifically designed for topical ophthalmic gene therapy. The authors first evaluated the transfection efficiency of the plasmid DNA-loaded NP in a human corneal epithelium cell model. Then they investigated the bioadhesion and internalization of the NP in the rabbit ocular epithelia and determined the in vivo efficacy of the nanocarriers in terms of their ability to transfect ocular tissues. The results showed that HA–CS NP and, in particular, those made of low molecular weight CS (10–12 kDa) led to high levels of expression of secreted alkaline phosphatase in the human corneal epithelium model. In addition, following topical administration to rabbits, the NP entered the corneal and conjunctival epithelial cells and become assimilated by the cells. More importantly, the NP provided an efficient delivery of the associated plasmid DNA inside the cells, reaching significant transfection levels.

12.7 CONCLUSION

Bioadhesive drug delivery systems have been widely studied in the past two decades and a number of interesting, novel drug delivery systems incorporating the use of bioadhesive polymers have been described. Primarily, the application of bioadhesion to drug delivery involves the process of mucoadhesion, a "wet-stick" adhesion requiring, primarily, spreading of the mucoadhesive on a mucin-coated epithelium, followed by interpenetration of polymer chains between the hydrated delivery system and mucin. An alternative mechanism can involve the use of specific bioadhesive molecules, notably lectins, primarily for oral drug delivery applications. Results, however, remain variable by this specific method, as with the use of bioadhesives, generally, for oral systemic drug delivery. The most notable applications of mucoadhesive systems to date remain those involving accessible epithelia, such as in ocular, nasal, buccal, rectal, or intravaginal drug delivery systems. The ancillary effects of certain mucoadhesive polymers, such as polycations, on promoting drug penetration of epithelial tissues, may be increasingly important in the future, particularly for the transmucosal delivery of therapeutic peptides and other biomolecules.

REFERENCES

Addy, M. (1994). Local delivery of antimicrobial agents to the oral cavity. *Advanced Drug Delivery Reviews* 13, 123–134.

Ahmed, I.S. and Ayres, J.W. (2007). Bioavailability of riboflavin from a gastric retention formulation. *International Journal of Pharmaceutics* 330, 146–154.

Ahuja, A., Khar, R.K., and Ali, J. (1997). Mucoadhesive drug delivery systems. *Drug Development and Industrial Pharmacy* 23, 489–515.

Alam, M.A., Ahmad, F.J., Khan, Z.I., Khar, R.K., and Ali, M. (2007). Development and evaluation of acid-buffering bioadhesive vaginal tablet for mixed vaginal infections. *AAPS PharmSciTech* 8, 109.

Anders, R. and Merkle, H.P. (1989). Evaluation of laminated mucoadhesive patches for buccal drug delivery. *International Journal of Pharmaceutics* 49, 231–240.

Andrews, G.P., Jones, D.S., Redpath, J.M., and Woolfson, A.D. (2004). Characterisation of protein-containing binary polymeric gel systems designed for the treatment of periodontal disease. *Journal of Pharmacy and Pharmacology* 56 (Suppl. S), S71–S72.

Andrews, G.P., Laverty, T.P., and Jones, D.S. (2009). Mucoadhesive polymeric platforms for controlled drug delivery. *European Journal of Pharmaceutics and Biopharmaceutics* 71, 505–518.

ASTM Designation D. (1984). American Society for Testing and Materials, West Conshohocken, PA, pp. 907–977.

Banchonglikitkul, C., Smart, J.D., Gibbs, R.V., Donovan, S.J., and Cook, D.J. (2002). An in vitro evaluation of lectin cytotoxicity using cell lines derived from the ocular surface. *Journal of Drug Targeting* 10, 601–606.

Bell, G.I., Dembo, M., and Bongrand, P. (1984). Cell adhesion: Competition between non-specific repulsion and specific bonding. *Biophysics Journal* 45, 1051–1064.

BlancoFuente, H., AnguianoIgea, S., OteroEspinar, F.J., and BlancoMendez, J. (1996). In-vitro bioadhesion of Carbopol hydrogels. *International Journal of Pharmacuetics* 142, 169–174.

Bouckaert, S. and Remon, J.P. (1993). In-vitro bioadhesion of a buccal, miconazole slow-release tablet. *Journal of Pharmacy and Pharmacology* 45, 504–507.

Bruschi, M.L., de Freitas, O., Lara, E.H.G., Panzeri, H., Gremiao, M.P.D., and Jones, D.S. (2008). Precursor system of liquid crystalline phase containing propolis microparticles for the treatment of periodontal disease: Development and characterization. *Drug Development and Industrial Pharmacy* 34, 267–278.

Cevher, E., Taha, M.A.M., Orlu, M., and Araman, A. (2008). Evaluation of mechanical and mucoadhesive properties of clomiphene citrate gel formulations containing carbomers and their thiolated derivatives. *Drug Delivery* 15, 57–67.

Charlton, S.T., Davis, S.S., and Illum, L. (2007). Nasal administration of an angiotensin antagonist in the rat model: Effect of bioadhesive formulations on the distribution of drugs to the systemic and central nervous systems. *International Journal of Pharmaceutics* 338, 94–103.

Ch'ng, H.S., Park, H., Kelly, P., and Robinson, J.R. (1985). Bioadhesive polymers as platforms for oral controlled drug delivery. II Synthesis and evaluation of some swelling, water-insoluble bioadhesive polymers. *Journal of Pharmaceutical Sciences* 74, 399–405.

Coucke, D., Schotsaert, M., Libert, C., Pringels, E., Vervaet, C., Foreman, P., Saelens, X., and Remon, J.P. (2009). Spray-dried powders of starch and crosslinked poly(acrylic acid) as carriers for nasal delivery of inactivated influenza vaccine. *Vaccine* 27, 1279–1286.

Craig, D.Q.M. and Tamburic, S. (1997). Dielectric analysis of bioadhesive gel systems. *European Journal of Pharmaceutics and Biopharmaceutics* 44, 61–70.

Cui, F.Y., He, C.B., He, M., Tang, C., Yin, L.C., Qian, F., and Yin, C.H. (2009). Preparation and evaluation of chitosan–ethylenediaminetetraacetic acid hydrogel films for the mucoadhesive transbuccal delivery of insulin. *Journal of Biomedical Materials Research Part A* 89A, 1063–1071.

Davis, S.S. (1985). The design and evaluation of controlled release systems for the gastro-intestinal tract. *Journal of Controlled Release* 2, 27–38.

de la Fuente, M., Seijo, B., and Alonso, M.J. (2009). Bioadhesive hyaluronan–chitosan nanoparticles can transport genes across the ocular mucosa and transfect ocular tissue. *Gene Therapy* 15, 668–676.

Deraguin, B.V. and Smilga, V.P. (1969). *Adhesion: Fundamentals and Practice*. McLaren, London U.K.

Donnelly, R.F., McCarron, P.A., Tunney M.M., and Woolfson, A.D. (2007). Potential of photodynamic therapy in treatment of fungal infections of the mouth. Design and characterisation of a mucoadhesive patch containing toluidine blue O. *Journal of Photochemistry and Photobiology B: Biology* 86, 59–69.

Donnelly, R.F., McCarron, P.A., Zawislak, A.A., and Woolfson, A.D. (2006). Design and physicochemical characterisation of a bioadhesive patch for dose-controlled topical delivery of imiquimod. *International Journal of Pharmaceutics* 307, 318–325.

Donnelly, R.F., McCarron, P.A., Zawislak, A., and Woolfson, A.D. (2009). *Photodynamic Therapy of Vulval Neoplasias and Dysplasias: Design and Evaluation of Bioadhesive Photosensitiser Delivery Systems*. VDM Verlag Dr. Müller, Saarbrücken, Germany.

Duchêne, D., Touchard, F., and Peppas, N.A. (1988). Pharmaceutical and medical aspects of bioadhesive systems for drug administration. *Drug Development and Industrial Pharmacy* 14, 283–318.

Gandhi, R.B. and Robinson, J.R. (1994). The oral cavity as a site for bioadhesive drug delivery. *Advanced Drug Delivery Reviews* 13, 43–74.

Good, W.R. (1983). Transdermal nitro-controlled delivery of nitroglycerin via the transdermal route. *Drug Development and Industrial Pharmacy* 9, 647–670.

Govil, S.K. (1988). Transdermal drug delivery systems. *Drug Delivery Devices* (P. Tyle, ed.), Marcel Dekker, New York.

Gudeman, L. and Peppas, N.A. (1995). Preparation and characterisation of pH-sensitive, interpenetrating networks of poly(vinyl alcohol) and poly(acrylic acid). *Journal of Applied Polymer Science* 55, 919–928.

Gurny, R., Meyer, J.M., and Peppas, N.A. (1984). Bioadhesive intraoral release systems: Design, testing and analysis. *Biomaterials* 5, 336–340.

Henriksen, I., Green, K.L., Smart, D., Smistad, G., and Karlsen, G. (1996). Bioadhesion of hydrated chitosans: An *in-vitro* and *in-vivo* study. *Int J Pharm* 145, 231–240.

Huntsberger, J.R. (1967). Mechanisms of adhesion. *Journal of Paint Technology* 39, 199–211.

Igawa, T., Maitani, Y., Machida, Y., and Nagai, T. (1990). Intranasal administration of human fibroblast interferon in mice, rats, rabbits and dogs. *Chem Pharm Bull (Tokyo)* 38(2), 549–551.

Ishida, M., Machida, Y., Nambu, N., and Nagai, T. (1981). New mucosal dosage forms of insulin. *Chemical and Pharmaceutical Bulletin* 29, 810–816.

Jiménez-Castellanos, M.R., Zia, H., and Rhodes, C.T. (1993). Mucoadhesive drug delivery systems. *Drug Development and Industrial Pharmacy* 19, 143–194.

Jones, D.S., Bruschi, M.L., de Freitas, O., Gremiao, M.P.D., Lara, E.H.G., and Andrews, G.P. (2009). Rheological, mechanical and mucoadhesive properties of thermoresponsive, bioadhesive binary mixtures composed of Poloxamer 407 and Carbopol 974P designed as platforms for implantable drug delivery systems for use in the oral cavity. *International Journal of Pharmaceutics* 372, 49–58.

Jones, D.S. and Woolfson, A.D. (1997). Measuring sensory properties of semi-solid products using texture profile analysis. *Pharmaceutical Manufacturing Review* 9, S3–S6.

Jones, D.S., Woolfson, A.D., and Brown, A.F. (1997). Textural, viscoelastic and mucoadhesive properties of pharmaceutical gels and polymers. *International Journal of Pharmaceutics* 151, 223–233.

Jones, D.S., Lawlor, M.S., and Woolfson, A.D. (2004). Formulation and characterisation of tetracycline-containing bioadhesive polymer networks designed for the treatment of periodontal disease. *Curr Drug Deliv* 1(1), 17–25.

Jones, D.S., Woolfson, A.D., Djokic, J., and Coulter, W.A. (1996). Development and physical characterisation of bioadhesive semi-solid, polymeric systems containing tetracycline for the treatment of periodontal diseases. *Pharmaceutical Research* 13, 1732–1736.

Jones, D.S., Muldoon, B.C., Woolfson, A.D., Andrews, G.P., and Sanderson, F.D. (2008). Physicochemical characterization of bioactive polyacrylic acid organogels as potential antimicrobial implants for the buccal cavity. *Biomacromolecules* 9(2), 624–633.

Kamath, K.R. and Park, K. (1992). Mucosal adhesive preparations. *Encyclopedia of Pharmaceutical Technology* (J. Swarbrick and J.C. Boylan, eds.), Marcel Dekker, New York, pp. 133.

Kim, J.K., Kim, M.S., Park, J.S., and Kim, C.K. (2009). Thermo-reversible flurbiprofen liquid suppository with HP-beta-CD as a solubility enhancer: Improvement of rectal bioavailability. *Journal of Inclusion Phenomena and Macrocyclic Chemistry* 64, 265–272.

Kinloch, A.J. (1980). The science of adhesion I. Surface and interfacial aspects. *Journal of Material Science* 15, 2141.

Larsson, K. (1980). Interfacial phenomena: Bioadhesion and biocompatibility. *Desalination* 35, 105–114.

Lehr, C.M. (1996). From sticky stuff to sweet receptors—Achievements, limits and novel approaches to bioadhesion. *European Journal of Drug Metabolism and Pharmacokinetics* 21, 139–148.

Lehr, C.M. and Poelma, F.G.J. (1991). An estimate of turnover time of intestinal mucus gel layer in the rat in situ loop. *International Journal of Pharmaceutics* 70, 235.

Leung, S.H.S. and Robinson, J.R. (1988). The contribution of anionic polymer structural features related to mucoadhesion. *Journal of Controlled Release* 5, 223–231.

Leung, S.H.S. and Robinson, J.A. (1992). Polyanionic polymers in bioadhesive and mucoadhesive drug delivery. *ACS Symposium Series* 480, 269–284.

Li, M.G., Lu, W.L., Wang, H.C., Zhang, X., Wang, X.Q., Zheng, A.P., and Zhang, Q. (2007). Distribution, transition, adhesion and release of insulin loaded nanoparticles in the gut of rats. *International Journal of Pharmaceutics* 329, 182–191.

Liu, Z.P., Lu, W.Y., Qian, L.S., Zhang, X.H., Zeng, P.Y., and Pan, J. (2005). In vitro and in vivo studies on mucoadhesive microspheres of amoxicillin. *Journal of Controlled Release* 102, 135–144.

Longer, M.A. and Robinson, J.R. (1986). Fundamental aspects of bioadhesion. *Pharm Int* 7, 114–117.

Machida, Y., Masuda, H., Fujiyama, N., Ito, S., Iwata, M., and Nagai, T. (1979). Preparation and phase-II clinical evaluation of topical dosage form for treatment of carcinoma colli containing bleomycin with hydroxypropylcellulose. *Chemical and Pharmaceutical Bulletin* 27, 93–100.

McBain, J.W. and Hopkins, D.G. (1925). On adhesives and adhesive action. *Journal of Physical Chemistry* 29, 188–204.

McCarron, P.A., Donnelly, R.F., Zawislak, A., and Woolfson, A.D. (2006). Design and evaluation of a water-soluble bioadhesive patch formulation for cutaneous delivery of 5-aminolevulinic acid to superficial neoplastic lesions. *European Journal of Pharmaceutical Sciences* 27, 268–279.

McCarron, P.A., Donnelly, R.F., Zawislak, A., Woolfson, A.D., Price J.H., and McClelland, R. (2005). Evaluation of a water-soluble bioadhesive patch for photodynamic therapy of vulval lesions. *International Journal of Pharmaceutics* 293, 11–23.

McCarron, P.A., Woolfson, A.D., Donnelly, R.F., Andrews, G.P., Zawislak, A., and Price, J.H. (2004). Influence of plasticiser type and storage conditions on the properties of poly(methyl vinyl ether-*co*-maleic anhydride) bioadhesive films. *Journal of Applied Polymer Science* 91, 1576–1589.

McInnes, F.J., O'Mahony, B., Lindsay, B., Band, J., Wilson, C.G., Hodges, L.A., and Stevens, H.N.E. (2007). Nasal residence of insulin containing lyophilised nasal insert formulations, using gamma scintigraphy. *European Journal of Pharmaceutical Sciences* 31, 25–31.

Medlicott, N.J., Rathbone, M.J., Tucker, I.J., and Holborow, D.W. (1994). Delivery systems for the administration of drugs to the periodontal pocket. *Advanced Drug Delivery Reviews* 13, 181–203.

Merrill, E.W. (1977). Properties of materials affecting the behaviour of blood at their surfaces. *Annals of the New York Academy of Science* 283, 6–16.

Mikos, A.G. and Peppas, N.A. (1986). Comparison of experimental techniques for the measurement of the bioadhesive forces of polymeric materials with soft tissues. *Proceedings of the International Symposium on Controlled Release Bioactive Materials* 13, 97.

Mortazavi, S.A. and Smart, J.D. (1994). Factors influencing gel-strengthening at the mucoadhesive–mucus interface. *Journal of Pharmacy and Pharmacology* 46, 86–90.

Nagai, T. (1985). Adhesive topical drug delivery systems. *Journal of Controlled Release* 2, 121–134.

Nagai, T. (1986a). Topical mucosal adhesive dosage forms. *Medical Research Reviews* 6, 227–242.

Nagai, T. (1986b). Bioadhesive and mucoadhesive drug delivery systems. *46th International Congress of Pharmaceutical Science of FIP*, September 1–5, Helsinki.

Nagai, T. and Konishi, R. (1987). Buccal/gingival drug delivery systems. *Journal of Controlled Release* 6, 353–360.

Nagai, T. and Machida, Y. (1985). Advances in drug delivery—Mucosal adhesive dosage forms. *Pharmacy International* 6, 196–200.

Nagai, T. and Machida, Y. (1990). Bioadhesive dosage forms for nasal administration. *Bioadhesive Drug Delivery Systems* (V. Lenaerts and R. Gurney, eds.), CRC Press, Boca Raton, FL.

Naisbett, B. and Woodley, J. (1994). The potential use of tomato lectin for oral drug delivery. *International Journal of Pharmaceutics* 107, 223–230.

Nicholls, T.J., Green, K.L., Rogers, D.J., Cook, J.D., Wolowacz, S., and Smart, J.D. (1996). Lectins in ocular drug delivery. An investigation of lectin binding sites on the corneal and conjunctival surfaces. *International Journal of Pharmaceutics* 138, 175–183.

Park, K. and Park, H. (1990). Test methods of bioadhesion. *Bioadhesive Drug Delivery Systems* (V. Lenaerts and R. Gurney, eds.), CRC Press, Boca Raton, FL.

Park, H. and Robinson, J.R. (1985). Physicochemical properties of water soluble polymers important to mucin/epithelium adhesion. *Journal of Controlled Release* 2, 47–57.

Peppas, N.A. and Buri, P.A. (1985). Surface, interfacial and molecular aspects of polymer bioadhesion on soft tissues. *Journal of Controlled Release* 2, 257–275.

Peppas, N.A., Little, M.D., and Huang, Y. (2000). Bioadhesive controlled release systems. *Handbook of Pharmaceutical Controlled Release Technology* (D.L. Wise, ed.), Marcel Dekker, New York, pp. 255–269.

Perioli, L., Ambrogi, V., Giovagnoli, S., Blasi, P., Mancini, A., Ricci, M., and Rossi, C. (2008). Influence of compression force on the behaviour of mucoadhesive buccal tablets. *AAPS PharmSciTech* 9, 274–281.

Ponchel, G. and Irache, J.M. (1998). Specific and non-specific bioadhesive particulate systems for oral delivery to the gastrointestinal tract. *Advanced Drug Delivery Reviews* 34, 191–219.

Ponchel, G., Touchard, F., Duchene, D., and Peppas, N.A. (1987). Bioadhesive analysis of controlled release systems I. Fracture and interpenetration analysis in poly(acrylic acid)-containing systems. *Journal of Controlled Release* 5, 129–141.

Pritchard, W.H. (1971). The role of hydrogen bonding in adhesion. *Aspects of Adhesion* 6, 11–23.

Radebaugh, G.W. (1992). Film coatings and film-forming materials: evaluation. *Encyclopedia of Pharmaceutical Technology* (J. Swarbrick and J.C. Boylan, eds.), Marcel Dekker, New York, pp. 1–28.

Rastogi, R., Sultana, Y., Aqil, M., Ali, A., Kumar, S., Chuttani, K., and Mishra, A.K. (2007). Alginate microspheres of isoniazid for oral sustained drug delivery. *International Journal of Pharmaceutics* 334, 71–77.

Reich, S., Levy, M., Meshorer, A., Blumental, M., Yalon, M., Sheets, J.W., and Goldberg, E.P. (1984). Intraocular-lens endothelial interface–adhesive force measurements. *Journal of Biomedical Materials Research* 18, 737–744.

Reinhart, C.P. and Peppas, N.A. (1984). Solute diffusion in swollen membranes II. Influence of crosslinking on diffusion properties. *Journal of Membrane Science* 18, 227–239.

Robinson, J.R. (1989). Ocular drug delivery mechanism(s) of corneal drug transport and mucoadhesive delivery systems. *STP Pharma Sciences* 5, 839–846.

Salman, H.H., Gamazo, C., Agueros, M., and Irache, J.M. (2007). Bioadhesive capacity and immunoadjuvant properties of thiamine-coated nanoparticles. *Vaccine* 48, 8123–8132.

Sayin, B., Somavarapu, S., Li, X.W., Thanou, M., Sesardic, D., Alpar, H.O., and Senel, S. (2008). Mono-*N*-carboxymethyl chitosan (MCC) and *N*-trimethyl chitosan (TMC) nanoparticles for non-invasive vaccine delivery. *International Journal of Pharmaceutics* 363, 139–148.

Sensoy, D., Cevher, E., Sarici, A., Yilmaz, M., Ozdamar, A., and Bergisadi, N. (2009). Bioadhesive sulfacetamide sodium microspheres: Evaluation of their effectiveness in the treatment of bacterial keratitis caused by *Staphylococcus aureus* and *Pseudomonas aeruginosa* in a rabbit model. *European Journal of Pharmaceutics and Biopharmaceutics* 72, 487–495.

Shemer, A., Amichai, B., Trau, H., Nathansohn, N., Mizrahi, B., and Domb, A.J. (2008). Efficacy of a mucoadhesive patch compared with an oral solution for treatment of aphthous stomatitis. *Drugs in R&D* 9, 29–35.

Sidhu, H., Price, J.H. McCarron, P.A., McCafferty, D.F., Woolfson, A.D., Biggart, D., and Thompson, W. (1997). A randomised control trial evaluating a novel cytotoxic drug delivery system for the treatment of cervical intraepithelial neoplasia. *British Journal of Obstetrics and Gynaecology* 104, 145–149.

Smart, J.D. (1991). An in vitro Assessment of some mucoadhesive dosage forms. *International Journal of Pharmaceutics* 73, 69–74.

Smart, J.D., Banchonglikitkul, C., Gibbs, R.V., Donovan, S.J., and Cook, D.J. (2003). Lectins in drug delivery to the oral cavity, in vitro toxicity studies. *STP Pharma Sciences* 13, 37–40.

Smart, J.D., Kellaway, I.W., and Worthington, H.E.C. (1984). An in vitro investigation of mucosa-adhesive materials for use in controlled drug delivery. *Journal of Pharmacy and Pharmacology* 36, 295–299.

Smart, J.D., Nicholls, T.J., Green, K.L., Rogers, D.J., and Cook, J.D. (1999). Lectins in drug delivery: A study of the acute local irritancy of the lectins from *Solanum tuberosum* and *Helix pomatia*. *European Journal of Pharmaceutical Sciences* 9, 93–98.

Uchida, T., Toida, Y., Sakakibara, S., Miyanaga, Y., Tanaka, H., Nishikata, M., Tazuya, K., Yasuda, N., and Matsuyama, K. (2001). Preparation and characterization of insulin-loaded acrylic hydrogels containing absorption enhancers. *Chemical and Pharmaceutical Bulletin* 49, 1261–1266.

Van Wachem, P.B., Beugeling, T., Feijen, J., Bantjes, A., Detmers, J.P., and Van Aken, W.G. (1985). Interaction of cultured human cells with polymeric surfaces of different wettabilities. *Biomaterials* 6, 403–408.

Veillard, M.M., Longer, M.A., Martens, T.W., and Robinson, J.R. (1987). Preliminary studies of oral mucosal delivery of peptide drugs. *Journal of Controlled Release* 6, 123–131.

Voorspoels, J., Remon, J.P., Eechaute, W., and DeSy, W. (1996). Buccal absorption of testosterone and its esters using a bioadhesive tablet in dogs. *Pharmaceutical Research* 13, 1228–1332.

Voyutskii, S.S. (1963). *Autoadhesion and Adhesion of High Polymers*. Wiley, New York.

Wake, W.C. (1982). *Adhesion and the Formulation of Adhesives*. Applied Science Publishers, London U.K.

Woolfson, A.D., McCafferty, D.F., McCallion, C.R., McAdams, E.T., and Anderson, J. (1995a). Moisture-activated, electrically-conducting bioadhesive hydrogels as interfaces for bio-electrodes: Effect of film hydration on cutaneous adherence in wet environments. *Journal of Applied Polymer Science* 58, 1291–1296.

Woolfson, A.D., McCafferty, D.F., McCarron, P.A., and Price, J.H. (1995b). A bioadhesive patch cervical drug delivery system for the administration of 5-fluorouracil to cervical tissue. *Journal of Controlled Release* 35, 49–58.

Woolfson, A.D., McCafferty, D.F., and Moss, G.P. (1998). Development and Characterisation of a moisture-activated bioadhesive drug delivery system for percutaneous local anaesthesia. *International Journal of Pharmaceutics* 169, 83–94.

Yang, X. and Robinson, J.R. (1988). Bioadhesion in mucosal drug delivery. *Biorelated Polymers and Gels* (T. Okano, ed.), Academic Press, London U.K.

13 Nanomedicines Coming of Age
Recent Developments in Nanoneuroscience and Nano-Oncology

Radoslav Savic, Jinzi Zheng, Christine Allen,
and Dusica Maysinger

CONTENTS

13.1 INTRODUCTION

In the last 10 years, nanoscience, nanotechnology, and nanomedicine took the spotlight including the very latest focus on RNA delivery and micro RNA detection, presently highly investigated areas in biology and medicine (Bagri et al. 2010, Service 2010, Wanunu et al. 2010). The numbers of publications dealing with various aspects of nano have increased by 16-fold in the 2000s compared to the 1990s. A simple search with a topic term "nano" retrieves 2929 results from the ISI Web of Knowledge [v.4.6] database for the period of 1990–1999, and 47,205 publications for the 2000–2010 period (search performed December 2010). Similarly, the U.S. National Nanotechnology Initiative, with 1.6 billion dollars in funding for 2010, was an agency in its infancy in the late 1990s. Already, in 2005, FDA approved albumin-bound form of paclitaxel (mean particle size of ~130 nm) against breast cancer (FDA 2005). This formulation improved tumor response rate (33% vs. 19%), increased the time to progression from 16.9 to 23.0 weeks (HR = 0.75), and eliminated the use of chemical solvents known to cause hypersensitivity reactions (Gradishar et al. 2005). Many other nanoformulations are still at an experimental stage, but it is certain that nanoformulated drugs and diagnostics will claim an important stake in healthcare in the coming decades (Kim et al. 2010a, Kojima 2010, Konstantatos and Sargent 2010, Wiley et al. 2010, Zelikin 2010).

The word nano comes from Greek *nanos*—dwarf, which, in drug delivery vehicle terms, encompasses the scale of 5–100 nm (Scheme 13.1a). The term nanoscience refers to manipulating materials at the atom/molecular scale (Feynman 1960) in order to create, e.g., 1–100 nm building blocks to be used in drug delivery and tissue engineering (Scheme 13.1b). Nanotechnology, on the other hand, involves creation of functioning devices using these nanosized building blocks (Staples et al. 2006). The devices themselves can be measured in inches (Scheme 13.1c). Definition of nanotechnology posted at the U.S. National Nanotechnology Initiative's website reads:

> Nanotechnology is the understanding and control of matter at dimensions between approximately 1 and 100 nanometers, where unique phenomena enable novel applications. Encompassing nanoscale science, engineering, and technology, nanotechnology involves imaging, measuring, modeling, and manipulating matter at this length scale.

The definition concludes: "Unusual physical, chemical, and biological properties can emerge in materials at the nanoscale. These properties may differ in important ways from the properties of bulk materials and single atoms or molecules."

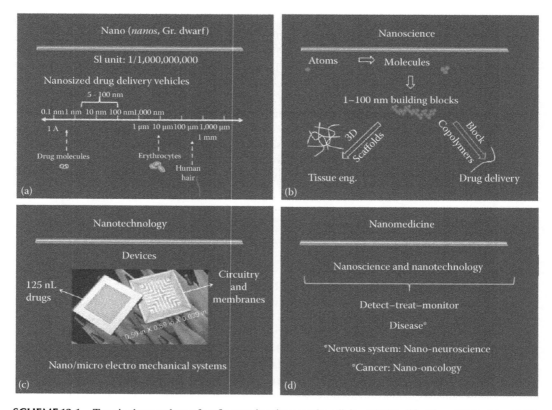

SCHEME 13.1   Terminology and use of prefix nano in science and medicine. ((a) Modified from Whitesides, G.M., *Nat. Biotechnol.*, 21, 1161, 2003; (c) minimally altered with kind permission from Springer Science + Business Media: Pharmaceutical Research. Staples, M. et al., *Pharm. Res.*, 23, 847, 2006. Source: MicroCHIPS, Inc.; photo credit: Dana Lipp Imaging.)

Nanomedicine refers to applications of nanoscience and nanotechnology against disease (Scheme 13.1d). A comprehensive review of all aspects of nano would exceed the limits of this chapter. Therefore, we will focus on discussing model, nanosized, delivery systems based on block copolymer micelles, and on new imaging tools—quantum dots (QDs). Current obstacles and opportunities needed for successful translation of these investigational delivery and imaging systems into the clinic will also be discussed. Specifically, we will use the brain and solid tumors as examples of translational research focus areas for the development of diagnostic, imaging, and therapeutic tools. Examples of the fate of these delivery and imaging vehicles in biological systems, including animals and humans, will be reviewed. First, a brief overview of drug development problems, costs, and search for solutions is given in the following section.

13.2   DRUG DEVELOPMENT AND TREATMENT ISSUES IN ONCOLOGY

Hundreds of millions of dollars are associated with the development of new drugs most of which face serious toxicities and pharmacokinetics/dynamics problems during development (Schemes 13.2 and 13.3). A typical development from conception to the market takes about 10–12 years per drug at a cost of about $ 800 million (Kola and Landis 2004).

Some of the approaches to improve the success of investigational drugs in oncology include suggestions of dosing adjustments (Sugiyma and Yamashita 2011) as well as changes in trial designs to accommodate genetic heterogeneity of the tumors, and to allow successful investigation of molecular-targeted therapeutics applied to selected populations of patients (Simon 2010). Oncology, in particular, is the single most difficult area with mere 5% success rate in bringing novel drugs to market (Kola and Landis 2004). Failure rates are high throughout all stages of development, and particularly concerning are the failures at the time of registration (almost 30%) when drug

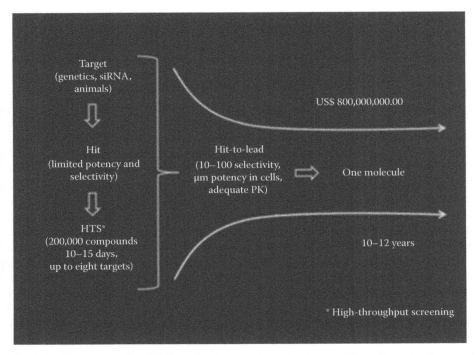

SCHEME 13.2   (See companion CD for color figure.) Drug development facts. Drug development begins with identification of a target, followed by an identification of a hit molecule and high-throughput screening to identify the lead. In the event of success, after an over a decade of work, a single molecule is identified at a cost of ~800 million dollars.

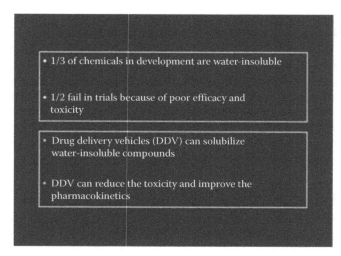

- 1/3 of chemicals in development are water-insoluble

- 1/2 fail in trials because of poor efficacy and toxicity

- Drug delivery vehicles (DDV) can solubilize water-insoluble compounds

- DDV can reduce the toxicity and improve the pharmacokinetics

SCHEME 13.3   (See companion CD for color figure.) Key issues behind the failures of new chemical entities during preclinical and clinical development.

development costs have been irretrievably incurred. Therefore, development of novel drug formulations that would overcome these failures is important (Scheme 13.3).

Here, we will briefly discuss selected new approaches, including available in-man results (from few patients to clinical trials), as they relate to drug delivery approaches to alleviate the issues of current and future drugs used in oncology.

13.2.1   DEVELOPMENT OF NOVEL DRUG FORMULATIONS

Ever since late Judah Folkman's pioneering work in angiogenesis, tumor blood supply has been a target of interest. For example, use of truncated tissue factor fused with endothelium-targeting NGR-motif-containing peptide occludes tumor blood vessels and its use is associated with reduced tumor growth in preclinical trials in mice, while tolerability has been established in five terminal-stage cancer patients (Bieker et al. 2009). The novelty of this approach includes the use of small peptides, rather than large antibodies, possibly allowing better penetrability into the tumor, reduced clearance by reticuloendothelial system (RES), and reduced unwanted toxicities. On the other hand, it is possible to cut off the tumor's blood supply by a surgical procedure—transarterial chemoembolization (TACE)—which is used in patients with intermediate-stage hepatocellular carcinoma. The procedure involves first delivering the drug into the tumor, followed by arterial embolization to induce ischemia and retain high local drug concentrations (Poggi et al. 2008). To further sustain the drug concentration over an extended period of time drug-eluting microspheres, instead of conventional drug emulsion, are used. The use of drug-eluting beads is superior in limiting the systemic toxicity of drugs and establishing better efficacy over conventional TACE (Varela et al. 2007). Radioactive microspheres have also been used in the treatment of solid tumors with moderate success (Gao et al. 2008). Similar to TACE, the latter approach requires surgical procedure involving catheterization of blood vessels.

Conversely, drug delivery systems such as block copolymer micelles are being developed to allow tumor-targeted drug delivery, sustained drug release, and limited unwanted side effects (Gindy and Prud'homme 2009, Kim et al. 2010b, Yokoyama 2010). The enhanced drug effectiveness is aimed to be achieved with the use of minimally invasive (e.g., intravenous) or noninvasive (e.g., oral) approaches to administration. Depending on the bioactive agent incorporated, the micelle delivery systems can be used to disrupt the tumor blood supply, induce apoptosis, deliver cell function–modifying nucleic acids, or conventionally attack the tumor with standard chemotherapeutics. The basic principles and the importance of block copolymer micelles are discussed in the following section.

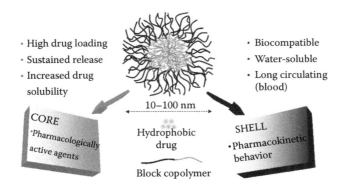

SCHEME 13.4 Schematic representation of block copolymer micelles and possible localizations of incorporated drugs. (Reproduced from Savic, R. et al., *J. Drug Target.*, 14, 343, 2006.)

13.2.2 Block Copolymer Micelles as Drug Delivery Systems

Block copolymer micelles can alleviate the unwanted toxicities, improve the pharmacokinetic profiles of drugs, and afford tumor accumulation of the drug-loaded nanoparticles by enhanced permeability through leaky tumor vasculature and by prolonged retention at the tumor site which is aided by insufficient lymphatic drainage (Maeda et al. 2009). Consequently, block copolymer micelles have been extensively investigated in the past two decades as formulations for anticancer drugs. Block copolymer micelles are self-assembled, usually spherical, delivery systems with 5–100 nm hydrodynamic radii (Scheme 13.4).

The type and length of each block of block copolymer dictate the physical–chemical properties of the delivery system. In general, the higher the ratio between the hydrophilic and hydrophobic parts of the block copolymer the greater the chance of spherical micelles being obtained. The loading of the micelle cores with drugs is less straightforward. It depends not only on the method used (e.g., dialysis vs. direct dissolution) but also on the compatibility of the drug and the polymer, as well as on the size of micelle cores (Savic et al. 2006b).

The principles of targeting drug delivery systems by covalently attached antibodies, or small peptides, have evolved over the years to include 2–3 orders of complexity. Micelle-based delivery systems using internalization facilitators (e.g., TAT peptides), and pH-sensitive polymers which disintegrate in lysosomes, following the micelle internalization, are but one such example (Lee et al. 2008a). These doxorubicin-loaded micelles were able to retard tumor growth in mice up to threefold, compared to free doxorubicin, over a period of 3 weeks. However, to date, the only micelles tested in clinical trials are passively targeted micelles. These micelles have no ligands attached to their coronas and they accumulate in tumors by traversing the leaky endothelium of tumor vasculature (enhanced permeability) and retention due to impaired lymphatic drainage.

13.2.3 Clinical Trials to Date Involving Block Copolymer Micelle-Based Therapeutics

The first suggestion of the use of block copolymer micelles as drug delivery systems occurred in the 1980s by Ringsdorf (Hirano et al. 1979, Gros et al. 1981), and several clinical trials have been conducted since then. In the clinical trial of methoxy-polyethylene glycol (PEG)–poly(D,L-lactide) micelle-incorporated paclitaxel, unfortunately, the elimination half-time and area under the curve were shorter and lower compared to free paclitaxel (Kim et al. 2004b), and there was no improvement in the response to treatment compared to paclitaxel. Similarly, Pluronic® L61 and Pluronic F127 micellar formulation of doxorubicin showed a delayed terminal clearance of the drug but no difference in area under the curve as compared to free doxorubicin. The response to treatment was observed in 3/21 patients but did not persist beyond 4 months after the last treatment (Danson et al. 2004). Doxorubicin–poly(aspartic acid)-*b*-PEG, investigated in patients with metastatic and recurrent solid tumors, had the 2-fold increase in the area under the curve, 1.5-fold increase in the volume of distribution, and 3.6-fold increase in the elimination half-time (Matsumura et al. 2004). However, liposomal formulation of doxorubicin at the same dose showed 556-fold increase of the area under the curve, 720-fold increase in the volume of distribution, and 57-fold increase in elimination half-time compared to doxorubicin (Gabizon et al. 1994). Together, these data suggest poor stability of micelles carrying the anticancer agents. While at least one micelle formulation is continuing in Phase II clinical trials (Lee et al. 2008b), there are no approved micelle formulations as of yet.

One of the greatest challenges remain the stability of micelle-formulated drugs and establishment of dosing regiments which will achieve antitumor effects, or extend the time to progression, with tolerable and manageable side effects. Delivery systems that lose the drug before they reach the target tissues and cells are useless. A simple schematic illustrates this important point (Scheme 13.5).

Since micelle integrity cannot be easily assessed in the presence of macromolecules and cells, it was necessary to overcome this obstacle. Recently, we designed and tested fluorogenic micelles (Savic et al. 2006a) to monitor the integrity of micelles in vitro and in vivo. Our results showed a

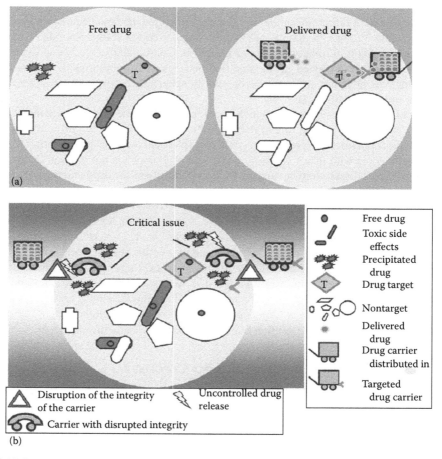

SCHEME 13.5 Importance of the integrity of micelles. The use of passively or actively targeted micelles is intended to bring the drug to the target tissues and cells (a). However, should the integrity of the delivery system be lost the micelles add no benefit. This holds true for both actively and passively targeted systems (b) as they are all in contact with physiological milieu (cells, enzymes, proteins, lipids) en route to target. (Reproduced from Savic, R. et al., *J. Drug Target.*, 14, 343, 2006.)

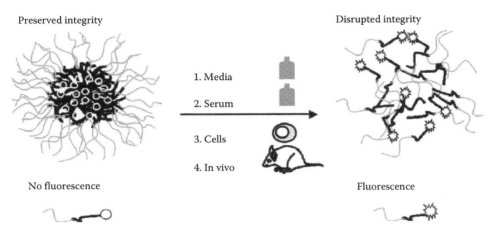

Preserved integrity   Disrupted integrity

1. Media

2. Serum

3. Cells

4. In vivo

No fluorescence   Fluorescence

SCHEME 13.6 Fluorogenic assessment of micelle integrity. Fluorogenic probe is covalently attached to the core-forming part of the block copolymer. With the micelle integrity intact, no fluorescence is detected. Loss of integrity is followed by time- and concentration-dependent increase in fluorescence. (From Savic, R. et al., *Langmuir*, 22, 3570, 2006; Reproduced from Savic, R. et al., *J. Drug Target.*, 14, 343, 2006.)

varied loss of integrity of micelles in different media, and largely preserved integrity in cells in vitro. We have also provided a proof of principle in vivo. The fluorogenic-based assessment of integrity (Scheme 13.6) lends itself to application of different polymer delivery systems, and, with the use of far-infrared probes, will yield quantitative in vivo data on the extent and time course of the loss of micelle integrity. A number of different approaches have been used to track the micelles, not necessarily their integrity, in vivo and in humans. These approaches are discussed in the following section.

13.3 IN VIVO IMAGING OF BLOCK COPOLYMER MICELLE SYSTEMS

The integration of imaging in pharmaceutical development provides a powerful means to gain insight into the in vivo transport and fate of nanosized drug delivery systems. Imaging enables anatomical and functional information to be acquired from animals and humans in a noninvasive manner. The imaging methods that are commonly used for clinical and research applications include computed tomography (CT), magnetic resonance (MR), single photon emission CT (SPECT), positron emission tomography (PET), ultrasound (US), and optical imaging. Each of the individual imaging methods has inherent strengths and weaknesses. Overall, the differences between the various imaging methods lie in their detection sensitivity, spatial and temporal resolution, complexity, and cost of use (Cassidy and Radda 2005, Zheng et al. 2008). Due to these differences, each of the individual modalities is commonly employed for a specific range of research and/or clinical applications.

Colloidal delivery systems can be easily labeled with probes, contrast agents, or radionuclides to support optical, CT, MR, US, SPECT, and/or PET imaging (Josephson et al. 2002, Kircher et al. 2003, Schellenberger et al. 2004, Huh et al. 2005, Sosnovik et al. 2005, Zheng et al. 2006, Zielhuis et al. 2006, Koyama et al. 2007, Rao 2008). These labeled systems can then be used as tools that self-report on their own in vivo pathway and fate at the macroscopic and microscopic scale depending on the imaging modality employed. Importantly, image-based assessment of the distribution of drug delivery systems can provide unique spatial and temporal information that is otherwise unobtainable using traditional methods (Zheng et al. 2009). In the future, these drug delivery systems may be used to pursue image-guided drug delivery for implementation of personalized medicine (Lammers et al. 2008b).

Similar to their impact in pharmaceutical applications nanosized drug delivery systems also have great potential for imaging applications (Islam and Harisinghani 2009). The colloidal size of the delivery systems results in prolonged circulation lifetimes in vivo which enables exploitation of the EPR effect and substantially improves their tumor accumulation in comparison to conventional low-molecular-weight contrast agents that are typically employed in medical imaging. In recent years, there has been significant interest in the development of macromolecular or colloidal imaging agents aimed at improving the sensitivity and specificity of image-based diagnostic tools for cancer and other diseases (Cassidy and Radda 2005, Rao 2008, Zheng et al. 2008, Islam and Harisinghani 2009).

Block copolymer micelles are a versatile nanotechnology platform that can be readily labeled with imaging moieties via conjugation of probes or radionuclides to their hydrophilic shell or encapsulation within their hydrophobic core (Blanco et al. 2009). The micelles can be prepared from biodegradable and/or biocompatible materials which further enhances the potential for translation of these agents into clinical development. Furthermore, the multi-compartment nature of the micelle (i.e., hydrophilic shell, hydrophobic core, hydrophilic–hydrophobic interface) enables simultaneous incorporation of both imaging moieties and therapeutic agents for pursuit of image-guided drug delivery. The following section reviews the various block copolymer micelle systems that have been labeled for imaging applications. Emphasis is placed on the new information that is obtained through image-based assessment of the distribution of these systems in vivo. In addition, potential for translation of these agents to clinical applications is highlighted.

13.3.1 Block Copolymer Micelle Systems That Support MR and CT Imaging

MR and CT are anatomical imaging techniques with sensitivities in the 1–10 µg/mL and 1–10 mg/mL ranges, respectively, and high spatial resolution (µm) (Krause 1999, Cassidy and Radda 2005). MR imaging (MRI) is advantageous in that it provides excellent visualization of soft tissue structures and requires no exposure to radiation; however, relatively long image acquisition time is required for high sensitivity, high resolution, and large field of view imaging, as well there is high cost and compatibility issues associated with its operation (Cassidy and Radda 2005, Zheng et al. 2008). Conversely, CT allows for quantitative, high temporal, and spatial resolution anatomical imaging; however, compared to MR, it provides relatively limited contrast differences between different types of soft tissues (particularly in the brain) and radiation exposure is involved (Zheng et al. 2008).

Several block copolymer micelle systems have been developed for use in MRI. These systems may be divided into two categories, namely, those labeled with gadolinium (Gd) and those incorporating superparamagnetic iron oxide (SPIO) particles. Gd is a positive or T_1-shortening agent that predominantly decreases the longitudinal or spin–lattice relaxation time constant T_1 resulting in an increase in signal intensity (positive contrast) on T_1-weighted MR images (Cassidy and Radda 2005). SPIO particles are negative or T_2-shortening agents that decrease the transverse or spin–spin relaxation time constant T_2 resulting in a reduction in signal intensity (negative contrast) on T_2-weighted images (Koenig and Brown 1994, Cassidy and Radda 2005).

Gd-based block copolymer micelle systems have been prepared by chelation of Gd to the hydrophilic shell of the micelles or incorporation of the metal atoms within the micelle core. Since Gd ions provide positive signal enhancement in T_1-weighted images by shortening the T_1 of the surrounding water protons, the degree of signal enhancement depends on access to water. Therefore, incorporation of Gd in the core of the micelle can result in suppression of the T_1-shortening ability of this agent. Yokoyama's group aimed to exploit this effect by designing a micelle system formed from a mixture of poly(ethylene glycol)-b-(polyaspartic acid) conjugated with DTPA for chelation of Gd (PEG-b-P(Asp(DTPA-Gd)) and polyallylamine or protamine (Nakamura et al. 2006). Measurement of the T_1 relaxation data and subsequent calculation of the R_1 relaxivity values revealed high relaxivity (i.e., T_1-shortening ability) values ranging from 10 to 11 mM^{-1}s^{-1} for the copolymer chains (i.e., PEG-b-P(Asp(DTPA-Gd)) and low relaxivity values for the intact micelles (2.1–3.6 mM^{-1}s^{-1}). Therefore, the incorporation of Gd into intact micelles via chelation to the P(Asp(DTPA)) core-forming component resulted in a decrease in the agent's T_1-shortening ability.

This micelle system could be used for tracking the state of the micelle (i.e., intact or dissociated) in vivo. Indeed, Nakamura et al. proposed that this system would provide minimal signal enhancement while intact and circulating in the bloodstream and high contrast enhancement when the micelles reached a tumor site and disassembled into single chains.

Yokoyama's group also designed a micelle system with Gd incorporated in the micelle core using PEG-b-poly(L-lysine-DOTA) as the polymeric building block (Shiraishi et al. 2009). Each copolymer chain was comprised of 118 ethylene glycol repeat units, 17 lysine units, 17 DOTA molecules conjugated to lysine repeat units, and 7 Gd ions chelated to DOTA groups. Micelles formed from this Gd-incorporated micelle system were 43 nm in diameter with secondary aggregates having a size of 226 nm. Evaluation of the pharmacokinetic profile of the PEG-b-poly(L-lysine-DOTA-Gd) micelles in mice bearing colon 26 tumor xenografts revealed that approximately 23% and 10% of the injected dose (0.05 mmol Gd/kg) remained in blood 24 and 48 h post i.v. administration, respectively. In contrast, the low-molecular-weight contrast agent that is commonly used clinically Gd-DTPA was found to be largely excreted only 1 h following administration (i.e., 1.4% in blood at 1 h). The PEG-b-poly(L-lysine-DOTA-Gd) micelles were found to localize at the tumor site to a significant extent (6% ID/g tumor) with 11% and 7% ID/g in the liver and spleen 24 h post-administration. Maximum signal enhancement of the tumor by the PEG-b-poly(L-lysine-DOTA-Gd) micelles was reached at 24 h postinjection. Therefore as shown in this study the key advantage of a colloidal platform, such as block copolymer micelles in comparison to low-molecular-weight contrast agents, is the extended circulation lifetime that is achieved which in turn enables passive targeting to solid tumors and sites of inflammation.

Block copolymer micelle systems have also been prepared with Gd chelated to the external shell (Zhang et al. 2008). For example, Zhang et al. used poly(L-glutamic acid)-b-polylactide (PG-b-PLA) with DTPA conjugated to the shell-forming block to prepare MRI visible micelles. Following chelation of Gd^{3+} ions to the micelles, the Gd content was reported to be 5% (w/w). The R_1 relaxivity of the PG(DTPA-Gd)-b-PLA micelles was found to be almost twofold higher than the R_1 relaxivity of Gd-DTPA. Therefore, chelation of Gd to the surface of the micelles not only results in higher R_1 relaxivity but also leaves the micelle core free for incorporation of therapeutic agents. However, the pharmacokinetics and biodistribution of this system were not reported; therefore, the influence of surface-chelated Gd on the circulation lifetime and tumor accumulation of the micelles in vivo is not known. It has been shown that in some cases the presence of charged groups at the surface of colloidal carriers can reduce their circulation lifetime and in turn decrease tumor accumulation (Zhang et al. 2004).

As mentioned earlier, another approach to the design of MR-visible block copolymer micelles relies on the encapsulation of SPIO particles within the micelles (Ai et al. 2005, Hong et al. 2008, Lu et al. 2009, Talelli et al. 2009). The general chemical formula for SPIO crystalline structures is $Fe_2^{3+}O_3M^{2+}O$, where M is a divalent metal (i.e., iron, nickel, manganese, cobalt, or magnesium) (Wang et al. 2001). In magnetite SPIO nanoparticles (SPIONs), the divalent metal is ferrous iron (Fe^{2+}). In the last several years, there has been significant interest in the development of SPIO-based contrast agents for MRI. Indeed, SPIO formulations have already been approved for clinical use while others are in late-stage clinical development. For example, AMI-25 (Feridex®) includes iron oxide crystals coated with dextran to form particles with a hydrodynamic diameter of 80 nm (Weissleder et al. 1989, Wang et al. 2001). AMI-25 is largely taken up by the RES with 83% of the dose accumulating in the liver and 6% in the spleen 1 h following administration (Weissleder et al. 1989). AMI-25 or Feridex is approved as an MRI contrast agent for detection of liver lesions.

To date, the majority of SPIO formulations developed include particles coated or associated with dextran or dextran derivatives (e.g., carboxydextran); however, other materials have also been tried including amphiphilic block copolymers. For example, Hennink's group reported SPIONs encapsulated within thermosensitive block copolymer micelles for image-guided drug delivery (Talelli et al. 2009). The copolymer employed was comprised of PEG as the hydrophilic block and poly[N-(2-hydroxypropyl)methacrylamide-dilactate] (p(HPMAm-Lac$_2$)) as the thermosensitive, core-forming block. Hydrophobic oleic acid–modified SPIONs (magnetite, Fe_3O_4) were encapsulated within the mPEG-*b*-p(HPMAm-Lac$_2$) copolymer micelles to produce particles of about 200 nm in diameter. MRI relaxivity measurements demonstrated that the superparamagnetic properties of the SPIONs were retained following micelle encapsulation.

Lu et al. recently reported superparamagnetic manganese ferrite nanocrystals ($MnO \cdot Fe_2O_3$; Mn-SPIO) encapsulated within mPEG-*b*-polycaprolactone (mPEG-*b*-PCL) copolymer micelles for liver imaging (Lu et al. 2009). The Mn-SPIOs of 7.8 nm in diameter were encapsulated within micelles to produce particles having an average hydrodynamic diameter of about 80 nm. Evaluation of the Mn-SPIO containing micelles using a 1.5 T clinical MR scanner revealed an R_2 relaxivity of 270 (Mn + Fe) $mM^{-1}s^{-1}$; this is significantly higher than that of mPEG-phospholipid micelles encapsulating a single Mn-SPIO nanocrystal (R_2 relaxivity = 66 (Mn + Fe) $mM^{-1}s^{-1}$). The clustering of numerous SPIONs within micelles is known to result in a dramatic increase in R_2 relaxivity. In this study, the MRI detection limit (i.e., defined as the nanocomposite concentration at which MRI signal intensity decreases to 50% of that of pure water in T_2-weighted images) was found to be much higher for the single Mn-SPION containing micelles, in comparison to the mPEG-*b*-PCL micelles containing Mn-SPIO clusters. The limit of detection for the single Mn-SPION containing micelles was 0.19 mM (Mn + Fe) and that for the Mn-SPION-mPEG-*b*-PCL micelles was 0.03 mM. Therefore, this data highlights the increase in sensitivity that is afforded by encapsulating multiple SPIO nanocrystals within a single micelle. The liver contrast enhancement of the Mn-SPION-mPEG-*b*-PCL micelles was assessed following the administration of 2.5 mg (Mn + Fe)/kg in mice. Overall, the liver signal enhancement decreased by ~80% 5 min following injection and returned to baseline after 48 h. The strong signal enhancement provided by this micelle-based contrast agent (imaging window of 36 h post-administration) suggests that it may be useful for the detection of liver lesions or diagnosis of liver diseases.

CT is another anatomical imaging technique that is commonly used for both disease diagnosis and image-guided interventions or treatment (e.g., surgery, radiation therapy). Contrast enhancement can be achieved in CT by increasing the difference in x-ray attenuation between two neighboring structures through an increase in the concentration of high atomic number (Z) material within a volume. Therefore, elements with a high Z, most commonly iodine (Z = 53) and barium (Z = 56), have been successfully employed as clinical CT contrast agents. However, other elements with even higher Z have also been explored in a research setting (i.e., gadolinium (Z = 64), gold (Z = 79), and bismuth (Z = 83)). CT has the unique advantage of possessing linear contrast agent concentration–imaging signal (i.e., HU) response profiles. The CT contrast agents used most commonly in the clinic are small molecules that are cleared rapidly from the circulation; therefore, one of the motivations for preparation of a nanotechnology-based CT agent has been to increase the circulation lifetime in order to afford passive targeting to tumor sites and/or to pursue longitudinal imaging sessions. In addition, a macromolecular contrast agent would greatly benefit blood pool imaging applications, as it minimizes the distribution of the agent in the extravascular space as well as significantly decreases the clearance of iodine via the kidneys. Iodine-associated renal toxicity is the major dose-limiting factor in contrast-enhanced CT imaging. On the other hand, due to the poor sensitivity of CT relative to other imaging modalities (i.e., mM range of contrast agent concentration is usually required), one of the major challenges in the design of a nanotechnology-based CT contrast agent is that high concentration of CT agent must be loaded within or conjugated to the nanoparticle for successful imaging.

Torchilin's group developed iodine-loaded MePEG-*b*-polylysine micelles as a CT contrast agent (Trubetskoy et al. 1997). The core-forming polylysine block of the copolymer was conjugated with an iodine-containing benzoic acid derivative. The MePEG-*b*-poly(ε,*N*-(triiodobenzoyl)-L-lysine) copolymer micelles were reported to have iodine contents up to 44% (w/w) and a diameter of about 80 nm. Micelle formulations containing up to 70 mg I/mL were prepared (Trubetskoy et al. 1997). Evaluation of the agent in rats and rabbits revealed prolonged residence times in the blood compartment and thus significant potential as a blood pool contrast agent (Torchilin et al. 1999, Torchilin 2002).

13.3.2 BLOCK COPOLYMER MICELLE SYSTEMS THAT SUPPORT RADIONUCLIDE IMAGING

The radionuclide imaging techniques SPECT and PET have high sensitivities that are in the ranges of 1 ng/mL–100 μg/mL and <1 pg/mL, respectively (Krause 1999). Use of these techniques to image block copolymer micelles in vivo requires labeling with a radioisotope (Park et al. 2002, Yang et al. 2007b, Matson and Grubbs 2008, Hoang et al. 2009). Each radioisotope is associated with a half-life that must be considered in planning the labeling process as well as experimental timelines. The radioisotopes most commonly used for SPECT imaging include gallium-67 (half-life = 78 h), indium-111 (68 h), iodine-123 (13.1 h), iodine-131 (8.08 h), technetium-99m (6.03 h), and thallium-201 (73 h); while the common isotopes for PET imaging are oxygen-15 (half-life = 2 min), nitrogen-13 (10 min), carbon-11 (20.4 min), fluorine-18 (110 min), bromine-76 (972 min), iodine-124 (6048 min), and copper-64 (762 min) (Fischman et al. 2002). The drawbacks associated with SPECT and PET imaging include the need to handle radioactivity, need to consider half-lives of isotopes in planning labeling process and experimental protocols, limited spatial resolution, lack of anatomic information, and high cost and/or limited availability of some isotopes and imaging equipment. Therefore despite the high sensitivity of the radionuclide imaging techniques, their use has only been reported in a limited number of studies for imaging block copolymer micelles in vivo (Yang et al. 2007b, Hoang et al. 2009). However, it should be noted that there are numerous reports on the use of SPECT for imaging liposomes and lipid micelles in vivo (Torchilin 2002).

In a recent study, Hoang et al. used microSPECT/CT imaging for tracking the in vivo pathway and fate of 111-indium (^{111}In)-labeled PEG-*b*-PCL micelles in healthy and tumor-bearing mice. The PEG-*b*-PCL micelles were labeled via conjugation of DTPA to the PEG terminus of a small population of the copolymer chains (1% of the total copolymer weight) followed by chelation of ^{111}In. Overall, the data obtained for the biodistribution of ^{111}In-micelles using noninvasive image-based region of interest analysis was found to be in good agreement with that obtained using a traditional method (i.e., excision of organs and gamma-counting). Moreover, as shown in Figure 13.1, analysis of the transversal slices of the tumor provided unique insight into the intratumoral distribution of the micelles in vivo. Specifically, the 58 nm PEG-*b*-PCL micelles were found to have a heterogeneous distribution within the MDA-MB-231 tumor xenografts with accumulation primarily at the tumor periphery. Importantly, information on the intratumoral distribution of micelles is unattainable by traditional evaluation of biodistribution and therefore this highlights the unique insight that can be gained via image-based assessment.

13.3.3 BLOCK COPOLYMER MICELLE SYSTEMS THAT SUPPORT OPTICAL IMAGING

In vivo optical imaging requires the detection of light photons that have been transmitted through tissues. The optimal window for tissue imaging is in the near-infrared (NIR) region, 700–1000 nm, since at these wavelengths most biological tissues exhibit low inherent scattering and minimal absorption. The two major optical techniques for in vivo imaging are NIR fluorescence and bioluminescence. The advantages associated with optical imaging include high sensitivity, accessibility (i.e., low cost), and lack of exposure to radiation, while the disadvantages include limited tissue penetration and ability to only obtain semiquantitative data. Organic fluorophores and QDs are the two main classes of contrast agents that are commonly used in fluorescence imaging.

In vivo optical imaging of block copolymer micelles has relied on both the incorporation of organic fluorophores and that of QDs into the nanoassemblies. This section will focus on the micelle–organic fluorophore combinations.

Recently, Kim et al. (2008) reported the use of optical imaging to assess the tumor accumulation of pH-sensitive and folate-conjugated micelles labeled with the NIR probe Cy5.5 in a KB epidermoid xenograft mouse model. The actively targeted multifunctional micelles incorporating both the NIR imaging probe Cy5.5 and the anticancer drug doxorubicin were formed from

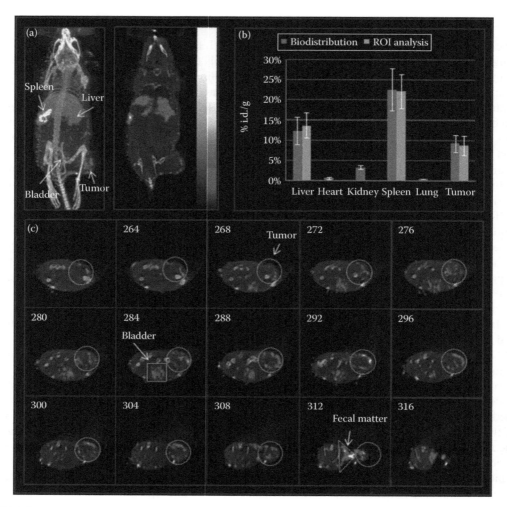

FIGURE 13.1 **(See companion CD for color figure.)** (a) Maximum intensity projection (MIP) and sagittal image of tissue accumulation of ^{111}In-PEG-b-PCL micelles 48 h postinjection in an athymic BALB/c mouse bearing an MDA-MB-231 tumor xenograft after i.v. administration of ^{111}In-micelles. Clear visualization of the liver, spleen, bladder, and tumor was observed. (b) Tissue distribution of ^{111}In-micelles acquired via conventional methodology and nanoSPECT/CT region of interest (ROI) analyses in MDA-MB-231 tumor-bearing mice at 48 h postinjection. (c) Transversal slices of tumor accumulation illustrating nonhomogeneous distribution of ^{111}In-micelles. (Reprinted with permission from Hoang, B. et al., *Mol. Pharm.*, 6, 581, 2009.)

PEG-b-poly(L-histidine) (PEG-b-poly-His) and folate-PEG-b-poly(L-lactic acid) (folate-PEG-b-PLLA) copolymers. The Cy5.5 probe was incorporated into the micelles by conjugation of Cy5.5 bis-NHS ester to primary amine groups of poly(benzyl-His). The Cy5.5 labeled poly(benzyl-His) was then mixed with folate-PEG-b-PLLA, PEG-b-poly-His (10/20/70 wt%/wt%/wt%), and doxorubicin for micelle preparation. Overall, the tumor accumulation evaluated by optical imaging ($\lambda_{ex} = 670$ nm, $\lambda_{em} = 700$ nm) of the pH-sensitive micelles was found to be 1.5-fold greater than that of pH-insensitive micelles (i.e., PEG-b-PLLA and folate-PEG-b-PLLA, 80/20 wt/wt%).

In another study, Rodriguez et al. (2008) entrapped the NIR probe indocyanine green (ICG) in poly(styrene-alt-maleic anhydride)-b-polystyrene (PSMA-b-PSTY) copolymer micelles. ICG is currently FDA-approved for use as an imaging agent in various applications including angiography, evaluation of cardiac output, and hepatic function. The use of ICG in additional preclinical and clinical applications is restricted by the probes limited aqueous, thermal, and photostability (Desmettre et al. 2000, Saxena et al. 2003). Importantly, entrapment of ICG in the PSMA-*b*-PSTY micelles was found to protect the probe, over a 3 week period, from aqueous and thermal (i.e., 37°C) degradation. The ICG was physically entrapped in PSMA-*b*-PSTY micelles via a solvent evaporation method following complexation of the probe to tetrabutylammonium iodide which resulted in the formation of a hydrophobic ICG-tetrabutylamine salt. The entrapped ICG was found to be well-retained within the micelles with only 11% of the total agent being released within 24 h of dialysis against water at 4°C. Retention of the ICG in the micelles in vivo would enable passive targeting of the agent to solid tumors or sites of inflammation via the EPR effect and thus the ICG-micelle formulation could have potential applications in cancer detection and diagnosis. This study demonstrates that formulation of a small molecule imaging agent in block copolymer micelles can provide a means to overcome the shortcomings of these agents (e.g., stability issues, rapid clearance) and in this way further exploit their clinical potential.

Finally, Yang et al. (2007b) created a dual-modality imaging agent using poly(PEG-methacrylate)-*b*-poly(triethoxysilyl propylmethacrylate) (PPEGMA-*b*-PESPMA) copolymer micelles by entrapment of the NIR probe Cy7-like hydrophobic dye 3-(triethoxysilyl)propyl-Cy7 into the core of the micelles and chelation of ^{111}In to the micelle surface (Yang et al. 2007b). The micelles were prepared by gradual addition of water to a THF solution of PPEGMA-*b*-PESPMA and 3-(triethoxysilyl)propyl-Cy7. Cross-linking of the core components of the micelles was then induced by addition of acetic acid. The metal chelator DTPA was then conjugated to the amine-terminated core cross-linked micelles for chelation of ^{111}In. The cross-linked micelles were reported to have an average size of 24 nm in diameter and a zeta potential of +1.2 with about 21 Cy7-like dye molecules and 19 DTPA chelating groups per micelle. Gamma and optical images (λ_{ex} = 730 nm, λ_{em} = 790 nm) were obtained in mice bearing MDA-MB468 tumor xenografts at time points following administration of the ^{111}In-cross-linked micelles containing Cy-7-like probe. Each mouse was administered a dose of 1.3×10^{14} particles (i.e., 150 µCi and 4.5 nmol Cy-7-like probe). The authors noted that optical imaging provided good detection of the superficial tumors and gamma scintigraphy provided visualization of the distribution of particles in internal organs as well. The nuclear images were reported to be in good agreement with data obtained via traditional assessment of biodistribution (i.e., excision of tissues and γ-counting). Indeed, the ^{111}In-cross-linked micelles containing Cy-7-like probe are only one example of numerous multimodality nanosystems that have been developed in recent years as a means to exploit the complementary nature of the different imaging modalities (Zheng et al. 2008).

Medical imaging has become recognized as a key enabler in pharmaceutical research and development. Certainly, many of the large pharmaceutical companies continue to invest in imaging as it has already made meaningful contributions in drug development. In a similar manner, the integration of imaging in drug delivery has the potential to revolutionize the field by providing new and unique insight on the pathway and fate of nanosystems in vivo. Image-based methods enable noninvasive assessment of a single animal over multiple time points. The pharmacokinetics and biodistribution, including tumor localization, of delivery systems can be determined quantitatively (Figure 13.2; Zheng et al. 2009). Furthermore information may be obtained on the in vivo intra-tissue or intra-organ distribution (i.e., whether homogeneous or heterogeneous) of the nanosystem (Figure 13.3). For example, image-based methodologies have shed new light on the in vivo tumor penetration and intratumoral distribution of drugs and delivery systems (Dreher et al. 2006). In addition, the merging of imaging and drug delivery opens the door for pursuing image-guided drug delivery for implementation of personalized medicine. Recent reports have demonstrated that imaging may be used not only to track the delivery system in vivo but also to report on in vivo drug release and response to therapy (Lammers et al. 2008a, Langereis et al. 2009). Indeed, numerous multifunctional nanosystems have been developed with imaging function, drug-loading capacity, and active targeting capability (Blanco et al. 2009). In the future, these "theranostic" systems could be used to tailor dose regimens on a patient-specific basis.

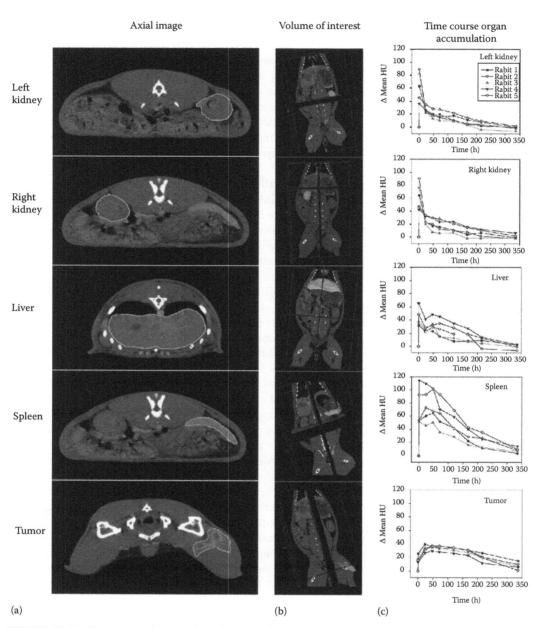

FIGURE 13.2 **(See companion CD for color figure.)** Visual illustration of (a) axial CT slices of the rabbit kidneys, liver, spleen, and tumor acquired at 48 h postinjection. These images are acquired at submillimeter resolution and they demonstrate potential for quantification of intra-organ heterogeneity. In this particular study, bulk organ analysis was performed on (b) the contoured organ/tissue volumes (in yellow). (c) The differential mean HU measured in each volume of interest (with respect to the preinjection data set) at selected time points. Each profile represents the values obtained for a given rabbit over 14 days. (Reprinted with permission from Zheng, J. et al., *Mol. Pharm.*, 6, 571, 2009.)

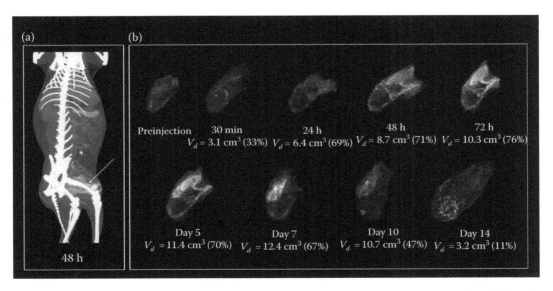

FIGURE 13.3 (a) Anterior views of 3D CT maximum intensity projections of a representative VX2 carcinoma-bearing male New Zealand white rabbit (3 kg) at 48 h post liposome administration. The arrow indicates the site of the tumor and the EPR effect is visualized through the opacification of the tumor area resulting from the accumulation of the iohexol and gadoteridol containing liposomes. (b) 3D maximum intensity projections of the segmented tumor volumes pre and up to 14 days post liposome injection. Note the percent volume of distribution (V_d) of liposomes in the tumor at the different time points. (Adapted from Zheng, J. et al., *Mol. Pharm.*, 6, 571, 2009.)

13.4 NANONEUROSCIENCE

Nanotechnology as an interdisciplinary area of science offers a number of complementary approaches to the traditional methods in neurology and neuroscience. Although still in its infancy, applications of nanotechnology to medicine and more specifically to neurology could improve early diagnosis and eventually therapeutic approaches in neurological disorders (Misgeld and Kerschensteiner 2006, Silva 2007, Flemming 2010). However, there are considerable challenges to deliver engineered nanomaterials to the central nervous system (CNS) because of the unique physiological and anatomical complexities and highly restricted access to the brain. One of the first challenges is to cross the blood–brain barrier (BBB) with minimal or no disruption of this barrier (Blakeley 2008). Nanoengineered molecular complexes might be well-suited to address this challenge because they can be designed to perform multiple functions in a coordinated manner. Great efforts are currently made to achieve this goal. Among the numerous nanomaterials in development as diagnostic nanosensors for clinical and preclinical studies are highly fluorescent nanoparticles, superparamagnetic nanoparticles, and fluorescently labeled polymeric materials. The following sections will highlight a type of fluorescent nanoparticles made of semiconducting materials, which have been extensively studied in different biomedical fields and basic neurological sciences in the past decade.

13.4.1 QDs (AND POLYMERS USED FOR THEIR COATING)

One of the most studied metallic nanoparticles for biomedical applications is the semiconductor QD. QDs are luminescent nanoparticles with unique optical properties that have been exploited for single cell and whole animal imaging. Excellent reviews summarizing the current status of QDs are available (Alivisatos et al. 2005, Medintz et al. 2008, Delehanty et al. 2009). A comparative analysis between fluorescent proteins, fluorescent dyes, and QDs was also reported (Giepmans et al. 2006).

Core composition of QDs for imaging cells and tissues in whole animals (and possibly in humans) must remain undisturbed for the duration of the imaging process. To achieve QD stability, several approaches were developed including the use of a ZnS coating and attachment of various polymeric materials as a surface layer. When coated with proteins or biocompatible polymers, QDs are generally less deleterious or even harmless to cells and organisms for at least several hours of the imaging process. The most effective surface modifications to prevent QD degradation and loss of luminescence include multilayer ZnS capping and surface conjugation with polymers such as PEG. Of course, the size and charge of the PEG chain is a critical determinant of QD internalization into cells, whole body distribution, and subcellular distribution. Several other polymers have been used to protect QD surfaces. For example, Nie and coworkers proposed the use of PEG-grafted polyethyleneimine as a stable QD-surface coating that is also able to disrupt endosomes, enhancing overall QD internalization (Duan and Nie 2007). Winnik et al. used a diblock copolymer poly(ethylene glycol-*b*-2-*N*,*N*-dimethylaminoethyl methacrylate) (PEG-*b*-PDMA) as a QD-surface coating. In this case, the DMA groups served as multidentate anchoring sites on the core of CdSe QD (Wang et al. 2007). Encapsulation of QDs within block copolymer shells and phospholipid micelles uses bifunctional amphiphilic molecules with distinct hydrophilic and hydrophobic segments. The hydrophobic units selectively interact and interdigitate with the native trioctylphosphine/trioctylphosphine oxide (TOP/TOPO) shell, while the hydrophilic units are exposed to an aqueous environment, thereby promoting QD dispersion in biological and other aqueous media. Di- and tri-block copolymers have been widely used as the amphiphilic coating shell for QDs conjugated with their original hydrophobic ligands (Wu et al. 2003, Ballou et al. 2004, Gao et al. 2004, Yu et al. 2007a). The method described by Wu, using 40% of octylamine-modified polyacrylic acid, is currently used for mass commercial production of QDs. The encapsulation of QDs by phospholipid derivatives results in the formation of micelles, as was proposed by Dubertret et al. (2002). This technique is relatively simple, as most phospholipid components are commercially available and the nanocrystals prepared using this protocol were reported to be stable in embryonic cells. Although no evidence is available yet, it is likely that each micelle contains more than one nanocrystal.

QDs for applications in neuroscience and other biological disciplines will require functional modifications of the nanocrystal surfaces. Delivery of QDs to a selected subcellular compartment requires QD surface modifications with different homing peptide sequences specific for the compartment. For example, studies by Derfus et al. (2003) and Chen and Gerion (2004) demonstrated the targeting of QDs to the mitochondria and the nucleus. The following sections describe some uses of different types of QDs for imaging purposes in different cell cultures and in whole animals.

13.4.2 QDs as Biomedical Tags (In Vitro and In Vivo Preclinical Studies)

QDs have been used both for single cell studies and in vivo studies in whole animals. A number of biological questions were addressed in these studies, including Where do QDs go? How long do they stay in the cell of interest? What is the rate of QD intracellular movement? Some of the studies addressing these questions are discussed in this section.

Imaging of living cells was rather successful and provided a wealth of knowledge and seminal information of selected QDs. These studies answered in part some of the key biological questions related to QD uptake (Zhang and Monteiro-Riviere 2009), subcellular distribution, and functional consequences. These studies focused primarily on identifying what factors (e.g., size, charge, or other surface properties) were primarily involved in determining the extent of uptake. Are QDs taken up retained or eliminated from the cells or do they accumulate intracellularly and ultimately get degraded? Studies by the Monteiro-Riviere group suggested several mechanisms of QD internalization into human epidermal keratinocytes. Their experiments include the use of different pharmacological inhibitors to block specific signal transduction pathways implicated in nanoparticle internalization (Zhang and Monteiro-Riviere 2009). They found that QDs with negatively charged surfaces were recognized by lipid rafts; however, the major QD-endocytic pathways primarily involve regulation by G-protein-coupled receptors and low-density lipoprotein scavenger receptors. Similar studies were not performed so far in any cell types of the nervous system. Studies in our and other laboratories show that metallic nanoparticles are uptaken by cells in the brain, and can even induce activation of the glial cells (Jackson et al. 2007, Maysinger et al. 2007). Glial cells, astrocytes, and especially microglia avidly respond to various insults to the nervous system (Block and Hong 2005, Hanisch and Kettenmann 2007) and they play a critical role in the overall functioning of the CNS at all stages of life, under both normal and pathological conditions (Nakanishi 2003, Pekny and Nilsson 2005). Additionally, microglia constitute a critically important "surveillance cell network" in the nervous system. They are highly dynamic and upon focal activation they switch from a patrolling to a shielding mode. In the recent study by Behrendt et al. (2009), we show that microglia are the major glial cell type which internalize InGaP/ZnS nanoparticles in mixed primary cortical cultures. In addition, we show that the distribution and uptake of InGaP/ZnS nanoparticles within individual organelles (lysosomes, mitochondria, and lipid droplets) is markedly affected in neural cells exposed to oleic acid. Since oleic acid is synthesized by glial cells, its disregulation may lead to an altered subcellular distribution of QDs in astrocytes and in neighboring cells. This finding points toward the importance of unsaturated fatty acids in uptake and subcellular distribution of nanoparticles. Oleic acid is a component of our daily alimentary products (e.g., olive oil) and its availability in different tissue compartments, including those of the CNS, could modulate the fate of nanoparticles to be used as delivery agents, imaging tools, or theranostics. Despite the convincing data, these results are limited to particular cell types and particular QDs; much more studies are required to have information for other types of cells and QDs.

13.4.3 QDs for Imaging and Biodistribution in Whole Animals

Studies related to QDs in whole animals are relatively limited (Table 13.1). One of the reasons for this is that imaging of whole animals using fluorescent nanoparticles in general remains difficult for two main reasons: (i) the strong autofluorescence of tissues, due to the presence of endogenous chromophores (e.g., collagens, porphyrins, and flavins), masks the fluorescence of the QDs and (ii) stability of the QD core and corona ligands are yet to be optimized to ensure biocompatibility. In most studies using CdSe/ZnS QDs, the issue of core stability within the time period tested did not seem to pose a problem. However, noncovalently attached surface molecules might have been partly lost en route toward the predicted destination. It is also unknown to which degree QDs interact with extracellular and cellular proteins and get quenched. A newer class of QDs with superior photophysical properties, which can at least in part overcome these limitations by emitting in NIR regions, were prepared and tested in rodents (Choi et al. 2009). Using NIR-emitting QDs avoids interfering noise from the animal's autofluorescence. An example of such QDs is QD705 (20–25 nm), which were injected in mice (intravenously). Results from these studies show that they are still present in the body after 24 h, accumulating in the liver and the spleen. Analyses of urine and feces did not reveal significant quantities in these specimens suggesting that most of the QD705 are retained in the body. Concordant with such a notion are results obtained by Fischer et al. (2006) using two types of QDs with different surface coatings. It seems that surface-modified QDs with PEG are biocompatible with both living cells in culture (Ryman-Rasmussen et al. 2007) and whole animals. PEG polymer is an attractive corona-forming candidate for micelles and QDs due to their physical, chemical, and biological properties, as well as the availability of the material (Bentzen et al. 2005, Boulmedais et al. 2006, Yu etal. 2006, Ballou et al. 2007). Reports on the distribution and pharmacokinetic properties (Gao et al. 2004, Kim et al. 2004a, So et al. 2006b) of cadmium selenide QDs show that these nanoparticles are sequestered in several organs after intravenous administration, but of particular note, they accumulate in lymph nodes (Frangioni et al. 2007) and solid tumors (Ballou et al. 2007).

Recent studies by Choi et al. (2007, 2009) showed that size and surface charge of QDs affect their renal clearance. Selected types of QDs were injected intravenously in rodents and their results showed that nanoparticles with a final hydrodynamic diameter <5.5 nm can be eliminated from the

TABLE 13.1

Examples of Selected In Vivo Studies Using QD

Particle Core	Surface Modification	Concentration (Diameter)	Findings	Reference
CdSe/ZnS	Peptides, PEG	1 mg/mL	PEG coating prevents nonselective accumulation of QDs in RES	Akerman et al. (2002)
CdSe/ZnS	Micelle	20 nM–1 µM (18–28 nm)	QDs visualized hundreds of microns beneath the skin of live mice	Larson et al. (2003)
CdTe/ CdSeTOPO	Oligomeric–polydentate phosphines	10–400 pmol (15–20 nm)	QDs are stable in serum. NIR QDs allow sentinel lymph node mapping and image-guided surgery of 1 cm deep lymph node	Kim et al. (2004a)
CdSe/ZnS	mPEG–750 QDs, mPEG–5000 QDs, cPEG–3400 QDs	1–2.5 pmol/µL	QDs accumulate primarily in the liver, but also in the spleen and bone. Fluorescence can be detected for up to 4 months. mPEG 5000 QDs have prolonged circulation and lower accumulation	Ballou et al. (2004)
CdSe/ZnS TOPO	COOH, PEG, PEG + prostate-specific membrane antigen	6 nmol (20–30 nm)	QDs accumulate in tumors after subcutaneous and systemic administration through both the EPR and antibody–antigen-specific binding	Gao et al. (2004)
CdSe/ZnS TOPO	DHLA	10 pmol	Tumor cells labeled with QDs did not impede tumor formation in mice	Voura et al. (2004)
CdS:Mn/ZnS	TAT peptide	10 mg/mL (3.1 nm)	Radio-opaque, paramagnetic QDs label brain tissue	Santra et al. (2005)
CdMgTe/Hg	BSA	100 µL of 1:1.5 QDs (5 nm)	NIR QDs used as photostable contrast agents	Morgan et al. (2005)
Cds/ZnS TOPO	PEG phospholipid micelles, PEG-PE-TAT	4 µM	QDs can be utilized to image and differentiate tumor vessels in GFP mice	Stroh et al. (2005)
CdSe/ZnS	Streptavidin	100 nM; 1 µM	QDs used for lineage tracing of embryonic zebra fish. QDs outline newly formed blood vessels by fluorescing in the endothelial wall	Rieger et al. (2005)
InAs/ZnSe	DHLA, DHLA-PEG	5.3 nm	QDs circulate in vivo for several minutes	Zimmer et al. (2006)
CdSe/ZnS	mPEG	2 µM (~35 nm)	QDs used to determine the upper size limit of particles that pass the BBB	Thorne and Nicholson (2006)
CdSe/ZnS	Arginine–glycine–aspartic acid	200 pmol (15–20 nm)	QDs targeted to a particular tumor marker in vitro, ex vivo, and in vivo	Cai et al. (2006)
CdTe/CdSe	Oligomeric phosphines and HSA800, or ICG:HSA	100 pmol–1 nmol (15 nm)	Type II NIR QDs used for mapping of sentinel lymph nodes	Tanaka et al. (2006)
CdTe/ZnS	Polymer with carboxy groups	8–16 pmol (6–12 nm)	QDs allow simultaneous imaging of two separate lymphatic flows and their drainage into lymph nodes	Hama et al. (2007)

TABLE 13.1 (continued)
Examples of Selected In Vivo Studies Using QD

Particle Core	Surface Modification	Concentration (Diameter)	Findings	Reference
CdSe/ZnS	Captopril	7 mg/kg	Analysis of pharmacokinetics and pharmacodynamics of a QD-tagged antihypertensive drug	Manabe et al. (2006)
CdSe/ZnS	Anti-HER 2 antibody	2 μM	Monitoring the trafficking of a tumor antibody	Tada et al. (2007)
CdSe/ZnS	Amino-PEG	3–17 nmol	QDs phagocytozed by macrophages and microglia which infiltrate gliomas	Jackson et al. (2007)
CdSe/ZnS	α-Fetoprotein antibody	0.4 nmol	QDs enable active tumor targeting and spectroscopic hepatoma imaging	Yu et al. (2007b)
CdSe/CdS	PEG	(37 nm)	Intradermally injected QDs move through, and degrade in, lymphatic ducts, lymph nodes, liver, kidney, spleen	Gopee et al. (2007)
CdSe/ZnS	Carboxy-PEG	8.2 nM	Long-term in vivo trafficking of selected human mesenchymal stem cells	Rosen (2007)
CdSe and CdTe	Carboxy-PEG	12–24 pmol (15–19 nm)	Five different QDs simultaneously reveal five separate lymphatic flows and their drainage to distinct lymph nodes	Kobayashi et al. (2007)
CdSe/ZnS	Phospholipid coat carboxy-conjugated	0.005 pg	QDs used to study in vivo fate, mapping, cellular migration, and differentiation of neural stem and progenitor cells (NSPCs)	Slotkin et al. (2007)
CdSe/ZnS and CdTe	PEGylated and non-PEGylated	30 nmol	Real-time assessment of QD-induced activation of astrocytes in the brains of mice with murine astrocyte promoter-driven luciferase expression	Maysinger et al. (2007)
CdSe/ZnS	In solid lipid nanoparticles	4–90 nm	QDs encapsulated within SLPs remain stable and fluorescent	Liu et al. (2007)
CdSe/ZnS	Amino-PEG-mAb (anti-PECAM, -VECAM, -ICAM, QD-IgG$_1$)	500 nM	QDs provide noninvasive optical imaging of retinal vasculature, and vascular events, by use of mAb-QDs targeted to the cell adhesion molecules	Jayagopal et al. (2007)
CdSe/ZnS	PEG	40 pmol	Nearly complete retention of QDs in mice after 28 days. Plasma $t_{1/2} = 18.5$ h. Continued depositions in the liver, kidneys, and spleen	Yang et al. (2007a)
CdSe/ZnS	Amino-PEG	5 pmol/g	QD imaging of RES, combined with bioluminescent tumor monitoring, may provide valuable information on localization of lesions	Inoue et al. (2007)

(continued)

TABLE 13.1 (continued)
Examples of Selected In Vivo Studies Using QD

Particle Core	Surface Modification	Concentration (Diameter)	Findings	Reference
CdSe/ZnS	PEG	10 nM	Successful in vivo multiplex imaging and trafficking of mouse embryonic stem cells labeled with QDs	Lin et al. (2007)
CdSe/ZnS	DOTA-QD-RGD with integrin $\alpha_v\beta_3$-specific radioligand	20 pmol (20 nm)	Quantization of tumor-targeting efficacy using a dual-function QD-based probe with PET and NIRF imaging. Dual-function PET/NIRF probe provides sufficient tumor contrast at lower concentrations than NIRF alone, thereby reducing potential toxicities	Cai et al. (2007)
CdSe/(ZnCdS)	Cysteine	3 μL (3.6 nm)	The small size of QD-Cys nanoparticles permits renal clearance	Choi et al. (2007)
CdHgTe	Thiol-capped	2–17.5 μg/g (7 nm)	Low doses of QDs not toxic after 3 months, high doses lethal. QDs accumulate in the liver; get excreted via intestine not kidneys	Chen et al. (2008)
CdSeTe/ZnS	EGF	10 mol	Measurable contrast enhancement of tumors is visible 4 h post-systemic administration; subsequent normalization occurs after 24 h	Diagaradjane et al. (2008)

body via urinary excretion. However, zwitterionic or neutral organic coatings on the QDs can prevent the adsorption of serum proteins and result in agglomeration of nanoparticles, thereby preventing their renal excretion. Another study by the same group (Choi et al. 2009) showed that in addition to size and charge, the length of polymer (PEG) chains on the surface of the QDs can significantly affect their biodistribution and clearance. Intermediate PEG chain lengths (2–22 units) resulted in the unexpected delivery of the nanoparticles to the liver, kidney, pancreas, and lymph nodes, but these QDs were cleared as predicted. QDs with short PEG chains (<2 units) were quickly taken up by the liver, whereas QDs with long PEG lengths (>22 units) can remain in the vasculature for an extended period of time.

Studies by Inoue et al. (2007) showed that long-term, repeated imaging of QDs in vivo is possible after intravenous injection. QD fluorescence was observed in the RES in mice despite a gradual decline of signal intensity due to repeated imaging. Their study also showed that QD-fluorescence imaging can be combined with bioluminescence imaging of luciferase-expressing tumor cells, providing a new and more accurate method to detect and locate tumor cells.

The concept of combining bioluminescence and fluorescence in living cells was first illustrated by So et al. (Feuk et al. 2006). These studies described the concept of self-illuminating QDs, and presented evidence for the application of these QDs in living cells and whole animals. QDs were first covalently bound with luciferase, and upon exposure to the luciferase substrate coelentrazine the bioluminescence energy released by substrate catabolism was transferred to the QDs, eliciting fluorescence emission. The protocol for making these self-illuminating QDs, provided by So et al. (2006a) is widely applicable to other conjugation products which can be utilized for different biosensing purposes.

QD distribution in the brain in real time has not been extensively explored. We studied glial cells in the nervous system because astrocytes and microglia are most likely the major players in delivering nanoparticles to the CNS (Perry et al. 1993, Banati 2003). The glial network significantly contributes to the viability and function of neurons and without their biochemical, morphological, and physiological communication, higher mental or emotional functions at the level of the whole body would not be achievable. Our studies involving the imaging of fluorescent CdSe/ZnS and InGaP/ZnS nanoparticles in cells in culture and in transgenic mice showed functional activation of glia in response to nanoparticles (Behrendt et al. 2009). Such studies are particularly useful for the assessment of nonfluorescent polymeric or metallic nanoparticles (e.g., cerium oxide, polymeric micelles, degenerate "naked" CdTe nanoparticles) in the CNS, which otherwise would not be easily detected by common imaging techniques. Studies by Maysinger et al. focused on astrocyte response to different types of QDs. Astrocytes are the principle macroglial cell type in the brain and their activation is one of the key components of the glial responses to stress and brain injuries. The passage from quiescent to reactive astrocytes is associated with strong up-regulation of the intermediate filament, glial fibrillary acidic protein (GFAP) (Ridet et al. 1997, Pekny and Nilsson 2005), and this up-regulation is considered to be a surrogate marker of neuronal stress and brain inflammatory response. Current methods of astrocyte and microglia detection are mainly based on immunocytochemistry. However, transgenic technology has created a transgenic mouse model carrying the luciferase gene under transcriptional control of the GFAP promoter (Zhu et al. 2004). Our group used this GFAP-transgenic mouse model in our studies and showed that the QD-induced up-regulation of GFAP can be observed noninvasively in live animals. The extent of GFAP up-regulation can be measured by an increase in luciferase expression, using biophotonic imaging and a high-resolution CCD camera (Kadurugamuwa et al. 2005). The principle of generating luciferase-expressing mice and imaging assays in real time is as follows: Luciferase promoter constructs for the GFAP protein (astroglial activation marker) and toll-like receptor (TLR2; microglia activation marker) are injected into donor oocytes, which are transplanted into pseudo mothers, used to breed transgenic reporter mice. Prior to imaging, mice receive intraperitoneal injections of D-luciferin, the substrate for luciferase. Given that luciferase expression is regulated by stressful stimuli specific to these glial cell types, the resultant conversion of the substrate to oxyluciferin, the bioluminescent product, is proportional to the degree of stress. An overview of studies with QDs in vivo aside of these briefly discussed earlier is provided in Table 13.1.

13.4.4 QD Interactions with Cells and Possible Untoward Effects

Nonfunctionalized nanocrystals can exploit cellular active transport machinery and be delivered into the cells and even the nucleus (Nabiev et al. 2007). This study and others showed that living human macrophages are able to rapidly internalize and accumulate QDs in distinct cellular compartments, the extent of which depends on QD size and charge. The most remarkable finding is that the smallest CdTe QDs specifically target histones in cell nuclei by a multistep process. The nuclear entry seems to be mediated via the nuclear pore complexes. These studies concur with previous findings by Lovric et al. (2005a) showing by confocal microscopy that this could indeed be the case. More recent studies (Choi et al. 2008) provided further evidence that minute amounts of small (<5 nm diameter) QDs can interfere with histones and induce epigenetic changes. Studies in this direction are warranted to uncover possible histone modifications with or without subsequent gene activations. This is an exciting and emerging new branch in nanotoxicology which we propose to call "nanoepigenetics" (Choi et al. 2008). When QDs are retained in cells or if they accumulate in the body over a long period of time, their coatings may be degraded, yielding "naked" QDs. We were particularly interested in the effects of "naked" QDs. Our studies showed that they can induce damage to the plasma membrane, mitochondrion, and nucleus, ultimately leading to cell death (Lovric et al. 2005b). Reactive oxygen species (ROS) were also found to be important players in mediating QD-induced cellular damage. QD-induced cytotoxicity can be reduced or even eliminated with covalent binding of protective polymers to the QD surface.

There are several primary modes of QDs uptake in glial cells: macropinocytosis, phagocytosis, and nonspecific entry via scavenger receptors. If QDs bear specific ligands for receptors recognized on glial surfaces, they can also be internalized by receptor-mediated endocytosis, e.g., via receptors for nerve growth factor, TRK receptors, which are expressed in many cancer cells, including neuroblastoma and glioblastoma cells. Brain tumors including glioblastoma and current challenges in treating them will be discussed in the following section.

13.4.5 Drug Delivery to the Impaired CNS

Glioblastomas are among the most common and lethal brain cancers. Therapeutic interventions are always combined with surgery to remove as much tumor as possible; however, tumors are often difficult to locate during surgery. To help delineate the margins of a glioma, surgeons are beginning to use optical probes. An advance in this regard was achieved by the utilization of iron oxide nanoparticles ($2r = 10\,$nm) together with an NIR imaging probe (e.g., fluorescent dye Cy5.5), and a targeting moiety chlorotoxin, a small peptide purified from giant Israeli scorpion that binds tightly to a membrane-bound matrix metalloprotein commonly overexpressed in gliomas and other brain tumors. These nanoparticles allowed for easy detection of the tumors using MRI and NIR spectroscopy. Single cell resolution can be achieved, and nanoparticles were not taken up by healthy cells, suggesting that similar constructs can be used for the delivery of chemotherapeutic agents and serve as a guide to surgery. None of such nanoparticles for cancer therapeutic interventions (i.e., obliteration of brain cancer cells) are yet approved. Development of theranostics for glioma is an active field and from some basic science research studies it seems a promising direction for nanoneuroscience.

A successful early diagnosis and treatment of neurological disorders does depend on not only new imaging tools and drugs but also on their crossing the BBB, successful biodistribution, and bioavailability. To reach this goal, new fields including nanoneuroscience, nanopharmacology, and nanotoxicology are now emerging and converging to target the problems of crossing the BBB and accessing the compromised brain regions including those bearing brain tumors (Beduneau et al. 2007, Gilmore et al. 2008, Juillerat-Jeanneret 2008, Bidros and Vogelbaum 2009, Sharma and Sharma 2009).

The BBB is a system of vascular cellular structures, mainly represented by tight junctions between endothelial cells, numerous enzymes, receptors, transporters, and efflux pumps that control and limit the access of molecules to the brain. Lipophilicity of the drugs is certainly facilitating drug distribution through the brain and chemical modifications of drugs that enhance lipid solubility are often employed to optimize drug effectiveness. In addition, hydrophilic compounds that are altered and functionalized to enter the brain via biological carriers must bear structural resemblance to normal transporter substances, allowing recognition by the transporter. Many anticancer agents are large hydrophobic molecules unable to freely cross the BBB and are also substrates for multidrug resistance efflux pumps expressed by the BBB vasculature and tumor cells. Thus innate drug resistance of the tumor cells is an additional problem which remains to be solved. Some success was achieved by using polymeric nanoparticles (e.g., micelles, liposomes) and re-engineering of biopharmaceuticals for delivery to the brain with molecular "Trojan horses" (Pardridge 2008, Boado et al. 2009). Among the molecular Trojan horses, endogenous peptides such as insulin or transferrin were exploited to ferry drugs across the BBB following attachment of the drug to them. Peptidomimetic monoclonal antibodies were also used as Trojan horses. For instance, murine monoclonal antibodies against transferrin receptors were used in rats, mice, and nonhuman primates. Humanized insulin receptor antibodies (HIRMAb) were found to be ninefold more effective than any known monoclonal antibody to the human transferrin receptor (Pardridge 1997). There are numerous therapeutic proteins that could be fused to HIRMAb to create new chemical entities and some of them including brain-derived neurotrophic factor (BDNF), glial-derived neurotrophic factor (GDNF) which are reviewed elsewhere (Boado et al. 2008). The bidirectional BBB transport of fusion protein antibodies against β-amyloid were utilized in different animal models of Alzheimer's disease. Initially, they caused encephalitis but considerable improvement with the fragments of purified antibodies was eventually achieved. The entry of the humanized antibody against insulin receptor linked to the Aβ antibody crossed the BBB via transport by the endogenous human insulin receptor. Once in the brain, the fusion antibody is free to bind to Aβ amyloid plaque and can promote disaggregation of the plaque. The fusion antibody is able to rapidly exit the brain via reverse transcytosis across the BBB because the BBB expresses an Fc receptor which functions as an IgG efflux system (Zhang and Pardridge 2001, Boado et al. 2008). Utilization of engineered molecular Trojan horses is promising; human Fc-soluble extracellular domain receptor fusion proteins such as etanercept are FDA-approved pharmaceuticals. They have low immunogenicity and minimal off target side effects in humans but future clinical studies are needed to confirm these in larger number of humans.

In order to bypass the BBB, nanoparticles carrying drugs affecting CNS functions can be used for direct nose-to-brain delivery (Mistry et al. 2009). Experiments in animal models have shown that nanosized drug delivery systems can enhance nose-to-brain delivery of drugs compared to equivalent dose of drugs in solution. However, one should not ignore several limitations with this route of drug delivery, including limited volume of administration, change in absorption in an inflamed nasal cavity, and possible nanoparticle-induced toxicity in the nasal cavity. Physiology of olfactory system and examples of intranasal delivery systems tested until recently were reviewed and their limitations and advances were discussed (Mistry et al. 2009).

BBB disruption strategies for enhancing drug delivery to the brain have not much advanced and they are generally too aggressive for the organism. For instance, osmotic BBB disruption has been used to improve therapeutic efficacy of chemotherapy in humans but the concern is the potential neurotoxicity from the high concentrations of chemotherapeutics reaching the normal noncancerous cells. Biochemical BBB disruption is somewhat less invasive than osmotic disruption. In this case, mediators of inflammation such as leukotrienes and histamine are used to induce transient vascular leakage and increased permeability of blood vessels. However, clinical studies using this approach were discontinued with carboplatin, dextran, and BCNU because of the side effects. Delivering drugs to brain tumors by systemic administration usually require high systemic drug concentrations and this causes considerable toxicity. Interstitial drug delivery was more appealing for the treatment of primary brain tumors because it provided the most direct method of overcoming barriers to the tumor. Approaches to local drug delivery include the use of implanted controlled-release polymer systems, catheter devices, and convection-enhanced delivery. These modalities of anticancer drug delivery were recently reviewed (Bidros and Vogelbaum 2009). After the first FDA-approved polymeric drug delivery system Gliadel (Eisai, Woodcliff Lake NJ), subsequent randomized trials in patients with newly diagnosed malignant gliomas were conducted with modest success. The problem is particularly serious in treating brain tumors with metastases. Malignant glioma is the most common type of primary brain tumors in adults (there are ~20,000 new cases every year; Bidros and Vogelbaum 2009). Gliobastoma multiforme progresses very rapidly and its recurrence after surgery results with an overall survival of <25 weeks. Consequently, treatments are being pursued and here are some facts reflecting the current status.

Several advances were made in delivering drugs with micelles. The major problem with block copolymer micelles is their stability. In contrast, some unimolecular micelles formed from dendrimers are stable upon dilution and do not disintegrate during circulation in the body. Dendrimer conjugates with anticancer agents were used to treat glioma (Wu et al. 2006, Yang and Kao 2006, Gilmore et al. 2008). Dendrimers of a fourth and fifth generation seem to be particularly interesting since they do not exert significant toxicity (Haensler and Szoka 1993, Chauhan et al. 2009). Interestingly, recent PAMAM dendrimers of the fourth generation exhibit an anti-inflammatory effect which is even stronger than with ibuprofen (Chauhan et al. 2009) (Kapoor et al. 2009). This finding is significant because brain tumors are often accompanied by inflammation and eventually as the chemotherapy, radiation, or surgery prove effective, the leakage of BBB should be reduced to prevent possible infections. Excellent reviews cover the chemistry of dendrimers and their current status as drug carriers (Franc and Kakkar 2008, 2009).

Since immune cells can enter the brain as a consequence of the establishment of chemokine gradient (Simard and Rivest 2004), monocytes were exploited for cell-mediated delivery. For instance, this mode of cell delivery was used by Batrakova et al. (2007) to deliver catalase in a self-assembled block ionomer complex made with PEI–PEG. This self-assembly system is a stable nanoconstruct 60–100 nm in size, which retained antioxidant activity. The nanoenzyme complex was taken up by bone marrow–derived monocytes and catalase was released for >24 h. Hence, such a cell–nanomaterial hybrid could be considered for the delivery of growth factors and other anti-inflammatory nanomedicines for neurodegenerative disorders.

In summary, intranasal route of administration avoids the BBB and is particularly attractive for highly potent therapeutics which should be administered in very small doses. However, this mode of drug delivery does not seem to be suitable for the treatment of brain tumors. Malignant gliomas remain one of the most lethal forms of brain cancers in humans and the prognosis is poor despite some advances in nanomedicines. In the light of the increasing number of humans with neurological disorders, interventions which will facilitate early diagnosis and prevention of neurodegenerative processes including the use of nanomedicines should be further investigated.

13.4.6 CURRENT STATUS AND FUTURE PROSPECTIVE FOR QD APPLICATIONS IN BIOMEDICINE

Stabilized QDs with high fluorescent yield and specific ligands on their surfaces may eventually become common nanomaterials for the development of highly accurate and sensitive tests for clinical diagnosis, including those for neurological disorders. Such tests could overcome the limitations of currently used organic fluorophores (e.g., bleaching and wide spectral emission bands). It is not likely that the QDs available now will be clinically used for therapeutic interventions, even in malignancies such as glioblastoma, in the near future, because there is still information missing to address the question of QD fate, and potential hazards from the core-forming metals. In the meantime, QDs will continue to be very useful tools to explore the mechanisms of cell–nanoparticle interactions in experimental animals and in living cells. Developments of novel biosensors employing QDs are promising for future diagnostic purposes in clinics.

13.5 CONCLUSIONS

Nanotechnology as an interdisciplinary area of science offers a number of complementary investigational and interventional approaches to the traditional ones used in medical sciences. Although still in its infancy, applications of nanotechnology to medicine, and more specifically to neurology, could improve early diagnosis and eventually therapeutic approaches in neurological disorders (Misgeld and Kerschensteiner 2006, Silva 2007). However, there are considerable challenges to deliver engineered nanomaterials to the CNS because of the unique physiological and anatomical complexities and highly restricted access to the brain. One of the first challenges is to cross the BBB with minimal or no disruption of this barrier. Nanoengineered molecular complexes might be well-suited to address this challenge because they can be designed to perform multiple functions in a coordinated manner. Great efforts are currently made to achieve this goal. Among numerous nanomaterials in development as diagnostic nanosensors for clinical and preclinical studies are the highly fluorescent nanoparticles, superparamagnetic nanoparticles, and fluorescently labeled polymeric materials.

On the other hand, the greatest impetus for and the marked advancement has been made in the application of nanoscience to the development of nanotherapeutics against cancer. The rationale includes the cost and time needed for development of new anticancer drugs and a staggeringly low (5%) success rate (Kola and Landis 2004). The failures are largely due to poor solubility of the compounds or their toxicity. Therefore, development of new formulations of approved drugs to improve

TABLE 13.2
Examples of Approved Nanosized Therapeutics

Trade Name	Active Ingredient	Application
Pegintron	Covalent conjugate of recombinant α-2b interferon with monomethoxy polyethylene glycol (PEG)	Treatment of chronic hepatitis C in patients with compensated liver disease
Genexol-PM[a]	Poly(ethylene glycol)-*block*-poly(D,L-lactide) micelle formulation of paclitaxel	Various malignancies (e.g., breast, lung, pancreas)
Abraxane	Albumin-bound nanoparticule formulation of paclitaxel	Treatment of metastatic breast cancer
Doxil	PEGylated liposomal formulation of doxorubicin	Ovarian cancer and HIV-associated Kaposi's sarcoma
DaunoXome	Liposomal formulation of daunorubicin	Treatment of advanced, HIV-associated Kaposi's sarcoma

Sources: Duncan, R., *Nat. Rev. Drug Discov.*, 2, 347, 2003; Davis, M.E. et al., *Nat. Rev. Drug Discov.*, 7, 771, 2008; Wagner, V. et al., *Nat. Biotechnol.*, 24, 1211, 2006.

[a] Approved in South Korea.

the efficacy and safety became one of the priorities in nano-oncology. In this regard, nanoparticles have been extensively investigated and reviewed (Savic et al. 2006b, Peer et al. 2007, Cho et al. 2008, Davis et al. 2008). Common to most nanoparticles is the change in the pharmacokinetics of the encapsulated drug resulting in reduced toxicity and increased accumulation in tumors, driven largely by an enhanced permeability and retention effect (Fang et al. 2003). Nanoparticles have also reached clinical applications including polymeric micelle formulation of paclitaxel (Genexol-PM) and nanosized liposomal formulations of doxorubicin (Doxil, DaunoXome) (Table 13.2).

Nanosystems such as block copolymer micelles are also becoming recognized as an important new class of contrast agents or probes for the detection and characterization of disease states. In contrast to the small molecule agents that have so far dominated the clinical arena, the physico-chemical properties of colloidal systems allow for an extended imaging window which in turn can enable acquisition of higher sensitivity and resolution image data sets. Furthermore, passive and active targeting to disease sites can be achieved with the potential to improve the specificity of image-based diagnosis and staging.

In summary, the original and innovative approaches that are being devised and applied in nano-medicine, some of which have been reviewed here, are postulated to provide breakthrough advances in disease diagnosis, treatment, and management in the coming decades.

REFERENCES

Ai, H., Flask, C., Weinberg, B. et al. 2005. Magnetite-loaded polymeric micelles as ultrasensitive magnetic-resonance probes. *Adv Mater* 17: 1949–1952.

Akerman, M. E., Chan, W. C., Laakkonen, P., Bhatia, S. N., and Ruoslahti, E. 2002. Nanocrystal targeting in vivo. *Proc Natl Acad Sci U S A* 99: 12617–12621.

Alivisatos, A. P., Gu, W., and Larabell, C. 2005. Quantum dots as cellular probes. *Annu Rev Biomed Eng* 7: 55–76.

Bagri, A., Mattevi, C., Acik, M., Chabal, Y. J., Chhowalla, M., and Shenoy, V. B. 2010. Structural evolution during the reduction of chemically derived graphene oxide. *Nat Chem* 2: 581–587.

Ballou, B., Ernst, L. A., Andreko, S. et al. 2007. Sentinel lymph node imaging using quantum dots in mouse tumor models. *Bioconjug Chem* 18: 389–396.

Ballou, B., Lagerholm, B. C., Ernst, L. A., Bruchez, M. P., and Waggoner, A. S. 2004. Noninvasive imaging of quantum dots in mice. *Bioconjug Chem* 15: 79–86.

Banati, R. B. 2003. Neuropathological imaging: In vivo detection of glial activation as a measure of disease and adaptive change in the brain. *Br Med Bull* 65: 121–131.

Batrakova, E. V., Li, S., Reynolds, A. D. et al. 2007. A macrophage–nanozyme delivery system for Parkinson's disease. *Bioconjug Chem* 18: 1498–1506.

Beduneau, A., Saulnier, P., and Benoit, J. P. 2007. Active targeting of brain tumors using nanocarriers. *Biomaterials* 28: 4947–4967.

Behrendt, M., Sandros, M., Mckinney, R. et al. 2009. Cell imaging in real time and organelle distribution of fluorescent InGaP/ZnS nanoparticles. *Nanomedicine* 4: 747–761.

Bentzen, E. L., Tomlinson, I. D., Mason, J. et al. 2005. Surface modification to reduce nonspecific binding of quantum dots in live cell assays. *Bioconjug Chem* 16: 1488–1494.

Bidros, D. S. and Vogelbaum, M. A. 2009. Novel drug delivery strategies in neuro-oncology. *Neurotherapeutics* 6: 539–546.

Bieker, R., Kessler, T., Schwoppe, C. et al. 2009. Infarction of tumor vessels by NGR-peptide-directed targeting of tissue factor: Experimental results and first-in-man experience. *Blood* 113: 5019–5027.

Blakeley, J. 2008. Drug delivery to brain tumors. *Curr Neurol Neurosci Rep* 8: 235–241.

Blanco, E., Kessinger, C. W., Sumer, B. D., and Gao, J. 2009. Multifunctional micellar nanomedicine for cancer therapy. *Exp Biol Med (Maywood)* 234: 123–131.

Block, M. L. and Hong, J. S. 2005. Microglia and inflammation-mediated neurodegeneration: Multiple triggers with a common mechanism. *Prog Neurobiol* 76: 77–98.

Boado, R. J., Zhang, Y., Wang, Y., and Pardridge, W. M. 2009. Engineering and expression of a chimeric trans-ferrin receptor monoclonal antibody for blood–brain barrier delivery in the mouse. *Biotechnol Bioeng* 102: 1251–1258.

Boado, R. J., Zhang, Y., Zhang, Y., Xia, C. F., Wang, Y., and Pardridge, W. M. 2008. Genetic engineering, expression, and activity of a chimeric monoclonal antibody–avidin fusion protein for receptor-mediated delivery of biotinylated drugs in humans. *Bioconjug Chem* 19: 731–739.

Boulmedais, F., Bauchat, P., Brienne, M. J. et al. 2006. Water-soluble pegylated quantum dots: From a composite hexagonal phase to isolated micelles. *Langmuir* 22: 9797–9803.

Cai, W., Chen, K., Li, Z. B., Gambhir, S. S., and Chen, X. 2007. Dual-function probe for PET and near-infrared fluorescence imaging of tumor vasculature. *J Nucl Med* 48: 1862–1870.

Cai, W., Shin, D. W., Chen, K. et al. 2006. Peptide-labeled near-infrared quantum dots for imaging tumor vasculature in living subjects. *Nano Lett* 6: 669–676.

Cassidy, P. J. and Radda, G. K. 2005. Molecular imaging perspectives. *J R Soc Interf* 2: 133–144.

Chauhan, A. S., Diwan, P. V., Jain, N. K., and Tomalia, D. A. 2009. Unexpected in vivo anti-inflammatory activity observed for simple, surface functionalized poly(amidoamine) dendrimers. *Biomacromolecules* 10: 1195–1202.

Chen, F. Q. and Gerion, D. 2004. Fluorescent CdSe/ZnS nanocrystal–peptide conjugates for long-term, non-toxic imaging and nuclear targeting in living cells. *Nano Lett* 4: 1827–1832.

Chen, H., Wang, Y., Xu, J. et al. 2008. Non-invasive near infrared fluorescence imaging of CdHgTe quantum dots in mouse model. *J Fluoresc* 18: 801–811.

Cho, K., Wang, X., Nie, S., Chen, Z. G., and Shin, D. M. 2008. Therapeutic nanoparticles for drug delivery in cancer. *Clin Cancer Res* 14: 1310–1316.

Choi, A. O., Brown, S. E., Szyf, M., and Maysinger, D. 2008. Quantum dot-induced epigenetic and genotoxic changes in human breast cancer cells. *J Mol Med* 86: 291–302.

Choi, H. S., Ipe, B. I., Misra, P., Lee, J. H., Bawendi, M. G., and Frangioni, J. V. 2009. Tissue- and organ-selective biodistribution of NIR fluorescent quantum dots. *Nano Lett* 9: 2354–2359.

Choi, H. S., Liu, W., Misra, P. et al. 2007. Renal clearance of quantum dots. *Nat Biotechnol* 25: 1165–1170.

Danson, S., Ferry, D., Alakhov, V. et al. 2004. Phase I dose escalation and pharmacokinetic study of pluronic polymer-bound doxorubicin (SP1049C) in patients with advanced cancer. *Br J Cancer* 90: 2085–2091.

Davis, M. E., Chen, Z. G., and Shin, D. M. 2008. Nanoparticle therapeutics: An emerging treatment modality for cancer. *Nat Rev Drug Discov* 7: 771–782.

Delehanty, J. B., Mattoussi, H., and Medintz, I. L. 2009. Delivering quantum dots into cells: Strategies, progress and remaining issues. *Anal Bioanal Chem* 393: 1091–1105.

Derfus, A. M., Chan, W. C. W., and Bhatia, S. N. 2003. Probing the cytotoxicity of semiconductor quantum dots. *Nano Lett* 4: 11–18.

Desmettre, T., Devoiselle, J. M., and Mordon, S. 2000. Fluorescence properties and metabolic features of indocyanine green (ICG) as related to angiography. *Surv Ophthalmol* 45: 15–27.

Diagaradjane, P., Orenstein-Cardona, J. M., Colón-Casasnovas, N. E. et al. 2008. Imaging epidermal growth factor receptor expression in vivo: Pharmacokinetic and biodistribution characterization of a bioconjugated quantum dot nanoprobe. *Clin Cancer Res* 14: 731–741.

Dreher, M. R., Liu, W., Michelich, C. R., Dewhirst, M. W., Yuan, F., and Chilkoti, A. 2006. Tumor vascular permeability, accumulation, and penetration of macromolecular drug carriers. *J Natl Cancer Inst* 98: 335–344.

Duan, H. and Nie, S. 2007. Cell-penetrating quantum dots based on multivalent and endosome-disrupting surface coatings. *J Am Chem Soc* 129: 3333–3338.

Dubertret, B., Skourides, P., Norris, D. J., Noireaux, V., Brivanlou, A. H., and Libchaber, A. 2002. In vivo imaging of quantum dots encapsulated in phospholipid micelles. *Science* 298: 1759–1762.

Duncan, R. 2003. The dawning era of polymer therapeutics. *Nat Rev Drug Discov* 2: 347–360.

Fang, J., Sawa, T., and Maeda, H. 2003. Factors and mechanism of "EPR" effect and the enhanced antitumor effects of macromolecular drugs including SMANCS. *Adv Exp Med Biol* 519: 29–49.

FDA. 2005. ABRAXANE for injectable suspension (paclitaxel protein-bound particles for injectable suspension). U.S. Food and Drug Administration.

Feuk, L., Carson, A. R., and Scherer, S. W. 2006. Structural variation in the human genome. *Nat Rev Genet* 7: 85–97.

Feynman, R. P. 1960. There's plenty of room at the bottom. *Eng Sci* 23: 22–36.

Fischer, H. C., Liu, L., Pang, K. S., and Chan, W. C. W. 2006. Pharmacokinetics of nanoscale quantum dots: In vivo distribution, sequestration, and clearance in the rat. *Adv Funct Mater* 16: 1299–1305.

Fischman, A. J., Alpert, N. M., and Rubin, R. H. 2002. Pharmacokinetic imaging: A noninvasive method for determining drug distribution and action. *Clin Pharmacokinet* 41: 581–602.

Flemming, A. 2010. Drug delivery: Nanobioconjugate shrinks brain tumours. *Nat Rev Drug Discov* 9: 917.

Franc, G. and Kakkar, A. 2008. Dendrimer design using Cu–I-catalyzed alkyne–azide "click-chemistry." *Chem Commun* 5267–5276.

Franc, G. and Kakkar, A. K. 2009. Diels-Alder "click" chemistry in designing dendritic macromolecules. *Chem Eur J* 15: 5630–5639.

Frangioni, J. V., Kim, S. W., Ohnishi, S., Kim, S., and Bawendi, M. G. 2007. Sentinel lymph node mapping with type-II quantum dots. *Methods Mol Biol* 374: 147–160.

Gabizon, A., Catane, R., Uziely, B. et al. 1994. Prolonged circulation time and enhanced accumulation in malignant exudates of doxorubicin encapsulated in polyethylene-glycol coated liposomes. *Cancer Res* 54: 987–992.

Gao, X., Cui, Y., Levenson, R. M., Chung, L. W., and Nie, S. 2004. In vivo cancer targeting and imaging with semiconductor quantum dots. *Nat Biotechnol* 22: 969–976.

Gao, W., Liu, L., Teng, G. J., Feng, G. S., Tong, G. S., and Gao, N. R. 2008. Internal radiotherapy using 32P colloid or microsphere for refractory solid tumors. *Ann Nucl Med* 22: 653–660.

Giepmans, B. N., Adams, S. R., Ellisman, M. H., and Tsien, R. Y. 2006. The fluorescent toolbox for assessing protein location and function. *Science* 312: 217–224.

Gilmore, J. L., Yi, X., Quan, L., and Kabanov, A. V. 2008. Novel nanomaterials for clinical neuroscience. *J Neuroimmune Pharmacol* 3: 83–94.

Gindy, M. E. and Prud'homme, R. K. 2009. Multifunctional nanoparticles for imaging, delivery and targeting in cancer therapy. *Expert Opin Drug Deliv* 6: 865–878.

Gopee, N. V., Roberts, D. W., Webb, P. et al. 2007. Migration of intradermally injected quantum dots to sentinel organs in mice. *Toxicol Sci* 98: 249–257.

Gradishar, W. J., Tjulandin, S., Davidson, N. et al. 2005. Phase III trial of nanoparticle albumin-bound paclitaxel compared with polyethylated castor oil-based paclitaxel in women with breast cancer. *J Clin Oncol* 23: 7794–7803.

Gros, L., Ringsdorf, H., and Schupp, H. 1981. Polymeric anti-tumor agents on a molecular and on a cellular-level. *Angew Chem Int Ed (English)* 20: 305–325.

Haensler, J. and Szoka, F. C., Jr. 1993. Polyamidoamine cascade polymers mediate efficient transfection of cells in culture. *Bioconjug Chem* 4: 372–379.

Hama, Y., Koyama, Y., Urano, Y., Choyke, P. L., and Kobayashi, H. 2007. Simultaneous two-color spectral fluorescence lymphangiography with near infrared quantum dots to map two lymphatic flows from the breast and the upper extremity. *Breast Cancer Res Treat* 103: 23–28.

Hanisch, U. K. and Kettenmann, H. 2007. Microglia: Active sensor and versatile effector cells in the normal and pathologic brain. *Nat Neurosci* 10: 1387–1394.

Hirano, T., Klesse, W., and Ringsdorf, H. 1979. Polymeric derivatives of activated cyclophosphamide as drug delivery systems in anti-tumor chemotherapy—Pharmacologically active polymers. 20. *Makromol Chem—Macromol Chem Phys* 180: 1125–1131.

Hoang, B., Lee, H., Reilly, R., and Allen, C. 2009. Non-invasive monitoring of the fate of ^{111}In-labeled block copolymer micelles by high resolution and high sensitivity microSPECT/CT imaging. *Mol Pharm* 6: 581–592.

Hong, G., Yuan, R., Liang, B., Shen, J., Yang, X., and Shuai, X. 2008. Folate-functionalized polymeric micelle as hepatic carcinoma-targeted, MRI-ultrasensitive delivery system of antitumor drugs. *Biomed Microdevices* 10: 693–700.

Huh, Y. M., Jun, Y. W., Song, H. T. et al. 2005. In vivo magnetic resonance detection of cancer by using multifunctional magnetic nanocrystals. *J Am Chem Soc* 127: 12387–12391.

Inoue, Y., Izawa, K., Yoshikawa, K., Yamada, H., Tojo, A., and Ohtomo, K. 2007. In vivo fluorescence imaging of the reticuloendothelial system using quantum dots in combination with bioluminescent tumour monitoring. *Eur J Nucl Med Mol Imaging* 34: 2048–2056.

Islam, T. and Harisinghani, M. G. 2009. Overview of nanoparticle use in cancer imaging. *Cancer Biomark* 5: 61–67.

Jackson, H., Muhammad, O., Daneshvar, H. et al. 2007. Quantum dots are phagocytized by macrophages and colocalize with experimental gliomas. *Neurosurgery* 60: 524–529; discussion 529–530.

Jayagopal, A., Russ, P. K., and Haselton, F. R. 2007. Surface engineering of quantum dots for in vivo vascular imaging. *Bioconjug Chem* 18: 1424–1433.

Josephson, L., Kircher, M. F., Mahmood, U., Tang, Y., and Weissleder, R. 2002. Near-infrared fluorescent nanoparticles as combined MR/optical imaging probes. *Bioconjug Chem* 13: 554–560.

Juillerat-Jeanneret, L. 2008. The targeted delivery of cancer drugs across the blood–brain barrier: Chemical modifications of drugs or drug–nanoparticles? *Drug Discov Today* 13: 1099–1106.

Kadurugamuwa, J. L., Modi, K., Coquoz, O. et al. 2005. Reduction of astrogliosis by early treatment of pneumococcal meningitis measured by simultaneous imaging, in vivo, of the pathogen and host response. *Infect Immun* 73: 7836–7843.

Kapoor, Y., Thomas, J. C., Tan, G., John, V. T., and Chauhan, A. 2009. Surfactant-laden soft contact lenses for extended delivery of ophthalmic drugs. *Biomaterials* 30: 867–878.

Kim, T. Y., Kim, D. W., Chung, J. Y. et al. 2004b. Phase I and pharmacokinetic study of Genexol-PM, a cremophor-free, polymeric micelle-formulated paclitaxel, in patients with advanced malignancies. *Clin Cancer Res* 10: 3708–3716.

Kim, D., Lee, E. S., Park, K., Kwon, I. C., and Bae, Y. H. 2008. Doxorubicin loaded pH-sensitive micelle: Antitumoral efficacy against ovarian A2780/DOXR tumor. *Pharm Res* 25: 2074–2082.

Kim, S., Lim, Y. T., Soltesz, E. G. et al. 2004a. Near-infrared fluorescent type II quantum dots for sentinel lymph node mapping. *Nat Biotechnol* 22: 93–97.

Kim, B. Y., Rutka, J. T., and Chan, W. C. 2010a. Nanomedicine. *N Engl J Med* 363: 2434–2443.

Kim, S., Shi, Y., Kim, J. Y., Park, K., and Cheng, J. X. 2010b. Overcoming the barriers in micellar drug delivery: Loading efficiency, in vivo stability, and micelle–cell interaction. *Expert Opin Drug Deliv* 7: 49–62.

Kircher, M. F., Mahmood, U., King, R. S., Weissleder, R., and Josephson, L. 2003. A multimodal nanoparticle for preoperative magnetic resonance imaging and intraoperative optical brain tumor delineation. *Cancer Res* 63: 8122–8125.

Kobayashi, H., Hama, Y., Koyama, Y. et al. 2007. Simultaneous multicolor imaging of five different lymphatic basins using quantum dots. *Nano Lett* 7: 1711–1716.

Koenig, S. H. and Brown, R. D. 1994. Relaxometry and MRI. In *NMR in Physiology and Biomedicine*, Gillies, R. J. (Ed.). San Diego, CA: Academic Press.

Kojima, C. 2010. Design of stimuli-responsive dendrimers. *Expert Opin Drug Deliv* 7: 307–319.

Kola, I. and Landis, J. 2004. Can the pharmaceutical industry reduce attrition rates? *Nat Rev Drug Discov* 3: 711–715.

Konstantatos, G. and Sargent, E. H. 2010. Nanostructured materials for photon detection. *Nat Nanotechnol* 5: 391–400.

Koyama, Y., Talanov, V. S., Bernardo, M. et al. 2007. A dendrimer-based nanosized contrast agent dual-labeled for magnetic resonance and optical fluorescence imaging to localize the sentinel lymph node in mice. *J Magn Reson Imaging* 25: 866–871.

Krause, W. 1999. Delivery of diagnostic agents in computed tomography. *Adv Drug Deliv Rev* 37: 159–173.

Lammers, T., Hennink, W. E., and Storm, G. 2008a. Tumour-targeted nanomedicines: Principles and practice. *Br J Cancer* 99: 392–397.

Lammers, T., Subr, V., Peschke, P. et al. 2008b. Image-guided and passively tumour-targeted polymeric nanomedicines for radiochemotherapy. *Br J Cancer* 99: 900–910.

Langereis, S., Keupp, J., Van Velthoven, J. L. et al. 2009. A temperature-sensitive liposomal 1H CEST and 19F contrast agent for MR image-guided drug delivery. *J Am Chem Soc* 131: 1380–1381.

Larson, D. R., Zipfel, W. R., Williams, R. M. et al. 2003. Water-soluble quantum dots for multiphoton fluorescence imaging in vivo. *Science* 300: 1434–1436.

Lee, K. S., Chung, H. C., Im, S. A. et al. 2008b. Multicenter phase II trial of Genexol-PM, a Cremophor-free, polymeric micelle formulation of paclitaxel, in patients with metastatic breast cancer. *Breast Cancer Res Treat* 108: 241–250.

Lee, E. S., Gao, Z., Kim, D., Park, K., Kwon, I. C., and Bae, Y. H. 2008a. Super pH-sensitive multifunctional polymeric micelle for tumor pH(e) specific TAT exposure and multidrug resistance. *J Control Release* 129: 228–236.

Lin, S., Xie, X., Patel, M. R. et al. 2007. Quantum dot imaging for embryonic stem cells. *BMC Biotechnol* 7: 67.

Liu, W., Choi, H. S., Zimmer, J. P., Tanaka, E., Frangioni, J. V., and Bawendi, M. 2007. Compact cysteine-coated CdSe(ZnCdS) quantum dots for in vivo applications. *J Am Chem Soc* 129: 14530–14531.

Lovric, J., Bazzi, H. S., Cuie, Y., Fortin, G. R., Winnik, F. M., and Maysinger, D. 2005a. Differences in subcellular distribution and toxicity of green and red emitting CdTe quantum dots. *J Mol Med* 83: 377–385.

Lovric, J., Cho, S. J., Winnik, F. M., and Maysinger, D. 2005b. Unmodified cadmium telluride quantum dots induce reactive oxygen species formation leading to multiple organelle damage and cell death. *Chem Biol* 12: 1227–1234.

Lu, J., Ma, S., Sun, J. et al. 2009. Manganese ferrite nanoparticle micellar nanocomposites as MRI contrast agent for liver imaging. *Biomaterials* 30: 291929–28.

Maeda, H., Bharate, G. Y., and Daruwalla, J. 2009. Polymeric drugs for efficient tumor-targeted drug delivery based on EPR-effect. *Eur J Pharm Biopharm* 71: 409–419.

Manabe, N., Hoshino, A., Liang, Y. Q., Goto, T., Kato, N., and Yamamoto, K. 2006. Quantum dot as a drug tracer in vivo. *IEEE Trans Nanobiosci* 5: 263–267.

Matson, J. B. and Grubbs, R. H. 2008. Synthesis of fluorine-18 functionalized nanoparticles for use as in vivo molecular imaging agents. *J Am Chem Soc* 130: 6731–6733.

Matsumura, Y., Hamaguchi, T., Ura, T. et al. 2004. Phase I clinical trial and pharmacokinetic evaluation of NK911, a micelle-encapsulated doxorubicin. *Br J Cancer* 91: 1775–1781.

Maysinger, D., Behrendt, M., Lalancette-Hebert, M., and Kriz, J. 2007. Real time imaging of astrocyte response to Q-dots: In vivo screening model system for biocompatibility of nanoparticles. *Nano Lett* 7: 2513–2520.

Medintz, I. L., Mattoussi, H., and Clapp, A. R. 2008. Potential clinical applications of quantum dots. *Int J Nanomed* 3: 151–167.

Misgeld, T. and Kerschensteiner, M. 2006. In vivo imaging of the diseased nervous system. *Nat Rev Neurosci* 7: 449–463.

Mistry, A., Stolnik, S., and Illum, L. 2009. Nanoparticles for direct nose-to-brain delivery of drugs. *Int J Pharm* 379: 146–157.

Morgan, N. Y., English, S., Chen, W. et al. 2005. Real time in vivo non-invasive optical imaging using near-infrared fluorescent quantum dots. *Acad Radiol* 12: 313–323.

Nabiev, I., Mitchell, S., Davies, A. et al. 2007. Nonfunctionalized nanocrystals can exploit a cell's active transport machinery delivering them to specific nuclear and cytoplasmic compartments. *Nano Lett* 7: 3452–3461.

Nakamura, E., Makino, K., Okano, T., Yamamoto, T., and Yokoyama, M. 2006. A polymeric micelle MRI contrast agent with changeable relaxivity. *J Control Release* 114: 325–333.

Nakanishi, H. 2003. Microglial functions and proteases. *Mol Neurobiol* 27: 163–176.

Pardridge, W. M. 1997. Drug delivery to the brain. *J Cereb Blood Flow Metab* 17: 713–731.

Pardridge, W. M. 2008. Re-engineering biopharmaceuticals for delivery to brain with molecular Trojan horses. *Bioconjug Chem* 19: 1327–1338.

Park, Y. J., Lee, J. Y., Chang, Y. S. et al. 2002. Radioisotope carrying polyethylene oxide–polycaprolactone copolymer micelles for targetable bone imaging. *Biomaterials* 23: 873–879.

Peer, D., Karp, J. M., Hong, S., Farokhzad, O. C., Margalit, R., and Langer, R. 2007. Nanocarriers as an emerging platform for cancer therapy. *Nat Nanotechnol* 2: 751–760.

Pekny, M. and Nilsson, M. 2005. Astrocyte activation and reactive gliosis. *Glia* 50: 427–434.

Perry, V. H., Andersson, P. B., and Gordon, S. 1993. Macrophages and inflammation in the central nervous system. *Trends Neurosci* 16: 268–273.

Poggi, G., Quaretti, P., Minoia, C. et al. 2008. Transhepatic arterial chemoembolization with oxaliplatin-eluting microspheres (OEM-TACE) for unresectable hepatic tumors. *Anticancer Res* 28: 3835–3842.

Rao, J. 2008. Shedding light on tumors using nanoparticles. *ACS Nano* 2: 1984–1986.

Ridet, A., Guillouf, C., Duchaud, E. et al. 1997. Deregulated apoptosis is a hallmark of the Fanconi anemia syndrome. *Cancer Res* 57: 1722–1730.

Rieger, S., Kulkarni, R. P., Darcy, D., Fraser, S. E., and Koster, R. W. 2005. Quantum dots are powerful multi-purpose vital labeling agents in zebrafish embryos. *Dev Dyn* 234: 670–681.

Rodriguez, V. B., Henry, S. M., Hoffman, A. S., Stayton, P. S., Li, X., and Pun, S. H. 2008. Encapsulation and stabilization of indocyanine green within poly(styrene-*alt*-maleic anhydride) *block*-poly(styrene) micelles for near-infrared imaging. *J Biomed Opt* 13: 014025.

Rosen, A. B. 2007. Everything is illuminated … with quantum dots. *Nanomed* 2: 951–954.

Ryman-Rasmussen, J. P., Riviere, J. E., and Monteiro-Riviere, N. A. 2007. Surface coatings determine cytotoxicity and irritation potential of quantum dot nanoparticles in epidermal keratinocytes. *J Invest Dermatol* 127: 143–153.

Santra, S., Yang, H., Holloway, P. H., Stanley, J. T., and Mericle, R. A. 2005. Synthesis of water-dispersible fluorescent, radio-opaque, and paramagnetic CdS:Mn/ZnS quantum dots: A multifunctional probe for bioimaging. *J Am Chem Soc* 127: 1656–1657.

Savic, R., Azzam, T., Eisenberg, A., and Maysinger, D. 2006a. Assessment of the integrity of poly(caprolactone)-*b*-poly(ethylene oxide) micelles under biological conditions: A fluorogenic-based approach. *Langmuir* 22: 3570–3578.

Savic, R., Eisenberg, A., and Maysinger, D. 2006b. Block copolymer micelles as delivery vehicles of hydrophobic drugs: Micelle–cell interactions. *Journal of Drug Targeting* 14: 343–355.

Saxena, V., Sadoqi, M., and Shao, J. 2003. Degradation kinetics of indocyanine green in aqueous solution. *J Pharm Sci* 92: 2090–2097.

Schellenberger, E. A., Sosnovik, D., Weissleder, R., and Josephson, L. 2004. Magneto/optical annexin V, a multimodal protein. *Bioconjug Chem* 15: 1062–1067.

Service, R. F. 2010. Nanotechnology. Nanoparticle Trojan horses gallop from the lab into the clinic. *Science* 330: 314–315.

Sharma, H. S. and Sharma, A. 2009. Conference Scene: New perspectives on nanoneuroscience, nanoneuropharmacology and nanoneurotoxicology. *Nanomedicine* 4: 509–513.

Shiraishi, K., Kawano, K., Minowa, T., Maitani, Y., and Yokoyama, M. 2009. Preparation and in vivo imaging of PEG–poly(L-lysine)-based polymeric micelle MRI contrast agents. *J Control Release* 136: 14–20.

Silva, G. A. 2007. What impact will nanotechnology have on neurology? *Nat Clin Pract Neurol* 3: 180–181.

Simard, A. R. and Rivest, S. 2004. Bone marrow stem cells have the ability to populate the entire central nervous system into fully differentiated parenchymal microglia. *FASEB J* 18: 998–1000.

Simon, R. 2010. Translational research in oncology: Key bottlenecks and new paradigms. *Expert Rev Mol Med* 12: e32.

Slotkin, J. R., Chakrabarti, L., Dai, H. N. et al. 2007. In vivo quantum dot labeling of mammalian stem and progenitor cells. *Dev Dyn* 236: 3393–3401.

So, M. K., Loening, A. M., Gambhir, S. S., and Rao, J. 2006a. Creating self-illuminating quantum dot conjugates. *Nat Protoc* 1: 1160–1164.

So, M. K., Xu, C., Loening, A. M., Gambhir, S. S., and Rao, J. 2006b. Self-illuminating quantum dot conjugates for in vivo imaging. *Nat Biotechnol* 24: 339–343.

Sosnovik, D. E., Schellenberger, E. A., Nahrendorf, M. et al. 2005. Magnetic resonance imaging of cardiomyocyte apoptosis with a novel magneto-optical nanoparticle. *Magn Reson Med* 54: 718–724.

Staples, M., Daniel, K., Cima, M. J., and Langer, R. 2006. Application of micro- and nano-electromechanical devices to drug delivery. *Pharm Res* 23: 847–863.

Stroh, M., Zimmer, J. P., Duda, D. G. et al. 2005. Quantum dots spectrally distinguish multiple species within the tumor milieu in vivo. *Nat Med* 11: 678–682.

Sugiyma, Y. and Yamashita, S. 2011. Impact of microdosing clinical study—Why necessary and how useful? *Adv Drug Deliv Rev* 63: 494–502.

Tada, H., Higuchi, H., Wanatabe, T. M., and Ohuchi, N. 2007. In vivo real-time tracking of single quantum dots conjugated with monoclonal anti-HER2 antibody in tumors of mice. *Cancer Res* 67: 1138–1144.

Talelli, M., Rijcken, C. J., Lammers, T. et al. 2009. Superparamagnetic iron oxide nanoparticles encapsulated in biodegradable thermosensitive polymeric micelles: Toward a targeted nanomedicine suitable for image-guided drug delivery. *Langmuir* 25: 2060–2067.

Tanaka, E., Choi, H. S., Fujii, H., Bawendi, M. G., and Frangioni, J. V. 2006. Image-guided oncologic surgery using invisible light: Completed pre-clinical development for sentinel lymph node mapping. *Ann Surg Oncol* 13: 1671–1681.

Thorne, R. G. and Nicholson, C. 2006. In vivo diffusion analysis with quantum dots and dextrans predicts the width of brain extracellular space. *Proc Natl Acad Sci U S A* 103: 5567–5572.

Torchilin, V. P. 2002. PEG-based micelles as carriers of contrast agents for different imaging modalities. *Adv Drug Deliv Rev* 54: 235–252.

Torchilin, V. P., Frank-Kamenetsky, M. D., and Wolf, G. L. 1999. CT visualization of blood pool in rats by using long-circulating, iodine-containing micelles. *Acad Radiol* 6: 61–65.

Trubetskoy, V. S., Gazelle, G. S., Wolf, G. L., and Torchilin, V. P. 1997. Block-copolymer of polyethylene glycol and polylysine as a carrier of organic iodine: Design of long-circulating particulate contrast medium for X-ray computed tomography. *J Drug Target* 4: 381–388.

Varela, M., Real, M. I., Burrel, M. et al. 2007. Chemoembolization of hepatocellular carcinoma with drug eluting beads: Efficacy and doxorubicin pharmacokinetics. *J Hepatol* 46: 474–481.

Voura, E. B., Jaiswal, J. K., Mattoussi, H., and Simon, S. M. 2004. Tracking metastatic tumor cell extravasation with quantum dot nanocrystals and fluorescence emission-scanning microscopy. *Nat Med* 10: 993–998.

Wagner, V., Dullaart, A., Bock, A. K., and Zweck, A. 2006. The emerging nanomedicine landscape. *Nat Biotechnol* 24: 1211–1217.

Wang, M. F., Felorzabihi, N., Guerin, G., Haley, J. C., Scholes, G. D., and Winnik, M. A. 2007. Water-soluble CdSe quantum dots passivated by a multidentate diblock copolymer. *Macromolecules* 40: 6377–6384.

Wang, Y. X., Hussain, S. M., and Krestin, G. P. 2001. Superparamagnetic iron oxide contrast agents: Physicochemical characteristics and applications in MR imaging. *Eur Radiol* 11: 2319–2331.

Wanunu, M., Dadosh, T., Ray, V., Jin, J., Mcreynolds, L., and Drndic, M. 2010. Rapid electronic detection of probe-specific microRNAs using thin nanopore sensors. *Nat Nanotechnol* 5: 807–814.

Weissleder, R., Stark, D. D., Engelstad, B. L. et al. 1989. Superparamagnetic iron oxide: Pharmacokinetics and toxicity. *AJR Am J Roentgenol* 152: 167–173.

Whitesides, G. M. 2003. The 'right' size in nanobiotechnology. *Nat Biotechnol* 21: 1161–1165.

Wiley, B. J., Qin, D., and Xia, Y. 2010. Nanofabrication at high throughput and low cost. *ACS Nano* 4: 3554–3559.

Wu, G., Barth, R. F., Yang, W., Kawabata, S., Zhang, L., and Green-Church, K. 2006. Targeted delivery of methotrexate to epidermal growth factor receptor-positive brain tumors by means of cetuximab (IMC-C225) dendrimer bioconjugates. *Mol Cancer Ther* 5: 52–59.

Wu, X., Liu, H., Liu, J. et al. 2003. Immunofluorescent labeling of cancer marker Her2 and other cellular targets with semiconductor quantum dots. *Nat Biotechnol* 21: 41–46.

Yang, R. S., Chang, L. W., Wu, J. P. et al. 2007a. Persistent tissue kinetics and redistribution of nanoparticles, quantum dot 705, in mice: ICP-MS quantitative assessment. *Environ Health Perspect* 115: 1339–1343.

Yang, H. and Kao, W. J. 2006. Dendrimers for pharmaceutical and biomedical applications. *J Biomater Sci Polym Ed* 17: 3–19.

Yang, Z., Zheng, S., Harrison, W. J. et al. 2007b. Long-circulating near-infrared fluorescence core-cross-linked polymeric micelles: Synthesis, characterization, and dual nuclear/optical imaging. *Biomacromolecules* 8: 3422–3428.

Yokoyama, M. 2010. Polymeric micelles as a new drug carrier system and their required considerations for clinical trials. *Expert Opin Drug Deliv* 7: 145–158.

Yu, W. W., Chang, E., Drezek, R., and Colvin, V. L. 2006. Water-soluble quantum dots for biomedical applications. *Biochem Biophys Res Commun* 348: 781–786.

Yu, W. W., Chang, E., Falkner, J. C. et al. 2007a. Forming biocompatible and nonaggregated nanocrystals in water using amphiphilic polymers. *J Am Chem Soc* 129: 2871–2879.

Yu, X., Chen, L., Li, K. et al. 2007b. Immunofluorescence detection with quantum dot bioconjugates for hepatoma in vivo. *J Biomed Opt* 12: 014008.

Zelikin, A. N. 2010. Drug releasing polymer thin films: New era of surface-mediated drug delivery. *ACS Nano* 4: 2494–2509.

Zhang, L. W. and Monteiro-Riviere, N. A. 2009. Mechanism of quantum dot nanoparticle cellular uptake. *Toxicol Sci* 110: 138–155.

Zhang, Y. and Pardridge, W. M. 2001. Mediated efflux of IgG molecules from brain to blood across the blood-brain barrier. *J Neuroimmunol* 114: 168–172.

Zhang, J. X., Zalipsky, S., Mullah, N., Pechar, M., and Allen, T. M. 2004. Pharmaco attributes of dioleoylphosphatidylethanolamine/cholesterylhemisuccinate liposomes containing different types of cleavable lipopolymers. *Pharmacol Res* 49: 185–198.

Zhang, G., Zhang, R., Wen, X., Li, L., and Li, C. 2008. Micelles based on biodegradable poly(L-glutamic acid)-*b*-polylactide with paramagnetic Gd ions chelated to the shell layer as a potential nanoscale MRI-visible delivery system. *Biomacromolecules* 9: 36–42.

Zheng, J., Jaffray, D., and Allen, C. 2008. Nanosystems for multimodality in vivo imaging. In *Multifunctional Pharmaceutical Nanocarriers*, Torchilin, V. (Ed.). New York: Springer.

Zheng, J., Jaffray, D., and Allen, C. 2009. Quantitative CT Imaging of the spatial and temporal distribution of liposomes in a rabbit tumor model. *Mol Pharm* 6: 571–580.

Zheng, J., Perkins, G., Kirilova, A., Allen, C., and Jaffray, D. A. 2006. Multimodal contrast agent for combined computed tomography and magnetic resonance imaging applications. *Invest Radiol* 41: 339–348.

Zhu, L., Ramboz, S., Hewitt, D., Boring, L., Grass, D. S., and Purchio, A. F. 2004. Non-invasive imaging of GFAP expression after neuronal damage in mice. *Neurosci Lett* 367: 210–212.

Zielhuis, S. W., Seppenwoolde, J. H., Mateus, V. A. et al. 2006. Lanthanide-loaded liposomes for multimodality imaging and therapy. *Cancer Biother Radiopharm* 21: 520–527.

Zimmer, J. P., Kim, S. W., Ohnishi, S., Tanaka, E., Frangioni, J. V., and Bawendi, M. G. 2006. Size series of small indium arsenide-zinc selenide core–shell nanocrystals and their application to in vivo imaging. *J Am Chem Soc* 128: 2526–2527.

14 Polymers for Myocardial Tissue Engineering

J.A. Roether, H. Jawad, R. Rai, N.N. Ali,
S.E. Harding, and A.R. Boccaccini

CONTENTS

14.1 INTRODUCTION

Cardiovascular disease (CVD) is the leading cause of death in the industrialized world (Lloyd-Jones et al. 2010). Heart attacks are the main cause of death in patients with CVD, although damage to heart muscle can also occur from infection, drugs, alcohol, chemotherapeutic agents, or because of congenital conditions. According to available statistical information (www.bhf.org.uk), approximately 30% of patients suffering from heart attacks each year die suddenly before reaching hospital. In the remaining patients who survive the initial acute event, the damage sustained to the heart may eventually develop heart failure. A heart attack, known as myocardial infarction (MI), occurs when one or more of the blood vessels supplying the heart suddenly occlude. These vessels are the coronary arteries, and when blocked abruptly, there is a sudden decrease in the supply of nutrients and oxygen to the portion of heart muscle supplied by the artery. If blood flow is not restored rapidly, the result is irreversible cell death within the affected part of the heart muscle.

The adult heart cannot adequately repair the damaged tissue, as the mature contracting cardiomyocytes (CMs) have severely limited ability to divide (Anversa et al. 2002), and the stem/progenitor population within the adult heart cannot provide the scale of replacement needed after substantial damage (Dimmeler et al. 2005). Thus the result of MI is the formation of scar tissue with different contractile, mechanical, and electrical properties to that of normal myocardium, which is unable to deliver sufficient blood to meet the body's metabolic requirements (Vunjak-Novakovic et al. 2010).

The replacement of contractile myocardium with noncontracting fibrous scar reduces the pumping efficiency of the ventricles, the heart's main pumping chambers. Various compensatory mechanisms are activated in response to the reduced cardiac output. These initially stabilize the damaged heart and maintain cardiac output at acceptable levels. Ultimately, these "compensatory systems" place extra burden on the weakened heart muscle. This leads to reduced cardiac function and the development of the clinical syndrome of heart failure (Leor et al. 2006). The deterioration of heart function accelerates as heart failure progresses. Ultimately at the end stage of heart failure, the only option is either mechanical ventricular assist devices (VADs) or heart transplantation. The high cost of VADs and the shortage of donor organs (Akins 2002) mean that many patients die while waiting for the operation.

Hence, there is great interest in developing new methods to repair and regenerate the infarcted area of the myocardium, which are summarized in Table 14.1 (Jawad et al. 2007). Repair of cardiac valves is also well advanced, and there are efforts to specifically promote revascularization, but this chapter will concentrate on the generation of new contractile muscle mass (although revascularization is often intimately connected with this endeavor). Replacement of scarred tissue with skeletal muscle cells, cells derived from bone marrow (mesenchymal and hematopoietic), or embryonic stem cells (ESCs) has been proposed (Laflamme and Murry 2005). For tissue regeneration to be successful, it is essential to generate appropriate numbers of cells to maintain the specific biological functions and for those cells to differentiate to the correct contractile phenotype. Some crucial cell functions in addition to active beating include interacting with neighboring cells/tissue, producing extracellular matrix (ECM) in the correct organization, as well as secreting signaling molecules.

Biomaterials play a key role in the success of several myocardial tissue regeneration strategies and the challenges in MTE have been discussed in the recent literature (Vunjak-Novakovic et al. 2010). Currently, the preferred method of introducing the cells into the nonviable myocardium is injection of cells in suspension either into the circulating blood or directly into the myocardium. However, the cell delivery route may be inefficient with substantial cell loss (Grossman et al. 2002, Hoffmann et al. 2005). The two different routes proposed for cell delivery to the damaged area, i.e., injection of cells directly into the infarcted region (Orlic et al. 2001) or via coronary circulation

---

TABLE 14.1
Advantages and Disadvantages of MTE Approaches

Approach	Advantages	Disadvantages
Cellular cardiomyoplasty (injection of cells only, either directly or intravenously)	Minimally invasive surgery if injection is intravenously	Lack of knowledge as to how cells contribute to myocardial regeneration or repair. Concern about direct injection only affecting endocardium and not epicardium. Concern regarding cell loss
In situ engineering: injection of cells and an injectable biomaterial	Biomaterial acts as a temporary supporting matrix while cells will simultaneously regenerate infarction	Involves open chest surgery and this suggestion is at infancy
Injection of a biomaterial alone	Matrix for homing autologous progenitor cells	Immunogenicity; as only natural polymers have been suggested
Left ventricular restraints ("wrapping" the ventricles with a biopolymer or patch)	Does not involve cell injection. Utilizes engineered biomaterial construct	Prevents remodeling, but does not repair or regenerate damaged area
Tissue engineering approaches using scaffold	Ensures cells are delivered to desired area with minimal cell loss. Provides mechanical support	Involves open chest surgery. More work is required to determine suitable cell type and biomaterial

Source: Adapted from Jawad, H. et al., *J. Tissue Eng. Regen. Med.*, 1, 327, 2007.

(Jackson et al. 2001), have both advantages and disadvantages. Directly injecting cells into the infarcted area ensures the delivery of cells directly to the damaged area, but is hampered by significant cell loss after needle withdrawal from the myocardium. The preferred method of introducing cells is by injection into the coronary vessels anticipating that they will home to the infarcted area. In an attempt to increase contact time of cells with the myocardium, blood flow in coronary vessels can be temporarily occluded by angioplasty balloon inflation. This technique is also inefficient at cell delivery, with only a small fraction of the delivered cells being retained in the heart (<5%) (Grossman et al. 2002, Hoffmann et al. 2005). However, the best route for cell delivery is still to be determined, and in particular to minimize cell loss.

These and related results have prompted search for alternative delivery techniques for the cells involving myocardial tissue engineering (MTE) concepts based on the application of a substrate or scaffold, for example, cells are delivered in a preformed cardiac patch or scaffold (Zimmermann et al. 2004, Jawad et al. 2007, Chen et al. 2008a,b, Vunjak-Novakovic et al. 2010, Yeong et al. 2010).

The aim of MTE is thus to repair or regenerate a damaged section of the heart following the classical TE approach of using a biomaterial-based construct. MTE involves the synthesis of a scaffold or patch made from a biomaterial which is combined with cells in vitro. The biomaterial–cell construct can be assessed ex vivo to determine the cell survival and their contractile function before it is implanted into the damaged (i.e., scarred) area of the heart where it acts as a support or tissue replacement. The main function of the biomaterial hence is to act as a vehicle for the delivery of cells to the damaged area and to improve the retention of the cells in the damaged region. Once cells are delivered to the desired region, the hope is that the cells integrate with the host tissue forming new myocardium and the biomaterial (patch) provides further mechanical support (Jawad et al. 2007, Chen et al. 2008a).

Biomaterials research is a broad subject area and improved biopolymers suitable for application as scaffolds in MTE are being continuously developed (Wang et al. 2002, Jawad et al. 2007, Chen et al. 2008a, Ott et al. 2008, Rosellini et al. 2008, Yeong et al. 2010, Park et al. 2011). To date, a great variety of synthetic and natural polymers, as well as composite materials, have been proposed for MTE (Jawad et al. 2008, Park et al. 2011). This chapter gives an overview of the biomaterials developed and applied in the most significant MTE strategies, summarizing results published in the specialized literature with focus in the last 10 years.

14.2   CONCEPT OF MYOCARDIAL TISSUE ENGINEERING (MTE)

MTE combines the principles of developmental biology, cardiac muscle cell biology, and materials science, and involves the use of preformed three-dimensional (3D) scaffolds, in the form of mesh, patch, or foam, being cultured with relevant cells, which is implanted into the infarcted region of the heart, as shown schematically in Figure 14.1 (Jawad et al. 2008). Many research groups worldwide are investigating MTE approaches and key challenges are discussed by Vunjak-Novakovic et al. (2010), Gaetani et al. 2011, Chiu et al. 2012. The goal of biomaterial-based MTE approaches is to design, fabricate, and characterize suitable biomaterials for the scaffolds or patches which are the substrates required to guide cardiac tissue development. As in all TE approaches, the main functions of an ideal biomaterial for MTE are to enhance cell attachment, growth, and differentiation. To initiate regeneration of the host tissue, it is essential that the biomaterial encourages in vivo revascularization and favors integration with the host tissue (Park et al. 2011). At the same time, the biomaterial should degrade at a predefined rate to enable its replacement with newly formed tissue by safely degrading at a similar rate to new tissue formation. The scaffold should be eventually removed from the body by natural metabolic pathways without producing toxic by-products. In relation to MTE and of relevance to this chapter, the basic requirements for myocardial-bioengineered constructs include robust yet flexible mechanical properties, ability to withstand contraction, electrophysiological stability, vascularization ability and, preferably, the construct should be biomimetic. Table 14.2 summarizes the essential and desirable properties for a successful MTE scaffold (Jawad et al. 2008).

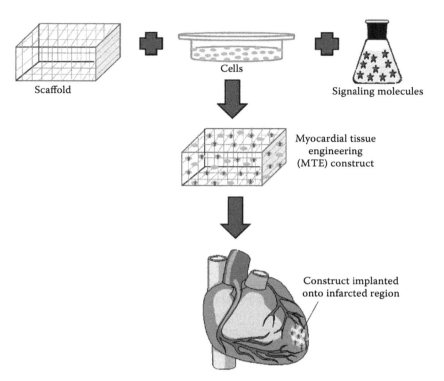

FIGURE 14.1   **(See companion CD for color figure.)** Schematic diagram illustrating the principle of MTE. A scaffold is made from a biopolymer. A selection of cells (cardiac or noncardiac) is expanded in vitro, with additional growth factors. Cells, with or without growth factors, are seeded onto the scaffold to form MTE construct and mechanical, electrical, and morphological properties optimized. This construct or patch is eventually sutured onto or into the infarcted region. (Adapted from Jawad, H. et al., *Br. Med. Bull.*, 87, 31, 2008.)

TABLE 14.2
Basic Requirements for Biomaterials for MTE

Biocompatible. Biomaterial must not be rejected or induces an inflammatory response in vivo.

Mechanical integrity. Biomaterial must enable handling during transplantation. More importantly, mechanical properties should match the host tissue it intends to replace and provide mechanical support during regeneration. In MTE, biomaterial is able to withstand, or even contribute to the continuous stretching/relaxing motion of the myocardium that occurs at each heartbeat.

Biodegradable. The degradation rate of the biomaterial should match the regeneration rate of the host tissue and the degradation by-products must be nontoxic and readily removed from the body.

Cell "friendly." Enhance cell adhesion and survival both in vivo and in vitro. Biomaterial must encourage cell proliferation and differentiation into cardiomyocytes as well as supporting vascular cells. Ideally, biomaterials could encourage cardiomyocyte alignment and maturation in vitro before implantation or in vivo, improving the contractile properties of the graft.

Electrical integration. Biomaterial must enable electrical integration of engineered graft with the native tissue to allow synchronized beating between the artificial construct and the heart. This requires matched excitability of host and grafted tissue, and support electrical of wave front propagation.

Vascularization. Biomaterial should encourage vascularization of the construct, to support survival of grafted cells.

Biomimetic. Reflect the extracellular matrix of the tissue it intends to replace.

Fabrication. Biomaterial must be easily accessible and designed with acceptable cost.

Source: Adapted from Jawad, H. et al., *Br. Med. Bull.*, 87, 31, 2008.

14.3 POLYMERS IN MTE

14.3.1 GENERAL CONSIDERATIONS

Polymers, both natural and synthetic (and combinations thereof), are the largest class of engineered biomaterials used today for TE, and they are available in a wide variety of compositions and properties (Nair and Laurencin 2007). Polymers are the materials of choice for applications in cardiovascular implants and in a wide range of MTE approaches (Chen et al. 2008a, Engelmayr et al. 2008, Rosellini et al. 2008, Freed et al. 2009, Tian et al. 2012, Guillemette et al. 2010, Karam et al. 2012). The successful development of a wide variety of synthetic biodegradable polymers has led to numerous applications in soft tissue engineering, considering that polymers can be tailor-made to match the properties of soft tissues (Nair and Laurencin 2007). A summary of polymers being proposed for MTE is presented in Table 14.3 (Jawad et al. 2008). Early studies on polymeric materials for MTE were based on hydrolytically degradable biocompatible polymers composed of polylactic acid (PLA), polyglycolic acid (PGA), and their copolymer polylactic-*co*-glycolic acid (PLGA) (Zammaretti and Jaconi 2004). With increasing research and experimental trials, it has been established that elastic properties of the biomaterial should match the elastic properties of the native heart as closely as possible to prevent the cells detaching from the bioengineered construct (McDevitt et al. 2002); thus, alternative polymers are continuously being developed.

Thermoplastic elastomers, specifically, nanostructured multiblock (segmented) copolymers, multiblock polyurethanes for example, have been recognized as key polymers in the medical field being

TABLE 14.3

Examples of Biopolymers, and the Processing Techniques Used, to Produce Engineered Constructs in MTE

Scaffold Material	Method Used for Construct Processing
Collagen	Commercially available scaffold
Collagen gel	Bioreactor
Gelatin mesh (Gelafoam)	Commercially available patches
	Biostretch/bioreactor
Collagen + glycosaminoglycan (GAG)	Cross-linking methods
Collagen type I matrix	Bioreactor
Collagen type I sponge + Matrigel	Bioreactor
Sodium alginate	Freeze-drying technique
Polyurethane	Solvent cast and spin coating
Elastomeric 1,3-trimethylene carbonate (TMC) and D,L-lactide (DLLA) and copolymers	Salt-leaching
Vicryl mesh (Dermagraft, Smith, and Nephew)	Commercially available
Poly(*N*-isopropylacrylamide) (PIPAAM)	Cell sheeting
Poly(ε-caprolactone)	Electrospinning
Polyglycolic acid (PGA)	Commercially available
Non-woven poly(lactide)- and poly(glycolide)-based (PLGA)	Electrospinning
Poly-*co*-caprolactone (PGCL)	Solvent casting and particle leaching
Poly(ester urethane) urea and collagen type I	Electrospinning
Polycaprolactone (PCL) mesh coated with collagen type I	Electrospinning
PGA, PLA, and PCL polymers with commercially available hydrophilic collagen sponge (Ultrafoam®)	Three polymers mixed and pores achieved using gas-forming method, followed by immersion in collagen
Collagen (type I and IV) and Matrigel matrix mixed with cells and seeded onto non-woven polymer mesh (poly(L-lactic acid) reinforced with PTFE)	Materials mixed in circular moulds

Source: Adapted from Jawad, H. et al., *Br. Med. Bull.*, 87, 31, 2008.

among the most bio- and blood-compatible known today (El Fray 2003). Moreover, poly(glycerol sebacate) (PGS), a chemically synthesized thermoelastomer, first developed by Wang et al. in 2002, has attracted interest for a number of medical applications particularly for MTE. PGS is a bioresorbable flexible polymer that undergoes surface hydrolytic degradation (Wang et al. 2002, Chen et al. 2008b, Rai et al. 2012). The starting materials for PGS are glycerol and sebacic acid. PGS has been shown to be safe in vivo (Tamada and Langer 1992, Wang et al. 2002). The low toxic nature of PGS can also be drawn from the approval of glycerol and polymers containing sebacic acid for medical applications by the US Food and Drug Administration (Wang et al. 2002). PGS has also been found to be hemocompatible which adds to its advantage as a cardiac patch material (Motlagh et al. 2006).

14.3.2 Methods Adopted in MTE

In this section, a brief insight is given on a series of techniques used to produce engineered constructs for MTE.

14.3.2.1 Electrospinning Methods

Collagen is the main constituent of the ECM, and since MTE constructs require an ECM-like structure and topography to mimic the in vivo scale of the collagen fibrils in the ECM, electrospinning has been suggested for this purpose because of the ability of the method to produce nanoscaled fibrous structures (Sill and von Recum 2008, Ravichandran et al. 2011). Electrospinning was developed first in 1934 and it has recently gained acceptance as a TE scaffold manufacturing method enabling development of fibrous scaffolds from synthetic, natural, or a combination of both types of polymers, with submicron pores and nanotopographic surfaces (Pham et al. 2006). Electrospinning involves an electrically charged jet of polymer solution produced by a high voltage. A constant pressure generated by a metering pump leads the polymer mixture to flow from the pipette, at a constant rate, onto a collector screen leaving a polymer fiber (fiber diameter can range from ~3 nm to ~5 μm). Electrospinning has gained acceptance for MTE applications and it has been suggested for several biomaterials.

For example, Ishii et al. (2006) cultured primary CMs harvested from neonatal rats onto biodegradable electrospun nanofibrous poly(ε-caprolactone) (PCL) meshes with average fiber diameter 250 nm, using the cell layering technique (Shimizu et al. 2002). The layering of individual grafts after 5–7 days of cell seeding enabled the construction of a 3D cardiac graft. It was found that beating CMs attached well to the meshes throughout the 14 day experimental period. Histological and immunohistochemistry tests confirmed morphological and electrical communications were established in constructs with up to five layers of meshes, proving strong adherence between the individual layers and the cells (Haraguchi et al. 2006). Wang et al. (2000) have assembled MTE scaffolds made of electrospun elastase-sensitive polyurethane urea nanofibers, encapsulating IGF-1 into poly(lactide-*co*-glycolide) (PLGA) microspheres. The IGF-1 that was released in culture with MSCs remained active for 4 weeks. Under hypoxia/nutrient starvation conditions, the IGF-1-loaded scaffolds were found to significantly improve MSC survival. However, long-term studies are yet to determine the cells' lifespan in vitro as well as in vivo. More recently a series of different studies based on electrospinning of a variety of biopolymers has brought new knowledge on the applicability of this technique for cardiac patch developments (for example: Ravichandran et al. 2011, Tian et al. 2012, Genovese et al. 2011).

14.3.2.2 Bioreactors

The general purpose of a bioreactor is to encourage growth and development of relevant cells or tissue on biomaterials as if under in vivo condition (Freed et al. 2006). The ability of producing a 3D myocardial tissue comprised of more than a few layers of muscle is the main advantage of using a bioreactor (Vunjak-Novakovic et al. 2010). Although the main purpose of using bioreactors is still considered to be scientific research, further improvements in in vitro design and quality control will eventually allow their application in MTE. Several different bioreactors have been suggested for cardiac constructs. These include static or mixed flask bioreactors where constructs are suspended in cultivation medium, rotating vessel bioreactors where constructs are suspended in medium that has a constant rotational flow, and finally perfusion cartridge bioreactors where constructs are perfused at interstitial velocities, comparable to blood flow in native tissue (Vunjak-Novakovic et al. 2006). One significant advantage of bioreactor systems used in MTE is the ability to cultivate cells on biomaterials producing contractile cell–polymer constructs (Vunjak-Novakovic et al. 2010). Moreover, several different geometries enabling various patterns of fluid dynamics have been suggested (Leor et al. 2005).

A limitation of bioreactors is their restricted ability to supply an adequate amount of nutrients to a tissue of thickness greater than approximately 100 μm or less than 10 cell layers thick (Leor et al. 2005). This limitation causes weak cellular integrity, short duration of contractility, as well as inhomogeneous cell seeding; which affects tissue function and cellular viability (Vunjak-Novakovic et al. 2006, 2010). Carrier et al. (2002) and Radisic et al. (2006a,b) demonstrated the importance of oxygen on engineered cardiac grafts to overcome the limitation of producing a 100 μm thick new tissue. It was found that by increasing the oxygen concentration supply to the in vitro construct, improved engineered cardiac muscle was produced. Additional studies are required to evaluate the specific properties of these engineered constructs, including the mechanical behavior, biodegradability, and eventually their in vivo performance. Primary neonatal rat ventricular cells were cultured on PGA scaffolds by Bursac et al. (1999) using bioreactors to form cardiac muscle constructs. The polymer scaffold provided the 3D substrate for cell attachment and for tissue formation, whereas the bioreactor was used to promote mass transfer of nutrients and gases to the forming tissue. More recently, Radisic et al. (2009) cultured neonatal rat heart cells on highly porous collagen scaffolds in a bioreactor with electrical field stimulation. An excitable tissue such as myocardium is able to propagate electrical impulses. Optical mapping was used to measure the electrical impulse propagation. The average conduction velocity of the electrically stimulated constructs was shown to be significantly higher than that of the non-stimulated constructs. The measured electrical propagation properties correlated to the contractile behavior and the compositions of tissue constructs. Electrical stimulation of the scaffolds during culture could significantly improve the amplitude of contractions, the tissue morphology, and the expression of connexin 43 compared to the non-simulated controls. The study hence provided evidence that electrical stimulation during bioreactor cultivation may improve electrical signal propagation in engineered constructs for MTE applications (Radisic et al. 2009).

14.3.2.3 Cell Sheeting (Temperature Sensitive)

This method reported by Shimizu et al. in 2002 (Shimizu et al. 2002) has been suggested for MTE. The concept involves temperature-responsive dishes made from a specific polymer, poly(N-isopropylacrylamide) (PIPAAM), which is temperature sensitive. At 37°C, the polymer is hydrophobic and cell adhesive; however, a 5°C reduction in temperature can cause the polymer to become non-cell adhesive as it hydrates and swells because its hydrophobic nature changes to hydrophilic. CMs seeded onto the polymer will produce individual spontaneously beating myocardial tissue sheets. The implantation of a 100 μm thick cardiac tissue made from six monolayered mesenchymal stem cell (MSC) sheets layered together was reported in 2006 (Miyahara et al. 2006). This method constitutes a very promising approach, as it does not need the involvement of enzymes during cell culture and has the possibility to produce individual cell sheets, forming 3D simultaneously beating myocardial tissue. Many other studies have illustrated the successful combination of cell sheeting and temperature-responsive culture substrates in MTE (Sekine et al. 2006, Sekiya et al. 2006, Sawa et al. 2012).

14.3.2.4 In Situ Engineering

This is a "scaffold-free" approach in MTE, which is included in this chapter for completeness. It involves the direct injection of the biomaterial and cell mixture into the infarcted region of the heart (Singelyn and Christman 2011). Unlike the use of prefabricated scaffolds or patches, injectable natural polymers are able to bond readily to the native tissue, as they can be easily shaped or cast to the heart's complex structure, simultaneously providing support for the cells. Different polymers have been suggested for in situ MTE, including alginate (Leor et al. 2000), fibrin glue (Christman et al. 2004a–c), collagen (Dai et al. 2005), and Matrigel (Kofidis et al. 2005a). Moreover, acellular alginate with bioactive molecules has been suggested with the intention that cardiac progenitor cells will home to the infarcted region (Leor et al. 2005). Biomaterials as carrier for peptides and growth factors (Iwakura et al. 2003) as well as "self-assembling" peptide nanofibers (Davis et al. 2005) have been also proposed. These strategies are still under consideration to produce new myocardial tissue, and experience with injection of cells directly into the heart has suggested that there is a low limit on the amount of new material that can be introduced into either the dense and continually compressing myocardium or the stiff and avascular scar.

Focus is now being placed on ECM proteins as crucial mediators in developing and maintaining the characteristics of 3D cardiac cell cultures (Akhyari et al. 2008). In studies carried out by Yu et al. (2009), alginate biopolymer was used as ECM and cell adhesion ligand, arginine–glycine–asparagine (RGD), as cell matrix mediator. The intramyocardial injection of RGD-modified alginate onto the infarcted rat model 5 weeks post MI demonstrated reshaping of the aneurismal LV, improved LV function, and induced angiogenesis (Yu et al. 2009). Microspheres have also been used as delivery vehicles for growth factors, drugs, and cells in tissue engineering (Buket et al. 2008, Patil and Sawant 2008). Alginate microspheres that have been found to be nontoxic, semipermeable, and able to provide immune protection for many cell types and recipients have also been studied for in situ myocardial engineering (Chang 1998, Yu et al. 2010). Human MSCs (hMSCs) have been encapsulated in RGD-modified alginate microspheres to be delivered onto the infarcted myocardium in rats. The in vitro studies of hMSCs demonstrated that the RGD-modified alginate can improve cell attachment, growth, and increase angiogenic growth factor expression. The RGD-alginate microspheres, shown in Figure 14.2, effectively arrested the negative remodeling of the LV and prevented the infarct wall thinning and chamber dilation after an infarct (Yu et al. 2010).

Fibrin glue, a biopolymer formed by polymerization of fibrinogen monomers, is another natural polymer being used for in situ MTE. This material has been reported to induce angiogenesis (Christman et al. 2004a) and it was used to engineer a contractile 3D cardiac tissue by seeding neonatal cardiac myocytes which were cultured in vivo in silicone chambers while being in close proximity to a vascular pedicle, therefore having an intrinsic vascular supply. Christman et al. (2004a,b) demonstrated the success of transplanting skeletal myoblasts in fibrin glue via direct injection into the infarcted region, where infarct size decreased and fibrin glue increased arteriole density and cell survival in vivo. Injecting fibrin alone was reported to preserve cardiac function

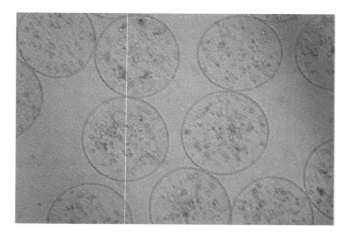

FIGURE 14.2   **(See companion CD for color figure.)** Human mesenchymal stem cells (hMSCs) encapsulated in RGD-modified alginate beads. (Reproduced from *Biomaterials*, 31, Yu, J., Du, K.T., Fang, Q., Gu, Y., Mihardja, S.S., Sievers, R.E., Wu, J.C., and Lee, R.J., The use of human stem cells encapsulated in RGD modified alginate microspheres in the repair of myocardial infarction in the rat, 7012–7020, Copyright 2010, with permission from Elsevier.)

and reduce cardiac remodeling where left ventricular geometry was unchanged (Christman et al. 2004c). Similar results with implantation of bone marrow mononuclear cells with and without fibrin matrix were reported by Ryu et al. (2005). The combination of bone marrow mononuclear cells and fibrin matrix produced approximately 350 microvessels/mm^2 density in the infarcted region, which was substantially higher than the density obtained without fibrin matrix and medium injection only. Injection of endothelial cells with fibrin glue also resulted in improved left ventricular function, myocardial blood flow, and neovascularization (Chekanov et al. 2003). A comprehensive review about injectable materials for MTE has been published recently (Singelyn and Christman 2011).

Collagen has also been suggested as an injectable biopolymer; however, conflicting results have been reported. Huang et al. (2005), for example, showed the effect of three different biopolymers injected into the infarcted region separately: fibrin, collagen, and Matrigel with endothelial cells. Results demonstrated increased capillary formation for all three biopolymers where collagen showed greater infiltration of myofibroblasts into the region in comparison to fibrin and Matrigel. Dai et al. (2005) reported an improvement in left ventricular function, increased scar thickness and remodeling, but no changes in regional blood flow, after injection of collagen into infarcted rat hearts. Implantation of a bioartificial liquid made of Matrigel and murine endothelial ESCs into the ischemic myocardium of mice improved cardiac function as well as attenuated left ventricular remodeling and expression of gap junctional protein connexin 43 at intercellular sites, indicating possible connections with the native myocardium (Kofidis et al. 2005a).

A different biomaterial was introduced in 2005 by Davis et al. "Self-assembling" peptide nanofibers were injected into the myocardium, creating microenvironments to recruit progenitor cells and induce vessel growth. Immunohistochemistry tests revealed progenitor vascular cells and cells expressing endothelial markers were homed to the nanofibers. Separately, neonatal CMs were injected with the peptides and they were reported to not only survive but also augment endogenous cell recruitment. Other studies reporting the success of injecting biomaterials with peptides and growth factors are those of Iwakura et al. (2003), Hsieh et al. (2006), and Wall et al. (2010). These approaches are relatively new in the MTE field and they have only been explored by limited number of researchers so far.

14.3.3 COMMON POLYMERS USED IN CARDIAC PATCH APPROACHES FOR MTE

A wide range of polymers has been suggested for MTE as shown in Table 14.3 (Jawad et al. 2008). This section will give an insight into the types of polymers reported specifically for developing cardiac patches for myocardial regeneration.

14.3.3.1 Synthetic Polymers

Several in vitro and in vivo methods are being proposed for making cardiac tissue from synthetic polymers. It has been demonstrated that the successive addition of spatially organized CMs onto microcontact laminin-patterned lines ($15 \times 10\,\mu m$) on polyurethane films can produce 3D-organized CM sheets (McDevitt et al. 2002). Pego et al. (2003) investigated the in vivo implantation of porous elastomeric 1,3-trimethylene carbonate (TMC) and D,L-lactide (DLLA) polymers and their copolymers made by a salt-leaching technique cultured with rat CMs. The construct was shown to disappear within 10 months and no severe tissue reaction was observed postimplantation. It was also shown that this copolymer has the ability to sustain the cyclic loading of the heart muscle under physiological conditions. Poly(lactic acid)/poly(glycolic-co-lactic acid) (PLLA/PLGA), two of the most popular biodegradable polymers for tissue engineering (Nair and Laurencin 2007), have been recently conveniently used to develop foams for heart muscle engineering (Lesman et al. 2010).

Dermagraft (Smith and Nephew, London, UK), which is a product composed of human dermal fibroblasts, has been cultured on Vicryl® mesh (90:10 poly(glycolide:lactide)) to develop angiogenic patches (Kellar et al. 2001). In vivo implantation was successful as the patch-stimulated angiogenesis within the region of cardiac infarction. It was also found that the patch does attenuate further loss of left ventricular function, suggesting that this effect may be related to myocardial revascularization.

The seeding of three different types of cells, i.e., fibroblasts, endothelial, and CM cells onto 2 mm poly(ethylene glycol) (PEG) disks, was reported by Iyer and Radisic (2006). Spontaneous beating was observed in both the tri-culture samples and PEG samples solely with CMs. Ke et al. (2005) have also investigated the effectiveness of grafting a commercially available PGA-biodegradable patch with mESCs (in an undifferentiated state). The patch was then sutured on the hearts of rats that were MI-induced, by ligation of the left coronary artery. Cardiac function and hemodynamics were evaluated 8 weeks post-patch implantation. A significant improvement in ventricular function and blood pressure with a survival rate of 82.1% was confirmed while, on the other hand, 46.2% of mice that did not receive the cardiac patch with ESCs died within 8 weeks. The potential of nano-structured elastomer substrates to deliver ESC-derived CMs (ESC-CM) to an infarcted area of the myocardium was investigated by Jawad et al. (2010). The polymers investigated were soft and strong poly(aliphatic/aromatic ester) multiblock thermoplastic elastomers with poly(ethylene terephthalate) (PET) hard segments and dimer fatty acid, i.e., dilinoleic acid (DLA) soft segments, respectively, with and without addition of 0.2 wt% TiO_2 nanoparticles to form nanocomposites. It was shown that addition of TiO_2 nanoparticles significantly altered surface roughness and enhanced adhesion and spreading of ESC-CM derived from mouse and human ESCs. The materials did not affect the functional activity of spontaneously beating hESC-CM, which was demonstrated by the unaltered rate of their beating. The cells demonstrated contractile activity on the materials for more than 2 months in culture. The approach of combining biomaterials and ESCs is in its beginning; however, ongoing research as the one reported in Jawad et al. (2010) will provide a more solid foundation to questions that need answering for advancement of the field.

Although a fairly new polymer in MTE applications, PGS is attracting increasing attention for development of cardiac patches (Chen et al. 2008b, 2010, Engelmayr et al. 2008, Freed et al. 2009, Park et al. 2011, Rai et al. 2012). Studies carried out to assess the mechanical properties of PGS films revealed that films synthesized in the temperature range 110°C–130°C had stiffness values ranging between several tens of kPa to ~1 MPa, thus covering the range of the passive stiffness of the heart muscle (Chen et al. 2008b). Pretreatment of PGS scaffold has also been studied by Radisic et al. (2008), whereby a porous PGS scaffold was pretreated with cardiac fibroblasts resulting in improved properties of the engineered heart tissues (EHT). During scaffold pretreatment when seeded with lower-density cardiac fibroblasts, these were able to recover from the isolation procedure and remodel the polymeric scaffold by depositing components of ECM and secreting soluble factors. Therefore, the scaffold already being conditioned to provide an environment similar to that of the native ventricle could support good growth and tissue assembly of the seeded myocytes (Radisic et al. 2008). Biomimetic approaches with PGS as matrix material have also been carried out in the framework of MTE (Radisic et al. 2006a,b, Engelmayr et al. 2008). Moreover, combination of PGS and PCL have been electrospun to produce 100–400 μm thick PGS–PCL (at 5:1, 3:1, 2:1, 0:1 ratio) scaffolds (Sant and Khademhosseini 2010). The diameter of fibers depended on the PGS–PLC ratio and electrospinning voltage ranging from ~2.8 to ~6.8 μm. Results demonstrated good cell proliferation and attachment on PGS–PLC scaffolds which also supported CMs growth.

CMs have high oxygen demand and rely on unobstructed oxygen supply when they are physiologically embedded in a delicate capillary network (Zimmermann 2008). A study was carried out to mimic this scenario using PGS scaffolds which could provide in vivo like oxygen supply to the cells within the engineered cardiac constructs consisting of a cell population of both myocytes and non-myocytes (fibroblasts). To mimic the capillary network, a highly porous PGS scaffold using salt-leaching technique was fabricated with the introduction of a parallel array of channels (Radisic et al. 2006b). Cardiac muscle fibers are highly branched and hierarchically surrounded and embedded in a 3D collagen network comprising distinct endomysial, perimysial, and epimysial levels of organization that resemble a honeycomb network. This complex structure imparts cardiac anisotropy, i.e., direction-dependent electrical and mechanical properties (Engelmayr et al. 2008, Zimmermann 2008, Freed et al. 2009). In this context, an interesting study was carried out by Engelmayr et al. (2008) in which they replicated such accordion-like honeycomb (ALH) structure using PGS in order to mimic human myocardium possessing anisotropic properties and promoting parallel heart cell alignment. Related multilayered elastomeric scaffolds based on PGS substrates with different pore architecture were also reported (Park et al. 2011). The scaffolds were microfabricated using excimer laser microablation in which modification of PGS to integrate a preferred (anisotropic) plane of flexibility into the scaffold material was carried out. An overview of the methods developed to fabricate the multilayered PGS scaffold is shown in Figure 14.3 (Park et al. 2011). The scaffold when seeded with neonatal heart cells demonstrated preferential heart cell alignment within 1 week of culture (Engelmayr et al. 2008). During this period, the scaffold was also able to retain anisotropic mechanical properties whilst withstanding in vitro fatigue cycle mimicking the dynamic physiological epicardial strains. The mechanical properties similar to native rat right ventricular myocardium were also achieved after optimization of polymer curing time (Engelmayr et al. 2008). Finite element (FE) simulations and a homogenization approach were carried out (Engelmayr et al. 2008) to retrospectively predict the anisotropic effective stiffness of the ALH PGS scaffold. This study showed that the FE model could be useful in designing variations in the ALH pore geometry that would then simultaneously provide proper cardiac anisotropy and reduced stiffness to enhance heart cell–mediated contractility (Engelmayr and Jean 2010). As the dynamic in vivo environment is different from in vitro, it becomes critical to assess the performance of any scaffold material in vivo, which ideally must be monitored serially and noninvasively. Pertaining to this, studies were carried out by Stuckey et al. (2010) in which magnetic resonance imaging was used to evaluate the in vivo performance of three patches made of PGS, poly(ethyleneterephathalate)/dimer

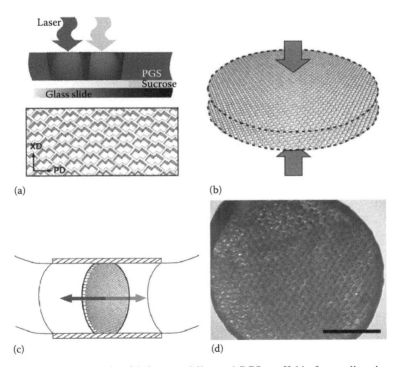

FIGURE 14.3 Method used to microfabricate multilayered PGS scaffolds for cardiac tissue engineering. (a) One-layered PGS scaffolds with accordion-like honeycomb pores made by laser microablation. (b) PGS membranes stacked and laminated to produce two-layered scaffolds. (c) Scaffold seeded with heart cells and cultured with bi-directional interstitial perfusion. (d) Phase contrast micrograph of a construct after 7 days, scale bar: 2mm. (Reproduced from *Biomaterials*, 32, Park, H., Larson, B. L., Guillemette, M. D. et al., The significance of pore microarchitecture in a multi-layered elastomeric scaffold for contractile cardiac muscle constructs, 1856–1864. Copyright 2011, with permission from Elsevier.)

fatty acid (PED), and TiO$_2$-reinforced PED (PED–TiO$_2$) (discussed in Jawad et al. 2010) grafted onto infarcted rat hearts. Patch-free rat infarcted heart was used as control. The results showed rapid in vivo degradation of PGS as opposed to its degradation in vitro. However, the PGS patch which was mechanically compatible to the rat heart was found to be successful in reducing hypertrophy giving it potential for limiting excessive post-infarct remodeling. (Stuckey et al. 2010).

14.3.3.2 Natural Biomaterials

A number of different natural polymers have been used in MTE as well as to regenerate other types of tissue, which are included in Table 14.3 (Jawad et al. 2008).

14.3.3.2.1 Collagen-Based Constructs

Zimmermann et al. (2000) have developed a convenient technique to engineer a cardiac muscle construct, which is known as EHT, which is based on the combination of neonatal CMs with an artificial ECM made mainly from collagen type I and Matrigel. The all natural EHT construct has several advantages. For example, EHT has been reported to contract for 8 weeks in vivo; newly formed myocardium of ~450 μm thickness has been observed; functional and morphological properties of differentiated heart muscle have been reported; and heart muscle shape and size can be manipulated accordingly. ETH has also some disadvantages. For example, upon implantation, left ventricular function was not improved according to echocardiography studies and the need for immunosuppression for the survival of the EHT in vivo has been reported (Zimmermann 2002). Some modifications have been implemented to the original EHT approach to improve in vivo performance; first culture took place under elevated ambient oxygen, second culture was carried out under auxotonic load (EHT contracts against load-adjusted coils), and finally culture medium was supplemented with insulin (Zimmermann et al. 2006). Twenty-eight days after implantation, electrical coupling to the native myocardium was observed as well as systolic wall thickening of the infarcted region. The results reveal that large contractile grafts may be constructed in vitro, implanted in vivo, and eventually support infarcted myocardial muscle. Zimmerman et al. (2006) have highlighted that further investigations on potential clinical use are required, considering cell sources, immunogenicity aspects of the EHTs, and graft size optimization. The latest progresses and challenges in this field have been highlighted recently (Zimmerman 2011).

In related developments, Guo et al. (2006) demonstrated that ESCs can be used for engineered muscle constructs. CMs derived from ESCs were mixed with type I collagen supplemented with Matrigel, and pipetted into circular casting moulds then incubated for 30–45 min to allow hardening of the mixture making the engineered cardiac tissue (ECT). After 7 days in culture, the ECT was placed into a stretch device to undergo unidirectional cyclic stretch for an additional 7 days and like rat neonatal CMs, produced contractile muscle strips. Li et al. (2000) seeded fetal rat ventricular cells onto commercially available 3D gelatin mesh (Gelafoam), forming cardiac tissue that contracted spontaneously both in vitro and in vivo for 5 weeks after implantation onto rat myocardial scar tissue, forming junctions with the recipient heart. However, postimplantation results showed no sign of cardiac function improvement. Zhong et al. (2005) developed a scaffold by electrospinning (see Section 14.3.2.1) collagen and glycosaminoglycan (GAG), both abundant proteins in the ECM of the body, forming a nanofibrous scaffold which exhibits a favorable environment for cell proliferation upon incorporation of collagen. Other studies have reported successful combination of collagen and GAG, where bone marrow–derived MSCs were implanted onto the biodegradable scaffold and onto the infarcted region resulting in neovascularization (Xiang et al. 2006).

Kofidis et al. (2005b) combined undifferentiated mESC with collagen type I matrix and reported the benefits of ESCs in a 3D collagen matrix when implanted in the infarcted area introduced by ligation of the left anterior descending artery in nude rat and surgically forming an intramural pouch. Stable intramyocardial grafts were observed without altering the myocardial geometry and increasing ventricular wall thickness (1.4 ± 0.1 mm) as well as expressing gap junctional protein connexin 43 in vivo. This study has thus highlighted the benefit of both the application of ESCs and the continually questioned immunogenicity of collagen when used in MTE.

Attempts have been made to improve the mechanical function of bioartificial tissues grown on natural polymers. For example, Akhyari et al. (2002) seeded heart cells on Gelafoam and applied cyclical mechanical stress at a frequency of 80 cycles/min for 14 days using a biostretch apparatus. The important effect of mechanical stress during development of bioartificial tissue was discussed, as the proliferation and distribution of cells improved upon incorporation of mechanical stress.

Collagen hemostatic scaffold, also called tissue fleece, was first suggested for MTE by Kofidis et al. (2003) who developed a contractile bioartificial myocardial tissue from collagen tissue fleece and CMs in vitro with homogeneous cellular distribution. In another study, Srinivasan and Sehgal (2010) used a microbial collagen extraction process to obtain pure collagen fibers from bovine tendon. In vitro and in vivo studies revealed that the biocompatible types I and III collagen fibers could be woven as a fleece to get a 3D scaffold with good mechanical properties suitable for cardiac patch applications. Gaballa et al. (2006) reported a reduction in cardiac remodeling and increased neoangiogenesis when grafted with acellular 3D collagen scaffold onto infarcted rat hearts. Further neoangiogenesis was reported when rats implanted with collagen scaffold received subcutaneous injection of granulocyte-colony-stimulating factor (G-CSF), which has been reported to mobilize progenitor cells. Although the collagen scaffold is applied to provide initial mechanical support and reduce remodeling, immunogenicity of the scaffold is still an obstacle. In this context, there are conflicting results as to whether G-CSF does actually mobilize resident cells to home to the infarct region (Quaini et al. 2002, Noral et al. 2003).

14.3.3.2.2 Alginate-Based Constructs

Alginate, a natural biomaterial, is a negatively charged linear copolymer produced by brown seaweed, although certain bacteria also produce alginates. Leor et al. (2000) grew fetal rat cardiac cells within 3D porous sodium alginate scaffolds made by freeze-drying technique. Biograft transplantation of the cardiac graft took place 7 days post MI. Visual and histological examinations revealed intensive neovascularization from neighboring coronary network and limited myofibers embedded in collagen matrix. Alginate scaffolds were shown to provide a supporting and conductive environment to encourage cardiac cell culturing as well as promising results for the regeneration and healing of the infarcted myocardium. However, because no experiments were carried out for alginate scaffolds without cell incorporation, it could not be determined whether it is the alginate alone or the combination of the two elements that encourages neovascularization. Blending of alginate with gelatine in films has also been considered as matrix material for MTE. Cell culture studies using C_2Cl_2 myoblasts showed best cell proliferation for blend films with 20:80 ratio of alginate to gelatin (Rosellini et al. 2008). A study carried out by Amir et al. (2009) considered a novel cardiac patch developed by seeding fetal CMs within macroporous alginate scaffolds followed by vascularization into the rat peritoneal cavity. The cardiac patch was then transplanted onto the damaged heart. The peritoneal-generated cardiac patch was tested in the heterotropic heart model which enables the replacement of a full thickness wall. The results showed that the patch lacked characteristic features of myocardium, such as the absence of striation. However, the patch was successful in sealing the defects in the LV wall and in maintaining the global contractility of the heart (Amir et al. 2009). An approach involving alginate/RGD microspheres (Wu et al. 2010) was discussed in Section 14.3.2.4.

14.3.3.2.3 Decellularized Matrices

Acellular tissue scaffolds provide natural ultrastructural, mechanical, and compositional platform for recellularization and tissue remodeling. The application of decellularized matrices in MTE is an expanding field, already well-investigated for cardiac valve engineering, and some typical examples are discussed in this section for completeness. Robinson et al. (2005) have reported successful implantation of an engineered cardiac patch made from four layer urinary bladder–derived ECM in the infarcted left ventricular walls of pigs. Results from 1 week and 1 month implantation revealed thrombus formation and inflammation. However, 3 months postimplantation showed fibrocellular tissue with contractile cells and biodegradation of the matrix when compared to the implantation of polytetrafluoroethylene (PTFE) where necrosis and calcification were observed. Kochupura et al. (2005) have found CM populations on a myocardial patch derived from ECM, which also provided mechanical benefit in the myocardium; however, this investigation was not carried out for infarcted hearts. More recently, decellularized cadaveric rat hearts were studied to engineer a bioartificial heart (Ott et al. 2008). The decellularized heart matrix was seeded with neonatal rat CMs and, to establish function, the seeded construct was maintained in a bioreactor that simulated cardiac physiology. Under physiological load and electrical stimulation by day 8, the construct could generate pump function (equivalent to about 2% adult or 25% of 16 week fetal heart function) (Ott et al. 2008). The authors suggested that with sufficient maturation the cultured organ could become transplantable either in part (as a ventricle for congestive heart failure) or as an entire donor heart at end-stage heart failure. This result thus opens avenues of using suitable parts of such cultured organs as cardiac patch and ultimately extending the approach with cadaveric human hearts. Wang et al. (2010) assessed the potential of decellularized porcine myocardium as a scaffold for thick cardiac patch tissue engineering. Bone marrow mononuclear cells seeded onto the decellularized scaffold were found to infiltrate and proliferate in the tissue constructs. The cells retained CM-like phenotype and showed indication of potential angiogenesis. The scaffolds showed stiffer mechanical response for both uniaxial and biaxial testing; however tissue extensibility and tensile modulus were found to recover in the constructs with increasing culture time (Wang et al. 2010). Hata et al. (2010) carried out studies to address the persistent problem of generating a functional myocardial patch that can maintain contractions in a thicker construct. By combining layered neonatal rat CM sheets with CM-seeded decellularized porcine small intestinal submucosa (SIS), a 3D myocardial patch of thicker construction with contraction in a defined direction was successfully generated (Hata et al. 2010).

14.3.3.3 Combinations of Natural and Synthetic Materials

As mentioned earlier, collagen is the major constituent of the ECM, and it enhances cell attachment when used as a scaffold. However, the mechanical properties of collagen are poor, being a disadvantage for its direct use in MTE. Combinations of synthetic polymeric materials (e.g., poly(ester urethane)urea) with collagen are being investigated to engineer suitable scaffolds using the electrospinning technique, in order to improve cellular adhesion (Stankus et al. 2004). Other combinations of natural and synthetic polymers have been discussed for hybridized cardiac patches (Pok and Jacot 2011, Ravichandran, et al. 2011, Rossellini et al. 2011).

In a related investigation, CMs from neonatal Lewis rats were cultured on electrospun, nanofibrous PCL meshes coated with type I collagen, forming cardiac nanofibrous meshes (CNM) (Shin et al. 2004). Formation of contractile cardiac grafts in vitro was achieved. The CNM consisted of a wire ring that acted as a passive load for the beating cells enabling contractions at natural frequency. Common to electrospun scaffolds, the advantages of this approach include nanoscale topography similar to that of the ECM, collagen coating permitting enhanced cellular adhesion and scaffold contraction in synchronization with beating cells.

The combination of polymeric synthetic materials with collagen using bioreactors has also been investigated to produce ECT. Biodegradable, hydrophobic PGA, poly(D,L-lactide-*co*-caprolactone), and commercially available hydrophilic collagen sponge (Ultrafoam™) have been mixed and made into a composite scaffold. Neonatal heart cells were seeded onto it, using bioreactor cartridges (Radisic et al. 2003). Krupnick et al. (2002) evaluated the transplantation of multipotent bone marrow–derived mesenchymal progenitor cells into infarcted syngenic rat hearts. Instead of directly injecting the cells in the infarcted region, they were seeded within a collagenous matrix made from collagen types I and IV and Matrigel. The mixture of cells and collagenase was then seeded onto a non-woven polymer mesh (poly(L-lactic acid) reinforced with PTFE) and implanted in vivo. Results were promising as minimal intra-cardiac inflammation occurred, cells retained fibroblastic shape, and immunohistochemistry revealed activity at the junction of native myocardium and the engineered construct in three of the four rats. Some transplanted cells stained for myosin indicated that cell differentiation occurred. Cardiac function appeared to be normal with no arrhythmias occurring. In a related investigation, Park et al. (2005) evaluated a composite scaffold of poly(D,L-lactide-*co*-caprolactone), PLGA, and type I collagen for cardiac tissue engineering. The scaffold contained open interconnected pores and an average void volume of 80% ± 5%. The composite scaffolds also enabled cell seeding of neonatal heart cells at high initial cell densities (1.35×10^3 cells/cm^3) because of its hydrophilic surface and provided efficient mass transfer to and from the cells because of the structural stability of the scaffold pores. The constructs also exhibited spontaneous localized contractions early in culture and were able to contract synchronously in response to electrical stimulation after 1–2 weeks of culture (Park et al. 2005).

14.4 DISCUSSION AND CONCLUDING REMARKS

This chapter has discussed the wide range of approaches being investigated for MTE involving the application of a great variety of biopolymers. The findings discussed, based on studies published in the last 10 years, contribute to analyze potential breakthroughs in the field of MTE, highlighting several important points as avenues for future research. MTE is an expanding field stimulating the development and characterization of many biomaterials, both synthetic and natural polymers (and composites). The analysis of the specialized literature has demonstrated the relevance of biopolymers and their importance in MTE. Several issues still need to be addressed for the success of MTE strategies based on biopolymers. For example, electrical coupling between cells is required to ensure the cells on the graft or patch beat in synchrony. Second, electrical coupling between the construct and native myocardium for simultaneous beating is still being investigated. It appears that cell sheeting methods could overcome this problem (Eschenhagen et al. 2006), where graft integration and no arrhythmias have been reported. It is also necessary to elucidate whether a dense "patch" is useful as opposed to a porous patch. It has been considered that for the cells to survive and carry out their full potential functions they must be embedded in a 3D scaffold containing pores. In this scenario, the required functions of the engineered construct must be assessed. If the construct will act solely as a "vehicle" to deliver cells to a damaged site degrading within a short period (e.g., within 3 months), a dense patch might be more suitable (Ke et al. 2005, Jawad et al. 2010). Alternatively, if the construct is to support the damaged area for a sustained period, then it is essential that it exhibits a porous structure (e.g., foam or mesh morphology) to ensure cell survival for the time they are in contact with the scaffold. In addition, the optimum design/shape of the construct and matrix material remains the subject of research, for example, are 3D structures like accordion honeycomb-type scaffolds, electrospan fibrous structures, or injectable gels more suitable in MTE strategies? Which specific biopolymer can withstand the complex requirements in each case? Are natural, synthetic, or combination of both types of polymers more suitable for MTE applications?

In addition, there is the need to identify which cell type is suitable for human use to regenerate the infarcted region of the heart, although this aspect is beyond the scope of the this chapter which has focused mainly on biomaterials. In conclusion, the ideal myocardial construct should mimic the morphological, physiological, and functional properties of the native cardiac muscle it intends to replace and remain viable after implantation. Biopolymers are an essential element of MTE strategies and their further development will undoubtedly contribute to future MTE breakthroughs, leading to improvement in function of the diseased myocardium as it integrates with the heart, with the ultimate goal of reducing morbidity and mortality of patients with heart failure.

ACKNOWLEDGMENT

The authors acknowledge financial support from the UK Biotechnology and Biological Sciences Research Council (BBSRC) (2006–2009) (grant number BB/D011027/1).

REFERENCES

Akhyari, P., Fedak, P. W. M., Weisel, R. D. et al. 2002. Mechanical stretch regimen enhances the formation of bioengineered autologous cardiac muscle grafts. *Circulation* 106: I137–I142.
Akhyari, P., Kamiya, H., Haverich, A., Karck, M., and Lichtenberga, A. 2008. Myocardial tissue engineering: The extracellular matrix. *European Journal of Cardio-Thoracic Surgery* 34: 229–241.

Akins, R. E. 2002. Can tissue engineering mend broken hearts? *Circulation Research* 90: 120–122.

Amir, G., Miller, L., Shahar, M. et al. 2009. Evaluation of a peritoneal generated cardiac patch in a rat model of heterotopic heart transplantation. *Cell Transplantation* 18: 275–282.

Anversa, P., Leri, A., Kajstura, J., and Nadal-Ginard, B. 2002. Myocyte growth and cardiac repair. *Journal of Molecular and Cellular Cardiology* 34: 91–105.

Buket B. F., Kose G. T., and Hasirci V. 2008. Sequential growth factor delivery from complexed microspheres for bone tissue engineering. *Biomaterials* 29: 4194–4204.

Bursac, N., Papadaki, M., Cohen, R. J. et al. 1999. Cardiac muscle tissue engineering: Toward an in vitro model for electrophysiological studies. *American Journal of Physiology-Heart C* 277: H433–H444.

Carrier, R. L., Rupnick, M., Langer, R., Schoen, F. J., Freed, L. E., and Vunjak-Novakovic G. 2002. Perfusion improves tissue architecture of engineered cardiac muscle. *Tissue Engineering* 8: 175–188.

Chang, T. M. 1998. Pharmaceutical and therapeutic applications of artificial cells including microencapsulation. *European Journal of Pharmaceutics and Biopharmaceutics* 45: 3–8.

Chekanov, V., Akhtar, M., Tchekanov, G. et al. 2003. Transplantation of autologous endothelial cells induce angiogenesis. *Pacing and Clinical Electrophysiology* 26: 496–499.

Chen, Q. Z., Bismarck, A., Hansen, U. et al. 2008b. Characterisation of a soft elastomer poly(glycerol sebacate) designed to match the mechanical properties of myocardial tissue. *Biomaterials* 29: 47–57.

Chen, Q. Z., Harding, S. E., Ali, N. N., Lyon, A., and Boccaccini, A. R. 2008a. Biomaterials in cardiac tissue engineering: Ten years of research survey. *Materials Science and Engineering R: Reports* 59:1–37.

Chen, Q. Z., Ishii, H., Thouas, G. A. et al. 2010. An elastomeric patch derived from poly(glycerol sebacate) for delivery of embryonic stem cells to the heart. *Biomaterials* 31: 3885–3893.

Chiu, L. L, Iyer, R. K., Reis, L. A., Nunes, S. S., and Radisic, M. 2012. Cardiac tissue engineering: current state and perspectives. *Frontiers in Bioscience* 17: 1533–1550.

Christman, K. L., Fang, Q., Yee, M. S., Johnson, K. R., Sievers, R. E., and Lee, R. J. 2004b. Enhanced neovasculature formation in ischemic myocardium following delivery of pleiotrophin plasmid in a biopolymer. *Biomaterials* 26: 1139–1144.

Christman, K. L., Fok, H. H., Sievers, R. E., Fang, Q., and Lee, R. J. 2004c. Fibrin glue alone and skeletal myoblasts in fibrin scaffold preserve cardiac function after myocardial infarction. *Tissue Engineering* 10: 403–409.

Christman, K. L., Vardanian, A. J., Fang, Q., Sievers, R. E., Fok, H. H., and Lee, R. J. 2004a. Injectable fibrin scaffold improves transplant survival, reduces infarct expansion, and induces neovasculature formation in ischemic myocardium. *Journal of the American College of Cardiology* 44: 654–660.

Dai, W., Wold, L. E., Dow, J. S., and Kloner, R. A. 2005. Thickening of the infarcted wall by collagen injection improves left ventricular function in rats. *Journal of the American College of Cardiology* 46: 714–719.

Davis, M. E., Motion, J. P. M., Narmonveda, D. A. et al. 2005. Injectable self-assembling peptide nanofibres create intramyocardial microenvironments for endothelial cells. *Circulation* 111: 442–450.

Dimmeler, S., Zeiher, A. M., and Schneider, M. D. 2005. Unchain my heart: The scientific foundations of cardiac repair. *Journal of Clinical Investigation* 115: 572–583.

El-Fray, M. 2003. Nanostructured elastomeric biomaterials for soft tissue reconstruction. 2003. Ref Type: Thesis/Dissertation.

Engelmayr, G. C., Cheng, M., Bettinger, C. J., Borenstein, J. T., Langer, R., and Freed, L. E. 2008. Accordion-like honeycombs for tissue engineering of cardiac anisotropy. *Nature Materials* 7: 1003–1010.

Engelmayr, G. C. J. and Jean, A. 2010. Finite element analysis of an accordion-like honeycomb scaffold for cardiac tissue engineering. *Journal of Biomechanics* 43: 3035–3043.

Eschenhagen, T., Zimmermann, W. H., and Kléber, A. G. 2006. Electrical coupling of cardiac myocyte cell sheets to the heart. *Circulation Research* 98: 705–712.

Freed, L. E., Engelmayr, G. C. J., Borenstein, J. T., Moutos, F. T., and Guilak, F. 2009. Advanced material strategies for tissue engineering scaffolds. *Advanced Materials* 21: 3410–3418.

Freed, L. D., Guilak, F., Guo, X. E. et al. 2006. Advanced tools for tissue engineering: Scaffolds, bioreactors and signalling. *Tissue Engineering* 12: 3285–3305.

Gaballa, M. A., Sunkomat, J. N., Thai, H., Morkin, E., Ewy, G., and Goldman, S. 2006. Grafting an acellular 3-dimensional collagen scaffold onto a non-transmural infarcted myocardium induces neo-angiogenesis and reduces cardiac remodelling. *Journal of Heart and Lung Transplantation* 25: 946–954.

Gaetani, R., Doevendans, P. A. F., Messina, E., and Sluijter, J. P. G. 2011. Tissue Engineering for Cardiac Regeneration. *Studies in Mechanobiology, Tissue Engineering and Biomaterials* 6: 1–27.

Genovese, J. A, Spadaccio, C., Rainer, A., and Covino, E. 2011. Electrospun nanocomposites and stem cells in cardiac tissue engineering. *Studies in Mechanobiology, Tissue Engineering and Biomaterials* 6: 215–242.

Grossman, P. M., Han, Z., Palasis, M. et al. 2002. Incomplete retention after direct myocardial injection. *Catheterization and Cardiovascular Interventions* 55: 392–397.

Guillemette, M. D., Park, H., Hsiao, J. C., Jain, S. R., Larson, B. L., Langer, R., and Freed, L. E. 2010. Combined technologies for microfabricating elastomeric cardiac tissue engineering scaffolds. *Macromolecular Bioscience* 10: 1330–1337.

Guo, X. M., Zhao, Y. S., Chang, H. X. et al. 2006. Creation of engineered cardiac tissue in vitro from mouse embryonic stem cells. *Circulation* 113: 2229–2237.

Haraguchi, Y., Shimizu, T., Yamato, M., Kikuchi, A., and Okano, T. 2006. Electrical coupling of cardiomyocyte sheets occurs rapidly via functional gap junction formation. *Biomaterials* 27: 4765–4774.

Hata, H., Bar, A., and Dorfman, S. 2010. Engineering a novel three dimensional contractile myocardial patch with cell sheets and decellularised matrix. *European Journal of Cardiothoracic Surgery* 38: 450–455.

Hofmann, M., Wollert, K. C., Meyer G. P. et al. 2005. Monitoring of bone marrow cell homing into the infarcted human myocardium. *Circulation* 111: 2198–2202.

Hsieh, P. C., Davis, M. E., Gannon, J., MacGillivray, C., and Lee, R. T. 2006. Controlled delivery of PDGF-BB for myocardial protection using injectable self-assembling peptide nanofibres. *Journal of Clinical Investigation* 116: 237–248.

Huang, N. F., Yu, J., Sievers, R., Li, S., and Lee, R. J. 2005. Injectable biopolymers enhance angiogenesis after myocardial infarction. *Tissue Engineering* 11: 1860–1866.

Ishii, O., Shin, M., Sueda, T., and Vacanti, J. P. 2006. In vitro tissue engineering of a cardiac graft using a degradable scaffold with an extracellular matrix-like topography. *Journal of Thoracic Cardiovascular Surgery* 130: 1358–1363.

Iwakura, A., Fujita, M., Kataoka, K. et al. 2003. Intramyocardial sustained delivery of basic fibroblast growth factor improves angiogenesis and ventricular function in a rat infarct model. *Heart Vessels* 18: 93–99.

Iyer, R. K. and Radisic, M. 2006. Microfabricated poly(ethylene glycol) templates for tri-culture in cardiac tissue engineering. *Journal of Molecular and Cellular Cardiology* 40: 877–882.

Jackson, K. A., Majka, S. M., Wang, H. et al. 2001. Regeneration of ischemic cardiac muscle and vascular endothelium by adult stem cells. *Journal of Clinical Investigation* 107: 1395–1402.

Jawad, H., Ali, N. N., Lyon, A., Chen Q. Z., Harding, S. E., and Boccaccini, A. R. 2007. Myocardial tissue engineering: A review. *Journal of Tissue Engineering Regenerative Medicine* 1: 327–342.

Jawad, H., El Fray, M., Boccaccini, A. R. et al. 2010. Nanocomposite elastomeric biomaterials for myocardial tissue engineering using embryonic stem cell-derived cardiomyocytes. *Advanced Engineering Materials (Advanced Biomaterials)* 12: B664–B674.

Jawad, H., Lyon, A. R., Harding, S. E., Ali, N. N., and Boccaccini, A. R. 2008. Myocardial tissue engineering. *British Medical Bulletin* 87: 31–47.

Karam, J. P., Muscari, C., and Montero-Menei, C. N. 2012. Combining adult stem cells and polymeric devices for tissue engineering in infarcted myocardium. *Biomaterials* 33: 5683–5695.

Ke, Q., Yang, Y., Rana, J. S., Chen, Y., Morgan, J. P., and Xiao, Y. F. 2005. Embryonic stem cells cultured in biodegradable scaffold repair infarcted myocardium in mice. *Acta Physiologica Sinica* 57: 673–681.

Kellar, R. S., Landeen, L. K., Shepherd, B. R., Naughton, G. K., Ratcliffe, A., and Williams, S. K. 2001. Scaffold-based, three-dimensional, human fibroblast culture provides a structural matrix that supports angiogenesis in infarcted heart tissue. *Circulation* 104: 2063–2068.

Kochupura, P. V., Azeloglu, E. U., Kelly, D. J. et al. 2005. Tissue-engineered myocardial patch derived from extracellular matrix provides regional mechanical function. *Circulation* 112: I144–I149.

Kofidis, T., Akhyari, P., Wachsmann, B. et al. 2003. Clinically established hemostatic scaffold (tissue fleece) as biomatrix in tissue- and organ-engineering research. *Tissue Engineering* 9: 517–523.

Kofidis, T., de Bruin, J. L., Hoyt, G. et al. 2005b. Myocardial restoration with embryonic stem cell bioartificial tissue transplantation. *Journal of Heart and Lung Transplantation* 24: 737–744.

Kofidis, T., Lebl, D. R., Martinez, E. C., Hoyt, G., Tanaka, M., and Robbins, R. C. 2005a. Novel injectable bio-artificial tissue facilitates targeted, less invasive, large-scale tissue restoration on the beating heart after myocardial injury. *Circulation* 112: 173–177.

Krupnick, A. S., Kreisel, D., Engels, F. H. et al. 2002. A novel small animal model of left ventricular tissue engineering. *Journal of Heart and Lung Transplantation* 21: 233–243.

Laflamme, M. A. and Murry, C. E. 2005. Regenerating the heart. *Nature Biotechnology* 23: 845–856.

Leor, J., Aboulafia-Etzion, S., Dar, A. et al. 2000. Bioengineered cardiac grafts—A new approach to repair the infarcted myocardium? *Circulation* 102: 56–61.

Leor, J., Amsalema Y., and Cohen, S. 2005. Cells, scaffolds, and molecules for myocardial tissue engineering. *Pharmacology & Therapeutics* 105: 151–163.

Leor, J., Rozen, L., Zuloff-Shani, A. et al. 2006. Ex vivo activated human macrophages improve healing, remodelling, and function of the infarcted heart. *Circulation* 114: 94–100.

Lesman, A., Habib, M., Caspi, O. et al. 2010. Transplantation of a tissue-engineered human vascularised cardiac muscle. *Tissue Engineering* (Part A) 16: 115–125.

Li, R. K., Yau, T. M., Weisel, R. D. et al. 2000. Construction of a bioengineered cardiac graft. *Journal of Thoracic Cardiovascular Surgery* 119: 368–375.

Lloyd-Jones D., Adams R. J., Brown, T. M. et al. 2010. Heart disease and stroke statistics—2010 Update: A report from the American Heart Association. *Circulation* 121: 46–215.

McDevitt, T., Woodhouse, K. A., Hauschka, S. D., and Murry, C. E. 2002. Spatially organized layers of cardiomyocytes on biodegradable polyurethane films for myocardial repair. *Journal of Biomaterials Research* 66A: 586–595.

Miyahara, Y., Nagaya, N., Kataoka, M. et al. 2006. Monolayered mesenchymal stem cells repair scarred myocardium after myocardial infarction. *Nature Medicine* 12: 459–465.

Motlagh, D., Yang, J., Lui, K. Y., Webb, A. R., and Ameer, G. A. 2006. Hemocompatibility evaluation of poly(glycerol-sebacate) in vitro for vascular tissue engineering. *Biomaterials* 27: 4315–4324.

Nair, L. S. and Laurencin, C. T. 2007. Biodegradable polymers as biomaterials. *Progress in Polymer Science* 32:762–798.

Noral, F., Merlet, P., Isnard, R. et al. 2003. Influence of mobilized stem cells on myocardial infarct repair in nonhuman primate model. *Blood* 102: 4361–4368.

Orlic, D., Kajstura, J., Chimenti, S. et al. 2001. Bone marrow cells regenerate infarcted myocardium. *Nature* 410: 701–705.

Ott, H. C., Matthiesen, T. S., Goh, S. K. et al. 2008. Perfusion-decellularised matrix: Using nature's platform to engineer a bioartificial heart. *Nature Medicine* 14(2): 213–221.

Park, H., Larson, B. L., Guillemette, M. D. et al. 2011. The significance of pore microarchitecture in a multi-layered elastomeric scaffold for contractile cardiac muscle constructs. *Biomaterials* 32: 1856–1864.

Park, H., Radisic, M., Lim, J. O., Chang, B. H., and Vunjak-Novakovic, G. 2005. A novel composite scaffold for cardiac tissue engineering. *In-Vitro Cellular and Developmental Biology. Animal* 41: 188–196.

Patil, S. B. and Sawant, K. K. 2008. Mucoadhesive microspheres: A promising tool in drug delivery. *Current Drug Delivery* 5: 312–318.

Pego, A. P., Van Luyn, M. J. A., Brouwer, L. A. et al. 2003. In vivo behavior of poly(1,3-trimethylene carbonate) and copolymers of 1,3-trimethylene carbonate with D,L-lactide or epsilon-caprolactone: Degradation and tissue response. *Journal of Biomedical Materials Research (Part A)* 67A: 1044–1054.

Pham, Q. P., Sharma, U., and Mikos, A. G. 2006. Electrospinning of polymeric nanofibres for tissue engineering applications: A review. *Tissue Engineering* 12: 1197–1211.

Pok, S. and Jacot, J. G. 2011. Biomaterials advances in patches for congenital heart defect repair. *Journal of Cardiovascular Translational Research* 4: 646–654.

Quaini, F., Urbanek, K., Beltrami, A. P. et al. 2002. Chimerism of the transplanted heart. *New England Journal of Medicine* 346: 5–15.

Radisic, M., Euloth, M., Yang, L., Langer, R., Freed, L. E., and Vunjak-Novakovic, G. 2003. High density seeding of myocyte cells for tissue engineering. *Biotechnology and Bioengineering* 82: 403–414.

Radisic, M., Fast, V. G., Sharifov, O. F., Iyer, R. K., Park, H., and Vunjak-Novakovic, G. 2009. Optical mapping of impulse propagation in engineered cardiac tissue. *Tissue Engineering A* 15: 851–860.

Radisic, M., Park, H., Chen, F. et al. 2006a. Biomimetic approach to cardiac tissue engineering: Oxygen carriers and channeled scaffolds. *Tissue Engineering* 12: 2077–2091.

Radisic, M., Park, H., Chen, F. et al. 2006b. Biomimetic approach to cardiac tissue engineering: Oxygen carriers and channeled scaffolds. *Tissue Engineering* 12: 2077–2091.

Radisic M., Park H., Martens T. P. et al. 2008. Pre-treatment of synthetic elastomeric scaffold by cardiac fibroblast improves engineered heart tissue. *Journal of Biomedical Materials Research* (Part A) 86: 713–724.

Rai, R., Tallawi, M., Grigore, A., and Boccaccini, A. R. 2012. Synthesis, properties and biomedical applications of poly(glycerol sebacate) (PGS): A review. *Progress in Polymer Science* 37: 1051–1078.

Ravichandran, R., Venugopal, J. R., Sundarrajan, S., Mukherjee, S., and Ramakrishna, S. 2011. Poly(glycerol sebacate)/gelatin core/shell fibrous structure for regeneration of myocardial infarction. *Tissue Engineering Part A* 17: 1363–1373.

Robinson, K. A., Li, J., Mathison, M. et al. 2005. Extracellular matrix scaffold for cardiac repair. *Circulation* 112: 135–143.

Rosellini, E., Cristallini, C., Barbani, N., and Giusti, P. 2011, Engineering multifunctional scaffolds for myocardial repair through nanofunctionalisation and microfabrication of novel polymeric biomaterials. *Studies in Mechanobiology, Tissue Engineering and Biomaterials* 6: 187–214.

Rosellini, E., Cristallini, C., Barbani, N., Vozzi, G., and Giusti, P. 2008. Preparation and characterisation of alginate/gelatin blend film for cardiac tissue engineering. *Journal of Biomedical Materials Research (Part A)* 91A: 447–453.

Ryu, J. H., Kim, I. K., Cho, S. W. et al. 2005. Implantation of bone marrow mononuclear cells using injectable fibrin matrix enhances neovascularization in infarcted myocardium. *Biomaterials* 26: 319–326.

Sant, S. and Khademhosseini, A. 2010. Fabrication and characterization of tough elastomeric fibrous scaffold for tissue engineering applications. *Conference Proceeding of the IEEE Engineering in Medicine and Biology Society* 1: 3546–3548.

Sawa, Y., Miyagawa, S., Sakaguchi, T., Fujita, T., Matsuyama, A., Saito, A., Shimizu, T., and Okano, T. 2012. Tissue engineered myoblast sheets improved cardiac function sufficiently to discontinue LVAS in a patient with DCM: Report of a case. *Surgery Today* 42: 181–184.

Sekine, H., Shimizu, T., Kosaka, S., Kobayashi, E., and Okano, T. 2006. Cardiomyocyte bridging between hearts and bioengineered myocardial tissues with mesenchymal transition of mesothelial cells. *Journal of Heart and Lung Transplantation* 25: 324–332.

Sekiya, S., Shimizu, T., Yamato, M., Kikuchi, A., and Okano, T. 2006. Bioengineered cardiac cell sheet grafts have intrinsic angiogenic potential. *Biochemical and Biophysical Research Communications* 341: 573–582.

Shimizu, T., Yamato, M., Isoi, Y. et al. 2002. Fabrication of pulsatile cardiac tissue grafts using a novel 3-dimensional cell sheet manipulation technique and temperature responsive cell culture surfaces. *Circulation Research* 90: e40–e48.

Shin, M., Ishii, O., Sueda, T., and Vacanti, J. P. 2004. Contractile cardiac grafts using a novel nanofibrous mesh. *Biomaterials* 25: 3717–3723.

Sill, T. J. and von Recum, H. A. 2008. Electrospinning: Applications in drug delivery and tissue engineering. *Biomaterials* 29: 1989–2006.

Singelyn, J. M. and Christman, K. L. 2011. Injectable materials for myocardial tissue engineering. *Studies in Mechanobiology, Tissue Engineering and Biomaterials* 6: 133–163.

Srinivasan, A. and Sehgal, P. K. 2010. Characterisation of biocompatible collagen fibres—A promising candidate for cardiac patch. *Tissue Engineering (Part C: Methods)* 16(5): 895–903.

Stankus, J. J., Guan, J., and Wagner, W. R. 2004. Fabrication of biodegradable elastomeric scaffolds with submicron morphologies. *Journal of Biomedical Materials Research* (Part A) 70A: 603–614.

Stuckey, D. J., Ishii, H., and Chen, Q. Z. 2010. Magnetic resonance imaging evaluation of remodeling by cardiac elastomeric tissue scaffold biomaterials in a rat model of myocardial infarction. *Tissue Engineering (Part A)* A16: 3395–3402.

Tamada, J. and Langer, R. 1992. The development of polyanhydrides for drug delivery applications. *Journal of Biomaterials Science Polymer Edition* 3: 315–353.

Tian, L., Prabhakaran, M. P., Ding, X., Kai, D., and Ramakrishna, S. 2012. Emulsion electrospun vascular endothelial growth factor encapsulated poly(l-lactic acid-co-ε-caprolactone) nanofibers for sustained release in cardiac tissue engineering. *Journal of Material Science* 47: 3272–3281.

Vunjak-Novakovic, G., Radisic, M., and Obradovic, B. 2006. Review: Cardiac tissue engineering: Effects of bioreactor flow environment on tissue constructs. *Chemical Technology and Biotechnology* 81: 485–490.

Vunjak-Novakovic, G., Tandon, N., Godier, A. et al. 2010. Challenges in cardiac tissue engineering. *Tissue Engineering (Part B)* 16: 169–187.

Wang, B., Borazjani, A., Tahai, M. et al. 2010. Fabrication of cardiac patch with decellularised porcine myocardial scaffold and bone marrow mononuclear cells. *Journal of Biomedical Materials Research (Part A)* A94: 1100–1110.

Wang, J. S., Shum-Tim, D., Galipeau, J., Chedrawy, E., Eliopoulos, N., and Chiu, R. C. 2000. Marrow stromal cells for cellular cardiomyoplasty: Feasibility and potential clinical advantages. *Journal of Thoracic Cardiovascular Surgery* 120: 999–1005.

Wall, S. T., Yeh, C. C., Tu, R. Y. K., Mann, M. J., and Healy, K. E. 2010. Biomimetic matrices for myocardial stabilization and stem cell transplantation. *Journal of Biomedical Materials Research (Part A)* 95A: 1055–1066.

Wang, Y., Ameer, G. A., Sheppard, B. J., and Langer, R. 2002. A tough biodegradable polymer. *Nature Biotechnology* 20: 602–606.

Xiang, Z., Liao, R., Kelly, M. S., and Spector, M. 2006. Collagen–GAG scaffolds grafted onto myocardial infarcts in a rat model: A delivery vehicle for mesenchymal stem cells. *Tissue Engineering* 12: 2467–2478.

Yeong, W. Y., Sudarmadji, N., Yu, H. Y. et al. 2010. Porous polycaprolactone scaffold for cardiac tissue engineering fabricated by selective laser sintering. *Acta Biomaterialia* 6: 2028–2034.

Yu, J., Du, K. T., Fang, Q., Gu, Y., Mihardja, S. S., Sievers, R. E., Wu, J. C., and Lee, R. J. 2010. The use of human stem cells encapsulated in RGD modified alginate microspheres in the repair of myocardial infarction in the rat. *Biomaterials* 31: 7012–7020.

Yu, J., Gu, Y., Du, K. T., Mihardja, S., Sievers, R. E., and Lee, R. J. 2009. The effect of injected RGD modified alginate on angiogenesis and left ventricular function in a chronic rat infarct model. *Biomaterials* 30: 751–756.

Zammaretti, P. and Jaconi, M. 2004. Cardiac tissue engineering: Regeneration of the wounded heart. *Current Opinions in Biotechnology* 15: 430–434.

Zhong, S., Teo, W. E., Zhu, X., Beuerman, R., Ramakrishna, S., and Yung, L. Y. 2005. Formation of collagen–glycosaminoglycan blended nanofibrous scaffolds and their biological properties. *Biomacromolecules* 6: 2998–3004.

Zimmermann, W. H. 2002. Cardiac grafting of engineered heart tissue in syngenic rats. *Circulation* 106: I151–I157.

Zimmermann, W. H. 2008. Tissue engineering: Polymers flex their muscles. *Nature Materials* 7: 932–933.

Zimmermann, W. H., 2011, Tissue engineered myocardium, *Studies in Mechanobiology, Tissue Engineering and Biomaterials* 6: 111–132.

Zimmermann, W. H., Fink, C., Kralisch, D., Remmers, U., Weil, J., and Eschenhagen, T. 2000. Three-dimensional engineered heart tissue from neonatal rat cardiac myocytes. *Biotechnology Bioengineering* 68: 106–114.

Zimmermann, W. H., Melnychenko, I., Eschenhagen, T. et al. 2004. Engineered heart tissue for regeneration of diseased hearts. *Biomaterials* 25: 1639–1647.

Zimmermann, W. H., Melnychenko, I., Wasmeier, G. et al. 2006. Engineered heart tissue grafts improve systolic and diastolic function in infarcted rat hearts. *Nature Medicine* 12: 452–458.

15 Acellular Tubular Grafts Constructed from Natural Materials in Vascular Tissue Engineering
From Bench to Bedside

M.J.W. Koens, A.G. Krasznai, J.A. van der Vliet,
T. Hendriks, R.G. Wismans, K.A. Faraj,
W.F. Daamen, and T.H. van Kuppevelt

CONTENTS

ABBREVIATIONS

aSIS	acellular small intestinal submucosa
BSE	bovine spongiform encephalopathy
CD31	cluster of differentiation molecule 31
CSH	construct–sleeve hybrid
DMSO	dimethyl sulfoxide
EC(s)	endothelial cell(s)
ECM	extracellular matrix
EDC	1-ethyl-3-(3-dimethylaminopropyl)carbodiimide
EDGE	ethylene glycol diglycidyl ether
EPC(s)	endothelial progenitor cell(s)
ePTFE	expanded polytetrafluoroethylene
FGF-2	fibroblast growth factor-2 (or basic fibroblast growth factor)
GAG(s)	glycosaminoglycan(s)
HUVEC(s)	human umbilical vein endothelial cell(s)
NCGT	negatively charged glutaraldehyde tanned
NHS	N-hydroxysuccimide
PAI-1	plasminogen activator inhibitor-1
PEO	polyethylene oxide
PCL	polycaprolactone
PGA	polyglycolic acid
PLGA	polylactic glycolic acid
PLLA	poly-L-lactic acid
PGI2	prostacyclin
RGD	abbreviation for integrin-binding recognition sequence arginine–glycine–aspartic acid
SDS	sodium dodecyl (lauryl) sulfate
SEM	scanning electron microscopy
SMC(s)	smooth muscle cell(s)
TEM	transmission electron microscopy
TGF-β	transforming growth factor-β
tPA	tissue plasminogen activator
UTS	ultimate tensile strength
VEGF	vascular endothelial growth factor
vWF	von Willebrand factor

15.1 INTRODUCTION

Tissue engineering is a rapidly evolving interdisciplinary scientific field that overlaps the areas of biology, medicine, (bio)chemistry, (bio)materials, engineering, and technology. It focuses upon regenerating or improving the biological function of organs, like skin, cartilage, bone, muscle, nerves, bladder, urethra, heart valves, and blood vessels.[1] The blood vessels have a pivotal role in (cardio)vascular tissue engineering.

Atherosclerotic disease causes high mortality in Western society.[2] In vascular surgery, large arteries, e.g., the aorta, can be replaced by large diameter grafts composed of synthetic polymers, like Dacron (polyethylene terephthalate) or expanded polytetrafluoroethylene (ePTFE). However, these polymers fail as grafts for small-diameter blood vessels (diameter ≤ 6 mm) due to (re)occlusion.[3,4] Here, the saphenous vein and mammary artery are still the golden standard replacements in peripheral and cardiac bypass surgery, respectively.[3,5] Many patients, however, lack high-quality and/or sufficient tissue for these autologous conduits, so, a need for alternatives persists.[3] Alternative grafts can be made from other synthetic polymers, natural materials, or a combination of both, producing hybrid grafts. Cells may also be added to the graft, either by static cell seeding, possibly followed by mechanical conditioning in a bioreactor, or by the body itself.[6] The landmark work of Weinberg and Bell[7] was one of the first to point out the intrinsic value of the arterial wall's native structure. It reported on a tissue-engineered blood vessel, constructed from collagen, and cultured vascular cells. Though the addition of cells to a graft may be beneficial, it is considered highly laborious. It also can hamper off-the-shelf availability in urgent clinical situations.[8]

This chapter focuses on the construction and application of acellular tubular grafts constructed from natural materials, as these are abundantly present in the native arterial wall. The following issues will be discussed. First, the (biochemical) structure and composition of the native blood vessel is addressed. Next, different methods for graft construction are discussed. In many cases, decellularization procedures are applied to remove cells and preserve the extracellular matrix (ECM) in its original conformation. Alternatively, vessels can be constructed originating from purified biomolecules. Third, in vitro criteria with respect to biocompatibility, biodegradability, and mechanical strength prior to in vivo experimentation will be discussed. Parameters that need to be studied in an in vivo setting will be discussed, including patency, cellular influx, (bio)degradability, and performance. Finally, clinically applied acellular natural grafts will be described and a future outlook will be given upon new directions of vascular graft development.

15.2 WHAT TO MIMIC: THE NATIVE ARTERIAL WALL

The vascular tree originates from the heart which is the center of the circulatory system. In a resting state, its left ventricle pumps blood into the aorta with a frequency of 70 beats per minute. In blood vessels, the systolic and diastolic pressures are approximately 120 and 80 mm Hg, respectively.[9] The resilient aortic wall can comprise up to 50% of elastin,[10] smoothening and guiding the blood flow into the vascular system. Typically, both elastic and muscular arteries harbor a three-layered architecture composing the intima, media, and adventitia (Figure 15.1).[11] The intima contains a confluent layer of endothelial cells (ECs) anchored to the basement membrane by cell–matrix adhesions. The basement membrane functions as a barrier and is mainly built up of a type IV collagen network containing laminin and proteoglycans. It rests upon the underlying lamina elastica interna that is connected to the media. The media mainly consists of smooth muscle cells (SMCs), elastic fibers, and type I collagen. This type of collagen is a relatively stiff protein and is abundantly present within

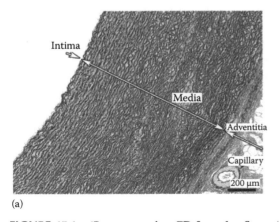
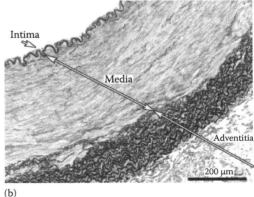

(a) (b)

FIGURE 15.1 (See companion CD for color figure.) Aorta (a) and arteria femoralis (b) from pig stained with Elastin von Gieson (paraffin, 5 μm), representing an elastic and muscular artery, respectively. Elastic fibers are black/blue and collagen is pink. Note the concentric layering of elastic fibers within the media of the aorta, and the dispersed medial elastic fibers of the femoral artery. Bars represent 200 μm.

the media of muscular arteries. With its muscular architecture, this type of artery orchestrates the continuous flow of nutrients and oxygen by luminal dilation and contraction.[11] An outer layer of type I collagen, the adventitia, enwraps the relatively weak concentric elastic rings in the tunica media of elastic arteries. The media and adventitia are separated by the lamina elastica externa. The adventitia contains, among others, blood vessels that supply blood to the arterial wall itself and fibroblasts.

15.3 CURRENT STATUS

15.3.1 SYNTHETIC GRAFTS

In the 1950s, pioneer Arthur Voorhees Jr. reported on the evaluation of synthetic grafts for large diameter aortic replacement.[12] In 1976, a graft of ePTFE was used for the first time in peripheral bypass surgery.[13] Today, Dacron and ePTFE are still extensively used in clinical practice for the replacement of atherosclerotic arteries.[14]

However, when considering small-diameter applications (inner diameter $\leq 6\,mm$), the application of synthetic grafts results in (re)stenosis due to thrombogenicity, intimal hyperplasia, compliance mismatch, and deficient recruitment of endothelial lining. Other frequent complications related to synthetic graft failure include infection, material deterioration, anastomotic aneurysm, and occlusion. Because of this, researchers continuously scope the applicability of other synthetic polymers, e.g., polyurethane,[15,16] such as polyether urethane urea (Tholaron®).[17] Also, the combination of existing grafts with other (bio)polymers such as polyethylene glycol,[18] phospholipids,[16,19,20] or poly(1,8-octanediol) citrate[21] is investigated. Additionally, surgical modifications to grafts are researched, such as Miller cuffs or Linton and Taylor patches.[14] Currently, nitric oxide–secreting and drug-eluting synthetic grafts rule as the next generation of state-of-the-art technology for synthetic grafts. Early results were encouraging in the case of drug-eluting stents, but late thrombosis rates were higher when eluting paclitaxel (a cytostatic agent) and sirolimus (an immunosuppressant agent) in comparison to bare metal stents.[22,23] Besides nitric oxide, ECM-based biomolecules are also used to functionalize synthetic grafts.

15.3.2 SYNTHETIC GRAFT FUNCTIONALIZATION WITH ECM-BASED BIOMOLECULES

Synthetic graft functionalization mostly includes biomolecular natural coatings to reduce thrombogenicity, preferably by inducing an endothelial lining. Modifications with natural materials, such as arginine–glycine–aspartic acid (RGD) peptides,[24,25] recombinant elastin-mimetic polymer,[26] or heparin,[3,27] are also investigated to prevent thrombosis.[14] Commercially available modified grafts include InterGard heparin (Intervascular) and Propaten (Gore-Tex), where heparin is covalently bound by end-point linkage to PTFE.[14] P15, a cell-binding domain derived from collagen, has also been coated to ePTFE grafts. It resulted in enhanced endothelialization and reduced intimal hyperplasia, hypothesizing that P15 stimulates EC adhesion.[28]

Another method to induce endothelialization includes the seeding of synthetic grafts with ECs.[14,29,30] EC sources may include artery, vein, and endothelial progenitor cells (EPCs) isolated from blood or bone marrow. Autologous EC seeding of ePTFE grafts improves patency rates in infrainguinal bypass patients.[30] Unfortunately, the cellular approach limits immediate graft use in emergency situations, especially in two-stage seeding procedures which include cell culture/expansion as well as cell harvesting (one-stage seeding). Currently, two-stage seeding in nonurgent situations and its routine handling in nontertiary hospitals is feasible.[29]

The development of synthetic grafts treated with biomolecules proves that natural materials are promising candidates to improve graft patency, to lower thrombogenic responses, and to limit SMC proliferation and subsequent intimal hyperplasia. Even more, the inclusion of natural materials has led the field of vascular tissue engineering into a new era, with a fully natural and acellular graft as its "holy grail."

15.4 NATURAL MATERIALS: SOURCE AND ORIGIN

15.4.1 GENERAL

The ECM constitutes the surrounding environment of cells, and supplies strength and physical support to tissues and organs, especially within connective tissues such as skin, bone, cartilage, arteries, and ligaments.[10] The ECM is also involved in providing cues to cells for proliferation, differentiation, and migration. This cell-embedding medium holds a great variety of excreted natural macromolecules. In blood vessels, type I collagen and elastin are main determinants for mechanical strength and elasticity, respectively. Additionally, glycosaminoglycans (GAGs) are essential for protein binding and hydration purposes.[3]

In vascular tissue engineering, different macromolecules and tissues have been used as source for tubular grafting (listed in Table 15.1). A large variety of sources have been used including pig,[31–47] dog,[48–53] cow,[54,55] rat,[56–58] rabbit,[59] sheep,[60,61] goat,[62] jellyfish,[63] or salmon.[64,65] Different macromolecular biomaterials[66] have been isolated for scaffolding purposes including collagen,[67] elastin,[68] fibrin,[69–71] silk,[72] and hyaluronan.[73,74]

15.4.2 COLLAGEN

Over 25 different types of fibrillar collagen exist, of which type I collagen is most abundantly present in organs. It is a relatively stiff protein that supplies strength and structural integrity to tissues.[75] The collagen fiber contains clustered collagen fibrils. These fibrils are built of collagen molecules of ~300 kDa each. The arrangement of these collagen molecules (quarter-staggered array) is responsible for the typical striated banding pattern (67 nm) of the collagen fibril (Figure 15.2a).[76] The collagen molecule itself is formed by three intertwined left-handed α (polypeptide) chains forming a right-handed triple helical structure (1.5×300 nm). Intermolecular cross-links result in the high stiffness and rigidity of fibrillar collagen.

TABLE 15.1
Overview of Source, Tissue, and Molecules Used for Vascular Grafting

Source	Tissue Used for Vascular Grafting
Pig	Carotid artery,[37,39,42,44] femoral artery,[33] ureter,[34] aorta from fetal pigs,[40] aortic wall[43]
Dog	Thoracic aorta,[48] abdominal aorta,[48] femoral iliac artery,[48] superior cava,[48] inferior cava,[48] femoral iliac vein,[48] external jugular vein,[49] ureter,[50] carotid artery,[52,53] external iliac artery,[52] abdominal aorta[52]
Cow	Carotid artery[54]
Rat	Aorta[58]
Rabbit	Carotid artery[51,59]
Sheep	Carotid artery,[60] abdominal aorta,[61] carotid artery,[61] jugular vein[61]
Goat	Carotid artery[62]
Source	**Isolated macromolecules**
Rat tail,[32,56,67] bovine Achilles tendon,[36,55,139] porcine carotid artery,[31] ascending porcine aorta,[41,45] porcine skin,[46] jellyfish,[63] salmon skin[64,65]	Type I collagen
Porcine carotid artery,[32,35,38] ascending porcine aorta,[38,41,45] equine ligamentum nuchae[55,139]	Elastin
Human fibrinogen, thrombin, and bovine aprotonin (Tissucol®)[70]	Fibrin
Cocoons of *Bombyx mori* silkworm[72]	Silk
HYAFF-11[73,74]	Hyaluronan
Swine small intestine[35,36]	Predominantly type I collagen

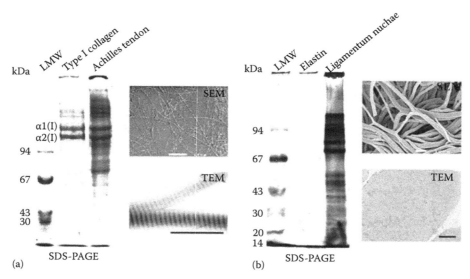

FIGURE 15.2 Analysis of highly purified type I collagen fibrils and elastin fibers. (a) SDS-PAGE analysis of type I collagen shows the typical α1 and α2 bands. Purity is compared to the source bovine Achilles tendon. Fibrils were visualized with scanning electron microscopy (SEM) and transmission electron microscopy (TEM). The latter depicts the typical striated banding pattern. (b) SDS-PAGE analysis of elastin. Due to the high insolubility of the cross-linked fibers, no protein enters the gel. Purity is compared to the source equine ligamentum nuchae. TEM indicates the absence of microfibrils. Bars represent 10 μm in SEM and 0.5 μm in TEM micrographs. (Reprinted with kind permission from Springer Science + Business Media: *Adv. Exp. Med. Biol.*, From molecules to matrix: Construction and evaluation of molecularly defined bioscaffolds, 585, 2006, 279–295, Geutjes PJ, Daamen WF, Buma P et al.)

15.4.3 ELASTIN

Elastic fibers provide elasticity to skin, lung, arteries, and ligaments.[77,78] The elastic fiber mainly consists of elastin and microfibrils. Highly purified elastin fibers are shown in Figure 15.2b. The protein elastin is a highly hydrophobic and amorphous protein. Its precursor, tropoelastin (72 kDa), forms a highly cross-linked network containing (iso)desmosine under the influence of the enzyme lysyl oxidase. This takes place upon a template of microfibrils, which are mainly composed of fibrillin. The high degree of cross-links together with the hydrophobic nature gives the elastic network its ability to recoil. This distinct feature makes elastin of great importance in the arterial system of mammals, which have high blood pressure circulatory systems.[77,79] Currently, an increasing amount of research focuses upon applying elastic fiber for vascular tissue engineering,[80,81] including the induction of elastin biosynthesis[68] and the study of cell–elastin interaction.[82,83] Elastin polypeptides have been applied as non-thrombogenic coatings.[84]

15.4.4 GLYCOSAMINOGLYCANS

GAGs are composed of repeating units of disaccharides, forming long unbranched polysaccharides. The disaccharide units are composed of two sugar moieties: a hexuronic acid/hexose and a hexosamine. The large diversity in GAGs is due to modification reactions like sulfation and epimerization, making GAGs highly heterogeneous and highly negatively charged molecules. Heparin has the highest negative charge density of any known natural molecule. GAGs may be grouped as heparan sulfate/heparin, keratan sulfate, chondroitin sulfate, dermatan sulfate, and hyaluronan. With the exception of hyaluronan, GAGs are generally bound to core proteins to form proteoglycans.[85]

GAGs have multiple functions in the body such as tissue hydration, binding of growth factors, and controlling signaling pathways. They are involved in a variety of cellular behaviors including cell adhesion, migration, and differentiation. GAGs have been investigated for (anti)thrombogenic effects[86] and are known to play key roles in atherosclerosis and restenosis.[87] The GAG heparin is well-known for its anticoagulant properties. Hyaluronan is thought to have a critical role in vascular stenosis and may contribute to atherosclerosis progression.[88] Exogenous hyaluronan oligosaccharides have an elastogenic effect upon vascular SMCs.[89] For more specific information on proteoglycans in atherosclerosis and tissue engineering, the reader is referred to an excellent review.[90]

15.5 GRAFT FABRICATION

15.5.1 TISSUE ENGINEERING APPROACH: ACELLULAR VERSUS CELLULAR

The application of PTFE and Dacron in surgical treatment serves the old paradigm of using highly inert materials for implantation purposes. Nowadays, the general idea has shifted to the research fields of tissue engineering and regenerative medicine, where the implant is provided as a so-called scaffold. This scaffold or graft may function as a template to steer and induce tissue-specific regeneration. To this end, a blood vessel scaffold, with or without cells, has to sustain the blood flow's mechanical forces instantly. Furthermore, it should gradually degrade and subsequently pass on the mechanical loading to newly formed arterial tissue, which in the end should form a new blood vessel (see Figure 15.3).[91]

The acellular versus cellular debate focuses upon whether or not this scaffold should be occupied with cells, and possibly mechanically conditioned, or should remain acellular at the moment of

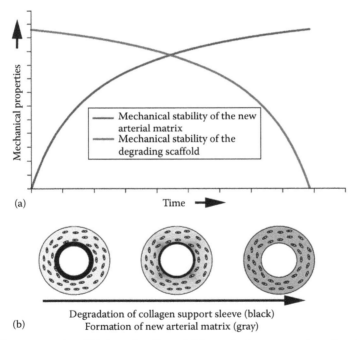

FIGURE 15.3 (See companion CD for color figure.) Theoretical representation of graft remodeling and mechanical loading. (a) Graph representing the delicate mechanical balance between scaffold degradation (red line) and the formation of new arterial tissue (blue line) plotted in time. (Reprinted with kind permission from Springer Science + Business Media: *Med. Biol. Eng. Comput.*, Achieving the ideal properties for vascular bypass grafts using a tissue engineered approach: A review, 45, 2007, 327–336, Sarkar, S., Schmitz-Rixen, T., Hamilton, G., and Seifalian, A.M.) (b) As a minimum threshold of strength is needed to sustain immediate blood flow, a supporting (collagen) sleeve (luminal black lining) can be added into a so-called construct–sleeve hybrid. In time, when the support sleeve degrades, strength is built up by newly formed tissue (depicted in gray). (Reprinted from *Biomaterials*, 24, Berglund, J.D., Mohseni, M.M., Nerem, R.M., and Sambanis, A., A biological hybrid model for collagen-based tissue engineered vascular constructs, 1241–1254. Copyright 2003, with permission from Elsevier.)

implantation. In the latter case, the cellular (re)population is consigned to the body itself. Whereas both strategies have their own advantages and disadvantages, the acellular approach appears more suitable with respect to direct application in case of emergencies, off-shelf availability, and economic feasibility. This chapter focuses upon the construction and validation of acellular vascular grafts.

15.5.2 GRAFT FABRICATION BY DECELLULARIZATION PROCEDURES

The rationale of decellularization procedures is based upon the physical and/or chemical removal of cells from tissues in order to obtain the ECM for scaffolding purposes.[92] A major advantage of this procedure is that the overall organization of the ECM fibers will remain intact.

Tissues can be chemically, enzymatically, or physically treated while retaining the ECM. Applied detergents are sodium dodecyl (lauryl) sulfate (SDS),[34,37,44,49,50,52,53,57] Triton X-100,[31,33,44,50–53,59] deoxycholate,[50,57] and IGEPAL.[33] Further tissue treatment may include enzymatic digestion including DNase, RNase, or trypsin in order to remove (nuclear) remnants.[40]

Generally, arteries are used in decellularization protocols. Ureters have also been decellularized to function as small-diameter blood vessels[34,50] because of their length and diameter and the absence of branches (avoiding ligations). A well-known example of decellularized tissue as a biomaterial is acellular small intestinal submucosa (aSIS), which is commercially available. aSIS has been successfully applied as grafts with high patency rates.[93,94] Vascular grafts from aSIS have been prepared by mandrel entrapment and suturing.[95] Huynh et al.[36] used aSIS for constructing an acellular collagen vascular graft by wrapping it with 10% overlay around a 4 mm mandrel, followed by luminal deposition of fibrillar collagen and subsequent EDC cross-linking. Hinds et al.[35] made grafts from NaOH-extracted elastin sleeves and enwrapped these with aSIS, where both layers were held together with fibrin glue. However, porcine DNA has been found in commercial aSIS (Depuy, J&J), suggesting that it may contain cellular debris, which may give rise to inflammatory responses.[96]

Generally, downsides of decellularization procedures include (1) the unknown molecular content of the obtained graft, (2) the possible toxicity of chemicals used,[97] and (3) the altered biomechanical properties.[44,59] It has been indicated that SDS treatment may lead to chemical or structural matrix alterations that negatively affect repopulation by cells.[97] Furthermore, possible application can be hampered as grafts, decellularized with Triton and trypsin, have markedly different geometrical and biomechanical properties with respect to original native arteries, such as a decreased wall thickness, increased elastic modulus and stiffness, and lower distensibility.[44,59] To overcome these concerns, one can consider creating defined tubular grafts from scratch.

15.5.3 FABRICATION FROM SCRATCH: MOLECULARLY DEFINED GRAFTS

Fabrication from scratch is defined here as the construction of a tubular graft with one or more purified component(s) of the ECM. For this, one needs characterized biomaterial(s) with a known molecular content. Such a component may already have a tubular shape. Fabrication of such grafts includes either (1) the creation of highly purified (elastin) sleeves from native arteries or (2) the use of purified materials applying tubular molding and casting techniques. Preferably, the defined graft is made to resemble the native arterial wall architecture. In the case of a small-diameter blood vessel, it holds that type I collagen and elastin should be used (Figure 15.1).

Type I collagen has proven to be a suitable material in vascular tissue engineering.[67] For example, tubular collagen films with and without seeded ECs and SMCs have been implanted into the inferior vena cava of Wistar rats and remained patent for 12 weeks displaying a continuous layer of ECs.[46] Elastin has also been employed in vascular applications. An elastin-mimetic protein polymer has been applied as a coating of PTFE[26] and human elastin polypeptides have been used to coat catheters.[84] Tubular elastin scaffolds, obtained by CNBr treatment of porcine carotid arteries, were placed in an acute rabbit carotid bypass model and exposed to blood flow for 30 min.[45] However, for long-term application and considering the high distensibility of elastin, additional structural integrity (e.g., conferred by collagen) is needed when applied in molecularly defined grafts. This can be obtained with layering techniques, thus mimicking the native layered architecture of blood vessels. Strengthening is essential as a vascular conduit instantly has to withstand blood pressure forces and gradually has to transfer these to newly formed tissue (Figure 15.3a). Collagen layers can also be cellularized. For instance, Berglund et al.[32] purified elastin by sequential steps of CNBr treatment and autoclaving cycles, and cast a cellularized collagen gel around it. The layering technique has also been exploited to make construct–sleeve hybrid (CSH) grafts, harboring a type I collagen support sleeve for (temporary) reinforcement (Figure 15.3b).[56] Besides sufficient mechanical strength, a vascular conduit needs to be nontoxic, biocompatible, non-thrombogenic, and noncalcifying.

Molecularly defined grafts from type I collagen and elastin can also be fabricated by molding and casting techniques. Nagai et al.[64] prepared tubular grafts by casting, molding, and cross-linking salmon collagen. The fabrication of collagen–GAG scaffolds can be highly controlled.[98,99] Type I collagen can be purified from pulverized bovine Achilles tendon by biochemical washings.[98] Elastin fibers[100] or elastin peptides,[101] purified from pulverized equine ligamentum nuchae, can be added to the collagenous scaffolds and stabilized by EDC/NHS cross-linking.[102] Additionally, using EDC/NHS cross-linking, heparin can be covalently bound to the collagen/elastin (Figure 15.4). Wissink et al.[103] found that EDC/NHS cross-linked collagen films support proliferation of ECs and confirmed this by determining EC-specific markers like tissue plasminogen activator (tPA), plasminogen activator inhibitor-1 (PAI-1), von Willebrand factor (vWF), and prostacyclin (PGI2). Buttafoco et al. made first step in using highly purified collagen and elastin to create flat[55] and tubular[104] scaffolds for blood vessel purposes. These scaffolding materials were created by freeze-drying and were able to sustain culture of SMCs.[105] To prevent graft loss, special attention is needed for mineralization, and so calcification studies include many. For instance, elastin tubes functionalized with agarose-imbedded FGF-2 did not calcify subcutaneously within adult rats while elastin tubes without FGF-2 did.[38] Subdermal implantation of elastin-based biomaterial with aluminum chloride merely delayed onset of calcification in BALB/c mice.[106]

Scaffold thickness and composition are key elements in graft survival. Walles et al.[61] using a subcutaneous rat model, found vascular matrices with a thickness of 0.88 mm to be repopulated with ECS and myofibroblasts, whereas thinner grafts degenerated within 8 weeks. These authors concluded that sufficient stabilizing fibers were needed to maintain scaffold integrity during the process of full in vivo recellularization. O'Brien et al.[107] analyzed the effect of pore size on cell adhesion in collagen–GAG scaffolds and found a linear relationship between cell attachment and specific surface area for pores.

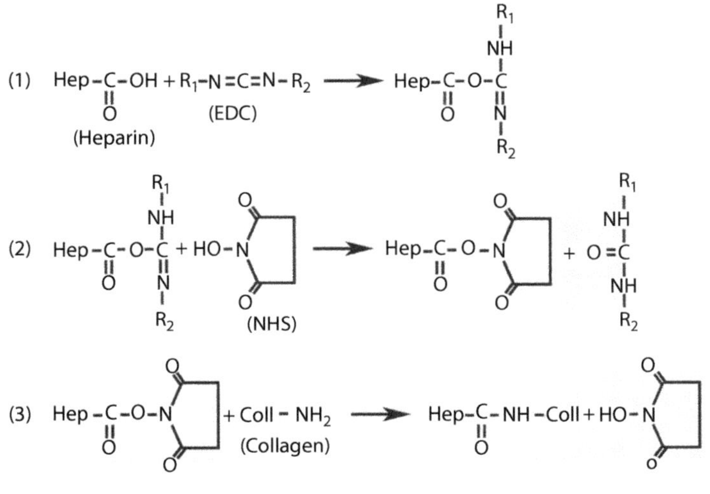

FIGURE 15.4 Cross-linking of heparin to collagen using EDC/NHS.

Tubular grafts have also been made solely from GAGs. Hyaluronan scaffolds (HYAFF-11™) have been molded by dipping dimethyl sulfoxide (DMSO)-dissolved esterified hyaluronan around a mandrel.[73,74] These grafts displayed 5 month patency and the ability to facilitate and promote the formation of circumferentially orientated elastin fibers.[74]

Electrospinning is a promising method to create molecularly defined grafts from scratch. It has large potential in the (vascular) tissue engineering field,[108] as electrospun scaffolds can be tailor-made for tissue-specific ECM applications.[109] The electrospinning process is displayed in Figure 15.5. Under influence of an electrical field, fibers can be drawn from a viscous polymer solution in a highly volatile solvent. For tubular grafting, a dry fiber can be collected upon a grounded and rotating mandrel. Fiber diameter can be controlled by a number of parameters like solution viscosity, needle thickness, collector distance, and voltage. Though pure electrospun collagen[110,111] or elastin fibers[111,112] can be made, a dissolved synthetic polymer may be used in the electrospinning fluid, e.g., polyethylene oxide (PEO),[72,113] polycaprolactone (PCL),[114] polyglycolic acid (PLGA),[115] polyglyconate,[116] polydioxanone,[117] or PGA.[116] The purpose of such addition is to increase time of fiber degradation, ensure viscosity, and increase fiber handability. Boland et al.[118] reported upon the fabrication of a three-layered electrospun vascular construct from collagen and elastin. Tubular grafts from electrospun blends of collagen and PCL have appropriate mechanical properties for implantation.[119] Stitzel et al.[115] prepared electrospun vascular grafts from a blend of type I collagen, elastin, and PLGA and showed tissue compatibility in a subdermal mice model. Tillman et al.[114] implanted collagen/PCL electrospun grafts in a rabbit aortoiliac bypass model and found patency up to 1 month. Additionally, absence of inflammatory infiltrate was noticed in contrast to previously performed studies with porcine decellularized grafts.[114]

Debate exists about the application of (pure) electrospun collagen fibers, as the fabrication process may cause such degradation that only gelatin remains.[120] A combination of electrospun man-made polymers with natural porous scaffolds may be an alternative approach for the fabrication of vascular grafts.[63] Therefore, nanotechnology has been introduced in (vascular) tissue engineering.[121]

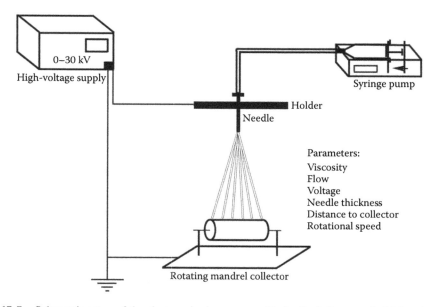

FIGURE 15.5 Schematic setup of the electrospinning process. Under the influence of a high-voltage electrical field, fibers are drawn from a viscous polymer solution that is delivered by a syringe pump to a needle. As a result, a drying fiber whips down manifesting a Taylor cone which is deposited at the collector. The parameters for the fabrication of electrospun tubular grafts include concentration of the polymer solution, viscosity, flow, voltage, needle thickness, distance to the collector, and the rotational speed of the collector.

15.5.4 Graft Cross-Linking and Stabilization

Cross-linking and stabilization are performed to strengthen the graft and to facilitate functionalization. This can be executed with a wide variety of chemical compounds, including glutaraldehyde,[56,93] polyepoxy compounds,[93] dye-mediated photooxidation,[42,93] ether,[55,122,123] J230,[55] acyl azide, and carbodiimides, like 1-ethyl-3-(3-dimethylaminopropyl)carbodiimide (EDC).[33,39,51,55,65] Though glutaraldehyde treatment has FDA approval, the necessity for alternatives remains as cytotoxicity and glutaraldehyde-induced calcification may occur.[43,93] One alternative is photooxidation. Meuris et al.[43] made a comparison between glutaraldehyde and photooxidative-treated tissues and found the latter to be potentially suitable as no calcification after implantation was found. McFetridge et al.[42] performed photooxidative cross-linking with methylene green on decellularized tissue and showed human umbilical vein ECs (HUVECs) adherence and maintenance over a short 3 day culture period.

Another frequently applied and low-toxic cross-linking method involves EDC, most generally used with the catalyst N-hydroxysuccimide (NHS) (Figure 15.4). EDC cross-linking can be controlled by the pH.[60] Also, EDC/NHS cross-linking is used in functionalizing grafts while cross-linking in the presence of a GAG. For instance, heparin can be used in providing an anticoagulant surface.[33,39,51] In the case of type I collagen, covalently bound heparin can act as a thromboresistant.[124] Furthermore, heparinization of grafts can be the onset to further graft functionalization by the addition of heparin-binding growth factors.

15.5.5 Graft Functionalization

15.5.5.1 Growth Factors

The incorporation of growth factors in vascular grafts may be the answer to the developing demand of smarter custom-made scaffolds with stimuli for growth and for creation of an optimal environment.[125] Growth factors can induce and modulate cellular influx after implantation. For vascular purposes, vascular endothelial growth factor (VEGF) and fibroblast growth factor 2 (FGF-2 or bFGF) are the most significant growth factors. VEGF-A is a 46 kDa disulfide-linked homodimer from the VEGF family and plays a key role in vasculogenesis and angiogenesis,[126] as does FGF-2 (16 kDa) from the FGF family, which is present in the subendothelial matrix of native blood vessels.[127]

It is known that porous collagen matrices can serve as carriers of growth factors.[128] The addition of VEGF and heparin increases the angiogenic potential[129] of collagen matrices. Release of immobilized VEGF from heparinized collagen matrices strongly depends upon proteinases.[130] Photoimmobilization of VEGF in photoreactive gelatin showed enhanced surface coverage by HUVECs grown in vitro.[131] The positive effect of VEGF upon endothelialization may help to increase patency.[132]

Kurane et al.[38] functionalized an elastin tube by addition of agarose with FGF-2. The sustained FGF-2 release enhanced cellular repopulation subdermally in rats.[38] Furthermore, a fibrin gel supplemented with VEGF and FGF stimulated angiogenesis in an arteriovenous loop model in rats.[70] VEGF and FGF-2 have a combined effect that leads to increased angiogenesis and blood vessel maturation when loaded to acellular heparinized collagen matrices in a subcutaneous model.[133] Other angiogenic agents have also been applied to acellular matrices, such as ginsenoside Rg1 which is retrieved from the plant *Panax ginseng*[134] and transforming growth factor-β (TGF-β) which synergistically enhanced elastin matrix regeneration with hyaluronan.[135]

Though the positive effects of growth factor application are evident, a cautionary note must be mentioned. Overstimulation of cells in vitro or in vivo may induce undesired effects like intimal hyperplasia.[136] However, when applied strategically, growth factors can be highly useful building blocks in the construction of tailor-made "smart" scaffolds.

15.5.5.2 Endothelialization

In vitro endothelialization, resulting in a confluent endothelium, can improve patency and decrease thrombogeneic responses.[132] EDC/NHS-cross-linked collagen sustains EC growth and proliferation and may therefore be a good candidate for in vivo application. Higher cell numbers were found on collagen with high cross-linking densities.[103] Yu et al.[123] pretreated ethylene glycol diglycidyl ether (EDGE)-decellularized thoracic arteries with lysine and type I collagen before endothelialization with HUVECs. Borschel et al.[57] showed that seeding of decellularized arteries with ECs increased patency rates in comparison to equivalent but non-seeded grafts. Acellular collagen matrices from porcine carotid arteries, also obtained by decellularization, were seeded with HUVECs and supported cellular expression of vasoactive agents.[31]

15.6 PERFORMANCE MONITORING

15.6.1 Graft Analyses and In Vitro Screening

It is important to realize that graft analyses and in vitro screening should precede in vivo experimentation. As described, type I collagen and elastin are the main extracellular building blocks in the native arterial architecture and, therefore, the main subjects of interest in this section. One of the first analyses of a biomaterial should be for successful cellular and nuclei removal. This can be done for instance by histological techniques and by DNA quantification.[41] Furthermore, morphometric analysis is useful in determining structure and porosity of the graft. The porosity of a graft is of main importance for in vitro and in vivo (re)population with cells. Pores need to be large enough (100–150 μm) for cells to enter and invade. The porosity and (ultra)structure can also be checked by electron microscopy.[76] Denaturation of proteins, as a side effect of heat or alkaline treatments, can be examined by, e.g., differential scanning calorimetry.[48] The purity of the "building blocks" needs to be assessed (Figure 15.2). Biochemical washings performed upon pulverized Achilles tendon resulted in purified insoluble type I collagen fibers with an intact striated banding pattern (Figure 15.2a).[98] Biochemical treatment of pulverized equine ligamentum nuchae resulted in pure elastin fibers devoid of microfibrils (Figure 15.2b).[101] Immunohistochemical techniques can be used to identify the location of a specific molecule within the graft. Addition of growth factors to molecularly defined grafts requires their quantification in the graft.

Both decellularization and purification protocols involve many chemical agents that can potentially induce cytotoxic effects. Therefore, in vitro cell seeding must be carried out to study the cell viability, adhesion, and proliferation of blood vessel cells like ECs and SMCs. ECs and SMCs can be isolated from arteries or veins and characterized by vWF/cluster of differentiation molecule 31 (CD31) and α-smooth muscle actin (αSMA) stainings, respectively. A strong linear correlation between cell attachment and scaffold surface area of collagen–GAG scaffolds has been reported.[107] Type I collagen membranes were seeded with ECs and SMCs and sustained co-culture for 21 days.[46] Berglund et al.[56] cast cellularized collagen gels around purified arterial elastin[32] and collagen support sleeves, thus obtaining layered constructs. Lu et al.[41] cultured 3T3 mouse fibroblasts upon pure elastin and collagen scaffolds. Wissink et al.[103] found secretion of the vasoactive agents PGI$_2$, tPa, and vWF by ECs seeded upon flat collagen films. Buijtenhuijs et al.[105] described αSMA-positive SMCs seeded on molecularly defined collagen–elastin scaffolds. Similar scaffolds were tubularized by a molding technique, seeded with SMCs, and cultured in a bioreactor for 7 days.[105] Rotational seeding was also used for the seeding of acellular collagen matrices with human saphenous vein ECs.[31]

Blood compatibility experiments including the thrombogenic and (anti)coagulant properties of the biomaterial may also be investigated. These can be measured by assays including platelet adhesion and activation, clotting time, thromboelastography, thrombin generation, and fibrin clotting. Boccafoschi et al.[67] found that acid-soluble type I collagen films slightly activated platelet aggregation, but did not enhance blood coagulation. Keuren et al.[86] tested polysaccharide surfaces for thrombogenicity and found that covalently bound heparin provides thromboresistance to collagen.[124] Recombinant human elastin polypeptides (EP-20-24-24) showed great potential as nonthrombogenic coating of synthetic polymer catheters both in vitro and in vivo in rabbits.[84]

Finally, an experimental graft has to possess sufficient mechanical strength to withstand blood pressure forces, 120 mm Hg in systole and 80 mm Hg in diastole. Native blood vessels have burst pressures in the range of 2000 mm Hg, well above physiological blood pressure values.[44,119] Additional mechanical determinants include compliance, modulus of elasticity, and suture retention strength. A minimal requirement for the suture retention strength is around 2 N.[36] Unidirectional testing can be used for calculation of the Young's modulus, elongation at break, and ultimate tensile strength (UTS). The human saphenous vein, the golden standard replacement graft in peripheral bypass surgery, has a burst pressure of 1680 ± 307 mm Hg,[64] porcine carotid arteries a burst pressure of 2000 mm Hg,[42] and ePTFE grafts[95] can have burst pressure values over 3500 mm Hg. Berglund et al.[56] measured peak bursting pressures for CSH grafts of 100 and 650 mm Hg for non-cross-linked and glutaraldehyde-fixed grafts, respectively. Burst pressure of aSIS vascular grafts range from 931[36] to 3517 mm Hg.[95] Human coronary arteries have a Young's modulus around 0.6 MPa.[137] In the case of decellularized arteries, there may be altered mechanical properties due to chemical treatment.[59] Decellularized rabbit carotid arteries displayed an increase in stiffness, a decreased extensibility, and a lower stress level in comparison to native ones.[59] Arterial replacement grafts need to have similar or matching mechanical properties as native vessels before in vivo application can take place, as mismatches may induce undesired long-term complications, like intimal hyperplasia, aneurysm formation, or even graft failure.[44]

15.6.2 ANIMAL MODELS

Numerous animal models have been developed to test biomaterials for (cardio)vascular tissue engineering purposes.[138] A broad range of implantation sites are used for acellular and natural tubular grafts, including subdermal implantation. Subdermal implantation is initially performed to study biocompatibility, cellular ingrowth, and biodegradability. Kurane et al.[38] found in vivo cellular repopulation in subdermally implanted tubular elastin scaffolds functionalized with FGF-2. The latter was found to suppress calcification of elastin. Daamen et al.[102] implanted scaffolds containing type I collagen and solubilized elastin in young Sprague Dawley rats, a sensitive calcification model, and found elastic fiber formation with no calcification. Liu et al.[40] placed decellularized aorta of fetal pigs subdermally in rats. The fetal tissue showed practically no immunological response and minimal calcification.

Vascular implantation sites include aorta, and coronary, carotid, and femoral arteries. Methods of graft preparation are listed in Table 15.2, together with some in vivo models and applications with natural materials. Liu et al.[58] placed strips of collagen and elastin matrices within the lumen of rat aorta and found that elastin matrices inhibited SMC proliferation and neointima formation in comparison to collagen matrices. Complete vascular replacement may initially include acute models, in which grafts are exposed to blood flow during a limited amount of time. Liao et al.[39] used a baboon external shunt model and studied radioactive platelet deposition during 60 min on decellularized and heparinized porcine carotid arteries. Hinds et al.[35] implanted composite elastin–aSIS grafts in domestic swine and compared these to ePTFE. In this acute model, the natural graft had significantly longer patency times than ePTFE, namely 5.23 h versus 4.15 h.

For long-term in vivo experiments, allogeneic[49,62] and xenogeneic[37] models have been used. Martin et al.[49] found satisfactory results in applying SDS-decellularized vein allografts with respect to strength and cellular repopulation within canine bilateral carotid interposition. Kim et al.[37] described histopathological changes of the graft in a xenogeneic pig-to-goat model. An acellular collagen graft from aSIS remained patent up to 90 days and became physiologically responsive when implanted end-to-side in carotid arteries of New Zealand rabbits.[36] In a similar animal model, bilateral surgery was performed with heparinized xenografts, indicating that heparinization reduces in vivo thrombogenicity.[51] Hyaluronan-based scaffolds (HYAFF-11) were placed end-to-end in rat abdominal aortas[73] and in porcine carotid arteries.[74] In the latter case, 70% of the grafts remained patent after 5 months, which was determined by Duplex scanning. The grafts were almost fully

TABLE 15.2
List of Acellular Natural Grafts Applied In Vivo

Grafts created by decellularization procedures

Source	Treatment	Method of Stabilization	Implantation Site	Outcome/Remarks
Bergmeister et al.[33] Porcine femoral artery	Triton X-100, deoxycholate, IGEPAL, EDTA, RNase, DNase	EDC cross-linking with and without heparinization	Interposition in rat infrarenal aorta	Transition from muscular to elastic artery
Borschel et al.[57] Rat iliac artery	Glycerine, SDS, and deoxycholate	—	Interposition in rat femoral arteries after EC recellularization	At 4 weeks patency was 89% for recellularized grafts versus 29% for acellular grafts
Hilbert et al.[62] Goat carotid artery	EDTA, n-octyl-β-D-glucospyranoside, DNase	Freeze-drying in the presence of a cryoprotectant	Interposition in goat carotid artery	All grafts remained patent at 7 months
Huynh et al.[36] Porcine small intestine	EDTA and salt solution (tubular entrapment and luminal deposition of fibrillated collagen)	EDC cross-linking	Interposition in rabbit carotid artery	In vivo cellularized vessels are responsive to vasoactive agents at 3 months
Kim et al.[37] Porcine carotid artery	Hypertonic solution and SDS. Kept frozen until use	—	Bilateral interposition in goat carotid artery	Grafts studied up to 1 year. No thrombi, but some stenosis found
Liao et al.[39] Porcine carotid artery	Unspecified hypotonic, enzymatic, and detergent treatment[151]	EDC cross-linking and heparinization	Eternal shunt in baboon	Reduced platelet deposition compared to non-heparin-coated vessels
Liu et al.[58] Rat aorta	NaOH treatment resulting in basal lamina, internal elastic lamina, or adventitial collagen at luminal side	—	Luminal implantation in rat infrarenal aorta	Elastic lamina lowers leukocyte adhesion and neointima formation
Martin et al.[49] Canine external jugular vein	SDS at 37°C	—	Bilateral interposition in canine carotid artery	Comparison to fresh allo- and autografts
Meuris et al.[43] Porcine aorta	Triton, RNase, and DNase	Glutaraldehyde and photooxidative fixation	Implantation of 1–2 cm² fragments in sheep jugular vein	Least calcification in photooxidized samples
Narita et al.[50] Canine ureter non-seeded and EC-seeded	Triton X-100/EDTA/RNase/DNase	—	Interposition in canine carotid artery	EC-seeded patent for at least 24 weeks. Non-seeded most occluded at 1 week

Source	Source material	Method	Stabilization	In vivo model	Results
Wang et al.[51]	Canine carotid artery	Hypo- and hypertonic, Tris, Triton X-100/EDTA, RNase, DNase	EDC cross-linking and heparinization	Bilateral bypass in rabbit carotid artery	Less occlusions in heparinized xenografts
Wilson et al.[52]	Canine carotid artery	Hypotonic solution and enzymatic digestion	—	End-to-side implantation in canine femoral, carotid, and infrarenal arteries in dogs	No antithrombotic drugs administered. Patency studied up to 6 years
Wilson et al.[53]	Canine carotid artery	Hypotonic solution, Triton X-100, SDS, and enzymatic digestion[52]	—	Bypass in canine coronary artery	4/9 grafts were patent
Grafts created from scratch					
Hinds et al.[35]	Elastin conduit isolated from porcine carotid artery with SIS	Acellular graft made by gluing elastin tube and SIS with fibrin	—	Interposition in porcine carotid artery	Grafts performed better than ePTFE in 6h study
Lepidi et al.[73]	Hyaluronan (HAYFF-11)	Tubular shaping by coating of mandrel	—	Interposition in rat aorta	Complete regeneration of vascular tube after 3 months
Nagai et al.[64]	Collagen from salmon skin	Casting technique to prepare a tube	EDC cross-linking	Interposition in rat abdominal aorta	Prior to implantation, grafts were coated with argatroban to prevent early thrombosis
Simionescu et al.[45]	Elastin and collagen isolated from ascending porcine aorta	Elastin grafts with maintained tubular shape		Interposition in rabbit carotid arteries	No leakage in 30min study, little platelet adhesion
Wu et al.[46]	Type I collagen from porcine skin	Vacuum suction to prepare membranes. These membranes were enrolled and sutured to prepare a tube	Glutaraldehyde followed by freeze-drying	Interposition in rat inferior vena cava	EC/SMC seeded graft: no thrombosis and intimal hyperplasia at 12 weeks. Acellular graft: platelet clot was observed in a cellular scaffold group; however, it was gradually dissolved some intimal hyperplasia (with the time and obviously reduced at the end of 12 weeks).
Tillman et al.[115]	Type I collagen and PCL	Tubular electrospinning in 1,1,1,3,3,3-hexafluoro-2-propanol	Glutaraldehyde cross-linking	Aortoiliac bypass in rabbit	7/8 implants remained patent for 1 month
Zavan et al.[74]	Hyaluronan (HAYFF-11)	Tubular shaping by coating of mandrel	—	Bilateral interposition in porcine carotids	Biosynthesis of organized layers of elastic fibers after 5 months

Note: Indicated are the source, type of treatment, method of stabilization, and in vivo model used.

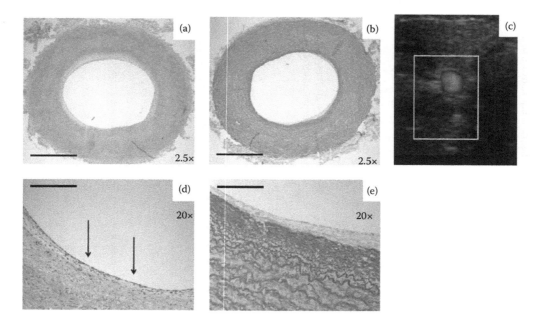

FIGURE 15.6 **(See companion CD for color figure.)** Histological sections of hyaluronan-based tubular scaffolds 5 months after implantation. No signs of intimal hyperplasia were observed (a and b). Hematoxylin staining was used for a general view (a and d) and Weigert staining was applied as specific stain for elastin (b and e). Histology shows the formation of concentric elastic layers of elastin and inner lining by flattened cells (d and e). Duplex scanning shows, like histology, patent grafts after 5 months (c). (Reprinted with permission from FASEB. Zavan, B. et al., *FASEB J.*, 22, 2853, 2008.)

degraded, remodeled into mature tissue and histology displayed CD31- and vWF-positive endothelial lining and formation of organized layers of elastin fibers (Figure 15.6). The ability to initiate the biosynthesis of elastin is considered to be the missing link in vascular tissue engineering.[68] Elastin's structural location, as displayed in Figure 15.1, is of great functional importance. Bergmeister et al.[33] implanted decellularized muscular arteries from pigs into rat aorta. The grafts displayed a transition from a muscular to an elastic phenotype.

Electrospun grafts from collagen in combination with PCL kept integrity for 1 month when implanted in a rabbit aortoiliac bypass model.[114] A hybrid tube of poly-L-lactic acid (PLLA) and PGA in combination with a collagen microsponge generated confluent endothelial lining in situ in a canine bilateral carotid artery interposition study.[139] Nonvascular conduits, obtained from decellularized ureter, were implanted in abdominal aorta of dogs and transformed into durable conduits.[144] Wilson et al.[52] used canine interposition in femoral and carotid arteries and followed patencies up to 6 years. Subsequently, this graft was also tested in another preclinical in vivo study, using in an allogeneic canine coronary artery bypass model.[53]

In the end, interpositioning and bypassing are the best animal models before future application in clinical human trials. The use of anticoagulant drugs is feasible and frequently performed, as this treatment can easily be adapted to a clinical setting. Additionally, anticoagulant medication gives the body time to integrate an acellular natural graft without the risk of thrombosis and graft occlusion.

15.7 CLINICAL USE OF ACELLULAR NATURAL GRAFTS

Several acellular natural grafts have reached clinical application. Benefits of the biografts include optimal availability and appropriate caliber. Artegraft, a collagen vascular graft, has been clinically applied since the 1970s. For this, carotid arteries of bovine spongiform encephalopathy (BSE)-free food cattle are the source. It is primarily used as AV shunt for hemodialysis purposes, femoro-popliteal bypass, and arterial replacement, bypass, or patch.[141] Cryolife® uses Synergraft® technology, a patented decellularization procedure, in processing bovine ureter as vascular conduit called SG 100.[140,142] The same company markets cryopreserved arteries and veins as well, by the names CryoArtery® and CryoVein®. Chemla and Morsy[142] ran a clinical trial and found that SG 100 has comparable achievements as ePTFE. In the 1980s, the new bovine collagen vascular graft Solcograft P was developed, but was taken off the market by the manufacturer due to failure in clinical trials, most probably due to aneurysm formation.[54,143] The negatively charged glutaraldehyde-tanned (NCGT) graft from Johnson & Johnson was demonstrated in a clinical trial to have a cumulative patency rate of 73% by Reddy et al.[144] Successful patencies of NCGT grafts are most probably caused by the surface charge and the relatively dense structure of NCGT grafts. This dense structure can cope with short- and long-term thrombosis.[145]

Hybrid grafts, that combine man-made and natural polymers, have also been used. Omniflow® II (Bio Nova international, Melbourne, Australia) is a vascular composite prosthesis that is created from stabilized sheep collagen with an integral endoskeleton of polyester mesh. The chemical stabilization of this prosthesis provides a superior structural stability and compliance. A multicenter randomized study was started in January 2009 to compare patency to PTFE. Other marketed hybrid grafts include Biograft, a glutaraldehyde-tanned human umbilical vein wrapped with Dacron from Synovis. The human application with this graft was first described in 1976 by Dardik et al.[146] In a 6 year study, only 17% of grafts resulted in aneurysm formation though dilatation of the graft was found.[147]

Contegra® pulmonary valve conduit (Medtronic) has been created from a bovine jugular vein with a trileaflet venous valve. An example of clinical application is an alternative during the Ross procedure, during which a diseased aortic valve is surgically replaced. According to Hickey et al.,[148] Contegra matches cryopreserved allograft performances in *truncus arteriosus* repair. This application of natural tubular grafts, especially in pediatric surgery, demonstrates their ability to function and remodel in specific tissue and grow along with the body itself.

15.8 CONCLUSION AND FUTURE OUTLOOK

The application of natural materials for vascular tissue engineering may counteract persisting clinical problems like intimal hyperplasia and (re)stenosis. Currently, the paradigm in medicine shifts from synthetic graft application to the use of biodegradable and porous biomaterials,[149] in combination with biological key effector molecules, which should trigger tissue-specific regeneration. Growth factors like VEGF and FGF-2 are promising candidates as both play key roles in angiogenesis, vasculogenesis, and embryological development.

Using Mother Nature as a blueprint, scaffolds or grafts should mimic the original composition and architecture of the organ or tissue in question. Considering the ECM, a multicomponent approach is most eminent with molecularly defined building blocks in order to construct characterized tailor-made scaffolds.[125] For this, vascular grafts may contain a molecular skeleton of type I collagen and elastin, to which bioactive molecules such as GAGs and growth factors can be added. To develop such "smart" scaffolds, casting and molding techniques or electrospinning may be applied using highly purified ECM components.

Microscale and high-throughput screening[150] of biomaterials in vitro and in vivo will most probably lead to the discovery of new graft compositions. In this way, development of even "smarter" scaffolds may bypass decellularized natural grafts, and become tailor-made off-the-shelf vascular grafts.

ACKNOWLEDGMENT

This work was financially supported by the Dutch Program for Tissue Engineering (DPTE 6735).

REFERENCES

1. Ikada Y. Challenges in tissue engineering. *J. R. Soc. Interface* 2006;3:589–601.
2. Gotlieb AI. Atherosclerosis and acute coronary syndromes. *Cardiovasc. Pathol.* 2005;14:181–184.
3. Devine C and McCollum C. Heparin-bonded Dacron or polytetrafluorethylene for femoropopliteal bypass: Five-year results of a prospective randomized multicenter clinical trial. *J. Vasc. Surg.* 2004;40:924–931.
4. Greenwald SE and Berry CL. Improving vascular grafts: The importance of mechanical and haemodynamic properties. *J. Pathol.* 2000;190:292–299.
5. Veith FJ, Moss CM, Sprayregen S, and Montefusco C. Preoperative saphenous venography in arterial reconstructive surgery of the lower extremity. *Surgery* 1979;85:253–256.
6. Thomas AC, Campbell GR, and Campbell JH. Advances in vascular tissue engineering. *Cardiovasc. Pathol.* 2003;12:271–276.
7. Weinberg CB and Bell E. A blood vessel model constructed from collagen and cultured vascular cells. *Science* 1986;231:397–400.
8. Rashid ST, Salacinski HJ, Fuller BJ, Hamilton G, and Seifalian AM. Engineering of bypass conduits to improve patency. *Cell Prolif.* 2004;37:351–366.
9. Guyton AC and Hall JE. *Textbook of Medical Physiology*. W.B. Saunders Company, Philadelphia, PA; 1996.
10. Ayad S, Boot-Handford RP, Humphries MJ, Kadler KE, and Shuttleworth CA. *The Extracellular Matrix Facts Book*. Academic Press, San Diego, CA; 1994.
11. Junqueira LC and Carneiro J. *Basic Histology: Text and Atlas*. McGraw–Hill, New York; 2005.
12. Voorhees AB, Jr., Jaretzki A, III, and Blakemore AH. The use of tubes constructed from vinyon "N" cloth in bridging arterial defects. *Ann. Surg.* 1952;135:332–336.
13. Campbell CD, Brooks DH, Webster MW, and Bahnson HT. The use of expanded microporous polytetrafluoroethylene for limb salvage: A preliminary report. *Surgery* 1976;79:485–491.
14. Kapadia MR, Popowich DA, and Kibbe MR. Modified prosthetic vascular conduits. *Circulation* 2008;117:1873–1882.
15. Seifalian AM, Salacinski HJ, Tiwari A et al. In vivo biostability of a poly(carbonate-urea)urethane graft. *Biomaterials* 2003;24:2549–2557.
16. Hong Y, Ye SH, Nieponice A et al. A small diameter, fibrous vascular conduit generated from a poly(ester urethane)urea and phospholipid polymer blend. *Biomaterials* 2009;30:2457–2467.
17. Farrar DJ. Development of a prosthetic coronary artery bypass graft. *Heart Surg. Forum* 2000;3:36–40.
18. Karrer L, Duwe J, Zisch AH et al. PPS–PEG surface coating to reduce thrombogenicity of small diameter ePTFE vascular grafts. *Int. J. Artif. Organs* 2005;28:993–1002.
19. Jordan SW, Faucher KM, Caves JM et al. Fabrication of a phospholipid membrane-mimetic film on the luminal surface of an ePTFE vascular graft. *Biomaterials* 2006;27:3473–3481.
20. Yoneyama T, Sugihara K, Ishihara K, Iwasaki Y, and Nakabayashi N. The vascular prosthesis without pseudointima prepared by antithrombogenic phospholipid polymer. *Biomaterials* 2002;23:1455–1459.
21. Yang J, Motlagh D, Allen J et al. Modulating ePTFE vascular graft host response via citric acid-based biodegradable elastomers. *Adv. Mater.* 2006;18:1493–1498.
22. Lagerqvist B, James SK, Stenestrand U et al. Long-term outcomes with drug-eluting stents versus bare-metal stents in Sweden. *N. Engl. J. Med.* 2007;356:1009–1019.
23. Spaulding C, Daemen J, Boersma E, Cutlip DE, and Serruys PW. A pooled analysis of data comparing sirolimus-eluting stents with bare-metal stents. *N. Engl. J. Med.* 2007;356:989–997.
24. Walluscheck KP, Steinhoff G, Kelm S, and Haverich A. Improved endothelial cell attachment on ePTFE vascular grafts pretreated with synthetic RGD-containing peptides. *Eur. J. Vasc. Endovasc. Surg.* 1996;12:321–330.
25. Krijgsman B, Seifalian AM, Salacinski HJ et al. An assessment of covalent grafting of RGD peptides to the surface of a compliant poly(carbonate-urea)urethane vascular conduit versus conventional biological coatings: Its role in enhancing cellular retention. *Tissue Eng.* 2002;8:673–680.
26. Jordan SW, Haller CA, Sallach RE et al. The effect of a recombinant elastin-mimetic coating of an ePTFE prosthesis on acute thrombogenicity in a baboon arteriovenous shunt. *Biomaterials* 2007;28:1191–1197.
27. Bosiers M, Deloose K, Verbist J et al. Heparin-bonded expanded polytetrafluoroethylene vascular graft for femoropopliteal and femorocrural bypass grafting: 1-year results. *J. Vasc. Surg.* 2006;43:313–318.
28. Li C, Hill A, and Imran M. In vitro and in vivo studies of ePTFE vascular grafts treated with P15 peptide. *J. Biomater. Sci. Polym. Ed.* 2005;16:875–891.

29. Deutsch M, Meinhart J, Zilla P et al. Long-term experience in autologous in vitro endothelialization of infrainguinal ePTFE grafts. *J. Vasc. Surg.* 2009;49:352–362.

30. Meinhart JG, Deutsch M, Fischlein T et al. Clinical autologous in vitro endothelialization of 153 infrainguinal ePTFE grafts. *Ann. Thorac. Surg.* 2001;71:S327–S331.

31. Amiel GE, Komura M, Shapira O et al. Engineering of blood vessels from acellular collagen matrices coated with human endothelial cells. *Tissue Eng.* 2006;12:2355–2365.

32. Berglund JD, Nerem RM, and Sambanis A. Incorporation of intact elastin scaffolds in tissue-engineered collagen-based vascular grafts. *Tissue Eng.* 2004;10:1526–1535.

33. Bergmeister H, Plasenzotti R, Walter I et al. Decellularized, xenogeneic small-diameter arteries: Transition from a muscular to an elastic phenotype in vivo. *J. Biomed. Mater. Res. B Appl. Biomater.* 2008;87B:95–104.

34. Derham C, Yow H, Ingram J et al. Tissue engineering small-diameter vascular grafts: Preparation of a biocompatible porcine ureteric scaffold. *Tissue Eng. Part A* 2008;14:1871–1882.

35. Hinds MT, Rowe RC, Ren Z et al. Development of a reinforced porcine elastin composite vascular scaffold. *J. Biomed. Mater. Res. A* 2006;77:458–469.

36. Huynh T, Abraham G, Murray J et al. Remodeling of an acellular collagen graft into a physiologically responsive neovessel. *Nat. Biotechnol.* 1999;17:1083–1086.

37. Kim WS, Seo JW, Rho JR, and Kim WG. Histopathologic changes of acellularized xenogenic carotid vascular grafts implanted in a pig-to-goat model. *Int. J. Artif. Organs* 2007;30:44–52.

38. Kurane A, Simionescu DT, and Vyavahare NR. In vivo cellular repopulation of tubular elastin scaffolds mediated by basic fibroblast growth factor. *Biomaterials* 2007;28:2830–2838.

39. Liao D, Wang X, Lin PH, Yao Q, and Chen C. Covalent linkage of heparin provides a stable anticoagulation surface of decellularized porcine arteries. *J. Cell Mol. Med.* 2009;13:2736–2743.

40. Liu GF, He ZJ, Yang DP et al. Decellularized aorta of fetal pigs as a potential scaffold for small diameter tissue engineered vascular graft. *Chin Med. J. (Engl.)* 2008;121:1398–1406.

41. Lu Q, Ganesan K, Simionescu DT, and Vyavahare NR. Novel porous aortic elastin and collagen scaffolds for tissue engineering. *Biomaterials* 2004;25:5227–5237.

42. McFetridge PS, Daniel JW, Bodamyali T, Horrocks M, and Chaudhuri JB. Preparation of porcine carotid arteries for vascular tissue engineering applications. *J. Biomed. Mater. Res. A* 2004;70:224–234.

43. Meuris B, Verbeken E, and Flameng W. Prevention of porcine aortic wall calcification by acellularization: Necessity for a non-glutaraldehyde-based fixation treatment. *J. Heart Valve Dis.* 2005;14:358–363.

44. Roy S, Silacci P, and Stergiopulos N. Biomechanical properties of decellularized porcine common carotid arteries. *Am. J. Physiol Heart Circ. Physiol.* 2005;289:H1567–H1576.

45. Simionescu DT, Lu Q, Song Y et al. Biocompatibility and remodeling potential of pure arterial elastin and collagen scaffolds. *Biomaterials* 2006;27:702–713.

46. Wu HC, Wang TW, Kang PL et al. Coculture of endothelial and smooth muscle cells on a collagen membrane in the development of a small-diameter vascular graft. *Biomaterials* 2007;28:1385–1392.

47. Dahl SL, Koh J, Prabhakar V, and Niklason LE. Decellularized native and engineered arterial scaffolds for transplantation. *Cell Transplant.* 2003;12:659–666.

48. Goissis G, Suzigan S, Parreira DR et al. Preparation and characterization of collagen–elastin matrices from blood vessels intended as small diameter vascular grafts. *Artif. Organs* 2000;24:217–223.

49. Martin ND, Schaner PJ, Tulenko TN et al. In vivo behavior of decellularized vein allograft. *J. Surg. Res.* 2005;129:17–23.

50. Narita Y, Kagami H, Matsunuma H et al. Decellularized ureter for tissue-engineered small-caliber vascular graft. *J. Artif. Organs* 2008;11:91–99.

51. Wang XN, Chen CZ, Yang M, and Gu YJ. Implantation of decellularized small-caliber vascular xenografts with and without surface heparin treatment. *Artif. Organs* 2007;31:99–104.

52. Wilson GJ, Yeger H, Klement P, Lee JM, and Courtman DW. Acellular matrix allograft small caliber vascular prostheses. *ASAIO Trans.* 1990;36:M340–M343.

53. Wilson GJ, Courtman DW, Klement P, Lee JM, and Yeger H. Acellular matrix: A biomaterials approach for coronary artery bypass and heart valve replacement. *Ann. Thorac. Surg.* 1995;60:S353–S358.

54. Schroder A, Imig H, Peiper U, Neidel J, and Petereit A. Results of a bovine collagen vascular graft (Solcograft-P) in infra-inguinal positions. *Eur. J. Vasc. Surg.* 1988;2:315–321.

55. Buttafoco L, Engbers-Buijtenhuijs P, Poot AA et al. First steps towards tissue engineering of small-diameter blood vessels: Preparation of flat scaffolds of collagen and elastin by means of freeze drying. *J. Biomed. Mater. Res. B Appl. Biomater.* 2006;77:357–368.

56. Berglund JD, Mohseni MM, Nerem RM, and Sambanis A. A biological hybrid model for collagen-based tissue engineered vascular constructs. *Biomaterials* 2003;24:1241–1254.

57. Borschel GH, Huang YC, Calve S et al. Tissue engineering of recellularized small-diameter vascular grafts. *Tissue Eng*. 2005;11:778–786.
58. Liu SQ, Tieche C, and Alkema PK. Neointima formation on vascular elastic laminae and collagen matrices scaffolds implanted in the rat aortae. *Biomaterials* 2004;25:1869–1882.
59. Williams C, Liao J, Joyce EM et al. Altered structural and mechanical properties in decellularized rabbit carotid arteries. *Acta Biomater*. 2009;5:993–1005.
60. Gratzer PF and Lee JM. Control of pH alters the type of cross-linking produced by 1-ethyl-3-(3-dimethylaminopropyl)-carbodiimide (EDC) treatment of acellular matrix vascular grafts. *J. Biomed. Mater. Res*. 2001;58:172–179.
61. Walles T, Herden T, Haverich A, and Mertsching H. Influence of scaffold thickness and scaffold composition on bioartificial graft survival. *Biomaterials* 2003;24:1233–1239.
62. Hilbert SL, Boerboom LE, Livesey SA, and Ferrans VJ. Explant pathology study of decellularized carotid artery vascular grafts. *J. Biomed. Mater. Res. A* 2004;69:197–204.
63. Jeong SI, Kim SY, Cho SK et al. Tissue-engineered vascular grafts composed of marine collagen and PLGA fibers using pulsatile perfusion bioreactors. *Biomaterials* 2007;28:1115–1122.
64. Nagai N, Nakayama Y, Zhou YM et al. Development of salmon collagen vascular graft: Mechanical and biological properties and preliminary implantation study. *J. Biomed. Mater. Res. B Appl. Biomater*. 2008;87:432–439.
65. Yunoki S, Mori K, Suzuki T, Nagai N, and Munekata M. Novel elastic material from collagen for tissue engineering. *J. Mater. Sci. Mater. Med*. 2007;18:1369–1375.
66. Stegemann JP, Kaszuba SN, and Rowe SL. Review: Advances in vascular tissue engineering using protein-based biomaterials. *Tissue Eng*. 2007;13:2601–2613.
67. Boccafoschi F, Habermehl J, Vesentini S, and Mantovani D. Biological performances of collagen-based scaffolds for vascular tissue engineering. *Biomaterials* 2005;26:7410–7417.
68. Patel A, Fine B, Sandig M, and Mequanint K. Elastin biosynthesis: The missing link in tissue-engineered blood vessels. *Cardiovasc. Res*. 2006;71:40–49.
69. Shaikh FM, Callanan A, Kavanagh EG et al. Fibrin: A natural biodegradable scaffold in vascular tissue engineering. *Cells Tissues Organs* 2008;188:333–346.
70. Arkudas A, Tjiawi J, Bleiziffer O et al. Fibrin gel-immobilized VEGF and bFGF efficiently stimulate angiogenesis in the AV loop model. *Mol. Med*. 2007;13:480–487.
71. Ahmed TA, Dare EV, and Hincke M. Fibrin: A versatile scaffold for tissue engineering applications. *Tissue Eng. Part B Rev*. 2008;14:199–215.
72. Soffer L, Wang X, Zhang X et al. Silk-based electrospun tubular scaffolds for tissue-engineered vascular grafts. *J. Biomater. Sci. Polym. Ed*. 2008;19:653–664.
73. Lepidi S, Grego F, Vindigni V et al. Hyaluronan biodegradable scaffold for small-caliber artery grafting: Preliminary results in an animal model. *Eur. J. Vasc. Endovasc. Surg*. 2006;32:411–417.
74. Zavan B, Vindigni V, Lepidi S et al. Neoarteries grown in vivo using a tissue-engineered hyaluronan-based scaffold. *FASEB J*. 2008;22:2853–2861.
75. Cen L, Liu W, Cui L, Zhang W, and Cao Y. Collagen tissue engineering: Development of novel biomaterials and applications. *Pediatr. Res*. 2008;63:492–496.
76. Geutjes PJ, Daamen WF, Buma P et al. From molecules to matrix: Construction and evaluation of molecularly defined bioscaffolds. *Adv. Exp. Med. Biol*. 2006;585:279–295.
77. Rosenbloom J, Abrams WR, and Mecham R. Extracellular matrix 4: The elastic fiber. *FASEB J*. 1993;7:1208–1218.
78. Daamen WF, Veerkamp JH, van Hest JC, and van Kuppevelt TH. Elastin as a biomaterial for tissue engineering. *Biomaterials* 2007;28:4378–4398.
79. Faury G. Function–structure relationship of elastic arteries in evolution: From microfibrils to elastin and elastic fibres. *Pathol. Biol. (Paris)* 2001;49:310–325.
80. Kielty CM, Stephan S, Sherratt MJ, Williamson M, and Shuttleworth CA. Applying elastic fibre biology in vascular tissue engineering. *Philos. Trans. R. Soc. Lond B Biol. Sci*. 2007;362:1293–1312.
81. Williamson MR, Shuttleworth A, Canfield AE, Black RA, and Kielty CM. The role of endothelial cell attachment to elastic fibre molecules in the enhancement of monolayer formation and retention, and the inhibition of smooth muscle cell recruitment. *Biomaterials* 2007;28:5307–5318.
82. Stephan S, Ball SG, Williamson M et al. Cell–matrix biology in vascular tissue engineering. *J. Anat*. 2006;209:495–502.
83. Robert L. Cell–elastin interaction and signaling. *Pathol. Biol. (Paris)* 2005;53:399–404.
84. Woodhouse KA, Klement P, Chen V et al. Investigation of recombinant human elastin polypeptides as non-thrombogenic coatings. *Biomaterials* 2004;25:4543–4553.

85. Toy EC, Seifert WE, Strobel HW, and Harms KP. *Case Files Biochemistry.* McGraw-Hill, New York; 2008.

86. Keuren JF, Wielders SJ, Willems GM et al. Thrombogenicity of polysaccharide-coated surfaces. *Biomaterials* 2003;24:1917–1924.

87. Wight TN and Merrilees MJ. Proteoglycans in atherosclerosis and restenosis: Key roles for versican. *Circ. Res.* 2004;94:1158–1167.

88. van den Boom M, Sarbia M, von Wnuck LK et al. Differential regulation of hyaluronic acid synthase isoforms in human saphenous vein smooth muscle cells: Possible implications for vein graft stenosis. *Circ. Res.* 2006;98:36–44.

89. Joddar B and Ramamurthi A. Elastogenic effects of exogenous hyaluronan oligosaccharides on vascular smooth muscle cells. *Biomaterials* 2006;27:5698–5707.

90. Ferdous Z and Grande-Allen KJ. Utility and control of proteoglycans in tissue engineering. *Tissue Eng.* 2007;13:1893–1904.

91. Sarkar S, Schmitz-Rixen T, Hamilton G, and Seifalian AM. Achieving the ideal properties for vascular bypass grafts using a tissue engineered approach: A review. *Med. Biol. Eng. Comput.* 2007;45:327–336.

92. Gilbert TW, Sellaro TL, and Badylak SF. Decellularization of tissues and organs. *Biomaterials* 2006;27:3675–3683.

93. Schmidt CE and Baier JM. Acellular vascular tissues: Natural biomaterials for tissue repair and tissue engineering. *Biomaterials* 2000;21:2215–2231.

94. Lantz GC, Badylak SF, Hiles MC et al. Small intestinal submucosa as a vascular graft: A review. *J. Invest. Surg.* 1993;6:297–310.

95. Roeder R, Wolfe J, Lianakis N et al. Compliance, elastic modulus, and burst pressure of small-intestine submucosa (SIS), small-diameter vascular grafts. *J. Biomed. Mater. Res.* 1999;47:65–70.

96. Zheng MH, Chen J, Kirilak Y et al. Porcine small intestine submucosa (SIS) is not an acellular collagenous matrix and contains porcine DNA: Possible implications in human implantation. *J. Biomed. Mater. Res. B Appl. Biomater.* 2005;73:61–67.

97. Gratzer PF, Harrison RD, and Woods T. Matrix alteration and not residual sodium dodecyl sulfate cytotoxicity affects the cellular repopulation of a decellularized matrix. *Tissue Eng.* 2006;12:2975–2983.

98. Pieper JS, Hafmans T, Veerkamp JH, and van Kuppevelt TH. Development of tailor-made collagen–glycosaminoglycan matrices: EDC/NHS crosslinking, and ultrastructural aspects. *Biomaterials* 2000;21:581–593.

99. Pieper JS, van Wachem PB, van Luyn MJA et al. Attachment of glycosaminoglycans to collagenous matrices modulates the tissue response in rats. *Biomaterials* 2000;21:1689–1699.

100. Daamen WF, Hafmans T, Veerkamp JH, and van Kuppevelt TH. Isolation of intact elastin fibers devoid of microfibrils. *Tissue Eng.* 2005;11:1168–1176.

101. Daamen WF, Nillesen ST, Wismans RG et al. A biomaterial composed of collagen and solubilized elastin enhances angiogenesis and elastic fiber formation without calcification. *Tissue Eng. Part A* 2008;14:349–360.

102. Daamen WF, van Moerkerk HT, Hafmans T et al. Preparation and evaluation of molecularly-defined collagen–elastin–glycosaminoglycan scaffolds for tissue engineering. *Biomaterials* 2003;24:4001–4009.

103. Wissink MJ, van Luyn MJ, Beernink R et al. Endothelial cell seeding on crosslinked collagen: Effects of crosslinking on endothelial cell proliferation and functional parameters. *Thromb. Haemost.* 2000;84:325–331.

104. Buttafoco L, Engbers-Buijtenhuijs P, Poot AA et al. Physical characterization of vascular grafts cultured in a bioreactor. *Biomaterials* 2006;27:2380–2389.

105. Buijtenhuijs P, Buttafoco L, Poot AA et al. Tissue engineering of blood vessels: Characterization of smooth-muscle cells for culturing on collagen-and-elastin-based scaffolds. *Biotechnol. Appl. Biochem.* 2004;39:141–149.

106. Hinds MT, Courtman DW, Goodell T et al. Biocompatibility of a xenogenic elastin-based biomaterial in a murine implantation model: The role of aluminum chloride pretreatment. *J. Biomed. Mater. Res. A* 2004;69:55–64.

107. O'Brien FJ, Harley BA, Yannas IV, and Gibson LJ. The effect of pore size on cell adhesion in collagen–GAG scaffolds. *Biomaterials* 2005;26:433–441.

108. Ashammakhi N, Ndreu A, Piras A et al. Biodegradable nanomats produced by electrospinning: Expanding multifunctionality and potential for tissue engineering. *J. Nanosci. Nanotechnol.* 2006;6:2693–2711.

109. Teo WE, He W, and Ramakrishna S. Electrospun scaffold tailored for tissue-specific extracellular matrix. *Biotechnol. J.* 2006;1:918–929.

110. Matthews JA, Wnek GE, Simpson DG, and Bowlin GL. Electrospinning of collagen nanofibers. *Biomacromolecules* 2002;3:232–238.

111. Li M, Mondrinos MJ, Gandhi MR et al. Electrospun protein fibers as matrices for tissue engineering. *Biomaterials* 2005;26:5999–6008.

112. Huang L, McMillan RA, Apkarian RP et al. Generation of synthetic elastin-mimetic small diameter fibers and fiber networks. *Macromolecules* 2000;33:2989–2997.

113. Buttafoco L, Kolkman NG, Engbers-Buijtenhuijs P et al. Electrospinning of collagen and elastin for tissue engineering applications. *Biomaterials* 2006;27:724–734.

114. Tillman BW, Yazdani SK, Lee SJ et al. The in vivo stability of electrospun polycaprolactone–collagen scaffolds in vascular reconstruction. *Biomaterials* 2009;30:583–588.

115. Stitzel J, Liu J, Lee SJ et al. Controlled fabrication of a biological vascular substitute. *Biomaterials* 2006;27:1088–1094.

116. Zhang X, Thomas V, and Vohra YK. In vitro biodegradation of designed tubular scaffolds of electrospun protein/polyglyconate blend fibers. *J. Biomed. Mater. Res. B Appl. Biomater.* 2009;89:135–147.

117. Sell SA, McClure MJ, Barnes CP et al. Electrospun polydioxanone–elastin blends: Potential for bioresorbable vascular grafts. *Biomed. Mater.* 2006;1:72–80.

118. Boland ED, Matthews JA, Pawlowski KJ et al. Electrospinning collagen and elastin: Preliminary vascular tissue engineering. *Front Biosci.* 2004;9:1422–1432.

119. Lee SJ, Liu J, Oh SH et al. Development of a composite vascular scaffolding system that withstands physiological vascular conditions. *Biomaterials* 2008;29:2891–2898.

120. Zeugolis DI, Khew ST, Yew ES et al. Electro-spinning of pure collagen nano-fibres—Just an expensive way to make gelatin? *Biomaterials* 2008;29:2293–2305.

121. Mironov V, Kasyanov V, and Markwald RR. Nanotechnology in vascular tissue engineering: From nanoscaffolding towards rapid vessel biofabrication. *Trends Biotechnol.* 2008;26:338–344.

122. Zeeman R, Dijkstra PJ, van Wachem PB et al. Crosslinking and modification of dermal sheep collagen using 1,4-butanediol diglycidyl ether. *J. Biomed. Mater. Res.* 1999;46:424–433.

123. Yu XX, Wan CX, and Chen HQ. Preparation and endothelialization of decellularised vascular scaffold for tissue-engineered blood vessel. *J. Mater. Sci. Mater. Med.* 2008;19:319–326.

124. Keuren JF, Wielders SJ, Driessen A et al. Covalently-bound heparin makes collagen thromboresistant. *Arterioscler. Thromb. Vasc. Biol.* 2004;24:613–617.

125. Bordenave L, Menu P, and Baquey C. Developments towards tissue-engineered, small-diameter arterial substitutes. *Expert Rev. Med. Devices* 2008;5:337–347.

126. Li B, Sharpe EE, Maupin AB et al. VEGF and PlGF promote adult vasculogenesis by enhancing EPC recruitment and vessel formation at the site of tumor neovascularization. *FASEB J.* 2006;20:1495–1497.

127. Vlodavsky I, Korner G, Ishai-Michaeli R et al. Extracellular matrix-resident growth factors and enzymes: Possible involvement in tumor metastasis and angiogenesis. *Cancer Metastasis Rev.* 1990;9:203–226.

128. Kanematsu A, Yamamoto S, Ozeki M et al. Collagenous matrices as release carriers of exogenous growth factors. *Biomaterials* 2004;25:4513–4520.

129. Steffens GC, Yao C, Prevel P et al. Modulation of angiogenic potential of collagen matrices by covalent incorporation of heparin and loading with vascular endothelial growth factor. *Tissue Eng.* 2004;10:1502–1509.

130. Yao C, Roderfeld M, Rath T et al. The impact of proteinase-induced matrix degradation on the release of VEGF from heparinized collagen matrices. *Biomaterials* 2006;27:1608–1616.

131. Ito Y, Hasuda H, Terai H, and Kitajima T. Culture of human umbilical vein endothelial cells on immobilized vascular endothelial growth factor. *J. Biomed. Mater. Res. A* 2005;74:659–665.

132. de Mel A, Jell G, Stevens MM, and Seifalian AM. Biofunctionalization of biomaterials for accelerated in situ endothelialization: A review. *Biomacromolecules* 2008;9:2969–2979.

133. Nillesen ST, Geutjes PJ, Wismans R et al. Increased angiogenesis and blood vessel maturation in acellular collagen–heparin scaffolds containing both FGF2 and VEGF. *Biomaterials* 2007;28:1123–1131.

134. Liang HC, Chen CT, Chang Y et al. Loading of a novel angiogenic agent, ginsenoside Rg1 in an acellular biological tissue for tissue regeneration. *Tissue Eng.* 2005;11:835–846.

135. Kothapalli CR, Taylor PM, Smolenski RT, Yacoub MH, and Ramamurthi A. Transforming growth factor beta 1 and hyaluronan oligomers synergistically enhance elastin matrix regeneration by vascular smooth muscle cells. *Tissue Eng. Part A* 2009;15:501–511.

136. Walpoth BH, Zammaretti P, Cikirikcioglu M et al. Enhanced intimal thickening of expanded polytetrafluoroethylene grafts coated with fibrin or fibrin-releasing vascular endothelial growth factor in the pig carotid artery interposition model. *J. Thorac. Cardiovasc. Surg.* 2007;133:1163–1170.

137. Abé H, Hayashi K, and Sato M. *Data Book on mechanical Properties of Living Cells, Tissues, and Organs*. Springer, New York; 1996.

138. Rashid ST, Salacinski HJ, Hamilton G, and Seifalian AM. The use of animal models in developing the discipline of cardiovascular tissue engineering: A review. *Biomaterials* 2004;25:1627–1637.

139. Yokota T, Ichikawa H, Matsumiya G et al. In situ tissue regeneration using a novel tissue-engineered, small-caliber vascular graft without cell seeding. *J. Thorac. Cardiovasc. Surg.* 2008;136:900–907.

140. Clarke DR, Lust RM, Sun YS, Black KS, and Ollerenshaw JD. Transformation of nonvascular acellular tissue matrices into durable vascular conduits. *Ann. Thorac. Surg.* 2001;71:S433–S436.

141. Gibson RA, Potter LA, Horn NJ, and VanDerVeer C. Artegraft®, available at http:\www.artegraft.com/bios.htm. Accessed June 2, 2009.

142. Chemla ES and Morsy M. Randomized clinical trial comparing decellularized bovine ureter with expanded polytetrafluoroethylene for vascular access. *Br. J. Surg.* 2009;96:34–39.

143. Guidoin R, Domurado D, Couture J et al. Chemically processed bovine heterografts of the second generation as arterial substitutes: A comparative evaluation of three commercial prostheses. *J. Cardiovasc. Surg. (Torino)* 1989;30:202–209.

144. Reddy K, Haque SN, Cohen L et al. A clinical experience with the NCGT graft. *J. Biomed. Mater. Res.* 1981;15:335–341.

145. Sawyer PN, Adamson R, Butt K et al. Long-term function of NCGT vascular conduits in a multicenter trial: Evaluation of physical chemical parameters. *Biomater. Med. Devices Artif. Organs* 1980;8:345–367.

146. Dardik H and Dardik II. Successful arterial substitution with modified human umbilical vein. *Ann. Surg.* 1976;183:252–258.

147. Strobel R, Boontje AH, and Van Den Dungen JJ. Aneurysm formation in modified human umbilical vein grafts. *Eur. J. Vasc. Endovasc. Surg.* 1996;11:417–420.

148. Hickey EJ, McCrindle BW, Blackstone EH et al. Jugular venous valved conduit (Contegra) matches allograft performance in infant truncus arteriosus repair. *Eur. J. Cardiothorac. Surg.* 2008;33:890–898.

149. Hollister SJ. Porous scaffold design for tissue engineering. *Nat. Mater.* 2005;4:518–524.

150. Khademhosseini A, Langer R, Borenstein J, and Vacanti JP. Microscale technologies for tissue engineering and biology. *Proc. Natl. Acad. Sci. U.S.A.* 2006;103:2480–2487.

151. Conklin BS, Richter ER, Kreutziger KL, Zhong DS, and Chen C. Development and evaluation of a novel decellularized vascular xenograft. *Med. Eng. Phys.* 2002;24:173–183.

16 pH-Responsive Polymers for Delivery of Nucleic Acid Therapeutics

Yu Nie and Ernst Wagner

CONTENTS

16.1 INTRODUCTION

Nonviral polymeric carriers have been recognized as promising systems for targeted delivery of therapeutic nucleic acid. Key features that can improve the bioefficiency of polymeric nanosystems include prolonged circulation in the bloodstream without undesired nontarget interactions, passive and active targeting into the disease site, cellular association and internalization into the target cells, and endosomal escape and subsequent intracellular transport. Because of these many different delivery tasks, nanocarriers are required which continuously optimize their various delivery functions in response to the various specific microenvironments. Nanocarriers can be chemically programmed for this purpose (Ganta et al. 2008; Oupicky and Diwadkar 2003; Roy and Gupta 2003; Wagner 2007). Chemical bonds and conformations can be integrated as sensors for environmental triggers which cleave the sensing bonds or alter the sensing conformation. Thus, the properties of the nanosystem should adopt favorably to the new task in the delivery cascade. Until now, various biological triggers have been exploited for tissue targeting and intracellular delivery, including pH (Meyer and Wagner 2006), temperature (Chilkoti et al. 2002; Zintchenko et al. 2006), and redox or special enzymatic microenvironment (Kommareddy and Amiji 2005; Saito et al. 2003). Another possible strategy is to utilize artificial physical triggers for targeting delivery of drugs and genes, like ultrasound, magnetic field, heat, or light (Berg et al. 2007; Duan et al. 2005; Hernot and Klibanov 2008).

Using the pH as trigger is highlighted in this chapter, as it offers the potential to discriminate on the one hand between different tissues, on the other hand between different intracellular compartments. The pH of diseased tissues in inflammation, infections, and cancer differs from that of the healthy tissues (Gerweck and Seetharaman 1996). As reported, the pH of inflamed tissue drops to 6.5, and most of the solid tumors have even lower extracellular pH of below 6.5 (Vaupel et al. 1989). Reasons are a faster proliferation of tumors cells, insufficient nutritional and oxygen supply, hypoxia, production of lactic acid, and also mislocation of vesicular proton ATPase to the cell surface, which consequently contribute to an acidic microenvironment. These biochemical properties can be exploited for the design of pH-triggered, disease-targeted drug, and gene delivery systems.

Furthermore, irrespective of normal or pathological conditions, cellular components display transmembrane pH gradients, which can also be benefit to the design of intracellular delivery. Organelle pH values (Gerweck and Seetharaman 1996) can be as low as 4.5 in the endolysosomal compartment. Likewise, the entry of nanocarriers to cells and subsequent transfer of the payload to subcellular organelles is associated with pH changes in the microenvironment. Both specific and nonspecific cell uptake of nanocarriers mainly occurs by endocytotic processes (Torchilin 2006; von Gersdorff et al. 2006; Wagner et al. 1990), which form the so-called endosome vesicles within the cells. And the endosomes mature into or fuse with lysosomes, which are of relatively low pH and rich in various degradative enzymes. The pH gradient of the endocytic pathway begins with the physiological pH of the cell surface (~7.4), drops to lower pH (5.5–6) in the endosomes, and approaches to the lysosomes (pH 4.5–5) (Asokan and Cho 2002). Endosomal acidic conditions and lysosomal enzymes are devastating to many biotherapeutics. Accumulation and degradation of therapeutic nucleic acids is a major bottleneck in the delivery process. To overcome such a dilemma, pH-responsive domains can be introduced into nanosystems to achieve efficient intracellular transport (Asokan and Cho 2002; Jabr-Milane et al. 2008; Khalil et al. 2006; Kulkarni et al. 2005; Mann et al. 2008; Medina-Kauwe et al. 2005; Meyer and Wagner 2006; Ropert 1999; Roy and Gupta 2003; Varshosaz 2008; Wagner 2007; Wolff and Rozema 2008).

16.2 PEG SHIELDING AND PH-TRIGGERED DESHIELDING OF POLYPLEXES

Polyplexes are nanoparticles generated by interacting negatively charged nucleic acid with cationic polymers (polycation [PC]) (Felgner et al. 1997). Effective and stable polyplexes are usually positively charged; such charges however can mediate undesired negative side effects. Shielding of polyplexes and other nonviral vectors with hydrophilic polymers like polyethylene glycol (PEG) reduces the positive surface charge and toxicity, prevents aggregation, protects from uptake by the reticuloendothelial system, increases circulation time, and hence, improves systemic targeted gene transfer (Ogris et al. 1999; Tam et al. 2000). However, at the same time, PEGylation lowers the transfection efficiency (Ogris et al. 2003; Oupicky et al. 2002) due to reduced cell surface and subsequently due to less endosomal lipid membrane interactions. Introduction of targeting ligands partly overcomes this hurdle, but also does not recover the activity of uncoated positively charged polyplexes; the vectors remain trapped in acidic vesicles after endocytosis. As reported (Meyer and Wagner 2006; Walker et al. 2005), electrostatic interactions between positively charged particles and the endosome membrane or additional endosomolytic moieties may lead to enhance gene transfer by membrane disruption through perturbation or fusion. But these different requirements constitute a dilemma that the surface charge and lytic activity must be shielded outside the cell but, after cellular uptake, the cationic surface charge and endosomal release functions should be re-exposed through deshielding of the particle. Therefore, it is vital that dynamic nonviral gene delivery systems have to be developed that undergo bioresponsive changes after reaching the target cell.

16.2.1 Deshielding of Polyplexes via Cleavage of pH-Labile PEG Linkages

Acid-labile linkers that are cleaved off in the endosomal milieu have been first utilized in conjugating chemotherapeutic agents such as doxorubicin to targeting moieties, including folic acid, antibodies, and antibody fragments (Srinivasachar and Neville 1989). Recently, such pH-sensitive bonds have also been used in gene delivery. Some of the highlighted examples are acid-labile acetals (Gillies et al. 2004; Knorr et al. 2007; Murthy et al. 2003a,b; Wong et al. 2008), ketals (Knorr et al. 2008a), hydrazones (HZN) (DeRouchey et al. 2008; Fella et al. 2008; Walker et al. 2005), diorthoesters (Choi et al. 2003; Li et al. 2005), and the vinyl ether (Shin et al. 2003; Xu et al. 2008), carboxylated dimethyl maleic acid (Rozema et al. 2007) linkers, which undergo cleavage at low pH (Figure 16.1).

Walker and colleagues (DeRouchey et al. 2008; Walker et al. 2005) developed HZN-linked pH-labile shielded polyplexes to undergo the "viruslike" changes outside and inside of targeting cells. In their research, a series of HZN bonds between the DNA-binding PC poly-L-lysine (PLL) and the shielding polymer PEG were selected. Only the acyl- and pyridylhydrazone bonds were sufficiently pH-sensitive, with the pyridylhydrazone displaying high stability at neutral pH. This PLL-pyridyl-HZN-PEG conjugates showed a half-life of 1.5 h at pH 7.4, while being hydrolyzed >90% in 10 min at pH 5 and 37°C. For polyplex formation, polyethylenimine (PEI) was induced for enhanced endosomal release and either transferrin-linked PEI (Tf-PEI) or EGF-linked PEI (EGF-PEI) was combined for targeting. Polyplexes with PLL-HZN-PEG remained shielded over 5 h, while completely deshielded in 1 h at pH 5. In cellular gene transfection experiments, targeted reversibly shielded polyplexes exhibited up to two orders of magnitude higher expression

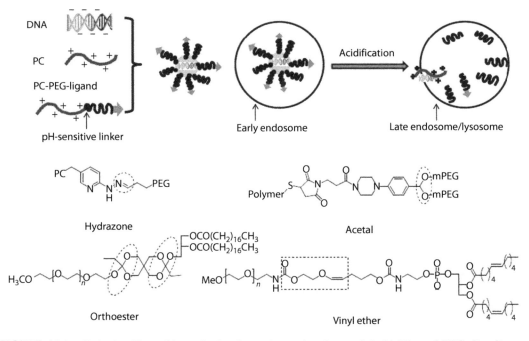

FIGURE 16.1 Cationic pH-sensitive polyplex formation and endosomal deshielding of PEG. Details are described in DeRouchey et al. (2008), Fella et al. (2008), Knorr et al. (2007), and Walker et al. (2005). Positive charges of polycation (PC) polyplexes can be shielded by polyethylene glycol (PEG) which in case of acid-labile linkages remains stable attached at neutral physiological conditions, while being removed at acidic endolysosomal pH-sensitive linkers. Hydrazones are described in DeRouchey et al. (2008), Fella et al. (2008), and Walker et al. (2005); acetals in Knorr et al. (2007); orthoester and vinyl ether are utilized in Choi et al. (2003) and Li et al. (2005), and Shin et al. (2003) and Xu et al. (2008), respectively.

than the stable shielding PEI-PEG polyplex on human leukemia K562, neuroblastoma Neuro-2A, hepatoma HUH-7, and renal carcinoma Renca-EGFR cells. The evaluation of systemic delivery in a subcutaneous HUH-7 tumor mouse model in vivo showed bioreversibly shielded polyplexes mediated about one logarithmic magnitude higher luciferase expression than stably shielded polyplexes. Subsequently, Fella et al. (2008) have improved the pH-sensitive PEG by bifunctional modification, which could be reacted with thiol group or amine group separately (Figure 16.1). DNA polyplexes were first formed with free PEI and EGF-PEG-PEI, and for post-PEGylation, a ω-2-pyridyldithio polyethylene glycol-butyraldehyde carboxypyridylhydrazone, N-hydroxysuccinimide ester (OPPS-PEG-HZN-NHS) was added for shielding. Both pH-sensitive and stable PEGylated polyplex remained shielded for 4 h in pH 7.4 buffer solution as confirmed by measurement of particle size and zeta potential, while at endosomal pH value of 5, the particles modified with PEG-HZN-NHS had deshielded. This property was also demonstrated by a more direct fluorescence correlation spectroscopy measurement with Alexa-488-labeled PEGylation reagents. Result from the different molecular masses, cleavable Alex-PEG at endosomal pH had shorter diffusion time (~200 μs), while uncleavable Alex-PEG remained on the polyplex had much longer time (>2000 μs). Luciferase gene transfections on EGF receptor overexpressing HUH7 cells showed up to 16-fold enhancement by the reversibly PEGylated polyplexes. Consistently, in vivo transgene expression in a subcutaneous HUH-7 SCID mouse model also enhanced tumor specificity after intravenous administration. Further extension and improvement of the pyridylhydrazone-based PEGylation has been performed by Nie et al. (2011) in the formulation of lipopolyplexes with both pH-reversible shielding and targeting on PEGylated cholesterol. Transfection results showed up to 40-fold higher protein expression in mouse neuroblastoma Neuro2A cells with the reversibly PEGylation compared to the stable ones. The effect of PEG deshielding on the intracellular fate was studied with confocal laser scanning microscopy.

Beside HZN, acetals are promising candidates for the development of acid-sensitive linkages. For example, Knorr et al. (2007) have synthesized an acetal-based PEGylation reagent (Figure 16.1). This pH-responsive compound had two acetal-linked PEG chains, with a maleimide (MAL) moiety for further reaction with thiol-functionalized PEI. Hydrolysis assay showed the half-life of PEG-acetal-PEI conjugate was about 3 min at endosomal pH, while 2 h at physiological pH conditions. DNA polyplexes were prepared containing free PEI, targeting conjugates (EGF-PEI or Tf-PEI) and shielding compounds (PEG-acetal-PEI or PEG-PEI). The reversibly shielded polyplex had an obvious contrast that being stable to salt-induced aggregation for 2 h at pH 7.4, while aggregated within 0.5 h at endosomal pH. With targeting, polyplexes shielded with the PEG-acetal-mediated enhanced luciferase gene expression in Renca-EGFR or K562 cells as compared to stably shielded control ones.

The function of endosomolytic moieties could be transiently shielded by PEG, and re-exposed through deshielding of the particle after uptake into the cells. In pioneering work, Hoffman and colleagues (Murthy et al. 2003a,b) designed encrypted polymers as multifunctional carriers, which incorporated a targeting agent that directs receptor-mediated endocytosis, a pH-responsive element that dynamically disrupts the endosome, and the therapeutic biomolecule, a peptide, or oligonucleotides (ODN) (Figure 16.2). The hydrophobic membrane-disruptive backbone was a terpolymer of butyl methacrylates, dimethylaminoethyl methacrylate (DMAEMA), and a styrene derivative with a reactive benzaldehyde group. PEG was anchored by acid-degradable acetal linkage as the solubilizing hydrophilic graft to "mask" the backbone. At the other terminus, PEG can be modified by targeting segment (mannose or lactose), cationic peptide (lysine), or fluorescein. Also, the therapeutic ODN can be electrostatically bound to the cationic terminals. Hydrolysis kinetics of acetal-linked PEG grafts had a half-life of 15 min at pH 5.4, while less than 10% degraded in 75 min at pH 7.4, which means two orders of magnitude improved PEG cleavage in acidic environment. In addition, antisense (AS)-ODN with the ability to inhibit the inducible nitric oxide synthesis in macrophages was used as therapeutic component. Obviously, AS-ODN polyplex with pH-responsive polymer showed 80% inhibition, while AS-ODN only reached 25% as control.

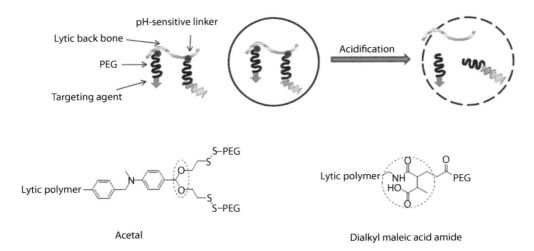

FIGURE 16.2 Endosomal deshielding of PEG in unmasking lytic polyplex domains. After acidification, the masked endosomolytic moieties are re-exposed through hydrolysis of the bond between lytic back bone and PEG. Descriptions are in Section 16.2.1, with pH-labile linkages acetal (Murthy et al. 2003a,b) and dialkyl maleic acid amide (Rozema et al. 2007). PEG, polyethylene glycol.

This strategy was extended by applying another useful pH-labile linkage between PEG and PC, such as the amide bond with dialkyl maleic acid. Using such a maleamate linkage, Rozema et al. (2007) developed a pH-reversible shielding endosomolytic vehicle for the siRNA delivery to hepatocytes in vitro and in vivo (Figure 16.2). Key features of this dynamic polyconjugate technology are similar to Murthy et al. (2003b), including a membrane-active polymer, reversibly masked with PEG. The siRNA was ferried specifically to hepatocytes in vivo after intravenous injection. The endosomolytic agent used in this study was an amphipathic poly(vinyl ether) (termed PBAVE). Bifunctional maleamate linkage was used to reversibly attach the shielding agent PEG and the hepatocyte-targeting ligand *N*-acetylgalactosamine to PBAVE. Therapeutic siRNA cargo was directly attached to PBAVE through a reversible disulfide linkage. Using this multifunctional delivery technology, two endogenous target genes (apolipoprotein B and peroxisome proliferator-activated receptor α) were effectively knocked down in vivo in the mouse liver.

16.2.2 pH-Sensitive Electrostatic Coating with PEG

Effective and stable polyplexes are usually positively charged on the surface; these preassembled DNA polyplexes can be ionically coated by negatively charged molecules. Plank and colleagues had utilized a so-called protective copolymers (PROCOPs), consisting of PEG, different anionic polyglutamate peptides, and the fusogenic negatively charged influenza-derived peptide (INF-7), which could mediate the electrostatic interaction with positively charged polyplex (Finsinger et al. 2000). With this electronic shielding by PEGs, PC (PEI, PLL, and liposomes)–DNA complexes were prevented from aggregation and undesired interactions. As designed, at acidic pH of the endosomes, glutamic acids (in the polyglutamate and INF-7 anchor) get protonated, leading to reduced electrostatic interactions between the PEG-coat and the particles, and the PROCOPs dissociating from the preassembled polyplexes. Results showed reduced opsonization and complement activation. At lower molar ratios of PC nitrogen atoms to DNA phosphate (N/P), an inhibitory influence on transfection of the PROCOPs was observed, compared with uncoated polyplexes. This was probably due to the loss of cell-binding capacity. However, at higher N/P ratios, the coated polyplex could improve the transfection because of reduced cytotoxicity. Moreover, membrane-disruptive INF-7-PROCOP significantly enhanced the transfection efficiency of PLL polyplexes on K562

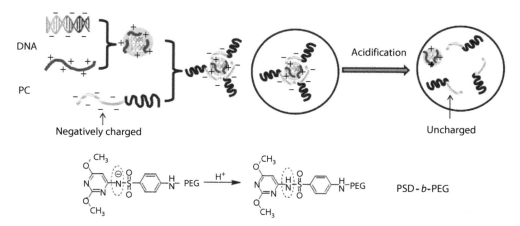

FIGURE 16.3 Ionic coating and decoating of PEG in cationic polyplex. Preformed cationic polyplexes can be coated by negative-charged molecules, whereas deshielding occurs at endosomal pH after charge neutralization of the coating polymer. Details are described in Section 16.2.2. Charge shift is illustrated by one example of poly(methacryloyl sulfadimethoxine)–poly(ethylene)glycol (PSD-*b*-PEG) in Sethuraman et al. (2006).

cells, because PLL itself does not have endosomolytic activity. Nevertheless, further stability of the ionic-coated PEG for in vivo gene delivery needs to be clarified. An illustration of the electrostatic decoating is given in Figure 16.3.

The electrostatic coating strategy was further developed (Sethuraman et al. 2006). A pH-sensitive diblock copolymer containing PEG and poly(methacryloyl sulfadimethoxine) (PSD-*b*-PEG) has been synthesized, which is negatively charged at pH 7.4, but electroneutral below pH 6.8. Thus, at physiological pH, PSD-*b*-PEG could coat polyplex (PEI-DNA) through electrostatic interaction. Whereas under acidic conditions, the anionic PSD moieties protonated and detach together with PEG from the polyplex. In contrast to the previous described system (Finsinger et al. 2000), which is based on the pH-sensitive carboxylic group with a pK_a around 5, the PSD system shows a sharp transition around pH 6.8. Cell viability and transfection data proved that PSD-*b*-PEG decreased the cytotoxicity of PEI at pH 7.4, whereas the transfection efficiency was recovered at pH 6.6. This highly defined shift may be suitable for the targeting to slightly acidic extracellular tumor matrix. In this respect, the strategy could also benefit from pre-administration of substances such as glucose for artificial decreasing of the pH of tumor area.

The main drawback of earlier system is that the "shield" PSD-*b*-PEG is not biodegradable. It is possible that the polymer might cause toxic effects if its accumulated dosage increases above a critical point. So later, Sethuraman et al. (2008) designed a new biodegradable pH-sensitive block copolymer poly(L-cystine bisamide-*g*-sulfadiazine)-*b*-PEG (PCBS-*b*-PEG), instead of PSD-*b*-PEG. A polymeric micelle consisting of poly(L-lactic acid)-*b*-PEG, together with trans-activator of transcription peptide (TAT), was introduced to preform positively charged polyplex and enhance the transfection efficiency.

16.2.3 Reversible Covalent Conjugation of Nucleic Acids with PEG

pH-labile bonds were used not only to link the cationic polymer to PEG, but also for direct conjugation of the polymer with the therapeutic nucleic acid. In 2003, Kataoka's group (Oishi et al. 2003) conjugated ODN to PEG through a pH-responsive β-propionate ester linkage (PEG-ODN) (Figure 16.4). This conjugate spontaneously associates with PEI to form a polyion complex (PIC) micelle. Hydrolysis experiments showed that the ester linkage in both PEG-ODN and PIC micelle had been cleaved at the endosomal pH (~5.5), suggesting that the ODN is released from the PIC micelle in the intracellular compartment.

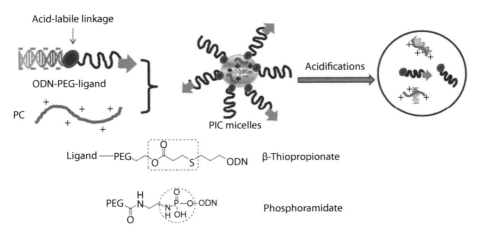

FIGURE 16.4 Reversible covalent PEGylation of PIC micelles and acid-labile linkages. For details, see Section 16.2.3. pH-responsive linker β-thiopropionate is described in Oishi et al. (2005a,b) and Oishi et al. (2003); phosphoramidate is described in Jeong et al. (2003). PEG, polyethylene glycol; ODN, oligonucleotide; PIC, polyion complex.

Later, Oishi et al. constructed a novel pH-responsive, targeted antisense ODN delivery system based on multimolecular assembly into PIC micelles, including PLL, and a targeting lacto-sylated PEG-AS-ODN conjugate (Lac-PEG-AS-ODN) containing the same acid-cleavable linker (β-propionate) between the PEG and ODN segments (Oishi et al. 2005a,b). In a dual luciferase reporter assay, the Lac-PIC micelles achieved a more efficient antisense effect, with 65% inhibition in HUH-7 cells, compared with ODN, Lac-PEG-ODN, or the lactose-free PIC micelle (45% inhibition). Clustered lactose moieties on the peripheries of PIC are supposed to involve an asialo-glycoprotein (ASGP) receptor–mediated endocytosis process, resulting in enhanced uptake than the untargeted ones. But pH-sensitive release of AS-ODN and PEG into the cellular plasma presented the most important contribution to the antisense effect, for a significant decrease of antisense effect (from 65% to 27% inhibition) was observed for the Lac-PIC micelle without β-propionate linkage. Cleaved PEG strand from the PIC micelle could increase the colloidal osmotic pressure, inducing swelling and disruption of the endosome. And intracellular delivery AS-ODN should be in a PEG-cleaved form to avoid the steric hindrance effect of PEG for the interaction between ODN and target RNA sequence. Consistent confirmation was also proved by using branched PEI (B-PEI) substituting the PLL for pH-sensitive PIC formulation, although with decrease in the antisense effect (9% or 42% inhibition). It was probably due to the "proton sponge and buffer" (see details at Section 16.4.1) effect of the B-PEI prevented the decrease of the pH, leading to inefficient cleavage of ODN and PEG in the endosome.

Jeong et al. (2003) reported PIC micelles composed of PEG-ODN conjugate containing another acid-cleavable phosphoramidate linkage (Figure 16.4) and the cationic fusogenic peptide KALA, which is known to disrupt endosomes and used to enhance gene transfection (Lee et al. 2001). From the image of confocal microscopy, it was observed that fluorescein (FITC)-labeled ODN in PIC micelles was distributed all over the cytoplasm of smooth muscle cells, whereas cells treated with FITC-ODN alone showed only negligible fluorescence in the cytoplasm. It was suggested that fuso-genic action of KALA contributes to efficient endosomal release compared with ODN alone. To test the efficiency of the PIC delivery system, a therapeutic AS-ODN having a complementary sequence to antisense c-myb was chosen for suppressing proliferation of muscle cells. Results showed that the PIC micelles had 70% inhibition efficiency. Moreover, different core-forming PCs (KALA, B-PEI, and protamine) with different functions were studied. All of the fusogenic KALA, endosomal-disruptive PEI, and simple nucleic acid–condensable protamine showed satisfied inhibition activity. This observation is in agreement with the former hypothesis of Oishi et al. (2005b) and Murthy et al. (2003c) who claimed that the acid-cleaved PEG chains were responsible for endosomal swelling and rupture, so that PEG-ODN/protamine micelles also mediate endosomal escape. In contrast, the use of PEI did not reduce the antisense activity of the delivery system as reported by Oishi et al. (2005b).

16.3 ACID-TRIGGERED POLYMER DEGRADATION

PEG shielding and pH-triggered deshielding have solved the dilemma in the way of shielding the positive charge and lytic activity of polyplexes outside the cell, and recovered them after cellular uptake. Here comes another contradictory requirement during endocytosis. When attaching to the cell or at early time of being taken up into endosomes, the binding between the PC and the nucleic acid has to be strong enough for protection of gene from enzymatic degradation. On the other hand, within the cell (or after endosomal release) the polyplex should finally dissociate for enabling the gene to move into the nucleus for transcription or target siRNA to mRNA for RNA interference (Knorr et al. 2008b; Roy and Gupta 2003). Moreover, high transfection efficiency is almost always associated with high vector toxicity. For example, PEI of molecular weight around 22–25 kDa compacts nucleic acid well and mediates gene transfer very effectively (Boussif et al. 1995; Zou et al. 2000) but causes considerable toxic effects (Chollet et al. 2002; Moghimi et al. 2005). In contrast, shorter PEI such as 800 Da oligoethylenimine (OEI) possesses negligible toxicity, but could only condense DNA at high N/P ratio and exhibit very poor gene transfer capacity. Therefore, one approach to fulfill the dual requirements is to design biodegradable PCs, which allow gene transfer as effective as their stable high-molecular-weight PEI counterparts, but with time or upon a specific stimulus, decompose into low-molecular-weight nontoxic degradation products. So researchers have inserted diverse degradable functions into their polymers, including disulfide bonds (Lee et al. 2007; Read et al. 2005), ester bonds (Russ et al. 2008; Thomas et al. 2005), imines (Kim et al. 2005), or phosphoesters (Wang et al. 2001; Zhao et al. 2003).

The Frechet's group has evaluated pH-sensitive acetal-linked degradable polymers for many years. They developed a new strategy (Goh et al. 2004) for the delivery of protein- or DNA-based vaccines (Goh et al. 2004; Murthy et al. 2003c; Paramonov et al. 2008) using acid-degradable polyacrylamide microparticles. Cross-linked with pH-sensitive linkers, the microparticles should degrade in the acidic environment, resulting in an increase in osmotic pressure within the compartment and leading to lysosomal disruption, thus providing cytoplasmic delivery of the encapsulated biomolecule intact. In fact, both protein and nuclear delivery was successful using this microbead system. pH-dependent release plots showed 10% burst release of DNA at physiological pH and no continuous increasing in the next 5 h, while under acidic conditions, full release of the DNA was achieved in 2 h. In vitro experiments were carried out by examining the immunostimulatory properties of the plasmid-loaded microparticles on RAW 264.7 macrophages; 0.1 μg DNA in pH-degradable microparticles resulted about four times higher response than 1 μg naked DNA.

Another degradable microparticle-based DNA delivery system was described by Little et al. (2004) for genetic vaccination composed of a degradable, pH-sensitive poly-β-amino ester (PBAE) and polylactic-co-glycolic acid (PLGA). PBAE not only increases the supercoiled content and overall effective loading of plasmid DNA but also significantly buffers the pH microenvironment created by ester bond degradation, rendering encapsulated plasmid more suitable for transfection. Release of plasmid from microparticles is controllable based on the amount of PBAE in the composition, which is optimal for 15% and 25%. These formulations are potent activators of dendritic cells in vitro. When used as vaccines in vivo, these microparticle formulations induced antigen-specific rejection of transplanted syngenic tumor cells (Little et al. 2005).

Acid-cleavable acetal-based cross-linkers have also recently been developed for improving PEI-based delivery systems. Knorr et al. (2008b) have designed two different acid-degradable nonviral gene carriers consisting of OEI polymerized either with acetone ketal cross-linker 2, 2-bis(N-maleimidoethyloxy)propane (MK) or the 4-methoxybenzaldehyde bisacrylate acetal cross-linker 1,

1-bis-(2-acryloyloxy ethoxy)-[4-methoxy-phenyl]methane) (BAA). Acid-insensitive polymers were synthesized as control, replacing the acetal-based linkers with stable ether or a hydrocarbon moiety. Polymers of OEI-MK and OEI-BAA degraded in a pH-dependent manner, with very fast degradation kinetics (approximately 3 min half-life at 37°C) in the acidic environment of pH 5 and much slower degradation (5 h for OEI-MK, 3.5 h for OEI-BAA) at physiological pH 7.4. In contrast, control polymers degraded very slowly at both conditions. The acetal-containing polymers showed clear advantages on in vitro cytotoxicity of polyplexes on B16 and Neuro2A cells, as well as in vivo liver histology after systemic administration of polymers in Balb/c mice. At low cation/plasmid (c/p) w/w ratios, the transfection efficiency of pH-sensitive polymers was slightly reduced, but became similar or superior to that of acid-stable polymers at higher c/p ratios. In the following, ketal-based PEG-OEI-MK (Knorr et al. 2008a) was synthesized by performing the polymerization of OEI with MK in the presence of mercapto-modified PEG. PEG-OEI-MK showed efficient DNA binding; the surface shielding of polyplexes remained stable at neutral pH 7.4, while deshielded and aggregated at pH 5 in 30 min. Introduction of PEG further reduced polymer toxicity and mediated a significant increased transfection efficiency compared to the non-PEGylated OEI-MK or the corresponding pH-stable ether-linked analog.

Jain and Frechet (2007) designed and synthesized another family of cationic, pH-sensitive poly(amidoamine)s. Acetal or ketal linkages were incorporated into the backbone. All of the polymers demonstrated a pH-dependent degradation profile with a very significant increase in hydrolysis rate as the pH was lowered from 7.4 to the pH value of 5.0 commonly found in lysosomes. The hydrolysis half-life of the poly(amidoamine)s varied from few hours to 80 days at pH 5.0 and from 6 to 161 days at pH 7.4, depending upon the structure of the components used to prepare each of the acid-degradable polymers. Moreover, an acetal-based acid-labile degradable hydrogel was combined together with other environment-specific stimuli for gene delivery (Namgung et al. 2009).

16.4 ENDOSOMAL ESCAPE

16.4.1 PROTON SPONGE EFFECT FOR ENDOSOMAL ESCAPE

Polyethyleneimine (PEI) (Akinc et al. 2005; Boussif et al. 1995; Sutton et al. 2006) and starburst dendrimers (Esfand and Tomalia 2001) are typical and efficient examples of pH-sensitive cationic polymers, which become protonated incrementally over a wide pH range. The application of these pH-sensitive cationic polymers to ODN and gene delivery has been studied extensively (Choi et al. 2004; Eichman et al. 2001).

Branched PEI is a positively charged polymer in physiological pH, containing primary, secondary, and tertiary amino groups frequently in a ratio of 1:2:1. These distinct amines have different pK_a, resulting in strong buffering capacity. It is thought that this "proton sponge" nature of PEI could also buffer the pH conditions in endosomes. PEI binds protons which are pumped into the endosome by natural vesicular H^+ ATPases, along with the concurrent influx of Cl^- to maintain electrical neutrality. As a consequence, the ionic strength inside the endosome increases, leading to the osmotic swelling. This, together with the highly positive charge of PEI, causes the lytic potential of the polymer, which disrupts the endosome and triggers the escape trafficking pathway from the acidic lysosomes (Demeneix and Behr 1996; Sonawane et al. 2003). Many research groups have explored the high efficiency mechanism of PEI-mediated gene delivery (Blessing et al. 2001; Lemkine and Demeneix 2001). Both PEI and its N-quaternized derivatives have been investigated by Akinc et al. (2005) for gene transfer efficiency. Consistent with the "proton sponge" theory, quaternized PEI polymers had much lower transfer efficiency because quaternization of amines prevents further protonation and therefore loses the buffering capacity. Therefore, the "proton sponge" hypothesis has also been applied to explain the relatively high transfection results of other amines buffering materials, like poly(amido amine) (PAMAM) dendrimers (Choi et al. 2004; Eichman et al. 2001) and imidazole-containing polymers (Asayama et al. 2007; Ihm et al. 2005).

Various synthetic poly(imidazole)s have been developed for gene delivery, with the repetition of the same basic heterocycle configuration as histidine, and suitable pK_a of 6.5–6.0 for adjusting the pH in the endosome. In fact, these imidazole-containing carriers really have shown high gene expression as the results, comparable with the effect of PEI in some cases. For example, poly(4-vinylimidazole) was explored for import of gene encoding yellow fluorescent protein and human osteoprotegerin (Ihm et al. 2005). In vivo transfection was explored by tail vein injections of rats, and results were evaluated both at the gene and protein levels in lung and spleen tissue, demonstrating similar efficiency as with PEI.

Another example of poly(1-vinylimidazole) (PVIm) (Asayama et al. 2007) with aminoethyl groups (PVIm-NH$_2$) was determined to have enhanced membrane-disruptive ability at endosomal pH. Agarose gel retardation assay proved that the introduced aminoethyl groups worked as anchor groups to retain DNA. Furthermore, the ternary complex of DNA, PVIm-NH$_2$, and PLL-lactose could specifically mediate the gene expression in HepG2 cells with ASGP receptors, while alkalization of PVIm-NH$_2$ could enhance the disruptive property by increasing the hydrophobic domain (Asayama et al. 2010).

16.4.2 PROTONATION OF POLYMERS TO TRIGGER ENDOSOMAL LYSIS

In nature, viruses are able to efficiently deliver their nucleic acid cargos from the endosome to the cytoplasm (Plank et al. 1994). Because they contain pH-sensitive fusogenic peptide, which exists in an ionized, hydrophilic state at physiological pH (pH 7.4), but becomes protonated and relatively hydrophobic at lower pH in endosome. Inspired by these phenomena, Cullis's group (Chen et al. 2004) and especially Hoffman's group (Cheung et al. 2001, 2005; Hoffman et al. 2002; Jones et al. 2003; Kyriakides et al. 2002; Murthy et al. 1999) followed a different approach to investigate a series of synthetic, pH-sensitive amphipathic polymers that show enhanced endosomal membrane disruption and endosomal release in cell culture studies (Figure 16.5). These polymers that mimic viral peptides contain a combination of acidic carboxylic acid groups and hydrophobic alkyl groups. This combination is useful for protonation of polymers at endosomal pH, resulting in increased hydrophobicity, which can lead to enhanced endosomal membrane disruption.

Poly(ethylacrylic acid) (PEAA) was first designed by Murthy et al. (1999) to efficiently disrupt lipid membranes of red blood cells in a pH-dependent mechanism, which is as efficient on a molar

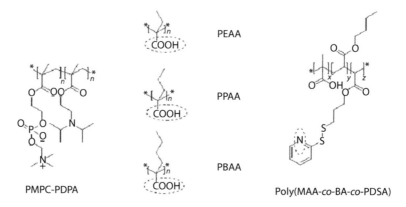

FIGURE 16.5 Protonable polymers to trigger endosomolysis. The negatively charged carboxylates of the polymers become neutralized by protonation at lower pH, resulting in a change in properties (from hydrophilic to hydrophobic), increased endosomal membrane interaction and disruption. PEAA, Poly(ethylacrylic acid); PPAA, poly(propyl acrylic acid); PBAA, poly(2-methylacrylic acid); poly(MAA-co-BA-PDSA), poly(methylacrylic acid-co-butylacrylic acid-co-pyridyl disulfide acrylate); PMPC-PDPA, poly(2-(methacryloyloxy)ethyl-phosphorylcholine)-co-poly(2-(diisopropylamino)ethyl methacrylate).

basis as the peptide melittin. Its hemolytic activity rises rapidly as the pH decreases from 6.3 to 5.0, and there is no hemolytic activity at pH 7.4. In the following, Hoffman's group has synthesized similar polymers belonging to the alkyl acrylic acid family, including poly(propyl acrylic acid) (PPAA) (Murthy et al. 1999), poly(butylacrylic acid) (PBAA), one glutathione-/pH-sensitive poly(methylacrylic acid-*co*-butylacrylic acid-*co*-pyridyl disulfide acrylate) (poly(MAA-*co*-BA-PDSA)). In addition, they also designed copolymers that combined acrylic acid with acrylate ester monomers that contained different alkyl substituents. All types of polymers were responsive to the lowered pH within endosomes, leading to disruption of the endosomal membrane and release of important biomolecular drugs such as DNA, RNA, peptides, and proteins to the cytoplasm before they are trafficked to lysosomes and degraded by lysosomal enzymes, but with different efficiencies.

PPAA was found as the most suitable one (Murthy et al. 1999). PPAA was approximately an order of magnitude more efficient in hemolysis than PEAA, as well as in hemolysis at a higher pH (~6.7–6.8). The addition of another methylene unit into the monomer unit (PBAA) shifted the sensitive pH range even further toward physiological pH (pH 7.4). The general shift in the pH profiles is consistent with the trend in carboxylate pK_a expected with longer and more hydrophobic alkyl groups. Random 1:1 copolymers of acrylic acid and alkyl acrylates can also act as pH-responsive membrane-disrupting agents. Although all of the copolymers displayed the ability to lyse RBC at lowered pH, none of them was as effective as PPAA (Hoffman et al. 2002; Murthy et al. 1999). It must be noted that the comparisons were made for similar molecular weights, usually about 40–60 kDa (Hoffman et al. 2002), because this pH shift profile are both concentration- and molecular weight–dependent. In general, as molecular weight or concentration increases the profiles are shifted toward higher pH transitions. To be similar as proton sponge effect, inducing hydrophobic moiety could also promote the endosomal lysis. For instance, diblock copolymer of cationic DMAEMA, zwitterionic PPAA, and hydrophobic BMA became sharply hemolytic at endosomal pH condition, with the best concentration of hydrophobic moiety at 48%, with no hemolysis at pH 7.4, 60% at pH 6.8 while 100% at pH 5.8 (Convertine et al. 2009). Ternary physical mixtures of the cationic lipid DOTAP, plasmid DNA, and PPAA were used for gene delivery (Cheung et al. 2001). Transfection of NIH3T3 fibroblasts showed marked enhancement of both gene expression levels and fraction of cells transfected compared to binary control mixtures of DOTAP and DNA. It was also observed that PPAA can significantly improve the serum stability of DOTAP/DNA vectors, which maintain high levels of transfection in media containing up to 50% serum (Cheung et al. 2001). This is mainly because the PPAA can protect integrity of lipoplexes from LDL, as well as retain high cellular uptake in the presence of BSA and HDL (Cheung et al. 2005). Later, Kyriakides et al. (2002) extended in vitro hemolysis and cell culture studies to an in vivo murine excisional wound healing model. A pilot study with a green fluorescent protein (GFP)–encoding plasmid indicated that injection of formulations containing PPAA into healing wounds resulted in increased GFP expression. Subsequently, by administering AS-DNA for the angiogenesis inhibitor thrombospondin-2 (TSP2), the lipoplex with PPAA can enhance in vivo transfections and that inhibition of TSP2 expression may lead to improved wound healing.

In vivo experiments on Chinese hamster ovary (CHO) cells confirmed the efficiency of PPAA for the ODN-lipoplex system (Lee et al. 2006). Incorporation of PPAA in DOTAP/ODN complexes improved two- to threefold the cellular uptake of fluorescently tagged ODN and also maintained high levels of uptake into cells upon exposure to serum. Antisense activity showed that addition of PPAA enhanced the efficiency in both serum-free, and to a lesser extent, serum-containing media.

Combining glutathione- and pH-sensitive properties has been done by Bulmus et al. (2003). Pyridyl disulfide acrylate was used as the functionalized monomer, incorporated into an amphiphilic copolymer consisting of methacrylic acid and butyl acrylate. This polymer showed no toxicity and enhanced cytoplasmic delivery of FITC-ODN in human leukemia monocytes THP-1.

Protonation of polymers can not only change polymer from hydrophilicity to hydrophobicity, but also cause the phase change of the polymers (or polyplex). Lomas et al. (2007) have used poly(2-(methacryloyloxy)ethyl-phosphorylcholine)-*co*-poly(2-(diisopropylamino)ethyl methacrylate) (PMPC-PDPA) diblock copolymers for gene delivery. The PMPC segment is highly biocompatible and nonfouling, while the PDPA part is pH-sensitive with pK_a of 5.8–6.6, depending on the ionic strength. It was reported that this diblock copolymer forms stable vesicles at physiological pH (namely biomimetic polymersomes) (Du et al. 2005). Upon acidification, tertiary amine groups on the PDPA chains can be protonated, switching from hydrophobicity at physiological pH to hydrophilicity (i.e., a weak cationic polyelectrolyte) in mildly acidic solution. This shift sequentially resulted in the rapid dissociation and disruption of the polymersome around pH 5–6 (Conner and Schmid 2003), forming molecularly dissolved monomers.

16.5 DELIVERY SYSTEM CONTAINING SEQUENCE-DEFINED POLYMERS OR PEPTIDES

Physicochemical characterization of polyplexes often suffers from lack of precise control during synthesis of polymers, resulting in heterogeneous polymer batches which also may vary in cellular toxicity and other biological properties. Therefore, recent efforts initiated to generate more precise polymer-like molecules (Schaffert and Wagner 2008). Dendrimers are one approach to overcome these limitations, but limit the carrier design space to hyperbranched structures. Solid-phase synthesis can be used to generate sequence-defined polymer units (Hartmann et al. 2006, 2008) or peptides (Mann et al. 2008) or combination of both (Wang et al. 2009a,b). These may present interesting alternative DNA delivery systems.

16.5.1 SEQUENCE-DEFINED POLYMER SYSTEMS

Solid-phase synthesis presents an interesting approach toward sequence-defined precise polymeric carriers. For example, Börner and colleagues synthesized monodisperse poly(amidoamine) sequencing for complexation of DNA (Hartmann et al. 2006, 2008). The group of Lu (Wang et al. 2007, 2008; Xu et al. 2010) assembled novel carriers by solid-phase synthesis for siRNA delivery. These contained protonatable amines of different pK_a, including histidine residues, cysteine residues for disulfide bond formation, and hydrophobic groups for stabilization of compact nanoparticles. The protonatable amino groups were designed to complex with siRNA and to introduce buffering capacity for pH-sensitive membrane disruption that will facilitate endosomal–lysosomal escape. Cysteine residues were designed to further stabilize the nanoparticles and siRNA complexes via reversible disulfide bond formation and to further chemically modify the nanoparticles. The incorporation of PEG at the nanoparticle surface shielded the net charges of the nanoparticles and minimized nonspecific cellular interaction.

Subsequently the investigators functionalized the siRNA nanoparticles with a bombesin peptide analog, which presents a targeting ligand to the bombesin receptor. This receptor is overexpressed on the cell surface of various cancer cells. The targeted siRNA nanoparticles resulted in specific gene silencing efficiency in CHO and U87 cells in vitro. Systemic administration of a therapeutic anti-HIF-1R siRNA with the bombesin-loaded carrier resulted in significant tumor growth inhibition in nude mice bearing human glioma U87 xenografts (Wang et al. 2009b).

16.5.2 pH-RESPONSIVE ENDOSOMOLYTIC PEPTIDES

Various peptides with specific sequence were explored to overcome different barriers during the whole trafficking of nucleic acid delivery (Mann et al. 2008). According to the function, these peptides can be categorized as (a) gene-binding cationic peptides, which could efficiently condense gene or form complexes with gene via electrostatic interaction and often be rich in lysine or arginine; (b) cell-targeting peptides with inherent important sequence in ligand or antibody; (c) cell-penetrating peptides (CPPs) or endosomolytic peptides, which efficiently help the vehicle to penetrate the phospholipid membrane, either outside of the cells or inside; and (d) peptides with specific nuclear localization sequences, which are supposed to transfer DNA to the cell nucleus via small pores on the surface. Only pH-responsive functional peptides are reviewed in the following text.

Viruses and bacteria have shown an innate ability to escape from the endocytic pathway, which benefit from the presence of an α-helical domain in the envelop proteins. They are sometimes pH-sensitive and responsible for the membrane-disruptive activity (Pecheur et al. 1999). A particularly well-studied example is the influenza virus homotrimer domain. The N-terminal sequence of the influenza virus hemagglutinin changes its conformation at endosomal pH, leading fusogenic α-helical sequence to insert into the bilayer lipid of the endosome, and eventually cause fusion (Wagner et al. 1992). Similarly, bacterial toxins such as the diphtheria toxin (Liu et al. 2005) and listeriolysin O (Dramsi and Cossart 2002) also act in a pH-dependent way. A conformational change exposes hydrophobic regions of toxins at low pH, which could fuse the membrane. Further development is to use the exact pH-sensitive sequence peptides, for retaining pH-sensitive endosomolytic property and reducing toxicity.

Fusogenic peptides mainly contain an amphipathic α-helical configuration that works for the interaction with the membrane of endosome or lysosome, leading to subsequent membrane disruption/pore formation (Mann et al. 2008). They are derived from fusogenic components of viral envelope proteins (e.g., HA-2, the N-terminal end of influenza virus hemagglutinin), or synthetically designed derivatives thereof such as GALA (Li et al. 2004), KALA, INF-7 (Plank et al. 1994; Wagner 1999).

For the use of peptides as directly DNA-binding delivery carriers, negatively charged glutamic acids in these peptides were replaced with positively charged lysine or arginine. Therefore, peptides like GALA and JTS-1 give rise to modified ones, like KALA and ppTG1, ppTG20 peptides (Wyman et al. 1997), showing high in vivo activity (Rittner et al. 2002).

Haas and Murphy (2004) have described synthesis and characterization of two derivatives of GALA, GALAdel3E, and YALA (Haas and Murphy 2004). GALAdel3E has deleted three centrally located glutamate residues from GALA, while YALA replaces one glutamate residue with the unusual amino acid 3,5-diiodotyrosine. Both derived peptides retain pH sensitivity, showing no ability to cause leakage of an encapsulated dye from unilamellar vesicles at neutral pH but substantial activity at acidic pH. Interestingly, the pH at which the peptides activate is shifted, with GALA becoming active at pH ~5.7, GALAdel3E at pH ~6.2, and YALA at pH ~6.7. Improved activity in the presence of cholesterol and onset of activity in the critical range between pH 6 and 7 may make these peptides useful in applications requiring intracellular delivery of genes.

Fusogenic peptides can also be used for AS-ODN delivery. As stated earlier (Jeong et al. 2003), KALA had been complexed with a pH-sensitive PEG-AS-ODN to form PIC micelles. On the one hand, cationic KALA peptide can ionically interact with the anionic ODN to form a polyelectrolyte core of PIC, and on the other hand, it also provides fusion activity to this delivery system. Furthermore, fusogenic peptides have been commonly used together with other component, working as auxiliary agents to enhance the efficacy of established delivery formulations. For instance, in order to promote the transfer effect of PLL, HA-2 was conjugated to PLL and then incorporated into transferrin-PLL/DNA polyplexes (Plank et al. 1992).

pH-specific fusogenic or lytic peptides do not have high efficiency at any cases. Various peptides have been compared for the promotion of polyplex-based gene transfer (Futaki et al. 2005; Lim et al. 2000; Wagner 1999). Results showed that the improvement also depends on the characteristics of the cationic carriers. Peptide HA-2 and other synthetic or natural sequences such as the amphipathic peptides GALA, KALA, EGLA, JTS1, and gramicidin S have been tested. Introduction of membrane-active peptides in ligand-PLL-mediated gene delivery could improve the transfection up to more than 1000-fold. Whereas other PCs like dendrimers or PEI as well as several cationic lipids are only slightly enhanced by endosomolytic peptides or adenoviruses, clearly demonstrating differences in the bottlenecks of the delivery processes. Electroneutral cationic lipid–DNA complexes however can be strongly improved by the addition of membrane-active peptides.

CPPs also have endosome-disrupting property (Mann et al. 2008). Recent discovery of these peptides with nature membrane penetration property has attracted a lot of attention for their promising prospect in assisting drug and protein delivery. Such CPPs include the derivatives from natural peptides (e.g., penetratin and TAT), chimeric peptides (e.g., transportan), and synthetic ones (e.g., oligoarginine) (Trehin and Merkle 2004). Most CPPs are amphiphilic or some are even totally hydrophobic, which provide well membrane lytic/fusogenic activities, but commonly have no pH-sensitive properties.

Apart from direct membrane lysis by fusogenic peptides, another approach to realize the escape from endosome is based on the incorporation of histidine-rich peptides, via the same "proton sponge effect" as described earlier. For this, the imidazole functionality in the side chain of the His residue is responsible, as it can be protonated at endosomal low pH and also may interact with negatively charged head groups of lipids. Histidylated polylysine (Bello Roufai and Midoux 2001) and gluconic acid–modified poly(His) (Pack et al. 2000) have also been shown to promote efficient gene transfer. H5WYG and E5WYG peptide are two kinds of variants of fusogenic HA-2, with histidine- and glutamic acid–rich sequence, respectively. The former one shows more efficient transfection because of early endosome (pH < 7) disruption as compared to the latter which is active only at the late endosome or lysosome (pH < 5.5) (Midoux et al. 1998).

Histidine has also been used to modify linear oligolysines for improved cytosolic delivery of ODNs (Pichon et al. 2000). Several branched lysine–histidine peptides show high transfection efficiency combining with liposomes (Chen et al. 2001). Leng and coworkers (Leng and Mixson 2005; Leng et al. 2005) generated various branched carriers consisting of lysine and histidine residues with different degrees of branching, length of terminal arms, and changed histidine–lysine sequences. They identified different carriers suitable for either DNA or siRNA delivery. Interestingly, carriers optimized for DNA gene transfer were not necessarily useful for delivery of siRNA.

Other strategies for pH-triggered activation of membrane-disruptive peptide were used. Lytic peptides like melittin and CPPs may display a particularly strong lytic activity that efficiently enhances the transport of gene carriers across membrane barrier (Boeckle et al. 2005; Ogris et al. 2001). But such a high lytic activity is harmful at physiological conditions, which may lead to toxic side effects before internalization into cells. Several approaches have been described to render melittin more pH-specific, for example by introduction of acidic residues (Boeckle and Wagner 2006; Boeckle et al. 2006) or site of linkage (Boeckle et al. 2005). Murata et al. (1987) showed that the lytic activity of melittin could be blocked irreversibly by acylation of lysines and terminal amino function in its sequences with succinic anhydride. This strategy was further developed by Rozema et al. (2003), who acylated melittin with a pH-reversible dimethylmaleic anhydride (DMMAn) derivative. DMMAn could shield the lytic activity around pH 7.4, while the maleamate mask is released, restoring the lytic activity of melittin at acidic environment. This approach also worked fine in the delivery of phosphorodiamidate morpholino ODNs. Masked melittin and the ODNs were co-incubated with HeLa cells for the transfection experiments. Co-administration resulted in a 5- to 12-fold increase of antisense efficiency in comparison to ODNs alone. These results are promising, but for in vivo delivery the membrane-disruptive peptide had to be associated with the vector. Inspired by the results, Meyer et al. (2007) covalently coupled pH-responsive DMMAn-modified melittin to PLL. Lytic activity of the conjugate was undetectable at neutral pH, while after deprotection at acidic pH of 5 it was activated. Acute toxicity was greatly reduced as compared to unmodified PLL-melittin and up to 1800-fold higher gene expression than unmodified PLL was obtained. Furthermore, PLL-PEG-DMMAn-Mel was successfully used for siRNA delivery (Meyer et al. 2008, 2009) (Figure 16.6).

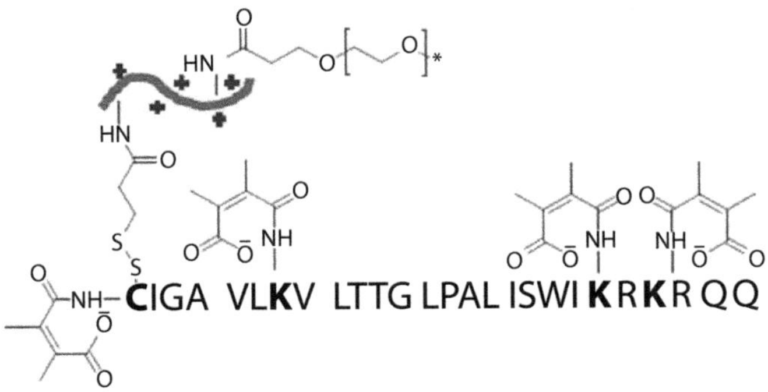

FIGURE 16.6 pH-responsive endosomolytic peptide. The lytic activity of the synthetic peptide melittin can be masked by blocking the lysine amine residues with dimethylmaleic anhydride (DMMAn). DMMAn is cleaved at endosomal pH, releasing lytic melittin. Optionally, as described in the text, DMMAn-modified melittin is conjugated to cationic polymers for nucleic acid complex formation.

16.6   CONCLUSION

Experiments evaluating the ability of pH-sensitive drug carriers to promote the cytoplasmic delivery of proteins, antisense ODN drugs, or DNA plasmid-based genes have met with considerable success, at least in vitro in cellular systems. Here the difference between extracellular neutral and endosomal acidic pH has been nicely utilized. The more demanding challenge of in vivo experiments can be attributed to a number of additional barriers. First, cytotoxicity associated with polymers appears to be a primary concern preventing their clinical application. Second, polymeric gene carriers are subjected to varying levels of deactivation by, e.g., plasma proteins, activation of the complement system, and unspecific interaction with nontarget cells and matrices. A continued thorough examination of carrier-activated clearance mechanisms will enable the design of better formulations with systemic longevity. The presence of numerous other barriers to macromolecular delivery at the tissue level cannot be ignored. A better integration between carrier design and in vivo experimentation will be critical for optimizing pH-responsive polymer-based carrier systems.

ACKNOWLEDGMENTS

Dr. Yu Nie was postdoctoral fellow at LMU and now is associate professor at National Engineering Research Center for Biomaterials, Sichuan University, China. She gratefully acknowledges the support from Alexander von Humboldt Foundation and Sino-German Center for Research Promotion. Funding for our work was provided by DFG WA1648/4-1, Excellence Cluster "Nanosystems Initiative Munich," and the SGC grant GZ756. We are very grateful to Olga Brück for secretarial support.

REFERENCES

Akinc, A., Thomas, M., Klibanov, A. M., and Langer, R. 2005. Exploring polyethylenimine-mediated DNA transfection and the proton sponge hypothesis. *J Gene Med.* 7: 657–663.

Asayama, S., Hakamatani, T., and Kawakami, H. 2010. Synthesis and characterization of alkylated poly(1-vinylimidazole) to control the stability of its DNA polyion complexes for gene delivery. *Bioconjug Chem.* 21: 646–652.

Asayama, S., Sekine, T., Kawakami, H., and Nagaoka, S. 2007. Design of aminated poly(1-vinylimidazole) for a new pH-sensitive polycation to enhance cell-specific gene delivery. *Bioconjug Chem.* 18: 1662–1667.

Asokan, A. and Cho, M. J. 2002. Exploitation of intracellular pH gradients in the cellular delivery of macromolecules. *J Pharm Sci.* 91: 903–913.

Bello Roufai, M. and Midoux, P. 2001. Histidylated polylysine as DNA vector: Elevation of the imidazole protonation and reduced cellular uptake without change in the polyfection efficiency of serum stabilized negative polyplexes. *Bioconjug Chem.* 12: 92–99.

Berg, K., Folini, M., Prasmickaite, L. et al. 2007. Photochemical internalization: A new tool for drug delivery. *Curr Pharm Biotechnol.* 8: 362–372.

Blessing, T., Kursa, M., Holzhauser, R., Kircheis, R., and Wagner, E. 2001. Different strategies for formation of pegylated EGF-conjugated PEI/DNA complexes for targeted gene delivery. *Bioconjug Chem.* 12: 529–537.

Boeckle, S., Fahrmeir, J., Roedl, W., Ogris, M., and Wagner, E. 2006. Melittin analogs with high lytic activity at endosomal pH enhance transfection with purified targeted PEI polyplexes. *J Control Release.* 112: 240–248.

Boeckle, S. and Wagner, E. 2006. Optimizing targeted gene delivery: Chemical modification of viral vectors and synthesis of artificial virus vector systems. *AAPS J.* 8: E731–E742.

Boeckle, S., Wagner, E., and Ogris, M. 2005. C- versus N-terminally linked melittin–polyethylenimine conjugates: The site of linkage strongly influences activity of DNA polyplexes. *J Gene Med.* 7: 1335–1347.

Boussif, O., Lezoualc'h, F., Zanta, M. A. et al. 1995. A versatile vector for gene and oligonucleotide transfer into cells in culture and in vivo: Polyethylenimine. *Proc Natl Acad Sci USA.* 92: 7297–7301.

Bulmus, V., Woodward, M., Lin, L. et al. 2003. A new pH-responsive and glutathione-reactive, endosomal membrane-disruptive polymeric carrier for intracellular delivery of biomolecular drugs. *J Control Release.* 93: 105–120.

Chen, T., McIntosh, D., He, Y. et al. 2004. Alkylated derivatives of poly(ethylacrylic acid) can be inserted into preformed liposomes and trigger pH-dependent intracellular delivery of liposomal contents. *Mol Membr Biol.* 21: 385–393.

Chen, Q. R., Zhang, L., Stass, S. A., and Mixson, A. J. 2001. Branched co-polymers of histidine and lysine are efficient carriers of plasmids. *Nucleic Acids Res.* 29: 1334–1340.

Cheung, C. Y., Murthy, N., Stayton, P. S., and Hoffman, A. S. 2001. A pH-sensitive polymer that enhances cationic lipid-mediated gene transfer. *Bioconjug Chem.* 12: 906–910.

Cheung, C. Y., Stayton, P. S., and Hoffman, A. S. 2005. Poly(propylacrylic acid)-mediated serum stabilization of cationic lipoplexes. *J Biomater Sci Polym Ed.* 16: 163–179.

Chilkoti, A., Dreher, M. R., Meyer, D. E., and Raucher, D. 2002. Targeted drug delivery by thermally responsive polymers. *Adv Drug Deliv Rev.* 54: 613–630.

Choi, J. S., MacKay, J. A., Szoka, F. C., Jr. 2003. Low-pH-sensitive PEG-stabilized plasmid-lipid nanoparticles: Preparation and characterization. *Bioconjug Chem.* 14: 420–429.

Choi, J. S., Nam, K., Park, J. Y. et al. 2004. Enhanced transfection efficiency of PAMAM dendrimer by surface modification with L-arginine. *J Control Release.* 99: 445–456.

Chollet, P., Favrot, M. C., Hurbin, A., and Coll, J. L. 2002. Side-effects of a systemic injection of linear polyethylenimine–DNA complexes. *J Gene Med.* 4: 84–91.

Conner, S. D. and Schmid, S. L. 2003. Regulated portals of entry into the cell. *Nature.* 422: 37–44.

Convertine, A. J., Benoit, D. S., Duvall, C. L., Hoffman, A. S., and Stayton, P. S. 2009. Development of a novel endosomolytic diblock copolymer for siRNA delivery. *J Control Release.* 133: 221–229.

Demeneix, B. A. and Behr, J. P. 1996. The proton sprenge: A trick the viruses didn't exploit. In *Artificial Self-Assembling Systems for Gene Delivery*, eds. Felgner, P. L., Heller, M. J., Lehn, P., Behr J. P., and Szoka, F. C. 146–151. Washington, DC: American Chemical Society.

DeRouchey, J., Schmidt, C., Walker, G. F. et al. 2008. Monomolecular assembly of siRNA and poly(ethylene glycol)–peptide copolymers. *Biomacromolecules.* 9: 724–732.

Dramsi, S. and Cossart, P. 2002. Listeriolysin O: A genuine cytolysin optimized for an intracellular parasite. *J Cell Biol.* 156: 943–946.

Du, J., Tang, Y., Lewis, A. L., and Armes, S. P. 2005. pH-sensitive vesicles based on a biocompatible zwitterionic diblock copolymer. *J Am Chem Soc.* 127: 17982–17983.

Duan, H., Wang, D., Sobal, N. S. et al. 2005. Magnetic colloidosomes derived from nanoparticle interfacial self-assembly. *Nano Lett.* 5: 949–952.

Eichman, J. D., Bielinska, A. U., Kukowska-Latallo, J. F., Donovan, B. W., and Baker, J. R., Jr. 2001. Bioapplications of PAMAM dendrimers. In *Dendrimers and Other Dendritic Polymers*, eds. Frechet, J. M. J. and Tomalia, D. A. 441–461. Hoboken, NJ: John Wiley & Sons.

Esfand, R. and Tomalia, D. A. 2001. Poly(amidoamine) (PAMAM) dendrimers: From biomimicry to drug delivery and biomedical applications. *Drug Discov Today.* 6: 427–436.

Felgner, P. L., Barenholz, Y., Behr, J. P. et al. 1997. Nomenclature for synthetic gene delivery systems. *Hum Gene Ther.* 8: 511–512.

Fella, C., Walker, G. F., Ogris, M., and Wagner, E. 2008. Amine-reactive pyridylhydrazone-based PEG reagents for pH-reversible PEI polyplex shielding. *Eur J Pharm Sci.* 34: 309–320.

Finsinger, D., Remy, J. S., Erbacher, P., Koch, C., and Plank, C. 2000. Protective copolymers for nonviral gene vectors: Synthesis, vector characterization and application in gene delivery. *Gene Ther.* 7: 1183–1192.

Futaki, S., Masui, Y., Nakase, I. et al. 2005. Unique features of a pH-sensitive fusogenic peptide that improves the transfection efficiency of cationic liposomes. *J Gene Med.* 7: 1450–1458.

Ganta, S., Devalapally, H., Shahiwala, A., and Amiji, M. 2008. A review of stimuli-responsive nanocarriers for drug and gene delivery. *J Control Release.* 126: 187–204.

von Gersdorff, K., Sanders, N. N., Vandenbroucke, R. et al. 2006. The internalization route resulting in successful gene expression depends on both cell line and polyethylenimine polyplex type. *Mol Ther.* 14: 745–753.

Gerweck, L. E. and Seetharaman, K. 1996. Cellular pH gradient in tumor versus normal tissue: Potential exploitation for the treatment of cancer. *Cancer Res.* 56: 1194–1198.

Gillies, E. R., Goodwin, A. P., and Frechet, J. M. 2004. Acetals as pH-sensitive linkages for drug delivery. *Bioconjug Chem.* 15: 1254–1263.

Goh, S. L., Murthy, N., Xu, M., and Frechet, J. M. 2004. Cross-linked microparticles as carriers for the delivery of plasmid DNA for vaccine development. *Bioconjug Chem.* 15: 467–474.

Haas, D. H. and Murphy, R. M. 2004. Design of a pH-sensitive pore-forming peptide with improved performance. *J Pept Res.* 63: 9–16.

Hartmann, L., Hafele, S., Peschka-Suss, R., Antonietti, M., and Borner, H. G. 2008. Tailor-made poly(amidoamine)s for controlled complexation and condensation of DNA. *Chemistry.* 14: 2025–2033.

Hartmann, L., Krause, E., Antonietti, M., and Borner, H. G. 2006. Solid-phase supported polymer synthesis of sequence-defined, multifunctional poly(amidoamines). *Biomacromolecules.* 7: 1239–1244.

Hernot, S. and Klibanov, A. L. 2008. Microbubbles in ultrasound-triggered drug and gene delivery. *Adv Drug Deliv Rev.* 60: 1153–1166.

Hoffman, A. S., Stayton P. S., Press, O. et al. 2002. Design of "smart" polymers that can direct intracellular drug delivery. *Polym Adv Technol.* 13: 992–999.

Ihm, J. E., Han, K. O., Hwang, C. S. et al. 2005. Poly(4-vinylimidazole) as nonviral gene carrier: In vitro and in vivo transfection. *Acta Biomater.* 1: 165–172.

Jabr-Milane, L., van Vlerken, L., Devalapally, H. et al. 2008. Multi-functional nanocarriers for targeted delivery of drugs and genes. *J Control Release.* 130: 121–128.

Jain, R. S., Standley, S. M., and Frechet, J. M. J. 2007. Synthesis and degradation of pH-sensitive linear poly(amidoamine)s. *Macromolecules.* 40: 452–457.

Jeong, J. H., Kim, S. W., and Park, T. G. 2003. Novel intracellular delivery system of antisense oligonucleotide by self-assembled hybrid micelles composed of DNA/PEG conjugate and cationic fusogenic peptide. *Bioconjug Chem.* 14: 473–479.

Jones, R. A., Cheung, C. Y., Black, F. E. et al. 2003. Poly(2-alkylacrylic acid) polymers deliver molecules to the cytosol by pH-sensitive disruption of endosomal vesicles. *Biochem J.* 372: 65–75.

Khalil, I. A., Kogure, K., Akita, H., and Harashima, H. 2006. Uptake pathways and subsequent intracellular trafficking in nonviral gene delivery. *Pharmacol Rev.* 58: 32–45.

Kim, Y. H., Park, J. H., Lee, M., Park, T. G., and Kim, S. W. 2005. Polyethylenimine with acid-labile linkages as a biodegradable gene carrier. *J Control Release.* 103: 209–219.

Knorr, V., Allmendinger, L., Walker, G. F., Paintner, F. F., and Wagner, E. 2007. An acetal-based PEGylation reagent for pH-sensitive shielding of DNA polyplexes. *Bioconjug Chem.* 18: 1218–1225.

Knorr, V., Ogris, M., and Wagner, E. 2008a. An acid sensitive ketal-based polyethylene glycol–oligoethylenimine copolymer mediates improved transfection efficiency at reduced toxicity. *Pharm Res.* 25: 2937–2945.

Knorr, V., Russ, V., Allmendinger, L., Ogris, M., and Wagner, E. 2008b. Acetal linked oligoethylenimines for use as pH-sensitive gene carriers. *Bioconjug Chem.* 19: 1625–1634.

Kommareddy, S. and Amiji, M. 2005. Preparation and evaluation of thiol-modified gelatin nanoparticles for intracellular DNA delivery in response to glutathione. *Bioconjug Chem.* 16: 1423–1432.

Kulkarni, R. P., Mishra, S., Fraser, S. E., and Davis, M. E. 2005. Single cell kinetics of intracellular, nonviral, nucleic acid delivery vehicle acidification and trafficking. *Bioconjug Chem.* 16: 986–994.

Kyriakides, T. R., Cheung, C. Y., Murthy, N. et al. 2002. pH-sensitive polymers that enhance intracellular drug delivery in vivo. *J Control Release.* 78: 295–303.

Lee, H., Jeong, J. H., and Park, T. G. 2001. A new gene delivery formulation of polyethylenimine/DNA complexes coated with PEG conjugated fusogenic peptide. *J Control Release.* 76: 183–192.

Lee, Y., Mo, H., Koo, H. et al. 2007. Visualization of the degradation of a disulfide polymer, linear poly(ethylenimine sulfide), for gene delivery. *Bioconjug Chem.* 18: 13–18.

Lee, L. K., Williams, C. L., Devore, D., and Roth, C. M. 2006. Poly(propylacrylic acid) enhances cationic lipidmediated delivery of antisense oligonucleotides. *Biomacromolecules.* 7: 1502–1508.

Lemkine, G. F. and Demeneix, B. A. 2001. Polyethylenimines for in vivo gene delivery. *Curr Opin Mol Ther.* 3: 178–182.

Leng, Q. and Mixson, A. J. 2005. Small interfering RNA targeting Raf-1 inhibits tumor growth in vitro and in vivo. *Cancer Gene Ther.* 12: 682–690.

Leng, Q., Scaria, P., Zhu, J. et al. 2005. Highly branched HK peptides are effective carriers of siRNA. *J Gene Med.* 7: 977–986.

Li, W., Huang, Z., MacKay, J. A., Grube, S., and Szoka, F. C., Jr. 2005. Low-pH-sensitive poly(ethylene glycol) (PEG)-stabilized plasmid nanolipoparticles: Effects of PEG chain length, lipid composition and assembly conditions on gene delivery. *J Gene Med.* 7: 67–79.

Li, W., Nicol, F., Szoka, F. C., Jr. 2004. GALA: A designed synthetic pH-responsive amphipathic peptide with applications in drug and gene delivery. *Adv Drug Deliv Rev.* 56: 967–985.

Lim, D. W., Yeom, Y. I., and Park, T. G. 2000. Poly(DMAEMA-NVP)-*b*-PEG-galactose as gene delivery vector for hepatocytes. *Bioconjug Chem.* 11: 688–695.

Little, S. R., Lynn, D. M., Ge, Q. et al. 2004. Poly-beta amino ester-containing microparticles enhance the activity of nonviral genetic vaccines. *Proc Natl Acad Sci USA.* 101: 9534–9539.

Little, S. R., Lynn, D. M., Puram, S. V., and Langer, R. 2005. Formulation and characterization of poly(beta amino ester) microparticles for genetic vaccine delivery. *J Control Release.* 107: 449–462.

Liu, T. F., Hall, P. D., Cohen, K. A. et al. 2005. Interstitial diphtheria toxin-epidermal growth factor fusion protein therapy produces regressions of subcutaneous human glioblastoma multiforme tumors in athymic nude mice. *Clin Cancer Res.* 11: 329–334.

Lomas, H., Canton, I., MacNeil, S., Du, J., Armes, S. P., Ryan, A. J., Lewis, A. L., and Battaglia, G. 2007. Biomimetic pH sensitive polymersomes for efficient DNA encapsulation and delivery. *Adv Mater.* 19: 4238–4243.

Mann, A., Thakur, G., Shukla, V., and Ganguli, M. 2008. Peptides in DNA delivery: Current insights and future directions. *Drug Discov Today.* 13: 152–160.

Medina-Kauwe, L. K., Xie, J., and Hamm-Alvarez, S. 2005. Intracellular trafficking of nonviral vectors. *Gene Ther.* 12: 1734–1751.

Meyer, M., Dohmen, C., Philipp, A. et al. 2009. Synthesis and biological evaluation of a bioresponsive and endosomolytic siRNA–polymer conjugate. *Mol Pharm.* 6: 752–762.

Meyer, M., Philipp, A., Oskuee, R., Schmidt, C., and Wagner, E. 2008. Breathing life into polycations: Functionalization with pH-responsive endosomolytic peptides and polyethylene glycol enables siRNA delivery. *J Am Chem Soc.* 130: 3272–3273.

Meyer, M. and Wagner, E. 2006. pH-responsive shielding of non-viral gene vectors. *Expert Opin Drug Deliv.* 3: 563–571.

Meyer, M., Zintchenko, A., Ogris, M., and Wagner, E. 2007. A dimethylmaleic acid–melittin–polylysine conjugate with reduced toxicity, pH-triggered endosomolytic activity and enhanced gene transfer potential. *J Gene Med.* 9: 797–805.

Midoux, P., Kichler, A., Boutin, V., Maurizot, J. C., and Monsigny, M. 1998. Membrane permeabilization and efficient gene transfer by a peptide containing several histidines. *Bioconjug Chem.* 9: 260–267.

Moghimi, S. M., Symonds, P., Murray, J. C. et al. 2005. A two-stage poly(ethylenimine)-mediated cytotoxicity: Implications for gene transfer/therapy. *Mol Ther.* 11: 990–995.

Murata, M., Nagayama, K., and Ohnishi, S. 1987. Membrane fusion activity of succinylated melittin is triggered by protonation of its carboxyl groups. *Biochemistry.* 26: 4056–4062.

Murthy, N., Campbell, J., Fausto, N., Hoffman, A. S., and Stayton, P. S. 2003a. Bioinspired pH-responsive polymers for the intracellular delivery of biomolecular drugs. *Bioconjug Chem.* 14: 412–419.

Murthy, N., Campbell, J., Fausto, N., Hoffman, A. S., and Stayton, P. S. 2003b. Design and synthesis of pH-responsive polymeric carriers that target uptake and enhance the intracellular delivery of oligonucleotides. *J Control Release.* 89: 365–374.

Murthy, N., Robichaud, J. R., Tirrell, D. A., Stayton, P. S., and Hoffman, A. S. 1999. The design and synthesis of polymers for eukaryotic membrane disruption. *J Control Release.* 61: 137–143.

Murthy, N., Xu, M., Schuck, S. et al. 2003c. A macromolecular delivery vehicle for protein-based vaccines: Acid-degradable protein-loaded microgels. *Proc Natl Acad Sci USA.* 100: 4995–5000.

Namgung, R., Nam, S., Kim, S. K. et al. 2009. An acid-labile temperature-responsive sol–gel reversible polymer for enhanced gene delivery to the myocardium and skeletal muscle cells. *Biomaterials.* 30: 5225–5233.

Nie, Y., Gunther, M., Gu, Z., and Wagner, E. 2011. Pyridylhydrazone-based PEGylation for pH-reversible lipopolyplex shielding. *Biomaterials.* 32: 858–869.

Ogris, M., Brunner, S., Schuller, S., Kircheis, R., and Wagner, E. 1999. PEGylated DNA/transferrin–PEI complexes: Reduced interaction with blood components, extended circulation in blood and potential for systemic gene delivery. *Gene Ther.* 6: 595–605.

Ogris, M., Carlisle, R. C., Bettinger, T., and Seymour, L. W. 2001. Melittin enables efficient vesicular escape and enhanced nuclear access of nonviral gene delivery vectors. *J Biol Chem.* 276: 47550–47555.

Ogris, M., Walker, G., Blessing, T. et al. 2003. Tumor-targeted gene therapy: Strategies for the preparation of ligand–polyethylene glycol–polyethylenimine/DNA complexes. *J Control Release.* 91: 173–181.

Oishi, M., Nagasaki, Y., Itaka, K., Nishiyama, N., and Kataoka, K. 2005a. Lactosylated poly(ethylene glycol)–siRNA conjugate through acid-labile beta-thiopropionate linkage to construct pH-sensitive polyion complex micelles achieving enhanced gene silencing in hepatoma cells. *J Am Chem Soc.* 127: 1624–1625.

Oishi, M., Nagatsugi, F., Sasaki, S., Nagasaki, Y., and Kataoka, K. 2005b. Smart polyion complex micelles for targeted intracellular delivery of PEGylated antisense oligonucleotides containing acid-labile linkages. *Chembiochem.* 6: 718–725.

Oishi, M., Sasaki, S., Nagasaki, Y., and Kataoka, K. 2003. pH-responsive oligodeoxynucleotide (ODN)–poly(ethylene glycol) conjugate through acid-labile beta-thiopropionate linkage: Preparation and polyion complex micelle formation. *Biomacromolecules.* 4: 1426–1432.

Oupicky, D. and Diwadkar, V. 2003. Stimuli-responsive gene delivery vectors. *Curr Opin Mol Ther.* 5: 345–350.

Oupicky, D., Ogris, M., Howard, K. A. et al. 2002. Importance of lateral and steric stabilization of polyelectrolyte gene delivery vectors for extended systemic circulation. *Mol Ther.* 5: 463–472.

Pack, D. W., Putnam, D., and Langer, R. 2000. Design of imidazole-containing endosomolytic biopolymers for gene delivery. *Biotechnol Bioeng.* 67: 217–223.

Paramonov, S. E., Bachelder, E. M., Beaudette, T. T. et al. 2008. Fully acid-degradable biocompatible polyacetal microparticles for drug delivery. *Bioconjug Chem.* 19: 911–919.

Pecheur, E. I., Sainte-Marie, J., Bienven e, A., and Hoekstra, D. 1999. Peptides and membrane fusion: Towards an understanding of the molecular mechanism of protein-induced fusion. *J Membr Biol.* 167: 1–17.

Pichon, C., Roufai, M. B., Monsigny, M., and Midoux, P. 2000. Histidylated oligolysines increase the transmembrane passage and the biological activity of antisense oligonucleotides. *Nucleic Acids Res.* 28: 504–512.

Plank, C., Oberhauser, B., Mechtler, K., Koch, C., and Wagner, E. 1994. The influence of endosome-disruptive peptides on gene transfer using synthetic virus-like gene transfer systems. *J Biol Chem.* 269: 12918–12924.

Plank, C., Zatloukal, K., Cotten, M., Mechtler, K., and Wagner, E. 1992. Gene transfer into hepatocytes using asialoglycoprotein receptor mediated endocytosis of DNA complexed with an artificial tetra-antennary galactose ligand. *Bioconjug Chem.* 3: 533–539.

Read, M. L., Singh, S., Ahmed, Z. et al. 2005. A versatile reducible polycation-based system for efficient delivery of a broad range of nucleic acids. *Nucleic Acids Res.* 33: e86.

Rittner, K., Benavente, A., Bompard-Sorlet, A. et al. 2002. New basic membrane-destabilizing peptides for plasmid-based gene delivery in vitro and in vivo. *Mol Ther.* 5: 104–114.

Ropert, C. 1999. Liposomes as a gene delivery system. *Braz J Med Biol Res.* 32: 163–169.

Roy, I. and Gupta, M. N. 2003. Smart polymeric materials: Emerging biochemical applications. *Chem Biol.* 10: 1161–1171.

Rozema, D. B., Ekena, K., Lewis, D. L., Loomis, A. G., and Wolff, J. A. 2003. Endosomolysis by masking of a membrane-active agent (EMMA) for cytoplasmic release of macromolecules. *Bioconjug Chem.* 14: 51–57.

Rozema, D. B., Lewis, D. L., Wakefield, D. H. et al. 2007. Dynamic polyconjugates for targeted in vivo delivery of siRNA to hepatocytes. *Proc Natl Acad Sci USA.* 104: 12982–12987.

Russ, V., Elfberg, H., Thoma, C. et al. 2008. Novel degradable oligoethylenimine acrylate ester-based pseudo-dendrimers for in vitro and in vivo gene transfer. *Gene Ther.* 15: 18–29.

Saito, G., Swanson, J. A., and Lee, K. D. 2003. Drug delivery strategy utilizing conjugation via reversible disulfide linkages: Role and site of cellular reducing activities. *Adv Drug Deliv Rev.* 55: 199–215.

Schaffert, D. and Wagner, E. 2008. Gene therapy progress and prospects: Synthetic polymer-based systems. *Gene Ther.* 15: 1131–1138.

Sethuraman, V. A., Lee, M. C., and Bae, Y. H. 2008. A biodegradable pH-sensitive micelle system for targeting acidic solid tumors. *Pharm Res.* 25: 657–666.

Sethuraman, V. A., Na, K., and Bae, Y. H. 2006. pH-responsive sulfonamide/PEI system for tumor specific gene delivery: An in vitro study. *Biomacromolecules.* 7: 64–70.

Shin, J., Shum, P., and Thompson, D. H. 2003. Acid-triggered release via dePEGylation of DOPE liposomes containing acid-labile vinyl ether PEG-lipids. *J Control Release*. 91: 187–200.

Sonawane, N. D., Szoka, F. C., Jr., and Verkman, A. S. 2003. Chloride accumulation and swelling in endosomes enhances DNA transfer by polyamine–DNA polyplexes. *J Biol Chem*. 278: 44826–44831.

Srinivasachar, K. and Neville, D. M., Jr. 1989. New protein cross-linking reagents that are cleaved by mild acid. *Biochemistry*. 28: 2501–2509.

Sutton, D., Durand, R., Shuai, X. T., and Gao, J. M. 2006. Poly(D,L-lactide-*co*-glycolide)/poly(ethylenimine) blend matrix system for pH sensitive drug delivery. *J Appl Polym Sci*. 100: 89–96.

Tam, P., Monck, M., Lee, D. et al. 2000. Stabilized plasmid-lipid particles for systemic gene therapy. *Gene Ther*. 7: 1867–1874.

Thomas, M., Ge, Q., Lu, J. J., Chen, J., and Klibanov, A. M. 2005. Cross-linked small polyethylenimines: While still nontoxic, deliver DNA efficiently to mammalian cells in vitro and in vivo. *Pharm Res*. 22: 373–380.

Torchilin, V. P. 2006. Recent approaches to intracellular delivery of drugs and DNA and organelle targeting. *Annu Rev Biomed Eng*. 8: 343–375.

Trehin, R. and Merkle, H. P. 2004. Chances and pitfalls of cell penetrating peptides for cellular drug delivery. *Eur J Pharm Biopharm*. 58: 209–223.

Varshosaz, J. 2008. pH-sensitive polymers for cytoplasmic drug delivery. *Expert Opin Ther Patents*. 18: 959–962.

Vaupel, P., Kallinowski, F., and Okunieff, P. 1989. Blood flow, oxygen and nutrient supply, and metabolic microenvironment of human tumors: A review. *Cancer Res*. 49: 6449–6465.

Wagner, E. 1999. Application of membrane-active peptides for nonviral gene delivery. *Adv Drug Deliv Rev*. 38: 279–289.

Wagner, E. 2007. Programmed drug delivery: Nanosystems for tumor targeting. *Expert Opin Biol Ther*. 7: 587–593.

Wagner, E., Plank, C., Zatloukal, K., Cotten, M., and Birnstiel, M. L. 1992. Influenza virus hemagglutinin HA-2 N-terminal fusogenic peptides augment gene transfer by transferrin–polylysine–DNA complexes: Toward a synthetic virus-like gene-transfer vehicle. *Proc Natl Acad Sci USA*. 89: 7934–7938.

Wagner, E., Zenke, M., Cotten, M., Beug, H., and Birnstiel, M. L. 1990. Transferrin–polycation conjugates as carriers for DNA uptake into cells. *Proc Natl Acad Sci USA*. 87: 3410–3414.

Walker, G. F., Fella, C., Pelisek, J. et al. 2005. Toward synthetic viruses: Endosomal pH-triggered deshielding of targeted polyplexes greatly enhances gene transfer in vitro and in vivo. *Mol Ther*. 11: 418–425.

Wang, J., Mao, H. Q., and Leong, K. W. 2001. A novel biodegradable gene carrier based on polyphosphoester. *J Am Chem Soc*. 123: 9480–9481.

Wang, X. L., Nguyen, T., Gillespie, D., Jensen, R., and Lu, Z. R. 2008. A multifunctional and reversibly polymerizable carrier for efficient siRNA delivery. *Biomaterials*. 29: 15–22.

Wang, X. L., Ramusovic, S., Nguyen, T., and Lu, Z. R. 2007. Novel polymerizable surfactants with pH-sensitive amphiphilicity and cell membrane disruption for efficient siRNA delivery. *Bioconjug Chem*. 18: 2169–2177.

Wang, X. L., Xu, R., and Lu, Z. R. 2009a. A peptide-targeted delivery system with pH-sensitive amphiphilic cell membrane disruption for efficient receptor-mediated siRNA delivery. *J Control Release*. 134: 207–213.

Wang, X. L., Xu, R., Wu, X. et al. 2009b. Targeted systemic delivery of a therapeutic siRNA with a multifunctional carrier controls tumor proliferation in mice. *Mol Pharm*. 6: 738–746.

Wolff, J. A. and Rozema, D. B. 2008. Breaking the bonds: Non-viral vectors become chemically dynamic. *Mol Ther*. 16: 8–15.

Wong, J. B., Grosse, S., Tabor, A. B., Hart, S. L., and Hailes, H. C. 2008. Acid cleavable PEG-lipids for applications in a ternary gene delivery vector. *Mol Biosyst*. 4: 532–541.

Wyman, T. B., Nicol, F., Zelphati, O. et al. 1997. Design, synthesis, and characterization of a cationic peptide that binds to nucleic acids and permeabilizes bilayers. *Biochemistry*. 36: 3008–3017.

Xu, Z., Gu, W., Chen, L. et al. 2008. A smart nanoassembly consisting of acid-labile vinyl ether PEG-DOPE and protamine for gene delivery: Preparation and in vitro transfection. *Biomacromolecules*. 9: 3119–3126.

Xu, R., Wang, X. L., and Lu, Z. R. 2010. New amphiphilic carriers forming pH-sensitive nanoparticles for nucleic acid delivery. *Langmuir*. 26: 13874–13882.

Zhao, Z., Wang, J., Mao, H. Q., and Leong, K. W. 2003. Polyphosphoesters in drug and gene delivery. *Adv Drug Deliv Rev*. 55: 483–499.

Zintchenko, A., Ogris, M., and Wagner, E. 2006. Temperature dependent gene expression induced by PNIPAM-based copolymers: Potential of hyperthermia in gene transfer. *Bioconjug Chem*. 17: 766–772.

Zou, S. M., Erbacher, P., Remy, J. S., and Behr, J. P. 2000. Systemic linear polyethylenimine (L-PEI)-mediated gene delivery in the mouse. *J Gene Med*. 2: 128–134.

17 Adhesive Biomaterials for Tissue Repair and Reconstruction

Sujata K. Bhatia

CONTENTS

17.1 INTRODUCTION: THE CLINICAL NEED FOR ADHESIVE BIOMATERIALS

Despite refinements in suturing and stapling techniques for tissue reattachment and wound closure, surgeons continue to struggle with difficult tissue repairs. Reliable closure of tissues and organs remains one of the great unsolved challenges of clinical medicine and physicians are constantly seeking better methods for achieving efficient and secure closure of both surgical and traumatic wounds. There is great demand for tissue adhesives to replace or augment sutures and staples for wound repair, as well as adhesive materials for hemorrhage control. An effective tissue adhesive has the potential to reduce operating times, lessen the bleeding and leakage from wounds, lower the incidence of complications, and reduce hospital stays. Such an adhesive will save time and money, as well as lessen morbidity and mortality in surgical patients.

The limitations of traditional approaches for wound closure are best illustrated by current rates of adverse events following common surgical procedures. For example, gastrointestinal anastomotic closures demonstrate a leakage rate of 3%–15%, resulting in dire clinical consequences (infection, hemorrhage) and a 2%–3% mortality rate for patients undergoing abdominal surgeries [1,2]. Leaking fluids can escape between suture or staple lines; fluids can also leak through needle holes and staple punctures created by the placement of sutures and staples. A slight seepage of intestinal contents can lead to local contamination and infection, abscess formation, and fulminant bacterial peritonitis. Likewise, a small oozing blood vessel can permit significant blood loss. Moreover, in patients undergoing a potentially curative surgery for colorectal cancer, the presence of gastrointestinal anastomotic leakage is associated with a lower 5 year survival rate (44.3% in patients with leakage versus 64.0% in patients without leakage), even when there are no immediate clinical consequences of wound leakage [3]. Failure to achieve robust tissue closure thus carries both short-term and long-term risks.

Even tiny surgical incisions present significant challenges with regard to wound closure. As a case in point, the optimal sealing of corneal incisions during cataract surgery is a complex issue for ophthalmic surgeons. Sutures have traditionally been utilized for closure of corneal wounds, but the use of sutures is associated with several disadvantages. Suture placement requires fine technical skill and a prolonged operative time and sutures inflict trauma to corneal tissue. In addition, suture materials can activate infection, inflammation, and neovascularization with resultant corneal scarring and defects in vision [4]. In addition, uneven tension on sutures can lead to asymmetric healing and astigmatism [5]. Postoperative integrity of sutures can also be compromised. There has recently been a progressive increase in the use of sutureless clear corneal incisions by cataract surgeons; however, sutureless corneal incisions have been associated with an increased incidence of acute endophthalmitis [6,7], likely due to bacterial contamination of the unsutured open wound.

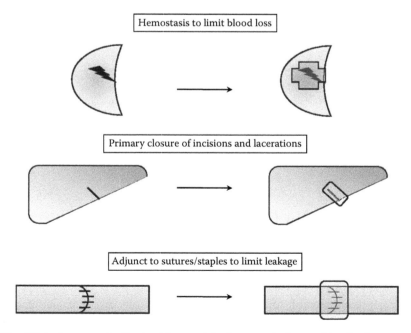

FIGURE 17.1 Clinical usage of adhesive biomaterials for wound closure; bioadhesives may function as hemostats, primary closure agents, and adjuncts to sutures and staples.

The earlier clinical cases reveal a pressing need in medicine for adhesive biomaterials for tissue reconstruction. Tissue adhesives are an attractive method for ensuring strong, leak-free repairs that preserve tissue and organ function, and subsequently improve the patient's quality-of-life and the patient's survival. An adhesive biomaterial for tissue reconstruction may function in multiple capacities (Figure 17.1): a hemostatic agent to limit blood loss, an adjunct to sutures and staples to limit leakage, or a primary closure agent to avoid the use of sutures or staples. In fact, polymeric adhesives can be applied to virtually every organ system in the body (Figure 17.2). Adhesive biomaterials have numerous potential clinical uses, ranging from cardiovascular medicine to orthopedics to neurosurgery; a detailed list of possible surgical applications for bioadhesives is given in Table 17.1.

The clinical imperative for tissue adhesives is clear and continued progress in this field is essential to patient's safety. This chapter reviews the current status of adhesive biomaterials for tissue reconstruction and highlights new developments in bioadhesive platforms. This chapter will begin with a discussion of practical considerations for the development of adhesive biomaterials, including recommendations for characterization of novel tissue adhesives. This chapter will then describe existing commercial adhesives for wound closure, including fibrin-based adhesives, cyanoacrylate-based adhesives, cross-linked protein-based adhesives, and PEG-based adhesives. This chapter will subsequently describe novel bioadhesives under development, including bio-inspired adhesives, polysaccharide-based adhesives, and dendrimer-based adhesives.

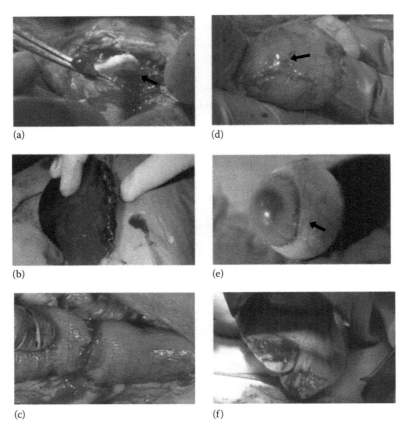

(a)   (d)

(b)   (e)

(c)   (f)

FIGURE 17.2 Application of polymeric bioadhesive to a variety of organs and tissue surfaces. A polysaccharide-based adhesive is shown applied to (a) vascular tissue; (b) liver; (c) large intestine; (d) bladder; (e) cornea; and (f) lung.

TABLE 17.1

Potential Clinical Applications for Polymeric Tissue Adhesives and Sealants

Neurosurgery
Repair of dural defects
Repair of central nervous system tissue
Spinal cord repair
Nerve grafting
Intervertebral disk surgery
Treatment of cerebrospinal fluid (CSF) leaks

Ophthalmic surgery
Clear corneal cataract surgery
Laser in situ keratomileusis (LASIK) surgery
Corneal ulcer treatment
Corneal transplantation
Conjunctival repair
Retinal attachment
Punctal plugging for treatment of dry eyes
Oculoplastics and blepharoplasty (eyelid lifts)
Vitrectomy closure
Attachment of extraocular muscles

Ear, nose, and throat (ENT) surgery
Control of epistaxis (nosebleeds)
Repair of vocal cord defects
Tympanoplasty for repair of perforated eardrum
Myringotomy (eardrum incision for drainage) with tube insertion
Sinus surgery
Nasal reconstructive surgery
Tonsillectomy surgery
Adenoidectomy surgery

Head and neck surgery
Salivary gland removal
Lymph node dissection
Treatment of chylous leakage after neck dissection

Interventional radiology
Therapeutic embolization
Femoral artery closure during interventional procedures

Vascular surgery
Arteriovenous fistula repair
Aortic aneurysm repair
Vascular anastomosis

Cardiovascular surgery
Cardiac valve repair
Repair of ventricular wall rupture
Coronary artery anastomosis during bypass surgery
Pacemaker and lead placement
Defibrillator and lead placement
Aortic anastomosis
Treatment of aortic dissection

TABLE 17.1 (continued)
Potential Clinical Applications for Polymeric
Tissue Adhesives and Sealants

Thoracic surgery
Lung lobectomy
Lung biopsy
Pneumothorax treatment

Gastrointestinal surgery
Gastrointestinal anastomosis
Peptic ulcer treatment
Treatment of esophageal rupture
Gallbladder or bile duct anastomosis
Gastric bypass surgery
Appendectomy
Cholecystectomy (gallbladder removal)
Pancreatic surgery
Gastrointestinal fistula repair
Rectal fistula repair
Treatment of diverticular bleeding
Hernia patch placement
Sealing of peritoneal dialysis catheter leakage
Prevention of intra-abdominal adhesions

Liver surgery
Liver resection
Liver transplantation

Gynecologic surgery
Hysterectomy
Myomectomy for uterine fibroid removal
Fallopian tube anastomosis
Vaginal fistula repair
Cervical surgery
Ovarian cyst removal
Breast biopsy
Mastectomy and lumpectomy
Management of preterm premature rupture of membranes

Urologic surgery
Nephrectomy
Kidney transplantation
Ureteral fistula repair
Ureteral anastomosis
Repair for stress urinary incontinence
Bladder closure
Radical prostatectomy
Vasectomy reversal surgery

Orthopedic surgery
Hip replacement surgery
Knee replacement surgery
Tendon reattachment
Cartilage repair

(continued)

TABLE 17.1 (continued)
Potential Clinical Applications for Polymeric
Tissue Adhesives and Sealants

Fracture repair

Bone grafting

Plastic and reconstructive surgery

Face lift surgery

Closure of skin incisions

Soft tissue augmentation

Trauma surgery

Closure of splenic lacerations and other solid organs

Closure of skin lacerations

Skin grafting for burn victims

17.2　PRACTICAL CONSIDERATIONS FOR ADHESIVE BIOMATERIALS

The concept of using an adhesive to join or rejoin tissues dates back to at least 1787, when it was noted that "many workmen glue their wounds with solid glue dissolved in water" [8]. The use of hide glue (similar to gelatin, which is itself derived from collagen protein) was most common, but other biological adhesives such as blood and egg white (albumin) have also been used for centuries. However, the search for the perfect operative sealant continues, as an ideal tissue adhesive must overcome difficult performance challenges.

A number of technical factors must be considered in the selection and development of novel adhesive polymers for tissue reconstruction. First, tissue adhesives must demonstrate adequate physical and mechanical properties. A useful tissue adhesive must exhibit strong adhesion to the target tissue, as well as sufficient cohesion to bind tissue sites together. The tissue adhesive should cure rapidly, preferably without requiring extra equipment to induce curing, and the material should have the ability to cure in a wet physiological environment. Second, in addition to meeting mechanical specifications, novel adhesive polymers must meet biocompatibility specifications. The tissue adhesive must be biocompatible to the target site, performing in its desired application without causing adverse effect. Both the polymer construct and its degradation products must be noncytotoxic, nonhemolytic, and noninflammatory; undesirable responses such as irritation and sensitization must be avoided. The polymer adhesive must not interfere with wound healing or induce fibrosis or a foreign body response; it is also necessary that the adhesive does not act as a hospitable environment for bacteria, so that it does not propagate an infection.

Third, adhesive polymers must meet the requirements of physiological metabolism. It is essential that the adhesive be biodegradable, with degradation achieved via either hydrolysis or enzymatic cleavage. The degradation time should be tuned such that the adhesive remains on the target site until physiological wound healing has taken place and degrades soon afterward to avoid polymer encapsulation by immune cells. Indeed, the success of the adhesive joint can be judged by how well the tissue is replaced. Moreover, the degradation products must be easily excreted by the kidneys. The molecular weight cutoff for kidney elimination of native globular proteins is considered to be 70,000, which is close to the molecular weight of serum albumin. Hydrophilic polymers may have a higher molecular volume than compact globular proteins; because of the larger effective size of polymers, the molecular weight cutoff for kidney excretion of polymers may be even more stringent. An additional consideration is that polymers with higher molecular weights will exhibit longer retention times in the blood.

Finally, tissue adhesives must satisfy commercial requirements. The ideal adhesive for clinical usage should be readily delivered through an easy-to-use device. The system should demonstrate adequate shelf stability and an optimal system should be storable at room temperature, requiring minimal advance preparation time. Production of the polymer adhesive must be scalable to allow cost-effective manufacture.

Throughout the development process, novel tissue adhesives must be assessed to ensure their suitability for clinical targets; characterization should include mechanical properties, physical/chemical properties, biological properties, shelf stability, and usability. A listing of recommended tests for adhesive biomaterials is presented in Table 17.2. Standardized testing protocols are available

---

TABLE 17.2

Methods for Characterizing Novel Polymeric Tissue Adhesives

Mechanical characterization

Adhesive strength
    Tensile strength
    Overlap shear strength
    Peel adhesion strength
    Impact strength
Cohesive tissue sealing ability
    Leak pressure test
    Burst pressure test

Physical/chemical characterization

Curing and reaction properties
    Tack-free time
    Total setting time
    Extent of reaction
    Residual monomer content
    Heat of polymerization
Degradation properties
    Degradation rate
    Degradation products
Swelling determination

Biological characterization

Sterility properties
    Bioburden test
    Bacterial endotoxin test
Tissue compatibility
    Cytotoxicity test
    Tissue irritation test
Hemolysis testing
Systemic effects
    Pyrogenicity test
    Sensitization test
Toxicokinetic testing
    Metabolic fate

Shelf life characterization

Accelerated shelf stability test
Viscosity determination

Delivery device characterization

Applicator functionality
Preparation time and ease-of-use

specifically for tissue adhesives; in particular, the American Society for Testing and Materials has developed several guidelines for measuring the strength of bioadhesives [9]. These include

- ASTM F2255-05 Standard Test Method for Strength Properties of Tissue Adhesives in Lap-Shear by Tension Loading
- ASTM F2256-05 Standard Test Method for Strength Properties of Tissue Adhesives in T-Peel by Tension Loading
- ASTM F2258-05 Standard Test Method for Strength Properties of Tissue Adhesives in Tension
- ASTM F2458-05 Standard Test Method for Wound Closure Strength in Tissue Adhesives and Sealants

These methods can provide a consistent means for comparing the performance of surgical adhesives on soft tissue.

The precise mechanical and chemical properties required of each adhesive biomaterial are determined to a large extent by the surgical target. Clinician input is an essential component of the design process, so that the surgeon's needs and the patient's needs can be translated into technical specifications. For example, an adhesive biomaterial for emergency hemostasis should demonstrate a very fast cure time and must be delivered over a wide surgical field; a hemostat may also exhibit a high degree of fluid absorption to stanch blood flow. In contrast, an adhesive for fine ophthalmic surgery or neurosurgery may need a slower cure time to allow adjustments of tissues and must be delivered through a smaller delivery device; such an adhesive might require low fluid absorption and low swell to avoid pressure on anatomical structures. The clinical target should continually guide the creation of a polymeric tissue adhesive.

17.3   FIBRIN-BASED TISSUE ADHESIVES

17.3.1   HISTORICAL DEVELOPMENT

Fibrin-based tissue adhesives are currently the principal biological sealant systems in clinical use. Fibrin glues take advantage of a physiological clotting cascade to form a synthetic fibrin clot on the surface of wounded tissue. As early as 1909, surgeons reported the hemostatic properties of fibrin powder applied in the operative field [10]. In 1915, the use of fibrin as a hemostatic agent during cerebral surgery was described [11]. Purified thrombin became available in 1938 and combinations of thrombin with fibrinogen were first utilized in 1944 to enhance adhesion of skin grafts to burned soldiers [12]; the use of fibrin sealants for peripheral nerve attachment was also first demonstrated in the 1940s [13]. The Cohn process for blood fractionation was developed during World War II and cryoprecipitation of fibrinogen was achieved in the 1960s, which led to the development of fibrin sealants in the 1970s. Commercially prepared fibrin sealants have been widely used in Europe since the 1970s. In 1972, a fibrin sealant containing concentrated fibrinogen was demonstrated to have utility for neural anastomoses [14]; by 1977, a two-suture microvascular anastomosis with fibrin glue was described [15]. However, these products were not available in the United States until more recently, due to concerns by the Food and Drug Administration (FDA) regarding viral transmission of HIV, hepatitis B, or hepatitis C from fibrin sealants prepared using pooled blood. Indeed, an increased risk of hepatitis B transmission had been previously demonstrated with fibrinogen prepared from pooled human plasma [16]. Since that time, viral elimination protocols have been implemented. On May 1, 1998, Tisseel® (Baxter Healthcare, Deerfield, IL) became the first fibrin sealant approved by the FDA for use in the United States.

17.3.2 Mechanism of Action

Fibrin-based tissue adhesives mimic the final steps of the physiological coagulation cascade (Figure 17.3) and produce a synthetic fibrin clot on the surface of wounded tissue. The key plasma proteins involved in the cascade are thrombin and fibrinogen. During physiological clotting, thrombin activates fibrinogen by cleaving fibrinopeptides A and B from fibrinogen. This cleavage step produces fibrin monomers, which then polymerize by hydrogen bonding and electrostatic interactions to form fibrin polymer, an unstable soft clot. Thrombin also activates clotting factor XIII to form factor XIIIa in the presence of calcium ions. Factor XIIIa catalyzes cross-linking of fibrin polymer to form a stable clot at the site of injury. Cross-linking occurs via the formation of amide links between glutamine and lysine residues in proteins [17–19]. Additional cross-linking occurs between fibrin and adhesive glycoproteins of the extracellular matrix, including collagen, fibronectin, and von Willebrand factor; fibrin also cross-links with cellular glycoproteins [20]. This covalent cross-linking helps to anchor the fibrin clot to the injury site.

Commercial fibrin-based adhesives are typically comprised of a two-component system, in which two solutions are mixed immediately before application to wounded tissue, to provide a controlled fibrin deposition. During mixing of the fibrin adhesive, a solution of thrombin and calcium is combined with a solution of fibrinogen and factor XIII to form a coagulum. The fibrinogen is at a much higher concentration than that in human plasma. On combining the two solutions, a reaction similar to that of the final stages of the blood clotting cascade occurs. In some preparations of fibrin

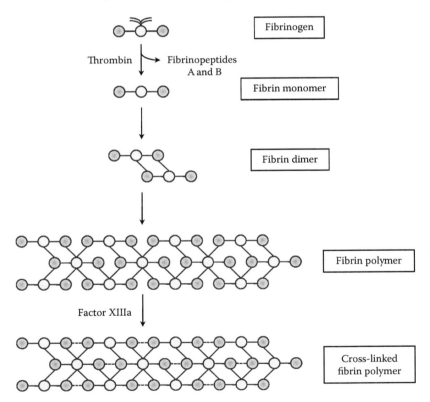

FIGURE 17.3 Mechanism of action of fibrin-based tissue adhesive. Thrombin cleaves fibrinopeptides A and B from fibrinogen to produce fibrin monomers. The monomers then polymerize to form fibrin polymer, an unstable soft clot. Thrombin-catalyzed fibrin polymerization is followed by Factor XIIIa-catalyzed fibrin cross-linking, resulting in the formation of a stable fibrin clot.

glue, the antifibrinolytic agent aprotinin is included in the tissue adhesive, presumably to prevent premature lysis of the clot. The exact composition of various commercial products differs and this alters the properties of the resulting fibrin clot [21]. Fibrinogen concentration contributes to the tensile strength of the fibrin glue, while thrombin concentration determines the curing time to achieve maximum adhesive strength. Tissue adhesives with high fibrinogen concentrations tend to produce stronger, but more slowly forming clots. On the other hand, those containing high thrombin concentrations clot rapidly, but the resulting clot is not as strong [22]. For example, a sealant formulation containing 4 IU/mL thrombin sets in approximately 30–60 s, while a sealant formulation containing 500 IU/mL thrombin sets in 10 s [23]. All fibrin glues produce clots that are biodegradable and bioabsorbable; the adherent fibrin clot on the tissue surface degrades naturally, within several days to weeks, by thrombolysis.

17.3.3 Clinical Applications

Fibrin-based tissue adhesives have been clinically utilized in a variety of organ systems and have been applied as hemostats, primary wound closure agents, and adjuncts to sutures and staples. A listing of commercially available fibrin tissue adhesives, along with their approved indications, is given in Table 17.3. Approved indications for fibrin adhesives include hemostasis in operations involving cardiopulmonary bypass, splenic trauma, and liver resection, as well as sealing of colonic anastomoses during the time of colostomy closure; in addition, one commercial fibrin sealant has gained approval for general hemostasis during surgery. However, fibrin sealants have found numerous uses in virtually every surgical discipline and have additionally been used off-label for drug delivery and tissue regeneration applications.

In vascular surgery, fibrin-based adhesives are an effective adjunct to sutures; fibrin adhesives prevent blood leakage from sewn vascular anastomoses and facilitate hemostasis in arteriovenous polytetrafluoroethylene (PTFE) vascular grafts [24,25]. In cardiac surgery, fibrin-based adhesives induce hemostasis during cardiopulmonary bypass procedures [26]; fibrin sealant is effective as a hemostatic agent to stop bleeding from adhesions at the time of reoperative cardiac surgery and from diffuse surfaces with capillary bleeding. Fibrin adhesives may also be used to seal sutured large vessel anastomoses, as well as woven Dacron cardiac prostheses [27]. Fibrin sealant additionally improves surgical results during acute aortic dissection [28]. The sealant is useful in both adult cardiac and pediatric cardiac surgical operations, including congenital heart surgery [29]. A fibrin sealant patch was reported to stop bleeding in the dramatic case of a free wall rupture of the left

TABLE 17.3
Commercially Available Fibrin-Based Tissue Adhesives

Commercial Product	Approved Indications	Constituents
Tisseel (Baxter Healthcare)	• Hemostasis during cardiopulmonary bypass surgery	Human fibrinogen
		Human thrombin
	• Treatment of splenic injuries	Human factor XIII
	• Adjunct in colostomy closure	Bovine aprotinin
Evicel® (J&J/Ethicon)	• General hemostasis in surgery	Human fibrinogen
		Human thrombin
CryoSeal® (Thermogenesis) Autologous fibrin sealant	• Hemostasis in liver resection surgery	Human fibrinogen
		Human fibronectin
		Human factor XIII
		Human factor VIII
		Human vWF
		Human thrombin

ventricle [30]. For thoracic surgery, there are reports of using fibrin glues to seal parenchymal air leaks to achieve pneumostasis and to close leaks of the bronchial tree [31].

In trauma settings, fibrin sealant has been used to control hemorrhage following solid organ injury in patients sustaining severe abdominal trauma [32] and is FDA-approved for repair of the liver and spleen. For gynecologic surgery, fibrin adhesives have been utilized to seal premature rupture of the membranes [33] and fibrin sealant has been used to enhance embryo transfer to improve adherence at the time of in vitro fertilization [34]. In gastrointestinal surgery, beyond its FDA-approved use for colonic anastomoses, fibrin adhesive has been applied to control bleeding from peptic ulcers [35] and for repair of anal fistulae [36]. Fibrin sealants have also been injected endoscopically in the gastrointestinal (GI) tract to control bleeding from gastric varices [37]. In orthopedic procedures, fibrin-based adhesives reduce bleeding during both hip replacement [38] and knee replacement [39].

Fibrin-based adhesives have found utility in plastic and reconstructive surgery, as hemostatic agents following burn debridement; adhesives for attachment of skin grafts; and hemostatics to reduce hematomas and bruising following facelift procedures [40,41]. In neurosurgery, fibrin sealants have been used for dural closure to prevent and treat cerebrospinal fluid leaks [42]. In the head and neck, fibrin-based adhesives reduce bleeding following tonsillectomy [43] and serve as a replacement for nasal packing to provide hemostasis in endonasal operations [44]. In ophthalmic surgery, fibrin-based adhesives have been employed for primary closure of conjunctival wounds to avoid suturing [45], as well as closure of corneal perforations and deep corneal ulcers [46].

Beyond their use in a variety of surgical applications, fibrin-based tissue adhesives have been demonstrated as carrier matrices for the controlled delivery of therapeutic drugs and biologics. One of the first drug delivery applications of fibrin glues was for antibiotics, to provide a local antimicrobial effect during wound healing [47,48]. Antibiotics that have been successfully delivered from fibrin glues include tetracycline [49], erythromycin, ciprofloxacin, vancomycin, gentamycin, cefazolin [50], ampicillin, carbenicillin, dibekacin, clindamycin, and cefotaxime. In addition, fibrin has been used for the delivery of chemotherapeutic agents such as doxorubicin [51] and cisplatin [52]; bone growth inducers such as bone morphogenetic protein and transforming growth factor-beta [53]; and nerve growth factors such as neurotrophin-4 to promote nerve regeneration [54].

17.3.4 LIMITATIONS AND COMPLICATIONS

Despite the versatility of fibrin-based tissue adhesives in surgical procedures, these adhesives are associated with important limitations. Although stringent precautionary measures are employed, fibrin glues containing blood components from pooled human plasma still pose a risk of viral transmission. There have been no known cases of hepatitis or HIV transmission with the use of commercial fibrin sealants. However, there are reports of transmission of symptomatic parvovirus B19 by fibrin sealant used during surgery [55,56].

An additional risk is raised by the formulation of fibrin glues containing bovine proteins; such foreign proteins can stimulate an immune response, with the resultant formation of antibodies to physiological clotting factors. Sensitization to bovine thrombin in fibrin sealants has been reported [57] and there are several cases of surgical patients developing antibodies to factor V and thrombin [58], following the application of fibrin glue containing bovine thrombin. The formation of inhibitor antibodies has been observed in peptic ulcer surgery [59], as well as adult and pediatric cardiac surgery [60]; this complication carries a bleeding risk and often requires plasmapheresis treatment. Serious clinical complications have also resulted from the use of currently approved fibrin glues containing bovine aprotinin. Patients who have been exposed to bovine aprotinin in fibrin sealant may develop antibodies to aprotinin and may exhibit an allergic or anaphylactic reaction upon reexposure to aprotinin. Cases of immediate allergic skin response [61] and severe anaphylaxis [62] have been reported as a result of bovine aprotinin in fibrin tissue adhesives. In one case, fatal intraoperative anaphylaxis occurred in response to local application of fibrin sealant [63]. The formation of aprotinin-specific antibodies following application of fibrin glues is not uncommon: in one study of children undergoing operations for congenital heart disease, 49% developed aprotinin-specific antibodies at 6 weeks after exposure to bovine aprotinin in fibrin glue, and 12% still had aprotinin-specific antibodies at 1 year after exposure to bovine aprotinin in fibrin glue [64].

The prevalence and seriousness of such adverse effects highlight the drawbacks of using human and bovine proteins in tissue adhesives. Moreover, a practical limitation of fibrin-based tissue adhesives is that they require approximately 20 min of preparation time [65], so the time of use must be correctly anticipated. This also places a constraint on the utility of fibrin sealants in bleeding emergencies. Finally, fibrin tissue adhesives demonstrate relatively poor adhesion to tissue in comparison to synthetic cyanoacrylate adhesives and gelatin–resorcinol–formaldehyde (GRF)/glutaraldehyde tissue adhesives [66]. For these reasons, there is clinical interest in developing both alternative biological sealants and synthetic sealants for tissue reconstruction.

17.4 CYANOACRYLATE-BASED TISSUE ADHESIVES

17.4.1 HISTORICAL DEVELOPMENT

Cyanoacrylate tissue adhesives are currently the principal synthetic polymer sealants in clinical usage. The preparation of alkyl alpha-cyanoacrylates was first described in 1949 for the production of "hard, clear, glass-like resins" [67,68] and the adhesive properties of alpha-cyanoacrylates were discovered in the 1950s [69]. Cyanoacrylates subsequently achieved widespread availability and household usage as "superglues," due to their fast-setting properties. The cyanoacrylate adhesive family has also found numerous applications in the automotive and construction industries. Cyanoacrylates were first used for medical applications in 1960, when the use of cyanoacrylate adhesives for small vessel surgery was first described [70]. In 1962, clinical applications of cyanoacrylate glues in massive liver resection surgery [71] and bronchial closure [72] were reported. In 1963, cyanoacrylates were utilized for nephrotomy closure [73], GI surgery [74], and ophthalmic closure [75,76]. Cyanoacrylate tissue adhesives have been commercially available since the 1980s in Europe and Canada, but were initially restricted from the medical market in the United States due to concerns over histotoxicity and carcinogenicity. In 1998, the FDA approved Dermabond® (J&J/Ethicon, Somerville, New Jersey) as the first synthetic cyanoacrylate tissue adhesive product for the U.S. market.

17.4.2 MECHANISM OF ACTION

Cyanoacrylates are distinct among adhesives in that they are single-component systems that polymerize at room temperature without the addition of a catalyst, evaporation of a solvent, heat, or pressure [77]. These adhesives require no external initiation for curing; cyanoacrylates can rely on small amounts of water to initiate the polymerization reaction and bonding can occur within seconds. The uniqueness of cyanoacrylate glues derives from the extreme reactivity of the cyanoacrylate monomer (Figure 17.4). The electron-withdrawing capacity of the nitrile and the alkoxylcarbonyl groups of the monomer results in the double bond being very polarized and amenable to nucleophilic attack. This permits a weak base, such as the water from tissue fluid, to initiate and complete the anionic polymerization process quickly. The polymerization reaction is exothermic and the rate of polymerization is inversely proportional to the amount of monomer. The reaction rate is also inversely proportional to the length of the alkyl side chain.

The basic cyanoacrylate monomer is a low-viscosity liquid. On contact with moist tissue, the cyanoacrylate polymerizes into a solid film that bridges apposed wound edges. Cyanoacrylate glues

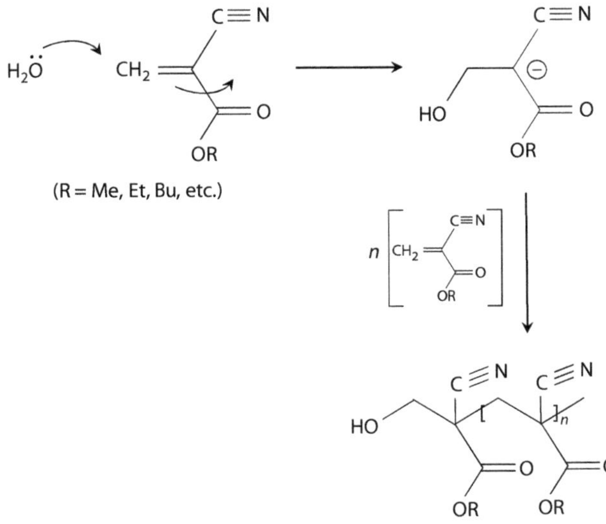

FIGURE 17.4 Mechanism of anionic polymerization of cyanoacrylate tissue adhesive. The ethylene bond of the cyanoacrylate monomer is polarized due to the electron-withdrawing capacity of the nitrile and alkoxylcarbonyl groups. A weak base, such as water from tissue fluid, can initiate and complete the anionic polymerization process.

achieve adhesion through two independent mechanisms: molecular interaction via covalent bonding to the tissue surface and mechanical interlocking of the poly(cyanoacrylate) with underlying tissues. Cyanoacrylates form primary chemical bonds with the tissue surface, by covalently bonding to functional groups in proteins, in particular amine groups. The adhesive creates physical or mechanical bonds by penetration of cyanoacrylate monomers into cracks and channels in the tissue surface. Strong mechanical interlocks form as the polymerization process commences and the adhesive sets; this results in strong mechanical bonds between closely approximated tissues. The combination of chemical and mechanical bonding establishes the bond strength of cyanoacrylate tissue adhesives. In general, the strength and physical properties of cyanoacrylate adhesives are determined by the length and complexity of the alkyl side chain. Short, straight-chain derivatives (such as butyl-cyanoacrylate) form tighter and stronger bonds in comparison to complex, long-chain derivatives (such as octyl-cyanoacrylate). The tight bonds formed by shorter-chain derivatives tend to be brittle, and fracture prematurely when used as a topical bridge; this results in lower tensile strength of the polymerized short-chain derivatives than that of long-chain derivatives [78]. Because of the brittle nature and lower bursting strength failure of shorter-chain butyl-cyanoacrylate glues, when device failure occurs resulting in wound dehiscence, it tends to be a cohesive failure such that the adhesive breaks in the middle. In contrast, when device failure occurs with the longer-chain octyl-cyanoacrylate glues, it tends to be an interfacial failure such that the adhesive peels away from the tissue surface [79].

Cyanoacrylate glues undergo hydrolytic degradation which takes place through nonenzymatic reactions; the main degradation products are formaldehyde and the corresponding alkyl cyanoacetate (Figure 17.5). Degradation takes place by breakdown of the polymer backbone and occurs because the methylene hydrogen in the polymer is highly activated inductively by the electron-withdrawing neighboring groups. The degradation rate of cyanoacrylate polymers decreases with an increase in length of the alkyl side chain, as a result of steric hindrance [80]. An alternative degradation mechanism has also been proposed [81], in which the cyanoacrylate polymer degrades by hydrolysis of the ester group to produce cyanoacrylic acid and alcohol (Figure 17.6). This reaction may occur in the physiological environment and may be catalyzed by enzymatic activity.

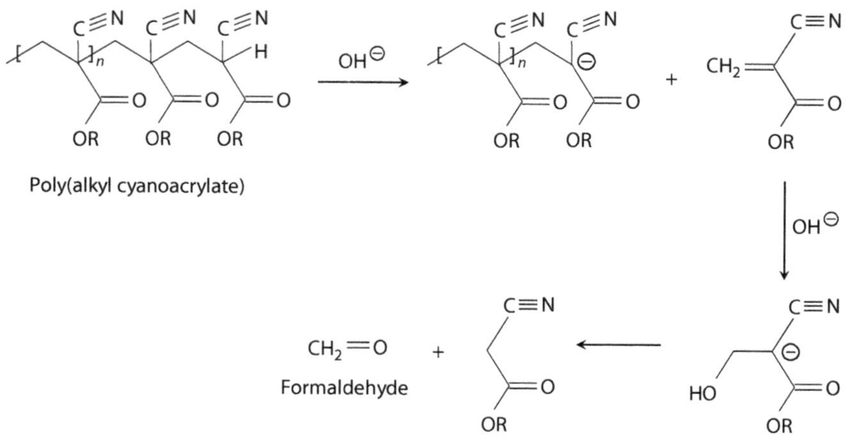

FIGURE 17.5 Mechanism of hydrolytic degradation of cyanoacrylate polymer. Water associated with tissue can induce cyanoacrylate hydrolysis, releasing formaldehyde and alkyl cyanoacetate as degradation products.

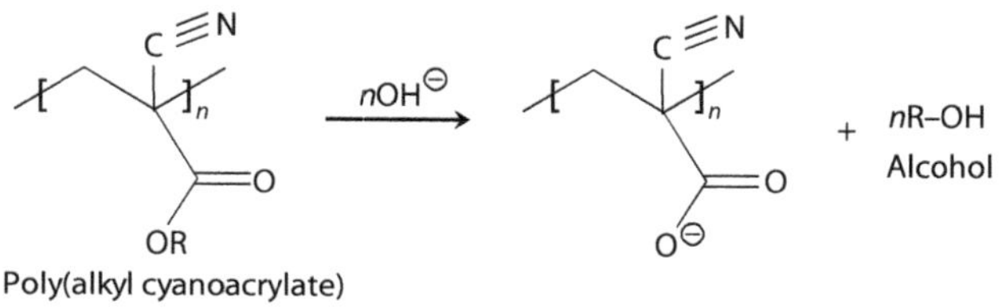

FIGURE 17.6 Alternative mechanism of degradation of cyanoacrylate polymer. The ester group is hydrolyzed to produce cyanoacrylic acid and alcohol.

17.4.3 CLINICAL APPLICATIONS

A listing of commercially available cyanoacrylate tissue adhesives, along with their approved indications, is given in Table 17.4. Cyanoacrylates are approved for external closure of topical skin incisions and trauma-induced skin lacerations; the commercial cyanoacrylate glue Omnex® (J&J/Ethicon, Somerville, New Jersey) is approved for internal hemostasis during vascular reconstruction surgery. When used for skin closure, the cyanoacrylates are recommended to be used with deep dermal sutures, because dehiscence can occur if the cyanoacrylates are the only means of skin attachment. In addition to their surgical adhesive indications, the 2-octyl-cyanoacrylate glues Dermabond and Indermil® (Covidien, Mansfield, Massachusetts) are approved for use as barriers against common bacterial microbes including certain staphylococci, pseudomonads, and *Escherichia coli*. However, cyanoacrylate-based tissue adhesives have found numerous off-label uses in a variety of surgical disciplines.

The chief clinical application of cyanoacrylates is for external skin closure; cyanoacrylates have played an important role in enabling fast, painless, and cosmetically acceptable closure of topical wounds. Because the polymerized cyanoacrylate is waterproof, it requires no dressing. When used as a skin adhesive, the cyanoacrylate remains in place for approximately 7–10 days, and then slowly sloughs off as skin cells regenerate and the natural healing process proceeds. Multiple clinical trials have confirmed the efficacy of cyanoacrylate glues in the management of lacerations and incisions [82,83]. A large, multicenter randomized controlled trial compared over 900 lacerations and surgical incisions closed with octyl-cyanoacrylate or a standard closure device (mostly sutures), and demonstrated that wound closure was faster using the cyanoacrylate adhesive, while the rates of wound infection, wound dehiscence, and optimal cosmetic results at 3 months were comparable [84]. In a clinical study of cyanoacrylate adhesives for topical closure in pediatric patients,

TABLE 17.4

Commercially Available Cyanoacrylate-Based Tissue Adhesives

Commercial Product	Approved Indications	Constituents
Indermil® (Covidien)	• Closure of topical skin incisions and trauma-induced skin lacerations • Additional approved claim as barrier to microbial penetration	n-Butyl-2-cyanoacrylate
Histoacryl® and Histoacryl® Blue (Tissueseal)	• Closure of topical skin incisions and trauma-induced skin lacerations	n-Butyl-2-cyanoacrylate
Dermabond (J&J/Ethicon)	• Closure of topical skin incisions and trauma-induced skin lacerations • Additional approved claim as barrier to microbial penetration	2-Octyl-cyanoacrylate
Omnex (J&J/Ethicon)	• Hemostasis for vascular reconstruction	2-Octyl-cyanoacrylate/butyl lactoyl cyanoacrylate

butyl-cyanoacrylate glues were utilized for the repair of scalp, face, and limb lacerations in over 1500 children, with excellent cosmetic results [85].

Cyanoacrylate-based tissue adhesives have also been used successfully for skin closure in hand surgery [86], and topical closure in head and neck surgery [87]. In craniofacial surgery, cyanoacrylates have been successful in external lip closure for sutureless repair of congenital cleft lip [88]. Cyanoacrylates have been used in neurosurgery for external wound closure in lumbar and cervical procedures [89], as well as scalp closure following craniotomies and shunt insertions [90]. In plastic and reconstructive surgery, cyanoacrylate glues have been applied topically during body contouring surgery [91], breast reduction surgery [92], and facial cosmetic surgery [93] including eyelid lifts [94]. In general surgery, cyanoacrylates have been demonstrated for topical closure of laparoscopic cholecystectomy incisions [95] and inguinal herniorrhaphy incisions [96]; in obstetrics, cyanoacrylates have shown efficacy for external repair of perineal lacerations and episiotomy incisions [97,98]. Cyanoacrylate adhesives have even found use in interventional cardiology, where the glues have been used to close wounds following cardiac device implantation [99], as well as cardiac surgery, where the glues demonstrate microbial barrier effects during skin closure of sternal incisions [100].

In vascular surgery, the cyanoacrylate glue Omnex has been utilized for its approved indication to control bleeding during vascular reconstruction. This cyanoacrylate-based adhesive has been demonstrated to facilitate hemostasis of polytetrafluoroethylene (PTFE) vascular grafts, Dacron vascular grafts, and autologous vascular grafts [101]. The efficacy of cyanoacrylate glues for vascular reconstruction has been reported for both arteriovenous shunts and femoral bypass grafts [102,103]. It should be noted that the n-butyl-cyanoacrylate product Trufill® n-BCA (J&J/Ethicon, Somerville, New Jersey) is also approved as a vascular occlusive agent for embolization of cerebral arteriovenous malformations, although this is not strictly a tissue reconstruction application.

Beyond their approved indications for external skin closure and vascular anastomotic closure, cyanoacrylate-based tissue adhesives have also been applied off-label in a variety of surgical fields. Cyanoacrylates have found utility in ophthalmology for treatment of corneal perforations [104] sealing of corneal incisions [105], and closure of scleral tunnel incisions [106]. Cyanoacrylate glues have been utilized in hernia repair surgery to secure polypropylene hernia meshes [107]. In gastroenterology, cyanoacrylates have been injected endoscopically to control bleeding from gastric varices [108]. The cyanoacrylate glues have been employed in urology to seal nephrostomy tube sites [109] and treat urinary fistulas [110]. In thoracic surgery, cyanoacrylate adhesives have been used in endoscopic repair of bronchial dehiscence following lung transplantation surgery [111]. In orthopedic surgery, there are case reports of cyanoacrylate glues for the fixation of osteochondral fractures of the knee [112], as well as fixation of talar osteochondral fractures [113]. Bone repair with cyanoacrylates has also been accomplished in head and neck surgery, where the glues have been delivered endoscopically for fixation of zygomatic fractures [114]. Cyanoacrylate glues have additionally been employed in tympanoplasty and myringoplasty surgeries for repair of the tympanic membrane [115].

17.4.4 LIMITATIONS AND COMPLICATIONS

Although cyanoacrylate-based tissue adhesives are unique single-component systems that are efficiently applied in a wide range of surgical procedures, these glues have significant and serious drawbacks because of limited biocompatibility. Cyanoacrylates are cytotoxic to connective tissue fibroblast cells [116] and cultured tendon cells [117]. In experimental implantation models, cyanoacrylates have been shown to induce acute inflammation and a prolonged foreign body giant cell response when applied subcutaneously [118]. When used in experimental ophthalmic surgery for scleral tunnel incisions, cyanoacrylate glues provoke a severe inflammatory response that inhibits collagen remodeling, and therefore interferes with wound healing [106]. In experimental models of nerve anastomosis, cyanoacrylates cause a foreign body inflammatory reaction and retractile fibrosis, reducing the nerve diameter by up to two-thirds [119]. In clinical neurosurgery, there are case reports of cyanoacrylates causing late complications following application to the frontal and lateral base of the skull; these serious complications include therapy-resistant fistulas, chronic sinusitis, and otogenic meningitis after n-butyl-2-cyanoacrylate glues were used for dural repair [120].

Both n-butyl-2-cyanoacrylate and 2-octyl-cyanoacrylate induce harmful histological changes in pancreatic tissue [121]; there is one clinical case report of an inflammatory tumor of the pancreas, induced by endoscopic injection of cyanoacrylate glue for gastric varices [122]. Necrosis associated with cyanoacrylate adhesives has been shown to contribute to thrombotic reactions [123]. Moreover, cyanoacrylates have been noted to inhibit new bone formation, cause a foreign body reaction, and impede fracture healing in experimental bone repair [124]. These effects raise questions regarding the routine use of cyanoacrylate adhesives for internal tissue bonding. Indeed, concerns have recently been raised regarding the hazards of cyanoacrylate use to both patients and healthcare workers; reported toxic effects of cyanoacrylates in the workplace include dermatologic, allergic, and respiratory conditions [125].

The cytotoxic, histotoxic, and inflammatory effects of cyanoacrylates may be due to the release of their degradation products, including formaldehyde. The production of formaldehyde inside the body represents a major issue regarding the suitability of cyanoacrylates for internal surgical wound closure. Formaldehyde has been classified as a human carcinogen by the International Agency for Research on Cancer, a panel of the World Health Organization. It is classified as a probable human carcinogen by the U.S. Environmental Protection Agency [126]. One experimental study in laboratory animals has shown that n-butyl-2-cyanoacrylate tissue adhesives induce the formation of cancerous soft tissue sarcomas [127]; this raises further questions about the potential carcinogenicity of cyanoacrylate glues. Given these considerations, it is prudent to limit the use of cyanoacrylates to their approved indications: external closure of topical skin wounds and hemostasis during vascular reconstruction.

Beyond their biological limitations, cyanoacrylates are also characterized by shortcomings in mechanical performance. After polymerizing, cyanoacrylate adhesives become brittle and are subject to fracturing when used in skin creases or long incisions. Cyanoacrylates are not as strong as 3/0 and some 4/0 sutures, so there is an increased risk of wound dehiscence when cyanoacrylates are used alone to close high-tension wounds [128]. A study comparing wound closure using n-butyl-2-cyanoacrylate versus sutures in children with groin incisions suggests that closure of these high-tension wounds with cyanoacrylates may result in a higher rate of wound dehiscence after repair [129]; dehiscence of the wound occurred in 26% of the cyanoacrylate-treated group and 0% of the suture-treated group. Cyanoacrylates are most appropriate for external closure of low-tension surgical incisions and traumatic lacerations whose edges are easily approximated.

17.5 CROSS-LINKED PROTEIN-BASED TISSUE ADHESIVES

Tissue adhesives based on cross-linked proteins were originally developed as less toxic alternatives to early cyanoacrylate glues. Cross-linked protein-based adhesives are typically composed of both natural and synthetic materials, and are fabricated by combining a natural protein with chemical cross-linking agents. The two major cross-linked protein-based sealants utilized in clinical medicine are gelatin–resorcinol–formaldehyde glues and albumin–glutaraldehyde glues; both of these adhesive platforms will be reviewed later. In general, the cross-linked protein-based tissue adhesives are characterized by very strong tissue adhesion but suboptimal biocompatibility. These adhesives have found specific utility in emergency cardiothoracic surgery, particularly for aortic dissections, by virtue of their superior adhesive properties. However, their toxicity profiles have limited these adhesives from achieving wider approval or usage.

17.5.1 Gelatin–Resorcinol–Formaldehyde Tissue Adhesives

The use of cross-linked gelatin as a tissue adhesive was first described in the 1960s by two different groups [130,131]. A gelatin–resorcinol mixture cross-linked with formaldehyde was chosen because of its high bond strength even in the presence of moisture. In 1966, it was reported that a gelatin–resorcinol mixture cross-linked with either formaldehyde (GRF glue) or a combination of formaldehyde and glutaraldehyde (GRFG glue) controlled hemorrhage from the liver and kidneys in an experimental model [132]. The GRF and GRFG glues were also demonstrated to give satisfactory tensile bond strengths in hepatic and renal tissues in an animal model, and favorable hemostatic effects in the aorta, atria, ventricles, and lungs [133]. In 1967, cross-linked gelatin tissue adhesives were successfully used in experimental GI surgery [134] and urinary tract surgery [135]. In 1968, gelatin cross-linked with formaldehyde was investigated as a potential tissue adhesive, and it was found that addition of resorcinol to the gelatin–formaldehyde mixture was necessary to improve bond strength [136]. GRF tissue adhesives were first utilized to treat aortic dissections in the late 1970s [137]. Although GRF tissue adhesives are now widely used in Japan and Europe to treat aortic dissections, these glues are not currently commercially available in the United States, due to concerns about possible toxicity and carcinogenicity of formaldehyde.

Curing of GRF tissue adhesives occurs by condensation reactions of formaldehyde with gelatin and resorcinol (Figures 17.7 and 17.8); condensation of formaldehyde with resorcinol yields a three-dimensional cross-linked resin. In use, the GRF glue is prepared by warming a 3:1 mixture of gelatin and resorcinol to 45°C, and applying the warmed mixture to the operative site; an 18%

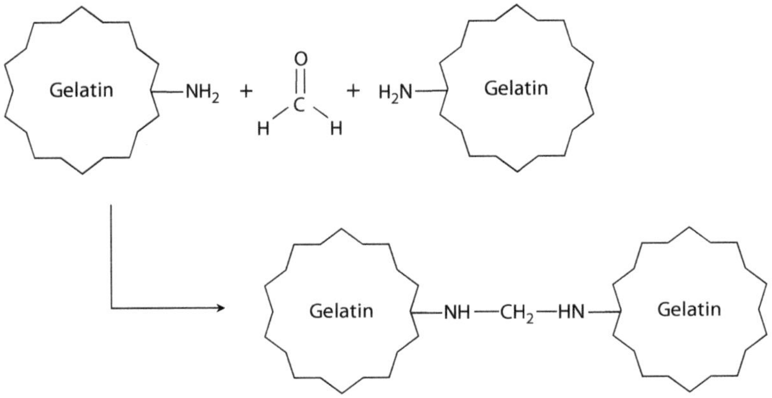

Cross-linked gelatin

FIGURE 17.7 Mechanism for curing of gelatin–resorcinol–formaldehyde tissue adhesives. Formaldehyde undergoes a condensation reaction with amine groups in gelatin protein to yield cross-linked gelatin.

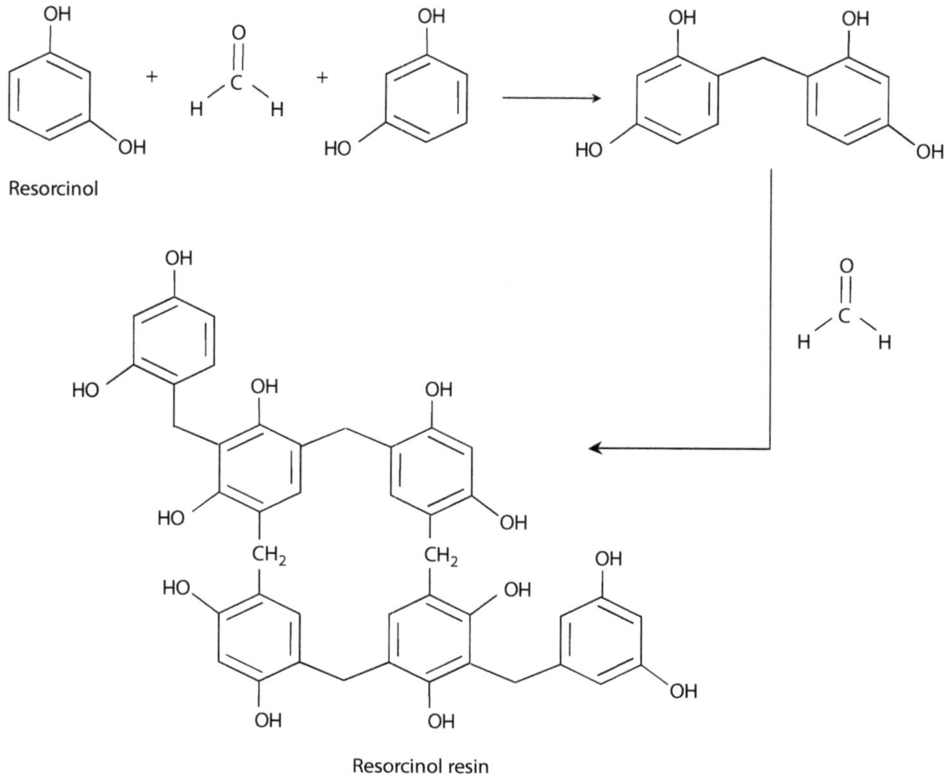

FIGURE 17.8 Additional mechanism for curing of gelatin–resorcinol–formaldehyde tissue adhesives. Formaldehyde undergoes a condensation reaction with resorcinol to yield a three-dimensional cross-linked resorcinol resin.

formaldehyde solution is then added to polymerize the tissue adhesive. Cross-linking of the GRF glue takes place in approximately 30 s. The curing profile of GRF adhesives can be altered by adjusting the ratios of the ingredients. The GRF glues bond well to wet tissues and form covalent linkages with functional groups on the tissue surface. Bonding strength may be enhanced by the penetration of ingredients into the tissue, as well as the hydrophobicity of the resultant resin.

Currently, the principal clinical application of GRF tissue adhesive is for treatment of acute aortic dissections [138]; several clinicians have noted the utility of GRF glue in difficult cases [139,140]. Because of its tissue bonding strength and cross-linked resin structure, the cured GRF polymer is exceptionally useful in reinforcing and reattaching the delicate structures of the dissected aortic wall. The GRF adhesive can be used to eliminate the abnormal dissection plane of the aortic vessel wall, as it glues together the layers of the aorta and strengthens the vessel wall to hold sutures more effectively. GRF glues have also been applied to remedy complications from aortic prostheses. In one clinical report, two patients presented with infected aortic bioprostheses complicated by annular abscesses; in each case, the aortic valve was replaced with a bioprosthesis and the annular abscesses were debrided and closed with the GRF glue, which completely sealed the abscess cavities [141]. GRF glues have additionally been employed in cardiac surgery for sutureless repair of left ventricular wall rupture [142,143] and ventricular septal defect [144]. In thoracic surgery, GRF adhesives have been applied following lung surgery to prevent air leakage [145]; there is also one case report of successful closure of a bronchopleural fistula with GRF glue [146]. In vascular surgery, GRF sealants have been demonstrated to facilitate hemostasis of ePTFE patch suture lines [147].

Despite its powerful ability to bind fragile tissue surfaces, GRF glue can cause substantial long-term damage and destruction to tissues. GRF adhesives are toxic to cardiovascular tissue, and devastating complications have occurred following the use of GRF glues in acute aortic dissection. Reported adverse outcomes include aortic root necrosis, aortic root redissection, pseudoaneurysm formation, and aortic regurgitation and insufficiency [148,149]. In one clinical case, application of GRF glue for aortic dissection resulted in severe coronary artery narrowing which required bypass surgery [150]. Given such potential outcomes, patients who receive GRF glues during surgery require careful long-term follow-up. Surgical application of GRF glues should be limited to special cases in which tissue integrity is poor, hemostasis is challenging, and high bonding strength is absolutely imperative.

A modified gelatin–resorcinol–aldehyde formulation has been developed to avoid the toxicity and potential carcinogenicity of formaldehyde [151,152]. In this modified tissue-adhesive, known as GR-DIAL, the formaldehyde component is replaced with two dialdehydes: glutaraldehyde (pentane-1,5-dial) and glyoxal (ethanedial). The GR-DIAL glue is commercially available in Europe under the trade name Gluetiss® (Geister Medizintechnik GmbH, Germany), and is indicated for treatment of aortic dissection.

17.5.2 ALBUMIN–GLUTARALDEHYDE TISSUE ADHESIVES

Albumin–glutaraldehyde sealants derive conceptually from GRF glues, as the albumin–glutaraldehyde agent not only serves as a tissue sealant, but also acts to strengthen friable tissues [153]. Glutaraldehyde-cross-linked albumin was investigated as a sealant for polyester vascular prostheses in the 1990s [154]; commercial albumin–glutaraldehyde tissue adhesives were introduced shortly thereafter. In 1997, the surgical adhesive BioGlue® (Cryolife, Atlanta, Georgia) received FDA approval in the United States under a humanitarian device exemption for use in acute aortic dissection. In 2001, BioGlue received FDA approval for general use as a hemostatic adjunct in cardiac and vascular surgery. The albumin–glutaraldehyde sealant has broader indications in Europe, whereas BioGlue received approval for vascular surgery in 1998, for pulmonary surgery in 1999, and for general surgical procedures in 2002. Under the CE mark, indicated soft tissues for BioGlue are cardiac, vascular, pulmonary, genitourinary, dural, alimentary (esophageal, gastrointestinal, and colorectal), and other abdominal tissues (pancreatic, splenic, hepatic, biliary). Additionally, BioGlue is allowed in Europe for the fixation of surgical meshes in hernia repair. In 2008, the albumin–glutaraldehyde glue received European approval for periosteal fixation following endoscopic browplasty or brow lift, a reconstructive plastic surgery procedure.

Curing of albumin–glutaraldehyde tissue adhesives occurs by condensation reactions between albumin and glutaraldehyde (Figure 17.9). Glutaraldehyde reacts with amine groups in the albumin protein, particularly amine groups within lysine amino acid residues, yielding a cross-linked albumin network. In practice, the adhesive is a two-component system, comprised of stoichiometrically equivalent doses of 45% bovine serum albumin and 10% glutaraldehyde (typically delivered in a 4:1 volume ratio). The two solutions are mixed as they are delivered, combining to form an adhesive polymer on the target tissue; no advance preparation is required. The glue begins to polymerize within 20 to 30 s, and reaches its bonding strength within 2 min. Albumin–glutaraldehyde tissue adhesives form covalent linkages with functional groups on the tissue surface. These glues also adhere to synthetic graft materials through mechanical bonding within the interstices of the graft matrix. Albumin–glutaraldehyde glues degrade very slowly, and one clinical report has demonstrated that the albumin–glutaraldehyde polymer persists at the repair site for up to 2 years after application [155].

Albumin–glutaraldehyde adhesives are primarily utilized in the clinic for repair of acute aortic dissection [156]. Like GRF glue, albumin–glutaraldehyde glue reinforces flimsy tissue into a tougher and more workable consistency; the treated tissue is firm and easy to suture. Albumin–glutaraldehyde tissue adhesives have also been used for a wide array of cardiac surgical procedures,

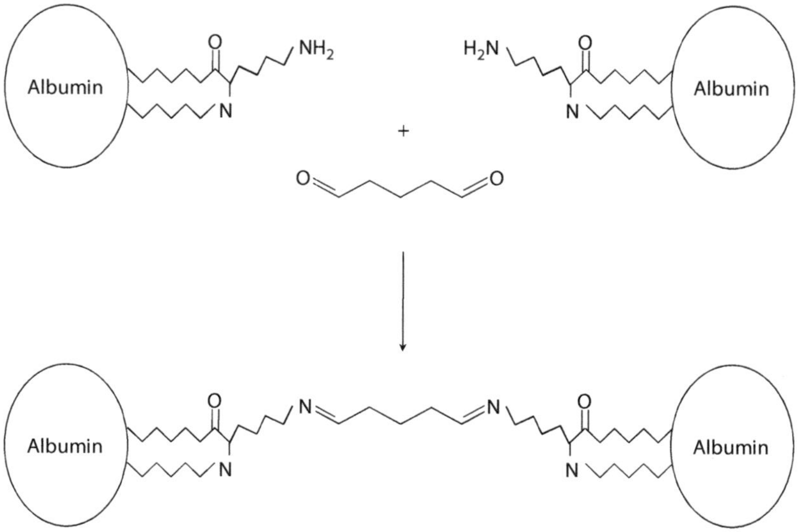

FIGURE 17.9 Mechanism for curing of albumin–glutaraldehyde tissue adhesives. Glutaraldehyde undergoes a condensation reaction with amine groups in albumin protein, particularly amines in lysine amino acid residues. The result is a cross-linked albumin network.

including coronary artery bypass grafting, valve procedures, ventricular aneurysm repair, closure of ventricular septal defect, and correction of congenital conditions [157]. In addition, the albumin–glutaraldehyde glues have been employed as hemostatic and structural adjuncts to reduce bleeding from cardiac and vascular anastomoses [158], and have been applied to attach Teflon patches during sutureless repair of ventricular free wall rupture [159]. Albumin–glutaraldehyde sealants have additionally found utility for hemostasis during left ventricular assist device implantation [160]. In thoracic surgery, albumin–glutaraldehyde adhesives have been used to seal lung lacerations, close bronchopleural fistulas, prevent lymph leakage, and prevent air leakage from suture or staple lines on pulmonary parenchyma [161,162].

In the abdominal cavity, albumin–glutaraldehyde sealant has been applied to seal the kidney [163] and spleen [164], and has been utilized during minimally invasive surgery to remove ovarian cysts [165]. In the lower GI tract, albumin–glutaraldehyde glue has been shown to reduce leakage from stapled anastomoses during hemorrhoid surgery [166], and has additionally been used to close anal fistulas [167]. In neurosurgery, albumin–glutaraldehyde sealant has been demonstrated to be effective in reducing cerebrospinal fluid leakage from dural defects and diaphragmatic defects [168]; the sealant has been used off-label for fixation of deep brain stimulation electrodes [169]. Albumin–glutaraldehyde glue has also been employed for tissue closure during middle ear surgery [170] and nasal surgery [171]. Finally, albumin–glutaraldehyde adhesives have been applied in facial plastic surgery for brow fixation following endoscopic brow lift procedures [172].

While albumin–glutaraldehyde tissue adhesives share many of the functional benefits of GRF tissue adhesives, they also share many of the same limitations. Although albumin–glutaraldehyde adhesives avoid the use of formaldehyde, significant toxicity can still result from the glutaraldehyde component of these sealants. Polymerized albumin/glutaraldhyde glue releases levels of glutaraldehyde that are capable of inducing cytotoxic effects in vitro to cultured human embryo fibroblasts and mouse myoblasts. The glue also releases amounts of glutaraldehyde sufficient to induce adverse effects in vivo, including high-grade inflammation, edema, and toxic necrosis in lung and liver tissue, as well as medium-grade inflammation in aortic tissue [173]. There are several case reports of late tissue toxicity and tissue scarring in patients who received albumin–glutaraldehyde glue; dense fibrosis and significant acute inflammation with foreign body giant cells were observed in these cases [174,175].

Use of albumin–glutaraldehyde tissue adhesives for aortic dissection repair is associated with a risk of aortic wall necrosis, leading to aortic root redissection [176]. Albumin–glutaraldehyde glue impairs aortic vascular tissue growth, and causes stricture when applied circumferentially around aortic anastomoses; for this reason, albumin–glutaraldehyde glue is contraindicated for cardiovascular anastomoses in pediatric patients [177]. Stenosis of the superior vena cava has additionally been reported following nearby application of albumin–glutaraldehyde glue [178]. Other reported adverse outcomes of albumin–glutaraldehyde adhesive application in cardiothoracic procedures include mediastinal cyst formation [179], lung fibrosis [180], and aortic pseudoaneurysm formation [181].

Migration and embolization of the albumin–glutaraldehyde polymer can also cause serious complications. Albumin–glutaraldehyde glue can leak through suture holes in aortic tissue and vascular grafts [182], and migrate to block distal blood vessels. In one clinical report, embolization of the glue led to blockage of the right and left coronary arteries, resulting in a fatal right ventricular heart attack [183]. In another report, two patients experienced acute loss of blood perfusion to the limbs, due to embolization of albumin–glutaraldehyde polymer after aortic dissection repair [184]. One patient required surgery to restore blood flow to the leg, and the other patient required surgery to restore blood flow to the arm. In addition, migration of the glue has resulted in cardiac valve malfunction. In one case report, one patient experienced blockage of a mechanical mitral valve leaflet following albumin–glutaraldehyde glue application [185], and in another case, one patient experienced malfunction of a prosthetic aortic valve due to albumin–glutaraldhehyde adhesive [186].

Moreover, glutaraldehyde may exhibit nerve toxicity, and adverse reactions have been noted following the use of albumin–glutaraldehyde sealant in neurosurgical procedures, as well as in the vicinity of peripheral nerves. Albumin–glutaraldehyde glue has been demonstrated in vivo to cause acute phrenic nerve injury leading to diaphragmatic paralysis, along with coagulation necrosis leading to cardiac conduction tissue damage [187]. A clinical study of 75 pediatric neurosurgical patients found a strong association between the use of albumin–glutaraldehyde adhesive and postoperative wound complications; a 10-fold increase in complications prompted the investigators to stop the use of albumin–glutaraldehyde glue in pediatric neurosurgery [188]. Observed reactions included both inflammatory and infectious complications, and the investigators proposed that the intense inflammatory response triggered by albumin–glutaraldehyde glue creates an ideal environment for bacterial growth. Indeed, albumin–glutaraldehyde sealant has even been associated with an increased risk of sepsis following treatment of anal fistulas [189]. Finally, the bovine albumin component of albumin–glutaraldehyde glue could act as an immunogen and sensitize patients to bovine products [190]. Taken together, these reports suggest that routine use of albumin–glutaraldehyde tissue adhesives is unwise. Tissue necrosis, neurotoxicity, infectious complications, and dense inflammatory adhesions are possible long-term outcomes that must be weighed against short-term benefits. Like GRF glues, albumin–glutaraldehyde glues should be limited to clinical situations in which strong tissue bonding is absolutely necessary, and no satisfactory alternatives exist.

17.6 POLYETHYLENE GLYCOL-BASED TISSUE ADHESIVES

New tissue adhesives based on polyethylene glycol (PEG) have been developed to provide highly biocompatible, bioresorbable, synthetic hydrogels for wound closure. PEG hydrogel-based tissue adhesives are advantageous relative to other synthetically derived sealants, including cyanoacrylates, GRF glues, and albumin–glutaraldehyde glues, because PEG is much more compatible with biological tissues, and PEG hydrogels are readily degraded by hydrolysis. In addition, synthetic PEG-based adhesives avoid the risks associated with biological fibrin glues, including viral transmission and sensitization; moreover, PEG-based sealants circumvent the potential immunogenicity of bovine albumin–glutaraldehyde glues. PEG polymers have a long history of clinical use as drug delivery agents for therapeutic proteins [191]. PEG conjugation reduces the immunogenicity of proteins and imparts "stealth" properties; PEG-modified proteins are nonimmunogenic, even with repeated infusions [192,193]. PEG polymers are thus well-known in clinical medicine as biocompatible materials. Three major PEG-based tissue adhesives that have been designed for clinical medicine are photopolymerizable PEG sealants, PEG–PEG sealants, and PEG–trilysine sealants; all three of these adhesive platforms will be reviewed later. In general, the PEG-based sealants are characterized by excellent biocompatibility, but are high swelling with very fast degradation profiles that limit their functionality in wound reinforcement.

17.6.1 Photopolymerizable PEG Tissue Adhesives

Hydrogel adhesives based on water-soluble, photopolymerizable macromers were developed in the early 1990s; these bioresorbable hydrogels are formed by photopolymerization of PEG-*co*-poly(α-hydroxy acid) diacrylate macromers [194]. In 1994, photopolymerized PEG sealants were reported to prevent postsurgical adhesion formation and allow intra-abdominal healing in experimental models [195,196]. In 1995, photopolymerizable PEG adhesives were demonstrated to seal human blood vessel anastomoses without inducing thrombogenicity [197]. By 1997, the photopolymerized hydrogel adhesives were found to be effective for sealing bronchial and parenchymal air leaks in experimental lung surgery [198]. On May 26, 2000, the FDA approved the commercial sealant FocalSeal®-L (Focal Incorporated, Lexington, Massachusetts) for sealing air leaks on the lungs following surgical removal of cancerous lung tumors. The photopolymerizable tissue adhesive also received CE Mark approval for sealing air leaks following lung surgery.

Each macromer of the commercial photopolymerizable adhesive consists of PEG modified with biodegradable and photoreactive elements. In each macromer, PEG is linked on both ends to hydrolyzable trimethylene carbonate or lactate oligomeric segments, and then end-capped with polymerizable acrylate groups (Figure 17.10). The macromers are amphiphilic in nature, with hydrophobic end regions on the central PEG chain, and form micellar structures in aqueous solution. The formation of such pre-organized configurations in aqueous solutions enables the macromers to undergo rapid photopolymerization and gelation. During clinical use, the macromers are applied to the target tissue in two parts: a primer solution to provide tissue bonding and a sealant solution to provide desired mechanical properties. Both components are introduced as aqueous solutions to the target site. The primer layer is first brushed onto the tissue surface to allow the low-viscosity solution to flow into tissue interstices. The sealant solution is then mixed with the primer solution using a brush

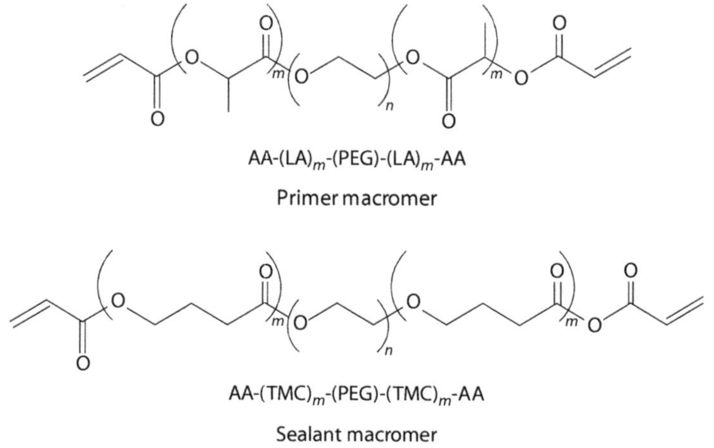

AA-(LA)$_m$-(PEG)-(LA)$_m$-AA

Primer macromer

AA-(TMC)$_m$-(PEG)-(TMC)$_m$-AA

Sealant macromer

FIGURE 17.10 Chemical structures of primer and sealant macromers used in photopolymerizable PEG tissue adhesives. In each macromer, polyethylene glycol (PEG) is linked on each end to either trimethylene carbonate (TMC) or lactate (LA)-hydolyzable segments, and end-capped with polymerizable acrylate groups (AA).

to provide a transition layer. The thicker sealant layer is then flowed in a continuous manner over the application area, and the macromers are photopolymerized.

To enable photopolymerization, the macromers are formulated in buffered saline solutions containing triethanolamine and eosin Y as the photoinitiator [199]. The polymerization is initiated using visible blue–green light illumination from a xenon arc lamp (470–520 nm) for 40 s at an intensity of 100 mW/cm^2. The macromers cross-link to form a clear, flexible, and adherent hydrogel network. Because the sealant is polymerized in situ, the polymer conforms to the tissue surface. The hydrogel expands upon contact with body fluids and reaches its equilibrium swell volume within 24 h; the hydrogel contains 95% water at equilibrium. Following implantation, the poly(L-lactide) and poly(trimethylene carbonate) segments of the hydrogel degrade by hydrolysis (Figure 17.11); the sealant thus degrades by dissolution rather than fragmentation. The biodegradation products are water-soluble and biocompatible; the components are sufficiently low in molecular weight to be cleared through the kidneys or locally metabolized.

Photopolymerizable PEG tissue adhesives have been utilized clinically in their approved application for sealing air leaks following pulmonary resection [200,201]; the adhesives have also been used to treat air leaks in patients suffering lung injury at cardiac reoperation [202]. In addition, there is a clinical report of successful use of photopolymerized PEG adhesives for repairing ventricular wall rupture following mitral valve replacement [203]. In experimental models, the sealants have demonstrated efficacy for repairing acute aortic dissection [204], coronary artery anastomoses [205], inguinal hernia [206], pancreatic–jejunal anastomoses [207], and intestinal anastomoses [208], as well as for prevention of peritendinous adhesions following flexor tendon repair surgery [209]. However, despite desirable characteristics of biocompatibility, biodegradability, and potential versatility in surgical applications, the photopolymerizable PEG adhesives have failed to achieve widespread clinical use. The requirement for additional equipment (a blue–green visible light lamp) in the operating room may limit the ease-of-use, and ultimately reduce the clinical acceptability of photopolymerizable sealants.

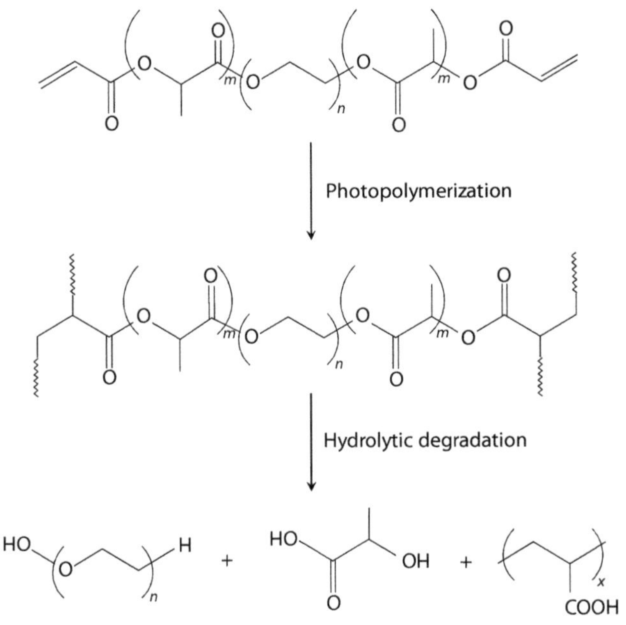

FIGURE 17.11 Photopolymerization and biodegradation reactions of poly(ethylene glycol)-*co*-poly(L-lactide) diacrylate monomer. The degradation products are soluble and biocompatible.

17.6.2 PEG–PEG Tissue Adhesives

A rapidly gelling synthetic PEG–PEG tissue sealant, formed by reacting two multifunctional four-arm star-branched PEG molecules, was first reported in 2001. The sealant, composed of tetra-succinimidyl-derivatized PEG and tetra-thiol-derivatized PEG, demonstrated adhesion to carotid arteries, collagen membranes, and PTFE grafts in vitro [210]. In subsequent clinical studies, the PEG–PEG tissue adhesive was successful in sealing suture lines of the aorta and coronary artery bypass grafts [211], as well as prosthetic vascular grafts [212]. On December 14, 2001, the commercial sealant CoSeal® (Cohesion Technologies, Palo Alto, California) received FDA approval as a hemostatic adjunct during vascular reconstruction surgery. The PEG–PEG sealant also received CE Mark approval for adjunctive hemostasis in vascular reconstructions.

The commercial PEG–PEG system is comprised of two PEG powder components: powdered pentaerythritol poly(ethylene glycol) ether tetra-succinimidyl glutarate and powdered pentaerythritol poly(ethylene glycol) ether tetra-thiol (Figure 17.12). The molecular weight of each four-arm star-branched PEG is approximately 10,000. Immediately prior to clinical use, the powder components are dissolved in an aqueous buffer; the two components are then mixed as they are delivered to the tissue site. Upon mixing, the functional groups on multiple arms of the PEGs react to form a covalently bonded three-dimensional matrix. The sulfur group of the multi-arm PEG thiol nucleo-philically attacks the carbonyl group attached to N-hydroxysuccinimide in the multi-arm PEG succinimidyl ester. The hydrogel is formed by the release of N-hydroxysuccinimide and concurrent formation of a thioester bond between the two substituted multi-arm PEGs (Figure 17.13). There is also a small amount of disulfide bond formation between thiol groups. The functionalized PEG end groups additionally react with functional groups (particularly amine groups) in the tissue matrix

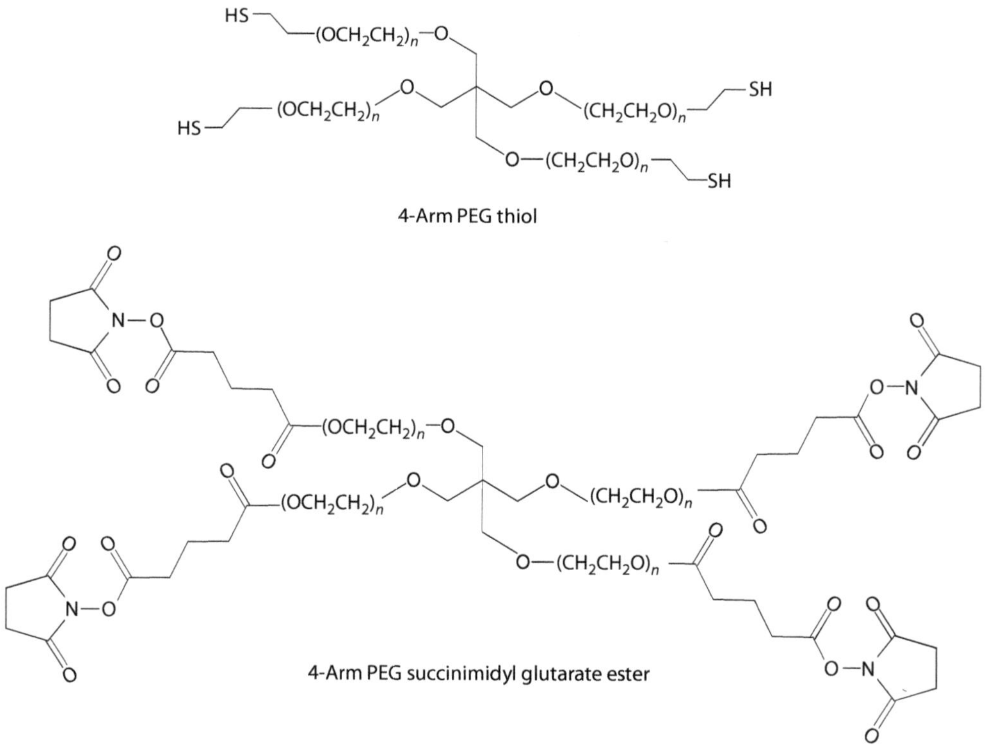

FIGURE 17.12 Chemical structures of multifunctional star-branched four-arm PEG polymers used in PEG–PEG tissue adhesives. The two components of the PEG–PEG system are pentaerythritol poly(ethylene glycol) ether tetra-thiol and pentaerythritol poly(ethylene glycol) ether tetra-succinimidyl glutarate.

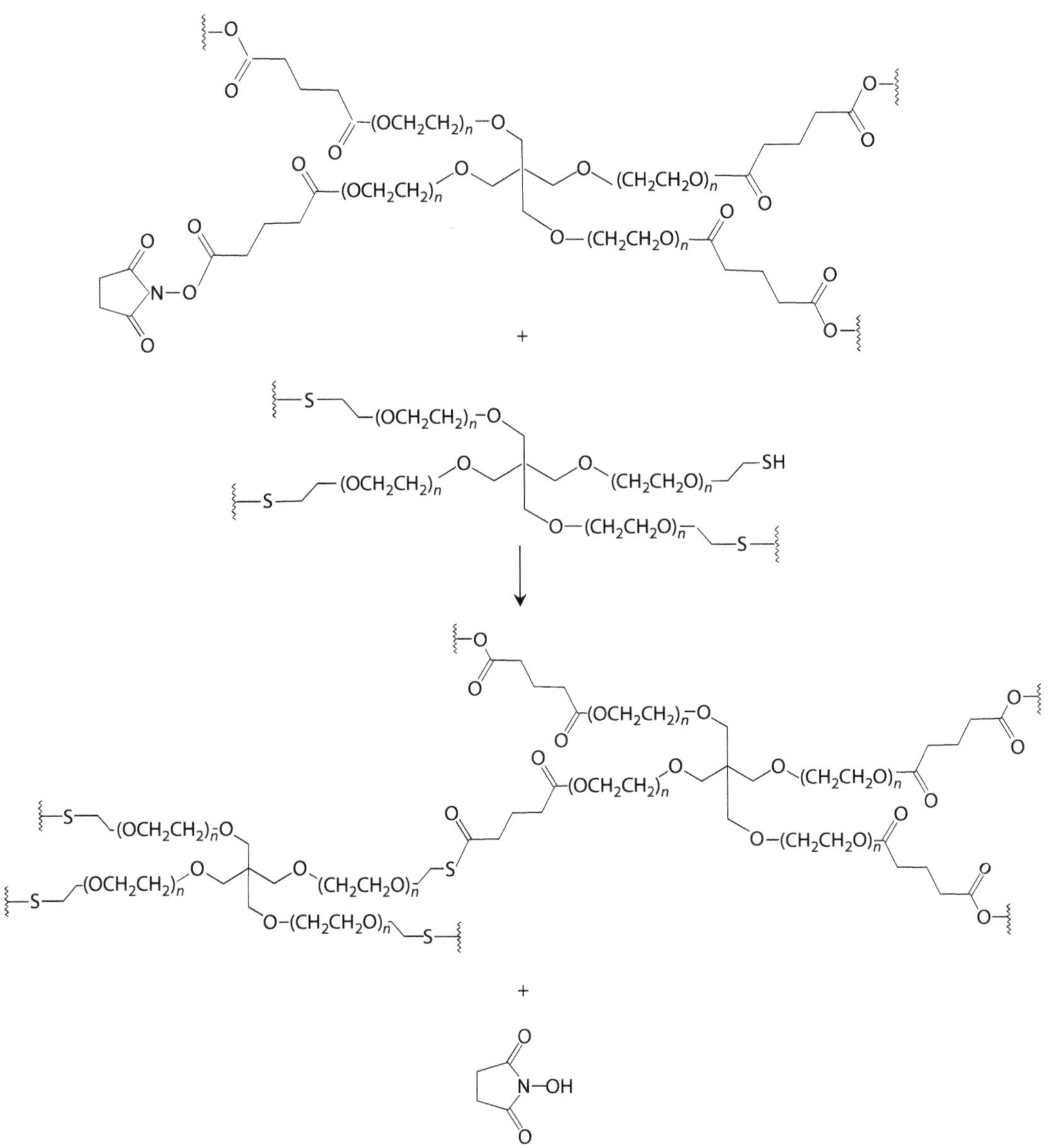

FIGURE 17.13 Reaction of multifunctional star-branched four-arm PEG polymers to form PEG–PEG tissue adhesives. The sulfur group of the multi-arm PEG thiol nucleophilically attacks the carbonyl group attached to N-hydroxysuccinimide in the multi-arm PEG succinimidyl ester. The hydrogel is formed by the release of N-hydroxysuccinimide and concurrent formation of a thioester bond between the two substituted multi-arm PEGs.

to form covalent bonds, providing a chemical linkage between the PEG–PEG hydrogel and the surrounding tissue. When applied to prosthetic vascular grafts, the PEG–PEG hydrogel partially penetrates the irregular graft surface and creates a mechanical bond.

The PEG–PEG tissue sealant provides sealing within 60 s; the mean time to complete anastomotic sealing during placement of prosthetic vascular grafts is 16.5 s. Following implantation, the PEG–PEG hydrogel absorbs water and swells up to four times its original volume within 24 h; application of the sealant should therefore be avoided near anatomic structures that are sensitive to compression. The hydrogel is biodegradable and contains two hydrolyzable bonds: the thioester between the two multi-arm PEGs and an O-ester that is within one of the PEGs and glutarate. The sealant is fully resorbed within 4 weeks.

Beyond its indicated use for adjunctive hemostasis of peripheral vascular anastomoses, the PEG–PEG system has demonstrated efficacy for minimizing anastomotic bleeding during aortic reconstruction [213]. In addition, PEG–PEG hydrogels have been successfully applied as sprayed barriers to reduce postoperative adhesion formation following uterine surgery [214], as well as adult and pediatric cardiac surgery [215,216]. The major limitations of the PEG–PEG sealant are its high degree of swell, and its relatively weak adhesion to tissue. On anastomotic closures, both cyanoacrylate glue and albumin–glutaraldehyde glue demonstrate greater mechanical integrity than the PEG–PEG sealant, and are capable of resisting higher loads before failure [217]. On the cut tissue surface of the kidney, fibrin sealant adheres more effectively than PEG–PEG sealant [218]. A practical limitation of the PEG–PEG system is the requirement for preparation and dissolution of the powdered PEGs into aqueous buffer prior to use; this may limit the convenient use of the sealant in the operating room.

17.6.3 PEG–Trilysine Tissue Adhesives

A novel hydrogel sealant, composed of a PEG ester and trilysine amine, was described in 2003 as an effective agent for dural closure to prevent cerebrospinal fluid leakage following neurosurgery [219]. In a preliminary clinical study conducted in 2005, the PEG–trilysine tissue adhesive demonstrated 100% closure of intraoperative cerebrospinal fluid leaks [220]. The commercial PEG–trilysine sealant DuraSeal® (Confluent Surgical, Waltham, Massachusetts) was granted FDA approval on April 7, 2005, as an adjunct to sutured dural repair during cranial surgery to provide watertight closure.

The commercial sealant is supplied as a two-component system, comprised of a PEG ester powder and a trilysine amine solution. The PEG component is dissolved in an aqueous solution immediately prior to clinical use. The two solutions mix as they are sprayed onto the dural tissue, and the components cross-link to form a watertight hydrogel seal. The sealant system also contains FD&C blue dye #1 to allow visualization of hydrogel coverage and thickness. The hydrogel absorbs water following implantation and swells by approximately 50% in volume; the sealant therefore should not be applied to confined bony structures where nerves are present, since neural compression can result due to hydrogel swelling. The tissue adhesive degrades hydrolytically within 4–8 weeks, and the degradation products are readily cleared by the kidneys. The PEG–trilysine system has continued to demonstrate 100% efficacy in stopping cerebrospinal fluid leakage in patients undergoing neurosurgical procedures [221].

17.7 EMERGING TISSUE ADHESIVES

A number of original adhesive platforms are currently being pursued for soft tissue repair and regeneration. Several of these innovative sealant technologies have demonstrated promise in experimental models of wound closure, but they have not yet reached clinical use. Three families of novel tissue adhesives are described later: naturally inspired tissue adhesives, polysaccharide-based tissue adhesives, and dendrimeric tissue adhesives. These emerging sealants are directed toward a variety of clinical applications, including external skin closure, gastrointestinal surgery, orthopedic surgery, and ophthalmic surgery.

17.7.1 Naturally Inspired Tissue Adhesives

Naturally occurring adhesive structures, such as frog glues, mussel proteins, and sticky gecko feet, have provided inspiration for unique new tissue sealants. For instance, the Australian frog *Notaden benetti* secretes an exudate which rapidly forms a tacky elastic solid. This protein-based material acts as a pressure-sensitive adhesive that functions in wet conditions, and covalent cross-linking does not seem to be necessary for the glue to set [222]. The frog glue demonstrated efficacy in

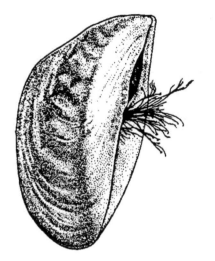

FIGURE 17.14 Adhesive byssal threads of the mussel (National Oceanic and Atmospheric Administration). The tough byssal threads are coated with mussel adhesive proteins to enable attachment to natural and manmade surfaces.

repairing torn meniscal tissue of the knee in an ex vivo model, and showed superior mechanical strength to both gelatin and fibrin glues [223]. This recently discovered biological glue may be considered for meniscal repairs in the future.

Marine and freshwater mussels also secrete specialized protein adhesives for rapid and durable attachment to wet surfaces. Mussels exude tough byssal threads (Figure 17.14) and coat these threads with adhesive proteins to attach to natural and manmade structures. Glues based on mussel adhesive proteins may thus be ideal for achieving adhesion to wet tissue substrates. Adhesive proteins extracted from *Mytilus edulis* mussel have demonstrated success in vitro for bonding porcine skin [224] and porcine small intestinal submucosa [225]; however, the mussel protein extracts required excessively long cure times. Synthetic polymers containing mussel protein functionality have been designed as a strategy for creating effective adhesives. The amino acid L-3,4-dihydroxylphenylalanine (DOPA) contributes to mussel protein solidification through oxidation and cross-linking reactions [226]; DOPA is formed in mussel proteins by posttranslational hydroxylation of the amino acid tyrosine (Figure 17.15). Biomimetic DOPA-functionalized PEG polymers have been shown to cross-link upon exposure to oxidizing reagents, and successfully bond porcine skin in vitro [227].

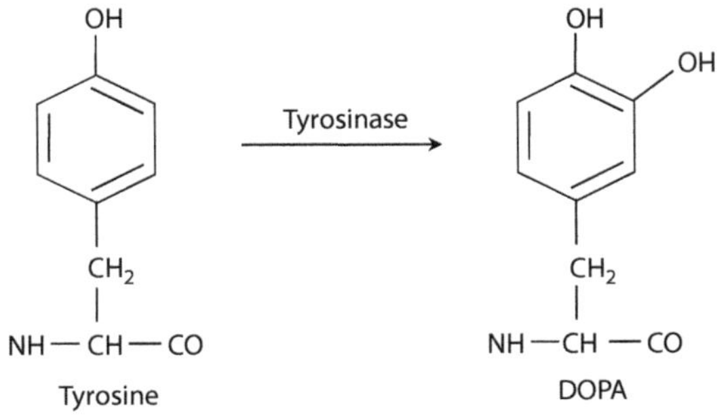

OH   OH

         Tyrosinase →

OH

CH$_2$ CH$_2$

NH — CH — CO   NH — CH — CO

Tyrosine   DOPA

FIGURE 17.15 Posttranslational modification of tyrosine residues in the *M. edulis* mussel adhesive protein. Hydroxylation of tyrosine residues creates L-3,4-dihydroxylphenylalanine (DOPA) residues, which are essential for adhesive protein cross-linking.

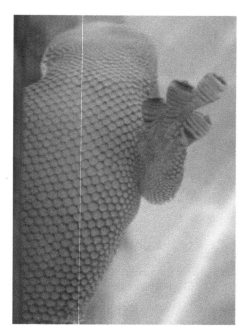

FIGURE 17.16   Adhesive footpads of the Madagascar gecko lizard (Montreal Biodome).

While frog glues and mussel adhesive proteins have inspired novel sealants based on their chemical compositions, the sticky footpad of the gecko lizard (Figure 17.16) has inspired new tissue adhesives based on its nanostructure. The gecko footpad is covered with a dense array of fibrils (setae), which maximize interfacial adhesion to surfaces [228]. Individual setae operate by van der Waals forces; the intermolecular attraction allows geckos to adhere to vertical and inverted surfaces. A biodegradable and biocompatible tissue adhesive has been designed to mimic the nanotopography of the gecko foot [229]. The gecko-inspired tissue adhesive is manufactured from a poly(glycerol-*co*-sebacate acrylate) (PGSA) elastomeric surface. The PGSA surface is etched into an array of nanoscale pillars to mimic the nanopatterns of the gecko foot, and the etched polymer is subsequently coated with oxidized dextran to allow tissue bonding. Applied as a tissue tape, the gecko-inspired adhesive has demonstrated efficacy in binding porcine intestinal tissue in vitro and rat abdominal tissue in vivo. The gecko-like tissue adhesive may provide the basis for an entirely new family of nanopatterned surgical adhesives.

17.7.2   POLYSACCHARIDE-BASED TISSUE ADHESIVES

Surgical glues composed of functionalized natural polysaccharides, including chondroitin and dextran, are showing early success as biocompatible sealants. The biopolymer chondroitin sulfate is a major component of cartilage extracellular matrix, and may provide an ideal foundation for designing biomaterials for cartilage repair. A photopolymerizable hydrogel composed of chondroitin sulfate functionalized with methacrylate and aldehyde groups has shown success in binding articular cartilage defects in vivo [230]. The chondroitin sulfate–methacrylate–aldehyde hydrogel is noncytotoxic, noninflammatory, and able to encapsulate cartilage cells, making it a promising platform for cartilage reconstruction. In addition, a photopolymerizable polysaccharide-based sealant composed of hyaluronic acid functionalized with methacrylate groups has shown efficacy in sealing experimental corneal incisions [231]. Finally, a polysaccharide-based sealant composed of dextran aldehyde and multi-arm PEG amine has been developed for wound closure (Figure 17.17).

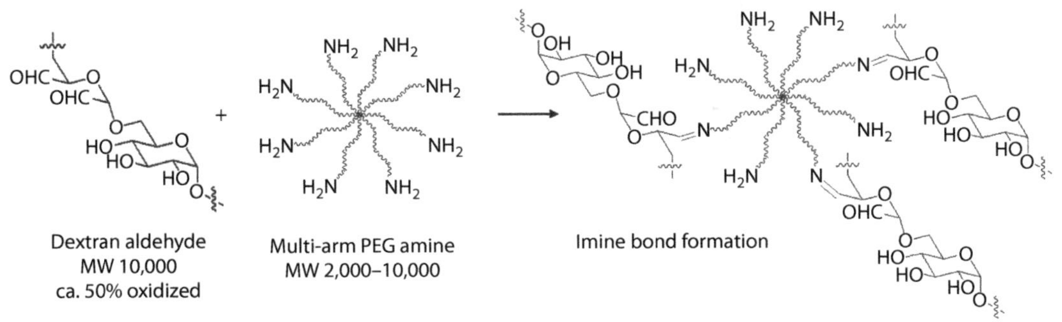

FIGURE 17.17 Foundation chemistry for dextran-based tissue adhesives. The oxidized polysaccharide dextran aldehyde reacts with an eight-arm star PEG amine to form a cross-linked hydrogel network.

The dextran-based tissue adhesive is noncytotoxic and noninflammatory [232], requires no external photoinitiator or other extra equipment, and has demonstrated efficacy in an ex vivo model of corneal closure [233].

17.7.3 DENDRIMERIC TISSUE ADHESIVES

Highly branched dendritic macromers have provided the basis for a new class of hydrogel sealants with unique physical and mechanical properties [234]. Dendrimers possess three main structural components: a central core, internal branching layers, and peripheral functional groups. Unlike linear polymers in which growth is accomplished by adding a single monomer, dendritic polymers grow by branching each monomer, leading to multiple additions. When used to construct hydrogel scaffolds, dendritic macromers allow increased cross-link density of the scaffold without significantly increasing the polymer concentration, as compared to linear polymer analogs; this approach leads to improved mechanical properties and minimal swelling of the hydrogel.

Biodendrimeric tissue adhesives, based on peptide dendrons functionalized with terminal cysteine residues, have been developed for ophthalmic applications. When the cysteine-terminated peptide dendrons are mixed with PEG dialdehyde, a hydrogel forms as a consequence of thiazolidene linkages between the two macromers [235]. These biodendrimeric sealants have been successful in sealing ex vivo corneal incisions and securing ex vivo corneal transplants [236]. A photo-cross-linkable biodendrimeric tissue adhesive has also been created from triblock copolymers; these hybrid dendritic–linear copolymers consist of a PEG core and methacrylated poly(glycerol succinic acid) dendrimer terminal blocks. The photopolymerized dendrimeric sealant is effective in sealing experimental full-thickness corneal lacerations [237] and securing ex vivo laser in situ keratomileusis (LASIK) flaps [238]. In addition, the photo-cross-linkable dendrimer adhesive attaches to experimental cartilage defects [239] and encapsulates chondrocyte cells [240]. Dendrimer-based sealants may thus be a promising new technology for cartilage repair.

17.8 CONCLUSIONS

Currently available tissue adhesives, including fibrin sealants, cyanoacrylate glues, cross-linked protein sealants, and PEG hydrogels, have been utilized in a wide variety of surgical procedures to aid in hemostasis and wound closure. The oldest and most widely used tissue adhesives make use of known biological pathways (fibrin sealants) or known industrial products (cyanoacrylate glues). Yet these traditional classes of sealants are associated with trade-offs between functionality and biocompatibility. Biological fibrin glues are tissue-friendly but more prone to tearing, due to their relatively poor tissue adhesion; these sealants carry additional risks of viral transmission and sensitization. Synthetically derived glues such as cyanoacrylates, GRF glues, and albumin–glutaraldehyde glues demonstrate extremely strong tissue adhesion and are more resistant to tearing, but they exhibit much higher tissue toxicity and poor biodegradability. PEG-based adhesives are tissue-friendly and readily biodegradable, but they do not adhere as strongly as cyanoacrylates and cross-linked protein glues, and they can swell considerably following implantation.

An optimal tissue sealant would exhibit all of the advantages of existing tissue adhesives, while avoiding the disadvantages and adverse effects. Such a sealant would combine biocompatibility and biodegradability with strong tissue adhesion and favorable physical properties. The perfect sealant must meet stringent mechanical and biological performance requirements; novel tissue adhesives are constantly being formulated to address these challenges. New technology platforms include naturally inspired adhesives, polysaccharide-based tissue adhesives, and dendrimeric adhesives. These emerging sealants demonstrate encouraging results in experimental models of wound closure, and they may soon realize clinical use. Continued development of adhesive biomaterials for tissue reconstruction will require a detailed understanding of both polymer properties and the biological environment. When polymer science and biology are brought to bear on tissue-adhesive design, the resulting biomaterials will enhance the lives of patients worldwide. The clinical imperative for improved tissue adhesives is clear, now more than ever.

REFERENCES

1. Bruce, J., Krukowski, Z. H., Al-Khairy, G., Russell, E. M., and Park, K. G. M. Systematic review of the definition and measurement of anastomotic leak after gastrointestinal surgery. *Br. J. Surg.*, 88, 1157–1168 (2001).
2. Pickleman, J., Watson, W., Cunningham, J., Fisher, S. G., and Gamelli, R. The failed gastrointestinal anastomosis: An inevitable catastrophe? *J. Am. Coll. Surg.*, 188, 473–482 (1999).
3. Walker, K. G., Bell, S. W., Rickard, M. J. F. X., Mehanna, D., Dent, O. F., Chapuis, P. H., and Bokey, E. L. Anastomotic leakage is predictive of diminished survival after potentially curative resection for colorectal cancer. *Ann. Surg.*, 240, 255–259 (2004).
4. Varley, G. A. and Meisler, D. M. Complications of penetrating keratoplasty: Graft infections. *Refract. Corneal Surg.*, 7, 62–66 (1991).
5. Binder, P. S. Selective suture removal can reduce postkeratoplasty astigmatism. *Ophthalmology*, 92, 1412–1416 (1985).
6. Nagaki, Y., Hayasaka, S., Kadoi, C., Matsumoto, M., Yanagisawa, S., Watanabe, K., Watanabe, K., Hayasaka, Y., Ikeda, N., Sato, S., Kataoka, Y., Togashi, M., and Abe, T. Bacterial endophthalmitis after small-incision cataract surgery. Effect of incision placement and intraocular lens type. *J. Cataract Refract. Surg.*, 29, 20–26 (2003).
7. Powe, N. R., Schein, O. D., Gieser, S. C., Tielsch, J. M., Luthra, R., Javitt, J., and Steinberg, E. P. Synthesis of the literature on visual acuity and complications following cataract extraction with intraocular lens implantation. Cataract Patient Outcome Research Team. *Arch. Ophthalmol.*, 112, 239–252 (1994).
8. Haring, R. Current status of tissue adhesives in Germany, in *Tissue Adhesives in Surgery*, T. Matsumoto, ed. New York: Medical Examination Publications (1972).
9. American Society for Testing and Materials. Annual book of ASTM standards. Volume 13.01 Medical and Surgical Materials and Devices; Anesthetic and Respiratory Equipment; Manufacture of Pharmaceutical Products. Philadelphia, PA: ASTM (2008).
10. Bergel, S. Uber Wirkungen des Fibrins. *Dtschr. Med. Wochenschr.*, 35, 633–665 (1909).
11. Grey, E. G. Fibrin as a haemostatic in cerebral surgery. *Surg. Gynecol. Obstet.*, 21, 452–454 (1915).
12. Cronkite, E. P., Lozner, E. L., and Deaver, J. Use of thrombin and fibrinogen in skin grafting. *JAMA*, 124, 976–978 (1944).
13. Young, J. Z. and Medawar, P. B. Fibrin suture of peripheral nerves. *Lancet*, 275, 126–132 (1940).
14. Matras, H., Dinges H. P., Lassman, H., and Mamoli, B. Zur nahtlosen interfaszikular-en Nerventransplantation im Tier experiment. *Wien Med. Wochenschr.*, 122, 517–523 (1972).
15. Matras, H., Chiari, F., Kletter, G., and Dinges, H. P. Zur klebung von mikrogefagefassanastomosen (Eine experimentelle studie). Proceedings. 13th Annual Meeting. *Dtsch. Gesfplast. Wiederhestell.*, 357–360 (1977).

16. Bove, J. R. Fibrinogen—Is the benefit worth the risk? *Transfusion*, 18, 129–136 (1978).
17. Martinowitz, U. and Saltz, R. Fibrin sealant. *Curr. Opin. Hematol.*, 3, 395–402 (1996).
18. Martinowitz, U. and Spotnitz, W. D. Fibrin tissue adhesives. *Thromb. Haemostas.*, 78, 661–666 (1997).
19. Martinowitz, U., Schulman, S., Horoszowski, H., and Heim, M. Role of fibrin sealants in surgical procedures on patients with hemostatic disorders. *Clin. Orthop. Relat. Res.*, 328, 65–75 (1996).
20. Radosevich, M., Goubran, H. A., and Burnouf, T. Fibrin sealant: Scientific rationale, production methods, properties, and current clinical use. *Vox Sang.*, 72, 133–143 (1997).
21. Buchta, C., Hedrich, H. C., Macher, M., Hocker, P., and Redl, H. Biochemical characterization of autologous fibrin sealants produced by CryoSeal® and Vivostat® in comparison to the homologous fibrin sealant product Tissucol/Tisseel®. *Biomaterials*, 26, 6233–6241 (2005).
22. Busuttil, R. W. A comparison of antifibrinolytic agents used in hemostatic fibrin sealants. *J. Am. Coll. Surg.*, 197, 1021–1028 (2003).
23. Detweiler, M. B., Detweiler, J. G., and Fenton, J. Sutureless and reduced suture anastomosis of hollow vessels with fibrin glue: A review. *J. Invest. Surg.*, 12, 245–262 (1999).
24. Schenk, W. G. 3rd, Burks, S. G., Gagne, P. J., Kagan, S. A., Lawson, J. H., and Spotnitz, W. D. Fibrin sealant improves hemostasis in peripheral vascular surgery: A randomized prospective trial. *Ann. Surg.*, 237, 871–876 (2003).
25. Schenk, W. G. 3rd, Goldthwaite, C. A. Jr., Burks, S., and Spotnitz, W. D. Fibrin sealant facilitates hemostasis in arteriovenous polytetrafluoroethylene grafts for renal dialysis access. *Am. Surg.*, 68, 728–732 (2002).
26. Rousou, J., Gonzalez-Lavin, L., Cosgrove, D., Weldon, C., Hess, P., Joyce, L., Bergsland, J., and Gazzaniga, A. Randomized clinical trial of fibrin sealant in patients undergoing resternotomy or reoperation after cardiac operations. *J. Thorac. Cardiovasc. Surg.*, 97, 194–203 (1989).
27. Koveker, G., De Vivie, E. R., and Helberg, K. D. Clinical experience with fibrin glue in cardiac surgery. *Thorac. Cardiovasc. Surg.*, 29, 287–289 (1981).
28. Seguin, J. R., Frapier, J.-M., Colson, R., and Chaptal, P. A. Fibrin sealants improve surgical results of type A acute aortic dissections. *Ann. Thorac. Surg.*, 52, 745–749 (1991).
29. Kjaergard H. K. and Fairbrother, J. E. Controlled clinical studies of fibrin sealant in cardiothoracic surgery—A review. *Eur. J. Cardiothorac. Surg.*, 10, 727–33 (1996).
30. Hvass, U., Chatel, D., Frikha, I., Pansard, Y., Depoix, J. P., and Julliard, J. M. Left ventricular free wall rupture. Long-term results with a pericardial patch and fibrin glue repair. *Eur. J. Cardiothorac. Surg.*, 9, 75–76 (1995).
31. Bayfield M. S. and Spotnitz, W. D. Fibrin sealant in thoracic surgery. Pulmonary applications, including management of bronchopleural fistula. *Chest Surg. Clin. N. Am.*, 6, 576–583 (1996).
32. Ochsner, M. G. Fibrin solutions to control hemorrhage in the trauma patient. *J. Long-Term. Effects Med. Implants*, 8, 161–173 (1998).
33. Sciscione, A. C., Manley, J. S., Pollock, M., Maas, B., Shlossman, P. A., Mulla, W., Lankiewicz, M., and Colmorgen, G. H. Intracervical fibrin sealants: A potential treatment for early preterm premature rupture of the membranes. *Am. J. Obstet. Gynecol.*, 184, 368–373 (2001).
34. Bar-Hava, I., Krissi, H., Ashkenazi, J., Orvieto, R., Shelef, M., and Ben-Rafael, Z. Fibrin glue improves pregnancy rates in women of advanced reproductive age and in patients in whom in vitro fertilization attempts repeatedly fail. *Fertil. Steril.*, 71, 821–824 (1999).
35. Lau, W. Y., Leung, K. L., Zhu, X. L., Lam, Y. H., Chung, S. C., and Li, A. K. Laparoscopic repair of perforated peptic ulcer. *Br. J. Surg.*, 82, 814–816 (1995).
36. Sentovich, S. M. Fibrin glue for anal fistulas: Long-term results. *Dis. Colon. Rectum*, 46, 498–502 (2003).
37. Heneghan, M. A., Byrne, A., and Harrison, P. M. An open pilot study of the effects of a human fibrin glue for endoscopic treatment of patients with acute bleeding from gastric varices. *Gastrointest. Endosc.*, 56, 422–426 (2002).
38. Wang G. J., Goldthwaite, C. A. Jr., Burks, S., Crawford, R., and Spotnitz, W. D. Orthopaedic Investigators Group. Fibrin sealant reduces perioperative blood loss in total hip replacement. *J. Long-Term Eff. Med. Implants*, 13, 399–411 (2003).
39. Wang, G. J., Goldthwaite, C. A. Jr., Burks, S. G., Spotnitz, W. D. Orthopaedic Research Group. Experience improves successful use of fibrin sealant in total knee arthroplasty: Implications for surgical education. *Long Term Eff. Med. Implants*, 13, 389–397 (2003).
40. Currie, L. J., Sharpe, J. R., and Martin, R. The use of fibrin glue in skin grafts and tissue-engineered skin replacements: A review. *Plast. Reconstr. Surg.*, 108, 1713–1726 (2001).
41. Marchac, D. and Sandor, G. Face lifts and sprayed fibrin glue: An outcome analysis of 200 patients. *Br. J. Plast. Surg.*, 47, 306–309 (1994).

42. Shaffrey, C. I., Spotnitz, W. D., Shaffrey, M. E., and Jane, J. A. Neurosurgical applications of fibrin glue: Augmentation of dural closure in 134 patients. *Neurosurgery*, 26, 207–210 (1990).

43. Moralee, S. J., Carney, A. S., Cash, M. P., and Murray, J. A. M. The effect of fibrin sealant haemostasis on post-operative pain in tonsillectomy. *Clin. Otolaryngol.*, 19, 526–528 (1994).

44. Vaiman, M., Eviatar, E., and Segal, S. Effectiveness of second-generation fibrin glue in endonasal operations. *Otolaryngol. Head Neck Surg.*, 126, 388–391 (2002).

45. Biedner, B. and Rosenthal, G. Conjunctival closure in strabismus surgery: Vicryl versus fibrin glue. *Ophthalmic Surg. Lasers*, 27, 967 (1996).

46. Sharma, A., Kaur, R., Kumar, S., Gupta, P., Pandav, S., Patnaik, B., and Gupta, A. Fibrin glue versus *N*-butyl-2-cyanoacrylate in corneal perforations. *Ophthalmology*, 110, 291–298 (2003).

47. Thompson, D. F. and Davis, T. W. The addition of antibiotics to fibrin glue. *South. Med. J.*, 90, 681–684 (1997).

48. van der Ham, A. C., Kort, W. J., Weijma, I. M., van den Ingh, H. F., and Jeekel, H. Effect of antibiotics in fibrin sealant on healing colonic anastomoses in the rat. *Br. J. Surg.*, 79, 525–528 (1992).

49. Woolverton, C. J., Fulton, J. A., Salstrom, S. J., Hayslip, J., Haller, N. A., Wildroudt, M. L., and MacPhee, M. Tetracycline delivery from fibrin controls peritoneal infection without measurable systemic antibiotic. *J. Antimicrob. Chemother.*, 48, 861–867 (2001).

50. Tredwell, S., Jackson, J. K., Hamilton, D., Lee, V., and Burt, H. M. Use of fibrin sealants for the localized, controlled release of cefazolin. *Can. J. Surg.*, 49, 347–352 (2006).

51. Kitazawa, H., Sato, H., Adachi, I., Masuko, Y., and Horikoshi, I. Microdialysis assessment of fibrin glue containing sodium alginate for local delivery of doxorubicin in tumor-bearing rats. *Biol. Pharm. Bull.*, 20, 278–281 (1997).

52. Miura, S., Mii, Y., Miyauchi, Y., Ohgushi, H., Morishita, T., Hohnoki, K., Aoki, M., Tamai, S., and Konishi, Y. Efficacy of slow-releasing anticancer drug delivery systems on transplantable osteosarcomas in rats. *Jpn. J. Clin. Oncol.*, 25, 61–71 (1995).

53. Kim, H. J., Kang, S. W., Lim, H. C., Han, S. B., Lee, J. S., Prasad, L., Kim, Y. J., Kim, B. S., and Park, J. H. The role of transforming growth factor-beta and bone morphogenetic protein with fibrin glue in healing of bone-tendon junction injury. *Connect. Tissue Res.*, 48, 309–315 (2007).

54. Yin, Q., Kemp, G. J., Yu, L. G., Wagstaff, S. C., and Frostick, S. P. Neurotrophin-4 delivered by fibrin glue promotes peripheral nerve regeneration. *Muscle Nerve*, 24, 345–351 (2001).

55. Hino, M., Ishiko, O., Honda, K. I., Yamane, T., Ohta, K., Takubo, T., and Tatsumi, N. Transmission of symptomatic parvovirus B19 infection by fibrin sealant used during surgery. *Br. J. Haematol.*, 108, 194–195 (2000).

56. Kawamura, M., Sawafuji, M., Watanabe, M., Horinouchi, H., and Kobayashi, K. Frequency of transmission of human parvovirus B19 infection by fibrin sealant used during thoracic surgery. *Ann. Thorac. Surg.*, 73, 1098–1100 (2002).

57. Enzmann, H. Sensibilisierung bei der Verwendung von Fibrinkleber. *Laryngol. Rhinol. Otol.*, 61, 302–304 (1982).

58. Berruyer, M., Amiral, J., Ffrench, P., Belleville, J., Bastien, O., Clerc, J., Kassir, A., Estanove, S., and Dechavanne, M. Immunization by bovine thrombin used with fibrin glue during cardiovascular operations. Development of thrombin and factor V inhibitors. *J. Thorac. Cardiovasc. Surg.*, 105, 892–897 (1993).

59. Caers, J., Reekmans, A., Jochmans, K., Naegels, S., Mana, F., Urbain, D., and Reynaert, H. Factor V inhibitor after injection of human thrombin (Tissucol) into a bleeding peptic ulcer. *Endoscopy*, 35, 542–544 (2003).

60. Muntean, W., Zenz, W., Finding, K., Zobel, G., and Beitzke, A. Inhibitor to factor V after exposure to fibrin sealant during cardiac surgery in a two-year-old child. *Acta Paediatr.*, 83, 84–87 (1994).

61. Beierlein, W., Scheule, A. M., Antoniadis, G., Braun, C., and Schosser, R. An immediate, allergic skin reaction to aprotinin after reexposure to fibrin sealant. *Transfusion*, 40, 302–305 (2000).

62. Kober, B. J., Scheule, A. M., Voth, V., Deschner, N., Schmid, E., and Ziemer, G. Anaphylactic reaction after systemic application of aprotinin triggered by aprotinin-containing fibrin sealant. *Anesth. Analg.*, 107, 406–409 (2008).

63. Oswald, A. M., Joly, L. M., Gury, C., Disdet, M., Leduc, V., and Kanny, G. Fatal intraoperative anaphylaxis related to aprotinin after local application of fibrin glue. *Anesthesiology*, 99, 762–763 (2003).

64. Scheule, A. M., Beierlein, W., Wendel, H. P., Eckstein, F. S., Heinemann, M. K., and Ziemer, G. Fibrin sealant, aprotinin, and immune response in children undergoing operations for congenital heart disease. *J. Thorac. Cardiovasc. Surg.*, 115, 883–889 (1998).

65. Conrad, K. and Yoskovitch, A. The use of fibrin glue in the correction of pollybeak deformity. *Arch. Facial Plast. Surg.*, 5, 522–527 (2003).

66. Albes, J. M., Krettek, C., Hausen, B., Rohde, R., Haverich, A., and Borst, H. G. Biophysical properties of the gelatin–resorcin–formaldehyde/glutaraldehyde adhesive. *Ann. Thorac. Surg.*, 56, 910–915 (1993).

67. Ardis, A. E. Preparation of monomeric alkyl alpha-cyanoacrylates. US Patent 2,467,926 (1949).

68. Ardis, A. E. Preparation of monomeric alkyl alpha-cyanoacrylates. US Patent 2,467,927 (1949).

69. Joyner, F. B. and Hawkins, G. F. Method of making alpha-cyanoacrylates. US Patent 2,721,858 (1955).

70. Carton, C. A., Kessler, L. A., Seidenberg, B., and Hurwitt, E. S. A plastic adhesive method of small blood vessel surgery. *World Neurol.*, 1, 356–362 (1960).

71. Marable, S. A. and Wagner, D. E. The use of rapidly polymerizing adhesives in massive liver resection. *Surg. Forum.*, 13, 264–266 (1962).

72. Healey, J. E. Jr., Sheena, K. S., Gallager, H. S., Clark, R. L., and O'Neill, P. The use of a plastic adhesive in the technique of bronchial closure. *Surg. Forum.*, 13, 153–155 (1962).

73. Mathes, G. L. and Terry, J. W. Jr. Non-suture closure of nephrotomy. *J. Urol.*, 89, 122–125 (1963).

74. Seidenberg, B., Garrow, E., Pimental, R., and Hurwitt, E. S. Studies on the use of plastic adhesive in gastro-intestinal surgery. *Ann. Surg.*, 158, 721–729 (1963).

75. Ellis, R. A. and Levine, A. M. Experimental sutureless ocular surgery. *Amer. J. Ophthalmol.*, 55, 733–741 (1963).

76. Bloomfield, S., Barnert, A. H., and Kanter, P. Use of Eastman-910 monomer as an adhesive in ocular surgery. II. Effectiveness in closure of limbal wounds in rabbits. *Amer. J. Ophthalmol.*, 55, 946–953 (1963).

77. Coover, H. W. Jr., Joyner, F. B., Shearer, N. H. Jr., and Wicker, T. H. Jr. Chemistry and performance of cyanoacrylate adhesives. *S.P.E. Tech. Papers*, 5, 92–97 (1959).

78. Quinn, J. V. Clinical approaches to the use of cyanoacrylate tissue adhesives, in *Tissue Adhesives in Clinical Medicine*, 2nd ed., J. V. Quinn, ed. Hamilton, Ontario, Canada: BC Decker Inc. (2005).

79. Singer, A. J., Zimmerman, T., Rooney, J., Cameau, P., Rudomen, G., and McClain, S. A. Comparison of wound bursting strength and surface characteristics of FDA approved tissue adhesives for skin closure. *J. Adhes. Sci. Technol.*, 18, 19–27 (2004).

80. Vezin, W. R. and Florence, A. T. In vitro heterogeneous degradation of poly(*n*-alkyl alpha-cyanoacrylates). *J. Biomed. Mater. Res.*, 14, 93–106 (1980).

81. Lenaerts, V., Couvreur, P., Christiaens-Leyh, D., Joiris, E., Roland, M., Rollman, B., and Speiser, P. Degradation of poly(isobutyl cyanoacrylate) nanoparticles. *Biomaterials*, 5, 65–68 (1984).

82. Quinn, J., Drzewiecki, A., Li, M., Stiell, I., Sutcliffe, T., Elmslie, T., and Wood, W. A randomized, controlled trial comparing a tissue adhesive with suturing in the repair of pediatric facial lacerations. *Ann. Emerg. Med.*, 22, 1130–1135 (1993).

83. Simon, H. K., McLario, D. J., Bruns, T. B., Zempsky, W. T., Wood, R. J., and Sullivan, K. M. Long-term appearance of lacerations repaired using a tissue adhesive. *Pediatrics*, 99, 193–195 (1997).

84. Singer, A. J., Quinn, J. V., Hollander, J. E., Clark R. E. TraumaSeal Study Group. Closure of lacerations and incisions with octyl-cyanoacrylate: A multi-center randomized clinical trial. *Surgery*, 131, 270–276 (2002).

85. Mizrahi, S., Bickel, A., and Ben-Layish, E. B. Use of tissue adhesives in the repair of lacerations in children. *J. Pediatr. Surg.*, 23, 312–313 (1988).

86. Sinha, S., Naik, M., Wright, V., Timmons, J., and Campbell, A. C. A single blind, prospective, randomized trial comparing *n*-butyl 2-cyanoacrylate tissue adhesive (Indermil) and sutures for skin closure in hand surgery. *J. Hand Surg. [Br.]*, 26, 264–265 (2001).

87. Laccourreye, P., Cauchois, R., Sharkawy, E. L., Menard, M., De Mones, E., Brasnu, D., and Hans, S. Octylcyanoacrylate (Dermabond) for skin closure at the time of head and neck surgery: A longitudinal prospective study. *Ann. Chir.*, 130, 624–630 (2005).

88. Spauwen, P. H., de Laat, W. A., and Hartman, E. H. Octyl-2-cyanoacrylate tissue glue (Dermabond) versus Monocryl 6 × 0 sutures in lip closure. *Cleft Palate Craniofac. J.*, 43, 625–627 (2006).

89. Hall, L. T. and Bailes, J. E. Dermabond for wound closure in lumbar and cervical neurosurgical procedures. *Neurosurgery*, 56, S147–S150 (2005).

90. Wang, M. Y., Levy, M. L., Mittler, M. A., Liu, C. Y., Johnston, S., and McComb, J. G. A prospective analysis of the use of octylcyanoacrylate tissue adhesive for wound closure in pediatric neurosurgery. *Pediatr. Neurosurg.*, 30, 186–188 (1999).

91. Nahas, F. X., Solia, D., Ferreira, L. M., and Novo, N. F. The use of tissue adhesive for skin closure in body contouring surgery. *Aesthetic Plast. Surg.*, 28, 165–169 (2004).

92. Scott, G. R., Carson, C. L., and Borah, G. L. Dermabond skin closures for bilateral reduction mammaplasties: A review of 255 consecutive cases. *Plast. Reconstr. Surg.*, 120, 1460–1465 (2007).

93. Toriumi, D. M., O'Grady, K., Desai, D., and Bagal, A. Use of octyl-2-cyanoacrylate for skin closure in facial plastic surgery. *Plast. Reconstr. Surg.*, 102, 2209–2219 (1998).

94. Greene, D., Koch, R. J., and Goode, R. L. Efficacy of octyl-2-cyanoacrylate tissue glue in blepharoplasty. A prospective controlled study of wound-healing characteristics. *Arch. Facial Plast. Surg.*, 1, 292–296 (1999).

95. Jallali, N., Haji, A., and Watson, C. J. A prospective randomized trial comparing 2-octyl cyanoacrylate to conventional suturing in closure of laparoscopic cholecystectomy incisions. *J. Laparoendosc. Adv. Surg. Tech.*, 14, 209–211 (2004).

96. Switzer, E. F., Dinsmore, R. C., and North, J. H. Jr. Subcuticular closure versus Dermabond: A prospective randomized trial. *Am. Surg.*, 69, 434–436 (2003).

97. Bowen, M. L. and Selinger, M. Episiotomy closure comparing enbucrilate tissue adhesive with conventional sutures. *Int. J. Gynaecol. Obstet.*, 78, 201–205 (2002).

98. Rogerson, L., Mason, G. C., and Roberts, A. C. Preliminary experience with twenty perineal repairs using Indermil tissue adhesive. *Eur. J. Obstet. Gynecol. Reprod. Biol.*, 88, 139–142 (2000).

99. Pachulski, R., Sabbour, H., Gupta, R., Adkins, D., Mirza, H., and Cone, J. Cardiac device implant wound closure with 2-octyl cyanoacrylate. *J. Interv. Cardiol.*, 18, 185–187 (2005).

100. Souza, E. C., Fitaroni, R. B., Januzelli, D. M., Macruz, H. M., Camacho, J. C., and Souza, M. R. Use of 2-octyl cyanoacrylate for skin closure of sternal incisions in cardiac surgery: Observations of microbial barrier effects. *Curr. Med. Res. Opin.*, 24, 151–155 (2008).

101. Brunkwall, J., Ruemenapf, G., Florek, H. J., Lang, W., and Schmitz-Rixen, T. A single arm, prospective study of an absorbable cyanoacrylate surgical sealant for use in vascular reconstructions as an adjunct to conventional techniques to achieve haemostasis. *J. Cardiovasc. Surg. (Torino)*, 48, 471–476 (2007).

102. Schenk, W. G. 3rd, Spotnitz, W. D., Burks, S. G., Lin, P. H., Bush, R. L., and Lumsden, A. B. Absorbable cyanoacrylate as a vascular hemostatic sealant: A preliminary trial. *Am. Surg.*, 71, 658–661 (2005).

103. Lumsden, A. B. and Heyman, E. R. Closure Medical Surgical Sealant Study Group. Prospective randomized study evaluating an absorbable cyanoacrylate for use in vascular reconstructions. *J. Vasc. Surg.*, 44, 1002–1009 (2006).

104. Taravella, M. J. and Chang, C. D. 2-Octyl cyanoacrylate medical adhesive in treatment of a corneal perforation. *Cornea*, 20, 220–221 (2001).

105. Chen, W. L., Lin, C. T., Hsieh, C. Y., Tu, I. H., Chen, W. Y., and Hu, F. R. Comparison of the bacteriostatic effects, corneal cytotoxicity, and the ability to seal corneal incisions among three different tissue adhesives. *Cornea*, 26, 1228–1234 (2007).

106. Kim, J. C., Bassage, S. D., Kempski, M. H., del Cerro, M., Park, S. B., and Aquavella, J. V. Evaluation of tissue adhesives in closure of scleral tunnel incisions. *J. Cataract Refract. Surg.*, 21, 320–325 (1995).

107. Jourdan, I. C. and Bailey, M. E. Initial experience with the use of *N*-butyl 2-cyanoacrylate glue for the fixation of polypropylene mesh in laparoscopic hernia repair. *Surg. Laparosc. Endosc.*, 8, 291–293 (1998).

108. Greenwald, B. D., Caldwell, S. H., Hespenheide, E. E., Patrie, J. T., Williams, J., Binmoeller, K. F., Woodall, L., and Haluszka, O. *N*-2-Butyl-cyanoacrylate for bleeding gastric varices: A United States pilot study and cost analysis. *Am. J. Gastroenterol.*, 98, 1982–1988 (2003).

109. Sofer, M., Greenstein, A., Chen, J., Nadu, A., Kaver, I., and Matzkin, H. Immediate closure of nephrostomy tube wounds using a tissue adhesive: A novel approach following percutaneous endourological procedures. *J. Urol.*, 169, 2034–2036 (2003).

110. Bardari, F., D'Urso, L., and Muto, G. Conservative treatment of iatrogenic urinary fistulas: The value of cyanoacrylic glue. *Urology*, 58, 1046–1048 (2001).

111. Maloney, J. D., Weigel, T. L., and Love, R. B. Endoscopic repair of bronchial dehiscence after lung transplantation. *Ann. Thorac. Surg.*, 72, 2109–2111 (2001).

112. Gul, R., Khan, F., Maher, Y., and O'Farrell, D. Osteochondral fractures in the knee treated with butyl-2-cyanoacrylate glue. A case report. *Acta Orthop. Belg.*, 72, 641–643 (2006).

113. Yilmaz, C. and Kuyurtar, F. Fixation of a talar osteochondral fracture with cyanoacrylate glue. *Arthroscopy*, 21, 1009 (2005).

114. Cheski, P. J. and Matthews, T. W. Endoscopic reduction and internal cyanoacrylate fixation of the zygoma. *J Otolaryngol.*, 26, 75–79 (1997).

115. Samuel, P. R., Roberts, A. C., and Nigam, A. The use of Indermil (*n*-butyl cyanoacrylate) in otorhinolaryngology and head and neck surgery. A preliminary report on the first 33 patients. *J. Layngol. Otol.*, 111, 536–540 (1997).

116. Ciapetti, G., Stea, S., Cenni, E., Sudanese, A., Marraro, D., Toni, A., and Pizzoferrato, A. Cytotoxicity testing of cyanoacrylates using direct contact assay on cell cultures. *Biomaterials*, 15, 63–67 (1994).

117. Evans, C. E., Lees, G. C., and Trail, I. A. Cytotoxicity of cyanoacrylate adhesives to cultured tendon cells. *J. Hand Surg. [Br.]*, 24, 658–661 (1999).

118. Toriumi, D. M., Raslan, W. F., Friedman, M., and Tardy, M. E. Variable histotoxicity of histocryl when used in a subcutaneous site: An experimental study. *Laryngoscope*, 101, 339–343 (1991).

119. Wieken, K., Angioi-Duprez, K., Lim, A., Marchal, L., and Merle, M. Nerve anastomosis with glue: Comparative histologic study of fibrin and cyanoacrylate glue. *J. Reconstr. Microsurg.*, 19, 17–20 (2003).

120. Chilla, R. Late histocryl-induced complications of dura surgery in the frontal and lateral base of the skull. *HNO*, 35, 250–251 (1987).

121. Lämsä, T., Jin, H. T., Sand, J., and Nordback, I. Tissue adhesives and the pancreas: Biocompatibility and adhesive properties of 6 preparations. *Pancreas*, 36, 261–266 (2008).

122. Sato, T., Yamazaki, K., Toyota, J., Karino, Y., Ohmura, T., and Suga, T. Inflammatory tumor in pancreatic tail induced by endoscopic ablation with cyanoacrylate glue for gastric varices. *J. Gastroenterol.*, 39, 475–478 (2004).

123. Papatheofanis, F. J. Cytotoxicity of alkyl-2-cyanoacrylate adhesives. *J. Biomed. Mater. Res.*, 23, 661–668 (1989).

124. Ekelund, A. and Nilsson, O. S. Tissue adhesives inhibit experimental new bone formation. *Int. Orthop.*, 15, 331–334 (1991).

125. Leggat, P. A., Smith, D. R., and Kedjarune, U. Surgical applications of cyanoacrylate adhesives: A review of toxicity. *ANZ. J. Surg.*, 77, 209–213 (2007).

126. U.S. EPA. Assessment of health risks to garment workers and certain home residents from exposure to formaldehyde. Office of Pesticides and Toxic Substances, U.S. Environmental Protection Agency, Washington, DC (1987).

127. Reiter, A. Induction of sarcomas by the tissue-binding substance Histocryl-blau in the rat. *Z. Exp. Chir. Transplant. Kunstliche Organe.*, 20, 55–60 (1987).

128. Singer, A. J., Quinn, J. V., and Hollander, J. E. The cyanoacrylate topical skin adhesives. *Am. J. Emerg. Med.*, 26, 490–496 (2008).

129. van den Ende, E. D., Vriens, P. W., Allema, J. H., and Breslau, P. J. Adhesive bonds or percutaneous absorbable suture for closure of surgical wounds in children. Results of a prospective randomized trial. *J. Pediatr. Surg.*, 39, 1249–1251 (2004).

130. Braunwald, N. S. and Tatooles, C. J. Use of a cross linked gelatin tissue adhesive to control hemorrhage from liver and kidney. *Surg. Forum*, 16, 345–346 (1965).

131. Falb, R. D. and Cooper, C. W. Adhesives in surgery. *New Scientist*, 308–309 (1966).

132. Tatooles, C. J. and Braunwald, N. S. The use of crosslinked gelatin as a tissue adhesive to control hemorrhage from liver and kidney. *Surgery*, 60, 857–861 (1966).

133. Braunwald, N. S., Gay, W., and Tatooles, C. J. Evaluation of crosslinked gelatin as a tissue adhesive and hemostatic agent: An experimental study. *Surgery*, 59, 1024–1030 (1966).

134. Bonchek, L. I. and Braunwald, N. S. Experimental evaluation of a cross-linked gelatin adhesive in gastrointestinal surgery. *Ann. Surg.*, 165, 420–424 (1967).

135. Bonchek, L. I., Fuchs, J. C., and Braunwald, N. S. Use of a cross-linked gelatin tissue adhesive in surgery of the urinary tract. *Surg. Gynecol. Obstet.*, 125, 1301–1306 (1967).

136. Cooper, C. W. and Falb, R. D. Surgical adhesives. *Ann. N.Y. Acad. Sci.*, 146, 214–224 (1968).

137. Laurian, C., Gigou, F., and Guilmet, D. Gelatin resorcin formaldehyde glue in vascular surgery. *Nouv. Presse Med.*, 6, 3221–3223 (1977).

138. Kunihara, T., Shiiya, N., Matsuzaki, K., Murashita, T., and Matsui, Y. Recommendation for appropriate use of GRF glue in the operation for acute aortic dissection. *Ann. Thorac. Cardiovasc. Surg.*, 14, 88–95 (2008).

139. Bachet, J., Goudot, B., Dreyfus, G. D., Brodaty, D., Dubois, C., Delentdecker, P., and Guilmet, D. Surgery for acute type A aortic dissection: The Hopital Foch experience (1977–1998). *Ann. Thorac. Surg.*, 67, 2006–2009 (1999).

140. Neri, E., Massetti, M., Capannini, G., Carone, E., and Sassi, C. Glue containment and anastomosis reinforcement in repair of aortic dissection. *Ann. Thorac. Surg.*, 67, 1510–1511 (1999).

141. Stassano, P., Rispo, G., Losi, M., Caputo, M., and Spampinato, N. Annular abscesses and GRF glue. *J. Card. Surg.*, 9, 357–360 (1994).

142. Iha, K., Arakaki, K., Horikawa, Y., Akasaki, M., Kuniyoshi, Y., and Koja, K. Sutureless technique for subacute left ventricular free wall rupture: A case report of an 85-year-old. *Ann. Thorac. Cardiovasc. Surg.*, 5, 265–268 (1999).

143. Okada, K., Yamashita, T., Matsumori, M., Hino, Y., Hanafusa, Y., Ozaki, N., Tsuji, Y., and Okita, Y. Surgical treatment for rupture of left ventricular free wall after acute myocardial infarction. *Interact. Cardiovasc. Thorac. Surg.*, 4, 203–206 (2005).

144. Isoda, S., Imoto, K., Uchida, K., Hashiyama, N., Yanagi, H., Tamagawa, H., and Takanashi, Y. Sandwich technique via right ventricle incision to repair postinfarction ventricular septal defect. *J. Card. Surg.*, 19, 149–150 (2004).
145. Takahashi, N., Tsunematsu, K., Sugawara, H., Kusajima, K., and Abe T. The estimation of the effectiveness of GRF glue in the respiratory. *Kyobu Geka*, 54, 141–145 (2001).
146. Hasumi, T., Yamanaka, S., Yamanaka, H., Endoh, C., and Suda, H. Clinical experience of gelatin–resorcin–formal (GRF) glue for acute empyema with bronchopleural fistula. *Kyobu Geka*, 56, 82–85 (2003).
147. Rittoo, D., Sintler, M., Burnley, S., Millns, P., Smith, S., and Vohra, R. Gelatine–resorcine–formol glue as a sealant of ePTFE patch suture lines. *Int. Angiol.*, 20, 214–217 (2001).
148. Suzuki, S., Imoto, K., Uchida, K., and Takanashi, Y. Aortic root necrosis after surgical treatment using gelatin–resorcinol–formaldehyde (GRF) glue in patients with acute type A aortic dissection. *Ann. Thorac. Cardiovasc. Surg.*, 12, 333–340 (2006).
149. Suehiro, K., Hata, T., Yoshitaka, H., Tsushima, Y., Matsumoto, M., Hamanaka, S., Mohri, M., Ohtani, S., Nagao, A., and Kojima, T. Late aortic root redissection following surgical treatment for acute type A aortic dissection using gelatin–resorcin–formalin glue. *Jpn. J. Thorac. Cardiovasc. Surg.*, 50, 195–200 (2002).
150. Martinelli, L., Graffigna, A., Guarnerio, M., Bonmassari, R., and Disertori, M. Coronary artery narrowing after aortic root reconstruction with resorcin–formalin glue. *Ann. Thorac. Surg.*, 70, 1701–1702 (2000).
151. Ennker, I. C., Ennker, J., Schoon, D., Schoon, H. A., Rimpler, M., and Hetzer, R. Formaldehyde-free collagen glue in experimental lung gluing. *Ann. Thorac. Surg.*, 57, 1622–1627 (1994).
152. Ennker, J., Ennker, I. C., Schoon, D., Schoon, H. A., Dörge, S., Meissler, M., Rimpler, M., and Hetzer, R. The impact of gelatin–resorcinol glue on aortic tissue: A histomorphologic evaluation. *J. Vasc. Surg.*, 20, 34–43 (1994).
153. Chao, H.-H. and Torchiana, D. F. BioGlue: Albumin/glutaraldehyde sealant in cardiac surgery. *J. Card. Surg.*, 18, 500–503 (2003).
154. Chafke, N., Gasser, B., Lindner, V., Rouyer, N., Rooke, R., Kretz, J. G., Nicolini, P., and Eisenmann, B. Albumin as a sealant for a polyester vascular prosthesis: Its impact on the healing sequence in humans. *J. Cardiovasc. Surg. (Torino)*, 37, 431–440 (1996).
155. Yuen, T. and Kaye, A. H. Persistence of Bioglue® in spinal dural repair. *J. Clin. Neurosci.*, 12, 100–101 (2005).
156. Raanani, E., Georghiou, G. P., Kogan, A., Wandwi, B., Shapira, Y., and Vidne, B. A. 'BioGlue' for the repair of aortic insufficiency in acute aortic dissection. *J. Heart Valve Dis.*, 13, 734–737 (2004).
157. Passage, J., Jalali, H., Tam, R. K., Harrocks, S., and O'Brien, M. F. BioGlue Surgical Adhesive—An appraisal of its indications in cardiac surgery. *Ann. Thorac. Surg.*, 74, 432–437 (2002).
158. Coselli, J. S., Bavaria, J. E., Fehrenbacher, J., Stowe, C. L., Macheers, S. K., and Gundry, S. R. Prospective randomized study of a protein-based tissue adhesive used as a hemostatic and structural adjunct in cardiac and vascular anastomotic repair procedures. *J. Am. Coll. Surg.*, 197, 243–252 (2003).
159. Leva, C., Bruno, P. G., Gallorini, C., Lazzarini, I., Musazzi, G., Vittonati, L., Rizzo, L., and Di Credico, G. Complete myocardial revascularization and sutureless technique for left ventricular free wall rupture: Clinical and echocardiographic results. *Interact. Cardiovasc. Thorac. Surg.*, 5, 408–412 (2006).
160. Goldstein, D. J. and Beauford, R. B. Left ventricular assist devices and bleeding: Adding insult to injury. *Ann. Thorac. Surg.*, 75, S42–S47 (2003).
161. Potaris, K., Mihos, P., and Gakidis, I. Preliminary results with the use of an albumin–glutaraldehyde tissue adhesive in lung surgery. *Med. Sci. Monit.*, 9, P179–P183 (2003).
162. Passage, J., Tam, R., Windsor, M., and O'Brien, M. Bioglue: A review of the use of this new surgical adhesive in thoracic surgery. *ANZ J. Surg.*, 75, 315–318 (2005).
163. Nadler, R. B., Loeb, S., Rubenstein, R. A., and Vardi, I. Y. Use of BioGlue in laparoscopic partial nephrectomy. *Urology*, 68, 416–418 (2006).
164. Biggs, G., Hafron, J., Feliciano, J., and Hoenig, D. M. Treatment of splenic injury during laparoscopic nephrectomy with BioGlue, a surgical adhesive. *Urology*, 66, 882 (2006).
165. Ehrlich, P. F., Teitelbaum, D. H., Hirschl, R. B., and Rescorla, F. Excision of large cystic ovarian tumors: Combining minimal invasive surgery techniques and cancer surgery—The best of both worlds. *J. Pediatr. Surg.*, 42, 890–893 (2007).
166. Anghelacopoulos, S. E., Tagarakis, G. I., Pilpilidis, I., Kartsounis, C., and Chryssafis, G. Albumin–glutaraldehyde bioadhesive ("Bioglue") for prevention of postoperative complications after stapled hemorrhoidopexy: A randomized controlled trial. *Wien. Klin. Wochenschr.*, 118, 469–472 (2006).
167. de la Portilla, F., Rada, R., León, E., Cisneros, N., Maldonado, V. H., and Espinosa, E. Evaluation of the use of BioGlue in the treatment of high anal fistulas: Preliminary results of a pilot study. *Dis. Colon Rectum*, 50, 218–222 (2007).

168. Dusick, J. R., Mattozo, C. A., Esposito, F., and Kelly, D. F. BioGlue for prevention of postoperative cerebrospinal fluid leaks in transsphenoidal surgery: A case series. *Surg. Neurol.*, 66, 371–376 (2006).

169. Bjarkam, C. R., Jorgensen, R. L., Jensen, K. N., Sunde, N. A., and Sørensen, J. C. Deep brain stimulation electrode anchoring using BioGlue((R)), a protective electrode covering, and a titanium microplate. *J. Neurosci. Methods*, 168, 151–155 (2008).

170. Sen, A., Green, K. M., Khan, M. I., Saeed, S. R., and Ramsden, R. T. Cerebrospinal fluid leak rate after the use of BioGlue in translabyrinthine vestibular Schwannoma surgery: A prospective study. *Otol. Neurotol.*, 27, 102–105 (2006).

171. Friedman, M. and Schalch, P. Middle turbinate medialization with bovine serum albumin tissue adhesive (BioGlue). *Laryngoscope*, 118, 335–338 (2008).

172. Sidle, D. M., Loos, B. M., Ramirez, A. L., Kabaker, S. S., and Maas, C. S. Use of BioGlue surgical adhesive for brow fixation in endoscopic browplasty. *Arch. Facial Plast. Surg.*, 7, 393–397 (2005).

173. Fürst, W. and Banerjee, A. Release of glutaraldehyde from an albumin–glutaraldehyde tissue adhesive causes significant in vitro and in vivo toxicity. *Ann. Thorac. Surg.*, 79, 1522–1528 (2005).

174. Calafiore, A. M., DiGiamarco, G., and Vitolla, G. Aortic valve exposure through a combined right atrial-ascending aortic approach in redo cases. *Ann. Thorac. Surg.*, 73, 318–319 (2002).

175. Erasmi, A. W. and Wohlschlager, C. Inflammatory response after BioGlue application. *Ann. Thorac. Surg.*, 73, 1025 (2002).

176. Kazui, T., Washiyama, N., Bashar, A. H., Terada, H., Suzuki, K., Yamashita, K., and Takinami, M. Role of biologic glue repair of proximal aortic dissection in the development of early and midterm redissection of the aortic root. *Ann. Thorac. Surg.*, 72, 509–514 (2001).

177. LeMaire, S. A., Schmittling, Z. C., Coselli, J. S., Undar, A., Deady, B. A., Clubb, F. J. Jr., and Fraser, C. D. Jr. BioGlue surgical adhesive impairs aortic growth and causes anastomotic strictures. *Ann. Thorac. Surg.*, 73, 1500–1505 (2002).

178. Economopoulos, G. C., Dimitrakakis, G. K., Brountzos, E., and Kelekis, D. A. Superior vena cava stenosis: A delayed BioGlue complication. *J. Thorac. Cardiovasc. Surg.*, 127, 1819–1821 (2004).

179. Szafranek, A., Podila, S. R., Al-Khyatt, W., and Kulatilake, E. N. Aseptic mediastinal cyst caused by BioGlue 7 months after cardiac surgery. *J. Thorac. Cardiovasc. Surg.*, 131, 1202–1203 (2006).

180. Haj-Yahia, S., Mittal, T., Birks, E., Carby, M., Petrou, M., Pepper, J., Dreyfus, G., and Amrani, M. Lung fibrosis as a potential complication of the hemostatic tissue sealant, biologic glue (Bioglue). *J. Thorac. Cardiovasc. Surg.*, 133, 1387–1388 (2007).

181. Ngaage, D. L., Edwards, W. D., Bell, M. R., and Sundt, T. M. A cautionary note regarding long-term sequelae of biologic glue. *J. Thorac. Cardiovasc. Surg.*, 129, 937–938 (2005).

182. LeMaire, S. A., Carter, S. A., Won, T., Wang, X. W., Conklin, L. D., and Coselli, J. S. The threat of adhesive embolization: BioGlue leaks through needle holes in aortic tissue and prosthetic grafts. *Ann. Thorac. Surg.*, 80, 106–111 (2005).

183. Mahmood, Z., Cook, D. S., Luckraz, H., and O'Keefe, P. Fatal right ventricular infarction caused by Bioglue coronary embolism. *J. Thorac. Cardiovasc. Surg.*, 128, 770–771 (2004).

184. Bernabeu, E., Castellá, M., Barriuso, C., and Mulet, J. Acute limb ischemia due to embolization of biological glue after repair of type A aortic dissection. *Interact. Cardiovasc. Thorac. Surg.*, 4, 329–331 (2005).

185. Devbhandari, M. P., Chaudhery, Q., and Duncan, A. J. Acute intraoperative malfunction of aortic valve due to surgical glue. *Ann. Thorac. Surg.*, 81, 1499–1500 (2006).

186. Shapira, Y., Raanani, E., and Sagie, A. "BioGlue" as a possible cause of acute blocked mechanical mitral valve leaflet. *J. Cardiovasc. Surg. (Torino)*, 47, 581–583 (2006).

187. Lemaire, S. A., Ochoa, L. N., Conklin, L. D., Schmittling, Z. C., Undar, A., Clubb, F. J. Jr., Li Wang, X., Coselli, J. S., and Fraser, C. D. Jr. Nerve and conduction tissue injury caused by contact with BioGlue. *J. Surg. Res.*, 143, 286–293 (2007).

188. Klimo, P. Jr., Khalil, A., Slotkin, J. R., Smith, E. R., Scott, R. M., and Goumnerova, L. C. Wound complications associated with the use of bovine serum albumin–glutaraldehyde surgical adhesive in pediatric patients. *Neurosurgery*, 60, 305–309 (2007).

189. Abbas, M. A. and Tejirian, T. Bioglue for the treatment of anal fistula is associated with acute anal sepsis. *Dis. Colon Rectum*, 51, 1155 (2008).

190. Van Belleghem, Y., Forsyth, R. G., Narine, K., Moerman, A., Taeymans, Y., and Van Nooten, G. J. Bovine glue (BioGlue) is catabolized by enzymatic reaction in the vascular dog model. *Ann. Thorac. Surg.*, 77, 2177–2181 (2004).

191. Caliceti, P. and Veronese, F. M. Pharmacokinetic and biodistribution properties of poly(ethylene glycol)–protein conjugates. *Adv. Drug. Deliv. Rev.*, 55, 1261–1277 (2000).

192. Abuchowski, A., van Es, T., Palczuk, N. C., and Davis, F. F. Alteration of immunological properties of bovine serum albumin by covalent attachment of polyethylene glycol. *J. Biol. Chem.*, 252, 3578–3581 (1977).

193. Abuchowski, A., McCoy, J. R., Palczuk, N. C., van Es, T., and Davis, F. F. Effect of covalent attachment of polyethylene glycol on immunogenicity and circulating life of bovine liver catalase. *J. Biol. Chem.*, 252, 3582–3586 (1977).

194. Sawhney, A. S., Pathak, C. P., and Hubbell, J. A. Bioerodible hydrogels based on photopolymerized poly(ethylene glycol)-*co*-poly(α-hydroxy acid) diacrylate macromers. *Macromolecules*, 26, 581–587 (1993).

195. Hill-West, J. L., Chowdhury, S. M., Sawhney, A. S., Pathak, C. P., Dunn, R. C., and Hubbell, J. A. Prevention of postoperative adhesions in the rat by in situ photopolymerization of bioresorbable hydrogel barriers. *Obstet. Gynecol.*, 83, 59–64 (1994).

196. Sawhney, A. S., Pathak, C. P., van Rensburg, J. J., Dunn, R. C., and Hubbell, J. A. Optimization of photopolymerized bioerodible hydrogel properties for adhesion prevention. *J. Biomed. Mater. Res.*, 28, 831–838 (1994).

197. Dumanian, G. A., Dascombe, W., Hong, C., Labadie, K., Garrett, K., Sawhney, A. S., Pathak, C. P., Hubbell, J. A., and Johnson, P. C. A new photopolymerizable blood vessel glue that seals human vessel anastomoses without augmenting thrombogenicity. *Plast. Reconstr. Surg.*, 95, 901–907 (1995).

198. Ranger, W. R., Halpin, D., Sawhney, A. S., Lyman, M., and Locicero, J. Pneumostasis of experimental air leaks with a new photopolymerized synthetic tissue sealant. *Am. Surg.*, 63, 788–795 (1997).

199. Sawhney, A. S. and Hubbell, J. A. In situ photopolymerized hydrogels for vascular and peritoneal wound healing, in *Tissue Engineering Methods and Protocols*, J. R. Morgan, M. L. Yarmush, eds. Totowa, NJ: Humana Press (1999).

200. Macchiarini, P., Wain, J., Almy, S., and Dartevelle, P. Experimental and clinical evaluation of a new synthetic, absorbable sealant to reduce air leaks in thoracic operations. *J. Thorac. Cardiovasc. Surg.*, 117, 751–758 (1999).

201. Wain, J. C., Kaiser, L. R., Johnstone, D. W., Yang, S. C., Wright, C. D., Friedberg, J. S., Feins, R. H., Heitmiller, R. F., Mathisen, D. J., and Selwyn, M. R. Trial of a novel synthetic sealant in preventing air leaks after lung resection. *Ann. Thorac. Surg.*, 71, 1623–1628 (2001).

202. Gillinov, A. M. and Lytle, B. W. A novel synthetic sealant to treat air leaks at cardiac reoperation. *J. Card. Surg.*, 16, 255–257 (2001).

203. Fasol, R., Wild, T., and El Dsoki, S. Left ventricular rupture after mitral surgery: Repair by patch and sealing. *Ann. Thorac. Surg.*, 77, 1070–1072 (2004).

204. Tanaka, K., Takamoto, S., Ohtsuka, T., Kotsuka, Y., and Kawauchi, M. Application of AdvaSeal for acute aortic dissection: Experimental study. *Ann. Thorac. Surg.*, 68, 1308–1312 (1999).

205. White, J. K., Titus, J. S., Tanabe, H., Aretz, H. T., and Torchiana, D. F. The use of a novel tissue sealant as a hemostatic adjunct in cardiac surgery. *Heart Surg. Forum*, 3, 56–61 (2000).

206. Kato, Y., Yamataka, A., Miyano, G., Tei, E., Koga, H., Lane, G. J., and Miyano, T. Tissue adhesives for repairing inguinal hernia: A preliminary study. *J. Laparoendosc. Adv. Surg. Tech. A*, 15, 424–428 (2005).

207. Argyra, E., Polymeneas, G., Karvouni, E., Kontorravdis, N., Theodosopoulos, T., and Arkadopoulos, N. Sutureless pancreatojejunal anastomosis using an absorbable sealant: Evaluation in a pig model. *J. Surg. Res.* 153, 282–286 (2009).

208. Sweeney, T., Rayan, S., Warren, H., and Rattner, D. Intestinal anastomoses detected with a photopolymerized hydrogel. *Surgery*, 131, 185–189 (2002).

209. Ferguson, R. E. and Rinker, B. The use of a hydrogel sealant on flexor tendon repairs to prevent adhesion formation. *Ann. Plast. Surg.*, 56, 54–58 (2006).

210. Wallace, D. G., Cruise, G. M., Rhee, W. M., Schroeder, J. A., Prior, J. J., Ju, J., et al. A tissue sealant based on reactive multifunctional polyethylene glycol. *J. Biomed. Mater. Res.*, 58, 545–555 (2001).

211. Marc Hendrikx, M., Mees, U., Hill, A. C., Egbert, B., Coker, G. T., and Estridge, T. D. Evaluation of a novel synthetic sealant for inhibition of cardiac adhesions and clinical experience in cardiac surgery procedures. *Heart Surg. Forum*, 4, 204–209 (2001).

212. Glickman, M., Gheissari, A., Money, S., Martin, J., and Ballard, J. L. CoSeal Multicenter Vascular Surgery Study Group. A polymeric sealant inhibits anastomotic suture hole bleeding more rapidly than gelfoam/thrombin: Results of a randomized controlled trial. *Arch. Surg.*, 137, 326–331 (2002).

213. Hagberg, R. C., Safi, H. J., Sabik, J., Conte, J., and Block, J. E. Improved intraoperative management of anastomotic bleeding during aortic reconstruction: Results of a randomized controlled trial. *Am. Surg.*, 70, 307–311 (2004).

214. Mettler, L., Hucke, J., Bojahr, B., Tinneberg, H. R., Leyland, N., and Avelar, R. A safety and efficacy study of a resorbable hydrogel for reduction of post-operative adhesions following myomectomy. *Hum. Reprod.*, 23, 1093–1100 (2008).

215. Konertz, W. F., Kostelka, M., Mohr, F. W., Hetzer, R., Hubler, M., Ritter, J., Liu, J., Koch, C., and Block, J. E. Reducing the incidence and severity of pericardial adhesions with a sprayable polymeric matrix. *Ann. Thorac. Surg.*, 76, 1270–1274 (2003).

216. Napoleone, C. P., Oppido, G., Angeli, E., and Gargiulo, G. Resternotomy in pediatric cardiac surgery: CoSeal initial experience. *Interact. Cardiovasc. Thorac. Surg.*, 6, 21–23 (2007).

217. Saunders, M. M., Baxter, Z. C., Abou-Elella, A., Kunselman, A. R., and Trussell, J. C. BioGlue and Dermabond save time, leak less, and are not mechanically inferior to two-layer and modified one-layer vasostomy. *Fertil. Steril.* 91, 560–565 (2009).

218. Bernie, J. E., Ng, J., Bargman, V., Gardner, T., Cheng, L., and Sundaram, C. P. Evaluation of hydrogel tissue sealant in porcine laparoscopic partial-nephrectomy model. *J. Endourol.*, 19, 1122–1126 (2005).

219. Preul, M. C., Bichard, W. D., and Spetzler, R. F. Toward optimal tissue sealants for neurosurgery: Use of a novel hydrogel sealant in a canine durotomy repair model. *Neurosurgery*, 53, 1189–1198 (2003).

220. Boogaarts, J. D., Grotenhuis, J. A., Bartels, R. H., and Beems, T. Use of a novel absorbable hydrogel for augmentation of dural repair: Results of a preliminary clinical study. *Neurosurgery*, 57 (1 Suppl), 146–151 (2005).

221. Cosgrove, G. R., Delashaw, J. B., Grotenhuis, J. A., Tew, J. M., Van Loveren, H., Spetzler, R. F., Payner, T., et al. Safety and efficacy of a novel polyethylene glycol hydrogel sealant for watertight dural repair. *J. Neurosurg.*, 106, 52–58 (2007).

222. Graham, L. D., Glattauer, V., Huson, M. G., Maxwell, J. M., Knott, R. B., White, J. W., Vaughan, P. R., et al. Characterization of a protein-based adhesive elastomer secreted by the Australian frog *Notaden bennetti*. *Biomacromolecules*, 6, 3300–3312 (2005).

223. Szomor, Z. L., Murrell, G. A. C., Appleyard, R. C., and Tyler, M. J. Meniscal repair with a new biological glue: An ex vivo study. *Tech. Knee Surg.*, 7, 261–265 (2008).

224. Ninan, L., Monahan, J., Stroshine, R. L., Wilker, J. J., and Shi, R. Adhesive strength of marine mussel extracts on porcine skin. *Biomaterials*, 24, 4091–4099 (2003).

225. Ninan, L., Stroshine, R. L., Wilker, J. J., and Shi, R. Adhesive strength and curing rate of marine mussel protein extracts on porcine small intestinal submucosa. *Acta Biomater.*, 3, 687–694 (2007).

226. Strausberg, R. L. and Link, R. Protein-based medical adhesives. *Trends Biotechnol.*, 8, 53–57 (1990).

227. Burke, S. A., Ritter-Jones, M., Lee, B. P., and Messersmith, P. B. Thermal gelation and tissue adhesion of biomimetic hydrogels. *Biomed. Mater.*, 2, 203–210 (2007).

228. Autumn, K., Liang, Y. A., Hsieh, S. T., Zesch, W., Chan, W. P., Kenny, T. W., Fearing, R., and Full, R. J. Adhesive force of a single gecko foot-hair. *Nature*, 405, 681–685 (2000).

229. Mahdavi, A., Ferreira, L., Sundback, C., Nichol, J. W., Chan, E. P., Carter, D. J., Bettinger, C. J. et al. A biodegradable and biocompatible gecko-inspired tissue adhesive. *Proc. Natl. Acad. Sci. USA*, 105, 2307–2312 (2008).

230. Wang, D. A., Varghese, S., Sharma, B., Strehin, I., Fermanian, S., Gorham, J., Fairbrother, D. H., Cascio, B., and Elisseeff, J. H. Multifunctional chondroitin sulphate for cartilage tissue-biomaterial integration. *Nat. Mater.*, 6, 385–392 (2007).

231. Miki, D., Dastgheib, K., Kim, T., Pfister-Serres, A., Smeds, K. A., Inoue, M., Hatchell, D. L., and Grinstaff, M. W. A photopolymerized sealant for corneal lacerations. *Cornea*, 21, 393–399 (2002).

232. Bhatia, S. K., Arthur, S. D., Chenault, H. K., and Kodokian, G. K. Interactions of polysaccharide-based tissue adhesives with clinically relevant fibroblast and macrophage cell lines. *Biotechnol. Lett.*, 29, 1645–1649 (2007).

233. Bhatia, S. K., Arthur, S. D., Chenault, H. K., Figuly, G. D., and Kodokian, G. K. Polysaccharide-based tissue adhesives for sealing corneal incisions. *Curr. Eye Res.*, 32, 1045–1050 (2007).

234. Grinstaff, M. W. Dendritic macromers for hydrogel formation: Tailored materials for ophthalmic, orthopedic, and biotech applications. *J. Polym. Sci. Part A: Polym. Chem.*, 46, 383–400 (2008).

235. Wathier, M., Jung, P. J., Carnahan, M. A., Kim, T., and Grinstaff, M. W. Dendritic macromers as in situ polymerizing biomaterials for securing cataract incisions. *J. Am. Chem. Soc.*, 126, 12744–12745 (2004).

236. Wathier, M., Johnson, C. S., Kim, T., and Grinstaff, M. W. Hydrogels formed by multiple peptide ligation reactions to fasten corneal transplants. *Bioconj. Chem.*, 17, 873–876 (2006).

237. Degoricija, L., Johnson, C. S., Wathier, M., Kim, T., and Grinstaff, M. W. Photo cross-linkable bio-dendrimers as ophthalmic adhesives for central lacerations and penetrating keratoplasties. *Invest. Ophthalmol. Vis. Sci.*, 48, 2037–2042 (2007).

238. Kang, P. C., Carnahan, M. A., Wathier, M., Grinstaff, M. W., and Kim, T. Novel tissue adhesives to secure laser in situ keratomileusis flaps. *J. Cataract Refract. Surg.*, 31, 1208–1212 (2005).
239. Degoricija, L., Bansal, P. N., Söntjens, S. H., Joshi, N. S., Takahashi, M., Snyder, B., and Grinstaff, M. W. Hydrogels for osteochondral repair based on photocrosslinkable carbamate dendrimers. *Biomacromolecules*, 9, 2863–2872 (2008).
240. Söntjens, S. H., Nettles, D. L., Carnahan, M. A., Setton, L. A., and Grinstaff, M. W. Biodendrimer-based hydrogel scaffolds for cartilage tissue repair. *Biomacromolecules*, 7, 310–316 (2006).

18 Polymeric Interactions with Drugs and Excipients

James C. DiNunzio and James W. McGinity

CONTENTS

18.1 INTRODUCTION

Numerous advances in polymer chemistry have brought to market a variety of polymeric materials which have had a substantial impact in numerous industrial fields. Particularly in the pharmaceutical field, these advances have revolutionized not only the type of materials incorporated into drug delivery systems, but also the way in which these systems are produced and the methods for engineering such systems. Examination of the pharmaceutical field shows that prior to the 1970s the vast majority of pharmaceutical systems intended for oral administration were designed to provide a rapid immediate release, with only a few simple matrix systems available at that time for controlled delivery applications. One also notes that the materials utilized for such applications were primarily of natural origin, stemming from cellulose derivatives. In more recent years, synthetic polymers have appeared on the pharmaceutical stage, allowing for significantly greater utility and versatility of polymeric materials for pharmaceutical applications. The emergence of these materials has also been coupled with tremendous advances in processing technology of pharmaceutical systems, particularly the emergence of nanotechnology and the development of amorphous solid dispersions, which have allowed polymeric materials to revolutionize the way in which drug delivery systems are designed, formulated, and manufactured.

As a direct result of the increasing complexity of pharmaceutical systems, polymeric interactions with drugs and other excipients in the formulations play a significant role in determining product attributes. Product properties such as chemical purity, drug release, and physical appearance can all be directly influenced by the type of polymeric material chosen for development of a pharmaceutical system, ultimately leading to success or failure of the product under development. Additionally, because of the regulatory requirements for pharmaceutical systems to provide well-controlled physical and chemical properties throughout the product shelf life, pharmaceutical scientists must consider not only the initial properties of the product but also the long-term properties to ensure accurate and reproducible performance throughout the product life. These properties can be impacted by phenomena occurring over extended timescales. As such, one must think about polymeric interactions with drugs and excipients in terms of the immediate effect and long-term stability of the system in order to develop a suitable pharmaceutical system.

In accordance with the Code of Federal Regulations, the United States Food and Drug Administration (FDA) has been established to determine the safety and efficacy of drug products intended for medicinal use within the country. Similar regulatory bodies exist in other regions of the world, which include the

European Medicines Agency (EMEA) and Japanese Ministry of Health, Labour and Welfare (MHLW). Recently, as the world marketplace has become more global in nature, the major regulatory bodies have worked to harmonize guidelines for regulatory practices under the International Conference on Harmonization (ICH). This consortium of regulatory agencies has worked to establish guidelines for the development of pharmaceutical and medical systems, outlining many of the synchronized regulatory policies to which pharmaceutical systems must comply. These policies cover the properties both for initial product release and for long-term stability. While many of these policies focus primarily on the properties of the active ingredient maintained within the product, growing regulatory discussion is also focusing on the properties of the polymers themselves. Additionally, these regulatory guidelines cover specific drug release properties which are frequently directly related to the chemical and physical stability of the active ingredient and polymers present, such as that found in a controlled release film-coated system. Since these regulations provide a basis for monitoring pharmaceutical product attributes in the context of providing optimized drug delivery systems dependent on the interaction with polymer systems, significant consideration must be given to the testing requirements provided.

Generally speaking, pharmaceutical systems intended for oral drug delivery can be subdivided into three major categories: traditional matrix systems, film-coated systems, and advanced systems. Traditional matrix systems encompass conventional tablet and capsule systems which have been available for over a century. These systems are generally characterized by a rapid drug dissolution profile and are commonly made using well-characterized excipients. A subcategory within this group is the controlled release matrix tablet, where a drug release rate-modifying system regulates the delivery of the active ingredient from the device. In these systems, specific interactions between the rate-controlling polymer and other components in the tablet may impact the observed release behavior from the system. Film-coated systems represent another distinct subclass of oral pharmaceutical systems and consist of dosage forms which have been coated for cosmetic and/or alternative functional reasons, such as immediate release tablets for taste masking, and multiparticulate systems which have been coated to achieve a functional release property, such as an enteric or sustained release. For such systems, particularly for those where the coating has been applied for release-modulating purposes, the interactions of the polymeric material providing the function with other materials present in the dosage form, as well as time, can play a significant role in determining product performance and even patient's safety. Environmental factors also provide significant interactions with the polymers chosen, allowing these interactions to regulate the release behavior from the system. Finally, a third class of oral dosage form is broadly classified as advanced dosage forms, which represent systems prepared using advanced pharmaceutical manufacturing technologies including hot-melt extrusion, spray–drying, and nanoparticle production. Unlike the previous two classifications of pharmaceutical systems, this group typically is employed in pharmaceutical production to enhance the oral bioavailability of the active pharmaceutical ingredient. In recent years, the advances in early-stage drug discovery technologies have led to an increasing number of candidate new chemical entities which exhibit poor aqueous solubility. These advanced technology platforms provide the ability to improve dissolution rate and solubility through increases in specific surface area and modification of thermodynamic properties of the formulation, and while effective at improving oral bioavailability, rely heavily on polymeric interactions to achieve and maintain such properties throughout the product shelf life. Divided into three major subsections based on the type of pharmaceutical production, this chapter describes the behavior of pharmaceutical polymers in terms of their interactions with active pharmaceutical ingredients, other commonly used pharmaceutical excipients, and various environmental factors to ultimately yield desired or undesired behavior of the system.

18.2 POLYMERIC INTERACTIONS IN TRADITIONAL MATRIX SYSTEMS

Traditional matrix pharmaceutical systems consist of a variety of specific dosage forms however, for oral delivery applications these typically include tablets and capsules. Fundamentally defined as an intimate mixture of drug with additional pharmaceutical adjuvants, these systems can be developed to provide either immediate release, controlled release, or some combination thereof, where drug release is controlled by two unique mechanisms: disintegration and dissolution. For the preparation of such systems, the basic underlying unit operations are similar and have been well-studied over the last century.

For the production of pharmaceutical tablets and capsules, the basic manufacturing scheme is similar regardless of the type of system to be prepared (i.e., immediate release or controlled release) and consists of two major unit operations: blending and compression or capsule filling. During the blending process, the active ingredient is combined with required adjuvants in order to facilitate downstream production of the finished dosage form, with such materials generally consisting of diluents, disintegrants, glidants, and lubricants. In the case of controlled release systems, rate-controlling polymers such as hypromellose (HPMC) may be added to the blend. While glidants and lubricants commonly used in pharmaceutical dosage forms are not polymeric in nature, many of the commonly used diluents and disintegrants can be polymeric materials. Following blending, the formulation is then prepared into either a tablet or capsule dosage form via compression or encapsulation, respectively. For compression, rotary compression is the most commonly used method of manufacture and requires a blend capable of meeting the requisite flow and compressibility characteristics. Underlying material properties and component interactions play a significant role in performance, frequently requiring the use of additives or materials exhibiting specific characteristics to enhance performance. Often times, one sees that specific trade-offs must be made to achieve the necessary balance between compressibility and flowability. Additionally, additives to enhance disintegration and dissolution may be incorporated into the formulation, further complicating the delicate balance in designing such systems. Similarly, for encapsulation adequate flow and compressibility properties are necessary to facilitate production. While both flow and compressibility characteristics are important for achieving acceptable production characteristics, the magnitude of each which must be achieved is not the same as for compression. Generally speaking, some degree of flowability and compressibility are necessary for a formulation to function on a dosator or tamping mechanism encapsulator; however, the degree to which these properties must be optimized is significantly less than that required for compression. As such, flow aids such as glidants and lubricants can be minimized in these formulations. Furthermore, because of the loose powder nature of the material superdisintegrant levels may be reduced, although cases have been observed where capsule fill volume and head space have impacted dissolution. Given this, it is clear that the interaction between polymeric components, such as diluents and disintegrants, and other materials can play a significant role on the observed properties of the dosage form.

For optimization of blend properties, the primary product attributes are primarily associated with the physical properties of the materials much more so than the chemical properties. These properties include flowability, compressibility, and uniformity and are substantially influenced by the material morphology and density. Such attributes have been shown to be impacted by material properties such as molecular weight. In a study by Shlieout et al. (2002), the researchers examined the impact of microcrystalline cellulose molecular weight on compressibility, a vital product attribute for the production of pharmaceutical tablets and capsules. Microcrystalline cellulose is a common polymeric pharmaceutical excipient used as both a diluent and disintegrant which is derived from wood pulp. By modifying the degree of polymerization, the researchers showed that flowability and compressibility properties were impacted due to changes induced to the underlying material morphology when celluloses were prepared with different molecular weights, such that increasing molecular weight resulted in decreased flowability with increased compressibility. As such, there is a clear trade-off for production characteristics based on the molecular weight of the material selected to balance compression and flow characteristics. From a formulation perspective, this means that based on the properties of polymeric additives selected, additional components may be required to achieve the requisite compressibility or flowability characteristics. Batch to batch variation of raw materials may also influence performance, depending on supplier quality requirements. The magnitude of these required compensations will be dependent on the inherent properties of other materials in the formulation as well, particularly the active ingredient and diluent which also generally make up large percentages of the blend formulation.

In addition to the optimization of blend properties to facilitate dosage form production, it is necessary to design a system capable of effective delivery of the active pharmaceutical ingredient. Polymeric interactions which can occur with other materials in the formulation can have a significant impact on oral bioavailability. Such polymeric materials, especially materials such as superdisintegrants and controlled release polymers, which play a critical role in bioavailability and onset of action, can also be subject to specific polymeric interactions with the gastrointestinal (GI) environment encountered upon administration. For example, numerous types of superdisintegrants are commercially available to assist in the disintegration of matrix tablet systems and the vast majority of these systems function by extensive water uptake which leads to a swelling of the polymeric network, thereby exerting a force on the tablet matrix capable of breaking apart interparticulate bonds. The rate at which water is taken up by the material plays a significant role in product performance. This behavior can be impacted by both environmental and compositional considerations. Bussemer et al. (2003) examined the swelling potential of several common disintegrants, evaluating performance using a novel swelling apparatus to measure the force exerted by the materials. Results from the study showed that the swelling energy was proportional to the rate of water uptake, with croscarmellose sodium providing superior performance over a variety of other materials including low substituted hydroxypropyl cellulose (HPC), sodium starch glycolate, crospovidone, and HPMC. Further insight was also provided on the environmental impact to the swelling potential of croscarmellose sodium due to the ionic nature of the polymer. Since croscarmellose sodium is an acidic polymer displaying carboxylic acid functional groups, these groups will remain largely uncharged in acidic conditions, however will become charged in more neutral conditions representative of the later stages of the GI tract. When ionized, the common charge across these groups functions as a repulsive force which complements swelling due to water uptake. Furthermore, the overall rate of water uptake will be increased due to the ionized nature of these materials. Increasing the ionic strength of the bulk fluid, however, showed that there was a decrease in swelling energy due to competition of the ions for free water. Even with this observed behavior however, croscarmellose sodium still provided significantly greater swelling energy than the other materials studied, which indicated that while such polymeric interactions exist they may not have a physiologically relevant impact on the drug product performance. The magnitude of such interactions must therefore be addressed on a case-by-case basis.

Similar polymeric interactions have also been observed in controlled release systems, particularly for controlled release systems based on diffusional release through swollen hydrophilic gels. These systems, which provide one of the main technological platforms for oral controlled release of pharmaceutical compounds, have been extensively characterized and also implemented successfully in numerous commercially available products, due in combination to both the well-understood science and simplicity of manufacture. In many cases, controlled release matrix tablets can be prepared simply by directly blending the active ingredients and other excipients with the controlled release polymer prior to tableting in order to prepare such a system. Other techniques may also be applied to the production of such systems; however, even many of the advanced production techniques such as coprocessing of the drug substance with the controlled release polymer fall well within the realm of traditional pharmaceutical manufacturing.

Controlled release hydrogel matrix tablet technology is based on the incorporation of a swellable polymeric material into the dosage form, which provides a platform for swelling and the development of a gel layer which regulates the release of drug from the device in concert with any competing erosion mechanisms (Siepmann et al. 1999, 2002; Siepmann and Peppas 2001; Siepmann and Siepmann 2008). Similar to the behavior of superdisintegrants discussed earlier, controlled release matrix polymers exhibit the ability to absorb water and create a gel layer. Unlike the superdisintegrant class of materials, swelling in these materials is not associated with the same mechanical force that is observed for the later. As noted by Bussemer et al. (2003), HPMC, a commonly used controlled release polymeric material, provided the lowest degree of swelling energy. This behavior allows such materials to form a mechanically intact gel barrier around the dosage form. Within

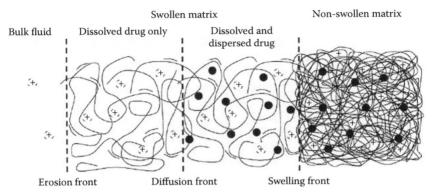

FIGURE 18.1 Schematic description of gel formulation in hydrophilic controlled release systems. (Reprinted from Siepmann, J. and Siepmann, F., *Int. J. Pharm.*, 364, 328, 2008. With permission.)

the gel, polymer chains are interconnected through a combination of intermolecular interactions, chemical bonds, and/or physical intermeshing to create a "weblike" structure capable of providing a diffusional barrier. The release of drug substance from such devices is controlled by a series of complex polymer interactions which contribute to the solubilization of the active ingredient, formation of the gel, and transport through the gel layer. Gel formation and resulting layer thickness are governed by three distinct mechanisms: solvent permeation, swelling, and erosion, as shown in Figure 18.1. Numerous researchers have developed empirical and first principle models describing release from such systems, with the most complete model to date being the sequential layer method. Simple models have been successfully shown to accurately describe drug release kinetics from such systems as well. Specifically, the Peppas equation illustrates the simple first-order release behavior of such systems which can actually be shown to account for the primary first principle behavior of Fickian diffusion based on drug concentration gradient from a simple geometric shape (Siepmann and Peppas 2001). In more complex models, such as the sequential layer method (Siepmann et al. 2002), Fickian release of drug from the dosage form is determined in a sequential, stepwise fashion accounting for water permeation, hydration, solubilization of the active ingredient, and erosion of the gel layer front. Using this model, Siepmann and coworkers were able to use the model to predict dissolution profiles from a variety of HPMC-based controlled release systems having varying polymeric molecular weights and different device geometries. Furthermore, the applicability of this model to a variety of drug substances having a range of physicochemical properties was also demonstrated, providing greater accuracy than could be achieved with simpler earlier generation exponential models.

Given the complex series of competing events which lead to drug release from controlled release matrix systems, it is not surprising that polymeric interactions can also significantly influence release behavior and ultimately pharmaceutical efficacy of the dosage form. Gel layers formed by hydration of the polymeric matrix influence the rate at which drug is released and the environmental factors can significantly influence product performance. It has been well-established that environmental factors such as pH, ionic strength, and temperature can influence the behavior of polymer gels, altering liquid uptake rate, intermolecular chain separation distances, and observed viscosity. Variations in gel behavior due to these interactions can lead to significant changes in release profile, altering the pharmaceutical efficacy of the product produced.

In a study by Johnson and coworkers (Johnson 1993), they examined the release behavior of phenylpropanolamine from HPC matrix systems having different particle size and molecular weight with respect to ionic strength. Their research demonstrated that at ionic strengths between 0.5 and 1.0 mol/L, disaggregation could occur resulting in dose dumping; however, they also noted that this was not likely to occur under normal physiological conditions where ionic strength is generally not more than 0.2 mol/L. Similar studies have also been undertaken to identify ionic strength–dependent

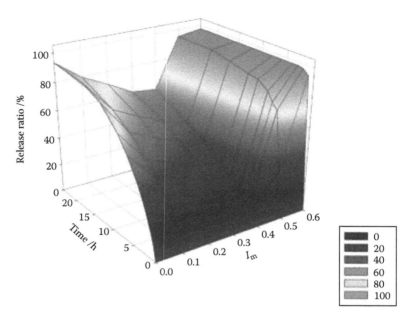

FIGURE 18.2   Dissolution profile of propanolol from HPMC K15M matrix tablets at varying ionic strength. (Reprinted from Xu, X.M. et al., *J. Appl. Polym. Sci.*, 102, 4066, 2006.)

phenomena in HPMC. Xu et al. (2006) reported that the ionic strength–dependent behavior of HPMC was also directly related to the ionic competition for free water within the system which could lead to salting out phenomena. Their results mirrored those reported by Johnson and coworkers, showing that at ionic strengths greater than 0.5 mol/L burst release of propanolol from HMPC K15 tablets resulted, while ionic strengths below 0.5 mol/L had only a slight impact on release properties (Figure 18.2). In addition to ionic strength, pH conditions experienced along the GI tract can also vary significantly. Although many of the common polymers used for hydrophilic controlled release applications are considered to be pH-independent, it is important to note that many pharmaceutical active ingredients are actually weak acids or weak bases capable of exhibiting pH-dependent solubility and ionization which can impact drug release. Changes in active ingredient solubility, particularly when being present at a high loading within the formulation, can also contribute to significant increases in erosion rates. Furthermore, the viscosity and strength of the gel layer formed may be affected due to the microenvironmental and regional pH of the GI tract. While HPMC has been shown to provide relatively independent release in the physiological pH ranges, controlled release tablets composed of carbomer and HPC have been shown to exhibit reduced gel viscosity properties based on pH and can be attributed to the chemical structure of the polymer. HPC, which has nearly 20 times the level of hydroxypropyl substituents than HPMC, providing greater potential for interaction with the environmental media. For carbomer, the carboxylic acids present in the polymer allow for ionization of the material based on environmental conditions, having a reported pK_a of 6.0. This ionization behavior allows for ionic interaction and also can impact intermolecular spacing within the gel based on the state of the material. Changes such as these can significantly impact release by reducing gel layer viscosity in the ionized state, leading to greater release rates in the later stages of the GI tract.

In order to better control the release characteristics from hydrogel-based controlled release matrix systems, pH-modifying agents can be incorporated into these devices to regulate microenvironmental pH. While some polymeric systems may show susceptibility to environmental changes resulting in variable release profiles, pH-modifying agents are generally incorporated in order to minimize a physical incompatibility of the active moiety or better control its pH-dependent solubility.

For example, oral dosage forms of fluvastatin (Kabadi and Vivilecchia 1994) and ifetroban (Nikfar et al. 1996) have been successfully stabilized from pH-dependent decomposition by the incorporation of pH-modifying additives into the formulation. With controlled release systems, the design of matrix systems containing active ingredients which exhibit pH-dependent solubility can be better-controlled using pH-modifying systems. As with active ingredients however, the pH-modifying agent is also subject to the same mass transport phenomena that occur with the active ingredient therefore requiring some consideration of the material properties such as solubility, as well as potential behavior in the polymer gel layer (Kranz et al. 2005). Although intuitively, one might think that the ideal modifying agent would have high solubility, this is in fact not the case. High solubility of the pH-modifying agent can lead to increased transport rates through the gel layer as a result of the concentration gradient, which may lead to a depletion of the component over time. Instead, incorporation of lower solubility material such as fumaric acid may provide a more stabilized release profile (Siepe et al. 2006). Solubility of these additives may also impact water uptake rate within the matrix, yielding a behavior where higher solubility materials facilitate water uptake. Using such an approach, controlled release matrix tablets for ZK 811 752, a novel weakly basic drug indicated in the treatment of autoimmune disease, were prepared by the addition of pH-modifying agents having several different polymeric base materials such as Kollidon® SR, ethyl cellulose, and HPMC (Kranz et al. 2005). This demonstrated the utility of such an approach, with numerous other examples also available in the literature.

Other interesting approaches to provide constant controlled release of drugs exhibiting a pH-dependent solubility have included the incorporation of pH-sensitive polymers in addition to the use of nonionic polymers to regulate drug release rate (Tatavarti et al. 2004). In such cases, the active ingredient exhibits a pH solubility profile opposite of that observed for the pH-sensitive polymer. This behavior creates a system which exhibits greater barrier layer permeability for a reduced concentration-driving force. Vice versa, one also notes a reduced barrier layer permeability corresponding to a higher driving force condition. In a patented composition, sodium alginate and HPMC matrices were prepared to regulate the controlled release of a pH-sensitive active ingredient (Howard and Timmins 1988). At low pH conditions where the active ingredient exhibited greater solubility, the sodium alginate remained undissolved and provided a greater barrier to drug diffusion through the gel layer. Similar approaches have also been successfully implemented using mixtures of ionic and nonionic polymers such as Eudragit® L100-55 and HPMC to provide constant release of pH-dependent active ingredients.

Given the variability of gel properties and the dependence of drug solubility to environmental conditions, coupled with the interaction behavior, it is not surprising that researchers have attempted to better control intermolecular interactions between the polymers and different components in both the environment and dosage form to provide a more controlled drug delivery platform. In recent years, molecular imprinting of polymers has been employed to yield materials having well-controlled physical and chemical properties capable of interacting with the different components of a formulation as well as the environment to regulate drug delivery (Byrne et al. 2002; Byrne and Salian 2008). Generally speaking, molecular imprinting is defined as the process of incorporating specific chemical moieties into a polymer to provide a specific response in the presence of stimuli, such as environmental factors (pH, ionic content, etc.) or specific chemical moieties (drug substance, indicator compounds, etc.). Additionally, the incorporated moieties may also help to refine the polymeric network properties, providing better definition of internal structure, porosity, and tortuosity, thereby regulating the diffusion properties through the materials. Numerous examples of molecularly imprinted systems have been described in the literature covering a range of applications from insulin-responsive systems to improved hydrogel diffusion control. In one recent example where molecular imprinting was used to provide stimuli-responsive control of insulin delivery, a pH-sensitive hydrogel-based polymer was imprinted with glucose oxidase (Byrne et al. 2008). When exposed to elevated glucose levels, the imprinted glucose oxidase converted glucose to gluconic acid resulting in an elevated pH which triggered swelling of the hydrogel allowing for the release of insulin. Other relevant examples include the development of hydrogels exhibiting a controlled cavity size capable of capturing specific chemical compounds to regulate transport. In a similar fashion, Venkatesh et al. (2008) described the use of imprinted hydrogels to control the transport of ketotifen fumarate through molecularly imprinted hydrogels due to both traditional diffusion and increased drug–polymer interactions resulting the presence of specifically designed chemical moieties intended to elicit specific hydrogen bonding interactions. While still commercially limited in terms of availability, the design of systems capable of providing stimuli-responsive controlled release with a greater degree of reproducibility represents the next generation in drug delivery systems. Control based on both chemical interactions and diffusional limitations also provides significantly greater control over mass transport through nanoscale domains, which represents a substantial advantage for the development of advanced micro- and nano-drug delivery systems.

18.3   POLYMERIC INTERACTIONS IN FILM-COATED SYSTEMS

The application of a coated layer to pharmaceutical systems has been extensively utilized over the last century in drug product development. The reasons for such coatings range from cosmetic to functional and can include systems utilizing small molecules, polymers, and different combinations thereof. Beginning with sugar-based coatings to achieve taste masking and provide a cover for cosmetic defects, pharmaceutical coating systems have grown extensively and now rely primarily on polymers applied using both aqueous and organic solvent–based systems to achieve a variety of functional and visual properties for the finished dosage form. Current coating technologies can be divided into two major categories based on the functionality of the system, specifically immediate release coating and modified release coatings. Immediate release coatings are used for both dosage forms and intermediates such as pellets and granules and provide functionality which does not significantly impact oral bioavailability and drug delivery properties. Such systems include cosmetic coatings, moisture protective coatings, and taste masking layers. Conversely, controlled release coatings provide a major function for drug delivery and are typically designed to directly impact oral bioavailability of the dosage form. Examples of controlled release coatings include enteric coatings for delayed release, permeable coating for sustained release, and semipermeable membranes which are designed to facilitate the development of osmotic systems.

In all cases, pharmaceutical coatings must be applied to a substrate which is typically a reservoir of the active ingredient to be delivered. Such coatings, regardless of application, can be applied by several basic unit operations which have been well-described in the literature today. The primary unit operations for such processes include perforated pan coating, fluid bed coating, and rotary coating and the method of which will depend primarily on the substrate size. Conducted in pan coating systems, perforated pan coating is primarily used for the coating of tablets which range in size from approximately 50 mg to upward of 2000 mg and beyond at batch sizes ranging from a few grams to hundreds of kilograms. In such processes, the materials are placed into the pan and rotated under a solution or suspension spray containing the coating system. Dried under heated air to remove the solvent, the coating forms a coherent layer on the substrate which is applicable for immediate and modified release systems, although most commonly used for immediate and delayed release coating due to spray limitations associated with some controlled release systems intended to provide sustained or pulsatile release. Fluid bed coating is commonly used for the coating of multiparticulate dosage forms (Jones 1994), although the technology has also been applied to tablets as well (Kucharski and Kmiec 1983). In the case of fluid bed coating, the material is placed inside a product container and large volumes of air are passed through to achieve a fluidized material state. Within the equipment, a Wurster insert separates the upbed and downbed, allowing material to pass upward through the collar and past the coating zone where the coating material is sprayed onto the substrate. Following the pass, material reaches the expansion chamber where the velocity slows before ultimately returning to the product bed via the downbed flow along the outside of the column. For such a system, significantly higher spray rates may be achieved due to the order of magnitude greater drying capacity present in the unit and substantially greater specific surface area of the substrate. Furthermore, multiparticulates present unique advantages over conventional dosage forms, specifically with the large distribution of active ingredient within the product, dose dumping becomes a less substantial concern, and patient-to-patient gastric emptying becomes more uniform (Ghebre-Shellasie 1989). Finally, rotor granulation is the third major technique for applying a coating of a pharmaceutical product. In this unit operation, a modified fluid bed apparatus is used to apply a coating typically onto substrates less than 2 mm in diameter and often times significantly smaller. Furthermore, many rotor granulation processes may be implemented without the need for starting seeds (Vecchio et al. 1994). Unlike a conventional fluid bed however, the rotor granulation unit consists of a traditional tower containing a rotating disk in place of the material screen which imparts an angular velocity to material in the product bed (Vertommen and Kinget 1997). As the bed of material circulates, it passes a coating zone where a liquid spray is applied to the substrate. Optionally, when configured for dry powder granulation, a second zone will also spray solid micronized material onto the wetted substrate to provide increased coating rates and reduced solvent burden. Due in part to the rotational energy applied to the materials, small powders are frequently coated within these systems for taste masking and controlled release applications.

Since the method of application and conditions during processing for controlled release coatings play a significant role in the performance and behavior of the system, it is important to consider the underlying physics involved in each unit operation. Coating operations, regardless of the specific manufacturing process employed, are thermodynamically similar and based on the drying capacity of the system, where drying capacity is defined as the amount of solvent which can be removed per volume of purge gas (Kucharski and Kmiec 1983). For most applications the purge gas used is air; however, when coatings are applied using potentially explosive organic solvent systems inerting purge gases such as nitrogen are used to prevent potential oxidation of components and also to apply an additional measure of safety. Considering the example of air as a purge gas and the use of an aqueous coating system, one notes that the gas enters the system with a specific temperature, mass flow rate, and quantity of moisture already present within the feed stream. These properties establish the maximum amount of moisture which can be removed from the system per unit time and indirectly determine the coating system spray rate. Additionally, examination of a psychrometric chart allows one to establish the behavior of such an air mass under adiabatic cooling conditions, such as those which can be assumed in most coating systems, in order to model and assist with scale-up of the process. Most importantly, these properties establish the concentration gradient between the bulk air mass traveling through the system and solvent layer formed over the surface of the substrate and ultimately determine the drying rate. Careful control of these conditions is necessary to ensure uniform spreading of materials across the substrate surface and achieve complete drying of the layer to avoid defects such as surface cracking and orange peeling.

Coating systems are not typically single component systems dispersed within a larger solvent phase, but rather they are commonly multicomponent systems, each having different solubilities within the solvent. Coating systems relying on organic solvents are commonly selected to achieve complete solubilization of the majority of materials within the coating system. During processing, the primary polymer for coating is dissolved, as are the majority of additives which may also be present, resulting in a generally simple mechanism for film formation. Upon being sprayed onto the substrate, droplets spread to cover the surface and drying occurs rapidly in many cases because of the commonly volatile nature of many solvents selected. As the solvent is removed, the solution becomes more and more concentrated until ultimately a precipitation point is achieved wherein the components precipitate out. Ensuring similar solubilities within the solvent and a rapid evaporation rate helps to minimize segregation which may occur during the coating process. Generally, because the components are dissolved in solution during the process, one commonly notes a more effective and uniform coating over the surface, with fewer aging-related events during shelf life compared to the aqueous counterpart.

In addition to solvent-based solubilized formulations, many coating systems may actually be colloidal dispersions, such as Aquacoat ECD which is an aqueous dispersion based on the water-insoluble ethyl cellulose. Such systems have gained significant popularity since their introduction in the 1970s (Banker and Peck 1981). Based on the suspension state of the particles within the system, as well as the number and type of other components, drying rates may have a critical effect on the behavior of the system. Many polymeric coatings, particularly those performed using aqueous systems, are considered to be pseudo-latex systems in which the polymeric particles are actually a true colloidal dispersion, generally prepared by dissolving the polymer within an organic solvent which is then emulsified and the organic phase evaporated to yield a composition containing discrete nanoscale particles (Carlin et al. 2008). Such systems have tremendous advantages for coating, specifically the ability to achieve a very high polymer content within the solvent without providing a substantial increase of viscosity. This allows for significantly greater solids application rates since there is less solvent per unit solids mass removed during coating. Even still, the drying rate within the system is critical because of the particulate nature of the polymer dispersed within the solvent. As shown in Figure 18.3, colloidal polymeric dispersions progress through a series of stages before achieving a final film coating. Following atomization, the spray containing the colloidal dispersion spreads across the surface of the substrate during which time substantial quantities of solvent are removed via evaporation. As the solvent is removed from the system, colloidal polymeric particles experience a gradual coalescence due to decreased void volume. As drying continues the colloidal particles begin to deform leading to the formation of an apparently continuous film which continues to form as water and polymer diffuse within the layer, ultimately yielding a final uniform film. Since the solvent plays a critical role in the coalescence and molecular mobility, the rate at which it is removed and extent to which it is removed play a significant role in the formulation of the dosage form. Additionally, incomplete drying may also result in changes in product performance over the storage period, commonly referred to as aging, which is a common problem among aqueous controlled release systems. With manufacturing playing such a critical role in product performance and product shelf life, optimization of the formulation and manufacturing must be investigated and controlled at an early stage to ensure successful development.

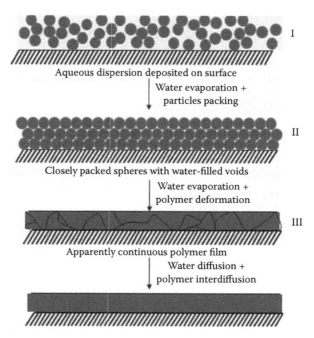

FIGURE 18.3 Stages of film formation from colloidal dispersions.

Most pharmaceutical systems are complex multicomponent systems containing a combination of polymers and additives all having an intended purpose. Such components include the primary polymer, an opacifying agent, plasticizer, antisticking agent, and stabilizing agent. Each material provides a specific function within the formulation and many of the additives facilitate a desired property of the formulation via a specific interaction with the polymer. For example, many polymeric coatings incorporate plasticizers into the formulation. These materials are added into the system to provide a specific interaction with the primary polymer, reducing the glass transition temperature of the polymer to promote film coalescence and uniformity. Components such as talc may also be included to minimize sticking of substrates during the coating process, with the levels of added material having a strong impact on both processing performance and dosage functionality in vitro and in vivo. Within this section, focus will be given to explaining polymeric interactions in terms of both controlled release and immediate release coatings, with a strong emphasis placed on controlled release systems.

Film-coated controlled release systems use a combination of erosion and diffusion to govern the release of active ingredient from the dosage form. Divided into two primary categories, delayed release and sustained release, the mechanism for each form of delivery is significantly different. For both forms of controlled release, successful and reliable implementation of these strategies is strongly dependent on polymeric interactions with other components as well as environmental factors.

For delayed release systems, which are designed to protect the contents from the acidic environment of the stomach, the rate of release is generally an erosion-dependent mechanism related to removal of the protective coating. Enteric polymers commonly exhibit acidic functional groups such as carboxylic acid, which remain unionized in the low pH gastric fluid rendering the coating essentially insoluble and impermeable. As the dosage form transits the GI tract, pH increases to above the pK_a of the polymer, resulting in substantial ionization of the polymer and dramatically increasing the solubility of the coating. Due to the increased solubility, rapid erosion of the coating occurs allowing for release of the once protected active ingredient. This interaction between the polymer and environmental pH has been exploited to develop numerous successful products, most notably the blockbuster product omeprazole which is subject to degradation in the presence of acidic gastric fluids (Larson et al. 1996).

Uniquely distinct from delayed release systems, sustained delivery systems provide extended durations of release and are commonly used to provide once daily dosing of drug substances which exhibit good absorption properties with a short half life. Additionally, this technique has gained substantial interest in recent years with the development of life cycle management programs for pharmaceutical compounds, which attempt to increase the exclusivity period of a compound by developing more advanced dosage forms which provides improved efficacy and/or reduced dosing. For such systems, release is controlled by a combination of erosion and diffusion, each providing unique release mechanisms and a combined effect which can also contribute to the rate of release. Within the matrix system, drug substance is solubilized at the substrate surface and within the core, establishing a high concentration domain with the film coating functioning as a barrier layer leading to the bulk liquid. This concentration gradient provides a driving force for mass transport through the film layer which can be described by Fick's law (Lecomte et al. 2005; Siepmann and Siepmann 2008). Simultaneously, erosion occurring at the surface may reduce the thickness of the film over time, leading to increased mass transfer rates. Unlike matrix systems which are governed by a combination of drug diffusion through the gel and barrier layer erosion, the diffusion properties through the barrier are primarily responsible for governing release of coated systems. From a mechanistic point of view, numerous factors can contribute to the observed diffusion rate through a polymeric film, and include compositional factors as well as environmental factors associated with the conditions during release and storage.

Numerous properties of a polymeric film contribute to the observed diffusion rate of material within the layer, with the type and level of plasticization contributing significantly to this behavior. Plasticizers are generally incorporated into coating systems, particularly aqueous pharmaceutical coatings to provide improved coalescence of colloidal polymeric particles and provide for increased molecular mobility by reducing the glass transition temperature of polymer. This reduction in glass transition temperature occurs due to an interaction between the polymer and plasticizer resulting in an increase in free volume (Aharoni 1998). Selection of the appropriate polymer–plasticizer combination is based on the chemical properties of each material and can generally be predicted based on the solubility parameter of the materials involved (Greenhalgh et al. 1999). Extensively studied, it has been well established that the selection of a specific plasticizer and given level within a formulation can have a significant impact on the glass transition temperature. Changes in material glass transition temperature can also have a substantial influence on the mechanical and storage properties of controlled release films. In a study conducted by Felton and McGinity (1997), their results highlighted the importance of plasticizer selection for achieving a reduction in glass transition temperature and also the impact of plasticization on the physicomechanical properties of films, indicating a correlation between adhesional strength of a film and the effectiveness of the plasticizer. Intuitively obvious as well is the impact of changing plasticizer levels within a film. With increasing levels of plasticizer, the polymer will exhibit a lower glass transition temperature and increased molecular mobility. During the coating process, this will result in a film exhibiting greater coalescence, providing a larger diffusional barrier and thereby slower drug release. There is an upper threshold to which plasticizer can be incorporated however, before negative effects are observed, which include the excessive mobility and visual defects of the coating.

Another substantial issue associated with polymeric coatings based on colloidal systems is the apparent instability of the release pattern over time. Due to a combination of molecular mobility and residual colloidal nature, many controlled release films may exhibit aging phenomenon which is a manifestation of continued film coalescence. As highlighted in a study by Amighi and Möes (1996), the rate of theophylline release from Eudragit RS 30D-coated multiparticulates decreased over storage time due to the continued coalescence of the film. Addition of increased levels of plasticizer also showed an increased rate of coalescence on storage due to the increased polymeric mobility resulting from the plasticizer–polymer interaction. Since this behavior can have a significant impact on the therapeutic efficacy of the product, care must be taken to minimize changes in release profile on stability. One technique to avoid such transitions over stability is the intentional curing of material as part of the manufacturing process by maintaining the material under elevated temperature conditions after application of the coating to further improve coalescence. Numerous studies have demonstrated the utility of curing, showing how this process can be utilized to prepare a pharmaceutically stable film (Amighi and Möes 1997).

Excipient interactions between formulations have also been shown to have a destabilizing effect on films during storage. The production of stable sustained release dosage form is predicated on the development of a homogeneous system or the ability to produce in a controlled and reproducible fashion a system with regular heterogeneity. Examples of each type include Eudragit L100-55 films designed to achieve uniform homogeneity and ethyl cellulose films containing low-molecular-weight polyvinylpyrrolidone (PVP) which are designed to have a uniform heterogeneity and drive pore formation to control release. In some cases however, excipients such as plasticizer and surfactants may be incorporated into the formulation which result in dynamic behavior on storage. In many cases, plasticizers may present issues with volatility resulting in a loss of the component from the film during storage. Such volatilization can change the mechanical properties of the formed film, resulting in a more brittle layer which may be subjected to mechanical breakage during the dissolution process, ultimately compromising the release rate (Anderson and Abdel-Aziz 1976). Surfactants, which are also commonly present in aqueous coating dispersions, may be subject to transformations during the storage period. Due to the relatively high level of the materials which are required to support emulsion polymerization, the surfactants may actually exceed the miscibility limit within the polymer. Furthermore, the relatively high melting point of many of these materials may drive recrystallization of the surfactant over the storage period. This behavior creates large crystalline domains within the film that are capable of dissolving more rapidly than the controlled release film, resulting in pore formation that drives a more rapid drug release. An extensively studied example of this behavior is nonoxynol 100 which is used as a surfactant during emulsion polymerization of Eudragit NE30D dispersions. Over time the formation of recrystallized material has been reported, which resulted in a more rapid release from film-coated systems.

Recently, formulation strategies have also been applied to develop more robust systems which can provide sustained release over pharmaceutically relevant timescales through the incorporation of stabilizing materials, ranging from insoluble additives and high glass transition components to immiscible polymers. Under a similar general theory, these materials are incorporated into the films to limit film mobility and prevent further coalescence on storage. By limiting the coalescence which occurs during storage, coated films will be able to maintain a requisite permeability and porosity to provide consistent drug release throughout the pharmaceutical life of the product.

Numerous studies have investigated the use of high glass transition temperature materials in combination with low glass transition temperature-controlled release films in order to stabilize the observed release characteristics. These systems are predicated on the miscibility of the two materials in order to ensure uniform distribution of materials within the film, allowing the higher glass transition material to stabilize the film. One successful example of this was the combination of Eudragit L100-55, a relatively high glass transition temperature polymer, with Eudragit RS 30D, a controlled release polymer having a low glass transition temperature and frequently reported to exhibit physical aging on storage (Wu and McGinity 2003). By incorporating L100-55 into the film, the researchers were able to stabilize the RS 30D and provide a consistent, albeit pH-dependent release of theophylline from the multiparticulate system.

The addition of insoluble additives, particularly at high levels within the formulation, can also be an effective mechanism for stabilizing controlled release films on stability. These materials present a scaffold that mechanically limits the motion of the film to coalesce during storage. Several common pharmaceutical materials such as colloidal silicon dioxide and talc can be used to exert such an effect. By combining extremely high levels of these materials in the formulation it is possible to limit the change in drug release over storage for a pharmaceutical composition (Maejima and McGinity 2001). While drug release is stabilized, it is also important to note that such high percentages can also drive incomplete film formation. Aesthetic properties such as surface roughness may also become exaggerated due to the excess of solid material within the formulation.

Functioning in a similar method to insoluble additives, immiscible materials may also be incorporated into the film formulation to stabilize release properties by providing a physical barrier to film transitions over storage. Numerous materials, including albumin and hydroxyethyl cellulose, have been investigated recently and have shown varying degrees of success in stabilizing films. Researchers have successfully demonstrated the use of albumin in stabilizing Eudragit RS/RL 30D films (Kucera et al. 2007). Interestingly, initial studies showed that albumin as a 10% additive actually destabilized the dispersion, resulting in a significant change of drug release rates on storage. By acidifying the coating system prior to addition of the albumin, researchers were able to demonstrate that this material could be successfully employed for the stabilization of pharmaceutical coatings. Furthermore, it was demonstrated that this behavior was due to the specific protein–polymer behavior at elevated pH conditions, resulting from interactions caused by the quaternary ammonia group of the polymer and the negatively charged nature of the protein above the isoelectric point. In another study examining the stabilizing properties of hydroxyethyl cellulose as a film stabilizer for Eudragit RS 30D films, researchers showed that incorporation of the material prevented changes in drug release, mechanical properties, and water vapor transmission rates over the storage period (Zheng et al. 2005). Utilizing atomic force microscopy the researchers were able to characterize the nanostructure of the films produced. As shown in Figure 18.4, films formed with hydroxyethyl

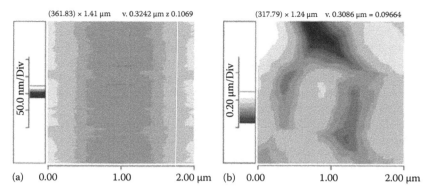

(361.83) × 1.41 μm   v. 0.3242 μm z 0.1069   (317.79) × 1.24 μm   v. 0.3086 μm = 0.09664

50.0 nm/Div   0.20 μm/Div

(a)   0.00   1.00   2.00 μm   (b)   0.00   1.00   2.00 μm

FIGURE 18.4 Atomic force microscopy images showing the difference in film morphology between unstabilized Eudragit RS 30D films and Eudragit RS 30D films containing hydroxyethyl cellulose. (Reprinted from Zheng, W. et al., *Eur. J. Pharm. Biopharm.*, 59, 147, 2005. With permission.)

cellulose showed discrete colloidal particles having a smooth surface while compositions without the stabilizer exhibited no discrete particles. This behavior was ultimately attributed to the coating of the hydrophobic colloidal particles with the hydrophilic stabilizer and that this barrier prevented the coalescence on storage.

Although the design of many multiparticulate systems limits the intermixing of drug substances with the functional coatings, many systems do provide an interfacial region which can result in potential drug–polymer interactions. These interactions can drive the formation of degradation products, thereby lowering the potency of the product and altering the impurity profile of the dosage form. Additionally, interactions may also compromise the drug release characteristics of the dosage form (Bruce et al. 2003). Studies have shown that active ingredients having alkaline properties can compromise enteric protection, while acidic moieties support the enteric protection by contributing to the microenvironmental pH (Dangel et al. 2000a,b). One also notes that the solubility of the active ingredient can have a significant impact on the performance of the system, not only due to the resulting surface–core concentration gradient, but also due to migration of the active ingredient into the film layer. In a study by Bodmeier and Paeratukul, it was shown that this behavior was due to a combination of aqueous solubility and drug–polymer affinity (Bodmeier and Paeratakul 1994). Such behavior indicated that there is not only an interfacial surface between the layers, but an interfacial volume in which the drug and polymer may be more intimately mixed. This allows for more complex interactions, including decomposition of the drug substance and partial plasticization of the film coating. Similar to the behavior of a traditional plasticizer, many drug substances have been shown to provide plasticizing effects on polymeric materials. In film-coated systems, partial penetration of the active ingredient into the coating can drive such behavior. Similarly, many drug substances will interact with the polymeric coatings to generate complex formation (Alvarez-Fuentes et al. 1994) and also drive degradation (Riedel and Leopold 2005; Stroyer et al. 2006). In general, these interaction behaviors can be limited through the use of subcoatings, which are designed to create a barrier between the layers, minimizing transport of drug into the polymer layer. Such behavior was illustrated by Bruce et al., where the application of a polymeric subcoating improved the enteric protection of Eudragit L100-55 films, as shown in Figure 18.5.

Noting the earlier discussion between suspension and solution application of polymeric films, one quickly realizes that there is an inherent difference in solubility for specific polymer–solvent combinations. In many cases, pharmaceutical polymers exhibit significantly higher solubilities in organic solvents. As a result, the presence of such materials in the GI tract can significantly impact drug delivery. While most people do not routinely imbibe many of the pharmaceutically acceptable solvents, alcohol is commonly consumed in many cultures throughout the world. The presence of alcohol can interact with many pharmaceutical compositions, including controlled release

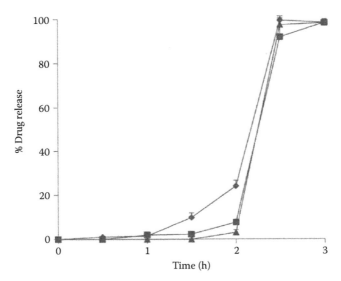

FIGURE 18.5 Drug release profile from Eudragit L100-55 film-coated chlorpheniramine maleate pellets containing 0% (♦); 3% (■); and 5% (▲) Opadry® AMB subcoating. Eudragit L100-55 film coating level is 10%. (Reprinted from Bruce, L.D. et al., *Drug Dev. Ind. Pharm.*, 29, 909, 2003. With permission.)

coatings, resulting in dose dumping. Dose dumping is defined as the rapid and uncontrolled release of drug from a dosage form. Dose dumping from controlled release systems is a major problem that may raise the likelihood of a fatal exposure due to rapid drug release from the system in combination with the generally higher dose present in the dosage form. In fact, several fatal cases have been reported over the years as a direct result of alcohol-induced dose dumping, with the most dramatic being those which resulted in the FDA-regulated removal of Palldone™ from the market (Fadda et al. 2008). The behavior of each specific pharmaceutical system varies significantly based on the overall composition and design of the product. Additionally, the concentration of alcohol within the GI tract plays a major role in the observed behavior, further complicating this situation. While many researchers have investigated ways to mitigate the effect of alcohol on controlled release systems, no well-accepted solution has been found to date. As such, products susceptible to dose dumping in the presence of alcohol generally carry a warning label to avoid consuming such products in conjunction with the medication. While an effective temporary solution which does improve safety and compliance, additional research in this field is required to achieve optimum product safety.

Controlled release systems based on modified release coatings are a successful platform for the delivery of pharmaceutical active ingredients, and the successful development and implementation of such products is based on a series of complicated polymeric interactions which help to provide product efficacy. Many of these interactions contribute to the overall success of the product, directly governing the rate of release. Other interactions, however, present a significant barrier to the development of a successful product. In recent years extensive research has been conducted to facilitate the development of successful systems, improving the reliability of release characteristics and minimizing the impact of aging on controlled release systems. Recent advances in immiscible material formulation and insoluble scaffold development present novel opportunities to improve long-term stability, while other areas of research specifically in the area of environmentally induced dose dumping still require significant contributions. Based on this, coated controlled release systems have been shown to be a viable platform for the production of pharmaceutical systems and will continue to play a critical role in the generation of advanced dosage forms designed to improve efficacy and safety.

18.4 POLYMERIC INTERACTIONS IN AMORPHOUS SOLID DISPERSIONS AND ADVANCED SYSTEMS

Advances in drug discovery technologies have led to an increasing number of new chemical entities which exhibit low aqueous solubility. This presents a significant limitation to oral absorption and results in a more challenging product development program. Over the last decade, there has been an explosion of new technologies and formulation strategies capable of addressing poor oral bioavailability due to low aqueous solubility, including crystal engineering, co-crystal formation, nanoparticle production, cyclodextrin complexation, and amorphous formation. Each of these processes seeks to maximize the dissolution rate through modification of the intermolecular interactions and/or reduction of particle size which often results in supersaturation, or the ability of a formulation to provide drug concentrations in excess of the equilibrium solubility. Ultimately, however, these formulations eventually return the drug concentration to its equilibrium solubility due to the thermodynamic instability associated with this state which can result in incomplete and variable oral bioavailability. An emerging field of research focuses on the maintenance of supersaturation to provide improved oral bioavailability of pharmaceutical compositions. This technique exploits unique interactions of drug and polymeric stabilizers to provide longer durations and elevated concentrations of drug in solution to thereby improve oral absorption. Recently, extensive research has been undertaken in this field to establish the utility supersaturating systems for oral bioavailability enhancement and also provide a more detailed scientific understanding of the underlying mechanisms for stabilization. In addition to new formulation technologies pioneered to address aqueous solubility limitations, processing technologies have also been developed to produce stable amorphous compositions providing dissolution and product stability benefits. While many technologies have focused on particle size reduction to provide increased surface area, a select group of processes have been developed to regulate the crystal structure of the active ingredient to alter the thermodynamic properties of the material and enhance dissolution rate. The successful implementation of such technologies for the production of pharmaceutical systems is highly dependent on polymeric interactions with other formulation components, specifically interactions which solubilize drug within the carrier, specific interactions associated with polymer melt behavior, stabilization properties of polymers which lead to amorphous stability, and the behavior of drug–polymer combinations leading to stabilized supersaturation. This section describes the underlying mechanisms for solubility enhancement from supersaturatable systems, as well as current applications of solid dispersion systems for bioavailability enhancement which are dependent on polymeric interactions.

Pharmaceutical materials can be described by a variety of physicochemical properties, including chemical structure, crystal structure, and particle morphology. The chemical structure of a compound refers to the particular atomic composition of the molecule, which can be combined in large clusters of molecules which present a macroscopic three-dimensional structure. Within these clusters, the intermolecular spacing is referred to as crystal structure. In many compounds, this arrangement presents both long and short range order, resulting in specific interactions between the molecules within the ensemble. Such structures are commonly referred to as crystal lattices and are present in seven basic forms, regardless of whether they are organic or ionic crystals. Out of these seven basic structures, organic molecules may also present different packing configurations within the crystal lattice, thereby expanding on the number of possible configurations for arrangement. For a given molecule, multiple configurations may be possible, and such materials are referred to as polymorphic forms. Each polymorphic form also presents a different long and short range order resulting in different intermolecular interactions and variable thermodynamic properties of the system, specifically free energy. As a general rule of thermodynamics, systems always attempt to achieve the lowest free energy state and the variation of free energy within the different crystal forms establishes a most stable polymorphic form and then metastable polymorphic forms having higher free energies (Hancock 2007). It is important to note that the specific crystal structure which is most stable may only be so under a given set of conditions, and deviation outside of those conditions would establish other forms as the most stable. This behavior is referred to as enantiotropic behavior, while a crystal form which is most stable over all conditions is referred to as monotropic (Brittain 2007). Free energy of a crystal form is also directly related to the apparent solubility of the compound, true density of the substance, and melting point of the material, such that higher free energy systems exhibit a lower melting point, reduced density, and greater apparent solubility. Application of polymorphic form selection has been well-illustrated throughout the history of the pharmaceutical industry for a variety of applications, including morphological control, stability, and solubility. While a theoretical potential exists for such system, they have generally shown only limited solubility improvements while being hampered by thermodynamic instability (Yu 2001). As a result, the pharmaceutical industry has not extensively embraced pharmaceutical polymorph screening for oral bioavailability enhancement.

Material properties of pharmaceutical compounds can be further manipulated to control properties, particularly solubility, by eliminating long and short range order associated with the crystal structure to create an amorphous material. Since amorphous forms lack coherent long range order associated with a crystal form, the material is presented in the highest free energy state and can provide the greatest apparent solubility and dissolution rate, often providing significant increases in such metrics (Yu 2001; Hancock 2007). Generation of an amorphous form also results in dramatic changes in the physical properties of such materials. Unlike the crystalline counterparts which are characterized by well-defined melting points, amorphous materials are actually considered super-cooled liquids below such transition points and exhibit a glass transition temperature where the equilibrium properties of the material, such as enthalpy and volume, deviate from that of the super-cooled glass, as shown in Figure 18.6. In such systems, properties such as solubility, molecular mobility, and vapor pressure are enhanced with respect to the crystal state. While improvements in solubility can offer the potential to enhance oral bioavailability, greater molecular mobility and elevated free energy can drive phase transformations and it is this instability that has been viewed by the pharmaceutical industry as one of the greatest drawbacks in the development of such systems.

Resulting from the solubility benefits provided by these systems, extensive research has been conducted investigating the fundamental mechanisms of recrystallization and polymorphic conversion, as well as techniques for improving the overall stability of such compositions. While recrystallization is predicted to occur as a result of the thermodynamic instability of the system, the rates of such transformations are determined by the kinetics associated with such processes, specifically activation energy and molecular mobility which may hinder such phenomena. Relaxation rates of pharmaceutical systems have been described using the Vogel–Tammann–Fulcher (VTF) equation which provides an Arrhenius-type relation for molecular relaxation as a function of temperature (Hancock 2007). Further understanding of molecular motion within such systems is also expressed using the Williams–Landel–Ferry

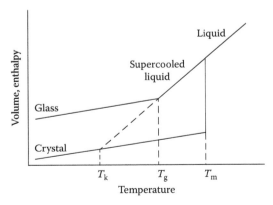

FIGURE 18.6 Schematic diagram illustrating the differences in enthalpy and volume for crystalline and glassy solid as a function of temperature. (Reproduced from Hancock, B.C. and Zografi, G., *J. Pharm. Sci.*, 86, 1, 1997. With permission.)

(WLF) equation which describes the temperature dependence of viscosity in such systems (Hancock 2007). In both relationships, the strong dependence of properties influencing molecular mobility on temperature indicates that controlling the environmental conditions can regulate recrystallization rates. As shown in Figure 18.6, as the supercooled liquid line is extrapolated past the glass transition temperature the line will intersect the enthalpy of the crystalline form which denotes the Kauzmann temperature where configurational entropy of the system reaches zero (Hancock 2007). Utilizing the WLF equation, an approximate temperature of 50 K below the glass transition temperature will provide a situation in which all molecular motion is essentially stopped and recrystallization is no longer possible (Hancock et al. 1995, 1998). Using this result, one approach to stabilizing such systems has been to store them under special conditions where this rule can be maintained. It is important to realize, however, that such behavior may be impacted by environmental conditions, such as moisture uptake, which can help to plasticize the material and provide greater molecular motion.

An alternative approach, which is much more common for pharmaceutical products today, is the development of amorphous solid dispersions, which combine the active ingredient and one more carrier materials which can stabilize the amorphous form through specific interactions or a general increase in glass transition temperature (Hancock et al. 1995). Many of the commonly used pharmaceutical coating polymers and matrix dosage form polymers have also shown utility in such applications. In the most common form, this material creates a solid solution in which the drug substance is homogeneously distributed within the carrier matrix, yielding a single glass transition temperature during solid-state characterization. The behavior of such miscible forms is governed by the Gordon–Taylor equation in a two-component situation which is actually a special case of the multicomponent Fox equation (Gordon and Taylor 1952). Interestingly, when the drug and carrier exhibit a synergistic interaction, such as hydrogen bonding, positive deviation in the Gordon–Taylor relationship is observed which results in elevated glass transition temperatures (Gupta et al. 2005). Several examples have been shown recently, highlighting the ability of pharmaceutical polymers such as PVP and polyvinyl acetate phthalate to interact with drugs such as celecoxib and itraconazole. In addition to the benefits provided by such systems for elevating the glass transition temperature, benefits in dissolution rate can also be achieved through the establishment of polymer dissolution rate-limited release systems, which are described in subsequent sections.

The production of amorphous active ingredients and amorphous solid solutions is generally accomplished by one of four basic techniques, including precipitation from solution, vapor condensation, milling/compaction, and supercooling of a melt (Hancock 2007). In three of the four schemes, the drug and optional carrier materials are dispersed in a molecular state and then applied through processing to render a solid amorphous form. For milling/compaction processes however, the amorphous form is produced by the introduction of high amounts of energy that drive the formation of molecular disorder within the system to form an amorphous material. While this method has been shown to provide the capability for amorphous material production, this class of unit operation is not commonly used for the production of such systems, but rather serves as a common example for the unintended conversion of crystalline material to amorphous material. For pharmaceutical production methods, amorphous systems are most commonly created using supercooling of a melt and precipitation from solution which are commonly referred to as fusion processing and solvent processing, respectively (Leuner and Dressman 2000). For such systems, the interactions between the polymers and other components such as drug substance are critical, helping to regulate uniformity and long-term amorphous stability of the composition.

Solvent processing techniques for the production of amorphous pharmaceutical systems are currently the most popular mode of manufacture for such products and consist of a variety of unit operations including fluid bed coating, spray drying, and particle engineering. While the specifics of each operation vary substantially, the basic principle is the same. Drug and optional carrier materials, most commonly polymers, are dissolved into an appropriate solvent, ranging from a super critical fluid to common organic solvent and mixed to develop a homogeneous solution. Polymers are frequently employed due to their stabilizing properties, high degree of miscibility with many common active ingredients, and relatively high solubility in volatile organic solvents. That solution is then applied by a process in order to rapidly remove the solvent on a timescale where molecular mobility cannot induce recrystallization. By inhibiting the molecular motion and providing similar solubility characteristics in conjunction with rapid solvent removal, it is possible to prevent recrystallization to develop an amorphous form, capable of providing enhanced oral bioavailability.

Fluid bed coating is the most common commercial method for amorphous form production and several currently marketed products are available based on this technology, including Sporanox® (Gilis et al. 1997) and Prograf® (Letko et al. 1999). In both cases, the drug and hydroxypropyl methylcellulose (HPMC) carrier are dissolved in an organic solvent system which is layered onto multiparticulate cores. While this technology platform is capable of rendering an amorphous form and also provides benefits in using existing production capabilities, it suffers from requirements for potentially toxic solvents, as well as the need for extended drying periods at elevated temperatures to ensure complete solvent removal. Another technology for solvent-based production of amorphous pharmaceutical systems is spray drying. Similar to fluid bed coating, the drug and appropriate carriers are dissolved in a common solvent and sprayed into a chamber to rapidly remove the liquid phase. Unlike fluid bed coating, the droplets are sprayed through an atomizing air nozzle and directly into a drying chamber, whereas in fluid bed coating, the droplets are sprayed onto the pellet surface forming a film on the substrate. By processing the droplets directly into the air stream, it is possible to control not only the amorphous nature of the particles but the particle morphology. Extensive studies have been conducted to identify the critical variables for controlling such properties and ultimately it has been shown to be related to the relationship of decreasing droplet size and diffusive molecular flow within the system, which can be described by the Péclet number (Vehring et al. 2007; Vehring 2008). Through careful optimization of the processing conditions it is possible to produce particles having maximized specific surface areas, capable of further improving dissolution, solubility, and oral bioavailability. This technology has also been extensively applied to the development of systems for pulmonary delivery, specifically the production of large porous particles (Shoyele and Cawthorne 2006; Vehring et al. 2007; Vehring 2008). In addition to the obvious drawbacks associated with solvent use during processing, the low bulk density of the resulting material can also have significant issues on downstream processing during compression and encapsulation (Çelik and Wendel 2005). When preparing solvent systems for such applications, the loading of material into solution may be a limiting factor due to viscosity or atomization considerations and as a result is generally less than 20% w/w during most conventional processing which results in an inherently low density (Çelik and Wendel 2005). While process optimization can facilitate the production of formulations with increased densities and many new spray-drying systems allow for production under pressurized environments, most products are still produced having bulk properties which can negatively impact downstream processing. This presents a need for additional post-processing to increase density using techniques such as roller compaction or slugging. Another critical aspect of spray-drying process control focuses around the optimization of drying rates to prevent phase separation (Patterson et al. 2007). During the drying process, solvent is removed which results in the formation of concentration gradients for the components dissolved within the particles. Varying drying rates with respect to the differing material solubility can lead to phase separation. Control of such issues during production is another critical area of development, since these issues may result in reduced dissolution rates and poor product stability.

Particle engineering technologies have also been researched extensively, providing another viable platform for the production of pharmaceutical materials having well-controlled physical and morphological properties (Hu et al. 2004; Bhardwaj et al. 2005; Jia 2005). Three major subclasses for such technologies are supercritical fluid technologies, anti-solvent processes, and cryogenic production, each providing its own unique advantages for the manufacture of such materials.

Supercritical fluid technologies utilize solvents maintained under conditions of temperature and pressure above the critical point, allowing for common materials such as carbon dioxide to be used as a solvent. As a result of the supercritical and often nontoxic nature of the solvent employed, removal is often rapid and complete, only requiring a return of the material to ambient conditions.

Unfortunately, supercritical fluids generally exhibit only minimal solubility for pharmaceutical materials, which can limit the applicability of this technology (Byrappa et al. 2008). As a result, many of the technologies using supercritical fluids for pharmaceutical applications rely on it for the extraction of residual solvents, spurring the development of anti-solvent processes. In anti-solvent processes for pharmaceutical applications, the drug and the carrier are dissolved in a solvent that is miscible with a second nontoxic solvent system. The drug–carrier solution is then added into the second solvent, where the miscible toxic solvent diffuses into the solvent, while the drug–carrier mixture which exhibits extremely low solubility precipitates to form a particle. The particles can then be isolated and further dried to remove any residual toxic solvent remaining. This technology has also been effectively used to produce amorphous systems. Kim et al. (2008) recently applied supercritical anti-solvent processing techniques to produce amorphous formulations of atorvastatin calcium to provide enhanced oral bioavailability. Their results showed that such technologies could be applied to the production of amorphous systems and provided significant improvements for oral bioavailability in a rat model. In another study by Vaughn et al. (2005, 2006a,b) using evaporative precipitation in aqueous solutions, an anti-solvent technique using an aqueous environment, substantially amorphous danazol particles were produced leading to in vitro supersaturation and enhanced oral bioavailability compared to a physical mixture of drug and carrier in a murine model. One negative attribute of such production techniques is the relatively long timescales for diffusion of the solvent in the bulk anti-solvent and the dynamics of drug and carrier concentration which occur as a result. As highlighted by Vaughn et al., such processes can result in partially crystalline material which can compromise the long-term stability and negatively impact dissolution rates of such systems. During optimization, it is essential to identify processing conditions and formulation variables such as stabilizers which can impact the amorphous nature of the finished product.

Cryogenic processes are a type of solvent process that utilizes significant changes in temperature to produce amorphous systems by exposing the drug–carrier-loaded solvent to significantly reduced temperatures that result in rapid freezing of the material. Rapid freezing is designed to occur on time scales comparable to the precipitation kinetics of the drug and carrier in solution, preventing phase separation. Additionally, drying conducted by lyophilization limits molecular motion as a result of the low temperatures used during processing. Furthermore, the absence of external heat required to drive off the solvent phase presents additional benefits for the production of amorphous particles containing heat-sensitive high-value compounds such as proteins and peptides. Based on this general principle, a variety of specific types of techniques have been utilized to produce amorphous forms, including spray freeze-drying (SFD) (Engstrom et al. 2007), spray freezing into liquids (SFL) (Hu et al. 2004), and thin film freezing (TFF) (Overhoff et al. 2009). Utilizing such technologies, Vaughn et al. (2006a,b) produced amorphous itraconazole particles dispersed in a matrix of polyethylene glycol (PEG) 800 and poloxamer 188 by SFL. Similarly, Overhoff et al. (2007) demonstrated the applicability of TFF for the production of amorphous danazol–PVP solid dispersion particles which exhibited high specific surface. Application of such technologies to the production of protein formulations was also demonstrated by Engstrom et al. (2007), again using TFF, where engineered particles provided high-specific surface area and also effectively maintain the activity of the model macromolecule.

An alternative to solvent-based processing, fusion processing involves the melting of drug and carrier material, followed by subsequent mixing and cooling to produce an amorphous solid dispersion. Although historically not as accepted as solvent processing, this class of manufacturing processes has become increasingly more common, with the commercially available Kaletra® formulation produced using hot-melt extrusion for oral bioavailability enhancement (Breitenbach 2006). Unlike solvent processing technologies, only a few variations of the process are currently used in pharmaceutical research and production, including hot spin mixing, fluidized bed melt granulation, and hot-melt extrusion (Leuner and Dressman 2000; Andrews 2007; Crowley et al. 2007; McGinity et al. 2007; Repka et al. 2007, 2008).

Fluidized bed melt granulation and hot spin mixing are two forms of fusion processing that have been reported in research literature, however have currently found little application in pharmaceutical production. In a recent study by Walker et al. (2007), fluidized bed melt granulation was utilized to produce amorphous compositions of ibuprofen by granulating for a predetermined time at an elevated temperature of 100°C. While applicable for this particular formulation, compositions containing high melting point drugs or high viscosity polymers may exhibit limitations under the current process design. Modifications to the process for incorporation of melt spraying may also be limited due to the high viscosity of polymer melts or temperatures required for flow and atomization. In another series of studies, hot spin mixing was utilized for the production of amorphous systems. This technology functioned by combining the drug and carrier into a rapidly spinning heated vessel, allowing the material to melt and then ejecting material into a cooling tower. While little information was provided on the exact setup of the process, it was shown to be an effective platform for the production of solid dispersions containing testosterone, dienogest, and progesterone (Dittgen et al. 1995a–c).

Hot-melt extrusion is the pharmaceutically preferred method of fusion-based solid dispersion production and is based on polymer melt extrusion which has been in use for over a century. First applied to pharmaceutical compositions in the 1970s, hot-melt extrusion has shown utility in a variety of applications including bioavailability enhancement (Breitenbach 2006), device manufacture (Rothen-Weinhold et al. 2000), and controlled release systems (McGinity and Zhang 2003; Fukuda et al. 2006; Lyons et al. 2006).

Based on the equipment utilized for polymer processing, melt extruders typically consist of one or more rotating screws which convey material through a heated barrel, providing intermixing of materials and generation of additional heat due to shear and friction, ultimately forcing material through a shaped die located at the end of the unit to produce a rodlike structure. A schematic diagram of an extruder is presented in Figure 18.7. Pharmaceutical extruders are derived from two basic designs based on the number of screws within the unit. Single screw extruders, as the name indicates, contain only one screw and are commonly used for pumping material through the extruder barrel. As a result of the single screw design, minimal flow perturbations are present along the flow pattern and less mixing is achieved than in other configurations. Additionally, these designs also may be subject to flow stagnation along interfacial surfaces, particularly along the melt–screw interface (Kim and Kwon 1996a,b). In some cases, specific mixing elements may be included on the screw to increase the frequency of flow discontinuities and improve convective mixing (Kim and Kwon 1996a,b). This can also result in prolonged residence times and increased potential for material degradation. Twin screw extruders are the other common form of pharmaceutical extruder and are designed with either corotating or counter rotating screws, each of which provides unique benefits to production. For counter rotating designs, increased convective mixing occurs as a result

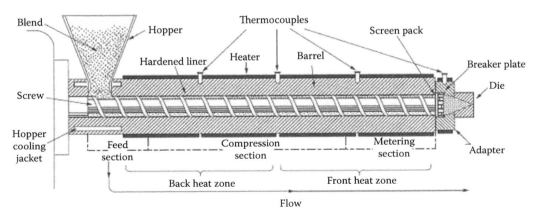

FIGURE 18.7 Schematic diagram of a hot-melt extruder illustrating critical equipment and process aspects. (Reproduced from Follonier, N. et al., *J. Control. Release*, 36, 243, 1995. With permission.)

of the opposite direction of motion which generates intersecting flow patterns inside the barrel. Corotating designs provide less convective mixing than the counter rotating equipment as a result of the similar flow direction at the screw intermeshing; however, this also provides for a material removal action along the screw which is commonly termed "self-wiping" (Thiele 2003). As a result, residence and holdup times inside twin screw extruders can be minimized with respect to the single screw contemporaries. Additional optimization of mixing and material flow can be controlled using different elements for the screw. Screw elements refer to interchangeable segments of the screw which are designed to provide different functions, such as kneading elements for enhanced mixing or conveying elements for greater material throughput. Process optimization of material feed rates, screw speeds, and zone temperatures can all contribute to residence times; however, these values generally range from 30 s to 10 min based on extruder configuration and scale.

Similar to solvent-based compositions, fusion-based compositions are intended to be produced as a single homogeneous phase to improve dissolution and reduce recrystallization potential due to molecular mobility. Proper preformulation identification of drug–polymer miscibility is essential for the development of a successful formulation. Solubility parameters provide an indication of interaction energies associated with the mixing of different materials based on the molecular structure of the components. Greenhalgh et al. (1999) successfully applied Hildebrand solubility parameters to predict the miscibility of ibuprofen in a variety of hydrophilic carriers and results indicated a strong correlation between observed behavior and that predicted by theory. Similar results have also been obtained for melt-extruded compositions containing indomethacin (Chokshi et al. 2005) and lacidipine (Forster et al. 2001). Other screening techniques have also proved vital in the development of melt-extruded systems. Applications of differential scanning calorimetry and hot stage microscopy have both been shown to provide valuable information about the behavior of drug–polymer combinations at elevated temperatures, specifically providing information about degradation, miscibility, drug solubility within the carrier, and recrystallization potential upon cooling (Forster et al. 2003; Van den Brande et al. 2004; Zhou et al. 2008). Extrapolation of results observed in these small-scale experiments can frequently be correlated to fusion processing behavior, although false negatives may occur due to the inability to approximate shear effects present in hot-melt extrusion. Even with careful examination of the formulation requirements, processing of certain types of materials may be difficult by melt extrusion, including the production of solid dispersions using polymers with high melt viscosity and the production of heat-sensitive components.

Unlike the solvent-based systems the use of thermal processes, hot-melt extrusion in particular, is highly dependent on the interaction between the drug and polymer to achieve the requisite processing characteristics. The ability to successfully manufacture compositions by hot-melt extrusion is directly related to the miscibility of the components within the system, melt behavior of the materials, and solubilization affinity of the carrier for the active ingredient. Formulations containing immiscible materials will form inherently unstable solid dispersions and may also yield difficulty during processing due to phase separation. Melt behavior may also drive processing difficulties. Many polymers exhibit a high molecular weight which is directly related to the observed melt viscosity of the material. In such cases, polymeric plasticizers must be included in the formulation or processing modifications must be modified to account for this behavior. Melt behavior also includes the chemical and physical stability of the materials. Many compounds, such as plasticizers, may have a low boiling point which results in venting of the specific material when maintained at temperatures in excess of the vaporization point. Solubilization capacity of the polymer is another critical attribute and is related to the miscibility of the system. It has been well-established in general science that "like dissolves like." In addition to being true for low-molecular-weight systems, it is also true to polymeric systems. Compositions having similar structures with respect to one another generally show some affinity to dissolve the other at elevated temperature. If the polymer shows similar chemical structures with respect to the active ingredient, which is frequently indicated with solubility parameters, solubilization of the drug within the polymer is maintained at temperatures above the glass transition temperature of the polymer. Such specific drug–polymer interactions can be exploited to facilitate lower temperature production and improved overall product quality.

The stability of such formulations is again a function of the glass transition temperature and specific interactions between the drug and carrier. Unlike many solvent processes, melt extrusion frequently requires the addition of a plasticizer to facilitate production by lowering the glass transition temperature to reduce melt viscosity (Crowley et al. 2007; McGinity et al. 2007). This reduction affects the material under both elevated and room temperature conditions, increasing the molecular mobility of the finished product. In a recent study by Miller et al. (2008a,b), a 20% plasticizer loading was required to achieve flow of melt-compounded itraconazole and Eudragit L100-55. Plasticization at this level reduced the glass transition temperature to approximately 50°C, making it unlikely that the formulation would have adequate shelf life upon storage. In another study by Bruce et al. (2007), melt-extruded compositions of guaifenesin and Acryl-Eze, a pre-plasticized form of Eudragit L100-55, exhibited extensive surface recrystallization, due in part to the reduce glass transition temperature and greater molecular mobility. It is important to note that in this case, the polymer was further plasticized by the presence of the drug substance within the solid dispersion. Due to the negative impact of the plasticizer on the solid-state properties of the finished product, studies have also been conducted using temporary plasticizers such as supercritical fluids. Verreck et al. examined the applicability of supercritical carbon dioxide as a temporary plasticizer for a variety of compositions to provide higher finished product glass transition temperatures and also process at reduced temperatures to improve the potency of heat-sensitive compositions (Verreck et al. 2006a,b, 2007). Another technique to facilitate processing of temperature-sensitive active ingredients is to control the feedstock characteristics of the active ingredient. Through polymorphic selection it is possible to identify active forms with lower melting points, ultimately achieving the lowest possible processing temperature through use of the amorphous form. Lakshman et al. (2008) successfully demonstrated the applicability of such an approach; however, many logistical issues may need to be addressed in order for such an approach to be commercially viable.

Even with the potential drawbacks associated with the addition of processing aids and degradation due to elevated temperatures, hot-melt extrusion has been extensively utilized for the production of amorphous solid dispersions for enhanced oral bioavailability, with some of the earliest uses of this technology for pharmaceutical applications being traced back to the 1970s (Leuner and Dressman 2000). To date, a multitude of publications have been presented documenting improvements in dissolution rate as well as bioavailability enhancement in animal models and human subjects. Hülsman et al. (2000) utilized hot-melt extrusion for the production of 17β-estradiol in a matrix of PVP and Gelucire® 44/14 which provided a 30-fold increase in dissolution rate. Nimodipine solid dispersion were also produced by melt extrusion and shown to provide improved in vitro dissolution rates as well as oral bioavailability in beagle dogs (Zheng et al. 2007a,b). In another example of melt extrusion technology R103757, an experimental compound exhibiting low aqueous solubility, was prepared as an amorphous solid dispersion in HPMC 2910 and provided improved in vitro dissolution rates (Verreck et al. 2004). Oral bioavailability studies in healthy volunteers showed that solid dispersions provided enhanced bioavailability, although the greatest improvement was provided by a cyclodextrin-based complex formulation. Itraconazole, another poorly water-soluble compound, has also been extensively studied using the melt extrusion platform, with results demonstrating that all solid dispersions could provide improved dissolution rates compared to the commercial multiparticulate formulation of Sporanox. Even with this dissolution improvement however, oral bioavailability studies in healthy volunteers showed no statistically significant improvement (Six et al. 2003, 2005). Subsequent studies by Miller et al. (2007, 2008a,b) indicated that the reason for this disparity was the formulation design which targeted supersaturation to the stomach instead of the upper small intestine. Kaletra is another example of a product produced using melt extrusion to provide improved oral bioavailability. This commercially marketed product, indicated in the treatment of human immunodeficiency virus, contains the poorly water-soluble drugs ritonavir and lopinavir (Breitenbach and Lewis 2002) and increases oral bioavailability lopinavir through a combination of solubility enhancement and preferential metabolism of ritonavir. Using this platform, a significant reduction in pill burden and removal of strict low temperature storage requirements were obtained when compared to the original soft gelatin capsule formulation. These examples highlight the capabilities of fusion-based solid dispersions to provide improved oral bioavailability.

The current development of formulations capable of achieving supersaturation has been focused primarily on the development of metastable polymorphic forms, nanomaterials, and solid dispersions, each of which is capable of achieving high solubilities due to inherent kinetic and thermodynamic properties of the system. Solid dispersions have recently gained significant popularity for the production of pharmaceuticals and can be defined as an intimate mixture of one or more active ingredients in an inert carrier or matrix at solid state prepared by thermal, solvent, or a combination of processing techniques (Chiou and Riegelman 1971). By developing these compositions it is possible to create formulations with the smallest possible drug domains, the individual molecules dispersed within a solid carrier, which are termed solid solutions. With solid dispersions, it is possible to maximize the solubility enhancement while increasing physical stability through proper selection of excipients when compared to polymorphic screening.

One of the most frequently cited properties of these systems is the dissolution rate, which is increased due to the enhancement of surface area. As particles decrease in size, the total surface area required for the same amount of mass increases significantly. Additionally, as described by the Ostwald–Freundlich equation, changes in metastable equilibrium solubility have also been reported due to decreases in particle size. By developing smaller particles, an enhancement in the overall magnitude of kinetic solubility can be achieved in addition to the rate at which that solubility is attained.

Crystal structure of the material is the other major property frequently cited in the literature for solubility enhancement. Pharmaceutical APIs may exist in a variety of crystal structures, commonly referred to as polymorphs, as well as amorphous forms which lack any type of long or short range order associated with a crystalline material. When examining the dissolution process, it can actually be viewed as two separate and discrete steps: dissociation of the solute molecules from the crystal lattice and solvation of solute molecules. Modifications of the crystal structure can be used to reduce the intermolecular interactions and facilitate the dissolution of the drug substance. Ultimately, these forms offer transient solid-state properties and will eventually transition to the thermodynamically stable form of the drug substance. Detailed polymorphic screening and formulation optimization can be used however to provide compositions which are stable for pharmaceutically relevant timescales.

As previously mentioned, an extensive portfolio of technologies has been developed to exploit these mechanisms of solubility enhancement. Particle size reduction processes, including "top-down" and "bottom-up" technologies, maximize surface area to exploit the kinetic and thermodynamic advantages offered by size reduction. Molecular complexation techniques and solid solutions reduce the intermolecular interaction to facilitate dissociation while providing the theoretically smallest possible structure for dissolution, i.e., the individual drug molecule. Additionally, formulation techniques such as the incorporation of hydrophilic polymers and surfactants have been shown to further enhance the dissolution rate through improved wetting of microscopic drug domains within the composition.

Traditional dissolution testing is conducted under sink conditions, meaning that the amount of material added to the vessel is three to five times less than that required to saturate the media within the vessel (Amidon et al. 1995). By operating under these conditions, it ensures a sufficient driving force for drug release throughout the testing phase in order to mimic conditions found in vivo during the dissolution and absorption process. Additionally, operating under sink conditions allows for convenient mathematical assumptions facilitating modeling of the release process. Furthermore, traditionally manufactured crystalline dosage forms lack the requisite thermodynamic and kinetic forces for supersaturation, eliminating the need for examination under these conditions. Solid dispersions are capable of supersaturating their environment, necessitating testing under these conditions. Typically, supersaturated dissolution testing is conducted under similar conditions to sink testing; however, the amount of drug added to the vessel is several fold the amount required for saturation of the media, allowing the formulation, provided it has the underlying properties, to supersaturate the media. Additionally, due to the presence of small particle precipitation which may occur during the testing period filter sizes are frequently smaller than those used under sink conditions. It is generally assumed that the particle size cutoff for cellular uptake is 200 nm, so frequently 0.2 μm polvinylidine fluoride (PVDF) or polytetrafluoroethylene (PTFE) filters are employed to reduce crystallization on the filter membrane while minimizing particle size. Utilizing this testing procedure it becomes possible to ascertain the dissolution rate kinetics associated with a formulation, as well as its ability to provide and maintain supersaturation over prolonged periods of time. In the following sections, examples of in vitro supersaturation are presented along with the resulting enhancement in bioavailability to illustrate the utility of solid dispersions and supersaturation for enhanced therapeutic performance.

According to the BCS system, class II compositions exhibit solubility-limited bioavailability making both the compositional equilibrium solubility and the rate at which it is achieved restrictive steps in the oral absorption process. In order to improve the bioavailability of these drugs many formulations, both investigative and commercial, have been developed to provide enhanced dissolution rates. By formulating the solid dispersion with hydrophilic excipients capable of rapid dissolution rates, the drug dissolution rate will become a function of the dissolution rate of the carrier polymer, allowing for enhanced dissolution rates and the potential for supersaturation. Several commonly used polymeric materials for this application include HPMC, PVP, low-molecular-weight PEG, vinylpyrrolidone vinylacetate (PVPVA), and Eudragit E100. Numerous publications are available in the literature focusing on the dissolution rate enhancement resulting from novel solid dispersion formulations to provide improved bioavailability or faster onset of action; however, only a paucity of these papers examined the ability of these compositions to supersaturate and correlated this behavior to performance in an animal model or human subjects. In recent years, the importance of supersaturation in achieving improved bioavailability has emerged as a critical design factor for formulation development.

Tacrolimus, which is currently marketed under the trade name Prograf, is produced as a solid dispersion using an organic solvent–based coating process (Yamashita et al. 2003). Utilizing a thermal solid dispersion process, it was possible to prepare compositions which exhibited substantial supersaturation and enhanced in vivo performance in an animal model that were similar to those produced by compositions using the solvent-based production process, as shown in Figure 18.8. The reason for the improved performance of formulations containing HPMC was attributed to the stabilizing interaction between the drug and polymer, which allowed the polymer to function as a recrystallization inhibitor. Similar approaches have also been taken to improve the oral bioavailability of itraconazole (Miller et al. 2008a,b) and nifedipine (Ho et al. 1996).

While the enhancement in dissolution rate can provide improved oral bioavailability by achieving the equilibrium solubility faster or providing higher metastable equilibrium solubility values, these formulations may not provide the greatest improvement in bioavailability. Weakly basic drugs, as stated previously, may be ionized at gastric pH and exhibit a higher solubility where hydrophilic polymer-based compositions will dissolve and release the drug. Upon transition to the upper small intestine, the pH rises and the drug may become partially or completely unionized driving a significant solubility reduction. Furthermore, most drugs are primarily absorbed in the upper small

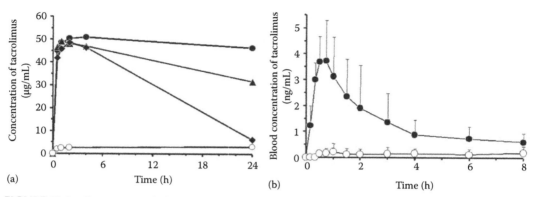

FIGURE 18.8 Supersaturated dissolution profiles of thermally processed tacrolimus solid dispersions (a) and in vivo plasma profile comparing solvent based and thermally processed solid dispersions (b). (•) Tacrolimus: HPMC, (▲) tacrolimus: PVP, (♦) tacrolimus: PEG 6000, (o) crystalline tacrolimus. (Reproduced from Yamashita, K. et al., *Int. J. Pharm.*, 267, 79, 2003. With permission.)

intestine, where the substantial surface area provided by the villi and microvilli facilitate transport across the membrane. Compositions which supersaturate the gastric environment for short durations may also be subject to partial or complete precipitation, achieving only equilibrium solubility prior to entering the upper small intestine and negating the tremendous advantages provided by solid dispersions. In these cases, it would be prudent to target supersaturation to the upper small intestine, which is commonly achieved by using pH-responsive carriers. These carrier materials are insoluble at gastric pH; however upon entering the upper small intestine, the pH change will trigger ionization of the carboxylic acid functional groups on the polymer chain resulting in dissolution. By providing a range of interactions between the drug and polymer, such formulations are capable of providing extended durations of supersaturation along with site targeting to minimize precipitation of the dissolved drug substance. These techniques have been demonstrated to be highly effective for a range of drugs, including itraconazole (DiNunzio et al. 2008), tacrolimus (Overhoff et al. 2008), and HO 221 (Kondo et al. 1993, 1994). These polymers, which not only maximize the rate of dissolution, but also provide significant stabilization and duration of supersaturation, are termed concentration-enhancing polymers. These materials are generally classified as high-molecular-weight hydrophilic polymers (HPMC E50, PVP K90, etc.) or enteric polymers. The behavior for stabilization, although not completely understood, is believed to be the result of steric hinderance associated with molecular weight as well specific drug–polymer interactions such as hydrogen bonding. Further improving stabilization behavior of enteric materials is the charge associated with the polymers when dissolved, which provides a repulsive force limiting colloidal growth.

The application of concentration-enhancing polymers was studied for oral bioavailability enhancement of celecoxib, a poorly water-soluble compound marketed under the trade name Celebrex® (Guzmán et al. 2007). Target concentration-enhancing polymers were identified using a high-throughput screening technique incorporating light scattering and yielded numerous potential candidates. Interestingly, the ability of these materials to inhibit precipitation was related to the CMC of the material. Addition of HPC was shown to enhance stabilization. When lead compositions consisting of TPGS and HPC, as well as formulations with Pluronic F127 and HPC, were dosed in a canine model, rapid and near complete absorption was observed. In a similar study by Overhoff et al. (2008), solid dispersions of tacrolimus were prepared using surfactant-based materials and shown to inhibit precipitation while also providing enhanced oral bioavailability in a rat model. Similar results, highlighted in Figure 18.9, were also demonstrated by DiNunzio et al., using engineered amorphous solid dispersions of enteric polymers to provide sustained durations of itraconazole supersaturation to enhance oral bioavailability (DiNunzio et al. 2008). Such techniques

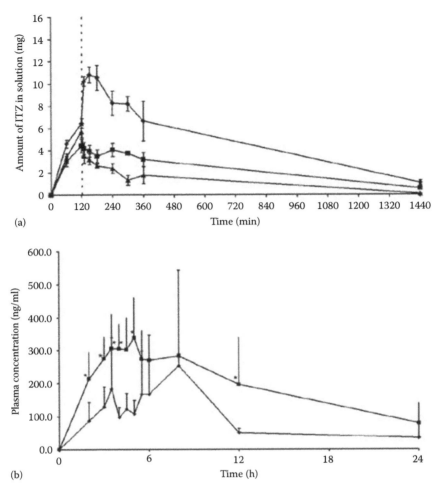

FIGURE 18.9 In vitro (a) and in vivo (b) behavior of enteric compositions. (a) Supersaturated dissolution profile of ITZ:CAP formulations. Key: 1:2 ITZ:CAP (◆), 1:1 ITZ:CAP (■), 2:1 ITZ:CAP (▲). Each vessel ($n = 3$) contained 37.5 mg ITZ equivalent corresponding to 10 times the equilibrium solubility of ITZ in the acid phase. Testing was conducted for 2 h in 750 mL of 0.1 N HCl followed by pH adjustment to 6.8 ± 0.5 with 250 mL of 0.2 M tribasic sodium phosphate solution. Dashed vertical line indicates the time of pH change. (b) In vivo plasma profile. Key: Sporanox pellets (◆), 1:2 ITZ:CAP (■). Formulations were administered by oral gavage at a dose of 15 mg ITZ/kg body weight per rat ($n = 6$). * indicates statistically significant concentration difference between test and reference formulation as determined by one-way ANOVA with Tukey post hoc testing. (Reprinted from DiNunzio, J.C. et al., *Mol. Pharm.*, 5, 968, 2008. With permission.)

have also led to the development of several patented compositions which have been demonstrated to improve oral bioavailability, as shown in Figure 18.10.

18.5 CONCLUSIONS

Polymeric materials play a critical role in the development of solid oral dosage forms and this contribution is likely to grow continuously in the foreseeable future as products continue to become more technologically advanced. Currently, polymers provide a basis for the development of controlled release systems and the stabilization of amorphous forms, presenting two unique delivery platforms which have contributed to the advancement of patient care. As scientists continue to develop a deeper

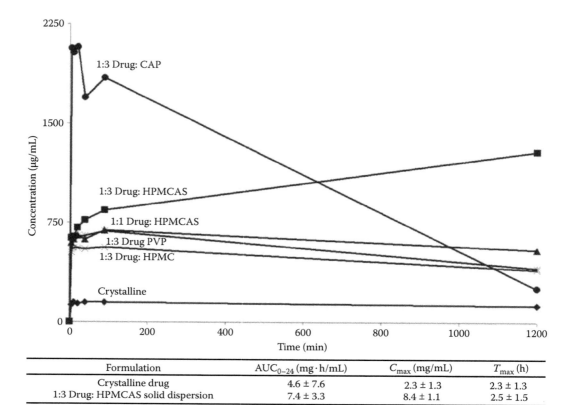

Formulation	AUC$_{0-24}$ (mg · h/mL)	C$_{max}$ (mg/mL)	T$_{max}$ (h)
Crystalline drug	4.6 ± 7.6	2.3 ± 1.3	2.3 ± 1.3
1:3 Drug: HPMCAS solid dispersion	7.4 ± 3.3	8.4 ± 1.1	2.5 ± 1.5

FIGURE 18.10 In vitro supersaturation profiles (top) and in vivo pharmacokinetic data from human trials (bottom) for developmental solid dispersions. (Adapted from data presented in Crew et al., Pharmaceutical compositions of a sparingly soluble glycogen phosphorylase inhibitor, US Patent 7,235,260 B2.)

and more complete understanding of the underlying interactions, and the impact of such behaviors on dosage form performance, pharmaceutical systems will continue to become more complex and more efficacious. Polymeric systems are also likely to provide significant contributions in many growing areas of pharmaceutical development, particularly for advanced device manufacturing and bionanotechnology applications. Many of the newest hybrid technologies exploit specific polymeric interactions in order to provide enhanced pharmaceutical efficacy. Continued learning and a greater understanding of these interactions will provide the basis for developing the next generation of drug delivery systems, providing improved treatment options for a number of disease states.

REFERENCES

Aharoni, S. M. (1998). Increased glass transition temperature in motionally constrained semicrystalline polymers. *Polymers for Advanced Technologies* 9(3): 169–201.

Alvarez-Fuentes, J., M. Fernandez-Arevalo et al. (1994). Morphine polymeric coprecipitates for controlled release: Elaboration and characterization. *Drug Development and Industrial Pharmacy* 20(15): 2409–2424.

Amidon, G. L., H. Lennernäs et al. (1995). A theoretical basis for a biopharmaceutic drug classification: The correlation of in vitro drug product dissolution and in vivo bioavailability. *Pharmaceutical Research* 12(3): 413–420.

Amighi, K. and A. J. Moës (1996). Influence of plasticizer concentration and storage conditions on the drug release rate from Eudragit® RS 30 D film-coated sustained-release theophylline pellets. *European Journal of Pharmaceutics and Biopharmaceutics* 42(1): 29–35.

Amighi, K. A. and A. J. Moës (1997). Influence of curing conditions in the drug release rate from Eudragit® NE 30 D film coated sustained release theopylline pellets. *STP Pharma Sciences* 7(2): 141–147.

Anderson, W. and S. A. M. Abdel-Aziz (1976). Ageing effects in cast acrylate–methacrylate film. *Journal of Pharmacy and Pharmacology* (Suppl 22): 28.

Andrews, G. P. (2007). Advances in solid dosage manufacturing technology. *Philosophical Transactions of the Royal Society of London A Mathematical and Physical Sciences* 365: 2935–2949.

Banker, G. and G. E. Peck (1981). The new, water based colloidal dispersions. *Pharmaceutical Technology* 5(4): 55–61.

Bhardwaj, V., S. Hariharan et al. (2005). Pharmaceutical aspects of polymeric nanoparticles for oral drug delivery. *Journal of Biomedical Nanotechnology* 1(3): 235–258.

Bodmeier, R. and O. Paeratakul (1994). The effect of curing on drug release and morphological properties of ethylcellulose pseudolatex-coated beads. *Drug Development and Industrial Pharmacy* 20(9): 1517–1533.

Breitenbach, J. (2006). Melt extrusion can bring new benefits to HIV therapy: The example of Kaletra tablets. *American Journal of Drug Delivery* 4(2): 61–64.

Breitenbach, J. and J. Lewis (2002). Two concepts, one technology: Controlled-release and solid dispersions with Meltrex. *Modified-Release Drug Delivery Technology*. M. J. Rathbone and J. H. M. S. Roberts, eds. New York: Informa Healthcare, 126, pp. 125–134.

Brittain, H. G. (2007). Polymorphism: Pharmaceutical aspects. *Encyclopedia of Pharmaceutical Technology*. J. Swarbrick and J. C. Boylan, eds. Hoboken, NJ: Informa Healthcare, pp. 2935–2945.

Bruce, C., K. A. Fegely et al. (2007). Crystal growth formation in melt extrudates. *International Journal of Pharmaceutics* 341(1–2): 162–172.

Bruce, L. D., J. J. Koleng et al. (2003). The influence of polymeric subcoats and pellet formulation on the release of chlorpheniramine maleate from enteric coated pellets. *Drug Development and Industrial Pharmacy* 29(8): 909–924.

Bussemer, T., N. A. Peppas et al. (2003). Evaluation of the swelling, hydration and rupturing properties of the swelling layer of a rupturable pulsatile drug delivery system. *European Journal of Pharmaceutics and Biopharmaceutics* 56(2): 261–270.

Byrappa, K., S. Ohara et al. (2008). Nanoparticles synthesis using supercritical fluid technology—Towards biomedical applications. *Advanced Drug Delivery Reviews* 60(3): 299–327.

Byrne, M. E., J. Z. Hilt et al. (2008). Recognitive biomimetic networks with moiety imprinting for intelligent drug delivery. *Journal of Biomedical Materials Research* 84A: 137–147.

Byrne, M. E., K. Park et al. (2002). Molecular imprinting within hydrogels. *Advanced Drug Delivery Reviews* 54(1): 149–161.

Byrne, M. E. and V. Salian (2008). Molecular imprinting within hydrogels II: Progress and analysis of the field. *International Journal of Pharmaceutics* 364(2): 188–212.

Carlin, B., J.-X. Li et al. (2008). Pseudolatex dispersions for controlled delivery applications. *Aqueous Polymeric Coaings for Pharmaceutical Dosage Forms*. J. W. McGinity and L. A. Felton, eds. Hoboken, NJ: Informa Healthcare, pp. 1–46.

Çelik, M. and S. C. Wendel (2005). Spray drying and pharmaceutical applications. *Handbook of Pharmaceutical Granulation Technology*. D. M. Parikh, ed. New York: Informa Healthcare, 154, pp. 129–158.

Chiou, W. L. and S. Riegelman (1971). Pharmaceutical applications of solid dispersions. *Journal of Pharmaceutical Sciences* 60(9): 1281–1302.

Chokshi, R. J., H. K. Sandhu et al. (2005). Characterization of physico-mechanical properties of indomethacin and polymers to assess their suitability for hot-melt extrusion process as a means to manufacture solid dispersion/solution. *Journal of Pharmaceutical Sciences* 94(11): 2463–2474.

Crowley, M. M., F. Zhang et al. (2007). Pharmaceutical applications of hot-melt extrusion: Part I. *Drug Development and Industrial Pharmacy* 33(9): 909–926.

Dangel, C., K. Kolter et al. (2000a). Aqueous enteric coatings with methacrylic acid copolymer type C on acidic and basic drugs in tablets and pellets. Part II: Dosage forms containing indomethacin and diclofenac sodium. *Pharmaceutical Technology* 24(4): 36–42.

Dangel, C., K. Kolter et al. (2000b). Aqueous enteric coatings with methacrylic acid copolymer type C. On acidic and basic drugs in tablets and pellets. Part I: Acetylsalicylic acid tablets and crystals. *Pharmaceutical Technology* 24(3): 64–70.

DiNunzio, J. C., D. A. Miller et al. (2008). Amorphous compositions using concentration enhancing polymers for improved bioavailability. *Molecular Pharmaceutics* 5(6): 968–980.

Dittgen, M., S. Fricke et al. (1995a). Hot spin mixing—A new technology to manufacture solid dispersions, part 1: Testosterone. *Pharmazie* 50(3): 225–226.

Dittgen, M., S. Fricke et al. (1995b). Hot spin mixing—A new technology to manufacture solid dispersions, part 3: Progesterone. *Pharmazie* 50(7): 507–508.

Dittgen, M., T. Gräser et al. (1995c). Hot spin mixing—A technology to manufacture solid dispersions, part 2: Dienogest. *Pharmazie* 50(6): 438–439.

Engstrom, J. D., D. T. Simpson et al. (2007). Stable high surface area lactate dehydrogenase particles produced by spray freezing into liquid nitrogen. *European Journal of Pharmaceutics and Biopharmaceutics* 65(2): 163–174.

Fadda, H. M., M. A. M. Mohamed et al. (2008). Impairment of the in vitro drug release behaviour of oral modified release preparations in the presence of alcohol. *International Journal of Pharmaceutics* 360(1–2): 171–176.

Felton, L. A. and J. W. McGinity (1997). Influence of plasticizers on the adhesive properties of an acrylic resin copolymer to hydrophilic and hydrophobic tablet compacts. *International Journal of Pharmaceutics* 154(2): 167–178.

Follonier, N. et al. (1995). Various way of modulating the release of diltiazem hydrochloride from hot-melt extruded sustained release pellets prepared using polymeric materials. *Journal of Controlled Release* 36(3): 243.

Forster, A., J. Hempenstall et al. (2001). Selection of excipients for melt extrusion with two poorly water-soluble drugs by solubility parameter calculation and thermal analysis. *International Journal of Pharmaceutics* 226(1–2): 147–161.

Forster, A., J. Hempenstall et al. (2003). Comparison of the Gordon–Taylor and Couchman–Karasz equations for prediction of the glass transition temperature of glass solutions of drug and polyvinylpyrrolidone prepared by melt extrusion. *Pharmazie* 58(11): 838–839.

Fukuda, M., N. A. Peppas et al. (2006). Properties of sustained release hot-melt extruded tablets containing chitosan and xanthan gum. *International Journal of Pharmaceutics* 310(1–2): 90–100.

Ghebre-Shellasie, I. (1989). Pellets: A general overview. *Pharmaceutical Pelletization Technology*. I. Ghebre-Shellasie, ed. New York: Marcel Dekker, pp. 6–7.

Gilis, P. M., V. F. V. De Condé et al. (1997). Beads having a core coated with an antifungal and a polymer. U.S.P.T. Office. US Patent 5,633,015.

Gordon, M. and J. S. Taylor (1952). Ideal copolymers and the second-order transitions of synthetic rubbers. I. Noncrystalline copolymers. *Journal of Applied Chemistry* 2: 493–500.

Greenhalgh, D. J., A. C. Williams et al. (1999). Solubility parameters as predictors of miscibility in solid dispersions. *Journal of Pharmaceutical Sciences* 88(11): 1182–1190.

Gupta, P., R. Thilagavathi et al. (2005). Role of molecular interaction in stability of celecoxib–PVP amorphous systems. *Molecular Pharmaceutics* 2(5): 384–391.

Guzmán, H. R., M. Tawa et al. (2007). Combined use of crystalline salt forms and precipitation inhibitors to improve oral absorption of celecoxib from solid oral formulations. *Journal of Pharmaceutical Sciences* 96(10): 2686–2702.

Hancock, B. C. (2007). Amorphous pharmaceutical systems. *Encyclopedia of Pharmaceutical Technology*. J. Swarbrick and J. C. Boylan, eds. Hoboken, NJ: Informa Healthcare, pp. 83–91.

Hancock, B. C., K. Christensen et al. (1998). Estimating the critical molecular mobility temperature (T_K) of amorphous pharmaceuticals. *Pharmaceutical Research* 15(11): 1649–1651.

Hancock, B. C., S. L. Shamblin et al. (1995). Molecular mobility of amorphous pharmaceutical solids below their glass transition temperatures. *Pharmaceutical Research* 12(6): 799–806.

Hancock, B.C. and Zografi, G. (1997). Characteristics and significance of the amorphous state in pharmaceutical systems. *Journal of Pharmaceutical Sciences* 86(1): 1.

Ho, H.-O., H.-L. Su et al. (1996). The preparation and characterization of solid dispersions on pellets using a fluidized-bed system. *International Journal of Pharmaceutics* 139(1–2): 223–229.

Howard, J. R. and P. Timmins (1988). Controlled release formulation. U.S.P.T. Office.

Hu, J., K. P. Johnston et al. (2004). Nanoparticle engineering processes for enhancing the dissolution rates of poorly water soluble drugs. *Drug Development and Industrial Pharmacy* 30(3): 233–245.

Hülsmann, S., T. Backensfeld et al. (2000). Melt extrusion. An alternative method for enhancing the dissolution rate of 17β-estradiol hemihydrate. *European Journal of Pharmaceutics and Biopharmaceutics* 49(3): 237–242.

Jia, L. (2005). Nanoparticle formulation increases oral bioavailability of poorly soluble drugs: Approaches, experimental evidences and theory. *Current Nanoscience* 1(3): 237–243.

Johnson, J. L. (1993). Influence of ionic strength on matrix integrity and drug release from hydroxypropyl cellulose compacts. *International Journal of Pharmaceutics* 90(1–2): 151–159.

Jones, D. (1994). Air suspension coating for multiparticulates. *Drug Development and Industrial Pharmacy* 20(20): 3175–3206.

Kabadi, M. B. and R. V. Vivilecchia (1994). Stabilized pharmaceutical compositions comprising an HMG-CoA reductase inhibitor compound. U.S.P.T. Office.

Kim, M.-S., S.-J. Jin et al. (2008). Preparation, characterization and in vivo evaluation of atorvastatin calcium nanoparticles using supercritical antisolvent (SAS) process. *European Journal of Pharmaceutics and Biopharmaceutics* 69(2): 454–465.

Kim, S. J. and T. H. Kwon (1996a). Enhancement of mixing performance of single-screw extrusion processes via chaotic flows: Part I. Basic concepts and experimental study. *Advances in Polymer Technology* 15(1): 41–54.

Kim, S. J. and T. H. Kwon (1996b). Enhancement of mixing performance of single-screw extrusion processes via chaotic flows: Part II. Numerical study. *Advances in Polymer Technology* 15(1): 55–69.

Kondo, N., T. Iwao et al. (1993). Pharmacokinetics of a micronized, poorly water-soluble drug, HO-221, in experimental animals. *Biological & Pharmaceutical Bulletin* 16(8): 796–800.

Kondo, N., T. Iwao et al. (1994). Improved oral absorption of enteric coprecipitates of a poorly soluble drug. *Journal of Pharmaceutical Sciences* 83(4): 566–570.

Kranz, H., C. Guthmann et al. (2005). Development of a single unit extended release formulation for ZK 811 752, a weakly basic drug. *European Journal of Pharmaceutical Sciences* 26(1): 47–53.

Kucera, S. A., N. H. Shah et al. (2007). The use of proteins to minimize the physical aging of Eudragit® sustained release films. *Drug Development and Industrial Pharmacy* 33: 717–726.

Kucharski, J. and A. Kmiec (1983). Hydrodynamics, heat and mass transfer during coating of tablets in a spouted bed. *The Canadian Journal of Chemical Engineering* 61(3): 435–439.

Lakshman, J. P., Y. Cao et al. (2008). Application of melt extrusion in the development of a physically and chemically stable-energy amorphous solid dispersion of a poorly water-soluble drug. *Molecular Pharmaceutics* 5(6): 994–1002.

Larson, C., N. Cavuto, J. et al. (1996). Bioavailability and efficacy of omeprazole given orally and by nanogastric tube. *Digestive Diseases and Sciences* 41(3): 475–479.

Lecomte, F., J. Siepmann et al. (2005). pH-Sensitive polymer blends used as coating materials to control drug release from spherical beads: Elucidation of the underlying mass transport mechanism. *Pharmaceutical Research* 22(7): 1129–1141.

Letko, E., K. Bhol et al. (1999). Tacrolimus (FK 506). *Annals of Allergy, Asthma & Immunology* 83(3): 179–190.

Leuner, C. and J. Dressman (2000). Improving drug solubility for oral delivery using solid dispersions. *European Journal of Pharmaceutical Sciences* 50(1): 47–60.

Lyons, J. G., D. M. Devine et al. (2006). The use of agar as a novel filler for monolithic matrices produced using hot melt extrusion. *European Journal of Pharmaceutics and Biopharmaceutics* 64(1): 75–81.

Maejima, T. and J. W. McGinity (2001). Influence of film additives on stabilizing drug release rates from pellets coated with acrylic polymers. *Pharmaceutical Development and Technology* 6: 211–221.

McGinity, J. W., M. A. Repka et al. (2007). Hot-melt extrusion technology. *Encyclopedia of Pharmaceutical Technology*. J. Swarbrick and J. C. Boylan, eds. Hoboken, NJ: Informa Healthcare: 2004–2020.

McGinity, J. W. and F. Zhang (2003). Melt-extruded controlled-release dosage forms. *Pharmaceutical Extrusion Technology*. I. Ghebre-Sellassie and C. Martin, eds. New York: Informa Healthcare: 133, pp. 183–208.

Miller, D. A., J. C. DiNunzio et al. (2008a). Enhanced in vivo absorption of itraconazole via stabilization of supersaturation following acidic-to-neutral pH transition. *Drug Development and Industrial Pharmacy* 34(8): 890–902.

Miller, D. A., J. C. DiNunzio et al. (2008b). Targeted intestinal delivery of supersaturated itraconazole for improved oral absorption. *Pharmaceutical Research* 25(6): 1450–1459.

Miller, D. A., J. T. McConville et al. (2007). Hot-melt extrusion for enhanced delivery of drug particles. *Journal of Pharmaceutical Sciences* 96(2): 361–376.

Nikfar, F., A. T. Serajuddin et al. (1996). Pharmaceutical compositions having good dissolution properties. U.S.P.T. Office.

Overhoff, K. A., J. D. Engstrom et al. (2007). Novel ultra-rapid freezing particle engineering process for enhancement of dissolution rates of poorly water-soluble drugs. *European Journal of Pharmaceutics and Biopharmaceutics* 65(1): 57–67.

Overhoff, K. A., K. P. Johnston et al. (2009). Use of thin film freezing to enable drug delivery: A review. *Journal of Drug Delivery Science and Technology* 19(2): 89–98.

Overhoff, K. A., J. T. McConville et al. (2008). Effect of stabilizer on the maximum degree and extent of supersaturation and oral absorption of tacrolimus made by ultra-rapid freezing. *Pharmaceutical Research* 25(1): 167–175.

Patterson, J. E., M. B. James et al. (2007). Preparation of glass solutions of three poorly water soluble drugs by spray drying, melt extrusion and ball milling. *International Journal of Pharmaceutics* 336(1): 22–34.

Repka, M. A., S. K. Battu et al. (2007). Pharmaceutical applications of hot-melt extrusion: Part II. *Drug Development and Industrial Pharmacy* 33(10): 1043–1057.

Repka, M. A., S. Majumdar et al. (2008). Applications of hot-melt extrusion for drug delivery. *Expert Opinion on Drug Delivery* 5(12): 1357–1376.

Riedel, A. and C. S. Leopold (2005). Degradation of omeprazole induced by enteric polymer solutions and aqueous dispersions: HPLC investigations. *Drug Development and Industrial Pharmacy* 31(2): 151–160.

Rothen-Weinhold, A., N. Oudry et al. (2000). Formation of peptide impurities in polyester matrices during implant manufacturing. *European Journal of Pharmaceutics and Biopharmaceutics* 49(3): 253–257.

Shlieout, G., K. Arnold et al. (2002). Powder and mechanical properties of microcrystalline cellulose with different degrees of polymerization. *AAPS PharmSciTech* 3(2): Article 11.

Shoyele, S. A. and S. Cawthorne (2006). Particle engineering techniques for inhaled biopharmaceuticals. *Advanced Drug Delivery Reviews* 58: 1009–1029.

Siepe, S., W. Herrmann et al. (2006). Microenvironmental pH and microviscosity inside pH-controlled matrix tablets: An EPR imaging study. *Journal of Controlled Release* 112(1): 72–78.

Siepmann, J. and N. A. Peppas (2001). Mathematical modeling of controlled drug delivery. *Advanced Drug Delivery Reviews* 48(2–3): 139–157.

Siepmann, J., K. Podual et al. (1999). A new model describing the swelling and drug release kinetics from hydroxypropyl methylcellulose tablets. *Journal of Pharmaceutical Sciences* 88(1): 65–72.

Siepmann, J. and F. Siepmann (2008). Mathematical modeling of drug delivery. *International Journal of Pharmaceutics* 364(2): 328–343.

Siepmann, J., A. Streubel et al. (2002). Understanding and predicting drug delivery from hydrophilic matrix tablets using the "sequential layer" model. *Pharmaceutical Research* 19(3): 306–314.

Six, K., H. Berghmans et al. (2003). Characterization of solid dispersions of itraconazole and hydroxypropyl-methylcellulose prepared by melt extrusion, part II. *Pharmaceutical Research* 20(7): 1047–1054.

Six, K., T. Daems et al. (2005). Clinical study of solid dispersions of itraconazole prepared by hot-stage extrusion. *European Journal of Pharmaceutical Sciences* 24(2–3): 179–186.

Stroyer, A., J. W. McGinity et al. (2006). Solid state interactions between the proton pump inhibitor omeprazole and various enteric coating polymers. *Journal of Pharmaceutical Sciences* 95(6): 1342–1353.

Tatavarti, A. S., K. A. Mehta et al. (2004). Influence of methacrylic and acrylic acid polymers on the release performance of weakly basic drugs from sustained release hydrophilic matrices. *Journal of Pharmaceutical Sciences* 93(9): 2319–2331.

Thiele, W. (2003). Twin-screw extrusion and screw design. *Pharmaceutical Extrusion Technology*. I. Ghebre-Sellassie and C. Martin, eds. New York: Informa Healthcare: 133, pp. 69–98.

Van den Brande, J., I. Weuts et al. (2004). DSC analysis of the anti-HIV agent loviride as a preformulation tool in the development of hot-melt extrudates. *Journal of Thermal Analysis and Calorimetry* 77(2): 523–530.

Vaughn, J. M., X. Gao et al. (2005). Comparison of powder produced by evaporative precipitation into aqueous solution (EPAS) and spray freezing into liquid (SFL) technologies using novel Z-contrast STEM and complimentary techniques. *European Journal of Pharmaceutics and Biopharmaceutics* (60): 81–89.

Vaughn, J. M., J. T. McConville et al. (2006a). Single dose and multiple dose studies of itraconazole nanoparticles. *European Journal of Pharmaceutics and Biopharmaceutics* (63): 95–102.

Vaughn, J. M., J. T. McConville et al. (2006b). Supersaturation produces high bioavailability of amorphous danazol particles formed by evaporative precipitation into aqueous solution and spray freezing into liquid technologies. *Drug Development and Industrial Pharmacy* 32(5): 559–567.

Vecchio, C., G. Bruni et al. (1994). Research papers: Preparation of indobufen pellets by using centrifugal rotary fluidized bed equipment without starting seeds. *Drug Development and Industrial Pharmacy* 20(12): 1943–1956.

Vehring, R. (2008). Pharmaceutical particle engineering via spray drying. *Pharmaceutical Research* 25(5): 999–1022.

Vehring, R., W. R. Foss et al. (2007). Particle formation in spray drying. *Journal of Aerosol Science* 38(7): 728–746.

Venkatesh, S., J. Saha et al. (2008). Transport and structural analysis of molecular imprinted hydrogels. *European Journal of Pharmaceutics and Biopharmaceutics* 69(3): 852–860.

Verreck, G., A. Decorte et al. (2006a). Hot stage extrusion of *p*-amino salicylic acid with EC using CO_2 as a temporary plasticizer. *International Journal of Pharmaceutics* 327(1–2): 45–50.

Verreck, G., A. Decorte et al. (2006b). The effect of pressurized carbon dioxide as a plasticizer and foaming agent on the hot melt extrusion process and extrudate properties of pharmaceutical polymers. *The Journal of Supercritical Fluids* 38(3): 383–391.

Verreck, G., A. Decorte et al. (2007). The effect of supercritical CO_2 as a reversible plasticizer and foaming agent on the hot stage extrusion of itraconazole with EC 20 cps. *Journal of Supercritical Fluids* 40(1): 153–162.

Verreck, G., R. Vandecruys et al. (2004). The use of three different solid dispersion formulations—melt extrusion, film-coated beads, and a glass thermoplastic system—to improve the bioavailability of a novel microsomal triglyceride transfer protein inhibitor. *Journal of Pharmaceutical Sciences* 93(5): 1217–1228.

Vertommen, J. and R. Kinget (1997). The influence of five selected processing and formulation variables on the particle size, particle size distribution, and friability of pellets produced in a rotary processor. *Drug Development and Industrial Pharmacy* 23(1): 39–46.

Walker, G. M., S. E. J. Bell et al. (2007). Co-melt fluidised bed granulation of pharmaceutical powders: Improvements in drug bioavailability. *Chemical Engineering Science* 62(1–2): 451–462.

Wu, C. and J. W. McGinity (2003). Influence of an enteric polymer on drug release rates of theophylline from pellets coated with Eudragit® RS 30 D. *Pharmaceutical Development and Technology* 8: 103–110.

Xu, X. M., Y. M. Song et al. (2006). Effect of ionic strength on the temperature-dependent behavior of hydroxypropyl methylcellulose solution and matrix tablet behavior. *Journal of Applied Polymer Science* 102(4): 4066–4074.

Yamashita, K., T. Nakate et al. (2003). Establishment of new preparation method for solid dispersion formulation of tacrolimus. *International Journal of Pharmaceutics* 267(1–2): 79–91.

Yu, L. (2001). Amorphous pharmaceutical solids: Preparation, characterization and stabilization. *Advanced Drug Delivery Reviews* 48(1): 27–42.

Zheng, W., D. Sauer et al. (2005). Influence of hydroxyethylcellulose on the drug release properties of theophylline pellets coated with Eudragit® RS 30 D. *European Journal of Pharmaceutics and Biopharmaceutics* 59(1): 147–154.

Zheng, X., R. Yang et al. (2007a). Part I: Characterization of solid dispersions of nimodipine prepared by hot-melt extrusion. *Drug Development and Industrial Pharmacy* 33(7): 791–802.

Zheng, X., R. Yang et al. (2007b). Part II: Bioavailability in beagle dogs of nimodipine solid dispersions prepared by hot-melt extrusion. *Drug Development and Industrial Pharmacy* 33(7): 783–789.

Zhou, D., G. G. Z. Zhang et al. (2008). Thermodynamics, molecular mobility and crystallization kinetics of amorphous griseofulvin. *Molecular Pharmaceutics* 5(6): 927–936.

Manufacturing
Multifunctional Scaffolds
for Tissue Engineering

Vincenzo Guarino, Antonio Gloria,
Roberto De Santis, and Luigi Ambrosio

CONTENTS

19.1 TISSUE REGENERATION PARADIGMS

The basic principle of tissue engineering entails the guided application and the control of cells, materials, and the microenvironment into which they are delivered [1]. In a tissue engineering process, tissues or organs may be created in vivo, in vitro, or ex vivo and implanted into the patient [2]. In this direction, the selection of three elements firstly concurs to satisfy the basic tissue engineering principles on fabricating biological tissues: (i) viable, responsive cells; (ii) a scaffold to support tissue formation; and (iii) a growth-inducing stimulus.

Firstly, the sites of tissue reconstruction have to contain a sufficient number of viable progenitor cells capable of producing the tissue of interest. If the number of cells is deficient, one must engineer the site to provide the necessary population of cells or provide some stimuli to recruit than to the reconstruction site [3].

Secondly, the reconstruction site should be filled with a 3D, porous scaffold that facilitates attachment, migration, and proliferation of cells. It should also stimulate the synthesis of extracellular matrix (ECM) proteins and the ingrowth of tissue throughout the site. Thirdly, the cells in the graft site must receive the appropriate stimuli (i.e., mechanical, biochemical) in the form of soluble, bioactive molecules (e.g., growth factors, cytokines, hormones) that will influence the cell phenotype and enhance the end result of tissue formation at the graft site [4,5].

Starting from these considerations, one of the most important challenges in tissue engineering is the appropriate design of open-pore biocompatible and biodegradable porous scaffolds which are able to provide a temporary substitute for the ECM. The relevance of the design of scaffolds to be used as ECM analogs should not be underestimated. Nearly 30 years ago, Bissell et al.

proposed dynamic reciprocity, which states that a tissue achieves a specific function in part through interactions of the cells with the ECM [6].

Subsequent work demonstrated that gene expression can be mediated by ECM binding to ECM receptors on the cell surface, which provides a link to the cytoskeleton and eventually the nuclear matrix [7]. The inclusion of neighboring cell interactions and soluble signals originating systemically or from cells in the immediate or distant vicinity provides a more complete model of tissue environment [8].

For these reasons, much attention has also been given to the simulation of the extracellular environment. Scaffolds for tissue regeneration occupy a fundamental role in tissue development, because they must support the proliferation and differentiation of cells as they mature into a functional tissue. In particular, the scaffold-assisted regeneration of specific tissues has been shown to be strongly dependent on morphological parameters such as surface-to-volume ratio and pore size and interconnectivity. Indeed, these micro-architectural features not only significantly influence cell morphology, cell binding, and phenotypic expression, but also control the extent and nature of nutrient diffusion and tissue ingrowth [9]. Furthermore, it has been suggested that the pore dimensions may directly affect some biological events; as a result, different tissues require optimal pore sizes for their regeneration [10]. Therefore, scaffolds with bimodal micron scale (l-bimodal) porosities may often be necessary for the regeneration of highly structured biological tissues, such as bone and cartilage [11]. On the other hand, transport issues, 3D cell colonization, and tissue ingrowth would be inhibited if the pores are not well-interconnected, even if the porosity of the scaffolds is high [12].

This chapter is aimed at discussing the basic functions and requirements of scaffolds in tissue engineering, underlining the ability of specific manufacturing techniques to impart all morphological and functional features in order to satisfy the specific demands of tissue regeneration.

19.2 BASIC GUIDELINES FOR SCAFFOLD DESIGN

Scaffold design is becoming essential to the success of scaffold-based tissue engineering strategies. It must combine several structural and functional properties through an appropriate selection of constituent materials, in order to adapt the scaffold features to the requirements of the specific application (Figure 19.1) on the micrometric and nanometric scale. In particular, an ideal scaffold should possess a repertoire of cues—chemical, biochemical, and biophysical—which are able to control and to promote specific events at the cellular (microscale) and macromolecular (nanoscale) level.

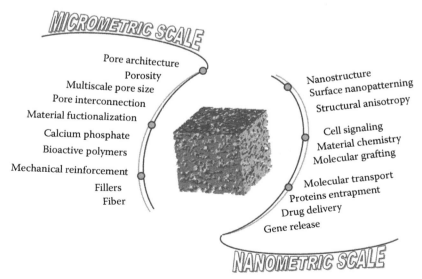

FIGURE 19.1 (See companion CD for color figure.) Key points of scaffold design on micro and nanometric scale.

Over the last two decades, the concept of cell guidance in tissue regeneration has been extensively discussed. In particular, cell guidance by the scaffold, from cell–material construct to the new engineered tissue, requires a complex balance between chemical, biochemical, and biophysical cues, able to mimic the spatial and temporal microenvironments of the natural ECM. A relevant part of this evolution concerns the development of novel scaffold materials, compatible with the cell guidance concept and resulting from contemporary advances in the fields of materials science and molecular biology [13]. In particular, the accurate design of materials chemistry allows the attainment of the optimum functional properties that a scaffold can achieve [14]. Material chemistry contributes to promote cell migration through the scaffold, so providing developmental signals to the cells, and directing the cell recruitment from the surrounding tissue. However, mass transport requirements for cell nutrition and metabolic waste removal, porous channels for cell migration, and surface characteristics for cell attachment impose the development of tailored porous structures by finely controlled processing techniques. A successful scaffold must also balance architectural features with biological function, allowing a sequential transition in which the regenerated tissue may assume a greater role as the scaffold degrades.

To act as an ECM substitute, a scaffold should impart a 3D geometry, show appropriate mechanical properties, enable cell attachment, and facilitate the advance of biological events involving cell activities during the formation of a functional tissue. At the microscopic level, a highly porous structure is absolutely essential to support the diffusion of nutrients and waste products through the scaffold. By this time, an optimal pore size in several matrices for cellular response in tissue regeneration (of bone, liver, nerve, cartilage) and for neovascularization [15,16] has been identified. Indeed, the optimal pore size tailored on the specific cell type has to be large enough to allow for cell migration and ECM formation, yet not be so small that pore occlusion occurs. This balance often presents a trade-off between a denser scaffold which provides better supporting function and a more porous scaffold which enables better interaction with the cell component.

Here, the key issue will be the identification and analysis of the materials and processing techniques which are able to develop micro- and nanostructured platforms. These are necessary to assure an optimal balance in terms of cell recognition, mass transport properties, and mechanical response in order to reproduce the morphological and functional features of new tissue at the microscopic and nanoscopic level.

19.2.1 Materials: Matrix Components and Signals

In the recent years, the use of biodegradable polymers for the administration of pharmaceuticals and biomedical devices has increased dramatically. The general criteria for selecting a biomaterial to be used as scaffold are to match the mechanical properties and the degradation rate as a specific demand of the application. Actually, the most important biomedical applications of biodegradable polymers are in the areas of controlled drug delivery systems [17], although the use of composite polymers with multiscale degradation kinetics is powerfully emerging as smart materials of temporary devices for tissue repair and regeneration [18].

To date, a large variety of biodegradable polymers have been tested in these applications, both natural and synthetic. Despite many advantages offered by materials from natural sources, notably biological recognition, synthetic polymers offer greater advantages than natural ones in that they can be tailored to give a wider range of properties. The use of synthetic material has been extensively exploited, for two important reasons. First, the immunogenic and purification issues relating to natural biomaterials are only partially overcome by recombinant protein technologies. Second, there is a relevant interest to control the material properties and to tailor performance in terms of tissue response, and synthetic materials satisfy this demand thanks to their highly chemically programmable and reproducible properties. However, some limitations in terms of cell recognition impose the need to improve their physical and chemical performance, by modification or combination with natural source materials to generate their semisynthetic counterparts [19].

In this survey, a successful alternative strategy consists of extending the biological performance of synthetic materials by chemically encoding some biomolecular cues into synthetic platforms (i.e., composites, peptide-grafted polymers, etc.). In this context, work on a new class of hybrid materials, capable of directly interacting with the cell, is currently in progress. This represents a significant challenge, since although there has been enormous progress over the last decades, several features of the processes that control cell guidance in three-dimensional (3D) materials still remain to be defined.

19.2.2 PROCESSING TECHNIQUES

To mimic the topological and microstructural characteristics of the ECM, a scaffold must present high degree of porosity, high surface-to-volume ratio, high degree of pore interconnection, appropriate pore size, and geometry control [14]. Although not all the details of scaffold design requirements have been completely defined, a large number of studies have explored the role of the topographical features of the scaffold on cellular responses [20]. These studies demonstrated that the microarchitecture of the scaffold may guide cell functions by regulating the interaction between cells, and the diffusion of nutrients and metabolic wastes throughout the 3D construct.

Several technologies for imprinting controlled porosity within polymeric matrices are progressively emerging. To date, a plethora of processing techniques are available in the literature for producing 3D scaffolds from various biodegradable polymers. These include fiber bonding [21,22], solvent casting and particulate leaching/porogen leaching [23,24], membrane lamination [25], phase inversion/particulate leaching [9,26], melt molding [14], solvent [27] or gel casting [28], thermally induced phase separation/sublimation/lyophilization/emulsion freeze-drying [29–32], gas foaming/high-pressure processing [33], co-continuous blend extrusion techniques [34], rapid prototyping (RP) techniques [35], and stereo-photolithographic [36] and inkjet techniques [37].

All these techniques could conceivably be developed to imprint the ordered porous arrays required for the control and design a micro- or nanometric "texture" within the polymer structure. In all cases, the architecture of the resultant material, in terms of pore dimensions and interconnectivity, can be finely controlled at the micron and submicron scale by modulating the process conditions. Moreover, several techniques can be subtly tuned to imprint, in turn, a highly ordered porosity pattern with tunable morphological orientation, and could be used in combination to realize 3D scaffolds with bimodal and highly oriented porosity and the desired high degree of pore interconnection [38,39].

In this chapter, some powerful fabrication techniques of 3D polymer and composite scaffolds are examined, emphasizing the potential of each technique to preferentially produce scaffold architectures on micron, submicron, or nanometer scale.

19.3 SCAFFOLD ARCHITECTURES ON MICROMETRIC SCALE

The architectural features of porous network, namely pore size and shape, pore wall morphology, porosity, surface area, and pore interconnectivity, are probably the most critical parameters as those have been shown to directly impact cell seeding, cell migration, tissue differentiation, transport of oxygen, nutrients and wastes, and new tissue formation in three dimensions. Several studies of this have indicated that the regeneration of specific tissues aided by synthetic materials is dependent on the porosity and pore size of the supporting 3D structure.

In particular, the active role of pore topological properties in cell adhesion and colonization has been demonstrated for composite scaffolds obtained via phase inversion/salt leaching techniques [26]. In this case, cells preferentially invade the macroporous systems, whereas the presence of small pores—due to solvent extraction coupled with the hydrophobic nature of the polymeric matrix—tends to hinder cell colonization. It has to be noted that cell proliferation and ECM deposition may progressively occlude the entire porosity of the scaffold and, consequently reduce the nutrient delivery to, and metabolic waste removal from, the interior of the construct. This aspect is relevant for in vitro scaffold-based tissue engineering strategies, as confirmed by large number of studies [40]. Recent studies have pointed out the importance of a uniform cell infiltration in 3D space as well as the adequate sustenance of cells assured by the presence of a micropore network to ensure the transport of fluids necessary for cell biosynthesis [41].

Since a tailored porous structure often fulfils the basic morphological and functional criteria for optimizing cell colonization, there has been a rapid rise in the use of associated manufacturing techniques, based upon intuitive theoretical principles and highly reproducible technologies including phase separation [42] gas foaming [43], solvent casting [44], or freeze-drying [45] in combination with traditional salt leaching techniques.

Because of their process versatility and low cost, these techniques are now commonly used to fabricate scaffolds. The lack of precise control over scaffold specifications such as pore size, shape, distribution, and interconnectivity as well as the overall scaffold shape remains a core limitation of these technologies.

Starting from several studies which underline the relevance of pore size in the ability of cells to adhere and proliferate on a scaffold [5], recent works have remarked on the relevance of solid-free form fabrication (SFF) techniques for obtaining porous scaffold with tailored pore geometry [35]. These techniques allow the reproducible fabrication of scaffolds directly from a computer-aided design (CAD) file. The ability to translate an electronic data set into a scaffold opens up the possibility of patient-specific scaffolds based on computed tomography (CT) or magnetic resonance imaging (MRI) data [46]. By these approaches, pore size is precisely controlled eliminating the variability in the pore size and structure which decouples the dependence of cell adhesion and proliferation on pore characteristics [47–49]. However, the porosity of the material, which is defined as the proportion of void space in a solid, is still a critical factor [47] on the definition of functional properties of the scaffold.

Besides, the graft site must bear loads at physiological levels just after implantation and, in some cases, internal fixation to provide adequate early stability cannot be used [48]. Moreover, a scaffold has to provide an adequate transfer of stresses in the implant site, assuring a valid match between the mechanical properties of a scaffold and the graft site morbidity so that the progression of tissue healing is not limited by mechanical failure of the scaffold prior to complete tissue regeneration. In this regard, the scaffold could supply these functions mechanically by the use of high strength materials such as cortical bone, metals ceramics, or carbon fiber–based polymers. However, the use of biodegradable polymers, characterized by high ductility but reduced stiffness, hinders the complete mimesis of the functionalities of hard tissues in load-bearing applications. It is essential that the implementation of scaffold manufacturing strategies leads to the design of micro-architectures of open-pore biodegradable scaffolds with high control of pore geometry and mechanical properties to finely guide specific biological interactions in order to satisfy the varied demand of tissue engineering applications.

19.3.1 Conventional Techniques

Judicious selection of processing techniques assures the development of scaffolds with tunable porosity as well as mechanical properties able to carry out the most appropriate biological response for hard tissue regeneration [35,50]. To date, particulate leaching is the most common methodology to impart pores with tailored features to the polymer matrix [51,52]. Generally, this method involves the mold casting of a polymer solution mixed with porogen, followed by porogen leaching-out, usually in water bath, to generate the pore network. Several different water-soluble particles including salts and carbohydrates, or hydrophobic systems (i.e., paraffin beads), have been largely used as the porogen agent [53,54]. In comparison with other techniques, particulate leaching guarantees simple control of pore structure in terms of pore size shape and spatial distribution with high definition of pore boundaries (Figure 19.2a and b). Beyond to offer a mechanical support, such hierarchical

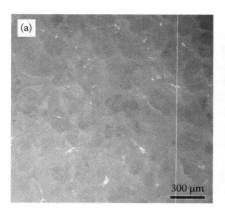 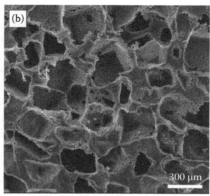

FIGURE 19.2 **(See companion CD for color figure.)** Porous scaffolds by phase inversion and salt leaching technique: images by (a) optical and (b) scanning electron microscopy.

porous architectures firstly create initial void spaces that are available for regenerating cells to form new tissues (including new blood vessels) as well as the pathways for mass transport via diffusion and/or convection [39].

A highly interconnected porosity does not still assure a homogeneous spatial colonization of cells [55] but firstly mediate all interactions between cells and biomaterials occurring at the interface, i.e., the entire internal pore walls of a 3D scaffold.

Savarino et al. have also demonstrated that PCL-based scaffold obtained by phase inversion and salt leaching technique, loaded with mesenchymal stem cells, may offer a valid alternative to autograft and allograft in order to control salient bone-forming properties without the disadvantages of traditional graft. Indeed, the use of biocompatible materials as cell carriers also reduces the interference with bone induction from inflammatory reactions. Moreover, the employment of biodegradable materials with long degradation kinetics such as PCL maintains bioactive elements at the site of implantation long enough for guiding the process of tissue growth and providing a temporal match between kinetic degradation and tissue maturation [56]. Furthermore, their capability of integrating insoluble (i.e., calcium phosphates, hydroxyapatite [HA]) and soluble (i.e., bone morphogenic proteins [BMP4, BMP7]) osteoconductive and osteoinductive signals have been proved to amplify the osteogenic potential for an efficacious bone regeneration.

For instance, to improve the osteoconductivity and bioactivity, biodegradable composite scaffolds have been prepared by adding HA particles within the polymeric matrix (Figure 19.3).

 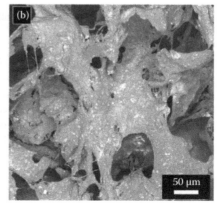

FIGURE 19.3 Bioactive composite scaffolds: (a) cross-sectional images of PCL/HA scaffolds and (b) homogeneous spatial distribution of micrometric HA particles along the scaffold trabecula.

The presence of bioactive solid signals such as HA in the polymer matrix may improve bone formation by osteoblasts mimicking the natural bone mineral phase [57]. Indeed, HA is known to be biocompatible, bioactive, i.e., ability to form a direct chemical bond with surrounding tissues, osteoconductive, nontoxic, non-inflammatory, and non-immunogenic agent [58].

The salt leaching-based techniques allow to easily integrate HA solid signals without relevant complications during the preparation procedure, using solvents that do not alter the osteoconductive potential of the HA filler, as confirmed by Ca/P ratios close to that of natural HAs. It is noticed that the presence of stoichiometric HA on the surface should play an important role in the osteoconduction and osteointegration capability of the proposed scaffolds to enhance bone cell response as widely reported in literature [59,60]. Moreover, an active role of the HA filler on the underlying in vitro degradation mechanisms by the simultaneous assessment of the influence of scaffold morphology and the physicochemical properties of the proposed scaffolds have been demonstrated. In detail, the presence of HA particles may control the nucleation and growth of pores, leading to the formation of low-size pores [26]. The slight reduction of pore size as a function of the increasing HA content, as well as the effective hydrophilicity of the scaffolds, drastically influences the kinetic of degradation mechanisms. Indeed, it has been verified that the surface-to-volume ratio of pores increases as a function of HA content, as a consequence of the relevant reduction of average pore sizes and, to a lesser extent, of the decrease of porosity. In contrast, the inclusion of micrometric HA also affects the polymer crystallinity, inducing a shielding effect of the polymer matrix against degradation. Properly, the increase of crystallinity of the polymer matrix in HA-loaded scaffolds hinders the degradation of the composites, preferentially deflecting the fluids at the polymer/ceramic interface, which are more susceptible to hydrolytic attack [60].

Beyond their osteoconductive enhancement, the inclusion of rigid bone-like particles within polymer matrix may improve the mechanical properties of the polymer, so strengthening its use as a substrate for hard tissue replacement. Indeed, HA clusters (Figure 19.3b) may act mainly as reinforcement agents but, upon increase of the HA content, they might act as structural defects which accelerate the crack formation and propagation in the composite skeleton with a risk of failure of the structure. However, the tendency of clustering due to the hydrophilic properties of the HA significantly limits the reinforcing potential of rigid particles used.

Alternatively, the integration of biodegradable PLA fibers into the polymer matrix has been efficaciously used to guarantee an adequate mechanical response for maintaining the spaces required for cellular ingrowth and matrix production [60]. In this case, the contextual addition of bioactive calcium phosphates particles, activated by hardening reaction via hydrolysis, generates needle-like HA crystals, which interact with the fiber-reinforced polymeric matrix [61] so improving the mechanical response up to an order of magnitude, with respect to the un-reinforced system. This combination of an interconnected macroporous structure with pore sizes suitable for the promotion of cell seeding and proliferation with adequate mechanical properties certainly contributes to design composite multifunctional scaffolds which are an excellent candidate for bone tissue engineering.

19.3.2 SFF METHODS

RP techniques have been applied to manufacturing components with complex geometries beyond the reach of conventional precise machining. The fabrication process is directed by CAD of a certain component. Methods including direct deposition [62], selective laser sintering (SLS) [63], 3D printing [64], SFF of drug delivery devices, and stereolithography (SLA) [65] have been developed recently, which build components in a laminated fashion.

As already highlighted, conventional manufacturing methods do not guarantee precise control over the internal architecture and interconnectivity, and scaffolds realized with these techniques can be shaped with custom-made molds.

Conversely, SFF allows to realize customized scaffolds with reproducible internal morphology and a higher degree of architectural control, thus increasing the mass transport of oxygen and nutrients throughout the scaffold [66,67].

In particular, SFF refers to a collective term for a group of technologies that are used to produce objects in a layer-by-layer manner from a 3D computer design of the object. Since SFF was initially developed for manufacturing prototype engineering parts, the name "rapid prototyping" is also widely used [68–71].

Over the past 20 years, several SFF technologies have been developed, and they mainly differ from each other on how the layers are laid down, solidified, and attached to the previous ones [69–71]. Although several variants of SFF technology exist, data input, data file preparation, and object building can be considered as three basic common steps [72].

All of the RP techniques are based on the use of CAD information that is converted to an STL type file format. This format is derived from the name stereolithography, the oldest of the RP technologies. The file format has been accepted as the golden standard of the industry. Basically, CAD data are converted into a series of cross-sectional layers. These computer-generated two-dimensional (2D) layers are then created as a solid model by a variety of processes. Starting from the bottom and proceeding upward, each layer is glued or otherwise bonded to the previous layer, thus producing a solid model of the object presented on the computer screen [67,73].

Furthermore, data obtained from CT or MRI medical scans can be used to realize customized CAD models. This means that the desired implant area of a patient can be scanned through CT or MRI, and then the data are imported into CAD software, thus enabling a surgeon to design an implant according to individual needs. After the information is transferred to a RP system, a biocompatible and biodegradable scaffold can be manufactured [67,74]. Briefly, these layer-by-layer manufacturing methods can be divided into three types: liquid-based, solid-based, and powder-based techniques. SLA is an example of liquid-based technologies, whereas fused deposition modeling represents a solid-based system. SLS and 3D printing are included in the category of powder-based methods. As for processing materials, the suitable choice spans from paper to several polymers, ceramics, and metals [75,76].

Implants used in craniomaxillofacial surgery can be considered as the first application of RP techniques in medicine. Even though a method for realizing patient skull models and prosthesis has been reported, the applications were strongly limited to surgical planning rather than actual cranioplasts manufacturing. However, for the first time these models enabled surgeons to plan the entire operation and to predict eventual outcomes [76,77]. Griffith and Halloran [78] described the fabrication of ceramic devices through SLA, where an ultraviolet (UV) photocurable monomer was loaded with suspensions of alumina, silicon nitride, and silica particles [67].

In particular, an UV laser beam was used to cure the monomer. The laser beam was guided according to the CAD cross-sectional data, thus obtaining a green body as a result of ceramic particles bonding. Successively, the polymer binder was removed through pyrolysis and the ceramic parts sintered [67]. Other workers used the same technique for manufacturing HA scaffolds to be used as orbital floor implants whereas ceramic scaffolds are usually limited to bone tissue engineering [66,67].

SLA has also been successfully considered to fabricate scaffolds for bone tissue engineering using photo-cross-linkable poly(propylene fumarate) (PPF). This material may be cross-linked through its carbon–carbon double bonds, and it can degrade in the body by simple hydrolysis of the ester bonds into nontoxic products [67,79].

In vitro and in vivo biological studies have been performed on 3D scaffolds to optimize pore size and porosity, and further investigations are needed [67,79]. However, the development of biomimetic scaffolds for bone tissue engineering results a growing field of research, and a minimum pore size of 100 μm has been suggested for mineralized tissue ingrowth.

Accordingly, in order to control the pore size, the need for novel scaffolds fabrication methods has emerged [67,80]. As previously underlined, the outer shape of a scaffold can be suitably designed according to the patient's defect. In this context, 3D printing has been used to manufacture custom-made scaffolds. This technique employs a conventional inject printing technology and its main advantage consists of realizing an implant directly from 3D data in one step without using an additional mold.

With regard to the basic process, a liquid binder is ejected from a printer head onto a thin layer of powder according to the sliced 2D profile of a CAD model. The binder plays an important role since it has to join adjacent powder particles together, thus obtaining the desired 3D structure.

It appears clear that for scaffolds manufacturing through 3D printing an important requirement results in the availability of biocompatible powder-binder systems. For example, bone-like calcium phosphates, which can be considered as highly biocompatible materials, have been adapted to 3D printing technology without using cytotoxic organic solvents. The porous structures obtained from 3D printing technology can be seeded with patient-derived cells and eventually implanted into the body. Besides ceramics, polymeric scaffolds can also be obtained from 3D printing technology [81,82].

However, it has been evidenced that SLS provides an efficient method to manufacture scaffolds matching the complex anatomical geometry of craniofacial or periodontal structures, and it may be also advantageous to realize bone tissue engineering scaffolds for sites such as the temporomandibular joint [67].

To manufacture scaffolds through SLS technique, any powdered biomaterial which will fuse but not decompose under a laser beam can be virtually considered, not requiring the use of any organic solvent. Basically, the laser selectively fuses the powdered material by scanning cross sections obtained from a 3D digital model of the part on the surface of a powder bed. Successively, the powder bed is lowered by one layer thickness and a new layer of material can be applied on the top. This process is then repeated until the object is obtained. Using the SLS technique, Williams and coworkers fabricated scaffolds from a biodegradable polymer already known for potential applications for bone and cartilage repair, polycaprolactone (PCL). Moreover, the SLS technique has been successfully used to realize scaffolds for human and porcine mandibular condyle, replicating the desired anatomy [83,84].

On the other hand, a group of researcher from the National University of Singapore and National University Hospital developed PCL scaffolds as bone patch to repair holes in the skull using the fused deposition modeling (FDM) [85]. The FDM process allows obtaining structures by extruding a polymeric fiber through a moving nozzle. The head extrudes thermoplastic polymers in a semiliquid state, thus obtaining ultrathin layers precisely placed. The extruded material then solidifies and adheres to the layer that has been previously created [67].

PCL scaffolds fabricated by FDM technique were used in a pilot study for cranioplasty, and the clinical outcome was very interesting [67].

Even though several polymers have been investigated for bone tissue engineering, none of them may be singularly used to satisfy all the requirements of a bone substitute. Composite materials may offer a solution to overcome the drawbacks related to individual materials, since bone matrix is an organic/inorganic composite of collagen and apatites [86].

Consequently, composite and polymer scaffolds with several compositions and complex internal architectures were fabricated through FDM and then investigated [67,87].

In the year 2000, a new RP technology based upon 3D dispensing of liquids and pastes in a liquid medium (also referred to as 3D plotting technology) was developed to manufacture objects with complex geometry and morphology according to computer design, thus producing architectures that are similar to the nonwoven ones [76,88,89].

Because of the possibility to vary the direction of the individual strands layer by layer, scaffolds with interconnecting pores can be designed, hence satisfying demands for cell attachment and cell growth.

Among all of the RP techniques, 3D plotting [76,88,89] and 3D fiber deposition [90,91] have been considered to make scaffolds for tissue engineering applications. In particular, 3D fiber deposition

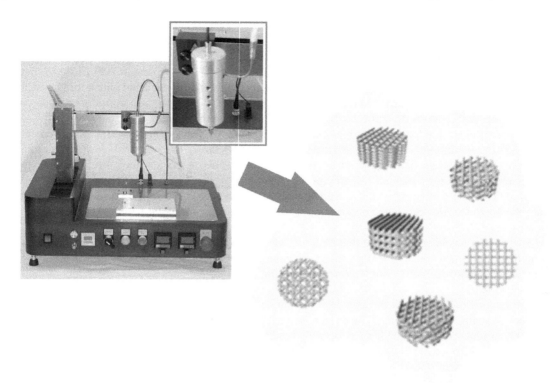

FIGURE 19.4 **(See companion CD for color figure.)** Example of 3D fiber deposition process: Bioplotter dispenser device and several scaffold architectures.

is a fused deposition technique in which a molten polymer is extruded and then deposited through a servo-mechanically controlled syringe that applies pressure (Figure 19.4), and it may be considered as a modified technique of 3D plotting for the extrusion of highly viscous polymers [92–94]. Based on CAD/CAM techniques, 3D fiber deposition allows the manufacture-customized scaffolds with defined internal structure, architecture, and 100% interconnectivity.

The Bioplotter represents the key element of the 3D fiber deposition technique. It is a dispensing machine developed by Landers and coworkers [76,88,89] to make scaffolds from hydrogel for soft tissue engineering, and it consists of a dispenser, equipped with a heating jacket, that is movable in 3D. The 3D Bioplotter is capable of extruding pastes, solutions, hot-melts, dispersions, polymers, monomers, or reactive solutions. It is a much more versatile system than FDM, and the basic process consists of extruding a material which is stored into a cartridge through a thin needle using an air-pressure control. The material can be dispensed in the presence of air or in a liquid, and, then, it solidifies [76]. The hardening process involves thermally induced solidification and solidification induced by a chemical reaction or by precipitation. The knowledge of the processing parameters plays a crucial role to develop 3D fiber-deposited scaffolds.

The ability to produce hydrogel scaffolds is probably one of the most attractive features of the 3D Bioplotter. For the first time, Lander and coworkers [76,88,89] proposed 3D bioplotting as a biofunctional and cell-compatible processing for hydrogels in the area of RP techniques, as they realized and characterized hydrogel-based scaffolds with desired external shape and well-defined internal pore structure. In particular, Lander and coworkers [76,88,89] also showed interesting results obtained from cell culture using two cell types (human osteosarcoma cell line and mouse connective tissue fibroblast) which were seeded on RP agar scaffolds coated with a mixture of hyaluronic and alginic acids, or with fibrin.

It is well-known that hydrogels (e.g., gelatin, agar, fibrin, or collagen) can be used as simple scaffold structures, like fibers, sheets, woven or nonwoven materials. Due to their flexibility, structural similarity to the ECM, and permeability to oxygen and metabolites, hydrogels can be considered as candidates for substituting soft tissues.

The earlier described processing conditions of SLA and SLS clearly prevent the possibility to use hydrogels, and hydrogel-based scaffolds have not been manufactured by using these RP techniques. Thus, the appearance of this novel RP technique was long awaited.

The 3D fiber deposition technique for manufacturing 3D poly(ethylene glycol terephthalate)–poly(butylene terephthalate) (PEGT/PBT) block copolymer scaffolds with 100% interconnectivity was presented by Woodfield et al. [90] for articular cartilage tissue engineering applications. This process allowed the design of scaffolds with specific characteristics in a layer-by-layer fashion, by controlling the deposition of molten copolymer fiber from a pressurized syringe placed onto the mobile arm of a 3D plotter. As for the mechanical properties of such 3D fiber-deposited PEGT/PBT scaffolds, dynamic–mechanical measurements provided values of storage modulus similar to those of native articular cartilage explants, by suitably varying PEGT/PBT composition, pore geometry, and porosity. Moreover, it was evidenced that 3D fiber-deposited PEGT/PBT scaffolds seeded with bovine articular chondrocytes resulted very promising for articular cartilage tissue engineering, since the presence of glycosaminoglycans and type II collagen, which represent articular cartilage ECM constituents, was highlighted throughout the interconnected pore structure. Furthermore, interesting results were also achieved with respect to the attachment of expanded human articular chondrocytes [1,90].

Generally, nutrient limitation (e.g., oxygen) may be considered as responsible for the onset of chondrogenesis solely within the peripheral boundaries of larger constructs. For this reason, Malda et al. studied the effect of the 3D fiber-deposited PEGT/PBT scaffold architecture on oxygen gradients in tissue-engineered cartilaginous constructs [91]. The oxygen gradients were assessed through microelectrode measurements, and then compared to those obtained from compression-molded and particle-leached scaffolds.

In their study, Malda et al. evidenced an enhancement of cell distribution and matrix deposition in 3D fiber-deposited scaffolds if compared to the compression-molded and particle-leached ones, but no significant effect of the 3D fiber-deposited scaffold architecture on oxygen gradients was observed. Accordingly, the results obtained clearly stressed the importance of a rationally designed scaffold for cartilage tissue engineering applications, suggesting that the well-organized 3D fiber-deposited scaffolds, that are characterized by a less tortuous and more open structure, can offer possibilities for regulation of nutrient supply [91].

Using the 3D fiber deposition technique, Moroni et al. [92–94] also fabricated scaffolds made up of poly(ethylene oxide terephthalate)–poly(butylene terephthalate) (PEOT/PBT) block copolymers. These polyether-ester multiblock copolymers belong to a class of thermoplastic elastomers which are characterized by good physical properties such as elasticity, strength, and toughness, in combination with easy processing. All of these properties are related to the phase-separated morphology copolymers, in which soft, hydrophilic PEO segments at handling temperatures are physically cross-linked by the presence of hard, semicrystalline PBT segments. Differently from chemically cross-linked materials, these cross-links are reversible and can be disrupted at temperatures above their glass transition or melting point, thus providing materials with good handling properties. By modulating the PEOT/PBT ratio, it is possible to obtain polymers with a broad range of hydrophilicity, degradation rate, swelling, and mechanical properties.

PEOT/PBT block copolymers have been widely investigated for in vitro and in vivo biocompatibility and used in the orthopedic surgery as bone fillers (PolyActive™, IsoTis OrthoBiologics S.A.). Since they are polyether-esters, degradation occurs in aqueous media through hydrolysis and oxidation, at a rate that varies from very low for high PBT contents to medium and high for larger contents of PEOT and longer PEO segments [95].

These 3D fiber-deposited scaffolds were realized by means of a Bioplotter device through heating PEOT/PBT copolymer granules with different compositions (Figure 19.4). By changing fiber diameter, spacing, sequence of fiber stacking (i.e., pattern), and layer thickness, pores were varied in shape and size [92–94].

As pore geometry, and, hence, porosity, is strongly related to fiber diameter and spacing, and layer thickness, the effect of deposition speed used during the process was also highlighted [94].

The influence of pore geometry and architecture on the mechanical properties of 3D fiber-deposited PEOT/PBT scaffolds was assessed through dynamic–mechanical analysis (DMA). DMA analysis evidenced an increase of the storage modulus (E') with decreasing fiber spacing, while a decrease of the modulus was evaluated by increasing porosity [94]. In addition, E' varied within a wide range of values for PEOT/PBT scaffolds characterized by the same composition and porosity but different architectures [94].

Benefiting from a rheological phenomenon known as "*viscous encapsulation,*" Moroni et al. manufactured hollow fibers with controllable cavity diameter and shell thickness directly integrated in a 3D fiber-deposited structure [96]. In summary, viscous encapsulation may be obtained when two components of a polymer blend possess a significant difference in viscosity in the molten state. When flowing through a narrow duct, such as the nozzle of an extruder or the needle used in the 3D bioplotting process, the polymer with lower viscosity tends to shift toward the walls of the nozzle during extrusion.

This phenomenon is related to the higher shear stresses present at the walls of the nozzle to which the lower-viscosity polymer adapts more easily as a consequence of its higher capacity to distort. The consequent separation of the components can lead to a stratification or a "canalization" effect, thus providing fibers with a shell–core structure. Hollow fibers can be obtained by removing the core polymer through selective dissolution [96].

In particular, using a Bioplotter device Moroni et al. [96] fabricated and characterized PEOT/PBT scaffolds with hollow fibers through the direct deposition of the viscous encapsulated fibers and the subsequent selective core dissolution. PEOT/PBT scaffolds with hollow fibers were obtained by soaking 3D shell–core scaffolds in a specific solvent (i.e., acetone) for the poly(butyl methacrylate-methyl methacrylate) (P(BMA/MMA)) or for the PCL core polymers. Accordingly, P(BMA/MMA) or PCL was selectively dissolved, leaving only the PEOT/PBT well-organized structure with hollow fibers. Results from this study have also evidenced that viscous encapsulation may be obtained for specific values of melting index ratios when polymers are extruded under proper rheological conditions [96]. The possibility to control the hollow cavity diameter and the shell thickness was also highlighted by varying the needle diameter, the polymers in the blend, and the blend composition [97].

Using the same basic principle without removing the core polymer, biphasic 3D fiber-deposited scaffolds for cartilage tissue engineering with a shell–core fiber structure were manufactured and studied [95]. In particular, biphasic shell–core PEOT/PBT 3D scaffolds were fabricated from PEOT/PBT copolymers using different compositions and, hence, different melting index. In this structure, the shell polymer played as a coating with specific physicochemical surface properties, while the core polymer provided suitable mechanical properties [95]. The core and shell polymers differed in terms of molecular weight of the initial poly(ethylene glycol) (PEG) segments used in the copolymerization and weight percentage of the PEOT domains.

3D fiber-deposited scaffolds entirely fabricated with the shell or with the core polymers were also investigated, and the results were then compared with those obtained from the biphasic shell–core scaffolds. With regard to the mechanical performances, the biphasic shell–core scaffolds also showed an improved storage modulus. On the other hand, cell cultures evidenced that all of the investigated scaffolds displayed similar amounts of entrapped chondrocytes and of ECM. However, during the culture period, chondrocytes maintained their specific morphology on the shell–core 3D fiber-deposited scaffolds, indicating a proper cell differentiation into articular cartilage and, hence, a promising solution for cartilage tissue engineering [95].

Although RP scaffolds display a well-organized and completely interconnected pore network, cell seeding efficiency still represents a critical factor for optimal tissue regeneration. If compared to a cell, the pore size of RP scaffolds is relatively large [96].

Accordingly, in order to obtain sufficient attached cells that are able to produce enough ECM, a great number of cells are needed. This may create expensive and extensive cell isolation, culture, and expansion processes, thus strongly limiting the clinical relevance of scaffolds obtained through RP techniques [96].

An interesting strategy can be to introduce micrometer and nanometer scale fibril networks into RP scaffolds. For this reason, Moroni et al. displayed the possibility to combine 3D fiber deposition technique with electrospinning [96], thus achieving multiscale networks in which the periodical macrofibers of 3D fiber-deposited scaffolds were integrated with the electrospun microfibers. In these integrated structures, the 3D fiber-deposited scaffold works as structural support with adequate mechanical properties, while the electrospun network functions as a sieve for cell entrapment and provides ECM-like cues to cells [96].

However, since most tissues and organs are multiphasic and contain multiple cell types, a further challenge in tissue engineering may be considered to design a suitable scaffold that should be able to support multilineage cell types. Over the past few years, great effort has been made to engineer tissues consisting of different cell types [98–100]. For example, Kyriakidou et al. (2008) proposed a cocultural endothelial and osteoblast-like cell model for enhanced bone tissue engineering [98]. To study the reciprocal cell interactions, stabilized osteoblast-like cells (MG63) and normal endothelial cells (human umbilical vein endothelial cells [HUVEC]) were co-seeded onto 3D fiber-deposited PCL scaffolds, and cultured by means of a rotary cell culture system [98]. The proposed cocultural model is sustained by the close mutual interaction of the two cell types, also highlighted by in vivo findings [101–105]. In particular, it has been found that osteogenesis and angiogenesis are mutually interdependent and that endothelial cells can accelerate bone formation through angiogenesis as well as in bone remodeling [104,105].

The 3D fiber-deposited PCL scaffolds used by Kyriakidou et al. in their study were manufactured with a Bioplotter device through melting the polymer granules, and characterized by a 0°/90° pattern [98]. Compression tests performed on these 3D PCL scaffolds evidenced a stress–strain curve [98] similar to that of a flexible foam [106] and a modulus of 134.6 ± 8.5 MPa. However, differently from the typical behavior of a flexible foam, the central part of the stress–strain curve did not present a plateau, but only a region with a lower stiffness [98].

On the other hand, 2D optical image and 3D micro-CT and imaging analyses (Figure 19.5a and b) evidenced the precise pore size, repeatable microstructure, and fully interconnected network of the 3D fiber-deposited PCL scaffolds [98]. A detail of structure confirms the high control of pore geometry (Figure 19.6a) as well as fiber roughness (Figure 19.6b).

As result of dynamic co-seeding onto the 3D PCL scaffolds, Kyriakidou et al. evidenced that osteoblasts increase the proliferation of endothelial cells, while endothelial cells enhance the growth

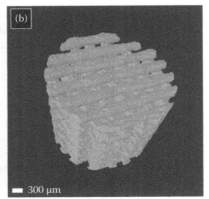

FIGURE 19.5 (See companion CD for color figure.) PCL scaffolds through 3D fiber deposition technique: 2D optical image (a) and 3D micro-CT reconstruction (b).

FIGURE 19.6   3D PCL scaffolds by Bioplotter dispenser device: (a) bulk image by backscattered electrons and (b) topographic image by secondary electrons source.

of osteoblasts decreasing their differentiation. The results emphasized that the dynamic seeding of osteoblasts and endothelial cells onto a well-organized 3D polymeric scaffold may be considered as a practical approach to obtain a functional hybrid in which angiogenesis, provided by neovascular organization of endothelial cells, could further support osteoblast's growth [98].

19.4   SCAFFOLD ARCHITECTURES ON NANOMETRIC SCALE

In biological and medical applications, the capability of controlling physical and chemical interactions at the level of natural building blocks, from proteins to cells, may favor a more efficient exploration, manipulation, and application of living systems and biological phenomena. In this context, nanotechnology has made great strides forward the creation of new materials, surfaces, and bulk architectures which find numerous applications in the biomedical area [107].

In particular, nanostructured biomaterials in the form of nanoparticles, nanofibers, nanosurfaces, and nanocomposites have gained increasing interest in regenerative medicine [108] because these materials often mimic the physical features of natural ECM at the nanoscale level.

Although different experts define *"nanomaterial"* in different ways, the most commonly accepted definition is that a nanomaterial is one with a basic structural unit in the range of 1–500 nm. When the conventional dimensions of a material are decreased into nanoscale, some unusual changes in physiochemical properties occur. This can be attributed to the size (i.e., size distribution), chemical composition (i.e., purity, crystallinity, electronic properties), surface structure (i.e., reactivity, surface groups), solubility, shape, and aggregation of nanometric materials compared to their micrometric equivalents [109]. Additionally, nanoscaled materials have more surface area, surface defects, density of atoms at the surface, and altered electron distributions, which change their surface properties (i.e., reactivity) with respect to conventional materials, further affecting the interactions between the surface and proteins.

In recent years, nanotechnology has been recognized as an important tool in scaffold design to reproduce the features of microenvironment-mediated signaling which determine tissue specificity and architecture of native tissues [110]. It is well-known that HA crystals, the main constituent of the inorganic phase of the bone ECM, are 2–5 nm thick while collagen type I, the main constituent of the organic phase of the bone, is 0.5 nm in diameter. This implies that cells are naturally accustomed to interacting with nanoscale features and surfaces. Indeed, in native tissues, nanoscale protein interactions are crucial for controlling cell functions such as proliferation, migration, and ECM production [111]. Protein adsorption characteristics are, in turn, dependent upon the surface features of the implanted biomaterials (roughness, charge, chemistry, wettability) [112]. Recent reports have also demonstrated that the unique properties of nanobiomaterials offer advantageous interactions with the proteins that control cellular function [113,114]. Particle or fiber dimension may strongly influence these surface properties and the corresponding protein interactions. It means that changes in material properties may manipulate protein characteristics, so altering protein adsorption behavior which further influences cell adhesion onto biomaterial surfaces [115].

It is also well-known that natural healing processes in highly dense tissues are limited by the mechanical deficiencies of the damaged tissue as well as by the hypocellularity which characterizes the peculiar vascular nature of the ECM. In particular, when healing does occur, the ordered structure of the native tissue is replaced with a disorganized fibrous scar with inferior mechanical properties, engendering sites that are prone to reinjury. Hence, the development of structurally motivated approaches based on organized nanofibrous assemblies rises in the attempt to reach a more aware engineering of such tissues.

In this context, several authors describes the possibility of patterning materials on nanometric scale able to control the adsorption or immobilization of adhesive proteins, or, to promote a selective adhesion of cells on substrates. However, it has been noticed that cellular responsiveness to surface biomaterials is also influenced by the nanoscale spatial distribution of macromolecular signals such as bioactive polymers or biomimetic peptides on the material surface [116]. To date, recent approach also involves the incorporation of soluble, biologically active factors to the bulk material or encapsulated within the biomaterial.

Currently, an important class of nanostructured biomaterials on which intensive research has been carried out is composed of nanofibrous materials, especially biodegradable polymer nanofibers, able to morphologically mimic the nanofibrillar structure of ECM [117]. In this regard, a powerful strategy to develop biomimetic nanofibrous material to be efficaciously used as ECM scaffold is represented by the electrospinning technique which allows arrangement of polymeric fibers at nanometer scale. This technique can also be used in combination with other techniques, such as phase separation, to obtain different and interesting geometries so satisfying several demands in tissue engineering applications.

19.4.1 Nanofibrous Platforms via Electrospinning Technique

Electrospinning is a simple and versatile technique for producing nonwoven, interconnected nanofiber mats that have potential applications in the field of engineering and medicine [118]. Traditionally, electrospinning has been used to generate platforms on submicron or nanometric scale, in order to overcome the limitation connected to traditional fiber drawing processes. Today, the current surge in nanotechnology has adopted electrospinning as an elegant technique to produce nanofibrous structures for biomedical applications such as wound healing, tissue engineering, and drug delivery. In comparison with conventional techniques, electrospinning has demonstrated its ability to fabricate nanofiber scaffolds that closely mimic the native ECM structure [119]. This technique exhibits a large versatility to produce scaffolds with a range of mechanical properties (either in plastic and elastic field), optimized porosity, and pore volume for tissue engineering applications [120,121].

Electrospinning is a nonmechanical, electrostatic technique which involves the use of a high-voltage electrostatic field to charge the surface of a polymer solution droplet and thus to induce the ejection of a liquid jet through a spinneret. In a typical process, an electrical potential is applied between a droplet of a polymer solution, or melt, held at the end of a capillary tube and a grounded target. This electrical field induces charge separation and hence causes charge repulsion within the polymer droplet. A polymer jet is initiated when the opposing electrostatic force overcomes the surface tension of the polymer solution. Just before the jet formation, polymer droplet under the influence of electric field assumes cone shape with convex sides and a rounded tip, which is known as the Taylor cone. Once the jet is launched, the dominating electric force promotes the elongation of the polymer droplet up to form a stretched fiber. On a microscopic scale, the Coulombic interaction within the jet and with external electric field manifests into an off-axial radial component of velocity that result in polymer jet bending and whipping instabilities. During the entire process, polymer jet under the influence of electrical force elongates thousands or even millions of times before the ultrathin solid fiber collects on the target.

Despite the electrospinning process having been known for a long time, the understanding of the physics of the process and a fundamental model to explain the process are still lacking. For this reason, a major problem in studying the electrospinning process is the optimization of materials and process parameters in order to correlate process features and fiber morphology.

A large number of polymeric biomaterials may be used to form nanofibrous scaffolds, both nonbiodegradable and/or biodegradable polymers, with the latter consisting of both natural and synthetic polymers. Nonbiodegradable polymers, such as polyurethane [122] and poly(ester urethane) [123], can be used to engineer tissues requiring substantial mechanical stability, such as ligament or muscle, but their long-lasting nature is likely to interfere with tissue turnover and remodeling.

In parallel to this, much attention has been devoted to biodegradable polymers in tissue engineering. By the combined effect of enzymatic and hydrolytic activities, polymer biodegradation generates void space within the scaffold that facilitates cellular processes, such as proliferation and the deposition of newly synthesized ECM [124]. The poly(α-hydroxy ester) polymer family is the most commonly used synthetic biodegradable polymers for nanofiber production and have shown promise for orthopedic applications [125]. For example, poly(ε-caprolactone)- and poly(L-lactic acid)-based nanofibrous scaffolds have been successfully used for cell-based engineering of cartilage and bone tissues in vitro. For example, PCL solutions in chloroform (10% w/w) allow to obtain electrospun nanofiber matrices (Figure 19.7a and b) with narrow size distribution and an average fiber diameter from 449 to 761 nm. However, intriguing by their high biological recognition, natural polymers, such as collagen, gelatin, elastin, silk fibroin, fibrinogen, hyaluronan, and chitosan, have also been fabricated into 3D, nanofibrous scaffolds for tissue engineering applications. Limited by their molecular weights and/or solubilities, there are many functional polymers that are not suitable for use with electrospinning [126]. One of most effective strategies for solving this problem is to blend them with polymers that are well-suited for electrospinning. Based on this approach, various polymer compositions such as synthetic–synthetic, synthetic–natural, and natural–natural polymer blends were recently electrospun to produce new scaffolds with required bioactivity and mechanical properties suitable for vascular, dermal, neural, and cartilage tissue engineering [127,128].

For instance, PCL/gelatin composite scaffolds were fabricated by dissolution in fluorinated solvents (i.e., TFE, HFP). Although synthetic polymers like PCL are intrinsically able to promote cell adhesion and to direct the growth of cells, however, cell affinity is partially compromised by their low hydrophilicity and lack of surface cell recognition [129]. The integration of a gelatin phase significantly improves the hydrophilic property, also supporting the more common strategies of

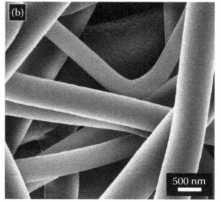

FIGURE 19.7 PCL electrospun membranes at low (a) and high (b) magnifications.

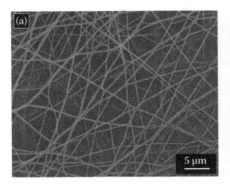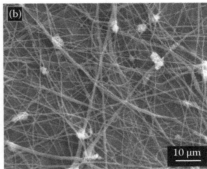

FIGURE 19.8   Bioactive electrospun membranes: (a) PCL and gelatine and (b) PCL and HA microparticles.

encoding cell-recognition domains [130]. Using such biocomposites (Figure 19.8a), the synthetic polymeric materials can function as a fibrous backbone providing mechanical properties while natural polymer component can act as promoters of cellular attachment, proliferation, and growth.

Further efforts have also been made to create polymer and ceramic electrospun scaffolds for bone tissue engineering applications [131,132]. Here, composite electrospun systems have been reported by a combination of a synthetic polymer, PCL, and HA particles on micrometric scale (Figure 19.8b). The integration of synthetic HA clays into a nanofibrous polymer matrix not only mimics the natural bone structure but also can potentially enhance mechanical and biological response of the scaffolds [133]. Moreover, the incorporation of nano-HA into the polymer matrix was considered to slow down the degradation process by neutralizing or buffering the pH changes caused by the typical acidic degradation products of polyesters [134].

Regardless of material systems used, it should be noted that all spinning parameters must be optimized for the specific polymer used in order to generate a homogeneous fiber array. Several processing variables generally affect the electrospinning process [135]: (i) polymer macromolecular features such as molecular weight distribution and chain architecture (i.e., branched, linear, etc.), (ii) polymer solution properties (i.e., viscosity, conductivity, electric permittivity, and surface tension), and (iii) process parameters such as electric potential, flow rate and concentration, distance between the capillary and collection screen, ambient parameters (temperature, humidity, and air velocity in the chamber) and, finally, the motion modality of the target screen.

For instance, changes in viscosity, due to polymer molecular weight as well as solution concentration, may drastically influence the mechanism of fiber formation in terms of size and fiber morphologies. It is essentially determined by concentration modification of polymer solution. In particular, highly concentrated polymer solutions, more viscous than less concentrated solutions, are preferred in order to obtain electrospun fibers. In contrast, less viscous solutions tend to produce beads due to activation of a new electrostatic mechanism namely electrospraying. In other words, electrospraying is an extension of the electrospinning process in the case of low concentration of solution. As a consequence, an accurate control of viscosity may be reached by switching from fibers to microspheres, moving from the electrospinning to the electrospraying technique. In the case of solutions with high viscosities, polymer chain entanglement can easily occur, which is capable of producing fibers at micrometer or nanometer sizes. At low viscosities, however, the applied electric field is able to form a conical jet and further break into droplets due to electrostatic force [136]. The size and distribution of droplets can be controlled by optimizing the processing parameters and the physical properties of the solution. SEM images of PCL particles obtained by electrospraying a solution directly onto the substrate were analyzed. Microspheres prepared from a 5 w/w% concentration solution of PCL in chloroform (Figure 19.9a) show a broad size distribution of particles with 2 μm as average diameter. However, a large number of microspheres on the nanoscale level have been detected (Figure 19.9b). The absence of surface porosity was attributed to the rapid chloroform evaporation at the early state of the ejecting passage from the nozzle to the

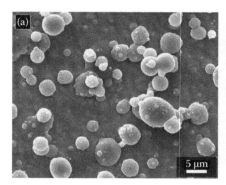
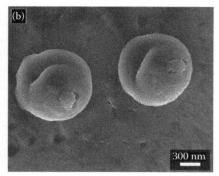

FIGURE 19.9 PCL beads on nanometric scale obtained via electrospraying technique.

collector due to the less interaction between the PCL and chloroform [137]. It has been demonstrated that the proper use of co-solvents (i.e., acetone) may induce a more controlled phase separation mechanism which is responsible for the formation of preordered porosity in order to produce denser or highly porous polymer particles with micro- or nanostructured surfaces [138].

Finally, a main aspect involving tissue replication concerns the mimicking of the structural organization firstly due to the spatial distribution of collagen fibers into the ECM. The majority of tissues present an ECM typically comprised of collagen assembled in a hierarchical fashion into dense bundles. In tissues that operate primarily in one direction, such as tendons and ligaments, collagen is organized along the prevailing line of action, and imbues such tissues with mechanical properties that are highly anisotropic (direction-dependent) and highest in the prevailing fiber orientation. In the case of the meniscus, which is exposed to a similarly complicated loading environment, the majority of collagen fibers are circumferentially organized, ranging from horn insertion site to horn insertion site. The need of regenerate these structurally complex tissues has driven the development of novel technologies which can impart multiscale rendering of fiber reinforcement from the nano- and micron scale through to the tissue level. The electrospinning technique is an optimal instrument to realize nanostructured platforms composed of ultrafine polymeric fibers able to spatially reproduce the structural anisotropy typical of fiber-reinforced tissues.

Efforts to mimic the natural ECM as closely as possible have led researchers to opt for various collector designs to align the fibers so that the cells seeded on the scaffold will have specific orientation and guided growth [139]. The effect of fiber orientation on the cell morphology has been published by various researchers [125,140,141]. In particular, bone has significant anisotropic mechanical properties, with highly oriented ECM and bone cells, so alignment of the ECM mimicking the nanofibrous scaffold is greatly preferred. The ability to orient also provides additional benefits, namely additional drawing of the fiber, thereby further decreasing the diameter and improving mechanical strength due to better fiber packing over nonwoven scaffolds. Recent studies have demonstrated that an increase in collector rotation speed, the nanofibers became more aligned and oriented perpendicular to the axis of rotation. This also resulted in a several-fold increase in the mechanical properties of the scaffold [142]. In other terms, the electrospinning is the unique approach for the multiscale rendering of fiber reinforcement from the nano- and micron levels with a tremendous potential for the repair or replacement of fiber-reinforced tissues.

19.5 SUMMARY AND FUTURE TRENDS

Tissue regeneration involves successful interplay between cells, biological signals, and biomaterials. It requires the fundamental understandings in both life sciences and materials sciences to develop successful regeneration technologies. With the advent of nanotechnology, enormous advances have been made in the field of biomaterials science and engineering. In this scenario, electrospinning has developed into one of the most elegant techniques to develop nanostructured scaffolds that could closely mimic the dimension of collagen fibrils in the ECM. Several studies have demonstrated the versatility of the process in controlling the structure and morphology of the fiber matrices and their favorable interactions with cells for tissue organization. Submicron or nanostructured tissue engineering scaffolds have the potential to direct cell destiny, as well as to regulate processes such as angiogenesis and cell migration. To induce cells to attach to certain areas or patterns of a particular biomaterial, one must alter the underlying substrate chemistry. The underlying surface chemistry has been modified to include hydrophilic or hydrophobic molecules or polymers [143], polysaccharides [144], stimuli-sensitive and responsive materials [145], and proteins or growth factors [146].

However, the biological processes that govern cell–cell and cell–matrix interactions by various biochemical cues present in the natural ECM are just as important as these structural features. Current studies focus on developing hierarchical structures with spatially presented biological cues from bioresorbable electrospun nanofibers could lead to the development of ideal scaffolds for tissue engineering applications [147]. Also, tissues are often a combination of small repeating units assembled over several scales. For instance, cortical bone and skeletal muscle are characterized by fascicles of repeating longitudinal units, respectively, osteons and muscle fibers (100–500 mm diameter). Such units form modules that operate on multiple lengths scale from the molecular and subcellular level through to the cell and tissue level [148]. In this regard and from an engineering point of view, it has recently been proposed to build tissues by assembling blocks mimicking those units in a bottom-up or modular approach [112,149].

The ability to engineer complex morphologies and the precise spatial distribution of bioactive signals, with molecular to macromolecular precision, has led to the creation of nanoscale materials and devices able to change the ability of a cell to attach, but to some extent, regulate cellular functions such as growth, differentiation, and apoptosis. Several approaches involving both top-down and bottom-up technologies are emerging to incorporate nanoscale control for tissue engineering scaffolds [150]. *Bottom-up approaches*, based on molecular self-assembly of small building blocks, have also been used to achieve 3D scaffold with nanoporosity, high pore interconnectivity, and outstanding biological recognition [151]. For instance, research into self-assembly of amphiphilic peptides has shown that they can self-assemble to form hydrogels for tissue engineering. Self-assembled scaffolds can be easily functionalized by incorporating peptide sequences that direct cell behavior directly onto the buildup molecule [152].

Conversely, *top-down approaches*, such as soft lithography, have greatly enhanced our ability to generate microscale and nanoscale features to fabricate tissue engineering scaffolds with control over features such as pore geometry, size, distribution, and spatial geometry, also minimizing the process costs [153].

To move to the next developmental phase of nanobiomaterial science, it is critical to understand the cellular and molecular basis governing the interaction between nanostructure and cells. Indeed, all these approaches appear very advantageous in terms of versatility and scalability to the fabrication of in vitro tissue models or implants. However, the fabrication of those tissue models still necessitates specific tools to create an initial architecture and systematically to manipulate their microenvironments in space and time [154]. In the near future, the introduction of strategies to create and manipulate local microenvironments will be absolutely necessary, in order to consciously investigate mechanisms of tissue morphogenesis, differentiation, and maintenance (i.e., cancer biology). Moreover, significant advancements are necessary to realize the full potential of nanobiomaterials in clinical use. Overall, current trends in nanotechnology foreshadow a bright future through the use of nanostructured materials to define new applications in tissue engineering.

REFERENCES

1. Vacanti, C.A. and Vacanti, J.P. 2000. The science of tissue engineering. *Orthop. Clin. North Am.* 31:351–356.
2. Musgrave, D.S., Fu, F.H., and Huard, J. 2002. Gene therapy and tissue engineering in orthopaedic surgery. *J. Am. Acad. Orthop. Surg.* 10:6–15.
3. Fleming, J.E., Cornell, C.N., and Muschler, G.F. 2000. Bone cells and matrices in orthopedic tissue engineering. *Orthop. Clin. North Am.* 31:357–374.
4. Laurencin, C.T., Attawia, M.A., Elgendy, H.E., and Herbert, K.M. 1996. Tissue engineered bone regeneration using degradable polymers: The formation of mineralized matrices. *Bone* 19:93S–99S.
5. Attawia, M.A., Herbert, K.M., and Laurencin, C.T. 1995. Osteoblast-like cell adherence and migration through 3 dimensional porous polymer matrices. *Biochem. Biophys. Res. Commun.* 213:639–644.
6. Bissell, M., Hall, H.G., and Pary, G. 1982. How does the extracellular matrix direct gene expression. *J. Ther. Biol.* 99:31–68.
7. Nickerson, J. 2001. Experimental observations of a nuclear matrix. *J. Cell Sci.* 114(3):463–474.
8. Nelson, C.M. and Bissell, M.J. 2006. Of extracellular matrix, scaffolds, and signaling: Tissue architecture regulates development, homeostasis, and cancer. *Annu. Rev. Cell Dev. Biol.* 22:287–309.
9. Yang, S., Leong, K., Du, Z., and Chua, C. 2001. The design of scaffolds for use in tissue engineering. Part I. Traditional factors. *Tissue Eng.* 7(6):679–689.
10. Zeltinger, J., Sherwood, J.K., Graham, D.A., Mueller, R., and Griffith, L.G. 2001. Effect of pore size and void fraction on cellular adhesion, proliferation, and matrix deposition. *Tissue Eng.* 7(5):557–572.
11. Salerno, A., Guarnieri, D., Iannone, M. et al. 2009. Engineered l-bimodal poly(e-caprolectone) porous scaffold for enhanced hMSC colonization and proliferation. *Acta Biol.* 5(4):1082–1093.
12. Moore, M.J., Jabbari, E., Ritman, E.L. et al. 2004. Quantitative analysis of interconnectivity of porous biodegradable scaffolds with micro-computed tomography. *J. Biomed. Mater. Res.* 71A(2):258–267.
13. Causa, F., Netti, P.A., and Ambrosio, L. 2007. A multi-functional scaffold for tissue regeneration: The need to engineer a tissue analogue. *Biomaterials* 28:5093–5099.
14. Hollister, S.J. 2005. Porous scaffold design for tissue engineering. *Nat. Mater.* 4:518–524.
15. O'Briena, F.J., Harley, B.A., Yannas, I.V., and Gibsona, L.J. 2005. The effect of pore size on cell adhesion in collagen–GAG scaffolds. *Biomaterials* 26:433–441.
16. Uebersax, L., Hagenmuller, H., Hofmann, S. et al. 2006. Effect of scaffold design on bone morphology in vitro. *Tissue Eng.* 12(12):3417–3429.
17. Langer, R. 1990. New methods of drug delivery. *Science* 249:1527–1533.
18. Holy, E., Fialkov, J.A., Davies, J.E., and Shoichet, M.S. 2003. Use of a biomimetic strategy to engineer bone. *J. Biomed. Mater. Res. Part A* 15:447–453.
19. Langer, R. and Tirrell, D.A. 2004. Designing materials for biology and medicine. *Nature* 1428(6982):487–492.
20. Karageorgiou, V. and Kaplan, D. 2005. Porosity of 3D biomaterial scaffolds and osteogenesis. *Biomaterials* 26(27):5474–5491.
21. Mikos, A.G., Bao, Y., Cima, L.G., Ingber, D.E., Vacati, J.P., and Langer, R. 1993. Preparation of poly(glycolic acid) bonded fiber structures for cell attachment and transplantation. *J. Biomed. Mater. Res.* 27:183–189.
22. Mooney, D.J., Mazzoni, C.L., Breuer, C. et al. 1996. Stabilized polyglycolic acid fibre-based tubes for tissue engineering. *Biomaterials* 17:115–124.
23. Ma, P.X. and Choi, J.W. 2001. Biodegradable polymer scaffolds with well defined interconnected spherical pore network. *Tissue Eng.* 7(1):23–33.
24. Murphy, W.L., Dennis, R.G., Kileny, J.L., and Mooney, D.J. 2002. Salt fusion: An approach to improve pore interconnectivity within tissue engineering scaffolds. *Tissue Eng.* 8(1):43–52.
25. Mikos, A.G., Sarakinos, G., Leite, S.M., Vacanti, J.P., and Langer, R. 1993. Laminated three-dimensional biodegradable foams for use in tissue engineering. *Biomaterials* 14(5):323–330.
26. Guarino, V., Causa, F., Netti, P.A. et al. 2008. The role of hydroxyapatite as solid signals on performance of PCL porous scaffolds for bone tissue regeneration. *J. Biomed. Mater. Res. Appl. Biomater.* 86B:548–557.
27. Athanasiou, K.A., Singhal, A.R., Agrawal, C.M., and Boyan, B.D. 1995. In vitro degradation and release characteristics of biodegradable implants containing trypsin inhibitor. *Clin. Orthop. Relat. Res.* 315:272–281.
28. Coombes, A.G.A. and Heckman, J.D. 1992. Gel casting of resorbable polymers 1. Processing and applications. *Biomaterials* 13(4):217–224.

29. Schugens, C., Maquet, V., Grandfils, C., Jerome, R., and Teyssie, P. 1996. Polylactide macroporous bio-degradable implants for cell transplantation. II. Preparation of polylactide foams by liquid–liquid phase separation. *J. Biomed. Mater. Res.* 30:449–461.

30. Schugens, C., Maquet, V., Grandfils, C., Jerome, R., and Teyssie, P. 1996. Biodegradable and macroporous polylactide implants for cell transplantation: 1. Preparation of macroporous polylactide supports by solid–liquid phase separation. *Polymer* 37(6):1027–1038.

31. Ma, P.X., Zhang, R., Xiao, G., and Franceschi, R. 2001. Engineering new bone tissue in vitro on highly porous poly(α-hydroxyl acids)/hydroxyapatite composite scaffolds. *J. Biomed. Mater. Res.* 54:284–293.

32. Zhang, R. and Ma, P.X. 1999. Poly(α-hydroxyl acids)/hydroxyapatite porous composites for bone–tissue engineering. I. Preparation and morphology. *J. Biomed. Mater. Res.* 44:446–455.

33. Holy, C.E., Dang, S.M., Davies, J.E., and Shoichet, M.S. 1999. In vitro degradation of a novel poly(lactide-*co*-glycolide) 75/25 foam. *Biomaterials* 20:1177–1185.

34. Washburn, N.R., Simon, C.G., Elgendy, H.M., Karim, A., and Amis, E.J. 2002. Co-extrusion of biocompatible polymers for scaffolds with co-continuous morphology. *J. Biomed. Mater. Res.* 60:20–29.

35. Hutmacher, D.W., Schantz, T., Zein, I., Ng, K.W., Teoh, S.H., and Tan, K.C. 2001. Mechanical properties and cell cultural response of polycaprolactone scaffolds designed and fabricated via fused deposition modelling. *J. Biomed. Mater. Res.* 55:203–216.

36. Grijpma, D.W., Hou, Q., and Feijen, J. 2005. Preparation of biodegradable networks by photo-crosslinking lactide, ϵ-caprolactone and trimethylene carbonate-based oligomers functionalized with fumaric acid monoethyl ester. *Biomaterials* 26:2795–2802.

37. Zhang, C., Wen, X., Vyavahare, N.R., and Boland, T. 2008. Synthesis and characterization of biodegradable elastomeric polyurethane scaffolds fabricated by the inkjet technique. *Biomaterials* 29:3781–3791.

38. Harris, L.D., Kim, B.S., and Mooney, D.J. 1998. Open pore biodegradable matrices formed with gas foaming. *J. Biomed. Mater. Res.* 42:396–402.

39. Guarino, V., Causa, F., Salerno, A., Ambrosio, L., and Netti, P.A. 2008. Design and manufacture of microporous polymeric materials with hierarchial complex structure for biomedical application. *Mater. Sci. Technol.* 24(9):1111–1117.

40. Richardson, T.P., Peters, M.C., Ennett, A.B., and Mooney, D.J. 2001. Polymeric system for dual growth factor delivery. *Nat. Biotech.* 19(11):1029–1034.

41. Silva, M.M.C.G., Cyster, L.A., Barry, J.J.A. et al. 2006. The effect of anisotropic architecture on cell and tissue infiltration into tissue engineering scaffolds. *Biomaterials* 27(35):5909–5917.

42. Lo, H., Ponticiello, M.S., and Leong, K.W. 1995. Fabrication of controlled release biodegradable foams by phase separation. *Tissue Eng.* 1(1):15–28.

43. Mooney, D.J., Baldwin, D.F., Suh, N.P., Vacanti, J.P., and Langer, R. 1996. Novel approach to fabricate porous sponges of poly(D,L-lactic-*co*-glycolic acid) without the use of organic solvents. *Biomaterials* 17(14):1417–1422.

44. Yang, Q., Chen, L., Shen, X., and Tan, Z. 2006. Preparation of polycaprolactone tissue engineering scaffolds by improved solvent casting/particulate leaching method. *J. Macromol. Sci. B: Phys.* 1525–609X, 45(6):1171.

45. Lv, Q. and Feng, Q. 2006. Preparation of 3D regenerated fibroin scaffolds with freeze drying method and freeze drying/foaming technique. *J. Mater. Sci. Mater. Med.* 17:1349–1356.

46. Zein, I. 2002. Fused deposition modelling of novel scaffold architectures for tissue engineering applications. *Biomaterials* 23:1169–1185.

47. Hutmacher, D.W. 2001. Scaffold design and fabrication technologies for engineering tissues. State of the art and future perspectives. *J. Biomater. Sci. Polym. Ed.* 12:107–124.

48. Leong, K.F., Chua, C.K., Leong, K.F., and Lim, C.S. 2003. Classification of rapid prototyping systems. In: Leong, K.F. et al., eds. *Rapid Prototyping, Principles and Applications.* Singapore: World Scientific Publishing, pp. 19–23.

49. Hutmacher, D.W., Sittinger, M., and Risbud, M.V. 2004. Scaffold-based tissue engineering: Rationale for computer-aided design and solid free-form fabrication systems. *Trends Biotechnol.* 22(7):334–362.

50. Guarino, V., Causa, F., and Ambrosio, L. 2007. Bioactive scaffolds for bone and ligament tissue. *Expert Rev. Med. Dev.* 4(3):406–418.

51. Hou, Q., Grijpma, D.W., and Feijen, J. 2003. Porous polymeric structures for tissue engineering prepared by a coagulation, compression moulding and salt leaching technique. *Biomaterials* 24:1937–1947.

52. Hou, Q., Grijpma, D.W., and Feijen, J. 2002. Preparation of porous poly(e-caprolactone) structures. *Macromol. Rapid Commun.* 23:247–252.

53. Guarino, V., Causa, F., and Ambrosio, L. 2007. Porosity and mechanical properties relationship in PCL based scaffolds. *J. Appl. Biomater. Biomech.* 5(3):149.

54. Spaans, C.J., de Groot, J.H., Belgraver, V.W., and Pennings, A.J. 1998. A new biomedical polyurethane with a high modulus based on 1,4-butanediisocyanate and e-caprolactone. *J. Mater. Sci. Mater. Med.* 9:675–678.

55. Ishaug-Riley, S.L., Crane-Kruger, G.M., Yaszemski, M.J., and Mikos, A.G. 1998. Three-dimensional culture of rat calvarial osteoblasts in porous biodegradable polymers. *Biomaterials* 19:1405.

56. Savarino, L., Baldini, N., Greco, M. et al. 2007. The performance of poly-ε-caprolactone scaffolds in a rabbit femur model with and without autologous stromal cells and bmp-4. *Biomaterials* 28:3101.

57. Koh, Y.H., Bae, C.J., Sun, J.J., Jun, I.K., and Kim, H.E. 2006. Macrochanneled poly(ε-caprolactone)/hydroxyapatite scaffold by combination of bi-axial machining and lamination. *J. Mater. Sci. Mater. Med.* 17:773–778.

58. Hench, L.L. 1991. Bioceramics: From concept to clinic. *J. Am. Ceram. Soc.* 74:1487–1510.

59. Vaccaro, A.R. 2002. The role of the osteoconductive scaffold in synthetic bone graft. *Orthopedics* 25(Suppl 5):s571–s578.

60. Guarino, V., Taddei, P., Di Foggia, M., Fagnano, C., Ciapetti, G., and Ambrosio, L. 2009. The influence of hydroxyapatite particles on in vitro degradation behaviour of PCL based composite scaffolds. *Tissue Eng. A* 15(11):3655–3668.

61. Guarino, V. and Ambrosio, L. 2008. The synergic effect of polylactide fiber and calcium phosphate particles reinforcement in poly e-caprolactone based composite scaffolds. *Acta Biomater.* 4:1778.

62. Vozzi, G., Flaim, C., Ahluwalia, A., and Bhatia, S. 2003. Fabrication of PLGA scaffolds using soft lithography and microsyringe deposition. *Biomaterials* 24:2533–2540.

63. Vail, N.K., Barlow, J.W., Beaman, J.J., Marcus, H.L., and Bourell, D.L. 1994. Development of a poly(methyl methacrylate-*co*-*n*-butyl methacrylate) copolymer binder system. *J. Appl. Polym. Sci.* 9:789–812.

64. Wu, B.M., Borland, S.W., Giordano, R.A., Cima, L.G., Sachs, E.M., and Cima, M.J. 1996. Solid free-form fabrication of drug delivery devices. *J. Control. Release* 40:77–87.

65. Cooke, M.N., Fisher, J.P., Dean, D., Rimnac, C., and Mikos, A.G. 2002. Use of stereolithography to manufacture critical-sized 3D biodegradable scaffolds for bone ingrowth. *J. Biomed. Mater. Res. Part B: Appl. Biomater.* 64B:65–69.

66. Sachlos, E. and Czernuske, J.T. 2003. Making tissue engineering scaffold work: Review on the application of SFF technology to the production of tissue engineering scaffolds. *Eur. Cell Mater.* 5:29–40.

67. Peltola, S.M., Melchels, F.P.W., Grijpma, D.K., and Kellomäki, M. 2008. A review of rapid prototyping techniques for tissue engineering purposes. *Ann. Med.* 40:268–280.

68. Chu, T.M.G. 2006. Solid freeform fabrication of tissue engineering scaffolds. In: Ma, P.X. and Elisseeff, J., eds. *Scaffolding in Tissue Engineering*. Boca Raton, FL: Taylor & Francis, pp. 139–153.

69. Yang, S., Leong, K.F., Du, Z., and Chua, C.K. 2002. The design of scaffolds for use in tissue engineering. Part II. Rapid prototyping techniques. *Tissue Eng.* 8:1–11.

70. Hutmacher, D.W. 2000. Scaffolds in tissue engineering bone and cartilage. *Biomaterials* 21:2529–2543.

71. Chua, C.K., Leong, K.F., and Lim, C.S. 2003. Rapid prototyping process chain. In: Chua, C.K., Leong, K.F., and Lim, C.S., eds. *Rapid Prototyping—Principles and Applications*. Singapore: World Scientific Publishing, pp. 25–33.

72. Fedchenko, R.P. and Jacobs, P.F. 1996. Introduction. In: Fedchenko, R.P. and Jacobs, P.F., eds. *Stereolithography and Other RP&M Technologies*. Dearborn, MI: Society of Manufacturing Engineers, pp. 1–26.

73. Jacobs, P.F. 1996. Special applications of RP&M. In: Fedchenko, R.P. and Jacobs, P.F., eds. *Stereolithography and Other RP&M Technologies*. Dearborn, MI: Society of Manufacturing Engineers, pp. 317–366.

74. Landers, R., Pfister, A., Hubner, U., John, H., Schmelzeisen, R., and Mulhaupt, R. 2002. Fabrication of soft tissue engineering scaffolds by means of rapid prototyping techniques. *J. Mater. Sci.* 37:3107–3116.

75. Hutmacher, D.W., Sittinger, M., and Risbud, M.V. 2004. Scaffold-based tissue engineering: Rationale for computer-aided design and solid free-form fabrication systems. *Trends Biotechnol.* 22:354–362.

76. Webb, P.A. 2000. A review of rapid prototyping (RP) techniques in the medical and biomedical sector. *J. Med. Eng. Technol.* 24:149–153.

77. Truscott, M., de Beer, D., Vicatos, G. et al. 2007. Using RP to promote collaborative design of customised medical implants. *J. Rapid Prototyping* 13:107–114.

78. Griffith, M.L. and Halloran, J.W. 1996. Free form fabrication of ceramics via stereolithography. *J. Am. Ceram. Soc.* 79(10):2601–2608.

79. Lee, K.W., Wang, S., Fox, B.C., Ritman, E.L., Yaszemski, M.J., and Lu, L. 2007. Poly(propylene fumarate) bone tissue engineering scaffold fabrication using stereolithography: Effects of resin formulations and laser parameters. *Biomacromolecules* 8:1077–1084.

80. Saiz, E., Gremillard, L., Menendez, G., Miranda, P., Gryn, K., and Tomsia, A.P. 2007. Preparation of porous hydroxyapatite scaffolds. *Mater. Sci. Eng. C* 27:546–550.

81. Leukers, B., Gülkan, H., Irsen, S.H., Milz, S., Tille, C., and Schieker, M. 2005. Hydroxyapatite scaffolds for bone tissue engineering made by 3D printing. *J. Mater. Sci. Mater. Med.* 16:1121–1124.

82. Khalyfa, A., Vogt, S., Weisser, J. et al. 2007. Development of a new calcium phosphate powder-binder system for the 3D printing of patient specific implants. *J. Mater. Sci. Mater. Med.* 18:909–916.

83. Williams, J.M., Adewunmi, A., Schek, R.M. et al. 2005. Bone tissue engineering using polycaprolactone scaffolds fabricated via selective laser sintering. *Biomaterials* 26:4817–4827.

84. Partee, B., Hollister, S.J., and Das, S. 2006. Selective laser sintering process optimization for layered manufacturing of CAPA6501 polycaprolactone bone tissue engineering scaffolds. *Trans. ASME J. Manuf. Sci. Eng.* 128:531–540.

85. Tan, L.L. 2004. Plugging bone the painless way. *Innovation* 4(3):20.

86. Liu, X. and Ma, P.X. 2004. Polymeric scaffolds for bone tissue engineering. *Ann. Biomed. Eng.* 32:477–486.

87. Tellis, B.C., Szivek, J.A., Bliss, C.L., Margolis, D.S., Vaidyanathan, R.K., and Calvert, P. 2007. Trabecular scaffolds created using micro CT guided fused deposition modelling. *Mater. Sci. Eng. C* 81B:30–39.

88. Landers, R. and Mulhaupt, R. 2000. Desktop manufacturing of complex object, prototypes and biomedical scaffolds by means of computer-assisted design combined with computer-guided 3D plotting of polymers and reactive oligomers. *Macromol. Mater. Eng.* 282:17–21.

89. Landers, R., Hubner, U., Schmelzeisen, R., and Mulhaupt, R. 2000. Rapid prototyping of scaffold derived from thermoreversible hydrogels and tailored for application in tissue engineering. *Biomaterials* 23:4437–4447.

90. Woodfield, T.B.F., Malda, J., de Wijn, J., Peters, F., Riesle, J., and Van Blitterswijk, C.A. 2004. Design of porous scaffolds for cartilage tissue engineering using a three-dimensional fibre-deposition technique. *Biomaterials* 25:4149–4161.

91. Malda, J., Woodfield, T.B., van der Vloodt, F. et al. 2004. The effect of PEGT/PBT scaffold architecture on the composition of tissue engineered cartilage. *Biomaterials* 26:63–72.

92. Moroni, L., de Wijn, J.R., and Van Blitterswijk, C.A. 2005. Three-dimensional fiber-deposited PEOT/PBT copolymer scaffolds for tissue engineering: Influence of porosity, molecular network mesh size and swelling in aqueous media on dynamic mechanical properties. *J. Biomed. Mater. Res. A* 75:957–965.

93. Moroni, L., de Wijn, J.R., and Van Blitterswijk, C.A. 2006. 3D fiber-deposited scaffolds for tissue engineering: Influence of pores geometry and architecture on dynamic mechanical properties. *Biomaterials* 27:974–985.

94. Moroni, L., Poort, G., van Keulen, F., de Wijn, J., and Van Blitterswijk, C.A. 2006. Dynamic mechanical properties of 3D fiber-deposited PEOT/PBT scaffolds: An experimental and numerical analysis. *J. Biomed. Mater. Res.* 78A:605–614.

95. Moroni, L., Hendriks, J.A.A., Schotel, R., de Wijn, J.R., and Van Blitterswijk, C.A. 2007. Design of biphasic 3-Dimensional fiber deposited scaffolds for cartilage tissue engineering applications. *Tissue Eng.* 13:361–371.

96. Moroni, L., Schotel, R., Hamann, D., de Wijn, J.R., and Van Blitterswijk, C.A. 2008. 3D fiber-deposited electrospun integrated scaffold enhance cartilage tissue formation. *Adv. Funct. Mater.* 18:53–60.

97. Moroni, L., Schotel, R., Sohier, J., de Wijn, J.R., and Van Blitterswijk, C.A. 2006. Polymer hollow-fiber three-dimensional matrices with controllable cavity and shell thickness. *Biomaterials* 27:5918–5926.

98. Kyriakidou, K., Lucarini, G., Zizzi, A. et al. 2008. Dynamic co-seeding of osteoblast and endothelial cells on 3D polycaprolactone scaffolds for enhanced bone tissue engineering. *J. Bioact. Compat. Polym.* 23:227–243.

99. Li, W., Tuli, R., Huang, X., Laquerriere, P., and Tuan, R.S. 2005. Multilineage differentiation of human mesenchymal stem cells in a three-dimensional nanofibrous scaffold. *Biomaterials* 26:5158–5166.

100. Unger, R.E., Sartoris, A., Peters, K. et al. 2007. Tissue-like self-assembly in cocultures of endothelial cells and osteoblasts and the formation of microcapillary-like structures on three-dimensional porous biomaterials. *Biomaterials* 28:3965–3976.

101. Decker, B., Bartles, H., and Decker, S. 1995. Relationships between endothelial cells, pericytes, and osteoblasts during bone formation in the sheep femur following implantation of tricalcium phosphate-ceramic. *Anat. Rec.* 242:310–320.

102. Villars, F., Bordenave, L., Bareille, R., and Amedee, J. 2000. Effect of human endothelial cells on human bone marrow stromal cell phenotype: Role of VEGF. *J. Cell Biochem.* 79:679–685.

103. Carrington, J.L. and Reddi, A.H. 1991. Parallels between development of embryonic and matrix-induced endochondral bone. *Bioassay* 13:403–408.
104. Collin-Osdoby, P. 1994. Role of vascular endothelial cells in bone biology. *J. Cell Biochem.* 55:304–309.
105. Wang, D.S., Miura, M., Demura, H., and Sato, K. 1997. Anabolic effects of 1,25-dihydroxyvitamin D3 on osteoblasts are enhanced by vascular endothelial growth factor produced by osteoblasts and by growth factors produced by endothelial cells. *Endocrinology* 138:2953–2962.
106. Gibson, L.J. and Ashby, M.F. 1997. *Cellular Solids: Structure and Properties.* Cambridge, U.K.: Cambridge University Press, p. 532.
107. Ma, P.X. 2008. Biomimetic materials for tissue engineering. *Adv. Drug Deliv. Rev.* 60:184–198.
108. Hasirci, V., Vrana, E., Zoplutuna, P. et al. 2006. Nanobiomaterials: A review of the existing science and technology, and new approaches. *J. Biomater. Sci. Polym. Ed.* 17:1241.
109. Nel, A., Xia, T., Madler, L. et al. 2006. Toxic potential of materials at the nanolevel. *Science* 311:622–627.
110. Veiseh, M., Turley, E.A., and Bissell, M.J. 2008. A top-down analysis of a dynamic environment: Extracellular matrix structure and function. In: Laurencin, C.T. and Nair, L.S., eds. *Nanotechnology and Engineering; The Scaffold.* Boca Raton, FL: CRC Press, Taylor & Francis Group, pp. 33–51.
111. Benoit, D.S.W. and Anseth, K.S. 2005. The effect on osteoblast function of colocalized RGD and PHSRN epitopes on PEG surfaces. *Biomaterials* 26:5209–5220.
112. Wilson, C.J., Clegg, R.E., Leavesley, D.I. et al. 2005. Mediation of biomaterial–cell interactions by adsorbed proteins: A review. *Tissue Eng.* 11:1–18.
113. Webster, T.J., Schadler, L.S., Siegel, R.W. et al. 2001. Mechanisms of enhanced osteoblast adhesion on nanophase alumina involve vitronectin. *Tissue Eng.* 7:291–301.
114. Yao, C., Perla, V., McKenzie, J. et al. 2005. Anodized Ti and Ti6A14V possessing nanometer surface features enhance osteoblast adhesion. *J. Biomed. Nanotechnol.* 1:68–77.
115. Mrksich, M., Dike, L.E., Tien, J., Ingber, D.E., and Whitesides, G.M. 1997. Using micro-contact printing to pattern the attachment of mammalian cells to self-assembled monolayers of alkanethiolates on transparent films of gold. *J. Am. Chem. Soc.* 235:305–313.
116. Neff, J.A., Tresco, P.A., and Caldwell, K.D. 1999. Surface modification for controlled studies of cell–ligand interactions. *Biomaterials* 20:2377–2393.
117. Wei, G. and Ma, P.X. 2008. Nanostructured biomaterials for regeneration. *Adv. Funct. Mater.* 18:3568–3582.
118. Huang, Z., Zhang, Y., Kotaki, M., and Ramakrishna, S. 2003. A review on polymer nanofibers by electrospinning and their applications in nanocomposites. *Compos. Sci. Technol.* 63:2223–2253.
119. Nair, L.S., Bhattacharyya, S., and Laurencin, C.T. 2004. Development of novel tissue engineering scaffolds via electrospinning. *Expert Opin. Biol. Ther.* 4:659–668.
120. McManus, M.C., Boland, E.D., Koo, H.P. et al. 2006. Mechanical properties of electrospun fibrinogen structures. *Acta Biomater.* 2:19–28.
121. Li, W.J., Laurencin, C.T., Caterson, E.J., Tuan, R.S., and Ko, F.K. 2002. Electrospun nanofibrous structure: A novel scaffold for tissue engineering. *J. Biomed. Mater. Res.* 60:613–621.
122. Lee, C.H., Shin, H.J., Cho, I.H. et al. 2005. Nanofiber alignment and direction of mechanical strain affect the ECM production of human ACL fibroblast. *Biomaterials* 26:1261–1270.
123. Riboldi, S.A., Sampaolesi, M., Neuenschwander, P. et al. 2005. Electrospun degradable polyesterurethane membranes: Potential scaffolds for skeletal muscle tissue engineering. *Biomaterials* 26:4606–4615.
124. Li, W.J., Mauck, R.L., and Tuan, R.S. 2005. Electrospun nanofibrous scaffolds: Production, characterization, and applications for tissue engineering and drug delivery. *J. Biomed. Nanotechnol.* 1:259–275.
125. Xu, C., Inai, R., Kotaki, M., and Ramakrishna, S. 2004. Electrospun nanofiber fabrication as synthetic extracellular matrix and its potential for vascular tissue engineering. *Tissue Eng.* 10:1160–1168.
126. Li, D. and Xia, Y. 2004. Electrospinning of nanofibres: Reinventing the wheels? *Adv. Mater.* 16(14):1151–1170.
127. Chong, E.J., Phan, T.T., Lim, I.J. et al. 2007. Evaluation of electrospun PCL/gelatin nanofibrous scaffold for wound healing and layered dermal reconstitution. *Acta Biomater.* 3:321–330.
128. Li, M., Guo, Y., Wei, Y., MacDiarmid, A.G., and Lelkes, P.I. 2006. Electrospinning polyaniline contained gelatin nanofibers for tissue engineering applications. *Biomaterials* 27:2705–2715.
129. Ciardelli, G., Chiono, V., Vozzi, G. et al. 2005. Blends of poly(ε-caprolactone) and polysaccharides in tissue engineering applications. *Biomacromolecules* 6:1961–1976.
130. Ghasemi-Mobarakeh, L., Prabhakaran, M.P., Morshed, M., Esfahani, M.H.N., and Ramakrishna, S. 2008. Electrospun poly(3-caprolactone)/gelatin nanofibrous scaffolds for nerve tissue engineering. *Biomaterials* 29:4532–4539.

131. Bhattacharyya, S., Nair, L.S., Singh, A. et al. 2005. Development of biodegradable polyphosphazene–nanohydroxyapatite composite nanofibers via electrospinning. *Mater. Res. Soc. Symp. Proc.* 845:91–96.

132. Deng, X.L., Xu, M.M., Li, D., Sui, G., Hu, X.Y., and Yang, X.P. 2007. Electrospun PLLA = MWNTs = HA hybrid nanofiber scaffolds and their potential in dental tissue engineering. *Key Eng. Mater.* 330–332:393–396.

133. Webster, T.J., Ergun, C., Doremus, R.H., Siegel, R.W., and Bizios, R. 2000. Specific proteins mediate enhanced osteoblast adhesion on nanophase ceramics. *J. Biomed. Mater. Res.* 51:475–483.

134. Zhang, N., Nichols, H.L., Tylor, S., and Wen, X. 2007. Fabrication of nanocrystalline hydroxyapatite doped degradable composite hollow fiber for guided and biomimetic bone tissue engineering. *Mater. Sci. Eng. C Biomim. Supramol. Syst.* 27:599–606.

135. Greiner, A. and Wendorff, J.H. 2007. Electrospinning: A fascinating method for the preparation of ultra-thin fibers. *Angew. Chem. Int. Ed.* 46:5670–5703.

136. Wu, Y.Q. and Clark, R.L. 2007. Controllable porous polymer particles generated by electrospraying. *J. Colloid Interf. Sci.* 310:529–535.

137. Bognitzki, M., Czado, W., Frese, T. et al. 2001. Nanostructured fibers via electrospinning. *Adv. Mater.* 13:70.

138. Zhou, X.D., Zhang, S.C., and Huebner, W. 2001. Effect of the solvent on the particle morphology of spray dried PMMA. *J. Mater. Sci.* 36:3759.

139. Xu, C.Y., Inai, R., Kotaki, M., and Ramakrishna, S. 2004. Aligned biodegradable nanofibrous structure: A potential scaffold for blood vessel engineering. *Biomaterials* 25:877–886.

140. Yang, F., Murugan, R., Wang, S., and Ramakrishna, S. 2005. Electrospinning of nano/micro scale poly(L-lactic acid) aligned fibers and their potential in neural tissue engineering. *Biomaterials* 26:2603–2610.

141. Zhong, S., Teo, W.E., Zhu, X., Beuerman, R.W., Ramakrishna, S., and Yung, L.Y.L. 2006. An aligned nanofibrous collagen scaffold by electrospinning and its effects on in vitro fibroblast culture. *J. Biomed. Mater. Res. A* 79:456–463.

142. Thomas, V., Jose, M.V., Chowdhury, S., Sullivan, J.F., Dean, D.R., and Vohra, Y.K. 2006. Mechano-morphological studies of aligned nanofibrous scaffolds of polycaprolactone fabricated by electrospinning. *J. Biomater. Sci. Polym. Ed.* V17:969–984.

143. Tan, J., Tien, J., and Chen, C. 2002. Microcontact printing of proteins on mixed self-assembled monolayers. *Langmuir* 18:519–523.

144. Griffith, L.G. and Lopina, S. 1998. Microdistribution of substratum-bound ligand affects cell function: Hepatocyte spreading on PEO-tethered galactose. *Biomaterials* 19:979–986.

145. Yamato, M., Konno, C., Utsumi, M., Kikuchi, A., and Okano, T. 2002. Thermally responsive polymer-grafted surfaces facilitate patterned cell seeding and coculture. *Biomaterials* 23:561–567.

146. Sorbías, H., Padeste, C., and Tiefenauer, L. 2002. Photolithographic generation of protein micropatterns for neuron culture applications. *Biomaterials* 23:893–900.

147. Christenson, E.M., Anseth, K.S., Van den Beucken, J.J.P. et al. 2007. Nanobiomaterial applications in orthopedics. *J. Orthop. Res.* 25:11–22.

148. Webster, T.J., Ergun, C., Doremus, R.H. et al. 2000. Enhanced functions of osteoblasts on nanophase ceramics. *Biomaterials* 21:1803–1810.

149. Li, W.J., Jiang, Y.J., and Tuan, R.S. 2006. Chondrocytes phenotype in engineered fibrous matrix is regulated by fiber size. *Tissue Eng.* 12:1775–1785.

150. Whitesides, G.M. 2006. The origins and the future of microfluidics. *Nature* 442:368–373.

151. Zhang, S. 2003. Fabrication of novel biomaterials through molecular self-assembly. *Nat. Biotechnol.* 21(10):1171–1178.

152. Whitesides, G.M. and Boncheva, M. 2002. Supramolecular chemistry and self-assembly special feature: Beyond molecules: Self-assembly of mesoscopic and macroscopic components. *Proc. Natl. Acad. Sci. USA* 99:4769–4774.

153. Vozzi, G., Flaim, C., Ahluwalia, A., and Bhatia, S. 2003. Microfabrication of biodegradable polymeric structure for guided tissue engineering. *Biomaterials* 24:2533–2540.

154. Rivron, N.C., Rouwkema, J., Truckenmuller, R., Karperien, M., De Boer, J., and Van Blitterswijk, C.A. 2009. Tissue assembly and organization: Developmental mechanisms in microfabricated tissues. *Biomaterials* 30:4851–4858.

20 Virus-Based Nanoparticles as Drug Delivery Systems

Eva Roblegg and Andreas Zimmer

CONTENTS

20.1 INTRODUCTION

Recently, drug delivery is one of the main challenges for the development of new pharmaceutical applications in the field of nanotechnology. Many drugs have physicochemical characteristics that are not favorable to pass biological/enzymatic barriers within the human body. Thus, the development of carriers for drug delivery has seen a significant increase in the last 20 years. Prodrugs, soft drugs, and codrugs are designed to transport efficient and selective drugs to the site of action and to achieve therapeutic efficiency as well as treatment safety [1]. Liposomes, dendrimers, smart polymers, as well as viral nanoparticles (VNPs) are used as tools for potent delivery of therapeutics [2]. In the first part of this chapter, we present the current state of progress regarding the main tools for potent drug delivery. In the second part of this chapter, we particularly focus on virus-based nanoparticles as drug delivery systems.

20.2 TOOLS FOR POTENT DRUG DELIVERY

20.2.1 LIPOSOMES

Liposomes, made out of the same material as the cell membrane (phospholipids), have shown to improve the solubility of amphiphilic drugs and to increase the transport into the cells and tissues. Liposomes are mainly prepared in a three-step method, starting with the dissolution of the lipids in an organic solvent, followed by solvent evaporation and finally by dispersion of the dried liquids in an aqueous medium. The size of liposomes can vary from 80 to 3000 nm reliant on the number of bilayers they consist of (i.e., multilamellar vesicles, small unilamellar vesicles, large unilamellar vesicles) [3]. Depending on the drug characteristics, hydrophobic substances can be anchored into the bilayer whereas hydrophilic substances can be encapsulated into the liposomal cavity. However, the delivery of the incorporated/anchored drug is limited, due to the fact that interactions with components of the human body (i.e., serum proteins, lipoproteins, and the mononuclear phagocytic system [MPS]) can take place, resulting in irreversible degradation and/or inactivation of the drug. To overcome this limitation, pH-sensitive liposomes, made of charged or neutral phospholipids, have been developed that are stable at physiological pH (7.4). In tumor cells, for instance, the polymer precipitates under acidic conditions and releases the drug [4,5]. Another possibility to make liposomes more sensitive has been the development of temperature-sensitive liposomes [6–9] which are not discussed in detail here. Many papers have demonstrated that rapid uptake of liposomes in vivo by cells of the MPS, which has restricted their therapeutic utility, can be overcome by conjunction with the hydrophilic polymer polyethylene glycol (PEG). Additionally, the presence of PEG at the surface of a liposomal carrier has been clearly shown to extend the circulation lifetime of the vehicle in the blood [10–16].

20.2.2 Dendrimers

Dendrimers are synthetic macromolecules with a tree-like structure and well-controllable sizes. The structure of these macromolecules can be divided into three separate architectural parts: (i) the core, (ii) the interior shells (branches), and (iii) the multivalent surface (terminal groups). Dendrimers can be synthesized according to the divergent method, where the dendrimer is built outwards of the core (repetition of a sequence of reactions) and to the convergent method, where the core is incorporated in the final step of the elaboration of the dendrimer [17,18]. The interior of the dendrimer is well-suited for host–guest interactions as well as for the encapsulation of guest molecules and therefore improves the stability and bioavailability of active substances. Another possibility for drug binding is the complexation of the drug with the dendrimer through electrostatic interactions. Recently, up to 78 molecules (e.g., DNA [19,20], oligonucleotides [21], paclitaxel [22], ibuprofen [23,24], doxorubicin [25]) can be complexed with polyamido-amine dendrimers (PAMAM) which form spherical polymers with high aqueous solubility and have the ability to target selectively cancer cells [23,26].

20.2.3 Smart Polymers

Over the last decade, smart polymeric materials are widely used for biological applications. Much progress has been made in the synthesis of these polymers that imitate naturally occurring polymers and show brilliant sensitivity to factors like pH, temperature, and mechanical stress. They are utilized to shuttle therapeutic substances past the endosomal membrane into the cytoplasm of targeted cells [27–32]. However, the key example for this technology is the designing of hydrogels which are three-dimensional networks of hydrophilic polymers that swell or expand by taking up water. Smart (or stimuli-responsive) polymers are reversible soluble–insoluble (SIS) in aqueous media or cross-linked to build up a hydrogel. SIS polymers can be synthetic, including methylmethacrylate polymers and poly(N-isopropylacrylamide), and/or natural like chitosan, alginate and carrageenan, proteins, and polysaccharides [33,34]. The type of cross-linking can either be chemical or physical. Chemical hydrogels are held together by chemical cross-linking via covalent bonds whereas physical hydrogels are obtained by non-covalent forces from spontaneous self-assembly. The encapsulation of the hydrophilic drug can be achieved by mixing the drug with the monomers followed by polymerization or by the swelling step in the aqueous medium. The drug is released out of the hydrogel by diffusion according to Fick's law [35]. By variation of the medium parameters (pH, temperature and light), hydrogels often compress or swell. Poly(N-isopropylacrylamide) is one of the most commonly studied smart polymers with variation of its water solubility as a function of the temperature. The hydrogel beads exhibit lower critical solution temperature (LCST) behavior in aqueous solution by expanding and swelling when cooled below the LCST (the hydrogen bonds between the polar parts and the polymer prevail) and shrinking and collapsing when heated above the LCST (hydrophobic interactions in the hydrogel predominate). The pore sizes of these temperature-sensitive hydrogel beads are influenced significantly by the temperature as well as by the gel composition, but not much by the cross-linker concentration [36].

High extensive effort in this field has been made by developing hydrogels that are blood sugar–sensitive. These pH-sensitive hydrogels show pulsatile and/or pulsed drug delivery characteristics and are able to release insulin when the glucose level is too high [37–40]. Goldraich and coworkers have described hydrogel matrices prepared of 2-hydroxyethyl methacrylate, *N,N*-dimethyl-aminoethyl methacrylate, tetraethylene glycol dimethacrylate, ethylene glycol, and water solutions containing glucose oxidase, bacitracin, or insulin. At lower pH or higher glucose concentration, the swelling of the hydrogel as well as the release rates of the insulin were faster and higher [41,42].

20.2.4 VIRUS-BASED NANOPARTICLES

A variety of viruses have been investigated for new applications including biomaterials, vaccines, tissue targeting, chemical tools, molecular electronic materials and imaging. Recombinant products, such as viral proteins and polymers, could be used as excipients to create tissue-specific nanoparticles and targeted nanodevices with multifunctional capabilities to make an early diagnosis possible and provide targeted treatment of the disease (such as cancer). Additionally, viral proteins are attractive as self-assembled and uniform structured in the nanoscale level with well-defined geometrics. Being protein-based, they have a natural biocompatibility and show relatively rigid structures which make it feasible to display molecules in precise spatial distribution [43,44]. Virus-like particles (VPLs) are defined as an assembly of structural proteins which are organized into capsomeres (substructures) that further organize into the capsid. They build an empty shell loaded with DNA or small molecules or proteins.

To date, more than 30 different viruses have been used for the preparation of VLPs including single and multiple capsid proteins and those with and without lipid envelopes [45]. VNPs currently under investigation include polyoma virus, cowpea mosaic virus (CPMV), cowpea chlorotic virus (CCMV), and bacteriophages (MS2, M13, and Qβ). In this study, we primarily focus on polyoma and briefly on CPMV.

20.3 VIRUS-BASED PARTICLES DERIVED FROM PAPOVAVIRIDAE

The family Papovaviridae is made up of the initials of the most important representatives: rabbit papilloma virus (pa), mouse polyoma virus (po), and vaculating virus (simian virus 40 [SV 40]) (va). These representatives are characterized by capsids of small, non-enveloped icosahedral viruses with genomes composed of a covalently circular closed, double-stranded DNA (dsDNA) molecule. The Papovaviridae family is divided into two subgroups: (i) the Papillomaviridae and (ii) the Polyomaviridae which differ in their genomic composition of the capsids [46–48]. Papillomaviridae are 55 nm sized virions, icosahedral with a molecular weight of the DNA of 5×10^6 Da. Polyomaviridae, classified as group 1 virus in the Baltimore classification scheme, contain a dsDNA, are smaller (45 nm), icosahedral, and have a molecular weight of the DNA of 3×10^6 Da [49] (see Figure 20.1). For the cell infection process, they use multiple and often distinct receptors and entry mechanisms. Liddington et al. [50] and Neu et al. [51] investigated the structures of two major members of this family and gave an insight into the mechanism of the recognition of the host cells. Table 20.1 gives a survey of the most important Papovaviridae. Table 20.2 represents the polypeptide classification, molecular weight, and percentage arrangement of Papillomaviridae.

Two major exponents of Polyomaviridae, both identified 1971 from immunosuppressed patients, use humans as their hosts. The JC virus, named after the patients' initials, was isolated in human

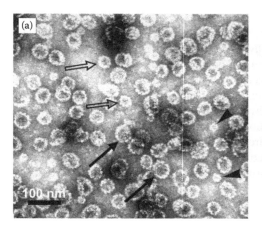
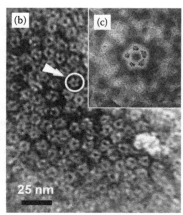

FIGURE 20.1 **(See companion CD for color figure.)** TEM images from polyomavirus capsids and pentamers. (a) Shows functional viral particles ~55 nm (closed arrows) and smaller particles which are assumed to be incomplete (20–30 nm, open arrows and dark tips). (b) Shows VP1 pentamers ~8.5 nm. (c) Visualizes the five VP1 proteins after image reconstruction from TEM analysis. (Modified from Henke, S. et al., *Pharm. Res.*, 17, 1062, 2000.)

TABLE 20.1
Overview of the Subfamily Polyomaviridae

Virus	Host	Genomic Quantity (bp)	Virus Capsid Protein		
			VP1	VP2	VP3
Polyomavirus	Mice	5392	42	35	22
JC virus	Man	5130	39	38	25
BK virus	Man	5133	40	39	26
Simian virus 40 (SV 40)	Rhesus monkey	5243	40	39	26
Lymphotropic papovavirus	Vervet monkey	5270	40	39	26
Bovine papovavirus	Cow	4967	40	39	26
Hamster papovavirus	Hamster	5366	41	38	24
Kirsten virus	Mice	4754	41	35	24
Rabbit papovavirus	Rabbit	n.d.	n.d.	n.d.	n.d.
Rat papovavirus	Rat	n.d.	n.d.	n.d.	n.d.
Simian Agent 12	Baboon	n.d.	n.d.	n.d.	n.d.
Budgerigar fledgling disease virus	Budgerigar	4980	38	38	26

Source: Modified from Cole, C., *Fundamental Virology*, B.N. Fields, ed., Lippincott-Raven, Philadelphia, PA, pp. 917–944, 1996.
n.d. = not determined.

fetal brain cultures and is the etiological agent of progressive leukoencephalopathy (PML) [52,53]. The BK virus, also named after the patients' initials, was isolated in VERO cells from the urine of a renal transplant recipient [54]. The BK virus causes polyomavirus–associated nephropathy (PVAN) in 1%–10% of renal transplant recipients after kidney allograft [55]. The SV 40 is thought to have a similar structure as the two viruses described earlier. For the first time, SV 40 was discovered as a contaminant of formalin-inactivated polioviruses, grown in rhesus monkey kidney cells [56]. The characterization of this virus as a papovavirus, able to transform cells in vitro and induce tumors, was a cause for great concern as several thousand individuals were inoculated with the

TABLE 20.2
Overview of the Subfamily Papillomaviridae

Virus	Polypeptide Classification	Molecular Weight (kDa)	Overall Protein (%)
Rabbit papilloma virus	VP1	85	11
	VP2	70	21
	VP3	60	48
	VP4	19	9
	VP5	15	7
	Other		4
Human papilloma virus	VP1	115	8
	VP2	100	1
	VP3	85	9
	VP4	63	60
	VP5	51	17
	VP6	14	1
	Other		2

Source: Modified from Melnick, J.L. et al., *Intervirology*, 3, 106, 1974.

TABLE 20.3
Overview of the Receptors and Entry Pathways Utilized by BK Virus, JC Virus, and SV 40

	BK Virus	JC Virus	SV 40
Receptors	$\alpha(2,3)$-linked sialic acid	Terminal $\alpha(2,3)$- or $\alpha(2,6)$-linked sialic acid	Sialic acid on ganglioside GM1
	Ganglioside GD1b, GT1b	Ganglioside GT1b	Ganglioside GM1
Co-receptors	Unknown	Serotonin receptor (5HT-2a)	MHC class I
Entry mechanism	Caveolae-mediated	Clathrin-dependent	Primarily caveolae-mediated
Cytoplasmic transport	Vesicles localize to the endoplasmic reticulum	Early endosomes to caveosomes	Fuse into caveosome and localize to the endoplasmic reticulum

Source: Modified from Neu, U. et al., *Virology*, 384, 389, 2009.

contaminated vaccine. The cancer risk for humans who received SV 40-contaminated vaccines did not apparently increase, but fragments of SV 40 DNA retrieved in specific human tumors [57,58]. The receptor, co-receptors, and entry mechanisms of the JC virus, BK virus, and SV 40 are summarized in Table 20.3.

20.3.1 STRUCTURE

Polyomaviruses and papillomaviruses belong to the papovavirus family, which is characterized by enveloped dsDNA viruses. One of the differences between these two subfamilies is the variation in their capsomer morphology and the intercapsomer associations, causing capsids assembly [59]. Murine polyomavirus, SV 40, and human polyomaviruses are the most commonly studied viruses due to their small genome size and will turn our main attention in this chapter. Polyomaviruses are made up of three viral-encoded proteins namely VP1, VP2, and VP3, found in different ratios. These proteins embed the viral chromatin composed of virus DNA and the cellular histones H2A, H2B, and H4 [60,61]. The recombinant VP1 protein is the major protein of polyoma virus [62].

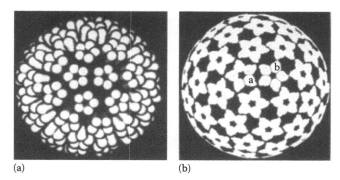

(a)   (b)

FIGURE 20.2   **(See companion CD for color figure.)** Computer graphical illustration of the polyomavirus capsid (VP1 pentamers). (a) Demonstrates the surface morphology of the capsid (diameter of the capsid: ~50 nm; diameter of the pentamers: up to 8.5 nm). (b) Displays the pentavalent pentamers (a) and the hexavalent pentamers (b). (Modified from Salunke, D.M. et al., *Cell*, 46, 895, 1986.)

The computer graphical illustration of VP1 pentamers is demonstrated in Figure 20.2. The outer virus shell of the capsid is made up of 72 pentameric VP1 proteins (see Figure 20.3); the pentamers are set up in 60 hexavalent pentamers and 12 pentavalent pentamers. Each pentavalent pentamer, in turn, is surrounded by further five pentamers, each hexavalent pentamer by six pentamers [63]. The N-terminal arms are responsible for the binding of polyomavirus DNA. Each VP1 pentamer binds either one of the minor coat proteins, VP2 or VP3. The C-terminus of VP1 and VP2 is linked into the axial loop of a VP1 pentamer and stabilized by hydrophobic interactions [60,64,65]. The genome of polyomaviruses, which is circular and covalent closed, is composed of a dsDNA molecule with 5000 bp in size [66]. The replication cycle is divided into early and late stages of infection [67]; thus, the early and late regions of the virus genome are responsible for this process. The virus infection in permissive cells takes place in three steps: (i) protein biosynthesis of the early proteins, (ii) replication of the virus DNA, and (iii) biosynthetization of the late proteins. The transcription processes bidirectional from the initial point and the early and late mRNAs are read off from the contrary DNA strands. The early region includes the sequence for proteins also identified as tumor

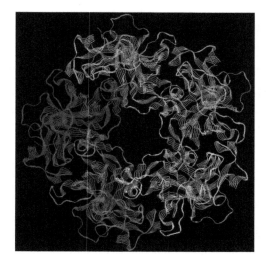

FIGURE 20.3   **(See companion CD for color figure.)** Computer graphical illustration of the VP1 pentamer structure. The outside of the VP1 pentamer complex is demonstrated based on the coordinates of Stehle and Harrison [148]. Each color illustrates a VP1 monomer. (Modified from Georgens, C. et al., *Curr. Pharm. Biotechnol.*, 6, 49, 2005.)

antigens (T-antigens). SV 40 and associated (related) viruses codify for the small T-antigens, placed between the nucleus and the cytoplasm, and for the large T-antigens, mainly located in the nucleus. The murine polyomavirus codifies for the middle T-antigens, found in the plasma membrane. These three types of antigens have different characteristics. The small T-antigens concern in the accumulation of viral DNA, the large T-antigens regulate the early transcription due to the viral DNA binding in the region of the early promoter, and the latter antigens (middle T) provide the basis for cell transformation [68]. In the late stage of the infection, the structural proteins VP1, VP2, and VP3 are synthesized. The accordant genes overlap on the genome and are translated from three different mRNAs [69]. VP1 virion was found to be composed of six distinct species, whereas polyoma capsid VP1 was found to contain only four species, all with the same amino acid sequence but generated by modifications of the initial translation product (i.e., phosphorylation, acetylation) [70]. Due to the synthesis of the structural proteins at the ribosomes of the endoplasmic reticulum (ER) in the cytoplasm, the single constituents are transported into the nucleus to accumulate. Gillock et al. demonstrated via nuclear localization signal the localization of the VP1 sequence within the first 11 amino acids of the N-terminus [71], for VP2 and VP3 12 amino acids of the C-terminus are responsible for the transportation into the nucleus [72–74]. It can be concluded that VP1 can migrate into the nucleus on its own and interacts with VP2 and VP3 in the cytoplasm to improve the transport of the minor proteins into the nucleus.

20.3.2 CELLULAR UPTAKE

The earlier studies of polyomavirus host–cell interactions appeared closely behind their initial discovery and isolation in mammalian tissue culture. Mattern et al. envisioned the infection of cells by the mouse polyomavirus as a three-step process including (i) virus adsorption, (ii) virus penetration, and (iii) virus uncoating. The whole virions were taken up into cells by phagocytosis as membrane-bound particles and aggregates [75,76]. In the 1970s, Mackay and Consigli used the plaque-assay technique as a tool to evaluate the optimal conditions for the adsorption of polyoma virions to host cells. Accordingly, they investigated the early events of infection by electron microscopy. The results demonstrated that when virions were used to infect either permissive or nonpermissive cells, identical early events of viral infection were observed, suggesting that these early events of infection are a property of the virion and not the host cell. Empty capsids and virions were endocytosed by two different pathways: virions were taken up by monopinocytic vesicles through the cell membrane, subsequently enclosed by the membrane and transported through the cytoplasm to the nucleus. The second pathway includes the absorption of empty capsids as phagocytic vesicles into the cell. They appeared to fuse with lysosomes, apparently destined for degradation [77,78]. Further investigations in the field of cellular uptake approved that an alternative factor controls the uptake of the viruses—the receptor-mediated endocytosis. For the receptor recognition sialic acid (N-actyl neuraminic acid), present on the cell surface of virtually every cell type in higher vertebrates [79], is required for the binding process to the host cell and mediates or modulates a wide variety of physiological and pathological processes [80–88].

20.3.3 PRODUCTION

20.3.3.1 Expression System and Purification

A bioprocess used for the production of virus-based particles is often associated with problems regarding the large-scale mammalian cell culture. Thus, other eukaryotic cells, for instance, insects [53], yeast [89], and *Escherichia coli* [90,91], are utilized for recombinant expression of the structural proteins VP1, VP2, and VP3.

Goldman et al. investigated the expression of the major protein VP1 of the human polyoma JC virus in the insect cell line Sf158. The highest expression was obtained 5 days after infection by using the recombinant baculoviruses [53]. Further investigations of An et al. described the cloning of the structural proteins of murine polyomavirus into the p2Bac dual multiple cloning site vector [92]. The corresponding proteins were expressed in insect cell line Sf9 by the co-transfection of Sf9 cells with p2Ba and the linear DNA of *Autographa californica* multiple nuclear polyhedrosis virus.

As mentioned earlier, another possibility for the expression of VP1 is the use of yeast. *Saccharomyces cerevisae* (INVSC1) was used for the expression of the VP1 protein. The VP1 gene was cloned into the yeast expressing plasmid and transfected into INVSC1 yeast cells [93].

The third procedure to express the major capsid protein VP1 utilizes the recombinant gene of *E. coli*. Here, we have to differ between VP1 used from (i) murine polyomavirus [63,94], (ii) avian polyomavirus, (iii) budgerigar fledgling disease virus, where the VP1 gene was inserted into a truncated form of the pFlag-1 vector and expressed in *E. coli* [91], and (iv) L1 capsid protein of papilloma virus type 11 (HPV 11) [90]. After the expression of the viral structural proteins, the cells are lysed and separated of cell constituents. To achieve the recombinant proteins in a pure form, different purification methods can be used including the cesium chloride (CsCl) density gradient centrifugation [71,95,96], the chitin affinity chromatography [97], affinity chromatography on immobilized glutathione (fusion proteins) [98], and dialysis and proceeding to a phosphocellulose (P-11) column [94]. The amount of the purified recombinant proteins can be evaluated by separation via sodium dodecyl sulfate–polyacrylamide gel electrophoresis (SDS-PAGE) [71,95,99].

20.3.3.2 Assembly and Loading

A variety of nanoassembled structures have been proposed due to the fact that viruses have well-characterized surface properties, regular geometrics, and dimensions in the nanoscale. VP1 monomers, achieved after expression and purification, immediately build up pentamers. These VP1 pentamers are linked to capsids and polymorphic aggregates at high ionic strength. The assembled structures were stabilized by calcium ions. These findings apply for mouse polyomavirus as well as for other members of this family [63,83,100,101]. Another possibility to stabilize the capsid is disulfide bonds. Recent studies by Chen et al. showed that VP1 of the JC virus expressed from yeast cells, assembled into a VLP, in which disulfide bonds were found. These disulfide bonds caused dimeric and trimeric VP1 linkage. Subsequently, the disulfide bonds were reduced and without disulfide bonds the capsid disassembled into capsomeres. The capsomeres were reassembled in the presence of calcium ions. The addition of diamide, to reconstruct the disulfide bonds, formed irregular aggregates instead of a VLP [93]. Further studies also showed that disulfide bonds help to stabilize the capsid between neighboring L1 molecules and build up the inter-L1 disulfide bond formation. However, the kinetics during morphogenesis have only been investigated in pseudoviruses but not in papillomaviruses [102–104].

In 1979, Barr et al. [105] described for the first time the preparation of polyoma-like particles, formed of polyoma DNA and purified empty capsids, incubated in cell-free system (sterile distilled water at 37°C) [66,101]. The uptake was performed by osmotic shock. The integrated dsDNA, with a molecular weight of 1.1×10^6 Da, was protected against the action of pancreatic DNase by the capsid [106]. The particles were stable in solutions of high ionic strength. VLPs as well referred as pseudoparticles are also able to incorporate exogenous DNA (till 3 kbp) through a process identified as pseudofection [107]. Slilaty and coworkers demonstrated to incorporate either linear, circular, or supercoiled polyoma DNA, as well as single-stranded DNA (ssDNA; antisense oligonucleotides), rRNA, and the synthetic homopolymers poly(dA), poly(dT), poly(dG), and poly(dC) into empty capsids [108–110]. To increase the transfer of DNA, fragments larger than 3 kbp DNA transfection for gene therapy were used by Soeda et al. [107]. The DNA was linked to polylysine and encapsulated into the pseudocapsid, to protect the DNA against degradation. Polylysine was found to increase the efficiency of short-term expression in both experiments, in vitro as well as in vivo. Electron microscopic investigations demonstrated that VP1 pseudocapsids created two different types of interactions with dsDNA. On the one side, they build up highly stable complexes depending on the free DNA ends; on the other side, they form weaker interactions with internal parts of the DNA chain [111].

It can be concluded that the "key" structural protein VP1 is able to bind DNA in a high affinity way (stronger for ssDNA; weaker for dsDNA), whereas VP2 and VP3 failed to bind. The domains, responsible for the binding process, are the first five amino acids of VP1 (i.e., Ala1-Pro-Lys-Arg-Lys-5) [112–114]. VP1 VLPs linked with heterologous DNA are more efficient in transporting the DNA into the tissue than DNA on its own [115]. They are shown to be excellent candidates for the incorporation of DNA and other biological molecules as drug delivery tools to target cells [116].

20.4 VIRUS-BASED PARTICLES DERIVED FROM CPMV

CPMV is a non-enveloped plant virus belonging to the Comoviridae family and has received recent attention as a nanoscale scaffold for the design of vaccines and therapeutics. The genome consists of two separate positive-sense RNAs, referred to as RNA-1 and RNA-2, which are encapsulated in separate particles. RNA-1 encodes the replicative mechanism of the virus (including polymerase and protease enzymes). The RNA-2 encodes the key capsid proteins, i.e., the large (L) with a molecular weight of 42×10^3 Da, the small (S) with 24×10^3 Da, and the movement protein (M) with 48×10^3 Da. The latter one enables spread of the virus between infected leaf cells during replication [117].

The infection can take place in two different ways: (i) naturally, by transmission via insects or (ii) experimentally, by inoculating the primary leaves of cowpea seedlings using a wound agent. The capsid is 31 nm in size and demonstrates an icosahedral pseudo $T = 3$ symmetry. The L and S proteins (60 of each per capsid) are assembled in an icosahedral surface lattice around the ssRNA. The designed β-barrel domains, formed by the L and S proteins, consist of external loops that are variable in their structure and allow the insertion of additional residues into the βb–βC loop of the capsid [118,119]. Regarding the natural chemical properties of CPMV, it could be demonstrated that this small virus is an excellent candidate for nanoassembly, is stable to acid (pH 1.0) as well as temperatures up to 60°C, and can also withstand organic solvents. Wang et al. demonstrated that the CPMV capsid consists of a lysine residue with improved reactivity in each asymmetric unit and 60 of such lysines per virus particle. Further results describe that under forcing conditions, up to four lysine residues per asymmetric unit can be addressed [120]. A great number of assemblies utilizing CPMV include for instance CPMV-Au [121], CPMV-cysteine mutants [122], CPMV chimeras [123,124], and PEGylated CPMV [125].

Another interesting aspect is that CPMV is convenient to link a variety of molecules to its coat protein and to modify the coat protein sequence by genetic means. It is already known that CPMV interacts with several mammalian cell lines and tissues in vivo [44]. Recent studies used transferrin, a circulatory iron carrier protein that provides the uptake of iron in active cells by interaction with the transferring receptor (TfR), conjugated to CPMV [126,127]. Iron is required by tumor cells and therefore the TfRs are up-regulated in the different tumor types compared to normal cells. As a consequence, transferrin, linked to CPMV, is a promising candidate for drug delivery to tumor cells [128,129]. Another ligand used for viral targeting to tumors is folic acid. This vitamin B, only obtained from exogenous sources, is important for the cell growth and the proliferation. The uptake is mediated by the folic acid receptor (FR). In contrast to normal tissue with low FR expression, tumor tissues exhibit an increase of the FR expression [130–133]. In addition, it is also important to increase the specificity and affinity of the targeted virus-based nanoparticles. Thus, the virus, such as CPMV, has to be PEGylated to reduce the background binding to mammalian cells [131]. Destito et al. attached a folic acid–PEG conjugate through an azide–alkyne cycloaddition reaction using copper as catalyst. The PEG-folate moiety approved CPMV-specific identification of tumor cells bearing the folate receptor [134].

In recent times, CPMV has been investigated as a nanoparticle delivery system for therapeutics targeting not only a number of tumors but also neurodegeneration. Newer studies by Shriver et al. demonstrate that CPMV particles, taken up during infection of mice with neurotropic mouse hepatitis virus (MHV), could be used as a vehicle to deliver therapeutics to the damaged central nervous system (CNS) during neurodegenerative and infectious diseases of the CNS [135]. Regarding the biodistribution, toxicity and pathology of CPMV in vivo, Singh et al. demonstrated CPMV as safe and nontoxic platform for in vivo applications [44].

20.5 OUTLOOK

Virus-based particles have undergone a rapid development as viral vectors for cancer treatment and other diseases. Due to their efficiency to stimulate cellular and humoral responses, they are interesting candidates as carrier molecules for drug delivery. However, detailed studies how each behaves in vitro as well as in vivo and the kind of immune response they generate are a great challenge in research.

Papillomavirus VLPs are the most studied viruses and are at the most sophisticated stage. They are used as cervical cancer vaccine targeting human papillomavirus 16/18 (HPV 16/18). Two preventive vaccines comprising recombinant HPV L1 VLPs have been approved by the FDA and licensed, i.e., Cervarix® (recombinant HPV genotype 16/18; GlaxoSmithKline) and Gardasil® (recombinant HPV genotype 6/11/16/18; Merck). These vaccines target two of the approximately 15 known oncogenic HPV types. Nevertheless, these two types are considered to cause 70% of cervical cancer cases [136,137]. Recent studies demonstrated the effectiveness of HPV L1 VLP vaccine. During the clinical phase III, 30,000 women worldwide were involved into the double-blinded trial, thereof 158 10–14 year old healthy girls and 458 15–25 year old young women received the vaccine at 0, 1, and 6 months. The anti-HPV antibody titers were evaluated. After 6 months, 100% seropositivity was achieved for HPV 16 as well as for HPV 18. The antibody titers were higher for the 10–14 year age group, although the vaccine was tolerated by all patients. Thus, it could be demonstrated that up to 4.5 years the HPV L1 VLP vaccine is safe, immunogenic, and induces protection against HPV 16 and HPV 18 and associated cervical lesions [138,139].

Murine VP1 VLPs generate an antibody response in normal and immunodeficient mice [115,140]. The antiviral IgM and IgG response (T-cell independent) was investigated on T-cell-deficient mice, which were infected with the structural protein VP1, VLPs, and live papillomavirus. It could be shown that the IgG production was only detectable by immunization with the live virus whereas the IgM response arose from each of the viral antigens [141]. These characteristics can be used to clone highly immunogenic VLPs for the development of new vaccines [142]. VP1 capsids are also potential candidates as drug carriers, especially for the delivery of foreign protein antigens. Recent studies describe that VLPs were well-suited for inducing a protective immune response on several routes of vaccine administration as these antigens generate humoral as well as cell-mediated immunity. The most important benefit of these carrier devices is presented by their capacity to directly target antigenic proteins or DNA vaccines to immature dendritic cells along their maturation pathway [143].

Another possibility for the improvement of VLPs used as drug delivery systems displays the employment of the structural protein VP2. VP2 can be used as molecular anchor where active agents are associated via cross-linker. Each pentamer of the VP1 is linked to a VP2 molecule in a stable and specific way, while the maximum loading capacity increases up to 72 VP2 proteins [65].

The plant comovirus CPMV is able to attach a variety of molecules to its coat protein and to modify the coat protein sequence by genetic means. Thus, an increasing range of bionanotechnology applications, including vaccines and antiviral therapeutics, are under investigation [144]. So far, most of the used viral particles are not examined due to their biodistribution, toxicity, and pathology. To address this question, Rae et al. studied the localization of CPMV VLPs after oral or intravenous inoculation to mice. CPMV was found in different tissues all over the body, including the spleen, kidney, liver, lung, stomach, small intestine, lymph nodes, brain and bone marrow. Additionally, the stability of CPMV virions demonstrated a high stability in the gastrointestinal tract which makes them feasible as orally bioavailable nanoparticles [144]. In further studies, Singh and coworkers investigated not only the biodistribution but also the toxicology and pathology of CPMV in mice.

Normally, particles, viruses included, are removed by the reticuloendothelial system of the liver and spleen [145,146]. In this study, the CPMV VLPs were cleared rapidly from plasma, after 30 min most of the injected CPMV VLPs were trapped in the liver and only some in the spleen. No cellular toxicity was noted and the hematology was largely normal. Thus, this plant virus seems to be a safe and nontoxic candidate for biomedical applications [44].

Due to the understanding of the structures and molecular basis of pathogenic viruses, their use as tools to create tissue-specific nanoparticles and targeted nanodevices with multifunctional capabilities makes an early diagnosis possible and provides targeted treatment of diseases. However, detailed studies of their in vivo behavior, including toxicity, as well as the knowledge about the immune responses they generate are important to evaluate the potential and the limitations of these promising biomedical nanotechnology applications.

REFERENCES

1. Serafin, A. and Stanczak, A., 2009. Different concepts of drug delivery in disease entities. *Mini Rev Med Chem*, 9: 481–497.
2. Portney, N.G. and Ozkan, M., 2006. Nano-oncology: Drug delivery, imaging, and sensing. *Anal Bioanal Chem*, 384: 620–630.
3. Kozubek, A., Gubernator, J., Przeworska, E., and Stasiuk, M., 2000. Liposomal drug delivery, a novel approach: PLARosomes. *Acta Biochim Pol*, 47: 639–649.
4. Drummond, D.C., Zignani, M., and Leroux, J., 2000. Current status of pH-sensitive liposomes in drug delivery. *Prog Lipid Res*, 39: 409–460.
5. Zignani, M., Drummond, D.C., Meyer, O., Hong, K., and Leroux, J.C., 2000. In vitro characterization of a novel polymeric-based pH-sensitive liposome system. *Biochim Biophys Acta*, 1463: 383–394.
6. Bikram, M. and West, J.L., 2008. Thermo-responsive systems for controlled drug delivery. *Expert Opin Drug Deliv*, 5: 1077–1091.
7. Han, H.D., Choi, M.S., Hwang, T., Song, C.K., Seong, H., Kim, T.W., Choi, H.S., and Shin, B.C., 2006. Hyperthermia-induced antitumor activity of thermosensitive polymer modified temperature-sensitive liposomes. *J Pharm Sci*, 95: 1909–1917.
8. Paasonen, L., Romberg, B., Storm, G., Yliperttula, M., Urtti, A., and Hennink, W.E., 2007. Temperature-sensitive poly(N-(2-hydroxypropyl)methacrylamide mono/dilactate)-coated liposomes for triggered contents release. *Bioconjug Chem*, 18: 2131–2136.
9. Ponce, A.M., Vujaskovic, Z., Yuan, F., Needham, D., and Dewhirst, M.W., 2006. Hyperthermia mediated liposomal drug delivery. *Int J Hyperthermia*, 22: 205–213.
10. Lasic, D. and Martin, F., 1995. *Stealth Liposomes*. CRC Press, Boca Raton, FL, pp. 103–114.
11. Lee, S.M., Chen, H., Dettmer, C.M., O'Halloran, T.V., and Nguyen, S.T., 2007. Polymer-caged lipsomes: A pH-responsive delivery system with high stability. *J Am Chem Soc*, 129: 15096–15097.
12. Allen, C., Dos Santos, N., Gallagher, R., Chiu, G.N., Shu, Y., Li, W.M., Johnstone, S.A. et al. 2002. Controlling the physical behavior and biological performance of liposome formulations through use of surface grafted poly(ethylene glycol). *Biosci Rep*, 22: 225–250.
13. Blume, G. and Cevc, G., 1990. Liposomes for the sustained drug release in vivo. *Biochim Biophys Acta*, 1029: 91–97.
14. Hashizaki, K., Taguchi, H., Itoh, C., Sakai, H., Abe, M., Saito, Y., and Ogawa, N., 2003. Effects of poly(ethylene glycol) (PEG) chain length of PEG-lipid on the permeability of liposomal bilayer membranes. *Chem Pharm Bull (Tokyo)*, 51: 815–820.
15. Sadzuka, Y., Kishi, K., Hirota, S., and Sonobe, T., 2003. Effect of polyethyleneglycol (PEG) chain on cell uptake of PEG-modified liposomes. *J Liposome Res*, 13: 157–172.
16. Woodle, M.C. and Lasic, D.D., 1992. Sterically stabilized liposomes. *Biochim Biophys Acta*, 1113: 171–199.
17. Soussan, E., Cassel, S., Blanzat, M., and Rico-Lattes, I., 2009. Drug delivery by soft matter: Matrix and vesicular carriers. *Angew Chem Int Ed Engl*, 48: 274–288.
18. Tekade, R., Kumar, P., and Jain, N., 2009. Dendrimeres in oncology: An expanding horizon. *Chem Rev*, 109: 49–87.
19. Bielinska, A.U., Chen, C., Johnson, J., and Baker, J.R., Jr., 1999. DNA complexing with polyamidoamine dendrimers: Implications for transfection. *Bioconjug Chem*, 10: 843–850.

20. Kukowska-Latallo, J.F., Chen, C., Eichman, J., Bielinska, A.U., and Baker, J.R., Jr., 1999. Enhancement of dendrimer-mediated transfection using synthetic lung surfactant exosurf neonatal in vitro. *Biochem Biophys Res Commun*, 264: 253–261.

21. Delong, R., Stephenson, K., Loftus, T., Fisher, M., Alahari, S., Nolting, A., and Juliano, R.L., 1997. Characterization of complexes of oligonucleotides with polyamidoamine starburst dendrimers and effects on intracellular delivery. *J Pharm Sci*, 86: 762–764.

22. Ooya, T., Lee, J., and Park, K., 2004. Hydrotropic dendrimers of generations 4 and 5: Synthesis, characterization, and hydrotropic solubilization of paclitaxel. *Bioconjug Chem*, 15: 1221–1229.

23. Kolhe, P., Misra, E., Kannan, R.M., Kannan, S., and Lieh-Lai, M., 2003. Drug complexation, in vitro release and cellular entry of dendrimers and hyperbranched polymers. *Int J Pharm*, 259: 143–160.

24. Kannan, S., Kolhe, P., Raykova, V., Glibatec, M., Kannan, R.M., Lieh-Lai, M., and Bassett, D., 2004. Dynamics of cellular entry and drug delivery by dendritic polymers into human lung epithelial carcinoma cells. *J Biomater Sci Polym Ed*, 15: 311–330.

25. Khopade, A.J. and Caruso, F., 2002. Stepwise self-assembled poly(amidoamine) dendrimer and poly(styrenesulfonate) microcapsules as sustained delivery vehicles. *Biomacromolecules*, 3: 1154–1162.

26. Kolhe, P., Khandare, J., Pillai, O., Kannan, S., Lieh-Lai, M., and Kannan, R.M., 2006. Preparation, cellular transport, and activity of polyamidoamine-based dendritic nanodevices with a high drug payload. *Biomaterials*, 27: 660–669.

27. Al-Tahami, K. and Singh, J., 2007. Smart polymer based delivery systems for peptides and proteins. *Recent Pat Drug Deliv Formul*, 1: 65–71.

28. El-Sayed, M.E., Hoffman, A.S., and Stayton, P.S., 2005. Smart polymeric carriers for enhanced intracellular delivery of therapeutic macromolecules. *Expert Opin Biol Ther*, 5: 23–32.

29. Fogueri, L.R. and Singh, S., 2009. Smart polymers for controlled delivery of proteins and peptides: A review of patents. *Recent Pat Drug Deliv Formul*, 3: 40–48.

30. Stayton, P.S., El-Sayed, M.E., Murthy, N., Bulmus, V., Lackey, C., Cheung, C., and Hoffman, A.S., 2005. 'Smart' delivery systems for biomolecular therapeutics. *Orthod Craniofac Res*, 8: 219–225.

31. Traitel, T., Goldbart, R., and Kost, J., 2008. Smart polymers for responsive drug-delivery systems. *J Biomater Sci Polym Ed*, 19: 755–767.

32. Williams, D., 2005. Environmentally smart polymers. *Med Device Technol*, 16: 9–10, 13.

33. Gupta, M.N., Kaul, R., Guoqiang, D., Dissing, U., and Mattiasson, B., 1996. Affinity precipitation of proteins. *J Mol Recognit*, 9: 356–359.

34. Roy, I. and Gupta, M.N., 2003. kappa-Carrageenan as a new smart macroaffinity ligand for the purification of pullulanase. *J Chromatogr A*, 998: 103–108.

35. Jeong, B., Kim, S.W., and Bae, Y.H., 2002. Thermosensitive sol–gel reversible hydrogels. *Adv Drug Deliv Rev*, 54: 37–51.

36. Park, T.G. and Hoffman, A.S., 1994. Estimation of temperature-dependent pore size in poly(*N*-isopropylacrylamide) hydrogel beads. *Biotechnol Prog*, 10: 82–86.

37. Kikuchi, A. and Okano, T., 2002. Pulsatile drug release control using hydrogels. *Adv Drug Deliv Rev*, 54: 53–77.

38. Mahkam, M., 2005. Using pH-sensitive hydrogels containing cubane as a crosslinking agent for oral delivery of insulin. *J Biomed Mater Res B Appl Biomater*, 75: 108–112.

39. Nakamura, K., Murray, R.J., Joseph, J.I., Peppas, N.A., Morishita, M., and Lowman, A.M., 2004. Oral insulin delivery using P(MAA-*g*-EG) hydrogels: Effects of network morphology on insulin delivery characteristics. *J Control Release*, 95: 589–599.

40. Zhang, Y. and Chu, C.C., 2002. In vitro release behavior of insulin from biodegradable hybrid hydrogel networks of polysaccharide and synthetic biodegradable polyester. *J Biomater Appl*, 16: 305–325.

41. Goldraich, M. and Kost, J., 1993. Glucose-sensitive polymeric matrices for controlled drug delivery. *Clin Mater*, 13: 135–142.

42. Roy, I. and Gupta, M.N., 2003. Smart polymeric materials: Emerging biochemical applications. *Chem Biol*, 10: 1161–1171.

43. Douglas, T. and Young, M., 2006. Viruses: Making friends with old foes. *Science*, 312: 873–875.

44. Singh, P., Prasuhn, D., Yeh, R.M., Destito, G., Rae, C.S., Osborn, K., Finn, M.G., and Manchester, M., 2007. Bio-distribution, toxicity and pathology of cowpea mosaic virus nanoparticles in vivo. *J Control Release*, 120: 41–50.

45. Noad, R. and Roy, P., 2003. Virus-like particles as immunogens. *Trends Microbiol*, 11: 438–444.

46. Cole, C., 1996. Polyomaviridae: The viruses and their replication. *Fundamental Virology*, ed. B.N. Fields. Lippincott-Raven, Philadelphia, PA, pp. 917–944.

47. Phelps, D.K., Speelman, B., and Post, C.B., 2000. Theoretical studies of viral capsid proteins. *Curr Opin Struct Biol*, 10: 170–173.
48. Gibson, W., 1974. Polyoma virus proteins: A description of the structural proteins of the virion based on polyacrylamide gel electrophoresis and peptide analysis. *Virology*, 62: 319–336.
49. Melnick, J.L., Allison, A.C., Butel, J.S., Eckhart, W., Eddy, B.E., Kit, S., Levine, A.J. et al. 1974. Papovaviridae. *Intervirology*, 3: 106–120.
50. Liddington, R.C., Yan, Y., Moulai, J., Sahli, R., Benjamin, T.L., and Harrison, S.C., 1991. Structure of simian virus 40 at 3.8-Å resolution. *Nature*, 354: 278–284.
51. Neu, U., Woellner, K., Gauglitz, G., and Stehle, T., 2008. Structural basis of GM1 ganglioside recognition by simian virus 40. *Proc Natl Acad Sci USA*, 105: 5219–5224.
52. Padgett, B.L., Walker, D.L., ZuRhein, G.M., Eckroade, R.J., and Dessel, B.H., 1971. Cultivation of papova-like virus from human brain with progressive multifocal leucoencephalopathy. *Lancet*, 1: 1257–1260.
53. Goldmann, C., Petry, H., Frye, S., Ast, O., Ebitsch, S., Jentsch, K.D., Kaup, F.J. et al. 1999. Molecular cloning and expression of major structural protein VP1 of the human polyomavirus JC virus: Formation of virus-like particles useful for immunological and therapeutic studies. *J Virol*, 73: 4465–4469.
54. Gardner, S.D., Field, A.M., Coleman, D.V., and Hulme, B., 1971. New human papovavirus (B.K.) isolated from urine after renal transplantation. *Lancet*, 1: 1253–1257.
55. Comoli, P., Hirsch, H.H., and Ginevri, F., 2008. Cellular immune responses to BK virus. *Curr Opin Organ Transplant*, 13: 569–574.
56. Sweet, B.H. and Hilleman, M.R., 1960. The vacuolating virus, S.V. 40. *Proc Soc Exp Biol Med*, 105: 420–427.
57. Girardi, A.J., Sweet, B.H., Slotnick, V.B., and Hilleman, M.R., 1962. Development of tumors in hamsters inoculated in the neonatal period with vacuolating virus, SV-40. *Proc Soc Exp Biol Med*, 109: 649–660.
58. Carroll-Pankhurst, C., Engels, E.A., Strickler, H.D., Goedert, J.J., Wagner, J., and Mortimer, E.A., Jr., 2001. Thirty-five year mortality following receipt of SV40-contaminated polio vaccine during the neonatal period. *Br J Cancer*, 85: 1295–1297.
59. Schwartz, R., Garcea, R.L., and Berger, B., 2000. "Local rules" theory applied to polyomavirus polymorphic capsid assemblies. *Virology*, 268: 461–470.
60. Sandalon, Z. and Oppenheim, A., 1997. Self-assembly and protein-protein interactions between the SV40 capsid proteins produced in insect cells. *Virology*, 237: 414–421.
61. Palkova, Z., Spanielova, H., Gottifredi, V., Hollanderova, D., Forstova, J., and Amati, P., 2000. The polyomavirus major capsid protein VP1 interacts with the nuclear matrix regulatory protein YY1. *FEBS Lett*, 467: 359–364.
62. Georgens, C., Weyermann, J., and Zimmer, A., 2005. Recombinant virus like particles as drug delivery system. *Curr Pharm Biotechnol*, 6: 49–55.
63. Salunke, D.M., Caspar, D.L., and Garcea, R.L., 1986. Self-assembly of purified polyomavirus capsid protein VP1. *Cell*, 46: 895–904.
64. Barouch, D.H. and Harrison, S.C., 1994. Interactions among the major and minor coat proteins of polyomavirus. *J Virol*, 68: 3982–3989.
65. Chen, X.S., Stehle, T., and Harrison, S.C., 1998. Interaction of polyomavirus internal protein VP2 with the major capsid protein VP1 and implications for participation of VP2 in viral entry. *EMBO J*, 17: 3233–3240.
66. Ou, W.C., Hseu, T.H., Wang, M., Chang, H., and Chang, D., 2001. Identification of a DNA encapsidation sequence for human polyomavirus pseudovirion formation. *J Med Virol*, 64: 366–373.
67. Li, T.C., Takeda, N., Kato, K., Nilsson, J., Xing, L., Haag, L., Cheng, R.H., and Miyamura, T., 2003. Characterization of self-assembled virus-like particles of human polyomavirus BK generated by recombinant baculoviruses. *Virology*, 311: 115–124.
68. Ogris, E., Mudrak, I., and Wintersberger, E., 1992. Polyomavirus large and small T antigens cooperate in induction of the S phase in serum-starved 3T3 mouse fibroblasts. *J Virol*, 66: 53–61.
69. Forstova, J., Krauzewicz, N., Wallace, S., Street, A.J., Dilworth, S.M., Beard, S., and Griffin, B.E., 1993. Cooperation of structural proteins during late events in the life cycle of polyomavirus. *J Virol*, 67: 1405–1413.
70. Bolen, J.B., Anders, D.G., Trempy, J., and Consigli, R.A., 1981. Differences in the subpopulations of the structural proteins of polyoma virions and capsids: Biological functions of the multiple VP1 species. *J Virol*, 37: 80–91.
71. Gillock, E.T., An, K., and Consigli, R.A., 1998. Truncation of the nuclear localization signal of polyomavirus VP1 results in a loss of DNA packaging when expressed in the baculovirus system. *Virus Res*, 58: 149–160.

72. Chang, D., Haynes, J.I., 2nd, Brady, J.N., and Consigli, R.A., 1992. Identification of a nuclear localization sequence in the polyomavirus capsid protein VP2. *Virology*, 191: 978–983.

73. Chang, D., Haynes, J.I., 2nd, Brady, J.N., and Consigli, R.A., 1992. The use of additive and subtractive approaches to examine the nuclear localization sequence of the polyomavirus major capsid protein VP1. *Virology*, 189: 821–827.

74. Chang, D., Haynes, J.I., Jr., Brady, J.N., and Consigli, R.A., 1993. Identification of amino acid sequences in the polyomavirus capsid proteins that serve as nuclear localization signals. *Trans Kans Acad Sci*, 96: 35–39.

75. Mattern, C.F., Takemoto, K.K., and Daniel, W.A., 1966. Replication of polyoma virus in mouse embryo cells: Electron microscopic observations. *Virology*, 30: 242–256.

76. Mattern, C.F., Takemoto, K.K., and DeLeva, A.M., 1967. Electron microscopic observations on multiple polyoma virus-related particles. *Virology*, 32: 378–392.

77. Mackay, R.L. and Consigli, R.A., 1976. Early events in polyoma virus infection: Attachment, penetration, and nuclear entry. *J Virol*, 19: 620–636.

78. Bolen, J.B. and Consigli, R.A., 1979. Differential adsorption of polyoma virions and capsids to mouse kidney cells and guinea pig erythrocytes. *J Virol*, 32: 679–683.

79. Varki, A., 2008. Sialic acids in human health and disease. *Trends Mol Med*, 14: 351–360.

80. Clayson, E.T., Brando, L.V., and Compans, R.W., 1989. Release of simian virus 40 virions from epithelial cells is polarized and occurs without cell lysis. *J Virol*, 63: 2278–2288.

81. Clayson, E.T. and Compans, R.W., 1989. Characterization of simian virus 40 receptor moieties on the surfaces of Vero C1008 cells. *J Virol*, 63: 1095–1100.

82. Nagata, Y., Yamashiro, S., Yodoi, J., Lloyd, K.O., Shiku, H., and Furukawa, K., 1992. Expression cloning of beta 1,4 *N*-acetylgalactosaminyltransferase cDNAs that determine the expression of GM2 and GD2 gangliosides. *J Biol Chem*, 267: 12082–12089.

83. Stehle, T., Yan, Y., Benjamin, T.L., and Harrison, S.C., 1994. Structure of murine polyomavirus complexed with an oligosaccharide receptor fragment. *Nature*, 369: 160–163.

84. Keppler, O.T., Stehling, P., Herrmann, M., Kayser, H., Grunow, D., Reutter, W., and Pawlita, M., 1995. Biosynthetic modulation of sialic acid-dependent virus–receptor interactions of two primate polyoma viruses. *J Biol Chem*, 270: 1308–1314.

85. Shi, W.X., Chammas, R., and Varki, A., 1996. Linkage-specific action of endogenous sialic acid *O*-acetyltransferase in Chinese hamster ovary cells. *J Biol Chem*, 271: 15130–15138.

86. Caruso, M., Belloni, L., Sthandier, O., Amati, P., and Garcia, M.I., 2003. Alpha4beta1 integrin acts as a cell receptor for murine polyomavirus at the postattachment level. *J Virol*, 77: 3913–3921.

87. Tsai, B., Gilbert, J.M., Stehle, T., Lencer, W., Benjamin, T.L., and Rapoport, T.A., 2003. Gangliosides are receptors for murine polyoma virus and SV40. *EMBO J*, 22: 4346–4355.

88. Neu, U., Stehle, T., and Atwood, W.J., 2009. The Polyomaviridae: Contributions of virus structure to our understanding of virus receptors and infectious entry. *Virology*, 384: 389–399.

89. Hale, A.D., Bartkeviciute, D., Dargeviciute, A., Jin, L., Knowles, W., Staniulis, J., Brown, D.W., and Sasnauskas, K., 2002. Expression and antigenic characterization of the major capsid proteins of human polyomaviruses BK and JC in *Saccharomyces cerevisiae*. *J Virol Methods*, 104: 93–98.

90. Li, M., Cripe, T.P., Estes, P.A., Lyon, M.K., Rose, R.C., and Garcea, R.L., 1997. Expression of the human papillomavirus type 11 L1 capsid protein in *Escherichia coli*: Characterization of protein domains involved in DNA binding and capsid assembly. *J Virol*, 71: 2988–2995.

91. Rodgers, R.E., Chang, D., Cai, X., and Consigli, R.A., 1994. Purification of recombinant budgerigar fledgling disease virus VP1 capsid protein and its ability for in vitro capsid assembly. *J Virol*, 68: 3386–3390.

92. An, K., Gillock, E.T., Sweat, J.A., Reeves, W.M., and Consigli, R.A., 1999. Use of the baculovirus system to assemble polyomavirus capsid-like particles with different polyomavirus structural proteins: Analysis of the recombinant assembled capsid-like particles. *J Gen Virol*, 80(Pt 4): 1009–1016.

93. Chen, P.L., Wang, M., Ou, W.C., Lii, C.K., Chen, L.S., and Chang, D., 2001. Disulfide bonds stabilize JC virus capsid-like structure by protecting calcium ions from chelation. *FEBS Lett*, 500: 109–113.

94. Leavitt, A.D., Roberts, T.M., and Garcea, R.L., 1985. Polyoma virus major capsid protein, VP1. Purification after high level expression in *Escherichia coli*. *J Biol Chem*, 260: 12803–12809.

95. Gillock, E.T., Rottinghaus, S., Chang, D., Cai, X., Smiley, S.A., An, K., and Consigli, R.A., 1997. Polyomavirus major capsid protein VP1 is capable of packaging cellular DNA when expressed in the baculovirus system. *J Virol*, 71: 2857–2865.

96. Chang, D., Fung, C.Y., Ou, W.C., Chao, P.C., Li, S.Y., Wang, M., Huang, Y.L., Tzeng, T.Y., and Tsai, R.T., 1997. Self-assembly of the JC virus major capsid protein, VP1, expressed in insect cells. *J Gen Virol*, 78(Pt 6): 1435–1439.

97. Schmidt, U., Rudolph, R., and Bohm, G., 2000. Mechanism of assembly of recombinant murine polyomavirus-like particles. *J Virol*, 74: 1658–1662.

98. Smith, D.B. and Johnson, K.S., 1988. Single-step purification of polypeptides expressed in *Escherichia coli* as fusions with glutathione *S*-transferase. *Gene*, 67: 31–40.

99. Garcea, R.L., Salunke, D.M., and Caspar, D.L., 1987. Site-directed mutation affecting polyomavirus capsid self-assembly in vitro. *Nature*, 329: 86–87.

100. Stehle, T., Gamblin, S.J., Yan, Y., and Harrison, S.C., 1996. The structure of simian virus 40 refined at 3.1 Å resolution. *Structure*, 4: 165–182.

101. Braun, H., Boller, K., Lower, J., Bertling, W.M., and Zimmer, A., 1999. Oligonucleotide and plasmid DNA packaging into polyoma VP1 virus-like particles expressed in *Escherichia coli*. *Biotechnol Appl Biochem*, 29(Pt 1): 31–43.

102. Buck, C.B., Thompson, C.D., Pang, Y.Y., Lowy, D.R., and Schiller, J.T., 2005. Maturation of papillomavirus capsids. *J Virol*, 79: 2839–2846.

103. Caldeira, J.C. and Peabody, D.S., 2007. Stability and assembly in vitro of bacteriophage PP7 virus-like particles. *J Nanobiotechnology*, 5: 10.

104. Li, M., Beard, P., Estes, P.A., Lyon, M.K., and Garcea, R.L., 1998. Intercapsomeric disulfide bonds in papillomavirus assembly and disassembly. *J Virol*, 72: 2160–2167.

105. Barr, S.M., Keck, K., and Aposhian, H.V., 1979. Cell-free assembly of a polyoma-like particle from empty capsids and DNA. *Virology*, 96: 656–659.

106. Bertling, W.M., Gareis, M., Paspaleeva, V., Zimmer, A., Kreuter, J., Nurnberg, E., and Harrer, P., 1991. Use of liposomes, viral capsids, and nanoparticles as DNA carriers. *Biotechnol Appl Biochem*, 13: 390–405.

107. Soeda, E., Krauzewicz, N., Cox, C., Stokrova, J., Forstova, J., and Griffin, B.E., 1998. Enhancement by polylysine of transient, but not stable, expression of genes carried into cells by polyoma VP1 pseudocapsids. *Gene Ther*, 5: 1410–1419.

108. Hunger-Bertling, K., Harrer, P., and Bertling, W., 1990. Short DNA fragments induce site specific recombination in mammalian cells. *Mol Cell Biochem*, 92: 107–116.

109. Gareis, M., Harrer, P., and Bertling, W.M., 1991. Homologous recombination of exogenous DNA fragments with genomic DNA in somatic cells of mice. *Cell Mol Biol*, 37: 191–203.

110. Slilaty, S.N., Berns, K.I., and Aposhian, H.V., 1982. Polyoma-like particle: Characterization of the DNA encapsidated in vitro by polyoma empty capsids. *J Biol Chem*, 257: 6571–6575.

111. Stokrova, J., Palkova, Z., Fischer, L., Richterova, Z., Korb, J., Griffin, B.E., and Forstova, J., 1999. Interactions of heterologous DNA with polyomavirus major structural protein, VP1. *FEBS Lett*, 445: 119–125.

112. Willwand, K. and Kaaden, O.R., 1988. Capsid protein VP1 (p85) of Aleutian disease virus is a major DNA-binding protein. *Virology*, 166: 52–57.

113. Chang, D., Cai, X., and Consigli, R.A., 1993. Characterization of the DNA binding properties of polyomavirus capsid protein. *J Virol*, 67: 6327–6331.

114. Soussi, T., 1986. DNA-binding properties of the major structural protein of simian virus 40. *J Virol*, 59: 740–742.

115. Heidari, S., Krauzewicz, N., Kalantari, M., Vlastos, A., Griffin, B.E., and Dalianis, T., 2000. Persistence and tissue distribution of DNA in normal and immunodeficient mice inoculated with polyomavirus VP1 pseudocapsid complexes or polyomavirus. *J Virol*, 74: 11963–11965.

116. Goldmann, C., Stolte, N., Nisslein, T., Hunsmann, G., Luke, W., and Petry, H., 2000. Packaging of small molecules into VP1-virus-like particles of the human polyomavirus JC virus. *J Virol Methods*, 90: 85–90.

117. van Lent, J., Storms, M., van der Meer, F., Wellink, J., and Goldbach, R., 1991. Tubular structures involved in movement of cowpea mosaic virus are also formed in infected cowpea protoplasts. *J Gen Virol*, 72 (Pt 11): 2615–2623.

118. Porta, C., Spall, V.E., Loveland, J., Johnson, J.E., Barker, P.J., and Lomonossoff, G.P., 1994. Development of cowpea mosaic virus as a high-yielding system for the presentation of foreign peptides. *Virology*, 202: 949–955.

119. Khor, I.W., Lin, T., Langedijk, J.P., Johnson, J.E., and Manchester, M., 2002. Novel strategy for inhibiting viral entry by use of a cellular receptor-plant virus chimera. *J Virol*, 76: 4412–4419.

120. Wang, Q., Kaltgrad, E., Lin, T., Johnson, J.E., and Finn, M.G., 2002. Natural supramolecular building blocks. Wild-type cowpea mosaic virus. *Chem Biol*, 9: 805–811.

121. Wang, Q., Lin, T., Tang, L., Johnson, J.E., and Finn, M.G., 2002. Icosahedral virus particles as addressable nanoscale building blocks. *Angew Chem Int Ed Engl*, 41: 459–462.

122. Wang, Q., Lin, T., Johnson, J.E., and Finn, M.G., 2002. Natural supramolecular building blocks. Cysteine-added mutants of cowpea mosaic virus. *Chem Biol*, 9: 813–819.

123. Lin, T., Porta, C., Lomonossoff, G., and Johnson, J.E., 1996. Structure-based design of peptide presentation on a viral surface: The crystal structure of a plant/animal virus chimera at 2.8 Å resolution. *Fold Des*, 1: 179–187.

124. Porta, C., Spall, V.E., Lin, T., Johnson, J.E., and Lomonossoff, G.P., 1996. The development of cowpea mosaic virus as a potential source of novel vaccines. *Intervirology*, 39: 79–84.

125. Raja, K.S., Wang, Q., Gonzalez, M.J., Manchester, M., Johnson, J.E., and Finn, M.G., 2003. Hybrid virus–polymer materials. 1. Synthesis and properties of PEG-decorated cowpea mosaic virus. *Biomacromolecules*, 4: 472–476.

126. Sen Gupta, S., Kuzelka, J., Singh, P., Lewis, W.G., Manchester, M., and Finn, M.G., 2005. Accelerated bioorthogonal conjugation: A practical method for the ligation of diverse functional molecules to a polyvalent virus scaffold. *Bioconjug Chem*, 16: 1572–1579.

127. Gomme, P.T., McCann, K.B., and Bertolini, J., 2005. Transferrin: Structure, function and potential therapeutic actions. *Drug Discov Today*, 10: 267–273.

128. Qian, Z.M., Li, H., Sun, H., and Ho, K., 2002. Targeted drug delivery via the transferrin receptor-mediated endocytosis pathway. *Pharmacol Rev*, 54: 561–587.

129. Hogemann-Savellano, D., Bos, E., Blondet, C., Sato, F., Abe, T., Josephson, L., Weissleder, R. et al. 2003. The transferrin receptor: A potential molecular imaging marker for human cancer. *Neoplasia*, 5: 495–506.

130. Antony, A.C., 1996. Folate receptors. *Annu Rev Nutr*, 16: 501–521.

131. Stephenson, S.M., Low, P.S., and Lee, R.J., 2004. Folate receptor-mediated targeting of liposomal drugs to cancer cells. *Methods Enzymol*, 387: 33–50.

132. Weitman, S.D., Weinberg, A.G., Coney, L.R., Zurawski, V.R., Jennings, D.S., and Kamen, B.A., 1992. Cellular localization of the folate receptor: Potential role in drug toxicity and folate homeostasis. *Cancer Res*, 52: 6708–6711.

133. Weitman, S.D., Lark, R.H., Coney, L.R., Fort, D.W., Frasca, V., Zurawski, V.R., Jr., and Kamen, B.A., 1992. Distribution of the folate receptor GP38 in normal and malignant cell lines and tissues. *Cancer Res*, 52: 3396–3401.

134. Destito, G., Yeh, R., Rae, C.S., Finn, M.G., and Manchester, M., 2007. Folic acid-mediated targeting of cowpea mosaic virus particles to tumor cells. *Chem Biol*, 14: 1152–1162.

135. Shriver, L.P., Koudelka, K.J., and Manchester, M., 2009. Viral nanoparticles associate with regions of inflammation and blood brain barrier disruption during CNS infection. *J Neuroimmunol*, 211: 66–72.

136. Inoue, M., 2008. [Protection against uterine cervical cancer by HPV vaccines]. *Uirusu*, 58: 155–163.

137. Karanam, B., Jagu, S., Huh, W.K., and Roden, R.B., 2009. Developing vaccines against minor capsid antigen L2 to prevent papillomavirus infection. *Immunol Cell Biol*, 87: 287–299.

138. Harper, D.M., Franco, E.L., Wheeler, C., Ferris, D.G., Jenkins, D., Schuind, A., Zahaf, T. et al. 2004. Efficacy of a bivalent L1 virus-like particle vaccine in prevention of infection with human papillomavirus types 16 and 18 in young women: A randomised controlled trial. *Lancet*, 364: 1757–1765.

139. Harper, D.M., Franco, E.L., Wheeler, C.M., Moscicki, A.B., Romanowski, B., Roteli-Martins, C.M., Jenkins, D., Schuind, A., Costa Clemens, S.A., and Dubin, G., 2006. Sustained efficacy up to 4.5 years of a bivalent L1 virus-like particle vaccine against human papillomavirus types 16 and 18: Follow-up from a randomised control trial. *Lancet*, 367: 1247–1255.

140. Vlastos, A., Andreasson, K., Tegerstedt, K., Hollanderova, D., Heidari, S., Forstova, J., Ramqvist, T., and Dalianis, T., 2003. VP1 pseudocapsids, but not a glutathione-*S*-transferase VP1 fusion protein, prevent polyomavirus infection in a T-cell immune deficient experimental mouse model. *J Med Virol*, 70: 293–300.

141. Szomolanyi-Tsuda, E., Le, Q.P., Garcea, R.L., and Welsh, R.M., 1998. T-cell-independent immunoglobulin G responses in vivo are elicited by live-virus infection but not by immunization with viral proteins or virus-like particles. *J Virol*, 72: 6665–6670.

142. Gedvilaite, A., Frommel, C., Sasnauskas, K., Micheel, B., Ozel, M., Behrsing, O., Staniulis, J., Jandrig, B., Scherneck, S., and Ulrich, R., 2000. Formation of immunogenic virus-like particles by inserting epitopes into surface-exposed regions of hamster polyomavirus major capsid protein. *Virology*, 273: 21–35.

143. Beyer, T., Herrmann, M., Reiser, C., Bertling, W., and Hess, J., 2001. Bacterial carriers and virus-like-particles as antigen delivery devices: Role of dendritic cells in antigen presentation. *Curr Drug Targets Infect Disord*, 1: 287–302.

144. Rae, C.S., Khor, I.W., Wang, Q., Destito, G., Gonzalez, M.J., Singh, P., Thomas, D.M. et al. 2005. Systemic trafficking of plant virus nanoparticles in mice via the oral route. *Virology*, 343: 224–235.
145. Fisher, K., 2006. Striking out at disseminated metastases: The systemic delivery of oncolytic viruses. *Curr Opin Mol Ther*, 8: 301–313.
146. Peiser, L., Mukhopadhyay, S., and Gordon, S., 2002. Scavenger receptors in innate immunity. *Curr Opin Immunol*, 14: 123–128.
147. Henke, S., Rohmann, A., Bertling, W.M., Dingermann, T., and Zimmer, A., 2000. Enhanced in vitro oligonucleotide and plasmid DNA transport by VP1 virus-like particles. *Pharm Res*, 17: 1062–1070.
148. Stehle, T. and Harrison, S.C., 1997. High-resolution structure of a polyomavirus VP1–oligosaccharide complex: Implications for assembly and receptor binding. *EMBO J*, 16: 5139–5148.

21 Polymeric Biomaterials in Pulmonary Drug Delivery

Nicole A. Beinborn and Robert O. Williams III

CONTENTS

21.1 INTRODUCTION

Inhalation drug therapy has received increased interest in recent years as an attractive route of administration for both local and systemic effects. The advantages of pulmonary drug delivery include noninvasiveness, high surface area available for absorption, high blood flow, and avoidance of first-pass metabolism. However, the lungs are designed to protect the body against the environment making targeted drug delivery to the lungs challenging for pharmaceutical scientists (Azarmi et al. 2008; Rytting et al. 2008; Scheuch et al. 2006).

Due to the sensitivity of lung tissue, a limited number of excipients are currently approved by the Food and Drug Administration (FDA) for inhalation. The polymeric biomaterials discussed in this chapter have proven biocompatibility and biodegradability when delivered by other routes of administration; however, there is concern when delivering these materials to the lungs. Do they damage the epithelial barrier making the lungs more susceptible to environmental hazards? How long will they remain in the lungs? Will they adversely affect lung function? All of these questions must be addressed in preclinical and clinical studies before we will see polymeric biomaterials as excipients in FDA-approved products for inhalation.

The efficacy of pulmonary drug delivery depends on the efficiency of drug deposition in the airways; therefore, the structure and function of the lungs play an important role. Many books and articles have been written in regard to the structure and function of the respiratory system. For that reason, we do not intend to give an extensive review of this topic. Instead, Section 21.2

will present basic information regarding the structure of the lungs and how it will influence pulmonary drug delivery using polymeric biomaterials. Section 21.3 provides an extensive review of various polymeric biomaterials currently being investigated for use in pulmonary drug delivery.

21.2 STRUCTURE AND FUNCTION OF THE LUNGS

Human lungs are incredibly complex organs. They are part of the respiratory system that allows for inhalation and gas exchange. The respiratory system is divided into two parts: the upper and lower airways. The upper airways consist of the nose to the larynx while lower airways are composed of trachea and all parts of the lung (Figure 21.1) (Stocks and Hislop 2002). The lower airways can be further subdivided into the conducting zone and the respiratory zone. The conducting zone consists of the first 16 generations of airways. Starting at the trachea, a series of branching occurs, with each generation smaller in diameter and shorter in length but greater in number and surface area than the preceding generation. The respiratory zone consists of the respiratory bronchioles, alveolar ducts, and alveoli, all of which participate in gas exchange (Figure 21.2) (Adjei et al. 2007). Each bifurcation in the lungs provides increased resistance to the delivery of drug particles. Therefore, it is imperative that drug particles or droplets for inhalation have a mass median aerodynamic diameter (MMAD) between 1 and 5 μm with low polydispersity in order to reach the alveoli (Adjei et al. 2007; Altiere and Thompson 2007; Stocks and Hislop 2002).

The entire respiratory tract is lined with epithelial cells, designed to protect the body from the external environment and substances that may be present in the inhaled air. Table 21.1 provides a

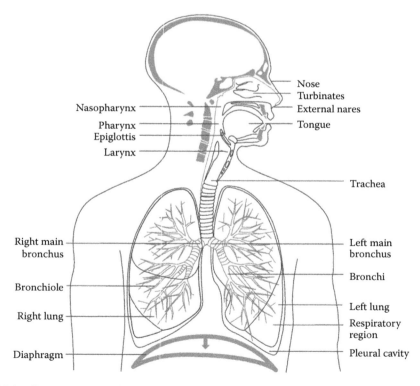

FIGURE 21.1 Gross structure of the respiratory system. (Reprinted from Stocks, J. and Hislop, A.A., Structure and function of the respiratory system: Developmental aspects and their relevance to aerosol therapy, in *Drug Delivery to the Lung*, H. Bisgaard, C. O'Callaghan, and G.C. Smaldone, eds., Vol. 162, Marcel Dekker, New York, 2002, pp. 47–104. With permission.)

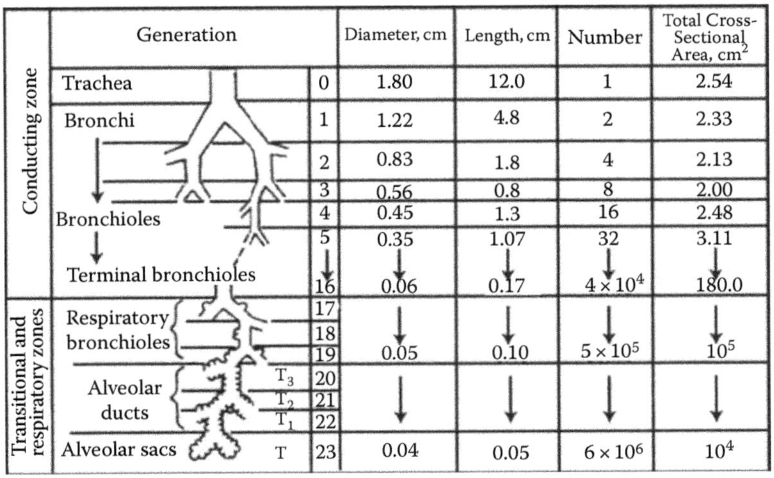

Generation		Diameter, cm	Length, cm	Number	Total Cross-Sectional Area, cm^2
Trachea	0	1.80	12.0	1	2.54
Bronchi	1	1.22	4.8	2	2.33
	2	0.83	1.8	4	2.13
	3	0.56	0.8	8	2.00
Bronchioles	4	0.45	1.3	16	2.48
	5	0.35	1.07	32	3.11
Terminal bronchioles	16	0.06	0.17	4×10^4	180.0
Respiratory bronchioles	17				
	18				
	19	0.05	0.10	5×10^5	10^5
Alveolar ducts	T$_3$ 20				
	T$_2$ 21				
	T$_1$ 22				
Alveolar sacs	T 23	0.04	0.05	6×10^6	10^4

Conducting zone (rows Trachea through Terminal bronchioles); Transitional and respiratory zones (rows Respiratory bronchioles through Alveolar sacs)

FIGURE 21.2 A schematic representation of airway branching in the human lung. (Reprinted from Adjei, A.L. et al., Bioavailability and pharmacokinetics of inhaled drugs, in *Inhalation Aerosols: Physical and Biological Basis for Therapy*, A.J. Hickey, ed., Vol. 221, Informa Healthcare, New York, 2007, pp. 187–218. With permission.)

TABLE 21.1
Airway Surface Epithelial Cells

Cell	Location	Principal Function
Airways		
Ciliated columnar	Trachea through respiratory bronchioles	Mucociliary escalator; glycoprotein secretion
Goblet	Trachea and bronchi	Mucus secretion; pro-genitor for ciliate cells
Basal[a]	Trachea and bronchi	Progenitor cells; aid in attachment of columnar cells to basement membrane
Clara	Bronchioles	Glycoprotein secretion; progenitor for ciliated and Clara cells; xenobiotic metabolism
Serous[b]	Sparse	Serous fluid secretion
Brush[c]	Sparse	Transition cell
Neuroendocrine (Klutschitsky; APUD)	Bronchi	Chemoreceptor, paracrine functions
Alveoli		
Type I pneumocyte		Gas exchange surface; fluid transport
Type II pneumocyte		Surfactant secretion; progenitor for type I cells
Type III pneumocyte[c] (alveolar brush)		Remains to be established

Source: Reprinted from Altiere, R.J. and Thompson, D.C., Physiology and pharmacology of the airways, in *Inhalation Aerosols: Physical and Biological Basis for Therapy*, A.J. Hickey, ed., Vol. 221, Informa Healthcare, New York, 2007, pp. 83–126. With permission.

[a] Function of basal cells not resolved.

[b] Found in fetal lung and several nonhuman mammalian species.

[c] Found in nonhuman mammalian species; rarely in human.

summary of the types of epithelial cells found in the airways, as well as their location and function (Altiere and Thompson 2007). The epithelial cell barrier must be overcome so that the drug can reach its site of action. Ciliated columnar epithelial cells are the major cells present from the trachea all the way down to the respiratory bronchioles. These cells make up what is known as the mucociliary escalator. Nonciliated goblet cells are present in the trachea and bronchi and nonciliated Clara cells are present in the bronchioles. Both are responsible for secreting a mucus-like substance. When particles deposit in tracheobronchial region, the cilia force mucus and the particles trapped in it upward until they are swallowed or expectorated. Some materials may affect the beat frequency of the cilia and therefore influence the rate of mucus clearance, which may be an important consideration for sustained-release action via pulmonary drug delivery to the airways. The alveolar surface (140–160 m² in surface area) is covered mostly by a single layer of type I cells. These cells are very thin with only 0.1–0.5 μm separating the air from the pulmonary capillary blood, although they have tight junctions that only permit penetration of substances less than 0.6 nm. Metabolically active type II cells are also present on the alveolar surface, which are responsible for both producing lung surfactant (10–20 nm in thickness) and type I cell renewal (Altiere and Thompson 2007; Stocks and Hislop 2002; Washington et al. 2001). Therefore, drug particles delivered to the alveoli must first dissolve in the lung surfactant before being able to pass through the epithelial cells to the systemic circulation.

Once particles reach the alveoli, the lung has yet another defense mechanism: alveolar macrophages. Alveolar macrophages reside in the lung surfactant and have a cell diameter of approximately 15–22 μm. Particles of 1–3 μm in diameter are readily phagocytosed, while larger particles and nanosized particles tend to evade alveolar macrophages due to their size (Yang et al. 2008). Once the foreign substance undergoes phagocytosis, the alveolar macrophage will migrate to the mucociliary escalator or the lymph tissue where it can be cleared from the respiratory tract (Altiere and Thompson 2007). The last major barrier for pulmonary drug delivery is metabolism. All metabolizing enzymes found in the liver are also present in the lungs (Bhat et al. 2007).

The pH of the lung fluid is approximately 7. Hence, it may be desirable for polymeric biomaterials intended for pulmonary drug delivery to be soluble or swellable at neutral pH. The soluble polymeric biomaterials discussed in this chapter are commonly used as absorption enhancers in the lungs while the swellable polymeric biomaterials are used primarily as mucoadhesive, sustained-release agents.

21.3 POLYMERIC BIOMATERIALS USED IN PULMONARY DRUG DELIVERY SYSTEMS

21.3.1 Poly(Lactic Acid-co-Glycolic Acid) and Polylactic Acid

Poly(lactic acid-co-glycolic acid) (PLGA) is a biodegradable, biocompatible copolymer consisting of lactic acid and glycolic acid monomers, while polylactic acid (PLA) is a biodegradable polymer composed only of lactic acid monomers. There has been significant preclinical research in recent years on the use of PLGA in pulmonary drug delivery formulations although PLGA is not yet approved by the FDA for inhalation. PLGA and PLA are currently being investigated for sustained-release delivery via the lungs. Formulations containing PLGA or PLA polymers may enable the dosing frequency of many drugs to be reduced, thus increasing patient compliance.

Suarez et al. produced PLGA microspheres containing rifampicin by spray drying that resulted in a longer residence time in the lungs. When the formulation was administered by insufflation to guinea pigs, 20 μg of drug was detected in the lungs after 72 h compared to only 6 μg with rifampicin alone. The increased concentration in the lungs from the PLGA microspheres proved to be superior in an animal model of tuberculosis as well. The microspheres could be administered by either insufflation with lactose as an excipient or nebulization of aqueous suspension (Suarez et al. 2001a,b). Taylor et al. engineered spray-dried, respirable particles (MMAD between 1 and 5 μm) comprised of ipratropium bromide and glycine. These particles were then spray-coated with PLA coatings of 1%, 5%, 10%, 15%, 30%, and 50% w/w to create sustained-release respirable particles. The pharmacodynamic effect of the PLA-coated ipratropium bromide particles was measured following dry powder insufflation in guinea pigs. The onset of drug action was 0.3 min for uncoated drug particles while the 5% and 30% w/w PLA-coated particles delayed the onset of action to 1.67 and 9.78 min, respectively. Furthermore, the duration of effect was increased from 11 min with uncoated drug particles to 13 and 54 min with 5% and 30% w/w PLA-coated particles, respectively (Taylor et al. 2006).

PLGA nanospheres containing insulin have been administered to guinea pigs via nebulization, resulting in significantly lower blood glucose levels over 48 h. Monodispersed nanospheres with a mean diameter of 400 nm were manufactured using an emulsion solvent diffusion method in water followed by lyophilization. The nanospheres were redispersed to form a suspension and aerosolized using an ultrasonic nebulizer producing a respirable fraction (droplet particle size < 7 μm) of approximately 70%. Sustained release of insulin in the lungs was achieved with the nebulized nanosphere suspension compared to nebulized aqueous solution of insulin (Kawashima et al. 1999). Pandey et al. investigated drug-loaded PLGA nanoparticles administered to guinea pigs via nebulization. The nanoparticle formulation contained rifampicin, isoniazid, and pyrazinamide for the treatment of tuberculosis. Pharmacokinetic analysis revealed an increase in bioavailability of 6.5, 19.1, and 13.4-fold for rifampicin, isoniazid, and pyrazinamide, respectively, compared to intravenous administration of the conventional drugs. The drugs were present at therapeutic concentrations in the lungs 10 days following a single dose of drug-loaded nanoparticles by nebulization, while the drugs were not detected in the lungs beyond 24 h after oral or aerosol administration of the conventional drugs. The efficacy against tuberculosis was equivalent for the nebulized drug-loaded nanoparticles administered once every 10 days for 6 weeks and the conventional oral formulation administered once daily for 46 days (Pandey et al. 2003). Similar sustained-release properties were observed with lectin-coated PLGA nanoparticles containing rifampicin, isoniazid, and pyrazinamide delivered to the lungs (Sharma et al. 2004). Surti et al. also reported an increase in bioavailability of PLGA and lectin-conjugated PLGA nanoparticles compared to free drug. The site enhancement factors were approximately four and six times that of the free drug for the PLGA formulation and the lectin-conjugated PLGA formulation, respectively (Surti et al. 2009).

Yamamoto et al. developed a dry powder inhaler formulation containing insulin-loaded PLGA nanospheres with sustained-release properties. The authors granulated the freeze-dried nanospheres with mannitol in a spray drying fluidized bed to form particles with acceptable inhalation properties (Yamamoto et al. 2007). Ohashi et al. used spray drying to produce rifampicin/PLGA nanoparticles, mean diameter of 213 nm, dispersed in mannitol microspheres with a mean diameter of 3.2 μm. A four-fluid nozzle was used to produce the formulation in one step. The mannitol microspheres containing rifampicin/PLGA nanoparticles were compared in vitro and in vivo to rifampicin/PLGA microspheres without mannitol (mean diameter = 2.1 μm). The in vitro aerosol characteristics using a Jethaler® inhalation device were synonymous with a fine particle fraction of approximately 35%; however, alveolar macrophage uptake was significantly greater for the formulation of mannitol microspheres containing rifampicin and PLGA. Since *Mycobacterium tuberculosis* is active in the alveolar macrophages, this formulation may be useful for targeted inhalation therapy of tuberculosis. The authors concluded that the mannitol quickly dissolved in the lungs, leaving the rifampicin/PLGA nanoparticles to disperse and deposit. The rifampicin/PLGA microspheres without mannitol were removed by mucociliary clearance but the rifampicin/PLGA nanoparticles were difficult to clear, resulting in increased lung retention and macrophage uptake (Ohashi et al. 2009). Similarly, Tomoda et al. formed nanocomposite particles for inhalation where trehalose dissolved quickly leaving

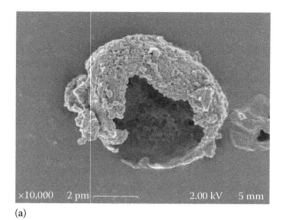

(a)

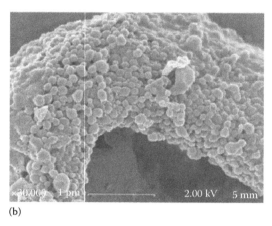

(b)

FIGURE 21.3 SEM image of spray-dried porous nanoparticle-aggregate formed from the PLGA nanoparticles. 10 × magnification (a) 30 × magnification (b). (With kind permission from Springer Science+Business Media: *Pharmaceut. Res.*, Formulation and pharmacokinetics of self-assembled rifampicin nanoparticle systems for pulmonary delivery, 26(8), 2009, 1847–1855, Sung, J.C., Padilla, D.J., Garcia-Contreras, L., VerBerkmoes, J.L., Durbin, D., Peloquin, C.A., Elbert, K.J., Hickey, A.J., and Edwards, D.A., Figure 1.)

behind the drug-loaded PLGA nanoparticles for uptake by epithelial cells (Tomoda et al. 2009). Sung et al. investigated porous nanoparticle-aggregate particles for lung delivery to guinea pigs. Rifampicin-containing PLGA nanoparticles were first prepared by an emulsion solvent evaporation technique. The nanoparticle suspension was then added to an aqueous leucine solution and spray dried to form aggregate particles with suitable aerosolization properties. Figure 21.3 shows the rifampicin-loaded PLGA nanoparticles incorporated into the spray-dried porous nanoparticle-aggregate particles. Concentrations of rifampicin in the lungs remained elevated for 8 h following insufflations of these porous nanoparticle-aggregate particles compared to porous particles without nanoparticles even though the porous particles without nanoparticles exhibited a higher fine particle fraction in vitro (68% versus approximately 40%). These results indicate the sustained-release properties of the PLGA contributed to the prolonged lung levels of rifampicin (Sung et al. 2009). Figures 21.4 and 21.5 are schematic representations of the fate of inhaled nanocomposite particles that represent the concept of all the formulations just described: dissolution and/or deaggregation of microparticles into primary nanoparticles with sustained action in lung tissue.

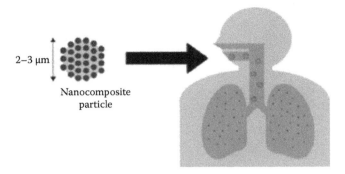

2–3 μm

Nanocomposite
particle

FIGURE 21.4 Fate of PLGA nanoparticles following inhalation of micron-sized nanocomposite particle. (Reprinted from *Colloids Surf. B Biointerfaces*, 71(2), Tomoda, K., Ohkoshi, T., Hirota, K., Sonavane, G.S., Nakajima, T., Terada, H., Komuro, M., Kitazato, K., and Makino, K., Preparation and properties of inhalable nanocomposite particles for treatment of lung cancer, 177–182. Copyright 2009, with permission from Elsevier.)

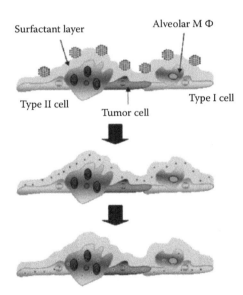

Surfactant layer

Alveolar M Φ

Type II cell

Tumor cell

Type I cell

FIGURE 21.5 Decomposition/dissolution of nanocomposite particle in lung lining fluid into PLGA nanoparticles followed by uptake of nanoparticles by epithelial cells and alveolar macrophages. (Reprinted from *Colloids Surf. B Biointerfaces*, 71(2), Tomoda, K., Ohkoshi, T., Hirota, K., Sonavane, G.S., Nakajima, T., Terada, H., Komuro, M., Kitazato, K., and Makino, K., Preparation and properties of inhalable nanocomposite particles for treatment of lung cancer, 177–182. Copyright 2009, with permission from Elsevier.)

PLGA has also been used to produce low-density, large porous particles for pulmonary delivery (Figure 21.6). Low-density, large porous particles can achieve higher respirable fractions with improved lung deposition. Furthermore, low-density, large porous particles, by virtue of their larger diameter, will evade alveolar macrophages, thus delivering another mechanism of sustained release in the lungs. Twelve milligrams of powder comprising large porous PLGA particles loaded into the Spinhaler® dry powder inhaler exhibited a fine particle fraction (MMAD < 4.7 μm) of 48%; a significant improvement over the fine particle fraction for small nonporous PLGA particles (16%). Animal studies were conducted to evaluate the large porous PLGA particles for both local and

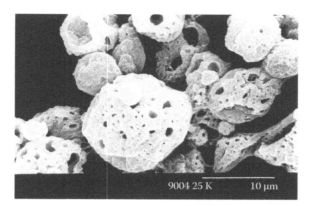

FIGURE 21.6 SEM image of low-density, large porous polymeric particles ($\rho = 0.1\,g/cm^3$ and d = 8.5 μm) produced by double-emulsion solvent evaporation technique. (Reprinted from Ben-Jebria, A. et al., *Aerosol Sci. Technol.*, 32(5), 421, 2000. Copyright 2000, Mount Laurel, NJ. With permission.)

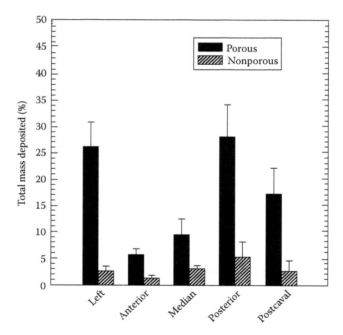

FIGURE 21.7 Particle mass deposited in different lobes of rat lungs. Large porous particles shown in Figure 21.6 compared to small nonporous particles (not shown). (Reprinted from Ben-Jebria, A. et al., *Aerosol Sci. Technol.*, 32(5), 421, 2000. Copyright 2000, Mount Laurel, NJ. With permission.)

systemic delivery in the lungs. Figure 21.7 shows the in vivo lung deposition of large porous particles compared to small nonporous particles. Systemic inhalation therapy with large porous particles containing PLGA and insulin or estradiol resulted in an extended period of absorption (approximately 100 h) and increased bioavailability compared to small nonporous PLGA particles. Local drug delivery of large porous PLGA particles comprising albuterol sulfate resulted in improved therapeutic effectiveness compared to small nonporous PLGA particles (Ben-Jebria et al. 2000). Koushik et al. reported similar outcomes with large porous PLGA particles containing deslorelin. The large porous PLGA particles achieved sustained release significantly longer than small conventional PLGA particles containing deslorelin and both formulations achieved sustained release significantly longer than deslorelin powder alone (Koushik et al. 2004). Edwards et al. determined that the size of large porous particles affected the bioavailability following inhalation. Testosterone was incorporated into large porous PLGA particles with two different mean diameters, 10 and 20 μm, and administered to rats as inhalation aerosols. The 20 μm diameter particles resulted in a plasma C_{max} of approximately 32 ng/mL, whereas, the 10 μm particles resulted in a plasma C_{max} of about 23 ng/mL (Edwards et al. 1997).

PLGA has been successful in achieving sustained drug delivery in the lungs in preclinical studies; however, PLGA does have some disadvantages. PLGA may cause some cytotoxicity and immunogenicity in lung tissue even though it has a long history of safety in humans via other routes of administration. Sivadas et al. compared the cytotoxicity of some of the biopolymers mentioned in this chapter on human airway epithelial Calu-3 cells. Polymeric microparticles of comparable size (aerodynamic diameter range of 1–3 μm) containing chitosan, alginate, PLGA, HPC, or sodium hyaluronate were produced by spray drying (Figure 21.8). PLGA was the only biopolymer to exhibit some cytotoxicity and immunogenicity (Sivadas et al. 2008). PLGA requires a long period for degradation (weeks to months), which may also decrease its safety in pulmonary drug delivery (Fu et al. 2002).

Talton et al. have reported a new technique to apply nano-thin coatings of PLGA or PLA onto dry powders for inhalation. Micronized corticosteroids, budesonide and triamcinolone acetonide, with a particle size of 1–5 μm were used. The pulsed laser technique successfully applied an ultrathin film of polymer coating (less than 1% by weight) onto the micronized drug substances. The in vitro aerosol characteristics were unchanged before and after the coating process. When tested for cytotoxicity using a murine macrophage cell line J774A.1, there was no difference in cell viability between the polymer-coated corticosteroid and the uncoated corticosteroid. Formulation of inhaled powders with these nano-thin polymer films may lead to sustained release with reduced polymer load and therefore reduced toxicity (Talton et al. 2000). Arya et al. showed that budesonide coated with a nano-thin film of PLA resulted in increased residence time in the lungs of neonatal rats following intratracheal instillation (Arya et al. 2006). Other researchers have developed a novel, fast-degrading PLGA derivative for pulmonary delivery: diethylaminopropylamine polyvinyl alcohol-grafted-poly(lactic-co-glycolic acid), also known as DEAPA-PVAL-g-PLGA (Oster et al. 2004). Figure 21.9 shows the structure of this PLGA derivative. Nanoparticles formed using the DEAPA-PVAL-g-PLGA biopolymer were less cytotoxic than PLGA nanoparticles in A549 cells (Dailey et al. 2006) and more stable during nebulization than PLGA nanoparticles (Dailey et al. 2003). In addition, nebulization of DEAPA-PVAL-g-PLGA nanoparticles loaded with 5(6)-carboxyfluorescein resulted in slower drug release and absorption compared to nebulization of 5(6)-carboxyfluorescein solution in an ex vivo lung model (Beck-Broichsitter et al. 2009).

21.3.2 Polyethylene Glycol

Polyethylene glycol (PEG) is a water-soluble polymer composed of repeating units of ethylene oxide. Polyethylene glycol is generally recognized as safe (GRAS) by the FDA and PEG 1000 is approved for inhalation at concentrations of 0.02% or less, according to the FDA's Inactive Ingredient Database (FDA CDER 2010). PEG 1000 is used as a lubricant in SYMBICORT® HFA Inhalation Aerosol drug product (AstraZeneca 2010). Inhalation of PEG has been shown to be relatively safe in preclinical studies. Klonne et al. evaluated the toxicity of PEG 3350 (20% w/w in water) in rats following aerosol exposure for 6 h per day, 5 days per week for 2 weeks. Only at the highest exposure level (1008 mg/m³) was there an observed increase in lung weight, serum neutrophils, and pulmonary macrophages. Overall, inhalation of PEG 3350 produced very little toxicity (Klonne et al. 1989).

Recently, PEG 2000 was used to form a PEGylated PAMAM dendrimer formulation for pulmonary delivery of low molecular weight heparin (LMWH). The pharmacokinetics and efficacy of LMWH-PEG-dendrimer micelle solution was compared to a plain LMWH solution following intratracheal administration to rats using a MicroSprayer®. The LMWH-PEG-dendrimer micelle

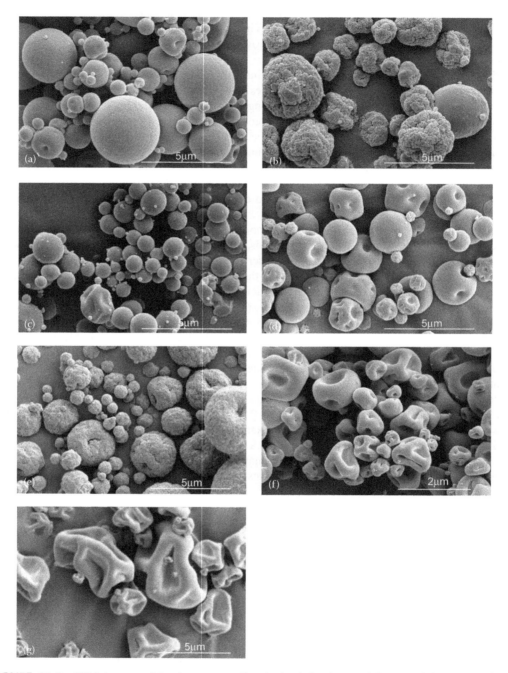

FIGURE 21.8 SEM images of bovine serum albumin loaded microparticles containing: (a) alginate; (b) chitosan; (c) PLGA; (d) gelatin; (e) sodium hyaluronate; (f) ovalbumin; and (g) HPC. Particles displayed very different morphology depending on the type of biopolymer used in the formulation. (Reprinted from *Int. J. Pharmaceut.*, 358(1–2), Sivadas, N., O'Rourke, D., Tobin, A., Buckley, V., Ramtoola, Z., Kelly, J.G., Hickey, A.J., and Cryan, S., A comparative study of a range of polymeric microspheres as potential carriers for the inhalation of proteins, 159–167. Copyright 2008, with permission from Elsevier.)

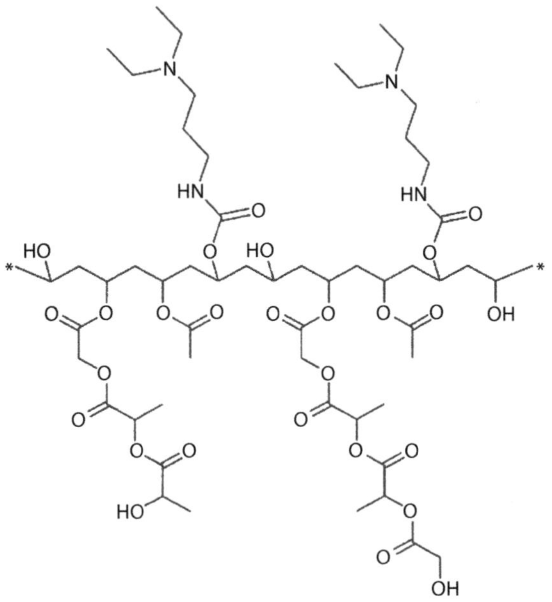

FIGURE 21.9 Structure of diethylaminopropylamine polyvinyl alcohol-grafted-poly(lactic-co-glycolic acid). (Reprinted from *Int. J. Pharmaceut.*, 367(1–2), Beck-Broichsitter, M., Gauss, J., Packhaeuser, C.B., Lahnstein, K., Schmehl, T., Seeger, W., Kissel, T., and Gessler, T., Pulmonary drug delivery with aerosolizable nanoparticles in an ex vivo lung model, 169–178. Copyright 2009, with permission from Elsevier.)

solution resulted in a slightly lower C_{max} and an equivalent t_{max} but the area-under-the-curve (AUC), bioavailability, and LMWH half-life were doubled, thus indicating sustained release. Furthermore, the LMWH-PEG-dendrimer micelle formulation administered every 48 h was as efficacious as the LMWH solution administered every 24 h (Bai and Ahsan 2009). Garcia-Contreras et al. compared the pharmacokinetic and pharmacodynamic effects of insulin-calcium phosphate-polyethylene glycol (insulin-CAP-PEG) particles in suspension and insulin solution administered to rats via intratracheal instillation and spray instillation. The bioavailability of the insulin-CAP-PEG particles was approximately twice that of the insulin solution. The sustained-release effect resulted in a longer duration of action evidenced by the extended hypoglycemic effect (Garcia-Contreras et al. 2003). Fu et al. investigated microparticulate aerosols composed of poly(sebacic anhydride-*co*-PEG). Microparticles with increased PEG content had lower densities, lower aerodynamic diameters, and faster degradation/dissolution rates. These polymers may be better than PLGA for pulmonary delivery since PLGA tends to degrade over a period of weeks leading to unwanted buildup of polymer in the lungs (Fu et al. 2002). Corrigan et al. co-spray-dried salbutamol sulfate with either PEG 4,000 or PEG 20,000 and evaluated the powders using a Rotahaler® DPI device. The powders were intended to avoid nonhomogeneity when mixing drug and carrier particles; however, all of the co-spray-dried powders studied (PEG concentrations of 5%, 20%, and 40%) resulted in low fine particle fractions (less than 10%) (Corrigan et al. 2006a).

PEG has also been investigated for use in carrier particles for dry powder inhalers. PEG 6000 with a mean particle size of 4 μm was blended with coarse lactose to obtain a 95% lactose/5% PEG 6000 blend. This formulation was used as carrier powder for bovine serum albumin-maltodextrin particles in a Rotahaler DPI device. The lactose/PEG blend increased the fine particle fraction of bovine serum albumin–maltodextrin particles from 24% to 34% compared to coarse lactose alone (Lucas et al. 1998). Gilani et al. spray-dried α-lactose monohydrate with different grades of PEG and compared the aerosolization properties of beclomethasone dipropionate from these carriers and Pharmatose® 325 M, a commercially available inhalation grade of lactose used in dry powder inhalers. The particle size of the spray- dried lactose/PEG (40:1) particles was dependent on the grade of PEG used in the formulation. The use of PEG 400 resulted in smaller particles ($d_{50\%}$ = 6.7 μm) than PEG 3000 ($d_{50\%}$ = 9.7 μm) and PEG 6000 ($d_{50\%}$ = 9.4 μm), all of which were smaller than Pharmatose 325 M ($d_{50\%}$ = 53.5 μm). All of the spray-dried lactose/PEG particles were spherical while the commercially available Pharmatose 325 M was irregularly shaped. The use of spray-dried lactose/PEG particles as carrier particles for beclomethasone dipropionate in the Spinhaler DPI device significantly improved the emitted dose and fine particle fraction compared to Pharmatose 325 M (emitted dose = 69%; FPF = 7%). Furthermore, carrier particles composed of PEG 3000 (emitted dose = 91%; FPF = 26%) or PEG 6000 (emitted dose = 92%; FPF = 25%) and lactose proved to be better than carrier particles composed of PEG 400 and lactose (emitted dose = 82%; FPF = 14%). The size of the carrier particles as well as the surface characteristics created good aerosolization from the DPI device and improved detachment of drug particles (Gilani et al. 2004).

21.3.3 Poly(Vinyl Pyrrolidone) and Poly(Vinyl Alcohol)

Poly(vinyl pyrrolidone) (PVP) is a 1-ethenyl-2-pyrrolidinone homopolymer and PVA is an ethanol homopolymer. Both are water-soluble polymers and available in various grades of increasing molecular weight and viscosity (Rowe et al. 2009). PVP K25 is approved for inhalation at concentrations of 0.0001% or less, according to the FDA's Inactive Ingredient Database, and is present in SYMBICORT HFA Inhalation Aerosol as a suspending agent (AstraZeneca 2010; FDA CDER 2010).

Researchers have investigated the use of PVP and PVA as stabilizing excipients in pressurized metered dose inhalers (pMDI) containing hydrofluoroalkane (HFA) propellants. Four formulations were spray dried to form drug microparticles: DNase I alone, DNase I with trehalose (1:1), DNase I with trehelose and PVA 80% hydrolyzed (1:1:1), and DNase I with trehalose, PVA 80% hydrolyzed, and PVP K15 (1:1:1:1). Only the formulation containing PVP was able to retain 100% biological activity of DNase I after the spray drying process. Additionally, the DNase I-trehalose-PVA-PVP formulation also produced the best physical stability in the HFA pMDI over a 24 week time period (Jones et al. 2006b). The authors performed a similar study with beclomethasone dipropionate. Spray-dried, polymer-coated drug particles were produced with PVA40, PVA 70, PVA80, low molecular weight PVA87-89, medium molecular weight PVA87-89, high molecular weight PVA87-89, PVA98, PVP K15, and PVP K90. All of the formulations yielded microparticles with a mean particle size of 3–5 μm except PVP K90 which resulted in particles larger than 5 μm. The combination of beclomethasone dipropionate, PVA 80, and PVP K15 showed the most improvement in physical stability of the inhalable microparticles in pMDIs containing HFA propellants. The stabilizing excipients decreased the amount of particle aggregation, and therefore increased the fine particle fraction emitted from the device. The fine particle fractions were 42% and 18% for the beclomethasone dipropionate-PVA 80-PVP K15 formulation and beclomethasone dipropionate without stabilizing excipients, respectively (Jones et al. 2006a).

A study by Buttini et al. demonstrated the capability of vinyl polymers, such as PVP and PVA, to modify the adhesion forces between budesonide and smooth lactose carrier particles in a dry powder inhaler. Polymer-coated budesonide microparticles were generated by spray drying. Fine particle fraction was increased from 29% to 53% when the budesonide particles were coated with 1% PVA80 or PVP K15 compared to unmodified budesonide particles with a similar particle size (Buttini et al. 2008). Liu et al. investigated the mucoadhesive properties of polymer-coated drug particles. Scutellarin was spray dried with PVP K15 (9:1), PVA80 (9:1), and PVP K15 and PVA80 (8:1:1). The particles containing scutellarin/PVP and scutellarin/PVA did not disperse well from a dry powder insufflator; however, the scutellarin/PVP/PVA formulation resulted in increased bioavailability compared to spray-dried scutellarin without excipients. Even though the particle sizes were similar, the fine particle fraction of the scutellarin/PVP/PVA formulation was much higher (61% versus 39%). In vitro mucociliary transport rates also indicated a significant difference between the scutellarin/PVP/PVA formulation (5 mm/min) and spray-dried scutellarin without excipients (16 mm/min). PVA80 showed greater mucoadhesiveness than PVP K15. Neither excipient exhibited cytotoxicity up to 5 mg/mL in Calu-3 and A549 lung cell lines (Liu et al. 2008).

21.3.4 HYDROXYPROPYL METHYL CELLULOSE, HYDROXYPROPYL METHYL CELLULOSE PHTHALATE, AND HYDROXYPROPYL CELLULOSE

HPMC is a semisynthetic polymer manufactured from the polymeric backbone of cellulose and is generally recognized as safe by the FDA. There are many commercially available grades of HPMC with different solution viscosities. Hydroxypropyl methyl cellulose (HPMC) is visco-elastic and water-soluble with sustained-release and mucoadhesive properties. Hydroxypropyl methyl cellulose phthalate (HPMCP) is a cellulose polymer in which some of the hydroxyl groups are replaced with methyl ethers, 2-hydroxypropyl ethers, or phthalyl esters. Several grades of HPMCP are commercially available; they are insoluble in acidic media but soluble above a certain pH. HPC is also a semisynthetic polymer made from cellulose that is generally recognized as safe by the FDA. There are many commercially available grades of HPC with different solution viscosities. HPC is soluble in cold water and it has sustained-release and mucoadhesive properties (Rowe et al. 2009).

In dry powder inhaler formulations, carrier particles such as lactose are often used to improve aerosolization of micronized drug substances from the dry powder inhalation device. Iida et al. investigated the effect of surface-coating these lactose carrier particles on aerosolization. Pharmatose® 200M was coated in a Wurster coater with an aqueous solution containing dissolved lactose (13%) and HPMC (2%). The lactose carrier particles were sieved into the range of 40–88 μm prior to blending with salbutamol sulfate–micronized drug substance. The surface-coating process resulted in more spherical particles with lower surface roughness and lower specific surface area. In vitro aerosolization testing using a Jethaler DPI device showed that surface-coating of lactose carrier particles resulted in a slight decrease in the amount of drug emitted from the device (94% versus 86%) but the respirable fraction of drug was significantly increased (15% versus 32%). The surface-coated lactose carrier particles had a lower adhesion force: fewer micronized drug particles adhered to the carrier particles during blending but those drug particles that did adhere were released more easily during aerosolization (Iida et al. 2005).

Westmeier et al. used an antisolvent precipitation process to produce drug particles containing both salmeterol xinafoate and fluticasone propionate for pulmonary delivery. A variety of biocompatible, polymeric excipients were tested for their influence on stabilization during the precipitation process, particle size, and aerosolization properties. A combination of 0.01% HPMC and 0.01% Polysorbate 80 with the two active ingredients produced the best particles for inhalation. When blended with carrier lactose, the fine particle fraction emitted from the Aerolizer® DPI device was 36%, an improvement over the commercially available DPI product Seretide® (fine particle fraction of approximately 20%) (Westmeier and Steckel 2008).

Nanospheres comprising hydroxypropylmethylcellulose phthalate (HPMCP-55) were prepared by emulsion solvent diffusion method in water and compared to PLGA nanospheres prepared by the same method. HPMCP-55 solubilizes above pH 5.5 while PLGA is insoluble in the lungs. The authors hypothesized that dissolution of the pH-dependent polymer, HPMCP-55, in the lungs could increase the viscosity of the mucus layer in the respiratory tract, decreasing the rate of clearance of the active drug substance in the lungs. Although the drug-loaded HPMCP-55 nanospheres produced a more effective therapeutic response in the lungs when administered intratracheally to rats, they were more cytotoxic to the A549 lung adenocarcinoma cell line than drug-loaded PLGA nanospheres (Yang et al. 2007).

Aerosol administration of microspheres containing various grades of HPC has demonstrated enhanced pulmonary adsorption in guinea pigs. Sakagami et al. studied fluorescein microspheres, with a mass median aerodynamic diameter of approximately 2 μm, containing HPC-H,

HPC-SL, HPC-L, or HPC-M at various ratios (1:1, 1:4, and 1:10). The 2% solution viscosities were 3–6, 6–10, 150–400, and 1000–4000 cps for the SL, L, M, and H grades of HPC, respectively. Fluorescein:HPC-H (1:4) microspheres resulted in the best bioavailability (88%) of all formulations tested. This was attributed to rapid dissolution due to amorphous drug substance and enhanced absorption due to mucoadhesion of the HPC microspheres. A drug to HPC ratio of 1:1 was not sufficient for mucoadhesion while a drug to HPC ratio of 1:10 did not further increase the bioavailability. The extent of pulmonary absorption was dependent on the polymer viscosity (H > M > L > SL) (Sakagami et al. 2001).

21.3.5 Chitosan

Chitosan is a polysaccharide composed of glucosamine and *N*-acetylglucosamine copolymers. It is formed by deacetylation of chitin, a natural polymer. Chitosan is mucoadhesive due to its cationic charge; it adheres to negatively charged mucosal surfaces. It is also biocompatible and biodegradable (Rowe et al. 2009).

Williams et al. investigated the use of chitosan microspheres to deliver therapeutic compounds from pMDI delivery systems. Chitosan microspheres containing fluorescein sodium with and without cross-linking excipients were prepared by spray drying. Noncross-linked chitosan microspheres and glutaraldehyde cross-linked chitosan microspheres exhibited a true density most similar to HFA 134a, the liquefied propellant; therefore, the authors expected the best suspension stability from these formulations. Both pMDI formulations had an acceptable respirable fraction of 18%, however, the noncross-linked chitosan microspheres are preferred due to a much higher drug-loading capacity (Williams et al. 1998).

Asada et al. formed spray-dried chitosan-theophylline particles with a mass mean aerodynamic diameter of 5 μm (Asada et al. 2004). Corrigan et al. formed spray-dried chitosan–salbutamol sulfate particles with a respirable fraction of approximately 30% using a Rotahaler DPI device (Corrigan et al. 2006b). Learoyd et al. produced spray-dried powders containing 4% w/w terbutaline, 36% w/w leucine as a dispersibility enhancer, 50% w/w chitosan as a sustained-release excipient, and 10% w/w lactose (Figure 21.10). A control formulation was used for comparison, which contained lactose in place of chitosan. When tested for aerosolization properties using a Spinhaler DPI device, all powders had mass median aerodynamic diameters between 1 and 3 μm with emitted doses greater than 90%. Figure 21.11 shows that the fine particle fraction decreased as the molecular weight (MW) of chitosan used in the formulation increased, but even spray-dried powder produced with the high MW chitosan had a fine particle fraction of 56%. The use of chitosan extended the duration of drug release in vitro: high MW > medium MW > low MW > control (Learoyd et al. 2008). All of these solid dispersions may be suitable for dry powder inhalers.

Grenha et al. used spray drying to produce mannitol microspheres containing insulin-loaded chitosan/tripolyphosphate nanoparticles for lung delivery. Microspheres prepared with mannitol/nanoparticle ratios of 80%/20% and 90%/10% resulted in dry powders with an aerodynamic diameter of 2–3 μm. in vitro testing showed quick dissolution of the mannitol releasing the chitosan/tripolyphosphate nanoparticles (Grenha et al. 2005). Confocal imaging of the formulations revealed that mannitol and the nanoparticles are homogenously distributed throughout the whole particle. The surface of the microspheres contained more mannitol than nanoparticles, as expected, due to the ratio of mannitol to nanoparticles in the formulation (Grenha et al. 2007b). Furthermore, the formulations exhibited low cytotoxicity in Calu-3 and A549 cells (Grenha et al. 2007a).

Chitosan microparticles and chitosan-coated PLGA microparticles of varying compositions were tested for nebulization efficiency using a compressor nebulizer system. For plain chitosan microparticles, nebulization efficiency decreased with increasing chitosan content due to increasing viscosity; however, nebulization efficiency increased with increasing chitosan content for chitosan-coated PLGA microparticles. The authors proposed that the chitosan stabilized the PLGA microparticles

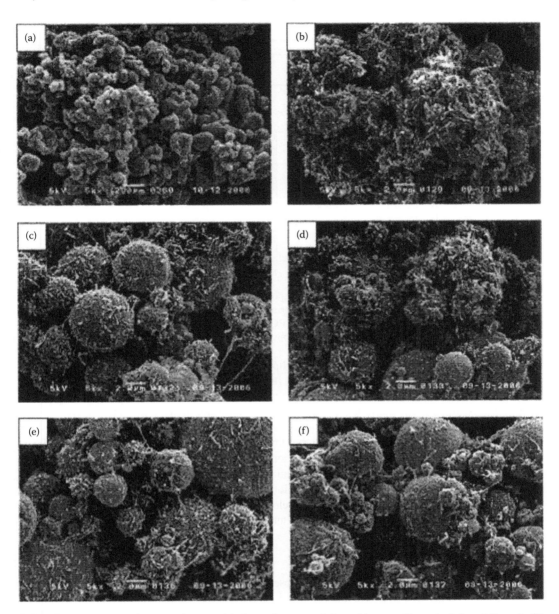

FIGURE 21.10 SEM images of spray-dried powders containing chitosan: (a) control (no chitosan), (b) low molecular weight (LMW) chitosan powder, (c) low/medium molecular weight (LMW/MMW) chitosan powder, (d) medium molecular weight (MMW) chitosan powder, (e) medium/high molecular weight (MMW/HMW) chitosan powder, and (f) high molecular weight (HMW) chitosan powder. Scale bar = 2 μm. (Reprinted from *Eur. J. Pharmaceut. Biopharmaceut.*, 68(2), Learoyd, T.P., Burrows, J.L., French, E., and Seville, P.C., Chitosan-based spray-dried respirable powders for sustained delivery of terbutaline sulfate, 224–234. Copyright 2008, with permission from Elsevier.)

during nebulization. Chitosan-coated PLGA microparticles also displayed better mucoadhesion properties than PLGA microparticles. As for cytotoxicity to human A549 alveolar cells, cell viability was affected by the particles in the following rank order: free drug > PLGA > chitosan-coated PLGA > chitosan (Manca et al. 2008). Zaru et al. investigated chitosan-coated liposomes for lung delivery via nebulization. Negatively charged liposomes were coated more efficiently with

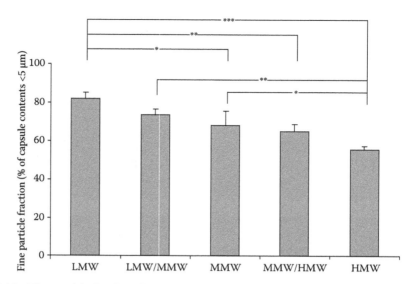

FIGURE 21.11 Fine particle fraction of spray-dried powders containing chitosan emitted from a Spinhaler®
DPI device. (*$p < 0.05$, **$p < 0.01$, ***$p < 0.001$.) (Reprinted from *Eur. J. Pharmaceut. Biopharmaceut.*, 68(2),
Learoyd, T.P., Burrows, J.L., French, E., and Seville, P.C., Chitosan-based spray-dried respirable powders for
sustained delivery of terbutaline sulfate, 224–234. Copyright 2008, with permission from Elsevier.)

the positively charged chitosan than uncharged liposomes, resulting in better mucoadhesive properties. All chitosan-coated liposomes displayed increased mucoadhesion compared to noncoated
liposomes (Zaru et al. 2009). Yamamoto showed that chitosan-surface-modified PLGA nanospheres
are eliminated from the lungs more slowly compared to unmodified PLGA nanospheres following nebulization to guinea pigs. The chitosan allowed the nanospheres to adhere to the mucus and
epithelial cell layer in the lungs, prolonging the time for drug absorption (Yamamoto et al. 2005).

Chitosan solution administered via nebulization to rats did not indicate pulmonary toxicity following a 70 mg/kg dose. Inflammatory and tissue stress biomarkers were essentially equivalent to
those exposed to distilled water or lactose solution. Chitosan may even display a protective effect
against oxidative stress, as indicated by the myeloperoxidase and glutathione disulfide biomarker
levels compared to distilled water and lactose solution (Valle et al. 2008).

21.3.6 CYCLODEXTRINS

Cyclodextrins are cyclic oligosaccharides composed of six to eight α-D-glucopyranose units linked
by 1,4 bonds. The general structure and the names and degree of substitution of various cyclodextrins are shown in Table 21.2. Cyclodextrins form complexes with drug molecules by partially or
completely encompassing the drug molecules in their hydrophobic core (Stella and He 2008).

In 1991, Marques et al. were the first to report the use of cyclodextrins for pulmonary delivery. Their first study investigated the absorption and pharmacokinetics of β-cyclodextrin,
dimethyl-β-cyclodextrin, and 2-hydroxypropyl-β-cyclodextrin following intratracheal instillation of
10 mg/kg doses in rabbits. While all three cyclodextrin derivatives resulted in similar bioavailability, 2-hydroxypropyl-β-cyclodextrin had a significantly longer mean absorption time (113 min versus 20–26 min for β-cyclodextrin and dimethyl-β-cyclodextrin) (Marques et al. 1991a). Following
dosing of salbutamol complexed with 2-hydroxypropyl-β-cyclodextrin via intratracheal instillation to rabbits, the time to reach C_{max} was increased (due to time for dissociation from the complex) compared to salbutamol alone; however, sustained release was not achieved. Bioavailability
of salbutamol from the 2-hydroxypropyl-β-cyclodextrin/salbutamol complex was 80% compared
to 100% for salbutamol alone (Marques et al. 1991b). A few years later, researchers began to

TABLE 21.2
Names and Structures of Various Cyclodextrins

General Structure

Name	R Group (May Have Differing Degrees of Substitution on the 2, 3, and 6 Positions)	n (# of α-D-Glucopyranose Units)
α-Cyclodextrin	H	6
Glucosyl-α-cyclodextrin	$C_6H_{11}O_5$ or H	6
Hydroxypropyl-α-cyclodextrin	$CH_2CH(OH)CH_3$ or H	6
Maltosyl-α-cyclodextrin	$C_{12}H_{21}O_{10}$ or H	6
β-Cyclodextrin	H	7
Dimethyl-β-cyclodextrin	CH_3 or H	7
Hydroxypropyl-β-cyclodextrin	$CH_2CH(OH)CH_3$ or H	7
Random methyl-β-cyclodextrin	CH_3 or H	7
Sulfobutylether-β-cyclodextrin	$(CH_2)_4SO_3Na$ or H	7
γ-Cyclodextrin	H	8
Hydroxypropyl-γ-cyclodextrin	$CH_2CH(OH)CH_3$ or H	8

investigate cyclodextrins as absorption enhancers in pulmonary drug delivery. Insulin solutions with and without cyclodextrins were administered via intratracheal instillation to rats. The relative efficacies of intratracheally administered insulin compared to an intravenous insulin dose were 8.2%, 10.5%, 16.9%, 23.9%, 34.7%, and 57.2% for insulin solution with no cyclodextrin, 5% hydroxypropyl-β-cyclodextrin, 5% γ-cyclodextrin, 1% β-cyclodextrin, 5% α-cyclodextrin, and 5% dimethyl-β-cyclodextrin, respectively (Shao et al. 1994). Hussain et al. also found that dimethyl-β-cyclodextrin enhanced pulmonary absorption of insulin following intratracheal administration with a MicroSprayer in rats (Hussain et al. 2003). Jalalipour et al. investigated dimethyl-β-cyclodextrin as both an absorption enhancer and a stabilizing agent for recombinant human growth hormone (rhGH) during spray drying. Particle aggregation was significantly reduced during spray drying when using increasing amounts of dimethyl-β-cyclodextrin; however, the aerosol characteristics varied greatly. Excipient-free rhGH exhibited a fine particle fraction of 33% while the addition of dimethyl-β-cyclodextrin at a molar ratio of 10:1 (dimethyl-β-cyclodextrin to rhGH) led to an increase in fine particle fraction to 53%. A further increase in dimethyl-β-cyclodextrin (100:1 and 1000:1) led to a decrease in fine particle fraction (33% and 23%, respectively). Interestingly, the formulations showed an absolute bioavailability of 29%, 25%, 77%, and 64% following dry powder inhalation of excipient-free rhGH, and powders containing 10:1, 100:1, and 1000:1 molar ratio of dimethyl-β-cyclodextrin to rhGH, respectively (Jalalipour et al. 2008b).

Fukaya et al. investigated four types of cyclodextrins with cyclosporine for dry powder inhalation: α-cyclodextrin, dimethyl-β-cyclodextrin, glucosyl-α-cyclodextrin, and maltosyl-α-cyclodextrin. Maltosyl-α-cyclodextrin was the most promising cyclodextrin for complexation with cyclosporine and dry powder inhalation due to very little inhibition of cilia, lower cytotoxicty, and good solubilization potential (80-fold increase in cyclosporine solubility). Aerosolized, micronized maltosyl-α-cyclodextrin/cyclosporine complex was approximately 10 times more effective than aerosolized, micronized cyclosporine in a mouse model of allergic asthma (Fukaya et al. 2003). In another study, doxycycline/hydroxypropyl-γ-cyclodextrin solution was nebulized to mice for the treatment of allergen-induced airway inflammation. The hydroxypropyl-γ-cyclodextrin was necessary to improve the solubility and stability of doxycycline in aqueous buffer solution. The authors found that exposure to aerosolized hydroxypropyl-γ-cyclodextrin had no effect on peribronchial inflammation (Gueders et al. 2008). Tolman et al. investigated the pharmacokinetics of an aerosolized solution containing voriconazole and sulfobutyl ether-β-cyclodextrin (Captisol®) in mice. The solubilization of voriconazole by sulfobutyl ether-β-cyclodextrin contributed to rapid absorption of the drug from the lungs to the systemic circulation (Tolman et al. 2009b).

Hydroxypropyl-β-cyclodextrin (HPβCD) has been used as an osmotic agent to form large porous particles containing PLGA and insulin. The use of HPβCD instead of salts in these formulations avoids insulin precipitation/aggregation/inactivation. Intratracheal delivery of the PLGA/HPβCD/insulin large porous particles via a dry powder insufflator confirmed the ability to achieve deep lung delivery. Fluorescence detection showed that insulin released from the particles and diffused into the alveolar tissues where it was absorbed (Ungaro et al. 2006, 2009). Dimethyl-β-cyclodextrin has been used to increase the solubility of insulin aqueous buffer solution leading to higher drug loading and increased stability in PLGA microspheres. The solubility of the sodium insulin in pH 7.4 phosphate-buffered saline was 48 mg/mL compared to 113 mg/mL for the 1:5 insulin:dimethyl-β-cyclodextrin complex. This resulted in PLGA microspheres with 15% drug loading compared to only 8% drug loading for uncomplexed insulin. Furthermore, the PLGA microspheres containing insulin:dimethyl-β-cyclodextrin complex had a longer and greater effect on plasma glucose levels in rats following intratracheal instillation. The sustained profile of insulin in the lungs was achieved due to the PLGA biodegradable polymer while the dimethyl-β-cyclodextrin complex probably enhanced the mucosal absorption of the insulin, increasing the bioavailability of insulin in the lungs (Aguiar et al. 2004; Rodrigues et al. 2003).

The safety of inhalation of cyclodextrins has been evaluated by various researchers. Lactate dehydrogenase levels in lung lavage fluid did not indicate any toxicity caused by pulmonary administration of 5% hydroxypropyl-β-cyclodextrin, 5% γ-cyclodextrin, 5% α-cyclodextrin, and 5% dimethyl-β-cyclodextrin solutions (Shao et al. 1994). The effects of 0.25% dimethyl-β-cyclodextrin solution were shown to be reversible following pulmonary administration since the respiratory epithelium returned to a normal state within 2 h after exposure (Hussain et al. 2003). Hence, acute lung exposure to cyclodextrins seems to promote drug absorption without long-term damage to lung tissues. Matilainen et al. studied the toxicity of α-cyclodextrin, β-cyclodextrin, γ-cyclodextrin, hydroxypropyl-α-cyclodextrin, hydroxypropyl-β-cyclodextrin, and randomly methylated-β-cyclodextrin on Calu-3 cells in vitro (Figure 21.12). β-Cyclodextrin and randomly methylated-β-cyclodextrin were the most cytotoxic

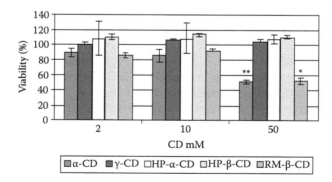

FIGURE 21.12 Viability of Calu-3 cells grown as cell layers after exposure to 2–50 mM α-cyclodextrin, γ-cyclodextrin, hydroxypropyl-α-cyclodextrin, hydroxypropyl-β-cyclodextrin, and randomly methylated-β-cyclodextrin. (*$p < 0.05$ and **$p < 0.01$ compared to control.) (Reprinted from *J. Control. Release*, 126(1), Matilainen, L., Toropainen, T., Vihola, H., Hirvonen, J., Järvinen, T., Jarho, P., and Järvinen, K., *In vitro* toxicity and permeation of cyclodextrins in Calu-3 cells, 10–16. Copyright 2008, with permission from Elsevier.)

while hydroxypropyl-α-cyclodextrin and hydroxypropyl-β-cyclodextrin were the safest toward the pulmonary cell line (Matilainen et al. 2008). Salem et al. compared the in vitro toxicity of a new cyclodextrin currently under development, Kleptose® Crysmeβ, to hydroxypropyl-γ-cyclodextrin, hydroxypropyl-β-cyclodextrin, and randomly methylated-β-cyclodextrin. The new cyclodextrin was more toxic to lung cells than hydroxypropyl-γ-cyclodextrin and hydroxypropyl-β-cyclodextrin but less toxic than randomly methylated-β-cyclodextrin (Salem et al. 2009). Evrard et al. evaluated the effect of nebulized solutions of hydroxypropyl-β-cyclodextrin, randomly methylated-β-cyclodextrin, or γ-cyclodextrin in mice. Three groups were exposed to the cyclodextrin aerosols for 30 min per day for 7 days via a whole body inhalation exposure chamber. A control group was exposed phosphate-buffered saline using the same procedure. There were no significant differences in lung histology, bronchial hyperresponsiveness, kidney histology, and blood urea levels for the cyclodextrin groups compared to the control group (Evrard et al. 2004). Tolman et al. evaluated an inhaled solution of Vfend®, which contained 100 mg/mL of sulfobutyl ether-β-cyclodextrin and 6.25 mg/mL of voriconazole. Rats were exposed to the aerosolized solution of voriconazole and sulfobutyl ether-β-cyclodextrin via a nose-only dosing chamber for 20 min twice daily for 21 days. Lung histopathology revealed an increased number of alveolar and respiratory duct macrophages, which resulted in an increased respiratory bronchiole index (RBI), but no evidence of an increased number of neutrophils, eosinophils, or lymphocytes compared to control. Furthermore, there was no evidence of ulceration, interstitial changes, or edema in the lungs and no significant changes in the liver, kidneys, or spleen compared to control (Tolman et al. 2009a).

A few clinical studies have involved the use of cyclodextrins for pulmonary delivery. In one pharmacokinetic study, healthy volunteers inhaled pMDI and DPI formulations containing FK224 and β-cyclodextrin (Nakate et al. 2003). Williamson et al. conducted a study in asthma patients, which compared the efficacy of Captisol-Enabled Budesonide Inhalation Solution twice daily for 2 weeks versus Pulmicort Respules® twice daily for 2 weeks. There were no significant differences in the outcome variables measured; however, a 120 μg dose of budesonide from the Captisol-Enabled Budesonide Inhalation Solution had comparable systemic exposure levels as a 500 μg dose of budesonide from the Pulmicort Respules (Williamson et al. 2009). The Captisol-Enabled Budesonide Inhalation Solution is currently in Phase II clinical trials according to Cydex Pharmaceuticals, Inc. Cydex is also expanding the regulatory safety data package for sulfobutyl ether-β-cyclodextrin (Captisol) with the FDA to include its use in pulmonary delivery (CyDex Pharmaceuticals 2009).

21.3.7 Hyaluronan and Hyaluronan Derivatives

Hyaluronan, also known as hyaluronic acid, is a naturally occurring biopolymer found in all tissues and body fluids, including the lungs. It is a linear polysaccharide composed of D-glucuronic acid and D-N-acetylglucosamine. Hyaluronan is mucoadhesive, dissolving slowly in water to form highly viscous solutions (Fraser et al. 1997; Surendrakumar et al. 2003).

Sodium hyaluronate has been shown to be a useful absorption enhancer for pulmonary delivery of peptide/protein drugs. Low-viscosity solutions of sodium hyaluronate containing insulin were administered to rats via intratracheal instillation. The effects of three different concentrations of sodium hyaluronate (0.05%, 0.1%, and 0.2% w/v) were studied and compared to control (insulin in aqueous solution without sodium hyaluronate). At physiological pH, 0.1% w/v sodium hyaluronate solution displayed the greatest absorption-enhancing effect of insulin, as measured by plasma glucose levels over 6 h. Pharmacological availability as a result of administration of the 0.05% w/v sodium hyaluronate solution was similar to control, and therefore, the level of sodium hyaluronate was not sufficient to significantly improve insulin absorption. Conversely, 0.2% w/v sodium hyaluronate solution was too viscous (14.3 cps compared to 4.6 cps for 0.1% w/w sodium hyaluronate solution) that insulin absorption was hindered due to increased time for diffusion of insulin through the formulation (Morimoto et al. 2001).

Sodium hyaluronate microparticles for pulmonary delivery have resulted in sustained release of insulin in dogs. Formulations were spray-dried to form amorphous, spherical microparticles with mass median aerodynamic diameters between 1 and 4 μm and administered via dry powder insufflation (Surendrakumar et al. 2003). Hyaluronan microspheres containing ofloxacin (HMO) achieved increased uptake by alveolar macrophages in vitro and high lung concentrations in rats. HMO (50% ofloxacin w/w) were prepared by spray drying and compared to spray-dried ofloxacin microspheres without hyaluronan (MO) and an ofloxacin solution (OS). An alveolar macrophage cell line phagocytosed significantly more drug from the HMO formulation than the MO and OS formulations due to increased mucoadhesiveness of the HMO formulation. This is beneficial since ofloxacin is used to treat pulmonary tuberculosis, which harbors in alveolar macrophages. Additionally, both microsphere formulations (HMO and MO) were administered to rats via dry powder insufflation while the solution (OS) was administered to rats intravenously and orally. Intratracheal administration of HMO resulted in significantly higher concentrations in the lungs and a lung–plasma partition coefficient of 13.1, which was 1.8-fold, 9.3-fold, and 10.8-fold greater than intratracheal administration of MO, oral administration of OS, and IV administration of OS, respectively. Greater lung accumulation and reduced systemic concentrations of drug indicate that pulmonary delivery of spray-dried hyaluronan microsphere-containing drug can increase the therapeutic effect of the drug while minimizing side effects (Hwang et al. 2008).

Aerosolized hyaluronic acid has also been investigated for therapeutic use in asthma patients. It was found to significantly reduce bronchial hyperreactivity when administered to asthma patients 30 min prior to exercise (Petrigni and Allegra 2006). Since hyaluronan is endogenous to the lungs and pulmonary administration of hyaluronan has shown therapeutic effects in the lungs, hyaluronan has a low risk of causing pulmonary toxicity (Petrigni and Allegra 2006; Rouse et al. 2007). Therefore, hyaluronan is a promising polymeric biomaterial for use in pulmonary delivery formulations.

21.3.8 ALGINATE AND CARRAGEENAN

Alginates and carrageenans are naturally occurring biopolymers extracted from seaweed. Alginates are composed of β-D-mannuronic acid and α-L-guluronic acid residues (Remminghorst and Rehm 2006). Carrageenans are composed of D-galactose residues linked by α-1,3 and β-1,4 bonds; they are classified according to substitutions that occur on free hydroxyl groups (Greer and Yaphe 1984).

Zahoor et al. produced sodium alginate nanoparticles containing three antitubercular drugs: isoniazid, pyrazinamide, and rifampicin. The nanoparticles had an average size of 235 nm with a drug to polymer ratio of 7.5:1. A compressor nebulizer system was used to aerosolize the nanoparticle suspension into droplets with a mass median aerodynamic diameter of 1.1 μm to guinea pigs. Following a single-dose nebulization of nanoparticles, all of the drugs were detected in the lungs and plasma for at least 10 days; whereas, all of the drugs were cleared from the animals within 24 h when free drugs were nebulized. In a chemotherapeutic model, only three doses (once every 15 days) of the nanoparticle suspension were required to achieve an equivalent therapeutic effect as daily doses of oral free drugs for 45 days, demonstrating the effectiveness of sodium alginate as a sustained-release biopolymer following pulmonary delivery (Zahoor et al. 2005).

Yamada et al. evaluated the use of sodium alginate and various grades of carrageenans as viscosity-increasing agents for theophylline and fluticasone propionate solutions administered intratracheally in rats. At a 2% w/v concentration in solution, sodium alginate was unable to prolong drug concentration in the lungs. However, 1% w/v iota-carrageenan was able to prolong theophylline concentrations in the lungs and serum and 0.5% w/v kappa-carrageenan was able to prolong fluticasone propionate concentrations in the lungs and serum resulting in increased bioavailability. Furthermore, 1% w/v solutions of iota- and kappa-carrageenans did not cause pulmonary inflammation in rats, measured by leukocyte cell levels, protein levels, and LDH enzymatic activity in the bronchoalveolar lavage fluid (Yamada et al. 2005).

21.3.9 GELATIN

Gelatin is a hydrolyzed form of collagen currently approved for use in oral and parenteral pharmaceutical products. It is not approved for inhalation other than for use as a capsule to hold dry powder for inhalation in some DPI devices (Rowe et al. 2009). The biodegradation of gelatin can be controlled by the degree of cross-linking (Tseng et al. 2008).

Yamada et al. incorporated 5% w/v gelatin into theophylline solutions for intratracheal administration in rats but there was no difference in pharmacokinetics compared to a solution without gelatin (Yamada et al. 2005). The same researchers had previously shown that 5% gelatin prolonged the absorption time of 5(6)-carboxyfluorescein solution from the lungs to the serum, although bioavailability was not increased (lower AUC) (Yamamoto et al. 2004).

Morimoto et al. investigated negatively and positively charged gelatin microspheres for pulmonary delivery of salmon calcitonin in rats. When administered as an intratracheal suspension, both the negatively and positively charged microspheres resulted in an increased pharmacodynamic effect compared to salmon calcitonin solution (Morimoto et al. 2000). Tseng et al. delivered gelatin nanoparticles containing cisplatin that were surface modified with epidermal growth factor to mice with induced tumors in pulmonary tissue. The functional groups present in gelatin (i.e., carboxyl, hydroxyl, and amino groups) allow for conjugations and surface modifications to enhance drug targeting. These nanoparticles were delivered via nebulization in a whole-body inhalation chamber. Significant accumulation of nanoparticles occurred in the lung tumors due to the targeted aerosol delivery. Additionally, the gelatin nanoparticles did not cause any pulmonary edema or neutrophil accumulation in the lungs compared to phosphate-buffered saline (Tseng et al. 2008, 2009). Brzoska et al. also found that gelatin-based nanoparticles were not cytotoxic and did not induce an inflammatory response in pulmonary epithelial cells (16HBE14o-). The nanoparticle concentrations tested in vitro were 5, 10, 50, and 100 μg/mL (Brzoska et al. 2004).

21.3.10 POLYMERIC SURFACTANTS: POLOXAMER, POLYSORBATE, AND POLYOXYETHYLENE ALKYL ETHERS

Poloxamers are nonionic polymeric surfactants composed of two polyoxyethylene chains and one polyoxypropylene chain. These copolymers are commonly named with the letter "P" (for poloxamer) followed by three digits; the first two digits of which, when multiplied by 100, correspond to the approximate average molecular weight of the polyoxypropylene portion of the copolymer and the third digit, when multiplied by 10, corresponds to the percentage by weight of the polyoxyethylene portion. Poloxamers are also known under the trade name Pluronic® (in the United States) and Lutrol® (in Europe). For the Pluronic trade name, coding of these copolymers starts with a letter to define its physical form at room temperature (L = liquid, P = paste, F = flake) followed by two or three digits. The first digit (or two digits in a three-digit number) in the numerical designation, represents the molecular weight of the polyoxypropylene portion, and the last digit indicates the weight percent of the polyoxyethylene portion (Rowe et al. 2009). Reverse poloxamers are composed of one polyoxyethylene chain flanked by two polyoxypropylene chains. They are indicated by an "R" in the nomenclature (Ridder et al. 2005). Polysorbates are polyoxyethylene sorbitan fatty acid esters. The number following the polysorbate is related to the type of fatty acid associated with the polyoxyethylene sorbitan part of the molecule. Monolaurate is indicated by 20, monopalmitate by 40, monostearate by 60, tristearate by 65, monooleate by 80, and trioleate by 85. Polysorbates are also known under the trade name Tween®. Polysorbate 80 is approved for use in oral suspensions for inhalation at concentrations of 0.04% or less, according to the FDA's Inactive Ingredient Database (FDA CDER 2010). Polyoxyethylene alkyl ether surfactants are polyoxyethylene glycol ethers of n-alcohols. Some of them are also known under the trade name Brij® and are available in various grades as well (Rowe et al. 2009).

Polymeric surfactants have been investigated in vivo as absorption enhancers. Polyoxyethylene oleyl ether (Brij 93) at a concentration of 0.5% exhibited excellent absorption enhancing effect for salmon calcitonin following intratracheal administration in rats. The area of calcium reduction in vivo was increased from 50% with no absorption enhancer to approximately 125% with Brij 93 (Kobayashi et al. 1994). Zancong et al. found similar results when using Brij 35 and Brij 78 to enhance absorption of insulin solution in rats (Zancong et al. 2000). Polyoxythylene sorbitan mono-oleate (Tween 80) and polyoxyethylene sorbitan trioleate (Tween 85) displayed moderate absorption-enhancing activity for salmon calcitonin following intratracheal administration in rats (Kobayashi et al. 1994). Absorption enhancers may have negative effects on the epithelial cell layer in the respiratory tract. Suzuki et al. investigated the toxic effects of intratracheal administration of 1% polyoxyethylene 9 lauryl ether. Lung lesions were observed 1 day after administration from the bronchi to the alveoli. In addition, edema, hemorrhage, and inflammatory cells were observed. Wound healing was observed 3 and 7 days following administration (Suzuki et al. 2000). Conversely, Vaughn et al. aerosolized itraconazole dispersions composed of itraconazole, polysorbate 80, and poloxamer 407 in a 1:0.75:0.75 ratio, excipient placebo (15 mg/mL polysorbate 80 and 15 mg/mL poloxamer 407), and saline control to mice twice daily for 12 days using a whole body inhalation chamber. There was no evidence of damage, inflammation, ulceration, or repair in all groups indicating these concentrations of polysorbate 80 and poloxamer 407 are safe for inhalation (Vaughn et al. 2007).

Nonionic polymeric surfactants may be useful as non-viral vectors in the delivery of genes by inhalation (Chao et al. 2007; Desigaux et al. 2005). Desigaux at al. showed a significant improvement in transgene expression the lung following intratracheal administration of DNA solutions with 3% and 5% w/v Lutrol compared to naked DNA (Desigaux et al. 2005).

Polymeric surfactants have been tested for suitability in nebulizer and pMDI formulations. McCallion et al. studied the influence of surfactant concentrations on aerosol characteristics emitted from various nebulizers. Tween 20 and Tween 80 solutions ranged in concentration from 0.0001 to 1.0% w/v. The general trend was an increase in MMAD as surfactant concentration increased; however, the total aerosol output also increased as surfactant concentration increased. The surfactant concentration influenced the surface tension of the solutions, positively affecting the aerosol generation rate but negatively impacting the size of aerosol droplets (McCallion et al. 1996). Brown et al. and Ridder et al. tested various surfactants for solubility in HFA 134a and HFA 227 for use in pMDI (Brown and George 1997; Ridder et al. 2005). Surfactants need to be soluble in pMDI propellants in order to aid in dispersing the pharmaceutical powders to form homogenous and stable suspensions. Brij 30 was soluble (>10 mg/mL); however, Brij 58 was insoluble. Tween 20 and Tween 40 were partially soluble (Brown and George 1997). Williams et al. found that Brij 97, Brij 98, and Tween 80 were all soluble in HFA 134a and therefore useful in dispersing bovine serum albumin in HFA 134a; however, the pMDI suspensions made with Tween 80 resulted in the highest dose delivered through valve and respirable fraction (Williams and Liu 1999). Pluronic F77 (poloxamer 217) was also found to influence the aerosol characteristics of a pMDI suspension comprising triamcinolone and a 50/50 mixture of HFA 134a and HFA 227. Cogrinding the triamcinolone with Pluronic F77 in a ratio of 3:1 prior to filling in the pMDI resulted in an emitted aerosol with lower MMAD and higher respirable fraction (Williams et al. 1999).

Polymeric surfactants have also been studied as stabilizing agents for dry powder formulations. Poloxamer 188 was used at a concentration of 2% w/w to stabilize dry powders for inhalation composed of thymopentin, mannitol, and leucine. The authors also found that inclusion of poloxamer 188 improved the flowability of the dry powders (Wang et al. 2009). Other researchers incorporated polysorbate 20, polysorbate 80, and poloxamer 407 into compositions containing itraconazole produced by evaporative precipitation into aqueous solution and spray freezing into liquid. The resulting amorphous nanoparticulate formulations were dispersed in normal saline and nebulized to mice in a whole-body inhalation chamber (McConville et al. 2006; Vaughn et al. 2006). Similarly, Tam et al. utilized polysorbate 80 to stabilize a cyclosporine A formulation prepared by controlled antisolvent precipitation into aqueous solution. The stabilized nanoparticulate dispersion was aerosolized to mice using a nose-only inhalation chamber, resulting in high lung levels of cyclosporine A (Tam et al. 2008). In another study, the addition of polysorbate 20 to solutions of recombinant human growth hormone decreased aggregation during spray drying but negatively affected the fine particle fraction of the dry powders (Jalalipour et al. 2008a). Singh et al. coated fluticasone propionate with poloxamer using a spray drying method to form microparticles suitable for pulmonary delivery when blended with carrier lactose. Following nose-only aerosolization to rabbits, the microparticles remained in the lungs for approximately 12 h (Singh et al. 2007).

Aerosol exposure studies for Pluronic L64 (poloxamer 184), Pluronic L31 (poloxamer 101), and Pluronic 17R1 were conducted by Ulrich et al. Rats were exposed to 100 mg/m³ aerosol concentrations for 6 h per day, 5 days per week for 2 weeks. There was no sign of toxicity except very slight alveolitis, which was resolved within 2 weeks after exposure (Ulrich et al. 1992).

21.4 CONCLUSION

Excipients used in pharmaceutical formulations for pulmonary delivery must be nontoxic and must not provoke an immune response. PLGA and HPMCP were the only biopolymers mentioned earlier that exhibited cytotoxicity and immunogenicity. Therefore, many polymeric biomaterials may be useful as excipients for pulmonary drug delivery. Scientists have begun to incorporate various polymeric biomaterials into pharmaceutical formulations intended for inhalation. With so much focus on preclinical and clinical research of these biopolymers in inhalation drug delivery in recent years, it is anticipated that the pharmaceutical industry will soon see more of these biopolymers in FDA-approved products.

REFERENCES

Adjei, A.L., Y. Qiu, and P.K. Gupta. 2007. Bioavailability and pharmacokinetics of inhaled drugs. In *Inhalation Aerosols: Physical and Biological Basis for Therapy*, ed. A.J. Hickey. New York: Informa Healthcare.

Aguiar, M.M.G., J.M. Rodrigues, and A.S. Cunha. 2004. Encapsulation of insulin-cyclodextrin complex in PLGA microspheres: a new approach for prolonged pulmonary insulin delivery. *Journal of Microencapsulation* 21 (5):553–564.

Altiere, R.J. and D.C. Thompson. 2007. Physiology and pharmacology of the airways. In *Inhalation Aerosols: Physical and Biological Basis for Therapy*, ed. A.J. Hickey. New York: Informa Healthcare.

Arya, V., I. Coowanitwong, B. Brugos, W.S. Kim, R. Singh, and G. Hochhaus. 2006. Pulmonary targeting of sustained release formulation of budesonide in neonatal rats. *Journal of Drug Targeting* 14 (10):680–686.

Asada, M., H. Takahashi, H. Okamoto, H. Tanino, and K. Danjo. 2004. Theophylline particle design using chitosan by the spray drying. *International Journal of Pharmaceutics* 270 (1–2):167–174.

AstraZeneca. 2010. (cited June 25, 2010). Available from http://www1.astrazeneca-us.com/pi/symbicort.pdf

Azarmi, S., W.H. Roa, and R. Loebenberg. 2008. Targeted delivery of nanoparticles for the treatment of lung diseases. *Advanced Drug Delivery Reviews* 60 (8):863–875.

Bai, S.H. and F. Ahsan. 2009. Synthesis and evaluation of pegylated dendrimeric nanocarrier for pulmonary delivery of low molecular weight heparin. *Pharmaceutical Research* 26 (3):539–548.

Beck-Broichsitter, M., J. Gauss, C.B. Packhaeuser et al. 2009. Pulmonary drug delivery with aerosolizable nanoparticles in an ex vivo lung model. *International Journal of Pharmaceutics* 367 (1–2):169–178.

Ben-Jebria, A., M.L. Eskew, and D.A. Edwards. 2000. Inhalation system for pulmonary aerosol drug delivery in rodents using large porous particles. *Aerosol Science and Technology* 32 (5):421–433.

Bhat, M., R.K. Wolff, J.K.H. Ma, and Y. Rojanasakul. 2007. Drug metabolism and enzyme kinetics in the lung. In *Inhalation Aerosols: Physical and Biological Basis for Therapy*, ed. A.J. Hickey. New York: Informa Healthcare.

Brown, A.R. and D.W. George. 1997. Tetrafluoroethane (HFC 134A) propellant-driven aerosols of proteins. *Pharmaceutical Research* 14 (11):1542–1547.

Brzoska, M., K. Langer, C. Coester, S. Loitsch, T.O. Wagner, and C. Mallinckrodt. 2004. Incorporation of biodegradable nanoparticles into human airway epithelium cells—in vitro study of the suitability as a vehicle for drug or gene delivery in pulmonary diseases. *Biochemical and Biophysical Research Communications* 318 (2):562–570.

Buttini, F., P. Colombo, M.P. Wenger, P. Mesquida, C. Marriott, and S.A. Jones. 2008. Back to basics: the development of a simple, homogenous, two-component dry-powder inhaler formulation for the delivery of budesonide using miscible vinyl polymers. *Journal of Pharmaceutical Sciences* 97 (3):1257–1267.

Chao, Y.C., S.F. Chang, S.C. Lu, T.C. Hwang, W.H. Hsieh, and J. Liaw. 2007. Ethanol enhanced in vivo gene delivery with non-ionic polymeric micelles inhalation. *Journal of Controlled Release* 118 (1):105–117.

Corrigan, D.O., O.I. Corrigan, and A.M. Healy. 2006a. Physicochemical and in vitro deposition properties of salbutamol sulphate/ipratropium bromide and salbutamol sulphate/excipient spray dried mixtures for use in dry powder inhalers. *International Journal of Pharmaceutics* 322 (1–2):22–30.

Corrigan, D.O., A.M. Healy, and O.I. Corrigan. 2006b. Preparation and release of salbutamol from chitosan and chitosan co-spray dried compacts and multiparticulates. *European Journal of Pharmaceutics and Biopharmaceutics* 62 (3):295–305.

CyDex Pharmaceuticals, Inc. 2009. (cited August 28, 2009). Available from http://www.cydexpharma.com/pdf/PR20090316%28AAAAI%29.pdf

Dailey, L.A., N. Jekel, L. Fink et al. 2006. Investigation of the proinflammatory potential of biodegradable nanoparticle drug delivery systems in the lung. *Toxicology and Applied Pharmacology* 215 (1):100–108.

Dailey, L.A., T. Schmehl, T. Gessler et al. 2003. Nebulization of biodegradable nanoparticles: impact of nebulizer technology and nanoparticle characteristics on aerosol features. *Journal of Controlled Release* 86 (1):131–144.

Desigaux, L., C. Gourden, M. Bello-Roufai et al. 2005. Nonionic amphiphilic block copolymers promote gene transfer to the lung. *Human Gene Therapy* 16 (7):821–829.

Edwards, D.A., J. Hanes, G. Caponetti et al. 1997. Large porous particles for pulmonary drug delivery. *Science* 276 (5320):1868–1871.

Evrard, B., P. Bertholet, M. Gueders et al. 2004. Cyclodextrins as a potential carrier in drug nebulization. *Journal of Controlled Release* 96 (3):403–410.

FDA Center for Drug Evaluation and Research. 2010. (cited October 22, 2010). Available from http://www.accessdata.fda.gov/scripts/cder/iig/index.cfm

Fraser, J.R.E., T.C. Laurent, and U.B.G. Laurent. 1997. Hyaluronan: Its nature, distribution, functions and turnover. *Journal of Internal Medicine* 242 (1):27–33.

Fu, J., J. Fiegel, E. Krauland, and J. Hanes. 2002. New polymeric carriers for controlled drug delivery following inhalation or injection. *Biomaterials* 23 (22):4425–4433.

Fukaya, H., A. Iimura, K. Hoshiko, T. Fuyumuro, S. Noji, and T. Nabeshima. 2003. A cyclosporin A/maltosyl-alpha-cyclodextrin complex for inhalation therapy of asthma. *European Respiratory Journal* 22 (2):213–219.

Garcia-Contreras, L., T. Morcol, S.J. Bell, and A.J. Hickey. 2003. Evaluation of novel particles as pulmonary delivery systems for insulin in rats. *AAPS PharmSci* 5 (2):E9.

Gilani, K., A.R. Najafabadi, M. Barghi, and M. Rafiee-Tehrani. 2004. Aerosolisation of beclomethasone dipropionate using spray dried lactose/polyethylene glycol carriers. *European Journal of Pharmaceutics and Biopharmaceutics* 58 (3):595–606.

Greer, C.W. and W. Yaphe. 1984. Enzymatic analysis of carrageenans—structure of carrageenan from eucheuma-nudum. *Hydrobiologia* 116 (September):563–567.

Grenha, A., C.I. Grainger, L.A. Dailey et al. 2007a. Chitosan nanoparticles are compatible with respiratory epithelial cells in vitro. *European Journal of Pharmaceutical Sciences* 31 (2):73–84.

Grenha, A., B. Seijo, and C. Remunan-Lopez. 2005. Microencapsulated chitosan nanoparticles for lung protein delivery. *European Journal of Pharmaceutical Sciences* 25 (4–5):427–437.

Grenha, A., B. Seijo, C. Serra, and C. Remunan-Lopez. 2007b. Chitosan nanoparticle-loaded mannitol microspheres: structure and surface characterization. *Biomacromolecules* 8 (7):2072–2079.

Gueders, M.M., P. Bertholet, F. Perin et al. 2008. A novel formulation of inhaled doxycycline reduces allergen-induced inflammation, hyperresponsiveness and remodeling by matrix metalloproteinases and cytokines modulation in a mouse model of asthma. *Biochemical Pharmacology* 75 (2):514–526.

Hussain, A., T.Z. Yang, A.A. Zaghloul, and F. Ahsan. 2003. Pulmonary absorption of insulin mediated by tetradecyl-beta-maltoside and dimethyl-beta-cyclodextrin. *Pharmaceutical Research* 20 (10):1551–1557.

Hwang, S.M., D.D. Kim, S.J. Chung, and C.K. Shim. 2008. Delivery of ofloxacin to the lung and alveolar macrophages via hyaluronan microspheres for the treatment of tuberculosis. *Journal of Controlled Release* 129 (2):100–106.

Iida, K., H. Todo, H. Okamoto, K. Danjo, and H. Leuenberger. 2005. Preparation of dry powder inhalation with lactose carrier particles surface-coated using a Wurster fluidized bed. *Chemical & Pharmaceutical Bulletin* 53 (4):431–434.

Jalalipour, M., K. Gilani, H. Tajerzadeh, A.R. Najafabadi, and M. Barghi. 2008a. Characterization and aerodynamic evaluation of spray dried recombinant human growth hormone using protein stabilizing agents. *International Journal of Pharmaceutics* 352 (1–2):209–216.

Jalalipour, M., A.R. Najafabadi, K. Gilani, H. Esmaily, and H. Tajerzadeh. 2008b. Effect of dimethyl-beta-cyclodextrin concentrations on the pulmonary delivery of recombinant human growth hormone dry powder in rats. *Journal of Pharmaceutical Sciences* 97 (12):5176–5185.

Jones, S.A., G.P. Martin, and M.B. Brown. 2006a. Manipulation of beclomethasone-hydrofluoroalkane interactions using biocompatible macromolecules. *Journal of Pharmaceutical Sciences* 95 (5):1060–1074.

Jones, S.A., G.P. Martin, and M.B. Brown. 2006b. Stabilisation of deoxyribonuclease in hydrofluoroalkanes using miscible vinyl polymers. *Journal of Controlled Release* 115 (1):1–8.

Kawashima, Y., H. Yamamoto, H. Takeuchi, S. Fujioka, and T. Hino. 1999. Pulmonary delivery of insulin with nebulized DL-lactide/glycolide copolymer (PLGA) nanospheres to prolong hypoglycemic effect. *Journal of Controlled Release* 62 (1–2):279–287.

Klonne, D.R., D.E. Dodd, P.E. Losco, C.M. Troup, and T.R. Tyler. 1989. 2-Week aerosol inhalation study on polyethylene glycol-(Peg)-3350 in F344 rats. *Drug and Chemical Toxicology* 12 (1):39–48.

Kobayashi, S., S. Kondo, and K. Juni. 1994. Study on pulmonary delivery of salmon calcitonin in rats: effects of protease inhibitors and absorption enhancers. *Pharmaceutical Research* 11 (9):1239–1243.

Koushik, K., D.S. Dhanda, N.P.S. Cheruvu, and U.B. Kompella. 2004. Pulmonary delivery of deslorelin: large-porous PLGA particles and HP beta CD complexes. *Pharmaceutical Research* 21 (7):1119–1126.

Learoyd, T.P., J.L. Burrows, E. French, and P.C. Seville. 2008. Chitosan-based spray-dried respirable powders for sustained delivery of terbutaline sulfate. *European Journal of Pharmaceutics and Biopharmaceutics* 68 (2):224–234.

Liu, X.B., J.X. Ye, L.H. Quan et al. 2008. Pulmonary delivery of scutellarin solution and mucoadhesive particles in rats. *European Journal of Pharmaceutics and Biopharmaceutics* 70 (3):845–852.

Lucas, P., K. Anderson, and J.N. Staniforth. 1998. Protein deposition from dry powder inhalers: fine particle multiplets as performance modifiers. *Pharmaceutical Research* 15 (4):562–569.

Manca, M.L., S. Mourtas, V. Dracopoulos, A.M. Fadda, and S.G. Antimisiaris. 2008. PLGA, chitosan or chitosan-coated PLGA microparticles for alveolar delivery? A comparative study of particle stability during nebulization. *Colloids and Surfaces B: Biointerfaces* 62 (2):220–231.

Marques, H.M.C., J. Hadgraft, I.W. Kellaway, and G. Taylor. 1991a. Studies of cyclodextrin inclusion complexes.3. The pulmonary absorption of beta-cyclodextrin, Dm-beta-cyclodextrin and hp-beta-cyclodextrin in rabbits. *International Journal of Pharmaceutics* 77 (2–3):297–302.

Marques, H.M.C., J. Hadgraft, I.W. Kellaway, and G. Taylor. 1991b. Studies of cyclodextrin inclusion complexes.4. The pulmonary absorption of salbutamol from a complex with 2-hydroxypropyl-beta-cyclodextrin in rabbits. *International Journal of Pharmaceutics* 77 (2–3):303–307.

Matilainen, L., T. Toropainen, H. Vihola et al. 2008. In vitro toxicity and permeation of cyclodextrins in Calu-3 cells. *Journal of Controlled Release* 126 (1):10–16.

McCallion, O.N.M., K.M.G. Taylor, M. Thomas, and A.J. Taylor. 1996. The influence of surface tension on aerosols produced by medical nebulisers. *International Journal of Pharmaceutics* 129 (1–2):123–136.

McConville, J.T., K.A. Overhoff, P. Sinswat et al. 2006. Targeted high lung concentrations of itraconazole using nebulized dispersions in a murine model. *Pharmaceutical Research* 23 (5):901–911.

Morimoto, K., H. Katsumata, T. Yabuta et al. 2000. Gelatin microspheres as a pulmonary delivery system: evaluation of salmon calcitonin absorption. *Journal of Pharmacy and Pharmacology* 52 (6):611–617.

Morimoto, K., K. Metsugi, H. Katsumata, K. Iwanaga, and M. Kakemi. 2001. Effects of low-viscosity sodium hyaluronate preparation on the pulmonary absorption of rh-insulin in rats. *Drug Development and Industrial Pharmacy* 27 (4):365–371.

Nakate, T., H. Yoshida, A. Ohike, Y. Tokunaga, R. Ibuki, and Y. Kawashima. 2003. Comparison of the lung absorption of FK224 inhaled from a pressurized metered dose inhaler and a dry powder inhaler by healthy volunteers. *European Journal of Pharmaceutics and Biopharmaceutics* 56 (3):319–325.

Ohashi, K., T. Kabasawa, T. Ozeki, and H. Okada. 2009. One-step preparation of rifampicin/poly(lactic-co-glycolic acid) nanoparticle-containing mannitol microspheres using a four-fluid nozzle spray drier for inhalation therapy of tuberculosis. *Journal of Controlled Release* 135 (1):19–24.

Oster, C.G., M. Wittmar, F. Unger, L. Barbu-Tudoran, A.K. Schaper, and T. Kissel. 2004. Design of amine-modified graft polyesters for effective gene delivery using DNA-loaded nanoparticles. *Pharmaceutical Research* 21 (6):927–931.

Pandey, R., A. Sharma, A. Zahoor, S. Sharma, G.K. Khuller, and B. Prasad. 2003. Poly (DL-lactide-*co*-glycolide) nanoparticle-based inhalable sustained drug delivery system for experimental tuberculosis. *Journal of Antimicrobial Chemotherapy* 52 (6):981–986.

Petrigni, G., and L. Allegra. 2006. Aerosolised hyaluronic acid prevents exercise-induced bronchoconstriction, suggesting novel hypotheses on the correction of matrix defects in asthma. *Pulmonary Pharmacology & Therapeutics* 19 (3):166–171.

Remminghorst, U. and B.H.A. Rehm. 2006. Bacterial alginates: from biosynthesis to applications. *Biotechnology Letters* 28 (21):1701–1712.

Ridder, K.B., C.J. Davies-Cutting, and I.W. Kellaway. 2005. Surfactant solubility and aggregate orientation in hydrofluoroalkanes. *International Journal of Pharmaceutics* 295 (1–2):57–65.

Rodrigues, J.M., K.D. Lima, C.E.D. Jensen, M.M.G. de Aguiar, and A.D. Cunha. 2003. The effect of cyclo-dextrins on the in vitro and in vivo properties of insulin-loaded poly(D,L-lactic-*co*-glycolic acid) micro-spheres. *Artificial Organs* 27 (5):492–497.

Rouse, J.J., T.L. Whateley, M. Thomas, and G.M. Eccleston. 2007. Controlled drug delivery to the lung: influence of hyaluronic acid solution conformation on its adsorption to hydrophobic drug particles. *International Journal of Pharmaceutics* 330 (1–2):175–182.

Rowe, R.C., P.J. Sheskey, and M.E. Quinn, eds. 2009. *Handbook of Pharmaceutical Excipients*, 6th edn. London, U.K.: Pharmaceutical Press and the American Pharmacists Association.

Rytting, E., J. Nguyen, X.Y. Wang, and T. Kissel. 2008. Biodegradable polymeric nanocarriers for pulmonary drug delivery. *Expert Opinion on Drug Delivery* 5 (6):629–639.

Sakagami, M., K. Sakon, W. Kinoshita, and Y. Makino. 2001. Enhanced pulmonary absorption following aerosol administration of mucoadhesive powder microspheres. *Journal of Controlled Release* 77 (1–2):117–129.

Salem, L.B., C. Bosquillon, L.A. Dailey et al. 2009. Sparing methylation of beta-cyclodextrin mitigates cyto-toxicity and permeability induction in respiratory epithelial cell layers in vitro. *Journal of Controlled Release* 136 (2):110–116.

Scheuch, G., M.J. Kohlhaeufl, P. Brand, and R. Siekmeler. 2006. Clinical perspectives on pulmonary systemic and macromolecular delivery. *Advanced Drug Delivery Reviews* 58 (9–10):996–1008.

Shao, Z.Z., Y.P. Li, and A.K. Mitra. 1994. Cyclodextrins as mucosal absorption promoters of insulin. 3. Pulmonary route of delivery. *European Journal of Pharmaceutics and Biopharmaceutics* 40 (5):283–288.

Sharma, A., S. Sharma, and G.K. Khuller. 2004. Lectin-functionalized poly(lactide-*co*-glycolide) nanoparticles as oral/aerosolized antitubercular drug carriers for treatment of tuberculosis. *Journal of Antimicrobial Chemotherapy* 54 (4):761–766.

Singh, D.J., J.J. Parmar, D.D. Hegde et al. 2007. Poloxamer coated fluticasone propionate microparticles for pulmonary delivery; in vivo lung deposition and efficacy studies. *Indian Journal of Pharmaceutical Sciences* 69 (5):714–715.

Sivadas, N., D. O'Rourke, A. Tobin et al. 2008. A comparative study of a range of polymeric microspheres as potential carriers for the inhalation of proteins. *International Journal of Pharmaceutics* 358 (1–2):159–167.

Stella, V.J. and Q. He. 2008. Cyclodextrins. *Toxicologic Pathology* 36 (1):30–42.

Stocks, J. and A.A. Hislop. 2002. Structure and function of the respiratory system: developmental aspects and their relevance to aerosol therapy. In *Drug Delivery to the Lung*, eds. H. Bisgaard, C. O'Callaghan, and G.C. Smaldone. New York: Marcel Dekker.

Suarez, S., P. O'Hara, M. Kazantseva et al. 2001a. Airways delivery of rifampicin microparticles for the treat-ment of tuberculosis. *Journal of Antimicrobial Chemotherapy* 48 (3):431–434.

Suarez, S., P. O'Hara, M. Kazantseva et al. 2001b. Respirable PLGA microspheres containing rifampicin for the treatment of tuberculosis: screening in an infectious disease model. *Pharmaceutical Research* 18 (9):1315–1319.

Sung, J.C., D.J. Padilla, L. Garcia-Contreras et al. 2009. Formulation and pharmacokinetics of self-assembled rifampicin nanoparticle systems for pulmonary delivery. *Pharmaceutical Research* 26 (8):1847–1855.

Surendrakumar, K., G.P. Martyn, E.C.M. Hodgers, M. Jansen, and J.A. Blair. 2003. Sustained release of insu-lin from sodium hyaluronate based dry powder formulations after pulmonary delivery to beagle dogs. *Journal of Controlled Release* 91 (3):385–394.

Surti, N., S. Naik, and A. Misra. 2009. Pharmacokinetic evaluation of wheat germ agglutinin-grafted nanopar-ticles of mometasone furoate. *Scientia Pharmaceutica* 77:123–321.

Suzuki, M., M. Machida, K. Adachi et al. 2000. Histopathological study of the effects of a single intratracheal instillation of surface active agents on lung in rats. *The Journal of Toxicological Sciences* 25 (1):49–55.

Talton, J., J. Fitz-Gerald, R. Singh, and G. Hochhaus. 2000. Nano-thin coatings for improved lung targeting of glucocorticoid dry powders: in-vitro and in-vivo characteristics. *Respiratory Drug Delivery* VII:67–74.

Tam, J.M., J.T. McConville, R.O. Williams, 3rd, and K.P. Johnston. 2008. Amorphous cyclosporin nano-dispersions for enhanced pulmonary deposition and dissolution. *Journal of Pharmaceutical Sciences* 97 (11):4915–4933.

Taylor, M.K., A.J. Hickey, and M. VanOort. 2006. Manufacture, characterization, and pharmacodynamic evaluation of engineered ipratropium bromide particles. *Pharmaceutical Development and Technology* 11 (3):321–336.

Tolman, J.A., N.A. Nelson, S. Bosselmann et al. 2009a. Dose tolerability of chronically inhaled voriconazole solution in rodents. *International Journal of Pharmaceutics* 379 (1):25–31.

Tolman, J.A., N.A. Nelson, Y.J. Son et al. 2009b. Characterization and pharmacokinetic analysis of aerosolized aqueous voriconazole solution. *European Journal of Pharmaceutics and Biopharmaceutics* 72 (1):199–205.

Tomoda, K., T. Ohkoshi, K. Hirota et al. 2009. Preparation and properties of inhalable nanocomposite particles for treatment of lung cancer. *Colloids and Surfaces B: Biointerfaces* 71 (2):177–182.

Tseng, C.L., W.Y. Su, K.C. Yen, K.C. Yang, and F.H. Lin. 2009. The use of biotinylated-EGF-modified gelatin nanoparticle carrier to enhance cisplatin accumulation in cancerous lungs via inhalation. *Biomaterials* 30 (20):3476–3485.

Tseng, C.L., S.Y. Wu, W.H. Wang et al. 2008. Targeting efficiency and biodistribution of biotinylated-EGF-conjugated gelatin nanoparticles administered via aerosol delivery in nude mice with lung cancer. *Biomaterials* 29 (20):3014–3022.

Ulrich, C.E., R.G. Geil, T.R. Tyler, G.L. Kennedy, Jr., and H.A. Birnbaum. 1992. Two-week aerosol inhalation study in rats of ethylene oxide/propylene oxide copolymers. *Drug and Chemical Toxicology* 15 (1):15–31.

Ungaro, F., R.D.D.V. Bianca, C. Giovino et al. 2009. Insulin-loaded PLGA/cyclodextrin large porous particles with improved aerosolization properties: In vivo deposition and hypoglycaemic activity after delivery to rat lungs. *Journal of Controlled Release* 135 (1):25–34.

Ungaro, F., G. De Rosa, A. Miro, F. Quaglia, and M.I. La Rotonda. 2006. Cyclodextrins in the production of large porous particles: Development of dry powders for the sustained release of insulin to the lungs. *European Journal of Pharmaceutical Sciences* 28 (5):423–432.

Valle, M.J.D.J., R.J. Dinis-Oliveira, F. Carvalho, M.L. Bastos, and A.S. Navarro. 2008. Toxicological evaluation of lactose and chitosan delivered by inhalation. *Journal of Biomaterials Science-Polymer Edition* 19 (3):387–397.

Vaughn, J.M., J.T. McConville, D. Burgess et al. 2006. Single dose and multiple dose studies of itraconazole nanoparticles. *European Journal of Pharmaceutics and Biopharmaceutics* 63 (2):95–102.

Vaughn, J.M., N.P. Wiederhold, J.T. McConville et al. 2007. Murine airway histology and intracellular uptake of inhaled amorphous itraconazole. *International Journal of Pharmaceutics* 338 (1–2):219–224.

Wang, L., Y. Zhang, and X. Tang. 2009. Characterization of a new inhalable thymopentin formulation. *International Journal of Pharmaceutics* 375 (1–2):1–7.

Washington, N., C. Washington, and C.G. Wilson. 2001. Pulmonary drug delivery. In *Physiological Pharmaceutics: Barriers to Drug Absorption*. New York: Taylor & Francis.

Westmeier, R. and H. Steckel. 2008. Combination particles containing salmeterol xinafoate and fluticasone propionate: Formulation and aerodynamic assessment. *Journal of Pharmaceutical Sciences* 97 (6):2299–2310.

Williams, R.O., M.K. Barron, M.J. Alonso, and C. Remunan-Lopez. 1998. Investigation of a pMDI system containing chitosan microspheres and P134a. *International Journal of Pharmaceutics* 174 (1–2):209–222.

Williams, R.O., 3rd and J. Liu. 1999. Formulation of a protein with propellant HFA 134a for aerosol delivery. *European Journal of Pharmaceutical Sciences* 7 (2):137–144.

Williams, R.O., M.A. Repka, and M.K. Barron. 1999. Application of co-grinding to formulate a model pMDI suspension. *European Journal of Pharmaceutics and Biopharmaceutics* 48 (2):131–140.

Williamson, P.A., D. Menzies, A. Nair, A. Tutuncu, and B.J. Lipworth. 2009. A proof-of-concept study to evaluate the anti-inflammatory effects of a novel soluble cyclodextrin formulation of nebulized budesonide in patients with mild to moderate asthma. *Annals of Allergy Asthma & Immunology* 102 (2):161–167.

Yamada, K., N. Kamada, M. Odomi et al. 2005. Carrageenans can regulate the pulmonary absorption of anti-asthmatic drugs and their retention in the rat lung tissues without any membrane damage. *International Journal of Pharmaceutics* 293 (1–2):63–72.

Yamamoto, H., W. Hoshina, H. Kurashima et al. 2007. Engineering of poly(DL-lactic-*co*-glycolic acid) nanocomposite particles for dry powder inhalation dosage forms of insulin with the spray-fluidized bed granulating system. *Advanced Powder Technology* 18 (2):215–228.

Yamamoto, H., Y. Kuno, S. Sugimoto, H. Takeuchi, and Y. Kawashima. 2005. Surface-modified PLGA nanosphere with chitosan improved pulmonary delivery of calcitonin by mucoadhesion and opening of the intercellular tight junctions. *Journal of Controlled Release* 102 (2):373–381.

Yamamoto, A., K. Yamada, H. Muramatsu et al. 2004. Control of pulmonary absorption of water-soluble compounds by various viscous vehicles. *International Journal of Pharmaceutics* 282 (1–2):141–149.

Yang, W., J.I. Peters, and R.O. Williams. 2008. Inhaled nanoparticles—a current review. *International Journal of Pharmaceutics* 356 (1–2):239–247.

Yang, M., H. Yamamoto, H. Kurashima, H. Takeuchi, and Y. Kawashima. 2007. In vitro and in vivo inhalation characterization of pH-dependent HP-55 nanospheres containing salmon calcitonin. *Asian Journal of Pharmaceutical Sciences* 2 (1):1–10.

Zahoor, A., S. Sharma, and G.K. Khuller. 2005. Inhalable alginate nanoparticles as antitubercular drug carriers against experimental tuberculosis. *International Journal of Antimicrobial Agents* 26 (4):298–303.

Zancong, S., C. Yi, Z. Qiang, W. Shuli, and W. Kui. 2000. Pulmonary delivery of insulin: absorption enhancement of insulin by various absorption promoters in rats. *Journal of Chinese Pharmaceutical Sciences* 9 (1):22–25.

Zaru, M., M.L. Manca, A.M. Fadda, and S.G. Antimisiaris. 2009. Chitosan-coated liposomes for delivery to lungs by nebulisation. *Colloids and Surfaces B: Biointerfaces* 71 (1):88–95.

22 Polymeric Gene Delivery Carriers for Pulmonary Diseases

Xiang Gao, Regis R. Vollmer, and Song Li

CONTENTS

22.1 INTRODUCTION

22.1.1 Nucleic Acid Therapeutics for Lung Diseases

Several genetic and acquired lung diseases represent significant public health problems that are often costly and difficult to treat with conventional therapeutic approaches. Examples include cystic fibrosis (CF), alpha1-antitrypsin deficiency (α1AT), pulmonary fibrosis, emphysema, epidemic viral infection, acute lung injuries (ALIs), and certain lung cancers. Nucleic acid-based macromolecular drugs capable of either enhancing or suppressing the activities of host gene expression relevant to a disease offer new strategies that may be more effective in managing these complicated lung diseases. These experimental therapeutic agents can be classified into several categories: genes that encode either wide-type (functional) proteins or mutant derivatives that act as inhibitors; triplex-forming agents and DNA decoys that specifically inhibit a particular gene expression at the transcriptional level; antisense oligonucleotides, small interfering RNAs, and micro RNAs that specifically inhibit particular gene expression at the posttranscriptional level. These nucleic acid-based therapeutic agents are fundamentally different from small-molecule drugs in mechanism of action, specificity, pharmacokinetics, and delivery method. First, these agents can be designed to be "disease gene-specific" based on our understanding of human diseases at the molecular level and the sequence information of a disease-related gene. In addition, nucleic acid therapeutics can be used to either positively or negatively regulate gene expression, depending upon the underlying mechanism of a specific disease. Thus, the agents provide versatility, which enable them to be used to treat a broad range of human diseases that are not amenable to treatment with small-molecule drugs. Rational approaches used in the discovery and development phases could significantly accelerate the process. Typically, genes and siRNA can provide therapeutic effects that last for a minimum of days instead of hours for most small-molecule drugs. However, low in vivo bioavailability severely limits the utility of these macromolecular agents. A successful implementation of most nucleic acid-based therapeutics usually relies on a delivery method or carrier that efficiently escorts these agents to intracellular locations of target cell populations. Several viral vectors have been developed with high gene delivery efficiency; Some of which can provide a sustained transcription of foreign genes, rendering them attractive as gene delivery systems. However, unexpected serious adverse effects of immunogenicity and oncogenicity were noticed from several clinical trials using these viral vectors (Marshall, 2002; Hacein-Bey-Abina et al., 2003), suggesting that further development of alternative safe and efficient delivery systems is necessary.

In this chapter, we will discuss the principles for developing safe and efficient polymers-based gene and siRNA transfer methods. We will also identify the barriers for in vivo gene delivery and present the strategies to overcome these challenges. Additionally, we will discuss the advances made in pulmonary gene delivery in experimental animal models and in humans. Finally, we will summarize the implications of novel therapeutic strategies of gene therapy for various inherited and acquired pulmonary diseases, including infectious, inflammatory, and malignant disorders.

22.1.2 Nonviral Nucleic Acid Carriers

Nonviral gene delivery systems use low immunogenic synthetic and natural polymers or lipids as the carrier to introduce nucleic acids into cells. Unprotected "naked" DNA and RNA molecules are inefficient by themselves for multiple reasons. Their cellular uptake is limited because they are hydrophilic polyanionic macromolecules that are impermeable to cell membranes. They are vulnerable to mechanic damage by shear force during aerosolization, a method frequently used to disperse liquid droplets evenly throughout the respiratory track. In addition, widespread nuclease activities in extracellular fluid, endosome/lysosome compartments, or cytoplasm can readily degrade these macromolecules.

Proper protection of nucleic acid molecules during the transfection process is a priority for most nonviral gene delivery systems. This can be readily achieved using cationic polymers that form condensed complexes with DNA. Alternatively, non-cationic polymers capable of physically entrapping DNA inside nanoparticles can also deliver intact DNA into cells.

The simplest form involves the direct formation of ionic complex of nucleic acid and a slight excess of cationic-charged carrier molecules. During this process, nucleic acid molecules in extended configuration are condensed into nanoparticles in which nucleic acids are well protected. The excessive carrier molecules provide positive surface charges to the nanoparticles which promote avid interaction with the negatively charged cell surface and trigger efficient cellular uptake. More sophisticated features can be incorporated into these nanoparticles using block polymers or multiple components to improve the surface properties and biocompatibility, and to add additional functionalities. These improved nanoparticles exhibit long circulation time in blood, targeting capacity to specific cell/tissue, endosomal release capability, disintegration feature in cells, and inclusion of nuclear trafficking signals to enhance the rate of transfection. Thus, modifications of nanoparticles to improve their effectiveness are crucial to maximize their in vivo applications.

22.2 CATIONIC DNA POLYMER-BASED DNA CONDENSATION AND TRANSFECTION AGENTS

22.2.1 OVERVIEW

Cationic polymers constitute an important class of nonviral gene delivery agents. Historically, DEAE-dextran was the first synthetic polymer used for transfection in vitro in 1965 (Pagano and Vaheri, 1965). The next significant step in 1987 was the design of asialoglycoprotein-poly-L-lysine conjugates, which led to the demonstration of the concept and feasibility of polymer-based, receptor-mediated gene delivery to hepatoma cells in vitro (Wu and Wu, 1987), and subsequently to liver hepatocytes in vivo (Wu and Wu, 1988). However, transfection efficiency achieved by these polymers alone was very low, unless they were combined with other strategies such as osmotic shock and/or the use of lysosomotropic agents. The failure of efficient gene transfer with these DNA/polymer complexes alone was due to their inability to escape from digestive endosome/lysosome compartments into the cytoplasm following intracellular delivery. This was clearly demonstrated by the markedly increased transfection efficiency when the polymer/DNA complexes were supplemented with inactivated adenovirus particles, or bacterial hemolysins that were capable of rupturing endosomal vesicles (Cotten et al., 1992; Walton et al., 1999). Synthetic materials with endosomal lytic properties and low immunogenicity, such as amphipathic peptides (Wagner, 1998; Lee et al., 2001; Ogris et al., 2001) and pH-sensitive polymers (Kiang et al., 2004), were also designed to enhance polymer-based transfection. The discovery that certain amine-rich cationic polymers, such as polyethyleneimines (PEIs) and partially degraded polyamidoamine dendrimers, have excellent transfection activities in vitro (Boussif et al., 1995; Tang et al., 1996; Ferrari et al., 1997) and in vivo (Ferrari et al., 1997; Goula et al., 1998; Densmore et al., 2000; Zou et al., 2000) marks a new era for polymer-based transfection. These polyamine compounds trigger the release of DNA/polymer complexes by increasing intravesicular osmotic pressure in endosome compartments due to protonation of uncharged amine groups and influx of counter ions accompanied with the protons. Cationic polymers with similar proton-binding capacity, but with improved biocompatibility and biodegradability over PEI, have been designed; some were proven to be excellent transfection agents (Arote et al., 2009; Gao et al., 2009; Zaliauskiene et al., 2010)

While most cationic polymeric transfection agents show variable in vitro transfection activity, only a few have demonstrated significant in vivo transfection efficiency when administered as

TABLE 22.1

List of Cationic Polymers That Showed Significant In Vivo Transfection Activities

Polymers	Biodegradable	In Vivo Model	Route	References
Branched PEI	No	Mouse	Aerosolization	Gautam et al. (2000)
Linear PEI	No	Mouse	i.v.	Goula et al. (1998)
		Rat	Airway	Uduehi et al. (2001)
		Rabbit	Airway	Ferrari et al. (1997)
PAMAM dendrimers	No	Mouse	i.v. airway	Kukowska-Latallo et al. (2000)
Polyaminoesters	Yes	Mouse	i.v., i.p.	Zugates et al. (2007)
Amphiphilic peptides	Yes	Mouse	i.v.	Rittner et al. (2002)
Chitosan	Yes	Mouse	Airway	Okamoto et al. (2003)
Aminated dextrans	Yes	Mouse	Airway	Abdullah et al. (2010)
PEI-PEG P123	Partially	Mouse	i.v.	Nguyen et al. (2000)
PEI-PEG	Partially	Mouse	Nasal	Kichler et al. (2002)
PLL-PEG	Yes	Mouse	Airway	Ziady et al. (2003b)
	Yes	Human	Nasal	Konstan et al. (2004)
Antibody-PEI	Partially	Mouse	i.v.	Li et al. (2000)
Antibody-PLL	Yes	Mouse	i.v.	Ferkol et al. (1995)

polyplexes intravenously or through airway either as bolus instillation or aerosolization. Table 22.1 lists major categories of cationic polymers that demonstrated significant levels of in vivo transfection efficiency.

22.2.2 POLYETHYLENIMINES: CHEMISTRY OF SYNTHESIS

Branched and linear PEIs are among the best cationic polymeric transfection agents. Branched PEI is typically synthesized in a ring opening reaction from azrine with tertiary, secondary, and primary amines as the branching, chain extension, and termination points, respectively. Kissel group identified the optimum conditions for synthesizing medium-range molecular weight, branched PEI (~12,000 Da) by acid-catalyzed ring-opening polymerization of aziridine at lower temperature over a long period. The polymer showed a lower degree of branching, and was less cytotoxic and up to two orders of magnitude higher in transfection efficiency than those obtained from a commercial PEI of ~600,000 Da. (Fischer et al., 1999).

Linear PEI, which has most nitrogen in the form of secondary amines, is generated by hydrolyzing protected precursor poly(2-ethyl-2-oxazoline), which is obtained by the cationic ring-opening polymerization of 2-ethyl-2-oxazoline. Using a specially designed initiator, one could generate a monofunctional linear polymer with a unique end group introduced on one of its termini. The Zuber group has reported the synthesis of linear PEI (~5000 Da) terminated at one end with a nucleophilic hydrazine residue, which can be used to selectively conjugate with an aldehyde-containing macromolecule, such as oxidized transferrin to form PEI–transferrin conjugate for targeted delivery of DNA (Pons et al., 2006). Through polymerization using a mono- or a bifunctional macroinitiator, followed by deprotection, PEI-macroinitiator diblocker, or PEI-macroinitiator-PEI triblocker, can be generated, respectively. Neutral polymer such as polyethylene glycol (PEG) can be incorporated to render the PEI–PEG conjugates serum resistant and less toxic while retaining excellent DNA condensation and transfection activity (Zhong et al., 2005).

The degree of hydrolyzation of the *N*-acetyl group from the precursor poly(2-ethyl-2-oxazoline) has a profound effect on the transfection efficiency of linear PEI. Fully deprotected linear PEI is 20 and 10,000 times more effective in transfection in vitro and in vivo than incompletely deprotected commercial product, respectively. Fully deacylated PEIs have been used for pulmonary delivery of siRNA via the vascular route, resulting in significant suppression of the expression of a reporter gene in mouse lungs. Delivery of a siRNA against the influenza viral nucleocapsid protein gene using these PEIs led to the reduction of virus titers in the lungs of influenza-infected animals to 5% of that in control group (Thomas et al., 2005).

22.2.3 CATIONIC POLYMER–DNA COMPLEX FORMATION

Upon mixing of positively charged cationic polymers and negatively charged DNA, polyelectrolyte complexes, also called polyplexes, are formed through a multistep process. The process is initiated due to strong electrostatic interactions between oppositely charged polyions that form the primary complexes. As the net charge densities of the polymer backbones in the primary complexes are reduced to a critical point, the polymer chains collapse to form unstable intercomplexes, usually accompanied with the formation of new bonds among polymer chains. This is followed by a slower aggregation process, which involves mainly hydrophobic interactions among intercomplexes (Tsuchida, 1994; Xavier and Jean-Francois, 1996).

The structural parameters of cationic polymer (length of side chain groups, basicity of cationic groups, polymer molecular weight, nature of counter ions, and charge spacing along the polymer backbone), polymer-to-DNA ratio, pH and ionic strength of the mixing medium, and concentrations of polymer and DNA have been shown to influence dramatically the biophysical properties of polyplexes. Properties such as surface charge density and particle size could influence cellular uptake rate. The overall stability of the complexes following intracellular uptake, which is determined by the polymer molecular weight, backbone charge spacing, length of side chain groups, and the basicity of charged groups, can affect the accessibility of polymer-bound DNA to the host transcriptional machinery and influence the outcome of transfection (Wolfert et al., 1999). Under conditions of low ionic strength and excess of charge ratios of cationic polymer over DNA, polyplexes of between 20 and 100 nm in diameter can be readily obtained. The presence of salt causes aggregation and the particles increase from 150 nm to ~1 μm in diameter. It has been shown that large complexes formed in the presence of salt were more active in transfection in vitro than the smaller ones formed at low ionic strength. However, smaller linear PEI polyplexes appear to be more active in vivo in the lungs when administered intravenously (Wightman et al., 2001). The greater efficacy of the smaller PEI polyplexes in vivo may be due to the differential effect of size on the cellular uptake of lipoplexes between in vitro and in vivo.

22.2.4 EFFECT OF FREE PEI

Polyplexes prepared with ratios of polymer and DNA that show efficient transfection activity should in theory contain free polymers because an excess of polymer is used in the preparation process. Boeckle et al. demonstrated free PEI can be separated from polyplexes using size-exclusion chromatography (SEC). The purified polyplexes had a composition of PEI nitrogen/DNA phosphate ratio of 2.5 regardless of the amount of excess PEI used for the preparation. The purified PEI polyplexes demonstrated low cellular and systemic toxicity when compared to unpurified polyplexes. However, the purified polyplexes were less potent in transfection in vitro and in vivo by systemic administration based on the same DNA dosage, suggesting that the free PEI is important for transfection activity but also contributes to toxicity. Purified polyplexes had to be applied at a dose that was nearly four times of that for unpurified complexes to achieve high transfection levels (Boeckle et al., 2004). Similar conclusions were reported when purification was accomplished with ultrafiltration or electrophoretic methods (Erbacher et al., 2004; Fahrmeir et al., 2007).

Standard approaches for characterizing polyplexes include analysis of size in solution by dynamic light scattering and morphology by EM; Zeta potential analysis for surface charge; assay for exclusion of intercalating dyes; cellular uptake using differentially labeled carrier and DNA molecules using fluorescence microscopy; transfection efficiency using plasmid DNA encoding firefly luciferase, β-galactosidase, or green fluorescent protein; and cytotoxicity on cells by MTT or lactate dehydrogenase release assays (Shcharbin et al., 2009, 2010).

22.2.5 POLYETHYLENIMINES-MEDIATED TRANSFECTION

22.2.5.1 Mechanism of Cellular Uptake

Cationic PEI condenses DNA into small particles called polyplexes, which have a net positive surface charge. These polyplexes efficiently bind to the cell surface through electrostatic interactions with negatively charged membrane components, such as proteoglycans. Polyplexes are taken up by cells through different endocytic pathways, depending upon the size of the polyplexes. For example, Grosse et al. showed that uptake of large PEI polyplexes by human airway epithelial cells was largely mediated via macropinocytosis. For intermediate (100–200 nm) and small polyplexes, clathrin-coated pits and caveolae play a more important role, respectively (Grosse et al., 2005). The impact of different endocytosis pathways on nonviral gene delivery is still not fully understood. As discussed before, in vitro transfection of large-sized polyplexes formed under physiologic ionic strength are more efficient than small-sized polyplexes formed at low salt, isotonic condition. However, a trafficking study showed that these polyplexes coexist in multiple types of endosomal vesicles, some of which are acidified and others are not. Morphological studies using differentially labeled cholera toxin B and plasmid DNA revealed the co-localization of polyplexes in caveolae-mediated vesicles. Disruption of caveolae-mediated endocytosis via cholesterol depletion (methyl-β-cyclodextrin) or specific inhibitor of this pathway (genistein) showed greater inhibitory effects on transfection than inhibitors that block other endocytic pathways. This seems to suggest (although not exclusively) that caveolae-mediated endocytosis, which does not advance to lysosomal compartments in these cell types, does play an important role in PEI-mediated transfection (van der Aa et al., 2007).

22.2.5.2 Endosomal Release Mechanism

Both branched and linear forms of PEI have densely packed amine groups along their backbones, most of which (~80%) are nonprotonated at the physiological pH due to a neighboring effect of adjacent amine groups that depresses the pK_a values (Suh et al., 1994; Tang, and Szoka, 1997). These uncharged amines act as a "proton sponge" and effectively slow down the acidification process inside endosomes by neutralizing protons that are pumped into the endosome via an active membrane transporter, H^+-ATPase. This results in an interruption of maturation process from endosome to lysosome (Boussif et al., 1995; Akinc et al., 2005). At the same time, the accumulation of chloride counter ions and the buildup of osmotic pressure within the endosome compartment occur, which causes the swelling and rupture of the endosomal membrane (Sonawane et al., 2003). Such activity seems to be also shared by other polymers containing significant numbers of titratable amine groups, noticeably partially degraded polyamidoamine dendrimers (Haensler and Szoka, 1993), polyamines with overall backbone structures similar to PEIs (Gao et al., 2009; Zaliauskiene et al., 2010), biodegradable poly-β-aminoester (Arote et al., 2009), and certain cationic acrylic polymers (Hu et al., 2009).

22.2.5.3 Nuclear Trafficking Barrier

It is well known that cells undergoing rapid cell division are much easier to transfect than resting cells. Cells undergo drastic physiologic changes during cell cycling. These changes include alterations in the levels of cell surface proteoglycans, variations in the rate of pinocytosis, and nuclear reorganization. More efficient cellular uptake of polyplexes and transfection occurred during the S or G2 phase, while expression was the lowest in the G1 phase (Brunner et al., 2000; Männistö et al., 2007). It has been generally believed that translocation of DNA across the nuclear membrane represents a major rate-limiting step in the transfection process. The translocation requires either the disassembly of the nuclear envelope during cell division or active nuclear transport via the nuclear pore complex (NPC). For resting cells, the NPC has a size exclusion limitation for molecules of 40 kDa in MW or 10 nm in size. Polyplexes that exceed the diffusion limit but smaller than 60 nm in diameter can be imported into nuclei through an active transporting system. The efficiency for nuclear translocation of larger particles (150 nm) is low (Chan et al., 2000; Chan and Jans, 2002). To facilitate translocation, several strategies have been developed using synthetic peptides or proteins that contain a class of peptide sequences called nuclear localization signals (NLSs). These sequences can be covalently conjugated to DNA directly or indirectly through a spacer, such as streptavidin. Zanta et al. synthesized a NLS–oligonucleotide stem-and-loop cap structure and linked it to a 3.3 kb linear expression cassette to produce linear dsDNA with a single SV40 T-antigene NLS coupled to one end. This construct resulted in much higher transfection efficiency than the control construct that lacks the NLS cap structure (Zanta et al., 1999). Alternatively, DNA-binding proteins carrying NLS (such as Tet-NLS) at locations that are not overlapped with the DNA-binding region can be used to interact with the DNA that contains the binding sequence for protein and facilitate the nuclear entry (Chan and Jans, 2001; Vaysse et al., 2004). Another strategy is to incorporate into DNA backbone of sequences that specifically bind to different endogenous nuclear proteins that carry NLSs. Examples include SV40 replication origin, NF-kB binding sequences, tet O, and glucocorticoid response element (Dean et al., 1999; Mesika et al., 2001; Vacik et al., 1999; Dames et al., 2007). Nuclear DNA-binding proteins such as histones, protamine, high mobility group (HMG)-box proteins, topoisomerase, and hexon protein from adenovirus capsid have been used to condense and transfect cells either alone or together with other helping agents (Bottger et al., 1988; Chen et al., 2000; Carlisle et al., 2001; Kaouass et al., 2006; Kogure et al., 2008).

22.2.5.4 Effect of MW on PEI-Mediated Transfection Efficiency and Toxicity

PEI is a nonbiodegradable cationic polymer. At the dosages that are effective in transfection, it often causes toxicity. The transfection efficiency and toxicity of PEI are determined by several parameters, including molecular weight (MW), architecture, amine contents, and the ratio of polymer to DNA used. There is a clear trend that high MW PEIs (>25,000 Da) are significantly toxic and less efficient in transfection while polymers with low MW (<5,000 Da) are neither toxic, nor effective, with PEIs of medium-to-low MW range (5,000–25,000 Da) being the optimal transfection agents (Fischer et al., (1999). Synthetic, high MW PEIs derived from low MW PEIs crosslinked by reducible (Gosselin et al., 2001; Lee et al., 2007) or degradable low MW linkers (Tang et al., 2006; Park et al., 2008; Xu et al., 2008) are effective in transfection with reduced toxicity.

22.2.5.5 Effect of PEI Derivatization on Transfection Efficiency and Toxicity

PEIs derivatized by alkylation and acylation with many types of groups were tested to assess the importance of various polymer properties in transfection such as proton sponge capacity, hydrophobic–hydrophilic balance, and lipophilicity. Conversion of primary and secondary amine groups to tertiary amines or quaternary ammoniums significantly reduced the transfection efficiency, reaffirming the critical role of primary and secondary amines as proton sponge groups. Substitution with many other groups had rather weak effects, with the exception that lipid or cholesteryl derivatization on low MW PEI resulted in more efficient transfection in the presence of serum (Han et al., 2001; Thomas and Klibanov, 2002). Upon hydrophobic modification, these conjugates self-assemble into nanoparticles which effectively turn low MW PEI into larger species. In an interesting study on siRNA delivery using PEI modified with the hydrophobic amino acid tyrosine, the conjugates self-assembled into nanoparticles capable of forming stable polyplexes with siRNA and the complexes were efficiently delivered into cells (Creusat et al., 2010). Unlike DNA, PEI does not form polyplexes with small siRNA that are stable enough in extracellular media for effective delivery. While PEI–tyrosine–siRNA polyplexes are stable in physiological media at neutral pH, the polyplexes disassemble into individual PEI–tyrosine conjugates at acidic pH, causing rupture of the endosomal membrane. Partial acetylation or succinylation that eliminates a portion of primary and secondary amines reduced toxicity and improved transfection efficiency of branched PEI (Forrest et al., 2004; Gabrielson and Pack, 2006; Zintchenko et al., 2008). Under proper conditions, surface modification of PEI modified with sugar residues (Grosse et al., 2008), monovalent PEG (Ogris et al., 1999; Sung et al., 2003), pluronic triblock polymers (P123) (Nguyen et al., 2000), or dextran (Erbacher et al., 1999) typically leads to improved stability and reduced toxicity while maintaining transfection efficiency. When compared with unmodified PEI, galactose–PEG–PEI polyplexes showed 4.5- and 11.6-fold increases of reporter gene expression in A549 cell line and in mouse lung, respectively (Chen et al., 2008a).

The protein transduction domain found in HIV Tat protein, TAT, is a short cationic arginine-rich peptide that acts as a ubiquitous ligand by interacting with anionic molecules on cell surface. A synthetic TAT-peptide mimetic, arginine 8-mer, was conjugated either directly or through various spacers with PEI and the resulting conjugates showed higher transfection efficiency in vitro and in vivo in mouse lungs. PEI–TAT derived from high MW PEI improved the internalization of PEI–TAT–DNA polyplexes, but the transfection efficiency was reduced compared to unmodified PEI. This might be due to altered kinetics of cellular transport and unpacking of the resulted polyplexes (Doyle and Chan, 2007). TAT-peptide conjugated at the terminus of PEG chain in branched PEI–PEG conjugate was also shown to have reduced transfection efficiency in vitro compared to unmodified PEI. However, significantly higher transfection efficiency and lower toxicity were found in mice after airway instillation; reporter gene expression was distributed throughout bronchial and alveolar tissue (Kleemann et al., 2005).

22.2.5.6 PEI-Mediated Transfection Assisted with Membrane-Lytic Peptides

Although PEI and its derivatives are active transfection agents, their performance can be further enhanced by the presence of polymers with additional functionalities. Significant improvement of transfection efficiency was achieved when a peptide or polymer with membrane lytic and/or nuclear localization activity was introduced via direct conjugation or co-formulated into the delivery system. Ogris et al. showed that cationic membrane-lytic peptide melittin–PEI conjugates have excellent activity to condense DNA into small particles. The membrane-lytic activity led to efficient release of DNA into cytoplasm, resulting in a significant increase of transfection activity over PEI in a number of established and primary cell types. A rapid onset of gene expression was detectable as early as 4 h. It reached maximal values after 12 h, instead of 24–32 h for regular PEI-mediated transfection. This suggests that melittin facilitates the release of DNA into the cytoplasm. Further studies using cytoplasmic microinjection of melittin–PEI polyplexes into fibroblasts produced nearly four times more reporter gene-expressing cells than for PEI polyplexes alone, suggesting that melittin motifs can also enhance cytoplasm–nucleus translocation of the polyplexes (Ogris et al., 2001). Moreover, PEI_{2000}–melittin conjugates mediated efficient transfection in primary and post-mitotic cells using a functional mRNA-based cytoplasmic expression system (Bettinger et al., 2001). The orientation of melittin linked to PEI was found to have a strong influence on the membrane-destabilizing activities and toxicity. N-terminally linked melittin–PEI conjugates showed more selective membrane-lytic activity at endosomal pH versus neutral pH and promoted better endosomal release of polyplexes and efficient gene delivery with less toxicity in different cell lines when compared to C-terminally linked melittin–PEI conjugates (Boeckle et al., 2005).

To further tune the membrane-lytic activity toward the endosomal pH over neutral pH to reduce the damage to plasma membrane, melittin derivatives were designed by replacing neutral glutamines (Gln-25 and Gln-26) with glutamic acid residues. The mutant peptides greatly improved the lytic activity of C-terminally linked melittin–PEI conjugates at the endosomal pH of 5 and demonstrated excellent transfection efficiency in several cell lines. These endosome-selective melittin–PEI conjugates were incorporated into EGF–PEG-shielded polyplexes and purified by SEC. The purified polyplexes mediated EGF-receptor-specific gene transfer at an efficiency up to 70-fold higher than polyplexes without melittin (Boeckle et al., 2006).

A fusion peptide, KALA-Antp, which contains the protein transduction domain (PTD) of the third alpha-helix of Antennapedia (Antp) homeodomain and the fusogenic peptide KALA, was designed to provide both cellular internalization and nuclear localization signal (Antp), and to promote the release of the peptide–DNA complexes and peptide–DNA–PEI complexes. Optimal KALA-Antp/DNA polyplexes were nearly 400–600-fold more efficient than Antp alone or poly-lysine-Antp in gene delivery. Incorporation of PEI into the system further increased the efficiency of KALA-Antp-mediated gene delivery (Min et al., 2010).

22.2.5.7 PEI-Mediated Transfection Assisted with Nuclear Proteins

To overcome the rate-limiting nuclear transport of exogenous DNA, high mobility group box 1 (HMGB1) was combined with branched or linear PEI (b-PEI or l-PEI) to form polymer/protein/DNA terplexes to provide a NLS to promote nuclear import. The pDNA/HMGB1/l-PEI ternary complexes had small sizes, low cytotoxicity, and up to 4.0-fold greater transfection than that for pDNA/l-PEI complexes and pDNA/b-PEI complexes, respectively (Shen et al., 2009). In a follow-up study, PEG-PEI was used to prepare terplexes; greater nuclear accumulation of pDNA was observed for HMGB1/PEG-PEI terplexes by confocal laser scanning microscopy, which was associated with enhanced transgene expression (Shen et al., 2010).

Adenovirus is known to be actively transported from the cytoplasm to the nucleus with its nucleocaspid structure largely remaining intact. Hexon, an adenovirus nucleocapsid protein that carries a NLS, was covalently linked to PEI (800 kDa) and the resulting conjugate was compared with PEI or PEI–albumin conjugate for transfection activity. PEI–hexon polyplexes gave 10-fold greater transgene expression than PEI/DNA or PEI–albumin polyplexes in HepG2 cells by transfection and in Xenopus laevis oocytes by microinjection. PEI–hexon polyplexes were more efficient in delivery at lower copy numbers than polyplexes containing a classical SV40 T-antigen NLS (Carlisle et al., 2001).

22.2.5.8 PEI Polyplexes Transfection In Vivo

Transgene can be delivered via multiple routes of administration in order to target pulmonary vascular endothelial cells or airway epithelial cells in different sections of the respiratory tree. Successful routes of administration include intravenous (i.v.), intranasal, intratracheal instillation, or aerosol delivery. Polyplexes formed between DNA and linear PEI are active in vivo when administered intravenously with a level of transfection similar or better than that of DNA/cationic lipid lipoplexes (Ferrari et al., 1997; Goula et al., 1998; Zou et al., 2000). The levels of cytokine induction by DNA/PEI complexes were much lower than that of DNA/lipid complexes (Hwang and Davis, 2001). In contrast, branched PEI has higher toxicity and lower transfection efficiency via vascular route, a finding partly explained by the fact that linear PEI can deposit more plasmid DNA in the lung vasculature than the branched PEI (Jeong et al., 2007). Initial optimization revealed that the dose of DNA, polymer-to-DNA ratio, and the solution used to prepare the polyplexes are important factors for successful in vivo transfection. As discussed earlier, studies showed that the degree of deprotection and MW of linear PEI had a major effect on the in vivo gene transfer. Fully deprotected linear PEIs of 20,000–40,000 Da are particularly efficacious (Thomas et al., 2005). Upon systemic administration, these polyplexes accumulate to a greater extent in the lungs. The expression of transgene in liver, heart, spleen, and kidney is orders of magnitude lower than in lungs (Goula et al., 1998). Pulmonary vascular endothelial cells are the major cell population that is transfected, although a number of other cell types are also positive for transgene expression, including type I and type II pneumocytes, and septal cells (Dif et al., 2006).

The polyplexes prepared from branched PEI are more stable and better suited for aerosolization to the airways. Transfection by aerosolization showed no sign of toxicity and the expression lasted for more than a week (Densmore et al., 2000; Gautam et al., 2000; Rudolph et al., 2000). Recently, aerosolization of PEI 25 kDa polyplexes in sheep showed equal potency as compared to the cationic lipid GL67, a well-studied transfection agent for CF gene therapy (McLachlan et al., 2007).

The polyplexes are sufficiently safe that they could be repeatedly administered to mice for up to 21 consecutive days without toxic side effects, suggesting a potential of this vector in gene therapy of CF (Ferrari et al., 1999). Transfection by inhalation of aerosolized polyplexes in distilled water, at a submicrogram dosage of DNA, was more (15-fold) efficient than direct instillation with a DNA dosage of 50 μg, suggesting that PEI is a very promising nonviral vector for pulmonary gene transfer (Rudolph et al., 2005).

Park et al. designed a degradable low MW PEI-poly(ethylene glycol) diacrylate copolymer (poly(ester amine)) and compared the gene expression of the polyplexes using i.v. injection and aerosolization in mice. The polymer was self-degraded in solution with a half-life time of approximately 20 h and exhibited high transfection efficiency with low toxicity. A single dose of polyplexes by aerosol administration to the lung resulted in much higher level of reporter gene expression both in lung and liver in poly(ester amine)-treated animals than in b-PEI transfected animals. Interestingly, gene expression was sustained at a high level for at least 7 days in poly(ester amine) transfected animals, while gene expression declined rapidly in the b-PEI transfected group (Park et al., 2008).

22.2.5.9 PEI Polyplexes Modified with Anionic Polymers for Intravenous and Airway Application

The need for surface modification for cationic polyplexes is twofold. Upon systemic administration, polyplexes of small particle size tend to aggregate to form larger complexes and accumulate in major tissues including lung and liver following i.v. administration, making targeted delivery to sites other than pulmonary vasculature difficult. Polyplexes delivered via airway can be readily trapped within anionic proteoglycan-enriched mucus hydrogel and cannot reach the epithelial cells below the mucus. In addition to surface conjugation of inert polymer such as PEG (5000 Da), pluronic triblock polymers (P127), or dextran to reduce the nonspecific interactions, biodegradable anionic polymers such as albumin, dextran sulfate, heparin, hyaluronic acid, alginate, and other synthetic or natural anionic polymers can be used to co-formulate with PEI and DNA to form ternary complexes to modify the surface charge and to overcome the toxicity and nonspecific interaction.

Wolff's group tested the strategy using polyanions with different charge densities to shield the cationic charges of polyplexes in order to reduce toxicity associated with i.v. administered polyplexes. The charge density, the distance between anionic groups to polymer backbone and the ratio between oppositely charged polymers determine the outcome of polyanion added polyplexes. Those with high charge density and shorter charge group to backbone distances have the tendency to disassemble DNA polyplexes by displacing DNA, while polyanions with a longer carboxyl/backbone distance effectively formed tertiary complexes (Trubetskoy et al., 1999).

Under optimal conditions, the inclusion of polyacrylic acid led to recharged polyplexes with anionic surface charges, some of which showed enhanced transfection efficiency and less toxicity in vivo (Trubetskoy et al., 2003).

Hyaluronic acid (HA), a natural anionic mucopolysaccharide, can be deposited onto the cationic surface of DNA/PEI complexes to recharge the surface potential and reduce nonspecific interactions with proteins. HA can also be used as a ligand to target-specific cell receptors. Furthermore, HA-coating enhanced the transcriptional activity of the plasmid/PEI complexes, probably through loosening the tight binding between DNA and PEI, which facilitated the approach of transcription factors. Amphoteric HA derivative having spermine side chains (Spn-HA) with a structure similar to HMG protein showed higher transcription-enhancing activity than HA. Plasmid/PEI/Spn-HA ternary complexes exhibited 29-fold higher transgene expression than naked plasmid/PEI complexes in CHO cells (Ito et al., 2006).

Sethuraman et al. designed poly(methacryloyl sulfadimethoxine)-*block*-PEG (PSD-*b*-PEG) copolymers as a pH-sensitive, conditional coating molecule for PEI polyplexes, taking the advantage that PSD groups possess negative charges at pH 7.4. Under weak acidic condition at pH 6.6, the charges are lost and the coating molecules dissociate from the polyplexes as indicated by the changes of particle size and zeta potential of DNA/PEI/PEG-*b*-PSD polyplexes. The polyplexes regain positive charges to allow transfection to take place (Sethuraman et al., 2006). This system has been developed for systemic gene delivery to tumors in vivo as the complexes are likely to be stable in blood circulation due to the shielding by PSD-*b*-PEG. Following reaching the acidic tumor microenvironment, degrafting of the PSD-*b*-PEG coating will lead to regain of surface positive charges, resulting in efficient interaction with tumor cells and their subsequent transfection.

A bifunctional shielding/targeting molecule with defined geometry was designed with multiple anionic succinyl groups located on one end of PEG and a RGD peptide targeting ligand on the other end. The PEG–succinyl derivative was used to coat the PEI polyplexes and protect these particles from albumin-induced aggregation. Improved in vitro transfection was achieved in Bl6 melanoma cells and high levels of transfection were obtained in vivo in tumor, lungs, and liver after i.v. administration (Sakae et al., 2008).

Nonspecific interaction of cationic polyplexes with thick anionic mucus secretions enriched in CF airway poses a significant problem to effective transfection of epithelial cells underneath the mucus layer. Human serum albumin was added to preformed PEI–DNA polyplexes and used to transfect confluent A549 and 9HTEo cells. The HSA terplexes increased luciferase expression in confluent cultures in a dose-dependent fashion up to 100 times as compared to PEI–DNA. No significant cytotoxicity was observed with either PEI or PEI-HSA–DNA terplexes. The terplexes were inhibited to a less extent by CF sputum than with PEI polyplexes (Carrabino et al., 2005).

22.3 POLYPEPTIDE GENE DELIVERY CARRIERS

Cationic homopolypeptides such as PLL, poly-L-ornithine, and poly-L-arginine carry abundant primary amine or guanidine groups that are expected to be fully protonated at the physiological pH ($pK_a \geq 10.5$). They are effective DNA condensing agents but with limited transfection capacity due to poor endosomal release and slow dissociation of the PLL–DNA complexes. PLLs with a dendritic configuration have better transfection activity than the linear counterparts (Ohsaki et al., 2002). A number of derivatives of linear PLL with modifications on the amine groups have been made, which showed improved transfection activity. Replacement of a portion of the primary amine groups with titratable imidazole groups ($pK_a \sim 6$) in the backbone of PLL resulted in polymers with proton absorption capacity and enhanced transfection activity, presumably due to increased endosomal release (Midoux and Monsigny, 1999). Hydrophobitized PLLs have been synthesized via conjugation with fatty acids with different chain lengths and levels of saturation. These PLL derivatives showed increased transfection efficiency as a single agent or in conjunction with LDL in primary cells (Kim et al., 1998; Abbasi et al., 2007).

A series of grafted diblock polymers with PEG and oligolysine represent the most impressive PLL derivatives in this category. Well-defined diblock polymers PEG-lysine$_{30}$ with a short 30 lysine cationic peptide segment and a bulky neutral PEG segment (10 kDa) can be manufactured under good manufacturing practice (GMP) condition (Ziady et al., 2003a). The unique geometric shape of the polymer conjugates enables the formation of small nanoparticles of ellipsoids or rods from individual plasmid DNA molecules with hairy PEG chains facing outward that effectively prevent secondary interaction between nanoparticles. As a result, complexes with small sizes of 10 nm in diameter can be readily formed at very high concentrations of polymer and DNA without significant aggregation or precipitation (Fink et al., 2006). The presence of PEG chains significantly improves the biocompatibility and particle stability and reduces toxicity. Both forms of nanoparticles are very effective transfection agents and result in widespread and high levels of gene expression in epithelial cells following administration to airways by direct instillation (Ziady et al., 2003b), or by intraocular injections (Farjo et al., 2006) without significant toxicity to the target tissues. At a dosage (100 µg) that gave maximal level of gene expression, only modest increases in neutrophil cell count, IL-6, and KC were noticed at 48 h in bronchoalveolar lavage (BAL) fluid (Ziady et al., 2003b). Cell surface nucleolin was shown to serve as the receptor for binding and internalization of PEG-lys$_{30}$/DNA complexes as indicated by surface plasmon resonance measurement for direct interaction between nucleolin and stable complexes and by fluorescence microscopic examination for colocalization. The role of nucleolin was further established in siRNA knockdown studies (Chen et al., 2008). Further studies showed that the initial binding is followed by internalization via lipid-raft, and finally transport to the nucleus as a nucleolin complex with the glucocorticoid receptor (Chen et al., 2011).

Most of cationic homopolyamino acids are not very effective transfection agents in vitro and in vivo due to poor endosome escape and slow complex dissociation. A high MW poly(oligo-D-arginine) was synthesized by polymerizing cys-nonargine-cys peptide units through terminal cysteine residues to form oligomers with reducible disulfide bonds. The high MW poly(oligo-D-arginine) is expected to break down to promote the dissolution of the polyplexes intracellularly once they are exposed to reducing environment in cytosol. In vivo transfection in lungs by intratracheal instillation of DNA/poly(oligo-D-arginine) polyplexes resulted in higher levels of gene expression than PEI polyplexes, which lasted for 1 week without toxicity. These data suggest poly(oligo-D-arginine) could be a promising nonviral gene carrier for lung diseases (Won et al., 2010).

Rittner et al. designed a new series of amphiphilic cationic peptides, ppTG1 and ppTG20 (20 amino acids), and tested their efficiency in vitro and in vivo as single-component gene transfer vectors. These peptides formed complexes with nucleic acids and mediated efficient transfection in several cell lines at low charge ratios, particularly at low DNA doses. Intravenous injection of a reporter plasmid/ppTG1 or ppTG20 polyplexes led to significant gene expression in the lung 24 h after injection. These amphipathic peptides possess membrane lytic activity toward liposomes, and their gene transfer activity is correlated with their propensity to form an α-helical conformation. Significant toxicity was noticed for these peptides in vivo (Rittner et al., 2002).

22.4 MOLECULAR CONJUGATES: TARGETED DELIVERY OF POLYPLEXES TO PULMONARY VASCULATURE ENDOTHELIAL CELLS

Targeted delivery using specific ligands allows the use of surface neutral polyplexes with improved biocompatibility. Several targeting strategies specific to pulmonary endothelial cells have been explored to deliver genes, oligonucleotides, and siRNA to the pulmonary vasculature (Kuruba et al., 2009). To achieve efficient systemic gene delivery to the lung with minimal toxicity, linear PEI was conjugated with antiplatelet/endothelial cell adhesion molecule (PECAM) antibody (Ab) via a disulfide linkage. Grafting of antibody onto PEI–DNA complexes led to a shielding effect that reduced the surface charge of polyplex particles. The resultant anti-PECAM Ab-PEI polyplexes transfected cultured primary mouse lung endothelial cells more efficiently than PEI polyplexes. Furthermore, the anti-PECAM Ab-PEI polyplexes were able to transfect mouse lung endothelial cells at lower N/P ratios, much more efficiently than PEI polyplexes and control IgG–PEI polyplexes, suggesting that the cellular uptake of anti-PECAM Ab-PEI polyplexes and subsequent gene expression were governed by a receptor-mediated process rather than a nonspecific charge interaction. Anti-PECAM Ab-PEI polyplexes also demonstrated significant improvement in the distribution of polyplexes in pulmonary vasculature and enhanced gene expression in lungs after i.v. administration. The antibody-directed transfection led to a decreased serum level of proinflammatory TNF-α compared to PEI polyplexes. These results indicate that targeted gene delivery to the lung endothelium is an effective strategy to enhance gene delivery to the pulmonary circulation while simultaneously reducing toxicity (Li et al., 2000).The same targeting antibody has been used to direct specific delivery of oligonucleotides (Wilson et al., 2005) and siRNA (Wilson et al., unpublished data) to the pulmonary vasculature using a neutral lipidic formulation. Several other antibodies have been used for targeted delivery to the pulmonary vascular endothelial cells of various types of therapeutic agents such as antioxidant enzymes, peptide fusion constructs, and adenoviral vectors (Muzykantov, 2005). For example, a bispecific antibody to angiotensin-converting enzyme (ACE) and adenovirus has been used successfully to increase pulmonary gene transfer by adenovirus by 20-fold (Reynolds et al., 2000).

Several short peptide sequences have been discovered via in vivo phage display biopanning, which show preferential binding to pulmonary vascular endothelial cells. The use of peptides as cell-specific targeting ligands has the advantages of ease of synthesis, improved stability, and reduced immunogenicity compared to antibodies. Rojotte et al. reported on GFE-1 (CGFECVRQCPERC) that binds to a membrane dipeptidase as a cell surface receptor (Rajotte et al., 1998; Rajotte and Ruoslahti, 1999). CGSPGWVRC was identified by Giordano et al. (target unknown) and used to direct in vivo delivery of an apoptosis-inducing fusion peptide to establish an emphysema disease model (Giordano et al., 2008). Work et al. discovered many peptide sequences that showed preferential binding to either the brain or pulmonary vasculature. Selected sequences were displayed on adenovirus particles to direct altered tropism to different vasculature beds in WKY rats. One of the peptides, VNTANST, was able to induce a 32-fold improvement in gene transduction to the pulmonary vasculature (Work et al., 2006). Greig et al. reported the use of two lung endothelium-specific sequences (CRPPR and CSGMARTKC) for pulmonary delivery of an antioxidant peptide gp91ds, which selectively inhibits assembly of NAD(P)H oxidase (Greig et al., 2010).

22.5 LIGANDS TARGETED TO AIRWAY EPITHELIAL CELLS

Several peptides and proteins have been reported capable of binding and being internalized by airway epithelial cells and gland epithelial cells (Table 22.2). Incorporation of these ligands into polyplexes could potentially improve the surface properties, promote their cellular entry, and reduce nonspecific interactions that may contribute to toxicity.

Lactoferrin (Lf) is one of the components of the mucosal innate defense system with antimicrobial activity (bactericide, fungicide). Lf interacts with anionic macromolecules, such as DNA and RNA, polysaccharides, and heparin. The Lf receptor binds and internalizes Lf. BEAS-2B human bronchial epithelial cells express high level of Lf receptor. PEI–Lf conjugates were synthesized and tested for the ability to enhance transfection by PEI polyplexes. At a low N/P ratio of 4, PEI–Lf polyplexes gave fivefold higher gene expression than PEI polyplexes. The transfection could be inhibited by an excess of free Lf. Lf–PEI polyplexes showed significantly lower cellular toxicity compared to PEI polyplexes (Elfinger et al., 2007).

TABLE 22.2
Peptide Ligand Facilitated Gene Delivery

Peptide	Sequence	Target Molecules/Cell Lines	References
RGD-peptide	ACRGDMFGCA	Integrin receptor/HUVECs	Schiffelers et al. (2004)
GFE	CGFECVRQCPERC	Dipeptidase/endothelial cells	Rajotte et al. (1998)
CGSPGWVRC	CGSPGWVRC	Unknown/mouse endothelial cells	Giordano et al. (2008)
VNTANST	VNTANST	Unknown/rat pulmonary vasculature	Work et al. (2006)
CD13 ligand	CNGRC	CD13/HUVECs, HT-1080, H1299	Moffatt et al. (2005)
CRPPR	CRPPR	Mouse pulmonary endothelial cells	Greig et al. (2010)
CSGMARTKC	CSGMARTKC	Mouse pulmonary endothelial cells	Greig et al. (2010)
IB1	CDSAFVTVDWGRSMSLC	Calu-3	Florea et al. (2003)
E	*K16*-GACSERSMNFCG	Airway epithelail cells	Tagalakis et al. (2008)
	(R/K)SM, L(P/Q)HK	1HAEo-cells	Writer et al. (2004)
	PSG(A/T)ARA		
THALWHT	THALWHT	Airway epithelial cells	Jost et al. (2001)
EHMALTYPFRPP	EHMALTYPFRPP	Lung cancer biopsy	Zang et al. (2009)
TAT	GRKKRRQRRRPPQ	16HBE, primary nasal epithelium	Rudolph et al. (2003)
TAT	GRKKKRRQRC	A549 lung cancer cells	Kleemann et al. (2005)
		Mouse lung airways	

The insulin receptor was known to internalize insulin upon binding. Insulin receptors exist on freshly isolated rabbit type II alveolar epithelial cells (pneumocytes). Addition of insulin to preformed PEI–DNA polyplexes increased particle sizes from 45 to 200–300 nm in diameter and reduced the surface charge, suggesting an association of insulin with polyplexes. The addition of insulin leads to enhanced transfection by approximately 16-fold in A549, the type II epithelial cells-derived adenocarcinoma cells that overexpress insulin receptor. In addition to increases in both the number of transfected cells and the level of gene expression per cell, this targeting system was associated with decreased cellular toxicity (Elfinger et al., 2009). In contrast, this system showed minimal effect on a normal bronchial epithelial cell-derived cell line.

IB-1 (CDSAFVTVDWGRSMSLC) is a peptide sequence that was identified using fully differentiated Calu-3 cell line as a screening target. Incorporation of biotinylated IB1/streptavidin complexes into PEI/DNA polyplexes led to a four- to sixfold enhancement in transfection of Calu-3 cells compared to PEI polyplexes (Florea et al., 2003). Tagalakis and colleagues identified a peptide K(16)GACSERSMNFCG that showed preferential binding toward airway epithelial cells over alveolar epithelial cells or macrophages (Tagalakis et al., 2008).

Writer et al. identified 14 peptide sequences by biopanning on 1HAEo⁻ cells, a well-characterized epithelial cell line. Thirteen of the peptides bound to 1HAEo⁻ cells with high affinity. Three clearly defined families of peptide were identified on the basis of sequence motifs, including (R/K)SM, L(P/Q)HK, and PSG(A/T)ARA. Peptides were incorporated into lipofectin/DNA complexes and shown to confer a high degree of transfection efficiency and specificity in 1HAEo⁻ cells. Improved transfection efficiency and specificity was also observed in human endothelial cells, fibroblasts, and keratinocytes (Writer et al., 2004).

THALWHT is a selective binding motif to airway epithelial cells that was identified by Jost and colleagues. Exposure of the epithelial cells to the labeled peptide led to specific binding followed by endocytosis as shown in a confocal microscopic study. A synthetic fusion peptide comprising a cyclic CTHALWHTC domain and a DNA-binding moiety enabled efficient targeted gene delivery into human airway epithelial cells. Competition assays with free THALWHT peptide confirmed the specificity of gene delivery (Jost et al., 2001). A similar approach has been used to select a peptide (EHMALTYPFRPP) that specifically binds to a lung cancer cell line and biopsies, but not normal cells. This peptide may be a potential ligand for targeted drug delivery for the treatment of lung cancer (Zang et al., 2009).

22.6 MOLECULAR CONJUGATES: TARGETED DELIVERY OF POLYPLEXES TO AIRWAY EPITHELIAL CELLS VIA BASOLATERAL TRANSCYTOSIS FOLLOWING INTRAVASCULAR ADMINISTRATION

The widespread and constant presence of a thick mucus layer on the surface of respiratory airways in patients with severe inflammatory diseases, such as CF, presents a serious challenge for an effective transfection of airway epithelial cells using positively charged particles, because these particles are most likely to be trapped and immobilized within the mucus hydrogel matrix and are unable to reach epithelial cells. A clever concept was proposed by Davis's group to overcome this difficulty by taking advantage of the efficient transcytosis process that is physiologically used to transport immunoglobulin Ig A and IgM from blood to the mucosal surface through epithelial cells. The transfecting agent first travels through the endothelial cell lining, then approaches epithelial cells basolaterally. The agent is subsequently endocytosed by epithelial cells through the polymeric immunoglobulin receptor (pIgR). Thus, the apical mucus layer is not a limiting factor by this route (Ferkol et al., 1995).

To achieve this aim, Fab fragments of anti-human secretory component (the extracellular portion of pIgR) antibody were prepared, linked to poly-L-lysine, and complexed to LacZ reporter plasmid. HT29.74 human colon carcinoma cells and primary cultures of human tracheal epithelial cells were induced to express the polymeric immunoglobulin receptor and transfected with the complexes.

A significant number of the respiratory epithelial cells expressed β-galactosidase activity after treatment. The transfection could be blocked by the addition of excess human secretory component to the culture medium at the time of transfection. The control polyplexes with an irrelevant Fab fragment were not effective, suggesting that this transfection system introduces DNA specifically into epithelial cells that express pIgR (Ferkol et al., 1993). To test the performance in vivo, Sprague Dawley rats were transfected by i.v. injection of anti-secretory component–polylysine polyplexes of small particle sizes, prepared using high ionic strength. Significant levels of luciferase activity were detected in extracts from liver and lung but not spleen or heart, tissues which do not express the receptor. Transfections using complexes with an irrelevant Fab fragment resulted in only background levels of luciferase activity. Histochemistry results showed that the reporter gene expression was localized to the airway epithelium and the submucosal glands. These data suggest that gene transfer targeting the lung epithelial cells can be achieved from blood side using polyplexes specific to pIgR system (Ferkol et al., 1995).

Other endothelial cell markers, such as aminopeptidase P, are capable of taking up ligand specifically through caveoli-mediated transcytosis, a process that has been studied and found to be very rapid and efficient in moving surface-bound antibody across endothelium into lung tissue, suggesting this can be potentially useful for drug and gene delivery to the epithelium from the blood (Oh et al., 2007).

22.7 PRECLINICAL AND CLINICAL THERAPEUTIC APPLICATIONS FOR LUNG DISEASES

22.7.1 Cystic Fibrosis

CF is a monogenic, autosomal recessive genetic disease caused by a defective cystic fibrosis transmembrane conductance regulator (CFTR) gene. CFTR acts as a chloride channel to concentrate chloride ions at the apical surface and thereby promote osmotic movement of water across the epithelial lining. The CFTR defect in CF patients causes dehydration of mucin into a thickened plug that cannot be removed readily by cilia movement. The thickened mucus layer provides a fertile environment for bacteria growth, which leads to chronic infection, inflammation, and progressive lung damage. In addition, blockade of pancreatic ducts also leads to dysfunction of this important digestive organ. Currently, there is no cure available, other than limited supportive care for these patients. The life expectancy of these patients is about mid-30s. Gene therapy provides a hope initially, but gene delivery technology remains largely inadequate. The hyperinflammatory nature of the disease and the rapid turnover of respiratory airway epithelial cells preclude the repeated use of immunogenic viral vectors. Several phase I/II clinical trials with nonviral cationic liposome/CFTR plasmid lipoplex formulations have been evaluated in humans, mostly by nasal application. A few trials have involved aerosolized delivery to the lungs. Modest success was achieved judging from partial correction of chloride channel activity, low and tolerable side effects, and evidence of low levels of transient transgene expression in some patients receiving the DNA-containing lipoplexes. The side effects are generally mild, ranging from none to "Flu-like" symptoms.

Up to this point, PEG-oligolysine compacted DNA polyplexes was the only polymer-mediated airway gene delivery method tested in a clinical trial based on the promising preclinical results in animals. Desirable characteristics included high transfection efficiency, low toxicity, as well as practical aspects, such as formulation stability during aerosolization and feasibility of production of DNA polyplex under GMP conditions. A double-blind, phase I dose escalation clinical trial was conducted in 12 subjects with CF. Patients received a nasal application of placebo (saline) or increasing dosages of the compacted DNA encoding a functional CFTR transgene. Nasal lavage and serum samples were collected at different time points. Cell counts, cytokine levels, the presence of plasmid, and the transgene expression levels were measured. No serious adverse events occurred, and there was no association of increased inflammatory mediators in serum or nasal samples with administration of compacted DNA. The compacted DNA nanoparticles can be safely administered into the nares of CF subjects, with evidence of relatively long-term presence of vector DNA in treated nasal epithelium. Despite being unable to detect transgene expression, there was evidence of partial to complete correction of the nasal potential difference in majority of the subjects (Konstan et al., 2004). It remains to be demonstrated that DNA polyplexes can be safely and efficiently delivered to the respiratory epithelium in the form of an aerosol and that such transfection can produce a level of gene expression that is sufficient to be therapeutically beneficial. Furthermore, toxicity of repeated administration of the transfection agents needs to be thoroughly evaluated.

Collectively, the results from clinical trials showed general promising safety and some efficacy as indicated in partial correction of CFTR channel activities. These benefits were detected in all trials using lipid or polymer-based gene delivery methods and the effects were shown to last as long as 1 week. There is a discrepancy on the threshold of CFTR mRNA transcription level that is required in the airway epithelium to correct chloride permeability. Researchers were encouraged by an estimation based on in vitro data, suggesting that transfection of exogenous CFTR cDNA into only 5%–10% of the cells is sufficient to correct the defective chloride secretion in CF, a level seemingly achievable by existing gene transfer methods (Yoshimura et al., 1991; Rosenfeld et al., 1994). However, recent studies indicated that significantly more cells need to be transfected with normal CFTR cDNA in order to normalize sodium absorption of the airway epithelium in CF patients (Johnson et al., 1992; Goldman et al., 1995). The consensus is that the expression levels achieved in humans are below the level required to achieve a significant clinical benefit. As discussed before, the presence of a thick mucus layer in CF patients clearly makes the delivery through airway route more difficult. Another issue is the difficulty in achieving transfection in serous cells of the secretory glands at a level that is sufficient to correct respiratory manifestations. This cell population is not readily accessible via airway route for any delivery systems (Engelhardt et al., 1992). New developments in transfection technologies with improved transfection efficiency, possibly through ligand-facilitated delivery, and/or alternative routes and prolonged expression of the transgene are needed to meet the needs in clinical trials.

22.7.2 Lung Cancers

Several strategies have been tested to treat lung cancers. These treatments are based on the expression of genes of therapeutic potential, such as a tumor suppressor gene (e.g., P53), or a gene that inactivates oncogenes (such as adenoviral E1a, FUS-1), a gene-directed enzyme–prodrug combination therapy (also known as suicide gene therapy), a cytokine that is effective in boosting immunotherapy (IL-12, IL-7, interferon-γ, GM-CSF), and molecules that attack tumor vasculature (such as soluble VEGF receptor). Although most of these studies involved the use of viral vectors, polymer-based gene delivery either through airways or systemic routes has also been vigorously studied. For example, administration of PEI:IL-12 expressing plasmid by aerosolization in a mouse osteosarcoma lung metastases model resulted in selective gene expression and protein production in the tumor area, reduced tumor burden, and metastases sizes. Aerosol-delivered PEI:IL-12 locally produced IL-12, which may avoid systemic toxicities associated with i.v. IL-12 dosing in patients. Because osteosarcoma develops metastases almost exclusively in the lung, aerosolization of PEI:IL-12 expressing vector may be therapeutically relevant (Jia et al., 2003).

The targeting of tumors via the systemic route using ligand-directed PEI/DNA compelxes has been reported. Moffatt et al. have developed a tumor-targeted PEI polyplex formulation capable of efficient tumor-specific delivery after i.v. administration to nude mice. A peptide sequence, CNGRC specific for aminopeptidase N (CD13), was used as the ligand, linked to polyplexes through a PEG linker. Intravenous administration of the CNGRC polyplexes to nude mice bearing subcutaneous tumors resulted in up to a 12-fold increase in reporter expression in tumors as compared with expression in either lungs or tumors from animals treated with unliganded PEI polyplexes. The expression of yellow fluorescence protein in subcutaneous human lung cancer model H1299 confirmed successful delivery of plasmid to both tumor cells and tumor endothelial cells (Moffatt et al., 2005). Further studies by the same group suggest that the in vivo expression levels can be enhanced by incorporating two types of NLSs to DNA constructs. A SV40 NLS was non-covalently linked to DNA through peptide nucleic acid chemistry; and a DNA nuclear targeting sequence capable of binding nuclear proteins was incorporated into plasmid backbone, resulting in an approximately 200-fold higher reporter gene expression in vitro and an approximately 20-fold enhanced selectivity in tumors. An EBV-based episomal vector was used to achieve a sustained p53 gene expression in tumor cells. Both tumor and tumor-associated endothelial cells but not normal cells were targeted. The expression of wild-type p53 causes tumor cell apoptosis and a significant tumor regression accompanied with 95% animal survival after 60 days. The strategies used exemplify the most sophisticated rational design for more efficient nonviral gene therapeutics (Moffatt et al., 2006).

22.7.3 OTHER LUNG DISEASES

22.7.3.1 Asthma

Allergic, eosinophilic inflammation of the airway is considered as a basic mechanism of an asthma attack. Induction and proliferation of type II helper T cells (Th2) that produce IL-4 and -5 is essential for the establishment of allergen-specific immune responses. It has been demonstrated that intratracheal administration of Th1 cytokines, such as IFN-γ or IL-12, can tilt the balance of Th1 and Th2 responses and suppress eosinophil recruitment into the airways and prevent allergic sensitizations in mice (Sur et al., 1996). Aerosolization of PEI-IL-12 expressing plasmid polyplexes may be beneficial for such treatment.

22.7.3.2 Pulmonary Fibrosis

Idiopathic pulmonary fibrosis (IPF) is a rapidly progressive disorder with poor prognosis. The 5 year survival rate is only about 50% after the initial diagnosis. The causes of IPF seem to have both genetic as well as environmental factors contributing to the disease progression. The apoptosis of alveolar epithelial cells via a Fas-mediated pathway, abnormal TGF-β and other growth factor-mediated signaling pathways, and excessive inflammatory response in the airways are involved in the onset of the disease (American Thoracic Society, 2000). Several experimental treatments, designed to protect respiratory epithelial cells from apoptotic cell death, suppress growth factor activation, or inhibit inflammatory response and the fibrosis process, have been conducted in experimental animal models of pulmonary fibrosis. Ad-mediated hepatic growth factor (HGF) expression systemically or locally in the lung also prevented collagen deposition following bleomycin treatment (Yaekashiwa et al., 1997). Airway administration of adenovirus expressing SMAD, a transcription factor that inhibits TGF signaling pathway, resulted in attenuation of bleomycin-induced lung fibrosis in mice (Nakao et al., 1999). In addition, overexpression of the bone morphogenetic protein receptor type 2 (BMPR2) resulted in upregulation of SMAD signaling, reduced cell proliferation, and attenuated hypoxic pulmonary hypertension (Reynolds et al., 2007).

Suppressing inflammatory aspect of IPF by shifting Th1/Th2 imbalance in the lung with Th1 cytokines, such as IFN-γ, represents another potential approach to treat this disease. Alveolar macrophages from patients with IPF are known to be defective in IFN-γ production. Systemic administrations of Th1 cytokines have been shown to prevent lung fibrosis induced by bleomycin in animal models (Gurujeyalakshmi and Giri, 1995). An added benefit with IFN-γ treatment is the suppression of expression of cytokines known to induce proliferation of fibroblasts, such as TGF-β and connective tissue growth factor (CTGF). Preliminary results from a small-sized clinical trial of 18 IPF patients treated with long-acting IFN-γ-1b and glucosteriods showed improved lung volumes and gas exchange parameters, as well as lowered TGF-β and CTGF mRNA transcripts (Ziesche et al., 1999). However, a subsequent double-blind, placebo-controlled trial involving s.c. administration of IFN-γ-1b in 330 IPF patients has not been able to confirm the initial observations in terms of both molecular changes (Strieter et al., 2004) and clinical benefits (Raghu et al., 2004). It remains to be tested if locally expressed IFN-γ through less-toxic gene transfer approaches would have any therapeutic values in IPF.

22.7.3.3 Acute Lung Injury

ALI is a severe pathologic response secondary to many initial assaults, including sepsis, multiple traumas, and pancreatitis. This condition involves activation and subsequent migration of inflammatory cells into the lung, predominately the interstitial space of the pulmonary vasculature. The activated inflammatory cells release free radicals and proteases that induce severe injuries in endothelial and parenchymal cells, causing vascular permeability changes (edema) and respiratory failure (Artigas et al., 1998). Strategies that antagonize pro-inflammatory cytokines, subdue cytokine-induced cell activation, and protect cell injuries have been tested to manage the disease progression. Soluble TNF-α receptor acts as an antagonist of the potent pro-inflammatory cytokine by competing with TNF-α receptor. Soluble TNF-α receptor or IL-10 (an anti-inflammatory cytokine) that was expressed from recombinant adenoviral vectors effectively protected mice from LPS-induced sepsis (Rogy et al., 1995). Antisense oligonucleotides against pro-inflammatory cytokines (IL-1 or TNF-α), IL-1 receptor, or a downstream adhesion molecule participating in inflammation (ICAM-1) have been shown to possess effective inhibitory effects in vivo.

Apoptosis of several cell types including lymphocytes, parenchyma cells (including intestinal and lung epithelial cells), and vascular endothelial cells is increased during ALI. Increased cell death through apoptosis contributes to the onset of organ failure and mortality associated with sepsis (Oberholzer et al., 2001). Gene silencing using siRNA specific to Fas gene has shown promising results for the prevention of cell death in different acute experimental models of liver and kidney injury (Song et al., 2003; Hamar et al., 2004). Matsuda et al. (2009) showed that systemic administration of siRNA targeting Fas-associated death domain (FADD), which recruits procaspase-8 into the death-inducing signaling complex, was protective in septic ALI and decreased mortality in a polymicrobial sepsis model induced by cecal ligation and puncture (CLP) in BALB/c mice. In vivo delivery of siRNA at 10 h after CLP prevented the ALI development, judging from the fact that parameters of blood-gas derangements, histologic lung damage, increased pulmonary inflammatory cells, and the survival of CLP mice were dramatically improved in siRNA-treated group. These results suggested the significance of the role of the death receptor apoptotic pathway, including FADD, in septic ALI and the potential therapeutic value of FADD siRNA for septic syndrome.

Conary et al. have shown that expression of protective lipid mediator in pulmonary vasculature by intravascular transfection of endothelial cells, a prostaglandin G/H synthase gene, is potentially promising strategy in an experimental model (Conary et al., 1994). Several studies have demonstrated the cytoprotective roles of HO-1 and/or its enzymatic reaction byproduct carbon monoxide (CO) in various acute lung injury models (Jin and Choi, 2005). HO-1 exerts its anti-oxidative, anti-inflammatory, and anti-apoptotic roles through several mechanisms. HO-1 reduces apoptotic cell death under oxidative stress through CO by activating p38 mitogen-activated protein kinase. HO-1 downregulates the production of pro-inflammatory cytokines, GM-CSF, IL-6, IL-1b, and TNF-α, while increasing anti-inflammatory IL-10 production (Morse, 2003; Otterbein et al., 2003). Sarady et al. have demonstrated that inhaled CO could increase the survival of rats exposed to lethal endotoxemia (Sarady et al., 2004). CO can inhibit iNOS expression in lung and alveolar macrophages and attenuate neutrophilic inflammation in response to LPS or hyperoxia (Otterbein et al., 2003). Preadministration of HO-1 expressing adenovirus intratracheally improved the survival rate of rats under hyperoxia (Miller et al., 2005).

22.8 CONCLUDING REMARKS

Although many examples of experimental gene therapy for various lung diseases discussed here were conducted with other gene transfer method, most notably viral vectors, opportunities as well as challenges clearly exist for the further development of polymer-based gene transfer methodologies and the testing of these therapeutic approaches using improved delivery vectors. Novel polymers with improved structural features that lead to enhanced transfection capacity and reduced toxicity are needed to maximize in vivo application. Improved target specificity and reduced nontarget uptake are crucial for systemic applications. Enhanced endosomal releasing capacity of the delivery system represents another important strategy to improve transfection efficiency. Better plasmid design that extends expression duration through carefully selecting the optimal promoter, inclusion of nuclear matrix association sequences, and the use of NLS also seem to be desirable attributes. One of the major advantages of polymeric systems is the flexibility in their design, which allows incorporation of various new functionalities to overcome various cellular and in vivo barriers. Our improving understanding of cell biology involved in transfection will certainly facilitate the effort of rational design of gene delivery vectors to maximize transfection efficiency and minimize toxicity.

ACKNOWLEDGMENT

This work was supported by National Institutes of Health Grants HL-68688 and HL-91828.

REFERENCES

van der Aa MA, Huth US, Häfele SY, Schubert R, Oosting RS, Mastrobattista E, Hennink WE, Peschka-Süss R, Koning GA, Crommelin DJ. (2007). Cellular uptake of cationic polymer-DNA complexes via caveolae plays a pivotal role in gene transfection in COS-7 cells. *Pharm Res* 24:1590–1598.

Abbasi M, Uludag H, Incani V, Olson C, Lin X, Clements BA, Rutkowski D, Ghahary A, Weinfeld M. (2007). Palmitic acid-modified poly-L-lysine for non-viral delivery of plasmid DNA to skin fibroblasts. *Biomacromolecules* 8:1059–1063.

Abdullah S, Wendy-Yeo WY, Hosseinkhani H, Hosseinkhani M, Masrawa E, Ramasamy R, Rosli R, Rahman SA, Domb AJ. (2010). Gene transfer into the lung by nanoparticle dextran-spermine/plasmid DNA complexes. *J Biomed Biotechnol* 2010:Article ID 284840.

Akinc A, Thomas M, Klibanov AM, Langer R. (2005). Exploring polyethylenimine-mediated DNA transfection and the proton sponge hypothesis. *J Gene Med* 7:657–663.

American Thoracic Society. (2000). Idiopathic pulmonary fibrosis: diagnosis and treatment. International consensus statement. American Thoracic Society (ATS), and the European Respiratory Society (ERS). *Am J Respir Crit Care Med* 161:646–664.

Arote RB, Lee ES, Jiang HL, Kim YK, Choi YJ, Cho MH, Cho CS. (2009). Efficient gene delivery with osmotically active and hyperbranched poly(ester amine)s. *Bioconjug Chem* 20:2231–2241.

Artigas A, Bernard GR, Carlet J, Dreyfuss D, Gattinoni L, Hudson L, Lamy M et al. (1998). The American-European Consensus Conference on ARDS, Part 2: Ventilatory, pharmacologic, supportive therapy, study design strategies, and issues related to recovery and remodeling. Acute respiratory distress syndrome. *Am J Respir Crit Care Med* 157:1332–1347.

Bettinger T, Carlisle RC, Read ML, Ogris M, Seymour LW. (2001). Peptide-mediated RNA delivery: A novel approach for enhanced transfection of primary and post-mitotic cells. *Nucleic Acids Res* 29:3882–3891.

Boeckle S, Fahrmeir J, Roedl W, Ogris M, Wagner E. (2006). Melittin analogs with high lytic activity at endosomal pH enhance transfection with purified targeted PEI polyplexes. *J Control Release* 112:240–248.

Boeckle S, von Gersdorff K, van der Piepen S, Culmsee C, Wagner E, Ogris M. (2004). Purification of polyethylenimine polyplexes highlights the role of free polycations in gene transfer. *J Gene Med* 6:1102–1111.

Boeckle S, Wagner E, Ogris M. (2005). C- versus N-terminally linked melittin-polyethylenimine conjugates: The site of linkage strongly influences activity of DNA polyplexes. *J Gene Med* 7:1335–1347.

Bottger M, Vogel F, Platzer M, Kiessling U, Grade K, Strauss M. (1988). Condensation of vector DNA by the chromosomal protein HMG1 results in efficient transfection. *Biochim Biophys Acta* 950:221–228.

Boussif O, Lezoualc'h F, Zanta MA, Mergny MD, Scherman D, Demeneix B, Behr JP. (1995). A versatile vector for gene and oligonucleotide transfer into cells in culture and *in vivo*: Polyethylenimine. *Proc Natl Acad Sci USA* 92:7297–7301.

Brunner S, Sauer T, Carotta S, Cotten M, Saltik M, Wagner E. (2000). Cell cycle dependence of gene transfer by lipoplex, polyplex and recombinant adenovirus. *Gene Ther* 7:401–407.

Carlisle RC, Bettinger T, Ogris M, Hale S, Mautner V, Seymour LW. (2001). Adenovirus hexon protein enhances nuclear delivery and increases transgene expression of polyethylenimine/plasmid DNA vectors. *Mol Ther* 4:473–483.

Carrabino S, Di Gioia S, Copreni E, Conese M. (2005). Serum albumin enhances polyethylenimine-mediated gene delivery to human respiratory epithelial cells. *J Gene Med* 7:1555–1564.

Chan CK, Jans DA. (2001). Enhancement of MSH receptor-and GAL4-mediated gene transfer by switching the nuclear import pathway. *Gene Ther* 8:166–171.

Chan CK, Jans DA. (2002). Using nuclear targeting signals to enhance non-viral gene transfer. *Immunol Cell Biol* 80:119–130.

Chan CK, Senden T, Jans DA. (2000). Supramolecular structure and nuclear targeting efficiency determine the enhancement of transfection by modified polylysines. *Gene Ther* 7:1690–1697.

Chen J, Gao X, Hu K, Pang Z, Cai J, Li J, Wu H, Jiang X. (2008a). Galactose-poly(ethylene glycol)-polyethylenimine for improved lung gene transfer. *Biochem Biophys Res Commun* 375:378–383.

Chen TY, Hsu CT, Chang KH, Ting CY, Whang-Peng J, Hui CF, Hwang J. (2000). Development of DNA delivery system using Pseudomonas exotoxin A and a DNA binding region of human DNA topoisomerase I. *Appl Microbiol Biotechnol* 53:558–567.

Chen X, Kube DM, Cooper MJ, Davis PB. (2008b). Cell surface nucleolin serves as receptor for DNA nanoparticles composed of pegylated polylysine and DNA. *Mol Ther* 16:333–342.

Chen X, Shank S, Davis PB, Ziady AG. (2011). Nucleolin-mediated cellular trafficking of DNA nanoparticle is lipid raft and microtubule dependent and can be modulated by glucocorticoid. *Mol Ther* 19:93–102.

Conary JT, Parker RE, Christman BW, Faulks RD, King GA, Meyrick BO, Brigham KL. (1994). Protection of rabbit lungs from endotoxin injury by in vivo hyperexpression of the prostaglandin G/H synthase gene. *J Clin Invest* 93:1834–1840.

Cotten M, Wagner E, Zatloukal K, Phillips S, Curiel DT, Birnstiel ML. (1992). High-efficiency receptor-mediated delivery of small and large (48 kilobase) gene constructs using the endosome-disruption activity of defective or chemically inactivated adenovirus particles. *Proc Natl Acad Sci USA* 89:6094–6098.

Creusat G, Rinaldi AS, Weiss E, Elbaghdadi R, Remy JS, Mulherkar R, Zuber G. (2010). Proton sponge trick for pH-sensitive disassembly of polyethylenimine-based siRNA delivery systems. *Bioconjug Chem* 21:994–1002.

Dean DA, Dean BS, Muller S, Smith LC. (1999). Sequence requirements for plasmid nuclear import. *Exp. Cell Res* 253:713–722.

Densmore CL, Orson FM, Xu B, Kinsey BM, Waldrep JC, Hua P, Bhogal B, Knight V. (2000). Aerosol delivery of robust polyethyleneimine-DNA complexes for gene therapy and genetic immunization. *Mol Ther* 1:180–188.

Dif F, Djediat C, Alegria O, Demeneix B, Levi G. (2006). Transfection of multiple pulmonary cell types following intravenous injection of PEI-DNA in normal and CFTR mutant mice. *J Gene Med* 8:82–89.

Doyle SR, Chan CK. (2007). Differential intracellular distribution of DNA complexed with polyethylenimine (PEI) and PEI-polyarginine PTD influences exogenous gene expression within live COS-7 cells. *Genet Vaccines Ther* 5:11.

Elfinger M, Maucksch C, Rudolph C. (2007). Characterization of lactoferrin as a targeting ligand for nonviral gene delivery to airway epithelial cells. *Biomaterials* 28:3448–3455.

Elfinger M, Pfeifer C, Uezguen S, Golas MM, Sander B, Maucksch C, Stark H, Aneja MK, Rudolph C. (2009). Self-assembly of ternary insulin-polyethylenimine (PEI)-DNA nanoparticles for enhanced gene delivery and expression in alveolar epithelial cells. *Biomacromolecules* 10:2912–2920.

Engelhardt, JF, Yankaskas JR, Ernst SA, Yang Y, Marino CR, Boucher RC, Cohn JA, Wilson JM. (1992). Submucosal glands are the predominant site of CFTR expression in the human bronchus. *Nat Genet* 2:240–248.

Erbacher P, Bettinger T, Belguise-Valladier P, Zou S, Coll JL, Behr JP, Remy JS. (1999). Transfection and physical properties of various saccharide, poly(ethylene glycol), and antibody-derivatized polyethylenimines (PEI). *J Gene Med* 1:210–222.

Erbacher P, Bettinger T, Brion E, Coll JL, Plank C, Behr JP, Remy JS. (2004). Genuine DNA/polyethylenimine (PEI) complexes improve transfection properties and cell survival. *J Drug Target* 12:223–236.

Fahrmeir J, Gunther M, Tietze N, Wagner E, Ogris M. (2007). Electrophoretic purification of tumor-targeted polyethylenimine-based polyplexes reduces toxic side effects *in vivo*. *J Control Release* 122:236–245.

Farjo R, Skaggs J, Quiambao AB, Cooper MJ, Naash MI. (2006). Efficient non-viral ocular gene transfer with compacted DNA nanoparticles. *PLoS ONE* 1:e38.

Ferkol T, Kaetzel CS, Davis PB. (1993). Gene transfer into respiratory epithelial cells by targeting the polymeric immunoglobulin receptor. *J Clin Invest* 92:2394–2400.

Ferkol T, Perales JC, Eckman E, Kaetzel CS, Hanson RW, Davis PB. (1995). Gene transfer into the airway epithelium of animals by targeting the polymeric immunoglobulin receptor. *J Clin Invest* 95:493–502.

Ferrari S, Moro E, Pettenazzo A, Behr JP, Zacchello F, Scarpa M. (1997). ExGen 500 is an efficient vector for gene delivery to lung epithelial cells in vitro and *in vivo*. *Gene Ther* 4:1100–1106.

Ferrari S, Pettenazzo A, Garbati N, Zacchello F, Behr JP, Scarpa M. (1999). Polyethylenimine shows properties of interest for cystic fibrosis gene therapy. *Biochim Biophys Acta* 1447:219–225.

Fink TL, Klepcyk PJ, Oette SM, Gedeon CR, Hyatt SL, Kowalczyk TH, Moen RC, Cooper MJ. (2006). Plasmid size up to 20 kbp does not limit effective in vivo lung gene transfer using compacted DNA nanoparticles. *Gene Ther* 13:1048–1051.

Fischer D, Bieber T, Li Y, Elsässer HP, Kissel T. (1999). A novel non-viral vector for DNA delivery based on low molecular weight, branched polyethylenimine: Effect of molecular weight on transfection efficiency and cytotoxicity. *Pharm Res* 16:1273–1279.

Florea BI, Molenaar TJ, Bot I, Michon IN, Kuiper J, Van Berkel TJ, Junginger HE, Biessen EA, Borchard G. (2003). Identification of an internalising peptide in differentiated Calu-3 cells by phage display technology; application to gene delivery to the airways. *J Drug Target* 11:383–390.

Forrest ML, Meister GE, Koerber JT, Pack DW. (2004). Partial acetylation of polyethylenimine enhances in vitro gene delivery. *Pharm Res* 21:365–371.

Gabrielson NP, Pack DW. (2006). Acetylation of polyethylenimine enhances gene delivery via weakened polymer/DNA interactions. *Biomacromolecules* 7:2427–2435.

Gao X, Kuruba R, Damodaran K, Day BW, Liu D, Li S. (2009). Polyhydroxylalkyleneamines: A class of hydrophilic cationic polymer-based gene transfer agents. *J Control Release* 137:38–45.

Gautam A, Densmore CL, Xu B, Waldrep JC. (2000). Enhanced gene expression in mouse lung after PEI-DNA aerosol delivery. *Mol Ther* 2:63–70.

Giordano RJ, Lahdenranta J, Zhen L, Chukwueke U, Petrache I, Langley RR, Fidler IJ, Pasqualini R, Tuder RM, Arap W. (2008). Targeted induction of lung endothelial cell apoptosis causes emphysema-like changes in the mouse. *J Biol Chem* 283:29447–29460.

Goldman MJ, Yang Y, Wilson JM. (1995). Gene therapy in a xenograft model of cystic fibrosis lung corrects chloride transport more effectively than the sodium defect. *Nat Genet* 9:126–131.

Gosselin MA, Guo W, Lee RJ. (2001). Efficient gene transfer using reversibly cross-linked low molecular weight polyethylenimine. *Bioconjug Chem* 12:989–994.

Goula D, Benoist C, Mantero S, Merlo G, Levi G, Demeneix BA. (1998). Polyethylenimine-based intravenous delivery of transgenes to mouse lung. *Gene Ther* 5:1291–1295.

Greig JA, Shirley R, Graham D, Denby L, Dominiczak AF, Work LM, Baker AH. (2010). Vascular-targeting anti-oxidant therapy in a model of hypertension and stroke. *J Cardiovasc Pharmacol* 56:642–650.

Grosse S, Aron Y, Thevenot G, Francois D, Monsigny M, Fajac I. (2005). Potocytosis and cellular exit of complexes as cellular pathways for gene delivery by polycations. *J Gene Med* 7:1275–1286.

Grosse S, Thévenot G, Aron Y, Duverger E, Abdelkarim M, Roche AC, Monsigny M, Fajac I. (2008). In vivo gene delivery in the mouse lung with lactosylated polyethylenimine, questioning the relevance of in vitro experiments. *J Control Release* 132:105–112.

Gurujeyalakshmi G, Giri SN. (1995). Molecular mechanisms of antifibrotic effect of interferon gamma in bleomycin-mouse model of lung fibrosis: Downregulation of TGF-beta and procollagen I and III gene expression. *Exp Lung Res* 21:791–808.

Hacein-Bey-Abina S, Von Kalle C, Schmidt M, McCormack MP, Wulffraat N, Leboulch P, Lim A. et al. (2003). LMO2-associated clonal T cell proliferation in two patients after gene therapy for SCID-X1. *Science* 302:415–419.

Haensler J, Szoka FC Jr. (1993). Polyamidoamine cascade polymers mediate efficient transfection of cells in culture. *Bioconjug Chem* 4:372–379.

Hamar P, Song E, Kökény G, Chen A, Ouyang N, Lieberman J. (2004). Small interfering RNA targeting Fas protects mice against renal ischemia-reperfusion injury. *Proc Natl Acad Sci USA* 101:14883–14888.

Han So, Mahato RI, Kim SW. (2001). Water-soluble lipopolymer for gene delivery. *Bioconjug Chem* 12:337–345.

Hu Y, Atukorale PU, Lu JJ, Moon JJ, Um SH, Cho EC, Wang Y, Chen J, Irvine DJ. (2009). Cytosolic delivery mediated via electrostatic surface binding of protein, virus, or siRNA cargos to pH-responsive core-shell gel particles. *Biomacromolecules* 10:756–765.

Hwang SJ, Davis ME. (2001). Cationic polymers for gene delivery: Designs for overcoming barriers to systemic administration. *Curr Opin Mol Ther* 3:183–191.

Ito T, Iida-Tanaka N, Niidome T, Kawano T, Kubo K, Yoshikawa K, Sato T, Yang Y, Koyama Y. (2006). Hyaluronic acid and its derivative as a multi-functional gene expression enhancer: Protection from non-specific interactions, adhesion to targeted cells, and transcriptional activation. *J Control Release* 112:382–388.

Jeong GJ, Byun HM, Kim JM, Yoon H, Choi HG, Kim WK, Kim SJ, Oh YK. (2007). Biodistribution and tissue expression kinetics of plasmid DNA complexed with polyethylenimines of different molecular weight and structure. *J Control Release* 118:118–125.

Johnson LG, Olsen JC, Sarkadi B, Moore KL, Swanstrom R, Boucher RC. (1992). Efficiency of gene transfer for restoration of normal airway epithelial function in cystic fibrosis. *Nat Genet* 2:21–25.

Jia SF, Worth LL, Densmore CL, Xu B, Duan X, Kleinerman ES. (2003). Aerosol gene therapy with PEI: IL-12 eradicates osteosarcoma lung metastases. *Clin Cancer Res* 9:3462–3468.

Jin Y, Choi AM. (2005). Cytoprotection of heme oxygenase-1/carbon monoxide in lung injury. *Proc Am Thorac Soc* 2:232–235.

Jost PJ, Harbottle RP, Knight A, Miller AD, Coutelle C, Schneider H. (2001). A novel peptide, THALWHT, for the targeting of human airway epithelia. *FEBS Lett* 489:263–269.

Kaouass M, Beaulieu R, Balicki D. (2006). Histonefection: Novel and potent non-viral gene delivery. *J Control Release* 113:245–254.

Kiang T, Bright C, Cheung CY, Stayton PS, Hoffman AS, Leong KW. (2004). Formulation of chitosan-DNA nanoparticles with poly(propyl acrylic acid) enhances gene expression. *J Biomater Sci Polym Ed* 15:1405–1421.

Kichler A, Chillon M, Leborgne C, Danos O, Frisch B. (2002). Intranasal gene delivery with a polyethylenimine-PEG conjugate. *J Control Release* 81:379–388.

Kim JS, Kim BI, Maruyama A, Akaike T, Kim SW. (1998). A new non-viral DNA delivery vector: The terplex system. *J Control Release* 53:175–182.

Kleemann E, Neu M, Jekel N, Fink L, Schmehl T, Gessler T, Seeger W, Kissel T. (2005). Nano-carriers for DNA delivery to the lung based upon a TAT-derived peptide covalently coupled to PEG-PEI. *J Control Release* 109:299–316.

Kogure K, Akita H, Yamada Y, Harashima H. (2008). Multifunctional envelope-type nano device (MEND) as a non-viral gene delivery system. *Adv Drug Deliv Rev* 60:559–571.

Konstan MW, Davis PB, Wagener JS, Hilliard KA, Stern RC, Milgram LJ, Kowalczyk TH et al. (2004). Compacted DNA nanoparticles administered to the nasal mucosa of cystic fibrosis subjects are safe and demonstrate partial to complete cystic fibrosis transmembrane regulator reconstitution. *Hum Gene Ther* 15:1255–1269.

Kukowska-Latallo JF, Raczka E, Quintana A, Chen C, Rymaszewski M, Baker JR Jr. (2000). Intravascular and endobronchial DNA delivery to murine lung tissue using a novel, nonviral vector. *Hum Gene Ther* 11:1385–1395.

Kuruba R, Wilson A, Gao X, Li S. (2009). Targeted delivery of nucleic-acid-based therapeutics to the pulmonary circulation. *AAPS J* 11:23–30.

Lee H, Jeong JH, Park TG. (2001). A new gene delivery formulation of polyethylenimine/DNA complexes coated with PEG conjugated fusogenic peptide. *J Control Release* 76:183–192.

Lee Y, Mo H, Koo H, Park JY, Cho MY, Jin GW, Park JS. (2007). Visualization of the degradation of a disulfide polymer, linear poly(ethylenimine sulfide), for gene delivery. *Bioconjug Chem* 18:13–18.

Li S, Tan Y, Viroonchatapan E, Pitt BR, Huang L. (2000). Targeted gene delivery to the lung by anti-PECAM antibody. *Am J Physiol* 278:504–511.

Männistö M, Reinisalo M, Ruponen M, Honkakoski P, Tammi M, Urtti A. (2007). Polyplex-mediated gene transfer and cell cycle: Effect of carrier on cellular uptake and intracellular kinetics, and significance of glycosaminoglycans. *J Gene Med* 9:479–487.

Marshall, E. (2002). Gene therapy a suspect in leukemia-like disease. *Science* 298:34–35.

Matsuda N, Yamamoto S, Takano K, Kageyama S, Kurobe Y, Yoshihara Y, Takano Y, Hattori Y. (2009). Silencing of fas-associated death domain protects mice from septic lung inflammation and apoptosis. *Am J Respir Crit Care Med* 179:806–815.

McLachlan G, Baker A, Tennant P, Gordon C, Vrettou C, Renwick L, Blundell R et al. (2007). Optimizing aerosol gene delivery and expression in the ovine lung. *Mol Ther* 15:348–354.

Mesika A, Grigoreva I, Zohar M, Reich Z. (2001). A regulated, NFkappaB-assisted import of plasmid DNA into mammalian cell nuclei. *Mol Ther* 3:653–657.

Midoux P, Monsigny M. (1999). Efficient gene transfer by histidylated polylysine/pDNA complexes. *Bioconjug Chem* 10:406–411.

Miller WH, Brosnan MJ, Graham D, Nicol CG, Morecroft I, Channon KM, Danilov SM, Reynolds PN, Baker AH, Dominiczak AF. (2005). Targeting endothelial cells with adenovirus expressing nitric oxide synthase prevents elevation of blood pressure in stroke-prone spontaneously hypertensive rats. *Mol Ther* 12:321–327.

Min SH, Kim DM, Kim MN, Ge J, Lee DC, Park IY, Park KC, Hwang JS, Cho CW, Yeom YI. (2010). Gene delivery using a derivative of the protein transduction domain peptide, K-Antp. *Biomaterials* 31:1858–1864.

Moffatt S, Wiehle S, Cristiano RJ. (2005). Tumor-specific gene delivery mediated by a novel peptide-polyethylenimine-DNA polyplex targeting aminopeptidase N/CD13. *Hum Gene Ther* 16:57–67.

Moffatt S, Wiehle S, Cristiano RJ. (2006). A multifunctional PEI-based cationic polyplex for enhanced systemic p53-mediated gene therapy. *Gene Ther* 13:1512–1523.

Morse D, Pischke SE, Zhou Z, Davis RJ, Flavell RA, Loop T, Otterbein SL, Otterbein LE, Choi AM. (2003). Suppression of inflammatory cytokine production by carbon monoxide involves the JNK pathway and AP-1. *J Biol Chem* 278:36993–36998.

Muzykantov VR. (2005). Biomedical aspects of targeted delivery of drugs to pulmonary endothelium. *Expert Opin Drug Deliv* 2:909–926.

Nakao A, Fujii, M, Matsumura, R, Kumano, K, Saito, Y, Miyazono, K, Iwamoto, I. (1999). Transient gene transfer and expression of Smad7 prevents bleomycin-induced lung fibrosis in mice. *J Clin Invest* 104:5–11.

Nguyen HK, Lemieux P, Vinogradov SV, Gebhart CL, Guérin N, Paradis G, Bronich TK, Alakhov VY, Kabanov AV. (2000). Evaluation of polyether-polyethyleneimine graft copolymers as gene transfer agents. *Gene Ther* 7:126–138.

Oberholzer C, Oberholzer A, Clare-Salzler M, Moldawer LL. (2001). Apoptosis in sepsis: A new target for therapeutic exploration. *FASEB J* 15:879–892.

Oh P, Borgström P, Witkiewicz H, Li Y, Borgström BJ, Chrastina A, Iwata K. et al. (2007). Live dynamic imaging of caveolae pumping targeted antibody rapidly and specifically across endothelium in the lung. *Nat Biotechnol* 25:327–337.

Okamoto H, Nishida S, Todo H, Sakakura Y, Iida K, Danjo K. (2003). Pulmonary gene delivery by chitosan-pDNA complex powder prepared by a supercritical carbon dioxide process. *J Pharm Sci* 92:371–380.

Ohsaki M, Okuda T, Wada A, Hirayama T, Niidome T, Aoyagi H. (2002). In vitro gene transfection using dendritic poly(L-lysine). *Bioconjug Chem* 13:510–517.

Ogris M, Brunner S, Schüller S, Kircheis R, Wagner E. (1999). PEGylated DNA/transferrin-PEI complexes: Reduced interaction with blood components, extended circulation in blood and potential for systemic gene delivery. *Gene Ther* 6:595–605.

Ogris M, Carlisle RC, Bettinger T, Seymour LW. (2001). Melittin enables efficient vesicular escape and enhanced nuclear access of nonviral gene delivery vectors. *J Biol Chem* 276:47550–47555.

Otterbein LE, Bach FH, Alam J, Soares M, Tao-Lu H, Wysk H, Davis RJ, Flavell RA, Choi AM. (2003). Carbon monoxide has anti-inflammatory effects involving the mitogen-activated protein kinase pathway. *Nat Med* 6:422–428.

Pagano JS, Vaheri A. (1965). Enhancement of infectivity of poliovirus RNA with diethylaminoethyl-dextran (DEAE-D). *Arch Gesamte Virusforsch* 17:456–464.

Park MR, Kim HW, Hwang CS, Han KO, Choi YJ, Song SC, Cho MH, Cho CS. (2008). Highly efficient gene transfer with degradable poly(ester amine) based on poly(ethylene glycol) diacrylate and polyethylenimine in vitro and *in vivo*. *J Gene Med* 10:198–207.

Pons B, Mouhoubi L, Adib A, Godzina P, Behr JP, Zuber G. (2006). omega-Hydrazino linear polyethylenimine: A monoconjugation building block for nucleic acid delivery. *Chembiochem* 7:303–309.

Raghu G, Brown KK, Bradford WZ, Starko K, Noble PW, Schwartz DA, King TE Jr, Idiopathic Pulmonary Fibrosis Study Group. (2004). A placebo-controlled trial of interferon gamma-1b in patients with idiopathic pulmonary fibrosis. *N Engl J Med* 350:125–133.

Rajotte D, Arap W, Hagedorn M, Koivunen E, Pasqualini R, Ruoslahti E. (1998). Molecular heterogeneity of the vascular endothelium revealed by in vivo phage display. *J Clin Invest* 102:430–437.

Rajotte D, Ruoslahti E. (1999). Membrane dipeptidase is the receptor for a lung-targeting peptide identified by in vivo phage display. *J Biol Chem* 274:11593–11598.

Reynolds AM, Xia W, Holmes MD, Hodge SJ, Danilov S, Curiel DT, Morrell NW, Reynolds PN. (2007). Bone morphogenetic protein type 2 receptor gene therapy attenuates hypoxic pulmonary hypertension. *Am J Physiol Lung Cell Mol Physiol* 292:L1182–L1192.

Reynolds PN, Zinn KR, Gavrilyuk VD, Balyasnikova IV, Rogers BE, Buchsbaum DJ, Wang MH et al. (2000). A targetable, injectable adenoviral vector for selective gene delivery to pulmonary endothelium *in vivo*. *Mol Ther* 2:562–578.

Rittner K, Benavente A, Bompard-Sorlet A, Heitz F, Divita G, Brasseur R, Jacobs E. (2002). New basic membrane-destabilizing peptides for plasmid-based gene delivery in vitro and *in vivo*. *Mol Ther* 5:104–114.

Rosenfeld MA, Rosenfeld SJ, Danel C, Banks TC, Crystal RG. (1994). Increasing expression of the normal human CFTR cDNA in cystic fibrosis epithelial cells results in a progressive increase in the level of CFTR protein expression, but a limit on the level of cAMP-stimulated chloride secretion. *Hum Gene Ther* 5:1121–1129.

Rogy MA, Auffenberg T, Espat NJ, Philip R, Remick D, Wollenberg GK, Copeland EM III, Moldawer LL. (1995). Human tumor necrosis factor receptor (p55) and interleukin 10 gene transfer in the mouse reduces mortality to lethal endotoxemia and also attenuates local inflammatory responses. *J Exp Med* 181:2289–2293.

Rudolph C, Lausier J, Naundorf S, Müller RH, Rosenecker J. (2000). In vivo gene delivery to the lung using polyethylenimine and fractured polyamidoamine dendrimers. *J Gene Med* 2:269–278.

Rudolph C, Plank C, Lausier J, Schillinger U, Müller RH, Rosenecker J. (2003). Oligomers of the arginine-rich motif of the HIV-1 TAT protein are capable of transferring plasmid DNA into cells. *J Biol Chem* 278:11411–11418.

Rudolph C, Schillinger U, Ortiz A, Plank C, Golas MM, Sander B, Stark H, Rosenecker J. (2005). Aerosolized nanogram quantities of plasmid DNA mediate highly efficient gene delivery to mouse airway epithelium. *Mol Ther* 12:493–501.

Sakae M, Ito T, Yoshihara C, Iida-Tanaka N, Yanagie H, Eriguchi M, Koyama Y. (2008). Highly efficient in vivo gene transfection by plasmid/PEI complexes coated by anionic PEG derivatives bearing carboxyl groups and RGD peptide. *Biomed Pharmacother* 62:448–453.

Sarady JK, Zuckerbraun BS, Bilban M, Wagner O, Usheva A, Liu F, Ifedigbo E, Zamora R, Choi AM, Otterbein LE. (2004). Carbon monoxide protection against endotoxic shock involves reciprocal effects on iNOS in the lung and liver. *FASEB J* 18:854–856.

Schiffelers RM, Ansari A, Xu J, Zhou Q, Tang Q, Storm G, Molema G, Lu PY, Scaria PV, Woodle MC. (2004). Cancer siRNA therapy by tumor selective delivery with ligand-targeted sterically stabilized nanoparticle. *Nucleic Acids Res* 32:e149.

Sethuraman VA, Na K, Bae YH. (2006). pH-responsive sulfonamide/PEI system for tumor specific gene delivery: An in vitro study. *Biomacromolecules* 7:64–70.

Shcharbin D, Pedziwiatr E, Blasiak J, Bryszewska M. (2010). How to study dendriplexes II: Transfection and cytotoxicity. *J Control Release* 141:110–127.

Shcharbin D, Pedziwiatr E, Bryszewska M. (2009). How to study dendriplexes I: Characterization. *J Control Release* 135:186–197.

Shen Y, Peng H, Deng J, Wen Y, Luo X, Pan S, Wu C, Feng M. (2009). High mobility group box 1 protein enhances polyethylenimine mediated gene delivery *in vitro*. *Int J Pharm* 375:140–147.

Shen Y, Peng H, Pan S, Feng M, Wen Y, Deng J, Luo X, Wu C. (2010). Interaction of DNA/nuclear protein/polycation and the terplexes for gene delivery. *Nanotechnology* 21:045102.

Sonawane ND, Szoka, FC Jr, Verkman AS. (2003). Chloride accumulation and swelling in endosomes enhances DNA transfer by polyamine-DNA polyplexes. *J Biol Chem* 278:44826–44831.

Song E, Lee SK, Wang J, Ince N, Ouyang N, Min J, Chen J. et al. (2003). RNA interference targeting Fas protects mice from fulminant hepatitis. *Nat Med* 9:347–351.

Strieter RM, Starko KM, Enelow RI, Noth I, Valentine VG. (2004). Effects of interferon-γ-1b on biomarker expression in idiopathic pulmonary fibrosis patients. *Am J Respir Crit Care Med* 170:133–140.

Suh J, Paik H-J, Hwang BK. (1994). Ionization of poly(ethylenimine) and poly(allylamine) at various pH's. *Bioorg Chem* 22:318–327.

Sung SJ, Min SH, Cho KY, Lee S, Min YJ, Yeom YI, Park JK. (2003). Effect of polyethylene glycol on gene delivery of polyethylenimine. *Biol Pharm Bull* 26:492–500.

Sur S, Lam J, Bouchard P, Sigounas A, Holbert D, Metzger WJ. (1996). Immunomodulatory effects of IL-12 on allergic lung inflammation depend on timing of doses. *J Immunol* 157:4173–4180.

Tagalakis AD, McAnulty RJ, Devaney J, Bottoms SE, Wong JB, Elbs M, Writer MJ et al. (2008). A receptor-targeted nanocomplex vector system optimized for respiratory gene transfer. *Mol Ther* 16:907–915.

Tang GP, Guo HY, Alexis F, Wang X, Zeng S, Lim TM, Ding J, Yang YY, Wang S. (2006). Low molecular weight polyethylenimines linked by beta-cyclodextrin for gene transfer into the nervous system. *J Gene Med* 8:736–744.

Tang MX, Redemann CT, Szoka FC Jr. (1996). In vitro gene delivery by degraded polyamidoamine dendrimers. *Bioconjug Chem* 7:703–714.

Tang MX, Szoka FC Jr. (1997). The influence of polymer structure on the interactions of cationic polymers with DNA and morphology of the resulting complexes. *Gene Ther* 4:823–832.

Thomas M, Klibanov AM. (2002). Enhancing polyethylenimine's delivery of plasmid DNA into mammalian cells. *Proc Natl Acad Sci* 99:14640–14645.

Thomas M, Lu JJ, Ge Q, Zhang CC, Chen JZ, Klibanov AM. (2005). Full deacylation of polyethylenimine dramatically boosts its gene delivery efficiency and specificity to mouse lung. *Proc Natl Acad Sci USA* 102:5679–5684.

Trubetskoy VS, Loomis A, Hagstrom JE, Budker VG, Wolff JA. (1999). Layer-by-layer deposition of oppositely charged polyelectrolytes on the surface of condensed DNA particles. *Nucleic Acids Res* 27:3090–3095.

Trubetskoy VS, Wong SC, Subbotin V, Budker VG, Loomis A, Hagstrom JE, Wolff JA. (2003). Recharging cationic DNA complexes with highly charged polyanions for in vitro and in vivo gene delivery. *Gene Ther* 10:261–271.

Tsuchida E. (1994). Formation of polyelectrolyte complexes and their structures. *J Macromol Sci Pure Appl Chem* A31:1–15.

Uduehi AN, Stammberger U, Frese S, Schmid RA. (2001). Efficiency of non-viral gene delivery systems to rat lungs. *Eur J Cardiothorac Surg* 20:159–163.

Vacik J, Dean BS, Zimmer WE, Dean DA. (1999). Cell-specific nuclear import of plasmid DNA. *Gene Ther* 6:1006–1014.

Vaysse L, Harbottle R, Bigger B, Bergau A, Tolmachov O, Coutelle C. (2004). Development of a self-assembling nuclear targeting vector system based on the tetracycline repressor protein. *J Biol Chem* 279:5555–5564.

Wagner E. (1998). Effects of membrane-active agents in gene delivery. *J Control Release* 53:155–158.

Walton CM, Wu CH, Wu GY. (1999). A DNA delivery system containing listeriolysin O results in enhanced hepatocyte-directed gene expression. *World J Gastroenterol* 5:465–469.

Wightman L, Kircheis R, Rössler V, Carotta S, Ruzicka R, Kursa M, Wagner E. (2001). Different behavior of branched and linear polyethylenimine for gene delivery in vitro and *in vivo*. *J Gene Med* 3:362–372.

Wilson A, Zhou W, Champion H, Alber S, Tang Z-L, Kennel S, Watkins S, Huang L, Pitt BR, Li S. (2005). Targeted delivery of oligodeoxynucleotides to mouse lung endothelial cells in vitro and *in vivo*. *Mol Ther* 12:510–518.

Won YW, Kim HA, Lee M, Kim YH. (2010). Reducible poly(oligo-D-arginine) for enhanced gene expression in mouse lung by intratracheal injection. *Mol Ther* 18:734–742.

Wolfert MA, Dash PR, Nazarova O, Oupicky D, Seymour LW, Smart S, Strohalm J, Ulbrich K. (1999). Polyelectrolyte vectors for gene delivery: Influence of cationic polymer on biophysical properties of complexes formed with DNA. *Bioconjug Chem* 10:993–1004.

Work LM, Büning H, Hunt E, Nicklin SA, Denby L, Britton N, Leike K et al. (2006). Vascular bed-targeted in vivo gene delivery using tropism-modified adeno-associated viruses. *Mol Ther* 13:683–693.

Writer MJ, Marshall B, Pilkington-Miksa MA, Barker SE, Jacobsen M, Kritz A, Bell PC et al. (2004). Targeted gene delivery to human airway epithelial cells with synthetic vectors incorporating novel targeting peptides selected by phage display. *J Drug Target* 12:185–193.

Wu GY, Wu CH. (1987). Receptor-mediated in vitro gene transformation by a soluble DNA carrier system. *J Biol Chem* 262:4429–4432.

Wu GY, Wu CH. (1988). Receptor-mediated gene delivery and expression *in vivo*. *J Biol Chem* 263:14621–14624.

Xavier C, Jean-Francois J. (1996). Adsorption of polyelectrolyte solutions on surface: A Debye-Huckel theory. *J Phys II France* 6:1669–1686.

Xu S, Chen M, Yao Y, Zhang Z, Jin T, Huang Y, Zhu, H. (2008). Novel poly(ethylene imine) biscarbamate conjugate as an efficient and nontoxic gene delivery system. *J Control Release* 130:64–68.

Yaekashiwa M, Nakayama S, Ohnuma K, Sakai T, Abe T, Satoh K, Matsumoto K. et al. (1997). Simultaneous or delayed administration of hepatocyte growth factor equally represses the fibrotic changes in murine lung injury induced by bleomycin. A morphologic study. *Am J Respir Crit Care Med* 156:1937–1944.

Yoshimura K, Nakamura H, Trapnell BC, Dalemans W, Pavirani A, Lecocq JP, Crystal RG. (1991). The cystic fibrosis gene has a "housekeeping"-type promoter and is expressed at low levels in cells of epithelial origin. *J Biol Chem* 266:140–144.

Zaliauskiene L, Bernadisiute U, Vareikis A, Makuska R, Volungeviciene L, Petuskaite A, Riauba L, Lagunavicius A, Zigmantas S. (2010). Efficient gene transfection using novel cationic polymers poly(hydroxyalkylene imines). *Bioconjug Chem* 21:1602–1611.

Zang L, Shi L, Guo J, Pan Q, Wu W, Pan X, Wang J. (2009). Screening and identification of a peptide specifically targeted to NCI-H1299 from a phage display peptide library. *Cancer Lett* 281:64–70.

Zanta MA, Belguise-Valladier P, Behr JP. (1999). Gene delivery: A single nuclear localization signal peptide is sufficient to carry DNA to the cell nucleus. *Proc Natl Acad Sci USA* 96:91–96.

Zhong Z, Feijen J, Lok MC, Hennink WE, Christensen LV, Yockman JW, Kim YH, Kim SW. (2005). Low molecular weight linear polyethylenimine-b-poly(ethylene glycol)-b-polyethylenimine triblock copolymers: Synthesis, characterization, and in vitro gene transfer properties. *Biomacromolecules* 6:3440–3448.

Ziady AG, Gedeon CR, Miller T, Quan W, Payne JM, Hyatt SL, Fink TL et al. (2003a). Transfection of airway epithelium by stable PEGylated poly-L-lysine DNA nanoparticles *in vivo*. *Mol Ther* 8:936–947.

Ziady AG, Gedeon CR, Muhammad O, Stillwell V, Oette SM, Fink TL, Quan W et al. (2003b). Minimal toxicity of stabilized compacted DNA nanoparticles in the murine lung. *Mol Ther* 8:948–956.

Ziesche R, Hofbauer E, Wittmann K, Petkov V. (1999). A preliminary study of long-term treatment with interferon gamma-1b and low-dose prednisolone in patients with idiopathic pulmonary fibrosis. *N Eng J Med* 341:1264–1269.

Zintchenko A, Philipp A, Dehshahri A, Wagner E. (2008). Simple modifications of branched PEI lead to highly efficient siRNA carriers with low toxicity. *Bioconjug Chem* 19:1448–1455.

Zou SM, Erbacher P, Remy JS, Behr JP. (2000). Systemic linear polyethylenimine (L-PEI)-mediated gene delivery in the mouse. *J Gene Med* 2:128–134.

Zugates GT, Peng W, Zumbuehl A, Jhunjhunwala S, Huang YH, Langer R, Sawicki JA, Anderson DG. (2007). Rapid optimization of gene delivery by parallel end-modification of poly(beta-amino ester)s. *Mol Ther* 15:1306–1312.

23 Biomedical Application of Membranes in Bioartificial Organs and Tissue Engineering

Thomas Groth, Xiao-Jun Huang, and Zhi-Kang Xu

CONTENTS

23.1 INTRODUCTION

Membranes are applied in biotechnology and medicine to control exchange of compounds between different phases or compartments or separation of particulate material from liquid phases (Bowry 2002; Klinkmann and Vienken 1995; Paul 1998; Ulbricht 2006). It must be emphasized that particularly in biomedical applications membranes based on polymeric materials are dominating, which is related to their ease of manufacturing, a wide range of product properties that allow an efficient adjustment of transport properties (molecular cut-off), sterilization procedure, good mechanical properties, biocompatibility, and product safety (Bowry 2002; Klinkmann and Vienken 1995; Ulbricht 2006). In this context it should be also noted that polymer membranes represent low-cost materials, which is an essential requirement for biomedical application due to the fact that they represent disposal materials. On the other hand, in biotechnology also other materials like metals or ceramics are applied, which have the advantage that much higher pressures and temperatures inherent to certain processes can be applied and these materials can be more easily reused after cleaning and sterilization procedures (Paul 1998).

Membranes for hemodialysis represent the most frequent membrane application in medicine to treat patients with stage-renal disease (ESRD) (Bowry 2002; Klinkmann and Vienken 1995; Paul 1998). Applications of membranes for blood detoxification such as hemodialysis to treat kidney failure are linked to transport of solutes to remove (toxic) metabolites, salts, and water from the organism. On the other hand, the undesired loss of substances such as proteins like albumin must be prevented (Bowry 2002; Klinkmann and Vienken 1995). Since ESRD is a frequent disease and patients require hemodialysis at least three times a week, membranes for ESRD treatment stand for the largest market segment in the field of membrane production (Paul 1998; Ulbricht 2006). However, membranes not only play a pivotal role in blood oxygenation applied in cardiopulmonary bypass (CPB) devices but also envisaged for the treatment of adult respiratory distress syndrome (ARDS), lung fibrosis, and other diseases that impair lung function. While CPB devices are large devices applied in the operation theater, also wearable oxygenators for the treatment of ARDS are under development to improve the quality of life of these patients (Rho et al. 2003). It should be noted that artificial organs like the artificial kidney (used as synonym for hemodialysis) replace only certain physical functions of the organ such as the filtration function of the glomerulus membrane; other functions like the readsorption or hormonal function of kidney cannot be replaced by technical devices (Uhlenbusch-Körwer et al. 2004). Artificial organs lack also the precise feedback mechanisms of natural organs, which can result in nonphysiological (pathological) conditions. As a consequence, hemodialysis is merely considered as a palliative and not curing therapy. In fact, outcome of long-term hemodialysis is rather poor due to the fact that patients begin to suffer from a variety of diseases, which results in increased morbidity and mortality of patients (Rocco et al. 2006). It has been also observed that acute failure of kidney, and also liver, which is treated by hemodialysis has a very poor outcome with a mortality as high as 50% (Kim et al. 2008). For this reason, combinations of technical with biological functions have been envisaged that shall reduce the drawbacks of artificial organs. A first example of this approach is the concept of bioartificial (or biohybrid) liver that combines hemodialysis with bioreactors containing hepatocytes that perform major detoxification functions (Yu et al. 2009). Also the poor outcome of treatment of patients with acute kidney failure by conventional hemodialysis has brought up the concept of a bioartificial kidney, a device having a hemofiltration filter corresponding to the glomerulus (filtration) unit of kidney and a bioreactor with kidney epithelial cells as tubular (re-adsorptive) unit of kidney (Tumlin et al. 2008). Membranes have also been applied to replace the hormone function of different glands like the pancreatic islets producing insulin. However, the membrane functions here merely focused on the isolation of cells of xenogenic or allogenic origin from the immune system of the patient. For this reason these applications are out of the focus of this review and will not be dealt with. For more information about this topic one can read more in the work of Colton (1995). Membranes play also a pivotal role in certain tissue engineering applications. Here often membranes play a role as mechanical support providing attachment sites for cells and tissue components, guiding tissue reformation during the healing, while the separation and exchange function are also important but play a minor role. An excellent survey on major therapeutic fields in which membranes are applied can be found elsewhere (Peinemann and Nunes 2012).

The current review is separated into two major parts. The first section will give a comprehensive overview on how typical polymer membranes for biomedical and technical applications are prepared and how essential membranes features like morphology, transport, and surface properties can be tailored. It will also show how surface modifications can be used to achieve membrane properties that allow adsorption of specific molecules, but providing also attachment sites for cells. The second section of this review will focus on a few selected biomedical applications of polymer membranes. Since conventional therapeutic applications of polymer membranes in the treatment of ESRD by hemodialysis, hemofiltration, or for blood oxygenation have been reviewed in many other articles, these applications will be not further discussed here. However, the review will give a few examples on the application of membranes in the field of bioartificial organs, with emphasis on liver replacement and tissue engineering of skin.

23.2   POLYMER MEMBRANE FORMATION AND SURFACE MODIFICATION

23.2.1   FORMATION OF POLYMER MEMBRANES

Asymmetric polymer membranes, which are widely used in various membrane-involving industries, are mostly fabricated by a process called phase inversion. It can be mainly achieved through the following four principal methods: immersion precipitation (Akthakul et al. 2005; Bazarjani et al. 2009; Bindal et al. 1996; Lin et al. 2002; Matsuyama et al. 2003a; Pitol et al. 2006; Wang et al. 2009; Wienk et al. 1996), vapor-induced phase separation (Khare et al. 2005; Li et al. 2008; Park et al. 1999; Tsai et al. 2006; Yip and McHugh 2006), thermally induced phase separation (Gu et al. 2008; Lloyd et al. 1990; Matsuyama et al. 2000; Rajabzadeh et al. 2009), and dry-casting (Altinkaya and Ozbas 2004; Kim et al. 2009). Among them, the immersion precipitation technique is practically applied to prepare polymeric membranes for applications in biotechnology and medicine, for example, hemodialysis membranes. Principally, a polymer solution is extruded through a spinneret or cast as a thin film on a support, and subsequently immersed in a non-solvent coagulation bath that leads to the formation of an asymmetric membrane with a greater or lesser porous structure. The diffusion of the non-solvent inside the polymer solution induces a phase separation between polymer and solvent, yielding the formation of pores. The structure of the polymer membrane obtained during this process is represented by a dense top layer surface and porous sublayer, which is important for the application of membranes.

Normally, the permeability and selectivity of these asymmetric membranes are mostly determined by its skin-layer morphologies like surface roughness, the nodule, and particularly pore size (Fritzsche et al. 1993; Kesting 1990). The porous sublayer is usually regarded as the supporting layer for asymmetric membrane giving some mechanical strength. Different porous sublayers such as macrovoid and sponge-like structures can often be tailored in asymmetric polymeric membranes related to their practical applications. For example, membranes with macrovoid structures are useful as reservoir systems in osmotic drug delivery and transdermal delivery device (Herbig et al. 1995; Wang et al. 1998). In contrast, the presence of macrovoids in asymmetric membranes is not favorable in high-pressure operations like ultra and nanofiltration, reverse osmosis, and gas separation because they may lead to mechanically instable weak spots (Paulsen et al. 1994; Smolders et al. 1992; Won et al. 1999).

Compared with the process of skin-layer formation, the mechanism of the porous sublayer formation is more complicated. Much attention has been paid and several explanations have been proposed for porous sublayers formation, for example, by Smolders et al. (1992), Zeman and Fraser (1993), and others (Altinkaya and Ozbas 2004; Andersson et al. 1980; Chen and Young 1991; Fritzsche et al. 1993; Frommer and Messalem 1973; Herbig et al. 1995; Kesting 1990; Kim et al. 2001, 2009; Machado et al. 1999; McKelvey and Koros 1996; Paulsen et al. 1994; Saier et al. 1974; Shimizu et al. 2002; Smolders et al. 1992; Vogrin et al. 2002; Wang et al. 1998; Won et al. 1999; Zeman and Fraser 1993). To obtain the favorable sponge-like cross-section of the membrane, a reduction in fluidity of the casting solution is often effective, which can be described as a dynamic way to inhibit the diffusion exchange between solvent and non-solvent in the coagulation bath in order to lower phase inversion rate. This usually can be achieved by increasing the casting solution viscosity that converts a fluid solution to a 3D network. Therefore, the polymeric asymmetrical porous membranes with a very large variety of structures (skin layer and cross-section) can be made by varying the membrane preparation conditions like polymer molecular weight, solution concentration, non-solvent additives, coagulation medium, temperature, and so on, which will be described in more detail in the following.

23.2.1.1 Effect of Polymer Molecular Weight and Solution Concentration

The higher molecular weight and higher solution concentration of the polymer decrease the fluidity of the casting solution and usually decelerate the exchange between solvent and non-solvent during the phase separation process, resulting in a decrease of pore size and pore numbers on the membrane surface (Kang and Lee 1991; Radovanovic et al. 1992; vandeWitte et al. 1996). Figure 23.1 shows as an example the effect of poly(acrylonitrile-*co*-2-hydroxyethyl methacrylate) (PANCHEMA) solution concentration on the structure of the skin layer. It was found that the pore size and number on the skin layer deceased as PANCHEMA solution concentration increased, which is ascribed to the increased viscosity at a high solution concentration resulting in a tight structure of the membrane surface. The water flux and bovine serum albumin rejection for these PANCHEMA membranes are shown in Figure 23.2. It can be seen that the water flux of the membrane decreased gradually, whereas the BSA rejection increased for the corresponding membranes as the result of decreased pore size and number on the skin layer.

23.2.1.2 Effect of Non-Solvent Additives

The fluidity of the casting solution and the polymer interchain entanglement can also be adjusted by the addition of non-solvent into the polymer solution. The non-solvent additives can reduce the gyration radius of polymer chains and make it easier to achieve polymer aggregation and phase separation of the casting solution due to its bad compatibility with the polymer. In order to control the membrane structure further, a low molecular weight component or a secondary polymer is frequently used as an

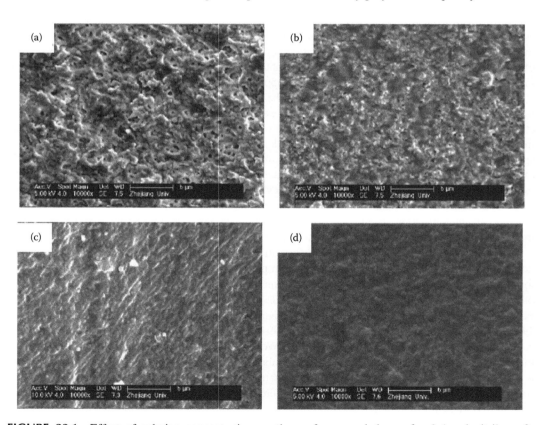

FIGURE 23.1 Effect of solution concentration on the surface morphology of poly(acrylonitrile-*co*-2-hydroxyethyl methacrylate) (PANCHEMA) membrane. The viscosity-average molecular weight (M_η) of PANCHEMA is 15.0×10^4 g/mol and 2-hydroxyethyl methacrylate (HEMA) content in PANCHEMA is 9.3 mol%. PANCHEMA concentration: (a) 8 wt.%, (b) 10 wt.%, (c) 12 wt.%, and (d) 15 wt.%.

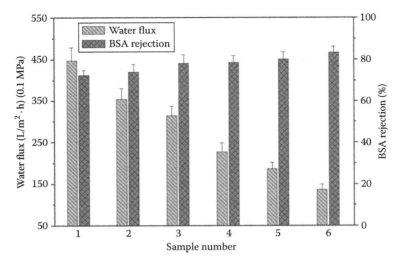

FIGURE 23.2   Effect of casting solution concentration on the water flux and bovine serum albumin (BSA) rejection of poly(acrylonitrile-*co*-2-hydroxyethyl methacrylate) (PANCHEMA) membranes. The viscosity-average molecular weight (M_η) of PANCHEMA is 15.0×10^4 g/mol and 2-hydroxyethyl methacrylate (HEMA) content in PANCHEMA is 9.3 mol%. Casting solution concentration: (1) 8 wt.%, (2) 10 wt.%, (3) 12 wt.%, (4) 13 wt.%, (5) 14 wt.%, (6) 15 wt.%.

additive in the membrane-forming system, because it offers a convenient and effective way to develop membranes with high performances. Particularly hydrophilic, water-soluble polymers like poly(vinyl pyrrolidone) (PVP) and poly(ethylene glycol) (PEG) have been widely used as additives to modify membrane structures and properties (Boom et al. 1992; Kang and Lee 2002; Kim and Lee 1998; Li and Jiang 2005; Matsuyama et al. 2003b; Wang et al. 2008; Zheng et al. 2006).

Figure 23.3 shows as an example the effect of PVP used as additive on the morphology of cross-section, external surface, and inner edge of PANCHEMA membranes. It can be seen that the PANCHEMA membranes have an asymmetric structure consisting of a dense skin layer and a porous sublayer.

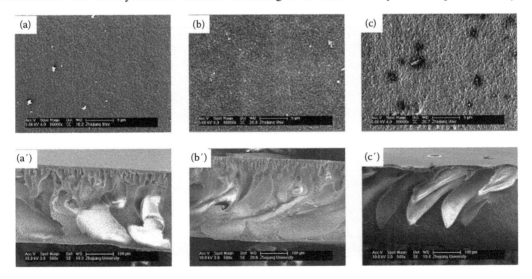

FIGURE 23.3   Effect of polyvinyl pyrrolidone (PVP) addition on the surface morphology and cross-section of poly(acrylonitrile-*co*-2-hydroxyethyl methacrylate) (PANCHEMA) membranes. The viscosity-average molecular weight (M_η) of PANCHEMA is 15.0×10^4 g/mol and 2-hydroxyethyl methacrylate (HEMA) content in PANCHEMA is 9.3 mol%. Polyvinyl pyrrolidone (PVP) concentration: (a) 0 wt.%, (b) 4 wt.%, and (c) 6 wt.%.

By the addition of 4.0 wt.% PVP in 12 wt.% of PANCHEMA solution, some pores are formed on the outer surface because PVP is water-soluble and leaches out during phase separation with water as non-solvent, making pores on the membrane surface. However, finger-like macrovoids beneath the membrane surface seems to be suppressed in the case of PVP. When keeping PANCHEMA concentration at 12 wt.% and adding PVP to the casting solution, the formation of macrovoids was gradually suppressed. A sponge-like substructure was obtained instead of the finger-like macrovoids with the increase of PVP concentration from 0 to 4 wt.% and then to 6 wt.%. Moreover, with the exception of roughness to some extent, a few pores were observed on the external surface with the increase of PVP concentration. These morphologies are consistent with the water permeation results shown in Figure 23.4. It is visible that the water flux increased for membranes increase of PVP concentration from 0 to 6 wt.% in the casting solution.

Compared with these water-soluble polymers, small molecular non-solvent additives like water-soluble salts, surfactants, ethanol, glycerin, and even water can also translate the finger-like structure of phase inversion membranes into a sponge-like structure (Lai et al. 1996, 1999; Lin et al. 1997; Rahimpour et al. 2007; Wang et al. 1998, 2000a–c). For example, Lee et al. (2002) added LiCl as inorganic salt additive to the casting solution containing poly(amic acid) and N-methyl pyrrolidone (NMP) to form a complex between LiCl and the polar solvent, resulting in sponge-like asymmetric membrane by increasing the solution viscosity. As it was reported by (Lai et al. 1999; Lin et al. 2010a; Wang et al. 1998), surfactants such as span series can delay demixing of the casting solution, which result also in a sponge-like structure. Furthermore, as a non-solvent, besides its application as major coagulation agent, water was also used as additive during polymer membrane preparation. It has obviously an impact on the gelation of polymers during the structure-forming process. The connecting junctions of the polymer are already generated in the initial liquid and form a 3D network that provides enough mechanical strength for the gel to be self-supporting (Li et al. 1996; Lin et al. 2010b; Wang et al. 2006). Li et al. (1996) studied the role of gelation during the structure-forming process of membrane by mixing water into polyethersulfone/DMAc system. Water was also used by Lai and coworkers to suppress the macrovoids in poly(methylmethacrylate) (PMMA) membranes through this gelation method from NMP solution (Lin et al. 2010b).

The effect of addition of water on the morphology of PANCHEMA asymmetric membranes is shown in Figure 23.5. When PANCHEMA concentration in the casting solution was kept at 12 wt.%, at increasing water content from 0 to 12 wt.% was gradually suppressing the appearance

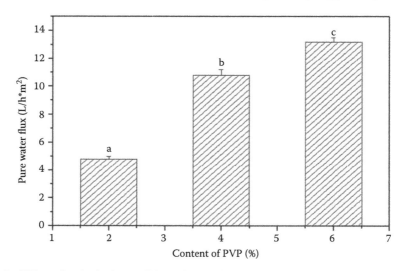

FIGURE 23.4 Effect of polyvinyl pyrrolidone (PVP) addition on the water flux of poly(acrylonitrile-*co*-2-hydroxyethyl methacrylate) (PANCHEMA) membranes. The viscosity-average molecular weight (M_η) of PANCHEMA is 15.0×10^4 g/mol and 2-hydroxyethyl methacrylate (HEMA) content in PANCHEMA is 9.3 mol%. Polyvinyl pyrrolidone (PVP) concentration: (a) 0 wt.%, (b) 4 wt.%, and (c) 6 wt.%.

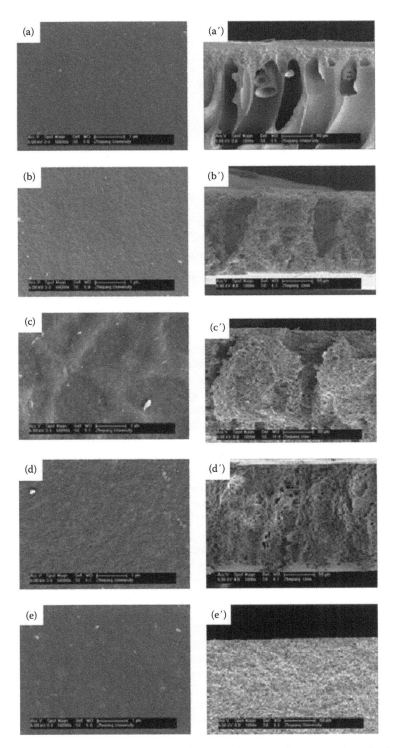

FIGURE 23.5 Effect of water addition on the surface morphology and cross-section of poly(acrylonitrile-*co*-2-hydroxyethyl methacrylate) (PANCHEMA) membranes. The viscosity-average molecular weight (M_η) of PANCHEMA is 15.0×10^4 g/mol and 2-hydroxyethyl methacrylate (HEMA) content in PANCHEMA is 9.3 mol%. PANCHEMA concentration: 12 wt.%. Water concentration in wt.%: (a) 0, (b) 2, (c) 4, (d) 8, (e) 12.

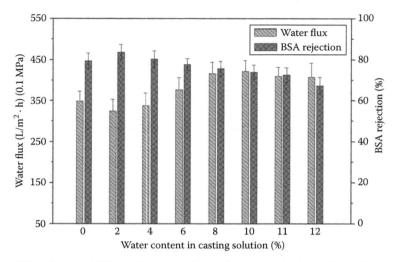

FIGURE 23.6 Effect of water addition on the permeation performances of poly(acrylonitrile-*co*-2-hydroxyethyl methacrylate) (PANCHEMA) membranes. The viscosity-average molecular weight (M_η) of PANCHEMA is 15.0×10^4 g/mol and 2-hydroxyethyl methacrylate (HEMA) content in PANCHEMA is 9.3 mol%. PANCHEMA concentration: 12 wt.%. Water concentration in polymer solution is 2, 4, 6, 8, 10, and 12 wt.%, respectively.

of macrovoids in the membranes. Hence, it can be stated that more water addition leads to a higher gelation rate during the membrane formation process. With the help of gels, the mobility of polymer chains is lowered, impeding the water penetration rate followed by the coalescence of the polymer-poor phase, which suppresses the formation of macrovoids.

The effects of water addition to the polymer solution before phase inversion on pure water flux and BSA rejection of the resulting PANCHEMA membranes are shown in Figure 23.6. It was observed that with increasing water content in the polymer solution, the pure water flux showed an increase with water content higher than 6 wt.%, while the BSA rejection decreased gradually with water contents higher than 4 wt.%. These results are also related to the surface morphology of the membranes shown in Figure 23.5. It was reported previously that the membrane performance was influenced by the presence of nodules on the surface. The surface nodule size actually increased as the non-solvent content in the casting solution increased (see Figure 23.5 as well). Moreover, a higher non-solvent concentration could make the casting solution easier to precede phase separation and thus result in a more porous membrane (Wang et al. 1998). In the example described herein, water as a non-solvent could also act in such competitive way that the water flux firstly decreased with increasing water concentration in the dope solution (indicating the dominating effect of the change in surface tension) and then increased with increasing water concentration further more (indicating the domination of the effect of phase separation). In addition, the increasing flux might be correlated to gelation. Different water concentrations result in different gelation rates leading to different states of polymer chain aggregation. Thus a lower water concentration might make the aggregation tighter because the polymer chains have more time for arrangement. The gelation could impede the water permeation rate after immersion, so the pre-solidified membrane structure could be retained. Overall, rapid gelation would result in less tight polymer aggregation. It can be seen in Figure 23.5 that the membrane prepared by the addition of 12 wt.% water also showed larger pores on the surface, which correlates with the least BSA rejection.

23.2.1.3 Effect of Casting Temperature

The mobility of polymer chains and the fluidity of the casting solution will increase at higher temperatures, which should speed up the exchange rate of solvent and non-solvent during the contacting of casting (polymer) solution and coagulant (non-solvent) (Saljoughi et al. 2009). It can be expected that macrovoids are formed under such rapid precipitation condition. From this observation it can

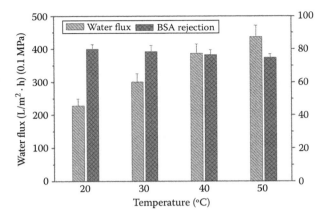

FIGURE 23.7   Effect of temperature on the permeation performances of poly(acrylonitrile-*co*-2-hydroxyethyl methacrylate) (PANCHEMA) membranes. The viscosity-average molecular weight (M_η) of PANCHEMA is 15.0×10^4 g/mol and 2-hydroxyethyl methacrylate (HEMA) content in PANCHEMA is 9.3 mol%, temperature is 20°C, 30°C, 40°C, and 50°C.

be deduced that a decreasing casting temperature should have a similar effect on skin and sublayer structures as increased polymer molecular weight and solution concentration. As it is shown in Figure 23.7, the water flux of the PANCHEMA membrane increased gradually as the temperature increased, whereas the BSA rejection decreased for the corresponding membranes.

23.2.1.4   Effect of Coagulant (Non-Solvent) Composition

The coagulant composition can greatly affect the structure and properties of the membrane. Solvents for the polymer are often added to the coagulation bath to control the phase inversion process (Nie et al. 2004; Pesek and Koros 1994; Yan and Lau 1998). It is easy to understand that the addition of a polymer solvent into the coagulation medium decreases its coagulation efficacy. During contact of polymer solution and coagulant, the ratio of the counter-diffusion rate of solvent and non-solvent is decreased, which results in a delayed demixing process. Macrovoids often result when demixing occurs rapidly, whereas delayed demixing conditions suppress or even eliminate macrovoid formation (Albrecht et al. 2001).

Figure 23.8 shows an example of the addition of solvents to the coagulation solution during poly(acrylonitrile-*co*-maleic acid) (PANCMA) hollow fiber membrane preparation. The SEM images in Figure 23.8 show that the thickness of the sublayer decreases with an increase of dimethylsulfoxide (DMSO) concentration in the coagulant solution. The size and number of the macrovoids in the sublayer also decreased, and their shape turned from large cavities to relatively regular finger-like voids. The impacts of DMSO concentration on the water flux and BSA rejection of the resulting PANCMA membranes are illustrated in Figure 23.9. It is demonstrated that with increasing DMSO concentration from 10 to 50 wt.%, the water flux of membranes decreased gradually from 135 to $100 \text{L/m}^2 \cdot \text{h} \cdot \text{atm}$, while the BSA rejection increased slightly from 95% to 96%. These results are consistent with the membrane structure mentioned earlier. It can be presumed that the increased DMSO concentration in the coagulant, the exchange rate of solvent and non-solvent by diffusion decreases, and a delayed demixing during phase inversion occurs. The polymer concentration in the polymer-rich phase decreases as well, which results in the delay of solidification. The polymer solution solidifies more slowly, which provides the polymer chains more time to become compacted on the inner surface layer before solidification. Hence the size and the number of macrovoids decrease and accompanied by the formation of a denser and thicker top layer on the inner surface of the hollow fiber membranes. Hence a decrease in water permeability and a slight increase in the rejection of BSA are the consequences.

In addition to the aforementioned parameters, other membrane preparation conditions like air gap distance, the temperature, and the flow rate of the coagulant solution can also be used to develop membranes with the optimized structure and performance.

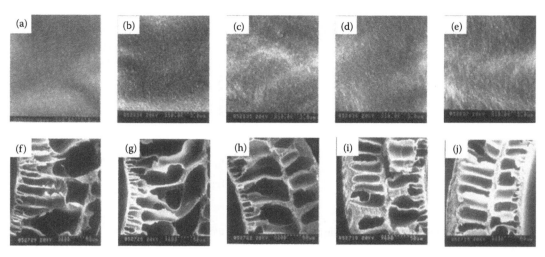

FIGURE 23.8 Scanning electron microscope (SEM) photographs of the membrane structure with different amounts of dimethyl sulfoxide (DMSO) in internal coagulant. Upper row: cross-section. Lower row: inner surface. The viscosity-average molecular weight (M_η) of poly(acrylonitrile-*co*-2-hydroxyethyl methacrylate) (PANCHEMA) is 2.04×10^5 g/mol and maleic acid content in copolymer was 5.13 mol%, DMSO/H$_2$O ratio in internal coagulant is (a) and (f): 0/100, (b) and (g): 20/80, (c) and (h): 30/70, (d) and (i): 40/60, (e) and (j): 50/50.

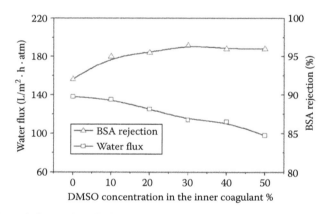

FIGURE 23.9 Effect of dimethyl sulfoxide (DMSO)/H$_2$O ratio in internal coagulant on the permeation performances of poly(acrylonitrile-*co*-2-hydroxyethyl methacrylate) (PANCHEMA) membranes. The viscosity-average molecular weight (M_η) of PANCHEMA is 15.0×10^4 g/mol and 2-hydroxyethyl methacrylate (HEMA) content in PANCHEMA is 9.3 mol%. PANCHEMA concentration: 12 wt.%. DMSO/H$_2$O ratio in internal coagulant is 0/100, 20/80, 30/70, 40/60, and 50/50, respectively.

23.2.2 Surface Modification of Polymer Membranes

Polymer separation membranes develop rapidly and have been applied in many fields of biotechnology and medicine. Many technologies are not only dependent on the transport properties of membranes, like flux of water and other solutes but also require a functional membrane surface. Therefore, surface modification of polymeric membranes has been of prime importance in various membrane-involving applications. It offers versatile means to minimize undesired properties of or introduce additional functions to polymer membranes. Herein, various methods of surface modification for polymeric membranes are also summarized, which include coating and self-assembly, chemical treatment, plasma treatment, and surface graft polymerization.

23.2.2.1 Surface Coating and Self-Assembly

Coating and self-assembly are physical modification methods. The principle and operation of these methods are very simple. For the surface coating, the surface modification is achieved by adsorption or coating suitable materials on the polymer membrane surface directly and the membrane surface can change, for example, from hydrophobic or non-biocompatible to hydrophilic and biocompatible. The key advantage of this technique is that the surface of polymeric membrane can be simply modified or tailored to acquire very distinct properties through the choice of a specific coating material while maintaining the bulk properties of the membrane polymer like elastic modulus. For example, Iwasaki et al. (2002) successfully prepared a blood-compatible gas-permeable membrane by coating a hydrophobic polyethylene (PE) porous membrane surfaces with copolymers of methacryloylphosphatidylcholine (MPC) and dodecyl methacrylate. The hydrophobic dodecyl alkyl chains of the phospholipid copolymer bound to the hydrophobic polymeric membrane and formed a stable add-on layer with improved blood compatibility, which was due to the hydrophilic nature of the phosphatidylcholine groups.

Other surface-coating methods are based on self-assembly like layer-by-layer (LBL) method, which represent new technique for membrane surface-engineering. Among the large variety of surface modification methods, this technique represents a simple, non-covalent, physical surface modification method, which can be used to produce well-defined coatings at a nanometer scale (Liu et al. 2010). The physical basis of this method is the attraction of oppositely charged polyelectrolytes in aqueous solutions by charged surfaces and the multiple adsorptions of polyanions and polycations to make multilayer coatings. The crucial feature of this method is the overcompensation of surface charge by excessive adsorption of the counter polyelectrolyte at every stage of multilayer film formation. The process is easy, and the procedure can be adapted to almost any surface as long as surface charges are present. For example, Liu et al. (2010) constructed polyelectrolyte multilayers (PEM) on the poly(ethylene imine) (PEI)-modified poly(L-lactide) (PLLA) surfaces by alternating immersing of polymer films in sulfated hyaluronan and chitosan solutions. It was found that surface chemistry had a significant effect on the shape and functional activity of an osteoblast-like cell line (MG 63) and the cells spread well with polarized shapes on the polyelectrolyte multilayers (PEM) of PLLA, while most cells were still round and did not have a regular shape on the blank PLLA. The morphology of MG 63 cells indicated that the formed polyion complex multilayer could significantly improve the biocompatibility of PLLA as shown in Figure 23.10.

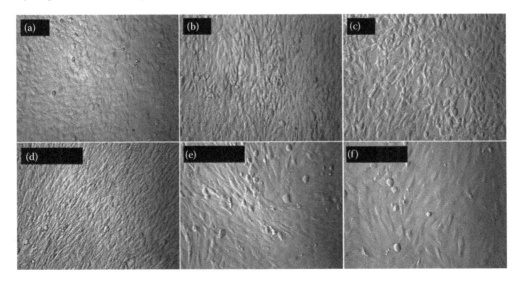

FIGURE 23.10 Effect of multilayer coatings of PLLA on growth of MG 63 osteoblasts. (a) Adsorption of polyethylene imine (PEI). (b) Chemical immobalization of PEI. (c) Plain PLLA. (d) Chitosan as terminal layer. (e) Sulfated hyaluronan as terminal layer. (f) Tissue culture polystyrene.

23.2.2.2 Chemical Treatment

Surface-modified membranes can also be obtained by the simple method of chemical treatment, namely, being immersed into reactive agents. Depending on the timescale of treatment only the surface is modified; therefore, polymer membranes maintain their original characteristics in terms of mechanical strength and thermal stability. Furthermore, various functional groups can be introduced onto the surface this way. Moreover, the modified membrane surface is relatively stable compared with some physical modification methods. Various chemical reactions including oxidation, hydrolysis, addition, and substitution reactions can be applied for the chemical surface modification of polymer membranes.

In the case of oxidation, oxygen-containing groups such as hydroxyl and carboxyl are introduced onto the membrane surface by the use of nitric acid, sulfuric acid, phosphoric acid, alone or in the combination with hydrogen peroxide, sodium hypochlorite, permanganate, chromate, or dichromate of potassium, transition metal nitrates, etc. Hydrolysis is one of the most common surface reactions and the degree of hydrolysis can be controlled by the reaction time. The nitrile group of polyacrylonitrile (PAN) can be easily hydrolyzed by NaOH or amine, and converted into carboxyl, acrylamide, or amide groups, which helps to improve the hydrophilicity and antifouling performance of membranes (Chiang and Hu 1990; Godjevargova and Dimov 1992; Oh et al. 2001a,b), which is shown in Figure 23.11. In addition, enzymes and affinity ligands can be easily immobilized on PAN membrane surface through these functional groups for further applications in biotechnological and biomedical applications.

Other modifications of membrane surfaces are useful to introduce various functional molecules to generate an affinity for specific molecules. For example, Huang et al. (2010) prepared a low-density lipoprotein (LDL) affinity membrane based on polysulfone (PSu) membranes through a series of surface activation reactions. The PSu membrane was firstly activated by a simple chloromethylation reaction, which was followed by covalent binding of a diamine to obtain amino groups on the surface of the otherwise inert PSU membrane. Subsequently, an affinity ligand like heparin could be easily immobilized via reaction with carbodiimide on the surfaces of PSU membrane. The heparin-modified PSu showed high affinity useful for selective adsorption of LDL, which is promising for the treatment of severe cases of hypercholesterolemia as novel kind of LDL apheresis therapy.

23.2.2.3 Plasma Treatment

Plasma that can be regarded as the fourth state of matter is composed of highly excited atomic, molecular, ionic, and radical species. Plasma provides a highly reactive chemical environment to

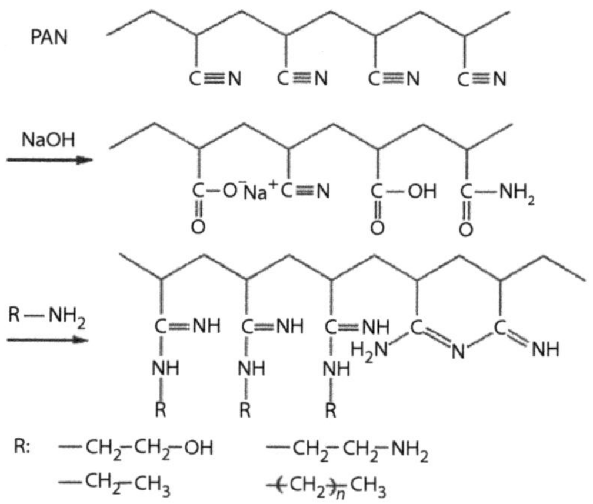

FIGURE 23.11 Reaction mechanism of polyacrylo nitrile (PAN) with NaOH and primary amines

polymers in which many plasma-surface reactions can occur. Surface modification with plasma is an effective and economical surface treatment technique for polymer membranes, too. Normally, inert gases like Argon are activated to form plasma, and accelerated toward the substratum during the plasma treatment. The energy of plasma is transferred to the surface atoms via elastic and inelastic collisions with the materials. Some surface atoms will acquire enough energy and escape from the substrate to the vacuum chamber. With sufficient time, surface contaminations will be cleaned off, which is also known as plasma cleaning of surfaces. Moreover, the high-energy state of plasma generates also radicals on the surface of polymers, which can be used for subsequent grafting reactions. Plasma treatment can also be applied to introduce other elements on the membrane surface. When polymeric membranes are exposed to plasma and if the plasma density and treatment time are proper, new functionalities can be created on the surface and also cross-linked polymer chains can be formed. In a typical plasma treatment process, hydrogen is first abstracted from the polymer chains, which creates radicals of the polymer chains. The polymer radicals can recombine with other simple radicals obtained in the plasma gas to form surface groups such as oxygen or nitrogen functionalities. Generally, formation of oxygen functionalities can convert the membrane surface from hydrophobic to hydrophilic, which improves the adhesion strength, biocompatibility, and other important properties.

For example, Yu et al. used different gases like nitrogen (Yu et al. 2007), ammonia (Yu et al. 2005), oxygen (Yu et al. 2008a), carbon dioxide (Yu et al. 2005), air (Yu et al. 2008b,c), and water (Yu et al. 2008d) as plasma gases to modify polypropylene microporous membranes for the application in bioreactors. After plasma treatment in these gases, the hydrophilicity of the membrane surface was improved because of the introduction of oxygen-, nitrogen-containing functional groups during the plasma treatment; carbon radicals will be formed at the polymer chain during plasma treatment. These carbon radicals will be successively oxidized into oxygen functional groups such as hydroxyl, carbonyl, carboxyl groups, etc., when the membrane is taken out from the plasma reactor into the surrounding air, which is called post-reaction (Wavhal and Fisher 2002). The modified membranes showed better filtration behavior in a membrane bioreactor than the unmodified membranes. The reduction of the initial pure water flux was lower, while the flux recovery after water and caustic cleaning were higher. Hence, the irreversible fouling resistance was decreased after plasma treatment. Furthermore, the chemical nature of the plasma gases has also a strong impact on surface modification reactions, which can change pore size and alter the character of the surface (Yu et al. 2008b,c). Among CO_2, H_2O, and N_2 plasma-treated membranes, the CO_2 plasma-treated membranes possessed the best antifouling characteristics, indicating that membrane surfaces having carboxyl groups are more resistant against fouling than surfaces with amido or hydroxyl groups. However, it must be noted that both the tensile strength and the tensile elongation at break decrease quickly with the increase of plasma treatment time. The decrease of tensile strength and the tensile elongation at break can be attributed to the scission of the molecular chains on the membrane surface. This is a negative effect due to the surface etching phenomenon of plasma treatment (Yu et al. 2008a). Taking the antifouling characteristics and the mechanical properties together, the plasma treatment time must be optimized to obtain the ideal membrane (Yu et al. 2008a).

One can conclude that plasma treatment can conveniently treat specimens with a complex geometry and handle a wide spectrum of materials. However, one major drawback of plasma treatment is the so-called hydrophobic recovery. It is commonly known that hydrophilicity gained by plasma modification is sometimes unstable. The hydrophilicity of surface is reduced and may disappear completely with the increase of storage time. This behavior is believed to be caused by gradual reorientation of the surface chain segments in response to interfacial forces when the membrane surface is exposed to the air or to other nonpolar media. This reorientation may lead to time dependency of surface properties of the plasma-treated membrane. Therefore, plasma polymerization and plasma-induced graft polymerization were developed to endow a membrane surface with a permanent effect. This alternative plasma treatment technique is utilized in the introduced first polar groups to initiate graft polymerization on membranes.

23.2.2.4 Surface Graft Polymerization

Surface grafting is a universal modification method to prepare "tailored" membrane surfaces with desired functions. The grafted polymer chains on the membrane surface play an important role in many membrane applications. During surface grafting, the modification is achieved by tethering suitable macromolecular chains on the membrane surface through covalent bonding. The key advantage of this technique is that the membrane surface can be modified or tailored to acquire distinctive properties through the choice of different grafting monomers or macromolecular chains, while maintaining the bulk properties of the membrane polymer. It also ensures an easy and controllable introduction of tethered chains with a high density and exact localization onto the membrane surface. Compared with physical modification methods such as coating and self-assembly, the covalent attachment of polymer chains onto the membrane surface avoids desorption and maintains a long-term chemical stability of the modified surface.

The grafting methods can be generally divided into two classes, i.e., "grafting-to" and "grafting-from" processes. In the case of "grafting-to" method, pre-formed polymer chains carrying reactive groups at the end or side chains are covalently coupled onto the membrane surface. The "grafting-from" method utilizes active species existing on the membrane surfaces to initiate the polymerization of monomers from the surface toward the outside bulk phase. These techniques can also be classified as chemical, radiation, photochemical, and plasma-induced according to different methods used for the generation of reactive groups.

23.2.2.4.1 Grafting Initiated by Chemical Means

In chemical graft process, the role of initiator is very important as it determines the path of the grafting process (Bhattacharya and Misra 2004). Usually, there are two types of initiators used in the solution phase. They are redox initiator and free-radical initiator, which are determined by the species of initiator. In these two paths, active sites are produced from the initiators and transferred to the substrate to react with monomer and then to form grafted copolymers. For example, Freger et al. (2002) used $K_2S_2O_8/Na_2S_2O_5$ as a redox initiator to modified commercial polyamide (PA) reverse osmosis (RO) and nanofiltration (NF) membranes with vinyl monomers. It was found that the polymerization could take place not only on the membrane surface but also inside the pores, particularly when a high grafting degree (GD) was achieved. In the case of free-radical initiator, the initiators like azo compounds, peroxides, hydroperoxides, and peroxide diphosphates can decompose into free radicals upon heating and then initiate graft polymerization. Son et al. (2006) used azobis (isobutyronitrile) (AIBN) as a heat-sensitive radical initiator that was first grafted on PE membranes at 70°C with divinylbenzene (DVB) as the cross-linker. It was found that the GD increased substantially with the concentration of AIBN and DVB.

Apart from the general free-radical mechanism, living graft polymerization is an alternative interesting technique for surface modification. It provides living polymeric chains with regulated molecular weights and low polydispersity, which means that a controlled and uniform polymer layer can be generated onto the membrane surface. Normally, atomic transfer radical polymerization (ATRP) initiators like 2-bromopropionyl groups are firstly generated by chemical reactions of active reagents on the membrane surface. For example, Yang (2007) employed surface-initiated ATRP to tailor the functionality of polypropylene microporous membranes (PPMM). The ATRP initiator on the poly(HEMA) (PHEMA)-functionalized PPMM surface (PPMM-g-PHEMA) was firstly generated by the reaction between 2-bromopropionyl bromide and the hydroxyl group (PPMM-g-PHEMA-Br). The generated 2-bromopropionyl groups were used as a surface initiator for the subsequent surface-initiated ATRP of a glycomonomer, D-gluconamidoethyl methacrylate (GAMA), to form comb-like glycopolymer brushes. In this way, membranes functionalized by poly(HEMA)-b-poly-(PGAMA) diblock copolymer brushes were prepared.

23.2.2.4.2 Surface Modification by Plasma-Induced Graft Polymerization

Another facile surface-treated technique is the plasma-induced grafting or graft polymerization. The major virtues of these techniques are based on a fast reaction, which normally takes a short time to achieve the required results. Since the plasma and membrane surface produces radicals only close to the surface of the membrane, plasma-grafting polymerization is restricted near the surface. It opens up vast possibilities for designing and developing membrane surfaces with tailored chemical functionality and suitable morphology. For example, the plasma-induced grafting of α-allyl glucoside (AG) on polypropylene microporous hollow fiber membranes (PPHFMs) was carried out by Kou et al. The sugar-containing monomer, α-allyl glucoside (AG), was grafted onto PPHFMs by N_2-plasma radiation as shown in Figure 23.12 (Kou et al. 2003). Due to the presence of plenty of hydroxyl groups, the water contact angle of the grafted membrane decreased significantly from 120° to 36° with the increase of the AG grafting degree from 0 to 3.46 wt.%. This means that a high surface hydrophilicity at a low grafting degree can be achieved by such reaction. In addition, the pure water flux of PPHFMs grafted with 2.50 wt.% AG reached tremendously 3.82 × 103 kg/(m²·h), which is 10 times higher than that of the ethanol-wetted original membranes. Such super hydrophilic PPHFMs may have important application prospects, e.g., for the production of potable water. Furthermore, the hydrated polyglucoside on the modified membrane surfaces may act as a protection layer to inhibit the nonspecific adsorption of proteins, which is demonstrated to some extent by the static adsorption and dynamic filtration of BSA. This indicates a good fouling resistance of such membrane. Furthermore, the PAG-modified PPHFMMs were also applied to assemble a biphasic enzyme membrane bioreactor (EMR), on which lipase from *Candida rugosa* was immobilized (Deng et al. 2005). It was found that, at optimal operational conditions, an apparent volumetric reaction rate of about 0.074 mmol/L·h could be obtained. This result indicates that the super hydrophilic and biocompatible layer formed by the pre-tethering polyglucoside can stabilize the conformation of the enzyme and thus improves the stability of the lipase.

23.2.2.4.3 Surface Modification by UV-Induced Graft Polymerization

When specific chemical groups on the membrane surface absorb light at a specific wave length, excited states can be obtained, which may result in reactive radicals useful to initiate grafting processes (Dyer 2006). The particular feature of this technique is that it works under ambient conditions and does not change the properties of the bulk properties of the polymer. The grafting process by a photochemical technique can proceed in two ways: with or without a sensitizer, which can promote the generation of reactive radicals.

The mechanism "without sensitizer" involves the generation of free radicals on the polymer backbone, which reacts with monomer free radicals to form a grafted copolymer. For example, poly(arylsulfone) (PAS), polysulfone (PSu), and poly(ether sulfone) (PES) membranes are photosensitive in the ultraviolet (UV) range (200–320 nm) (Kaeselev et al. 2001; Kuroda et al. 1990; Nayak et al. 2006; Pieracci et al. 2002a,b). Hence, they do not need a photoinitiator for radical production, i.e., they self-initiate and can produce sufficient radicals for vinyl grafting of various functional groups. Compared with PAS and PSu, PES is far more sensitive to UV-induced graft polymerization, and thus requires less energy to

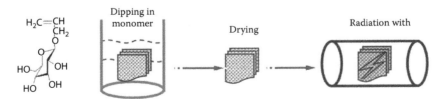

FIGURE 23.12 Schematic representative of N_2-plasma-induced graft polymerization of α-allyl glucoside on polypropylene microporous hollow fiber membranes (PPHFMs).

attain a desired degree of grafting than PSA and PSu (Kaeselev et al. 2001). Furthermore, Pieracci et al. reported that the UV wavelength used for irradiating PES membranes has a strong influence on performance of the modified PES membrane (Pieracci et al. 2002a,b). UV lamps with an emission wavelength maximum of 300 nm and two specially selected UV light filters, benzene and aromatic polyester films, were used to filter out the 254 nm wavelength. As a result a considerable reduction of protein rejection was observed. Membranes modified with 300 nm using the benzene filter had a higher surface wettability than the base membrane, which resulted in a lower irreversible flux loss.

Nayak et al. (2006) prepared an optically reversible switching membrane by photo grafting spiropyran monomers onto PES membrane surfaces. Spiropyran monomer was pre-adsorbed on the PES membrane, which was then exposed to 300-nm UV radiation. ATR-FTIR spectra of the modified PES membrane showed an increased peak at about $1663 \, cm^{-1}$ whereas the peak at $1720–1725 \, cm^{-1}$ decreased, when it was exposed to visible light for 5 min afterwards. In contrast, when the modified PES membrane was exposed to 254 nm UV radiation for 1 h after initial illumination with 300-nm UV light, the peak at about $1663 \, cm^{-1}$ decreased whereas that at $1720–1725 \, cm^{-1}$ increased. This was ascribed to a light-switchable photo chromic group in spiropyran, in which a colored polar "open" merocyanine form and a white nonpolar "closed" form can be switched by exposing to UV and visible light, respectively. Therefore, the vinyl spiropyran-grafted PES membrane showed different colors after exposure to UV or visible light.

On the other hand, in the mechanism "with sensitizer," the sensitizer forms free radicals, which can abstract hydrogen atoms from the base polymer, producing the radical sites required for grafting. Until now, by UV-induced graft polymerization with photoinitiator, a number of vinyl monomers were successfully used to modify and functionalize polymer membranes. Among them, BP and its derivatives were mostly used as photoinitiator. Generally, the graft polymerization of monomers not only takes place on the membrane surface but also at the wall of membrane pores. Xu and coworkers have carried out several work on surface modification by UV-induced graft polymerization with photoinitiator. Yu et al. (2006, 2007) reported that acrylic acid and acrylamide, respectively, could be grafted on PPMM with BP as photoinitiator, which improved the antifouling characteristics of membranes significantly. Yang et al. (2005b, 2006a,b, 2007) developed a series of methods to construct glycosylated membrane surface using UV-induced graft polymerization. A vinyl glycomonomer of 2-gluconamidoethyl methacrylate (GAMA) bearing linear glucose residues was firstly synthesized and then glycosylation of PPMM surface was carried out by UV-induced graft polymerizations in the presence of benzophenone as shown in Figure 23.13. It was found that the rise of PGAMA grafting degree from 2.23 to 6.03 wt.%, the water flux was increased from 466 ± 26 to $764 \pm 28 \, kg/m^2 \cdot h$. Furthermore, the GAMA polymer layer on surface was able to prevent the adsorption of BSA effectively, indicating good fouling resistance of such a membrane. When the original PPMM membrane was used in a dynamic protein solution permeation processes with BSA as the model protein, it was found that the flux decreased with the permeation of BSA solution due to the deposition of proteins on the membrane surface and in the pores. After cleaning in NaOH solution, the flux exhibited some recovery. However, flux recovery was enhanced significantly with the increase of grafting degree. Even at a lower grafting degree a relatively high flux recovery ratio (>80%) was achieved.

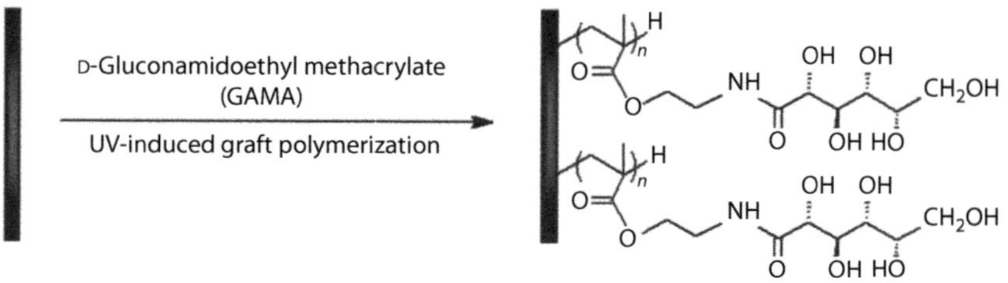

FIGURE 23.13 Schematic representation of ultraviolet (UV)-induced graft polymerizations of D-gluconamidoethyl methacrylate (GAMA) onto polypropylene microporous membrane (PPMM) surface.

Hu (2006, 2007, 2008, 2009) further developed a process to modify PP membrane with HEMA by UV-induced graft polymerization using BP and FeCl3 as photoinitiator, which was followed by reaction with acetylated saccharides such as α-glucose pentaacetate, β-galactose pentaacetate, and lactose octaacetate, to construct a glycosylated membrane surface. The glycosylated membranes were then used to selectively isolate lectins from protein solutions. It was found that the tentacle-like poly(HEMA) chains on the membrane with pendent saccharide ligands not only increases the specific area of membrane surface but were also beneficial to adsorb proteins in multilayers, which enhanced the binding capacity of protein on the affinity membrane. On the other hand, the glycosylated membrane surfaces became also highly hydrophilic, which greatly inhibited the nonspecific adsorption of bovine serum albumin or peanut agglutinin on these surfaces. Therefore, glycosylated membranes may have great potentials for application in selective protein isolation.

23.2.2.4.4 Other Methods

Except the previously mentioned methods, some techniques, such as high-energy radiation, ion beam radiation, laser radiation, and ozone treatment, also may be chosen to initiate graft polymerization on polymer membranes. For example, Liu et al. (2004) applied γ-ray pre-irradiation to graft NVP on PPMM surface. Compared with UV-induced graft polymerization, in which the grafting degree was never higher than 2.0 wt.% with the monomer concentration ranging from 10 to 70 vol.%, a high grafting degree of 64.7 wt.% could be obtained when the dose of γ-ray was 15 kGy. Wang et al. developed an ozone-induced surface graft polymerization method, in which peroxides generated by the ozone treatment could also be decomposed by redox reaction with $FeCl_2$ and then initiate HEMA graft polymerization at mild temperatures (Wang et al. 2000). The grafting degree increased from 4% to 23% with increasing the ozone treatment time.

Overall, a large variety of methods exists to modulate inner structure and surface properties of polymer membranes, which affects not only their transport properties but also permeation and adsorption of proteins. Particularly, the latter may also have a great impact on the so-called fouling of membranes in biotechnological applications, when a protein film is formed, which eventually may lead to clogging of pores with reduction of flux through the membrane. This can also pave the way or be accompanied by adhesion of bacteria in certain biotechnological applications as a mostly undesired effect. On the other hand, control over transport properties, adsorption of proteins, and adhesion of cells are also the key to membrane applications in biomedical applications, which will be outlined with a few examples in the following sections of this review.

23.3 APPLICATION OF MEMBRANES IN BIOARTIFICIAL ORGANS

23.3.1 Organ Failure and Bioartificial Organ Technology

The term bioartificial organ was introduced by Chick et al. in 1975 to define a device containing living cells separated from the organism by a synthetic membrane to substitute the function of organ and tissues (Chick et al. 1975). Today both terms, bioartificial organ and biohybrid organ, are used. Bioartificial organs (BAO) may be represented not only by small implantable devices for the treatment of chronic disorder, such as diabetes (Hunkeler 1999), dwarfism (Chang et al. 1993), and Parkinson's disease (Aebischer et al. 1991; Lindner and Emerich 1998) but also by larger extracorporal systems like bioartificial liver (Stadbauer and Jalan 2007) and kidney systems (Saito 2003). Since most of the BAO developed and tested to date are based on allogenic or even xenogenic cells, such applications require usually that cells are located in a compartment separate from the surrounding tissue to maintain the structural integrity of the immobilized cell mass and to achieve their isolation from the immune system of the recipient. Membranes have been identified as useful tool enabling not only protection from host immune system but also sufficient exchange of substances and oxygenation of immobilized cells (Colton 1995).

TABLE 23.1
Survey on Membrane Application in Bioartificial Organs Development and Application

Organ	Polymer	Membrane Type	Reference
Liver	Poly(acrylonitrile-*co*-*N*-vinylpyrrolidone)	Flat membrane	Krasteva et al. (2002)
	Cellulose nitrate/cellulose acetate	Hollow fiber membrane	Rozga and Demetriou (1995)
	Polyamide, polysulfone, and polypropylene	Flat membrane	Catapano et al. (1996), DeBartolo et al. (1999)
	Polypropylene and polysulfone	Hollow fiber membrane	Gerlach (1996)
	Poly(tetrafluor ethylene)	Flat membrane	Bader et al. (2000)
Kidney	Polysulfone	Hollow fiber membrane	Humes et al. (1999)
	Poly(acrylo nitrile), polysulfone	Hollow fiber membrane	Aebischer et al. (1987a), Fey-Lamprecht et al. (2003)
Pancreas	Alginate	Microcapsules	Calafiore et al. (2001)
	Poly(acrylonitrile-sodiummethallylsulfonate) copolymer	Hollow fiber membrane	Rivereau et al. (1997)
Parathyroid	Alginate-poly-L-lysine	Microcapsules	Fu and Sun (1989)

The number of cells required to replace a specific tissue or organ covers a large range, which is determined by the specific activity (e.g., secretion of hormone) per cell and quantity of the agent or activity required by the body (Colton 1995). Since the quantity of hormones needed by the body is relatively low, also a lower amount of cells of about 10^6–10^7 is necessary for replacement of most endocrine functions. This corresponds to a cell volume of about $10\,\mu L$. By contrast, implantation of Langerhans islets cells to treat diabetes needs already 10^9 cells, which matches to about $1\,mL$ volume. However, a full replacement of liver function, for example, requires between 10^{10} and 10^{11} cells, which resembles more than $100\,mL$ cell suspension (Colton 1995). This makes also understandable that the replacement of endocrine functions can be still achieved by an implantable device while liver or kidney replacement therapies still require extracorporal organs.

A large number of different setups of implantable devices have been developed, such as microcapsules, hollow fiber, or planar diffusion chambers with flat membranes (Colton 1995). A majority of devices is based on encapsulation of cells into hollow spheres, which can be implanted, for example, into the peritoneal cavity. Microcapsules have many advantages with regard to mass transfer properties due to their large surface to volume ratio. However, their application can be problematic because they are not retrievable and become quickly surrounded by connective tissue. Hollow fiber membranes have been used to encapsulate cells with endocrine functions to overcome this limitation. For example, hollow fibers were placed into blood vessels or in the lumbal region of the backbone, which represent locations from where they can also be retrieved. While the current chapter focuses on the application of membranes in BAO for detoxification, others authors have highlighted the application of cells in implantable devices for the replacement of endocrine functions in more detail (Chang et al. 1993; Colton 1995; Hunkeler 1999). Table 23.1 shows a short survey on implantable BAO with the type of organ, geometry of the device, and type of membrane taken (Aebischer et al. 1987a; Bader et al. 2000; Calafiore et al. 2001; Catapano et al. 1996; De Bartolo et al. 1999; Fey-Lamprecht et al. 2003; Fu and Sun 1989; Gerlach 1996; Humes et al. 1999; Krasteva et al. 2002; Rivereau et al. 1997; Rozga and Demetriou 1995).

23.3.2 BIOARTIFICIAL ORGANS FOR BLOOD DETOXIFICATION

The separation of toxins from blood is crucial for the treatment of end-stage renal disease (ESRD), acute renal, and also fulminate liver failure. The current majority of extracorporal support and replacement therapies of these organs are based on physical processes that comprise dialysis, filtration, and adsorption (Bowry 2002; Klinkmann and Vienken 1995; Uhlenbusch-Körwer et al. 2004).

However, kidney and liver represent organs with multiple functions. Kidney has basically a filtration function, and also a re-adsorptive and endocrine function that cannot be replaced by artificial organ technology completely (Lamb and Delaney 2009). Likewise, the numerous tasks of the liver, such as conversion of toxins generated by the metabolism, removal of xenobiotics, protein synthesis, etc. cannot be substituted by hemofiltration or adsorption technologies alone (Atillasoy and Berk 1995). Hence it is not surprising that the conventional organ replacement therapies are associated with an increased morbidity and mortality, which is known from long-term treatment with hemodialysis in ESRD and acute renal or in fulminant hepatic failure (Atillasoy and Berk 1995; Ring-Larsen and Palazzo 1981; Stadbauer and Jalan 2007). To date, organ transplantation is the most successful therapy for acute and chronic failure of internal organs like heart, kidney, liver, lung, etc. However, there is an increasing gap between the number of donor organs available and the patients on the waiting list. Because of the shortage in donor organs, the advent of BAO may be a solution of the described problem because they may support failing organs, which can lead to organ regeneration in acute liver and renal failure or may bridge the time to find a suitable transplant for the patient.

Kidney and liver consist of connective tissue, endothelial, and epithelial cells. Epithelial cells which represent the functional units, such as distal and proximal tubule cells in the kidney or hepatocytes in the liver, have a polar organization where the cell is separated into an apical and a basolateral region. Main morphological characteristic of epithelial cells is the presence of tight junctions and microvilli on the apical cell part, which are both related to transepithelial transport processes (Fromm and Hierholzer 1997). Tight junctions separate the basal and apical region preventing or controlling the exchange of solutes along concentration gradients. Microvilli on the apical cell surface increase the surface area for the uptake of substances and are an important feature of many epithelia as well. Typical epithelia have a planar structure with an underlying complex of basal membranes composed of different extracellular matrix proteins, such as collagen IV, fibronectin, laminin, etc. (Lodish et al. 1995). Figure 23.14 shows the typical morphology of a kidney epithelial cell with the presence of a tight junction between neighboring cells and microvilli on the apical region obtained by transmission electron microscopy.

Bioartificial liver and kidney resemble a combination of immobilized epithelial cells in/on a suitable type of carrier, which can also be a polymer membrane in a special bioreactor design (see for example Allen et al. 2001). Organ cells in bioartificial organs not only have to make an

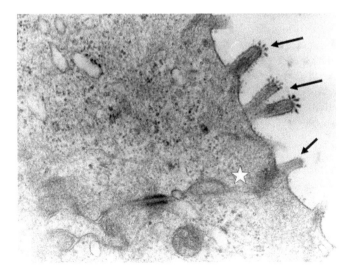

FIGURE 23.14 Transmission electron micrograph (TEM) of kidney epithelial cells with typical morphological features such as microvilli on the apical cell surface (arrows) and tight junctions (asterisk) between adjacent cells.

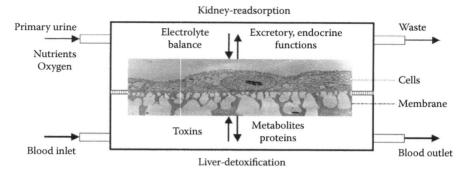

FIGURE 23.15 Schematic illustration of bioartificial organs for blood detoxification underlining the critical role of membranes as support for epithelial cells, as surface in contact with blood components, and as barrier controlling the exchange of solutes between the different compartments.

intimate contact with the surface of the membrane but also to develop close cell–cell connections, which is a perquisite for their survival and a high functional activity. On the other hand, the blood which has to be detoxified will contact the other side of the membrane and may not become activated by the synthetic material. Finally, as a third important requirement, the transport properties of the membrane must be adjusted to allow a sufficient exchange of oxygen, small molecular weight solutes such as toxins, smaller proteins, and must also protect the immobilized organ cells from the immune system of the recipient (Colton 1995). Figure 23.15 shows a simplified setup of a bioartificial organ to emphasize the complex requirements on membranes in this kind of application.

While bioartificial liver support systems (BAL) will be discussed in more detail in Section 3.2, approaches to replace failing kidneys by BAO will be described here only briefly. Aebischer et al. suggested for the first time the combination of hollow fiber membranes with kidney tubular cells to transport substances across the cell-attached membranes like in the kidney tubules (Aebischer et al. 1987a,b). This concept was named bioartificial kidney (BAK). The principal setup of a BAK for preclinical and clinical studies has been only realized by Humes et al. (1999, 2003; Humes 2000; Tiranathanagul et al. 2006). It consists basically of a commercial hemofiltration device, which generates an ultrafiltrate like the primary urine in the kidney glomerulus. This "primary urine" is then transferred to the proximal tubule device, which consists of a hollow fiber bioreactor containing kidney epithelial cells. These cells are able to perform active transport as in the renal tubule along with a multitude of cellular metabolic reactions. As a result, some reabsorbate is generated, which moves back across the membrane into the blood of the patient. The readsorption process is important for return of amino acids, electrolytes, and water. On the other hand, also excretion of certain metabolites like $\beta2$ microglobulin can be achieved. Preclinical and clinical studies with BAK have shown significantly improved survival in acute renal failure, which is normally below 50% (Humes et al. 2003; Tiranathanagul et al. 2006).

23.3.3 Bioartificial Liver Support Systems

Fulminant hepatic failure after intoxication or viral infections is distinguished by defective blood protein synthesis, gluconeogenesis, urogenesis, by impaired plasma detoxification, neurological complications (often associated with cerebral edema), and finally multiorgan failure (Atillasoy and Berk 1995). Contemporary therapies are based on hemodialysis, hemofiltration, and apheresis that have unfortunately a very poor outcome with low survival rates around 20% (Atillasoy and Berk 1995). Therefore, orthotopic liver transplantation is still the gold standard in medical therapy with survival rates higher than 90% after 1 year (Cacciarelli et al. 1997). However, access to donor organs is limited due to not only lack of immediate availability of organs but also immunological mismatch between donor and recipient. Therefore, development of BAL devices has been promoted to bridge the time to transplantation or to allow regeneration of the damaged liver tissue (Tzanakakis et al. 2000).

BAL devices are primarily based on hepatocytes cultured in a bioreactor module. Different setups of BAL devices have been developed, which have been reviewed in more detail by others (Legallais et al. 2001; Park et al. 2005; Tzanakakis et al. 2000). Particularly, membrane-based systems may offer certain advantages because they (1) separate donor hepatocytes from the immune system of the recipient, (2) provide a substratum for cell attachment, and (3) have a large surface to volume ratio to make up a compact bioreactor design (Legallais et al. 2001). However, there are certain critical issues of membrane properties that must be considered. In particular, the ability of membranes to allow adequate adhesion for hepatocytes and the ability for bidirectional mass transfer have not been considered in the past to necessary extent. Transport properties of membranes are important not only for transfer of oxygen and nutrients but also for transfer of albumin-bound toxins, which normally reach hepatocytes through the fenestrated endothelium in the liver (Tzanakakis et al. 2000).

It should be noted that the majority of membranes applied in BAL systems have been developed for other biomedical applications such as hemodialyzers or oxygenators (Legallais et al. 2001; Park et al. 2005; Tzanakakis et al. 2000). These kinds of membranes were chosen because of their relatively good blood compatibility (Park et al. 2009). The molecular cut-off of these membranes is selected to avoid exposure of cells in the bioreactor to the immune system of recipient such as immunoglobulins (ca. 160 kDa) or factors of the complement system (>200 kDa) and leukocytes to avoid host vs. graft rejection. However, it is desirable that the permeation of serum albumin (ca. 70 kDa) as a carrier of many lipophilic toxins and metabolites is still possible. Therefore, some groups have selected membranes with a molecular cut-off of around 100 kDa, while others used conventional membranes for hemodialysis with a cut-off below 60 kDa (Legallais et al. 2001; Park et al. 2005, 2009; Tzanakakis et al. 2000). On the other hand, some authors argued that membranes with a lower cut-off should be used to prevent the entrance of cellular proteins from donor cells into the blood of patient. This seems to be particularly important when porcine or hepatocytes derived from human cancer cell lines like the C3A cells are applied in the bioreactors (Arkadopoulos et al. 1998; Rozga and Demetriou 1995). However, there are also examples that membranes with microfiltration properties can be used with porcine cells, showing that immune isolation is not an absolute requirement (Dixit 1994; Gerlach 1996). Table 23.2 shows examples of bioartificial liver support systems that were preclinically or clinically studied (Arkadopoulos et al. 1998; Dixit 1994; Gerlach 1996).

It is surprising to learn that the majority of membranes used in bioartificial liver systems have not been optimized regarding the promotion of attachment and function of hepatocytes. This is certainly due to the fact that bioengineers and clinicians used available polysulfone hemodialyzers, polypropylene oxygenators, or similar devices based on existing membrane technology for their purpose. Obvious disadvantages of these membranes regarding the poor functionality and survival of hepatocytes in the original devices have been tried to overcome by pre-coating with extracellular matrix

TABLE 23.2
Examples of Membrane-Based Bioartificial Liver Systems

Devices/Company	Cell Type	Polymer	Cut-Off/Pore Size	Reference
ELAD/Amphioxus Cell Technology	C3A hepatoblastoma cells	Cellulose acetate	70 kDa	Arkadopoulos et al. (1998)
HepatAssist/Circe Biomedical	Porcine hepatocytes	Polysulfone	0.2 μm	Dixit (1994)
LSS/Charité Humboldt University	Porcine or human hepatocytes	Combination of Polyamide	100 kDa	Gerlach (1996)
		Polypropylene	Oxygenation	
		Polysulfone	80 kDa	
BLSS/Excorp Biomedical, Inc.	Porcine hepatocytes	Polysulfone	100 kDa	Mazariegos et al. (2002)

proteins like collagen (Legallais et al. 2001) or even Matrigel—an ill-defined complex mixture of matrix components obtained from a mouse tumor cell line (Te Velde et al. 1995). Nevertheless, it was observed that the majority of BAL devices can only be used for treatment sessions of 6–8 h, each patient receiving multiple sessions. The main cause for this restriction is that the primary hepatocytes have an unstable function ex vivo. They lose viability and essential enzymatic activity with culture time and die after short time ex vivo (Legallais et al. 2001; Park et al. 2005, 2009). In addition to hydrophobic membranes mentioned earlier, also very hydrophilic membranes based on cellulose and its derivatives have shown to be less suitable for primary hepatocytes (Legallais et al. 2001). To overcome these obvious limitations, protein pre-coating has been introduced, which improved the functional activity and survival of hepatocytes (Legallais et al. 2001). However, protein pre-coating changes the transport properties of membranes and has other undesired effects, such as increase of costs, possible immunological problems, and also safety of the treatment.

Therefore, it can be anticipated that the development of membranes with polymer compositions that are tailored for hepatocytes could improve the length of treatment time for patients with acute liver failure and make production of the devices also easier. Catapano and coworkers have demonstrated in several papers that hepatocyte adhesion and urea synthesis were enhanced with increasing wetting and surface roughness of surfaces and polypropylene membranes (Catapano et al. 1996, 2001; De Bartolo et al. 1999). Also the chemical composition of polymer membranes has been used to control the functional activity of hepatocytes in vitro finding that hepatocyte adhesion and function was better maintained on more hydrophilic than hydrophobic membranes (De Bartolo et al. 2002, 2004). As mentioned before, other polymer membranes as those developed for hemodialysis or blood oxygenation have been tested rarely in this regard. One example here is poly(ether ketone) (however, also coated with proteins) as a substratum for hepatocytes (De Bartolo et al. 2004).

To adjust the chemical composition of membranes to the requirements of specific cells, our group developed a number of membrane types which are based on copolymers of acrylonitrile (Groth et al. 2002). In contrast to most of the previous works, we developed a broad range of copolymers that were studied with regard to their effects on hepatocyte adhesion, growth, and function (Krasteva et al. 2002, 2005). The monomers applied here were acrylonitrile, hydrophilizing, nonionic N-vinylpyrrolidone (NVP 5, 20, and 30 mol%), aminoethylmethacrylate or aminopropylmethacrylate (AEMA or APMA, both about 1 mol%), and sodium methallylsulfonate (NaMAS, about 2 mol%). Copolymers were compared with the poly(acrylo nitrile) (PAN) homopolymer. Membranes were prepared by phase inversion and possessed porosities in the ultrafiltration range. Most of the studies were performed with C3A hepatoblastoma cells, investigating their adhesion, growth, and functional activity. This human liver-derived cell line has been applied in clinical studies with the ELAD device (Arkadopoulos et al. 1998). In our studies it was observed that particularly hydrophilic copolymers, which contained 20 mol% N-vinylpyrrolidone (NVP 20) supported moderate cell attachment to the membrane and supported cell–cell adhesions of C3A cells (Krasteva et al. 2002, 2005). Figure 23.16 shows the impact of chemical composition of the different copolymer membranes on adhesion and morphology of C3A hepatoblastoma cells after 24 h. For example, an increasing content in NVP from 5 to 30 mol% leads to a concomitant decrease in cell number and also spreading. It is interesting to note that the decrease in cell spreading is accompanied by improved cell–cell contacts. Figure 23.18 shows comparative immune fluorescence studies of C3A cells on PAN homopolymer and the PAN copolymer with 20 mol% NVP after 72 h of culture on the membranes. It is visible that cells on PAN (upper lane of Figure 23.17) form spread aggregates. Actin, the cytoskeletal protein, forms partly long stress fibers, which are similar to those found in fibroblasts. C3A cells on PAN possess also numerous short focal adhesion plaques visualized by staining of vinculin, which indicate also stronger adherence of cells on this substratum. In contrast, formation of cell–cell contacts seems to be at least to some part inhibited because distribution of E-Cadherin—a junctional protein in epithelial cells—is irregular. In comparison to PAN, C3A cells on the more hydrophilic NVP 20 copolymer have a quite condensed actin structure, where no stress fibers are visible. Also focal adhesions are smaller and more dot-like in these cells, while

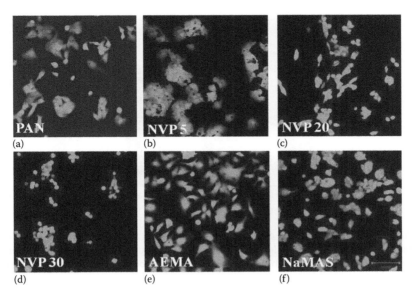

FIGURE 23.16 **(See companion CD for color figure.)** Adhesion of human C3A hepatoblastoma cells on membranes made from different acrylonitrile copolymers. Cells were cultured in the presence of 10% fetal bovine serum for 24 h, stained with fluorescein diacetate, and visualized with confocal laser scanning microscopy. (a) PAN—poly(acrylo nitrile); (b) NVP 5—poly(acrylo nitrile–*N*-vinylpyrrolidone) copolymer with 5 mol% *N*-vinylpyrrolidone (NVP); (c) NVP 20—same copolymer with 20 mol% NVP; (d) NVP 30—same copolymer with 30 mol% NVP; (e) AEMA—poly(acrylonitrile–aminoethylacrylate) copolymer with 1 mol% AEMA, and (f) NaMAS—poly(acrylo nitrile–sodium methallylsulfonate) copolymer with 2 mol% NaMAS.

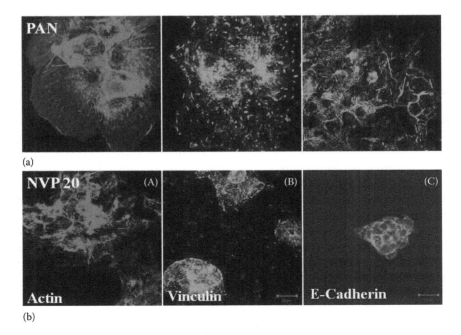

FIGURE 23.17 **(See companion CD for color figure.)** Immunofluorescence studies of C3A hepatoblastoma cells after 72 h culture on (a) polyacrylonitrile (PAN) and (b) poly(acrylo nitrile–*N*-vinylpyrrolidone) copolymer with 20 mol% *N*-vinylpyrrolidone (NVP 20) membranes. Cells were fixed, permeabilized, and stained for actin (A), vinculin (B), and E-Cadherin (C).

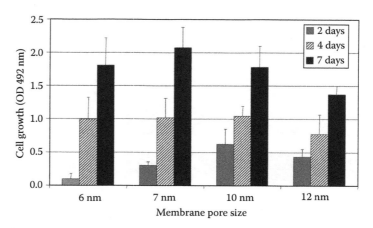

FIGURE 23.18 Comparison of growth of C3A hepatoblastoma cell attachment and growth in dependence on mean pore diameter of poly(acrylo nitrile–*N*-vinylpyrrolidone) copolymer with 20 mol% *N*-vinylpyrrolidone (NVP 20) membranes. Please note that initial number of cells after 2 days is higher on membranes with larger than smaller pores. However, cell numbers are higher after 7 days on membranes with smaller pores indicating better growth on smoother surfaces.

E-Cadherin visualizes the well-developed junctions between the neighboring cells, which is evident by the regular distribution of E-Cadherin identifying the individual cells in the aggregate. This was accompanied by enhanced functional activity of cells in terms of certain cytochrome P450 enzymes important for the detoxification activity of hepatocytes (Krasteva et al. 2005). Results of this study underline the significant role of regulating the adhesion of epithelial cells (Tzoneva et al. 2007).

In further studies also the effect of membrane porosity in the ultrafiltration range of NVP 20 membranes on C3A cell was investigated (Krasteva et al. 2004). Different average pore sizes of the membranes were adjusted ranging from 6 to 12 nm. Indeed, an inverse relationship between initial cell attachment and growth was found as shown in Figure 23.18. While the initial attachment was higher on membranes with larger pores, the cell growth was higher on membranes with smaller pores. The functional activity, however, measured by the activity of liver-specific P450 enzymes was maximal at intermediate pore size (Krasteva et al. 2004). Also the hemocompatibility of membranes used in bioartificial liver systems is of great interest since blood must be in intimate contact with the membrane to allow mass transfer of toxins, metabolites, and smaller proteins. While mass transfer properties are certainly an issue, the activation of blood coagulation, complement activation, and adhesion of blood cells like blood platelets and leukocytes are important prerequisites for the clinical application. Therefore, further studies on the blood compatibility of these membranes were conducted, finding that introduction of *N*-vinylpyrrolidone improves the hemocompatibility of membranes greatly, when compared to polyacrylonitrile as standard membrane for hemodialyzers (Groth et al. 2005). This was an encouraging finding and indicated the potential of such type of hydrophilic membrane to be applied in BAL systems for clinical applications.

Additional studies were carried to investigate the effect of the same copolymers on the survival and functional activity of primary rat hepatocytes (Grant et al. 2005). Here, an additional comonomer acrylamido-2-methyl-propansulfonic acid (AMPS) was introduced and used also to synthesize polyacrylonitrile copolymers for membrane formation. Besides the positive impact of NVP-containing copolymers on primary hepatocytes, particularly the poly(AN–AMPS) copolymer was found to be the most compatible option for maintaining primary hepatocyte morphology, function, and survival. It maintained viable, functional cells for at least 16 days in culture (Grant et al. 2005). Since copolymers of AN and AMPS had rubber-like properties even at low AMPS content of 2 mol%, different blends from polyacrylonitrile homopolymer and AMPS–AN copolymers were prepared in additional studies. Investigations were carried out here with human C3A hepatoblastoma cells again (Kostadinova et al. 2009). Results of this study showed that certain blend ratios provided both good mechanical properties for membrane formation and also excellent biocompatibility visible by good attachment, growth, and also functional activity of cells (Kostadinova et al. 2009). Overall, tailoring membrane properties regarding wettability, chemical composition, and porosity has a great potential to improve further the functional activity and to prolong the survival of hepatocytes in bioartificial liver systems.

It must be noted here that despite the worldwide activities on the development of BAL, no system has been implemented into routine clinical practice except some clinical studies. (1) Main obstacles to be overcome here are an unlimited source of functionally active cells. Recently, also embryonic stem cells have been used to replace primary liver cells from human outdated liver transplants or porcine hepatocytes (Shamra et al. 2010). Also in this case, appropriate liver-specific biocompatibility of membranes will be still an issue. (2) Moreover, storage or cryopreservation of BAL bioreactors to be available on demand (off-the-shelf availability) is still an unsolved issue. (3) Finally, quite recently more efficient cell-free detoxification procedures based on adsorption or diffusion/convection to remove toxins from blood of patients with failing livers like the PROMETHEUS™ system from Fresenius AG or the MARS™ system from Gambro AG have shown good success in clinical settings, which makes also some of the efforts in the past obsolete (Rifai and Mann 2006; Stange et al. 1999). More new cell-free systems like the Microsphere-Derived Detoxification System (MDS) developed by the group of Falkenhagen bear great promises for future clinical application in hepatic failure and sepsis (Falkenhagen et al. 1999; von Appen et al. 1996). Indeed, small-scale BAL may have also interesting applications in testing hepatotoxicity of pharmaceuticals in an organ-like setting that differs greatly from the conventional 2D tissue culture setups.

23.4 APPLICATION OF MEMBRANES IN TISSUE ENGINEERING

23.4.1 Introduction to Tissue Engineering

Tissue engineering has been introduced by Langer and Vacanti as a new discipline aiming at the restoration of tissue structure and function (Langer and Vacanti 1993). This approach is based on three components. (1) First, some kind of scaffold acts as guiding structure for tissue regeneration, providing an adhesive support for cells. The scaffold material must be degradable and will dissolve in a timescale needed to be replaced by newly formed tissue (e.g., new bone). The scaffold should also have a certain porosity to allow the colonization with cells, neovascularization, and also transfer of oxygen, nutrients, and waste products (Cohen et al. 1993). (2) Furthermore, the scaffold can be loaded with bioactive (signaling) molecules, which stimulate anchoring, growth, and differentiation of cells (Drotleff et al. 2004). However, this is not always necessary. The material or its surface may also have some inherent bioactivity, which stimulates cells (e.g., calcium phosphates for osteoblasts) (Ramay and Zhang 2004). (3) Finally, the scaffold must be colonized by tissue-specific cells. Cells applied in tissue engineering are preferentially autologous obtained by biopsies from patients and seeded on the scaffold. Another possibility is that cells colonize the scaffold by migration and growth from surrounding body fluids or tissues (Minuth et al. 2004).

23.4.2 Application of Membranes in Tissue Engineering of Skin

Adequate repair of deep skin defects such as burns or chronic skin ulcers in elderly or diabetic patients represents still a great challenge in medicine. Contemporary therapies aim to restore the normal skin structure, which is mainly composed of the underlying dermal part with fibroblast and blood vessels embedded in extracellular matrix material overlaid by the epidermis as a nonvascularized multilayered epithelium (Boyce and Warden 2002; Chester and Papini 2004). Successful repair of deep skin defects requires the restoration of the dermal part as a prerequisite for proper regeneration of the epidermis (Horch et al. 2001). Application of autologous split skin grafts that contain both components is still the medical gold standard. However, it is not sufficient when the wound size is too large or the constitution of the patient does not permit a further skin injury at the donor site (Boyce and Warden 2002; Chester and Papini 2004). First clinical trials to repair skin defects by tissue engineering were based on the regeneration of the epidermis only. Rheinwald and Green (1975) developed decades ago a technique to prepare cultured epidermal autografts (CEA) in vitro, which are derived from skin biopsies of the patient. The cells grow in vitro up to confluence as a multilayered keratinocyte sheet, which has to be separated from the culture flask by enzymatic treatment. Unfortunately, these CEA are quite fragile because they lack any mechanical support. Their production is quite labor-consuming and requires about three weeks. It must be noted as well that the healing of CEA is often not as good as that of split skin grafts. Even months after transplantation, the epidermis can still get lost due to blistering (Horch et al. 2001).

Current medical practice in the treatment of deep skin defects includes debridement of the wounded area and temporary coverage by wound dressings. Indeed, the wound dressing must be porous enough to allow oxygenation and removal of wound exudates. On the other hand, they must also protect the patient from dehydration and infection of the wound (Balasubramani et al. 2001; Jones et al. 2002). It has been acknowledged that polymer membranes meet the majority of these requirements. Moreover, it would be a great advantage if keratinocytes could be cultured on membranes to generate a combination of conventional wound dressings with a carrier for cell transplantation. So far a limited number of approaches have been put forward to apply membranes in tissue engineering of epidermis. One good example for this approach is the polyurethane membrane HydroDerm® developed by HD Innovative Technologies Ltd., which has been developed primarily as a wound dressing material (Wright et al. 1998). Polyurethanes have been known for their good biocompatibility in different biomedical applications. Also their mechanical properties can be varied depending on their chemical composition from stiff to elastic materials (Santerre et al. 2005). Therefore, they were studied regarding their applicability as tissue engineering scaffold for regeneration of epidermis (Rennekampff et al. 1996). In the study described here, keratinocytes were seeded on the polyurethane membrane in vitro. It was found that keratinocyte growth was delayed when compared to standard tissue culture polystyrene. However, the formation of a multilayered epidermis was observed that indicates support of cell differentiation by the membrane (Rennekampff et al. 1996; Wright et al. 1998). Preclinical studies with this constructs placing the membrane with cells of early pre-confluent stages on the wound of an animal model showed a partial reformation of the epidermis. It is important to recognize that epidermis is a polarized tissue with basal layer of living keratinocytes that differentiate to form a multilayered epithelium with an acellular uppermost layer. This means also that the epidermal reformation requires transmigration of keratinocytes from the membrane placed on the wound to the underlying dermis followed by reattachment and formation of a new basal layer (Grant et al. 2001). In the study described earlier, a delayed growth of keratinocytes was observed that might be attributed to the chemical composition of the polyurethane membrane, which was not especially designed to promote keratinocyte attachment and growth (Rennekampff et al. 1996; Wright et al. 1998).

Besides synthetic polymers, natural polymers or combinations with polyesters have also been applied to make membranes for the transplantation of keratinocytes. The obvious advantage of natural polymers or polyesters, such as poly(ε-caprolactone) (PCL), is their degradability, which avoids the necessity to remove the material at later stages of treatment. Hyaluronic acid (HA) is a major component of the extracellular matrix in connective tissues and plays a promoting role in wound healing (Price et al. 2007). However, HA is water-soluble and forms hydrogels, which have poor mechanical properties. To overcome this disadvantage, HA has been esterified with benzyl groups and cross-linked, which decreases its solubility in water and increases the mechanical stability (Caravaggi et al. 2003). The modified HA has been used to generate different scaffold materials for engineering of skin, cartilage, and other tissues. In this context, esterified hyaluronic acid (HYAFF) has also been used as a carrier system for keratinocytes. The commercial product is produced by Fidia Advanced Biopolymers and denoted as HYAFF-11 or Laserskin™. The Laserskin represents a membrane that contains additionally larger pores generated by a laser. Keratinocytes are seeded on both sides of Laserskin and may migrate through the pores to cover the underlying dermal structures of the skin. Laserskin is applied mainly for the treatment of nonhealing skin ulcers with good success (Caravaggi et al. 2003). Also membranes made of blends from collagen and PCL have been used for tissue engineering of epidermis and dermis. One of the examples is a membrane based on a collagen scaffold, which is prepared by a freeze drying process and subsequent impregnation with PCL solution (Dai et al. 2004). The resulting scaffold represents a microporous membrane with larger pores in the range of about 40 μm. It was shown that both dermal fibroblasts and keratinocytes were able to grow on this scaffold. A specific co-culture model was developed where the membrane separates fibroblasts building the dermal component and keratinocytes making the epidermis (Dai et al. 2004, 2005). Such a construct can be used in the treatment of deep skin defects when not only epidermis but also the dermal part of the skin must be replaced.

23.4.2.1 Application of PAN Copolymers or Blends for Tissue Engineering of Skin

To address the specific needs of cells from epithelial origin like hepatocytes, kidney epithelial cells, and keratinocytes, the chemical composition of polymer membranes or membrane surfaces based on copolymers of acrylonitrile introduced in (Groth et al. 2002, Krasteva et al. 2004, 2005) or on surface-modified polyetherimide membranes (Altankov et al. 2005) was tailored. All membrane types described herein were prepared by phase inversion and possessed porosities in the ultrafiltration range. Interestingly, membranes prepared from copolymers of acrylonitrile and AEMA or APMA that possess primary amino groups promoted growth of keratinocytes in vitro to an unexpected extend. Figure 23.19 shows a comparison of scanning electron micrographs of

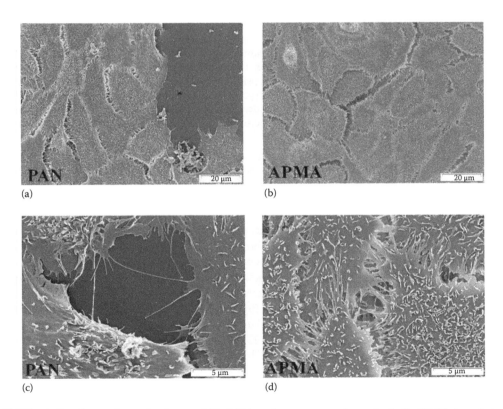

(a)   (b)

(c)   (d)

FIGURE 23.19 Comparison of cultures of HaCaT keratinocytes on polyacrylonitrile (PAN) (a and c) and aminopropylmethacrylate (APMA) (b and d) copolymer membranes after 7 days of culture visualized with scanning electron microscopy at low (a and b) and high (c and d) magnification. Note that the amine-functionalized copolymer APMA supports more rapid monolayer formation (b) with closer cell–cell contacts (d) compared to PAN where no monolayer of cells was established at the same time (a and c).

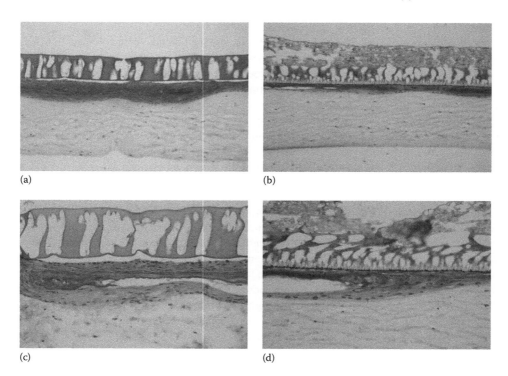

(a)   (b)

(c)   (d)

FIGURE 23.20 Organotypic culture of HaCaT keratinocytes seeded on aminopropylmethacrylate (APMA) (a and c) and poly(acrylo nitrile–N-vinylpyrrolidone) copolymer with 5 mol% N-vinylpyrrolidone (NVP 5) (c and d) membranes, which were after 1 day placed upside-down on a collagen gel with embedded fibroblasts for 14 days. Organotypic cultures were fixed and cryosections were stained with hematoxylin/eosin. It is shown that APMA supports the formation of a neoepidermis (a and c) indicated by the thick blue layer beneath the membrane attached on the collagen gel compared to NVP 5 (b and d). (a and b) Magnification 10 times and (c and d) magnification 20 times.

keratinocytes cultures on PAN (Figure 23.19a and c) and APMA (Figure 23.19b and d) at two different magnifications. It is visible that cells on APMA cover the entire membrane surface and have close intercellular contacts. In contrast, keratinocytes cultured on PAN did not reach a confluent monolayer in the time frame of the investigation. Also intercellular contacts between the cells were less well expressed. Moreover, it was possible to establish an organotypic co-culture model based on collagen gels with embedded fibroblasts as dermal equivalent on which the membranes with keratinocytes were placed upside down. Here it was possible to observe that a part of keratinocytes was able to transmigrate from the APMA membrane to the surface of the dermal equivalent starting there to grow and differentiate, which means to build up a neo-epidermis-like structure (Grant et al. 2001; Horch et al. 2001). Figure 23.20 shows a cross-section through membranes with neo-epidermis and dermal equivalent stained with hematoxylin–eosin. It is shown that the APMA copolymer membrane (Figure 23.20a small and c high magnification) promotes the formation of a neo-epidermis in contrast to a hydrophilic NVP 5 copolymer membrane (Figure 23.20c small and d high magnification).

23.4.2.2 Poly(ether imide) Blends or Surface-Modified PEI as Potential Materials for Skin Tissue Engineering

To explore further the possibility that polymer membranes with amino groups may support growth and differentiation of keratinocytes, we developed a blend made of poly(ether imide) (PEI) and poly(benzimidazole), which was used to prepare membranes by phase inversion method

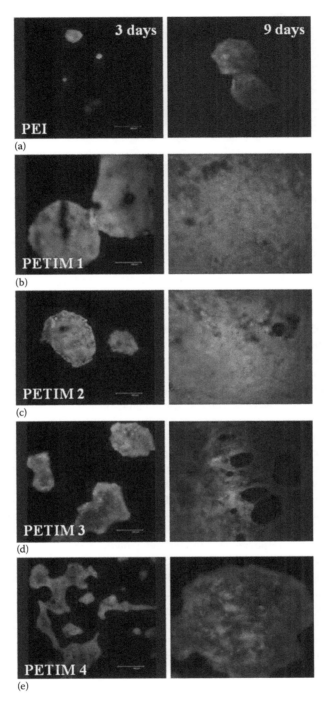

FIGURE 23.21 **(See companion CD for color figure.)** Growth of human HaCaT keratinocytes on poly(ether imide) (PEI) membranes covalently modified with different poly(ethylene imines). Cells were cultured in the presence of 10% fetal bovine serum for 3 or 9 days, stained with fluorescein diacetate, and visualized with confocal laser scanning microscopy. (a) PEI—plain poly(ether imide) membranes; (b) PETIM 1—PEI membrane modified with poly(ethylene imine) (PETIM) of weight-average molecular weight (M_w) of 800 g/mol; (c) PETIM 2—PEI membranes modified with PETIM of M_w 2,000 g/mol; (d) PETIM 3—PEI membranes modified with PETIM of M_w 25,000 g/mol; and (e) PETIM 4—PEI membranes modified with PETIM of M_w 750,000 g/mol.

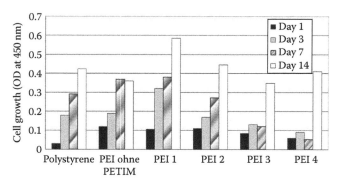

FIGURE 23.22 Quantitative estimation of human HaCaT keratinocytes growth on poly(ether imide) (PEI) membranes covalently modified with different poly(ethylene imines) after 1, 3, 7, and 14 days in the presence of 10% fetal bovine serum. A modified lactate dehydrogenase test was used, lysing adherent viable cells with Triton X-100 at the times indicated. PEI—plain poly(ether imide) membranes; PETIM 1—PEI membrane modified with poly(ethylene imine) (PETIM) of weight-average molecular weight (M_w) of 800 g/mol; PETIM 2—PEI membranes modified with PETIM of M_w 2,000 g/mol; PETIM 3—PEI membranes modified with PETIM of M_w 25,000 g/mol, and PETIM 4—PEI membranes modified with PETIM of M_w 750,000 g/mol.

(Altankov et al. 2005). PEI has the advantage of good membrane formation properties and high thermal stability (Kneifel and Peinemann 1992), which makes this polymer interesting for biomedical applications (Seifert et al. 2002). The blend membrane was more hydrophilic than the pure PEI membrane, less porous, and had an increased surface (zeta) potential. It was also observed that blending PEI with PBI promoted cell–cell contacts. Here, a clear tendency for homotypic cellular interaction of keratinocytes was observed that was much stronger on the blend membrane (Altankov et al. 2005). Thus, membranes based on blends of PEI with PBI could provide a tissue-compatible scaffold that may be used as a temporary carrier of keratinocytes as well.

During our studies we discovered also the possibility to functionalize PEI by a simple wet chemical procedure using different diamines or polyamines (Albrecht et al. 2003). It was observed that covalent binding of poly(ethylene imine) (PETIM) of lower molecular weight may provide advantages for the culture of keratinocytes, such as a rapid establishment of a pre-confluent cell layer on the surface and subsequent release of the cell layer from the membrane support (Trimpert et al. 2006). These studies were recently extended studying the effect of a wide range of molecular weights of PETIM on keratinocytes attachment and growth. The following molecular weights of poly(ethylene imines) were applied: 800 g/mol (PETIM 1), 2,000 g/mol (PETIM 2), 25,000 g/mol (PETIM 3), and 750,000 g/mL (PETIM 4). These modifications were compared to the pure PEI membrane and in addition also to tissue culture polystyrene as standard material with good cell compatibility. Figure 23.21 shows confocal images of HaCaT keratinocytes stained with fluorescein diacetate cultured for 3 and 9 days. It is evident that growth of cells is inhibited on the pure PEI membrane, while immediate growth of cells can be observed on PETIM 1 and PETIM 2 with formation of confluent monolayers after 9 days. By contrast, higher molecular weight poly(ethylene imine) such as PETIM 3 and PETIM 4 leads to delayed growth because after 9 days no complete monolayer formation can be observed. These qualitative observations were also confirmed by quantitative cell growth measurements shown in Figure 23.22, showing that highest number of cells was achieved with PETIM 1, which has the smallest molecular weight. The results of all studies mentioned earlier underline the beneficial role of amino groups on adhesion growth and differentiation of keratinocytes and formation of neo-epidermis-like structures. While first organotypic culture models have shown advantageous effect of the polymer compositions and also first animal experiments have been carried out, clinical studies are still required to study the applicability of these materials for treatment of skin defects.

23.5 SUMMARY AND CONCLUSIONS

Polymeric membranes have the great advantage that already during the membrane formation process specific membrane properties like permeability and also surface topography can be tailored according to the requirements of the application. Moreover, also a given membrane type may be further adapted to the desired purpose by a large variety of surface modification techniques that may also pave the way for a multitude of biomedical and clinical applications. As stated herein, polymer membranes held great promises for different kinds of applications in medicine apart from the long-time established use in hemodialysis and blood oxygenation. Particularly in the field of bioartificial organs and tissue engineering, membranes may be applied as guiding structures providing also mechanical support to cells. For helping cell attachment, growth, functionality, and survival ex vivo, particularly the chemical composition of the membrane surface and also their surface topography play a crucial role. In addition, membranes may also be useful as reservoir system to accumulate growth factors and cytokines in the wounded area to promote healing of defects. Overall, membranes can represent an essential component in many tissue engineering applications and for BAO technology.

ACKNOWLEDGMENTS

The authors gratefully acknowledge the collaboration with Dr. Wolfgang Albrecht, Dr. Gregor Boese, Dr. Michael Schossig, Dr. Barbara Seifert, and Prof. Dr. Dieter Paul from Institute of Chemistry, GKSS Centre Geesthacht. Also Prof. George Altankov from Institute of Biomedical Engineering, Barcelona, and Dr. Natalia Krasteva from Bulgarian Academy of Sciences made significant contributions to the work cited herein. Furthermore, Dr. Hans-Jürgen Stark and Prof. Dr. Norbert Fusenig from German Cancer Research Centre, Heidelberg, made significant contributions to the work on tissue engineering of skin. We are particularly indebted to our colleague, Dr. Günter Malsch, who deceased on December 24, 2007. Without his work and experience, a significant part of the work on polyacrylonitrile membranes reviewed in this chapter would not have been possible. Also the financial support of some studies reviewed herein by BMBF (grant FKZ 01GN0119), Commission of European Union (BE 97-4326), the DAAD (funding of bilateral collaboration between Zhejiang University Hangzhou and Martin Luther University Halle Wittenberg), and the temporary support to X-J Huang from Fresenius Medical Care Deutschland GmbH's Nephrocore Program for Young Scientists is gratefully acknowledged.

REFERENCES

Aebischer, P., Ip, T.K., Panol, G., and Galletti, PM. 1987a. The bioartificial kidney—Progress towards an ultra-filtration device with renal epithelial cells. *Life Support Syst* 5: 159–168.

Aebischer, P., Miracoli, T., and Galletti, P.M. 1987b. Renal epithelial cells grown on semipermeable hollow fibers as a potential ultrafiltrate processor. *ASAIO Trans* 33: 96–102.

Aebischer, P., Tresco, P.A., Winn, S.R., Greene, L.A., and Jaeger, C.B. 1991. Long-term cross-species brain transplantation of a polymer-encapsulated dopamine-secreting cell-line. *Exp Neurol* 11: 269–275.

Akthakul, A., Scott, C.E., Mayes, A.M. et al. 2005. Lattice Boltzmann simulation of asymmetric membrane formation by immersion precipitation. *J Membr Sci* 249: 213–226.

Albrecht, W., Seifert, B., Weigel, T. et al. 2003. Amination of poly(ether imide) membranes using di- and multivalent amines. *Macromol Chem Phys* 204: 510–521.

Albrecht, W., Weigel, T., Schossig-Tiedemann, M. et al. 2001. Formation of hollow fiber membranes from poly(ether imide) at wet phase inversion using binary mixtures of solvents for the preparation of the dope. *J Membr Sci* 192: 217–230.

Allen, J.W., Hassanein, T., and Bhatia, S.N. 2001. Advances in bioartificial liver devices. *Hepatology* 34: 447–455.

Altankov, G., Albrecht, W., Richau, K., Groth, T., and Lendlein, A. 2005. On the tissue compatibility of poly(ether imide) membranes: An in vitro study on their interaction with human dermal fibroblasts and keratinocytes. *J Biomater Sci Polym Ed* 16: 23–42.

Altinkaya, S.A. and Ozbas, B. 2004. Modeling of asymmetric membrane formation by dry-casting method. *J Membr Sci* 230: 71–89.

Andersson, B., Sundby, C., and Albertsson, P.A. 1980. A mechanism for the formation of inside-out membrane-vesicles—Preparation of inside-out vesicles from membrane-paired randomized chloroplast lamellae. *Biochim Biophys Acta* 599: 391–402.

Arkadopoulos, N., Detry, O., Rozga, J., and Demetriou, A.A. 1998. Liver assist systems: State of the art. *Int J Artif Organs* 21: 781–787.

Atillasoy, E. and Berk, P.D. 1995. Fulminant hepatic failure—Pathophysiology, treatment and survival. *Ann Rev Med* 46: 181–191.

Bader, A., De Bartolo, L., and Haverich, A. 2000. High level benzodiazepine and ammonia clearance by flat membrane bioreactors with porcine liver cells. *J Biotechnol* 81: 95–105

Balasubramani, M., Kumar, T.R., and Babu, M. 2001. Skin substitutes: A review. *Burns* 27: 534–544.

Bazarjani, M.S., Mohammadi, N., and Ghasemi, S.M. 2009. Ranking the key parameters of immersion precipitation process and modeling the resultant membrane structural evolution. *J Appl Polym Sci* 113: 1529–1538.

Bhattacharya, A. and Misra, B.N. 2004. Grafting: A versatile means to modify polymers—Techniques, factors and applications. *Prog Polym Sci* 29: 767–814.

Bindal, R.C., Hanra, M.S., and Misra, B.M. 1996. Novel solvent exchange cum immersion precipitation technique for the preparation of asymmetric polymeric membrane. *J Membr Sci* 118: 23–29.

Boom, R.M., Wienk, I.M., Vandenboomgaard, T. et al. 1992. Microstructures in phase inversion membranes. 2. The role of a polymeric additive. *J Membr Sci* 73: 277–292.

Bowry, S.K. 2002. Dialysis membranes today. *Int J Artif Organs* 25: 447–460.

Boyce, S.T. and Warden, G.D. 2002. Principles and practices for treatment of cutaneous wounds with cultured skin substitutes. *Am J Surg* 183: 445–456.

Cacciarelli, T.V., Esquivel, C.O., Moore, D.H. et al. 1997. Factors affecting survival after orthotopic liver transplantation in infants. *Transplantation* 64: 242–248.

Calafiore, R., Luca, G., Calvitti, M. et al. 2001. Cellular support systems for alginate microcapsules containing islets, as composite bioartificial pancreas. In: *Bioartificial Organs III: Tissue Sourcing, Immunoisolation, and Clinical Trials*. Hunkeler, D., Cherrington, A., Prokop, A., and Rajotte, R. (Eds.). New York Academy of Sciences, New York, Vol. 944, pp. 240–251.

Caravaggi, D., De Giglio, R., Pritelli, C. et al. 2003. HYAFF 11-based autologous dermal and epidermal grafts in the treatment of noninfected diabetic plantar and dorsal foot ulcers—A prospective, mulicenter, controlled, randomized clinical trial. *Diabetes Care* 26: 2853–2859.

Catapano, G., De Bartolo, L., Vico, V., and Ambrosio, L. 2001. Morphology and metabolism of hepatocytes cultured in Petri dishes on films and in non-woven fabrics of hyaluronic acid esters. *Biomaterials* 22: 659–665.

Catapano, G., DiLorenzo, M.C., DellaVolpe, C., DeBartolo, L., and Migliaresi, C. 1996. Polymeric membranes for hybrid liver support devices: The effect of membrane surface wettability on hepatocyte viability and functions. *J Biomater Sci Polym Ed* 7: 1017–1027.

Chang, P.L., Shen, N., and Westcott, A.J. 1993. Delivery of recombinant gene-products with microencapsulated cells in-vivo. *Hum Gene Ther* 4: 433–440.

Chen, L.W. and Young, T.H. 1991. Effect of nonsolvents on the mechanism of wet-casting membrane formation from eval copolymers. *J Membr Sci* 59: 15–26.

Chester, D.L. and Papini, R.P.G. 2004. Skin and skin substitutes in burn management. *Trauma* 6: 54–87.

Chiang, W.Y. and Hu, C.M. 1990. Studies of reactions with polymers. 6. The modification of pan with primary amines. *J Polym Sci Polym Chem* 28: 1623–1636.

Chick, W.L., Like, A.A., Lauris, V. et al. 1975. Hybrid artificial pancreas. *Trans Am Soc Artif Intern Organs* 21: 8–15.

Cohen, S., Bano, M., Cima, L. et al. 1993. Design of synthetic polymeric structures for cell transplantation and tissue engineering. *Clin Mater* 13: 3–10.

Colton, C.K. 1995. Implantable biohybrid artificial organs. *Cell Transplant* 4: 415–436.

Dai, N.T., Williamson, M.R., Khammo, N., Adams, E.F., and Coombes, A.G.A. 2004. Composite cell support membranes based on collagen and polycaprolactone for tissue engineering of skin. *Biomaterials* 25: 4263–4271.

Dai, N.T., Yeh, M.K., Liu, D.D. et al. 2005. A co-cultured skin model based on cell support membranes. *Biochem Biophys Res Commun* 329: 905–908.

De Bartolo, L., Catapano, G., Della Volpe, C., and Drioli E. 1999. The effect of surface roughness of microporous membranes on the kinetics of oxygen consumption and ammonia elimination by adherent hepatocytes. *J Biomater Sci Polym Ed* 10: 641–655.

De Bartolo, L., Gugliuzza, A., Morelli, S., Cirillo, B., Gordano, A., and Drioli, E. 2004. Novel PEEK-WC membranes with low plasma protein affinity related to surface free energy parameters. *J Mater Sci Mater Med* 15: 877–883.

De Bartolo, L., Morelli, S., Bader, A., and Drioli, E. 2002. Evaluation of cell behaviour related to physico-chemical properties of polymeric membranes to be used in bioartificial organs. *Biomaterials* 23: 2485–2497.

Deng, H.T., Xu, Z.K., Dai, Z.W. et al. 2005. Immobilization of Candida rugosa lipase on polypropylene micro-filtration membrane modified by glycopolymer: Hydrolysis of olive oil in biphasic bioreactor. *Enzyme Microb Technol* 36: 996–1002.

Dixit, V. 1994. Development of a bioartificial liver using isolated hepatocytes. *Artif Organs* 18: 371–384.

Drotleff, S., Lungwitz, U., Breunig, M. et al. 2004. Biomimetic polymers in pharmaceutical and biomedical sciences. *Eur J Pharm Biopharm* 58: 385–407.

Dyer, D.J. 2006. Photoinitiated synthesis of grafted polymers. *Adv Polym Sci* 197: 47–65.

Falkenhagen, D., Strobl, W., Vogt, G. et al. 1999. Fractionated plasma separation and adsorption system: A novel system for blood purification to remove albumin bound substances. *Artif Organs* 23: 81–86.

Fey-Lamprecht, F., Albrecht, W., Groth, T., Weigel, T., and Gross, U. 2003. Morphological studies on the culture of kidney epithelial cells in a fibre-in-fibre bioreactor design using hollow fibre membranes. *J Biomed Mater Res* 65A: 144–157.

Freger, V., Gilron, J., and Belfer, S. 2002. TFC polyamide membranes modified by grafting of hydrophilic polymers: An FT-IR/AFM/TEM study. *J Membr Sci* 209: 283–292.

Fritzsche, A.K., Arevalo, A.R., Moore, M.D. et al. 1993. The surface-structure and morphology of polyacrylonitrile membranes by atomic-force microscopy. *J Membr Sci* 81: 109–120.

Fromm, M., and Hierholzer, K., 1995, 1997. *Epithelien*. In; Physiologie des Menschen. *Kapitel 34*, R.F. Schmidt and G. Thews (Eds.)., 26 and 27. Auflage, Springer, Berlin, Heidelberg, New York, pp. 719–736.

Frommer, M.A. and Messalem, R.M. 1973. Mechanism of membrane formation. 6. Convective flows and large void formation during membrane precipitation. *Ind Eng Chem Prod Res Dev* 12: 328–333.

Fu, X.W. and Sun, A.M. 1989. Microencapsulated parathyroid cells as a bioartificial parathyroid—In vivo studies. *Transplantation* 47: 432–435.

Gerlach, J.C. 1996. Development of a hybrid liver support system: A review. *Int J Artif Organs* 19: 645–654.

Gerlach, J.C., Schnoy, N., Vienken, J., Smith, M., and Neuhaus P. 1996. Comparison of hollow fibre membranes for hepatocyte immobilisation in bioreactors. *Int J Artif Organs* 19: 610–616.

Godjevargova, T. and Dimov, A. 1992. Permeability and protein adsorption of modified charged acrylonitrile copolymer membranes. *J Membr Sci* 67: 283–287.

Grant, M.H., Morgan, C., Henderson, C. et al. 2005. The viability and function of primary rat hepatocytes cultured on polymeric membranes developed for hybrid artificial liver devices. *J Biomed Mater Res Part A* 73A: 367–375.

Grant, I., Ng, R.L.H., Woodward, B., Bevan, S., Green, C., and Martin, R. 2001. Demonstration of epidermal transfer from a polymer membrane using genetically marked porcine keratinocytes. *Burns* 27: 1–8.

Groth, T., Seifert, B., and Albrecht, W. et al. 2005. Development of polymer membranes with improved haemo-compatibility for biohybrid organ technology. *Clin Hemorheol Microcirc* 32: 129–143.

Groth, T., Seifert, B., Malsch, G. et al. 2002. Interaction of human skin fibroblasts with moderate wettable polyacrylonitrile-copolymer membranes. *J Biomed Mater Res* 61A: 290–300.

Gu, M.H., Zhang, J., Xia, Y. et al. 2008. Poly(vinylidene fluoride) crystallization behavior and membrane structure formation via thermally induced phase separation with benzophenone diluent. *J Macromol Sci B* 47: 180–191.

Herbig, S.M., Cardinal, J.R., Korsmeyer, R.W. et al. 1995. Asymmetric-membrane tablet coatings for osmotic drug-delivery. *J Control Release* 35: 127–136.

Hoenich, N.A., Woffindin, C., Mathews, J.N.S., and Vienken, J. 1995. Biocompatibility of membranes used in the treatment of renal failure. *Biomaterials* 16: 587–592.

Horch, R.E., Munster, A.M., and Achinger, B.A. 2001. *Cultured Human Keratinocytes and Tissue Engineered Skin Substitutes*. Georg Thieme Verlag, Stuttgart, Germany.

Hu, M.X., Wan, L.S., Fu, Z.S. et al. 2007. Construction of glycosylated surfaces for poly(propylene) beads with a photoinduced grafting/chemical reaction sequence. *Macromol Rapid Commun* 28: 2325–2331.

Hu, M.X., Wan, L.S., Liu, Z.M. et al. 2008. Fabrication of glycosylated surfaces on microporous polypropylene membranes for protein recognition and adsorption. *J Mater Chem* 18: 4663–4669.

Hu, M.X., Wan, L.S., and Xu, Z.K. 2009. Multilayer adsorption of lectins on glycosylated microporous polypropylene membranes. *J Membr Sci* 335: 111–117.

Hu, M.X., Yang, Q., and Xu, Z.K. 2006. Enhancing the hydrophilicity of polypropylene microporous membranes by the grafting of 2-hydroxyethyl methacrylate via a synergistic effect of photoinitiators. *J Membr Sci* 285: 196–205.

Huang, X.J., Guduru, D., Xu, Z.K. et al. 2010. Immobilization of heparin on polysulfone surface for selective adsorption of low-density lipoprotein (LDL). *Acta Biomater* 6: 1099–1106.

Humes, H.D. 2000. Bioartificial kidney for full renal replacement therapy. *Semin Nephrol* 20: 71–82.

Humes, H.D., Buffington, D.A., Lou, L. et al. 2003. Cell therapy with a tissue-engineered kidney reduces the multiple-organ consequences of septic shock. *Crit Care Med* 31: 2421–2428.

Humes, H.D., MacKay, S.M., Funke, A.J., and Buffington, D.A. 1999. Tissue engineering of a bioartificial renal tubule assist device: In vitro transport and metabolic characteristics. *Kidney Int* 55: 2502–2514.

Hunkeler, D. 1999. Bioartificial organs—Risks and requirements. In: *Bioartificial Organs II: Technology, Medicine, and Materials*. Hunkeler, D., Prokop, A., Cherrington, A.D., Rajotte, R.V., and Sefton, M. (Eds.). *Ann NY Acad Sci* 875: 1–6.

Iwasaki, Y., Uchiyama, S., Kurita, K. et al. 2002. A nonthrombogenic gas-permeable membrane composed of a phospholipid polymer skin film adhered to a polyethylene porous membrane. *Biomaterials* 23: 3421–3427.

Jones, I., Currie, L., and Martin, R. 2002. A guide to biological skin substitutes. *Br J Plast Surg* 55: 185–193.

Kaeselev, B., Pieracci, J., and Belfort, G. 2001. Photoinduced grafting of ultrafiltration membranes: Comparison of poly(ether sulfone) and poly(sulfone). *J Membr Sci* 194: 245–261.

Kang, Y.S., Kim, H.J., and Kim, U.Y. 1991. Asymmetric membrane formation via immersion precipitation method.1. Kinetic effect. *J Membr Sci* 60: 219–232.

Kang, J.S. and Lee, Y.M. 2002. Effects of molecular weight of polyvinylpyrrolidone on precipitation kinetics during the formation of asymmetric polyacrylonitrile membrane. *J Appl Polym Sci* 85: 57–68.

Kesting, R.E. 1990. The four tiers of structure in integrally skinned phase inversion membranes and their relevance to the various separation regimes. *J Appl Polym Sci* 41: 2739–2752.

Khare, V.P., Greenberg, A.R., and Krantz, W.B. 2005. Vapor-induced phase separation—Effect of the humid air exposure step on membrane morphology Part I. Insights from mathematical modeling. *J Membr Sci* 258: 140–156.

Kim, J.H. and Lee, K.H. 1998. Effect of PEG additive on membrane formation by phase inversion. *J Membr Sci* 138: 153–163.

Kim, J.H., Min, B.R., Won, J. et al. 2001. Phase behavior and mechanism of membrane formation for polyimide/DMSO/water system. *J Membr Sci* 187: 47–55.

Kim, J.K., Taki, K., Nagamine, S. et al. 2009. Preparation of a polymeric membrane with a fine porous structure by dry casting. *J Appl Polym Sci* 111: 2518–2526.

Kim, H.W., Yu, M.H., Lee, J.H. et al. 2008. Experiences with acute kidney injury complicating non-fulminant hepatitis A. *Nephrology* 13: 451–458.

Klinkmann, H. and Vienken, J. 1995. Membranes for dialysis. *Nephrol Dial Transplant* 10: 39–45.

Kneifel, K. and Peinemann, K.V. 1992. Preparation of hollow fiber membranes form polyetherimide for gas separation. *J Membr Sci* 65: 295–307.

Kostadinova, A., Seifert, B., Albrecht, G. et al. 2009. Novel polymer blends for the preparation of membranes for biohybrid liver systems. *J Biomater Sci Polym Ed* 20: 821–839

Kou, R.Q., Xu, Z.K., Deng, H.T. et al. 2003. Surface modification of microporous polypropylene membranes by plasma-induced graft polymerization of alpha-allyl glucoside. *Langmuir* 19: 6869–6875.

Krasteva, N., Harms, U., Albrecht, W. et al. 2002. Membranes for biohybrid liver support systems—Investigations on hepatocyte attachment, morphology and growth. *Biomaterials* 23: 2467–2478.

Krasteva, N., Seifert, B., Albrecht, W. et al. 2004. Influence of polymer membrane porosity on C3A hepatoblastoma cell adhesive interaction and function. *Biomaterials* 25: 2467–2476.

Krasteva, N., Seifert, B., Hopp, M. et al. 2005. Membranes for biohybrid liver support—The behaviour of C3A hepatoblastoma cells is dependent on the composition of acrylonitrile copolymers. *J Biomater Sci Polym Ed* 16: 1–22.

Kuroda, S., Mita, I., Obata, K. et al. 1990. Degradation of aromatic polymers. 4. Effect of temperature and light-intensity on the photodegradation of polyethersulfone. *Polym Degrad Stabil* 27: 257–270.

Lai, J.Y., Lin, F.C., Wang, C.C. et al. 1996. Effect of nonsolvent additives on the porosity and morphology of asymmetric TPX membranes. *J Membr Sci* 118: 49–61.

Lai, J.Y., Lin, F.C., Wu, T.T. et al. 1999. On the formation of macrovoids in PMMA membranes. *J Membr Sci* 155: 31–43.

Lamb, E. and Delaney, M. 2009. ACB Venture In: *Kidney Disease and Laboratory Medicine*. Lapsey, M. and Harris, B. (Eds.). ACB Venture Publications, London, U.K., 226pp.

Langer, R. and Vacanti, J.P. 1993. Tissue engineering. *Science* 260: 920–926.

Lee, H.J., Won, J., Lee, H. et al. 2002. Solution properties of poly(amic acid)-NMP containing LiCl and their effects on membrane morphologies. *J Membr Sci* 196: 267–277.

Legallais, C., David, B., and Dore, E. 2001. Bioartificial livers (BAL): Current technological aspects and future developments. *J Membr Sci* 181: 81–95.

Li, S.G., vanden Boomgaard, T., Smolders, C.A. et al. 1996. Physical gelation of amorphous polymers in a mixture of solvent and nonsolvent. *Macromolecules* 29: 2053–2059.

Li, Z.S. and Jiang, C.Z. 2005. Investigation of the dynamics membrane formation by of poly(ether sulfone) membrane formation by precipitation immersion. *J Polym Sci Polym Phys* 43: 498–510.

Li, J.F., Xu, Z.L., and Yang, H. 2008. Microporous polyethersulfone membranes prepared under the combined precipitation conditions with non-solvent additives. *Polym Adv Technol* 19: 251–257.

Lin, D.J., Chang, C.L., Chen, T.C. et al. 2010a. Microporous PVDF membrane formation by immersion precipitation from water/TEP/PVDF system. *Desalination* 145: 25–29.

Lin, F.C., Wang, D.M., Lai, C.L. et al. 1997. Effect of surfactants on the structure of PMMA membranes. *J Membr Sci* 123: 281–291.

Lin, K.Y., Wang, D.M., Lai, J.Y. 2010b. Nonsolvent-induced gelation and its effect on membrane morphology. *Macromolecules* 35: 6697–6706.

Lindner, M.D. and Emerich, D.F. 1998. Therapeutic potential of a polymer-encapsulated L-DOPA and dopamine-producing cell line in rodent and primate models of Parkinson's disease. *Cell Transplant* 7: 165–174.

Liu, Z.M., Lee, S.Y., Sarun, S. et al. 2010. Biocompatibility of poly(L-lactide) films modified with poly(ethylene imine) and polyelectrolyte multilayers. *J Biomater Sci Polym Ed* 21: 893–912.

Liu, Z.M., Xu, Z.K., Wang, J.Q. et al. 2004. Surface modification of polypropylene microfiltration membranes by graft polymerization of N-vinyl-2-pyrrolidone. *Eur Polym J* 40: 2077–2087.

Lloyd, D.R., Kinzer, K.E., and Tseng, H.S. 1990. Microporous membrane formation via thermally induced phase-separation. 1. Solid liquid-phase separation. *J Membr Sci* 52: 239–261.

Lodish, H., Baltimore, D., Berk, A. et al. 1995. *Molecular Cell Biology*. W. H. Freeman and Company, Scientific American Books, New York.

Machado, P.S.T., Habert, A.C., and Borges, C.P. 1999. Membrane formation mechanism based on precipitation kinetics and membrane morphology: Flat and hollow fiber polysulfone membranes. *J Membr Sci* 155: 171–183.

Matsuyama, H., Maki, T., Teramoto, M. et al. 2003a. Effect of PVP additive on porous polysulfone membrane formation by immersion precipitation method. *Sep Sci Technol* 38: 3449–3458.

Matsuyama, H., Nakagawa, K., Maki, T. et al. 2003b. Studies on phase separation rate in porous polyimide membrane formation by immersion precipitation. *J Appl Polym Sci* 90: 292–296.

Matsuyama, H., Yuasa, M., Kitamura, Y. et al. 2000. Structure control of anisotropic and asymmetric polypropylene membrane prepared by thermally induced phase separation. *J Membr Sci* 179: 91–100.

Mazariegos, G.V., Patzer II, J.F., Lopez, R.C. et al. 2002. First clinical use of a novel bioartificial liver support. *Am J Transplant* 2: 260–266.

McKelvey, S.A. and Koros, W.J. 1996. Phase separation, vitrification, and the manifestation of macrovoids in polymeric asymmetric membranes. *J Membr Sci* 112: 29–39.

Minuth, W.W., Strehl, R., and Schumacher, K. 2004. Tissue factory: Conceptual design of a modular system for the in vitro generation of functional tissues. *Tissue Eng* 10: 285–294.

Nayak, A., Liu, H.W., and Belfort, G. 2006. An optically reversible switching membrane surface. *Angew Chem Int Ed* 45: 4094–4098.

Nie, F.Q., Xu, Z.K., Ming, Y.Q. et al. 2004. Preparation and characterization of polyacrylonitrile-based membranes: Effects of internal coagulant on poly(acrylonitrile-co-maleic acid) ultrafiltration hollow fiber membranes. *Desalination* 160: 43–50.

Oh, N.W., Jegal, J., and Lee, K.H. 2001a. Preparation and characterization of nanofiltration composite membranes using polyacrylonitrile (PAN). I. Preparation and modification of PAN supports. *J Appl Polym Sci* 80: 1854–1862.

Oh, N.W., Jegal, J., and Lee, K.H. 2001b. Preparation and characterization of nanofiltration composite membranes using polyacrylonitrile (PAN). II. Preparation and characterization of polyamide composite membranes. *J Appl Polym Sci* 80: 2729–2736.

Park, H.C., Kim, Y.P., Kim, H.Y. et al. 1999. Membrane formation by water vapor induced phase inversion. *J Membr Sci* 156: 169–178.

Park, J.K. and Lee, D.H. 2005. Bioartificial liver systems: Current status and future perspective. *J Biosci Bioeng* 99: 311–319.

Park, J.K., Lee, S.K., Lee, D.H., and Kim, Y.V. 2009. Bioartificial liver. In: *Fundamental of Tissue Engineering and Regenerative Medicine*, Meyer, U., Handschel, J., Meyer, T., and Wiesmann, H.P. (Eds.). Springer, Berlin, Germany, pp. 397–407.

Paul, D. 1998. Polymer membranes in separation processes. *Chem Unserer Zeit* 32: 197–205.

Paulsen, F.G., Shojaie, S.S., and Krantz, W.B. 1994. Effect of evaporation step on macrovoid formation in wet-cast polymeric membranes. *J Membr Sci* 91: 265–282.

Peinemann, K.V. and Nunes, S. 2012. *Membrane Technology*. Wiley-VCH, Weinheim, Germany.

Pesek, S.C. and Koros, W.J. 1994. Aqueous quenched asymmetric polysulfone hollow fibers prepared by dry wet phase-separation. *J Membr Sci* 88: 1–19.

Pieracci, J., Crivello, J.V., and Belfort, G. 2002a. Increasing membrane permeability of UV-modified poly(ether sulfone) ultrafiltration membranes. *J Membr Sci* 202: 1–16.

Pieracci, J., Crivello, J.V., and Belfort, G. 2002b. UV-assisted graft polymerization of N-vinyl-2-pyrrolidinone onto poly(ether sulfone) ultrafiltration membranes using selective UV wavelengths. *Chem Mater* 14: 256–265.

Pitol, L., Torras, C., Avalos, J.B. et al. 2006. Modelling of polysulfone membrane formation by immersion precipitation. *Desalination* 200: 427–428.

Price, R.D., Myers, S., Leigh, I.M., and Navsaria, H.A. 2005. The role of hyaluronic acid in wound healing—Assessment of clinical evidence. *Am J Clin Dermatol* 6: 393–402.

Radovanovic, P., Thiel, S.W., and Hwang, S.T. 1992. Formation of asymmetric polysulfone membranes by immersion precipitation. 2. The effects of casting solution and gelation bath compositions on membrane-structure and skin formation. *J Membr Sci* 65: 231–246.

Rahimpour, A., Madaeni, S.S., and Mansourpanah, Y. 2007. The effect of anionic, non-ionic and cationic surfactants on morphology and performance of polyethersulfone ultrafiltration membranes for milk concentration. *J Membr Sci* 296: 110–121.

Rajabzadeh, S., Maruyama, T., Ohmukai, Y. et al. 2009. Preparation of PVDF/PMMA blend hollow fiber membrane via thermally induced phase separation (TIPS) method. *Sep Purif Technol* 66: 76–83.

Ramay, H.R.R. and Zhang, M. 2004. Biphasic calcium phosphate nanocomposite porous scaffolds for load-bearing bone tissue engineering. *Biomaterials* 25: 5171–5180.

Rennekampff, H.O., Hansbrough, J.F., Kiessig, V., Abiezzi, S., and Woods, V. 1996. Wound closure with human keratinocytes cultured on a polyurethane dressing overlaid on a cultured human dermal replacement. *Surgery* 120: 16–22.

Rheinwald, J.G. and Green, H. 1975. Serial cultivation of strains of human epidermal keratinocytes—Formation of keratinizing colonies from single cells. *Cell* 6: 331–344.

Rho, Y.R., Choi, H., Lee, J.C. et al. 2003. Applications of the pulsatile flow versatile ECLS: In vivo studies. *Int J Artif Organs* 26: 428–435.

Rifai, K. and Mann, M.P. 2006. Clinical experience with Prometheus. *Ther Apher Dial* 10: 232–236.

Ring-Larsen, H. and Palazzo, U. 1981. Renal failure in fulminant hepatic failure and terminal cirrhosis: A comparison between incidence, types, and prognosis. *Gut* 22: 585–591.

Rivereau, A., Darquy, S., Chaillous, L. et al. 1997. Reversal of diabetes in non-obese diabetic mice by xenografts of porcine islets entrapped in hollow fibres composed of polyacrylonitrile sodium methallylsulphonate copolymer. *Diabetes Metab* 23: 205–212.

Rocco, M.V., Frankenfield, D.L., Hopson, S.D., and McClellan, W.M. 2006. Relationship between clinical performance measures and outcomes among patients receiving long-term haemodialysis treatment. *Ann Intern Med* 145: 7512–7519.

Rozga, J. and Demetriou, A. 1995. Artificial liver. Evolution and future perspectives. *ASAIO J* 4: 831–837.

Saier, H.D., Strathmann, H., and Mylius, U.V. 1974. Mechanism of asymmetric membrane formation. *Angew Makromol Chem* 40: 391–404.

Saito, A. 2003. Development of bioartificial kidneys. *Nephrology* 8: S10–S15.

Saljoughi, E., Amirilargani, M., and Mohammadi, T. 2009. Effect of poly(vinyl pyrrolidone) concentration and coagulation bath temperature on the morphology, permeability, and thermal stability of asymmetric cellulose acetate membranes. *J Appl Polym Sci* 111: 2537–2544.

Santerre, J.P., Woodhouse, K., Laroche, G., and Labow, R.S. 2005. Understanding the biodegradation of polyurethanes: From classical implants to tissue engineering materials. *Biomaterials* 26: 7457–7470.

Seifert, B., Mihanetzis, G., Groth, T. et al. 2002. Polyetherimide: A new membrane-forming polymer for biomedical applications. *Artif Organs* 26: 189–199.

Sharma, R., Greenhough, S., Medine, C.N., and Hay, D.C. 2010. Three-dimensional culture of human embryonic stem cell derived hepatic endoderm and its role in bioartificial liver construction. *J Biomed Biotechnol* 2010: 2362–2371

Shimizu, H., Kawakami, H., and Nagaoka, S. 2002. Membrane formation mechanism and permeation properties of a novel porous polyimide membrane. *Polym Adv Technol* 13: 370–380.

Smolders, C.A., Reuvers, A.J., Boom, R.M. et al. 1992. Microstructures in phase-inversion membranes. 1. formation of macrovoids. *J Membr Sci* 73: 259–275.

Son, H.D., Cho, M.S., Nam, J.D. et al. 2006. Depression of methanol-crossover using multilayer proton conducting membranes prepared by layer-by-layer deposition onto a porous polyethylene film. *J Power Sources* 163: 66–70.

Stadbauer, V. and Jalan, R. 2007. Acute liver failure: Liver support therapies. *Curr Opin Crit Care* 13: 215–221.

Stange, J., Mitzner, B., and Risler R. 1999. Molecular adsorbent recycling system (MARS): Clinical results of a new membrane-based blood purification system for bioartificial liver support. *Artif Organs* 23: 319–323.

Te Velde, A.A., Ladiges, N.C.J.J., Flendrig, L.M., and Chamuleaux, R.A.F.M. 1995. Functional activity of isolated pig hepatocytes attached to different extracellular matrix substrates. Implication for application of pig hepatocytes in a bioartificial liver. *J Hepatol* 23: 184–195.

Tiranathanagul, K., Brodie, J., and Humes, H.D. 2006. Bioartificial kidney in the treatment of acute renal failure associated with sepsis. *Nephrology* 11: 285–291.

Trimpert, C., Boese, G., Albrecht, W. et al. 2006. Poly(ether imide) membranes modified with poly(ethylene imine) as potential carriers for epidermal substitutes. *Macromol Biosci* 6: 274–284.

Tsai, H.A., Kuo, C.Y., Lin, J.H. et al. 2006. Morphology control of polysulfone hollow fiber membranes via water vapor induced phase separation. *J Membr Sci* 278: 390–400.

Tumlin, J., Wali, R., Williams, W. et al. 2008. Efficacy and safety of renal tubule cell therapy for acute renal failure. *J Am Soc Nephrol* 19: 1034–1040.

Tzanakakis, E.S., Hess, D.J., Sielaff, T.D., and Hu, W.S. 2000. Extracorporeal tissue engineered liver-assist devices. *Ann Rev Biomed Eng* 2: 607–632.

Tzoneva, R., Faucheux, N., and Groth, T. 2007. Wettability of substrata controls cell-substrate and cell-cell-adhesion. *Biochim Biophys Acta Gen Issues* 1770: 1538–1547.

Uhlenbusch-Körwer, I., Bonnie-Schorn, E., Grassmann, A., and Vienken, J. 2004. *Understanding Membranes and Dialysers*. Pabst Science Publishers, Lengerich, Germany.

Ulbricht, M. 2006. Advanced functional polymer membranes. *Polymer* 47: 2217–2262.

vandeWitte, P., Dijkstra, P.J., vandenBerg, J.W.A. et al. 1996. Phase separation processes in polymer solutions in relation to membrane formation. *J Membr Sci* 117: 1–31.

Vogrin, N., Stropnik, C., Musil, V. et al. 2002. The wet phase separation: The effect of cast solution thickness on the appearance of macrovoids in the membrane forming ternary cellulose acetate/acetone/water system. *J Membr Sci* 207: 139–141.

von Appen, K., Weber, C., Losert, U. et al. 1996. Microsphere-based detoxification system: A new method in convective blood purification. *Artif Organs* 20: 420–425.

Wang, Y., Kim, J.H., Choo, K.H. et al. 2000. Hydrophilic modification of polypropylene microfiltration membranes by ozone-induced graft polymerization. *J Membr Sci* 169: 269–276.

Wang, D.L., Li, K., Teo, W.K. 2000a. Preparation of annular hollow fiber membranes. *J Membr Sci* 166: 31–39.

Wang, D.L., Li, K., Teo, W.K. 2000b. Highly permeable polyethersulfone hollow fiber gas separation membranes prepared using water as non-solvent additive. *J Membr Sci* 176: 147–158.

Wang, D.L., Li, K., Teo, W.K. 2000c. Porous PVDF asymmetric hollow fiber membranes prepared with the use of small molecular additives. *J Membr Sci* 178: 13–23.

Wang, D.M., Lin, F.C., Chen, L.Y. et al. 1998. Application of asymmetric TPX membranes to transdermal delivery of nitroglycerin. *J Control Release* 50: 187–195.

Wang, D.M., Lin, F.C., Chiang, J.C. et al. 1998. Control of the porosity of asymmetric TPX membranes. *J Membr Sci* 141: 1–12.

Wang, D.M., Lin, F.C., Wu, T.T. et al. 1998. Formation mechanism of the macrovoids induced by surfactant additives. *J Membr Sci* 142: 191–204.

Wang, Z.G., Xu, Z.K., and Wan, L.S. 2006. Modulation the morphologies and performance of polyacrylonitrile-based asymmetric membranes containing reactive groups: Effect of non-solvents in the dope solution. *J Membr Sci* 278: 447–456.

Wang, X.Y., Zhang, L., Sun, D.H. et al. 2009. Formation mechanism and crystallization of poly(vinylidene fluoride) membrane via immersion precipitation method. *Desalination* 236: 170–178.

Wavhal, D.S. and Fisher, E.R. 2002. Modification of porous poly(ether sulfone) membranes by low-temperature CO_2-plasma treatment. *J Polym Sci Polym Phys* 40: 2473–2488.

Wienk, I.M., Boom, R.M., Beerlage, M.A.M. et al. 1996. Recent advances in the formation of phase inversion membranes made from amorphous or semi-crystalline polymers. *J Membr Sci* 113: 361–371.

Won, J., Park, H.C., Kim, U.Y. et al. 1999. The effect of dope solution characteristics on the membrane morphology and gas transport properties: PES/gamma-BL/NMP system. *J Membr Sci* 162: 247–255.

Wright, K.A., Nadire, K.B., Busto, P., Tubo, R., McPherson, J.M., and Wentworth, B.M. 1998. Alternative delivery of keratinocytes using a polyurethane membrane and the implications for its use in the treatment of full-thickness burn injury. *Burns* 24: 7–17.

Yan, J.S. and Lau, W.W.Y. 1998. Effect of internal coagulant on morphology of polysulfone hollow fiber membranes. I. *Sep Sci Technol* 33: 33–55.

Yang, Q., Hu, M., X., Dai, Z.W. et al. 2006. Fabrication of glycosylated surface on polymer membrane by UV-induced graft polymerization for lectin recognition. *Langmuir* 22: 9345–9349.

Yang, Q., Tian, J., Dai, Z.W. et al. 2006. Novel photoinduced grafting-chemical reaction sequence for the construction of a glycosylation surface. *Langmuir* 22: 10097–10102.

Yang, Q., Tian, J., Hu, M.X. et al. 2007. Construction of a comb-like glycosylated membrane surface by a combination of UV-induced graft polymerization and surface-initiated ATRP. *Langmuir* 23: 6684–6690.

Yang, Q., Tian, J., and Xu, Z.K. 2007. Photo-induced graft polymerization of acrylamide on polypropylene membrane surface in the presence of dibenzyl trithiocarbonate. *Chin J Polym Sci* 25: 221–226.

Yang, Q., Xu, Z.K., Dai, Z.W. et al. 2005a. Surface modification of polypropylene microporous membranes with a novel glycopolymer. *Chem Mater* 17: 3050–3058.

Yang, Q., Xu, Z.K., and Ulbricht, M. 2005b. Surface modification of polypropylene microporous membrane by the immobilization of dextran. *Chem J Chin Univ* 26: 189–191.

Yip, Y. and McHugh, A.J. 2006. Modeling and simulation of nonsolvent vapor-induced phase separation. *J Membr Sci* 271: 163–176.

Yu, H.Y., He, X.C., Liu, L.Q. et al. 2007. Surface modification of polypropylene microporous membrane to improve its antifouling characteristics in an SMBR: N_2 plasma treatment. *Water Res* 41: 4703–4709.

Yu, H.Y., He, X.C., Liu, L.Q. et al. 2008a. Surface modification of poly(propylene) microporous membrane to improve its antifouling characteristics in an SMBR: O_2 plasma treatment. *Plasma Process Polym* 5: 84–91.

Yu, H.Y., Hu, M.X., Xu, Z.K. et al. 2005. Surface modification of polypropylene microporous membranes to improve their antifouling property in MBR: NH_3 plasma treatment. *Sep Purif Technol* 45: 8–15.

Yu, H.Y., Liu, L.Q., Tang, Z.Q. et al. 2008b. Mitigated membrane fouling in an SMBR by surface modification. *J Membr Sci* 310: 409–417.

Yu, H.Y., Liu, L.Q., Tang, Z.Q. et al. 2008c. Surface modification of polypropylene microporous membrane to improve its antifouling characteristics in an SMBR: Air plasma treatment. *J Membr Sci* 311: 216–224.

Yu, C.B., Pan, X.P., and Li, L.J. 2009. Progress in bioreactors of bioartificial livers. *Hepatobiliary Pancreat Dis* 8: 134–140.

Yu, H.Y., Tang, Z.Q., Huang, L. et al. 2008d. Surface modification of polypropylene macroporous membrane to improve its antifouling characteristics in a submerged membrane-bioreactor: H_2O plasma treatment. *Water Res* 42: 4341–4347.

Yu, H.Y., Xie, Y., Hu, M.X. et al. 2005. Surface modification of polypropylene microporous membrane to improve its antifouling property in MBR: CO_2 plasma treatment. *J Membr Sci* 254: 219–227.

Yu, H.Y., Xu, Z.K., Lei, H. et al. 2007. Photoinduced graft polymerization of acrylamide on polypropylene microporous membranes for the improvement of antifouling characteristics in a submerged membrane-bioreactor. *Sep Purif Technol* 53: 119–125.

Yu, H.Y., Xu, Z.K., Yang, Q. et al. 2006. Improvement of the antifouling characteristics for polypropylene microporous membranes by the sequential photoinduced graft polymerization of acrylic acid. *J Membr Sci* 281: 658–665.

Zeman, L. and Fraser, T. 1993. Formation of air-cast cellulose-acetate membranes.1. Study of macrovoid formation. *J Membr Sci* 84: 93–106.

Zheng, Q.Z., Wang, P., and Yang, Y.N. 2006. Rheological and thermodynamic variation in polysulfone solution by PEG introduction and its effect on kinetics of membrane formation via phase-inversion process. *J Membr Sci* 279: 230–237.

24 Controlled Release Systems for Bone Regeneration

Hossein Hosseinkhani

CONTENTS

24.1 OVERVIEW

Bone defects and fracture nonunion are common problems, affecting as many as 1000 patients in the world every year, and are difficult to heal using current therapies. Previously, these cases have been treated by surgery, using techniques such as autologous bone grafting or artificial bone grafting. However, autologous bone grafts have a number of problems including donor-site problems, the limitations of harvested bone, or the weak strength of graft-bone, while artificial bone grafts also have associated problems caused by the use of biomaterials, including immunogenicity, biodegradation, or strength limitations. Bone regeneration is an attractive research field of tissue engineering because of its high clinical requirement. It is widely recognized that various osteogenic growth factors regulate the proliferation and differentiation of osteogenic cells and enhance bone formation. However, the use of osteogenic growth factor alone requires large amounts of protein because of its short half-life. Furthermore, the response to osteogenic growth factor varies between human species and primates need larger amounts of osteogenic growth factor than rodents. Aging has also been reported to lead to a reduction in response. To overcome these problems and to reduce the amounts of osteogenic growth factor required, developments in new types of materials by use of drug delivery systems (DDS) and combined treatments with other reagents that can enhance bone regeneration are challenging. Thus, if one can accelerate bone regeneration using osteogenic growth factors in a suitable manner, this regeneration technology will provide a new clinical procedure bone repair and be substituted for autogenous and allogenous bone grafts or biomaterial implants. This chapter reviews the basic principle of controlled release systems and the recent developments of new materials for their potential applications in regenerative medicine therapy for bone regeneration. This chapter emphasizes that controlled release technology in combination with principle of tissue engineering represents a viable strategy for the development of certain engineered tissue replacements and tissue regeneration systems to enhance bone regeneration.

24.2 INTRODUCTION

Bone reconstruction is a clinically important procedure to treat bone defects and has been widely tried by different methods. Basically, bone has the inherent ability to spontaneously repair itself for the bone fracture of small size. However, such a self-repairing cannot always be expected for large-size defects that are caused by trauma, tumor resection, spinal arthodesis, and congenital abnormalities. This situation often happens clinically and therapeutic demand has been being increased recently. Autograft, which is considered to be a gold standard as bone substitutes, is applied to the defect site because it provides a suitable environment for cell attachment, proliferation, and differentiation for bone regeneration. However, it has several disadvantages, such as the limited donor supply, potential complications such as chronic pain at the donor supply and at the donor sites. On the other hand, allograft is being performed clinically, but the rate of graft integration into the natural bone is lower than that of autograft. In addition, it is necessary to consider a risk of disease transmission and postoperative complications due to the tissue rejection in case of an allograft. Therefore, under these circumstances, as the substitute for bone grafts, the biomaterials of metals and ceramics have been investigated and developed. Although the above problems may be cleared, they have other disadvantages, such as the lack of biodegradability under physiological conditions and the limited processability. Especially, metals show poor integration property to the bone tissue at the implantation site compared with the autograft and allograft although they provide mechanical support. Different from artificial biomaterials, one of the important advantages of the bone graft is its ability to positively accelerate osteoconduction and osteoinduction. With the aim of addressing the issues mentioned earlier, bone tissue engineering has been attracted much attention as a new therapeutic technology [1,2]. The basic idea is to provide key cells the local environment suitable to promote their proliferation and differentiation for the induction of tissue regeneration.

Generally, there are three factors necessary for tissue engineering, such as cells, the scaffold for cell proliferation and differentiation, and growth factors. Without any treatment, a large-sized body defect will be naturally occupied with the fibrous tissue. If this tissue occupation takes place, the defect will never be regenerated and repaired by the right tissue to be expected. For successful bone regeneration, it is indispensable to efficiently build up the regeneration environment by making use of various biomaterials or their combination with key cells and growth factors. Figure 24.1 shows a schematic illustration of tissue regeneration based on the scaffold for cell proliferation and differentiation, and growth factors. It is well recognized that the extracellular matrix of natural scaffold provides not only a physical support for cells but also plays an important role in the cell proliferation and differentiation or cell-mediated morphogenesis. The scaffold of biomaterials has been designed and prepared for the local environment of tissue regeneration. The scaffolds should be biodegradable to disappear in the body accompanied with bone regeneration at the defect. In addition, the degradation products should not be toxic and must be naturally excreted by metabolic pathways.

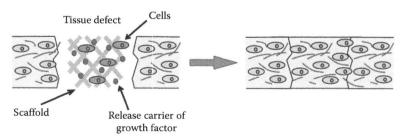

FIGURE 24.1 (See companion CD for color figure.) Three key factors—scaffold for cell proliferation, and differentiation, and growth factors—for successful tissue regeneration.

If the surrounding tissue of bone defect has a high potential toward regeneration, bone tissue will be newly formed in the scaffold implanted into bone defect by seeded cells or cells infiltrated from the surrounding tissues. However, when the regeneration potential is very low, bone regeneration will not always be expected similarly. In this situation, it is one of the practically possible ways to utilize growth factors to accelerate the induction of bone regeneration. The recent research developments in basic biology and medicine reveal that growth factors play an important role in proliferation and differentiation of cells both in vitro and in vivo [3]. However, only by the directed injection of growth factor in the solution form into the target site to be regenerated, we cannot always expect the growth factor–induced tissue regeneration. This is because the growth factor generally has a very short half-life in the body, and is rapidly excreted from the site injected or deactivated by the attack of enzymes and antibodies. Consequently, the high-dosage administration and repeated regimens are necessary for bone regeneration. However, they often cause adverse effects. As one trail to efficiently enhance the in vivo biological efficacy of growth factor, it is practically possible to make use of technology and methodology of DDS. For example, a growth factor incorporated into a carrier matrix for controlled release and applied to the site has to be regenerated. It is likely that this release system efficiently enhances the local consideration of growth factor over a certain time period, resulting in promoted tissue regeneration.

24.3   SCAFFOLDING BIOMATERIALS

Regenerative medicine is an interdisciplinary field that combines engineering and live sciences in order to develop techniques that enable the restoration, maintenance, or enhancement of living tissues and organs. Its fundamental aim is the creation of natural tissue with the ability to restore missing organ or tissue function, which the organism has not been able to regenerate in physiological conditions. By doing so, it aspires to improve the health and quality of life for millions of people worldwide and to give solution to the present limitations: rejections, low quantity of donors, etc. [4]. Tissue engineering needs scaffolds to serve as a substrate for seeding cells and as a physical support in order to guide the formation of the new tissue. The majority of the techniques that are used utilize three-dimensional polymeric scaffolds, which are composed of natural or synthetic polymers. Synthetic materials are attractive because their chemical and physical properties (e.g., porosity, mechanical strength) can be specifically optimized for a particular application. The polymeric scaffold structures are endowed with a complex internal architecture, channels, and porosity that provide sites for cell attachment and maintenance of differentiated function without hindering proliferation. Ideally, a polymeric scaffold for tissue engineering should have the following characteristics: (1) To have appropriate surface properties promoting cell adhesion, proliferation, and differentiation, (2) To be biocompatible, (3) To be highly porous, with a high surface area/volume ratio, with an interconnected pore network for cell growth and flow transport of nutrients and metabolic waste, and (4) To have mechanical properties sufficient to withstand any in vivo stresses. The last requirement is difficult to combine with the high porosity in volume of the material. That is why it is necessary to use polymeric matrices with special or reinforced properties, especially if the polymer is a hydrogel.

The polymeric scaffold design depends on the regarded applications, but in any case, it must achieve structures with the aforementioned characteristics, which are necessary to their correct function. To achieve it with success is conditional on two factors: materials used, both the porogen, and the reticulate polymer, which is infiltrated in the porogen to become a scaffold; and the structural architecture, both external and internal, basically shown by its porosity (high surface area/volume ratio), geometry, size pore, and having in mind that the structures must be easily processed into three-dimensional. On the basis of the extensive range of polymeric materials, different processing techniques have been developed to design and fabricate 3D scaffolds for tissue engineering implants [4]. They include: (a) phase separation, (b) gas foaming, (c) fiber bonding, (d) photolithography, (e) solid free form (SFF), and (f) solvent casting in combination with particle leaching. However, none of the techniques have achieved a suitable model of three-dimensional architecture so that the scaffolds can fulfill their aims in the way as required using equipments with high cost even, for the reasons that are going to be discussed. So, using phase separation, a porous structure can be easily obtained by adjusting thermodynamic and kinetic parameters. However, because of the complexity of the processing variables involved in phase-separation technique the pore structure cannot be easily controlled. Moreover, it is difficult to obtain large pores and may exhibit a lack of interconnectivity. Gas foaming has the advantage of room temperature processing but produces a largely nonporous outer skin layer and a mixture of open and closed pores within the center, leaving incomplete interconnectivity. The main disadvantage of the gas foaming method is that it often results in a nonconnected cellular structure within the scaffold. Fiber bonding provides a large surface area for cell attachment and a rapid diffusion of nutrients in favor of cell survival and growth. However, these scaffolds, as the ones used to construct a network of bonded polyglycolic acid (PGA), lacked the structural stability necessary for in vivo use. In addition, the technique does not lend itself to easy and independent control of porosity and pore size. Photolithography has also been employed for patterning, to obtain structures with high resolution, although this resolution may be unnecessary for many applications of patterning in cell biology. In any case, the disadvantage of this technique is that the high cost of the equipment need limits their applicability. SFF scaffold manufacturing methods provide excellent control over scaffold external shape, and internal pore interconnectivity and geometry, but offer limited microscale resolution. Moreover, the minimum size of global pores is 100 µm. Additionally, SFF requires complex correction of scaffold design for anisotropic shrinkage during fabrication. Moreover, it needs high cost equipments. Finally, solvent casting in combination with particulate leaching method, which involves the casting of a mixture of monomers and initiator solution and a porogen in a mold, polymerization, followed by leaching-out of the porogen with the proper solvent to generate the pores, is inexpensive but still has to overcome some disadvantages in order to find engineering applications, namely the problem of residual porogen remains, irregular shaped pores, and insufficient interconnectivity.

The proposed scaffolds may find applications as structures that facilitate either tissue regeneration or repair during reconstructive operations. So far, three-dimensional materials with a porous structure have been designed for the cell scaffold from glycolide–lactide copolymer nonwoven fabrics, collagen sponges, calcium phosphate ceramics, and PEG-based hydrogels [5]. Among them, hydroxyapatite (HAp) and β-tricalcium phosphate (β-TCP) have been intensively investigated as the scaffold materials for bone tissue engineering because it is well recognized that they are compatible to natural bone tissue and osteoconductive [5,6]. However, HAp is not practically degraded under physiological conditions and remains inside the bone tissue regenerated. Therefore, as one trial to control the in vivo degradability, the HAp is combined with organic materials, such as collagen and glycolide–lactide copolymers [7]. The combination is effective in manipulating the degradation and mechanical properties for the HAp scaffolds. On the other hand, β-TCP is advantageous from the viewpoint of the biodegradability, although brittle is compared with HAp. Several requirements should be considered in the design of three-dimensional scaffolds for bone tissue engineering. Firstly, the scaffold should have sufficient porosity for cell proliferation, differentiation, and ingrowth, resulting in promoted bone regeneration. Porosity higher than 90% is important for scaffolds. The pore size is one key factor for the scaffold design. It is reported that the pore size ranging from 150 to 400 µm is preferable for the bone regeneration [8]. High interconnectivity between pores is also desirable for homogenous cell seeding and distribution, oxygen or nutrient supply, and excretion of metabolic waste from the cell–scaffold constructs. It has been widely accepted that the cell–scaffold interaction is greatly influenced by their porous structure. In addition, the scaffold requires suitable surface properties because the cell behavior is greatly influenced by roughness, topography, wettability, charge, and chemical composition of scaffold surface. Figure 24.2 indicates tissue regeneration based on the principle of tissue engineering.

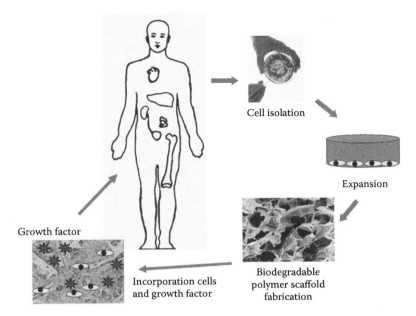

Cell isolation

Expansion

Growth factor

Incorporation cells
and growth factor

Biodegradable
polymer scaffold
fabrication

FIGURE 24.2 **(See companion CD for color figure.)** Schematic illustration of tissue regeneration based on the principle of tissue engineering.

24.4 GROWTH FACTORS

Tissue engineering is designed to regenerate natural tissues or to create biological substitutes for defective or lost organs by making use of cells. Considering the usage of cells in the body, it is no doubt that a sufficient supply of nutrients and oxygen to the transplanted cells is vital for their survival and functional maintenance. Without a sufficient supply, only a small number of cells preseeded in the scaffold or migrated into the scaffold from the surrounding tissue would survive. Rapid formation of a vascular network at the transplanted site of cells must be a promising way to provide cells with the vital supply. This process of generating new microvasculature, termed neovascularization, is a process observed physiologically in development and wound healing. It is recognized that basic fibroblast growth factor (bFGF) function to promote such an angiogenesis process. The growth factors stimulate the appropriate cells (e.g., endothelial cells), already present in the body, to migrate from the surrounding tissue, proliferate, and finally differentiate into blood vessels [9]. However, one cannot always expect the sustained angiogenesis activity when these proteins are only injected in the solution form probably because of their rapid diffusional excretion from the injected site. One possible way for enhancing the in vivo efficacy is to achieve its controlled release over an extended time period by incorporating the growth factor in a polymer carrier. If this carrier is biodegraded and harmonized with tissue growth, it will work as a scaffold for tissue regeneration in addition to a carrier matrix for the growth factor release. The use of angiogenic factors is a popular approach to induce neovascularization. Among them, bFGF plays a multifunctional role in the stimulation of cell growth and tissue repair. However, it has a very short half-life when injected and is unstable in solution. To overcome these problems, bFGF was encapsulated within alginate, gelatin, agarose/heparin, collagen, and poly(ethylene-*co*-vinyl acetate) carriers [10–12]. According to the results of these studies, it is conceivable to incorporate the angiogenic factor to a sustained releasing system, and using it prior to the implantation. Some studies have demonstrated that bFGF achieved promoted angiogenesis when used in combination with delivery matrices and scaffold [13].

There are other growth factors currently used in tissue regeneration. Hepatocyte growth factor (HGF) is originally discovered as a protein factor to accelerate hepatocyte proliferation. Previous studies have demonstrated that HGF has great potential for proliferation, differentiation, mitogenesis, and morphogenesis of various cells [14]. Therefore, HGF can be used for various tissue engineering applications where angiogenesis is needed. However, since growth factors such as HGF have very short half-time therefore when they inject into the body, they lose their biological activities due to rapid digestion. Sustained release technology has been used widely for different drugs and proteins to overcome this problem.

Osteoinductive properties of bone morphogenetic protein (BMP) family have attracted much attention in terms of bone regeneration because BMP has high potentials to stimulate the differentiation of stem cells into ostogenic lineage. Bone morphogenetic proteins belong to the transforming growth factor-β superfamily and play an important role in osteogenesis and bone metabolism. It is strong enough to induce bone formation even at ectopic sites, such as subcutis and muscle. There are at least 15 types of BMP currently reported, and some recombinant human BMPs (rhBMP) are available at large amounts through the route of recombinant DNA technology. Among them, rhBMP-2 and rhBMP-7 (OP-1) have already been clinically applied to repair the critical-sized bone defect and accelerate healing [15]. Osteogenic growth factors such as transforming growth factor-β ((TGF-β), bone morphogenetic proteins (BMPs), and bFGF can induce bone formation in both ectopic and orthotopic sites in vivo. Table 24.1 summarizes characteristics of the growth factors used in tissue engineering.

Since recombinant human BMP-2 (rhBMP-2) has become available, many animal studies on the induction of bone formation by implantation of rhBMP-2 using various carriers have been performed [16]. However, the use of BMP alone requires large amounts of protein because of its short half-life. Furthermore, the response to BMPs varies between animal species and primates need

TABLE 24.1
Characteristics of the Growth Factors Used in Tissue Engineering

Growth Factor	Isoelectric Point (IEP)	Molecular Weight (kDa)	Biological Substances for Growth Factor Binding	Functions of Growth Factor
Basic fibroblast growth factor (bFGF)	9.6	16	Heparin or heparan sulfate	Stimulating the cells involved in the healing process (bone, cartilage, nerve, etc.) Angiogenesis
Transforming growth TGF-β1 (TGF-β1)	9.5	25	Heparin or heparan sulfate Collagen type IV Latency associated protein Latent TGF-β1 binding protein	Enhancing the wound healing, stimulating the osteoblast proliferation to enhance bone formation
Bone morphogenetic protein-2 (BMP-2)	8.5	32	Collagen type IV	Stimulating the mesenchymal stem cells to osteoblast lineage and inducing the bone formation both at bone and ectopic sites
Vascular endothelial growth factor (VEGF)	8.5	38	Heparin or heparan sulfate	Stimulating the endothelial cell growth, angiogenesis, and capillary permeability
Hepatocyte growth factor (HGF)	5.5	100	Heparin or heparan sulfate	Stimulating of matrix remodeling and epithelial regeneration (liver, spleen, kidney, etc.)

larger amounts of BMP (up to milligram quantities) than rodents. Aging has also been reported to lead to a reduction in response. To overcome these problems and to reduce the amounts of BMP required, developments of new types of scaffold and combined treatments with other reagents that can enhance bone regeneration are in progress.

24.5 CONTROLLED RELEASE TECHNOLOGY

The controlled release of drugs such as proteins, growth factors, genes, and siRNAs is still one of the main objectives of drug delivery systems. The success of the controlled release of a drug for the required duration of time with the optimum release mode depends on various factors, such as the physicochemical properties of the drug and the drug–carrier matrix in addition to the dosage form and administration route. Successful drug delivery will have enormous academic, clinical, and practical impacts on gene therapy, cell and molecular biology, pharmaceutical and food industries, and bio-production. The objective of drug delivery involves the controlled release of drug, the prolongation of drug life span, the acceleration of drug absorption, and the drug targeting. To achieve these objectives, several research trials have been extensively performed by the use of several materials and methods. For example, drug has been chemically coupled with various water-soluble polymers to enlarge the apparent molecular size, which allows the drug to prolong the serum half-life period, in addition to encapsulation into nano-order drug carriers including polymeric nanospheres, polymer micelles, lipid emulsion, and liposomes [17–22]. Drug is often modified with the polymers and polymer micelles to cover the molecular surface by a hydrophilic layer, which results in reduced exclusion of the drug by the mononuclear phagocyte system (MPS) present in the liver and spleen from blood circulation [23–26]. This polymer modification also allows the drug to prolong the in vivo life span. When modified with a drug carrier of polymers or encapsulated into a particulate carrier, drug will be actively targeted to a tissue and organ according to the tissue-specific affinity of carrier itself [27–30] as well as the phagocytic cells based on the high susceptibility of carrier to cellular uptake [31,32]. Once the combination of drug and carrier matrix is fixed, the molecular compatibility between the drug and the matrix, in other words, their miscibility becomes a most influential factor for the sustained release of drug from the matrix. If the compatibility is poor, the drug molecules will not be dispersed homogeneously in the matrix phase, creating microscopic phase separation, resulting in the initial burst effect. On the other hand, good compatibility between the drug and the matrix will lead to a long-term sustained release with zero-, first-order, or diffusion-controlled kinetics. Various biodegradable or nonbiodegradable polymeric materials have been developed in recent years for medical, pharmaceutical, drug delivery, and tissue engineering applications [33]. The objective of controlled release technology involves the controlled release of drug, the prolongation of drug life span, the acceleration of drug absorption, and the drug targeting.

24.6 CONTROLLED RELEASE SYSTEMS FOR BONE REGENERATION

Several preclinical studies have revealed that BMP administrated in the solution forms does not always induce the expected efficiency in bone regeneration. BMP at physiologically high doses is often required to achieve bone formation [34]. To tackle these problems, various biodegradable carriers, including collagen, lactide–glycolide copolymers, β-tricalcium phosphate, and ethylene glycol–lactic acid copolymers have been employed for the carrier matrices for BMP release [35–38]. An osteogenetic product composed of BMP-7 and collagen of the release carrier is commercially available [39]. It is strongly indicated from these findings that the combinations of BMP with the release materials is absolutely needed to achieve the in vivo BMP-induced bone formation. However, little has been investigated on the ability of BMP for bone regeneration from viewpoint of the in vivo release profile.

In our previous research, we have already prepared a hydrogel from gelatin with different levels of biodegradability and succeeded in augmenting the biological effects of bFGF, TGF-β1, and HGF [40–42]. Gelatin was selected as the carrier material for growth factor release because it is

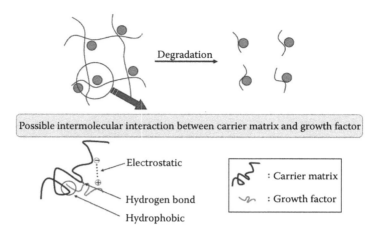

FIGURE 24.3 **(See companion CD for color figure.)** Mechanism on the controlled release of growth factor.

commercially available with various physicochemical properties and has been extensively used for industrial, pharmaceutical, and medical purposes. The biosafety of gelatin has been proved through the long clinical use [43]. Another unique advantage is the electrical nature of gelatin, which allows a growth factor with an electrical charge to physically immobilize into the gelatin-based hydrogel without its denaturation [44]. It is demonstrated that the time period of growth factor remaining in the hydrogel is in good accordance with that of hydrogel remaining. Only when the hydrogel is enzymatically degraded to generate water-soluble gelatin fragments, the growth factor immobilized can be released from the hydrogel [45]. Figure 24.3 shows a schematic illustration of the mechanism of growth factor release from the carrier matrix based on the degradation of the matrix. The in vivo degradability of gelatin hydrogels depends on their water content, which can be modified by changing the preparation conditions. Moreover, it should be noted that gelatin hydrogels can be formulated into different shapes, such as disks, tubes, sheets, and microspheres.

Recently, Tabata's groups have prepared a biodegradable hydrogel from gelatin with an isoelectric point (IEP) of 9.0 for the controlled release of BMP-2 based on hydrogel biodegradation. In this study, ectopic bone formation by the BMP release system was investigated and compared with the in vivo profile of BMP release. Hydrogels with different water contents were prepared through glutaraldehyde cross-linking of gelatin with IEP of 9.0 under varied reaction conditions. Following subcutaneous implantation of the gelatin hydrogels incorporating BMP-2 into the back of mice, the in vivo period of BMP-2 prolonged with a decrease in the water content of hydrogels used, although every time period was much longer than that of BMP-2 solution injection. Ectopic bone formation studies demonstrated that alkaline phosphate activity (ALP) and osteocalcin content (OCN) around the implanted site of BMP-2-incorporated gelatin hydrogels were significantly high compared with those around the injected site of BMP-2 solution. The values became maximum for the gelatin hydrogel incorporating BMP-2 with a middle period of BMP-2 retention, while bone formation was histologically observed around the hydrogel incorporating BMP-2. The ALP activity was significantly higher than that of the gelatin hydrogel incorporating BMP-2. Controlled release technology of BMP-2 for a certain time period was essential to induce the potential activity for bone formation [46]. The same materials were used to regenerate bone at a skull defect of nonhuman primates. A critical-sized defect (6 mm in diameter) was prepared at the skull bone of cynomolgus monkeys skeletally matured while gelatin hydrogels incorporating various doses of BMP-2 were applied into the defects. When the bone regeneration was evaluated by soft x-ray examinations, the gelatin hydrogen incorporating BMP-2 exhibited significantly high osteoinduction activity. The gelatin hydrogel enabled BMP-2 to induce the bone regeneration induction in nonhuman primates [47]. Table 24.2 summarizes research reports of BMP-2 release for ectopic bone formation.

TABLE 24.2
Research Reports of BMP-2 Release for Ectopic Bone Formation

	Material	Implanted Site	Animal
Inorganic material	β-TCP (tricalcium phosphate)	Muscle	Mouse
	HAp (hydroxyapatite)	Muscle	Rat
	Plaster of paris	Muscle	Mouse
Organic material	IBM (insoluble bone matrix)	Muscle	Mouse
	Type I collagen	Muscle	Rat
	Fibrin glue	Muscle	Mouse
	PLA (polylactide)	Muscle	Mouse
	PLGA (poly(lactide-co-glycolide))	Muscle	Mouse
	PLA-PEG (poly(lactide-coethylene glycole)	Muscle	Mouse
	Gelatin hydrogel	Muscle	Rat

Our recent study has indicated that a 3-D network of self-assembled nanofibers was formed by mixing of bFGF suspension with aqueous solution of peptide amphiphile as an injectable carrier for controlled release of growth factors and used it for feasibility of prevascularization by the bFGF release from the 3-D networks of nanofibers in improving efficiency of tissue regeneration [48]. The bFGF incorporated in self-assembled peptide could be delivered to living tissues by simply injection of a liquid (i.e., peptide amphiphile solutions) and bFGF solution. The injected solutions would form a solid scaffold at the injected site of the tissue and the released bFGF induced significant angiogenesis around the injected site, in marked contrast to bFGF injection alone or PA injection alone. This release system also was enabling to induce significant bone formation when peptide amphiphile solutions and TGF-β were subcutaneously injected to the back of rat as shown in Figure 24.4.

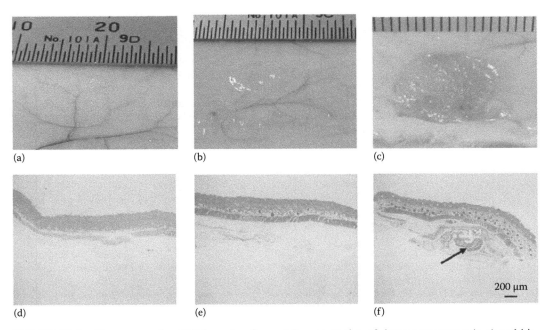

FIGURE 24.4 **(See companion CD for color figure.)** Representative of tissue appearance (a–c) and histological cross-sections (d–f) of ectopically formed bone after subcutaneous injection of TGF-β (a and d), peptide solution (b and e), and peptide solution with TGF-β (c and f). The concentration of TGF-β is 10 μg. Each specimen is subjected to H-E staining. Arrow indicates the newly formed bone.

The injected solutions of peptide and TGF-β formed solid gel and the sustained release of TGF-β induced significant ectopic bone compared with TGF-β injection. As a flexible delivery system, these scaffolds can be adapted for sustained release of many different growth factors and biomolecules. Significantly ectopic bone formation also was observed using these systems when peptide amphiphile solutions and BMP-2 were subcutaneously injected to the back of rat [49].

Controlled release of plasmid DNA encoded osteogenic growth factors has been attempted to induce significant bone regeneration. Our recent studies have indicated that incorporation of plasmid DNA encoded BMP-2 into collagen sponge significantly enhanced in vivo ectopic bone formation [50–54]. Controlled release of plasmid DNA-BMP-2 from cationized gelatin enhanced gene transfection of mesenchymal stem cells (MSC) and formed genetically engineered MSC. Homogeneous bone formation was histologically observed throughout the genetically engineered MSC 4 weeks after subcutaneous implantation of scaffolds into the back of rats. The bone mineral density (BMD) of new bone formed at the implanted sites of scaffolds seeded with the genetically engineered MSC was significantly higher compared with scaffolds seeded with only naked plasmid DNA.

24.7 CLOSING REMARKS

This chapter gives us a wide knowledge about the basic principle underlying controlled release technology for bone regeneration. It gives a general overview of different kinds of release systems. The research papers quoted in this chapter clearly inform the reader about the potential applications of polymeric materials that can be applied for fabricating wider range of novel biomaterials for the use in controlled release systems. With appropriate references and examples, this chapter opens up the reader's mind incorporating a wider range of knowledge about the growth factor delivery systems. These macroscopic structures have inspired the researchers to use them in various areas of science such as biotechnology, nanotechnology, and medicine. However, it is necessary to create new and other alternative methods if we face any problems using the current technology in delivering biomolecules drug; such as protein, growth factors, and DNA.

REFERENCES

1. Langer, R. and Vacanti, J.P. 1993. Tissue engineering. *Science* 260: 920–926.
2. Salgado, A.J., Coutinho, O.P., and Reis, R.L. 2004. Bone tissue engineering: State of the art and future trends. *Macromol. Biosci.* 4: 743–765.
3. Taipale, J. and Keski-Oja, J. 1997. Growth factors in the extracellular matrix. *FASEB J.* 11: 51–59.
4. Thomson, R.C., Wake, M.C., Yaszemski, M.J. et al. 1995. Biodegradable polymer scaffolds to regenerate organs. *Adv. Polym. Sci. (Biopolymers II)* 122: 245–273.
5. Damien, C.J. and Parsons, J.R. 1991. Bone graft and bone graft substitutes: A review of current technology and applications. *J. Appl. Biomater.* 2: 187–208.
6. Bruder, S.P. and Caplan, A.J. 2000. Bone regeneration through cellular engineering, *Principle of Tissue Engineering*, 2nd edn. Academic Press, New York.
7. Xu, H.H.K. and Simon, C.G. 2004. Self-hardening calcium phosphate composite scaffold for tissue engineering. *J. Orthop. Res.* 22: 535–543.
8. Karageorgiou, V. and Caplan, D. 2005. Porosity of 3D biomaterial scaffolds and osteogenesis. *Biomaterials* 26: 5474–5491.
9. Pankajakshan, D., Krishnan, V.K., and Krishnan, L.K. 2007. Vascular tissue generation in response to signaling molecules integrated with a novel poly(epsilon-caprolactone)-fibrin hybrid scaffold. *J. Tissue Eng. Regen. Med.* 1: 389–397.
10. Iwakura, A., Tabata, Y., Tamura, N. et al. 2001. Gelatin sheets incorporating basic fibroblast growth factor enhances healing of devascularized sternum in diabetic rats. *Circulation* 104: 1325–1329.
11. Edelman, E.R., Mathiowitz, E., and Langer, R. 1991. Controlled and modulated release of basic fibroblast growth factor. *Biomaterials* 12: 619–626.
12. Ware, J.A. and Simons, M. 1997. Angiogenesis in ischemic heart disease. *Nature Med.* 3: 158–164.
13. Tabata, Y., Nagano, A., and Ikada, Y. 1995. Biodegradation of hydrogel carrier incorporating fibroblast growth factor. *Tissue Eng.* 5: 127–138.

14. Rubin, R.S., Chan, A.M., and Bottaro, D.P. 1991. A broad-spectrum human lung fibroblast-derived mitogen is a variant of hepatocyte growth factor. *Proc. Natl Acad. Sci. USA* 88: 415–419.

15. Lieberman, J.R., Daluski, A., and Einhorn, T.A. 2002. The role of growth factors in the repair of bone: biology and clinical applications. *J. Bone Joint Surg. Am.* 84: 1032–1044.

16. Fujimura, K., Bessho, K., Kusumoto, K. et al. 1995. Experimental studies on bone inducing activity of composites of atelopeptide type I collagen as a carrier for ectopic osteoinduction by rhBMP-2. *Biochem. Biophys. Res. Commun.* 208: 316–322.

17. Guiot, P. and Couvreur, P. 1986. *Polymeric Nanospheres and Microspheres*. CRC Press, Boca Raton, FL.

18. Zuber, G., Dauty, E., Nothisen, M. et al. 2001. Towards synthetic viruses. *Adv. Drug. Deliv. Rev.* 52: 245–253.

19. Demeneix, B., Hassani, Z., and Behr, J.P. 2004. Towards multifunctional synthetic vectors. *Curr. Gene Ther.* 4: 445–455.

20. Takenaga, M., Igarashi, R., Tsuji, H. et al. 1993. Enhanced antitumor activity and reduced toxicity of 1,3-bis(2-chloroethyl)-1-nitrosourea administered in lipid microspheres to tumor-bearing mice, *Jpn. J. Cancer Res.* 84: 1078–1085.

21. Liu, D., Mori, A., and Huang, L. 1992. Role of liposome size and RES blockade in controlling biodistribution and tumor uptake of GM1-containing liposomes. *Biochim. Biophys. Acta* 1104: 95–101.

22. Liu, D., Mori, A., and Huang, L. 1991. Large liposomes containing ganglioside GM1 accumulate effectively in spleen. *Biochim. Biophys. Acta* 1066: 159–165.

23. Molineux, G. 2003. Pegylation: Engineering improved biopharmaceuticals for oncology. *Pharmacotherapy* 23: 3S–8S.

24. Francis, G.E., Fisher, D., Delgado, C. et al. 1998. PEGylation of cytokines and other therapeutic proteins and peptides: The importance of biological optimisation of coupling techniques. *Int. J. Hematol.* 68: 1–18.

25. Roberts, M.J., Bentley, M.D., and Harris, J.M. 2002. Chemistry for peptide and protein PEGylation. *Adv. Drug. Deliv. Rev.* 54: 459–476.

26. Harris, J.M. and Chess, R.B. 2003. Effect of pegylation on pharmaceuticals. *Nat. Rev. Drug. Discov.* 2: 214–221.

27. Wu, G.Y. and Wu, C.H. 1988. Evidence for targeted gene delivery to Hep G2 hepatoma cells in vitro. *Biochemistry* 27: 887–892.

28. Nakazono, K., Ito, Y., Wu, C.H. et al. 1996. Inhibition of hepatitis B virus replication by targeted pretreatment of complexed antisense DNA in vitro. *Hepatology* 23: 1297–1303.

29. Lu, X.M., Fischman, A.J., Jyawook, S.L. et al. 1994. Antisense DNA delivery in vivo: Liver targeting by receptor-mediated uptake. *J. Nucl. Med.* 35: 269–275.

30. Wu, G.Y. and Wu, C.H. 1988. Receptor-mediated gene delivery and expression in vivo. *J. Biol. Chem.* 263: 14621–14624.

31. Tabata, Y. and Ikada, Y. 1991. Drug delivery systems for antitumor activation of macrophages. *Crit. Rev. Ther. Drug Carrier Syst.* 7: 121–148.

32. Ahsan, F., Rivas, I.P., Khan, M.A. et al. 2002. Targeting to macrophages: Role of physicochemical properties of particulate carriers—liposomes and microspheres—On the phagocytosis by macrophages. *J. Control. Release* 79: 29–40.

33. Nalwa, H.S. 2005. *Handbook of Nanostructured Biomaterials and Their Applications in Nanobiotechnology, Vol. 1: Biomaterials*. American Scientific Publishers, Los Angeles, CA.

34. Valentin-Opran, A., Wozney, J., and Csimma, C. 2002. Clinical evaluation of recombinant human bone morphogenetic protein-2. *Clin. Orthop. Rel. Res.* 395: 110–120.

35. Geiger, M., Li, R.H., and Friess, W. 2003. Collagen sponges for bone regeneration with rhBMP-2. *Adv. Drug Deliv. Rev.* 55: 1613–1629.

36. Boyan, B.D., Lohmann, C.H., Somers, A. et al. 1999. Potential of porous poly-D,L-lactide-*co*-glycolide particles as a carrier for recombinant human bone morphogenetic protein-2 during osteoinduction in vivo. *J. Biomed. Mater. Res.* 46: 51–59.

37. Urist, M.R., Lietze, A., and Dawson, E. 1984. β-Tricalcium phosphate delivery systems for bone morphogenetic protein. *Clin. Orthop. Rel. Res.* 187: 277–280.

38. Sito, N., Okada, H., and Horiuchi, H. 2001. A biodegradable polymer as a cytokine delivery system for inducing bone formation. *Nat. Biotech.* 19: 332–335.

39. Seeherman, H. and Wozney, J.M. 2005. Delivery of bone morphogenetics proteins for orthopedic tissue regeneration. *Cytokine Growth Factor Rev.* 16: 329–345.

40. Tabata, Y. and Ikada, Y. 1999. Vascularization effect of basic fibroblast growth factor released from gelatin hydrogels with different biodegradabilities. *Biomaterials* 20: 2169–2175.

41. Yamamoto, M., Tabata, Y., Hong, L. et al. 2000. Bone regeneration by transforming growth factor β1 released from a biodegradable hydrogel. *J. Control. Release* 64: 133–142.
42. Ozeki, M., Ishii, T., and Tabata, Y. 2001. Controlled release of hepatocyte growth factor from gelatin hydrogels based on hydrogel degradation. *J. Drug Targeting* 9: 461–471.
43. Zekorn, D. 1969. Modified gelatin as plasma substitutes. *Bibl. Haematol.* 33: 30–60.
44. Tabata, Y. and Ikada, Y. 1998. Protein release from gelatin matrices. *Adv. Drug Delivery Rev.* 31: 287–301.
45. Tabata, Y. 2003. Tissue regeneration based on growth factor release. *Tissue Eng.* 9: S5–S15.
46. Yamamoto, M., Takahashi, Y., and Tabata, Y. 2003. Controlled release by biodegradable hydrogels enhances the ectopic bone formation of bone morphogenetic protein. *Biomaterials* 24: 4375–4383.
47. Takahashi, Y., Yamamoto, M., and Tabata, Y. 2007. Skull bone regeneration in non-human primates by controlled release of bone morphogenetic protein-2. *Tissue Eng.* 13: 293–300.
48. Hosseinkhani, H., Hosseinkhani, M., and Tabata, Y. 2006. Enhanced angiogenesis through controlled release of basic fibroblast growth factor from peptide amphiphile for tissue regeneration. *Biomaterials* 27: 5836–5844.
49. Hosseinkhani, H., Hosseinkhani, M., Khademhosseini, A. et al. 2007. Bone regeneration through controlled release of bone morphogenetic protein-2 from 3-D tissue engineered nano-scaffold. *J. Control. Release* 117: 380–386.
50. Hosseinkhani, H., Hosseinkhani, M., Khademhosseini, A. et al. 2008. DNA nanoparticles encapsulated in 3-D tissue engineered scaffold enhance osteogenic differentiation of mesenchymal stem cells. *J. Biomed. Mater. Res. Part A* 85: 47–60.
51. Hosseinkhani, H., Yamamoto, M., Inatsugu, Y. et al. 2006. Enhanced ectopic bone formation using a combination of plasmid DNA impregnation into 3-D scaffold and bioreactor perfusion culture. *Biomaterials* 27: 1387–1398.
52. Hosseinkhani, H., Azzam, T., Kobayashi, H. et al. 2006. Combination of 3-D tissue engineered scaffold and non-viral gene enhance in vitro DNA expression of mesenchymal stem cells. *Biomaterials* 27: 4269–4278.
53. Hosseinkhani, H., Inatsugu, Y., Hiraoka, Y. et al. 2005. Impregnation of plasmid DNA into three-dimensional scaffolds and medium perfusion enhance in vitro DNA expression of mesenchymal stem cells. *Tissue Eng.* 11: 1459–1475.
54. Hosseinkhani, H., Inatsugu, Y., Inoue, S. et al. 2005. Perfusion culture enhances the osteogenic differentiation of rat mesenchymal stem cells in collagen sponge rein forced with poly (glycolic acid) fiber. *Tissue Eng.* 11: 1476–1488.

25 Controlled Release Systems Targeting Angiogenesis

*Stéphanie Deshayes, Karine Gionnet, Victor Maurizot,
and Gérard Déléris**

CONTENTS

ABBREVIATIONS

AMD   age-related macular degeneration
BMP   bone morphogenetic protein
BMSC   bone marrow stromal cell
BNCT   boron neutron capture therapy
CAM   chicken chorioallantoic membrane
Doxo   doxorubicin
DSPE   distearoylphosphatidylethanolamine
ECM   extracellular matrix

---

* Deceased.

EGF	epidermal growth factor
EPR	enhanced permeation and retention
FDA	food and drug administration
FGF-1/aFGF	fibroblast growth factor-1/acidic fibroblast growth factor
FGF-2/bFGF	fibroblast growth factor-2/basic fibroblast growth factor
FGFR	fibroblast growth factor receptor
FITC	fluorescein isothiocyanate
Flk-1	fetal liver kinase-1
Flt-1	fms-like tyrosine kinase-1
G-CSF	granulocyte colony-stimulating factor
GM-CSF	granulocyte-macrophage colony-stimulating factor
HGF/SF	hepatocyte growth factor/scatter factor
HPMA	N-(2-hydroxypropyl) methacrylamide
HUVEC	human umbilical vein endothelial cell
IFN-α	interferon-α
IFN-β	interferon-β
IGF-1	insulin growth factor-1
IGF-2	insulin growth factor-2
Ig-like	immunoglobulin-like
IL	interleukin
IP-10	interferon-γ-inducible protein
KDR	kinase insert domain containing receptor
KGF	keratinocyte growth factor
LDL	low density lipoprotein
MMP	matrix metalloproteinase
MMPI	matrix metalloproteinase inhibitor
MRI	magnetic resonance imaging
NGF	nerve growth factor
PAGA	poly[R-(4-aminobutyl)-L-glycolic acid]
PAMAM	polyamidoamine
PCL	poly(ε-caprolactone)
PDGF	platelet derived growth factor
PEDF	pigment epithelium-derived factor
PEG	poly(ethylene glycol)
PEI	poly(ethylenimine)
PF-4	platelet factor-4
PGA	poly(glycolic acid)
PLA	poly(lactic acid)
PLGA	poly(lactic-co-glycolic acid)
PlGF	placenta growth factor
PLL	poly(l-lysine)
PNIPAM	poly(N-isopropyl acrylamide)
PPAA	poly(propylacrylic acid)
PPI	polypropyleneimine
PVP	poly(N-vinyl pyrrolidone)
RES	reticuloendothelial system
RTK	receptor-type tyrosine kinase
SCID	severe combined immunodeficient
sFlt-1	soluble fms-like tyrosine kinase-1
Shh	sonic hedgehog
SPIO	superparamagnetic iron oxide

TAF	tumor angiogenesis factor
TFPI-2	tissue factor pathway inhibitor-2
TGF-α	transforming growth factors-α
TGF-β	transforming growth factors-β
TIMP	tissue inhibitor of metalloproteinase
TNF-α	tumor necrosis factor-α
tPA	tissue plasminogen activator
TSP-1	thrombospondine-1
TSP-2	thrombospondine-2
uPA	urokinase plasminogen activator
VCAM	vascular cell adhesion molecule
VEGF	vascular endothelial growth factor
VEGI	vascular endothelial growth inhibitor
VPF	vascular permeability factor
WSLP	water-soluble lipopolymer

25.1 INTRODUCTION

Angiogenesis, the formation of new blood vessels from existing ones, is an important event in several biological processes and may be used as a specific target for fighting diseases characterized by either poor vascularization or abnormal expansion of vasculature (Pandya et al., 2006). According to pathology, the therapy consists in inhibiting or promoting angiogenesis (Carmeliet, 2003). Various drugs such as therapeutic proteins, growth factors, or encoding genes have been designed to target pathological angiogenesis (Polverini, 1995; Choksy and Chan, 2006; Folkman, 2006). However, these biomacromolecules are rapidly deactivated by enzymes or immune system and hence require a vehicle when administered as a drug (Allen and Cullis, 2004; Faraji and Wipf, 2009). Controlling the drug release to a targeted site remains a formidable challenge in medicine, and numerous systems have undergone extensive investigations. Among them, polymers are promising candidates for the controlled release due to their architecture variability (Jagur-Grodzinski, 2009). Polymers provide several unique properties and capabilities including drug bioavailability enhancement, drug protection, pharmacokinetic improvement, prolonged blood circulation time with a reduction in immunological body response due to a "stealth" character, a passive or active targeting, and the possibility to form a multifunctional system carrying several biomolecules (different drugs and/or imaging agent) (Khandare and Minko, 2006).

This chapter seeks to highlight recent advances regarding the polymeric systems that may be used for targeting angiogenesis in different pathologies. A description of angiogenesis with associated mediators will precede a short overview of current pro- or anti-angiogenesis therapies. The importance of polymeric carriers in drug delivery and the different ways to reach the target site will be then demonstrated. The different polymeric architectures used for these applications will be lastly described.

25.2 ANGIOGENESIS

25.2.1 DEFINITION AND PROCESS

The emergence of the blood vascular network is one of the first events of the embryonic development which is the result of two distinct processes: vasculogenesis and angiogenesis. During vasculogenesis, mesodermal cells differentiate into endothelial cell precursors (angioblasts), which proliferate and coalesce into a primitive network of homogeneously sized vessels known as primary capillary plexus. This initial capillary meshwork is then remodeled by angiogenesis into a mature and functional vascular bed comprising arteries, capillaries, and veins (Risau and Flamme, 1995; Jain, 2003). It was in 1787 that the angiogenesis term was introduced by Dr. John Hunter, a British

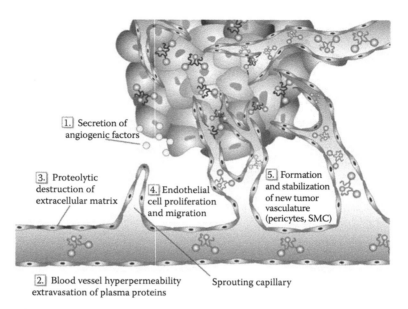

1. Secretion of angiogenic factors

3. Proteolytic destruction of extracellular matrix

4. Endothelial cell proliferation and migration

5. Formation and stabilization of new tumor vasculature (pericytes, SMC)

2. Blood vessel hyperpermeability extravasation of plasma proteins

Sprouting capillary

FIGURE 25.1 **(See companion CD for color figure.)** Schematic representation of the angiogenesis process by sprouting. (From Satchi-Fainaro, R. et al., *Adv. Polym. Sci.*, 193, 1, 2006. With permission.)

surgeon, and described the history (genesis) of the blood vessels (angio). Angiogenesis, a normal physiological process, occurs during the embryonic development, wound healing, nerve regeneration, and female ovulatory cycle. In pathological states, angiogenesis is observed during solid tumor growth and metastasis, age-related macular degeneration (AMD), cardiac ischemia, atherosclerosis, osteoporosis, diabetic retinopathy, and chronic inflammatory disorders (Folkman, 1995).

Three distinct mechanisms have been defined for angiogenesis: sprouting, intussusceptive microvascular growth, and bridging (Carmeliet, 2000). The latter is characterized by the formation of vascular conduits separated inside of vessels, leading to vascular bridges. This mechanism is misunderstood and will not be discussed in this chapter.

Sprouting angiogenesis. During sprouting angiogenesis (Figure 25.1), endothelial cells are first activated by specific growth factors that bind to specific receptors. Subsequently, enzymes (proteases) are released and degrade the extracellular matrix (ECM) and basement membrane. Then, endothelial cells invade the surrounding matrix and proliferate. Through differentiation and polarization, endothelial cells organize in tubular structures with a new basement lamina, sprouts connect to close vessels and form a lumen. The stabilization of these immature vessels is established by the recruitment of mural cells and the generation of ECM (Hillen and Griffioen, 2007).

Intussusceptive angiogenesis. In contrast to sprouting angiogenesis, intussusception (growth within itself) is a relatively new concept discovered by Peter H. Burri et al. during their investigation about the transformation of vascular network in the rats and human lungs (Caduff et al., 1986; Burri and Djonov, 2002). Intussusceptive microvascular growth divides existing vessel lumens by formation and insertion of endothelial columns into the lumen. This process does not require relevant proliferation of endothelial cells but rather rearrangement of existing structures (Djonov et al., 2003). This mechanism predominates in lungs which contain intrinsic endothelial precursors and are initially vascularized by vasculogenesis.

25.2.2 Endogenous Mediators of Angiogenesis

Angiogenesis is regulated by a balance of positive and negative factors (Table 25.1). In most physiological status, angiogenesis is "turned off" by the production of more inhibitors than stimulators.

TABLE 25.1

Endogenous Mediators of Angiogenesis

Stimulators	Inhibitors
Degradation of ECM	
uPA	α-2-macroglobuline
tPA	PEX
MMP	TIMP
	TFPI-2
	RECK
	N-Tes
Proliferation, migration, and differentiation of endothelial cells	
VEGF	TSP-1, -2
PIGF	Troponine I
FGF-1, -2	IFN-α, -γ
HIV-*tat*	PEDF
PDGF	IP-10
HGF/SF	PF-4
TGF-α, -β	IL-4, -12
EGF	VEGI
IGF-1	PAI-1
TNF-α	Retinoic acid
IL-8	Ang-2
IL-3	2-Methoxyoestradiol
COX-2	sFlt-1
Angiogenin	Angiostatin
Proliferin	Endostatin
Erythropoietin	Prolactin
G-CSF	Tumstatin
GM-CSF	
BMP	
Cell adhesion	
Integrins $\alpha_V\beta_3$, $\alpha_V\beta_5$	Arresten
VCAM-1	
E-selectin	

Abbreviations are listed at the beginning of this chapter.

When pro-angiogenic factors are produced in excess, the balance is tipped in favor of blood vessel growth. Every step of angiogenesis, such as ECM degradation, cell activation, and cell adhesion, is mediated by specific molecules.

25.2.2.1 ECM Degradation

The degradation process requires the cooperative action of plasminogen activators and matrix metalloproteinases (MMPs). Plasmin is a 560 amino acid proteolytic enzyme which degrades several compounds of the ECM. Plasmin is produced by plasminogen activation via uPAs (urokinase plasminogen activators) and tPAs (tissue plasminogen activators) mediation (Mignatti and Rifkin, 1996; Mazar, 2008). MMPs represent a family of 23 endopeptidases implicated in the proteolytic degradation of the ECM. MMPs are either released in the matrix or associated with the cell membrane and categorized as the membrane-type MMPs. All MMPs share a basic structural organization with a "pre" region to direct their secretion from cells, a "pro" region to maintain latency, and an active catalytic region that contains the zinc-binding active site (Overall, 2002). The catalytic activity of MMPs is mediated by physiological inhibitors such as α-2-macroglobuline (Sottrup-Jensen, 1989), PEX (a fragment of MMP-2 which comprises the *C*-terminal hemopexin-like domain) (Brooks et al., 1998), and tissue inhibitors of metalloproteinases (TIMPs) (Liekens et al., 2001; Handsley and Edwards, 2005).

25.2.2.2 Activation of Endothelial Cells

25.2.2.2.1 Stimulation

After the degradation of ECM, migration and proliferation of endothelial cells occur and are stimulated by a variety of growth factors. The investigation of growth factors started in 1948 when I. C. Michaelson first suggested that a soluble "factor X" produced by hypoxic retina stimulated neovascularization (Michaelson, 1948). In 1971, J. Folkman isolated tumor angiogenesis factor (TAF) which promoted the vascularization on chicken chorioallantoic membrane (CAM) (Folkman et al., 1971). In the 1980s, numerous other factors have been identified including epidermal growth factor (EGF), platelet-derived growth factor (PDGF), transforming growth factors (TGF-α and-β), insulin growth factors (IGF-1 and -2), and tumor necrosis factor-α (TNF-α) which exhibited a pro-angiogenic activity in vascular proliferation. In addition, D. Gospodarowicz isolated fibroblast growth factors (FGF-1 and FGF-2) which stimulated the proliferation and differentiation of endothelial cells and smooth muscle cells (Gospodarowicz et al., 1978). In 1983, D. R. Senger and H. F. Dvorak reported the discovery of a vascular permeability factor (VPF) from tumor cells that promoted accumulation of ascites (Senger et al., 1983; Dvorak, 2006). The major finding was the identification of vascular endothelial growth factor (VEGF) by N. Ferrara (Ferrara and Henzel, 1989). After comparison between VPF and VEGF, the two proteins proved to be identical. VEGF and FGF have been reported as the most potent regulators of angiogenesis.

Vascular endothelial growth factor (VEGF). VEGF family is composed of several members including VEGF-A, B, C, D, and placenta growth factor (PlGF), a functional homolog of VEGF-B. Those growth factors are dimeric glycoproteins with a molecular mass of 34–45 kDa. VEGF-A was the first discovered member having a mitotic specific activity in endothelial cells (Ferrara and Henzel, 1989). It possesses a structural homology to PDGF (Tischer et al., 1989). VEGF-A exists in several isoforms due to alternative splicing of mRNA with 121, 145, 165, 189, and 206 amino acids by monomer ($VEGF_{121}$, $VEGF_{145}$, $VEGF_{165}$, $VEGF_{189}$, $VEGF_{206}$, respectively) (Tischer et al., 1991). VEGF exerts its action via binding to two main receptor-type tyrosine kinases (RTKs): fms-like tyrosine kinase-1 (Flt-1 also known as VEGFR-1) (de Vries et al., 1992) and kinase insert domain containing receptor (KDR also known as VEGFR-2) for its human form or fetal liver kinase-1 (Flk-1) for its murine form (Terman et al., 1992). These receptors embody seven immunoglobulin-like extracellular domains, one intracellular kinase insert sequence, and a single transmembrane region (Hubbard, 1999). VEGF binding to immunoglobulin-like (Ig-like) domains induces receptor dimerization, followed by the protein kinase activation and tyrosine phosphorylation. This results in the activation of signal transduction cascade leading to cell proliferation and migration, cell survival, and vascular permeability induction (Fuh et al., 1998; Cross et al., 2003).

Fibroblast growth factor (FGF). FGF-2 (also known as bFGF for basic fibroblast growth factor) was the first discovered member of FGF family, rapidly followed by its homologous FGF-1 (also known as aFGF for acidic fibroblast growth factor) (Slavin, 1995). Both belong to a wide family of more than 20 members. FGFs are small polypeptides which share similar sequence homology and common properties: they bind, with high affinity, to heparin and heparan sulfate proteoglycans at cell membrane. Four distinct high-affinity receptors were identified: FGFR1, FGFR2, FGFR3, and FGFR4. These receptors have three extracellular Ig-like domains, and their interaction with one of the growth factors activates a tyrosine kinase activity. FGFs are mitogen and they promote proliferation of endothelial cells and their physical organization into tube-like structures (Bikfalvi et al., 1997; Powers et al., 2000).

25.2.2.2.2   Inhibition

The activation of endothelial cells can be downregulated by a number of endogenous factors among thrombospondine-1 (TSP-1), angiostatin (plasminogen fragment), endostatin (*C*-terminal fragment of collagen XVIII) (Sim et al., 1998), platelet factor 4 (PF-4) (Watson et al., 1994), and interferon-α (IFN-α) (Gresser, 1989). TSP-1 produced by normal cells is considered as the main physiological inhibitor of angiogenesis. It was shown that TSP-1 production is regulated by tumor suppressor gene p53 (Grossfeld et al., 1997). Angiostatin was initially isolated from mice bearing Lewis lung carcinoma and was identified as a 38 kDa internal proteolytic cleavage product of plasminogen (amino acids 98–440) that includes the first four kringles of the molecule. Angiostatin was demonstrated to be a specific inhibitor of endothelial cell proliferation (O'Reilly et al., 1994). An endogenous anti-angiogenic isoform variant of $VEGF_{165}$ was identified and was called $VEGF_{165b}$. It reduces proliferation and migration of endothelial cells induced by VEGF (Bates et al., 2002; Woolard et al., 2004).

25.2.2.3   Cell Adhesion

The process of cell invasion is governed by cell adhesion molecules involved in their binding to other cells or ECM. Cell adhesion molecules can be classified into four families according to their biochemical and structural features: selectins, Ig superfamily, cadherins, and integrins (Liekens et al., 2001). Integrins consist of two noncovalently bound transmembrane α and β subunits and are cell surface receptors that are responsible for anchoring cells to ECM (Eliceiri and Cheresh, 1999). $\alpha_v\beta_3$ integrin was found to be especially important during angiogenesis. It is a receptor for a number of protein with an exposed Arg-Gly-Asp (RGD) sequence, including fibronectin, vitronectin, laminin, vWF (von Willebrand Factor), fibrinogen, and collagen (Hynes, 1992).

25.2.3   Angiogenesis-Dependent Diseases and Therapy

Some pathologies such as cardiovascular diseases, arteriogenesis, osteoporosis, or nerve damages require the vascular promotion through therapeutic angiogenesis, while other diseases including cancer need the disruption of angiogenesis. According to the case, angiogenic activators or inhibitors may be used for the treatment.

25.2.3.1   Therapeutic Angiogenesis for Tissue Engineering

Therapeutic angiogenesis is an investigational method to stimulate new vessel formation via administration of pro-angiogenic growth factors. The concept of therapeutic angiogenesis has been first introduced in 1994 by Takeshita et al. at Boston using an ischemia model in a rabbit (Takeshita et al., 1994). VEGF and FGF are the most studied pro-angiogenic agents which are used either directly as recombinant proteins or via a viral vector containing a coding gene in clinical trials.

Schumacher et al. reported the first study of therapeutic angiogenesis in a human coronary artery disease (Schumacher et al., 1998). This study revealed neovascularization from the injection of FGF-1 into the heart of patients, when in the control group no vascularization was observed. A clinical trial showed the persistence of the neovascularization (Pecher and Schumacher, 2000), but on the other hand, the death rate was similar for both groups. Numerous clinical trials have since been conducted, with generally doubtful results. Simons et al. performed a large study of recombinant FGF-2 for the treatment of chronic myocardial ischemia. This study called FIRST (FGF-2 Initiating Revascularization Support Trial) confirmed that administration of recombinant FGF is safe and feasible, but revealed that its efficiency as agent for therapeutic angiogenesis was questionable (Simons et al., 2002). Another trial called VIVA (VEGF in Ischemia for Vascular Angiogenesis) was realized in double-blind with a control group in order to evaluate the safety and efficiency of a recombinant human protein ($rhVEGF_{165}$). Patients with inoperable coronary artery disease and stable angina were randomized to receive placebo or low dose of $rhVEGF_{165}$

(17 ng/kg/min) or high dose of rhVEGF$_{165}$ (50 ng/kg/min). Sixty days after treatment, the VIVA trial showed a good tolerance and safety of recombinant VEGF; but no significant difference between the three groups was observed. After 120 days, only patients treated with the high dose demonstrated a significant improvement in angina class (Henry et al., 2003). To improve the efficiency of treatments, VEGF has been delivered via a viral vector in a large clinical trial using an adenovirus (adVEGF$_{121}$). This study called REVASC (Randomized Evaluation of VEGF for Angiogenesis in Severe Coronary disease) was done with 67 patients that received adVEGF$_{121}$ or placebo by direct intramyocardial injections. This trial demonstrated sustained and continuous clinical improvement from 3 to 6 months and served as a proof-of-concept study for angiogenic myocardial gene delivery in patients with severe coronary disease (Stewart, 2002).

Besides cardiovascular diseases, bone formation is also dependent on VEGF-induced angiogenesis (Uchida et al., 2003). Bone loss is frequently caused by accidents or diseases such as osteoporosis, leading to an upregulation of VEGF during the first days, followed by an extensive decrease. Street et al. showed that VEGF promoted bone repair in a damaged tissue (Street et al., 2002). VEGF acts synergistically with osteogenic factors such as bone morphogenetic proteins (BMPs). BMPs are members of TGF-β and are implicated in the development of various tissues, most notably bone. They are growth factors and cytokines inducing osteoblast differentiation and stimulating angiogenesis through the production of VEGF-A (Deckers et al., 2002). Peng et al. have elaborated retroviral vectors expressing human BMP-4 and VEGF to assess the therapeutic benefit of synergistic delivery (Peng et al., 2002). In the presence of BMP-4, VEGF promoted recruitment of mesenchymal stem cells for bone formation and improved cell survival. Since then, other studies have showed the crucial role of those factors in bone healing, in particular by gene therapy (Sugiyama et al., 2005; Yang et al., 2005a; Shi et al., 2008).

Angiogenesis has recently been found to be implicated in degenerative nerve diseases such as Alzheimer's or Parkinson's diseases. VEGF is a neurotrophic factor which stimulates axonal outgrowth and neuronal survival in vitro (Sondell et al., 1999). Nerve growth factor (NGF) has also been reported to modulate angiogenesis in nerve damages (Nico et al., 2008). NGF was the first described neurotrophic factor that stimulates the growth and differentiation of nerve cells. A host of clinical trials has been performed using single growth factors in neurologica pathologies (Jönhagen et al., 1998; Apfel et al., 2000; McArthur et al., 2000; Ropper et al., 2009). Despite initial encouraging reports with these agents, early attempts to develop clinical treatments were unsuccessful (Thoenen and Sendtner, 2002).

Overall, current approaches to therapeutically administer growth factors, either as recombinant protein or by gene transfer, might provide novel safe and feasible treatments for patients suffering from tissue damages (cardiovascular, nerve, bone). However, the efficiency of therapeutic angiogenesis applying regenerative factors by single injection or via viral vectors remains uncertain. One reason for this may be the short elimination half-life of VEGF and FGF that is approximately 1 h (Lazarous et al., 1996; Khosravi et al., 2007). Furthermore, only 3%–5% of the growth factor dose remains into the heart after intracoronary administration (Lazarous et al., 1997). Consequently, inappropriate local and temporal availability of growth factors explains these disappointing results. Polymeric vehicles may represent an ideal approach to deliver growth factors in a better controlled and sustained manner to the target site.

25.2.3.2 Anti-Angiogenesis Therapy and Cancer

In the 1970s, J. Folkman hypothesized that tumor growth is angiogenesis-dependent (Folkman, 1971). At the early stage of tumor proliferation, tumor is in a primitive state and only receives nutrients by passive diffusion and grows slowly because cell apoptosis compensates for cell division. This state is called quiescent phase. From a size of 1–2 mm, its further growth requires the elaboration of vascular network to supply nutrients and oxygen. This neovascularization phase is essential because it avoids hypoxia and tumor cell apoptosis. The transition between both phases is named angiogenic "switch" which can be represented as a balance where neovascularization takes place when pro-angiogenic factors outweigh anti-angiogenic factors. After the switch, new blood vessels develop from preexisting ones, and tumor cells invade neovessels and form tumor embolism. Cancer cells are released into the blood circulation inducing the formation of metastasis. The angiogenic switch can be triggered by various signals including genetic mutations, hypoxia, immune or inflammatory response, and other stress (Bergers and Benjamin, 2003).

Nowadays, anti-angiogenic therapy is considered one of the most promising approaches for controlling cancer. The first angiogenic inhibitors were discovered in the 1980s from experiments performed in Folkman laboratory (Taylor and Folkman, 1982; Crum et al., 1985). These drugs are based on the concept that removal of the angiogenic blood vessels will limit the supply of nutrients and oxygen to cancer cells. By the mid-1990s, new drugs with anti-angiogenic activity entered clinical trials and various strategies for disrupting angiogenesis have been investigated. For instance, it is possible to interfere with angiogenic process inhibiting ECM degradation in order to block the invasion and migration of endothelial cells. This strategy consists in controlling MMP activity via small synthetic inhibitors (MMPIs) targeting the catalytic Zn^{2+} including batimastat (Botos et al., 1996) marismastat (Sparano et al., 2004), prinomastat (Shalinsky et al., 1999), or neovastat (Falardeau et al., 2001). However, the development of MMPIs was stopped due to the systemic toxicity, lack of correlation between activity of MMPIs and MMP levels in plasma, and poor efficacy (Hoekstra et al., 2001; Overall and Kleifeld, 2006). Other research groups have focused on targeting integrins which are barely detectable from normal vessels but are overexpressed on tumor vessels. Thus, several antagonist peptides containing RGD sequence as cilengitide (Eskens et al., 2003) and antibodies as LM609 (Brooks et al., 1995; Gutheil et al., 2000) or MEDI-522 (a human form of LM609) (Hersey et al., 2005) have been developed to block $\alpha_v\beta_3$ and $\alpha_v\beta_5$ integrins and have been demonstrated to inhibit the proliferation of endothelial cells. The most used strategy consists in the inhibition of growth factors or their receptors. VEGF-trap (AVE0005), a protein carrying extracellular domains of soluble receptors, has been developed to capture VEGF with an affinity 100 times higher than with monoclonal antibodies. VEGF-trap suppresses tumor growth and vascularization in vivo, decreasing metastasis (Holash et al., 2002; Konner and Dupont, 2004). Several angiogenic inhibitors including bevacizumab, sunitinib (Goodman et al., 2007), and sorafenib (Kane et al., 2006) have been approved for cancer therapy by Food and Drug Administration (FDA) in the United States and by 28 other countries. Bevacizumab (Avastin®) was the first anti-angiogenic factor approved for the treatment of metastatic colorectal cancer in the United States in 2004 (Hurwitz et al., 2004; Hurwitz and Saini, 2006). This molecule is a high-affinity humanized monoclonal antibody directed against VEGF. It has been demonstrated to decrease tumor perfusion, microvessel density, interstitial pressure, and the number of endothelial progenitor cells.

Another angiogenesis-dependent disease is age-related macular degeneration (AMD). AMD is a pathology leading to visual loss and it is characterized by choroidal neovascularization (Ferris III et al., 1984; Bressler et al., 1988). Hypoxia was shown to play a crucial role in the development of choroidal neovascularization. Indeed, diffusion of oxygen from the choroid to the retina is decreased. Under these hypoxic conditions, retinal pigment epithelium cells upregulate VEGF-A secretion to their basal side (Witmer et al., 2003). Thus, VEGF plays a predominant role in the progression of AMD and a possible way to treat AMD is to use anti-angiogenic molecules. Although studies in vivo using animal choroidal neovascular models have demonstrated favorable results for several anti-angiogenic drugs such as IFN-β and thalidomide, these drugs were not effective in humans with a high incidence of side effects (Ciulla et al., 1998).

The clinical results of these strategies have generally caused a limited enthusiasm owing to a lack of effective intracellular delivery, a lack of tumor target specificity, development of drug resistance, and side effects (Izzedine et al., 2007). Indeed, anti-angiogenic treatments are often restricted by adverse systemic toxicity, which limits the dose of drug that can be administered. A great deal of research is concentrated to develop controlled release polymeric systems in order to improve the efficacy and reduce the toxicity of current anti-angiogenic therapies.

25.3 POLYMERIC DELIVERY SYSTEMS FOR PRO- AND ANTI-ANGIOGENIC THERAPY

A host of pro- or anti-angiogenic drugs like growth factors, recombinant proteins, peptides, oligonu-cleotides are available for application in therapies such as cardiovascular, degenerative nerve diseases, atherosclerosis, osteoporosis, and cancers. However, these biomacromolecules lack specificity for the malignant site and are fragile, thus they are required to be administered in a high and repeated dose at a given time. This sometimes results in damaging side effects. As a result, research activity has been focused on design of systems delivering drugs continually at therapeutic doses for prolonged time periods. Various drug delivery systems such as viral vectors for gene therapy, liposomes, inorganic nanoparticles, or polymers have been designed to adjust drug release rates, target desirable site, increase in drug systemic half-life, bioavailability, and therapeutic index. Further, an appropriate design of polymeric system can provide more benefits including biocompatibility, biodegradability, increased solubility, "stealth" character ("invisible" to macrophages), protection from proteases and nucleases in systemic circulation, relatively easy and inexpensive formulation (Narayani, 2007). Polymeric carriers can carry physically entrapped or chemically conjugated drugs. Then, the active molecule is released by polymer degradation or diffusion through the polymeric matrix. Consistent with this, numerous synthetic or natural polymers have been investigated (Table 25.2) in the following:

Cationic polymers such as poly(ethylenimine) (PEI) and poly(L-lysine) (PLL) are used to associate with genetic material through electrostatic interactions, they are called polyplexes.

Poly(ethylene glycol) (PEG) and its derivatives are hydrophilic polymers, completely water-soluble, nontoxic, and uncharged. They have been extensively used for their "stealth" character avoiding the reticuloendothelial system (RES). Indeed, these polymers can prevent the interactions between the carrier and the opsonins and thus the recognition by the immune system (Peracchia et al., 1999).

Polyesters such as poly(ε-caprolactone) (PCL), poly(lactic acid) (PLA), poly(glycolic acid) (PGA), and their block copolymer poly[R-(4-aminobutyl)-L-glycolic acid] (PAGA), poly(lactic-co-glycolic acid) (PLGA) are hydrophobic polymers. They are cleaved by hydrolysis to form natural metabolites which are removed from the body by the citric acid cycle. For PLA, several distinct forms exist owing to the chiral nature of lactic acid: poly(D,L-glycolic acid), poly(L-glycolic acid), and poly(D-glycolic acid). These different forms of PLA present different degrees of crystallinity. Indeed, poly(D,L-glycolic acid) is amorphous and poly(L-glycolic acid) and poly(D-glycolic acid) are semi-crystalline. As a result, biodegradability of poly(D-glycolic acid) is slower than for poly(L-glycolic acid) due to the higher crystallinity of poly(D-glycolic acid).

Polysaccharides such as chitosan and dextran are not only biocompatible but also biodegradable and have been widely used for drug delivery.

Poly(amino acids) present a good biocompatibility, biodegradability, low toxicity; however, their use as carriers for drug delivery is limited by the ease of their enzymatic degradation and by possible immunogenic reactions (Park et al., 2008; Jagur-Grodzinski, 2009).

25.3.1 Ways of Targeting Angiogenic Sites

In anticancer therapies, most drugs have no tumor selectivity and are randomly distributed in the body. Two strategies to improve the efficiency of tumor targeting can be considered using polymeric nanoparticles: active and passive targeting.

25.3.1.1 Passive Targeting

Recent studies have shown that nanoparticles accumulate passively in tumors even in the absence of targeting ligands, suggesting the existence of a passive retention mechanism (Duncan, 2003). This phenomenon was first identified by Maeda et al. and called the enhanced permeation and

TABLE 25.2

Chemical Structures of the Most Used Polymers as Drug Delivery Systems

Polymers and Abbreviations	Chemical Structures	Polymers and Abbreviations	Chemical Structures
Poly(ethylenimine), PEI		Poly(lactic acid), PLA	
Poly(L-lysine), PLL		Poly(glycolic acid), PGA	
Poly(ethylene glycol), PEG		Poly(lactic-co-glycolic acid), PLGA	
Poly(N-vinyl pyrrolidone), PVP		Poly(ε-caprolactone), PCL	
Poly(N-isopropyl acrylamide), PNIPAM		Chitosan	
Poly(propylene oxide), PPO		Dextran	
α-Poly(glutamic acid)		γ-Poly(glutamic acid)	

DA = degree of acetylation.

retention (EPR) effect (Matsumura and Maeda, 1986). Tumor vessels are generally characterized by abnormalities such as high proliferation of endothelial cells, tortuosity increase, pericyte deficiency, aberrant basement membrane formation. This defective vascular structure is the result of the rapid vascularization to supply nutrients necessary to tumor survival, leading to a lymphatic drainage decrease and increasing vessel permeability to macromolecules (Park et al., 2008). Further studies have investigated the relation between EPR effect and the molecular weight of polymers conjugated with drugs. Small polymers (M_w < 40,000 Da) are subject to rapid renal clearance, when higher molecular weight polymers, owing to their size, cannot escape via the kidney and persist in the blood circulation for long periods (Seymour et al., 1995). Consistent with this, numerous studies have been performed with drug-entrapped polymeric nanoparticles. For instance, hydrophobically modified glycol chitosan nanoparticles were loaded with paclitaxel, an anti-cancerous drug, and have shown to passively accumulate in tumor tissue due to an EPR effect (Kim et al., 2006b).

25.3.1.2 Active Targeting

Specific receptors are overexpressed on cells of angiogenic site while they are found at low levels on normal cells. Active targeting of polymeric carriers that carry drugs can be achieved by chemical attachment of a ligand that strongly interacts with these upregulated receptors, leading to drug accumulation and better cellular uptake of the drug by receptor-mediated endocytose. The use of a targeting moiety significantly decreases adverse side effects by allowing the drug to be delivered to the specific site of action. In the field of angiogenesis, the most studied target is integrin which is a receptor highly expressed on activated endothelial cells and new-born vessels, but absent in normal endothelial cells (Zhaofei et al., 2008). RGD sequence being found in many adhesive glycoproteins of ECM, a variety of cyclic RGD peptides and peptide templates have been investigated as targeting agents and confer greater stability and selectivity over linear peptides. The cyclic pentapeptides cyclo(Arg-Gly-Asp-*D*Phe-Val) or c(RGDfV) and cyclo(Arg-Gly-Asp-*D*Phe-Lys) or c(RGDfK) were first designed and synthesized by Kessler and co-workers (Haubner et al., 1996). c(RGDfK) is interesting because it can be functionalized with various linker molecules through the amino group on the lysine residue. A

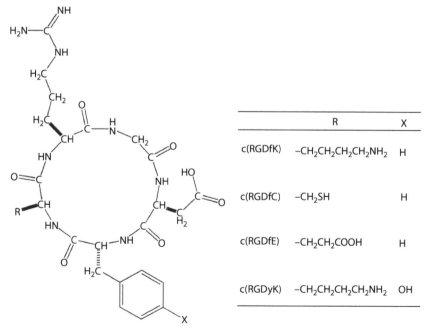

	R	X
c(RGDfK)	$-CH_2CH_2CH_2CH_2NH_2$	H
c(RGDfC)	$-CH_2SH$	H
c(RGDfE)	$-CH_2CH_2COOH$	H
c(RGDyK)	$-CH_2CH_2CH_2CH_2NH_2$	OH

FIGURE 25.2 Chemical structures of cyclic RGD pentapeptides: c(RGDfK), c(RGDfC), c(RGDfE) and c(RGDyK) used as targeting agents.

peptide based on c(RGDfV), named cilengitide, has been assessed for the treatment of gliomas and is currently in a clinical phase II trial (Stupp et al., 2007; Reardon et al., 2008). In addition, cyclo(Arg-Gly-Asp-DPhe-Cys) or c(RGDfC), cyclo(Arg-Gly-Asp-DPhe-Glu) or c(RGDfE) and cyclo(Arg-Gly-Asp-DTyr-Lys) or c(RGDyK) are also available for potential targeting (Figure 25.2) (Haubner et al., 2005). Bicyclic RGD peptides such as H-Glu[cyclo(Arg-Gly-Asp-DPhe-Lys)]₂ also known as E-[c(RGDfK)₂] and RGD4C peptide (KACDCRGDCFCG) were reported to possess high affinity for α_vβ₃ integrin (Figure 25.3) (Koivunen et al., 1995; Chen et al., 2004). Furthermore, these cyclic pentapeptides also exhibited high metabolic stability in vivo (Wester and Kessler, 2005). A host of studies have been made linking those peptides to polymeric carrier (Nasongkla et al., 2004; Line et al., 2005; Shukla et al., 2005; Kim et al., 2006c; Mitra et al., 2006a), and numerous examples will be detailed later.

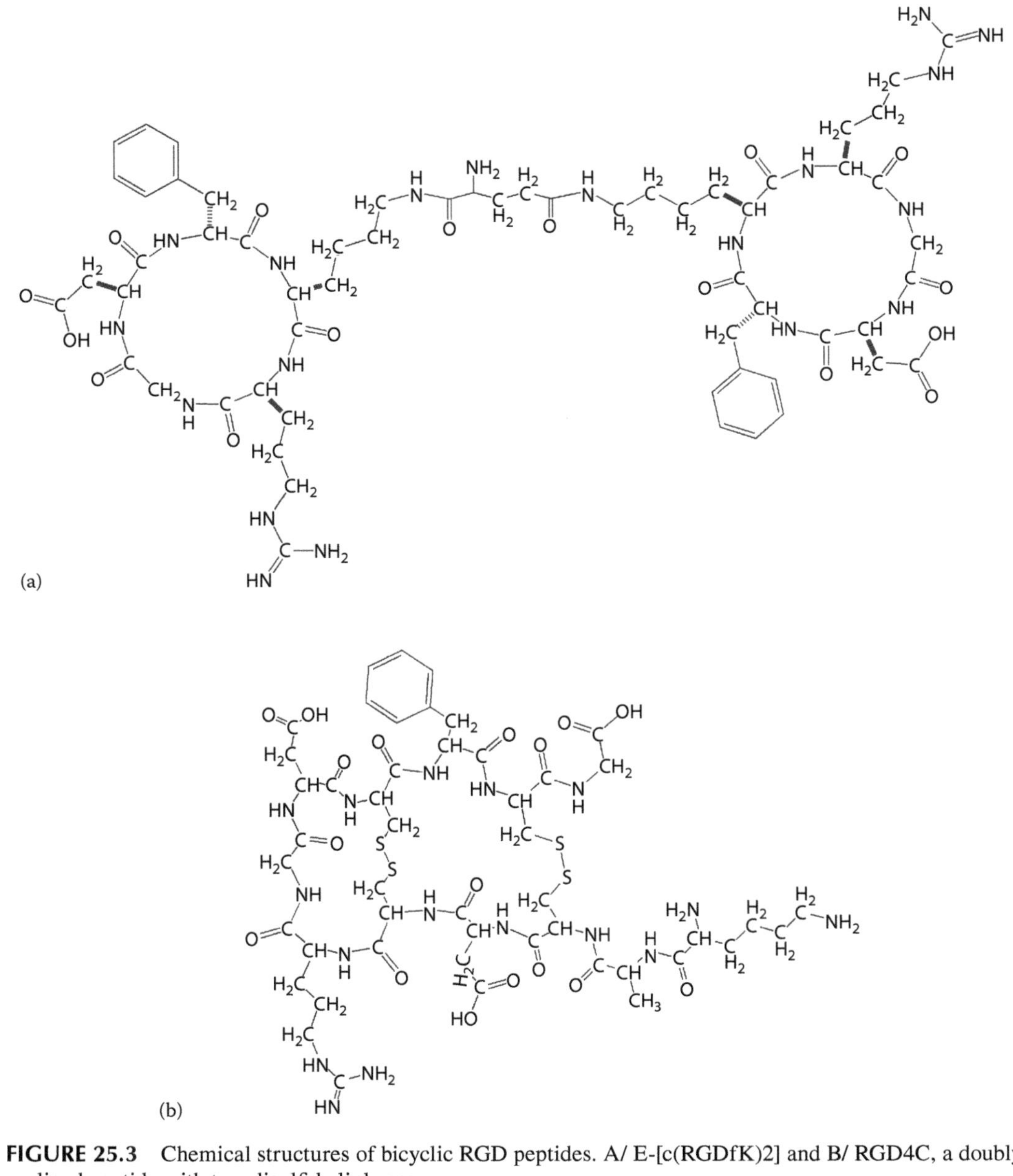

(a)

(b)

FIGURE 25.3 Chemical structures of bicyclic RGD peptides. A/ E-[c(RGDfK)2] and B/ RGD4C, a doubly cyclized peptide with two disulfide linkages.

25.3.2 POLYMERIC ARCHITECTURES

The term "polymer therapeutics" includes a large number of polymeric systems with various architectures such as polyplexes (DNA–polymer complexes) that are being developed as non-viral vectors, polymer–protein conjugates, polymer–peptide conjugates, polymer–drug conjugates, micro- or nanocapsules, polymeric micelles, dendrimers, and hydrogels (Figure 25.4).

25.3.2.1 Polymeric Gene Delivery (Polyplexes)

Direct injection of free DNA into tissues has been shown to produce surprisingly high levels of gene expression. However, naked DNA has short lifetime due to in vivo enzymatic degradation, and its use is limited owing to low efficiency of transfection. Two different kinds of vectors have then been developed: viral and non-viral vectors. Viral vectors are classified into five families: adenovirus, adeno-associated virus, lentivirus, retrovirus, and herpes simplex-1 virus (Tandle et al., 2004). The main difference between these vectors is the insertion or not of genetic material into genome. Although viral vectors are currently the most effective way to transfer genes to cells, their use is limited by difficulties in encapsulating the large genetic material, as well as by toxic effects due to their high immunogenicity and oncogenic potential (Lehrman, 1999; Sun et al., 2003). Non-viral vectors could circumvent some of these problems and are increasingly being considered for in vivo gene delivery. Non-viral vectors are useful in therapeutic angiogenesis not only to promote angiogenesis, for example, in cardiovascular diseases (Choi, 2007) or tissue engineering but also to inhibit angiogenesis such as in cancer pathologies (Tandle et al., 2004). Among non-viral vectors, cationic polymers offer numerous benefits. Indeed, they present a low immunogenicity and

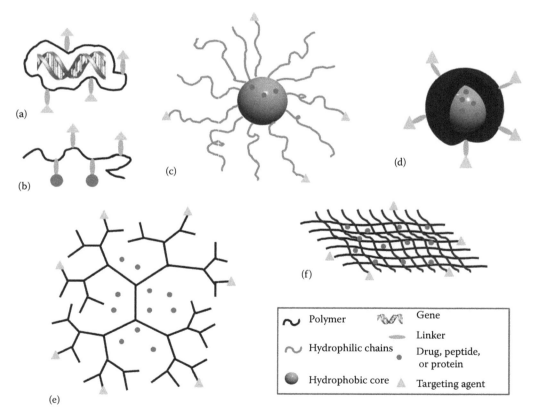

FIGURE 25.4 **(See companion CD for color figure.)** Different architectures of polymeric carrier for drug delivery. (a) Polyplex, (b) polymer-drug conjugate, (c) micelle, (d) micro- or nanocapsule, (e) dendrimer, and (f) hydrogel.

low toxicity, simplicity of use, and ease of large-scale production (Niidome and Huang, 2002). Nevertheless, a major problem when using cationic polymers for gene delivery is that serum components interfere with the formation of complexes and thereby adversely affect transfection (Hatefi et al., 2006). Incorporation of PEG into polymeric gene carriers allowed to overcome this drawback and has been shown to stabilize cationic polymer/DNA complexes (Ogris et al., 1999).

Poly(ethylenimine) (PEI) has proved to be an effective candidate as gene carrier and was extensively used to complex DNA. PEI has a high transfection level due to proton-sponge effect over a broad pH range (Boussif et al., 1995; Huang et al., 2005a). Huang et al. have described the condensation of DNA encoding for BMP-4 in PEI for bone tissue engineering (Huang et al., 2005b). PEI scaffold incorporating condensed DNA was implanted into rat cranium critical-sized defects and has demonstrated a significant bone regeneration and formation compared with control samples. However, high molecular weight PEI (25 kDa), which is commonly used for gene delivery, has high toxicity (Benns et al., 2001). Although low molecular weight PEI (1.8 kDa) has low cytotoxicity, it also has low transfection efficiency. Therefore, other formulations have been designed including water-soluble lipopolymer (WSLP) (Lee et al., 2003; Yockman et al., 2008). These systems consist of plasmid DNA condensed with cationic polymer and lipoprotein (cholesterol). The cholesterol moiety allows a better condensation by forming stable micellar complexes owing to its hydrophobic character. Moreover, cholesterol can increase the transfection efficiency of polymer carriers through selective cell uptake mediated by lipoprotein lipase (Yokoyama et al., 2007). WSLP was used to condensate the gene encoding for $hVEGF_{165}$ and was administrated in vivo in rabbit myocardium. WSLP demonstrated higher transfection efficiency than PEI with a lower cytotoxicity. These results suggest that WSLP could be applied to in vivo gene therapy for the treatment of ischemic heart disease (Lee et al., 2003). In addition, PEI polyplexes can be conjugated with PEG chains to reduce interaction with blood components avoiding, in this way, aggregation and extending circulation in blood. It is known that surface modification with PEG prevents the recognition of polyplexes by RES and the uptake by macrophages in the liver and spleen, yielding "stealth" polyplexes (Merdan et al., 2005). PEG being biocompatible also reduces cytotoxicity of PEI. Kim and co-workers have worked on soluble Flt-1 (sFlt-1) gene delivery using PEI-g-PEG conjugate in order to inhibit angiogenesis (Kim et al., 2005). Indeed, sFlt-1 is a potent and selective inhibitor trapping VEGF (Kendall et al., 1996). Further, PEG-g-PEI copolymer was modified with a RGD-motif-containing peptide to target integrins. The carrier might condensate gene encoding sFlt-1 and displayed good transfection efficiency through integrins, inhibiting VEGF-driven proliferation of endothelial cells by 63%. This polyplex appeared as a promising gene carrier in anti-angiogenic therapies. To improve again the transfection efficiency, a similar system was elaborated where genetic material was covalently linked via a disulfide bridge and not condensed by electrostatic interactions. PEG/PEI micelles were used to carry small interfering RNA (siRNA) silencing the VEGF gene in human prostate carcinoma cells. siRNA–PEG/PEI micelles have shown higher transfection efficiency than siRNA/PEI complexes, and siRNA could be released in a reductive cytosolic environment by cleavage of the disulfide linkage (Kim et al., 2006a).

Poly(L-lysine) (PLL), a linear polypeptide, is one of the polymers that have been thoroughly investigated as a non-viral gene delivery vector. It yields with DNA polyelectrolyte complexes mediated by interactions between the cationic charges of protonated primary amino groups of PLL and negative charges of phosphate moieties on the DNA backbone. In a similar manner to PEI, the high molecular weight of PLL (>4 kDa) tends to improve the stability of polyplexes; on the other hand, it increases their toxicity (Wolfert and Seymour, 1996). To obviate this problem, PEG chains have been grafted on PLL through peptide bound avoiding aggregation with proteins and red blood cells (Männistö et al., 2002). PEGylation is expected to improve the stability, the pharmacokinetic and transfection efficiency of PLL/DNA complexes. Kataoka et al. synthesized micellar polyplexes of PEG–PLL copolymer associated with plasmid DNA (pDNA) and conjugated to RGD peptide (Oba et al., 2007). pDNA did not encode for a specific gene, but was only used to monitor transfection efficiency. c(RGDfK)-PEG-PLL/pDNA polyplexes contributed to an increased transfection efficiency for the cultured HeLa cells possessing $\alpha_v\beta_3$ and $\alpha_v\beta_5$ integrins, compared with polyplexes without targeting agent. This polyplex attached with RGD peptide as ligand appeared useful for anti-angiogenic therapy in tumors. An approach similar to WSLP has been developed with PLL, namely TerplexDNA consisting in pDNA, stearyl-PLL, and low density lipoprotein (LDL) (Kim et al., 1998; Affleck et al., 2001; Yu et al., 2001). The insertion of the stearyl groups into the core of lipoprotein is mediated via hydrophobic interactions. Incorporation of LDL into the polymeric carrier likely enhances gene delivery through a rise of the LDL receptor-mediated endocytosis pathway. This vehicle was used to increase myocardial transfection for the treatment of cardiovascular disease. Transfection rate in the rabbit myocardium was found 20–100-fold higher with TerplexDNA than with naked plasmid. TerplexDNA gene carrier presented negligible cytotoxicity and appeared as a promising alternative to viral vectors and liposomes in gene delivery (Affleck et al., 2001).

In addition to high transfection, an ideal vector should be excreted from the body after delivery of its genetic material to avoid accumulation in cells. In this sense, biodegradable polymers have emerged, such as poly(lactic-co-glycolic acid) (PLGA), poly[R-(4-aminobutyl)-L-glycolic acid] (PAGA), polysaccharides, poly(amino acids), or modified PEI. PLGA is the most common FDA-approved polymer used for its biodegradability and biocompatibility. PLGA undergoes degradation by the hydrolysis of ester linkages to yield lactic and glycolic acid (Houchin and Topp, 2008). The degradation kinetics depends on the ratio of lactic to glycolic acid units and on molecular weight. The higher the glycolic acid content, the faster the degradation rate (Robert et al., 1977; Sanders et al., 1984). This hydrolysis allows to control gene or drug delivery. Kang et al. have employed PLGA nanospheres to entrap a gene coding for VEGF (pDNA) inducing neovascularization in mouse ischemic limbs (Kang et al., 2008). PLGA nanospheres offer several advantages such as a high stability, easy uptake into the cells by endocytosis, and the targeting ability to specific tissue or organs by coupling with ligand materials at the surface. The pDNA encapsulation yield in PLGA nanospheres reached 87% and pDNA was released over 11 days in vitro. Transfection efficiency in human dermal fibroblasts was higher with PLGA compared with PEI. In addition, PLGA nanospheres allowed higher cell viability. The PLGA/pDNA nanospheres were transplanted into skeletal muscles of normal mice. At 12 days after administration in the mouse limb ischemic model, the VEGF concentration is higher with PLGA nanospheres compared with the naked pDNA. PLGA nanospheres displayed slower gene release than PEI favoring sustained gene expression in the target cells or tissues. VEGF gene delivery using PLGA nanospheres resulted in more extensive neovascularization into the mice's ischemic limbs.

Chitosan, deacetylated chitin found in crustacean shells, is a semi-natural, biocompatible, biodegradable, cationic polysaccharide. With its positive charges, chitosan may be used to condensate DNA. However, these simple (chitosan–DNA) polyplexes do not have high transfection efficiency compared with PEI–DNA and PLL–DNA polyplexes. Therefore, chitosan has undergone several modifications, such as modulation of its degree of deacetylation, alkylation, quaternization, PEGylation. Thus, chitosan may be adapted for numerous applications in gene delivery. Dass and colleagues reported the synthesis of chitosan microparticles containing a plasmid expressing human pigment epithelium-derived factor (PEDF) by complex coacervation in an orthotopic metastatic model of osteosarcoma (Dass et al., 2007). PEDF is a potent endogenous anti-angiogenic factor inducing endothelial cell apoptosis, as well as decreasing the expression of significant pro-angiogenic factors such as VEGF (Dawson et al., 1999). In vivo studies revealed that these microparticles inhibited primary tumor growth at the tibial site. In addition to the activity at bone site, microparticles reduced establishment of lung (predominant site for osteosarcoma cell spread) metastases.

Transfection efficiency also depends on endocytosis, release of endocytosed DNA from the endosome into the cytoplasm, and transport of the DNA into the nucleus. In particular, gene delivery carriers suffer from their trafficking to lysosomes where they are degraded. The drug release in the cytosol is crucial in order to avoid degradation by lysosomal enzymes. Since the pH of an endosome is lower than that of the cytosol by 1–2 pH units, Kyriakides and colleagues have designed pH-sensitive and membrane-disruptive polymers (Kyriakides et al., 2002). These polymers were expected to disrupt lipid bilayer membranes of endosome at pH 6.5 and below, but be non-lytic at pH 7.4 in the cytosol. In this way, poly(propylacrylic acid) (PPAA) was used as pH-sensitive polymer for its ability to hemolyze red blood cells suspended in acidic buffers. Therefore, ternary complexes have been designed after addition of PPAA to lipoplexes (liposome/DNA complexes) to enhance their transfection efficiency. Lipoplexes were made of cationic lipid dioleyltrimethylammonium propane (DOTAP) and sense or antisense DNA encoding for the angiogenesis inhibitor thrombospondin-2 (TSP-2). The in vivo transfection efficiency and consequently the wound healing were improved.

25.3.2.2 Polymer–Drug Conjugates

The polymer–drug conjugates are nano-sized buildings that covalently link a bioactive moiety with a polymer to ensure its efficient delivery and its availability within a specific period of time. They are the simplest polymeric systems used in drug delivery and allow to control the amount of conjugated molecule.

The first polymeric anti-angiogenic conjugate was caplostatin. This is a water-soluble and biocompatible copolymer, named N-(2-hydroxypropyl) methacrylamide (HPMA) linked to TNP-470 via a peptide (Gly-Phe-Leu-Gly) (Figure 25.5) (Satchi-Fainaro et al., 2004). TNP-470, an analog of fumagillin, is a potent inhibitor of angiogenesis; however, when used free, it shows a relatively high toxicity with neural side effects such as dizziness, decreased concentration, short-term memory loss, confusion, and depression. Moreover, it presents a poor oral availability and short half-life. HPMA-TNP-470 conjugate allowed to improve the solubility of drug in water and reduce drug toxicity. Drug is released via peptide cleavage by the lysosomal enzyme cathepsin B present in endothelial cells. This conjugate inhibited endothelial cell proliferation and endothelial sprouting in the same proportion than free TNP-470. Thus, copolymer did not impair drug activity and showed a preferential accumulation in tumor blood vessels by EPR effect.

The same copolymer was used to deliver radiotherapeutic agents and actively target tumor vasculature. Radioisotopes can eradicate tumor from the emission of energy and are prone to have an effect on the peripheral rim of the tumor, which cannot be reached by anti-angiogenic molecules since it is nourished by mature blood vessels. Side chains with chelators for ^{99m}Tc and ^{90}Y were incorporated to HPMA, which was further conjugated to a doubly cyclized targeting peptide (RGD4C) (Figure 25.6). Radiolabeled HPMA copolymer–RGD4C conjugate has been demonstrated to interrupt tumor growth with no significant radiation-induced toxicity to other organs (Mitra et al., 2006b).

Another peptide, able to target tumor angiogenic vasculature, was assessed from a phage-displayed peptide library and its sequence was determined as APRPG. This peptide was conjugated to PEG hydrophobized with distearoylphosphatidylethanolamine (DSPE) phospholipids yielding DSPE-PEG-APRPG. Phospholipids self-assemble in vesicular nanostructures to form liposomes which may encapsulate hydrophilic drugs in their inner hydrophilic compartment and hydrophobic moieties within the hydrophobic bilayer membrane. However, their loading capacity is limited due to membrane destabilization effects (Liu et al., 2006). In order to obtain "stealth" liposomes, PEG chains were attached leading to a carrier with a longer blood exposure and an EPR effect. The biodistribution of this conjugate was investigated in tumor-bearing model mice. The results showed that DSPE-PEG-APRPG had prolonged circulation times in blood, avoiding RES and accumulated in tumor through passive targeting and also active targeting thanks to APRPG ligand (Maeda et al., 2004). A subsequent study was performed with a radiolabel, [2-^{18}F]2-fluoro-2-deoxy-D-glucose ([2-^{18}F]FDG) encapsulated into DSPE-PEG-APRPG in order to assess the in vivo trafficking of this drug carrier in tumor with positron emission tomography (Maeda et al., 2006).

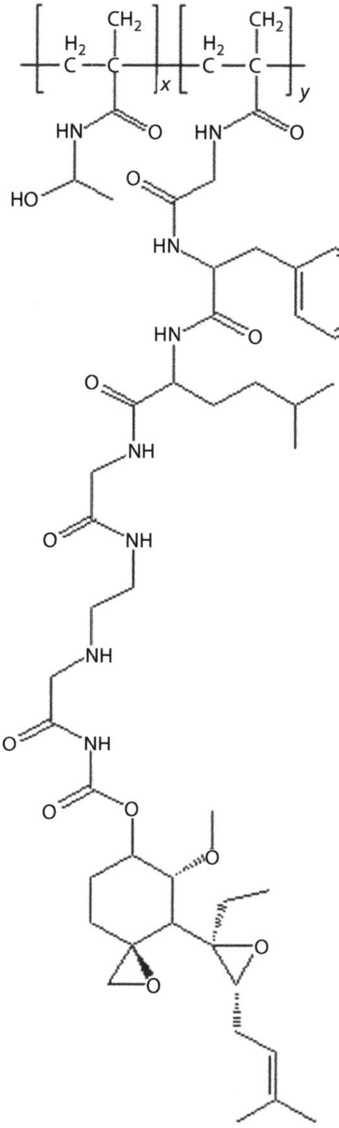

FIGURE 25.5 Chemical structure of caplostatin. (Adapted from Satchi-Fainaro, R. et al., *Nat. Med.*, 10, 255, 2004.)

Besides drugs and peptides, it is also possible to conjugate proteins with polymers. Wall et al. have attached a recombinant form of the protein Sonic hedgehog (Shh) to linear polymers poly(acrylic acid) and hyaluronic acid (Wall et al., 2008). Shh is a known angiogenic growth factor (Pola et al., 2001). The Shh–polymer conjugates with high ratio of protein increased angiogenesis determined by the in vivo CAM assay, which exhibits a great potential for tissue engineering. Linear polymers did not alter the potency of protein while protecting it and enhancing its circulation half-life.

25.3.2.3 Micro- and Nanoparticles

Micro- and nanoparticles gained considerable attention in recent years due to their potential applications for targeted drug delivery (Faraji and Wipf, 2009). They provide several unique benefits. Indeed, their properties can be tuned by changing their size in order to fit to the aimed application.

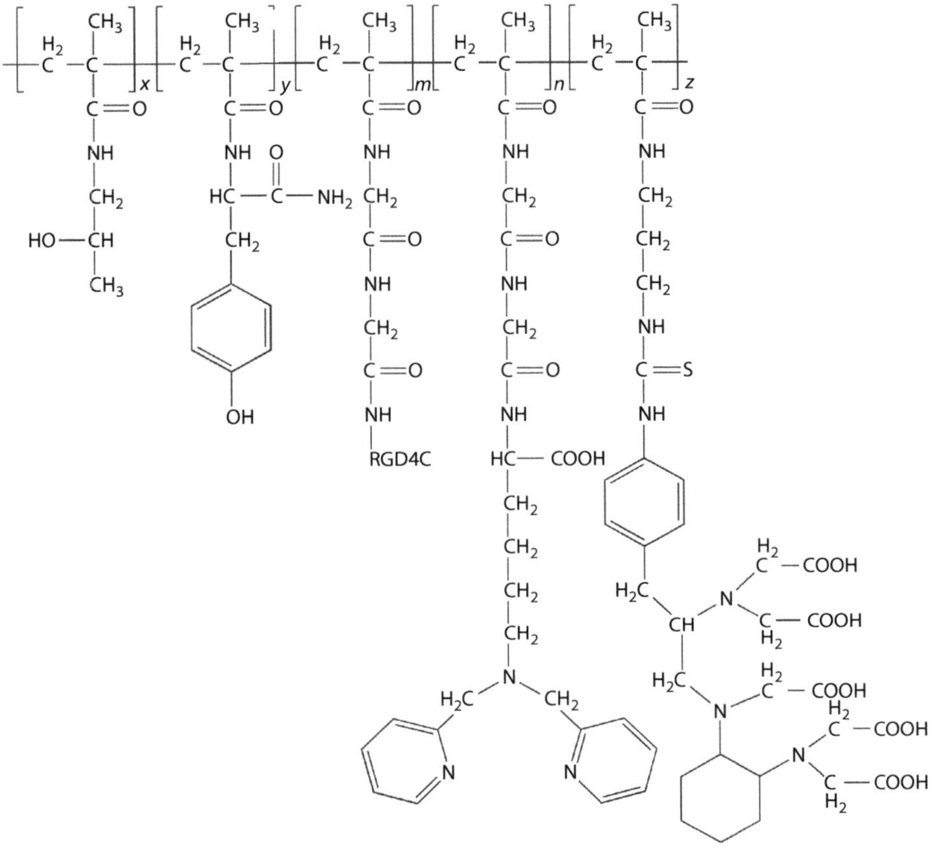

FIGURE 25.6 Structure of HPMA copolymer-RGD4C conjugate. (Adapted from Mitra, A. et al., *Nucl. Med. Bio.*, 33, 43, 2006b.)

Moreover, nanoparticles have a high surface area to carry a large number of biological molecules. This allows to reduce the number of injections in patients.

25.3.2.3.1 Pro-Angiogenesis

In therapeutic angiogenesis, VEGF has a short half-life once introduced into a host environment (Lazarous et al., 1996). Thus, growth factors need to be protected to extend its blood circulation time. In this way, aliphatic polyesters (PLA, PGA, and PLGA) are the most widely used carriers for growth factor delivery and commonly processed into microspheres. Growth factors are generally incorporated into polymers via emulsion techniques involving organic solvents and high temperature. The drawback is that the growth factors are not stable in those conditions. To circumvent this problem, a gas foaming particulate technique has been established to allow scaffold fabrication without the use of organic solvents or high temperature (Harris et al., 1998). Thus, 3D highly porous matrices from PLGA particles were obtained under high-pressure CO_2 gas. Semi-crystalline homopolymers such as PLA (poly(L-lactic acid) and poly(D-lactide acid)) or PGA did not form porous matrices, while amorphous copolymers (PLGA) can yield matrices with porosity up to 95% (Sheridan et al., 2000). The controlled release of growth factors is regulated by the hydrolytic degradation of PLGA. The rate of release depends on the molecular weight and chemical structure of polymer (Jain, 2000). Multiple growth factors (VEGF, PDGF, FGF) have been incorporated into PLGA scaffolds for tissue engineering (Sheridan et al., 2000; Cleland et al., 2001; Richardson et al., 2001; Ennett et al., 2006).

In coronary artery diseases, microcapsules may be delivered by injection to either the apex or the base of the heart wall, while porous scaffolds may be useful as patches and wraps (Fischbach and Mooney, 2006). Ennett and co-workers have made a comparison between different polymer constructs (Ennett et al., 2006). Growth factors were incorporated into scaffolds by two approaches. The first approach involved simply mixing VEGF with polymeric particles before processing the polymer into a porous scaffold. The second approach involved pre-encapsulating VEGF in PLGA microspheres, and then fabricating scaffolds from these particles. In both cases, VEGF has been successfully released from polymer in vitro, but with two distinct kinetics. The rate of delivery was faster when incorporated in the pores of scaffolds than pre-encapsulated in microspheres where VEGF is more deeply embedded. Thus, it was possible to control kinetic of protein release according to the used process. In vivo, the scaffolds with VEGF directly incorporated have been implanted into the subcutaneous tissue of mice. The study of protein distribution showed that VEGF remained localized to their target tissue with little systemic exposure. Moreover, VEGF release allowed the enhancement of microvessel density relative to the control conditions in two mice models, severe combined immunodeficient (SCID) and C57Bl/6J mice. The formulation of drug delivery system will depend on the aim.

Microparticles could be also used for bone or nerve regeneration. Richardson et al. have developed a polymeric system that allowed the tissue-specific delivery of multiple growth factors (VEGF$_{165}$ and PDGF-BB) (Richardson et al., 2001). This vehicle was a porous polymer scaffold based on PLGA. Multiple delivery promoted higher vessel maturation than individual growth factors. Other approaches have been employed to improve bone regeneration. For instance, Murphy et al. have coated the surface of 3D porous PLGA scaffold with a biomineral layer in order to provide an osteoconductive behavior to their vehicle (Murphy et al., 2004). The mineral grown on PLGA surfaces was a carbonate apatite, the major mineral component of vertebrate bone. VEGF$_{165}$ was incorporated into mineralized scaffold in order to be delivered in a localized and sustained manner for regeneration of bone and vascular tissue. This multifunctional scaffold induced a high density of blood vessels and a 62% increase in bone regeneration into critical-sized cranial defects in Lewis rats in comparison with control scaffolds. However, VEGF release is delayed due to the mineralization process and has to be further controlled by the polymer composition. VEGF has been also combined with BMP to enhance bone regeneration. Gelatin microparticles were incorporated into a porous polymer scaffold made from poly(propylene fumarate) and used to deliver synergistically VEGF and BMP-2 for bone regeneration in a rat cranial critical size defect (Patel et al., 2008). With regard to nerve regeneration, polymeric conduits termed guidance channels and bridges have been employed to deliver NGF. These conduits were made of PLGA microspheres and porogen (NaCl). Microspheres retained the bioactivity of nerve factor and allowed to stimulate neurite outgrowth from primary dorsal root ganglion (Yang et al., 2005b). Other studies showed the efficiency of PLGA microspheres as NGF delivery system (Cao and Shoichet, 1999; Rosner et al., 2003).

25.3.2.3.2 Anti-Angiogenesis

Polymeric particles may also be employed for delivery of anti-angiogenic drugs. PLGA-based nanospheres have been loaded with two anti-angiogenic drugs, endostatin and paclitaxel. E-selectin, a surface membrane receptor overexpressed by the vascular endothelium in many diseases including malignant tumors, was used as a specific target of tumor angiogenesis. A synthetic analog of Sialyl Lewis^x, specific agent for E-selectin, was grafted on nanospheres. The selective cellular uptake of nanoparticles was investigated in human umbilical vascular endothelial cell (HUVEC) cultures. An enhanced anti-proliferative effect on HUVECs and a higher anti-angiogenic activity on rat aorta ring cultures were observed for the loaded drugs compared to the free molecules (Hammady et al., 2009).

25.3.2.4 Micelles

Polymeric micelles have been first introduced as drug carriers by Ringsdorf in 1984 (Bader et al., 1984). They are formed from self-assembly of amphiphilic block copolymers in aqueous solution. They usually have spherical shape and nanometric size (10–100 nm). Their hydrophobic core is covered with a hydrophilic shell and allows to encapsulate and protect hydrophobic molecules. Micellar systems present a relatively long blood lifetime owing to their hydrophilic surface which is known to avoid protein adsorption and opsonization by RES (Bader et al., 1984). It has been reported that nanoparticles with hydrophobic surfaces tend to be quickly cleared by liver and spleen (Klibanov et al., 1990). The most commonly used hydrophilic corona are PEG derivatives with molecular weight of 2–15 kDa. PEG are biocompatible with a high chain mobility in aqueous environment and a large excluded volume, thus decreasing interactions with constituents of biological fluids (Blanco et al., 2009). They prevent nanoparticles from aggregation and are easy to modify. Other hydrophilic polymers such as poly(N-vinyl pyrrolidone) (PVP) or poly(N-isopropyl acrylamide) (PNIPAM) have also been used to form the micelle corona. Hydrophobic cores are typically polyesters (PLA, PLGA, PCL), but also polyethers such as poly(propylene oxide) or polypeptides as poly(aspartic acid) (Otsuka et al., 2003).

Hydrophobic drugs present some limitations to be directly used as bioactive molecules: relative toxicity, low water solubility, fast phagocytic, and renal clearance. In this way, Nasongkla et al. have loaded, a hydrophobic antitumor agent, the doxorubicin (Doxo) inside the solid core of PCL–PEG micelles (Nasongkla et al., 2004). In order to target tumor vasculature, they used RGD peptide as ligand on the surface of micelles. PEG chains enclosed maleimide groups at one extremity, allowing coupling with thiol functions of the ligand. The typical Doxo loading content in the micelles was 3.10 wt%. The maximal uptake with cRGD-containing micelles in tumor endothelial cells which overexpress $\alpha_v\beta_3$ integrins was 30-fold higher than with nonfunctionalized micelles. Thus, the crucial role of targeting ligand to be specific tumor endothelial cells was demonstrated. Confocal laser scanning microscopy images of tumor cells suggest that micelles are entrapped in the endosomal compartments. In order to improve the controlled release of Doxo, Nasongkla and co-workers have designed a multifunctional micellar platform responding with a change in conformation to pH stimuli (Nasongkla et al., 2006). Micelles were made of amphiphilic block copolymers of maleimide-terminated poly(ethylene glycol)-block-poly(D,L-lactide acid) (MAL-PEG-PLA). A cyclic RGD motif used as targeting ligand was attached to the micelle surface through a covalent thiol–maleimide linkage and a cluster of superparamagnetic iron oxide (SPIO) nanoparticles used as magnetic resonance imaging (MRI) contrast agent was loaded inside the micelle core. The MRI signal magnitude was measured in endothelial cells as a function of the peptide loading content, indicating the good specificity of nanoparticles for integrin receptors. Interestingly, higher concentrations of released Doxo were found in cell nuclei compared to that in the previous study with PCL–PEG. Two reasons can explain this result. First, contrary to semi-crystalline PCL, poly(D,L-lactide) is an amorphous polymer which makes easier the release of drug. In addition, the release from PEG–PLA was found to depend on pH due to the presence of the ionizable ammonium groups on Doxo (pKa ~ 7). Indeed, the release is faster inside endosome when pH is acidic than in the physiological environment where pH is neutral.

Polysaccharides can also present amphiphilic behavior after modification with hydrophobic moieties and they self-associate in water with core-shell structure via intra- and inter-molecular associations. For example, glycol chitosan was hydrophobically modified with 5β-cholanic acid to encapsulate RGD peptide yielding micellar RGD–HGC nanoparticles with a high loading efficiency (>85%). In this study, RGD peptide was used as anti-angiogenic molecule. Thus, self-assembled glycol chitosan nanoparticles have been explored as carrier for RGD peptide delivery in cancer therapy. RGD–HGC nanoparticles showed a sustained and controlled release of the peptide by diffusion for up to 7 days. The inhibition of cell adhesion in vitro by RGD–HGC nanoparticles is similar to free RGD, suggesting that the carrier does not decrease the activity of the peptide. Moreover, bFGF-induced angiogenesis was inhibited in vivo by RGD–HGC nanoparticles with a hemoglobin content reduced to 72% relative to 51% by free RGD. Intratumoral injection allowed to decrease tumor growth and microvessel density in comparison with free RGD peptide because nanoparticles provided high local drug concentrations within the solid tumors (Kim et al., 2008).

Similar to caplostatin, Benny et al. have designed a polymeric platform with a potent anti-angiogenic activity, named lodamin (Benny et al., 2008). This system was made of monomethoxy-PEG-PLA micelles conjugated to TNP-470 located in the core (mPEG-PLA-TNP-470). After 48 h of lodamin incubation in HUVECs, cell proliferation was inhibited up to 95%. Furthermore, lodamin showed an inhibition of bFGF and VEGF-induced angiogenesis and a preferential accumulation in tumor presumably through the EPR effect, without neurotoxicity in mice. This platform emerged as a promising therapy in the treatment of solid tumors and metastasis in mice.

Two MMP-specific peptides, GPLGV and GPLGVRG (P5D and P7D, respectively), have been conjugated to Doxo and they were found to be cleaved by MMP-2 secreted by cancer cells, with a lower cytotoxicity compared with free Doxo. However, they do not possess a prolonged residence in blood. To overcome this problem, they have been modified with PEG. The conjugation of hydrophobic Doxo and hydrophilic PEG gave an amphiphilic character to PEGylated peptide–Doxo conjugate and contributed to form micelles. The longevity of peptide–Doxo conjugate was improved with micellar formulation. Doxo was loaded in micelles in addition to Doxo conjugated to peptides, leading up to 72% of inhibition of tumor growth relative to the control in vivo. Doxo-loaded PEGylated peptide–Doxo conjugate micelles appeared as a potent anticancer platform that allows to reduce the toxicity of drug while retaining its activity (Lee et al., 2007).

Micelles could also be used for the treatment of AMD. Kataoka and colleagues have developed a polyion complex micelle from charged block copolymers (Ideta et al., 2004). These micelles consisted of an outer PEG shell and an inner core formed by poly(aspartic acid) encapsulating fluorescein isothiocyanate-labeled poly(L-lysine) (FITC-PLL) through electrostatic interactions. Polyion complex micelles were found to preferentially accumulate to choroidal neovascular lesions due to a targeting by EPR effect. Their blood lifetime could be controlled by the charge ratio and the length of the poly-amino acid. In addition, polyion complex micelles decreased adverse effects of FITC-PLL.

Polymeric micelles have been widely studied to target angiogenic site and deliver drug for the treatment of cancer or AMD. In contrast, they have been rarely used for the delivery of growth factors in therapeutic angiogenesis.

25.3.2.5 Dendrimers

The dendrimer term was first introduced in 1985 by Tomalia (Tomalia et al., 1985) and comes from the Greek dendron meaning tree. Indeed, monomers lead to a monodisperse polymer such as a tree. Two major synthetic strategies are used to construct dendritic structures, namely the divergent and convergent approaches. The first one is a stepwise layer-by-layer approach which assembles the molecule from nucleus toward the periphery and the convergent strategy starts from exterior to nucleus. The intensified interest in dendrimers is due to their unique characteristics including structural uniformity, multivalency, high degree of branching, well-defined architecture, and highly variable chemical composition. A classical dendrimer is composed of three distinct parts: (1) a focal core where dendrons are attached and which can encapsulate various chemical species, (2) inner blocks with several layers composed of repeating units which can provide a flexible space and can encapsulate various small molecules, (3) multivalent surface with numerous functions which can interact with external environment (Jang et al., 2009). The dendrimers are related to biomacromolecules such as proteins because of their 3D structure, their size, and their surface. Over the past 20 years, numerous dendrimers have been developed based on inspiration from biological systems. Among them, polyamidoamine (PAMAM) and polypropyleneimine (PPI) dendrimers have been extensively investigated (Figure 25.7).

In order to target angiogenic vasculature, dendrimers could be coupled with RGD peptide. Shukla et al. described the synthesis of PAMAM-RGD dendrimers labeled with a fluorescent dye (Alexa Fluor 488) (Shukla et al., 2005). Amine terminated PAMAM dendrimers bind to the cells in a nonspecific manner owing to positive charge on the surface. To improve the efficiency of targeting and reduce the nonspecific interactions, one part of amine functions was modified with acetic anhydride and another part was functionalized with RGD peptide (Figure 25.8). The researchers have studied the uptake of dendrimers in HUVECs expressing high cell surface $\alpha_v\beta_3$ receptor, and they showed that an excess of peptide impeded the internalization of dendrimers indicating that uptake was mediated by integrin receptors.

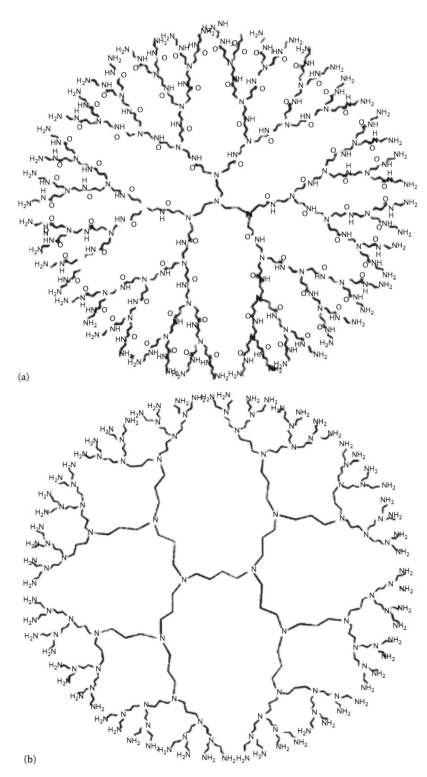

(a)

(b)

FIGURE 25.7 Chemical structures of dendrimers. (a) PAMAM and (b) PPI. (From Jang, W.D., et al., *Prog. Poly. Sci.*, 34, 1, 2009. With permission.)

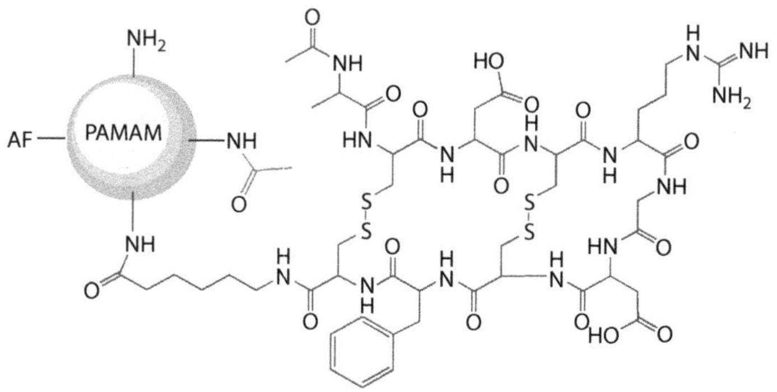

FIGURE 25.8 **(See companion CD for color figure.)** Molecular structure of PAMAM dendrimers modified with acetic anhydride and functionalized with cyclic RGD peptide (RGD4C) and Alexa Fluor 488 (AF) dye. (Adapted from Shukla, R. et al., *Chem. Commun.*, 46, 5739, 2005.)

The dendrimers may also play a crucial role in the field of gene therapy. Vincent et al. reported the inhibition of tumor growth and angiogenesis via PAMAM dendrimers associated with 36-mer anionic oligonucleotide (ON36) to deliver genes encoding angiostatin (Kringle 1–3) and MMP inhibitors (TIMP-2) (Vincent et al., 2003). DNA/dendrimer complexes possess a global positive charge, which allows them to bind to negatively charged receptors on the cell surface and optimize the entry of DNA into the cell. Angiostatin transfection was demonstrated to suppress the proliferative effect of bFGF on endothelial cells, but not modify the proliferation of cancer cells. In addition, TIMP-2 gene transfer inhibited both cancer cell invasion and endothelial cell migration. The synergistic delivery of angiostatin and TIMP-2 genes by dendrimers allowed to delay capillary tube formation and inhibit tumor growth in vivo indicating the great interest of a combined treatment.

Other kinds of dendrimers were elaborated in gene therapy. A lipophilic amino acid dendrimer (Figure 25.9) was designed to deliver an anti-VEGF oligonucleotide (ODN-1) into the eyes of rats and inhibit laser-induced choroidal neovascularization. ODN-1 is complementary to a section of sequence in the 5′ untranslated region of VEGF and inhibits VEGF expression both in vitro and in vivo (Garrett et al., 2001). In a first study, ODN-1 was delivered via dendrimers into the retinal tissue of rats; the results suggested that the dendrimers acted as protective agents against the effects of nucleases while preserving the ODN-1 bioactivity (Marano et al., 2004). Then, the study was extended to assess the longevity of dendrimer/ODN-1 conjugates. The data revealed that this dendritic system inhibited the development of choroidal neovascularization for 4–6 months by up to 95%, which represents a great improvement compared to classical ODN-based therapies requiring treatments every 28 days. Moreover, no evidence of adverse effects was found on the rodent eye after a long treatment period (Marano et al., 2005).

Dendrimers were also used in boron neutron capture therapy (BNCT) combined to an anti-angiogenic targeting. BNCT is a radiotherapy based on the injection of stable isotope, ^{10}B, followed by its irradiation to produce α-particles and ^{7}Li nuclei which have high linear energy transfer. This

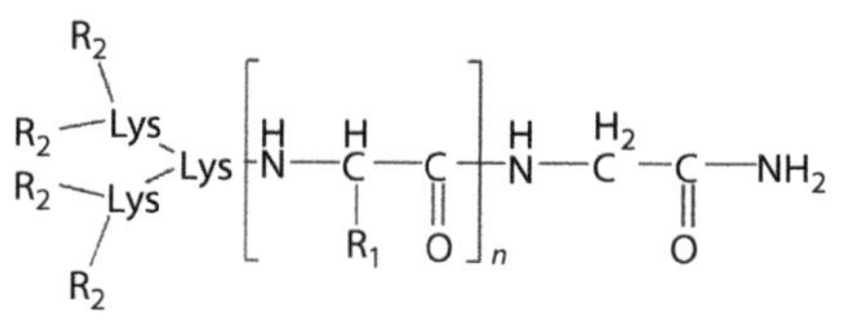

$R_1 = C_{12}H_{25}, \ R_2 = Lys(NH_2)_2, \ n = 2$

FIGURE 25.9 Chemical structure of dendrimer used for ODN-1 delivery into the retinal tissue of rats. (Adapted from Marano, R.J. et al., *Gene Ther.*, 12, 1544, 2005.)

technique is very attractive since the radiation damage only occurs over a short range and thus normal tissues can be spared whether the therapy is targeted to malignant site. Boronated EGF was conjugated to PAMAM dendrimers arranged in a starburst pattern (BSD-EGF) to target brain tumors and deliver radiotherapeutic agent. Indeed, the gene of EGF receptor is amplified in human glioblastoma but remains downregulated in normal cells and thus it is considered as a potential target for the tumor brain therapy. Dendrimers were labeled with ^{131}I to obtain ^{131}I-BSD-EGF and were administrated by intratumoral or intravenous injection in rats bearing intra cerebral implants of the $C6_{EGFR}$ glioma. In contrast to intravenous injection, intratumoral injection allowed to selectively deliver BSD-EGF to glioma avoiding the dendrimer clearance by the liver or the spleen (Yang et al., 1997).

25.3.2.6 Hydrogels

Hydrogels are defined as tridimensional networks made of water-insoluble polymeric chains able to form a gel, sometimes found in a colloidal state, with a huge adsorbent power. Indeed, hydrogels may contain up to 99% water. They possess a flexibility degree similar to natural tissues due to their high water content. These low-density materials can be classified on the basis of the mechanisms by which the cross-links are produced within the networks. Physical gels are formed by molecular self-assembly due to secondary forces, such as ionic or hydrogen bonds, while chemical gels are formed by covalent bonds (Silva et al., 2009). The most used polymers as hydrogels for medicinal applications are from natural origin including chitosan, dextran, xanthan, agarose, alginate, gelatin, cellulose, and hyaluronic acid.

In the angiogenic field, hydrogels find their main application in the delivery of growth factors such as VEGF or FGF in order to induce angiogenesis for tissue engineering (Dogan et al., 2005; Peattie et al., 2006; Silva et al., 2009). They are used as implants or injected and are ideally processed in a way to mimic ECM. From a structural perspective, ECM is a gel composed of various protein fibrils and fibers interwoven within a hydrated network of glycosaminoglycan chains (Silva et al., 2009). Growth factors may be embedded into hydrogels either by direct loading, electrostatic interaction, or covalent binding. Kanematsu et al. developed natural and synthetic collagenous matrices and added growth factors into it by direct loading (rehydration of the lyophilized matrix in an aqueous solution of growth factors), such as bFGF, HGF, PDGF-BB, VEGF, IGF, or heparin binding-EGF (Kanematsu et al., 2004). A natural bladder matrix was compared with a synthetic sponge matrix of porcine type 1 collagen. The growth factor release profiles from the collagen sponge correlated well to that of natural matrix, suggesting these hydrogels may be applied to every field tissue reconstruction in which collagenous matrices are employed. However, every growth factor has demonstrated distinct release profiles indicating different mechanisms of binding with the matrix. To alter the release profiles, cross-linking density, swelling ability, and degradation property can be adjusted. Dogan and colleagues have prepared dextran hydrogels with epichlorohydrin acting as the cross-linker (Dogan et al., 2005). The high swelling ability of the hydrogel allowed high loading capacity of bFGF and EGF. Although an initial burst of proteins was observed during the first 10 h, in vitro release kinetics of EGF and bFGF showed a longer period (up to 350 h for EGF and 650 h for bFGF) of sustained release. In in vivo studies, bFGF containing hydrogel significantly enhanced neovascularization in rat skin defects and EGF-loaded systems accelerated wound healing. Nevertheless, these hydrogel systems were not biodegradable and needed to be removed through a second surgery. Efforts have been provided to elaborate biodegradable hydrogel carriers.

Since, many efforts have been provided to elaborate biodegradable hydrogel carriers. Nakajima and co-workers designed biodegradable gelatin hydrogel to release bFGF for ischemic limb and heart injury (Nakajima et al., 2004). bFGF was incorporated by electrostatic intermolecular interaction without losing its bioactivity. The bFGF-gelatin hydrogel system was either intramuscularly injected in rabbits with limb ischemia or subepicardially injected into heart infarcts in rats. In both cases, the bFGF controlled release increased vessel density and thus improved blood perfusion.

A complete angiogenic response requires the contribution of multiple factors. In this sense, Peattie et al. have employed two growth factors: VEGF and keratinocyte growth factor (KGF) preloaded into hyaluronan hydrogels (Peattie et al., 2006). KGF is a member of FGF family; it acts specifically on epithelial cells, stimulates keratinocytes, and hastens wound healing (Gillis et al., 1999). Hyaluronan is a natural, non-immunogenic glycosaminoglycan present in ECM and possesses unique properties: an ease of production and modification, a hydrophilic and non-adhesive character, and an enzymatic degradation. Hydrogel films were prepared from thiolated derivatives of hyaluronan and cross-linked by disulfide bridging with poly(ethylene glycol)-diacrylate (PEGDA) to engineer robust materials (Figure 25.10). These hydrogels were implanted in an ear pinna model of mice and were able to provide sustained and localized release of the growth factors in vivo, while retaining their biological activity. The combined delivery of VEGF and KGF enabled to obtain the greatest density of microvessels relative to the control groups. Even a low dose of growth factors produced a strong angiogenic response.

In a similar combined approach, Hao and colleagues have investigated the multiple delivery of growth factors (VEGF$_{165}$ and PDFG-BB) with alginate hydrogels (Hao et al., 2007). The main benefit of this system was that it did not require incision because hydrogel could be injected. Alginates are polysaccharides isolated from algae. They are biocompatible, but suffer from a low degradation. To obviate this limitation, alginate was modified and partially oxidized to provide controlled degradation. PDGF-BB stimulates smooth muscle cell recruitment to newly formed vessels and is partially responsible for enhancement of vessel functionality with maturation. The release of VEGF$_{165}$ and PDFG-BB was not simultaneous. Indeed, the release of PDGF-BB was delayed compared to that of VEGF. Alginate hydrogels have been demonstrated biodegradable and their combined release induced greater remodeling of the vasculature than single growth factor delivery in a myocardial infarction model.

Hydrogel may also be used as microparticles. Thus, Moribe and colleagues have described agarose hydrogel microparticles as reservoir for bFGF (Moribe et al., 2008). Agarose is a neutral unbranched polysaccharide isolated from red-purple seaweed. Agarose hydrogels possess a low immunogenicity, they are not toxic and can be degraded by hydrolysis under acidic conditions. These microparticles were prepared by emulsification/gelation method. The loading efficiency of bFGF was more than 99% in agarose hydrogels. bFGF was stably retained into the particles through electrostatic interaction and show slow release. The in vivo experiments were assessed in a mouse model of limb ischemia and showed a relevant limb blood flow and a high increase of capillary density in the agarose groups compared to the groups without bFGF.

Generally, growth factors are released from hydrogel matrices by passive diffusion, but the release may be controlled with a stimuli-responsive system. Lee et al. have designed a new biomaterial which responds to mechanical signals (Lee et al., 2000). This biomaterial is composed of VEGF-alginate hydrogels and, when exposed to mechanical stress, VEGF release rate was increased. Indeed, VEGF was linked to hydrogel in a reversible manner where the release could be regulated by the amplitude of the compression. After implantation of these VEGF-loaded hydrogels

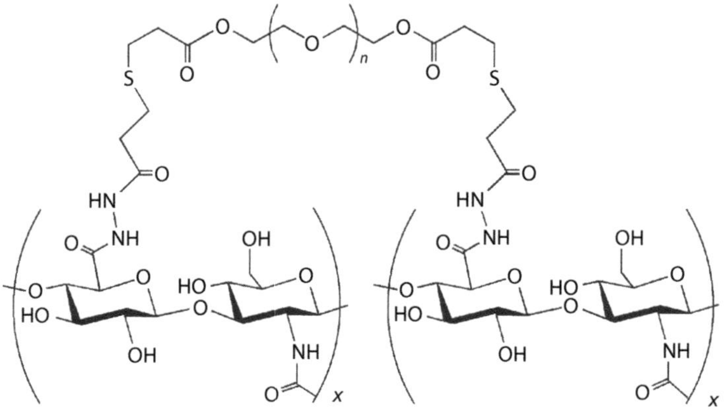

FIGURE 25.10 Chemical structure of crosslinked hyaluronan-PEGDA hydrogel. (Adapted from Peattie, R.J. et al., *Biomaterials*, 27, 1868, 2006.)

into the dorsal region of SCID mice, cyclic mechanical stimulation showed a significant increase in blood vessel density.

In an elegant strategy performed by Simmons et al., alginate hydrogels were applied for the delivery of bone progenitor cells along with combinations of growth factors for bone regeneration (Simmons et al., 2004). The incorporation of those cells allowed to avoid the use of large growth factor concentrations required to obtain beneficial effects. Alginates modified with RGD peptides were first combined with bone marrow stromal cells (BMSCs). BMSCs have been chosen because they are a promising, clinically significant autologous progenitor cell source. The alginate–BMSC mixtures were then combined with two growth factors: BMP-2 and TGF-β3. This dual growth factor delivery system allowed bone formation after implantation in SCID mice.

Hydrogel may also be prepared from synthetic polymer such as PEG, poly(vinyl alcohol), poly(acrylic acid), poly(propylene fumarate-co-ethylene glycol), and polypeptides. In addition to these usual polymers, HPMA polymer was also processed as biocompatible hydrogel (Figure 25.11a). This hydrogel was functionalized, by covalent linkage via a diglycidyl spacer, with an RGD-containing peptide to promote nerve regeneration. This scaffold was implanted at the lesion site of spinal cords in rats and was demonstrated to induce axonal growth through cord segment reconstruction (Figure 25.11b) (Woerly et al., 2001).

PEG hydrogels have been studied as growth factor delivery systems in tissue engineering because they act as a resorbable polymeric barrier preventing thrombosis. However, tissue engineering involves migration and proliferation of endothelial cells through angiogenesis, and PEG gels are highly non-adhesive toward proteins and cells and, without modification, would be unable to support the reattachment and proliferation of migrating cells (West and Hubbell, 1996). Zisch et al. have modified PEG hydrogel with the cell adhesive peptide motif RGD for cell recognition, while retaining its non-adhesive qualities toward platelets and thrombus formation. Furthermore, a 16-amino acid oligopeptide containing an MMP cleavage substrate was used as the cross-linking oligomer in order to make the hydrogels sensitive to proteolytic degradation by MMPs. Then, they have incorporated VEGF by covalent linkage and entrapped TGF-β1 into the RGD–PEG hydrogel network to regulate endothelial cell-mediated proteolysis. The covalent linkage provides retention of the factor in the matrix until VEGF may be released by plasmin or MMPs derived from cells invading the hydrogel. In short, migrating endothelial cells attached on PEG surfaces mediated by interaction between RGD peptide and integrin receptors. This cell attachment led to the production of MMPs degrading the PEG matrix and releasing growth factors. Then, VEGF promoted the

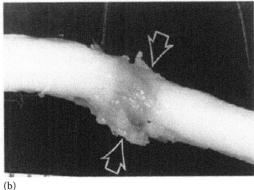

(a)   (b)

FIGURE 25.11 (a) Scanning electron microscopy micrographs of the swollen HPMA hydrogel after freeze-drying, showing the structure and spatial arrangement of the macromolecular network. (b) Macrophotographs showing the reconstruction of a spinal cord reconstruction 5 weeks after implantation of the HPMA-RGD hydrogel. Note in (b) the fusion of the polymer gel (arrows) to the ends of the cord. (From Woerly, S. et al., *Biomaterials*, 22, 1095, 2001. With permission.)

production of latent MMP-2, while TGF-β1 induced the activation of MMP-2 increasing the proteo-lytic degradation of hydrogel. Then, hydrogel matrix can be proteolytically remodeled into native and vascularized tissue by cell-associated MMPs. Functionalized PEG-peptide hydrogels appeared promising to restore the native vessel wall after balloon angioplasty for treatment of coronary artery disease (Zisch et al., 2003; Seliktar et al., 2004).

25.4 CONCLUSION

Angiogenesis is a biological event leading to the formation of new blood vessels from preexisting ones and occurs in various diseases. Angiogenesis is regulated by a balance of inductors and inhibitors. Vascular endothelia growth factor (VEGF) plays a pivotal role in this process. Targeting angiogenesis represents a promising strategy for treatment of angiogenesis-dependent diseases. Some pathologies such as cardiovascular diseases, arteriogenesis, osteoporosis, or nerve damages require the vascular promotion through therapeutic angiogenesis, while other diseases including cancer need the disruption of angiogenesis. Several clinical trials have been reported using pro- or anti-angiogenic molecules. However, they showed a limited success owing to significant side effects, a low drug bioavailability, a rapid clearance, and a lack of selectivity for malignant site. The controlled release of angiogenic agents from a carrier has thus been proposed as a promising approach in various therapies in order to avoid multiple dosing and to prolong the residence time of drug. Numerous controlled release systems have been developed over the past few years. Among them, polymers gained tremendous interest owing to their intrinsic features. They are known to their ability to be modulated according to the target application. They can be natural or synthetic, biocompatible and/or biodegradable, and they present relatively easy and inexpensive formulation. Their blood circulation time can be more prolonged by modifying their surface with judicious polymers, such as poly(ethylene glycol), yielding "stealth" systems, invisible to macrophages. They provide drug protection in different manners according to polymeric formulation. Indeed, the drug can be either covalently linked in polymer–drug conjugates, loaded into micro- or nanoparticles, encapsulated inside hydrophobic micelle core, incorporated into dendrimers, entrapped within hydrogel matrices, or in the case of genetic material, it can be condensed in polyplexes by electrostatic interactions. Polymers can be functionalized with a targeting ligand in order to increase their selectivity and specificity toward the malignant site. In angiogenic therapies, the most widely used agent is the RGD motif-containing peptide which targets integrin receptors at endothelial cell surface. Furthermore, many efforts have been made to improve the control of drug release such as the design of variable polymer architecture in response to stimuli. A variety of polymers including poly(lactic acid) and poly(lactic-co-glycolic acid) have been approved by Food and Drug Administration and are already used routinely in medicine. In the field of angiogenesis, polymers not only can deliver growth factors such as VEGF to induce angiogenesis for tissue engineering, but they can also carry anti-angiogenic molecules such as small drugs, peptides, recombinant proteins, or oligonucleotides to inhibit angiogenic process for the treatment of cancer or age-related macular degeneration.

In the future, controlled release systems should be more improved with greater efficacy and minimal side effects. "Intelligent" polymers with high precision of targeting and perfect controlled release will be in need. Research activity should help in the understanding of safety and pharmacokinetics of these systems and should focus on the design of novel combinations of polymers.

REFERENCES

Affleck, D. G.; Yu, L.; Bull, D. A.; Bailey, S. H., and Kim, S. W. 2001. Augmentation of myocardial transfection using TerplexDNA: A novel gene delivery system. *Gene Therapy* 8: 349–353.

Allen, T. M. and Cullis, P. R. 2004. Drug delivery systems: Entering the mainstream. *Science* 303: 1818–1822.

Apfel, S. C.; Schwartz, S.; Adornato, B. T. et al. 2000. Efficacy and safety of recombinant human nerve growth factor in patients with diabetic polyneuropathy: A randomized controlled trial. *JAMA: The Journal of the American Medical Association* 284: 2215–2221.

Bader, H.; Ringsdorf, H., and Schmidt, B. 1984. Water soluble polymers in medicine. *Angewandte Makromolekulare Chemie* 123: 457–485.

Bates, D. O.; Cui, T.-G.; Doughty, J. M. et al. 2002. VEGF$_{165b}$, an inhibitory splice variant of vascular endothelial growth factor, is down-regulated in renal cell carcinoma. *Cancer Research* 62: 4123–4131.

Benns, J. M.; Maheshwari, A.; Furgeson, D. Y.; Mahato, R. I., and Kim, S. W. 2001. Folate-PEG-folate-graft-polyethylenimine-based gene delivery. *Journal of Drug Targeting* 9: 123–139.

Benny, O.; Fainaru, O.; Adini, A. et al. 2008. An orally delivered small-molecule formulation with antiangiogenic and anticancer activity. *Nature Biotechnology* 26: 799–807.

Bergers, G. and Benjamin, L. E. 2003. Tumorigenesis and the angiogenic switch. *Nature Reviews Cancer* 3: 401–410.

Bikfalvi, A.; Klein, S.; Pintucci, G., and Rifkin, D. B. 1997. Biological roles of fibroblast growth factor-2. *Endocrine Reviews* 18: 26–45.

Blanco, E.; Kessinger, C. W.; Sumer, B. D., and Gao, J. 2009. Multifunctional micellar nanomedicine for cancer therapy. *Experimental Biology and Medicine* 234: 123–131.

Botos, I.; Scapozza, L.; Zhang, D.; Liotta, L. A., and Meyer, E. F. 1996. Batimastat, a potent matrix metalloproteinase inhibitor, exhibits an unexpected mode of binding. *Proceedings of the National Academy of Sciences of the United States of America* 93: 2749–2754.

Boussif, O.; Lezoualc'h, F.; Zanta, M. A. et al. 1995. A versatile vector for gene and oligonucleotide transfer into cells in culture and in vivo: Polyethylenimine. *Proceedings of the National Academy of Sciences of the United States of America* 92: 7297–7301.

Bressler, N. M.; Bressler, S. B., and Fine, S. L. 1988. *Age-Related Macular Degeneration.* New York: Elsevier.

Brooks, P. C.; Silletti, S.; von Schalscha, T. L.; Friedlander, M., and Cheresh, D. A. 1998. Disruption of angiogenesis by PEX, a noncatalytic metalloproteinase fragment with integrin binding activity. *Cell* 92: 391–400.

Brooks, P. C.; Strömblad, S.; Klemke, R. et al. 1995. Antiintegrin $\alpha_v\beta_3$ blocks human breast cancer growth and angiogenesis in human skin. *The Journal of Clinical Investigation* 96: 1815–1822.

Burri, P. H. and Djonov, V. 2002. Intussusceptive angiogenesis—The alternative to capillary sprouting. *Molecular Aspects of Medicine* 23: 1–27.

Caduff, J. H.; Fischer, L. C., and Burri, P. H. 1986. Scanning electron microscope study of the developing microvasculature in the postnatal rat lung. *The Anatomical Record* 216: 154–164.

Cao, X. and Shoichet, M. S. 1999. Delivering neuroactive molecules from biodegradable microspheres for application in central nervous system disorders. *Biomaterials* 20: 329–339.

Carmeliet, P. 2000. Mechanisms of angiogenesis and arteriogenesis. *Nature Medicine* 6: 389–395.

Carmeliet, P. 2003. Angiogenesis in health and disease. *Nature Medicine* 9: 653–660.

Chen, X.; Liu, S.; Hou, Y. et al. 2004. MicroPET imaging of breast cancer α_v-integrin expression with ^{64}Cu-labeled dimeric RGD peptides. *Molecular Imaging and Biology* 6: 350–359.

Choi, D. 2007. Polymer based cardiovascular gene therapy. *Biotechnology and Bioprocess Engineering* 12: 39–42.

Choksy, S. A. and Chan, P. 2006. Therapeutic angiogenesis. *British Journal of Surgery* 93: 261–263.

Ciulla, T. A.; Danis, R. P., and Harris, A. 1998. Age-related macular degeneration: A review of experimental treatments. *Survey of Ophthalmology* 43: 134–146.

Cleland, J. L.; Duenas, E. T.; Park, A. et al. 2001. Development of poly-($_{D,L}$-lactide-coglycolide) microsphere formulations containing recombinant human vascular endothelial growth factor to promote local angiogenesis. *Journal of Controlled Release* 72: 13–24.

Cross, M. J.; Dixelius, J.; Matsumoto, T., and Claesson-Welsh, L. 2003. VEGF-receptor signal transduction. *Trends in Biochemical Sciences* 28: 488–494.

Crum, R.; Szabo, S., and Folkman, J. 1985. A new class of steroids inhibits angiogenesis in the presence of heparin or a heparin fragment. *Science* 230: 1375–1378.

Dass, C. R.; Contreras, K. G.; Dunstan, D. E., and Choong, P. F. M. 2007. Chitosan microparticles encapsulating PEDF plasmid demonstrate efficacy in an orthotopic metastatic model of osteosarcoma. *Biomaterials* 28: 3026–3033.

Dawson, D. W.; Volpert, O. V.; Gillis, P. et al. 1999. Pigment epithelium-derived factor: A potent inhibitor of angiogenesis. *Science* 285: 245–248.

Deckers, M. M. L.; van Bezooijen, R. L.; van der Horst, G. et al. 2002. Bone morphogenetic proteins stimulate angiogenesis through osteoblast-derived vascular endothelial growth factor A. *Endocrinology* 143: 1545–1553.

Djonov, V.; Baum, O., and Burri, P. H. 2003. Vascular remodeling by intussusceptive angiogenesis. *Cell and Tissue Research* 314: 107–117.

Dogan, A. K.; Gumusderelioglu, M., and Aksoz, E. 2005. Controlled release of EGF and bFGF from dextran hydrogels in vitro and in vivo. *Journal of Biomedical Materials Research Part B: Applied Biomaterials* 74B: 504–510.

Duncan, R. 2003. The dawning era of polymer therapeutics. *Nature Reviews Drug Discovery* 2: 347–360.

Dvorak, H. F. 2006. Discovery of vascular permeability factor (VPF). *Experimental Cell Research* 312: 522–526.

Eliceiri, B. P. and Cheresh, D. A. 1999. The role of αv integrins during angiogenesis: Insights into potential mechanisms of action and clinical development. *Journal of Clinical Investigation* 103: 1227–1230.

Ennett, A. B.; Kaigler, D., and Mooney, D. J. 2006. Temporally regulated delivery of VEGF in vitro and in vivo. *Journal of Biomedical Materials Research Part A* 79A: 176–184.

Eskens, F. A. L. M.; Dumez, H.; Hoekstra, R. et al. 2003. Phase I and pharmacokinetic study of continuous twice weekly intravenous administration of Cilengitide (EMD 121974), a novel inhibitor of the integrins $\alpha_v\beta_3$ and $\alpha_v\beta_5$ in patients with advanced solid tumours. *European Journal of Cancer* 39: 917–926.

Falardeau, P.; Champagne, P.; Poyet, P.; Hariton, C., and Dupont, E. 2001. Neovastat, a naturally occurring multifunctional antiangiogenic drug, in phase III clinical trials. *Seminars in Oncology* 28: 620–625.

Faraji, A. H. and Wipf, P. 2009. Nanoparticles in cellular drug delivery. *Bioorganic & Medicinal Chemistry* 17: 2950–2962.

Ferrara, N. and Henzel, W. J. 1989. Pituitary follicular cells secrete a novel heparin-binding growth factor specific for vascular endothelial cells. *Biochemical and Biophysical Research Communications* 161: 851–858.

Ferris III, F. L.; Fine, S. L., and Hyman, L. 1984. Age-related macular degeneration and blindness due to neovascular maculopathy. *Archives of Ophthalmology* 102: 1640–1642.

Fischbach, C. and Mooney, D. 2006. Polymeric systems for bioinspired delivery of angiogenic molecules. *Advances in Polymer Science* 203: 191–221.

Folkman, J. 1971. Tumor angiogenesis: Therapeutic implications. *New England Journal of Medicine* 285: 1182–1186.

Folkman, J. 1995. Angiogenesis in cancer, vascular, rheumatoid and other disease. *Nature Medicine* 1: 27–30.

Folkman, J. 2006. Angiogenesis. *Annual Review of Medicine* 57: 1–18.

Folkman, J.; Abernathy, C., and Williams, G. 1971. Isolation of a tumor factor responsible for angiogenesis. *The Journal of Experimental Medicine* 133: 275–288.

Fuh, G.; Li, B.; Crowley, C.; Cunningham, B., and Wells, J. A. 1998. Requirements for binding and signaling of the kinase domain receptor for vascular endothelial growth factor. *Journal of Biological Chemistry* 273: 11197–11204.

Garrett, K. L.; Shen, W.-Y., and Rakoczy, P. E. 2001. In vivo use of oligonucleotides to inhibit choroidal neovascularisation in the eye. *The Journal of Gene Medicine* 3: 373–383.

Gillis, P.; Savla, U.; Volpert, O. V. et al. 1999. Keratinocyte growth factor induces angiogenesis and protects endothelial barrier function. *Journal of Cell Science* 112: 2049–2057.

Goodman, V. L.; Rock, E. P.; Dagher, R. et al. 2007. Approval summary: Sunitinib for the treatment of imatinib refractory or intolerant gastrointestinal stromal tumors and advanced renal cell carcinoma. *Clinical Cancer Research* 13: 1367–1373.

Gospodarowicz, D.; Brown, K. D.; Birdwell, C. R., and Zetter, B. R. 1978. Control of proliferation of human vascular endothelial cells. Characterization of the response of human umbilical vein endothelial cells to fibroblast growth factor, epidermal growth factor, and thrombin. *The Journal of Cell Biology* 77: 774–788.

Gresser, I. 1989. Antitumor effects of interferon. *Acta Oncologica* 28: 347–353.

Grossfeld, G. D.; Ginsberg, D. A.; Stein, J. P. et al. 1997. Thrombospondin-1 expression in bladder cancer: association with p53 alterations, tumor angiogenesis, and tumor progression. *Journal of the National Cancer Institute* 89: 219–227.

Gutheil, J. C.; Campbell, T. N.; Pierce, P. R. et al. 2000. Targeted antiangiogenic therapy for cancer using vitaxin: A humanized monoclonal antibody to the integrin $\alpha_v\beta_3$. *Clinical Cancer Research* 6: 3056–3061.

Hammady, T.; Rabanel, J.-M.; Dhanikula, R. S.; Leclair, G., and Hildgen, P. 2009. Functionalized nanospheres loaded with anti-angiogenic drugs: Cellular uptake and angiosuppressive efficacy. *European Journal of Pharmaceutics and Biopharmaceutics* 72: 418–427.

Handsley, M. M. and Edwards, D. R. 2005. Metalloproteinases and their inhibitors in tumor angiogenesis. *International Journal of Cancer* 115: 849–860.

Hao, X.; Silva, E. A.; Mansson-Broberg, A. et al. 2007. Angiogenic effects of sequential release of VEGF-A$_{165}$ and PDGF-BB with alginate hydrogels after myocardial infarction. *Cardiovascular Research* 75: 178–185.

Harris, L. D.; Kim, B.-S., and Mooney, D. J. 1998. Open pore biodegradable matrices formed with gas foaming. *Journal of Biomedical Materials Research* 42: 396–402.

Hatefi, A.; Megeed, Z., and Ghandehari, H. 2006. Recombinant polymer-protein fusion: A promising approach towards efficient and targeted gene delivery. *The Journal of Gene Medicine* 8: 468–476.

Haubner, R.; Gratias, R.; Diefenbach, B. et al. 1996. Structural and functional aspects of RGD-containing cyclic pentapeptides as highly potent and selective integrin $\alpha_v\beta_3$ antagonists. *Journal of the American Chemical Society* 118: 7461–7472.

Haubner, R.; Weber, W. A.; Beer, A. J. et al. 2005. Noninvasive visualization of the activated $\alpha_v\beta_3$ integrin in cancer patients by positron emission tomography and [^{18}F]galacto-RGD. *PLoS Medicine* 2: e70.

Henry, T. D.; Annex, B. H.; McKendall, G. R. et al. 2003. The VIVA trial: Vascular endothelial growth factor in ischemia for vascular angiogenesis. *Circulation* 107: 1359–1365.

Hersey, P.; Sosman, J.; O'Day, S. et al. 2005. A phase II, randomized, open-label study evaluating the anti-tumor activity of MEDI-522, a humanized monoclonal antibody directed against the human alpha v beta 3 ($\alpha_v\beta_3$) integrin, +/-Dacarbazine (DTIC) in patients with metastatic melanoma (MM). *Journal of Immunotherapy* 28: 643–644.

Hillen, F. and Griffioen, A. 2007. Tumour vascularization: Sprouting angiogenesis and beyond. *Cancer and Metastasis Reviews* 26: 489–502.

Hoekstra, R.; Eskens, F. A. L. M., and Verweij, J. 2001. Matrix metalloproteinase inhibitors: Current developments and future perspectives. *Oncologist* 6: 415–427.

Holash, J.; Davis, S.; Papadopoulos, N. et al. 2002. VEGF-Trap: A VEGF blocker with potent antitumor effects. *Proceedings of the National Academy of Sciences of the United States of America* 99: 11393–11398.

Houchin, M. L. and Topp, E. M. 2008. Chemical degradation of peptides and proteins in PLGA: A review of reactions and mechanisms. *Journal of Pharmaceutical Sciences* 97: 2395–2404.

Huang, Y.-C.; Riddle, K.; Rice, K. G., and Mooney, D. J. 2005a. Long-term in vivo gene expression via delivery of PEI-DNA condensates from porous polymer scaffolds. *Human Gene Therapy* 16: 609–617.

Huang, Y.-C.; Simmons, C.; Kaigler, D.; Rice, K. G., and Mooney, D. J. 2005b. Bone regeneration in a rat cranial defect with delivery of PEI-condensed plasmid DNA encoding for bone morphogenetic protein-4 (BMP-4). *Gene Therapy* 12: 418–426.

Hubbard, S. R. 1999. Structural analysis of receptor tyrosine kinases. *Progress in Biophysics and Molecular Biology* 71: 343–358.

Hurwitz, H.; Fehrenbacher, L.; Novotny, W. et al. 2004. Bevacizumab plus irinotecan, fluorouracil, and leucovorin for metastatic colorectal cancer. *The New England Journal of Medicine* 350: 2335–2342.

Hurwitz, H. and Saini, S. 2006. Bevacizumab in the treatment of metastatic colorectal cancer: Safety profile and management of adverse events. *Seminars in Oncology* 33: S26–S34.

Hynes, R. O. 1992. Integrins: Versatility, modulation, and signaling in cell adhesion. *Cell* 69: 11–25.

Ideta, R.; Yanagi, Y.; Tamaki, Y. et al. 2004. Effective accumulation of polyion complex micelle to experimental choroidal neovascularization in rats. *FEBS Letters* 557: 21–25.

Izzedine, H.; Brocheriou, I.; Deray, G., and Rixe, O. 2007. Thrombotic microangiopathy and anti-VEGF agents. *Nephrology Dialysis Transplantation* 22: 1481–1482.

Jagur-Grodzinski, J. 2009. Polymers for targeted and/or sustained drug delivery. *Polymers for Advanced Technologies* 20: 595–606.

Jain, R. A. 2000. The manufacturing techniques of various drug loaded biodegradable poly(lactide-co-glycolide) (PLGA) devices. *Biomaterials* 21: 2475–2490.

Jain, R. K. 2003. Molecular regulation of vessel maturation. *Nature Medicine* 9: 685–693.

Jang, W.-D.; Selim, K. M. K.; Lee, C.-H., and Kang, I.-K. 2009. Bioinspired application of dendrimers: From bio-mimicry to biomedical applications. *Progress in Polymer Science* 34: 1–23.

Jönhagen, M. E.; Nordberg, A.; Amberla, K. et al. 1998. Intracerebroventricular infusion of nerve growth factor in three patients with Alzheimer's disease. *Dementia and Geriatric Cognitive Disorders* 9: 246–257.

Kane, R. C.; Farrell, A. T.; Saber, H. et al. 2006. Sorafenib for the treatment of advanced renal cell carcinoma. *Clinical Cancer Research* 12: 7271–7278.

Kanematsu, A.; Yamamoto, S.; Ozeki, M. et al. 2004. Collagenous matrices as release carriers of exogenous growth factors. *Biomaterials* 25: 4513–4520.

Kang, S.-W.; Lim, H.-W.; Seo, S.-W. et al. 2008. Nanosphere-mediated delivery of vascular endothelial growth factor gene for therapeutic angiogenesis in mouse ischemic limbs. *Biomaterials* 29: 1109–1117.

Kendall, R. L.; Wang, G., and Thomas, K. A. 1996. Identification of a natural soluble form of the vascular endothelial growth factor receptor, FLT-1, and its heterodimerization with KDR. *Biochemical and Biophysical Research Communications* 226: 324–328.

Khandare, J. and Minko, T. 2006. Polymer–drug conjugates: Progress in polymeric prodrugs. *Progress in Polymer Science* 31: 359–397.

Khosravi, A.; Cutler, C. M.; Kelly, M. H. et al. 2007. Determination of the elimination half-life of fibroblast growth factor-23. *Journal of Clinical Endocrinology and Metabolism* 92: 2374–2377.

Kim, S. H.; Jeong, J. H.; Lee, S. H.; Kim, S. W., and Park, T. G. 2006a. PEG conjugated VEGF siRNA for anti-angiogenic gene therapy. *Journal of Controlled Release* 116: 123–129.

Kim, J.-H.; Kim, Y.-S.; Kim, S. et al. 2006b. Hydrophobically modified glycol chitosan nanoparticles as carriers for paclitaxel. *Journal of Controlled Release* 111: 228–234.

Kim, J.-H.; Kim, Y.-S.; Park, K. et al. 2008. Self-assembled glycol chitosan nanoparticles for the sustained and prolonged delivery of antiangiogenic small peptide drugs in cancer therapy. *Biomaterials* 29: 1920–1930.

Kim, J.-S.; Maruyama, A.; Akaike, T., and Kim, S. W. 1998. Terplex DNA delivery system as a gene carrier. *Pharmaceutical Research* 15: 116–121.

Kim, W. J.; Yockman, J. W.; Jeong, J. H. et al. 2006c. Anti-angiogenic inhibition of tumor growth by systemic delivery of PEI-g-PEG-RGD/pCMV-sFlt-1 complexes in tumor-bearing mice. *Journal of Controlled Release* 114: 381–388.

Kim, W. J.; Yockman, J. W.; Lee, M. et al. 2005. Soluble Flt-1 gene delivery using PEI-g-PEG-RGD conjugate for anti-angiogenesis. *Molecular Therapy* 11: S86.

Klibanov, A. L.; Maruyama, K.; Torchilin, V. P., and Huang, L. 1990. Amphipathic polyethyleneglycols effectively prolong the circulation time of liposomes. *FEBS Letters* 268: 235–237.

Koivunen, E.; Wang, B., and Ruoslahti, E. 1995. Phage libraries displaying cyclic peptides with different ring sizes: Ligand specificities of the RGD-directed integrins. *Nature Biotechnology* 13: 265–270.

Konner, J. and Dupont, J. 2004. Use of soluble recombinant decoy receptor vascular endothelial growth factor trap (VEGF Trap) to inhibit vascular endothelial growth factor activity. *Clinical Colorectal Cancer* 4(Suppl. 2): S81–S85.

Kyriakides, T. R.; Cheung, C. Y.; Murthy, N. et al. 2002. pH-sensitive polymers that enhance intracellular drug delivery in vivo. *Journal of Controlled Release* 78: 295–303.

Lazarous, D. F.; Shou, M.; Scheinowitz, M. et al. 1996. Comparative effects of basic fibroblast growth factor and vascular endothelial growth factor on coronary collateral development and the arterial response to injury. *Circulation* 94: 1074–1082.

Lazarous, D. F.; Shou, M.; Stiber, J. A. et al. 1997. Pharmacodynamics of basic fibroblast growth factor: Route of administration determines myocardial and systemic distribution. *Cardiovascular Research* 36: 78–85.

Lee, G. Y.; Park, K.; Kim, S. Y., and Byun, Y. 2007. MMPs-specific PEGylated peptide-DOX conjugate micelles that can contain free doxorubicin. *European Journal of Pharmaceutics and Biopharmaceutics* 67: 646–654.

Lee, K. Y.; Peters, M. C.; Anderson, K. W., and Mooney, D. J. 2000. Controlled growth factor release from synthetic extracellular matrices. *Nature* 408: 998–1000.

Lee, M.; Rentz, J.; Han, S. O.; Bull, D. A., and Kim, S. W. 2003. Water-soluble lipopolymer as an efficient carrier for gene delivery to myocardium. *Gene Therapy* 10: 585–593.

Lehrman, S. 1999. Virus treatment questioned after gene therapy death. *Nature* 401: 517–518.

Liekens, S.; De Clercq, E., and Neyts, J. 2001. Angiogenesis: Regulators and clinical applications. *Biochemical Pharmacology* 61: 253–270.

Line, B. R.; Mitra, A.; Nan, A., and Ghandehari, H. 2005. Targeting tumor angiogenesis: Comparison of peptide and polymer–peptide conjugates. *The Journal of Nuclear Medicine* 46: 1552–1560.

Liu, J.; Lee, H.; Huesca, M.; Young, A., and Allen, C. 2006. Liposome formulation of a novel hydrophobic aryl-imidazole compound for anti-cancer therapy. *Cancer Chemotherapy and Pharmacology* 58: 306–318.

Maeda, N.; Miyazawa, S.; Shimizu, K. et al. 2006. Enhancement of anticancer activity in antineovascular therapy is based on the intratumoral distribution of the active targeting carrier for anticancer drugs. *Biological and Pharmaceutical Bulletin* 29: 1936–1940.

Maeda, N.; Takeuchi, Y.; Takada, M.; Namba, Y., and Oku, N. 2004. Synthesis of angiogenesis-targeted peptide and hydrophobized polyethylene glycol conjugate. *Bioorganic & Medicinal Chemistry Letters* 14: 1015–1017.

Männistö, M.; Vanderkerken, S.; Toncheva, V. et al. 2002. Structure-activity relationships of poly(L-lysines): Effects of PEGylation and molecular shape on physicochemical and biological properties in gene delivery. *Journal of Controlled Release* 83: 169–182.

Marano, R. J.; Toth, I.; Wimmer, N.; Brankov, M., and Rakoczy, P. E. 2005. Dendrimer delivery of an anti-VEGF oligonucleotide into the eye: A long-term study into inhibition of laser-induced CNV, distribution, uptake and toxicity. *Gene Therapy* 12: 1544–1550.

Marano, R. J.; Wimmer, N.; Kearns, P. S. et al. 2004. Inhibition of in vitro VEGF expression and choroidal neovascularization by synthetic dendrimer peptide mediated delivery of a sense oligonucleotide. *Experimental Eye Research* 79: 525–535.

Matsumura, Y. and Maeda, H. 1986. A new concept for macromolecular therapeutics in cancer chemotherapy: Mechanism of tumoritropic accumulation of proteins and the antitumor agent Smancs. *Cancer Research* 46: 6387–6392.

Mazar, A. P. 2008. Urokinase plasminogen activator receptor choreographs multiple ligand interactions: Implications for tumor progression and therapy. *Clinical Cancer Research* 14: 5649–5655.

McArthur, J. C.; Yiannoutsos, C.; Simpson, D. M. et al. 2000. A phase II trial of nerve growth factor for sensory neuropathy associated with HIV infection. *Neurology* 54: 1080–1088.

Merdan, T.; Kunath, K.; Petersen, H. et al. 2005. PEGylation of poly(ethylene imine) affects stability of complexes with plasmid DNA under in vivo conditions in a dose-dependent manner after intravenous injection into mice. *Bioconjugate Chemistry* 16: 785–792.

Michaelson, I. C. 1948. The mode of development of the vascular system of the retina with some observations on its significance for certain retinal diseases. *Transactions of the Ophthalmological Society of the United Kingdom* 68: 137–180.

Mignatti, P. and Rifkin, D. B. 1996. Plasminogen activators and matrix metalloproteinases in angiogenesis. *Enzyme Protein* 49: 117–137.

Mitra, A.; Coleman, T.; Borgman, M. et al. 2006a. Polymeric conjugates of mono- and bi-cyclic $\alpha_v\beta_3$ binding peptides for tumor targeting. *Journal of Controlled Release* 114: 175–183.

Mitra, A.; Nan, A.; Papadimitriou, J. C.; Ghandehari, H., and Line, B. R. 2006b. Polymer–peptide conjugates for angiogenesis targeted tumor radiotherapy. *Nuclear Medicine and Biology* 33: 43–52.

Moribe, K.; Nomizu, N.; Izukura, S. et al. 2008. Physicochemical, morphological and therapeutic evaluation of agarose hydrogel particles as a reservoir for basic fibroblast growth factor. *Pharmaceutical Development and Technology* 13: 541–547.

Murphy, W. L.; Simmons, C. A.; Kaigler, D., and Mooney, D. J. 2004. Bone regeneration via a mineral substrate and induced angiogenesis. *Journal of Dental Research* 83: 204–210.

Nakajima, H.; Sakakibara, Y.; Tambara, K. et al. 2004. Therapeutic angiogenesis by the controlled release of basic fibroblast growth factor for ischemic limb and heart injury: Toward safety and minimal invasiveness. *Journal of Artificial Organs* 7: 58–61.

Narayani, R. 2007. Polymeric delivery systems in biotechnology: A mini review. *Trends in Biomaterials and Artificial Organs* 21: 14–19.

Nasongkla, N.; Bey, E.; Ren, J. et al. 2006. Multifunctional polymeric micelles as cancer-targeted, MRI-ultrasensitive drug delivery systems. *Nano Letters* 6: 2427–2430.

Nasongkla, N.; Shuai, X.; Ai, H. et al. 2004. cRGD-functionalized polymer micelles for targeted doxorubicin delivery. *Angewandte Chemie International Edition* 43: 6326–6327.

Nico, B.; Mangieri, D.; Benagiano, V.; Crivellato, E., and Ribatti, D. 2008. Nerve growth factor as an angiogenic factor. *Microvascular Research* 75: 135–141.

Niidome, T. and Huang, L. 2002. *Gene Therapy Progress and Prospects: Nonviral Vectors.* Hampshire, U.K.: Nature Publishing Group.

Oba, M.; Fukushima, S.; Kanayama, N. et al. 2007. Cyclic RGD peptide-conjugated polyplex micelles as a targetable gene delivery system directed to cells possessing $\alpha_v\beta_3$ and $\alpha_v\beta_5$ integrins. *Bioconjugate Chemistry* 18: 1415–1423.

Ogris, M.; Brunner, S.; Schüller, S.; Kircheis, R., and Wagner, E. 1999. PEGylated DNA/transferrin-PEI complexes: Reduced interaction with blood components, extended circulation in blood and potential for systemic gene delivery. *Gene Therapy* 6: 595–605.

O'Reilly, M. S.; Holmgren, L.; Shing, Y. et al. 1994. Angiostatin: A novel angiogenesis inhibitor that mediates the suppression of metastases by a Lewis lung carcinoma. *Cell* 79: 315–328.

Otsuka, H.; Nagasaki, Y., and Kataoka, K. 2003. PEGylated nanoparticles for biological and pharmaceutical applications. *Advanced Drug Delivery Reviews* 55: 403–419.

Overall, C. M. 2002. Molecular determinants of metalloproteinase substrate specificity: Matrix metalloproteinase substrate binding domains, modules, and exosites. *Molecular Biotechnology* 22: 51–86.

Overall, C. M. and Kleifeld, O. 2006. Towards third generation matrix metalloproteinase inhibitors for cancer therapy. *British Journal of Cancer* 94: 941–946.

Pandya, N. M.; Dhalla, N. S., and Santani, D. D. 2006. Angiogenesis—A new target for future therapy. *Vascular Pharmacology* 44: 265–274.

Park, J. H.; Lee, S.; Kim, J.-H. et al. 2008. Polymeric nanomedicine for cancer therapy. *Progress in Polymer Science* 33: 113–137.

Patel, Z. S.; Young, S.; Tabata, Y. et al. 2008. Dual delivery of an angiogenic and an osteogenic growth factor for bone regeneration in a critical size defect model. *Bone* 43: 931–940.

Peattie, R. A.; Rieke, E. R.; Hewett, E. M. et al. 2006. Dual growth factor-induced angiogenesis in vivo using hyaluronan hydrogel implants. *Biomaterials* 27: 1868–1875.

Pecher, P. and Schumacher, B. A. 2000. Angiogenesis in ischemic human myocardium: Clinical results after 3 years. *The Annals of Thoracic Surgery* 69: 1414–1419.

Peng, H.; Wright, V.; Usas, A. et al. 2002. Synergistic enhancement of bone formation and healing by stem cell–expressed VEGF and bone morphogenetic protein-4. *The Journal of Clinical Investigation* 110: 751–759.

Peracchia, M. T.; Fattal, E.; Desmaële, D. et al. 1999. Stealth® PEGylated polycyanoacrylate nanoparticles for intravenous administration and splenic targeting. *Journal of Controlled Release* 60: 121–128.

Pola, R.; Ling, L. E.; Silver, M. et al. 2001. The morphogen Sonic hedgehog is an indirect angiogenic agent upregulating two families of angiogenic growth factors. *Nature Medicine* 7: 706–711.

Polverini, P. J. 1995. The pathophysiology of angiogenesis. *Critical Reviews in Oral Biology & Medicine* 6: 230–247.

Powers, C. J.; McLeskey, S. W., and Wellstein, A. 2000. Fibroblast growth factors, their receptors and signaling. *Endocrine-Related Cancer* 7: 165–197.

Reardon, D. A.; Fink, K. L.; Mikkelsen, T. et al. 2008. Randomized phase II study of cilengitide, an integrin-targeting arginine-glycine-aspartic acid peptide, in recurrent glioblastoma multiforme. *Journal of Clinical Oncology* 26: 5610–5617.

Richardson, T. P.; Peters, M. C.; Ennett, A. B., and Mooney, D. J. 2001. Polymeric system for dual growth factor delivery. *Nature Biotechnology* 19: 1029–1034.

Risau, W. and Flamme, I. 1995. Vasculogenesis. *Annual Review of Cell and Developmental Biology* 11: 73–91.

Robert, A. M.; John, M. B., and Duane, E. C. 1977. Degradation rates of oral resorbable implants (polylactates and polyglycolates): Rate modification with changes in PLA/PGA copolymer ratios. *Journal of Biomedical Materials Research* 11: 711–719.

Ropper, A. H.; Gorson, K. C.; Gooch, C. L. et al. 2009. Vascular endothelial growth factor gene transfer for diabetic polyneuropathy: A randomized, double-blinded trial. *Annals of Neurology* 65: 386–393.

Rosner, B. I.; Siegel, R. A.; Grosberg, A., and Tranquillo, R. T. 2003. Rational design of contact guiding, neurotrophic matrices for peripheral nerve regeneration. *Annals of Biomedical Engineering* 31: 1383–1401.

Sanders, L. M.; Kent, J. S.; McRae, G. I. et al. 1984. Controlled release of a luteinizing hormone-releasing hormone analogue from poly(d,l-lactide-co-glycolide) microspheres. *Journal of Pharmaceutical Sciences* 73: 1294–1297.

Satchi-Fainaro, R.; Duncan, R., and Barnes, C. 2006. Polymer therapeutics for cancer: Current status and future challenges. *Advances in Polymer Science* 193: 1–65.

Satchi-Fainaro, R.; Puder, M.; Davies, J. W. et al. 2004. Targeting angiogenesis with a conjugate of HPMA copolymer and TNP-470. *Nature Medicine* 10: 255–261.

Schumacher, B.; Pecher, P.; von Specht, B. U., and Stegmann, T. 1998. Induction of neoangiogenesis in ischemic myocardium by human growth factors: First clinical results of a new treatment of coronary heart disease. *Circulation* 97: 645–650.

Seliktar, D.; Zisch, A. H.; Lutolf, M. P.; Wrana, J. L., and Hubbell, J. A. 2004. MMP-2 sensitive, VEGF-bearing bioactive hydrogels for promotion of vascular healing. *Journal of Biomedical Materials Research Part A* 68A: 704–716.

Senger, D. R.; Galli, S. J.; Dvorak, A. M. et al. 1983. Tumor cells secrete a vascular permeability factor that promotes accumulation of ascites fluid. *Science* 219: 983–985.

Seymour, L. W.; Miyamoto, Y.; Maeda, H. et al. 1995. Influence of molecular weight on passive tumour accumulation of a soluble macromolecular drug carrier. *European Journal of Cancer* 31: 766–770.

Shalinsky, D. R.; Brekken, J.; Zou, H. et al. 1999. Broad antitumor and antiangiogenic activities of AG3340, a potent and selective MMP inhibitor undergoing advanced oncology clinical trials. *Annals of the New York Academy of Sciences* 878: 236–270.

Sheridan, M. H.; Shea, L. D.; Peters, M. C., and Mooney, D. J. 2000. Bioabsorbable polymer scaffolds for tissue engineering capable of sustained growth factor delivery. *Journal of Controlled Release* 64: 91–102.

Shi, Z.; Huang, X.; Wang, K. et al. 2008. Construction of recombinant adeno-associated virus vector coexpressing $hVEGF_{165}$ and $hBMP_7$ gene. *Journal of Nanjing Medical University* 22: 205–210.

Shukla, R.; Thomas, T. P.; Peters, J. et al. 2005. Tumor angiogenic vasculature targeting with PAMAM dendrimer–RGD conjugates. *Chemical Communications* 46: 5739–5741.

Silva, A. K. A.; Richard, C.; Bessodes, M.; Scherman, D., and Merten, O.-W. 2009. Growth factor delivery approaches in hydrogels. *Biomacromolecules* 10: 9–18.

Sim, B. K. L.; MacDonald, N. J., and Gubish, E. R. 1998. Angiostatin and endostatin: Endothelial cell-specific endogenous inhibitors of angiogenesis and tumor growth. *Cancer and Metastasis Reviews* 2: 37–48.

Simmons, C. A.; Alsberg, E.; Hsiong, S.; Kim, W. J., and Mooney, D. J. 2004. Dual growth factor delivery and controlled scaffold degradation enhance in vivo bone formation by transplanted bone marrow stromal cells. *Bone* 35: 562–569.

Simons, M.; Annex, B. H.; Laham, R. J. et al. 2002. Pharmacological treatment of coronary artery disease with recombinant fibroblast growth factor-2: Double-blind, randomized, controlled clinical trial. *Circulation* 105: 788–793.

Slavin, J. 1995. Fibroblast growth factors: At the heart of angiogenesis. *Cell Biology International* 19: 431–444.

Sondell, M.; Lundborg, G., and Kanje, M. 1999. Vascular endothelial growth factor has neurotrophic activity and stimulates axonal outgrowth, enhancing cell survival and Schwann cell proliferation in the peripheral nervous system. *Journal of Neuroscience* 19: 5731–5740.

Sottrup-Jensen, L. 1989. α2-macroglobulins: Structure, shape, and mechanism of proteinase complex formation. *Journal of Biological Chemistry* 264: 11539–11542.

Sparano, J. A.; Bernardo, P.; Stephenson, P. et al. 2004. Randomized phase III trial of marimastat versus placebo in patients with metastatic breast cancer who have responding or stable disease after first-line chemotherapy: Eastern cooperative oncology group trial E2196. *Journal of Clinical Oncology* 22: 4683–4690.

Stewart, D. J. 2002. A phase 2, randomized, multicenter, 26-week study to assess the efficacy and safety of BIOBYPASS (Ad$_{gv}$VEGF$_{121.10}$) delivered through minimally invasive surgery versus maximum medical treatment in patients with severe angina, advanced coronary artery disease, and no options for revascularization. *Circulation* 106: 2986.

Street, J.; Bao, M.; deGuzman, L. et al. 2002. Vascular endothelial growth factor stimulates bone repair by promoting angiogenesis and bone turnover. *Proceedings of the National Academy of Sciences of the United States of America* 99: 9656–9661.

Stupp, R.; Goldbrunner, R.; Neyns, B. et al. 2007. Phase I/IIa trial of cilengitide (EMD121974) and temozolomide with concomitant radiotherapy, followed by temozolomide and cilengitide maintenance therapy in patients (pts) with newly diagnosed glioblastoma (GBM). *Journal of Clinical Oncology: 2007 ASCO Annual Meeting Proceedings Part I* 25:18S, abstract 2000.

Sugiyama, O.; Sung An, D.; Kung, S. P. K. et al. 2005. Lentivirus-mediated gene transfer induces long-term transgene expression of BMP-2 in vitro and new bone formation in vivo. *Molecular Therapy* 11: 390–398.

Sun, J. Y.; Anand-Jawa, V.; Chatterjee, S., and Wong, K. K. 2003. Immune responses to adeno-associated virus and its recombinant vectors. *Gene Therapy* 10: 964–976.

Takeshita, S.; Zheng, L. P.; Brogi, E. et al. 1994. Therapeutic angiogenesis. A single intraarterial bolus of vascular endothelial growth factor augments revascularization in a rabbit ischemic hind limb model. *The Journal of Clinical Investigation* 93: 662–670.

Tandle, A.; Blazer, D. G., and Libutti, S. K. 2004. Antiangiogenic gene therapy of cancer: Recent developments. *Journal of Translational Medicine* 2: 1–20.

Taylor, S. and Folkman, J. 1982. Protamine is an inhibitor of angiogenesis. *Nature* 297: 307–312.

Terman, B. I.; Dougher-Vermazen, M.; Carrion, M. E. et al. 1992. Identification of the KDR tyrosine kinase as a receptor for vascular endothelial cell growth factor. *Biochemical and Biophysical Research Communications* 187: 1579–1586.

Thoenen, H. and Sendtner, M. 2002. Neurotrophins: From enthusiastic expectations through sobering experiences to rational therapeutic approaches. *Nature Neuroscience* 5: 1046–1050.

Tischer, E.; Gospodarowicz, D.; Mitchell, R. et al. 1989. Vascular endothelial growth factor: A new member of the platelet-derived growth factor gene family. *Biochemical and Biophysical Research Communications* 165: 1198–1206.

Tischer, E.; Mitchell, R.; Hartman, T. et al. 1991. The human gene for vascular endothelial growth factor. Multiple protein forms are encoded through alternative exon splicing. *Journal of Biological Chemistry* 266: 11947–11954.

Tomalia, D. A.; Baker, H.; Dewald, J. et al. 1985. A new class of polymers: Starburst-dendritic macromolecules. *Polymer Journal* 17: 117–132.

Uchida, S.; Sakai, A.; Kudo, H. et al. 2003. Vascular endothelial growth factor is expressed along with its receptors during the healing process of bone and bone marrow after drill-hole injury in rats. *Bone* 32: 491–501.

Vincent, L.; Varet, J.; Pille, J.-Y. et al. 2003. Efficacy of dendrimer-mediated angiostatin and TIMP-2 gene delivery on inhibition of tumor growth and angiogenesis: In vitro and in vivo studies. *International Journal of Cancer* 105: 419–429.

de Vries, C.; Escobedo, J. A.; Ueno, H. et al. 1992. The fms-like tyrosine kinase, a receptor for vascular endothelial growth factor. *Science* 255: 989–991.

Wall, S. T.; Saha, K.; Ashton, R. S. et al. 2008. Multivalency of sonic hedgehog conjugated to linear polymer chains modulates protein potency. *Bioconjugate Chemistry* 19: 806–812.

Watson, J. B.; Getzler, S. B., and Mosher, D. F. 1994. Platelet factor 4 modulates the mitogenic activity of basic fibroblast growth factor. *Journal of Clinical Investigation* 94: 261–268.

West, J. L. and Hubbell, J. A. 1996. Separation of the arterial wall from blood contact using hydrogel barriers reduces intimal thickening after balloon injury in the rat: The roles of medial and luminal factors in arterial healing. *Proceedings of the National Academy of Sciences of the United States of America* 93: 13188–13193.

Wester, H.-J. and Kessler, H. 2005. Molecular targeting with peptides or peptide–polymer conjugates: Just a question of size? *Journal of Nuclear Medicine* 46: 1940–1945.

Witmer, A. N.; Vrensen, G. F. J. M.; Van Noorden, C. J. F., and Schlingemann, R. O. 2003. Vascular endothelial growth factors and angiogenesis in eye disease. *Progress in Retinal and Eye Research* 22: 1–29.

Woerly, S.; Pinet, E.; de Robertis, L.; Van Diep, D., and Bousmina, M. 2001. Spinal cord repair with PHPMA hydrogel containing RGD peptides (NeuroGel™). *Biomaterials* 22: 1095–1111.

Wolfert, M. A. and Seymour, L. W. 1996. Atomic force microscopic analysis of the influence of the molecular weight of poly(L)lysine on the size of polyelectrolyte complexes formed with DNA. *Gene Therapy* 3: 269–273.

Woolard, J.; Wang, W.-Y.; Bevan, H. S. et al. 2004. VEGF$_{165b}$, an inhibitory vascular endothelial growth factor splice variant: Mechanism of action, in vivo effect on angiogenesis and endogenous protein expression. *Cancer Research* 64: 7822–7835.

Yang, W.; Barth, R. F.; Adams, D. M., and Soloway, A. H. 1997. Intratumoral delivery of boronated epidermal growth factor for neutron capture therapy of brain tumors. *Cancer Research* 57: 4333–4339.

Yang, Y.; De Laporte, L.; Rives, C. B. et al. 2005b. Neurotrophin releasing single and multiple lumen nerve conduits. *Journal of Controlled Release* 104: 433–446.

Yang, M.; Ma, Q.-J.; Dang, G.-T. et al. 2005a. Adeno-associated virus-mediated bone morphogenetic protein-7 gene transfer induces C2C12 cell differentiation into osteoblast lineage cells. *Acta Pharmacologica Sinica* 26: 963–968.

Yockman, J. W.; Choi, D.; Whitten, M. G. et al. 2008. Polymeric gene delivery of ischemia-inducible VEGF significantly attenuates infarct size and apoptosis following myocardial infarct. *Gene Therapy* 16: 127–135.

Yokoyama, M.; Seo, T.; Park, T. et al. 2007. Effects of lipoprotein lipase and statins on cholesterol uptake into heart and skeletal muscle. *Journal of Lipid Research* 48: 646–655.

Yu, L.; Nielsen, M.; Han, S.-O., and Wan Kim, S. 2001. TerplexDNA gene carrier system targeting artery wall cells. *Journal of Controlled Release* 72: 179–189.

Zhaofei, L.; Fan, W., and Xiaoyuan, C. 2008. Integrin $\alpha_v\beta_3$-targeted cancer therapy. *Drug Development Research* 69: 329–339.

Zisch, A. H.; Lutolf, M. P.; Ehrbar, M. et al. 2003. Cell-demanded release of VEGF from synthetic, biointeractive cell-ingrowth matrices for vascularized tissue growth. *The FASEB Journal* 17: 2260–2262.

26 Bioceramics for Development of Bioartificial Liver

Tomokazu Matsuura and Mamoru Aizawa

CONTENTS

26.1 INTRODUCTION

The cell culture carrier (scaffold) for loading cells into a column is the most important biomaterial for the construction of a bioartificial liver (BAL). Different kinds of materials are used for the construction of BALs of the extracorporeal and implant types. First, for the cases of a BAL of the extracorporeal type, the column volume necessary for human extracorporeal circulation is believed to be 100 mL at the minimum. BAL also has the role of a bioreactor for substance production, and should be available for the production of plasma proteins, physiologically active trace substances such as G-CSF and erythropoietin, and vaccines. The industrial bioreactor for these purposes requires liters of column volume, thus the scaffold used for filling the column should be rigid and have sufficient strength to withhold the weight of the column volume. In addition, as a matter of course, the scaffold should be porous in order to allow attachment of the cells over an increased area and flow passage for cell survival. In this chapter we explain the use of porous hydroxyapatite beads (APACERAM, HOYA PENTAX) [1]. Next, for aBAL of the implant type, the material should be safe when implanted into the living body, and ideally a dissoluble material would be preferable for exogenous construction of the liver organoid. We describe a hydroxyapatite fiber scaffold (AFS) in detail as a material that fulfills the aforementioned requirements [2,3]. We also describe the extracorporeal circulation experiments conducted using the extracorporeal BAL constructed with the radial-flow bioreactor (RFB) filled with porous hydroxyapatite beads, as well as with the liver organoid produced by co-culturing AFS with immortalized hepatocytes, hepatic stellate cells, and sinusoidal endothelial cells [4–6].

26.2   BIOMATERIALS FOR THE DEVELOPMENT OF BIOARTIFICIAL LIVER

26.2.1   GENERAL INFORMATION ABOUT BIOMATERIALS

26.2.1.1   Microcarrier Packed in Bioartificial Liver Module

The best reference model for constructing BAL is, of course, the actual liver tissue. However, unfortunately, the 3D architecture of the liver is not yet fully understood. A helpful model is the concept of "primary lobule" of the liver, which was proposed in reference to the angioarchitectonics of the liver by Matsumoto et al. [7] (Figure 26.1). In this model, the blood from the peripheral portal vein around the lobule flows radially into the central vein. In the lobule, the blood flows slowly in the peripheral region, but different streams converge approximately at 1/3 distance from the periphery of the lobule, and then flow into the central region. The radius of the lobule is about 200 μm, and seven to eight of these primary lobules cluster together to form the second lobule, and further assemblage of these secondary lobules, in turn, builds up the 3D architecture of the liver. This sophisticated organization of the liver is reasonable also in terms of the fluid dynamics. In order to maintain a high density of cells, nutrients and oxygen have to be evenly supplied to them; however, the migration length of these substances by diffusion is considered to be limited to 200 nm. While

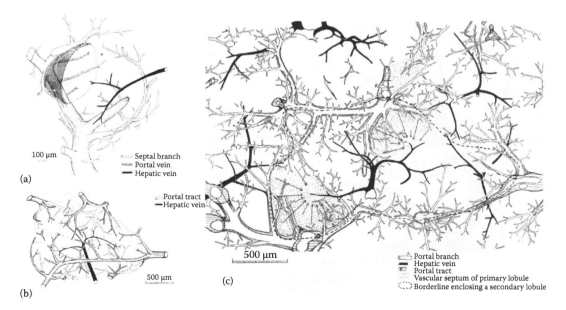

FIGURE 26.1   Matsumoto's primary lobule. The radius of the lobules in the liver about 200 μm (a). Seven to eight of these primary lobules cluster together to form the second lobule (b). Secondary (classical) lobule is constituted with primary lobules (c). (From Matsumoto, T. et al., *Jikei. Med. J.*, 26, 1, 1979. With permission.)

the sinusoid structure of the hepatic lobules meets this condition, the nutrients and oxygen are first consumed at the inflow sites of the blood at the periphery of hepatic lobule, and the concentrations of the nutrients and partial pressure of oxygen in the blood gradually decrease as the blood reaches the central efflux site of the lobule. This unevenness is corrected by the radial structure of the lobule, because it allows the blood to flow faster as it comes closer to the central part of the lobule, that is, the flow volume per unit time compensates for the decrease in the amount of nutrients and oxygen delivered, thereby allowing the cells at the central part also to be maintained.

The ideal method of constructing a BAL would be to reproduce the unique lobular structure of the liver. In fact, a study is underway in which an attempt is being made to construct a BAL by dissolving the cellular components of the liver from a dead body to obtain only the matrix and then culturing fresh cells on it to obtain the BAL [8]. When the liver is subjected to staining by the silver impregnation method, the frame architecture forming the lobule becomes apparent (Figure 26.2). Construction of a 3D BAL requires cells to be arranged on a 3D frame and maintained as a tissue by circumfusion culture, and further for the artificial liver to exert hepatic functions. Therefore, construction of a BAL requires suitable scaffolds for the respective bioreactors. Conversely, development of a suitable bioreactor for the respective scaffolds is necessary.

We mainly use porous beads as scaffold, but depending on the purpose or application of the study, further improvements of the material and configuration will be needed. The advantages of a porous material are (1) a scaffold with pores of approximately 100–200 μm in diameter can provide a larger cell attachment area for a defined volume of the bioreactor, enabling high-density and

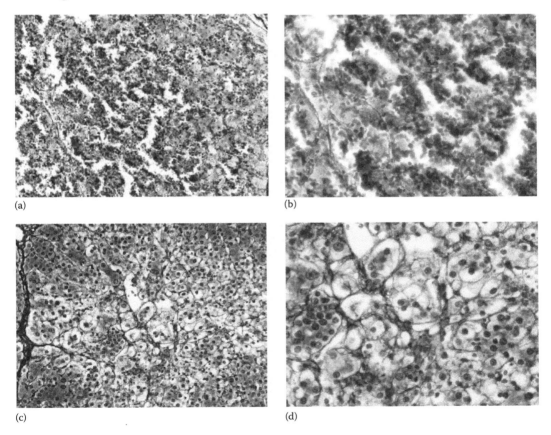

(a)　　　　　　　　　　　(b)

(c)　　　　　　　　　　　(d)

FIGURE 26.2 (**See companion CD for color figure.**) Silver staining of pig livers on fulminant hepatic failure model. Even after severe acute liver damage in pig liver by toxins, silver-stained collagen fiber remained as frame (a, b). Regenerative liver cells were reconstructed along frame (c, d).

large-scale culture; (2) circumfusion culture in a bioreactor allows uniform cultivation of cells on the outer and inner surfaces of the beads and pores, respectively; (3) small pores with a diameter of several dozen micrometers enhance permeation of the perfusion solution into the attached cells. Rigid porous beads are convenient for culturing a large amount of cells in a very high dense state using a large-size bioreactor, and are the appropriate filling carrier for the extracorporeal type of BAL and for a substance-producing bioreactor. On the other hand, soft beads are not suitable for a large column because the beads in the lower layers of the bioreactor will be squashed owing to the force of gravity, which would result in incomplete perfusion. However, soft beads would be a convenient filling for a small bioreactor for use in morphological observational research, and thus might be worth using. When we construct a BAL with RFB, we use porous hydroxyapatite beads or porous cellulose beads, and AFS as the filling carriers.

26.2.1.2 Biomaterials for Filling Microcarriers

In general, biomaterials have been used in a human body. The definition will be "a biomaterials is a nonviable material used in a medical device, intended to interact with biological systems" according to Williams [9]. The biomaterials were classified into three kinds of materials: polymer, metal, and ceramics. In other case, the biomaterials may be consisting of the composite of polymer, metal, and ceramics. Polymers as biomaterials are well applied to replace soft tissues, while metals and ceramics are mainly used to substitute with hard tissue, such as bone.

In the development of BAL, scaffolds for cells are needed. The scaffolds play a role of a matrix for which hepatocytes attached to proliferate. Generally, such scaffolds are prepared from polymer materials. For example, Funatsu and Nakazawa [10] have used polyurethane foam in their BAL development. Akaike et al. [11] have used the polymer with side chain of galactose, PVLA (poly[N-p-vinylbenzyl-4-O-β-D-galactopyranosyl-D gluconamide]). The PVLA can specifically recognize hepatocytes through asialoglycoprotein receptor (ASGPR). In addition to those described earlier, the BAL system developed in the United States, "HeaptAssist," was used as hollow fibers for dialysis of a renal disease as a matrix for cells [12]. Recently, we have reported that novel bioceramics with biocompatibility can be applied as the carrier of the BAL [13]. The bioceramics will be described in the next section.

26.2.2 Bioceramics

There are bioceramics in a kind of biomaterials; this is based on ceramics, i.e., inorganic material. In general, comparing biocompatibility to tissues with three kinds of biomaterials, the order is of ceramics > polymer, metal. It is well known that bioceramics has the most excellent biocompatibility to both soft and hard tissue. The bioceramics are classified into three kinds: (1) bioactive ceramics, (2) bioinert ceramics, and (3) biodegradable ceramics.

One of the most famous bioactive ceramics will be a hydroxyapatite ($Ca_{10}(PO_4)_6(OH)_2$; HAp) [14]. When the HAp ceramics are implanted into host's hard tissue, newly formed bone is chemically and directly bonded with the biological apatite formed on the surface of the HAp ceramics. This phenomenon is defined as "bioactivity" in the bioceramics field. On the other hand, bioinert ceramics is very stable in a body, and has high strength for clinical application. There are alumina (α-Al_2O_3) and tetragonal zirconia (t-ZrO_2) ceramics in typical bioinert ceramics. In addition to these, one of the most important biodegradable ceramics will be a β-tricalcium phosphate (β-$Ca_3(PO_4)_2$; β-TCP) [15]. When this β-TCP ceramics are implanted into host's hard tissue, the β-TCP ceramics are finally replaced with newly formed. Porous β-TCP ceramics with 60% of porosity, Osferion, are manufactured from Japanese company, Olympus Terumo Biomaterial (OTB).

The aforementioned bioceramics, such as HAp and β-TCP ceramics, are being applied with various material shapes: (1) dense ceramics, (2) porous ceramics, (3) cement (paste-like artificial bone), and (4) granules (beads). Among these shapes, porous ceramics is well used as artificial bone graftings, together with scaffolds for tissue engineering of hard tissue. The shapes of beads (or granule) are also used as carriers for hepatocytes of the BAL. We will show (1) apatite beads and (2) apatite-fiber scaffolds (AFS) created by ourselves which can be used in the carriers of the BAL system as follows.

26.2.2.1 Porous Hydroxyapatite Beads

Hydroxyapatite beads are used for clinical application as a filling material for defects of the bone in the fields of orthopedics, plastic surgery, cerebral surgery, and dentistry. The size of the porous hydroxyapatite beads (PENTAX, Tokyo, Japan) used by us ranges from 800 to 1000 μm in diameter, and contains an adequate number of pores that are at least 200 μm in diameter. The porosity of the beads is 85% and multiple pores measuring 50–200 μm in diameter are linked together inside the beads (Figure 26.3). The beads have sufficient strength to be used in a bioreactor with a volume of approximately 100 mL at the minimum.

Porous hydroxyapatite beads can be used for the construction of an extracorporeal BAL that might require scaling-up. For developing industrial BALs that produce plasma proteins, etc., further scaling-up up to 5 L is needed; therefore, the strength and other relevant properties of porous hydroxyapatite beads should be investigated.

26.2.2.2 Apatite Fiber Scaffold

One of the bioactive ceramics, HAp, has been widely applied as a biomaterial and as an adsorbent for chromatography [16]. Figure 26.4 illustrates scheme of crystal growth of the HAp crystal. By controlling the morphology of HAp crystals, novel properties may be produced by enabling controlled orientation of the crystal planes, as HAp crystal has two crystal planes with different charges: positive on the $a(b)$-planes and negative on c-planes [17].

We have successfully synthesized apatite fiber with long axis size of 60–100 μm [18], as shown in Figure 26.5a. It was confirmed from the results of a high-resolution transmission electron microscopy (HR-TEM) using a shadow imaging technique that the apatite fibers were of single crystals with the c-axis orientation parallel to the long axis of the fiber. Selected area electron diffractions

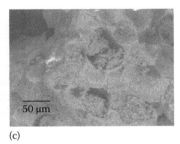

(a)   (b)   (c)

FIGURE 26.3 (See companion CD for color figure.) Commercially available apatite beads with high porosity: (a) overview, (b, c) particle morphology (SEM image).

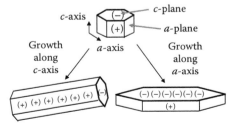

FIGURE 26.4 (See companion CD for color figure.) Scheme of crystal growth of apatite crystal.

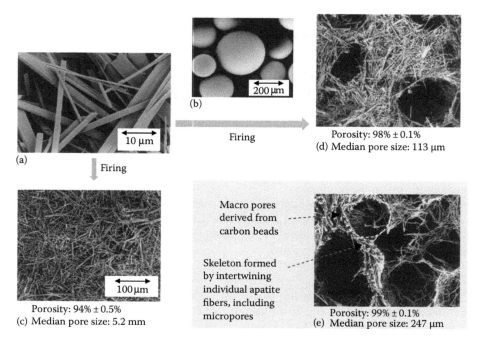

FIGURE 26.5 **(See companion CD for color figure.)** Fabrication of apatite-fiber scaffolds (AFS) and their microstructure: (a) apatite fiber, (b) carbon beads, (c) AFS0, (d) AFS1000, and (e) AFS2000.

(SAED) were performed at five points along the long axis of the apatite fiber. The diffraction pattern from the five points showed clear spots corresponding to the apatite structure with high crystallinity. As the five diffraction patterns showed the identical features along the long axis of the fiber, we have concluded that the apatite fibers were not polycrystalline, but single crystals. The HR-TEM observations combined with the XRD results suggest that the single crystal apatite fibers may grow along the c-axis to develop the $a(b)$-plane of the hexagonal HAp. Thus, the present apatite fibers have a positive charge on the surface.

We have developed novel scaffolds for tissue engineering, apatite-fiber scaffolds (AFS) [19,20]. The AFSs are fabricated as follows. Firstly, the HAp fibers as shown in Figure 26.3a were suspended with spherical carbon beads (Nika beads; Nihon Carbon Company, as shown in Figure 26.5b with a diameter of ~150 μm in the mixed solvent (ethanol/water = 1/1 (v/v)). The carbon beads are added to the HAp fiber in the following carbon/HAp (w/w) ratios: 20/1, 10/1, 1/1, and 0/1. The green compacts for the scaffold are fabricated by pouring and vacuum pumping the aforementioned mixed suspension containing ~1 mass% of the HAp (5 cm^3) into the vinyl chloride mold of 16.5 mm in internal diameter. The resulting compacts are fired at 1300°C for 5 h in a steam atmosphere to develop the structure of the scaffold. Hereafter, we name the scaffolds derived from carbon/HAp = 20/1, 10/1, and 0/1 (w/w) "AFS2000," "AFS1000," and "AFS0," respectively.

Figure 26.5c–e shows microstructure of the resulting AFS0, AFS1000, and AFS2000, respectively. The SEM observation shows that the AFS2000 was composed of large pores of 100–300 μm in diameter and smaller pores formed by intertwining of individual fibers and that the pores were interconnected in the structure. As compared with microstructure of the carbon-free derived scaffold (AFS0), the pore sizes of the scaffolds, AFS1000 and AFS2000, were significantly enlarged with increasing amounts of the carbon beads added.

The pore-size distribution of the scaffolds is measured using a mercury porosimeter. In the case of the AFS0 scaffold, the pore size distribution is very limited, in the range of about 5 μm. On the other hand, the pore size distribution of the AFS2000 scaffold shifts toward larger pore sizes, that is, in

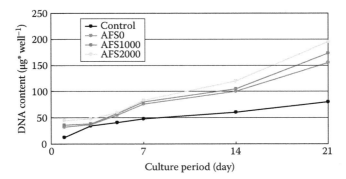

FIGURE 26.6 (See companion CD for color figure.) Proliferation of FLC-4 cells cultured in/on the AFS on the basis of DNA determination. Cell suspension (0.25 mL) with density of 5.0×10^5 (/cm^3) was seeded on Control (48 well plate for cell culture, polystyrene), AFS0 (7.5 mmf × 1 mm), AF1000 (7.5 mmf × 2.5 mm, and AFS2000 (7.5 mmf × 3.5 mm).

the range of 100–500 μm. In fact, the median pore sizes of the AFS0, AFS1000, and AFS2000 scaffolds are of 5.2, 112.8, and 247.3 μm, respectively. These large pores are generated by releasing the carbon beads during firing. The AFS1000 and AFS2000 scaffolds developed using the carbon beads have a large pore size and high porosity, as compared to the carbon-free derived scaffold (AFS0). In particular, the AFS2000 scaffold may contain pores of dimensions suitable for cell in-growth.

The present AFSs have been applied as a scaffold for tissue engineering of bone. The AFSs are biologically evaluated using two kinds of cells, osteoblastic cell (MC3T3-E1) and rat bone marrow cells [19]. In both cases, the cells cultured on/in the scaffolds show excellent cellular responses, such as good initial cell-attachment efficiency, good cell proliferation, and enhanced differentiation into osteoblasts. The AFS2000 scaffold is particularly able to support 3D cell proliferation. The bone marrow cells cultured in the scaffold are calcified creating a bone-like structure. We make it clear that the present AFSs promote bone formation in vivo using rat model [20,21]. Especially, the AFS2000 scaffolds with high porosity and large pore size may be effective as the matrix for tissue engineering leading to bone regeneration.

In addition to these, the AFSs can be used for 3D cell culture of hepatocyte. In order to apply the aforementioned AFS to the regeneration of real organs (e.g., liver), we have performed the culture of FLC-4 cells [22] as a model of hepatocyte. Figure 26.6 illustrates proliferation of the FLC-4 cells cultured in/on the AFSs on the basis of DNA determination. The cells are well-proliferated in/on all the specimens examined; the cells in the AFS group are much grown than that on the control. Especially, the cells in AFS2000 proliferate in a similar manner to those on the AFS0 and AFS1000 from 1 day up to 7 days and more efficiently after that to 21 days. This result of good cell proliferation may be due to the 3D structure of the AFS2000. Figure 26.7 shows the histological observation of the AFS2000 cultured with FLC-4 cells for 10 days. We can see that the cells align along the skeleton of macropores in the AFS2000. Still after cell culture for a long time, the macropores are maintained inside of the AFSs. These macropores are effective for supplying the medium to the cells grown in the AFSs.

The present AFS can be easily shaped to be cartridge type. The AFS with the shape of cartridge type is settled into a bioreactor in a RFB system. Figure 26.8 illustrates overview of the RFB system settled with the AFS2000. As the FLC-4 cells are 3Dcultured using the RFB and the AFS, the tissue-engineered liver organoid is constructed as described later.

26.2.3 Porous Cellulose Beads

Porous cellulose beads are soft and were originally developed for efficiently culturing a large amount of adherent cells in a high-dense state in a suspended cell culture system. We have been performing

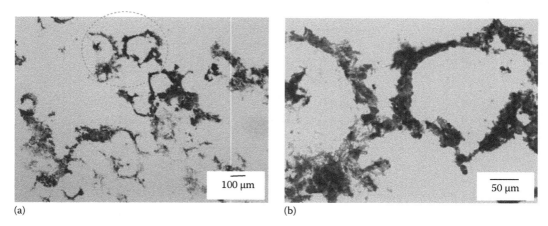

(a)

(b)

FIGURE 26.7 (See companion CD for color figure.) Histological observation of the AFS1000 cultured FLC-4 cells for 10 days (cross-section; hematoxylin-eosin stain): (a) low magnification, (b) high magnification of pink region as shown in dotted circle.

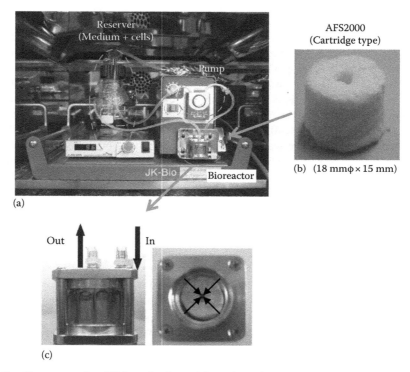

FIGURE 26.8 (See companion CD for color figure.) Overview of a simple radial flow bioreactor (RFB) system settled the AFS2000. Five milliliter volume of RFB (a). AFS2000 column for RFB (b). AFS column in RFB (c).

high-density circumfusion culture using RFB filled with the microcarrier for industrial production, supplied by Asahi Kasei Medical Co. Ltd. This carrier originally consists of macroporous cellulose microspheres for suspended cell culture, thus the particles are approximately 200 μm in diameter and the effective surface area in the dry state is 180 m²/g. The surface of the beads is processed to bear a cationic charge for facilitating cell attachment (Figure 26.9).

Porous cellulose beads are soft and prone to being squashed when used as a scaffold for RFB, which would make scaling-up of the bioreactor impossible. However, porous cellulose beads are

(a) (b) (c)

FIGURE 26.9 **(See companion CD for color figure.)** Cellulose beads with high porosity: (a) overview, (b, c) particle morphology (SEM image).

suitable for use as the scaffold for RFB of 5 mL volume for experimental use. In particular, they are suitable for morphological observations due to their softness, thus they allow easy access to histological observation, immunostaining, and electron microscopic observation. We could culture the cells of Huh-7, a human hepatocellular carcinoma cell line, three-dimensionally for 2 months in RFB filled with porous cellulose beads, whose particle diameter was adjusted to approximately 1 mm (Figure 26.10).

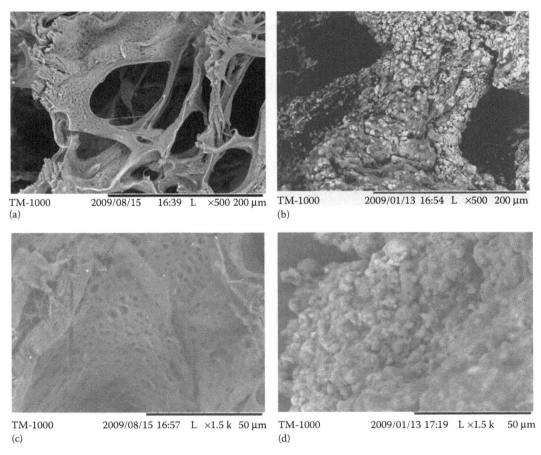

TM-1000   2009/08/15   16:39 L   ×500 200 μm
(a)

TM-1000   2009/01/13 16:54 L   ×500 200 μm
(b)

TM-1000   2009/08/15 16:57 L ×1.5 k 50 μm
(c)

TM-1000   2009/01/13 17:19 L ×1.5 k   50 μm
(d)

FIGURE 26.10 Huh-7 human hepatocellular carcinoma cells were cultured on the surface of and in the hole of cellulose beads in RFB. SEM image of cellulose beads (a, c). SEM image of Huh-7 cells on cellulose beads (b, d).

26.3   BIOREACTORS FOR BIOARTIFICIAL LIVER

26.3.1   General Information About Bioreactors

At the present time, the hollow-fiber type of module is mainly used for the construction of a BAL of the extracorporeal type (e.g., ELAD, HepatAssist, BLSS, MELS) [23]. Membrane materials such as cellulose acetate-polysulfone, cellulose nitrate, polyamide, and polyether sulfone are used most often. In this type of module, cells are attached around the hollow fiber and the whole blood is refluxed in the lumen. This type of bioreactor is more useful for realizing the function of extracorporeal circulation rather than the biological functions; since the cells are 3D cultured in a high-dense state, it is uncertain whether the original morphology and function of the liver can be reproduced with this type of module.

In order to design a bioreactor in which the liver structure is assembled exogenously and the liver function is reproduced, the system used should allow highly dense cell culture, as well as the possibility of both scaling-up and scaling-down of the culture. In order to scale up highly dense cell culture, the circumfusion culture method is required for supplying cells with sufficient amounts of oxygen and nutrients. In addition, the unevenness of the concentrations of oxygen and nutrients between the inflow and the efflux sites in the bioreactor should be improved; that is, even if the concentrations of oxygen and nutrients are maintained at high levels at the inflow site of the media, they are rapidly consumed when the media flows through the bioreactor, as the cells are cultured in a high-dense state; as a result, the cells at the efflux site cannot receive sufficient oxygen and nutrients. Moreover, the uneven distribution of the cells in the bioreactor caused by gravity should also be corrected.

The rotary cell culture system (RCCS) developed by NASA has succeeded in correcting the aforementioned uneven distribution of the cells due to gravity during the construction of a small organoid [24]. However, it is impossible to sufficiently scale-up this system for constructing an extracorporeal type of BAL.

26.3.2   Radial-Flow Bioreactor

We have developed a BAL using the RFB (ABLE-BIOTT Co., Ltd., Tokyo, Japan) [4–6,25–32]. RFB is a cylindrical, packed-column-type bioreactor filled with a cell culture carrier and used for highly dense cell culture (Figure 26.11). The culture media is radially refluxed from the inner periphery of the cylinder toward the central part, which eliminates the uneven supply of nutrients and oxygen, enabling highly dense cell culture.

(a)   (b)

FIGURE 26.11   (See companion CD for color figure.) Overview of a fully equipped radial-flow bioreactor (RFB) system settled hydroxyapatite beads. Fifteen milliliter volume of RFB (a). Fully equipped RFB system (b).

26.4 APPLICATIONS

26.4.1 EXTRACORPOREAL-TYPE BIOARTIFICIAL LIVER

26.4.1.1 Background

Availability of a BAL is hoped to contribute to intensive care of patient with acute hepatic failure as a bridge to sustain life until liver transplantation or until recovery from hepatic failure. In acute hepatic failure, irreversible brain damage is a particular threat to survival. Hepatic encephalopathy is associated with ammonia, cytokines such as tumor necrosis factor (TNF-α), endogenous benzodiazepine-like substances, and postulated but unidentified hepatic coma agents increased in blood of acute hepatic failure patients [33]. Established blood purification therapies using extracorporeal circulation, plasmapheresis, and hemodialysis therapy, have been applied to removal of toxins causing hepatic coma [34]. However, plasmapheresis sometimes can exacerbate brain edema, and the patient could develop progressive brain death despite hemodialysis therapy. Given limited success with conventional extracorporeal circulation therapies, a more physiologic treatment such as the BAL is desired.

A clinically effective BAL requires development of a high-density cell culture module and a highly functioning liver cell line. Preclinical investigation in appropriate animal models of human hepatic failure then is necessary. Furthermore, we need to learn more about the mechanism that induces hepatic encephalopathy, brain edema, and brain death in acute hepatic failure, and how BAL therapy can ameliorate this encephalopathy. We sought to develop a high-performance BAL to avoid lethal hepatic encephalopathy in acute hepatic insufficiency and establish the BAL as an extracorporeal circulation therapy able to surpass conventional blood purification procedures. Our extracorporeal BAL support system used a highly functional human hepatocellular carcinoma (HCC) cell line (FLC-4) cultured in a RFB [4]. One important problem when cells are cultured densely is the delivery of sufficient oxygen and nutrients even when these are plentiful at the inflow site. In our RFB, adequate delivery was accomplished by monitoring and adjusting the flow rate at the center of the outflow site. As a result, we could incubate cells successfully at a density of 10^8/mL.

We presently tested the BAL in mini-pigs with acute hepatic failure induced by α-amanitin, a mushroom-derived poison, while monitoring the electroencephalogram (EEG) to assess the effectiveness of BAL against hepatic encephalopathy.

26.4.1.2 Materials and Methods

26.4.1.2.1 Mini-Pigs and Monitoring

Male mini-pigs (CSK-MS) weighing 10–15 kg were a kind gift from Chugai Pharmaceutical (Tokyo, Japan). Prior to experiments, they were maintained for 1–4 weeks at Laboratory Animal Facilities of the Jikei University School of Medicine, receiving a standard chow. The study was approved by the institution's committee concerning animal experimentation.

Pigs were given inhalation anesthesia 3%–4% isoflurane during bilateral implantation of six stainless steel epidural electrodes at the frontal poles, central regions, and parietal regions. In addition, we fixed electrocardiographic (ECG) electrodes to the right and left chest wall and girdled the animal with a belt incorporating a transducer to detect respiratory movements. We placed the pig in a hammock and recorded EEGs at the time of awakening, using a bipolar lead array. We simultaneously recorded the ECG, and pneumogram. The EEG recording was subjected to power spectrum analysis.

26.4.1.2.2 Acute Hepatic Failure Model

During inhalation anesthesia with 3%–4% isoflurane, 0.05 mg/kg of α-amanitin (Calbiochem, Darmstadt, Germany), and 1 µg/kg of lipopolysaccharide (LPS; Sigma, St. Louis, MO) dissolved in 10 mL of saline was administered via the splenic vein. For maintaining a secure access route and for extracorporeal circulation, we cannulated a cervical artery and a cervical vein. The animal was restrained in a hammock again while recording EEG, ECG, and pneumogram. We also periodically collected blood for hematologic, biochemical, and coagulation testing. A saline solution with

5% glucose was given by intravenous drip to animals with hepatic failure. When needed a 50% glucose solution and 7% sodium bicarbonate were injected as venous blood glucose and arterial blood acid–base parameters were monitored.

26.4.1.2.3 Human Hepatocellular Carcinoma Cell (HCC) Line

A human HCC cell line, FLC-4, which we established at the Jikei University School of Medicine, was used as a functioning liver cell culture in the BAL [29]. FLC-4 cells were maintained in serum-free ASF 104 medium (Ajinomoto, Tokyo, Japan) at 37°C in an atmosphere of 95% air and 5% CO_2. Cells were separated for using a trypsin solution (25 UPS units/mL; Difco Laboratories, Detroit, MI) including 0.02% EDTA.

26.4.1.2.4 Bioartificial Liver Using the Radial-Flow Bioreactor

The RFB (Biott, Tokyo, Japan) is a cell filling type bioreactor of 15 mL capacity that filled up a cylindrical module with porous hydroxyapatite beads (PENTAX, Tokyo, Japan) of diameter approximately 1 mm [25]. Culture apparatus consists of the RFB, a reservoir adjusting culture fluid, and an automatic controller to adjust dissolved oxygen content and pH of the culture fluid. We injected 10^8 of FLC-4 cells into the reservoir, which contained ASF 104 culture medium with 2% fetal bovine serum (FBS). Flow of culture medium was initiated, and cell culture was continued for 10 days after adhesion of cells to porous hydroxyapatite beads in the RFB. We used the RFB nearly filled with cells as a BAL.

26.4.1.2.5 Extracorporeal Circulation Using the Radial-Flow Bioreactor

Arterial blood was extracted at 20–30 mL/min, and plasma was separated at 10–15 mL/min by a plasma separator (Plasmaflo, OP-02W; Asahi Kasei Medical, Tokyo, Japan) and allowed to circulate through the BAL after passage through an oxygenator (silicone rubber tube module M40-3000; Nagayanagi, Tokyo, Japan) (Figure 26.12). The entire device was maintained constantly at 37°C. Together with the separated blood cells, the purified plasma was returned to the animal via the cervical vein. The extracorporeal circulation time in this experiment was 4–6 h. Intravenous treatment with 50% glucose and 7% sodium bicarbonate solution continued after the BAL extracorporeal circulation. Heparin was injected as an anticoagulant. At the initiation of

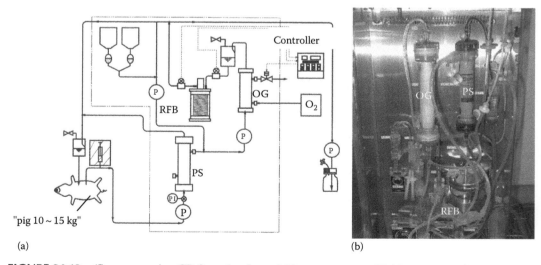

(a)   (b)

FIGURE 26.12 **(See companion CD for color figure.)** The extracorporeal BAL system. (a) Schematic diagram. From blood obtained from an artery, and plasma is separated from cells at 10–15 mL/min by the plasma separator (PS) and flows into the BAL after taking up oxygen from the oxygenator (OG). The entire device is maintained constantly at 37°C. The purified plasma is mixed with blood cells and returned to a vein. (b) Photograph of the system. (From Kanai, H. et al., *Artif. Organs*, 31, 148, 2007. With permission.)

extracorporeal circulation, an intravenous injection of 1000 units was given. Heparin then was infused into the withdrawn arterial blood at 500 unit/h.

26.4.1.2.6 Blood and Biochemical Tests
We collected blood at various time points for blood cell count, prothrombin time, and biochemical tests (aspartate aminotransferase, alanine aminotransferase, total bilirubin, total cholesterol, and ammonia).

26.4.1.2.7 Pathologic Analysis
Pathologic examination was performed in animals killed after experiments as well as in those dying during experiments. Pig livers and brains were fixed in 10% neutral buffered formalin for histological tissue examinations. After plasma perfusion, we also removed the hydroxyapatite beads with FLC-4 cells remaining in 1.4% glutaraldehyde, and performed scanning electron microscopy (SEM) after fixation.

For immunohistochemistry, formalin-fixed, paraffin-embedded tissue sections of the livers and brains were deparaffinized and subjected to microwave oven heating in 10 mM citrate buffer (pH 6.0) for 10 min at 90°C as an antigen-retrieval procedure. After incubation with the primary antibody [anti-Mib-1 IgG or anti-glial fibrillary acidic protein (GFAP) IgG], secondary antibodies were applied, and followed by avidin–viotin peroxydase complex. Color was developed by 3.3′-diaminobenzidine (DAB).

26.4.1.3 Results
26.4.1.3.1 FLC-4 Cells Cultured on Hydroxyapatite Beads
When we observed cells by pre-perfusion SEM, FLC-4 cells were adherent single-layer to the porous hydroxyapatite beads in a cubic 3D array (Figure 26.13). Cell-to-cell adhesion was noted, and microvilli were numerous on the perfused cell surface. Thus cells maintained polarity and an orderly array when cultured in high density. Upon post-perfusion SEM, we observed a fibrin network on the surfaces of FLC-4 cells adherent to the hydroxyapatite beads. However, microvilli on the perfused cell surface were well-preserved.

26.4.1.3.2 Biochemical, EEG, and Neuropathologic Abnormalities in Acute Hepatic Failure
Figure 26.14 demonstrates changes in biochemical tests and EEG during acute hepatic failure, caused by the administration of α-amanitin and LPS to the animal via the splenic vein. In this hepatic failure model, serum transaminase elevation was most prominent for AST. In addition, total protein and total cholesterol decreased. Hyperammonemia was also a typical finding. A normal EEG initially was seen when a mini-pig awakened from anesthesia (fast-wave pattern about 8–15 Hz). Then, the EEG slowed in proportion to progression of hepatic dysfunction. The EEG change reflected cerebral edema, which ultimately was associated with ECG. The EEG became flat as a result of cerebral edema, then breathing ceased, followed later by cardiac arrest. We observed Alzheimer type II astrocytes in the swollen brain of the animal dying 48 h after toxin administration (Figure 26.15).

26.4.1.3.3 Courses of Individual Animals
26.4.1.4 Controls Study (Animals 1 and 2)
We first established extracorporeal circulation through a BAL system without FLC-4 cells in two animals with α-amanitin/LPS-induced hepatic dysfunction. Animal 1 died 2 h after initiation of extracorporeal circulation. Animal 2 also died at a completion point of extracorporeal circulation in time of 6 h (Figure 26.16).

26.4.1.5 BAL Therapy Initiated 12–20 h after Injection of α-Amanitin and LPS (Animals 3–5)
We thus used the BAL system with FLC-4 cells to treat hepatic dysfunction in three animals, obtaining EEG improvement and survival in all (Figure 26.17).

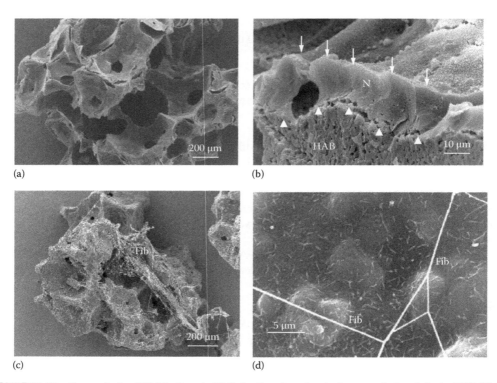

(a)

(b)

(c)

(d)

FIGURE 26.13 Pre-perfusion SEM findings in FLC-4 cells cultured on hydroxyapatite beads in the RFB. FLC-4 cells form a single layer on porous hydroxyapatite beads in a cubic array (a, b). A fibrin network is observed on surfaces of FLC-4 cells after perfusion with plasma (c, d). FLC-4 cells remain attached to the beads, and microvilli are intact on the perfused aspects of cells. (From Kanai, H. et al., *Artif. Organs*, 31, 148, 2007. With permission.)

	0 h	13 h	21 h
AST (IU/L)	65	232	15,410
ALT (IU/L)	39	53	606
LDH (IU/L)	759	1350	11,480
T.Bil (mg/dL)	0.2	1.7	2.8
T.Chol (mg/dL)	76	46	16
NH_3 (μg/dL)	90	685	2,990
PT (%)	90	20	11

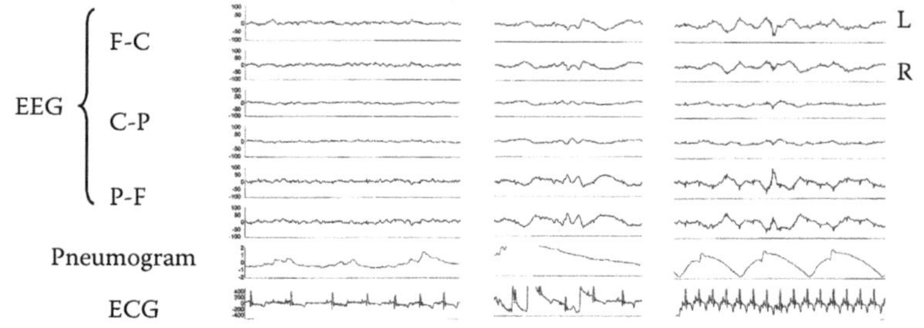

FIGURE 26.14 As hepatic failure progresses blood transaminases increase, particularly AST. Prothrombin time and especially total cholesterol decrease. Hyperammonemia typically is present. A normal EEG at the time of awakening from anesthesia shows 8–15 Hz activity, but the waves show slowing as hepatic dysfunction progresses. As cerebral edema develops, slowing worsens and heart rate increases (ECG), sometimes producing artifact in the EEG. AST, aspartate amino transferase; ALT, alanine amino transferase; LDH, lactate dehydrogenase; T.Bil, total bililbin; T.Chol, total cholesterol; PT, prothrombin time.

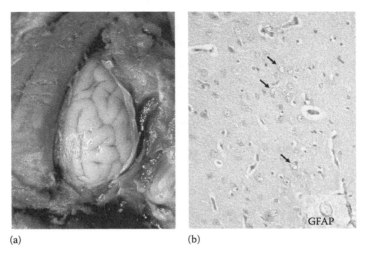

(a)   (b)

FIGURE 26.15   (See companion CD for color figure.) Lethal cerebral edema in a pig with acute hepatic failure. (a) Macroscopically brain swelling is evident. (b) Large Alzheimer type II astrocytes are prominent (arrows) in the brain. These cells show GFAP immunoreactivity with cytoplasmic staining (inset). (b) Hematoxylin and eosin staining.

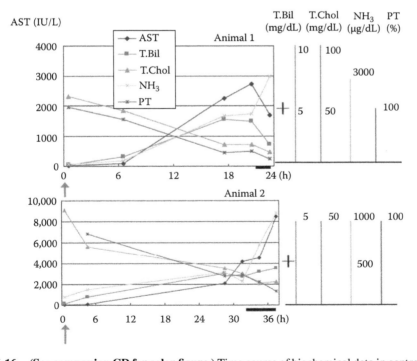

FIGURE 26.16   (See companion CD for color figure.) Time course of biochemical data in controls (animals 1 and 2) developing hepatic failure after administration of α-amanitin and LPS. Plasma was perfused through the RFB *without* FLC-4 cells. Neither animal recovered at any time from fatal hepatic failure. Red arrow indicates time of α-amanitin and LPS administration. Black bar indicates extracorporeal BAL perfusion without cells. (From Matsuura, T. et al. *Clin. Gastroenterol.*, 23, 1821, 2008. With permission.)

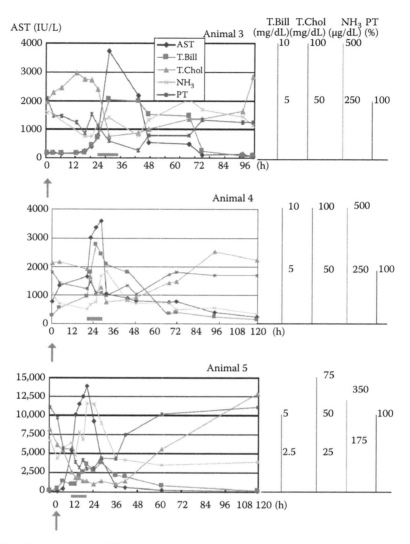

FIGURE 26.17 **(See companion CD for color figure.)** Time course of biochemical data in three animals (animals 3, 4, and 5) developing acute hepatic failure after administration of α-amanitin and LPS. Plasma was perfused through the RFB containing FLC-4 cells beginning 12 or more hours after toxin administration. All three animals survived. Red arrow indicates time of α-amanitin and LPS administration. Green bar indicates the extracorporeal BAL support with FLC-4 cells. (From Matsuura, T. et al. *Clin. Gastroenterol.*, 23, 1821, 2008. With permission.)

26.4.1.6 BAL Therapy Initiated 6 h after Injection of α-Amanitin and LPS (Animal 7)

After α-amanitin and LPS administration, about half of the animals died suddenly within 12 h, before BAL extracorporeal circulation therapy could be initiated (e.g., animal 6). In these instances, even when blood biochemical data were not seriously abnormal, the animals died rapidly with progression of cerebral edema. In animal 7, however, BAL extracorporeal circulation therapy was started 6 h after α-amanitin and LPS administration, when no remarkable biochemical derangement was evident in plasma. The animal died accidentally by pulling out an arterial catheter 21 h after administration of α-amanitin and LPS. Although blood analysis results were worsening prior to death with AST of 19,000 IU/L, and PT of <10% of normal activity, EEG activity had not slowed, and no cerebral edema could be found at autopsy. Body weights of animals 6 and 7 both were 12 kg, but brain weights for animals 6 (hepatic failure death) and 7 (BAL treatment) were 80 g versus 55 g.

26.4.1.6.1 Power Spectrum Analysis of the Electroencephalogram

We assessed EEG changes in all animals by power spectrum analysis. Figure 26.17 (animal 4) is typical for survival with BAL therapy. Normal anterior and posterior rhythms disappeared for 10 h after administration of α-amanitin and LPS, being replaced by slow waves. Normal rhythms reappeared about 7 h after conclusion of the BAL therapy (Figure 26.18). In animal 7, early BAL therapy achieved no improvement of plasma biochemical data, which were not markedly abnormal. However, early EEG slowing did not occur in this animal receiving toxins. This suggested that BAL therapy was neuroprotective in the presence of hepatic dysfunction.

26.4.1.6.2 Pathological Analysis

Intense hemorrhagic necrosis was apparent in the liver of an animal that died of hepatic dysfunction caused by α-amanitin and LPS, to the extent that viable-appearing hepatocytes were difficult to find (Figure 26.19). Lobular arrangement of hepatocytes was fairly well preserved in the liver of an animal surviving with BAL therapy. Two percent of hepatocytes were Mib-1-positive in the normal liver of a 3 months old control animal, while no Mib-1-positive hepatocytes were found in an animal rapidly dying of hepatic dysfunction. On the other hand, the animal surviving with RFB therapy showed 50% of Mib-1 positivity among hepatocytes.

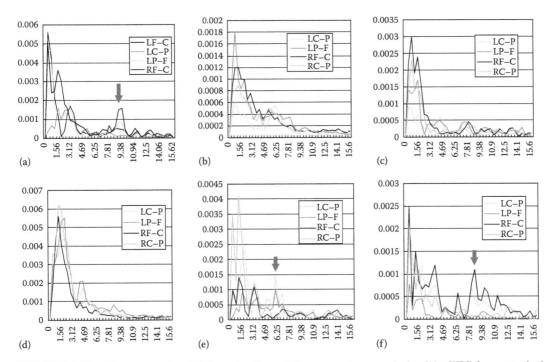

FIGURE 26.18 (See companion CD for color figure.) Power spectrum analysis of the EEG from a typical animal (animal 4) that recovered with BAL therapy. X bar: Frequency (Hz), Y bar: Normalized power spectral density. Normal fast rhythm disappeared anteriorly and posteriorly for 10 h after administration of α-amanitin and LPS, was replaced by slow waves. Fast rhythm reappeared 7 h after conclusion of BAL therapy. (a) Before administration of toxins. (b) At 10.5 h after administration of toxins. (c) At 17.5 h after administration of toxins (5.5 h after extracorporeal circulation therapy initiation). (d) At 19.5 h after administration of toxins (1.5 h after extracorporeal circulation therapy completion). (e) At 25 h after administration of toxins (7.0 h after extracorporeal circulation therapy completion). (f) At 47 h after administration of toxins (29 h after extracorporeal circulation therapy completion).

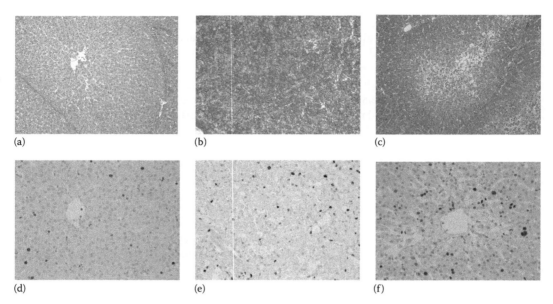

FIGURE 26.19 **(See companion CD for color figure.)** Microscopic findings in the liver of a normal animal (a, d), an animal dying of acute hepatic failure induced by α-amanitin and LPS (b, e), and an animal that recovered with BAL therapy (c, f). (a–c) Masson trichrome staining. (d–f) Mib-1 immunohistochemical staining. In panel (e), most Mib-1-positive cells are macrophages, not hepatocytes. Strongly Mib-1-positive hepatocytes are prominent in panel (f).

26.4.1.7 Discussion

Extracorporeal artificial livers have been developed in Europe and the United States. Liu et al. [35] analyzed 12 published reports of clinical results with such extracorporeal livers. In 2001, Allen et al. [36] summarized results with two types of dialysis and six of biologic models tested since 1990 in Europe and the United States. There were two of these trials advancing to Phase III. In the trial of BAL, a result of clinical test of many institutions mainly on Demetriou was reported recently [37]. They compared results with such treatments to those in a patient group receiving only conventional combined-modality therapy for, BAL-treated patients showed significantly better survival than conventionally treated patients at 30 days. Such a report was the first to suggest that a BAL treatment provided clinically useful life support to patients with acute hepatic failure. Why the BAL therapy was beneficial, however, was not clear. Prevention of brain death from hepatic coma, most important cause of death in acute hepatic failure, is the principal goal of BAL extracorporeal circulation therapy.

Agents potentially responsible for hepatic coma include not only ammonia and manganese compounds but also an assumed unknown substance with molecular weight of 5–20,000 Da [38]. In the brain affected by hepatic encephalopathy, changes in energy metabolism and neural mechanisms related to glutamine, serotonin, catecholamine, γ-aminobutylic acid (GABA), and the endogenous opioid system are thought to occur. In addition, changes occur in synthesis of peripheral-type benzodiazepine receptors and of neurogenic steroids. Increased nitric oxide also affects the onset of hepatic encephalopathy [33]. Astrocytes form the blood brain barrier (BBB), and sustain nerve cells as they function [39,40]. Increases of the postulated hepatic coma agent in blood induce early functional impairment in astrocytes. However, removal of unidentified hepatic coma agents is difficult using conventional blood-purifying treatments. We therefore feel urgency about developing modalities such as the BAL, in which human liver cells (highly functioning HCC cell lines) are cultured at high density in a RFB. The human HCC cell that we chose for BAL therapy (FLC-4) produces large amounts of human albumin and expresses human isoforms of cytochrome P 450 (CYP) [25,35]. In addition, L-type amino acid transporter (LAT) 3, a new organic solute transporter, was cloned from FLC-4 [41].

In the present preclinical study, our ultimate aim was to find a way to prevent or reverse potentially lethal hepatic encephalopathy in acute hepatic failure using BAL as a bridge to either recovery or transplantation. We used an acute hepatic failure model involving a relatively large animal to test the effectiveness of densely cultured FLC-4 cells in a module for extracorporeal circulation. Devising an acute hepatic failure model in a relatively large animal is extremely difficult. We used the method of Takada et al. [42], who induced acute hepatic failure by injecting of α-amanitin and LPS via the portal vein. However, courses in our model were not uniform, with half of the animals dying of fulminant hepatic failure before initiation of extracorporeal circulation therapy. As for deciding when to start BAL extracorporeal circulation therapy, increases of ammonia and transaminase in blood sample were not reliable indicators. Even when transaminase and ammonia concentrations in blood were not markedly elevated, many animals that had been injected with α-amanitin and LPS died of hemorrhagic necrosis of the liver marked cerebral edema. Among blood tests, the best index of hepatic failure was a decrease in cholesterol concentration. We therefore administered BAL for 4–6 h, when the EEG showed slowing and plasma cholesterol decreased, about 20 h after toxin administration. Three animals with acute hepatic failure showed considerable normalization of slow-wave activity in the EEG after extracorporeal circulation therapy using FLC-4 cultured in the RFB, with ultimate survival.

The reason for survival of animals in acute hepatic failure treated with the BAL is thought to be prevention of rapidly progressive cerebral edema. The remarkable capacity for hepatic regeneration in mini-pigs 3–4 months old helped them to recover from hemorrhagic hepatic necrosis resulting from the single toxin administration. Only one pig (animal 7) received early BAL therapy, at 6 h after administration of toxins. This animal did not show recovery on plasma biochemical data or liver pathology at 21 h after administration of toxins, when dislodgment of an arterial catheter caused death. However, no slow waves were observed, and brain edema was not detected at autopsy. Therefore, BAL therapy appeared to protect against brain edema despite hepatic dysfunction. Results of this experiment suggest that BAL can ameliorate hepatic encephalopathy by removal of suspected and/or unknown hepatic coma agents.

We next examined whether BAL removed agents injurious to astrocytes from plasma. Damage to cerebral astrocytes is considered an early event leading to cerebral edema in hepatic encephalopathy. We measured plasma concentration of S-100 β protein, which reflect brain astrocyte damage, in plasma from animals with hepatic failure [4]. Just prior to death, those concentrations were extremely high. On the other hand, surviving animals receiving BAL treatment showed attenuated elevations of S-100 β protein during BAL therapy. This result suggested that brain damage in hepatic encephalopathy was inhibited by BAL treatment.

26.4.2 Implant-Type Bioartificial Liver

In order to construct the implant-type BAL, a cylindrical AFS cartridge was prepared and filled into an RFB of 5 mL volume (Figure 26.8). Then, we investigated whether a liver organoid could be produced exogenously by co-culturing AFS with the hepatocytes, stellate cells, and endothelial cells. Mouse immortal cells were used for co-culturing, and a liver organoid with capillary vessel-like structure was successfully prepared (Figures 26.20 and 26.21). In addition, small pores similar to liver sinusoids were reproduced among the endothelial cells in the lumen region of the capillary vessels. The liver organoids could be transplanted and survived in the omentum and under the kidney capsule in nude mice [43].

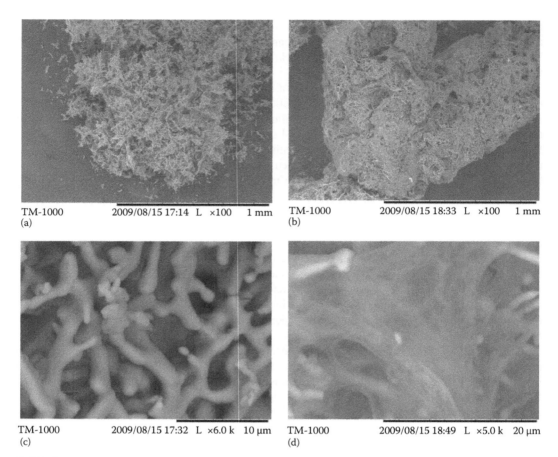

TM-1000   2009/08/15 17:14 L ×100   1 mm   TM-1000   2009/08/15 18:33 L ×100   1 mm
(a)   (b)

TM-1000   2009/08/15 17:32 L ×6.0 k 10 μm   TM-1000   2009/08/15 18:49 L ×5.0 k   20 μm
(c)   (d)

FIGURE 26.20 Liver organoid co-cultured immortalized mouse hepatocyte, hepatic stellate cells, and sinusoidal endothelial cells in RFB. Packed AFS (a, c). Liver organoid (b) and endothelial cells (d).

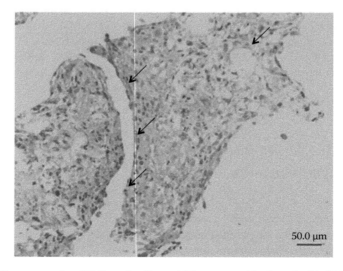

FIGURE 26.21 **(See companion CD for color figure.)** Liver organoid cultured with AFS in RFB (HE staining). Vessel-like structure was observed in liver organoid (arrow).

26.5 CONCLUSIONS

Three-dimensional highly-dense cell culture using RFB filled with a rigid scaffold such as porous hydroxyapatite beads enables scaling-up of the culture and construction of an extracorporeal type of BAL. The liver organoid can be reproduced by filling the column with soluble AFS and culturing the hepatocytes, stellate cells, and endothelial cells inside. These reproduced tissues can be used potentially for organ implantation in the future. BAL can be applied not only clinically but also to the testing of drug metabolism for pharmaceutical development and for the production of virus particles for the manufacture of vaccine and physiologically active substances. The development and selection of a suitable BAL for each purpose are important challenges for the future.

ACKNOWLEDGMENTS

We thank Prof. K. Ohkawa, Prof. N. Kimura, Dr. T. Iwaki, Prof. H. Hano, and Dr. T. Fukuda at the Jikei University School of Medicine for experimental support on this study. This study was supported in part by grants-in-aids from the University Start-Ups Creation Support System, The Promotion and Mutual Aid Corporation for Private Schools of Japan (1999–2001), The Japan Health Sciences Foundation (KHC1202, 2009–2011), the Program for Promotion of Fundamental Studies in Health Sciences of the National Institute of Biomedical Innovation (NIBIO) (2007–2011), and "Academic Frontier" Project for Private Universities: matching fund subsidy from MEXT (Ministry of Education, Culture, Sports, Science and Technology) (2006–2010).

REFERENCES

1. Nakatsu, M., Okihana, H., Sakamoto, M., Matsumoto, T., Ogalva, T. 2005. Bone formation of highly porous hydroxy apatite with different pore size ranges. *Bull Ceram Soc Jpn* 40: 828–830.
2. Honda, M., Fujimi, T.J., Kanzawa, K., Izawa, K., Tsuchiya, T., Aizawa, M. 2006. Osteogenic differentiation in a three-dimensional apatite-fiber scaffolds. *Arch BioCeram Res* 6:171–174.
3. Hiramoto, A., Matsuura, T., Aizawa, M. 2006. Three-dimensional cell culture of hepatocytes using apatite-fiber scaffold and application to a radial-flow bioreactor. *Arch BioCeram Res* 6:220–223.
4. Kanai, H., Marushima, H., Kimura, F., Iwaki, T., Ohkawa, K., Yanaga, K. et al. 2007. An extracorporeal bioartificial liver in treatment of pigs with experimental hepatic encephalopathy. *Artif Organs* 31:148–151.
5. Saito, M., Matsuura, T., Masaki, T., Maehashi, H., Shimizu, K., Hataba, Y. et al. 2006. Reconstruction of liver organoid using a bioreactor. *World J Gastroenterol* 12:1881–1888.
6. Saito, M., Matsuura, T., Nagatsuma, K., Tanaka, K., Maehashi, H., Shimizu, K. et al. 2007. The functional interrelationship between gap junctions and fenestrae in endothelial cells of the liver organoid. *J Membr Biol* 217:115–121.
7. Matsumoto, T., Komori, R., Magara, T., Ui, T., Kawakami, M., Hano, H. 1979. A study on the normal structure of the human liver, with special reference to its angioarchitecture. *Jikei Med J* 26:1–40.
8. Sato-Gutierrez, A., Uygun, B.F., Yagi, H., Uygun, K., Yarmush, M.L. 2009. Liver engineering using natural livers as scaffolds. *Organ Biol* 16:80.
9. Williams, D.F. 1986. Definitions in biomaterials. *Proceedings of a Consensus Conference of the European Society for Biomaterials*. Vol.4 (March):3–5. New York: Elsevier.
10. Funatsu, K., Nakazawa, K. 2002. Novel hybrid artificial liver using hepatocyte organoids. *Int J Artif Organs* 25:77–82.
11. Kobayashi, K., Kobayashi, A., Akaike, T. 1994. Culturing hepatocytes on lactose-carrying polystyrene layer via asialoglycoprotein receptor-mediated interactions. *Methods Enzymol* 247:409–418.
12. Stevens, A.C., Demetriou, A. et al. 2001. An interin analysis of a phase II/III prospective randomized, multicenter, controlled trial of the hepatassist bioartificial liver support system for the treatment of fulminant hepatic failure. *Hepatology* 34:299A.
13. Aizawa, M., Hiramoto, A., Maehashi, H., Matsuura, T. 2008. Reconstruction of liver organoid using an apatite-fiber scaffold, a radial-flow bioreactor, and FLC-4 cells of hepatocyte model. *Key Eng Mater* 361–363: 1165–1168.
14. Hench, L.L. 1998. Bioceramics. *J Am Ceram Soc* 81:1705–1728.

15. LeGeros, R.Z., LeGeros, J.P. 1993. Dense hydroxyapatite. In *An Introduction to Bioceramics*, eds. Hench, L.L. and Wilson, J., pp. 139–180. Hackensack, NJ: World Scientific.

16. Ota, K., Monma, H., Kawasaki, T. 1999. Characterization of interaction between hydroxyapatite and organic macromolecules by liquid chromatography. *Inorg Mater* 6:224–230.

17. Kawasaki, T.J. 1991. Hydroxyapatite as a liquid chromatographic packing. *J Chromatogr* 544:147–184.

18. Aizawa, M., Porter, A.E., Best, S.M., Bonfield, W. 2005. Ultrastructural observation of single-crystal apatite fibres. *Biomaterials* 26:3427–3433.

19. Aizawa, M., Shinoda, H., Uchida, H., Okada, I., Fujimi, T.J., Kanzawa, N. et al. 2004. in vitro biological evaluations of three-dimensional scaffold developed from single-crystal apatite fibres for tissue engineering of bone. *Phosphorus Res Bull* 17:268–273.

20. Morisue, H., Matsumoto, M., Chiba, K., Matsumoto, H., Toyama, Y., Aizawa, M. et al. 2006. A novel hydroxyapatite fiber mesh as a carrier for recombinant human bone morphogenetic protein-2 enhances bone union in rat posterolateral fusion model. *Spine* 31:1194–2000.

21. Morisue, H., Matsumoto, M., Chiba, K., Matsumoto, H., Toyama, Y., Aizawa, M. et al. 2008. In vivo bone formation using three-dimensional scaffolds developed from a single crystal apatite fiber. *J Biomed Mater Res A*, 90A: 811–818.

22. Matsuura, T. 1997. Cell sources for bioartificial liver—Hepatocytes and sinusoidal cells. *Tissue Culture Eng* 23:288–291.

23. van de Kerkhove, M.P., Hoekstra, R., Chamuleau, R.A.F.M., van Gulik, T.M. 2004. Clinical application of bioartificial liver support systems. *Ann Surg* 240:216–230.

24. Nickerson, C.A., Ott, C.M., Wilson, J.W., Ramamurthy, R., Pierson, D.L. 2004. Microbial responses to microgravity and other low-shear environments. *Microbiol Mol Biol Rev* 68:345–361.

25. Matsuura, T., Kawada, M., Hasumura, S., Nagamori, S., Obata, T., Yamaguchi, M. et al. 1998. High density culture of immortalized liver endothelial cells in the radial-flow bioreactor in the development of an artificial liver. *Int J Artif Organs* 21:229–234.

26. Kawada, M., Nagamori, S., Aizaki, H., Fukaya, K., Niiya, M., Matsuura, T. et al. 1998. Massive culture of human liver cancer cells in a newly developed radial flow bioreactor system: Ultrafine structure of functionally enhanced hepatocarcinoma cell lines. *In Vitro Cell Dev Biol* 34:109–115.

27. Nagamori, S., Hasumura, S., Matsuura, T., Aizaki, H., Kawada, M. 2000. Developments in bioartificial liver research: Concepts, performance, and applications. *J Gastroenterol* 35:493–503.

28. Iwahori, T., Matsuura, T., Maehashi, H., Sugo, K., Saito, M., Hosokawa, M. et al. 2003. CYP3A4 inducible model for in vitro analysis of human drug metabolism using a bioartificial liver. *Hepatology* 37:665–673.

29. Aizaki, H., Nagamori, S., Matsuda, M., Kawakami, H., Hashimoto, O., Matsuura, T. et al. 2003. Production and release of infections hepatitis C virus from human liver cell cultures in the three-dimensional radial-flow bioreactor. *Virology* 314:16–25.

30. Hascilowicz, T., Kosuge, M., Matsuura, T., Matsufuji, S., Murai, N. 2005. Two-dimensional protein analysis of functional liver cells for bioartificial liver. *Jikei Med J* 52:109–114.

31. Matsuura, T. 2006. Bioreactors for 3-dimensional high-density culture of human cells. *Human Cell* 19:11–16.

32. Kosuge, M., Takizawa, M., Maehashi, H., Matsuura, T., Matsufuji, S. 2007. A comprehensive gene expression analysis of human hepatocellular carcinoma cell lines as components of bioartificial liver using radial flow bioreactor. *Liver Int* 27:101–108.

33. Hazell, A.S., Butterworth, R.F. 1999. Hepatic encephalopathy: An update of pathophysiologic mechanisms. *Proc Soc Exp Biol Med* 222:99–112.

34. Yoshiba, M., Inoue, K., Sekiyama, K., Koh, I. 1996. Favorable effect of new artificial liver support on survival of patients with fulminant hepatic failure. *Artif Organs* 20:1169–1172.

35. Liu, J.P., Gluud, L.L., Als-Nielsen, B., Gluud, C. 2004. Artificial and bioartificial support systems for liver failure. *Cochrane Database Syst Rev* 1:CD003628.

36. Allen, J.W., Hassanein, T., Bhatia, S.N. 2001. Advances in bioartificial liver devices. *Hepatology* 34:447–455.

37. Demetriou, A.A., Brown Jr., R.S., Busuttil, R.W., Fair, J., McGuire, B.M., Rosenthal, P. et al. 2004. Prospective, randomized, multicenter, controlled trial of a bioartificial liver in treating acute liver failure. *Ann Surg* 239:660–670.

38. Yamazaki, Z., Fujimori, Y., Sanjo, K., Kojima, Y., Sugiyama, M., Wada, T. et al. 2000. New artificial liver support system (plasma perfusion detoxification) for hepatic coma. *Ther Apher* 4:23–25.

39. Matsushita, M., Yamamoto, T., Gemba, H. 1999. The role of astrocytes in the development of hepatic encephalopathy. *Neurosci Res* 34:271–280.

40. Jalan, R., Shawcross, D., Davies, N. 2003. The molecular pathogenesis of hepatic encephalopathy. *Int J Biochem Cell Biol* 35:1175–1181.
41. Babu, E., Kanai, Y., Chairoungdua, A., Kim, D.K., Iribe, Y., Tangtrongsup, S. et al. 2003. Identification of a novel system L amino acid transporter structurally distinct from heterodimeric amino acid transporters. *J Biol Chem* 278:43838–43845.
42. Takada, Y., Ishiguro, S., Fukunaga, K., Gu, M., Taniguchi, H., Seino, K. et al. 2001. Increased intracranial pressure in a porcine model of fulminant hepatic failure using amatoxin and endotoxin. *J Hepatol* 34:825–831.
43. Saito, R., Ishii, Y., Ito, R., Nagatuma, K., Tanaka, K., Saito, M. et al. 2011. Transplantation of liver organoids in the omentum and kidney. *Artif Organs* 35:80–83.

27 Materials Biofunctionalization for Tissue Regeneration

Laura Cipolla, Laura Russo, Nasrin Shaikh, and Francesco Nicotra

CONTENTS

ABBREVIATIONS

APTES	3-aminopropyl-trimethoxysilane
bFGF	basic fibroblast growth factor
BMP	bone morphogenetic protein
CDI	carbonyl diimidazole
DCC	dicyclohexyl-carbodiimide
DNA	deoxyribonucleic acid
ECM	extracellular matrix
EDC	1-ethyl-3-(3-dimethylaminopropyl)-carbodiimide
EGF	epidermal growth factor
FGF	fibroblast growth factor
FGF-R	FGF receptor
FN	fibronectin
HA	hyaluronic acid
HB-EGF	heparin-binding epidermal growth factor
Lac	lactose
LN	laminin
NGF	neuronal growth factor
NT	neurotensin
OGP	osteogenic growth peptide
PBS	phosphate buffered saline
PDGF	platelet-derived growth factor
PDGF-BB	platelet-derived growth factor isoform BB

PEG   polyethyleneglycol
PLGA   poly-L-glycolic acid
PLLA   poly-L-lactic acid
RGD   Arg-Gly-Asp
SAM   self-assembled monolayer
SF   silk fibroin
TGF   transforming growth factor
VEGF   vascular endothelial growth factor
VN   vitronectin
WSC   water-soluble carbodiimide

27.1   INTRODUCTION

Recently, tissue engineering has attracted many scientists and surgeons with a hope to treat patients in a minimally invasive way (Vacanti and Langer 1999). Tissue engineering involves the isolation of specific cells through a small biopsy from a patient, their growth on a 3D biomimetic scaffold under precisely controlled culture conditions, the delivery of the construct to the desired site in the patient's body, and the stimulation of new tissue formation into the scaffold that can be degraded over time (Lee and Mooney 2001). Tissue engineering also offers unique opportunities to investigate aspects of the structure–function relationship associated with new tissue formation in the laboratory and to predict the clinical outcome of the specific medical treatment. In order to achieve successful regeneration of damaged organs or tissues, several critical elements should be considered including biomaterial scaffolds that serve as a mechanical support for cell growth (Sakiyama-Elbert and Hubbel 2001, Shin et al. 2003, Langer and Tirrell 2004, Peppas and Langer 2004, Hubbell 1995, Lutolf and Hubbell 2005), progenitor cells that can be differentiated into specific cell types, and inductive growth factors that can modulate cellular activities (Putnam and Mooney 1996, Heath 2005). The biomaterial plays an important role in most tissue engineering strategies (Chaikof 2002, Griffith and Naughton 2002, Langer and Tirrell 2004). For example, biomaterials can serve as a substrate on which cell populations can attach and migrate, being implanted with a combination of specific cell types as a cell delivery vehicle, or utilized as a drug carrier to activate specific cellular function in the localized region.

In its simplest form a tissue engineering scaffold provides mechanical support, shape, and cell-scale architecture for neo-tissue construction in vitro or in vivo as seeded cells expand and organize. Most degradable biomaterials used to date comprise a class of synthetic polyesters such as poly(L-lactic acid) (PLLA) and poly (L-glycolic acid) (PLGA), and/or natural biological polymers such as alginate, chitosan, collagen, and fibrin (Langer and Tirrell 2004). A multitude of fabrication techniques have been devised and afford an abundance of potential shapes, sizes, porosities, and architectures (Weigel et al. 2006). Composites of these synthetic and natural polymers, alone or with bioactive ceramics such as hydroxyapatite or bioglasses, can be designed to yield materials with a range of strengths and porosities, particularly for the engineering of hard tissues (Boccaccini and Blaker 2005).

The development of biomaterials for tissue engineering applications has recently focused on the design of biomimetic materials, fully integrating principles from cell and molecular biology. Materials equipped with molecular cues mimicking the structure or function of natural extracellular microenvironments are able to interact with surrounding tissues by biomolecular recognition (Healy 1999, Hubbell 1999, Sakiyama-Elbert et al. 2001). The design of biomimetic materials is an attempt to make the materials such that they are capable of eliciting specific cellular responses and directing new tissue formation mediated by specific interactions, which can be manipulated by altering design parameters.

Extensive studies have been performed to render materials biomimetic. The surface modification of biomaterials with bioactive molecules is an obvious way to make biomimetic materials.

27.2 EXTRACELLULAR MATRIX: THE BASIS FOR FUNCTIONAL AND STRUCTURAL "BIO-INSPIRATION"

An important area of tissue engineering is to develop improved scaffolds that more nearly recapitulate the biological properties of native extracellular matrix (ECM). Reconstructing mature ECM and understanding its complex functions in mature or regenerating tissues are formidable tasks. The native ECM, in addition to contributing to mechanical integrity, has important signaling and regulatory functions in the development, maintenance, and regeneration of tissues. ECM components, in synergy with soluble signals provided by growth factors and hormones, participate to the tissue-specific control of gene expression through a variety of transduction mechanisms. Furthermore, the ECM is itself a dynamic structure that is actively remodeled by the cells with which it interacts (Birkedal-Hansen 1995). Tissue dynamics, that is, its formation, function, and regeneration after damage, as well as its function in pathology, is the result of an intricate temporal and spatial coordination of numerous individual cell fate processes, each of which is induced by a myriad of signals originating from the extracellular microenvironment (Kleinman et al. 2003). The extracellular microenvironment, which surrounds cells and comprises molecular signals, is a highly hydrated network hosting the following three main effectors:

- Insoluble hydrated macromolecules (fibrillar proteins such as collagens, noncollagenous glycoproteins such as elastin, laminin, or fibronectin, and hydrophilic proteoglycans with large glycosaminoglycan side chains)
- Soluble macromolecules (growth factors, chemokines, and cytokines)
- Proteins on surfaces of neighboring cells

Thus, the ultimate decision of a cell to differentiate, proliferate, migrate, apoptose, or perform other specific functions is a coordinated response to the molecular interactions with these ECM effectors. It is noteworthy that the flow of information between cells and their ECM is highly bidirectional as, for example, observed in processes involving ECM degradation and remodeling (Lutolf and Hubbell 2005).

The exogenous ECMs are designed to bring the desired cell types into contact in an appropriate 3D environment, and also provide mechanical support until the newly formed tissues are structurally stabilized.

Biomaterials scientists have sought to approximate its functions using several different approaches. In the absence of methods for de novo construction of a true ECM from purified components, decellularized tissues or organs can serve as sources of biological ECM (Hodde 2002).

During normal development tissue morphogenesis is heavily influenced by the interaction of cells with the extracellular matrix. Yet simple polymers, while providing architectural support for neo-tissue development, do not adequately mimic the complex interactions between adult stem and progenitor cells and the ECM that promote functional tissue regeneration. In soft tissues, they consist mainly of collagens, proteoglycans, and glycosaminoglycans; the collagen fibers provide tensile strength while a hydrated gel of proteoglycans fills the extracellular space, creating a space for the tissue while allowing the diffusion of nutrients, metabolites, and growth factors. The ECM may also serve as a storage depot for growth factors and provide these factors in a controlled manner to cells adjacent to the ECM.

Design criteria for synthetic ECMs may vary considerably depending on the specific strategy utilized to create a new tissue (Lutolf and Hubbell 2005), while significant progress has been made in the use of naturally derived ECM molecules (Byung-Soo et al. 1998). Native ECM can also be approximated by the use of some components of ECM, either alone or in simple combinations. Structural proteins such as collagen, laminin, elastin, and fibronectin have been used as matrices for tissue engineering and as vehicles for cell delivery. Collagen has found widespread use as a scaffold and carrier for cells in tissue engineering and regenerative medicine, particularly in soft tissue applications such as skin (Patino et al. 2002). Carbohydrate polymers have been utilized not only in hydrogels for drug delivery but also in tissue engineering (Pouyani and Prestwich 1994, Morra 2005, Kogan et al. 2007, Nagahama et al. 2008). The linear glycosaminoglycan hyaluronic acid (HA), composed of repeating disaccharide units of glucuronic acid and N-acetylglucosamine, is widely distributed in the ECM and plays an important role in vertebrate tissue morphogenesis (Spicer and Tien 2004). HA has been approved for use in human patients both as viscous fluid and as sheet formulations, and is indicated for knee pain and surgical adhesions, respectively. The activity of HA, like that of other relatively simple carbohydrate matrix components, may be enhanced by modification to promote cell migration, spreading, and multiplication. Other carbohydrate polymers such as chitosan (Shi et al. 2006) and alginate (Totowa 2004), derived from the exoskeleton of shell-fish and brown algae, respectively, have been used in several biomedical applications. Chitosan is a polycationic material produced by the deacetylation of chitin; it readily forms hydrogels that have been used in a number of gene and drug delivery applications.

The relatively high degree of evolutionary conservation of many ECM components allows the use of xenogeneic materials. The pioneering identification of small fragments of ECM components (Ruoslahti 1996) involved in the interaction with cellular receptor, initiates the preparation of ligand-functionalized materials. Since these findings, bioactive molecules normally involved in ECM–cell and cell–cell interactions, have been used for surface modification in several studies.

27.3 DESIGNING SMART BIOMATERIALS

The ECM performs two main functions: (i) structural support for the cell environment; (ii) signaling, in order to elicit specific biological responses. The bioactive molecules for material functionalization should mimic these two roles of ECM, and it is really a hard task to design biomaterials possessing both structural and signaling features. The ideal biomaterial must be non-immunogenic, biocompatible, and biodegradable and must allow its functionalization with bioactive molecules eliciting the desired biological response. In designing such novel biomaterials, frequently referred to "smart biomaterials" (Anderson et al. 2004a), researchers have sought not merely to create bioinert materials but rather materials that can respond to the cellular environment around them to improve device integration and tissue regeneration (Sakiyama-Elbert et al. 2001). Different classes of biomolecules have been used to obtain a bioactive scaffold, including proteins, their engineered mutant variants or peptidic epitopes, polysaccharides or tailored glycidic structures, semi-synthetic materials incorporating protein domains.

Once the bioactive molecules have been identified, different approaches can be used in order to direct the desired biological activity into the biomaterial: (i) delivery of the bioactive molecules; (ii) immobilization (incorporation) of the bioactive molecules into the scaffold.

The first approach is based on the incorporation of soluble bioactive molecules such as growth factors or plasmid DNA into biomaterial carriers so that the bioactive molecule can be released from the material and trigger or modulate new tissue formation (Babensee et al. 2000, Richardson et al. 2001a, Whitaker et al. 2001). Natural ECMs modulate tissue dynamics through their ability to locally bind, store, and release soluble bioactive ECM effectors such as growth factors and direct them to the right place at the right time (Ramirez and Rifkin 2003). Many growth factors bind ECM via electrostatic interactions with heparan sulfate proteoglycans.

It raises their local concentration to levels appropriate for signaling, localizes their morphogenetic activity, protects them from enzymatic degradation, and in some cases may increase receptor–ligand interactions. As growth factors are required in only very tiny quantities to elicit proper biological response, the main focus in designing smart biomaterials has been to control their local concentration. Several strategies to engineer growth factor release from biomaterials have been presented over the past years, and some initial success has been reported in animal models for the regeneration of bone and skin as well as the induction of vascularization (Boontheekul and Mooney 2003, Chen and Mooney 2003, Zisch et al. 2003b). Since many cellular processes involved in morphogenesis require a complex network of several signaling pathways and usually more than one growth factor, recent efforts have also focused on schemes for sequential delivery of multiple growth factors (Richardson et al. 2001b). Incorporation of cysteine-tagged functional domains of fibronectin into thiol-modified hyaluronic acid gels was found to stimulate spreading and proliferation of human fibroblasts in vitro and to promote recruitment of dermal fibroblasts in an in vivo cutaneous wound model. Similarly, recombinant technology has been employed to generate a series of elastin–mimetic protein triblock copolymers (Nagapudi et al. 2005).

The second approach involves incorporation of cell-binding ligands into biomaterials via chemical or physical modification. The cell-binding ligands include native ECM proteins as well as short peptide sequences derived from intact ECM proteins that can incur specific interactions with cell receptors. For example, the immobilization of signaling peptides renders the surface of biomaterials cell adhesive that inherently have been non-adhesive to cells (Shin et al. 2002). The incorporation of peptide sequences into materials can also make the material degradable by specific protease enzymes (West and Hubbell 1999) or induce cellular responses that may not be present in a local native tissue (Suzuki et al. 2000). For example, the use of biological feedback mechanisms in growth factor delivery has been explored (Zisch et al. 2003a). In this case, a growth factor is bound to the matrix and released upon cellular demand through cell-mediated localized proteolytic cleavage from the matrix (Zisch et al. 2001). This approach substantially mimics the mechanism by which these factors are released in vivo from stores in the natural ECM by invading cells in tissue repair. Immobilized growth factors may be able to modulate subsequent cell functions, such as proliferation, differentiation, and activity, on biomaterial surfaces. Epidermal growth factor (EGF) was coupled to a polystyrene plate–induced phosphorylation of the EGF receptor (Ito et al. 1998). Immobilized EGF was as effective as the soluble growth factor in stimulating DNA synthesis in hepatocytes (Kuhl and Griffith-Cima 1996).

Similarly, immobilized insulin, although technically not a growth factor, stimulated cell growth better than the soluble form (Liu et al. 1992). Polymeric substrates have been extensively studied because they possess abundant reactive functional groups for use in immobilizing biomolecules (Puleo et al. 2002).

27.4 POLYVALENT INTERACTIONS IN BIOLOGICAL SYSTEM

While designing functionalized biomaterials, one should take into account that many molecular interactions are polyvalent in nature (Mammen et al. 1998). Polyvalent interactions are characterized by the simultaneous binding of multiple ligands on one biological entity (a molecule, a surface) to multiple receptors on the other. Polyvalent interactions can be collectively much stronger than corresponding monovalent interactions, and they can provide the basis for mechanisms of both agonizing and antagonizing biological interactions that are fundamentally different from those available in monovalent systems.

Polyvalent molecules can interact with cells (and bacteria) and cause responses ranging from growth and differentiation to clearance and death. Some of these responses are only possible if the molecule is polyvalent, and would not occur with even a very tightly binding monovalent molecule.

Cell growth, differentiation, migration, and apoptosis are regulated in part by polypeptide growth factors or cytokines. Many growth factors and cytokines are polyvalent, and exert their effects through binding and dimerizing (or oligomerizing) cell-surface receptors. Evidence of such a mechanism has been accumulated over the past few years (Heldin 1995). Association of a polyvalent ligand to multiple diffusing receptors in a cellular membrane is often involved.

One polyvalent interaction most widely exploited in the design of biomaterial for tissue engineering is that between multiple copies of Arg-Gly-Asp (RGD) presented by fibronectin (FN) and laminin (LN), and the integrin class of receptor (Ruoslahti and Pierschbacher 1987).

Applications of polyvalent molecules in tissue engineering should be possible, and will open new interesting doors in the biomaterial field.

A second type of application consists in agonizing interactions that are not naturally polyvalent with synthetic polyvalent molecules. For example, one may design a molecule that presents multiple copies of the desired ligand; since the affinity of this ligand will most likely be a function of the density and accessibility to receptors, the interaction with the receptor will be much stronger.

The degree of flexibility, the extent of hydration, and the size of the scaffold will play important roles in the design of successful polyvalent biomaterials. This suggests also the importance of incorporating stably bound ligands that remain intact and biologically active under physiologic shear forces.

The creation of chemically defined synthetic ECM analogs, in which ligand type, concentration, and spatial distribution can be modulated upon a passive background, may also help in deciphering the complexity of signaling in cell–ECM interactions. In addition, quantitative information on the ligand density required for a particular cellular response is needed for the design of functionalized material (Massia and Hubbell 1991). Studies on the influence of the density of adhesion ligand on cell migration (Di Milla et al. 1991, Burgess et al. 2000, Schense and Hubbell 2000, Gobin and West 2002, Lutolf et al. 2003a), studies on cell response to the nanoscale spatial organization of adhesion ligands (Irvine et al. 2002, Park et al. 2002) and ligand gradients (Brandley and Schnaar 1989), and finally studies on the coregulation of signals (Maheshwari et al. 1999, Koo et al. 2002) provide the basis for the design of well-controlled material matrices.

27.5 BIOMATERIAL FUNCTIONALIZATION METHODS

The surface modification of biomaterials with bioactive molecules is a simple way to make smart materials. However, modification of metallic biomaterials used for orthopedic and dental implants is particularly challenging since they possess a paucity of reactive functional groups. The use of superficial treatment, like plasma modification, is a strategy employed for immobilization of bioactive molecules on a "bioinert" material. One- and two-step carbodiimide strategies were used to immobilize lysozyme, a model biomolecule, and bone morphogenetic protein-4 (BMP-4), an osteo-inductive protein, on the aminated surfaces by plasma of titanium alloy (Kuhl 1996).

To date, different strategies can be used for the introduction of biomimetic elements into synthetic materials.

1. Physical adsorption (van der Waals, electrostatic, affinity, adsorbed and cross-linked)
2. Physical entrapment attachment (barrier system, hydrogel, dispersed matrix system)
3. Covalent surface immobilization, taking advantage of different natural or unnatural functional groups present both on the biomolecules and on the material surfaces (chemoselective ligation, via amino functionalities, heterobifunctional linkers, etc.)

The major methods of immobilizing a bioactive compound to a polymeric surface are adsorption via electrostatic interactions, ligand–receptor pairing (as in biotin–avidin), and covalent attachment (Figure 27.1).

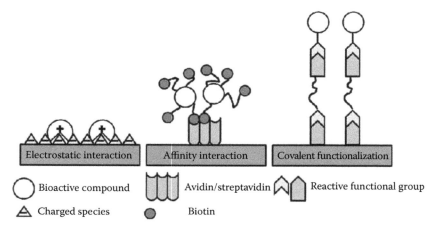

FIGURE 27.1 Different strategies for the introduction of biomimetic elements into synthetic materials.

Non-covalent adsorption is sometimes desirable, as in certain drug delivery applications. Covalent immobilizations offer several advantages by providing the most stable bond between the compound and the functionalized polymer surface. A covalent immobilization can be used to extend the half-life of a biomolecule, prevent its metabolism, or allow continued bioactivity of indwelling devices (Goddarda 2007).

Bioactive molecules (growth factors, ECM proteins, etc.) that are free in solution, as opposed to immobilized to the matrix, may induce significantly different biological responses. Growth factors are routinely added to cultures in vitro, and have been incorporated and released from polymeric systems with retention of bioactivity, as shown for neurotrophins (Whittlesey and Shea 2004) BMPs, (Simmons et al. 2004, Kroese-Deutman et al. 2005) and VEGF (Zisch 2003). In vivo, these soluble factors can be released from the delivery site, and the relevant parameter is the duration over which therapeutic concentrations can be maintained. Alternatively, bioactive molecules can be linked covalently to the scaffold, eventually in a reversible way or exploiting a degradable linking tether. For growth factor immobilization to fibrin, cell migration results in cell-activated plasmin degradation that can catalyze release of the factor. These scaffolds have been termed "cell-responsive" (Lutolf et al. 2003b) due to release of the factor upon cellular demand. Once released, these soluble factors can bind their receptors and initiate a signaling cascade.

Alternatively, immobilized biomolecules can ligate their receptors directly from the material surface; however, this type of interaction may not exactly replicate the signaling of soluble factors, as growth factor internalization can stimulate signaling pathways different from those activated at the surface (Haugh et al. 1999, Wiedlocha and Sorensen 2004). For example, neuronal growth factor NGF induces neurite outgrowth by signaling at the plasma membrane, yet promotes neuron survival when internalized (Zhang et al. 2000, Ye et al. 2003, Bronfman et al. 2003). Surface immobilization has been successfully used to attach several factors such as EGF (Kuhl and Griffith-Cima 1996), BMP-7 (Kirkwood et al. 2003), BMP-2 (Karageorgiou et al. 2004), VEGF (Zisch et al. 2003), NGF (Sakiyama-Elbert and Hubbell 2000a,b, Lee et al. 2003), and NT-3 (Taylor and McDonald 2004) to a variety of natural and synthetic biomaterials. Signaling by these immobilized or locally released bioactive ligands may be more potent than signaling by soluble versions added directly to culture media (Seliktar et al. 2004). These studies also demonstrate that the immobilization strategy must consider protein structure and active region topology, when designing suitable delivery systems in order to maximize bioactivity. Ultimately, some factors may be best delivered in a continuous manner, while others benefit from direct attachment to the biomaterial substrate (Dinbergs et al. 1996).

Different methods have been developed for covalent functionalization of biomolecules to diverse biomaterials. Many of these methods are illustrated in Figures 27.2 and 27.3 (Hersel et al. 2003).

For covalent functionalization to an inert solid polymer, the surface must first be chemically modified to provide reactive groups ($-OH$, $-NH_2$, COOH, $-SH$) for a second functionalization step. When the material does not contain reactive groups, they can be generated by chemical and physical modification on the polymer surfaces in order to permit covalent attachment of biomolecules. With this goal, a wide number of surface modification techniques have been developed, including plasma, ionic radiation graft polymerization, photochemical grafting, chemical modification, and chemical derivatization.

For example, peptides can react via the N-terminus with different groups on polymers (Figure 27.2). This is usually done by reacting an activated carboxylic acid group with the nucleophilic N-terminus of peptides. The carboxylic group can be activated with different peptide coupling reagent, e.g., 1-ethyl-3-(3-dimethylaminopropyl)-carbodiimide (EDC, also referred to as water-soluble carbodiimide, WSC), dicyclohexyl-carbodiimide (DCC), or carbonyl diimidazole (CDI).

In a more recent approach named chemoselective ligation (Figure 27.3), selected pairs of functional groups are used to form stable bonds without the need of an activating agent and without interfering with other functionalities usually encountered in biomolecules. Chemoselective ligation proceeds usually under mild conditions and results in good yields.

A biomolecule may also be attached, with these coupling methods, via a spacer group, in order to give better access to the target receptor. One useful and biocompatible spacer is PEG

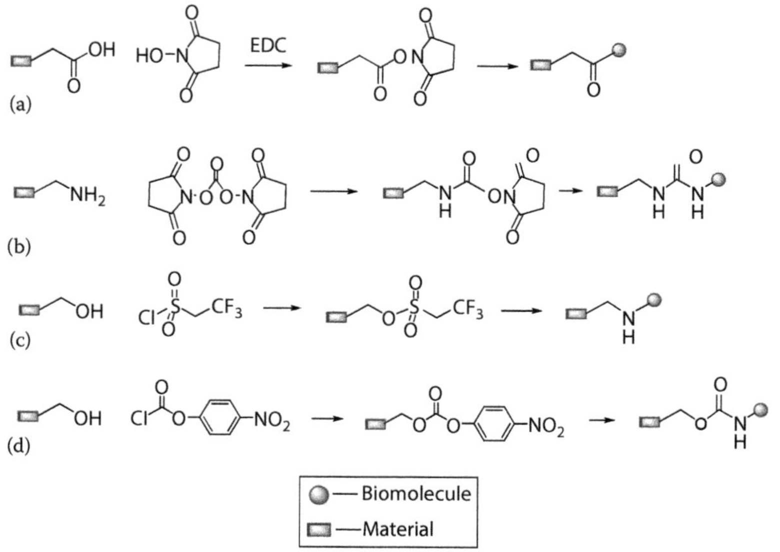

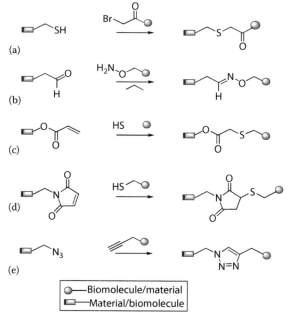

FIGURE 27.2 Coupling methods to different groups on materials. (a) Carboxyl groups, preactivated via EDC and NHS, (b) amino groups, preactivated with DSC, (c) hydroxyl groups, preactivated as tresylate, (d) hydroxyl groups, preactivated as p-nitrophenyl carbonate.

a. Thiol and bromoacetyl-biomolecule
b. Aldehyde and aminooxy-biomolecule
c. Acrylate and aminoxy-biomolecule
d. Maleinimde and thiol biomolecule
e. Azide and propargyl-biomolecule

FIGURE 27.3 Chemoselective ligation methods. (a) Thiol and bromoacetyl-biomolecule, (b) aldehyde and aminooxy-biomolecule, (c) acrylate and aminoxy-biomolecule, (d) maleinimide and thiol biomolecule, and (e) azide and propargyl-biomolecule.

that has been differently functionalized at two extremities (La Ferla et al. 2009). Metal or ceramic surface may also be silanized, exploiting functionalized triethoxysilanes (Weetall 1993, Dubruel et al. 2006).

27.6   THE ROLE OF LIGAND DISTRIBUTION

The strength of cell attachment and the cell migration rate depend on distribution of the ligand on the material. These biological responses are dependent upon several factors such as the receptor–ligand affinity, the density of ligand, and the spatial distribution of ligand.

Two relevant factors should be considered in designing smart biomaterials: (1) the spatial distribution or concentration of the ligand incorporated into biomimetic materials and (2) the spacer through which the ligand is conjugated.

Early studies have been performed to investigate the relationship between cell behavior and the concentration of immobilized whole ECM proteins such as FN and vitronectin (VN). Humphries et al. examined the cell attachment on a surface modified with FN and demonstrated that a minimal surface concentration of 1300 fmol/cm^2 was required to obtain sufficient cell spreading. Further studies using FN and VN showed that fibroblasts spread well when the surface concentration was greater than 400 fmol/cm^2 for VN and 100 fmol/cm^2 for FN (Danilov and Juliono 1989, Underwood and Bennett 1989). In addition to native ECM proteins, a number of model substrates covalently grafted with peptide sequences have been developed to study the concentration-dependent cellular functions (Massia and Hubbell 1990, 1991, Hubbell et al. 1992, Drumheller and Hubbell 1994, Hern and Hubbell 1998). A fibroblast surface density of at least 1 fmol/cm^2 was necessary for cell spreading (Neff et al. 1999), and one order of magnitude higher surface density of peptide (10 fmol/cm^2) was required for cytoskeletal organization. These values are considerably low compared to the concentration of peptide needed for sufficient spreading of fibroblasts on a surface coated with a native ECM protein. Fibroblasts were seeded on a model surface, in which RGD was grafted at different densities onto a cross-linked poly(acrylic acid) hydrogel via a PEG spacer (Drumheller et al. 1994). In this study, the required surface concentration of RGD was 12 pmol/cm^2 for cell spreading and 66 pmol/cm^2 for focal contact formation. These results indicate that cell attachment and spreading are dependent not only on peptide density but also on the hydrophilic/hydrophobic nature of the material. High peptide density enhances cell attachment, but it may also impede cell migration and proliferation. Neff et al. observed maximum proliferation of fibroblasts on peptide-modified polystyrene at an intermediate surface concentration of RGD (1.33 pmol/cm^2) (Neff et al. 1999). Similarly, cell migration showed biphasic trends in response to ligand density (Palecek et al. 1997) in which as ligand density increased or decreased, cell migration decreased. Recently, the relationship between peptide density and various cellular responses was extensively studied using PEG-based hydrogels (Mann and West 2002). The results of this study showed that concentrations of incorporated peptides (2.8–7 mmol/mL) cell attachment and spreading were promoted without hampering migration, proliferation, and matrix production of smooth muscle cells.

To maintain the biological activity of synthetic peptide sequences upon immobilization, the modified peptide should be flexible and experience minimal steric hindrance. Bioinert chains such as PEG with defined molecular weights (Drumheller et al. 1994, Shin et al. 2002) and repetitive sequences of nonspecific peptides, e.g., GGGG (Massia and Hubbell 1991) or Pro-Val-Glu-Leu-Pro (PVELP) (Sakiyama-Elbert et al. 2001), have been placed between main polymer chains or solid phase surfaces and signaling peptides. Hern et al. showed that cells adhered to the RGD modified surface with a PEG spacer (MW 3400) at low surface density (0.01 pmol/cm^2), provided that a PEG spacer is used (MW 3400). Without the spacer, limited cell adhesion was observed even at a higher surface density (1 pmol/cm^2) (Drumheller et al. 1994). In another study, different molecular weights of PEG were used to investigate the effect of the linker length on osteoblast adhesion (Kantlehner et al. 1999). It was determined that the effective distance between the peptide and the substrate is 3.5 nm, which is equivalent to the approximate length of PEG of MW 3500.

Although numerous studies attempt to draw general principles on the level of modification that induces optimal cellular responses, the desirable peptide density and length of spacer may vary depending on the specific cell types or intrinsic properties of biomaterials.

27.7 BIOMOLECULES FOR BIOMATERIAL DESIGN: PROTEIN AND PEPTIDES

ECM proteins such as fibronectin (FN), vitronectin (VN), and laminin (LN) used for tissue engineering applications are summarized in Table 27.1.

The use of proteins for biomedical application has some disadvantages. They must be isolated from other organisms and purified, thus they may elicit undesirable immune responses and increase infection risks. In addition, proteins are subject to proteolytic degradation and need to be refreshed continuously. Long-time applications of these materials would be impossible and costly. Furthermore, only a part of the proteins have proper orientation for cell adhesion due to their stochastic orientation on the surface (Elbert and Hubbell 2001). In addition, the texture of the surface determined by charge, wettability, and topography may influence the conformation and/or the orientation of the proteins. On hydrophobic surfaces proteins tend to maximize interaction with hydrophobic amino acid side chains; this causes denaturation or at least a different presentation of cell-binding motifs (Elbert and Hubbell 1996).

The finding of signaling domains (short peptides) of ECM proteins that primarily interact with cell membrane receptors has opened the way to the use of such short peptide fragments for surface modification (Seeger and Klingman 1985, Humphries et al. 1986). Some synthetic peptide sequences used in tissue engineering applications are summarized in Table 27.2.

TABLE 27.1
Growth Factors and Others ECM Proteins Used in Tissue Engineering

Proteins	Applications	Reference
FGF-2	Tissue development and remodeling, cellular proliferation, angiogenesis	Kanematsu et al. (2004), Cote et al. (2004)
PDGF-BB	Blood vessel maturation	Kanematsu et al. (2004)
VEGF	Angiogenesis	Kanematsu et al. (2004), Cleland et al. (2001), Murphy et al. (2000)
HB-EGF	Wound healing	Kanematsu et al. (2004)
TGF-β	Chondrogenesis, smooth muscle matrix deposition	Lee et al. (2004) Mann et al. (2001)
NGF	Promote neurite extension and axonal guidance in CNS and PNS injuries, neural regeneration	Sakiyama-Elbert and Hubbell (2000), Lee et al. (2003) Saltzman et al. (1999), Bloch et al. (2001), Yang et al. (2005a), Camarata et al. (1992), Cao and Schoichet (1999), Hadlock et al. (1999), Menei et al. (1993), Pean et al. (1998, 1999)
bFGF	Angiogenesis, wound healing Neural regeneration, angiogenesis	Sakiyama-Elbert and Hubbell (2000a), Jeon et al. (2005) Perets et al. (2003), Aebischer et al. (1989)
BMP-4/VEGF	Osteoinduction and angiogenesis	Huang et al. (2005)
FGF-1	Promote wound healing while inducing angiogenesis	Royce et al. (2004)
BMP-2	Bone regeneration	Rai et al. (2005), Mardegan Issa et al. (2008), Yang et al. (2004), Autefage et al. (2009), Piskounova et al. (2009), Li et al. (2009)
BMP	Osteogenesis, osteoinduction	Puleo et al. (2002), Hu et al. (2003)
Fusion protein	Cell adhesion, promoting tubular network formation, structural support, and affinity to collagen	Nakamura et al. (2008)

TABLE 27.2

Examples of Synthetic Peptide Sequences Derived from Extracellular Matrix Proteins Used in Tissue Engineering Application

Synthetic Sequence	Origin	Function	Reference
RGD and peptides including the RGD sequence	Fibronectin, vitronectin	Cell adhesion	Hersel et al. (2003), Chollet et al. (2009), Hennessy et al. (2008), Yang et al. (2005b), Smith et al. (2005), Anderson (2004b)
KQAGDV	Smooth muscle	Cell adhesion	Mann et al. (2001b)
YIGSR	Laminin B1	Cell adhesion	Ranieri et al. (1995), Bellamkonda et al. (1995), Massia et al. (1993)
REDV	Fibronectin	Endothelial cell adhesion	Hubbell et al. (1991), Panitch et al. (1999)
IKVAV	Laminin	Neurite extension	Ranieri et al. (1995), Heller et al. (2005), Schense et al. (2000)
RNIAEIIKDI	Laminin B2	Neurite extension	Schense et al. (2000)
KHIFSDDSSE	Neural cell adhesion molecules	Astrocyte adhesion	Kam et al. (2002)
VPGIG	Elastin	Enhance elastic modulus of artificial ECM	Panitch et al. (1999)
FHRRIKA	Heparin-binding domain	Improve osteoblastic mineralization	Rezania and Healy (1999), Schuler et al. (2009)
KRSR	Heparin-binding domain	Osteoblast adhesion	Dee et al. (1998)
NSPVNSKIPKACCVPTELSAI	BMP-2	Osteoinduction	Suzuki et al. (2000)
APGL	Collagenase	Mediated degradation	West and Hubbell (1999)
VRN	Plasmin	Mediated degradation	West and Hubbell (1999)
AAAAAAAA	Elastase	Mediated degradation	Mann et al. (2001a)
Osteogenic peptide TP508	BMP-2	Osteogenesis	Hedberg et al. (2002)
Peptide P24	BMP-2	Osteogenesis	Yuan et al. (2007), Duan et al. (2007)
BMP-7-derived peptides	BMP-7 (OP-1)	Osteoblastic differentiation and mineralization	Sakiyama-Elbert and Hubbell (2000)
Collagen-I derived peptides	Collagen-I	MSC adhesion differentiation and osteogenesis	Hennessy et al. (2009), Thorwarth et al. (2005)
EGF-derived peptides	EGF	Cell migration and proliferation	Miller et al. (2009)
GLRSKSKKFRRPDIQYPDATDEDITSHM	Collagen-binding motif (CBM) from osteopontin	Osteogenesis	Lee et al. (2007)
CDPGYIGSR, CSRARKQAASIKVAVSAD	Laminin-I	Nerve regeneration	Itoh et al. (2003)
Thrombin peptide TP508	Non-proteolytic receptor-binding domain of thrombin	Osteogenesis	Hedberg et al. (2002)
Osteogenic growth peptide (OGP) and fragments	OGP	Proliferation, bone formation, hematopoiesis	Hedberg et al. (2002), Bab and Chorev (2002), Hui et al. (2007), Chen et al. (2007)

Most of the problems arising from the use of proteins can, therefore, be overcome with small immobilized peptides. They exhibit higher stability toward sterilization conditions, heat treatment and pH-variation, storage and conformational shifting, as well as easier characterization and cost-effectiveness. Because of lower space requirement, peptides can be packed with higher density on surfaces; this provides a chance to compensate the possible lower cell adhesion activity exploiting the cluster effect. In addition, extracellular matrix proteins may contain different cell recognition motives, whereas small peptides represent only one of them. Therefore, they can selectively address one particular type of cell adhesion receptors. Sometimes linear peptides undergo slow enzymatic degradation; to circumvent this problem, small cyclic peptides can be employed (Aumailley et al. 1991) as well as tailored peptidomimetics (Sulyok et al. 2001). Finally, short peptide sequences can be massively synthesized in laboratories more economically.

The most commonly used peptide for surface modification is Arg-Gly-Asp (RGD), the signaling domain derived from fibronectin (FN) and laminin (LN). Additionally, other peptide sequences (Table 27.2) such as Tyr-Ile- Gly-Ser-Arg (YIGSR), Arg-Glu-Asp-Val (REDV), and Ile-Lys-Val-Ala-Val (IKVAV) have been immobilized on various scaffolds. A number of materials including glass, quartz, metal oxide, and polymers have been modified with these peptides and characterized for cellular interaction with their surfaces.

27.8 BIOMOLECULES FOR BIOMATERIAL DESIGN: GLYCIDIC STRUCTURES

The traditional view of carbohydrate polymers as energy source (starch and glycogen) and structural materials (cellulose, collagen, and proteoglycans) has expanded. Today, carbohydrates are known to have a wide variety of biological functions. For example, the sulfated polysaccharide, heparin, plays an essential role in blood coagulation (Linhardt and Toida 1997). Another related polysaccharide, hyaluronan, which acts as a lubricant in joints, has been used to protect the corneal endothelium during ophthalmologic surgery (Balazs and Laurent 1998). In addition to hyaluronan's lubricating and cushioning properties, polysaccharide and glycoprotein participate in a number of recognizing processes. Synthetic carbohydrate–based polymers are increasingly being explored as biodegradable, biocompatible, and biorenewable materials for use as water absorbent, chromatographic supports, and medical devices. Moreover, synthetic polymers bearing sugar residues also offer a good surface for cell attachment and might thus be applied to cell-recognizing events in tissue engineering. Different types of synthetic polymers bearing sugar residues are known (Wang and Dordick 2002): linear polymers, comb polymers, dendrimers, and cross-linking hydrogels represent the four major classes (Figure 27.4). One of the most relevant glycidic structure for biomaterial applications is hyaluronic acid.

It is well known that the glycidic part of glycolipids and glycoproteins strongly interact with lectins present at cell surfaces. In particular, it is known that β-galactose residues are recognized by asialoglycoprotein receptors on the surfaces of hepatocytes (Ashwell and Harford 1982, Varki 1993) and promote hepatocyte attachment (Weigel et al. 1979, Kobayashi et al. 1986, 1994, Oka and Weigel 1986, Yang et al. 2002, Yoon et al. 2002). Therefore, it is expected that the β-galactose

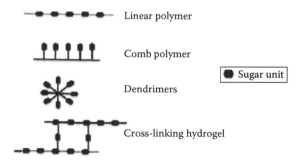

FIGURE 27.4 Major class of synthetic sugar containing polymers.

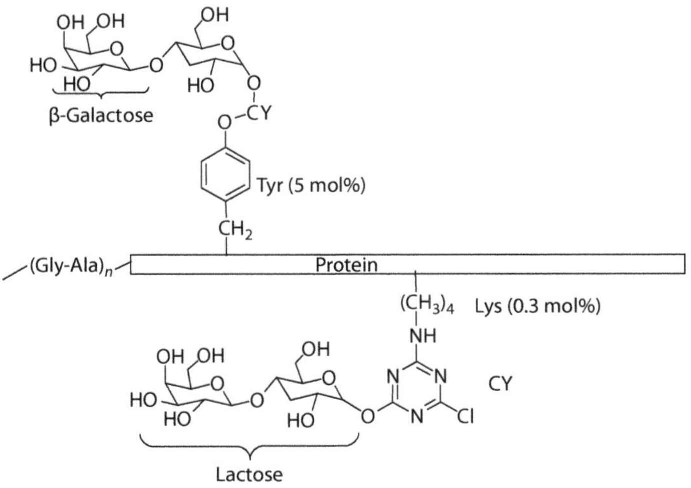

β-Galactose

Tyr (5 mol%)

CH_2

(Gly-Ala)$_n$

Protein

$(CH_3)_4$ Lys (0.3 mol%)

NH

CY

Lactose

FIGURE 27.5 Covalent linking of lactose to silk fibroin.

residues incorporated into silk fibroin may act as an efficient ligand for hepatocyte attachment to silk fibroin scaffolds.

Oligosaccharides were covalently coupled to silk fibroin using cyanuric chloride generating the glycoconjugate described in Figure 27.5, in which the sugar is linked to the tyrosine and cystine residues via a cyanuric chloride linker (Gotoh et al. 1996, 2002).

Although hepatocyte attachment on silk fibroin-coated dishes was lower than that on uncoated polystyrene dishes, the attachment on the dishes coated with 0.1%(w/v) and 1%(w/v) solutions of lactose-cyanuric chloride-silk fibroin (Lac–CY–SF) was comparable to that on collagen-coated dishes. After 2 days of culture, hepatocytes on the conjugate-coated dishes showed a small amount of dispersion compared to those on collagen-coated dishes. These results support the concept that the Lac–CY–SF glycoconjugates can serve as a scaffold for hepatocyte attachment.

Sato et al. exploited the anionic polysaccharide heparin to extend the scope of micropatterned carbohydrate displays by applying electrostatic interaction. They also used micropatterned carbohydrate displays on silicon substrates by a combination of photolithography and self-assembly (Miura et al. 2004). These techniques have interesting potential applications in tissue engineering. Top-down photolithography and bottom-up molecular self-assembly are the two main approaches to biomacromolecular micropatterning. Photolithography has been widely used in electronic devices and most practically applied for precise micropatterning of proteins for cell cultivation (Love et al. 2006).

Heparin is a glycosaminoglycan polysaccharide with a variable number of sulfate groups, and its binding to basic fibroblast growth factor (bFGF) is known to modulate cell proliferation and differentiation (Lindahl et al. 1994), (Sato et al. 2007). The interaction of heparin with bFGF induces the dimerization of bFGF, which is required for binding of bFGF to FGF receptor (FGF-R) on cell surfaces (Figure 27.6), as illustrated later.

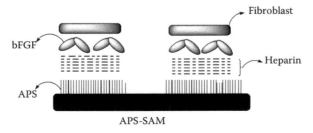

FIGURE 27.6 Adhesion of fibroblast cells.

The self-assembled monolayer (SAM) of 3-aminopropyl-trimethoxysilane (APTES) as ammonium-terminated template was prepared on silicon and glass substrates, and micropatterned by photolithography. Self-assembling of heparin onto the APTES-SAM through electrostatic interaction and then molecular recognition of bFGF to heparin and of fibroblast cells to bFGF have been achieved (Sato et al. 2007). Adsorption of heparin was not observed on a silicon substrate without APTES-SAM. Heparin adsorbed on APTES-SAM remained on the substrate after rinsing with PBS buffer. The electrostatic interaction was essential for anchoring heparin on APS-SAM. The time-course of heparin adsorption to APS-SAM does not fit with a simple Langmuir model, suggesting that, in addition to electrostatic interaction, some other interaction such as hydrogen bonding may contribute a little to the adsorption of heparin to APS-SAM (Foster 1961, Verli and Guimaraes 2004).

The aforementioned carbohydrate display using heparin has an attractive potential for tissue engineering, since it allows a facile and effective strategy for biomacromolecular patterning with significant bioactive molecules.

27.9 CONCLUSION

The modification of biomaterials with bioactive molecules is extremely useful to design biomimetic scaffolds that can provide biological cues to elicit specific cellular responses and direct new tissue formation. New generations of "smart biomaterials" are being developed in the past few years as 3D extracellular microenvironments mimicking the regulatory characteristics of ECM, both for therapeutic applications and for basic biological studies. The surface and bulk modification of materials with biomolecules with definite concentration and spatial distribution can modulate cellular functions such as adhesion, proliferation, and migration.

Modern synthetic biomaterials still represent oversimplified mimics of natural ECMs lacking the essential natural temporal and spatial complexity. Despite the recent advances toward the development of biomimetic materials for tissue engineering applications, several challenges still remain including the design of adhesion molecules for specific cell types as needed for guided tissue regeneration and the synthesis of materials exhibiting the mechanical responsiveness of living tissues. A growing symbiosis of materials engineering and cell biology may ultimately result in synthetic materials that contain the necessary signals to recapitulate developmental processes in tissue- and organ-specific differentiation and morphogenesis.

REFERENCES

Aebischer, P.; Salessiotis, A. N.; Winn, S. R. 1989. Basic fibroblast growth factor released from synthetic guidance channels facilitates peripheral nerve regeneration across long nerve gaps. *J. Neurosci. Res.* 23:282–289.

Anderson, D.G.; Burdick, J.A.; Langer, R. 2004a. Materials science: Smart biomaterials. *Science* 305:1923–1924.

Anderson, E.H.; Ruegsegger, M.A; Murugesan, G.; Kottke-Marchant, K.; Marchant, R.E. 2004b. Extracellular matrix-like surfactant polymers containing arginine-glycine-aspartic acid (RGD) peptides. *Macromol. Biosci.* 4:766–775.

Ashwell, G.; Harford J. 1982. Carbohydrate-specific receptors of the liver. *Annu. Rev. Biochem.* 51:531–554.

Aumailley, M.; Gurrath, M.; Muller, G.; Calvete, J.; Timpl, R.; Kessler, H. 1991. Arg-Gly-Asp constrained within cyclic pentapeptides, strong and selective inhibitors of cell adhesion to vitronectin and laminin fragment P1. *FEBS Lett.* 291:50–54.

Autefage, H.; Briand-Mésange, F.; Cazalbou, S. et al. 2009. Adsorption and release of BMP-2 on nanocrystalline apatite-coated and uncoated hydroxyapatite/tricalcium phosphate porous ceramics. *J. Biomed. Mater. Res. B* 91B:706–715.

Bab, I.; Chorev, M. 2002. Osteogenic growth peptide: From concept to drug design. *Biopolymers (Pept. Sci.)* 66:33–48.

Babensee, J.E.; McIntire, L.V.; Mikos, A.G. 2000. Growth factor delivery for tissue engineering. *Pharm. Res.* 17:497–504.

Balazs, E.A.; Laurent, T.C. 1998. New applications of hyaluronan. In *The Chemistry, Biology and Medical Applications of Hyaluronan and Its Derivatives*. Laurent, T.C., Ed., London, U.K.: Portland Press, pp. 325–336.

Bellamkonda, R.; Ranieri, J.P.; Aebischer, P. 1995. Laminin oligopeptide derivatized agarose gels allow three dimensional neurite extension in vitro. *J. Neurosci. Res.* 41:501–509.

Birkedal-Hansen, H. 1995. Proteolytic remodeling of extracellular matrix. *Curr. Opin. Cell Biol.* 7:728–735.

Bloch, J.; Fine, E.G.; Bouche, N.; Zurn, A.D.; Aebischer, P. 2001. Nerve growth factor- and neurotrophin-3-releasing guidance channels promote regeneration of the transected rat dorsal root. *Exp. Neurol.* 172:425–432.

Boccaccini, A.R.; Blaker, J.J. 2005. Bioactive composite materials for tissue engineering scaffolds. *Expert Rev. Med. Devices* 2:303–317.

Boontheekul, T.; Mooney, D.J. 2003. Protein-based signaling systems in tissue engineering. *Curr. Opin. Biotechnol.* 14:559–565.

Brandley, B.K.; Schnaar, R.L. 1989. Tumor cell haptotaxis on covalently immobilized linear and exponential gradients of a cell adhesion peptide. *Dev. Biol.* 135:74–86.

Bronfman, F.C.; Tcherpakov, M.; Jovin, T.M.; Fainzilber, M. 2003. Ligand-induced internalization of the p75 neurotrophin receptor: A slow route to the signaling endosome. *J. Neurosci.* 23:3209–3220.

Burgess, B.T.; Myles, J.L.; Dickinson, R.B. 2000. Quantitative analysis of adhesion-mediated cell migration in three-dimensional gels of RGD-grafted collagen. *Ann. Biomed. Eng.* 28:110–118.

Byung-Soo, K.; Mooney, D.J.; Rowley, J.A.; Madlambayan, G. 1998. Development of biocompatible synthetic extracellular matrices for tissue engineering. *Trends Biotechnol.* 16:224–230.

Camarata, P.J.; Suryanarayanan, R.; Turner, D.A.; Parker, R.G.; Ebner, T.J. 1992. Sustained release of nerve growth factor from biodegradable polymer microspheres. *Neurosurgery* 30:313–319.

Cao, X.; Schoichet, M.S. 1999. Delivering neuroactive molecules from biodegradable microspheres for application in central nervous system disorders. *Biomaterials* 20:329–339.

Chaikof, E.L. 2002. Biomaterials and scaffolds in reparative medicine. *Ann. N. Y. Acad. Sci.* 961:96–105.

Chen, Z.-X.; Chang, M.; Peng, Y.-L. et al. 2007. Osteogenic growth peptide C-terminal pentapeptide [OGP(10–14)] acts on rat bone marrow mesenchymal stem cells to promote differentiation to osteoblasts and to inhibit differentiation to adipocytes. *Regul. Pept.* 142:16–23.

Chen, R.R.; Mooney, D.J. 2003. Polymeric growth factor delivery strategies for tissue engineering. *Pharm. Res.* 20:1103–1112.

Chollet, C.; Chanseau, C.; Remy, M. et al. 2009. The effect of RGD density on osteoblast and endothelial cell behaviour on RGD-grafted polyethylene terephthalate surfaces. *Biomaterials* 30:711–720.

Cleland, J.L.; Duenas, E.T.; Park, A. et al. 2001. Development of poly-(D,Llactide-co-glycolide) microsphere formulations containing recombinant human vascular endothelial growth factor to promote local angiogenesis. *J. Control. Release* 72:13–24.

Cote, M.F.; Laroche, G.; Gagnon, E.; Chevallier, P.; Doillon, C.J. 2004. Denatured collagen as support for a FGF-2 delivery system: Physicochemical characterizations and in vitro release kinetics and bioactivity. *Biomaterials* 25:3761–3772.

Danilov, Y.N.; Juliono, R.N. 1989. (Arg-Gly-Asp)n-albumin conjugates as a model substratum for integrin-mediated cell adhesion. *Exp. Cell Res.* 182:186–196.

Dee, K.C.; Anderson, T.T.; Bizios, R. 1998. Design and function of novel osteoblast-adhesive peptides for chemical modification of biomaterials. *J. Biomed. Mater. Res.* 40:371–377.

Di Milla, P.A.; Barbee, K.; Lauffenburger, D.A. 1991. Mathematical model for the effects of adhesion and mechanics on cell migration speed. *Biophys. J.* 60:15–37.

Dinbergs, I.D.; Brown, L.; Edelman, E.R. 1996. Cellular response to transforming growth factor-beta1 and basic fibroblast growth factor depends on release kinetics and extracellular matrix interactions. *J. Biol. Chem.* 271:29822–29829.

Drumheller, P.D.; Elbert, D.L.; Hubbell, J.A. 1994. Multifunctional poly(ethylene glycol) semi-interpenetrating polymer networks as highly selective adhesive substrates for bioadhesive peptide grafting. *Biotechnol. Bioeng.* 43:772–780.

Drumheller, P.D.; Hubbell, J.A. 1994. Polymer networks with grafted cell adhesion peptides for highly biospecific cell adhesive substrates. *Anal. Biochem.* 222:380–388.

Duan, Z.; Zheng, Q.; Guo, X.; Yuan, Q.; Chen, S. 2007. Experimental research on ectopic osteogenesis of BMP2-derived peptide P24 combined with PLGA copolymers. *J. Huazhong Univ. Sci. Technol.* 27:179–182.

Dubruel, P.; Vanderleyden, E.; Bergadà, M. et al. 2006. Comparative study of silanisation reactions for the biofunctionalisation of Ti-surfaces. *Surf. Sci.* 600:2562–2571.

Elbert, D.L.; Hubbell, J.A. 1996. Surface treatments of polymers for biocompatibility. *Annu. Rev. Mater. Sci.* 26:365–394.

Elbert, D.L.; Hubbell, J.A. 2001. Conjugate addition reactions combined with free-radical cross-linking for the design of materials for tissue engineering. *Biomacromolecules* 2:430–441.

Foster, A.B. 1961. Chemistry of carbohydrate. *Annu. Rev. Biochem.* 30:45–70.

Gobin, A.S.; West, J.L. 2002. Cell migration through defined, synthetic ECM analogs. *FASEB J.* 16:751–753.

Goddarda, J.M. 2007. Polymer surface modification for the attachment of bioactive compounds. J.M. Goddarda and J.H. Hotchkiss. *Prog. Polym. Sci.* 32:698–725.

Gotoh, Y.; Minoura, N.; Miyashita, T. 2002. Preparation and characterization of conjugates of silk fibroin and chitooligosaccharides. *Colloid Polym. Sci.* 280:562–568.

Gotoh, Y.; Tsukada, M.; Aiba, S.-I.; Minoura, N. 1996. Chemical modification of silk fibroin with N-acetyl-chito-oligosaccharides. *Int. J. Biol. Macromol.* 18:19–26.

Griffith, L.G.; Naughton, G. 2002. Tissue engineering—Current challenges and expanding opportunities. *Science* 295:1009–1014.

Hadlock, T.; Sundback, C.; Koka, R.; Hunter, D.; Cheney, M.; Vacanti, J. 1999. A novel, biodegradable polymer conduit delivers neurotrophins and promotes nerve regeneration. *Laryngoscope* 109:1412–1416.

Haugh, J.M.; Schooler, K.; Wells, A.; Wiley, H.S.; Lauffenburger, D.A. 1999. Effect of epidermal growth factor receptor internalization on regulation of the phospholipase C-gamma1 signaling pathway. *J. Biol. Chem.* 274:8958–8965.

Healy, K.E. 1999. Molecular engineering of materials for bioreactivity. *Curr. Opin. Solid State Mater. Sci.* 4:381–387.

Heath, C.A. 2000. Cells for tissue engineering. *TIBTECH* 18:17–19.

Hedberg, E.L.; Tang, A.; Crowther, R.S.; Carney, D.H.; Mikos, A.G. 2002. Controlled release of an osteogenic peptide from injectable biodegradable polymeric composites. *J. Control. Release* 84:137–150.

Heldin, C.H. 1995. Dimerization of cell surface receptors in signal transduction. *Cell* 80:213–223.

Heller, D.A.; Garga, V.; Kelleher, K.J. et al. 2005. Patterned networks of mouse hippocampal neurons on peptide-coated gold surfaces. *Biomaterials* 26:883–889.

Hennessy, K.M.; Clem, W.C.; Phipps, M.C. et al. 2008. The effect of RGD peptides on osseointegration of hydroxyapatite biomaterials *Biomaterials* 29:3075–3083.

Hennessy, K.M.; Pollot, B.E.; Clem, W.C. et al. 2009. The effect of collagen I mimetic peptides on mesenchymal stem cell adhesion and differentiation, and on bone formation at hydroxyapatite surfaces. *Biomaterials* 30:1898–1909.

Hern, D.L.; Hubbell, J.A. 1998. Incorporation of adhesion peptides into nonadhesive hydrogels useful for tissue resurfacing. *J. Biomed. Mater. Res.* 39:266–276.

Hersel, U.; Dahmen, C.; Kessler, H. 2003. RGD modified polymers: Biomaterials for stimulated cell adhesion and beyond. *Biomaterials* 24:4385–4415.

Hodde, J. 2002. Naturally occurring scaffolds for soft tissue repair and regeneration. *Tissue Eng.* 8:295–308.

Hu, Y.; Zhang, C.; Zhang, S.; Xiong, Z.; Xu, J. 2003. Development of a porous poly(L-lactic acid)/hydroxyapatite/collagen scaffold as a BMP delivery system and its use in healing canine segmental bone defect. *J. Biomed. Mater. Res. A* 67:591–598.

Huang, Y.C.; Kaigler, D.; Rice, K.G.; Krebsbach, P.H.; Mooney, D.J. 2005. Combined angiogenic and osteogenic factor delivery enhances bone marrow stromal cell-driven bone regeneration. *J. Bone Miner. Res.* 20:848–857.

Hubbell, J.A. 1995. Biomaterials in tissue engineering. *Bio/Technology* 13:565–576.

Hubbell, J.A. 1999. Bioactive biomaterials. *Curr. Opin. Biotechnol.* 10:123–129.

Hubbell, J.A.; Massia, S.P.; Desai, N.P.; Drumheller, P.D. 1991. Endothelial cell-selective materials for tissue engineering in the vascular graft via a new receptor. *Biotechnology* 9:568–572.

Hubbell, J.A.; Massia, S.P.; Drumheller, P.D. 1992. Surface-grafted cell-binding peptide in tissue engineering of the vascular graft. *Ann. N. Y. Acad. Sci.* 665:253–258.

Hui, Z.; Yu, L.; Xiaoli, Y. 2007. C-terminal pentapeptide of osteogenic growth peptide regulates hematopoiesis in early stage. *J. Cell Biochem.* 101:1423–1429.

Humphries, M.J.; Akiyama, S.K.; Komoriya, A.; Olden, K.; Yamada, K.M. 1986. Identification of an alternatively spliced site in human plasma fibronectin that mediates cell type specific adhesion. *J. Cell. Biol.* 103:2637–2647.

Irvine, D.J.; Hue, K.A.; Mayes, A.M.; Griffith, L.G. 2002. Simulations of cell-surface integrin binding to nanoscale-clustered adhesion ligands. *Biophys. J.* 82:120–132.

Ito, Y.; Chen, G.; Imanishi, Y. 1998. Micropatterned immobilization of epidermal growth factor to regulate cell function. *Bioconj. Chem.* 9:277–282.

Itoh, S.; Yamaguchi, I.; Suzuki, M. et al. 2003. Hydroxyapatite-coated tendon chitosan tubes with adsorbed laminin peptides facilitate nerve regeneration in vivo. *Brain Res.* 993:111–123.

Jeon, O.; Ryu, S.H.; Chung, J.H.; Kim, B.S. 2005. Control of basic fibroblast growth factor release from fibrin gel with heparin and concentrations of fibrinogen and thrombin. *J. Control. Release* 105:249–259.

Kam, L.; Shain, W.; Turner, J.N.; Bizios, R. 2002. Selective adhesion of astrocytes to surfaces modified with immobilized peptides. *Biomaterials* 23:511–515.

Kanematsu, A.; Yamamoto, S.; Ozeki, M. et al. 2004. Collagenous matrices as release carriers of exogenous growth factors. *Biomaterials* 25:4513–4520.

Kantlehner, M.; Finsinger, D.; Meyer, J. et al. 1999. Selective RGD-mediated adhesion of osteoblast at surfaces of implant. *Angew. Chem. Int. Ed.* 38:560–562.

Karageorgiou, V.; Meinel, L.; Hofmann, S.; Malhotra, A.; Volloch, V.; Kaplan, D. 2004. Bone morphogenetic protein-2 decorated silk fibroin films induce osteogenic differentiation of human bone marrow stromal cells. *J. Biomed. Mater. Res. A* 71:528–537.

Kirkwood, K.; Rheude, B.; Kim, Y.J.; White, K.; Dee, K.K.C. 2003. In vitro mineralization studies with substrate-immobilized bone morphogenetic protein peptides. *J. Oral Implantol.* 29:57–65.

Kleinman, H.K.; Philp, D.; Hoffman, M.P. 2003. Role of the extracellular matrix in morphogenesis. *Curr. Opin. Biotechnol.* 14:526–532.

Kobayashi, A.; Akaike, T.; Kobayashi, K.; Sumitomo, H. 1986. Enhanced adhesion and survival efficiency of liver cells in culture dished coated with a lactose-carrying styrene homopolymer. *Makromol. Chem. Rapid Commun.* 7:645–650.

Kobayashi, A.; Goto, M.; Kobayashi, K.; Akaike. T. 1994. Receptor-mediated regulation of differentiation and proliferation of hepatocytes by synthetic polymer model of asialoglycoprotein. *J. Biomater. Sci. Polym. Ed.* 6:325–342.

Kogan, G.; Ladislav, S.; Stern, R.; Gemeiner, P. 2007. Hyaluronic acid: A natural biopolymer with a broad range of biomedical and industrial applications. *Biotechnol. Lett.* 29:17–25.

Koo, L.Y.; Irvine, D.J.; Mayes, A.M.; Lauffenburger, D.A.; Griffith, L.G. 2002. Co-regulation of cell adhesion by nanoscale RGD organization and mechanical stimulus. *J. Cell Sci.* 115:1423–1433.

Kroese-Deutman, H.C.; Ruhe, P.Q.; Spauwen, P.H.; Jansen, J.A. 2005. Bone inductive properties of rhBMP-2 loaded porous calcium phosphate cement implants inserted at an ectopic site in rabbits. *Biomaterials* 26:1131–1138.

Kuhl, P.R.; Griffith-Cima, L.G. 1996. Tethered epidermal growth factor as a paradigm for growth factor-induced stimulation from the solid phase. *Nat. Med.* 2:1022–1027.

La Ferla, B.; Zona, C.; Nicotra, F. 2009. Easy silica gel supported desymmetrization of PEG. *Synlett* 14:2325–2327.

Langer, R.; Tirrell, D.A. 2004. Designing materials for biology and medicine. *Nature* 428:487–492.

Lee, J.-Y.; Choo, J.-E.; Choi, Y.-S. et al. 2007. Assembly of collagen-binding peptide with collagen as a bioactive scaffold for osteogenesis in vitro and in vivo. *Biomaterials* 28:4257–4267.

Lee, J.E.; Kim, K.E.; Kwon, I.C. et al. 2004. Effects of the controlled released TGF-beta 1 from chitosan microspheres on chondrocytes cultured in a collagen/chitosan/glycosaminoglycan scaffold. *Biomaterials* 25:4163–4173.

Lee, K.Y.; Mooney, D.J. 2001. Hydrogels for tissue engineering. *Chem. Rev.* 101:1869–1879.

Lee, A.C.; Yu, V.M.; Lowe, J.B. 2003. Controlled release of nerve growth factor enhances sciatic nerve regeneration. *Exp. Neurol.* 184:295–300.

Li, B.; Yoshii, T.; Hafeman, A.E.; Nyman, J.S.; Wenke, J.C.; Guelcher, S.A. 2009. The effects of rhBMP-2 released from biodegradable polyurethane/microsphere composite scaffolds on new bone formation in rat femora. *Biomaterials* 30:6768–6779.

Lindahl, U.; Lidholt, K.; Spillmann, D.; Kjellén, L. 1994. More to "heparin" than anticoagulation. *Thromb. Res.* 75:1–32.

Linhardt, R.J.; Toida, T. 1997. Heparin oligosaccharides: New analogues development and applications. In *Carbohydrates in Drug Design*. Witczak, Z.J. and Nieforth, K.A., Eds. New York: Marcel Dekker.

Liu, S.Q.; Ito, Y.; Imanishi, Y. 1992. Cell growth on immobilized cell growth factor. I. Acceleration of the growth of fibroblast cells on insulin-immobilized polymer matrix in culture medium without serum. *Biomaterials* 13:50–58.

Love, C.; Ronan, J.L.; Grotenbreg, G.M.; van der Veen, A.G.; Ploegh, H.L. 2006. A microengraving method for rapid selection of single cells producing antigen-specific antibodies. *Nat. Biotechnol.* 24:703–707.

Lutolf, M.P.; Hubbell, J.A. 2005. Synthetic biomaterials as instructive extracellular microenvironments for morphogenesis in tissue engineering. *Nat. Biotechnol.* 23:47–55.

Lutolf, M.P.; Lauer-Fields, J.L.; Schmoekel, H.G. 2003a. Synthetic matrix metalloproteinase-sensitive hydrogels for the conduction of tissue regeneration: Engineering cell-invasion characteristics. *Proc. Natl. Acad. Sci. USA* 100:5413–5418.

Lutolf, M.P.; Raeber, G.P.; Zisch, A.H.; Tirelli, N.; Hubbel, J.A. 2003b. Cell responsive synthetic hydrogels. *Adv. Mater.* 15:888–892.

Maheshwari, G.; Wells, A.; Griffith, L.G.; Lauffenburger, D.A. 1999. Biophysical integration of effects of epidermal growth factor and fibronectin on fibroblast migration. *Biophys. J.* 76:2814–2823.

Mammen, M.; Choi, S.-K.; Whitesides, G.M. 1998. Polyvalent interactions in biological systems: Implications for design and use of multivalent ligands and inhibitors. *Angew. Chem. Int. Ed.* 37:2754–2794.

Mann, B.K.; Gobin, A.S.; Tsai, A.T.; Schmedlen, R.H.; West, J.L. 2001a. Smooth muscle cell growth in photopolymerized hydrogels with cell adhesive and proteolytically degradable domains: Synthetic ECM analogs for tissue engineering. *Biomaterials* 22:3045–3051.

Mann, B.K.; Schmedlen, R.H.; West, J.L. 2001b. Tethered-TGFbeta increases extracellular matrix production of vascular smooth muscle cells. *Biomaterials* 22:439–444.

Mann, B.K.; West, J.L. 2002. Cell adhesion peptides alter smooth muscle cell adhesion, proliferation, migration, and matrix protein synthesis on modified surfaces and in polymer scaffolds. *J. Biomed. Mater. Res.* 60:86–93.

Mardegan Issa, J.P.; do Nascimento, C.; Iyomasa, M.M. et al. 2008. Bone healing process in critical-sized defects by rhBMP-2 using poloxamer gel and collagen sponge as carriers. *Micron* 39:17–24.

Massia, S.P.; Hubbell, J.A. 1990. Covalent surface immobilization of Arg-Gly-Asp- and Tyr-Ile-Gly-Ser-Arg-containing peptides to obtain well-defined cell-adhesive substrates. *Anal. Biochem.* 187:292–301.

Massia, S.P.; Hubbell, J.A. 1991. An RGD spacing of 440 nm is sufficient for integrin alpha v beta 3-mediated fibroblast spreading and 140 nm for focal contact and stress fiber formation. *J. Cell Biol.* 114:1089–1100.

Massia, S.P.; Rao, S.; Hubbell, J.A. 1993. Covalently immobilized laminin peptide Tyr-Ile-Gly-Ser-Arg (YIGSR) supports cell spreading and co-localization of the 67-Kilodalton laminin receptor with α-actinin and vinculin. *J. Biol. Chem.* 268:8053–8059.

Menei, P.; Daniel, V.; Montero-Menei, C.; Brouillard, M.; Pouplard-Barthelaix, A.; Benoit, J.P. 1993. Biodegradation and brain tissue reaction to poly(D,L-lactide-co-glycolide) microspheres. *Biomaterials* 14:470–478.

Miller, D.S.; Chirayil, S.; Ball, H.L.; Luebke, K.J. 2009. Manipulating cell migration and proliferation with a light-activated polypeptide. *ChemBioChem* 10:577–584.

Miura, Y.; Sato, H.; Ikeda, T.; Sugimura, H.; Takai, O.; Kobayashi. K. 2004. Micropatterned carbohydrate displays by self-assembly of glycoconjugate polymers on hydrophobic templated on silicon. *Biomacromolecules* 5:1708–1713.

Morra, M. 2005. Engineering of biomaterials surfaces by hyaluronan. *Biomacromolecules* 6:1205–1223.

Murphy, W.L.; Peters, M.C.; Kohn, D.H.; Mooney, D.J. 2000. Sustained release of vascular endothelial growth factor from mineralized poly(lactide-co-glycolide) scaffolds for tissue engineering. *Biomaterials* 21:2521–2527.

Nagahama, H.; New, N.; Jayakumar, R.; Koiwa, S.; Furuike, T.; Tamura, H. 2008. Novel biodegradable chitin membranes for tissue engineering applications. *Carbohydr. Polym.* 73:295–302.

Nagapudi, K.; Brinkman, W.T.; Thomas, B.S. et al. 2005. Viscoelastic and mechanical behavior of recombinant protein elastomers. *Biomaterials* 26:4695–4706.

Nakamura, M.; Mie, M.; Mihara, H.; Nakamura, M.; Kobatake, E. 2008. Construction of multi-functional extracellular matrix proteins that promote tube formation of endothelial cells. *Biomaterials* 29:2977–2986.

Neff, J.A.; Tresco, P.A.; Caldwell, K.D. 1999. Surface modification for controlled studies of cell–ligand interactions. *Biomaterials* 20:2377–2393.

Oka, J.A.; Weigel, P.H. 1986. Binding and spreading of hepatocytes on synthetic galactose culture surfaces occur as distinct and separable threshold responses. *J. Cell Biol.* 103:1055–1060.

Palecek, S.P.; Loftus, J.C.; Ginsberg, M.H.; Luffenburger, D.A.; Horwitz, A.F. 1997. Integrin-ligand binding properties govern cell migration speed through cell-substratum adhesiveness. *Nature* 385:537–540.

Panitch, A.; Yamaoka, T.; Fournier, M.J.; Mason, T.L.; Tirrell, D.A. 1999. Design and biosynthesis of elastin-like artificial extracellular matrix proteins containing periodically spaced fibronectin CS5 domains. *Macromolecules* 32:1701–1703.

Park, K.I.; Teng, Y.D.; Snyder, E.Y. 2002. The injured brain interacts reciprocally with neural stem cells supported by scaffolds to reconstitute lost tissue. *Nat. Biotechnol.* 20:1111–1117.

Patino, M.G.; Neiders, M.E.; Andreana, S.; Noble, B.; Cohen, R.E. 2002. Collagen as an implantable material in medicine and dentistry. *J. Oral Implantol.* 28:220–225.

Pean, J.M.; Boury, F.; Venier-Julienne, M.C.; Menei, P.; Proust, J.E.; Benoit, J.P. 1999. Why does PEG 400 co-encapsulation improve NGF stability and release from PLGA biodegradable microspheres? *Pharm. Res.* 16:1294–1299.

Pean, J.M.; Venier-Julienne, M.C.; Boury, F.; Menei, P.; Denizot, B.; Benoit, J.P. 1998. NGF release from poly(D,L-lactide-coglycolide) microspheres: Effect of some formulation parameters on encapsulated NGF stability. *J. Control. Release* 56:175–187.

Peppas, N.A.; Langer, R. 2004. New challenges in biomaterials. *Science* 263:1715–1720.

Perets, A.; Baruch, Y.; Weisbuch, F.; Shoshany, G.; Neufeld, G.; Cohen, S. 2003. Enhancing the vascularization of three-dimensional porous alginate scaffolds by incorporating controlled release basic fibroblast growth factor microspheres. *J. Biomed. Mater. Res. A* 65:489–497.

Piskounova, S.; Forsgren, J.; Brohede, U.; Engqvist, H.; Strømme, M. 2009. In vitro characterization of bioactive titanium dioxide/hydroxyapatite surfaces functionalized with BMP-2. *J. Biomed. Mater. Res. B* 91B:780–787.

Pouyani, T.; Prestwich, G.D. 1994. Functionalized derivatives of hyaluronic acid oligosaccharides: Drug carriers and novel biomaterials. *Bioconj. Chem.* 5:339–347.

Puleo, D.A.; Kissling, R.A.; Sheu, M.-S. 2002. A technique to immobilize bioactive proteins, including bone morphogenetic protein-4 (BMP-4), on titanium alloy. *Biomaterials* 23:2079–2087.

Putnam, A.J.; Mooney, D.J. 1996. Tissue engineering using synthetic extracellular matrices. *Nat. Med.* 2:824–826.

Rai, B.; Teoh, S.H.; Hutmacher, D.W.; Cao, T.; Ho, K.H. 2005. Novel PCL-based honeycomb scaffolds as drug delivery systems for rhBMP-2. *Biomaterials* 26:3739–3748.

Ramirez, F.; Rifkin, D.B. 2003. Cell signaling events: A view from the matrix. *Matrix Biol.* 22:101–107.

Ranieri, J.P.; Bellamkonda, R.; Bekos, E.J.; Vargo Jr., T.G.; Aebischer, J.A.G. 1995. Neuronal cell attachment to fluorinated ethylene propylene films with covalently immobilized laminin oligopeptides YIGSR and IKVAV. II. *J. Biomed. Mater. Res.* 29:779–785.

Rezania, A.; Healy, K.E. 1999. Biomimetic peptide surfaces that regulate adhesion, spreading, cytoskeletal organization, and mineralization of the matrix deposited by osteoblast-like cells. *Biotechnol. Prog.* 15:19–32.

Richardson, T.P.; Murphy, W.L.; Mooney, D.J. 2001a. Polymeric delivery of proteins and plasmid DNA for tissue engineering and gene therapy. *Crit. Rev. Eukaryot. Gene Expr.* 11:47–58.

Richardson, T.P.; Peters, M.C.; Ennett, A.B.; Mooney, D.J. 2001b. Polymeric system for dual growth factor delivery. *Nat. Biotechnol.* 19:1029–1034.

Royce, S.M.; Askari, M.; Marra, K.G. 2004. Incorporation of polymer microspheres within fibrin scaffolds for the controlled delivery of FGF-1. *J. Biomater. Sci. Polym.* 15:1327–1336.

Ruoslahti, E. 1996. RGD and other recognition sequences for integrins. *Annu. Rev. Cell Dev. Biol.* 12:697–715.

Ruoslahti, E.; Pierschbacher, M.D. 1987. New perspectives in cell adhesion: RGD and integrins. *Science* 238:491–497.

Sakiyama-Elbert, S.E.; Hubbell, J.A. 2000a. Development of fibrin derivatives for controlled release of heparin-binding growth factors. *J. Control. Release* 65:389–402.

Sakiyama-Elbert, S.E.; Hubbell, J.A. 2000b. Controlled release of nerve growth factor from a heparin-containing fibrin-based cell ingrowth matrix. *J. Control. Release* 69:149–158.

Sakiyama-Elbert, S.E.; Hubbell, J.A. 2001. Functional biomaterials: Design of novel biomaterials. *Annu. Rev. Mater. Res.* 31:183–201.

Sakiyama-Elbert, S.E.; Panitch, A.; Hubbell, J.A. 2001. Development of growth factor fusion proteins for cell-triggered drug delivery. *FASEB J.* 15:1300–1302.

Saltzman, W.M.; Mak, M.W.; Mahoney, M.J.; Duenas, E.T.; Cleland, J.L. 1999. Intracranial delivery of recombinant nerve growth factor: Release kinetics and protein distribution for three delivery systems. *Pharm. Res.* 16:232–240.

Sato, H.; Miura, Y.; Saito, N.; Kobayashi, K.; Takai, O. 2007. A micropatterned carbohydrate display for tissue engineering by self-assembly of heparin. *Surf. Sci.* 601:3871–3875.

Schense, J.C.; Bloch, J.; Aebischer, P.; Hubbell, J.A. 2000. Enzymatic incorporation of bioactive peptides into fibrin matrices enhances neurite extension. *Nat. Biotechnol.* 18:415–419.

Schense, J.C.; Hubbell, J.A. 2000. Three-dimensional migration of neurites is mediated by adhesion site density and affinity. *J. Biol. Chem.* 275:6813–6818.

Schuler, M.; Hamilton, D.W.; Kunzler, T.P. et al. 2009. Comparison of the response of cultured osteoblasts and osteoblasts outgrown from rat calvarial bone chips to nonfouling KRSR and FHRRIKA-peptide modified rough titanium surfaces. *J. Biomed. Mater. Res. B* 91B:517–527.

Seeger, J.M.; Klingman, N. 1985. Improved endothelial cell seeding with cultured cells and fibronectin-coated grafts. *J. Surg. Res.* 38:641–647.

Seliktar, D.; Zisch, A.H.; Lutolf, M.P.; Wrana, J.L.; Hubbell, J.A. 2004. MMP-2 sensitive, VEGF-bearing bioactive hydrogels for promotion of vascular healing. *J. Biomed. Mater. Res. A* 68:704–716.

Shi, C.; Zhu, Y.; Ran, X.; Wang, M.; Su, Y.; Cheng, T. 2006. Therapeutic potential of chitosan and its derivatives in regenerative medicine. *J. Surg. Res.* 133:185–192.

Shin, H.; Jo, S.; Mikos, A.G. 2002. Modulation of marrow stromal osteoblast adhesion on biomimetic oligo(poly(ethylene glycol) fumarate) hydrogels modified with Arg-Gly-Asp peptides and a poly(ethylene glycol) spacer. *J. Biomed. Mater. Res.* 61:169–179.

Shin, H.; Jo, S.; Mikos, A.G. 2003. Biomimetic materials for tissue engineering. *Biomaterials* 24:4353–4364.

Simmons, C.A.; Alsberg, E.; Hsiong, S.; Kim, W.J.; Mooney, D.J. 2004. Dual growth factor delivery and controlled scaffold degradation enhance in vivo bone formation by transplanted bone marrow stromal cells. *Bone* 35:562–569.

Smith, E.; Yang, J.; McGann, L.; Sebald, W.; Uludag, H. 2005. RGD-grafted thermoreversible polymers to facilitate attachment of BMP-2 responsive C2C12 cells. *Biomaterials* 26:7329–7338.

Spicer, A.P.; Tien, J.Y. 2004. Hyaluronan and morphogenesis. *Birth Defects Res. C Embryo. Today* 72:89–108.

Sulyok, G.A.G.; Gibson, C.; Goodman, S.L.; Holzemann, G.; Wiesner, M.; Kessler, H. 2001. Solid-phase synthesis of a nonpeptide RGD mimetic library: New selective $\alpha v \beta 3$ integrin antagonists. *J. Med. Chem.* 44:1938–1950.

Suzuki, Y.; Tanihara, M.; Suzuki, K.; Saitou, A.; Sufan, W.; Nishimura, Y. 2000. Alginate hydrogel linked with synthetic oligopeptide derived from BMP-2 allows ectopic osteoinduction in vivo. *J. Biomed. Mater. Res.* 50:405–409.

Taylor, S.J.; McDonald, J.W. 2004. Controlled release of neurotrophin-3 from fibrin gels for spinal cord injury. *J. Control. Release* 98:281–294.

Thorwarth, M.; Schultze-Mosgau, S.; Wehrhan, F. et al. 2005. Bioactivation of an anorganic bone matrix by P-15 peptide for the promotion of early bone formation. *Biomaterials* 26:5648–5657.

Totowa, N.J. 2004. Alginates in tissue engineering. In *Biopolymer Methods in Tissue Engineering*. Hollander, A.P. and Hatton, P.V., Eds., Book Series: Methods in Molecular Biology, Vol. 238. Totowa, NJ: Humana Press Inc., pp. 77–86

Umezawa, Y.; Aoki, H. 2004. Ion channel sensors based on artificial receptors. *Anal. Chem.* 76:320–326.

Underwood, P.A.; Bennett, F.A. 1989. A comparison of the biological activities of the cell-adhesive proteins vitronectin and fibronectin. *J. Cell Sci.* 93:641–649.

Vacanti, J.P.; Langer R. 1999. Tissue engineering: The design and fabrication of living replacement devices for surgical reconstruction and transplantation. *Lancet* 354(suppl. I):32–34.

Varki, A. 1993. Biological roles of oligosaccharides: All of the theories are correct. *Glycobiology* 3:97–130.

Verli, H.; Guimaraes, J.A. 2004. Molecular dynamics simulation of a decasaccharide fragment of heparin in aqueous solution. *Carbohydr. Res.* 339:281–290.

Wang, Q.; Dordick, J.S. 2002. Synthesis and application of carbohydrate containing polymers. *Chem. Mater.* 14:3232–3244.

Weetall, H.H. 1993. Preparation of immobilized proteins covalently coupled through silane coupling agents to inorganic supports. *Appl. Biochem. Biotechnol.* 41:157–188.

Weigel, T.; Schinkel, G.; Lendlein, A. 2006. Design and preparation of polymeric scaffolds for tissue engineering. *Expert Rev. Med. Devices* 3:835–851.

Weigel, P.H.; Schnaar, R.L.; Kuhlenschmidt, M.S. et al. 1979. Adhesion of hepatocytes to immobilized sugars. *J. Biol. Chem.* 254:10830–10838.

West, J.L.; Hubbell, J.A. 1999. Polymeric biomaterials with degradation sites for proteases involved in cell migration. *Macromolecules* 32:241–244.

Whitaker, M.J.; Quirk, R.A.; Howdle, S.M.; Shakesheff, K.M. 2001. Growth factor release from tissue engineering scaffolds. *J. Pharm. Pharmacol.* 53:1427–1437.

Whittlesey, K.J.; Shea, L.D. 2004. Delivery systems for small molecule drugs, proteins, and DNA: The neuroscience/biomaterial interface. *Exp. Neurol.* 190:1–16.

Wiedlocha, A.; Sorensen, V. 2004. Signaling, internalization, and intracellular activity of fibroblast growth factor. *Curr. Top. Microbiol. Immunol.* 286:45–79.

Yang, Y.; De Laporte, L.; Rives, C.B. et al. 2005a. Neurotrophin releasing single and multiple lumen nerve conduits. *J. Control. Release* 104:433–446.

Yang, J.; Goto, M.; Ise, H.; Cho, C.-S.; Akaike, T. 2002. Galactosylated alginate as a scaffold for hepatocytes entrapment. *Biomaterials* 23:471–479.

Yang, X.B.; Whitaker, M.J.; Sebald, W. et al. 2004. Human osteoprogenitor bone formation using encapsulated bone morphogenetic protein 2 in porous polymer scaffolds. *Tissue Eng.* 10:1037–1045.

Yang, F.; Williams, C.G.; Wang, D. et al. 2005b. The effect of incorporating RGD adhesive peptide in polyethylene glycol diacrylate hydrogel on osteogenesis of bone marrow stromal cells. *Biomaterials* 26:5991–5998.

Ye, H.; Kuruvilla, R.; Zweifel, L.S.; Ginty, D.D. 2003. Evidence in support of signaling endosome-based retrograde survival of sympathetic neurons. *Neuron* 39:57–68.

Yoon, J.J.; Nam, Y.S.; Kim, J.H., Park, T.G. 2002. Surface immobilization of galactose onto aliphatic biodegradable polymers for hepatocyte culture. *Biotechnol. Bioeng.* 78:1–10.

Yuan, Q.; Lu, H.; Tang, S. et al. 2007. Ectopic bone formation in vivo induced by a novel synthetic peptide derived from BMP-2 using porous collagen scaffolds. *J. Wuhan Univ. Technol.-Mater. Sci. Ed.* 701–715.

Zhang, Y.; Moheban, D.B.; Conway, B.R.; Bhattacharyya, A.; Segal, R.A. 2000. Cell surface Trk receptors mediate NGF-induced survival while internalized receptors regulate NGF-induced differentiation. *J. Neurosci.* 20:5671–5678.

Zisch, A.H.; Lutolf, M.; Ehrbar, P.M. et al. 2003a. Cell-demanded release of VEGF from synthetic, biointeractive cell ingrowth matrices for vascularized tissue growth. *FASEB J.* 17:2260–2262.

Zisch, A.H.; Lutolf, M.P.; Hubbell, J.A. 2003b. Biopolymeric delivery matrices for angiogenic growth factors. *Cardiovasc. Pathol.* 12:295–310.

Zisch, A.H.; Schenk, U.; Schense, J.C.; Sakiyama-Elbert, S.E.; Hubbell, J.A. 2001. Covalently conjugated VEGF–fibrin matrices for endothelialization. *J. Control. Release* 72:101–113.

28 Polymers-Based Devices for Dermal and Transdermal Delivery

Donatella Paolino, Margherita Vono, and Felisa Cilurzo

CONTENTS

28.1 INTRODUCTION

The skin is the largest organ of our body and represents a unique barrier that protects us from external agents (Hadgraft 2004). It is made up of three main layers, the epidermis (divided into the stratum corneum and the viable epidermis), the dermis, and the hypodermis. In particular, the stratum corneum is characterized by a thickness of 10–20 μm and contains between 10 and 15 layers of corneocytes, which are continually removed and regenerated, while the viable epidermis consists of multiple layers of keratinocytes at various stages of differentiation.

A drug molecule can cross the intact *stratum corneum* by means of three routes: via skin appendages (shunt routes, generally it has a little influence on flux, it has been proposed that this route is important for large polar molecules and ions); through the intercellular lipid domains (drug diffuses through the continuous lipid matrix, and is generally accepted as the most common path for small uncharged molecules penetrating the skin) or by a transcellular route (through corneocytes, containing highly hydrated keratin alternated to lipid envelope connecting the cells to the interstitial lipids, for this reason drugs can penetrate *via* the transcellular route only after a number of partitioning and diffusion steps; this pathway is thought to be the predominant for highly hydrophilic drugs during steady-state flux).

The percutaneous drug permeation is evaluated by means of Fick's Law (Equation 28.1):

$$J = \frac{dQ}{dt} = \frac{K_s \times D}{h} \times C \times A \tag{28.1}$$

where

dQ/dt is the amount of drug diffused per unit of time or drug flux (J)
K_s is the partition coefficient
D is the diffusion coefficient
h is the thickness of the stratum corneum
C is the concentration of the active compound
A is the skin surface area utilized for drug administration

28.2 INNOVATIVE DEVICES FOR DERMAL AND TRANSDERMAL DELIVERY

Although the skin can be considered an advantageous route for drug delivery, thanks to the suitability that gives us to overcome the first pass metabolism, to have lower fluctuations in plasma drug levels, to allow targeting of the active ingredient for a local effect, and to have good patient compliance, its barrier nature makes very difficult for most drugs to penetrate into and permeate through it.

During the past decades new strategies have been developed in order to increase the percutaneous permeation. These strategies may be of chemical, physical nature, or based on innovative formulations.

Chemical enhancers are substances that interact with intercellular lipids improve the diffusion coefficient of the substance in the stratum corneum. They may (a) increase the diffusibility of the substance inside the barrier, (b) increase the solubility in the vehicle, or (c) improve the partition coefficient. These substances frequently do not have a specific action. Enhancers of this type are Azone, Dermac SR-38, and oleic acid. Unfortunately, in some cases these substances have an irritating effect and must be carefully evaluated in various preparations.

Several techniques of physical nature have been developed and the principal ones are (1) sonophoresis (the low-frequency ultrasound induces alterations in the structure of the stratum corneum and enhances permeability) may be applied to enhance penetration of drugs and other active principles in dermatology; (2) ionophoresis, which is able to increase the penetration of ionizable substances using the force of a weak electric field; electroporation, where a reversible electrical field, with short and high voltage pulses causes the formation of nonlamellar lipid phases and of subcellular pores.

Unfortunately, these cited techniques increase the percutaneous permeation of all the components and this may lead to an increase of side effect; moreover, they may cause irritative effects to the skin, for this reason in the past years innovative nanocarriers have been developed.

Nanocarriers are colloidal systems having structures below a particle or droplet size of 500 nm. Innovative nanocarriers for drug delivery through the skin are liposomes and their derivatives and cyclodextrins.

Liposomes were introduced in the early 1980s as potential transdermal drug delivery systems but from then on, many studies have been published with conflicting results. Recently, it became evident that, in most cases, classic liposomes are of little or no value as carriers for transdermal drug delivery as they do not deeply penetrate skin, but rather remain confined to upper layers of the stratum corneum. Therefore, liposomes can be indicated for the dermal drug delivery and they act like penetration enhancers loosening the lipid structure of the stratum corneum and promoting impaired barrier function of these layers to the drug, with less well-packed intercellular lipid structure form, and with subsequent increasing skin partitioning of the drug (Elsayed et al. 2007).

The potential application of liposomes in skin delivery, compared with conventional nonvesicle formulations, has been proposed because of the extremely promising properties of vesicular systems in terms of drug penetration enhancement (Betz et al. 2005), improved pharmacological effects, decreased side effects, controlled drug release, and drug photoprotection (Arsic et al. 1999). The penetration behavior of drug-loaded liposomes through the skin is based on the technological parameters of colloidal devices. In particular, preparation methods, vesicle-membrane structure (uni- or multilamellar), lipid composition, mean sizes, and physicochemical properties of drugs influence the enhancement of penetration when liposomal formulations are used as topical devices (Montenegro et al. 2006).

Intensive research in the field of vesicular system for dermal and transdermal drug delivery led to the introduction and development of a new class of highly deformable (elastic or ultraflexible) liposomes that have been investigated for the first time by Cevc and were termed Transfersomes® (Cevc 2003). These ultradeformable liposomes are very stable vesicles (Celia et al. 2009) and were reported to penetrate intact skin, carrying therapeutic concentrations of drugs, but only when applied under non-occluded conditions (Cevc and Blume 1992). The main difference in composition between liposomes and ultradeformable vesicles is the presence of an edge activating that is often a single chain surfactant, having a high radius of curvature, that destabilizes lipid bilayers of the vesicles and increases deformability of the bilayers. Examples of edge activators are sodium cholate, sodium deoxycholate, Span 60, Span 65, Span 80, Tween 20, Tween 60, Tween 80, and dipotassium glycyrrhizinate. Many studies have reported that ultradeformable liposomes were able to penetrate intact skin in vivo, transferring therapeutic amounts of drugs, including macromolecules, with an efficiency comparable with subcutaneous administration. The mechanisms proposed for the passage of ultradeformable vesicles through the skin are the following: the driving force for the vesicles entering the skin is xerophobia, i.e., the tendency to avoid dry surroundings; thus, they can cross the intact skin spontaneously, under the influence of the naturally occurring, in vivo transcutaneous hydration gradient (Cevc and Blume 2001), without permanent disintegration.

A relative new vesicular carrier for dermal and transdermal drug delivery is represented by ethosomes®, developed by Prof. Touitou (Touitou at al. 1996) that are lipidic systems composed of phospholipids, ethanol, and water. Ethosomes® are soft, malleable vesicles able to improve the delivery of drugs into the deeper layers of the skin.

Ethosomes® have been shown to exhibit high encapsulation efficiency for a wide range of molecules including lipophilic drugs, thanks to multilamellarity of ethosomal vesicles (Touitou et al. 2000), as well as by the presence of ethanol in ethosomes which allows for better solubility of many drugs. This carrier has been tested widely both in vitro (Touitou et al. 2001) and in vivo (Paolino et al. 2005, Aibinder et al. 2010) and its efficacy has been demonstrated. In contrast to ultradeformable vesicular systems such as transfersomes for which skin permeation enhancement is observed only under non-occlusive conditions, for ethosomes occlusion does not affect the skin permeation profile.

Another alternative to conventional liposomes is represented by niosomes that are non-ionic surfactant vesicles formed through self-assembly in an aqueous media. In respect to liposomes, niosomes are characterized by low cost and great stability. Niosomes may be made up of a variety of amphiphiles bearing sugar, polyoxyethylene, polyglycerol, crown ether, and amino acid hydrophilic head groups; and these amphiphiles typically possess one to two hydrophobic alkyl, perfluoroalkyl, or steroidal groups (Manconi et al. 2003), and bolaform surfactants (Paolino et al. 2007, 2008). The topical application of niosomes can increase the residence time of the drug in the stratum corneum, allowing the epidermis to receive a local therapeutic effect, thus reducing the systemic absorption of the drug compound and consequently the appearance of side effects. Moreover, the topical application reduces transepidermal water loss and increases smoothness through the replenishment of lost skin lipids. The structure of non-ionic surfactant vesicles offers great opportunities for drug delivery because their size, shape, surface charge, and composition can be altered and adapted to the characteristics of the drug.

Another carrier used to enhance percutaneous permeation of drugs is ciclodextrins that can be used to have both local and systemic effect. Cyclodextrins are cyclic oligosaccharides, consisting of ($\alpha - 1,4$)-linked α-D-glucopyranose units; they are characterized by the presence of a lipophilic central cavity and a hydrophilic outer surface. The glucopyranose units are in the form of a chair and, for this reason, the cyclodextrins may be represented as a truncated cone. The OH groups are oriented with the primary hydroxyl groups of the various units of glucose on the narrow side of the cone and the secondary OH groups at the larger edge. The lipophilic character of the central cavity is determined by skeletal carbons and ethereal oxygens. It has been largely demonstrated that CDs can markedly enhance the dermal delivery of lipophilic drugs (Ventura et al. 2006). The effects of cyclodextrins on the permeation rates of drugs through the skin may be determined by the increase of thermodynamic activity of drugs in a vehicle, the fast dissolution rate of the included drug and to a destabilizing action exerted by the macrocycle on the biomembrane (Ventura et al. 2005, 2006), the extraction of skin component, and the partition coefficient of the drug between skin and vehicle.

In the past decades the use of polymers has represented an important step forward in the field of dermal and transdermal application of drug.

28.3 TRANSDERMAL DRUG DELIVERY SYSTEMS

A transdermal drug delivery system is a device made of one or more polymers embedded with drug, able to deliver the drug through the skin over a controlled period of time. The first patch was introduced in 1979 and it contained scopolamine for the treatment of motion sickness.

The transdermal drug delivery system is based on the concept that the limiting step is the diffusion from the devices; this mechanism is driven by the gradient between the high concentration in the delivery system and the zero concentration present in the skin. The drug permeation across the skin obeys Fick's first law (Equation 28.1).

Transdermal drug delivery systems are generally classified into three types: reservoir systems, matrix systems, microreservoir systems (Figure 28.1).

In the reservoir systems the drug reservoir is embedded between a backing layer and a membrane (Figure 28.1a). The drug release occurs only through the membrane, which is rate-controlling and may be microporous or nonporous. The drug can be added to the polymer matrix as solution, suspension, or gel or dispersed. To adhere to the skin, the outer surface of the polymeric membrane can be filmed with a thin layer of adhesive polymer.

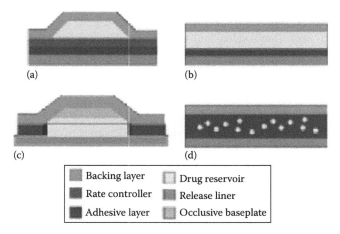

(a)   (b)

(c)   (d)

Backing layer	Drug reservoir
Rate controller	Release liner
Adhesive layer	Occlusive baseplate

FIGURE 28.1 **(See companion CD for color figure.)** Schematic representation of transdermal drug delivery systems: (a) reservoir System; (b) matrix-dispersion system; (c) peripheral adhesive design; (d) microreservoir system.

In the matrix systems, the drug reservoir is formed by dispersing the drug in an adhesive polymer and then the medicated polymer adhesive is added by solvent casting or by melting the adhesive (in the case of hot-melt adhesives) onto a backing layer. On top of the reservoir, layers of unmedicated adhesive polymer are applied (Figure 28.1b). A variant of matrix systems is peripheral design, which is characterized by the dispersion of the drug in the polymer matrix; this one is then fixed onto a drug-impermeable backing layer. The adhesive is spread along the circumference of the patch (Figure 28.1c).

A combination of reservoir and matrix-dispersion systems is represented by microreservoir systems where the drug reservoir is formed by first suspending the aqueous solution of the drug in a lipophilic polymer to form thousands of microscopic spheres of drug reservoirs. The dispersion even if thermodynamically unstable is stabilized by immediately cross-linking of the polymer (Figure 28.1d). Therefore, the application of a transdermal drug delivery system requires the controlled drug release as the first characteristic, and for this reason the choice of materials and techniques of preparation is fundamental. Commonly controlled drug release is guaranteed by the polymeric material that is able to release the drug in a predesigned controlled manner from the polymer into the blood stream.

28.4 POLYMERS IN DERMAL AND TRANSDERMAL DELIVERY

Polymers, today, are the backbone of dermal and transdermal drug delivery system. Polymers used in dermal and transdermal drug delivery must be characterized by biocompatibility and chemical compatibility with drugs and other components present in the formulation. Furthermore from the biological point of view, polymers must be inert, nontoxic, noncarcinogenic and teratogenic, nonallergenic, and sterilizable. In addition, they should have good chemical and physical stability, sufficient resistance to mechanical stress, and must not release impurities and/or residues of polymerization that can cause the formation of thrombi or alter the characteristics of biological fluids, particularly when used for the preparation of carrier for transdermal delivery.

In transdermal delivery polymers, in relation to their properties, may be used in various ways, in particular: matrix formers (used in the formulation of polymer matrix), rate-controlling membranes (used in reservoir-type transdermal drug delivery, where the drug diffuses through the membrane in a controlled manner), pressure-sensitive adhesives (PSAs, where the patch adheres to the skin by application of light force), backing layers and release liners (i.e., protective liner that is removed immediately before the application of the patch to the skin).

In this chapter the most used polymer in dermal and transdermal delivery are treated, and polymers are classified in natural (polysaccharides, gelatin, rosin, gums, and their derivatives) and synthetic (polyvinyl alcohol, silicon, polyethylene, polypropylene, polyacrylate, polyvinylpyrrolidone, polymethylmethacrylate).

28.5 NATURAL POLYMERS

28.5.1 CELLULOSE AND ITS DERIVATIVES

Cellulose is a polysaccharide with the formula $(C_6H_{10}O_5)$ consisting of a linear chain of several hundred to over 10,000 $\beta(1 \rightarrow 4)$ linked D-glucose units (Figure 28.2).

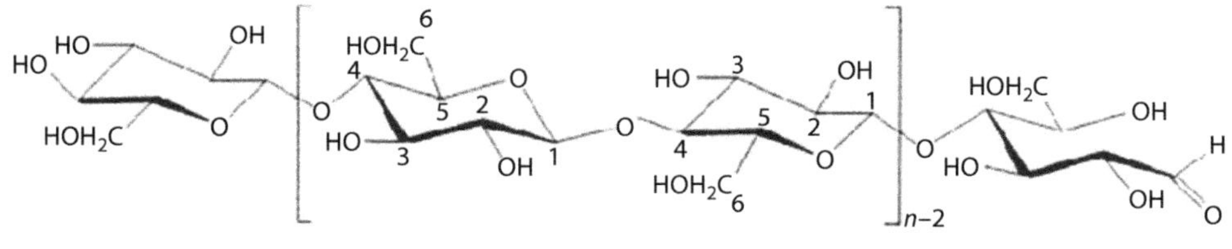

FIGURE 28.2 Structure of cellulose.

Cellulose is characterized by special properties as biocompatibility, biodegradation, and extreme-resistant in a wide temperature range.

The hydroxyl groups of $\beta - 1$, 4-glucan cellulose are placed at positions C2 and C3 (secondary, equatorial) as well as C6 (primary). The CH_2OH side group is arranged in a trans-gauche (t_g) position relative to the O5–C5 and C4–C5 bonds. As a result of the supramolecular structure of cellulose, the solid state is represented by areas of both high order (crystalline, usually in the range of 40%–60%) and low order (amorphous). The morphology of cellulose has an important effect on its reactivity; in fact, the hydroxyl groups located in the amorphous regions are highly accessible and react readily, whereas those in crystalline regions with close packing and strong interchain bonding can be completely inaccessible. The principal modifications of cellulose are esterifications and etherifications or oxidation at the hydroxyl groups of cellulose. These kinds of substitution drastically influence the original properties of cellulose and can lead to water-soluble and organic solvent-soluble cellulose derivatives.

Oxycellulose is cellulose in which some of the terminal primary alcohol groups of the glucose residues have been converted to carboxyl group. This cellulose derivative readily disperses in water and forms thixotropic dispersions that may be used in pharmaceutical field in the preparation of topical formulations (as emulsions) characterized by bioadhesion.

Cellulose ethers are widely used as pharmaceutical excipients in dermal formulations.

Sodium carboxymethyl cellulose is a polyanionic polysaccharide derivative of cellulose, employed in pharmaceutical and cosmetic field as emulsifying or gelling agent. In a recent approach, sodium carboxymethyl cellulose has been proposed as temporary dermal filler (Falcone et al. 2009). In that approach it was demonstrated that the clinical persistence of the sodium carboxymethyl cellulose dermal fillers correlates linearly with the concentration of polymer.

Methylcellulose is a neutral, odorless, tasteless, and inert polymer. In water it swells and forms a viscous colloidal solution; it is insoluble in organic solvents. Its solutions are stable in the range of pH 2–12 with no alteration of viscosity. It is principally used as thickening agent in emulsions and gels even in the presence of non-neutral pH. Functionalization with methacrylate groups and photo-crosslinked hydrogels has been proposed for potential application in plastic and reconstructive surgery (Stalling et al. 2009). These hydrogels tested in vitro exhibited no significant changes in cell viability, indicating that the materials are noncytotoxic; when implanted subcutaneously they were able to maintain their integrity in vivo after a long period (80 days), with no observed inflammatory reactions; these data suggest that hydrogels may be of use in soft tissue reconstruction.

Ethylcellulose is non-ionic, pH insensitive, insoluble in water but soluble in many polar organic solvents. It is used as thickener, stabilizer, and suspending agent for dermal applications of various surfactant-free water-in-oil or oil-in-water emulsions (Melzer et al. 2003). Recently it was used as hydrophobic film former (Limpongsa and Umprayn 2008).

Hydroxypropyl methyl cellulose is a water-soluble cellulose ether, and it has been explored as a gelling agent (Kikwai et al. 2005) and as a matrix in the design of films (Chandak and Verma 2008, Patel et al. 2009). Hydroxypropyl methyl cellulose has been shown to yield clear films because of the adequate solubility of the drug in the polymer. The drug release from hydroxypropyl methyl cellulose matrices follows two mechanisms, drug diffusion through the swelling gel layer and release by matrix erosion of the swollen layer (Ranga Rao et al. 1990). Due to its high hydrophilicity, matrices of hydroxypropyl methyl cellulose without rate-controlling membranes exhibit a burst effect during dissolution testing because the polymer is hydrated easily and swelled, leading to the fast release of the drug.

28.5.2 Collagen

Collagen is a protein largely present in mammals and it gives strength to tissue. The molecule of collagen molecule consists of three intertwined protein chains that form a helical structure. These molecules polymerize together to form collagen fibers of varying length.

Collagen has been proposed in a variety of formulations, including porous sponges, gels, and sheets, and can be cross-linked with several substances to make it stronger or to alter its degradation rate. Collagen sponges were proposed for the topical application of recombinant basic fibroblast growth factor for the treatment of chronic traumatic ulcers and were able to increase the incidence of complete wound closure, to shorten the complete healing time, and to improve the healing quality of chronic traumatic ulcers (Yao et al. 2006). The controlled release of the encapsulated bioactive may be obtained by cross-linking collagen sponges (Yang 2008).

For the dermal and transdermal applications of the drugs it has been used for the delivery by means of patch (Erdogan and van Gulik 2008).

A recent interesting application of collagen is as dressing material (Singh et al. 2011). This retrospective study showed that no significant better results in terms of completeness of healing of burn and chronic wounds between collagen dressing and conventional dressing were found; however, collagen dressing may avoid the need of skin grafting and provide additional advantage of patients' compliance and comfort.

28.5.3 ROSIN

Rosin is a natural oleoresin obtained from pine trees. It is a low molecular weight (MW = 400) polymer containing ~90% of tricyclic diterprene carboxylic acids (abietic and pimaric, Figure 28.3). This polymer and its derivatives are hydrophobic, biocompatibile, and biodegradable. In particular, biocompatibility of rosin was demonstrated by the absence of necrosis or abscess formation in the surrounding tissues when injected into the dermis (Pathak et al. 1990).

This resin is widely used in cosmetic and pharmaceutical fields especially for its excellent film-forming property. Moreover, the biodegradability of rosin makes possible the implant into the body without the need to remove the empty resin at the end of the drug release (Sinha et al. 2003).

The transdermal patches prepared with rosin in mixture with polyvinyl pyrrolidone (at different concentrations) were characterized by interesting pharmacokinetic and pharmacodynamic properties, and in particular the drug release occurs with a first-order kinetic (no erosion occurs) with a sustained/controlled release profile (Prashant et al. 2005).

28.5.4 CARRAGEENANS

Carrageenans are a family of linear sulfated polysaccharides extracted from red seaweeds. It is a high molecular weight polysaccharide with 15%–40% of ester-sulfate content. It is formed by alternate units of ß-D-galactose and 3,6-anhydro-α-D-galactose joined by α-1,3 and β-1,4-glycosidic linkage (Figure 28.4). The three most important carrageenans are called ι-(mono-sulfate), κ-(di-sulfate),

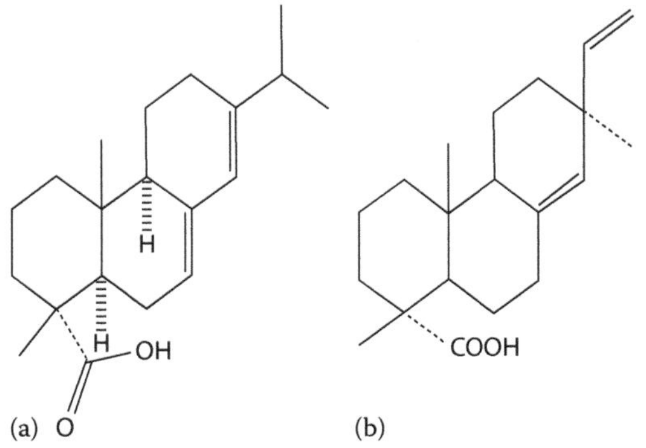

FIGURE 28.3 Structures of (a) abietic acid and (b) pimaric acid.

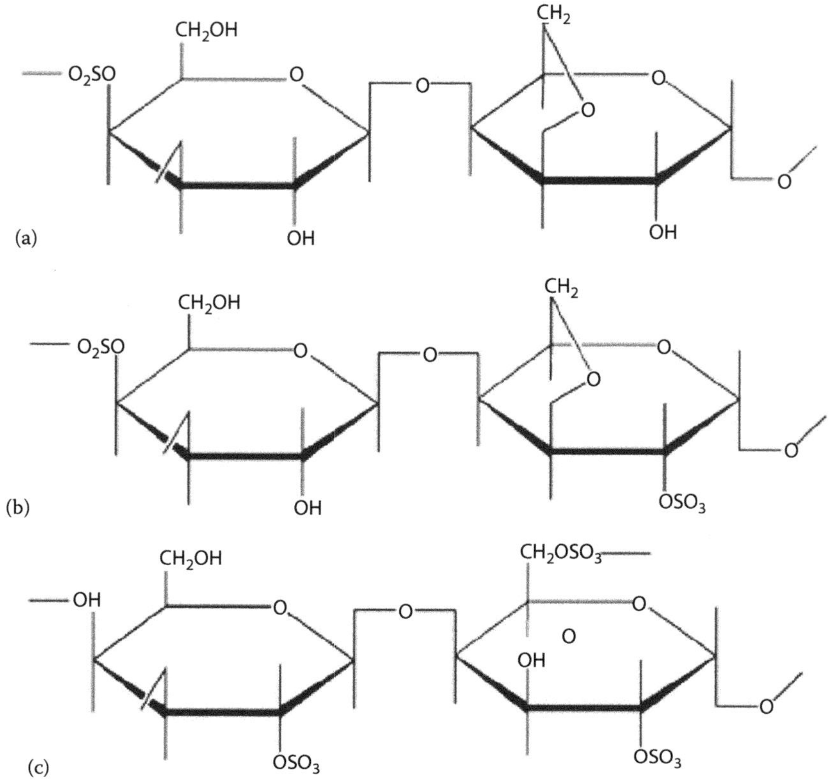

FIGURE 28.4 Structures of Carrageenans: (a) ι-(mono-sulfate) (b) κ-(di-sulfate) (c) λ-carrageenan.

and λ-carrageenan (tri-sulfate). The aqueous solutions of first two types of carrageenans form thermoreversible gels upon its cooling and for this characteristic are used for dermal application of drugs as gel-forming systems, whereas λ-carrageenan is a thickener agent used for the formulation of emulsions for topical use. κ-carrageenan, in particular, has been used to prepare hydrogels (also in blend with other natural polymers, i.e., agar and gelatin); in this case, release rate of the loaded molecules depends on the hydrophobicity; more hydrophobic is the drug, stronger is the adsorption in the chains leading to a lower diffusion coefficient (Sjoberg et al. 1999).

28.5.5 CHITOSAN

Chitosan is a highly basic, linear polysaccharide composed of randomly distributed β – (1 – 4)-linked D-glucosamine (deacetylated unit) and N-acetyl-D-glucosamine (acetylated unit) (Figure 28.5). Chitosan has positive charge under acidic conditions derived from protonation of its free amino groups; this explains the chitosan insolubility in neutral and basic environments; instead, in acidic environments, protonation of the amino groups leads to an increase in solubility. Because of its bioadhesive properties, biocompatibility, biodegradability, and nontoxicity, it is largely used in transdermal drug delivery. Principal fields of applications of chitosan in transdermal delivery are for the formation of bioadhesive films, and for skin generation in wound dressing.

Chitosan has been proposed for the preparation of easy-to-handle polymeric film bioadhesive for the incapsulation of several drugs. Transdermal delivery systems using chitosan membranes with different cross-link densities as drug release have been performed loading both lipophilic and hydrophilic drugs. The drug release was found depending on the cross-link density within the

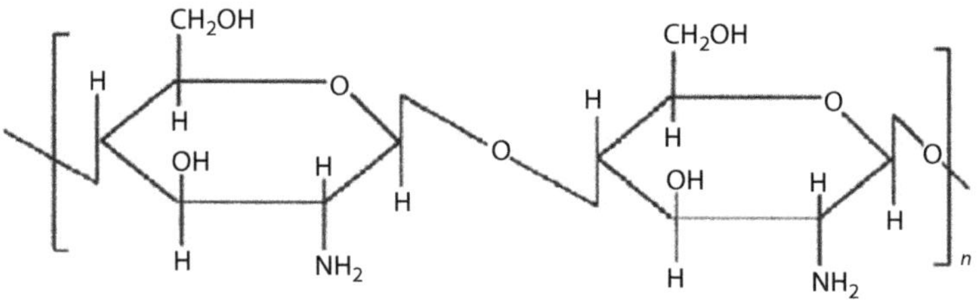

FIGURE 28.5 Structure of chitosan.

membranes; and in particular the higher the cross-link density, the lower the amount of released drug because of the decreased permeability coefficient of membranes (Thacharodi and Panduranga Rao 1993, 1995).

Several preparations have been proposed as a function of molecular weight and degree of deacetylation (Nunthanid et al. 2001). Both molecular weight and degree of deaceylation affected the film properties; in particular, the increase in molecular weight of chitosan increased the tensile strength and elongation as well as moisture absorption of the films; the increase in the degree of deacetylation of chitosan increased the tensile strength of the films depending on its molecular weight; the higher the degree of deacetylation of chitosan, the more brittle and the less moisture absorption the films became.

Chitosan films were used for the topical application of tramadol for treatment of moderate or severe chronic pain (Ammar et al. 2009a). Monolithic tramadol matrix film of chitosan showed best excellent properties in terms of flexibility, elasticity, smoothness, adhesion. Unfortunately, these chitosan films showed a high moisture uptake capacity, insufficient strength, and rapid release and permeation of loaded drug. These aspects were overcome by using chitosan in mixture with other polymers (Ammar et al. 2009b).

The efficacy of chitosan films as transdermal devices were also evaluated in vivo (Ammar et al. 2008). Chitosan films were proposed as transdermal delivery system for glimepiride, an antidiabetic drug. Release studies revealed adequate release rates from chitosan films, high flux values were obtained from films comprising, and in vivo studies on diabetic rats revealed a significant therapeutic efficacy prolonged for about 48 h.

The antibacterial and wound-healing properties of chitosan are well known (Raafat and Sahl 2009), for this property chitosan has been largely studied as a supporting material for tissue engineering especially for skin regeneration following serious burns. The application of physical hydrogels made up of chitosan and water was proposed. These hydrogels were constituted by two layers: the external was a rigid protective gel, able to ensure good mechanical properties and gas exchange and an internal soft and flexible layer able to adapt itself to the geometry of the wound and to ensure a superficial contact. These hydrogels were well-tolerated and able to promote the tissue regeneration; in fact, they induced the migration of the inflammatory cells favoring the vascularization of the newly formed tissue. Moreover, they were able to stimulate the biosynthesis of collagen under the granulation tissue and also the formation of dermal–epidermal junction was stimulated. At the end of the experiment (100 days), the newly formed tissue was really similar to the native skin (Boucard et al. 2007).

Recently, 3D chitosans produced as nanofibrillar scaffolds or as sponges were evaluated (Tchemtchoua et al. 2011). The experimental evidences showed that nanofibrillar structure strongly improved cell adhesion and proliferation in vitro; and when implanted in vivo in mice, the nanofibrillar scaffold was colonized by mesenchymal cells and blood vessels and an accumulation of collagen fibrils was also observed. Moreover, it was demonstrated that the use of chitosan nanofibrils as a dressing covering full-thickness skin wounds in mice induced a faster regeneration of both the epidermis and the dermis compartments. In contrast, chitosan administered as sponges induced a foreign body granuloma.

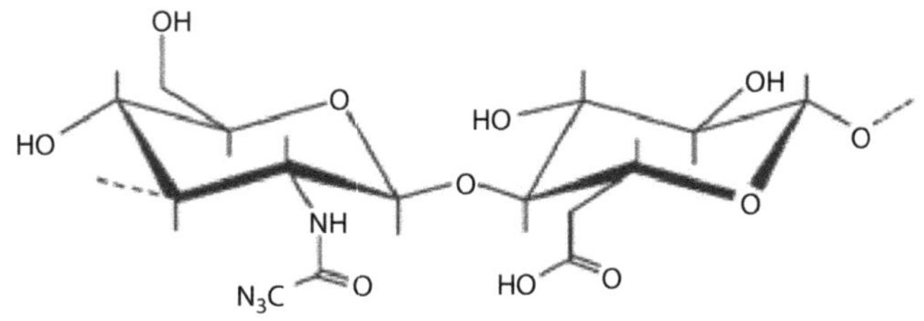

FIGURE 28.6 Structure of hyaluronic acid.

28.5.6 HYALURONIC ACID

Hyaluronic acid is a water-soluble polysaccharide that is widely distributed throughout the cell membranes of connective tissues in human and other animals. It is formed by multiple disaccharide units of D-glucuronic acid and N-acetyl-D-glucosamine (Figure 28.6). Hyaluronic acid for commercial use is isolated from several animal sources, as the synovial fluid or the skin, or from bacteria through a process of fermentation or direct isolation. As a function of the source and of the isolation process, it is possible to obtain numerous molecular weight grades. Biocompatibility, non-immunogenicity, biodegradability, and viscoelasticity of this polymer have been largely demonstrated, and for this hyaluronic acid is an ideal biomaterial for cosmetic, medical, and pharmaceutical applications.

Hyaluronic acid in aqueous solution undergoes a transition from Newtonian to non-Newtonian characteristics with increasing molecular weight, concentration, or shear rate (Gribbon et al. 2000). It has been demonstrated that a solution of hyaluronic acid possesses an increasing viscoelasticity, increasing its molecular weight and concentration of hyaluronic acid (Gibbs et al. 1968, Kobayashi et al. 1994). Its viscoelasticity in aqueous solution is pH-dependent and affected by the ionic strength of its environment (Mo et al. 1999).

This polymer is able to trap a great amount of water (1000 times its weight in water), and for this reason has been extensively utilized in cosmetic products because of its viscoelastic properties and excellent biocompatibility (Friedman et al. 2002). Largely used as a component of commercial dermal fillers to reduce facial lines and wrinkles (Duranti el at. 1998, Narins et al. 2003), its principal main side effect is the allergic reaction, possibly due to impurities present (Lupton and Alster 2000).

Hyaluronic acid has been also proposed for the formulation of cosmetic products as moisturizing or sunscreens, and in fact is able to protect the skin against ultraviolet irradiation due to its free radical scavenging properties (Trommer et al. 2003).

More recently, hyaluronic acid has been investigated as a vehicle able to localize the drug in the epidermis, and a formulation, marketed as Solaraze®, was already approved in the United States, Canada, and most European countries for the topical treatment of actinic keratoses (Jarvis and Figgitt 2003). In this application hyaluronic acid, as largely demonstrated (Brown et al. 1995), enhances significantly the partitioning of diclofenac into human skin and its retention and localization in the epidermis with respect to other pharmaceutical formulations (aqueous solution, or other gelling agents as polysaccharides [Brown and Martin 2001]).

28.6 SYNTHETIC POLYMERS

28.6.1 ETHYLENE VINYL ACETATE

Ethylene vinyl acetate is the copolymer of ethylene and vinyl acetate (Figure 28.7). This polymer has good barrier properties, hot-melt adhesive water proof properties, resistance to UV radiation, and odorless (RIF).

Ethylene vinyl acetate has been proposed as rate-controlling membrane (Charoo et al. 2005). Generally, the polymer contains a percentage of vinyl acetate from 10% to 70%; however, adjusting

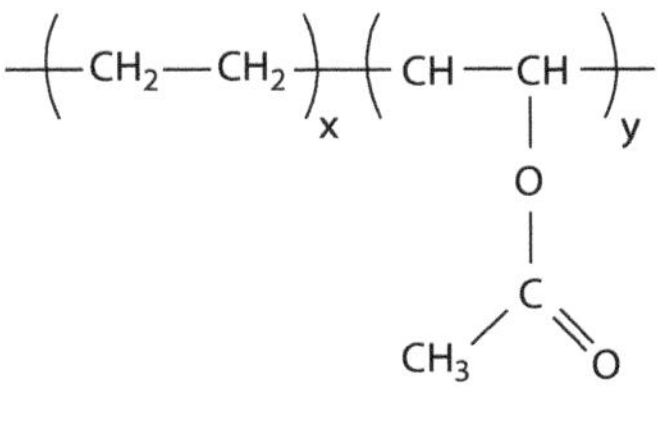

FIGURE 28.7 Structure of ethylene and vinyl acetate copolymer.

the amount of vinyl acetate, it is possible to change the membrane permeability. In fact, the copolymerization of the ethylene (which is crystalline) with vinyl acetate (not isomorphous), the degree of crystallinity, and the crystalline melting point decreases and amorphousness increases, so the permeability increases since an amorphous region is more permeable.

Moreover, the copolymerization also causes an increase in polarity and consequently an increase in solubility and in the diffusivity of polar compounds in the polymer. Generally, for a polar compound, the maximum flux of permeation is obtained at 60% of vinyl acetate content (Kandavilli et al. 2002).

28.6.2 POLYISOBUTYLENE

Polyisobutylene, produced by low-temperature cationic polymerization, is a polymer with no asymmetric carbons (Figure 28.8). In its unstrained state, the polymer is in an amorphous state, and its T_g is ~70°C (Wood 1976). Its physical state changes at the increase of molecular weight. Low molecular weight polymers are viscous liquids, and at the increase of molecular weight, they become more viscous to reach elastomeric solids. Un-crosslinked polymers exhibit a high degree of self-adhesion and for this reason they have been largely used as pressure-sensitive adhesive. (Qvist et al. 2002, Schulz et al. 2010).

28.6.3 POLYURETHANES

Polyurethanes are a class polymers derived from condensation of polyisocyanates and polyols having an intramolecular urethane bond or carbamate ester bonds. They may be synthesized from polyether polyol and are termed *polyether urethanes*, or from polyester polyol and are termed *polyester urethanes*. Initially most used were polyether type cause of their high resistance to hydrolysis (Boretos et al. 1971); however, polyester polyurethanes due to their biodegradability have been evaluated (Kambe et al. 1999). This class of polymers is generally used as rate-controlling membranes. The permeability of polyurethanes is generally optimized modifying the hydrophilic–hydrophobic ratio in these polymers (Lyman et al. 1967). Polyurethane membranes are used for hydrophilic polar compounds.

28.6.4 POLYVINYL ALCOHOL

Polyvinyl alcohol is a water-soluble, odorless and tasteless, translucent, white or cream-colored, granular powder synthetic polymer (Figure 28.9). Polyvinyl alcohol has a melting point of

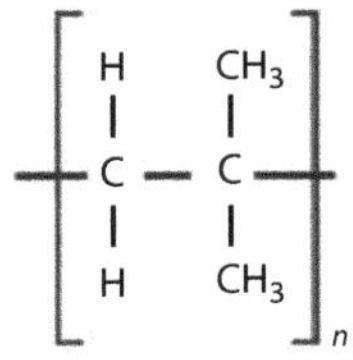

FIGURE 28.8 Structure of polyisobutylene.

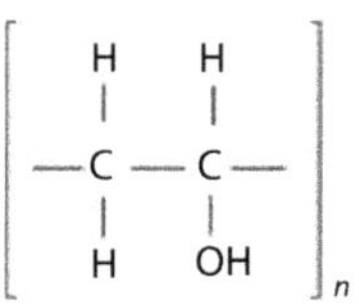

FIGURE 28.9 Structure of polyvinyl alcohol.

180°C–190°C. It is really biocompatibile, and for this it is well-studied in several medicine fields. It is able to form hydrogels that may be obtained by using several methods: chemical cross-linking (Ossipov et al. 2006); using glutaraldehyde as the cross-linking agent (Purss et al. 2005); cross-linking by γ-radiation (Benamer et al. 2006), by UV radiation (Martens and Anseth 2000) or by the use of successive freezing/thawing cycles (Peppas et al. 1997). An interesting field of application of polyvinyl alcohol hydrogels is in wound healing. In fact, they were used to deliver successfully a lot of drugs such as growth factors (Bourke et al. 2003), clindamycin (Kim et al. 2008), gentamicin (Hwang et al. 2010). Moreover, polyvinyl alcohol was used as film-forming polymer for the formulation of transdermal film (Aqil et al. 2004, Nicoli et al. 2006).

28.6.5 POLYVINYL PYRROLIDONE

This polymer has been proposed in mixture with ethylcellulose as matrix-type transdermal patches (Arora et al. 2002, Alam et al. 2009, De et al. 2009) to improve the characteristics of the patches. Polyvinyl pyrrolidone enhanced drug release, in vitro percutaneous permeation, and in vivo activity of loaded drugs.

Recently, it was found that polyvinyl pyrrolidone is able to inhibit the crystallization of drug (hydrophilic or hydrophobic) in acrylate and silicone adhesives. The incorporation of polyvinyl pyrrolidone in patches allowed the incorporation of drugs in amounts higher than their saturation solubility in pure adhesives (Jain et al. 2010).

28.6.6 POLYACRYLATES AND POLYMETHACRYLATES

Polyacrylates are a class of polymers derived from the polymerization of acrylic acid, methacrylic acid, and their esters (Figure 28.10). Acrylates, cause of vinyl group, polymerize readily and are characterized by a long stability. In fact, they possess a great resistance to both acidic and alkaline hydrolysis and are insensitive to UV. The mechanical properties of acrylic polymers improve as the molecular weight increases. These polymeric matrices have been largely used in pharmaceutical field, especially for the formulation of transdermal patch as PSA.

The characteristics of the polymer matrix largely depend on the properties of the used monomer, molecular weight, and reticulation degree. These polymers possess T_g really low; however, it is possible to add plasticizers to lower it further. The principal utilized method to produce polymeric film is the film casting technique. To obtain the required characteristics, several substances are added to acrylic polymers as organic tackifiers, stabilizers, antioxidants, plasticizers.

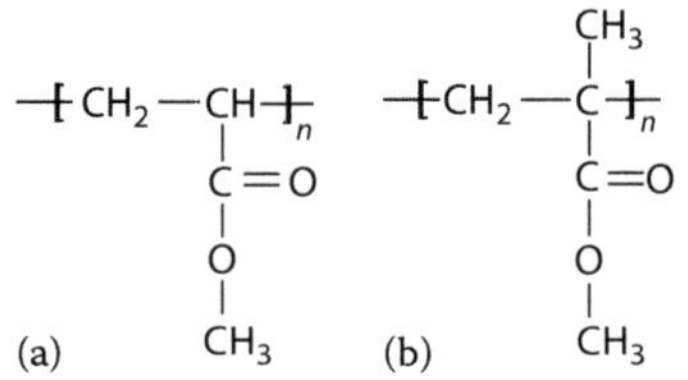

FIGURE 28.10 Structure of (a) poly methyl acrylate and (b) poly methyl methacrylate.

A lot of substances have been carried out by means of transdermal drug delivery systems based on acrylates polymers and a lot of quoted reviews have been written (Argoff 2011, Cachia et al. 2011, Dhillon 2011, Greenspoon et al. 2011, Rapoport et al. 2010); however, these kinds of polymers are able to ensure both adhesive properties and a controlled release of the loaded drug.

28.6.7 Silicones

Silicone in transdermal drug delivery is used principally as PSA (Pfister et al. 1992). It is prepared by condensing a polymer (low-viscosity dimethylsiloxane) and a resin (a 3D silicate structure that is end capped with trimethyl siloxy groups and contains residual silanol functionality). The resultant silicone has a viscoelastic behavior: the viscous component is due to the dimethylsiloxane (fluid, at low viscosity) and is able to guarantee the wetting and spreadability properties; the elastic behavior is due to the resin, which acts as a tackyfying and reinforcing component. To obtain the optimum in terms of tack, adhesion, and peel release, the resin content has to be adjusted.

Unlike acrylic polymers, medical-grade silicone adhesives do not contain organic tackifiers, stabilizers, antioxidants, plasticizers, catalysts, or other potentially toxic substances because silicone PSAs are stable throughout a wide range of temperatures ($-73°C$ to $+250°C$).

28.7 CONCLUSIONS

Transdermal drug delivery systems represent a fascinating field of the pharmaceutical research that has been largely studied in the past four decades. This field of application is really intricate, in fact to date only a few products (~40) have been launched on the market.

To obtain more products with appropriate characteristics is probably necessary to use different approaches from those presented in this chapter, as some researchers are already doing (Liu et al. 2011).

REFERENCES

Ainbinder, D., Paolino, D., Fresta, M., and Touitou, E. 2010. Drug delivery applications with ethosomes. *J. Biomed. Nanotechnol.* 6:558–568.

Alam, M.I., Baboota, S., Kohli, K., Ali, J., and Ahuja, A. 2009. Development and evaluation of transdermal patches of celecoxib. *PDA J. Pharm. Sci. Technol.* 63:429–437.

Ammar, H.O., Ghorab, M., El-Nahhas, S.A., and Kamel, R. 2009a. Polymeric matrix system for prolonged delivery of tramadol hydrochloride, part I: Physicochemical evaluation. *AAPS PharmSciTech.* 10(1):7–20.

Ammar, H.O., Ghorab, M., El-Nahhas, S.A., and Kamel, R. 2009b. Polymeric matrix system for prolonged delivery of tramadol hydrochloride, part I: Physicochemical evaluation. *AAPS PharmSciTech.* 10(3):1065–1070.

Ammar, H.O., Salama, H.A., El-Nahhas, S.A., and Elmotasem, H. 2008. Design and evaluation of chitosan films for transdermal delivery of glimepiride. *Curr. Drug Deliv.* 5(4):290–298.

Aqil, M., Sultana, Y., Ali, A., Dubey, K., Najmi, A.K., and Pillai, K.K. 2004. Transdermal drug delivery systems of a beta blocker: Design, in vitro, and in vivo characterization. *Drug Deliv.* 11:27–31.

Argoff, C.E. 2011. Recent developments in the treatment of osteoarthritis with NSAIDs. *Curr Med. Res. Opin.* 27:1315–1327.

Arora, P. and Mukherjee, B. 2002. Design, development, physicochemical, and in vitro and in vivo evaluation of transdermal patches containing diclofenac diethylammonium salt. *J. Pharm. Sci.* 91:2076–2089.

Arsic, I., Vidovic, A., and Vuleta, G. 1999. Influence of liposomes on the stability of vitamin A incorporated in polyacrylate hydrogel. *Int. J. Cosm. Sci.* 21:219–225.

Benamer, S., Mahlous, M., Boukrif, A., Masouri, B., and Larbi, Y.S. 2006. Synthesis and characterisation of hydrogels based on poly(vinyl pyrrolidone). *Nucl. Instrum. Methods Phys. Res. Sect. B.* 248:284–290.

Betz, G., Aeppli, A., Menshutina, N., and Leuenberger, H. 2005. In vivo comparison of various liposome formulations for cosmetic application. *Int. J. Pharm.* 296:44–54.

Boretos, J.W., Detmer, D.E., and Donachy, J.H. 1971. Segmented polyurethane: A polyether polymer, II: Two years' experience. *J. Biomed. Mat. Res.* 5:373–387.

Boucard, N., Viton, C., Agay, D., Mari, E., Roger, T., Chancerelle Y., and Domard A. 2007. The use of physical hydrogels of chitosan for skin regeneration following third-degree burns. *Biomaterials* 28(24):3478–3488.

Bourke, S.L., Al-Khalili, M., Briggs, T., Michniak, B.B., Kohn, J., and Poole-Warren, L.A. 2003. A photo-crosslinked poly(vinyl alcohol) hydrogel growth factor release vehicle for wound healing applications. *AAPS PharmSci.* 5(4):101–111, doi:10.1208/ps050433.

Brown, M.B. and Martin, G.P. 2001. Comparison of the effect of hyaluronan and other polysaccharides on drug skin partitioning. *Int. J. Pharm.* 225:113–121.

Brown, M.B., Marriott, C., and Martin, G.P. 1995. The effect of hyaluronan on the in vitro deposition of diclofenac within the skin. *Int. J. Tissue React.-Exp. Clin. Asp.* 17:133–140.

Cachia, E. and Ahmedzai, S.H. 2011. Transdermal opioids for cancer pain. *Curr. Opin. Support Palliat. Care* 5:15–19

Celia, C., Trapasso, E., Cosco, D., Paolino, D., and Fresta, M. 2009. Turbiscan Lab® expert analysis of the stability of ethosomes® and ultradeformable liposomes containing a bilayer fluidizing agent. *Colloids Surf. B Biointerfaces* 72:155–160.

Cevc, G. 2003. Transdermal drug delivery of insulin with ultradeformable carriers. *Clin. Pharmacokinet.* 42(5):461–474.

Cevc, G. and Blume, G. 1992. Lipid vesicles penetrate into intact skin owing to the transdermal osmotic gradients and hydration force. *Biochim. Biophys. Acta.* 104:226–232.

Cevc, G. and Blume, G. 2001. New, highly efficient formulation of diclofenac for the topical, transdermal administration in ultradeformable drug carriers, Transfersomes. *Biochim. Biophys. Acta.* 1514:191–205.

Chandak, A.R. and Verma, P.R. 2008. Design and development of hydroxypropyl methylcellulose (HPMC) based polymeric films of methotrexate: Physicochemical and pharmacokinetic evaluations. *Yakugaku Zasshi.* 128(7):1057–1066.

Charoo, N.A., Anwer, A., Kohli, K., Pillai, K.K., and Rahman, Z. 2005. Transdermal delivery of flurbiprofen: Permeation enhancement, design, pharmacokinetic, and pharmacodynamic studies in albino rats. *Pharm. Dev. Technol.* 10(3):343–351.

De P., Damodharan N., Mallick S., and Mukherjee B. 2009. Development and evaluation of nefopam transdermal matrix patch system in human volunteers. *PDA J. Pharm. Sci. Technol.* 63:537–546.

Dhillon, S. 2011. Rivastigmine transdermal patch: A review of its use in the management of dementia of the Alzheimer's type. *Drugs.* 71:1209–1231.

Duranti, F., Salti, G., Bovani, B., Calandra, M., and Rosati, M.L. 1998. Injectable hyaluronic acid gel for soft tissue augmentation – A clinical and histological study. *Dermatol. Surg.* 24:1317–1325.

Elsayed, M.M.A., Abdallah, O.Y., Naggar, V.F., and Khalafallah, N.M. 2007. Lipid vesicles for skin delivery of drugs: Reviewing three decades of research. *Int. J. Pharm.* 332:1–16.

Erdogan, D. and van Gulik, T.M. 2008. Evolution of fibrinogen-coated collagen patch for use as a topical hemostatic agent. *J. Biomed. Mater. Res. B Appl. Biomater.* 85(1):272–278.

Falcone, S.J. and Berg, R.A. 2009. Temporary polysaccharide dermal fillers: A model for persistence based on physical properties. *Dermatol. Surg.* 35(8):1238–1243.

Friedman, P.M., Mafong, E.A., Kauvar, A.N.B., and Geronemus, R.G. 2002. Safety data of injectable nonanimal stabilized hyaluronic acid gel for soft tissue augmentation. *Dermatol. Surg.* 28:491–494.

Gibbs, D.A., Merrill, E.W., Smith, K.A., and Balazs, E.A. 1968. Rheology of hyaluronic acid. *Biopolymers* 6:777–791.

Greenspoon, J., Herrmann, N., and Adam, D.N. 2011. Transdermal rivastigmine: Management of cutaneous adverse events and review of the literature. *CNS Drugs.* 25:575–583.

Gribbon, P., Heng, B.C., and Hardingham, T.E. 2000. The analysis of intermolecular interactions in concentrated hyaluronan solutions suggest no evidence for chain–chain association. *Biochem. J.* 350:329–335.

Hadgraft, J. 2004. Skin deep. *Eur. J. Pharm. Biopharm.* 58:291–299.

Hwang, M.R., Kim, J.O., Lee, J.H., Kim, Y.I., Kim, J.H. et al. 2010. Gentamicin-loaded wound dressing with polyvinyl alcohol/dextran hydrogel: Gel characterization and in vivo healing evaluation. *AAPS PharmSciTech.* 11:1092–1103.

Jain, P. and Banga, A.K. 2010. Inhibition of crystallization in drug-in-adhesive-type transdermal patches. *Int. J. Pharm.* 394:68–74.

Jarvis, B. and Figgitt, D.P. 2003. Topical 3% diclofenac in 2.5% hyaluronic acid gel: A review of its use in patients with actinic keratoses. *Am. J. Clin. Dermatol.* 4(3):203–213.

Kambe, T.N. et al. 1999. Microbial degradation of polyurethane, polyester polyurethanes, and polyether polyurethanes. *Appl. Microbiol. Biotechnol.* 51:134–140.

Kandavilli, S., Nair, V., and Panchagnula, R. 2002. Polymers in transdermal drug delivery systems. *Pharm. Technol.* 2002(May):62–80.

Kikwai, L., Babu, R.J., Prado, R., Kolot, A., Armstrong, C.A., Ansel, J.C., and Singh, M. 2005. In vitro and in vivo evaluation of topical formulations of spantide II. *AAPS PharmSciTech.* 6(4):E565–E572.

Kim, J.O., Choi, J.Y., Park, J.K., Kim, J.H., Jin, S.G., Chang, S.W. et al. 2008. Development of clindamycin-loaded wound dressing with polyvinyl alcohol and sodium alginate. *Biol. Pharm. Bull.* 31:2277–2282.

Kobayashi, Y., Okamoto, A., and Nishinari, K. 1994. Viscoelasticity of hyaluronic-acid with different molecular-weights. *Biorheology.* 31:235–244.

Limpongsa, E. and Umprayn, K. 2008. Preparation and evaluation of diltiazem hydrochloride diffusion-controlled transdermal delivery system. *AAPS PharmSciTech.* 9(2):464–470.

Lyman, D.J. and Loo, B.H. 1967. New synthetic membranes for dialysis, IV: A copolyether-urethane membrane system. *J. Biomed. Mater. Res.* 1:17–26.

Liu, X., Liu, H., Liu, J., He, Z., Ding, C. et al. 2011. Preparation of a ligustrazine ethosome patch and its evaluation in vitro and in vivo. *Int. J. Nanomed.* 6:241–247.

Lupton, J.R. and Alster, T.S. 2000. Cutaneous hypersensitivity reaction to injectable hyaluronic acid gel. *Dermatol. Surg.* 26:135–137.

Manconi, M., Valenti, D., and Sinico, C., 2003. Niosomes as carriers for tretinoin II. Influence of vesicular incorporation on tretinoin photostability. *Int. J. Pharm.* 260:261–272

Martens, P. and Anseth, K.S. 2000. Characterization of hydrogels formed from acrylate modified poly(vinyl alcohol) macromers. *Polymer.* 41:7715–7722.

Melzer, E., Kreuter, J., and Daniels, R. 2003. Ethylcellulose: A new type of emulsion stabilizer. *Eur. J. Pharm. Biopharm.* 56(1):23–27.

Mo, Y., Takaya, T., Nishinari, K. et al. 1999. Effects of sodium chloride, guanidine hydrochloride, and sucrose on the viscoelastic properties of sodium hyaluronate solutions. *Biopolymers.* 50:23–34.

Montenegro, L., Paolino, D., Drago, R., Pignatello, R., Fresta, M., and Puglisi, G. 2006. Influence of liposome composition on in vitro permeation of diosmine through human stratum corneum and epidermis. *J. Drug. Del. Sci. Tech.* 16:133–140.

Narins, R.S., Brandt, F., Leyden, J. et al. 2003. A randomized, double-blind, multicenter comparison of the efficacy and tolerability of Restylane versus Zyplast for the correction of nasolabial folds. *Dermatol. Surg.* 29:588–595.

Nicoli, S., Penna, E., Padula, C., Colombo, P., and Santi, P. 2006. New transdermal bioadhesive film containing oxybutynin: In vitro permeation across rabbit ear skin. *Int. J. Pharm.* 325:2–7.

Nunthanid, J., Puttipipatkhachorn, S., Yamamoto, K., and Peck, G.E. 2001. Physical properties and molecular behavior of chitosan films. *Drug. Dev. Ind. Pharm.* 27(2):143–157.

Ossipov, D.A. and Hilborn, J. 2006. Poly(vinyl alcohol)-based hydrogels formed by "click chemistry". *Macromolecules.* 39:1709–1718.

Paolino, D., Cosco, D., Muzzalupo, R., Trapasso, E., Picci, N., and Fresta, M. 2008. Innovative bola-surfactant niosomes as topical delivery systems of 5-fluorouracil for the treatment of skin cancer. *Int. J. Pharm.* 353:233–242.

Paolino, D., Lucania, G., Mardente, D. et al. 2005. Ethosomes for skin delivery of ammonium glycyrrhizinate: In vitro percutaneous permeation through human skin and in vivo anti-inflammatory activity on human volunteers. *J. Control Release.* 106:99–110.

Paolino, D., Muzzalupo, R., Ricciardi, A., Celia, C., Picci, N., and Fresta, M. 2007. In vitro and in vivo evaluation of Bola-surfactant containing niosomes for transdermal delivery. *Biomed. Microdevices* 9:421–433.

Patel, D.P., Setty, C.M., Mistry, G.N., Patel, S.L., Patel, T.J., Mistry, P.C. et al. 2009. Development and evaluation of ethyl cellulose-based transdermal films of furosemide for improved in vitro skin permeation. *AAPS PharmSciTech.* 10(2):437–442.

Pathak, Y.V. and Dorle, A.K. 1990. Rosin and rosin derivatives as hydrophobic matrix materials for controlled release of drugs. *Drug Dev. Ind. Pharm.* 6:223–227.

Peppas, N.A. and Mongia, N.K. 1997. Ultrapure poly(vinyl alcohol) hydrogels with mucoadhesive drug delivery characteristics. *Eur. J. Pharm. Biopharm.* 43:51–58.

Pfister, W.R., Woodard, J.T., and Grigoras, S. 1992. Developing drug-compatible adhesives for transdermal drug delivery devices. *Pharm. Technol.* 16:42–58.

Prashant, M.S., Suniket, V.F., and Avinash, K.D. 2005. Evaluation of polymerized rosin for the formulation and development of transdermal drug delivery system: A technical note. *AAPS PharmSciTech.* 6(4):E649–E654.

Purss, K.H., Qiao, G.G., and Solomon, D.H. 2005. Effect of "glutaraldehyde" functionality on network formation in poly(vinyl alcohol) membranes. *Appl. Polym. Sci.* 96:780–792.

Qvist, M.H., Hoeck, U., Kreilgaard, B., Madsen, F., and Frokjaer, S. 2002. Release of chemical permeation enhancers from drug-in-adhesive transdermal patches. *Int. J. Pharm.* 231(2):253–263.

Raafat, D. and Sahl, H.G. 2009. Chitosan and its antimicrobial potential-a critical literature survey. *Microb. Biotechnol.* 2(2):186–201.

Ranga Rao K.V., Padmalatha D., and Buri P. 1990. Influence of molecular size and water solubility of the solute on its release from swelling and erosion controlled polymeric matrices. *J. Control. Release.* 12:133–141.

Rapoport, A.M., Freitag, F., and Pearlman, S.H. 2010. Innovative delivery systems for migraine: The clinical utility of a transdermal patch for the acute treatment of migraine. *CNS Drugs.* 24:929–940.

Schulz, M., Fussnegger, B., and Bodmeier, R. 2010. Drug release and adhesive properties of crospovidone-containing matrix patches based on polyisobutene and acrylic adhesives. *Eur. J. Pharm. Sci.* 41(5):675–684.

Singh, O., Gupta, S.S., Soni, M., Moses, S., Shukla, S., and Mathur, R.K. 2011. Collagen dressing versus conventional dressings in burn and chronic wounds: A retrospective study. *J. Cutan. Aesthet. Surg.* 4:12–16.

Sinha, V.R. and Trehan, A. 2003. Biodegradable microspheres for protein delivery. *J. Control. Release.* 90:261–280.

Sjoberg, H., Persson, S., and Caram-Lelham, N. 1999. How interactions between drugs and agarose–carrageenan hydrogels influence the simultaneous transport of drugs. *J. Control Release.* 59:391–400.

Stalling, S.S., Akintoye, S.O., and Nicoll, S.B. 2009. Development of photocrosslinked methylcellulose hydrogels for soft tissue reconstruction. *Acta Biomater.* 5(6):1911–1918.

Tchemtchoua, V.T., Atanasova, G., Aqil, A., Filée, P., Garbacki, N., Vanhooteghem, O. et al. 2011. Development of a chitosan nanofibrillar scaffold for skin repair and regeneration. *Biomacromolecules.* 12(9):3194–3204.

Thacharodi, D. and Panduranga Rao, K. 1993. Release of nifedipine through crosslinked chitosan membranes. *Int. J. Pharm.* 96:33–39.

Thacharodi, D. and Panduranga Rao, K. 1995. Development and in vitro evaluation of chitosan based transdermal drug delivery systems for controlled delivery of propranolol hydrochloride. *Biomaterials.* 16:145–148.

Touitou, E. 1996. Compositions for applying active substances to or through the skin. US patent US5540934.

Touitou, E., Dayan, N., Bergelson, L. et al. 2000. Ethosomes—Novel vesicular carriers for enhanced delivery: Characterization and skin penetration properties. *J. Control Release.* 65:403–418.

Touitou, E., Godin, B., Dayan, N. et al. 2001. Intracellular delivery mediated by an ethosomal carrier. *Biomaterials.* 22:3053–3059.

Trommer, H., Wartewig, S., Bottcher, R. et al. 2003. The effects of hyaluronan and its fragments on lipid models exposed to UV irradiation. *Int. J. Pharm.* 254:223–234.

Ventura, C.A., Giannone, I., Paolino, D., Pistarà, V., Corsaro, A., and Puglisi, G. 2005. Preparation of celecoxib-dimethyl-beta-cyclodextrin inclusion complex: Characterization and in vitro permeation study. *Eur. J. Med. Chem.* 40:624–631.

Ventura, C.A., Tommasini, S., Falcone, A., Giannone, I., Paolino, D., Sdrafkakis, V. et al. 2006. Influence of modified cyclodextrins on solubility and percutaneous absorption of celecoxib through human skin. *Int. J. Pharm.* 314:37–45.

Wood, L.A. 1976. Physical constants of different rubber. *Rubber Chem. Technol.* 49(1976):189, doi:10.5254/1.3534955.

Yang, C.H. 2008. Evaluation of the release rate of bioactive recombinant human epidermal growth factor from crosslinking collagen sponges. J. Mater Sci. Med. 19:1433–1440.

Yao, C., Yao, P., Wu, H, and Zha, Z. 2006. Acceleration of wound healing in traumatic ulcers by absorbable collagen sponge containing recombinant basic fibroblast growth factor. *Biomed. Mater.* 1(1):33–37.

Index

Milton Keynes UK
Ingram Content Group UK Ltd.
UKHW050306111024
449327UK00043B/2021